AMERICAN GEOLOGICAL LITERATURE, 1669 to 1850

AMERICAN GEOLOGICAL LITERATURE, 1669 to 1850

Robert M. Hazen
and
Margaret Hindle Hazen

Dowden, Hutchinson & Ross, Inc.
Stroudsburg   Pennsylvania

Copyright © 1980 by **Dowden, Hutchinson & Ross, Inc.**
Library of Congress Catalog Card Number: 79-25898
ISBN: 0-87933-371-5

All rights reserved. No part of this book may be reproduced or transmitted in any form or by any means—graphic, electronic, or mechanical, including photocopying, recording, taping, or information storage and retrieval systems—without written permission of the publisher.

82 81 80   1 2 3 4 5
Manufactured in the United States of America.

LIBRARY OF CONGRESS CATALOGING IN PUBLICATION DATA
Hazen, Robert M 1948–
   American geological literature, 1669 to 1850.
   Includes index.
   1. Geology—United States—Early works to 1800—
Bibliography. 2. Geology—United States—Bibliography,
I. Hazen, Margaret Hindle, joint author. II. Title.,
Z6034.U49H39 [QE77]   016.5573   79-25898
ISBN 0-87933-371-5

Distributed world wide by Academic Press,
a subsidiary of Harcourt Brace Jovanovich,
Publishers.

CONTENTS

Foreword	vii
Preface	ix
Introduction	1
Secondary References	12
Periodicals Consulted	17
Guide to the Use of the Bibliography	40
List of Abbreviations	42
Bibliography	45
Index	393
About the Authors	431

FOREWORD

Geologists and historians are indebted to Robert and Margaret Hazen for this scholarly and enormously detailed bibliography. It is noteworthy that some of the great geological bibliographies have been produced by professional scientists—Agassiz, de Margerie, Pestana, and Sarjeant, to name a few. Now we have another to add to the list.

The existing bibliographies of American geology provide an entry to some of the early geological literature—those of Meisel, Darton, and Nickles (and to a certain extent Sabin, Evans, and Howes)—but these are far from complete for the earliest publications, especially those not in book form. The searching out of reviews, broadsides, and other similar ephemeral and "off beat" publications has obviously been a tremendous task; the inclusion of these is in itself an important contribution.

The recent work of the late R. P. Stearns makes us familiar with early American science published overseas, but that actually published in the United States, or in what was to become the United States, has been little known. And yet the indigenous publications may have contributed to the development of geology in the United States as much as did the foreign publications. American reviews of foreign publications are often of great value, and here we find them listed. Certainly the American editions of foreign works were of great importance. Here inquirers can find in one place what editions of Buckland, Cuvier, and Lyell were published in the United States.

A good specialized bibliography is a "thing of beauty and a joy forever." It is in the closest bookshelf at the hand of the scholar. Working geologists have Nickles and the state survey bibliographies for their areal studies, but for complete reference to their own areas, they will find this bibliography useful. Geologist historians as well as historians of science need the Hazen bibliography nearby along with other critical apparatus. Now we will have Hazen and Hazen to lead us through the jungle of the earliest works on American geology and of ephemera, reviews, and pamphlet reports of the first six decades of the Republic.

GEORGE W. WHITE
University of Illinois

PREFACE

That the literature of geology is extensive and diverse constitutes a major problem for historians of geology who must become familiar with numerous publications dealing with such varied subjects as fossils, soils, mines, and topography. Because many of the early American publications relating to these topics have been long neglected, research can be difficult and time-consuming.

To date, several important compilations have simplified the bibliographic search for students of North American geology. Foremost among these tools are the well-known bibliographies of Darton (1896), Nickles (1923), Meisel (1924–1927), and Pestana (1972). The usefulness of these reference works is limited, however, by the restrictions placed on content by their compilers. Darton, Nickles, and Meisel, for example, listed only what they considered to be major scientific works and Pestana collected references for one type of publication only (U. S. government documents.)

The present bibliography has a different scope. Conceived as a reference tool for historians of American geology, it includes geology-related books, reviews, maps, broadsides, pamphlets, journal aritcles, and other nonnewspaper sources published before 1851 in what is now the United States.

The bibliography was limited to pre-1851, nonnewspaper sources for reasons of time and space. The extension of the work through 1860 would have doubled the number of entries, and examination of the two million pre-1851 newspaper issues was impossible. The bibliography was limited to works published in the United States partly because access to many important foreign sources was limited. Any attempt to include foreign-published material would have fallen far short of a complete record. Furthermore it seemed to us that a realistic guide to the intellectual development of geologists in America would be the corpus of works published in America. Of course, some foreign texts and articles captured the attention of American scientists, but most of these were abstracted, reviewed, or even republished in the United States and thus appear in the bibliography.

We compiled the bibliography from three main groups of sources: geology bibliographies, periodicals, and library catalogs. The first step was to extract all relevant references from the large number of existing geological and scientific bibliographies (see Secondary References). Most important were the sources relating to North American geology listed in Darton and Nickles, as well as the more general science bibliography of Meisel. To these we added entries from more specialized works, such as Pestana's *Bibliography of Congressional Geology* (1972), bibliographies of the published works of individual geologists (such as *Samuel Latham Mitchill* by C. R. Hall, 1934), and more than fifty state and local geology bibliographies. Finally Charles Evans's *American Bibliography* (1903–1934) provided a valuable guide to the pre-1801 American literature, whereas bibliographies by Sabin (1961), Shaw and Shoemaker (1958–1965), and Shoemaker (1964–) contained references for the period through 1850. In all, we examined approximately eighty bibliographies, and the references discovered provided a core of four thousand entries for the bibliography.

Among the core references, fewer than fifty American-published, pre-1851 periodicals were cited. To ensure that additional periodicals containing geological literature were not omitted, we conducted a systematic search of the *Union List of Serials* (Titus, 1965), and all pre-1851 journals were noted. More than two thousand titles were discovered, and although more than half of these were unavailable in complete runs, nearly a thousand complete early American journals were to be found in U. S. libraries. Of these, seven hundred were accessible to us. Included in this number were all major popular, literary, review, eclectic, juvenile, agricultural, medical, military, mechanical, religious, and scientific periodicals published in the United States through 1850. Each of these journal runs was examined page by page for all earth-science notes, reviews, and articles. References extracted include both descriptive and analytical accounts of earth materials, geological processes (such as earthquakes and volcanoes), mining techniques and safety, geomorphology, and cosmology. Journal sources added almost ten thousand references to the core list.

The list of fourteen thousand references was further expanded by entries found during a search of the Library of Congress printed catalogs and the card catalogs of selected research libraries. The most comprehensive source was the *National Union Catalog* (Mansell, 1968–), which was complete through the letter *V* (volume 640) at the time of compilation. All authors of earth-science articles were searched in Mansell for additional entries. For authors *W* through *Z*, we used other catalogs, including those of the Library of Congress and Harvard University. In addition, the card catalog of the Baker Library of the Harvard Business School provided an important source of data on mining and railroad company reports. Nearly a thousand additional book and pamphlet references were added in systematic searches of these library catalogs.

We consolidated the total of fifteen thousand references to the 11,133 entries of this edition in three ways. Books that had multiple printings, or in some cases multiple editions, are listed under a single entry. Journal articles that appeared in two or more different periodicals are combined into a single numbered entry. Finally, in several periodicals, such as *Niles' Weekly Register* and *Scientific American*, which contained hundreds of short (one- to nine-line) notes on earthquakes, volcanoes, mining, and other geology-related subjects, we have combined several brief notes on the same topic into a single entry.

The bibliography is incomplete for several types of published sources. Perhaps the least complete portion is the listing of mining and railroad company reports. This literature (including annual stockholders' reports, share advertisements, geological reports, articles of incorporation, and bylaws) must have been published for many of the thousands of operating mines in nineteenth-century America; however, only about three hundred such reports were found in our search. It is anticipated that many more will be found by other researchers. It is also expected that additional earth-science articles will be found in the thirteen hundred periodicals that we did not examine. However, we have included all scientific journals, and most of the omitted journals had only one or two volumes published. Therefore it is assumed that the majority of such articles have been found. Our listing of works by several authors whose names begin with letters W through Z is probably incomplete because Mansell's *National Union Catalog* had not yet completed publication. As new volumes of Mansell appear, additional titles will undoubtedly be added.

Although no attempt was made to examine systematically the vast newspaper literature of the United States, a few important original articles have been included. In addition, many such articles that were republished in journals also appear in this bibliography. It should be noted, however, that newspapers often contain valuable information on the earth sciences, especially on transitory phenomena such as earthquakes, volcanoes, and the discovery of unusual fossils or minerals. Newspapers from mining districts may also contain production reports and other elusive statistics (see, for example, the *Pottsville Miner's Journal*, Pottsville, Pennsylvania).

Throughout the compilation of this bibliography, we have depended on the aid, advice, and encouragement of numerous individuals. Brooke Hindle, senior historian at the National Museum of History and Technology, Smithsonian Institution, inspired our interest in the history of science and offered guidance in all stages of the project. George W. White at the Department of Geology, University of Illinois, generously gave both advice and encouragement and helped to define the scope and potential applications of this work. John D. Haskell, editor of *A New England Bibliography*, significantly eased our task and saved us from many pitfalls by sharing his experiences as a bibliographer.

Special thanks are due to Michele L. Aldrich, Bailey Bishop, Donald W. Fisher, John C. Greene, Clifford Nelson, Harold Pestana, Alonzo W. Quinn, and John Sinkankas, who discovered and reported unusual references. James X. Corgan has been of great help in searching out geological articles in the early periodicals of Tennessee. W. A. S. Sargeant, Henry Faul, Dennis Dean, and Cecil J. Schneer detected errors and suggested revisions based on a preliminary version of the bibliography. We are grateful to the numerous librarians who aided in the search for rare journals and texts. The majority of the references were examined in one of the following libraries: Department of Geological Sciences Library, Museum of Comparative Zoology Library, and Widener Library, Harvard University; Baker Library, Harvard Business School; Countway Library, Harvard Medical School; Boston Public Library; Boston Athenaeum; Massachusetts Historical Society; American Antiquarian Society; New York Public Library; Library of Congress; University of Maryland Library; National Library of Medicine; Department of Agriculture Library; and the U. S. Geological Survey Library.

Cindi Finger accomplished the formidable task of typing and editing the reference and index data with efficiency, accuracy, and enthusiasm. The bibliography was typed onto magnetic disks and is currently in storage at the Geophysical Laboratory of the Carnegie Institution of Washington, D. C. The production of the bibliography was greatly simplified by Dr. Larry W. Finger, Carnegie Institution of Washington, who devised and implemented procedures for the word processing, editing, indexing, and printing of the bibliography and index portions of this volume. The bibliography and index were generated on a letter quality computer printer.

Users of this bibliography who discover new references or who detect errors are requested to contact the authors at Geophysical Laboratory, 2801 Upton St. N. W., Washington, D. C. 20008.

ROBERT M. HAZEN
MARGARET HINDLE HAZEN

AMERICAN GEOLOGICAL LITERATURE,
1669 to 1850

INTRODUCTION

A rich literary heritage has been left to historians of American geology. Commencing in 1669 with a Cambridge publication describing an eruption of Mount Etna, American writings on geology increased and diversified over the years. Because the earth—its topography, composition, and riches—was of paramount importance to America's inhabitants, the literature of geology as a whole reflects the basic economic, intellectual, and social interests of an expanding nation.

Growth, a key concept in American history, is also a striking aspect of the development of American geological literature. Figures 1 (1660–1800) and 2 (1800–1850) illustrate the number of bibliographic references versus year of publication. In spite of year-to-year fluctuations, a steady increase in publication rate is shown, and the total number of publications approximately doubled every ten years. (Note that the vertical scale in figure 2 is ten times that of figure 1.) For the year 1850, almost nine hundred references are recorded in the bibliography. Several events that were significant for the publication of geology in the United States are reflected in figures 1 and 2. In 1727 and 1755, major earth tremors occurred in New England, and many "earthquake sermons" and descriptive accounts followed the shocks. Other important dates include the first years of publication of the American Philosophical Society *Transactions* (1769–1771), the American Academy of Arts and Sciences *Memoirs* (1785), *The Medical Repository* (1798), *the American Mineralogical Journal* (1810-1814), and the *American Journal of Science* (1818). However, the dates of publication of these journals show on the graph as relatively small peaks when superimposed on the continuous increase of geological publications.

Predictably the number of places publishing American geo-

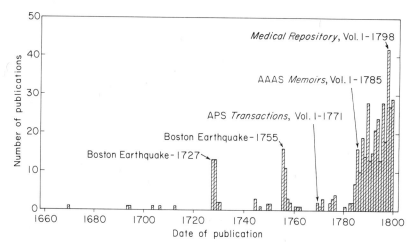

Figure 1 The growth of earth-science publications in the United States from 1660 to 1800.

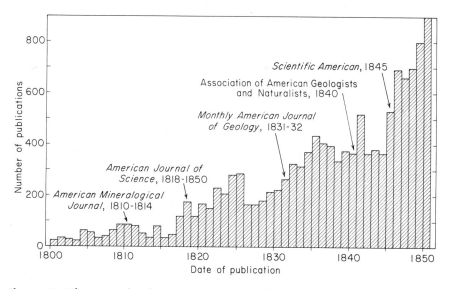

Figure 2 The growth of earth-science publications in the United States from 1800 to 1850. (Note that the vertical scale in figure 2 is ten times that in figure 1.)

logical literature increased markedly over time. In the years before 1769, the first year of publication of the American Philosophical Society's *Transactions*, New England printers, especially those in Boston, produced the vast majority of American earth-science works. This regional predominance can be explained largely by the combination of New England's religious climate and its susceptibility to earthquakes, which led to the extensive publication of earthquake descriptions and sermons. Only New York City and Philadelphia contributed additional earth-science works in this early period, and imprints from these cities are rare. With the establishment of the Republic following the Revolution, geological works were published in increasing numbers throughout the United States. In the late eighteenth century, the number of geological tracts published in New York City and Philadelphia expanded rapidly, and other cities, including Baltimore, Washington, and Charleston, began publishing in the field for the first time. As westward expansion commenced and settlers crossed the Appalachians, new centers of culture and commerce boasted printers who produced, among other publications, geological books and articles. In the 1820s and 1830s, printers in Lexington and Louisville, Kentucky, and numerous Ohio towns were regularly publishing geological literature; by the 1840s, Chicago and New Orleans imprints were established. Although the increasing number of towns publishing geological literature generally reflects patterns of settlement and the growth of cultural centers in early America, occasionally an area of special geological interest gave rise to publications on a specific topic. Printers in Saratoga Springs, New York, for example, published numerous descriptions of the mineral springs in the area. Thus the settlement of America and the associated discoveries of new geological features promoted the development of new centers for the dissemination of geological literature.

The subject matter of the early writings on geology was varied. Individual authors had different interests and training, and these are mirrored in their publications: physicians wrote on mineral waters, theologians discussed the divine origin of earthquakes, and chemists analyzed the soil. The following sections discuss several topics that appear frequently in the literature of geology. These topics reflect modern as well as historical subdivisions. For convenience in citing references, Hazen and Hazen (HH) numbers have been provided.

MINING IN AMERICA

Before the Revolutionary War, there was little effort in America to publish essays on mining techniques or ore processing. In fact, Jared Eliot's *An Essay on . . . Making Very Good, If Not the Best Iron, from Black Sea Sand* (1762, HH3475) and a Pennsylvania German edition of Alvaro Barba's seventeenth-century treatise on mining (1763, HH1483) are the only pre-1776 contributions. During the two decades following the war, several more mining pamphlets were published. Undoubtedly most of the knowledge of mining and smelting was passed by example from one worker to the next. Still these early sources provide some insight to young America's mineral industry.

U.S. mining became big business in the early nineteenth century, and corporate publications include a wealth of details on mine geology, techniques, and production statistics. Although approximately three hundred mining company reports are included in the bibliography, these must constitute only a modest fraction of the total body of these ephemeral publications. The Lehigh Coal and Navigation Company alone issued more than thirty reports between 1824 and 1850. Most pre-1850 mining efforts, and consequently most mining publications, were concentrated in five regions: the gold mines of Virginia, the Carolinas, and Georgia; the lead mines of Missouri; the copper mines of the Lake Superior district; the coal fields of Pennsylvania and Maryland; and the gold mines of California. Each of the regions was the subject of privately sponsored mineral surveys by prominent geologists of the day. Benjamin Silliman, C. T. Jackson, James Hall, and the Rogers brothers were all authors of mine surveys. In addition, a new breed of mining geologist, including R. C. Taylor, W. R. Johnson, and Henry King, produced detailed studies of mineral regions.

Additional valuable sources of data for historians of American mining are the commercial and mining periodicals of the nineteenth century. *De Bow's Review, Hunt's Merchant's Magazine, Hazard's Register,* and *Fisher's National Magazine* contain regular summaries of world mine production. Even more detailed statistics, as well as mining news and advertisements, are to be found by the 1840s in the *American Railroad Journal* and the *Mining Journal and American Railroad Gazette*. These mining publications constitute a large body of little-known literature that reflects the growing wealth and westward expansion of the nation.

AGRICULTURAL GEOLOGY

Early America was predominantly an agrarian society, and the application of geological principles to agriculture received widespread attention. The principal contribution of earth scientists to farming was in the area of soil analysis and improvement. Two early essays on the use of gypsum to enhance crop yield were George Logan's *Agricultural Experiments on Gypsum* (1797, HH6466) and Richard Peters's *Agricultural Enquiries on Plaister of Paris* (1797, HH8247), both published in Philadelphia. Efforts to approach agriculture in more scientific ways were furthered by published transactions of the Connecticut Society for Promoting Agriculture (1802), the Massachusetts Agricultural Society (1793–1850), the New York Society for the Promotion of the Useful Arts (1792–1819), and the Philadelphia Society for Promoting Agriculture (1808–1826).

These earliest American publications on scientific agriculture reached a very limited audience. The average farmer was not interested in urban intellectual societies or in scholarly essays. A far wider readership was reached by the first of the popular agricultural journals, the *American Farmer* (1819–1850), which publicized the need for planned agricultural development of American land. Many other popular periodicals for farmers followed, including Edwin Ruffin's influential *Farmer's Register* (1835–1844), which contained nearly two hundred notes and articles on soil analysis and the use of fertilizers in its eleven volumes. By 1850 more than a dozen farming journals, such as the *Michigan Farmer*, the *Boston Cultivator*, the *Maine Farmer*, and the *Delaware Register and Farmer's Magazine*, served the special local interests of farmers in America by emphasizing the importance of soil chemistry.

That Americans attached great importance to scientific agriculture is perhaps best revealed by the number of agricultural textbooks and primers available in 1850. Foremost among these were the works of J. F. W. Johnston, a British agricultural chemist and geologist, who made popular lecture tours through the United States in the 1840s. Johnston's *Lectures on Agricultural Chemistry and Geology* (1842–1843, HH5818-5819) was reprinted several times and was reviewed in more than a dozen agricultural periodicals of the day. Even schoolchildren were exposed to Johnston's ideas on soil chemistry through his *Catechism of Agricultural Chemistry and Geology* (1845, HH5825). These texts were supplemented by M. M. Rodger's *Scientific Agriculture* (1848, HH8829), Andrew Ure's *Dictionary of Arts* (1842, HH10532), and other practical guidebooks for American farmers. Through such publications, nineteenth-century geologists effectively aided America's principal economic activity.

RELIGION AND GEOLOGY

Colonial American clergymen, responsible for the moral education and religious guidance of their congregations, frequently linked worldly events with God's purposes. It is not surprising, therefore, that the New England ministers proposed religious explanations for the Boston earthquakes of 1705, 1727, and 1755. The earliest American-published earthquake sermons ascribed to divine judgment the origins of shocks in London (Doolittle, 1693, HH3201) and Jamaica (Corbin, 1703, HH2641). A similar explanation was given for the New England earthquakes by virtually all authors of the more than forty pamphlet sermons that followed these events. The gravity with which New Englanders considered these shocks is reflected in a typical sermon title: *The Judgements of Providence in the Hand of Christ: His Voice to Us in the Terrible Earthquake. And, the Earth Devoured by the Curse* (Colman, 1727, HH2486).

Religious beliefs continued to figure prominently in discussions of geological phenomena. In the nineteenth century, the work of many geologists posed serious threats to a purely biblical interpretation of the earth's history. For example, Charles Lyell and other uniformitarianists supported the concept of a very ancient earth, in direct conflict with the scriptural account in Genesis. In addition, evidence of continental glaciation put forth by Louis Agassiz and others explained many topographical features previously attributed to the action of the Noachian deluge and thus called into question the very occurrence of the Flood. Discoveries of extinct fossil animals by Cuvier and his colleagues further eroded confidence in a highly ordered, recently created world. These challenges to long-held beliefs were met in various ways. Many attempts were made to rationalize geological observations within the context of religious beliefs. Edward Hitchcock wrote on "The Connection between Geology and natural religion" (1835, HH5107), and Benjamin Silliman published *Consistency of the Discoveries of Modern Geology, with the Sacred History of the Creation and the Deluge* (1833, HH9712). But, religious journals were quick to criticize any publications—even those that tried to accommodate new scientific discoveries to the Bible—that refuted basic theological doctrine. The *Christian Review, Biblical Repository,* and *Methodist Review* all contained such articles. For many years, as the science of geology grew in America, numerous debates occurred that contrasted religious ideas with geological theories.

MEDICINE AND GEOLOGY

Many of the earliest natural scientists in America were educated as medical doctors, some of whom applied their geological discoveries to the improvement of health. Their research often focused on the origin, composition, and beneficial effects of mineral waters and springs. As early as 1792, John Rouelle produced *A Complete Treatise on the Mineral Waters of Virginia* (HH9040), and in 1793 there appeared the controversial *Dissertation on the Mineral Waters of Saratoga* by Valentine Seaman (HH9383).* Through the efforts of these authors and others, the mineral fountains of western Virginia and Saratoga Springs, New York, became the focus of major health resorts, where guides to the use of medicinal springs were popular publications; one was P. H. Nicklin's *Letters Descriptive of the Virginia Springs* (1835, HH7865). Although mineral springs in New York and Pennsylvania attracted the most attention, articles on natural medicinal wells in Kentucky, New Jersey, Ohio, Pennsylvania, and other states reflected the widespread interest in these useful phenomena.

Medical topography, the influence of local rock formations on the prevalence of disease, occupied the research of several prominent early American scientists and physicians. Samuel Latham Mitchill was one of the first Americans to champion the theory that regions with limestone bedrock provided healthier places to live than those underlain by shale. His "Outlines of medical geography" (1799, HH7276) was followed by a series of essays on the medical topography of various parts of New York state. Mitchill's ideas were applied to the western communities of Cincinnati, Ohio, and Louisville, Kentucky, by Daniel Drake ("Medical Topography," 1816, HH3219). As late as 1831, the concept of the "Connexion of disease with the rock formations of a country" (HH1022) was still publicized, indicating a long-lasting partnership between medicine and geology.

SCIENTIFIC GEOLOGY

The science of geology, including both theoretical and field study, is by far the best-documented and most-studied aspect of American geological history. However, even in this area of investigation, there exists a large number of little-known references. Few scientific publications appeared in the United States before 1800, but three important works have received little analysis. The first theoretical earth-science study by an American was John

* Seaman participated in a prolonged debate with David Hosack on the efficacy of New York medicinal waters. This controversy culminated seventeen years later with Hosack's "Answer to Dr. Seaman's "Examination" of a review of his Dissertation on the mineral waters of Saratoga" (1810, HH5282).

Winthrop's *Lecture on Earthquakes* (1755, HH11002), which sought a physical explanation for observations made during the 1755 Boston quake. S. L. Mitchill's *Observations . . . to Which Are Added Geological Remarks on the Maritime Parts of New York* (1787, HH7264), is not cited in standard Mitchill bibliographies, even though it contains what may be the earliest use of the word *geology* in the United States. A third neglected eighteenth-century source is the unsigned *Compendius System of Mineralogy and Metallurgy* (1794, HH2502), based on the system of Swedish chemist Axel Cronstedt. This text was the first systematic treatise on earth materials published in America.

Throughout the first half of the nineteenth century, the most significant contributions to the earth sciences were made in Europe. Americans subsequently learned of foreign advances through correspondence, lecture tours by Europeans, and republication of European texts. U.S. publishers were quick to issue American editions of most major European geological works, including those of Bakewell, Buckland, Cuvier, and Lyell. Just as important were the numerous abstracts of European scientific journals to be found in *Medical Repository*, the *American Journal of Science*, the *Museum of Foreign Literature and Science*, and dozens of other periodicals. Little attention has been directed toward the transfer of European ideas through these American publications.

Magnetic studies are another type of scientific publication to receive little attention from historians. As early as 1771, John Lorimer noted the need for systematic observations of magnetic dip and intensity (HH6501). By the 1840s, through the efforts of such workers as W. C. Bond, A. D. Bache, John Locke, and Joseph Lovering, accurate magnetic charts were made available to scientists and travelers alike.

Prominent geologists occasionally published reports of original research in unlikely or obscure periodicals. For reasons not immediately apparent, H. H. Hayden chose to publish his longest geological article in the *Methodist Recorder* ("Inquiry into some of the geological phenomena to be found in various parts of America, and elsewhere," 1824, HH4880). Similarly Amos Eaton published in the *Massachusetts Agricultural Recorder* ("A geological and agricultural survey of Rensselaer County," 1824, HH3339), James Eights in *Zodiac* ("Synopsis of the rocks of New York," 1835, HH3463), and Ebenezer Emmons in the *Monthly Chronicle* ("Iron ore at Duane, New York," 1841, HH3503) and the *American Magazine and Repository of Useful Knowledge* ("Geological observations," 1842, HH3509). Other lesser-known journals with contributions by well-known geologists include *Knickerbocker* (H. R. Schoolcraft), *Columbian Magazine* (S. L. Mitchill), and the *Journal of Science and the Arts* (W. Maclure). Such obscure articles must be considered in any thorough assessment of the contributions of these geologists.

BOOK REVIEWS

Several hundred American earth-science reviews are important pre-1851 sources for the historian of geology. They call attention to the geological publications of the day and provide useful summaries of the contents of these publications. More importantly, these reviews, though frequently written anonymously, afford a useful measure of the popularity and acceptability of early geological theories.

Virtually every major geological publication in the United States was reviewed at least once. (Book reviews are noted following book entries in this bibliography.) The 1804 edition of Volney's *View of the Soil and Climate of the United States* (HH10653) was the subject of some of the earliest reviews of geological ideas (see *Literary Miscellany*, 1805, HH6326). Other pre-1820 works to receive notice include William Maclure's *Observations on the Geology of the United States* (HH6615, 6616 and 6620), the three editions of which prompted eight reviews, and Parker Cleaveland's *Elementary Treatise on Mineralogy and Geology* (1816, HH2420), which was praised in a dozen periodicals. A few influential or controversial works, such as Robert Chamber's *Vestiges of the Natural History of Creation* (1845, HH2292), provoked as many as twenty reviews.* Charles Lyell, an extremely popular English author, was the most reviewed geologist in American periodicals with nearly forty discussions of his several books published before 1850.

As the rate of geological publication increased, reviewers found it more convenient to consider several recent texts together in one essay. All of the review journals (*North American Review, American Whig Review, Southern Review, New York Review,* and others) as well as many literary periodicals followed this procedure, providing useful comparisons of differing geological theses. Typical was the 1830 article in the *Southern Review* (HH9969) that analyzed Bakewell's *Introduction to Geology,* Silliman's *Geological Lectures,* Ure's *New System of Geology,* Brande's *Outlines of Geology,* and the rock classification scheme of De La Beche.

* Chambers's elaborate but unsupported theory of evolution was severely criticized by scientists and theologians alike.

THE POPULARIZATION OF GEOLOGY

Geology was of importance to many Americans; farmers, miners, physicians, and theologians all had reason to study aspects of the earth sciences in their daily lives. Yet geology appealed to an even wider audience as a result of the compelling curiosity with which people view their world. The imagination has always been excited by natural phenomena, especially earthquakes and volcanoes, among the most awe-inspiring events of nature. The first earth-science publication in North America was *A True and Exact Relation of the Late Prodigious Earthquake and Eruption of Mount Aetna* by Heneage Finch (1669, HH3825). In fact, this pamphlet was one of the first 150 publications of any kind in North America (Evans, 1903-1934). Similar descriptive accounts of earthquakes and volcanoes were published frequently throughout the next 180 years.

Hundreds of popular American journals contained short articles and notes on subjects of geological interest. Discoveries of new mines or unusual fossils were publicized widely, as were the occurrences of volcanoes and earthquakes. The basic principles of geology were also described in the popular media, where the theories of Lyell, Cuvier, Agassiz, and others received widespread circulation. Well-known geologists sometimes wrote expressly for the public at large. Gerard Troost, state geologist of Tennessee, published several such articles, including three versions of "Bones of the gigantic mastodon" (1833, HH10359-10361). G. A. Mantell's *Wonders of Geology* was an immensely popular book both in the original English edition and in the American reprint (1839, HH6732). Perhaps the most widely read American published geology texts were from the eloquent pen of Charles Lyell, whose *Elements of Geology* (1839, HH6533), *Eight Lectures on Geology* (1842, HH6540), and two travel accounts of the United States stand today as classics of popular geological literature.

By the 1820s, geological publications had reached a new and younger audience. Young people's periodicals, including the *Juvenile Miscellany*, the *Juvenile Rambler*, and *Scientific Tracts and Family Lyceum*, contained elementary descriptions of rocks, minerals, and geological concepts, and several primers in the earth sciences further testified to the growing popularity of geology. Delvalle Lowry's *Conversations on Mineralogy* (1822, HH6514), Josiah Holbrook's *First Lessons in Geology* (1833, HH5218), and S. G. Goodrich's *The Child's Geology* (1832, HH4442) are three typical examples of this juvenile literature. Even standard school geographies added sections on the earth sciences. *A Sys-*

tem of Universal Geography by W. C. Woodbridge (1824, HH11038) is notable for containing a woodcut version of William Maclure's geological map of the United States.

By the mid-nineteenth century, published information about geological phenomena was widely disseminated. Analysis of the geological literature that reached the public at large reveals much about the nature and growth of earth-science inquiry in the United States.

MISCELLANEOUS REFERENCES

Several types of miscellaneous references enrich the American earth-science literature. Activities of prominent geologists were often newsworthy. Consequently notices of lectures, field excursions, and publications were commonly printed. Obituaries memorialized the contributions of many earth scientists, both American and foreign. Catalogs of libraries and fossil and mineral collections provide insights into the nature of study matrials available to researchers, and society minutes and notices record the degree of interaction among contemporary workers. Lithographs of mines, minerals, and natural wonders and steel-plate engravings of prominent geologists give historians a source of data on how the earth and its investigators were perceived. Perhaps the oddest examples of earth science literature were the several geological poems of the pre-1851 years. Poetry on the formation of caves and rocks, the perfection of mineral crystals, and "The Geologist's Wife" (1847, HH6395) are to be found in early American periodicals.

CONCLUSION

From the first account in 1669 to the hundreds of 1850 publications, the changing intellectual and social climate, westward expansion, and growing prosperity of the United States are mirrored in the extensive earth-science literature. The lives of most Americans were affected by geological developments in scientific, industrial, agricultural, medical, religious, or popular contexts, and this, too, is reflected in the diversity of publications on geology. It is hoped that consideration of the complete scope of American earth-science publications will lead to further understanding of the founding and development of geology in the United States.

SECONDARY REFERENCES

The following secondary sources contain information on geologic literature published in America that has been incorporated into this bibliography.

Aalto, K. R. 1969. Specialization and professionalization of pre-Civil War North American geology. *J. Geol. Education* **17**:91–94.

Barbour, E. H., and C. A. Fisher. 1902. The geological bibliography of Nebraska. *Nebraska State Board of Agriculture Ann. Rept. 1901*, pp. 248–266.

Bates, R. L., and M. R. Burks. 1945. Geologic literature of New Mexico. *New Mexico State Bureau of Mines and Mineral Resources Bulletin 27*, 147p.

Bell, W. J. 1939. A box of old bones: a note on the identification of the mastodon, 1766–1806. *Am. Phil. Soc. Proc.* **93**:169–177.

Bennett, W. A. G. 1939. Bibliography and index of geology and mineral resources 1814–1836. *Washington Dept. of Conservation and Development, Division of Geology Bulletin 35*, 140p.

Bigelow, E. L. 1966. *A Taconic Bibliography; a List of References on the Taconic Problem.* Burlington: Vermont Academy of the Arts and Sciences, 6p.

Bovee, G. G. 1918. Bibliography and index of Wyoming geology 1823–1916. *Wyoming Geological Survey Bulletin 17*, pp. 317–446.

Branner, J. C. 1894. Bibliography of the geology of Arkansas. *Arkansas Geological Survey Ann. Rept. 1891*, vol. 2, pp. 319–340.

Bristol, R. P. 1970. *Supplement to Charles Evans' American Bibliography.* Charlottesville: University Press of Virginia, 636p.

Brown, S. B. 1901. A bibliography of works upon the geology and natural resources of West Virginia from 1764 to 1901. *West Virginia Geological Survey Bulletin 1*, 85p.

Budge, C. E. 1946. Bibliography of the geology and natural resources of North Dakota, 1814–1944. *North Dakota Research Foundation Bulletin 1*, 214p.

Cassidy, M. 1962. *A Partial Bibliography of the Geology of Massachusetts Through 1958.* Cambridge, Mass.: Harvard University, Department of Geological Sciences, 90p.

Clarke, J. M. 1921. *James Hall of Albany, Geologist and Paleontologist, 1811–1898.* Albany: E. E. Rankin, 565p.

Cockrill, E. 1911. Bibliography of Tennessee geology and related subjects, with subject index. *Tennessee Geological Survey Bulletin 1B,* 117p.

Corgan, J. 1975. The Naturalist, 1846. Typed manuscript, 3p.

Cramer, H. R. 1959. Annotated bibliography of Georgia geology through 1959. *Georgia Geological Survey Bulletin 79,* 368p.

Cramer, H. R. 1961. Annotated bibliography of Pennsylvania geology to 1949. *Pennsylvania Topographical and Geological Survey Bulletin G34,* 435p.

Darton, N. H. 1896. Catalogue and index of contributions to North American geology, 1732–1891. *U.S. Geological Survey Bulletin 127,* 1045p.

Dixon, D. E. 1926. *Bibliography of the Geology of Oregon.* University of Oregon Publications, Geology Series, vol. 1. Eugene: University of Oregon, 125p.

Ellis, M. 1903. Index to publications of the New York State Natural History Survey and New York State Museum, 1837–1902. *New York State Museum Bulletin 66,* 653p.

Evans, C. 1903–1934. *American Bibliography,* vols. 1–12. (Chicago: Blakely Press); Shipton, C. K. 1955. *Charles Evans' American Bibliography,* vol. 13. Worcester, Mass.: American Antiquarian Society.

Fitzpatrick, T. J. 1911. *Rafinesque: A Sketch of His Life, with Bibliography.* Des Moines, Iowa: Historical Department of Iowa, 241p.

Fuller, G. N., ed. 1928. *Geological Reports of Douglass Houghton, First State Geologist of Michigan, 1837–1845.* Lansing, Mich.: Michigan Historical Commission, 700p.

Fulton, J. F., and E. H. Thomson. 1947. *Benjamin Silliman, 1779–1864, Pathfinder in American Science.* New York: Henry Schuman, 294p.

Gianella, V. P. 1945. Bibliography of geologic literature of Nevada and bibliography of geologic maps of Nevada areas. *Nevada Geology and Mining Series 43,* 205p.

Grametbaur, A. B. 1946. Bibliography and index of the geology of New Jersey. *New Jersey Department of Conservation and Development Bulletin 59,* 142p.

Greene, J. C., and J. G. Burke. 1978. The science of minerals in the age of Jefferson. *Am. Philos. Soc. Trans.* **68**(4):1–113.

Gregor, D. K. 1945. Bibliography of Missouri geology. *Missouri Geological Survey and Water Resources 31* (2d series), 294p.

Gregory, H. E. 1907. Bibliography of the geology of Connecticut. *Connecticut Geological Survey Bulletin 8,* 123p.

Hall, C. R. 1934. *A Scientist of the Early Republic: Samuel Latham Mitchill, 1764–1831.* New York: Columbia University Press, 162p.

Harper, R. M. 1935. Bibliography of Alabama geology. *Alabama Geological Survey Bulletin 42,* pp. 59–108.

Haskell, D. C. 1942. *The United States Exploring Expedition, 1838–1842 and Its Publications, 1844–1874.* New York: New York Public Library, 188p.

Hazen, R. M. ed. 1979. *North American Geology: Early Writings.* Stroudsburg, Pa.: Dowden, Hutchinson & Ross, 357p.

Ireland, H. A. 1943. History of the development of geologic maps. *Geol. Soc. America Bull.* **54**:1227–1280.

Jillson, W. R. 1923. Geological research in Kentucky: A summary account of the several geological surveys of Kentucky including a complete list of their publications. *Kentucky Geological Survey 15,* Series 6, 228p.

Jillson, W. R. 1966. *A Bibliography of the Mineral Resources of Kentucky, 1818–1965.* Frankfort, Ky.: Roberts Printing Co., 66p.

Jones, O. M. 1914. Bibliography of Colorado geology and mining, with subject index from the earliest exploration to 1912. *Colorado Geological Survey Bulletin 7,* 493p.

Kemble, E. C. 1962. *A History of California Newspapers, 1846–1858.* Los Gatos, Calif.: Talisman Press, 398p.

Keyes, C. R. 1893. Bibliography of Iowa geology. *Iowa Geological Survey Annual Report 1892,* pp. 209–464.

Keys, C. 1925. William Maclure, father of modern geology. *Pan-American Geologist* **44**:81–94.

Laney, F. B., and K. H. Wood. 1909. Bibliography of North Carolina geology, mineralogy and geography, with a list of maps. *North Carolina Geological Survey Bulletin 18,* 428p.

Leighton, H. 1909. One-hundred years of New York state geologic maps: 1809–1909. *New York State Museum Bulletin 113,* pp. 115–155.

Long, H. K. 1971. *A Bibliography of Earth Science Bibliographies.* Washington, D.C.: American Geological Institute, 19p.

McAllister, E. M. 1941. *Amos Eaton: Scientist and Educator, 1776–1842.* Philadelphia: University of Pennsylvania Press, 587p.

Martin, H. M., and M. T. Straight. 1956. An index to the geology of Michigan, 1823–1955. *Michigan Geological Survey Publication 50,* 461p.

Mather, K. F., and S. L. Mason. 1970. *A Source Book in Geology, 1400–1900.* Cambridge, Mass.: Harvard University Press, 702p.

Mathews, E. B. 1897. Bibliography and cartography of Maryland, including publications on physiography, geology, mineral resources. *Maryland Geological Survey Report 1,* pp. 229–332.

Meisel, M. 1924, 1926, 1927. *A Bibliography of American Natural History: The Pioneer Century, 1769–1865,* vols. 1–3. New York: Premier Publishing Co.

Melone, T. G., and L. W. Weis. 1951. Bibliography of Minnesota geology. *Minnesota Geological Survey Bulletin 34,* 124p.

Merrill, G. P. 1904. Contributions to the history of American geology. *U.S. National Museum Report,* Part 2, pp. 187–733.

Merrill, G. P. 1920. Contributions to a history of American state geological and natural history surveys. *U.S. National Museum Bulletin 109,* 549p.

Merrill, G. P. 1924. *The First One Hundred Years of American Geology.* New Haven: Yale University Press, 773p.

Moore, J. P. 1947. William Maclure: Scientist and humanitarian. *Am. Philos. Soc. Proc.* **91**:234–249.

Moore, R. T., and E. D. Wilson. 1965. Bibliography of the geology and mineral resources of Arizona, 1848–1964. *Arizona Bureau of Mines Bulletin 173*, 321p.

National Union Catalog: Pre-1956 Imprints. 1968–. 640 volumes to date. London: Mansell.

Nevers, G. M., and R. D. Walker. 1894. Geological literature of Indiana. *Indiana Academy of Science Proceedings 1893*, pp. 156–191.

Nickles, J. M. 1923. Geologic literature on North America, 1785–1918. *U.S. Geological Survey Bulletin 746*. 1167p.

Parmelee, G. 1960. *A South Dakota Bibliography—Science and Applied Science.* Sioux Falls: South Dakota Library Association, 24p.

Perkins, G. H. 1902. List of reports on the geology of Vermont. *Vermont State Geologist Report 3*, pp. 14–21.

Pestana, H. R. 1972. *Bibliography of Congressional Geology.* New York: Hafner Publishing Co., 285p.

Petty, J. J. 1950. Bibliography of the geology of the state of South Carolina. University of South Carolina Publications, Series 11, *Physical Sciences Bulletin 1*, 86p.

Quinn, A., and D. H. Swann. 1950. *Bibliography of the Geology of Rhode Island.* Providence: Rhode Island Port and Industrial Commission, 26p.

Rice, W. N. 1907. The contributions of America to geology. *Science* **2**: 161–175.

Roberts, J. K. 1942. Annotated geological bibliography of Virginia. *University of Virginia Bibliographic Series 2*, 726p.

Ruhle, J. L. 1965. Geological literature of the coastal plain of Virginia, 1783–1962. *Virginia Division of Mineral Resources Information Circular 9*, 95p.

Sabin, J. 1961. *A Dictionary of Books Relating to America, from Its Discovery to the Present Time*, reprint ed., 29 volumes. Amsterdam: N. Israel.

Schneer, C. J. 1969a. Ebenezer Emmons and the foundations of American geology. *Isis* **60**(204):439–450.

Schneer, C. J. 1969b. *Towards a History of Geology.* Cambridge, Mass.: MIT Press, 469p.

Sellards, E. H. 1908. Bibliography of Florida geology. *Florida Geological Survey Annual Report 1*, pp. 73–108.

Sellards, E. H. 1933. Bibliography and subject index of Texas geology. *Texas University, Bureau of Economic Geology Publication 3232*, pp. 819–996.

Shaw, R. R., and R. H. Shoemaker. 1958–1965. *American Bibliography: A Preliminary Checklist*, 19 volumes, Addenda. New York: Scarecrow Press.

Shoemaker, R. H. 1964–. *Checklist of American Imprints, 1820–.* New York: Scarecrow Press, in progress.

Simpson, G. G. 1942. The beginnings of vertebrate paleontology in North America. *Am. Philos. Soc. Proc.* **86**:130–188.

Simpson, G. G. 1943. The discovery of fossil vertebrates in North America. *J. Paleontology* **17**:26–38.

Swem, E. G. 1919. An analysis of Ruffin's farmer's register, with a bibliography of Edmund Ruffin. *Virginia State Library Bulletin* **11**(3-4): 39–144.

Titus, E. B., ed. 1965. *Union List of Serials in Libraries of the United States and Canada*, 3d ed., 5 volumes. New York: H. W. Wilson, 4649p.

Trout, L. E., and G. H. Myers. 1915. Bibliography of Oklahoma geology with subject index. *Oklahoma Geological Survey Bulletin 25*, 105p.

Twinen, J. C. 1932. *Bibliography on the Geology of Maine, 1836-1930*. Augusta: Maine Geological Survey, 92p.

Ver Wiebe, W. A., and B. Cooper. 1938. Bibliography of Kansas geology, 1823-1938. *Municipal University of Wichitaw Bulletin 13*, 76p.

Vogdes, A. W. 1904. A bibliography relating to the geology, paleontology, and mineral resources of California. *California State Mine Bureau Bulletin 30*, 121p.

Von Zittel, K. 1962. *History of Geology and Paleontology*. New York: Hafner Publishing Co., 562p.

Waldron, C. R., and R. H. Earhard. 1942. Bibliography of the geology and mineral resources of Montana. *Montana Bureau of Mines and Geology Memoirs 21*, 356p.

Watkins, D. G. 1953. Bibliography of Ohio geology, 1879-1950. *Ohio Geological Survey Bulletin 52*, 103p.

Wells, J. W. 1959. Notes of the earliest geological maps of the United States, 1756-1832. *J. Washington Acad. Sci.* **49**:198–205.

Wells, J. W. 1963. Early investigations of the Devonian system in New York, 1656-1836. *Geological Society of America Special Paper 74*, 74p.

White, G. W. 1951. Lewis Evans' contributions to early American geology, 1743-1755. *Illinois Acad. Sci. Trans.* **44**:152–158.

White, G. W. 1973a. The history of geology and mineralogy as seen by American writers, 1803-1835: A bibliographic essay. *Isis* **64**(222): 197–214.

White, G. W. 1973b. History of investigation and classification of Wisconsinan drift in north-central United States. *Geological Society of America Memoir 136*, pp. 3–34.

Willman, H. B. 1968. Bibliography and index of Illinois geology through 1965. *Illinois State Geological Survey Bulletin 92*, 373p.

Wright, C. 1941. The religion of geology. *New England Quarterly* **14**: 335–358.

Xerox University Microfilms. 1975. *American Periodicals, 1700-1900: A Consolidated Bibliography*. Ann Arbor, Mich.: Xerox University Microfilms, 30p.

PERIODICALS CONSULTED

Below are listed the title, place of publication, and dates of publication (through 1850) of each periodical consulted. Journals in which no earth-science material was found are preceded by an asterisk (*).

*The ————— —————, Washington, D.C., 1826.
The academician, containing the elements of scholastic science . . ., New York, 1818–1819.
Academy of Natural Sciences of Philadelphia, see Philadelphia Academy of Natural Sciences.
*The adviser, or Vermont evangelical magazine, Middlebury, Vt., 1809.
The advocate of science, and annals of natural history, Philadelphia, 1834–1835.
*The aeronaut: A periodical paper by an association of gentlemen, New York, 1816–1819.
The Aesculapian register, Philadelphia, 1824.
*The African observer, Philadelphia, 1828.
The African repository and colonial journal, Washington, D.C., 1825–1850.
Agricultural intelligencer, and mechanic register, Boston, 1820.
The agricultural museum, Georgetown, Washington, D.C., 1811–1812.
The Agriculturist, Nashville, Tenn., 1840–1845.
*Albany bouquet; and literary spectator, Albany, 1835–1836.
Albany Institute, Transactions, Albany, 1830–1850.
Amaranth, a semi-monthly publication devoted to polite literature, science, poetry and amusement, Ashland, Ohio, 1847.
American Academy of Arts and Sciences, Memoirs of the, Boston, 1785–1850.
American Academy of Arts and Sciences, Proceedings of the, Boston, 1846–1850.
American Agricultural Association, Transactions of the, New York, 1846.
The American agriculturist, New York, 1842–1850.

The American almanac and repository of useful knowledge, Boston, 1829–1850.
*The American and Foreign Christian Union, New York, 1850.
The American annual register, New York, 1825–1833.
The American annual register, or, historical memoirs of the United States for the year 1796, Philadelphia, 1797.
*American Antiquarian Society. Proceedings of the, Worcester, Mass., 1812–1850.
American Antiquarian Society. Transactions and collections of the (i.e., Archaeologia Americana), Worcester, Mass., 1820, 1836.
American apollo, see Massachusetts Historical Society, Collections of the.
*American archives, Washington, D.C., 1837–1853.
American Association for the Advancement of Science, Proceedings of the, Philadelphia, 1849, Boston, 1850, Charleston, S.C., 1850.
American Association for the Promotion of Science, Literature and the Arts, Journal of the, New York, 1830–1831.
The American athenaeum; or, repository of the arts, sciences, and belles letters, New York, 1825.
The American Biblical repository, see The Biblical repository.
*The American botanical register, Washington, D.C., 1825–1830.
*American critic and general review, Washington, D.C., 1820.
American cultivator, see Boston cultivator.
The American eclectic; or, selections from the periodical literature of all foreign countries, New York, 1841–1842.
The American farmer, Baltimore, 1819–1850.
The American farmer's magazine, Philadelphia and New York, 1848–1850.
The American gardener's magazine and register, Boston, 1835–1850.
The American historical magazine and literary record, New Haven, 1836.
American Institute of New York, Annual reports of the, New York, 1847–1850.
American Institute of New York, Journal of the, New York, 1835–1850.
American Institute of New York, Transactions of the, New York, 1839–1850.
The American journal of agriculture and science, see American quarterly journal of agriculture and science.
*The American journal of foreign medicine, Boston, 1827.
The American journal of improvements in the useful arts and mirror of the Patent Office, Washington, D.C., 1846–1847.
The American journal of improvements in the useful arts and mirror of the Patent Office in the United States, Washington, D.C., 1828.
American journal of pharmacy, Philadelphia, 1829–1850.
The American journal of science, New Haven, 1818–1850.
The American journal of the medical sciences, Philadelphia, 1828–1850.
The American literary magazine, Albany, 1848, Hartford, 1848–1849.
The American magazine, Containing a miscellaneous collection of original and other valuable essays, in prose and verse, and calculated for both instruction and amusement, New York, 1787–1788.

The American magazine, a monthly miscellany, New York, 1815–1816.
The American magazine and historical chronicle, Boston, 1743–1746.
The American magazine and monthly chronicle for the British colonies, Philadelphia, 1757–1758.
The American magazine and repository of useful literature, Devoted to science, literature, and arts, Albany, 1841–1842.
The American magazine of useful and entertaining knowledge, Boston, 1834–1837.
The American magazine of wonders, and marvelous chronicle . . . , New York, 1809.
*The American magazine, or a monthly view of the political state of the British colonies, Philadelphia, 1741.
The American magazine, or general repository, Philadelphia, 1769.
American mechanics' magazine; containing useful original matter, on subjects connected with manufactures, the arts and sciences, New York, 1825–1826.
The American medical and philosophical register; or annals of medicine, natural history, agriculture, and the arts, New York, 1810–1814.
The American medical recorder, Philadelphia, 1818–1829.
The American medical review and journal, Philadelphia, 1824–1826.
*The American metropolitan magazine, New York, 1849.
The American mineralogical journal, Being a collection of facts and observations tending to elucidate the mineralogy and geology of the United States of America, New York, 1810–1814.
The American mining journal, see The mining journal and American railroad gazette.
*American monitor; or, republican magazine, Boston, 1785.
*The American monthly magazine, Philadelphia, 1824.
*The American monthly magazine, Boston, 1829–1831.
The American monthly magazine, New York, 1833–1838.
The American monthly magazine and critical review, New York, 1817–1819.
The American monthly review, Cambridge, 1832, Boston, 1833.
The American monthly review; or, literary journal, Philadelphia, 1795.
The American moral and sentimental magazine, New York, 1797–1798.
American museum, and repository of arts and sciences, Washington, D.C., 1822.
*American museum of literature and the arts, Baltimore, 1838–1839.
The American museum; or, annual register of fugitive pieces, ancient and modern, Philadelphia, 1799.
The American museum, or universal magazine, Philadelphia, 1787–1792.
*The American people's journal of science, literature, and art, New York, 1850.
American Philosophical Society, Proceedings of the, Philadelphia, 1838–1850.
American Philosophical Society, Transactions of the, Philadelphia, 1769–1850.

The American pioneer, a monthly periodical devoted to the objects of the Logan Historical Society, Cincinnati, 1842–1843.
The American quarterly journal of agriculture and science, Albany, 1845–1846, New York, 1846–1848.
American quarterly observer, Boston, 1833–1834.
The American quarterly register and magazine, Philadelphia, 1848–1850.
The American quarterly review, Philadelphia, 1827–1837.
American rail-road journal, New York, 1831–1850.
The American register, or, general repository of history, politics, & science, Philadelphia, 1807–1811.
The American register; or, summary review of history, politics, and literature, Philadelphia, 1817.
*The American remembrancer, Philadelphia, 1795.
The American repertory of arts, sciences, and manufactures, New York, 1840–1842.
The American repertory of arts, sciences, and useful literature, Philadelphia, 1830–1832.
*The American repository of useful information, Philadelphia, 1795–1797.
The American review, see The American Whig review.
The American review, and literary journal, New York, 1801–1802.
*The American review of history and politics, and general repository of literature and state papers, Philadelphia, 1811–1812.
The American universal magazine, see The universal magazine.
The American weekly messenger; or, register of state papers, history and politics, Philadelphia, 1813–1814.
The American Whig review, New York, 1845–1850.
The analectic magazine, Philadelphia, 1813–1820.
*The analyst; or, mathematical museum, Philadelphia, 1808–1809.
Anglo-American, a journal of literature, news, politics, the drama, fine arts, etc., New York, 1843–1847.
*Anglo-American magazine, Boston, 1843.
Annals of nature, Lexington, Ky., 1820.
The annual of scientific discovery; or, year-book of facts in science and art, Boston, 1850.
The antiquarian and general review, Schenectady, N.Y., 1845–1848.
Archaeologia Americana, see American Antiquarian Society. Transactions and collections of the.
Archives of useful knowledge, Philadelphia, 1810–1813.
Arcturus, a journal of books and opinions, New York, 1841–1842.
Army and Navy chronicle, Washington, D.C., 1835–1842.
Army and Navy chronicle, and scientific repository, Washington, D.C., 1843–1844.
Arthur's magazine, Philadelphia, 1844–1846.
Association of American Geologists and Naturalists, Proceedings of the annual meetings, New York and New Haven, 1840–1846. The proceedings of the fourth annual meeting were not found in a separate edition. They are located in American Journal of Science, volume 45.

*The astronomical journal, Cambridge, 1849–1850.
*The atheneum, New Haven, 1814.
Atlantic journal and friend of knowledge, Philadelphia, 1832–1833.
The atlantic magazine, New York, 1824–1825.
The Baltimore literary monument, Baltimore, 1839.
*The Baltimore magazine for July, 1807, Baltimore, 1807.
The Baltimore medical and philosophical lyceum, Baltimore, 1811.
The Baltimore medical and physical recorder, Baltimore, 1808–1809.
Baltimore medical and surgical journal, Baltimore, 1833–1834.
*The Baltimore monthly journal of medicine and surgery, Baltimore, 1830–1831.
*The Baltimore monthly visitor, Baltimore, 1842.
*The Baltimore monument; devoted to polite literature, science, and the arts, Baltimore, 1836–1838.
*The Baltimore philosophical journal and review, Baltimore, 1823.
The Baltimore repertory, on papers on literary and other topics, Baltimore, 1811.
*The Baltimore weekly magazine, Baltimore, 1800–1801.
The biblical journal, Boston, 1842.
Biblical repertory, Princeton, N.J., 1825–1850.
The biblical repository, Andover, Mass., 1831–1850.
The Boston cultivator, Boston, 1839–1850.
Boston journal of natural history, containing papers and communications read to the Boston Society of Natural History, Boston, 1834–1850.
The Boston journal of philosophy and the arts, exhibiting a view of the progress of discovery in natural philosophy, mechanics, chemistry, geology, and mineralogy; natural history, comparative anatomy, and physiology; geography, statistics, and the fine and useful arts, Boston, 1823–1826.
The Boston literary magazine, Boston, 1832–1833.
The Boston lyceum, Boston, 1827.
The Boston magazine, Boston, 1783–1786.
The Boston magazine, Boston, 1802–1806.
The Boston mechanic and journal of the useful arts and sciences, Boston, 1832–1836.
Boston medical intelligencer, Boston, 1823–1828.
The Boston miscellany of literature and fashion, Boston, 1842–1843.
The Boston monthly magazine, Boston, 1825–1826.
The Boston quarterly review, Boston, 1838–1842.
Boston Society of Natural History, Proceedings of the, Boston, 1841–1850.
The Boston spectator; devoted to politicks and belles-lettres, Boston, 1814–1815.
*The Boston weekly magazine, Boston, 1743.
The Boston weekly magazine, devoted to morality, literature, biography, history, the fine arts, agriculture, &c. &c., Boston, 1802–1806.
The Boston weekly magazine, Boston, 1815–1824.
Boston weekly magazine, Boston, 1838–1841.

Bowen's Boston news-letter and city record, Boston, 1825–1827.
Brownson's quarterly review, Boston, 1844–1850.
Buchanan's journal of man, Cincinnati, 1849–1850.
*The bulletin of medical science, Philadelphia, 1843–1846.
*The bureau; or repository of literature, politics, and intelligence, Philadelphia, 1812.
Burton's gentleman's magazine, Philadelphia, 1837–1840.
*The cabinet; a repository of polite literature, Boston, 1811.
The cabinet of natural history, and American rural sports with illustrations, Philadelphia, 1830–1833.
The cabinet of religion, education, literature, science, and intelligence, New York, 1830–1831.
Cambridge miscellany, Cambridge, 1842.
Campbell's foreign semi-monthly magazine; or, select miscellany of the periodical literature of Great Britain, Philadelphia, 1842–1844.
*Carey's library of choice literature containing the best works of the day, in biography, history, travels, novels, poetry, &c. &c., Philadelphia, 1835–1836.
The Carolina journal of medicine, science and agriculture, Charleston, S.C., 1825.
*The casket, Hudson, N.Y., 1811–1812.
The casket, devoted to literature, science, the arts, news, &c., Cincinnati, 1846.
The casket, or flowers of literature, see Graham's American monthly magazine.
The central New York farmer, Rome, N.Y., 1842–1844.
The Charleston medical journal and review, Charleston, S.C., 1846–1850.
*The Charleston medical register for the year MDCCCII, Charleston, S.C., 1803.
*Cheap repository, Philadelphia, 1800.
Chester County Cabinet of Natural Science, Papers of the, Westchester, Pa., 1826–1849.
*The children's magazine, calculated for the use of families and schools, Hartford, 1789.
*The Christian cabinet, Philadelphia, 1802.
The Christian review, Boston, 1836–1850.
The Christian review: Nashville, Tenn., 1847.
*The Christian world, New York, 1850.
The Christian's magazine; designed to promote the knowledge and influence of evangelical truth and order, New York, 1806–1811.
*Christian's magazine, reviewer, and religious intelligencer, Exeter, N.H., 1805–1808.
The Christian's, scholar's, and farmer's magazine, Elizabethtown, N.J., 1789–1791.
Cincinnati Historical Society, Annals of the, Cincinnati, 1845.
The Cincinnati literary gazette, Cincinnati, 1824–1825.
The Cincinnati mirror (and western gazette of literature, science, and the arts), Cincinnati, 1831–1836.

The Cincinnati miscellany, or antiquities of the west, Cincinnati, 1845–1846.
Civil engineer and herald of internal improvement, Columbus, Ohio, 1828.
Collections, topographical, historical, and bibliographical, relating principally to New Hampshire, Concord, N.H., 1822–1824.
*College of Physicians of Philadelphia, Summary of the transactions of the, Philadelphia, 1841–1850.
*College of Physicians of Philadelphia, Transactions of the, Philadelphia, 1793.
Columbian Chemical Society of Philadelphia, Memoirs of the, Philadelphia, 1813.
*Columbian historian, New Richmond, Ohio, 1824–1825.
The columbian magazine, or, monthly miscellany, Philadelphia, 1786–1790. See also Universal asylum and columbian magazine, for continuation.
*The columbian museum, or, universal asylum, Philadelphia, 1793.
*The columbian phoenix and Boston review, Boston, 1800.
*Colvin's weekly register, Washington, D.C., 1808.
*The comet, Boston, 1811–1812.
The commercial review of the South and West, see De Bow's review.
*The common school journal, Boston, 1839–1850.
*The common school journal of the state of Pennsylvania, Philadelphia, 1844.
*The companion and weekly miscellany, Baltimore, 1804–1806.
*Concord Society of Natural History, Journal of the, Concord, N.H., 1848.
Connecticut Academy of Arts and Sciences, Memoirs of the, New Haven, 1810–1816.
The Connecticut magazine, Bridgeport, Ct., 1801.
Connecticut Society for Promoting Agriculture, Transactions of the, New Haven, 1802.
The correspondent, New York, 1827–1829.
The corsair, New York, 1839–1840.
The cultivator (a monthly publication, designed to improve the soil and the mind), Albany, N.Y., 1834–1850.
The daguerreotype, A magazine of foreign literature and science; compiled chiefly from the periodical publications of England, France, and Germany, Boston, 1847–1849.
De Bow's review of the southern and western states, New Orleans, 1846–1850.
The Delaware register and farmer's magazine, Dover, Del., 1838–1839.
The Delaware register; or, farmers', manufacturers' and mechanics' advocate, containing a variety of original and selected articles, Wilmington, Del., 1828–1829.
*The dessert to the true American, Philadelphia, 1798–1799.
The dial, Boston, 1840–1844.
The dollar magazine; a literary, political, advertising, and miscellaneous newspaper, Philadelphia, 1833.

*The dollar magazine; a monthly gazette of current literature, music and art, New York, 1841–1842.
*Dorchester Antiquarian and Historical Society, Collections of the, Boston, 1844.
Dwight's American magazine, and family newspaper, New York, 1845–1850.
*The eastern magazine, Bangor, Me., 1835–1836.
*Eccentricities of literature and life; or the recreative magazine, Boston, 1822.
The eclectic, and medical botanist; devoted principally to improvements in the botanic practice of medicine, Columbus, Ohio, 1832–1833.
The eclectic journal of science, Columbus, Ohio, 1834–1835.
The eclectic magazine of foreign literature, science and art, New York, 1844–1850.
The eclectic museum of foreign literature, science and art, New York, 1843–1844.
The eclectric repertory and analytic review, medical and philosophical, Philadelphia, 1810–1819. *See also* The journal of foreign medical sciences and literature.
The emporium of arts & sciences, Philadelphia, 1812–1814.
*The entertaining and marvelous repository, Boston, 1842.
Essex County Natural History Society, Journal of the, Salem, Mass., 1836–1839.
Essex Institute, Proceedings of the, Salem, Mass., 1848–1850.
The evening fireside, Philadelphia, 1804–1806.
The examiner, and journal of political economy, Philadelphia, 1834–1835.
*The examiner; containing political essays on the most important events of the time, New York, 1813–1816.
The expositor; a weekly journal of foreign and domestic intelligence, literature, science, and the fine arts, New York, 1838–1839.
The expositor and universalist review, Boston, 1831–1840.
The family lyceum, Boston, 1832–1833.
The family magazine, or general abstract of useful knowledge, New York, 1833–1841.
The family visitor, Cleveland, Ohio, 1850.
The farmer and mechanic, devoted to agriculture, mechanics, manufactures, science and the arts, New York, 1847–1850.
*The farmers' and planters' friend, Philadelphia, 1821.
The farmers' cabinet; devoted to agriculture, horticulture, and rural economy, Philadelphia, 1836–1848.
The farmer's mechanic's, manufacturer's, and sportsman's magazine, New York, 1826–1827.
The farmer's monthly visitor, Concord, N.H., 1839–1849.
The farmers' register, Shellbanks, Va., 1833–1835, Petersburg, Va., 1835–1843.
*The farrier's magazine, or the archives of veterinary science, Philadelphia, 1818.
Fessenden's silk manual and practical farmer, Boston, 1835–1837.

The financial register of the United States, Philadelphia, 1838.
Fisher's national magazine and industrial record, New York, 1845–1846.
*The fly; or juvenile miscellany, Boston, 1805.
Foederal American monthly, see The knickerbocker: or, New-York monthly magazine.
Franklin Institute journal, Philadelphia, 1826–1850.
The free enquirer, New Harmony, Ind., 1830–1835. See also New-Harmony Gazette.
Free-masons magazine and general miscellany, Philadelphia, 1811–1812.
The friend, a periodical work, Albany, 1815–1816.
The friend, A religious and literary journal, Philadelphia, 1827–1850.
The friendly visitor, New York, 1825.
The friends' intelligencer; devoted to religion, morals, literature, the arts, science, &c., Philadelphia, 1838–1839.
*Friends' miscellany, Philadelphia, 1831–1850.
*The general magazine, and historical chronicle, for all the British plantations in America, Philadelphia, 1741.
*The general magazine and impartial review, Baltimore, 1798.
The general repository and review, Cambridge, 1812–1813.
The Genesee farmer and gardener's journal, Rochester, N.Y., 1831–1839.
The Genesee farmer and gardener's journal, Rochester, N.Y., 1840–1850.
The gentleman and lady's town and country magazine, Boston, 1784.
The gentleman's magazine and monthly American review, see Burton's gentleman's magazine.
The gentlemen and ladies town and country magazine, Boston, 1789–1790.
Geographical, historical, and statistical repository, Philadelphia, 1824.
Geological Society of Pennsylvania, Transactions of the, Philadelphia, 1834–1835.
*The Georgia analytical repository, Savannah, Ga., 1802–1803.
*Georgia Historical Society, Collections of the, Savannah, Ga., 1840.
The globe, New York, 1819.
Graham's American monthly magazine, Philadelphia, 1826–1850.
The Green Mountain gem, Bradford, Vt., 1843–1849.
The Green Mountain repository, for the year 1832, Burlington, Vt., 1832.
*Greenbank's periodical library, Philadelphia, 1833.
The guardian, a family magazine, Columbia, Tennessee, 1841–1850.
*The guardian of health, Boston, 1846–1850.
*The half-yearly abstract of the medical sciences, New York, 1845–1850.
*The harbinger of the Mississippi Valley, Frankfort, Ky., 1832.
*Harper's new monthly magazine, New York, 1850.
*Hartford County Agricultural Society, Transactions of the, Hartford, Ct., 1842–1845.
Hartford Natural History Society, see Natural History Society of Hartford.
*The Harvard lyceum, Cambridge, 1810–1811.
*The Harvard register, Cambridge, 1827–1828.
Hawaiian spectator, Honolulu, 1838–1839.
Hazard's register of Pennsylvania, Philadelphia, 1828–1835.

Hazard's United States commercial and statistical register, Philadelphia, 1839–1842.
The Hesperian, Columbus, Ohio, 1838–1839.
*The historical register of the United States, Washington, D.C., 1812–1814.
Historical Society of Michigan, Historical and scientific sketches of Michigan, Detroit, 1834.
Holden's dollar magazine, New York, 1848–1850.
The horticulturist and journal of rural arts and rural taste, Albany, 1846–1850.
Hovey's magazine of horticulture, see American gardener's magazine and register.
*Humphrey's journal of the daguerreotype & photographic arts, New York, 1850.
The Huntingdon literary museum, and monthly miscellany, Huntingdon, Pa., 1810.
Hunt's merchants' magazine, and commercial review, New York, 1839–1850.
*The idle man, New York, 1821–1822.
The Illinois medical and surgical journal, Chicago, 1844–1847.
The Illinois monthly magazine, Vandalia, Ill., 1830–1832.
Illustrated family magazine, Boston, 1845–1846.
The independent, New York, 1848–1850.
The independent republican; and miscellaneous magazine, Newburyport, Mass., 1805.
The indicator, Amherst, Mass., 1848–1850.
*The intellectual regalia; or ladies' tea tray, Philadelphia, 1814–1815.
The international monthly magazine of literature, science, and art, New York, 1850.
The investigator; religious, moral, scientific, etc., Washington, D.C., 1845–1846.
The iris: Devoted to science, literature, and the arts, Richmond, Va., 1848.
The iris, or literary messenger, New York, 1840–1841.
The Jeffersonian, Albany, 1838–1839.
The Jewish intelligencer, a monthly publication, New York, 1836–1837.
Journal of agriculture, New York, 1845–1848.
The journal of foreign medical sciences and literature, Philadelphia, 1821–1824. See also The eclectic repertory and analytic review.
The journal of health and monthly miscellany, Boston, 1846–1850.
The journal of science and the arts, New York, 1817–1818.
*Journal of the medical society of Maine, Hallowell, Me., 1834.
*Journal of the times, Baltimore, 1818–1819.
*The juvenile gazette, Providence, R.I., 1819.
Juvenile magazine, or miscellaneous repository, Philadelphia, 1802–1803.
The juvenile miscellany, Boston, 1826–1834.
The juvenile port-folio and literary miscelleny, Philadelphia, 1812–1816.
Juvenile rambler; or family and school journal, Boston, 1832–1833.
*The juvenile repository, Boston, 1811.

*The key, Fredericktown, Md., 1798.
The knickerbocker; or New-York monthly magazine, New York, 1833–1850.
*Ladies literary portfolio, Philadelphia, 1828–1829.
*The ladies' magazine, Boston, 1828–1836.
*Ladies' miscellany, Salem, Mass., 1828–1831.
*Ladies museum, Philadelphia, 1800.
*Ladies visitor, Boston, 1806–1807.
*Ladies visitor, Marietta, Pa., 1819–1820.
*The lady & gentleman's pocket magazine of literary and polite amusement, New York, 1796.
The lady's magazine and repository of entertaining knowledge, Philadelphia, 1792–1793.
*Lady's western magazine and garland of the valley, Cincinnati, 1849.
Lancaster hive, Lancaster, Pa., 1803–1805.
*The lancet, Newark, N.J., 1803.
The literary and philosophical repertory, Middlebury, Vt., 1812–1816.
Literary and Philosophical Society of New York, Transactions of the, Albany, 1815, 1825.
Literary and scientific repository, and critical review, New York, 1820–1822.
The literary and theological review, New York, 1834–1839.
The literary casket: Devoted to literature, the arts, and science, Hartford, Ct., 1826–1827.
*The literary companion, New York, 1821.
The literary focus, a monthly periodical, Oxford, Ohio, 1827–1828.
The literary gazette, Concord, N.H., 1834–1835.
The literary gazette and American athenaeum, New York, 1825–1827.
The literary gazette; or, journal of criticism, science, and the arts, Philadelphia, 1821.
*Literary gems, New York, 1833.
The literary inquirer; a semi-monthly journal devoted to literature and science, Buffalo, N.Y., 1833–1834.
*The literary journal, Schenectady, N.Y., 1834–1835.
The literary journal, and weekly register of science and the arts, Providence, R.I., 1833–1834.
*Literary magazine, Boston, 1835.
Literary magazine, and American register, Philadelphia, 1803–1807.
*The literary miscellany, Philadelphia, 1795.
The literary miscellany, Cambridge, 1805–1806.
The literary miscellany, or monthly review, New York, 1811.
The literary museum, or monthly magazine, Westchester, Pa., 1797.
Literary port folio, Philadelphia, 1830.
The literary record and journal of the Linnaean Association of Pennsylvania College, Gettysburg, Pa., 1844–1848.
The literary register, a weekly paper, Oxford, Ohio, 1828–1829.
The literary tablet; or general repository of various entertainment, Hanover, N.H., 1804–1807.

The literary union, Syracuse, N.Y., 1849–1850.
The literary world, New York, 1847–1850.
The living age, Boston, 1844–1850.
*The Louisville journal of medicine and surgery, Louisville, Ky., 1838.
The Lowell offering, Lowell, Mass., 1840–1844.
*Lynn Natural History Society, Publications of the, Lynn, Mass., 1846.
Maclurian Lyceum of the Arts and Sciences, Contributions of the, Philadelphia, 1827–1829.
Madison County Agricultural Society, Transactions of the, Hamilton, N.Y., 1845.
The magazine of useful and entertaining knowledge, New York, 1830–1831.
The Maine farmer and journal of the useful arts, Winthrop, Me., 1833–1850.
Maine Historical Society, Collections of the, Portland, Me., 1831–1850.
The Maine monthly magazine, Bangor, Me., 1836–1837.
Maryland Academy of Science and Literature, Transactions of the, Baltimore, 1837.
The Maryland medical and surgical journal, Baltimore, 1840–1843.
*The Maryland medical recorder, Baltimore, 1829–1832.
Massachusetts agricultural repository and journal, Boston, 1815–1832.
Massachusetts Agricultural Societies, Transactions of the, Boston, 1846–1850.
Massachusetts Agricultural Society, Papers on agriculture, Boston, 1793–1811.
Massachusetts Historical Society, Collections of the, Boston, 1792–1850.
*Massachusetts Horticultural Society, Transactions of the, Boston, 1829–1850.
The Massachusetts magazine, Boston, 1789–1796.
*Massachusetts Medical Society, Medical Communications of the, Boston, 1790–1850.
The Massachusetts quarterly review, Boston, 1847–1850.
Massachusetts Society for the Promotion of Agriculture, see Massachusetts Agricultural Society.
*The Massachusetts watchman, and periodical journal, Palmer, Mass., 1809.
*The mathematical correspondent, Reading, Pa., 1804–1807.
*The mathematical diary, New York, 1825, 1828, 1832.
*The mathematical miscellany, New York, 1836–1839.
The mechanic, see The young mechanic.
Mechanic apprentice, Boston, 1845–1846.
The mechanics' magazine, and journal of public internal improvement, Boston, 1830.
Mechanics' magazine, and register of inventions and improvements, New York, 1833–1837.
Mechanics' magazine, museum, register, journal and gazette, see American mechanics' magazine.
The mechanics' mirror, Albany, 1846.
The medical and agricultural register, Boston, 1806–1807.

*The medical and surgical register, New York, 1818.
The medical and surgical reporter, Burlington, N.J., and Philadelphia, 1847–1850.
The medical examiner, Philadelphia, 1838–1850.
*The medical intelligencer and literary advertiser, Philadelphia, 1842.
The medical magazine, Boston, 1832–1835.
*The medical news and library, Philadelphia, 1843–1850.
The medical repository, New York, 1798–1818.
Medical Society of the State of New-York, Transactions of the, New York, 1807–1831, Albany, 1833–1850.
The medley or monthly miscellany, Louisville, Ky., 1803.
The merchants' magazine and commercial review, see Hunt's merchants' magazine.
Merrimack magazine and ladies literary cabinet, Newburyport, Mass., 1805–1806.
The Merrimack magazine, and monthly register of politics, agriculture, literature, and religion, Haverhill, Mass., 1825.
Merrimack miscellany, Newburyport, Mass., 1805.
Merry's museum, New York, 1841–1850.
The Methodist magazine, Philadelphia, 1797–1798.
The Methodist magazine, New York, 1818–1850.
The Methodist quarterly review, see The Methodist magazine (New York)
The Methodist recorder, Baltimore, 1824–1825.
The Michigan farmer, and western horticulturist, Jackson, Mich., 1843–1850.
Michigan Historical Society, see Historical Society of Michigan.
*Michigan State Agricultural Society, Transactions of the, Lansing, Mich., 1850.
*The microscope, New Haven, 1820.
The microscope and general advertiser, Louisville, Ky., and New Albany, Ind., 1824–1825.
*The military and naval magazine of the United States, Washington, D.C., 1833–1836.
The Minerva; or, literary, entertaining, and scientific journal, New York, 1822–1825.
The mining journal and American railroad gazette, New York, 1847–1849.
The miscellaneous magazine, Trenton, N.J., 1824.
*Miss Leslie's magazine, Philadelphia, 1843.
Missouri Historical and Philosophical Society, Annals, Jefferson City, Mo., 1848.
The monthly American journal of geology and natural science, Philadelphia, 1831–1832.
The monthly anthology, and Boston review, Boston, 1803–1811.
The monthly chronicle of events, discoveries, improvements and opinions, Boston, 1840–1842.
*The monthly flora, or botanical magazine, New York, 1846.

The monthly journal of agriculture, New York, 1845–1848.
*The monthly journal of foreign medicine, Hartford, Ct., 1823–1824.
The monthly journal of medical literature, and American medical students' gazette, Lowell, Mass., 1832.
Monthly journal of medicine, Hartford, Ct., 1823–1825.
The monthly magazine and American review, New York, 1800–1802.
*The monthly magazine and literary journal, Winchester, Va., 1812–1813.
*The monthly magazine of religion and literature, Gettysburg, Pa., 1840–1841.
The monthly miscellany of religion and letters, Boston, 1839–1843.
*The monthly miscellany or Vermont magazine, Burlington, Vt., 1794.
*The monthly recorder, New York, 1813.
*The monthly register, magazine, and review of the United States, Charleston, S.C., 1805–1807.
*The monthly religious magazine, Boston, 1844–1850.
The monthly repository and library of entertaining knowledge, New York, 1830–1834.
The monthly review and miscellany of the United States, Charleston, S.C., 1805–1806.
The monthly scientific journal, New York, 1818.
*The mother's magazine, Utica, N.Y., 1833–1850.
The museum of foreign literature and science, Philadelphia and New York, 1822–1845.
National Institute for the Promotion of Science, Bulletin of the proceedings of the, Washington, D.C., 1841–1846.
*The national magazine and republican review, Washington, D.C., 1839.
*National magazine; or, a political, historical, biographical, and literary repository, Richmond, Va., 1799.
National magazine; or, cabinet of the United States, Washington, D.C., 1801–1802.
*The national magazine or lady's emporium, Baltimore, 1830–1831.
The national pilot, see The pilot.
The national recorder, see The Saturday magazine.
Natural History Society of Hartford, Transactions of the, Hartford, Ct., 1836.
The Naturalist, Nashville, Tenn., 1850.
The naturalist, and journal of natural history, agriculture, education, and literature, Nashville, Tenn., 1846.
The naturalist, containing treatises on natural history, chemistry, domestic and rural economy, manufactures and arts, Boston, 1830–1832.
The Naval magazine, New York, 1836–1837.
New Castle County Agricultural Society and Mechanics Institute, Transactions of the, Wilmington, Del., 1845.
*New-England Association of Farmers, Mechanics and Other Working Men, Proceedings of the, Boston, 1832.
The New England farmer, Boston, 1822–1845.

The New England farmer; a semi-monthly journal, devoted to agriculture, horticulture, and their kindred arts and sciences, Boston, 1848–1850.
The New-England journal of medicine and surgery, and collateral branches of science, Boston, 1812–1826.
*The New England literary herald, Boston, 1809.
The New-England magazine, Boston, 1831–1835.
*The New-England magazine of knowledge and pleasure, Boston, 1758–1759.
The New England medical eclectic, and guide to health, Worcester, Mass., 1846.
*The New England medical review and journal, Boston, 1827.
*New England offering, Lowell, Mass., 1848–1850.
*The New England quarterly magazine; comprehending literature, morals, and amusement, Boston, 1802.
*The New-England telegraph, and eclectic review, North Wrentham, Mass., 1835.
The New Englander, New Haven, 1843–1850.
New-Hampshire Historical Society, Transactions of the, Concord, N.H., 1824.
*The New-Hampshire journal of medicine, Concord, N.H., 1850.
*The New-Hampshire magazine, Manchester and Great Falls, N.H., 1843–1844.
*The New Hampshire repository, Gilmanton, N.H., 1845.
The New-Harmony Gazette, New Harmony, Ind., 1825–1827. See also The free enquirer.
*New Haven County Medical Society, Cases and observations, New Haven, 1788.
New Haven gazette, and the Connecticut magazine, New Haven, 1786–1789.
The New Jersey and Pennsylvania agricultural monthly intelligencer, and farmer's magazine, Camden, N.J., 1825.
*New Jersey Historical Society, Collections of the, New York, 1846–1850.
*New Jersey Historical Society, Proceedings of the, Newark, N.J., 1845–1850.
New Jersey magazine, and monthly advertiser, New Brunswick, N.J., 1786–1787.
*The New Jersey medical reporter, and transactions of the New Jersey Medical Society, Burlington, N.J., 1847–1850.
*The New Jersey monthly magazine, Newark, N.J., 1825.
*The new mirror, New York, 1843–1844.
New monthly magazine and literary journal, Boston and Philadelphia, 1821–1834.
*The New-Orleans medical journal, New Orleans, 1844–1850.
The New Orleans miscellany, New Orleans, 1847–1848.
*The new star, Concord, N.H., 1797.
*The new star; a republican paper, Concord, N.H., 1797.
New-York Historical Society, Collections of the, New York, 1809–1850.

*New-York Historical Society, Proceedings of the, New York, 1843–1849.
The New York journal of medicine and surgery, New York, 1839–1841.
The New York journal of medicine, and the collateral sciences, New York, 1843–1850.
New York Literary and Philosophical Society, Transactions of the, see Literary and Philosophical Society of New York, Transactions of the.
New-York literary journal, and belles-lettres repository, New York, 1819–1821.
New York Lyceum of Natural History, Annals of the, New York, 1823–1850.
*The New York magazine and general repository of useful knowledge, New York, 1814.
New York magazine; or, literary repository, New York, 1790–1797.
New York medical and philosophical journal and review, New York, 1809–1811.
New York medical and physical journal, New York, 1822–1830.
*The New York medical gazette, New York, 1842.
*The New-York medical journal, New York, 1830–1831.
*New York medical magazine, New York, 1814–1815.
New York Medical Society, see Medical Society of the State of New York.
*New-York medico-chirurgical bulletin, New York, 1831–1832.
The New York monthly chronicle of medicine and surgery, New York, 1824–1825.
The New York monthly magazine, New York, 1824.
*The New York register of medicine and pharmacy, New York, 1850.
The New York review, New York, 1837–1842.
The New York review or atheneum magazine, New York, 1825–1826.
New York Society for the Promotion of Agriculture, Arts and Manufactures, see New York Society for the Promotion of the Useful Arts.
New York Society for the Promotion of the Useful Arts, Transactions of the, Albany, 1792–1819.
New-York State Agricultural Society, Journal of the, Albany, 1850.
New York State Agricultural Society, Transactions of the, Albany, 1841–1850.
New York State, Board of Agriculture, Journal, see Plough boy.
New York State, Board of Agriculture, Memoirs of the, Albany, 1821–1826.
The New York state mechanic, a journal of the manual arts, trades, and manufactures, Albany, 1841–1843.
New York State, State Museum, Annual Reports, Albany, 1847–1850.
The New York weekly magazine, New York, 1795–1797.
The New York weekly review, New York, 1850.
The New-Yorker, New York, 1836–1841.
The nightingale, or a melange de litterature, Boston, 1796.
Niles' weekly register, Baltimore, 1812–1850. Also, supplements to volumes 5, 7, 8, 9, 15, 16, 23, 38, 43; also, addenda to volume 41.

*The nineteenth century, Philadelphia, 1848–1849.
North American archives of medical and surgical science, Baltimore, 1834–1836.
The North American medical and surgical journal, Philadelphia, 1826–1831.
The North American, or, weekly journal of politics, science, and literature, Baltimore, 1827.
North American review, Boston, 1815–1850.
*The North Carolina magazine, political, historical, and miscellaneous, n.p., 1813.
The north-western medical and surgical journal, Chicago and Indianapolis, 1848–1850.
The northern light, Albany, 1841–1844.
The oasis, a monthly magazine, devoted to literature, science and the arts, Oswego, N.Y., 1837–1838.
The Oberlin quarterly review, Oberlin, Ohio, 1845–1849.
*The observer, Baltimore, 1806–1807.
*The observer, New York, 1809.
*The observer, New York, 1810–1811.
The observer and record of agriculture, science, and art, Philadelphia, 1838–1839.
The Ohio cultivator, Columbus, Ohio, 1845–1850.
*Ohio medical and surgical journal, Columbus, Ohio, 1848–1850.
The Ohio medical repository, Cincinnati, 1826–1827.
Ohio miscellaneous museum, Lebanon, Ohio, 1822.
The Ohio school journal, Cleveland, Ohio, 1846, Columbus, Ohio, 1847–1848.
Ohio State Board of Agriculture, Annual Reports, Columbus, Ohio, 1847–1850.
The olio, a literary and miscellaneous paper, New York, 1813–1814.
*The olive branch and Christian inquirer, New York, 1828.
*Oliver's magazine, Boston, 1841.
Omnium gatherum, Boston, 1810.
*The ordeal; a critical journal of politicks and literature, Boston, 1809.
Parley's magazine, New York and Boston, 1833–1844.
The parlour companion, Philadelphia, 1817–1819.
The Parthenon, or, literary and scientific museum, New York, 1827.
*The patriarch; or, family library magazine, New York, 1841.
Pennsylvania Agricultural Society, Memoirs of the, Philadelphia, 1823–1824.
Pennsylvania Geological Society, see Geological Society of Pennsylvania.
*Pennsylvania Historical Society, Bulletin, Philadelphia, 1848.
Pennsylvania Historical Society, Memoirs, Philadelphia, 1826–1850.
The Pennsylvania magazine; or, American monthly museum, Philadelphia, 1775–1776.
*The penny post, n.p., 1769.

Philadelphia Academy of Natural Sciences, Journal of the, Philadelphia, 1817–1850.

Philadelphia Academy of Natural Sciences, Proceedings of the, Philadelphia, 1841–1850.

Philadelphia Academy of Natural Sciences, Report of the transactions of the, Philadelphia, 1824–1829.

Philadelphia College of Pharmacy journal, see American journal of pharmacy.

The Philadelphia journal of the medical and physical sciences, Philadelphia, 1820–1827.

The Philadelphia magazine and review, Philadelphia, 1799.

The Philadelphia magazine and weekly repository, Philadelphia, 1818.

The Philadelphia medical and physical journal, Philadelphia, 1804–1808.

Philadelphia medical museum, Philadelphia, 1805–1809, 1810–1811.

*The Philadelphia monthly journal of medicine and surgery, Philadelphia, 1827–1828.

The Philadelphia monthly magazine, Philadelphia, 1798.

The Philadelphia museum, or register of natural history and the arts, Philadelphia, 1824.

The Philadelphia register and national recorder, see The Saturday magazine.

The Philadelphia repertory, Philadelphia, 1810–1812.

The Philadelphia repository, and weekly register, Philadelphia, 1800–1806.

Philadelphia Society For Promoting Agriculture, Memoirs of the, Philadelphia, 1808–1826.

The philosophical medical journal, or family physician, New York, 1844–1848.

*Physico-medical Society of New York, Transactions of the, New York, 1817.

The pictorial national library, a monthly miscellany of the useful and entertaining in science, art and literature, Boston, 1848–1849.

The pilot, New Haven, 1821–1824.

*The pioneer, Boston, 1843.

The plaindealer, New York, 1836–1837.

The plough boy, Albany, 1819–1823.

The plough, the loom and the anvil, Philadelphia, 1848–1850.

The political magazine; and miscellaneous repository, Ballston, N.Y., 1800.

The port folio, Philadelphia, 1801–1827.

The portfolio, and companion to the select circulating library, Philadelphia, 1835–1836.

The portico, a repository of science and literature, Baltimore, 1816–1818.

*Portland magazine, Portland, Me., 1805.

The Portland magazine, devoted to literature, Portland, Me., 1834–1835.

*The Portsmouth weekly magazine, a repository of miscellaneous literary matter, in prose and verse, Portsmouth, N.H., 1824.

The Poughkeepsie casket, Poughkeepsie, N.Y., 1836–1841.
The prairie farmer, Chicago, 1843–1850.
*The present, Boston, 1843–1844.
*The Princeton magazine, Princeton, N.J., 1850.
The Princeton review, see Biblical repertory.
*Providence Franklin Society, Proceedings of the, Providence, R.I., 1846.
The quarterly journal and review, Cincinnati, 1846.
*The quarterly theological review, Philadelphia, 1818–1819.
Recreative magazine, Boston, 1822.
*The reflector, Boston, 1821.
*The register and library of medical and chirurgical science, Washington, D.C., 1834–1836.
*The religious magazine, and family lyceum, Boston, 1833–1838.
The religious magazine, or spirit of foreign theological journals and reviews, Philadelphia, 1828–1830.
*Rensselaer County Agricultural Society, Transactions of the, for the year 1846, Troy, N.Y., 1847.
Repository and ladies weekly museum, see The Philadelphia repository and weekly register.
*Repository of knowledge, historical, literary, miscellaneous, and theological, Philadelphia, 1801.
A republican magazine, Fairhaven, Vt., 1798.
*Rhode Island Historical Society, Collections of the, Providence, R.I., 1827–1850.
*Rhode Island literary repository, Providence, R.I., 1814–1815.
Richmond Lyceum, Journal, Richmond, Va., 1838–1839.
*Roberts' semi-monthly magazine, for town and country, Boston, 1841–1842.
Robinson's magazine, a weekly repository of original papers, Baltimore, 1818–1819.
*Rochester magazine and theological review, Rochester, N.Y., 1824.
The royal American magazine, Boston, 1774–1775.
The rural casket, Poughkeepsie, N.Y., 1798.
The rural magazine, Newark, N.J., 1798–1799.
The rural magazine and farmer's monthly museum, Hartford, Ct., 1819.
The rural magazine and literary evening fireside, Philadelphia, 1820.
The rural magazine; or, Vermont repository, Rutland, Vt., 1795–1796.
The rural visiter (sic.); a literary and miscellaneous gazette, Burlington, N.J., 1810–1811.
Sandwich Island gazette, & journal of commerce, Hawaii, 1836–1839.
*Sargent's new monthly magazine, New York, 1843.
Sartain's union magazine, New York and Philadelphia, 1847–1850.
The Saturday magazine, Philadelphia, 1819–1822.
The scholar's journal, Westfield, 1829.
The scholar's quarterly journal, Westfield, 1828.
The school journal, and Vermont agriculturist, Windsor, Vt., 1847–1850.
*The school magazine, Boston, 1829.

Scientific American, New York, 1845–1850.
*Scientific and literary journal for the diffusion of useful knowledge, Boston, 1837.
The scientific journal, Perth Amboy, N.J., 1818–1819.
Scientific mechanic, inventors' advocate, patent office reporter, and expositor of arts and trade, New York, 1847.
Scientific tracts and family lyceum, Boston, 1834–1835.
Scientific tracts for the diffusion of useful knowledge, Boston, 1836.
The select journal of foreign periodical literature, Boston, 1833–1834.
Select reviews of literature, and spirit of the foreign magazines, Philadelphia, 1809–1812.
*Sentimental and literary magazine, New York, 1797.
*The sidereal messenger, Cincinnati, 1846–1848.
*Smith's weekly volume, Philadelphia, 1845–1846.
Smithsonian Institution, Annual reports of the Board of Regents, Washington, D.C., 1847–1850.
Smithsonian Institution, Contributions, Washington, D.C., 1848–1850.
Society for the Promotion of Useful Arts, New York, 1792–1819.
*Society of Literary & Scientific Chiffonniers, Transactions of the, New York, 1844.
Something, Boston, 1809.
*Something new, Harvard, Mass., 1832.
South Carolina weekly museum, Charleston, S.C., 1797.
Southern agriculturist, horticulturist, and register of rural affairs, Charleston, S.C., 1828–1846.
Southern and western monthly magazine and review, Charleston, S.C., 1845.
The southern journal of medicine and pharmacy, see The Charleston medical journal and review.
*Southern literary journal and magazine of arts, Charleston, S.C., 1835–1838.
The southern literary messenger, Richmond, Va., 1834–1850.
*The southern magazine and monthly review, Petersburg, Va., 1841.
The southern medical and surgical journal, Augusta, Ga., 1836–1850.
The southern planter, Richmond, Va., 1841–1850.
The southern quarterly review, New Orleans, 1842–1850.
The southern review, Charleston, S.C., 1828–1832.
Southron, or Lily of the Valley, Gallatin, Tenn., 1841.
Southwestern journal and monthly review, Natchez, Miss., 1837–1838.
*The spirit of the age, New York, 1849–1850.
The spirit of the Pilgrims, Boston, 1828–1833.
*The spirit of the public journals, Baltimore, 1806.
*The stranger, a literary paper, Albany, 1814.
Stryker's American register, Philadelphia, 1848.
*The tablet, Boston, 1795.
*The talisman, New York, 1827–1829.
*The theological and literary journal, New York, 1848–1850.

*The thistle, Boston, 1807.
*The Thompsonian botanic watchman, Albany, 1834–1835.
*The times, Boston, 1807–1808.
*The toilet; a weekly collection of literary pieces, Charlestown, Mass., 1801.
*The town, New York, 1807.
The Transylvania journal of medicine and the associated sciences, Lexington, Ky., 1828–1839.
*Transylvania medical journal, Lexington, Ky., 1849–1850.
The Transylvanian, or Lexington literary journal, Lexington, Ky., 1829.
The union agriculturist and western prairie farmer, Chicago, 1841–1842.
The union magazine, Philadelphia, 1847–1850.
*The Unitarian, and foreign religious miscellany, Boston, 1847.
*The Unitarian essayist, Meadville, Pa., 1831–1832.
*The United States Christian magazine, New York, 1796.
The United States democratic review, Washington and New York, 1837–1850.
The United States literary gazette, Boston, 1824–1826.
*The United States magazine, A repository of history, politics and literature, Philadelphia, 1779.
The United States magazine, and democratic review, Washington, D.C., 1837–1850.
*The United States magazine, and literary and politcal repository, New York, 1823.
*The United States magazine, or, general repository, Newark, N.J., 1794.
The United States medical and surgical journal, New York and Philadelphia, 1834–1836.
*The United States Military Philosophical Society, Extracts, Washington, D.C., 1806–1809.
*The United States Naval chronicle, Washington, D.C., 1824.
The United States review and literary gazette, Boston, 1827.
The universal asylum and columbian magazine, Philadelphia, 1790–1792.
The universal magazine, Philadelphia, 1797–1798.
The Universalist quarterly and general review, Boston, 1844–1850.
The useful cabinet, Boston, 1808.
*The village museum, Cortland Village, N.Y., 1820.
The villager, a literary paper, Greenwich Village, N.Y., 1819.
Virginia Historical and Philosophical Society, Collections of the, Richmond, Va., 1833.
The Virginia Historical register and literary companion, Richmond, Va., 1848–1850.
The Virginia literary museum and journal of belles lettres, arts, sciences &c., Charlottesville, Va., 1829–1830.
The visitor, Richmond, Va., 1809.
*The Washington quarterly magazine, Washington, D.C., 1823–1824.
*The watchman, New Haven, 1819.
The weekly inspector, New York, 1806–1807.

The weekly magazine, Philadelphia, 1798–1799.
The weekly register, see Niles' weekly register.
The weekly visitor, or, ladies miscellany, New York, 1802–1812.
The western academician and journal of education and science, Cincinnati, 1837–1838.
Western and southern medical recorder, Lexington, Ky., 1841–1843.
Western farmer, see Michigan farmer.
The western farmer and gardener, Indianapolis, 1839–1845.
The western gleaner, or repository for arts, sciences and literature, Pittsburgh, 1813–1814.
*The western horticultural review, Cincinnati, 1850.
The western journal and civilian, see The western journal of agriculture.
The western journal of agriculture, manufactures, mechanic arts, internal improvement, commerce, and general literature, St. Louis, 1848–1850.
The western journal of medicine and surgery, Louisville, Ky., 1840–1850.
The western journal of the medical and physical sciences, Cincinnati, 1827–1838.
*The western ladies' casket, Connersville, Ind., 1824.
*The western literary emporium, Cincinnati, 1848–1849.
The western literary journal and monthly review, Cincinnati, 1836.
The western literary journal and monthly review, Cincinnati, 1844–1845.
*Western literary magazine, Pittsburgh, 1840–1844.
The western literary magazine, and journal of education, science, arts, and morals, Columbus, Ohio, 1849.
The western medical and physical journal, original and eclectic, Cincinnati, 1827–1828.
The western medical gazette, Cincinnati, 1832–1835.
The western messenger; devoted to religion and literature, Louisville, Ky., 1835–1841.
Western Minerva, or American annals of knowledge and literature, Lexington, Ky., 1821.
The western miscellany, Dayton, Ohio, 1848–1849.
The western monthly magazine, a continuation of the Illinois monthly magazine, Cincinnati, 1833.
The western monthly review, Cincinnati, 1827–1830.
*The western quarterly journal of practical medicine, Cincinnati, 1837.
The western quarterly reporter of medical, surgical, and natural science, Cincinnati, 1822–1823.
The western quarterly review, Cincinnati, 1849.
The western register, a monthly journal of commerce, navigation, science and the arts, St. Louis, 1849.
*The western review, Columbus, Ohio, 1846.
Western review and miscellaneous magazine, Lexington, Ky., 1819–1821.
*The Worcester Agricultural Society. Transactions, addresses, etc., Worcester, Mass., 1819–1850.
*Worcester County Horticultural Society, Transactions of the, Boston, 1847.

*Worcester County Mechanics' Association, Reports, Worcester, Mass., 1848–1850.

Worcester magazine and historical journal, Worcester, Mass., 1825–1826.

The working farmer devoted to agriculture, embracing the rights of agriculturists, New York, 1849–1850.

The Yale literary magazine, New Haven, 1836–1850.

The yankee, Boston, 1828–1829.

The young American's magazine of self-improvement, Boston, 1847.

The young ladies' journal of literature and science, Baltimore, 1830–1831.

The young mechanic, Boston, 1832–1834.

The young people's book; or, magazine of useful and entertaining knowledge, Philadelphia, 1841–1842.

*Youth's cabinet, Boston, 1815.

Youth's companion, Boston, 1827–1850.

The youth's instructor and guardian, New York, 1823–1832.

The youth's medallion, Boston, 1841–1842.

Youth's repository of Christian knowledge, New Haven, 1813.

The zodiac, a monthly periodical, devoted to science, literature and the arts, Albany, 1835–1837.

GUIDE TO THE USE OF THE BIBLIOGRAPHY

Each reference contains several elements, which are illustrated and described below.

                                                                  1
          Johnston, James Finley Weir (1796-1855)   5
5816   1841, Chemical geology: Am. Eclec., v. 2, no. 2, 395-396.
                Abstracted from Lit. Gaz.
2   3   4
           6

1. **Author:** References are listed alphabetically by author and chronologically for all entries of a given author. If journal articles were not signed, frequently the case, the periodical title has been substituted for the author entry. Unsigned books and pamphlets have been listed by title rather than by author.

2. **Reference number:** Each entry has a reference number for ease of indexing, cross-referencing, and referral.

3. **Date of publication:** Doubtful dates are followed by a question mark. Approximate dates are preceded by a c. Works published in parts over more than one year are handled as follows: 1821–1823 designates 1821 *through* 1823; 1821/1823 designates 1821 *and* 1823.

4. **Title:** If a specific title is lacking or if the title is insufficient to define the subject of the reference, additional information may appear in brackets.

5. **Source:** Source information includes journal title, volume, and pagination for periodical articles; place of publication,

publisher or printer, and pagination for books and pamphlets; place of engraving, engraver or lithographer, and size in centimeters for maps and lithographs. Journal titles and places of publication have been abbreviated following conventions in the List of Abbreviations on page 42. If the publisher is not known, then *n.p.* appears in the place of this information.

6. **Additional information:** Many references contain additional information to clarify the subject of the entry. In addition, cross-references to closely related works, such as reviews or prospecti or data on the original source of extracted articles may also be given.

LIST OF ABBREVIATIONS

The following abbreviations have been used for journal titles and for places of book and pamphlet publication.

Acad.	Academy	Ct.	Connecticut
Adv.	Advancement	Cult.	Cultivator
Ag.	Agriculture		
Agri.	Agricultural	Del.	Delaware
Alb.	Albany, New York	Dem.	Democratic
Am.	American	Disc.	Discovery
Ann.	Annual		
Anth.	Anthology	Eclec.	Eclectic
Arch.	Archives	Emp.	Emporium
Assoc.	Association	Ent.	Entertaining
Ath.	Athenaeum	Exam.	Examiner
Atl.	Atlantic		
		Fam.	Family
Balt.	Baltimore, Maryland	Farm.	Farmer
Bib.	Biblical	Farm's	Farmers
Bos.	Boston, Massachusetts	Fin.	Financial
Bot.	Botanical	For.	Foreign
Cab.	Cabinet	Gaz.	Gazette
Cas.	Casket	Gen.	General
Char.	Charleston, South Carolina	Geog.	Geographical
		Geol.	Geological
Chron.	Chronicle	Geol's	Geologists
Cin.	Cincinnati, Ohio		
Co.	County	Hart.	Hartford, Connecticut
Coll.	Collections	HD	House Document
Colum.	Columbian	HED	House Executive Document
Com.	Commercial		
Crit.	Critical	Hist.	Historical

Hort.	Horticulturist	NE	New England
Horti.	Horticultural	NH	New Hampshire
		NHav.	New Haven, Connecticut
Ill.	Illinois		
Illus.	Illustrated	NJ	New Jersey
Ind.	Independent	No.	Northern
Inst.	Institute	NY	New York (also New York City)
Intell.	Intelligencer		
Inter.	International		
		Obs.	Observer
J.	Journal		
JD	Joint Document	Penn.	Pennsylvania
Juv.	Juvenile	Phil.	Philosophy
		Phila.	Philadelphia, Pennsylvania
Know.	Knowledge		
		Philos.	Philosophical
Lit.	Literary	Phys.	Physical
Liter.	Literature	Phys'n	Physician
Lyc.	Lyceum	Plant.	Planter
		Pol.	Political
Mag.	Magazine	Prac.	Practical
Manu.	Manufacture	Proc.	Proceedings
Manu's	Manufactures	Prom.	Promotion
Mass.	Massachusetts		
Md.	Maryland	Q.	Quarterly
Me.	Maine		
Mech.	Mechanic	Rec.	Record
Mech's	Mechanics	Rec'r	Recorder
Med.	Medical	Reg.	Register
Mem.	Memoirs	Rel.	Religious
Merch.	Merchants	Rep.	Repertory
Mess.	Messenger	Repos.	Repository
Meth.	Methodist	Rept.	Reports
Mich.	Michigan	Repub.	Republican
Min.	Mineralogical	Rev.	Review
Mir.	Mirror	Rr.	Railroad
Misc.	Miscellaneous		
Mon.	Monthly	S.	Southern
Mtn.	Mountain	SC	South Carolina
Mus.	Museum	Sch.	School
		Sci.	Science
		Scien.	Scientific
N.	North	Sci's	Sciences
Nat.	Natural	SD	Senate Document
Nat'l	National	SED	Senate Executive Document
Nat's	Naturalists		
Natur.	Naturalist	SMD	Senate Miscellaneous Document
NC	North Carolina		

Soc.	Society	Use.	Useful
Surg.	Surgery		
		V.	Volume
Tenn.	Tennessee	Va.	Virginia
Topo.	Topographical	Vis.	Visitor
Trans.	Transactions	Vt.	Vermont
Un.	Union	Wash.	Washington, D.C.
Univ.	Universal	West.	Western
US	United States	Wk.	Weekly

BIBLIOGRAPHY

Abbot, J. H. (See 8236 by M. H. Perley)

Abbot, Joel (1766-1826)
1   1803, Speculations on magnetism: Med. Repos., v. 6, 317-319.
2   1814, Essay on the central influence of magnetism: Phila., Printed for the author, 24, 1 p.

Abbott, Austin
3   1820, Hudson Association For Improvement in Science. Extract of a letter: Am. J. Sci., v. 2, no. 2, 375-376.

Abdulla, Syed
4   1817, [Account of meteorite in Bombay, India]: J. Sci. Arts, v. 1, no. 1, 117-118.
    From Royal Soc. Edinburgh Proc.

Abel, Henry I.
5   1838, Geographical, Geological and Statistical Chart of Wisconsin and Iowa. Designed especially for the use of emigrants and travellers: Phila., H. I. Abel, Xylographic Press, Broadside 64 x 49 cm.
    Includes 1836 map by Aug. Mitchell. Also 1839 edition.

Abert, James William (1820-1897)
6   1846, Journal ... from Bent's Fort to St. Louis, in 1845: US SED 438, 29-1, v. 8 (477), 2-75, 11 plates, 2 maps.
    For review see 889. Alternate title: Report of an expedition ... on the upper Arkansas and through the country of the Camanche Indians, in the fall of the year 1845.
7   1848, ... Examination of New Mexico in the years 1846 to 1847: US SED 23, 30-1, v. 4 (506), 3-132, 24 plates, 1 map.
    Published with reports by W. H. Emory (see 3578) and J. W. Bailey (see 1439). For review with excerpts see 949. Also published as US HED 41, 30-1, v. 4 (517), 417-548, 24 plates, 1 map.

Abert, John James (1788-1863)
8   1837, [Testimonial letter in "Reply of G. W. Featherstonhaugh" (see 7630)]: Naval Mag., v. 2, 574-577.

Abich, Otto Hermann Wilhelm (1806-1886)
9   1846, Mount Ararat: Am. J. Sci., v. 2 (ns), no. 2, 291 (11 lines).
    Abstracted from l'Institut.
10 1848, On the volcanic plateaux of the lower Caucasus: Am. J. Sci., v. 5 (ns), no. 3, 423-428.
    Abstracted from Q. J. Geol. Soc.

Academy of Natural Sciences of Philadelphia
11 1828, Act of incorporation and by-laws ... : Phila., J. R. A. Skerrett, 12 p.

An account of the earthquakes which occurred in the United States, North America, on the 16th of December 1811, the 23rd of January, and the 7th of February, 1812
12 1812, Phila., R. Smith Jr., 84 p.

An account of the late dreadful earthquake and fire, which destroyed the city of Lisbon, the metropolis of Portugal. ... :
13 1756, Bos., Green and Russell, 23 p.
    Evans number 7602. First edition. Also "second edition" and "third edition", 23 p. (1756).

Accum, Friedrich Christian (1769-1838)
14 1808, System of theoretical and practical chemistry: Phila., Kimber and Conrad, 2 v.
    Contains an appendix by Thomas Cooper.
15 1809, A practical essay on the analysis of minerals, exemplifying the best methods of analysing ores, earths, stones, inflammable fossils and mineral substances in general: Phila., Kimber and Conrad, and B. and T. Kite, xxiv, [25]-236 p.
16 1814, System of theoretical and practical chemistry: Phila., Kimber and Conrad, 2v.
    Second American edition.
17 1817, A practical essay on chemical reagents, or tests: Phila., M. Carey and Son, 204 p.
    For review with excerpts see 8544.

Acosta, Col.
18 1850, Volcano in New Granada: Ann. Scien. Disc., v. 1, 234.
    Abstracted from Comptes Rendus.

Adam, M.
19 1849, On the identity of osmelite and pectolite: Am. J. Sci., v. 8 (ns), no. 1, 123 (5 lines).
    Abstracted from Annuaire de Chimie.

Adams, Benjamin
20 1820, Account of a great and very extraordinary cave in Indiana [in Jeffersonville]: Am. Antiquarian Soc. Trans. and Coll., v. 1, 434-436. Also in: Ohio Misc. Mus., v. 1, no. 4, 172-175 (1822).

Adams, Charles Baker (1814-1853)
    See also 7183 with J. W. Mighels.
21 1843, [Review with excerpts of Final Report on the geology of Massachusetts and notice of Elementary geology by Edward Hitchcock (see 5135 and 5132)]: N. Am. Rev., v. 56, 435-451.
22 1845, Circular [on Vermont geology]: Burlington, Vt., Chauncey Goodrich, 3 p.
    Also in 23.

Adams, C. B. (cont.)
23 1845, First annual report on the geology of the state of Vermont: Burlington, Vt., Chauncey Goodrich, 92 p., illus., folded colored map.
    Contains "Circular" (see 22), and reports or letters by E. H. Drury (see 3229), S. R. Hall and Z. Thompson (see 4734), E. Hitchcock (see 5170), J. Robbins (see 8809), E. M. Snow (see 9918), and S. W. Thayer (see 10220). Also contains "Bill" (see 10614).
    For review see 845, 3534.
24 1845, Some reminiscences of the geology of Jamaica: Assoc. Am. Geol's Nat's Proc., v. 6, 32-33.
25 1846, Notice of a small ornithichnite [from Wethersfield, Connecticut]: Am. J. Sci., v. 2 (ns), no. 2, 215-216, 2 figs.
26 1846, Second annual report on the geology of the state of Vermont: Burlington, Vt., Chauncey Goodrich, 2, 267, 2 p., illus.
    Contains reports or letters by C. L. Allen (see 163), S. R. Hall (see 4735), E. Hitchcock (see 5174), D. Olmsted (see 8033), Z. Thompson (see 10238), and J. W. Bailey (see 1436). Contains D. Olmsted obituary (see 2999).
    For review see 881 and 3560.
27 1847, Marl. [In answer to "S. E." (see 9251)]: Sch. J. and Vt. Agriculturist, v. 1, 42-43.
28 1847, Notice of an example of apparent drift furrows dependent on structure: Am. J. Sci., v. 3 (ns), no. 3, 433-434.
29 1847, [Observation on a polished rock]: Am. Q. J. Ag. Sci., v. 6, 215-216.
    See also article by W. C. Redfield 8719.
30 1847, On claystone concretions: Am. Q. J. Ag. Sci., v. 6, 207 [i.e. 255].
    In Assoc. Am. Geol's Nat's Proc.
31 1847, On the Taconic rocks: Am. Q. J. Ag. Sci., v. 6, 212 [i.e. 260].
    With discussion by E. Emmons. In Assoc. Am. Geol's Nat's Proc.
32 1847, Third annual report on the geology of the state of Vermont: Burlington, Vt., Chauncey Goodrich, 32 p., illus.
    Contains reports or letters by J. Hall (see 4710), S. R. Hall (see 4736), and T. S. Hunt (see 5408).
33 1848, Claystones [from Vermont]: Am. J. Sci., v.5 (ns), no. 1, 110 (7 lines).
    In Assoc. Am. Geol's Nat's Proc.
34 1848, Fourth annual report on the geological survey of Vermont: Burlington, Vt., Chauncey Goodrich, 8 p.
    Contains chemical analyses by T. S. Hunt (see 5409).
35 1848, Observations on a polished rock [from Winooski, Vermont]: Am. J. Sci., v. 5 (ns), no. 1, 110 (8 lines).
    In Assoc. Am. Geol's Nat's Proc.
36 1848, On the Taconic rocks [from Addison, Vermont]: Am. J. Sci., v. 5 (ns), no. 1, 108-110.
    In Assoc. Am. Geol's Nat's Proc.
37 1848, [Unusual glacial striae from the Onion River Valley]: Lit. World, v. 2, 229 (7 lines).
    Signed by E. B. (i.e. C. B.) Adams. In Assoc. Am. Geol's Nat's Proc.

Adams, Henry W.
38 1849, Aqueous agencies producing geological changes: Am. Lit. Mag., v. 4, no. 5, 649-658.
39 1849, Igneous agencies producing geological changes: Am. Lit. Mag., v. 4, no. 6, 713-722.

Adams, J.
40 1825, Remarks on art. VI. vol. V. no. 1. of this journal, and on a passage in Dr. Dwight's Travels, vol. III. p. 245, relating to some phenomena of moving rocks: Am. J. Sci., v. 9, no. 1, 136-144.
    Discussion of works by T. Dwight (see 3297) and C. A. Lee (see 6118).

Adams, John (Reverend)
41 1824, New locality of amethyst [at Bristol, Rhode Island]: Am. J. Sci., v. 8, no. 1, 199.

Adams, John (1735-1826)
42 1822, Honorable notice of Mr. Schoolcraft's memoir of a fossil tree: Am. J. Sci., v. 5, no. 1, 23-24.
    Notice of article by H. R. Schoolcraft (see 9263).

Adams, Michael
43 1808, Some account of a journey to the frozen sea, and of the discovery of the remains of a mammoth: Phila. Med. Phys. J., v. 3, pt. 1, 120-137.
    Translated from French. Also in: NY Med. Philos. Rev., v. 2, 248-258 (1810); Select Reviews, v. 3, 198-203 (1810); Emp. Arts Sci's, V. 2, 219-233 (1813). Title varies slightly.

Adams, W. A.
44 1843, Foot marks and other artificial impressions on rocks: Am. J. Sci., v. 44, no. 1, 200-202.
    On the origin of fossil human foot prints, from the limestone of St. Louis.

An address to the inhabitants of Rhode-Island on the subject of their coal mines:
45 1825, NY, J. Seymour, 16 p.

Adie, Alexander J.
46 1837, On the expansion of different kinds of stone from an increase of temperature, with a description of the pyrometer used in making the experiments: Franklin Inst. J., v. 20 (ns), 153-154; 200-201, 4 figs.
    From Royal Soc. Edinburgh Trans.

Adrain, Robert (1775-1843)
See also 5936 by T. Keith.
47   1818, Investigation of the figure of the earth, and of the gravity in different latitudes: Am. Philos. Soc. Trans., v. 1 (ns), 119-135.
48   1818, Research concerning the mean diameter of the earth: Am. Philos. Soc. Trans., v. 1 (ns), 353-366.

The Advocate of Science, and Annals of Natural History (Philadelphia, 1834-1835)
In addition to the five articles listed below, the single volume of this periodical contains many short notices on mines, earthquakes, volcanoes and recent scientific publications.
49   1834, Explosions in coal mines; the safety lamp: v. 1, 123-177 [i.e. 129].
50   1834, Country museums: v. 1, 178-180.
51   1834, The safey lamp: v. 1, 180-188, fig.
52   1834, [Review of Synopsis of the organic remains ..., by S. G. Morton (see 7470)]: v. 1, 198-199.
53   1834, Arctic expedition in search of a mineral treasure: v. 1, 292-294.

The African Repository and Colonial Journal (Washington, 1825-1850)
54   1846, Gold mine!: v. 22, 135 (10 lines).
Notice of mine discovery in Caldwell, Liberia.

Agassiz, Louis (1807-1873)
See also discussions by Agassiz of articles in the Assoc. Am. Geol's Nat's Proc. by N. D. Gale (4203), A. Guyot (4611, 4612); H. D. Rogers (8939, 8941); J. Hall (4730, 4731); R. W. Gibbes (4358, 4360); G. J. Chase (2327); and J. C. Warren (10701).
55   1835, Fossil fishes [from the Devonian of Great Britain]: Am. J. Sci., v. 28, no. 1, 74 (3 lines).
In British Assoc. Proc. See also article by R. I. Murchison (7508).
56   1835, Organic remains in the limestone of Burdic House: Am. J. Sci., v. 28, no. 1, 74 (13 lines).
In British Assoc. Proc.
57   1836, [Excerpts from Recherches sur les poissons fossiles]: Am. J. Sci., v. 30, no. 1, 33-53.
Translated and excerpted by R. Jameson (see 5712).
58   1838, Changes of temperature which our globe has undergone: Franklin Inst. J., v. 22 (ns), 342-348, fig.
From Edinburgh New Philos. J.
59   1838, Prof. Agassiz on the Echinodermata: Am. J. Sci., v. 34, no. 1, 212-213.
60   1841, Bird on the continent: Am. Eclec., v. 1, no. 2, 388 (8 lines).
Abstracted from British Assoc. Proc. On fossil birds from the Glaris slate.
61   1841, The former existence of glaciers in Scotland: Am. J. Sci., v. 41, no. 1, 191-193.
62   1841, Glaciers and boulders in Switzerland: Am. J. Sci., v. 41, no. 1, 59-60.
63   1841, Proceedings of the Geological Society, June 10th, 1840: Am. J. Sci., v. 41, no. 1, 190-191.
Abstracted from Annals and Mag. Nat. History. On Agassiz's glacial theory.
64   1842, The ice-period. A period of the history of our globe: Am. Eclec., v. 3, no. 2, 307-326; no. 3, 521-544.
Translated from the French by J. H. Agnew.
65   1846, Fins of fishes afford important characteristics: Am. J. Sci., v. 1 (ns), no. 3, 440 (4 lines).
66   1846, M. Agassiz on the geological development of animal life: Am. J. Sci., v. 1 (ns), no. 2, 280.
Abstracted from Annals and Mag. Nat. History.
67   1847, Analogy between the fossil flora of the European Miocene and the living flora of America: Am. J. Sci., v. 4 (ns), no. 3, 424-425.
Abstracted from London Ath.
68   1847, An introduction to the study of natural history in a series of lectures delivered in the hall of the College of Physicians and Surgeons, N. Y....: NY, Greeley and McElrath, 58 p., illus.
69   1847, [On the characters of Pygorhynchus from Georgia]: Bos. Soc. Nat. History Proc., v. 2, 193.
70   1847, [On transporting power of water currents]: Am. Q. J. Ag. Sci., v. 6, 208-209 [i.e. 256-257].
From Assoc. Am. Geol's Nat's Proc.
71   1847, Professor Agassiz lectures -The glaciers of Switzerland: Farm. and Mech., v. 1 (ns), 595-597, 15 figs.
72   1848, [Comparison of Zeuglodon and Dorudon]: Phila. Acad. Nat. Sci's Proc., v. 4, no. 1, 4-5.
73   1848, [Correlation of European and American strata]: Lit. World, v. 2, 225 (9 lines).
Abstracted from Assoc. Am. Geol's Nat's Proc.
74   1848, Geology - the glaciers of Switzerland: Genesee Farm. and Gardener's J., v. 9, 292-294, 14 woodcuts.
Abstract of Agassiz's Brooklyn Institute lecture.
75   1848, and A. A. Gould, Principles of zoology; touching the structure, development, distribution, and natural arrangement of the races of animals, living and extinct with numerous illustrations...for the use of schools and colleges: Part 1: Comparative physiology: Bos., Gould, Kendall and Lincoln, xix, 216 p., illus., plate, map.
76   1848, [Remarks on American fossil fish]: NY J. Medicine and Collateral Sci's, v. 10, 339.
Abstracted from NY Lyc. Nat. History Minutes.
77   1848, [Zoological classification, and geological succession]: Bos. Soc. Nat. History Proc., v. 3, 65.
78   1849, On the origin of the actual outlines of Lake Superior: Am. Assoc. Adv. Sci. Proc., v. 1, 79.
79   1849, [Remarks on crocodiles of the green sand of New Jersey and on Atlantochelys]: Phila. Acad. Nat. Sci's Proc., v. 4, no. 7, 169.

Agassiz, L. (cont.)
80 1849, The terraces and ancient river bars, drift, boulders, and polished surfaces of Lake Superior: Am. Assoc. Adv. Sci. Proc., v. 1, 68-70.
81 1850, The erratic phenomena about Lake Superior: Am. J. Sci., v. 10 (ns), no. 1, 83-101, fig.
82 1850, [Importance of fossil Leiodon from Alabama]: Am. Assoc. Adv. Sci. Proc., v. 3, 74 (4 lines).
83 1850, Lake Superior: its physical character, vegetation, and animals, compared with those of other and similar regions. With a narrative of the tour by J. Elliott Cabot, and contributions by other scientific gentlemen: Bos., Gould, Kendall and Lincoln, x, [2], 9-428 p., illus., maps.
For review with excerpts see 6367 and 8594. For review see 1006, 2359, 5488, and 10528. For notice see 1002.
84 1850, [On the age of the Connecticut sandstone]: Bos. Soc. Nat. History Proc., v. 3, 336-337; 341.
See also article by C. T. Jackson (5673).
85 1850, On the differences between progressive, embryonic, and prophetic types in the succession of organized beings through the whole range of geological times: Am. Assoc. Adv. Sci. Proc., v. 2, 432-438.
86 1850, On the fossil remains of an elephant found in Vermont: Am. Assoc. Adv. Sci. Proc., v. 2, 100-101.
With discussion by J. C. Warren and H. D. Rogers.
87 1850, [A memoir of Hugh Miller]: in Footprints of the Creator by H. Miller (see 7186), Preface.
88 1850, [On Massachusetts drift]: Bos. Soc. Nat. History Proc., v. 3, 183.
89 1850, ["Rain-drops" formed by surf action]: Bos. Soc. Nat. History Proc., v. 3, 201 (3 lines).
90 1850, Recommendations: Ann. Scien. Disc., v. 1, iii.
Testimonial letter for the Annual of Scientific Discovery.
91 1850, [Remarks on G. Troost's list of Tennessee crinoids (see 10409)]: Am. Assoc. Adv. Sci. Proc., v. 2, 62-63; 64.

Agricola (pseudonym; see 2244, 6603, 7825).

Agricultural Intelligencer, and Mechanic Register (Boston, 1820)
92 1820, Silver mine [at Zanesville, Ohio]: v. 1, 27, 55, 134.
93 1820, [Notice of Index of geology..., second edition, by A. Eaton (see 3325)]: v. 1, 134.
94 1820, A valuable discovery: v. 1, 134 (12 lines).
Notice of a freestone quarry at Savannah, Georgia.

The Agricultural Museum (Georgetown, D. C., 1811-1812)
95 1811, Mineral black [from Nazareth, Pennsylvania]: v. 1, 40 (9 lines).
Abstracted from Bos. Sentinel.
96 1811, Mines of Perkyomen [i.e. Perkiomen, Pennsylvania]: v. 1, 40.
Abstracted from Aurora.
97 1811, Salt works on the Kanawha River: v. 1, 227-229.
Abstracted from Richmond Enquirer.
98 1811, [Coal vein at Chesterfield, Virginia]: v. 1, 279 (9 lines).
Abstracted from Richmond Enquirer.

The Agriculturist (Nashville, Tennessee, 1840-1845)
99 1840, Tennessee marble quarries: v. 1, 121.

Aigster, F.
100 1813, Mineralogy in the vicinity of Pittsburgh: Med. Repos., v. 16, 211-213.

Aikin, Arthur (1773-1854)
101 1815, Manual of mineralogy: Phila., S. W. conrad, viii, [9]-275 p.
For review with excerpts see 3405. For review see 7345.
102 1826, Practical observations and results, in the manufacture of iron from the ore into bloom, in Shropshire: Franklin Inst. J., v. 2, 74-78.
From Technical Repos.

Aikin, John (1747-1822)
103 1797, Account of the lead mines in Derbyshire, (England) with the manner of working them: NY Mag., v. 2 (ns), 240-243.
From Description of Manchester.
104 1797, Account of the rock-salt works at Northwich: NY Mag., v. 2 (ns), 404-405.
From Description of Manchester.

Aikin, William E. A. (1807-1888)
105 1834, Some notices of the geology of the country between Baltimore and the Ohio River, with a section illustrating the superposition of the rocks: Am. J. Sci., v. 26, no. 2, 219-232, plate.
106 1842, Report of a geological examination of some tracts of land in Alleghany Co., Md., and Hampshire Co., Va.: Balt., n.p.
Not seen. In Philadelphia Academy of Natural Sciences Library.

Ainsworth, William Francis (1807-1896)
107 1832, Notice of Graham Island: Naturalist, v. 2, no. 4, 97-101, plate.
108 1832, Notice of the volcanic island thrown up between Pantellaria and Sciacca: Am. J. Sci., v. 21, no. 2, 399-404, 2 figs.
109 1837, Scientific results of the voyage to the Pacific and Behring's Strait, of H. B. M. Ship Blossom: Naval Mag., v. 2, 153-168.
From Royal Geog. Soc. London J.
110 1849, Ancient and modern monster reptiles: Daguerreotype, v. 3, 468-479.
From New Mon. Mag.

Akerly, Samuel (1785-1845)
111 1806, Mineralogical description of the country near the Wall-kill and the Shawangunk Mountains, in New York: Med. Repos., v. 9, 324-327.

Akerly, S. (cont.)
112 1810, A geological account of Dutchess Co. in New York: Am. Min. J., v. 1, no. 1, 11-16.
113 1811, On the improbability of finding coal on Long-Island or in the vicinity of New-York: Am. Min. J., v. 1, no. 2, 85 [i.e. 84]-86.
114 1814, On the geology and mineralogy of the island of New York: Am. Min. J., v. 1, no. 4, 191-198.
115 1820, An essay on the geology of the Hudson River and the adjacent regions; illustrated by a geological section of the country from the neighborhood of Sandy-Hook, in New Jersey, northward through the highlands in New-York toward the Catskill Mountains: NY, A. T. Goodrich and Co., 69 p., plate.
    For review see 7106.

Alabama. State of
116 1848, [Act appointing M. Tuomey state geologist]: Alabama General Assembly Documents, 1848.
    Not seen.

Albany Institute (Albany, New York)
117 1830, Catalogue of the library of the Albany Institute: Alb. Inst. Trans., v. 1, pt. 2, 7-24.

Alden, Timothy (1771-1839)
118 1804, On earthquakes [in New Hamphsire]: Mass. Hist. Soc. Coll., v. 9, 232-234.
119 1816, Letter from T. Alden on earthquakes: Mass. Hist. Soc. Coll., v. 4 (2s), 70-73.
120 1820, Antiquities and curiosities of western Pennsylvania: Am. Antiquarian Soc. Trans. and Coll., v. 1, 308-313.

Alessi, Guiseppe
121 1835, Origin of amber: Am. J. Sci., V. 28, no. 2, 293 (6 lines).
    Ascribed to M. T. Aessi (i.e. G. Alessi). In "Notice of the Transactions of the Geological Society of France" by C. U. Shepard (see 9497). Also in Advocate of Sci. and Annals Nat. History, v. 1, 283-286 (1835); and Am. J. Pharmacy, v. 6, 340-344 (1835).

Alexander, And.
122 1822, To convert earths or minerals into paints: Am. Mus. and Repos. Arts, v. 1, 01 [i.e. 10].

Alexander, Caleb (1755-1828)
123 1785, An account of eruptions, and the present appearances in West River Mountain [Chester County, New Hampshire]: Am. Acad. Arts Sci's Mem., v. 1, 316-317.

Alexander, Sir James Edward (1803-1885)
124 1833, Notice regarding the asphaltum or pitch lake of Trinidad: Franklin Inst. J., v. 11 (ns), 337-338; also in: Am. Farm., v. 15, 135 (1833); The Friend, v. 6, 192 (1833).
    From Edinburgh New Philos. J.

Alexander, John Henry (1812-1867)
    See also reports by J. T. Ducatel and J. H. Alexander (3231, 3232, 3235, 3236, 3238, 3239, and 3245).
125 1836?, [Acts of incorporation of the] George's Creek Coal & Iron Mining Company in Maryland: Balt.
126 1836, and P. T. Tyson, [Geology of the George's Creek coal basin]: in George's Creek Mining Co. An act..., (see 4330), 9-22, 6 folding plates.
127 1840, Report on the manufacture of iron; addressed to the governor of Maryland: Annapolis, Md., W. McNeir, printer to the senate, xxiv, [17]-269, 3 plates.
128 1840, Report on the manufacture of iron: Balt., F. Lucas, jr., xxiv, [17]-269.
    For review see 908.
129 1840, Second report on the manufacture of iron; addressed to the governor of Maryland: Annapolis, Md., W. McNeir, xviii, [9]-189.
    Pages [9]-26 are "An elementary treatise on iron making" by S. B. Rogers.
130 1848, Crystallography: memorandum: Am. J. Sci., v. 5 (ns), no. 1, 136.

Alger, Francis (1807-1863)
    See also reports on Nova Scotia geology with C. T. Jackson (5538; 5539; 5541, 5542, 5578)
131 1827, Notes on the mineralogy of Nova Scotia: Am. J. Sci., v. 12, no. 2, 227-232.
132 1840, Interesting minerals [from Nova Scotia, for trade]: Am. J. Sci., v. 38, no. 2, 380.
133 1840, Notice of minerals from New Holland: Am. J. Sci., V. 39, no. 1, 157-164, 7 figs.
    Also in: Bos. J. Nat. History, v. 3, no. 3, 306-318, 7 figs.
134 1840, Proposed exchange of minerals: Franklin Inst. J., v. 25 (ns), 143; 287.
135 1843, Beaumontite and lincolnite identical with heulandite: Bos. Soc. Nat. History Proc., v, 1, 145-146; also in: Am. J. Sci., v. 46, no. 2, 233-236, 2 figs. (1844); also in: Bos. J. Nat. History, v. 4, no. 4, 422-426, 2 figs. (1844)
136 1844, [Additions] containing the latest discoveries in American and foreign mineralogy: in An elementary treatise on mineralogy by W. Phillips (see 8325).
137 1844, Introduction: In An elementary treatise on mineralogy by W. Phillips (see 8325), i-cl.
138 1845, Formula of the masonite of Dr. Jackson: Am. J. Sci., v. 48, no. 1, 218-219.
139 1845, On the zinc mines of Franklin, Sussex County, New Jersey: Am. J. Sci., v. 48, no. 2, 252-264, fig.
    Also reprint edition, 15 p. With analysis of zincite by A. A. Hayes (see 4900).
140 1846, Dysluite identical with automolite: Am. J. Sci., v. 1 (ns), no. 1, 121.

Alger, F. (cont.)
141   1846, Notices of new localities of rare minerals, and reasons for uniting several supposed distinct species: Bos. J. Nat. History, v. 5, no. 3, 297-309; also in: Am. Q. J. Ag. Sci., v. 3, 69 (1846)
      Also reprint edition, 13 p.
142   1846, [Notes on certain minerals--phacolite, yttrocerite, ottrelite, polyadelphite]: Bos. Soc. Nat. History Proc., v. 2, 87-89.
143   1846, Phacolite observed in New York: Am. J. Sci., v. 1 (ns), no. 1, 121 (8 lines).
144   1846, Yttro-cerite: Am. J. Sci., v. 1 (ns), no. 1, 121 (3 lines).
145   1847, [Sapphire from Cherokee County, Georgia]: Bos. Soc. Nat. History Proc., v.2, 251 (8 lines).
146   1849, [Description of algerite from Franklin, NJ]: Bos. J. Nat. History, v. 6, no. 1, 118-120; also in: Am. J. Sci., v. 8 (ns), no. 1, 103-104 (1849).
      In articles by T. S. Hunt (see 5411).
147   1849, Examination of a mineral from Cherokee County, in Georgia: Bos. J. Nat. History, v. 6, 123-124.
148   1849, [Hydrate of magnesia from Hoboken, New Jersey]: Bos. Soc. Nat. History Proc., v. 3, 167.
149   1850, [American sandstone as building materials]: Bos. Soc. Nat. History Proc., v. 3, 241.
150   1850, [Barite from Nova Scotia]: Bos. Soc. Nat. History Proc., v. 3, 229 (7 lines).
151   1850, Crystallized gold from California: Am. J. Sci., v. 10 (ns), no. 1, 101-106, 10 figs; also in: Am. Rr. J., v. 23, 453-454, 10 figs (1850); also in: Bos. Soc. Nat. History Proc., v. 3, 266-267, fig. (1850).
      Also 7 p. reprint edition.
152   1850, [Gold from Vermont]: Bos. Soc. Nat. History Proc., v. 3, 382 (6 lines).
153   1850, [On the optical properties of kyanite]: Ann. Scien. Disc., v. 1, 131.
154   1850, [On a deposit of phosphorite in Hurdsville, Morris County, New Jersey]: Bos. Soc. Nat. History Proc., v. 3, 376-378.
155   1850, [On a singular cavity in a quartz crystal from Waterbury, Vermont]: Bos. Soc. Nat. History Proc., v. 3, 273-274, fig.
156   1850, On rutilated quartz crystals from Vermont and phenomena connected with them: Am. J. Sci., v. 10 (ns), no. 1, 12-19; also in: Am. Assoc. Adv. Sci. Proc., v. 2, 426-432, 28 figs. (1850).
157   1850, [Venniform mica from Waterbury, Vermont]: Bos. Soc. Nat. History Proc., v. 3, 249-250 (7 lines).
158   1850, [Zincite and franklinite from Franklin, New Jersey]: Bos. Soc. Nat. History Proc., v. 3, 321.

Allain (Mr.)
159   1850, and Bartenbach, On the extraction of gold from the copper ores of Chessy and Sain-Bel: Am. J. Sci., v. 9, no. 2, 297-298.
      From Comptes Rendus.

Allan, Robert (1806-1863; see 8325)

Allan, Thomas (1777-1833)
160   1814, The Faroe Islands: Am. Min. J., v. 1, no. 4, 259-261.
      Abstract from a letter.
161   1817, A geological account of the lead mine of Dufton in Westmoreland: J. Sci. Arts, v. 2, 198-200.

Allan, W.
162   1818, [Geology of Nice, Italy]: J. Sci. Arts, v. 5, no. 1, 173.
      From Royal Soc. Edinburgh Proc.

Allen, Charles Linnaeus (1828-1909)
163   1846, Letter [on a natural spring]: in second Vermont report by C. B. Adams (see 26), 261.

Allen, James (1691-1747)
164   1727, Thunder and earthquake, a loud and awful call to reformation. Consider'd in a sermon preached at Brooklyn, November the first; upon a special fast, occasion'd by the earthquake, which happen'd in the evening after the 29th day of October 1727...: Bos., Gamaliel Rogers for Joseph Edwards, [4], 49, [1] p.

Allen, James (1800?-1846)
165   1846, Journal of a march into the Indian country in the northern part of Iowa Territory in 1844, by Company I, 1st Regiment of Dragoons: US HED 168, 29-1, v. 6 (485), 7-18.
      See also letter with H. R. Schoolcraft (9272).

Allen, James M.
166   1842, The Onondaga salines: No. Light, v. 1, 153-155.

Allen, John H.
167   1846, Some facts respecting the geology of Tampa Bay, Florida: Am. J. Sci., v. 1 (ns), no. 1, 38-42.

Allen, Jonathan A.
168   1821, On the question whether there are any traces of a volcano in the West River Mountain: Am. J. Sci., v. 3, no. 1, 73-76.

Allen, Paul (1775-1826; see 6212 by M. Lewis)

Allen, Richard L. (1808-1873)
169   1838, Analysis of the mineral waters of Avon, New York: Not seen.
170   1844, A historical, chemical, and theraputical analysis of the principal mineral fountains at Saratoga Springs; together with general directions for their use: Saratoga Springs, NY, B. Huling, iv, 71 p.

Allen, R. L. (cont.)
171   1848, Analysis of the principal mineral fountains at Saratoga Springs, with general directions for their use(second edition): Saratoga Springs, NY, B. Huling, 72 p.

Allen, T.
172   1848, An account of the inflammable gas wells on the banks of the Kanawha River, in Virginia, as they appeared in June, 1847: Am. Philos. Soc. Proc., v. 4, 366-368.

Allen, Zachariah (1795-1882)
173   1833, The practical tourist, or sketches of the useful arts, and of society, scenery, &c. in Great Britain, France and Holland: Am. J. Sci., v. 23, no. 2, 213-225.

Allis, M.
174   1848, and M. Gillespie, Bromine from the bittern of salt works: Am. J. Sci., v. 6 (ns), no. 2, 293-294.

Allsop, Thomas
175   1826, On the mineral history of gold, and the separation of it from earthy and stony bodies: Franklin Inst. J., v. 2, 79-81; 168-171; 196-201.

Alpha (pseudonym; see 4025)

Amaranth, a Semi-monthly Publication Devoted to Polite Literature, Science, Poetry and Amusement (Ashland, Ohio, 1847)
176   1847, Origin of our planet: v. 1, no. 2, 6.

American Academy of Arts and Sciences, Boston, Massachusetts
   For yearly lists of officers and members see also the Academy's Proceedings and Memoirs.
177   1785, Presents made to the American Academy of Arts and Sciences; with the names of the donors: Am. Acad. Arts Sci's Mem., v. 1, xxxi-xxxii.
178   1804, Donations made to the American Academy of Arts and Sciences, since the publication of the first volume of their Memoirs: Am. Acad. Arts Sci's Mem., v. 2, pt. 2, 145-161.
179   1814, Donations made to the American Academy of Arts and Sciences since the publication of the 2d volume of their Memoirs: Am. Acad. Arts Sci's Mem., v. 3, pt. 2, [539]-541.
180   1821, Catalogue of books and other articles presented to the Academy since the publication of the third volume of Memoirs: Am. Acad. Arts Sci's Mem., v. 4, pt. 2, 417-419.
181   1846, Officers of the American Academy for the current year: Am. J. Sci., v. 2 (ns), no. 1, 144 (5 lines).
182   1848, Donations to the library: Am. Acad. Arts Sci's Proc., v. 1, 3-4; 36-38; 155-157; 181-182; 343-346.

American Agricultural Association of New York
183   1849, Analysis of soils: Hort. and J. Rural Arts, v. 3, no. 11, 534-535.

The American Agriculturist (New York, 1842-1850)
184   1842, [Review of Elements of agricultural chemistry and geology by J. F. W. Johnston (see 5818)]: v. 1, 188.
185   1842, [Review of Final report on the geology of Massachusetts by E. Hitchcock (see 5135)]: v. 1, 188-189.
186   1842, [Review of Fourche Cove report by W. B. Powell (see 8564)]: v. 1, 284.
187   1843-1850, [Brief notes on the analysis and improvement of soils]: v. 2, 168 (1843); v. 3, 188 (1844); v. 4, 227 (1845); v. 5, 91 (1846); v. 6, 190-191, 346-347 (1847); v. 7, 108-109, 233-234, 265-266 (1848); v. 8, 158-159, 267 (1849); v. 9, 220 (1850).
188   1846, [Review of Vestiges..., by R. Chambers (see 2292 et seq.)]: v. 5, 325.
189   1850, Agricultural geology: v. 9, 83-84; 125-126; 171-172; 268-270; 334-335.

The American Almanac and Repository of Useful Knowledge (Boston, 1829-1850)
190   1830, Shape of the earth and its size: v. for the year 1831, 101-103.
191   1830, Density of the earth: v. for the year 1831, 103-104.
192   1830, Temperature of the interior of the earth: v. for the year 1831, 104.
193   1832, Meteoric stones: v. for the year 1833, 71-75.
194   1836, Observations on the use of anthracite coal: v. for the year 1837, 61-69.
195   1839, Gold and silver: v. for the year 1840, 96-98.
196   1849, [Owen, Foster, and Whitney listed as United States geologists]: v. for the year 1850, 108.
197   1850, [Coal statistics from the United States]: v. for the year 1851, 188-195.
198   1850, [Owen, Foster, and Whitney listed as United States geologists]: v. for the year 1851, 111; 112.

American Association for the Advancement of Science
199   1849, American Association for the Advancement of Science: Am. J. Sci., v. 8 (ns), no. 2, 311-316; no. 3, 444.
200   1849, Communication relative to the geological survey of the state [of Vermont]: in Vt. House J. for 1849, 345-348.
201   1849, [Resolution supporting] geological surveys of the United States: Am. J. Sci., v. 8 (ns), no. 3, 450 (12 lines).
202   1850, American Association for the Advancement of Science: Am. J. Sci., v. 10 (ns), no. 1, 136-137 (10 lines).
203   1850, Charleston meeting of the American Association for the Advancement of Science: Am. J. Sci., v. 9 (ns), no. 3, 453.
204   1850, Meeting of the American Association for the Advancement of Science at New Haven: Am. J. Sci., v. 10 (ns), no. 2, 287-293.
205   1850, Proceedings of the American Association for the Advancement of Science, second meeting: Am. J. Sci., v. 9 (ns), no. 3, 454 (13 lines).

American Association for the Advancement of Science (cont.)
206   1850, Proceedings of the American Association for the Advancement of Science; third meeting, held at Charleston, S. C.: Am. J. Sci., v. 10 (ns), no. 2, 293-296.

American Association for the Promotion of Science
207   1848, American Association for the Promotion of Science: Am. J. Sci., v. 6 (ns), no. 2, 294.
208   1848, Notice of the meeting of the American Association for the Promotion of Science, held at Philadelphia, September 20 - 25, 1848: Am. J. Sci., v. 6 (ns), no. 3, 393-401.
209   1849, The American Association for the Promotion of Science: Am. J. Sci., v. 8 (ns), no. 1, 145-146.
210   1849, Memoirs read before the American Association for the Promotion of Science: Am. J. Sci., v. 7 (ns), no. 1, 151.

The American Athenaeum; or, Repository of the Arts, Sciences, and Belles Letters (New York, 1825)
211   1825, Earthquake [at Belida, Algiers]: v. 1, 196 (7 lines).
212   1825, Paleontography: v. 1, 231.
       On the distribution of fossil elks.
213   1825, [Review of Outline of geology by W. T. Brande]: v. 1, 244.
214   1825, The grotto of Balagansk, in Siberia: v. 1, 270.
215   1825, Vesuvius of West Pennsylvania: v. 1, 270.
       On a burning coal mine.
216   1825, The hot springs of Ouachita: v. 1, 352-353.

American Cultivator (see Boston Cultivator)

The American Eclectic; or, Selections from the Periodical Literature of All Foreign Countries (New York, 1841-1842)
217   1841, Animal remains: v. 1, no. 1, 152-155.
       On frozen fossil animals from Siberia.
218   1841, Use of geology to farmers: v. 1, no. 1, 178.
219   1841, Natural exhalation of carburetted hydrogen gas [at Glamorganshire]: v. 1, no. 1, 182.
220   1841, [Review of The Silurian system by R. I. Murchison]: v. 1, no. 1, 182.
221   1841, Erratic block of granite: v. 1, no. 1, 184 (4 lines).
222   1841, Convulsion and land-slip; near Axmouth, Devon: v. 1, no. 1, 184-186.
       Abstracted from Year Book of Facts.
223   1841, Land-slip in Russia: v. 1, no. 1, 186 (6 lines).
224   1841, Moses and the geologists [Reviews of The certainties of geology by W. S. Gibson and Scriptural geology, or an essay on the high antiquity ascribed to the organic remains imbedded in stratified rocks by G. Young]: v. 2, no. 1, 87-98.
225   1841, New volcanic islands: v. 2, no. 2, 386.
226   1841, [On fossil bones of the mammoth, Mastodon, Glyptodon, Dinotherium]: v. 2, no. 2, 387-388.
227   1841, Berghmehl [Sweden]: v. 2, no. 2, 391 (4 lines).
       Fossil infusoria found.
228   1841, Geology and physical geography: v. 2, no. 3, 587.
       From British Assoc. Proc. On the opinions of W. Phillips and W. Buckland.
229   1842, [Review of Elementary geology by E. Hitchcock (see 5134)]: v. 3, no. 3, 576-581.
       With excerpts from "Preface" by J. P. Smith (see 9896). From London Eclec. Rev.
230   1842, [Review of The combustion of coal by C. W. Williams]: v. 3, no. 3, 604-605.
       From Polytechnic J.
231   1842, [Review of Elements of mineral chemistry by F. Hoeffer]: v. 3, no. 3, 608-609.
       From Le Semeur.
232   1842, The ice period: - the economy of glaciers and the theory of their former extension over large portions of the earth's surface: v. 4, no. 1, 1-38.
       From Edinburgh Rev. Review of theories by L. Agassiz, F. J. Hugi, J. Charpentier, L. A. Necker, and others.
233   1842, Colliers and collieries: v. 4, no. 2, 316-351.

The American Farmer (Baltimore, 1819-1850)
234   1820, Geological: v. 2, 94-95.
       On recent fossil discoveries in America.
235   1820, Table of contents of the most remarkable medical springs in Europe: v. 2, 96 (table).
236   1821, The Island of Madeira, it wines, vines, geological phenomena, &c. &c. &c.: v. 3, 35-36.
237   1823-1830, [Short notes on coal and coal mines]: v. 4, 396 (1823); v. 8, 270 (1826); v. 10, 14 (1828); v. 12, 101 (1830).
238   1824-1849, [Short notes on soils and fertilization]: v. 5, 380-381 (1824); v. 14, 220 (1832); v. 15, 1, 103, 212, 236, 314-315, 322-323 (1833); v. 16, 10-12, 154 (1834); v. 2 (2s), 149, 218 (1835); v. 4 (2s), 162-163 (1837); v. 2, (3s), 6 (1840); v. 4 (3s), 402-403 (1843); v. 6 (3s), 6 (1844); v. 1 (4s), 14, 185-186 (1845); v. 2 (4s), 141-142 (1846); v. 5 (4s), 49, 116-117 (1849).
239   1824, Mineral waters [of New York]: v. 6, 134.
240   1824, Volcanoes: v. y, 191.
       Table of number of active volcanoes.
241   1824, Treatise on soils and manures: v. 6, 217-219; 225-227; 234-235; 241-243; 257-259; 314-316.
242   1826, Illinois lead mines: v. 8, 320 (9 lines).
243   1827, Mineral spring in New York: v. 9, 167-168.
244   1828, Temperature of the earth: v. 10, 231.
245   1829-1837, [Short notes on gold mining in the South]: v. 11, 95, 200, 222-223 (1829); v. 12, 134 (1830); v. 13, 55 (1831); v. 3 (2s), 317-318; v. 4 (2s), 62, 147 (1837).
246   1829, Mammoth: v. 11, 207.
       Fossil bones discovered at West Conococheague, Pennsylvania.

American Farmer (cont.)
247   1830, Mineral wealth of the West: v. 11, 374.
248   1832, Kanawha alum salt: v. 14, 271.
249   1832, Marl again [Review of Essay..., by E. Ruffin (see 9088)]: v. 14, 282.
      Signed by "M."
250   1834, Good news for the lower counties: v. 16, 137.
      Notice of J. T. Ducatel's geological researches in Maryland.
251   1834, Dr. Ducatel's survey of the upper counties: v. 16, 145.
252   1834, Geological surveys: v. 16, 191.
      Current surveys noted.
253   1834, Geological treat: v. 16, 207.
      On B. Silliman's geological lectures.
254   1834, Popular science: v. 16, 255.
      On the popularity of geology.
255   1835, Twelve reasons why geology should be introduced into common schools as a branch
      of study: v. 16, 351.
      From Scien. Tracts.
256   1835, Geology: v. 16, 377 (13 lines).
      On fossil palmwood from Paris, France.
257   1835, The importance of geological and mineral surveys: v. 2 (2s), 124-125.
258   1835, New and important discovery: v. 2 (2s), 150.
      On verd antique marble from Havrestraw, New York.
259   1836, Profesor Ducatel's report [notice of 3239, 3240]: v. 2 (2s), 355.
260   1836, American oil well [at Burkesville, Kentucky]: v. 3 (2s), 23.
261   1836, Lead mines in St. Lawrence County, New York: v. 3 (2s), 127.
262   1836, Geological definitions: v. 3 (2s), 215.
      Glossary of geological terms.
263   1837, The geological and topographical survey of Maryland [review of Report..., by J.
      T. Ducatel and J. H. Alexander (see 3245)]: v. 3 (2s), 359.
264   1837, Geological: v. 4 (2s), 59.
      On the importance of the Alabama Survey.
265   1837, Geology of Virginia: v. 4 (2s), 142.
      On the personnel and activities of the Virginia geological survey
266   1837, Plumbago [from Barrowdale, England and Raleigh, North Carolina]: v. 4 (2s), 167.
267   1837, Geological survey of Maryland: v. 4 (2s), 227-228.
268   1837, Valuable discovery: v. 4 (2s), 391.
      Concerns iron ore in Virginia.
269   1837, Mining region of Missouri: v. 4 (2s), 391.
270   1838, [Notice of 1837 report by J. T. Ducatel (see 3246)]: v. 5 (2s), 2.
271   1842, Structure of the earth [notice and review of W. L. Horton's essays to be
      published (see 5278)]: v. 3 (3s), 329, 358.
      See also reply by W. L. Horton (5279).
272   1842, The causes why some shell and petrified fish have been found upon the highest
      mountains of the world as well as some fossil remains at any depth under the
      earth's surface: v. 4 (3s), 134.
273   1843/1845, [Review of Lectures..., by J. F. W. Johnston (see 5819)]: v. 5 (3s), 85; 92;
      v, 1, (4s), 146.
274   1844, The earth a drop of melted lava: v. 6 (3s), 30.
      On the origin of the earth.
275   1845, [Prof. Johnston's lectures in Scotland]: v. 6 (3s), 382-383; 389-390.
276   1845, [Review of Travels..., by C. Lyell (see 6557)]: v. 1 (4s), 146.
      Signed by "B."

The American Gardener's Magazine and Register (Boston, 1835-1850)
277   1835, Boston Journal of Natural History [review of volume 1, nos. 1-2]: v. 1, 308-309.
      Signed by "B." Includes review of geological articles.
278   1836, Boston Journal of Natural History [review of volume 1, no. 3]: v. 2, 231-232.
      Signed by "R." Includes review of geological articles.
279   1837, Boston Journal of Natural History [review of volume 1, no. 4]: v. 2, 381-387.
      Signed by "J. L. R." Includes review of geological articles.
280   1844, [Review of Geographical and topographical descriptions of Wisconsin &c. by I. A.
      Lapham (see 6046)]: v. 10, 307-308.
281   1850, [Review of Scientific agriculture, or the elements of chemistry, geology, botany,
      and meteorology, applied to practical agriculture by M. M. Rogers]: v. 16, 80.

American Geological Society
282   1819-1825, American Geological Society: Am. J. Sci., v. 1, no. 4, 442 (1819); v. 2, no.
      1, 139-144; no. 2, 372-373 (1820); v. 3, no. 2, 360-362; v. 4, no. 2, 191-192
      (1821); v. 5, no. 2, 403-404 (1822); v. 6, no. 2, 377; v. 7, no. 2, 358; v. 8, no.
      2, 392-393 (1824); v. 9, no. 1, 178-179; v. 9, no. 2, 387; v. 10, no. 1, 201-203
      (1825).
283   1822, Several other catalogues of rocks and minerals presented to the American
      Geological Society: Am. J. Sci., v. 5, no. 2, 265-272.
284   1824, American Geological Society: Bos. J. Phil. Arts, v. 2, no. 2, 203-204.
285   1824, [Honorary members of the American Geological Society]: Am. J. Sci., v. 7, no. 2,
      393 (4 lines).

The American Historical Magazine and Literary Record (New Haven, 1836)
286   1836, Gold mines: v. 1, no. 1, 20-22.
      Signed by "S."
287   1836, Greenfield marble quarry: v. 1, no. 5, 199.

American Institute of New York
288   1835, [Memorial on a state geological survey]: Am. Rr. J., v. 4, 324-325.

American Institute of New York. Journal of the (New York, 1835-1839)
289   1835-1838, [Short notes on mines and mining]: v. 1, 152, 325, 392 (1835); v. 2, 40,
      222, 272, 608 (1836); v. 3, 147, 159, 222 (1837); v. 4, 112, 152 (1838).
290   1836, Gold mining [in North Carolina]: v. 1, 241-242.

American Institute of New York (cont.)
291   1836, Geological report of the secretary of state [of New York]: v. 1, 267-268.
292   1836, Volcanoes and earthquakes: v. 1, 378-380.
293   1836, Marble, granite, slate, and cement [from New England]: v. 2, 151-152.
294   1837, Barnard slate quarry [in Maine]: v. 2, 274-275.
295   1837, Salt: v. 2, 301-303.
          Manufacturing techniques used in New York.
296   1837, Geological survey [of New York]: v. 2, 483-487.
297   1837, Granite: v. 2, 504.
          Comparative densities of different types.
298   1837, Caledonia Springs [in Upper Canada]: v. 2, 606-607.
299   1837, Coal basins of Pennsylvania: v. 3, 220-221.
300   1837, Cabinet of mineralogy: v. 3, 235-236.
          Description of American Institute's mineral collection.
301   1838, The coal mines of Cumberland [Maryland]: v. 3, 557-558.
302   1838, Geological survey of the state of New York: v. 3, 598-602.
          From second annual New York report (see 7737).
303   1838, Gigantic mastodon: v. 4, 152-153.
          Fossil bones discovered in Crawford County, Ohio.
304   1839, Earthquake at Martinique: v. 4, 383-386.
          Account of February 1839 earthquake.
305   1839, Geology of the state of New York: v. 4, 550-551.
          From third annual New York report (see 7740).
306   1839, The Current River Copper Mining Company [in Shannon County, Missouri]: v. 4, 597-598.
307   1839, Geological surveys: v. 4, 598.
          Benefits of state geological surveys.
308   1839, Iron ore [from Essex County, New York]: v. 4, 613-614.

American Journal of Improvements in the Useful Arts and Mirror of the Patent Office (Washington, 1846-1847)
309   1846, [Brief notes on American mining]: v. 1, no. 2, 38-40.
310   1846, Gold washing: v. 1, no. 4, 82-83.
          Techniques described.

American Journal of Improvements in the Useful Arts and Mirror of the Patent Office of the United States (Washington, 1828)
311   1828, Oil stone [from Perry County, Ohio]: v. 1, 212 (3 lines).
312   1828, Notice regarding steatite or soapstone, and its principal uses: v. 1, 428-429.
313   1828, Temperature of a burning coal mine: v. 1, 451 (8 lines).

American Journal of Pharmacy (Philadelphia, 1829-1850)
314   1833, Analysis of coal: v. 5, 292-293.
          Signed by "T."
315   1835, Origin of sulphur: v. 1 (ns), 260 (7 lines).
316   1837, Iodine in minerals and plants: v. 3 (ns), 177 (9 lines).
317   1838, Silex: v. 3 (ns), 263-264.
          On the structure of semi-opal.

The American Journal of Science (New Haven, 1818-1850)
          The American Journal of Science, also known as Silliman's Journal, was from its founding in 1818 America's leading scientific periodical. Much of the unsigned material listed below (see 318 to 1021) is undoubtedly by Silliman or his assistant editors Benjamin Silliman, jr. (volume 34, 1838, through 1850) and James Dwight Dana (volume 1, new series, 1846, through 1850).
318   1818, Petrified wood from Antigua: v. 1, no. 1, 56-57.
319   1818, Porcelain and porcelain clays: v. 1, no. 1, 57-58.
320   1818, Native sulphur from Java: v. 1, no. 1, 58-59.
321   1818, Productions of Wier's Cave, in Virginia: v. 1, no. 1, 59.
322   1818, Notice of Professor Mitchill's edition of Cuvier's Essay on the Theory of the Earth [notice of 2754]: v. 1, no. 1, 68-69.
323   1818, Notice of Eaton's Index to the geology of the northern states, with a transverse section of them from Catskill Mountain to the Atlantic [notice of 3324]: v. 1, no. 1, 69-70.
324   1819, [Introductory remarks to "Hints on some of the outlines of geological arrangement ... " by W. Maclure (see 6622)]: v. 1, no. 3, 209-213.
325   1819, Localities of minerals and animal remains and acknowledgments of specimens received: v. 1, no. 3, 237-243.
          Contains letters by G. Chase (see 2324) and Mr. Mather (see 6865).
326   1819, Remarks on "Asbestos in anthracite" by J. W. Webster (see 10726): v. 1, no. 3, 244 (11 lines).
327   1819, Biographical notice of the late Archibald Bruce, M. D. Professor of Materia Medica and Mineralogy in the Medical Institution of the State of New-York, and Queen's College, New-Jersey; and a member of various learned societies in America and Europe. With a portrait: v. 1, no. 3, 299-304, plate.
328   1819, Dr. J. W. Webster's lectures: v. 1, no. 3, 304.
329   1819, Dr. Webster's cabinet: v. 1, no. 3, 305.
330   1819, Supposed identity of copal and amber: v. 1, no. 3, 306.
331   1819, The neocronite. - (A supposed new mineral.): v. 1, no. 3, 306-307.
332   1819, Cleaveland's Mineralogy [review of 2420]: v. 1, no. 3, 308-309.
333   1819, A new alkali: v. 1, no. 3, 309.
          On the discovery of lithium by J. A. Arfvedson.
334   1819, Red Rain: v. 1, no. 3, 309-310.
          On a meteorite fall at Naples, Italy.
335   1819, Augite: v. 1, no. 3, 310 (3 lines).
336   1819, New minerals: v. 1, no. 3, 310 (3 lines).
          Notice of scorodite and tungstate of lead from Europe.
337   1819, New metal: v. 1, no. 3, 310 (3 lines).
          On the discovery of selenium by J. J. Berzelius.
338   1819, Pure alumine [from France]: v. 1, no. 3, 310 (2 lines).
339   1819, Collections of American minerals: v. 1, no. 3, 310-311 (8 lines).
340   1819, C. S. Rafinesque, Esq.: v. 1, no. 3, 311.

American Journal of Science (cont.)
341   1819, Discovery of American tungsten and tellurium [from New Stratford, Connecticut]: v. 1, no. 3, 312.
342   1819, Additional note concerning the tungsten and tellurium [from New Stratford, Connecticut]: v. 1, no. 3, 316.
343   1819, Brucite: v. 1, no. 4, 439 (3 lines).
344   1820, Curious geological facts: v. 2, no. 1, 144-146.
      In part from Q. Rev. On fossils from Europe.
345   1820, Note [to "Fossil bones found in red sandstone" by N. Smith (see 9899)]: v. 2, no. 1, 147.
346   1820, American verd antique marble [from New Haven]: v. 2, no. 1, 165-166.
347   1820, Map shewing the relative height of the principal mountains on the globe: v. 2, no. 1, 168-169.
      Signed by "A. B."
348   1820, Cabinet of minerals, for sale: v. 2, no. 1, 169.
      Notice of G. Gibb's collection.
349   1820, Exploring expedition [to the Great Lakes]: v. 2, no. 1, 178.
350   1820, Lignite: v. 2, no. 2, 341 (9 lines).
      Abstracted by J. Griscom from Brande's J.
351   1820, Death of M. F. de St. Fond: v. 2, no. 2, 352.
      Abstracted by J. Griscom from Brande's J.
352   1820, Preparing for publication "A mineralogical dictionary": v. 2, no. 2, 352-353.
      Abstracted by J. Griscom from Tilloch's Philos. Mag.
353   1820, Dr. John Murray [obituary]: v. 2, no. 2, 355-356.
      Abstracted by J. Griscom from Tilloch's Philos. Mag.
354   1820, Breccia of Mont D'or [France]: v. 2, no. 2, 356-357.
      Abstracted by J. Griscom from Tilloch's Philos. Mag.
355   1820, A new metal, (aurum millium): v. 2, no. 2, 363 (8 lines).
356   1820, Cleaveland's mineralogy [notice of 2422]: v. 2, no. 2, 375 (6 lines).
357   1820, Sulphate of magnesia [from New York]: v. 2, no. 2, 375 (2 lines).
358   1821, Notice of "Geological essays, or an inquiry into some of the geological phenomena, to be found in various parts of America and elsewhere, by Horace H. Hayden" [notice of 4877]: v. 3, no. 1, 47-57.
359   1821, Notice of "A view of the lead mines of Missouri including some observations on the mineralogy, geology, geography, antiquities, soil, climate, population, and productions of Missouri and Arkansaw, and other sections of the western country, accompanied by three engravings; by Henry R. Schoolcraft" [review of 9255]: v. 3, no. 1, 59-72.
360   1821, Upon the fusion of various refractory bodies by Hare's blowpipe: v. 3, no. 1, 87-93, fig.
      Translated by B. Silliman from Annales de Chimie.
361   1821, Micaceous iron ore [from New Stratford, Connecticut] for oligiste of Hauy: v. 3, no. 2, 232.
362   1821, Limpid quartz, from Fairfield, New-York: v. 3, no. 2, 233-234.
363   1821, Geological survey of the county of Albany, &c., [by A. Eaton and T. R. Beck (notice of 3328)]: v. 3, no. 2, 239-240.
364   1821, Alum in decomposed mica slate [from Connecticut]: v. 3, no. 2, 240-241.
365   1821, Remarkable locality of garnet [from Connecticut]: v. 3, no. 2, 241.
366   1821, Curious variety of carbonate of lime resembling agaric mineral [from Vermont]: v. 3, no. 2, 242-243.
      Signed by "J. R."
367   1821, Pumice stone floating on the Mississippi: v. 3, no. 2, 247.
368   1821, Remarkable petrifaction: v. 3, no. 2, 282-283.
      From Thompson's Annals. On a fossil tree from Glasgow, Scotland.
369   1821, Fluor spar of Illinois: v. 3, no. 2, 367-368.
370   1821, Royal Society of London: v. 3, no. 2 378-379.
      From Thompson's Annals. Wollaston elected president.
371   1821, New works on petrifaction: v. 4, no. 1, 31-32.
      From Edinburgh Philos. J. Notice of works by Baron von Schlotheim, G. K. von Sternberg, and Emmerling.
372   1821, Other mineral localities, &c.: v. 4, no. 1, 50-55.
373   1821, Geological survey of the county of Rensselaer, state of New York [review of survey by A. Eaton and T. R. Beck (see 3328): v. 4, no. 1, 189.
374   1822, New system of mineralogy [notice of Treatise on Mineralogy by Brewster]: v. 4, no. 2, 245-246.
375   1822, Discovery of the fossil elk of Ireland, in the Isle of Man: v. 4, no. 2, 246.
      From Edinburgh Philos. J. abstracted by J. W. Webster.
376   1822, Notice regarding the working and polishing of granite in India: v. 4, no. 2, 246-249.
      Abstracted by J. W. Webster from Edinburgh Philos. J.
377   1822, Notice of a new work [review with excerpts of A description of the Island of St. Michael..., by J. W. Webster (see 10730)]: v. 4, no. 2, 251-266.
378   1822, Account of the earthquake at Kutch, on the 16th of June, 1819. Drawn up from published and unpublished letters from India: v. 4, no. 2, 315-319.
      From Edinburgh Philos. J.
379   1822, Test for barytes and strontian: v. 4, no. 2, 372-373.
      From Brande's J. abstracted by J. Griscom.
380   1822, French voyage of discovery: v. 4, no. 2, 391-392.
      Abstracted by J. Griscom from Annales de Chimie.
381   1822, [Activities of A. Brongniart and G. Cuvier]: v. 4, no. 2, 394-396.
382   1822, Prices of some minerals in London: v. 5, no. 1, 169-170.
      Signed by "F. L.".
383   1822, Human bones in a fossil state [from Saxony]: v. 5, no. 1, 171-173.
      Abstracted by J. Griscom from Bibliotheque Universelle.
384   1822, [Brief geological notes abstracted from Revue Encyclopedique by J. Griscom]: v. 5, no. 1, 175-177, 193.
385   1822, Geological survey of North Carolina [notice of D. Olmsted survey (see 8029 and 8030)]: v. 5, no. 1, 202.
386   1822, Professor Eaton's geological and agricultural survey of Rensselaer County. Dr. J. H. Steel's report of the geological structure of the county of Saratoga [reviews of 3331 and 10002]: v. 5, no 1, 203.

American Journal of Science (cont.)

387 1822, Yellow mineral from Sparta, N. Jersey, imbedded in white granular limestone: v. 5, no. 1, 203.
388 1822, Miscellaneous localities of minerals: v. 5, no. 2, 254-256.
389 1822, Mineralogy [notice of mineralogy text by F. S. Beudant]: v. 5, no. 2, 384 (4 lines).
390 1822, Mr. Brongniart's notice of American specimens of organized remains: v. 5, no. 2, 397.
391 1822, Professor Berzelius: v. 5, no. 2, 397-398.
      His opinions on Swedish minerals.
392 1822, Second edition of Cleaveland's Mineralogy [notice of 2422]: v. 5, no. 2, 404.
393 1822, Formation of calcareous spar: v. 5, no. 2, 405.
      Artificial calcite crystals grown.
394 1823, Vesuvius: v. 6, no. 2, 385-386.
      Abstracted by J. Griscom from Bibliotheque Universelle. On recent eruptions of Mt. Vesuvius.
395 1823, New journal [notice of Boston journal of philosophy and the arts]: v. 6, no. 2, 379 (3 lines).
396 1823, Ittro cerite [from Franklin, New Jersey]: v. 6, no. 2, 379 (2 lines).
397 1823, Finch's Geology [notice of 3833]: v. 7, no. 1, 178 (3 lines).
398 1823, Secondary granite [from Italy]: v. 7, no. 1, 186 (13 lines).
      Abstracted from Edinburgh Philos. J.
399 1823, A new fluid, with remarkable physical properties discovered in cavities of minerals: v. 7, no. 1, 186-187.
      Abstracted from Edinburgh Philos. J.
400 1823, Mineralogie appliquée aux arts [notice of text by M. Brard]: v. 7, no. 1, 188 (10 lines).
      From Revue Encyclopédique. Abstracted by J. Griscom.
401 1824, Review of "Outlines of the geology of England and Wales; with an introductory compendium of the general principles of that science: and comparative view of the structure of foreign countries. By Rev. W. D. Conybeare and Wm. Phillips" [(see 2597)]: v. 7, no. 2, 203-240.
      See also "Remarks additional" (417).
402 1824, Boston journal of philosophy and the arts, conducted by Drs. Webster and Ware, and Mr. D. Treadwell [notice and review]: v. 7, no. 2, 359.
403 1824, Cleavelandite: v. 7, no. 2, 390-391.
      Abstracted by J. Griscom from Annals of Phil. On localities of cleavelandite.
404 1824, Polishing of granite: v. 8, no. 1, 185 (7 lines).
405 1824, Prof. Hall's catalogue of minerals [notice of 4650]: v. 8, no. 1, 193.
406 1824, [Notice of A catalogue of American minerals by Dr. S. Robinson (see 8822)]: v. 8, no. 1, 200 (14 lines).
407 1824, Notice of the malleable iron of Louisiana: v. 8, no. 2, 218-225.
      Signed by "C. H.".
408 1824, Facts tending to illustrate the formation of crystals in geodes: v. 8, no. 2, 282-287.
409 1824, Notice of a geological and agricultural survey of the district adjoining the Erie Canal, in the state of New York - taken under the direction of the Hon. Stephen Van Rensselaer. Part I. By A. Eaton [notice of 3338]: v. 8, no. 2, 358-362.
410 1824, Biographical notice of Haüy: v. 8, no. 2, 362-371.
      Abstracted by J. Griscom from Revue Encyclopédique.
411 1824, Single blocks of stone: v. 8, no. 2, 375-376.
      Abstracted by J. Griscom from Revue Encyclopédique. On red granite from Russia.
412 1824, French periodical journals: v. 8, no. 2, 385-389.
413 1824, Copper of Great Britain and Ireland: v. 8, no. 2, 390 (4 lines).
      Abstracted from Edinburgh Philos. J.
414 1824, Medicinal properties of the waters of the Mississippi - mineral impregnation of the well water of Henderson, Ohio: v. 8, no. 2, 396-397.
415 1824, A system of universal geography [review of text by W. C. Woodbridge and E. Willard (see 10962 and 11038)]: v. 8, no. 2, 397-398.
416 1825, On bowlders and rolled stones: v. 9, no. 1, 28-39.
      Signed by "N."
417 1825, Remarks additional to the review of Conybeare and Phillips's Geology of England and Wales, (vol. VII, no. 2 of this journal.) with reference to the communication of Prof. Eaton in the last no. of the work, page 261: v. 9, no. 1, 146-154.
      See also 401 and 2597.
418 1825, Mr. Hitchcock's geological sketch of the country on Connecticut River (see 5081): v. 9, no. 1, 179 (8 lines).
419 1825, Russia - gold mines: v. 9, no. 1, 183.
      Abstracted by J. Griscom from Cronstadt Gaz.
420 1825, Effects of an earthquake on the vegetation of wheat: v. 9, no. 1, 208 (7 lines).
      Abstracted by J. Griscom from Edinburgh Philos. J.
421 1825, Notice of Prof. Eaton's geological survey of the district adjoining the Erie Canal [notice of 3338]: v. 9, no. 2, 355-356.
      Signed by "C. D." (i.e. Chester Dewey?).
422 1825, Ferussac's Bulletin universel des sciences, &c. &c.: v. 9, no. 2, 374-375.
423 1825, Marine fossil plants [review of text by A. Brongniart]: v. 9, no. 2, 375-376 (14 lines).
      Signed by "C. H."
424 1825, Prehnite - olivine: v. 9, no. 2, 378-379.
      Signed by "C. H."
425 1825, Dr. Robinson's catalogue of American minerals [review of 8825]: v. 9, no. 2, 396-397.
426 1825, Dr. Van Rensselaer's geology [review of 10551]: v. 9, no. 2, 397 (7 lines).
427 1825, Col. George Gibbs: v. 9, no. 2, 397 (4 lines).
      Notice of his Siberian mineral collection.
428 1825, Lehigh coal: v. 9, no. 2, 397-398.
      New coal vein discovered at Lehigh, Pennsylvania.
429 1825, Mineralogy of the coast of Labrador, and of the shores of the St. Lawrence: v. 9, no. 2, 398-399.
      Abstracted from the Quebec Gaz.

American Journal of Science (cont.)
430   1825, Mineralogical notice: v. 9, no. 2, 400-401.
         Notice of Dr. Morton's mineral collection for sale.
431   1825, Amethyst of Rhode-Island: v. 9, no. 2, 401-402.
432   1825, Soapstone of Middlefield, Mass.: v. 10, no. 1, 19.
         "Correction of a passage in Dr. Dwight's Travels" (see 3297).
433   1825, Large mass of amber found in the Island of New-Providence: v. 10, no. 1, 171.
         Abstracted by J. Griscom from Bibliothèque Universelle.
434   1825, [Revue encyclopédique]: v. 10, no. 1, 178-179.
         Am. J. Sci., reviewed by Revue Encyclopédique.
435   1825, English locality of metallic lead: v. 10, no. 1, 191.
         Abstracted from Brewster's J.
436   1826, Notice of scientific societies in the United States: v. 10, no. 2, 369-376.
         Signed by "S. E. D."
437   1826, Subterranean sounds [at Meleda Island]: v. 10, no. 2, 377.
         Abstracted by J. Griscom from Bibliothèque Universelle.
438   1826, Suggestions as to the origin of fountains: v. 10, no. 2, 394-395.
         Signed by "Z."
439   1826, Lead mines of the United States: v. 10, no. 2, 398 (7 lines).
440   1826, Mineralogy [notice of Manual of Mineralogy, by J. L. Comstock (see 2505)]: v. 10, no. 2, 400 (4 lines).
441   1826, Proofs that general and powerful currents have swept and worn the surface of the earth: v. 11, no. 1, 100-104.
         Signed by "A."
442   1826, Collections of foreign minerals: v. 11, no. 1, 197 (12 lines).
         Minerals for sale by Moldenhauer in Germany.
443   1826, Reliquiae diluvianae. Hayden's geological essays. Eaton's survey [notice of 3338 and 4877]: v. 11, no. 1, 197-198.
444   1826, Collection of minerals: v. 11, no. 2, 385 (11 lines).
         Notice of collection of United States minerals for sale.
445   1826, Double refraction: v. 11, no. 2, 385 (8 lines).
         On the optical properties of calcite.
446   1827, Native gold [from New Fane, Vermont]: v. 12, no. 1, 171 l (8 lines).
         Abstracted from Bos. J. Phil. Arts.
447   1827, New works on mineralogy and geology [notice of Elements of mineralogy by J. L. Comstock (see 2506) and Manual of mineralogy and geology by E. Emmons (see 3491)]: v. 12, no. 1, 173 (9 lines).
448   1827, Geological survey of Pennsylvania: v. 12, no. 1, 173-176.
449   1827, Mineralogy of Nova Scotia: v. 12, no. 1, 176 (15 lines).
         Abstracted from Bos. J. Phil. Arts.
450   1827, Collection of aerolites and meteoric iron: v. 12, no. 1, 183.
         Notice of Mr. Heuland's collection in London.
451   1827, Mineralogy of Vesuvius: v. 12, no. 1, 185-187.
         Abstracted by J. Griscom from Bibliothèque Universelle.
452   1827, Georama: v. 12, no. 1, 190-191.
         Abstracted by J. Griscom from Revue Encyclopédique. On a geological teaching aid.
453   1827, Necrology: v. 12, no. 1, 192-193.
         Obituary of Scipio Breislak. From Revue Encyclopédique.
454   1827, Russian mines: v. 12, no. 1, 197 (10 lines).
         Abstracted by J. Griscom from Revue Encyclopédique.
455   1827, Remarks on Prof. Eaton's proposed improvements in the manufacture of compass needles [remarks on 3343]: v. 12, no. 2, 232-234.
         Signed by "Surveyor."
456   1827, Physical and medical journal of Cincinnati [notice of]: v. 12, no. 2, 382.
457   1827, New work on geology [notice of Introduction to the study of geology by J. Finch (see 3833)]: v. 12, no. 2, 383 (8 lines).
458   1827, [Mineralogical notes abstracted from Ferussac's Bulletin by J. Griscom]: v. 12, no. 2, 384-386.
         On nitre and gold from Russia, opal from Hungary, and menardite from Spain.
459   1827, [French opinions of American journals]: v. 12, no. 2, 394-395.
         Abstracted by J. Griscom from Revue Encyclopédique.
460   1827, Enormous fossil vertebra [from Dorset, England]: v. 13, no. 1, 186-187.
         Abstracted by C. U. Shepard from Annales de Chimie.
461   1827, New edition of Cleaveland's Mineralogy [notice of 2426]: v. 13, no. 1, 198-199.
462   1827, Cabinet of minerals for sale: v. 13, no. 1, 199 (10 lines).
463   1827, The late Dr. Robinson's collection of minerals: v. 13, no. 1, 199 (10 lines).
464   1828, Annunciation of the second part of Professor A. Eaton's report of the geological survey of the Erie Canal [notice of 3338]: v. 13, no. 2, 383-385.
         Published with Eaton's "Tabular view of North American rocks" (see 3352).
465   1828, Remarks on Mr. Barnes' notice respecting magnetic polarity: v. 14, no. 1, 121-124.
         Signed by "Surveyor."
466   1828, Geological notice [notice of "Geological nomenclature" by A. Eaton (see 3348)]: v. 14, no. 1, 190.
467   1828, Valuable collection in geology and mineralogy: v. 14, no. 1, 197-198.
         Notice of G. W. Featherstonhaugh's mineral collection.
468   1828, Asbestos [from New Milford, Connecticut]: v. 14, no. 1, 199 (6 lines).
469   1828, [Notice of minerals abstracted from Ferussac's Bulletin by J. Griscom]: v. 14, no. 1, 204-205.
         On gold and platinum from Russia and sapphire from Naxos.
470   1828, Notice of the reports on the geology of North Carolina, conducted under the direction of the Board of Agriculture; by Denison Olmsted [review with excerpts of 8029 and 8030]: v. 14, no. 2, 230-251.
471   1828, Notice regarding steatite or soapstone, and its principal uses: v. 14, no. 2, 376-377.
472   1828, Singular organic relic [from New Haven, Connecticut]: v. 14, no. 2, 393-394.
473   1828, Chesterfield [Massachusetts] tourmaline: v. 14, no. 2, 400 (4 lines).
474   1828, Comparative durability of marble and granite: v. 15, no. 1, 168.
         Abstracted by J. Griscom from Edinburgh Philos. J.

American Journal of Science (cont.)
475 1828, Sideroscope: v. 15, no. 1, 177-178.
         Abstracted by J. Griscom from Ferussac's Bulletin.
476 1828, Crystal bed [from Russia]: v. 15, no. 1, 186 (14 lines).
         Abstracted by J. Griscom from Annales Patriotiques.
477 1829, Plaster casts of fossil bones of the mastodon: v. 15, no. 2, 400.
478 1829, William Maclure, Esq.: v. 15, no. 2, 400-401.
         Describes Maclure's activities from 1828-1829.
479 1829, Cleaveland's Mineralogy, 3rd edition [notice of 2426]: v. 15, no. 2, 401 (6 lines).
480 1829, Polar explorations: v. 16, no. 1, 124-151.
481 1829, Remarks on the specimens transmitted by Mr. Goodrich [from Hawaii]: v. 16, no. 2, 347-350.
482 1829, Carpenter's Saratoga powders, for making Congress Spring or Saratoga waters: v. 16, no. 2, 369-370.
483 1829, Notice of a projected improvement in the method of blasting rocks, making tunnels through mountains, &c. with the results of some preliminary experiments: v. 16, no. 2, 372-373, fig.
484 1829, Ohio oil stone: v. 16, no. 2, 374 (13 lines).
485 1829, Report of the Chester County cabinet: v. 16, no. 2, 374-375.
486 1829, Chalcedony: v. 16, no. 2, 375-376.
487 1829, Group of crystals of common salt [from Curacao]: v. 16, no. 2, 377.
488 1829, Fibrous gypsum of Onondaga County, New York: v. 16, no. 2, 377.
489 1829, Natural history in Canada: v. 16, no. 2, 378.
         Notice of societies.
490 1829, Cabinet of the late William Phillips: v. 16, no. 2, 379-380.
491 1829, Remains of the mammoth [from Behring's Strait]: v. 16, no. 2, 382-383.
         Abstracted from the Scotsman.
492 1829, [Mineralogical notes abstracted from Ferussac's Bulletin]: v. 16, no. 2, 384-390.
         On analysis of pluranium from Russia, and potash, iodine, and platinum.
493 1829, A congress of savans [in Berlin]: v. 16, no. 2, 386-387.
         Abstracted from Revue Encyclopédique.
494 1829, Manufactory of diamonds: v. 16, no. 2, 394 (11 lines).
         Abstracted from Annals de Chimie.
495 1829, Sketch of the geology of the Arctic regions, and the Steppes of Russia, with notices of Siberia, Kamschatka, and the Kurile Islands: v. 17, no. 1, 1-34.
496 1829, Geology: v. 17, no. 1, 197 (9 lines).
         Abstracted from Diario de Fisica. Describes alluvial deposits in the Gulf of Bengal.
497 1829, Foreign memoirs and pamphlets: v. 17, no. 1, 201-202.
         Notice of works by C. Lyell, A. Brongniart, R. Murchison, and J. Fitton.
498 1829, Gold in Maryland: v. 17, no. 1, 202 (7 lines).
499 1829, Acidulous sulphate of iron: v. 17, no. 1, 203 (11 lines).
         Signed by "Correspondent."
500 1829, Iron mines [of Orange County, New York]: v. 17, no. 1, 203-204.
501 1830, Philosophical transactions of the Royal Society of London, for the year 1829 [notice and excerpts]: v. 17, no. 2, 361-368.
502 1830, Use of the blowpipe in chemistry and mineralogy. [In German.] By Jacob Berzelius [notice of 1668]: v. 17, no. 2, 411 (2 lines).
503 1830, The diamond: v. 17, no. 2, 372-373.
         Abstracted by J. Griscom from Bibliothèque Universelle.
504 1830, Lyceums [in Massachusetts]: v. 17, no. 2, 415 (8 lines).
505 1830, Baron Humboldt's expedition [to Siberia]: v. 17, no. 2, 405-406.
         Abstracted from Courier des Electeurs.
506 1830, A map of the U. States' lead mines on the Upper Mississippi, by R. W. Chandler of Galena [review of 2306]: v. 17, no. 2, 416 (9 lines).
507 1830, New treatise on mineralogy. Preparing for press [notice of <u>Treatise</u> <u>of</u> <u>mineralogy</u> by C. U. Shepard (see 9489)]: v. 17, no. 2, 416 (10 lines).
508 1830, Iron works of Sweden: v. 18, no. 1, 173.
         Abstracted from Revue Encyclopédique.
509 1830, Gold and platina [of Russia]: v. 18, no. 1, 190.
         Abstracted from Annals of Phil.
510 1830, Tennessee meteorite: v. 18, no. 1, 200 (9 lines).
511 1830, Rensselaer School flotilla: v. 18, no. 1, 200-201.
512 1830, Dr. Morton's paper [notice of <u>Organic</u> <u>remains</u> by S. G. Morton (see 7470)]: v. 18, no. 1, 201 (5 lines).
513 1830, Natural, statistical, and civil history of the state of New York, in three volumes, 8vo. by James MacAuley [review of 6588]: v. 18, no. 1, 206-207.
514 1830, On the carbonization of lignite: v. 18, no. 2, 371-372.
515 1830, Cannel coal in Ohio: v. 18, no. 2, 376-377.
516 1830, Plumbago of Sturbridge, Mass.: v. 18, no. 2, 377-378.
517 1830, Notice of the circumstances attending the fall of the Tennessee meteorites, May 9, 1827: v. 18, no. 2, 378-379.
         Abstracted from Nashville Banner.
518 1830, A history of the county of Berkshire, Mass. in two parts, the first being a general view of the whole county, and the second an account of the several towns; by gentlemen of the county, clergymen and laymen [review of 5069]: v. 18, no. 2, 387 (12 lines).
519 1830, Collection of New England rocks with their imbedded minerals, for sale: v. 18, no. 2, 390 (9 lines).
520 1830, Cloth of Amianthus: v. 18, no. 2, 401 (10 lines).
         Abstracted by J. Griscom from Annales d'Hygiène Publique.
521 1830, On the treatment of siliceous minerals by carburetted alkalies: v. 18, no. 2, 404.
         Abstracted by J. Griscom from Bibliothèque Universelle.
522 1830, Daily magnetic variation: v. 19, no. 1, 189 (13 lines).
         Abstracted from Q. J. of Sci.
523 1830, Science in Madrid: v. 19, no. 1, 194-195.
         Abstracted from Revue Encyclopédique.
524 1830, Russian diamond mines: v. 19, no. 1, 199 (14 lines).
         Abstracted from Revue Encyclopédique.

American Journal of Science (cont.)
525 1831, Meteoric iron in Bohemia: v. 19, no. 2, 384-386.
     Abstracted from Brewster's J.
526 1831, Burning coal mine at New Sauchie [Devon, England] : v. 19, no. 2, 386.
527 1831, On the connection of the solfaterra with Vesuvius: v. 19, no. 2, 387.
     Abstracted from Brewster's J.
528 1831, A notice of the mammoth [from Mourum] : v. 19, no. 2, 388 (11 lines).
     Abstracted from Bibliothèque Universelle.
529 1831, Mauch Chunk anthracite mines: v. 20 no. 1, 163 (13 lines).
530 1831, Encyclopaedia Americana: v. 20, no. 1, 167 (12 lines).
     Notice of geology articles (see 3586).
531 1831, Literary and scientific societies of Canada: v. 20, no. 1, 168 (13 lines).
532 1831, The last annual meeting of the naturalists and physicians of Germany: v. 20, no. 1, 175.
     Abstracted by J. Griscom from Revue Encyclopédique.
533 1831, Arsenic in sea salt: v. 20, no. 1, 193 (9 lines).
     Abstracted from Philos. Mag.
534 1831, Polarizing rock: v. 20, no. 1, 198-199.
     Abstracted by C. U. Shepard from Schweigger's Jahrbuch.
535 1831, Notices of eminent men deceased in Great Britain: v. 20, no. 2, 300-308.
     Obituaries of J. S. Miller, Chenevix, Smithson, and the Duke of Athol.
536 1831, Interesting discovery of fossil animals [from France] : v. 20, no. 2, 382 (11 lines)
     Abstracted from Philos. J. by J. Griscom.
537 1831, Projected branch mint of North Carolina: v. 20, no. 2, 401-404.
538 1831, Marl for manure: v. 20, no. 2, 410-411.
539 1831, New monthly journal [review of journal by Featherstonhaugh (see 3746)] : v. 20, no. 2, 412 (6 lines).
540 1831, Alum in mica slate [from Colchester, Connecticut] : v. 20, no. 2, 418.
541 1831, Physical geography: v. 21, no. 1, 127-132.
     Translated by J. Griscom from Bibliothèque Universelle. On mountain ranges of Europe.
542 1831, Addresses of the Rev. Adam Sedgwick at the anniversary meetings of the Geological Society of London: v. 21, no. 1, 186.
543 1831, Science in the West: v. 21, no. 1, 187-188.
     On scientific societies of western United States, especially the Detroit Lyceum.
544 1831, Science in the South: v. 21, no. 1, 188 (7 lines).
     On scientific societies of southern United States, especially the Tuscaloosa Lyceum.
545 1832, Professor Eaton's geological text-book [notice of 3374] : v. 21, no. 2, 389 (4 lines).
546 1832, New work on mineralogy [notice of Treatise on mineralogy by C. U. Shepard (see 9489)] : v. 21, no. 2, 389-390 (9 lines).
547 1832, The journal of a naturalist [review of book by J. L. Knapp (see 5977)] : v. 21, no. 2, 390-391.
548 1832, Remarks upon the natural resources of the western county: v. 22, no. 1, 122-125.
     Signed by "K."
549 1832, Mineralogy and geology of Nova Scotia, by C. T. Jackson and F. Alger [review of 5538, 5539] : v. 22, no. 1, 167-169.
550 1832, Statistics of iron in the United States: v. 22, no. 1, 179-180.
551 1832, Cabinet of minerals, &c.: v. 22, no. 1, 180.
     Notice of L. Feuchtwanger's mineral collection for sale.
552 1832, Notice of Eaton's geological text book, second edition, 8vo., pp. 140. [notice and review of 3374] : v. 22, no. 2, 391.
553 1832, Treatise on mineralogy; by Charles Upham Shepard [review with excerpts of 9489] : v. 22, no. 2, 395-402.
554 1832, West Chester County Cabinet of Natural Science: v. 22, no. 402 (12 lines).
555 1832, Note on the progressive increase of temperature as we descend beneath the surface of the earth: v. 23, no. 1, 14-18.
     Signed "J."
556 1832, Fossil shells of the Tertiary formations of North America, illustrated by figures drawn on stone, from nature; by T. A. Conrad [review of 2545] : v. 23, no. 1, 204-205.
557 1833, Professor Jacob Green's monograph of the trilobites of North America, with colored models of the species [review with excerpts of 4535] : v. 23, no. 2, 395-398.
558 1833, Prof. Hitchcock's report on the geology of Massachusetts [review of 5101] : v. 23, no. 2, 389-390.
559 1833, Fossil shells of the Tertiary formations of the United States, by T. A. Conrad [review of 2545] : v. 23, no. 2, 405.
560 1833, Anthracite in Wrentham, Mass.: v. 23, no. 2, 405 (14 lines).
561 1833, Lehigh Coal and Navigation Company: v. 24, no. 1, 173-174.
     Notice of their explorations for coal in Pennsylvania.
562 1833, Delaware Academy of Natural Sciences, and the address of Dr. Henry Gibbons, at Wilmington: v. 24, no. 1, 177 (13 lines).
563 1833, Flint's History and geography of the Valley of the Mississippi, &c. 2nd edition [review of 3887] : v. 24, no. 1, 179-181.
564 1833, New works in England: v. 24, no. 1, 212.
     Recent English geology texts listed.
565 1833, Miscellaneous notices, in a letter to the editor, dated from an American national ship, off Cape de Gott, in Spain, August 11, 1832: v. 24, no. 2, 237-246.
     A description of an eruption of Mt. Vesuvius, and Grotto del Cane.
566 1833, Use of mica in chemical analyses on a small scale: v. 24, no. 2, 373 (7 lines).
     Abstracted by J. Griscom from Archives de Brandes.
567 1833, Mines of Freyberg in Saxony: v. 24, no. 2, 376-377.
     Abstracted by J. Griscom from Annales des Mines.
568 1833, Extract from the MS. of an unpublished narrative of travels and observations in South America: v. 24, no. 2, 382-383.
     On the volcano Cotopaxi. Signed by "T."

American Journal of Science (cont.)

569   1833, Prof. Hitchcock's reports on the geology of Massachusetts [notice of 5105]: v. 24, no. 2, 396-397.
570   1833, Manual of mineralogy and geology; by Ebenezer Emmons [review of 3492]: v. 24, no. 2, 397.
571   1833, Geological text-book; second edition [review of text by A. Eaton (see 3374)]: v. 24, no. 2, 399 (10 lines).
572   1833, Hezekiah Howe & Co., New Haven: v. 24, no. 2, 400.
       Notice of geological publications for sale.
573   1833, Marine shells in the coal formation [of Yorkshire, England]: v. 25, no. 1, 199.
574   1833, Col. George Gibbs [obituary]: v. 25, no. 1, 214-215.
575   1833, Dr. William Meade [obituary]: v. 25, no. 1, 215-216.
576   1834, Contributions to geology: by Isaac Lea [review with excerpts of 6093]: v. 25, no. 2, 413-423.
577   1834, Recent scientific publications in the United States: v. 25, no. 2, 425-428.
578   1834, Physical discovery. - (Retrospective.) - The magnetic needle made to indicate the true North, and rendered more steady by a newly invented magnetic process: v. 26, no. 1, 90-92.
579   1834, British and American journals of science: v. 26, no. 1, 174-175.
580   1834, New observations upon the action of sulphate of lime: v. 26, no. 1, 181-182.
       Translated "by a lady" from Annales de l'Institut Royal Horticole de Fromont. On the use of gypsum. Signed by "S. B."
581   1834, Recent scientific publications in the United States [notice of Outlines of geology by J. L. Comstock (see 2511), and Geology of the Scriptures by G. Fairholme (see 3628)]: v. 26, no. 1, 208-209.
582   1834, Cabinet of the late Dr. William Meade: v. 26, no. 1, 209-210.
583   1834, Outlines of geology, &c., by Dr. J. L. Comstock [review of 2511]: v. 26, no. 1, 212-213.
584   1834, Prof. Hitchcock's report on the geology, mineralogy, botany and zoology of Massachusetts: 1833 [review of 5105]: v. 26, no. 1, 213.
585   1834, Second American edition of Bakewell's Geology [review of 1457]: v. 26, no. 1, 213-214.
586   1834, Mr. C. U. Shepard's private school of mineralogy and other branches of natural history: v. 26, no. 1, 215.
587   1834, Ligneous stems of American coal-fields desired: v. 26, no. 1, 215-221.
       Notice of H. T. M. Witham's experiments on thin sections of fossil plants.
588   1834, Mantell's geology of the south east of England: v. 26, no. 1, 216-217.
       Notice of G. A. Mantell's research and publications.
589   1834, Septaria of extraordinary size and beauty [from Lyme Regis, England]: v. 26, no. 1, 217-218.
590   1834, On porcelain and earthenware: v. 26, no. 2, 233-261, 3 plates.
       On the mineralogy and localities of porcelin clay, 245-249.
591   1834, Recent scientific publications in the United States: v. 26, no. 2, 397.
592   1834, Elements of geology [review of text by W. W. Mather (see 6875)]: v. 26, no. 2, [406] (10 lines).
593   1834, Vertebral bone of a mastodon [from Britain, Connecticut]: v. 27, no. 1, 165-166.
594   1834, Fossil tooth [from Chatauque Lake, New York]: v. 27, no. 1, 166-168.
595   1834, Ledererite not a new mineral: v. 27, no. 1, 171-172 (12 lines).
       Abstracted from London and Edinburgh Philos. Mag.
596   1834, Annual report of the Regents of the University to the legislature of New York: v. 27, no. 1, 177-178.
       On mineralogy classes taught in New York University.
597   1834, Treatise on mineralogy, by Charles U. Shepard [notice of 9502]: v. 27, no. 1, 200 (6 lines).
598   1835, Notice of the Transactions of the Geological Society of Pennsylvania - August, 1834, Part I [notice of 4314]: v. 27, no. 2, 347-355.
599   1835, Dr. Morton's Synopsis of the organic remains of the Cretaceous group of the United States [review of 7470]: v. 27, no. 2, 377-381.
600   1835, Large mass of native copper [from Lake Michigan]: v. 27, no. 2, 381.
601   1835, Great mass of meteoric iron from Louisiana: v. 27, no. 2, 382.
602   1835, Soapstone or steatite, of Middlefield [Massachusetts]: v. 27, no. 2, 382-383.
603   1835, Prof. Hitchcock's geology of Massachusetts [review of 5105]: v. 27, no. 2, 383 (13 lines).
604   1835, Modern trilobites of New South Scotland: v. 27, no. 2, 395 (10 lines).
605   1835, Recent scientific publications in the U. States: v. 27, no. 2, 395-396.
606   1835, Progressive increase of the internal heat of the crust of the globe: v. 27, no. 2, 397-399.
       Abstracted by J. Griscom from Bibliothèque Universelle. See also 9726.
607   1835, Water obtained by boring: v. 27, no. 2, 399-400.
       Abstracted by J. Griscom from Bulletin d'Encouragement.
608   1835, Mr. Murchison on the geology of Wales: v. 27, no. 2, 412 (5 lines).
       Notice of R. I. Murchison's research and publication on Wales.
609   1835, Mr. De la Beche's Manual [notice of 3037]: v. 27, no. 2, 412 (3 lines).
610   1835, Researches on theoretical geology, by Mr. T. De la Beche [review of 3039]: v. 27, no. 2, 412 (7 lines).
611   1835, Mr. Lyell and his geology [review of 6532]: v. 27, no. 2, 412-413.
612   1835, Manual of mineralogy, by Rob. Allan [notice of]: v. 27, no. 2, 413 (4 lines).
613   1835, Notices of Egypt - in a letter to the editor from an American gentleman, dated on the Nile, July 30th, 1834: v. 28, no. 1, 23-33, fig.
614   1835, Annual yield of cementation - copper of the Rio Tinto Mine in Spain: v. 28, no. 1, 144.
       Abstracted by C. U. Shepard from Annales des Mines.
615   1835, On the roasting of copper ores: v. 28, no. 1, 145 (7 lines).
       Abstracted by C. U. Shepard from Annales des Mines.
616   1835, Ancient mineralogy, or an inquiry respecting the mineral substances mentioned by the ancients - their uses, &c.; by N. F. Moore [review of 7419]: v. 28, no. 1, 188.
617   1835, Recherches sur les poissons fossiles, par L'Agassiz. - Great work of Prof. Agassiz on fossil fishes: v. 28, no. 1, 193-194.

American Journal of Science (cont.)
618   1835, Specimens from Mr. Mantell: v. 28, no. 1, 197-198.
619   1835, Vesuvius and Etna: v. 28, no. 1, 199-200.
       Abstracted from Galignani's Messenger.
620   1835, Information respecting the variation of the magnetic needle: v. 28, no. 1, 200.
621   1835, Earthquake in Chile, Feb. 20, 1835: v. 28, no. 2, 336-340.
       Abstracted from New Bedford Gazette.
622   1835, Earthquake at Florence: v. 28, no. 2, 340 (9 lines).
       Abstracted from London Ath.
623   1835, Report on the fresh water limestone of Burdic House, near Edinburgh, by Dr. S. Hibbert: v. 28, no. 2, 365-366.
624   1835, Ichthyosaurus fossil fish, wood, &c.: v. 28, no. 2, 369-370.
625   1835, Treatise on mineralogy, consisting of descriptions of the species, with five hundred wood cuts. By C. U. Shepard [review of 9502]: v. 28, no. 2, 374-378.
626   1835, Geological report on the elevated country between Missouri and Red River, from an examination in 1834, by G. W. Featherstonhaugh [review of 3778]: v. 28, no. 2, 379.
627   1835, Report on the new map of Maryland [review of 3235]: v. 28, no. 2, 379-380.
628   1835, Geological Survey of Connecticut: v. 28, no. 2, 381-382.
629   1835, On the proofs of a gradual rising of the land in certain parts of Sweden. By Charles Lyell: v. 28, no. 2, 387-388.
       Abstracted from London and Edinburgh Philos. J.
630   1835, Discovery of saurian bones in the magnesian conglomerate near Bristol [England]: v. 28, no. 2, 389.
       Abstracted from London and Edinburgh Philos. J.
631   1835, Diamonds at Algiers: v. 28, no. 2, 394 (6 lines).
       Abstracted from Jameson's Edinburgh New Philos. J.
632   1835, Allanite of Greenland: v. 28, no. 2, 394.
       Abstracted from Jameson's Edinburgh New Philos. J.
633   1835, Needle ore: v. 28, no. 2, 395 (12 lines).
       Abstracted from Jameson's Edinburgh New Philos. J.
634   1835, Platina and gold of the Uralian Mountains: v. 28, no. 2, 395 (7 lines).
       Abstracted from Jameson's Edinburgh New Philos. J.
635   1835, For sale - the cabinet of minerals of the late Dr. Young, of Edenville, New York: v. 28, no. 2, 400 (7 lines).
636   1835, List of new publications since the commencement of the present year: v. 29, no. 1, 161-168.
637   1836, Visit to the quicksilver mines of Idria [Spain]; in a letter from an officer in the American Navy: v. 29, no. 2, 219-222.
638   1836, The salt mountains of Ischil [Germany]; in a letter from an officer in the American Navy: v. 29, no. 2, 225-229, fig.
639   1836, Fossil forest [near Glasgow]: v. 29, no. 2, 352.
       From British Assoc. Proc.
640   1836, The two north magnetic poles: v. 29, no. 2, 352 (4 lines).
       From British Assoc. Proc.
641   1836, Lyell's Geology, 4th London edition, four volumes, 8vo. 1835 [review of 6532]: v. 29, no. 2, 358-359.
642   1836, Specimens from Dr. Mantell: v. 29, no. 2, 362-363.
       Notice of donation of plaster fossil casts.
643   1836, Recherches sur les poissons fossiles, par L'Agassiz, &c. [review of]: v. 29, no. 2, 363.
644   1836, Extracts of a letter to the editor, from a gentleman in England, dated Scarborough, Oct. 12, 1835, with a notice of a Plesiosaurus and other fossils, and of remarkable human remains: v. 29, no. 2, 364-364 [i.e. 366].
645   1836, Volcanic eruption: v. 29, no. 2, 364 (12 lines).
       From J. de la Haye. On the November 1833 eruption of volcano Bochet Kaba.
646   1836, Collection of saurian remains, made by Mr. Hawkins [in England]: v. 29, no. 2, 367 (6 lines).
647   1836, Depth of mines: v. 29, no. 2, 374.
       Abstracted from Mining Rev.
648   1836, Topaz in Ireland: v. 29, no. 2, 374 (3 lines).
       Abstracted from Dublin Geol. Soc. Trans.
649   1836, Roasting of copper ores: v. 29, no. 2, 374 (6 lines).
       Abstracted from "Ann. du Comptoir des Mines de Fer in Sweden."
650   1836, Transactions of the Geological Society of Pennsylvania, vol. I, part II. [review of 4314]: v. 29, no. 2, 391 (12 lines).
651   1836, Fossil flora of North America: v. 29, no. 2, 393-394 (7 lines).
       Notice of C. S. Rafinesque's work on fossil plants.
652   1836, Diamonds in North America: v. 29, no. 2, 394 (5 lines).
       Abstracted from Geol. Soc. Penn. Trans.
653   1836, List of new publications: v. 29, no. 2, 396-400.
654   1836, Earthquake and rising of the sea coast of Chili [sic], in November, 1822: v. 30, no. 1, 110-113.
       See also 8825.
655   1836, Remarks on a "Critical examination of some passages in Gen. I. by M. Stuart [review of 10057]: v. 30, no. 1, 114-130.
       Signed by "K."
656   1836, Essay on calcareous manures, by Edmund Ruffin. Second edition. Shellbanks, Va., 1835. 8vo. pp. 116. On the use of lime as a manure, by M. Puvis, translated for the Farmer's Register. [reviews with excerpts from 9088 and 8612]: v. 30, no. 1, 138-163.
657   1836, Medal of the Royal Society conferred on Mr. Lyell: v. 30, no. 1, 174.
658   1836, [Notes on meteorites and minerals abstracted from the Philosophical Magazine]: v. 30, no. 1, 175-177.
       Signed by "O. P. H."
659   1836, Fossil wax [from Moldavia]: v. 30, no. 1, 185-186.
       Abstracted from Atheneum.
660   1836, Iron [from Mt. Etna]: v. 30, no. 1, 186 (5 lines).
661   1836, Medical and physical researches, or original memoirs in medicine, surgery, physiology, geology, zoology and comparative anatomy, illustrated by plates containing 160 figures; by R. Harlan [review of 4805]: v. 30, no. 1, 188.

American Journal of Science (cont.)

662   1836, List of new publications: v. 30, no. 1, 198-202.
663   1836, New scientific journal: v. 30, no. 2, 382 (4 lines).
        Abstracted from London and Edinburgh Philos. Mag. Notice of the London Geol. J.
664   1836, Fertilizing properties of limestone: v. 30, no. 2, 383-384.
        Signed by "a gentleman in Geneseo, NY."
665   1836, Report on the new map of Maryland, 1835: v. 30, no. 2, 393-394.
        Notice of progress on several state surveys. See also 3240.
666   1836, Notice of Dr. Hildreth's article on the coal deposits of the Ohio, &c. [review of 5053]: v. 30, no. 2, 399-400.
667   1836, Dr. J. L. Comstock, M. D. [notice of his Outlines of geology, second edition (see 2511)]: v. 30, no. 2, 400 (5 lines).
668   1836, Ashmolean Society: v. 31, no. 1, 164-165.
        Communicated by J. Barrett from the Oxford Herald. On fossil footprints from Saxony.
669   1836, Bird tracks at Middletown, Conn. in the New Red Sandstone: v. 31, no. 1, 165 (11 lines).
670   1836, Remarks on the lavas, &c. of Mexico and South America, in a letter to the editor, dated January 24, 1836: v. 31, no. 1, 176-177 (14 lines).
671   1836, Plumbago and black lead pencils: v. 31, no. 1, 177-178.
        Abstracted from Rec. of Gen. Sci.
672   1837, New work on mineralogy [notice of Treatise on mineralogy by J. D. Dana (see 2798)]: v. 31, no. 2, 413 (10 lines).
673   1837, Mineralogical and geological collections: v. 31, no. 2, 418-419.
        Notice of specimens for sale by W. W. Mather and F. Hall.
674   1837, Geology and mineralogy considered with reference to natural theology, by the Rev. William Buckland [review of 2160]: v. 31, no. 2, 419-420.
675   1837, Supposed volcano at sea [near Galapagos]: v. 32, no. 1, 195-196.
        From Bos. Daily Advertiser.
676   1837, Twelve lectures on the connection between science and revealed religion; delivered in Rome, by Nicholas Wiseman, D. D. [review of 11007]: v. 32, no. 1, 209-210.
677   1837, Dr. Buckland's new work [notice of Geology and mineralogy ... (see 2160)]: v. 32, no. 1, 210 (5 lines).
678   1837, Lyell's Geology, 5th edition [notice of 6532]: v. 32, no. 1, 210 (11 lines).
679   1837, East Indian geology: v. 32, no. 1, 216 (7 lines).
        Notice of researches by J. McClelland in East India.
680   1837, Fossil remains of the elephant, Elephas primigeneus: v. 32, no. 2, 377-379.
681   1837, A system of mineralogy: including an extended treatise on crystallography: with an appendix containing the application of mathematics to crystallographic investigation, and a mineralogical bibliography .., by James Dwight Dana [review of 2798]: v. 32, no. 2, 387-392.
        Signed by "T."
682   1837, Meteorite [from East Bridgewater, Massachusetts]: v. 32, no. 2, 395.
        Abstracted from Bos. Daily Advertiser.
683   1837, The American edition of Dr. Buckland's late work of geology and mineralogy, considered with reference to natural theology [review of 2160]: v. 32, no. 2, 397-398.
684   1837, Earthquake [at Hartford, Connecticut]: v. 32, no. 2, 399 (6 lines).
685   1837, Geological survey of Connecticut [notice of 9512]: v. 32, no. 2, 399 (8 lines).
686   1837, Lyell's Geology. First American from the fifth and last London edition, 2 vols. [review of 6532]: v. 33, no. 1, 182-183.
687   1837, Newly discovered ichnolites [from Middletown, Connecticut]: v. 33, no. 1, 201-202.
        Signed by "R."
688   1837, Hot Springs of Arkansas, &c.: v. 33, no. 1, 202 (11 lines).
689   1837, Lethaea Geognostica, oder Abbildung und Bescreibung der fur die Gebirgs - Formationen bezeichnendsten Versteinerungen: von Dr. H. G. Bronn [notice and review of]: v. 33, no. 1, 204 (15 lines).
690   1838, Prairies of Ohio: v. 33, no. 2, 230-236.
        From West. Mon. Mag.
691   1838, Geology of the desert between Suez and Cairo: v. 33, no. 2, 288.
        From British Assoc. Proc.
692   1838, Impressions of feet in rocks [from St. Louis, Missouri]: v. 33, no. 2, 398.
        Signed by an "English correspondent."
693   1838, Heat of the earth: v. 34, no. 1, 36-37.
        From British Assoc. Proc.
694   1838, Outlines of geology, prepared for the use of the Junior Class of Columbia College; by Jas. Renwick [review of 8751]: v. 34, no. 1, 183-185.
695   1838, Geological reports: v. 34, no. 1, 185-198.
        Review of progress of the several state geological surveys.
696   1838, Fossil fishes [from the Connecticut River Valley]: v. 34, no. 1, 198-200.
        Signed by "D. L. H."
697   1838, Sienitic granite, near Christiana, Norway: v. 34, no. 1, 204.
698   1838, New locality of tourmaline [Orford, New Hampshire]: v. 34, no. 1, 204 (6 lines).
699   1838, Aerolites [from Brazil]: v. 34, no. 1, 209-210 (11 lines).
        From London Ath.
700   1838, Prof. Agassiz' great work on fossil fishes: v. 34, no. 1, 212 (14 lines).
701   1838, Second report of the geology of Maine [notice of report by C. T. Jackson (see 5560)]: v. 34, no. 1, 219 (3 lines).
702   1838, Third American edition of Bakewell's Geology, from the fifth and last of the author in London [notice of 1460]: v. 34, no. 1, 219 ((14 lines).
703   1838, First annual report on the geological survey of the state of Ohio; by W. W. Mather [review of 689]: v. 34, no. 2, 347-364.
704   1838, Additional notices of trilobites [from Bourbon County, Kentucky and Otisco, New York]: v. 34, no. 2, 380-381.
705   1838, Wonders of geology, in two vols. 12mo., with numerous plates and wood cuts; by Gideon Mantell [review of 6732]: v. 34, no. 2, 387-392.
706   1838, Gold in Georgia: v. 34, no. 2, 397 (6 lines).
707   1838, Geological and other reports: v. 34, no. 2, 402-404.

American Journal of Science (cont.)

708   1838, Second part to Shepard's Descriptive mineralogy [review of 9502]: v. 35, no. 1, 187 (5 lines).

709   1838, A treatise on gems, in reference to their practical and scientific value. ... By Lewis Feuchtwanger [notice of 3805]: v. 35, no. 1, 189-190.

710   1838, Green feldspar and galena [from Massachusetts]: v. 35, no. 1, 192 (5 lines). Abstracted from Bos. J. Phil. Arts.

711   1838, Annals of natural history, or magazine of zoology, botany, and geology [review of]: v. 35, no. 1, 194-195.

712   1839, Fossil shells and bones [from Wilmington Island, Georgia]: v. 35, no. 2, 380. Signed by "T. R. D."

713   1839, Dr. Mantell's Wonders of geology [notice of 6732]: v. 35, no. 2, 384 (10 lines).

714   1839, Geological surveys [in the United States]: v. 35, no. 2, 384 (7 lines).

715   1839, Mr. Bakewell's Geology [review of 1460]: v. 35, no. 2, 385.

716   1839, Elements of geology; by Charles Lyell [review of 6533]: v. 35, no. 2, 385.

717   1839, The science of geology, from the Glasgow treatises, with additions [review of 9292]: v. 35, no. 2, 387 (6 lines).

718   1839, Dr. Charles T. Jackson's reports on the geology of Maine [(notice of 5558 and 5560)]: v. 35, no. 2, 387-388 (15 lines).

719   1839, Marble and serpentine in Vermont: v. 35, no. 2, 390.

720   1839, N. Dunn's Chinese collection at Philadelphia, communicated: v. 35, no. 2, 391-400.
Notice of Chinese minerals on p. 399

721   1839, Prof. Agassiz and his works: v. 35, no. 2, 400.

722   1839, Citations from and abstract of, the Geological reports on the state of New York, for 1837-8, being State Document no. 200; communicated by Gov. W. L. Marcy [review with excerpts of 7737]: v. 36, no. 1, 1-49.

723   1839, Dr. Jackson's Reports on the geology of the state of Maine, and on the public lands belonging to Maine and Massachusetts [reviews with excerpts of 5556 and 5566]: v. 36, no. 1, 143-156.

724   1839, Obituary notice of the Hon. Stephen Van Rensselaer: v. 36, no. 1, 156-164.

725   1839, Fossil fishes of the red sandstone: v. 36, no. 1, 186-187 (13 lines). Signed by "R."

726   1839, The mammoth, (mastodon? Eds.): v. 36, no. 1, 198-200.
Discovery of fossil bones from Gasconada County, Missouri.

727   1839, Third American from the fifth English edition of Bakewell's Geology [review of 1460]: v. 36, no. 1, 201-202.

728   1839, African meteorite: v. 36, no. 2, 393 (16 lines).
Abstracted from London Nautical Mag.

729   1839, New works received: v. 36, no. 2, 399-400.

730   1839, Great scheme for magnetical observations: v. 37, no. 1, 198.
Abstracted from London and Edinburgh Philos. Mag.

731   1839, Notices of geological and other physical facts and of antiquities in Asia, from Sir Robert K. Porter's Travels in Georgia, Persia, Armenia, Ancient Babylonia, &c. &c., during the years 1817, '18, '19, and '20, with numerous engravings of portraits, costumes, antiquities, &c. [review with excerpts of 8521]: v. 37, no. 2, 347-356.

732   1839, Rose mica lepidolite: v. 37, no. 2, 361 (18 lines).
Abstracted from London and Edinburgh Philos. Mag.

733   1839, Fossil-tree at Granton, near Edinburgh: v. 37, no. 2, 363 (6 lines).
Abstracted from Jameson's J.

734   1839, Geological surveys [in the United States]: v. 37, no. 2, 375-383.

735   1839, Explosions in American coal mines: v. 37, no. 2, 387-389.

736   1839/1840, British Antarctic expedition: v. 37, no. 2, 397; v. 38, no. 1, 204.

737   1839, British annual and epitome of the progress of science for 1839. Edited by R. D. Thomson [review of]: v. 37, no. 2, 400.

738   1840, Observations météorologiques et magnétiques faites dans l'entendue de l'empire de Russie redigées et publiées aux frais du gouvernement, par A. T. Kupffer [review of]: v. 38, no. 2, 380-381.

739   1840, Notice of the Wonders of geology by Gideon Algernon Mantell [review of 6732]: v. 39, no. 1, 1-18.

740   1840, Notice of a report of a geological, mineralogical and topographical examination of the coal field of Carbon Creek, with an analysis of the minerals, accompanied by maps, profiles and sections; by Walter R. Johnson [review with excerpts of 5772]: v. 39, no. 1, 137-149.

741   1840, Magnetic observations: v. 39, no. 1, 193-194 (12 lines).

742   1840, Petroleum oil well [from Burksville, Kentucky]: v. 39, no. 1, 195.
From the "N. O. Bulletin."

743   1840, Earthquake in Connecticut, &c.: v. 39, no. 2, 335-342.

744   1840, Hitchcock's Geology [review of 5132, with excerpts from "Preface" by J. P. Smith (see 9896)]: v. 39, no. 2, 391-392.

745   1840, Leonhard's Geology [review of v. 2 of Géologie des gens du monde by K. C. von Leonhard]: v. 39, no. 2, 393-394.

746   1841, African meteorite of Cold Bokkeveld: v. 40, no. 1, 199-201.
Abstracted from South African Com. Advertiser.

747   1841, [Notes on elaterite and minerals of France abstracted from London Athenaeum]: v. 40, no. 1, 215-217.

748   1841, Necrology: v. 40, no. 1, 218-220.
Obituaries of W. Smith, J. Esmark and F. Mohs.

749   1841, Prof. Agassiz and his works: v. 41, no. 1, 194-195.

750   1841, Artesian boring at Paris: v. 41, no. 1, 209-210.
Notice of various strata encountered in boring.

751   1841, Address delivered at the annual meeting of the Boston Natural History Society, May 5th, 1841; by J. E. Teschemacher. [review of 10169]: v. 41, no. 2, 370-371.

752   1841, Report on the manufature of iron, addressed to the Governor of Maryland, by J. W. Alexander [review of 127]: v. 41, no. 2, 376-378.

753   1841, Final report on the geology of Massachusetts; by Edward Hitchcock [review of 5135]: v. 41, no. 2, 384-385.

754   1841, First annual report on the geology of the state of New Hampshire; by Dr. Charles T. Jackson, state geologist [review of 5576]: v. 41, no. 2, 383-384.

American Journal of Science (cont.)

755   1841, [On recent American geological surveys]: v. 41, no. 2, 385.
756   1841, A sketch of the geology of Surrey [review of work by G. A. Mantell]: v. 41, no. 2, 386-387.
757   1841, Popular lectures on geology, treated in a very comprehensive manner; by K. C. Von Leonhard [review of 6192]: v. 41, no. 2, 387 (16 lines).
758   1841, Notice of the relation between the Holy Scriptures and some parts of geological science; by John Pye Smith [review of]: v. 41, no. 2, 387-389.
759   1841, Scientific visit of Charles Lyell, Esq. of London, to the United States: v. 41, no. 2, 403.
760   1842, Lectures on the applications of chemistry and geology to agriculture; by James F. W. Johnston [review of 5819]: v. 42, no. 1, 187-191.
761   1842, Principles of geology; or modern changes of the earth and its inhabitants, considered as illustrative of geology; by Charles Lyell and Elements of geology; by Charles Lyell [reviews of 6533 and 6543]: v. 42, no. 1, 191-192.
762   1842, Notes on the use of anthracite in the manufacture of iron, with some remarks on its evaporating power; by Walter R. Johnson. [review of 5777]: v. 42, no. 1, 192.
763   1842, Another meteorite in France: v. 42, no. 1, 203, (4 and 12 lines). Abstracted from the New York Obs.
764   1842, Fall of a meteoric stone at Grüneberg in Silesia: v. 42, no. 1, 203 (15 lines). Abstracted from Poggendorff's Annalen.
765   1842, Monographie d'echinodermes vivans et fossiles, par L. Agassiz [review and abstract of]: v. 42, no. 2, 378-379.
766   1842, Infusorial animals [from Prussia]: v. 42, no. 2, 388 (11 lines). Abstracted from Lit. Gaz.
767   1842, Meteorite of Château-Renard [France]: v. 42, no. 2, 403-404. Abstracted from l'Institut.
768   1842, Memoir on a portion of the lower jaw of the Iguanodon and of the remains of the Hylaesaurus and other saurians, discovered in the strata of Tilgate Forest in Sussex, (England,) by Gideon Algernon Mantell [review of]: v. 43, no. 1, 189-190.
769   1842, Practical geology and mineralogy, with instructions for the qualitative anaysis of minerals; by Joshua Trimmer [review of 10330]: v. 43, no. 1, 191.
770   1842, A muck manual for farmers; by Samuel L. Dana [review of 2885]: v. 43, no. 1, 192-197.
771   1842, New works in science: v. 43, no. 1, 205.
772   1842, Soirées of the President of the Geological Society of London: v. 43, no. 1, 206 (9 lines).
773   1842, Testimonial to Dr. Mantell: v. 43, no. 1, 206-207 (12 lines).
774   1842, Thoughts on a pebble, or a first lesson in geology; by the author of the Wonders of geology [review of book by G. A. Mantell]: v. 43, no. 2, 382-383.
775   1842, Monographie d'echinodermes vivans et fossiles, par L. Agassiz [notice of v. 3 and 4]: v. 43, no. 2, 390.
776   1842, New scientific journals: v. 43, no. 2, 392.
777   1842, Terrestrial origin of the alleged meteoric rain in Hungary: v. 43, no. 2, 401. Abstracted from Poggendorff's Annalen.
778   1843, Report on the geology of Connecticut; by James G. Percival [review of 8230]: v. 44, no. 1, 187-188.
779   1843, Lectures on the application of chemistry and geology to agriculture. Part II, on the inorganic elements of plants. By J. F. W. Johbston [review of 5819]: v. 44, no. 1, 189.
780   1843, Murchison's Silurian system [review of]: v. 44, no. 1, 193 (14 lines). Notice of its sale in America.
781   1843, Phillip's Mineralogy, new American edition [review of 8325]: v. 44, no. 1, 199 (3 lines).
782   1843, Obituary [of Louis Lederer]: v. 44, no. 1, 216 (17 lines).
783   1843, Earthquakes in the United States: v. 44, no. 2, 419 (7 lines).
784   1843, Great earthquakes in the West Indies, Feb. 8: v. 44, no. 2, 419 (9 lines).
785   1843, New York State reports: v. 44, no. 2, 420 (8 lines). Notice of New York geological survey reports.
786   1843, Monograghies d'echinodermes vivans et fossiles; par Louis Agassiz [review of]: v. 45, no. 2, 399-400. Signed by "J. W."
787   1843, Death of Mr. Bakewell: v. 45, no. 2, 403-404.
788   1843, Death of Prof. Hall: v. 45, no. 2, 404 (10 lines).
789   1843, Death of Mr. J. N. Nicollet: v. 45, no. 2, 404.
790   1844, Notice of travels in the Alps of Savoy and other parts of the Pennine Chain, with observations on the phenomena of glaciers; by James D. Forbes [review of]: v. 46, no. 1, 172-192.
791   1844, Mr. Alger's edition of Phillip's mineralogy [review of 8325]: v. 46, no. 1, 203 (13 lines).
792   1844, Remarkable fulgurite [from Dresden]: v. 46, no. 1, 210-211.
793   1844, Death of Richard Harlan: v. 46, no. 1, 216 (4 lines).
794   1844, Review of Alger's Phillips' Mineralogy, and Shepard's Treatise on Mineralogy [review with excerpts of 8325 and 9542]: v. 47, no. 2, 333-351.
795   1844, New books received: v. 47, no. 2, 415.
796   1845, New York geological survey: v. 48, no. 1, 210-211.
797   1845, Prof. J. F. W. Johnston's works on agricultural chemistry [notices of 5825 and 5818]: v. 48, no. 1, 212-213.
798   1845, Mr. Phillip's mineralogical collection: v. 48, no. 1, 219 (6 lines). Signed by "A". Collection donated to Liverpool Medical Inst.
799   1845, Final report on the geology and mineralogy of the state of New Hampshire ... by Charles T. Jackson [notice of 5611]: v. 48, no. 2, 393-394.
800   1845, The Taconic system, based on observations in New York, Massachusetts, Maine, Vermont, and Rhode Island; by Prof. Ebenezer Emmons [notice of 3520]: v. 48, no. 2, 394 (12 lines).
801   1845, Report on American coals to the Navy Department of the United States. By Walter R. Johnson [notice of 5796]: v. 48, no. 2, 394-395.
802   1845, Vestiges of the natural history of Creation [notice of text by R. Chambers (see 2292 et seq.)]: v. 48, no. 2, 395 (13 lines).

American Journal of Science (cont.)

803   1845, Geological observations on the region near Centerville, Alabama: v. 48, no. 2, 399-400 (15 lines).
804   1845, School for the study of elementary and analytical chemistry, by Dr. C. T. Jackson: v. 48, no. 2, 403-404.
805   1845, Geological survey of Canada: v. 48, no. 2, 404 (6 lines). Abstracted from "Chron. Guard." (Toronto). Notice of appropriations.
806   1845, A geological survey of Vermont: v. 48, no. 2, 404 (3 lines).
807   1845, Prof. W. R. Johnson on the heating power of various coals [review of 5796]: v. 49, no. 1, 166-171.
808   1845, Handwörterbuch der Topographischen Mineralogie von Gustav Leonhard [review of]: v. 49, no. 1, 188-189.
809   1845, Owen's Illustrated catalogue [review of Illustrated catalogue of fossil mammals by R. Owen]: v. 49, no 1, 190 (11 lines).
810   1845, Vestiges of the natural history of Creation [review of text by R. Chambers (see 2292 et seq.)]: v. 49, no. 1, 191.
811   1845, Dr. G. A. Mantell on the geological structure of the country seen from Leith Hill in the county of Surrey [review of]: v. 49, no. 1, 191.
812   1845, Large skeleton of the Zeuglodon of Alabama: v. 49, no. 1, 218. Notice of A. Koch's collection. From Mobile Daily Advertiser.
813   1845, Bones of the extinct gigantic birds of New Zealand, called Moa: v. 49, no. 1, 219 (6 lines).
814   1845, Plumbago formed by pressure: v. 49, no. 1, 227 (8 lines).
815   1845, Report to the Navy Department of the United States on American coals, applicable to steam navigation and to other purposes; by Prof. Walter R. Johnson [review of 5796]: v. 49, no. 2, 310-336.
816   1845, Travels in North America in the years 1841-42, with geological observations on the United States, Canada, and Nova Scotia; by Charles Lyell [review of 6557]: v. 49, no. 2, 368-373.
817   1845, On the geological constitution of the Altai; by M. P. de Tchihatcheff [review of]: v. 49, no. 2, 378-379. From Comptes Rendus.
818   1845, Whitney's translation of Berzelius on the blowpipe [review of 1668]: v. 49, no. 2, 379 (16 lines).
819   1845, New books received: v. 49, no. 2, 386 (17 lines).
820   1845, Lieut. Wright's Treatise on Mortars [review of 11067]: v. 49, no. 2, 379-384.
821   1845, Dissertation on a natural system of chemical classification; by Oliver Wolcott Gibbs [review of 4384]: v. 49, no. 2, 384-386.
822   1845, Lithographic stones [from France]: v. 49, no. 2, 401 (7 lines).
823   1845, Notice in a letter from London to the senior editor, dated August 30, 1845: v. 49, no. 2, 405-406. On fossils from railroad excavations in England.
824   1846, Native lead [from Kerry County, Ireland]: v. 1 (ns), no. 1, 120 (5 lines).
825   1846, Observations made at the magnetical and meteorological observatory at Toronto in Canada, 1840, 1841, and 1842. 4to. [notice of]: v. 1 (ns), no. 1, 137-138.
826   1846, Proceedings connected with the magnetical and meterological conference held at Cambridge, (Eng.) in June, 1845: v. 1 (ns), no. 1, 142.
827   1846, At the Italian Scientific Association for 1844: v. 1 (ns), no. 1, 146 (6 lines). Abstracted from l'Institut. Includes notice of Hoffmann's geological map of Sicily.
828   1846, Rail road excavations in England: v. 1 (ns), no. 1, 146 (13 lines).
829   1846, Die Rizopodi caratteristeci dei terreni Sopracretacei, or on the characteristic Rhizopodi of the Supracretaceous deposits; by G. Michelotti. Modena, 1841: v. 1 (ns), no. 1, 149 (5 lines). Notice of G. Michelotti's work.
830   1846, Dr. Mantell: v. 1 (ns), no. 1, 149 (12 lines). Notice of new editions of his works.
831   1846, Mr. Murchison: v. 1 (ns), no. 1, 149 (2 lines). Notice of his researches in Russia.
832   1846, Report of the exploring expedition to the Rocky Mountains in the year 1842, and to Oregon and California in 1843-44; by Brevet Capt. J. C. Fremont [notice of 4113 and 4114]: v. 1 (ns), no. 1, 149-150.
833   1846, A history of fossil insects in the Secondary rocks of England, &c.; by the Rev. Peter Brodie [notice of]: v. 1 (ns), no. 1, 150 (5 lines).
834   1846, Dr. William Horton [obituary]: v. 1 (ns), no. 1, 152 (11 lines).
835   1846, William C. Woodbridge [obituary]: v. 1 (ns), no. 1, 152.
836   1846, Sequel to the vestiges of Creation [review of text by R. Chambers (see 2291)]: v. 1 (ns), no. 2, 250-254.
837   1846, The mastodon of Newburg, N. Y., discovered in August, 1845: v. 1 (ns), no. 2, 268-270.
838   1846, Lower green-sand fossils: v. 1 (ns), no. 2, 280 (4 lines). Abstracted from London Q. J. Geol. Soc.
839   1846, Fossil vertebrae of shark: v. 1 (ns), no. 2, 286 (3 lines). Abstracted from Edinburgh Royal Soc. Trans.
840   1846, Antarctic continent: v. 1 (ns), no. 2, 307 (7 lines). On magnetic field.
841   1846, Nuttall's cabinet of minerals for sale: v. 1 (ns), no. 2, 309 (6 lines).
842   1846, Cabinet of minerals of the late Dr. J. P. Young, of Edenville, N. Y., for sale: v. 1 (ns), no. 2, 309 (6 lines).
843   1846, Paléontologie universelle des coquilles et des mollusques [notice of works by A. D'Orbigny]: v. 1 (ns), no. 2, 309-310.
844   1846, Essai d'une carte géologique du globe terrestre; by A. Boué [notice of]: v. 1 (ns), no. 2, 310 (7 lines).
845   1846, First annual report of the geology of the state of Vermont, 1845; by C. B. Adams, State Geologist, Burlington [review of 26]: v. 1 (ns), no. 2, 310 (12 lines).
846   1846, New York scientific reports: v. 1 (ns), no. 2, 310-311.
847   1846, Reports on the Bear Mountain Railroad [notice of report by J. Hall (see 5753)]: v. 1 (ns), no. 2, 311 (6 lines).

American Journal of Science (cont.)
848   1846, Artificial asbestus: v. 1 (ns), no. 3, 429 (10 lines).
           Abstracted from Glascow Philos. Soc. Proc.
849   1846, Gradual rise of Newfoundland above the sea: v. 1 (ns), no. 3, 434.
           Abstracted from Jameson's J.
850   1846, Cataract Cave, Schoharie [New York]: v. 1 (ns), no. 3, 434-435.
           Abstract, communicated by A. Eggleston.
851   1846, Geological survey of Vermont: v. 1 (ns), no. 3, 435.
852   1846, Pterodactyl in the English Chalk: v. 1 (ns), no. 3, 436 (6 lines).
853   1846, Prof. Louis Agassiz of Neufchatel, Switzerland: v. 1 (ns), no. 3, 451-452.
           United States visit announced.
854   1846, The collection of fossils of Herr Munster of Bayreuth: v. 1 (ns), no. 3, 452 (5 lines).
855   1846, Roderick Impey Murchison: v. 1 (ns), no. 3, 452 (7 lines).
           Notice of honors from England and Russia.
856   1846, [Bibliography, with recent geological literature]: v. 1 (ns), no. 3, 455-456.
857   1846, Acknowledgments of books, etc., received: v. 1 (ns), Appendix, 1-15.
858   1846, Diamonds in North Carolina: v. 2 (ns), no. 1, 119 (4 lines).
859   1846, The naturalist, and journal of agriculture, horticulture, education and literature [notice of]: v. 2 (ns), no. 1, 147 (11 lines).
860   1846, A history of British fossil mammals and birds; by Richard Owen [review of]: v. 2 (ns), no. 1, 148-149.
861   1846, Thoughts on animalcules, or a glimpse of the invisible world revealed by the microscope; by Gideon Algernon Mantell [and] Notes of a microscopic examination of chalk and flint; by the same author [reviews of]: v. 2 (ns), no. 1, 149-150.
862   1846, The geology of Russia in Europe, and the Ural Mountains; by Sir R. I. Murchison [review of]: v. 2 (ns), no. 1, 152-153.
863   1846, List of works: v. 2 (ns), no. 1, 153-156; no. 2, 302-304; no. 3, 446-448.
           Notice of recently published scientific works.
864   1846, Notice of an earthquake and a probable subsidence of the land in the district of Cutch, near the mouth of the Koree, or eastern branch of the Indus, in June, 1845: v. 2 (ns), no. 2, 270.
           Abstracted from a letter, reported in J. Geol. Soc. London.
865   1846, Geological chart of M. Boué [review of]: v. 2 (ns), no. 2, 272.
           Abstracted from Geol. Soc. France Bulletin.
866   1846, [Notes on volcanoes and Australian copper mines abstracted from London Athenaeum]: v. 2 (ns), no. 2, 290-291.
867   1846, Obituary. - Denison Olmsted, jr.: v. 2 (ns), no. 2, 297.
868   1846, M. D'Orbigny: v. 2 (ns), no. 2, 299-300.
           Reviews of several of his recent publications.
869   1846, Leçons de géologie pratique, tome premier [review of work by E. de Beaumont]: v. 2 (ns), no. 2, 300 (12 lines).
870   1846, Work of M. Agassiz: v. 2 (ns), no. 2, 300-302.
           Reviews of several of his recent works.
871   1846, Vanadate of copper [from Russia]: v. 2 (ns), no. 3, 414 (4 lines).
           Abstracted from l'Institut.
872   1846, Phyllite [from Massachusetts]: v. 2 (ns), no. 3, 422 (6 lines).
873   1846, Prof. Louis Agassiz: v. 2 (ns), no. 3, 440 (6 lines).
           Notice of his arrival in Boston in 1846.
874   1846, A monograph on fossil crinoidea; by Thos. Austin, Esq., F. G. S., and Thos. Austin, jr., Esq., A. B. J. [notice of]: v. 2 (ns), no. 3, 446 (8 lines).
875   1847, Benjamin Silliman M. D., L. L. D.: v. 50, front. steelplate engraving.
876   1847, Ilmenium: v. 3 (ns), no. 1, 116-117.
           Notice of new element in yttro-tantalite.
877   1847, On the origin of the coal of Silesia: v. 3 (ns), no. 1, 119 (13 lines).
           Abstracted from London Ath. by Prof. Goppert.
878   1847, Fall of meteorites [in Italy]: v. 3 (ns), no. 1, 141-142.
           Abstracted from l'Institut.
879   1847, Artesian well in the Duchy of Luxemburg: v. 3 (ns), no. 1, 142-143 (4 lines).
           Abstracted from London Ath.
880   1847, Native gold [from South Australia]: v. 3 (ns), no. 1, 143 (2 lines).
           Abstracted from London Ath.
881   1847, Second annual report on the geology of Vermont; by Prof. C. B. Adams [review of 29]: v. 3 (ns), no. 1, 144 (10 lines).
882   1847, Geological observations on South America [review of work by C. Darwin]: v. 3 (ns), no. 1, 146 (11 lines).
883   1847, The London geological journal and record of discoveries in British and foreign paleontology [review of v. 1]: v. 3 (ns), no. 1, 146 (7 lines).
884   1847, Instruction in chemical analysis (quantitative); by Dr. C. Remigius Fresenius, edited by J. Lloyd Bullock [review of]: v. 3 (ns), no. 1, 149.
885   1847, Taschenbuch für Freunde der Geologie von Karl Cäsar v. Leonhard [review of]: v. 3 (ns), no. 1, 149.
886   1847, [Bibliography, with recent geological literature]: v. 3 (ns), no. 1, 150-152; no. 2, 311-312; no. 3, 455-456.
887   1847, [Notices on submarine volcanoes abstracted from London Athenaeum]: v. 3 (ns), no. 2, 273.
888   1847, Monument of the late Thomas Say: v. 3 (ns), no. 2, 298-299, fig.
889   1847, Journal of Lieut. J. W. Abert, from Bent's Fort to St. Louis, in 1845 [review of 6]: v. 3 (ns), no. 2, 301-302.
890   1847, Russian geology [notice of work by Helmersen]: v. 3 (ns), no. 3, 430.
           Abstracted from l'Institut.
891   1847, Chalk and coal fires: v. 3 (ns), no. 3, 451 (6 lines).
           Abstracted from London Ath. On combination of coal and chalk as a fuel.
892   1847, New appointments to professorships in Harvard and Yale: v. 3 (ns), no. 3, 451 (8 lines).
           Notice of appointments of B. Silliman, jr. and E. N. Horsford.
893   1847, Obituary. - Dr. Amos Binney: v. 3 (ns), no. 3, 451-452.
894   1847, Paleontology of New York; by James Hall [notice of 4712]: v. 3 (ns), no. 3, 454 (7 lines).
895   1847, Coal and iron in India: v. 4 (ns), no. 1, 109-110.
           Abstracted from Mining J.

American Journal of Science (cont.)

896   1847, Volcanic eruption at the Cape Verds: v. 4 (ns), no. 1, 146 (12 lines).
Abstracted from Salem Reg.
897   1847, Elementary geology; by Edward Hitchcock [review of 5177]: v. 4 (ns), no. 1, 146-147.
898   1847, Dr. Mantell's geology of the Isle of Wight [notice of]: v. 4 (ns), no. 1, 147 (4 lines).
899   1847, Principles of geology - or, the modern changes of the earth and its inhabitants, considered as illustrative of geology; by Charles Lyell [review of 6543]: v. 4 (ns), no. 1, 147-148.
900   1847, [Bibliography, with recent geological literature]: v. 4 (ns), no. 1, 148-150; no. 2, 303-304; no. 3, 455-456.
901   1847, Reducing copper ores by electricity: v. 4 (ns), no. 2, 276-277.
From London Mining J.
902   1847, Smelting copper ore: v. 4 (ns), no. 2, 292.
903   1847, Obituary. - Ithamar B. Crawe, M. D.: v. 4 (ns), no. 2, 300.
Signed by "C. D.", of Rochester, New York (i.e. Chester Dewey?).
904   1847, Geology: Introductory, descriptive, and practical. II. The ancient world, or picturesque sketches of creation. By D. T. Ansted [notice and review of 1280]: v. 4 (ns), no. 2, 300-302.
905   1847, Exploration of the volcano Rucu-Pichincha, (Quito;) by MM. Seb. Wisse et Garcio Moreno, during August, 1845: v. 4 (ns), no. 3, 417-419.
Abstracted from the Q. J. Geol. Soc.; signed by "J. C. M."
906   1847, Smithsonian Institution: v. 4 (ns), no. 3, 438-440.
Announcement of formation. Abstracted from Lit. World.
907   1847, American science in Turkey: v. 4 (ns), no. 3, 449 (13 lines).
Notice of J. Lawrence's appointment as geologist to the Sultan of Turkey.
908   1847, Prof. Agassiz: v. 4 (ns), no. 3, 449 (5 lines).
Notice of his work at Harvard College.
909   1847, Large crystal of columbite: v. 4 (ns), no. 3, 499 (5 lines).
910   1847, The London geological journal and record of discoveries in British and foreign paleontology [review of first volume]: v. 4 (ns), no. 3, 450-451.
911   1847, Lexicon scientiarum - a dictionary of terms used in the various branches of anatomy, astronomy, botany, geology, geometry, hygiene, mineralogy, natural philosophy, physiology, zoology, &c., for the use of all who read or study in college, school, or private life. By Henry McMurtrie, M. D. [review of 6646]: v. 4 (ns), no. 3, 454-455.
912   1848, Mineral localities in New York: v. 5 (ns), no. 1, 132-133.
Signed by "F. B. H."
913   1848, A Thanksgiving present to Prof. Agassiz: v. 5 (ns), no. 1, 139.
From a "New York paper", on gift of appreciation from the New York College of Physicians.
914   1848, Paleontology of New York, vol. I; containing descriptions of the organic remains of the lower division of the New York system, (equivalent of the Lower Silurian rocks of Europe,) by James Hall [review of 4712]: v. 5 (ns), no. 1, 149-150.
915   1848, A. B. Gray, Esq.; report on the mineral lands of Lake Superior [notice of 4498]: v. 5 (ns), no. 1, 151 (9 lines).
916   1848, [Bibliography, with recent geological literature]: v. 5 (ns), no. 1, 151-152; no. 2, 304-306; no. 3, 460.
917   1848, A memoir of Dr. Douglass Houghton [obituary]: v. 5 (ns), no. 2, 217-227.
918   1848, Effect of the fusion of siliceous minerals on their specific gravity: v. 5 (ns), no. 2, 258.
Abstracted from Berzelius Jahresbericht.
919   1848, Manganocalcite of Breithaupt: v. 5 (ns), no. 2, 268 (4 lines).
Abstracted from l'Institut.
920   1848, Sulphuret of cobalt [from India]: v. 5 (ns), no. 2, 268 (4 lines).
Abstracted from Chemical Gaz.
921   1848, Coal in the East Indies: v. 5 (ns), no. 2, 268 (5 lines).
Abstracted from l'Institut.
922   1848, Meteorite of July 4, 1847 [in Bohemia]: v. 5 (ns), no. 2, 285 (5 lines).
923   1848, Original columns of the Giant's Causeway, and model of this wonderful group of natural pillars: v. 5 (ns), no. 2, 294.
924   1848, Italian Congress of Science: v. 5 (ns), no. 2, 297 (6 lines).
925   1848, History of Vermont, natural, civil, and statistical, in three parts, with a new map of the state and 200 engravings; by Zaddock Thompson [review of 10237]: v. 5 (ns), no. 2, 301.
926   1848, Daubeny on active and extinct volcanoes [notice of new edition]: v. 5 (ns), no. 2, 303 (9 lines).
927   1848, Dr. Mantell's Wonders of geology [notice of 6732]: v. 5 (ns), no. 2, 303 (8 lines).
928   1848, New volcano [at Amargura Island, the Friendly Group]: v. 5 (ns), no. 3, 422 (11 lines).
Abstracted from London Ath.
929   1848, Earthquake and eruption in Ternate: v. 5 (ns), no. 3, 422-423.
Abstracted from J. Indian Archipelago.
930   1848, Falling in of a mountain in Timor: v. 5 (ns), no. 3, 423 (6 lines).
Abstracted from J. Indian Archipelago.
931   1848, The late Professor Vanuxem: v. 5 (ns), no. 3, 445-446.
From N. Am. (Philadelphia). Obituary of L. Vanuxem.
932   1848, A description of active and extinct volcanoes, of earthquakes, and of thermal springs; with remarks on the causes of these phenomena, the character of their respective products, and of their influence on the past and present condition of the globe. By Charles Daubeny [review of second edition]: v. 5 (ns), no. 3, 446-447.
933   1848, Wonders of geology, or a familiar exposition of geological phenomena; by Gideon Algernon Mantell [review of 6732]: v. 5 (ns), no. 3, 447-449.
934   1848, Geology and mineralogy of the Malay Peninsula: v. 6 (ns), no. 1, 129-132.
From Mining J.
935   1848, Burra Burra Copper Mine, South Australia: v. 6 (ns), no. 1, 134.
From Mining J.

American Journal of Science (cont.)

936 1848, Histoire des progrès de la géologie de 1834 à 1845; par Le Vicomte D'Archiac [review of]: v. 6 (ns), no. 1, 134-135.
937 1848, Coal in Chili [sic]: v. 6 (ns), no. 1, 146.
     Abstracted from Mining J.
938 1848, Statistics of coal [review of text by R. C. Taylor (see 10147)]: v. 6 (ns), no. 1, 150-151.
939 1848, Observations on the Temple of Serapis at Pozzuoli near Naples; by Charles Babbage [review of]: v. 6 (ns), no. 1, 152.
940 1848, [Bibliography, with recent geological literature]: v. 6 (ns), no. 1, 155-156; no. 2, 303-304; no. 3, 455-456.
941 1848, On the Indian Archipelago: v. 6 (ns), no. 2, 157-170.
     Abstracted from Indian Archipelago and Eastern Asia J.
942 1848, Produce of gold in the Ural and Siberia in the year 1846: v. 6 (ns), no. 2, 275.
     Abstracted from Q. J. Geol. Soc.
943 1848, Cabinet and observatory at Amherst College, Mass.: v. 6 (ns), no. 2, 293.
944 1848, F. Markoe's mineralogical cabinet: v. 6 (ns), no. 2, 297 (6 lines).
     Mineral collection sold to West Point.
945 1848, Meteorite of Arkansas: v. 6 (ns), no. 2, 297 (5 lines).
946 1848, Manual of mineralogy ... by James D. Dana [notice and review of 2832]: v. 6 (ns), no. 2, 302.
947 1848, Obituary. - Death of J. Richardson: v. 6 (ns), no. 2, 297 (3 lines).
948 1848, Letters on geology; by David Christy [review of 2365]: v. 6 (ns), no. 2, 298-299.
949 1848, New Mexico and California [reviews with excerpts of works by J. W. Abert (see 7), W. H. Emory (see 3578), and A. Wislizenus (see 11009)]: v. 6 (ns), no. 3, 376-392, fig.
950 1848, Lapis lazuli and mica [from Russia]: v. 6 (ns), no. 3, 425 (7 lines).
951 1848, Yield of lead in Great Britain: v. 6 (ns), no. 3, 445 (8 lines).
     Abstracted from Geol. Soc. Great Britain Mem.
952 1848, Lithographic limestone [from Deccan, India]: v. 6 (ns), no. 3, 446 (3 lines).
     Abstracted from London Ath.
953 1848, Museum of Economic Geology, England: v. 6 (ns), no. 3, 446-447 (5 lines).
954 1848, Interesting collection for sale: v. 6 (ns), no. 3, 447-448.
     Mineral collection offered.
955 1848, Recherches sur les animaux fossiles; par L. de Koninck. Liège, 1847 [review of]: v. 6 (ns), no. 3, 448.
956 1848, Berzelius [obituary]: v. 6 (ns), no. 3, 448-451.
     From London Ath.
957 1849, Platinum [from France]: v. 7 (ns), no. 1, 137 (5 lines).
958 1849, Mines in Australia: v. 7 (ns), no. 1, 138 (7 lines).
     Abstracted from London Ath.
959 1849, Death of Robert Gilmor, Esq., of Baltimore: v. 7 (ns), no. 1, 142-143.
960 1849, Dr. A. Goldfuss [obituary]: v. 7 (ns), no. 1, 143 (2 lines).
961 1849, The philosophy of geology; by A. C. G. Jobert [review of]: v. 7 (ns), no. 1, 150.
962 1849, Observations on belemnites and other fossil remains of Cephalopoda, discovered by Mr. Reginald Neville Mantell [review of]: v. 7 (ns), no. 1, 150 (11 lines).
963 1849, A memoir on the structure of the maxillary and dental organs of the Iguanodon; by G. A. Mantell, &c. [review of]: v. 7 (ns), no. 1, 150-151 (7 lines).
964 1849, A memoir on the Dinorthis, part III [notice of work by R. Owen]: v. 7 (ns), no. 1, 151 (10 lines).
965 1849, [Bibliography, with recent geological literature]: v. 7 (ns), no. 1, 151-152; no. 2, 311-312; no. 3, 457-458.
966 1849, Gold mining in Virginia: v. 7 (ns), no. 2, 295-299.
     Abstracted from Mining J.
967 1849, Coal in the Straits of Magellan: v. 7 (ns), no. 2, 304 (5 lines).
968 1849, Native silver in Norway: v. 7 (ns), no. 2, 305 (6 lines).
969 1849, Naphtha spring near Alfreton, England: v. 7 (ns), no. 2, 305 (6 lines).
970 1849, Mineral coal of Vancouver's Island, North West America: v. 7 (ns), no. 3, 436 (9 lines).
     Abstracted from London Ath.
971 1849, Gold of Africa: v. 7 (ns), no. 3, 436-437.
     Abstracted from London Ath.
972 1849, Vesuvius: v. 7 (ns), no. 3, 437 (5 lines).
     Notice of volcanic activity.
973 1849, Verhandlungen der Russisch - Kaiserlichen Mineralogischen Gesellschaft zu St. Petersburg. Jahr. 1847, St. Petersburg, 1848 [notice of]: v. 7 (ns), no. 3, 457.
     Table of contents translated.
974 1849, Observations at the magnetic and meteorological observatory, at the Girard College, Philadelphia, made under the direction of A. D. Bache [review of 1386]: v. 7 (ns), no. 3, 455.
975 1849, Julius T. Ducatel [obituary]: v. 8 (ns), no. 1, 146-149.
976 1849, The fossil footmarks of the United States and the animals that made them; by Edward Hitchcock [review of 5183]: v. 8 (ns), no. 1, 151-152.
977 1849, Report on the geological survey of Canada, for the year 1847-48; by W. E. Logan [review of]: v. 8 (ns), no. 1, 154-155.
978 1849, Système Silurien du centre de la Bohême; by Joachim Barrande [notice of]: v. 8 (ns), no. 1, 158 (9 lines).
979 1849, Manual of mineralogy or the natural history of the mineral kingdom; by James Nicol [review of]: v. 8 (ns), no. 1, 159.
980 1849, The earth and man ... by Arnold Guyot [review of 4609]: v. 8 (ns), no. 1, 158 (10 lines).
981 1849, [Bibliography, with recent geological literature]: v. 8 (ns), no. 1, 159-160; no. 2, 310; no. 3, 457-458.
982 1849, Baierine [from Bavaria]: v. 8 (ns), no. 2, 274 (5 lines).
     Abstracted from l'Institut. Mineral compared with columbite.
983 1849, Gold at Port Phillip, south shore of Australia: v. 8 (ns), no. 2, 290-291.
984 1849, The tin mines of Banca [China]: v. 8 (ns), no. 2, 291-292.
     Abstracted from Mining J.

American Journal of Science (cont.)
985   1849, Platinum and diamonds in California: v. 8 (ns), no. 2, 294 (9 lines).
986   1849, A memoir on the geological action of the tidal and other currents of the ocean; by Charles Henry Davis [review of 2947]: v. 8 (ns), no. 2, 305-306 (14 lines).
987   1849, Ko Doü Dzu Roku, or, A memoir on smelting copper, illustrated with plates: v. 8 (ns), no. 3, 336-348.
   From the Chinese Repos.
988   1849, Geographical survey of Tennessee [review of report by G. Troost (see 10404)]: v. 8 (ns), no. 3, 419-420.
989   1849, Meteorite of Arva: v. 8 (ns), no. 3, 439-440.
990   1849, California gold: v. 8 (ns), no. 3, 449 (3 lines).
991   1849, Coal in Egypt: v. 8 (ns), no. 3, 449 (4 lines).
992   1849, Geology of the United States exploring expedition under Wilkes, U. S. N.; by James D. Dana [notice of 2841]: v. 8 (ns), no. 3, 454-455.
993   1850, Gold on the farm of Samuel Elliot, Montgomery County, Md., thirty miles from Baltimore: v. 9 (ns), no. 1, 126.
   From Am. Philos. Soc. Proc.
994   1850, Cabinet of geology and mineralogy for sale: v. 9 (ns), no. 1, 147 (11 lines).
   Notice of sale of L. Vanuxem's collection.
995   1850, Contributions to the history of British fossil mammals (first series); by Richard Owen, F. R. S., &c. [review of]: v. 9 (ns), no. 1, 149-151.
996   1850, Foster's complete geological chart [review of 3920]: v. 9 (ns), no. 1, 151-152.
997   1850, [Bibliography, with recent geological literature]: v. 9 (ns), no. 1, 152; no. 2, 312; no. 3, 457-458.
998   1850, Anniversary of the Royal Society of London: v. 9 (ns), no. 2, 304.
   Notice of awards presented to R. I. Murchison and G. A. Mantell.
999   1850, Mastodon angustidens [from Turin, Italy]: v. 9 (ns), no. 2, 304.
1000 1850, Report of a geological reconnoissance of the Chippewa Land District of Wisconsin..., by David Dale Owen [review of 8078]: v. 9 (ns), no. 2, 306-307.
1001 1850, Foster's geological chart [review of 3920]: v. 9 (ns), no. 2, 309.
1002 1850, The annual of scientific discovery: v. 9 (ns), no. 2, 309.
   New publication noted. Also notice of L. Agassiz's Lake study (see 83).
1003 1850, The lagoons of Tuscany [Italy]: v. 9 (ns), no. 3, 431-434.
   From Geol. Soc. France Bulletin.
1004 1850, The annual of scientific discovery, or year book of facts in science and arts, &c. [notice of]: v. 9 (ns), no. 3, 454.
1005 1850, The physical atlas of natural phenomena; for the use of colleges, academies and families; by Alexander Keith Johnston [review of 5811]: v. 9 (ns), no. 3, 454-455.
1006 1850, Lake Superior, its physical character, vegetation and animals, compared with those of other and similar regions [review of book by L. Agassiz (see 83)]: v. 9 (ns), no. 3, 455-456.
1007 1850, A natural scale of heights by the application of which the measure of different countries are reduced to a common measure known to all geographers, constructed by Miss Colthurst [notice of]: v. 9 (ns), no. 3, 456.
1008 1850, The East; sketches of travel in Egypt and the Holy Land; by the Rev. J. A. Spencer, M. A. [notice of]: v. 9 (ns), no. 3, 456.
1009 1850, Man primeval, or the constitution and primitive conditions of the human being, &c.; by John Harris, D. D. [review of 4832]: v. 9 (ns), no. 3, 456.
1010 1850, Review of the geological report on the Chippewa Land District of Wisconsin and part of Iowa, made in the year 1847, under the direction of D. D. Owen, M. D. [review of 8078]: v. 10 (ns), no. 1, 1-12.
1011 1850, Fowlerite: v. 10 (ns), no. 1, 121.
1012 1850, Prof. C. U. Shepard, on meteorites: v. 10 (ns), no. 1, 127-128.
   Notice of his recent work on meteorites of South Carolina.
1013 1850, Mr. Edward J. Chapman has been appointed to the Professorship of Mineralogy recently instituted in University College, London: v. 10 (ns), no. 1, 137 (2 lines).
1014 1850, A system of mineralogy..., by James D. Dana [review of 2865]: v. 10 (ns), no. 1, 138-143.
1015 1850, An elementary course of geology, mineralogy and physical geography; by David T. Ansted [review of]: v. 10 (ns), no. 1, 144-145.
1016 1850, Index Paleontologicus, oder übersicht der bis jetzt bekannten Fossilen Organismen, unter mitwirkung der H. H. Prof. H. R. Göppert und Herm. von Meyer, bearbeitet von Dr. H. G. Bronn [review of]: v. 10 (ns), no. 1, 145-146.
1017 1850, Minerals and their uses, in a series of letters to a lady; by J. R. Jackson [review of]: v. 10 (ns), no. 1, 147 (6 lines).
1018 1850, Third annual report of the Regents of the University, of the condition of the state cabinet of natural history and the historical and antiquarian collection annexed thereto [notice of 7757]: v. 10 (ns), no. 1, 147-148.
1019 1850, [Bibliography, with recent geological literature]: v. 10 (ns), no. 1, 149-150; no. 2, 303-304.
1020 1850, Dr. Troost [obituary]: v. 10 (ns), no. 2, 298.
1021 1850, First biennial report on the geology of Alabama; by M. Tuomey, geologist of the state [review of 10440]: v. 10 (ns), no. 2, 299-300.

The American Journal of the Medical Sciences (Philadelphia, 1828-1850)
1022 1831, Connexion of diseases with the rock formations of a country: v. 8, 257 (17 lines).
1023 1831, On baths and mineral waters [notice of work by J. Bell (see 1622)]: v. 9, 263 (5 lines).
1024 1833, Hot springs of Virginia: v. 13, 273-274.
   Signed by "W. E. H."
1025 1842, Trimmer's practical geology and mineralogy [notice of 10330]: v. 3 (ns), 522 (6 lines).

The American Magazine, a Monthly Miscellany (New York, 1815-1816)
1026 1815, Diamonds: v. 1, no. 3, 114-115.
   Largest known specimens described.
1027 1816, Fuel: v. 1, no. 8, 281-282.
   Sources of fuel on the East Coast of the United States.

American Magazine, a Monthly Miscellany (cont.)
1028   1816, State of Ohio: v. 1, no. 12, 425-428.
        Signed "H." Coal and minerals noted.

The American Magazine and Historical Chronicle (Boston, 1743-1746)
1029   1744, An account of a subterranean city near Mount Vesuvius, which had been overturned
        near the said Mount, and buried many years since by the eruptions and earthquake,
        and which had been discovered through a well: Number for May, 1744, 379-380.
1030   1744, A curious description of a burning vulcano, in the city of Kerry: Number for
        August, 1744.
        Signed by "Bishop of K--m--e."

The American Magazine and Monthly Chronicle for the British Colonies (Philadelphia, 1757-1758)
1031   1757, The philosophy of earthquakes: v. 1, no. 1, 23-24.
        By "a member of the Roy. Acad. Berlin."

The American Magazine and Repository of Useful Literature (Albany, 1841-1842)
1032   1841, [Short geological notices on fossils and volcanoes]: v. 1, 30-31.
1033   1841, Hitchcock's Geology [review of 5132]: v. 1, 33.
        Signed by "E. N. H."
1034   1841, Visit to the volcano Kirauea [in Hawaii]: v. 1, 40-41.
        From Atheneum.
1035   1841, Mammoth caves of Kentucky: v. 1, 86-90; 130-133; 184-190, map.
        Signed by "J. S. W."
1036   1841, [Short notices on fossils, Virginia salt, Tennessee meteorite, and magnetic
        observations]: v. 1, 93-94.
1037   1841, Geological report of the state of New York, 1841 [notice of 7746]: v. 1, 94.
1038   1841, Coral islands: v. 1, 136.
        On origin of coral islands.
1039   1841, [Geological notices on gold in North Carolina, sulfur from Dubuque, and Charles
        Lyell's visit]: v. 1, 158.
1040   1842, Mineral spring [at Ballston, New York]: v. 2, 31.
1041   1842, The natural bridge [in Virginia]: v. 2, 76-78, woodcut.

The American Magazine of Useful and Entertaining Knowledge (Boston, 1835-1837)
1042   1835-1837, [Brief notes on ores and mining]: v. 1, no. 1, 28, 33, 42; no. 2, 136; no.
        3, 160; no. 8, 328; no. 12, 499 (1835); v. 2, no. 5, 211-212; no. 6, 258; no. 9,
        358, 362, 364, 393-394; no. 10, 410; no. 12, 508 (1836); v. 3, no. 1, 6; no. 2, 56,
        76, no. 4, 158; no. 5, 180, fig., 202; no. 6, 229, 236, fig. (1837).
1043   1835-1837, [Brief notes on fossils]: v. 1, no. 2, 110; no. 10, 422 (1835); v. 2, no. 6,
        241; no. 9, 391-392; no. 10, 407-408 (1836); v. 3, no. 1, 15; no. 7, 280, fig.; no.
        8, 334, fig. (1837).
1044   1835-1836, [Brief notes on mineral waters]: v. 1, no. 3, 160; no. 11, 489-490 (1835);
        v. 2, no. 1, 9-10, 33; no. 8, 340; no. 12, 518 (1836).
1045   1835-1837, [Brief notes on volcanoes and earthquakes]: v. 1, no. 4, 186; no. 9, 401;
        no. 12, 520 (1835); v. 2, no. 2, 60; no. 5, 205-206 (1836); v. 3, no. 2, 81, fig.;
        no. 3, 102; no. 6, 227, fig., 245, fig.; no. 7, 268, fig. (1837).
1046   1835-1837, [Brief notes on meteorites]: v. 1, no. 8, 352-354 (1835); v. 2, no. 2, 52;
        no.12, 515 (1836); v. 3, no. 1, 7-8; no. 9, 384 (1837).
1047   1835, Geography: v. 1, no. 10, 427.
        Relationship between geology and geography.
1048   1835, Compression of the earth: v. 1, no. 10, 430, fig.
1049   1835, Physical changes on the globe: v. 1, no. 1,, 459-461.
        On the process of erosion.
1050   1836-1837, [Brief notes on caves]: v. 2, no. 2, 75; no. 12, 518 (1836); v. 3, no. 4,
        131; no. 5, 184 (1837).
1051   1836, Former temperature of the earth: v. 2, no. 8, 325-326.
1052   1836, Coal: v. 2, no. 12, 498-499, woodcut.
        Pennsylvania mining techniques discussed.
1053   1836, Caverns: v. 2, no. 12, 507-508, 3 figs.
        Notice of recently discovered caves in Tennessee, Kentucky, and Virginia.
1054   1836, Hot springs of Iceland: v. 2, no. 12, 512-513, fig.
1055   1837, Original condition of the earth: v. 3, no. 1, 10.
1056   1837, First principles of geology: v. 3, no. 4, 142, 3 figs.
        Elementary concepts of stratigraphy.
1057   1837, Dip and strike of geological strata: v. 3, no. 4, 165, fig.
        Definitions.
1058   1837, Immense boulders: v. 3, no. 5, 184, 2 figs.
        Notice of rocking stones from Massachusetts.
1059   1837, Rent rock near Brewster [Massachusetts]: v. 3, no. 5, 200, fig.
1060   1837, Springs and artesian wells: v. 3, no. 6, 212-214, fig.
1061   1837, Salt manufacture on the Kenawha River [Virginia]: v. 3, no. 8, 346-347, fig.
1062   1837, Iceland: v. 3, no. 10, 400-402, fig.
        Geological notes on Iceland, with figure of a basalt cave.
1063   1837, Earthquake at Lisbon, in 1755: v. 3, no. 12, 452-455.
        Eyewitness account quoted.

The American Magazine of Wonders, and Marvelous Chronicle (New York, 1809)
1064   1809, The wonderful fountain: v. 1, 23.
        On a geyser in Iceland.
1065   1809, A curiosity in Virginia which seems unparalled any where: v. 1, 304-305.
        On a burning spring in Fincastle County.
1066   1809, Wonderful spring in Iceland described: v. 1, 309-310.
        Geyser described.
1067   1809, Account of a burning island that rose out of the sea: v. 1, 316-318.
        Volcanic island in Santerini Bay, c. 1800.
1068   1809, Description of asphaltite lake, or Dead Sea [in Jordan]: v. 1, 381-382.
1069   1809, Remarkable eathquakes, and other consequent and wonderful phenomena: v. 2, 50-54.

The American Magazine of Wonders (cont.)
1070 1809, Aerial stones: v. 2, 121-126.
1071 1809, The nature of fossil asbestos: v. 2, 249.
1072 1809, An account of a burning well [in Shropshire, England]: v. 2, 250-251.
1073 1809, An account of the death of a woman, killed by a sudden imperceptible eruption from the earth [in France]: v. 2, 259-260.
1074 1809, Dreadful earthquakes and eruptions [on Terceira Island, the Azores]: v. 2, 264-266.

The American Magazine, or General Repository (Philadelphia, 1769)
1075 1769, On the increase of continents: v. 1, February, 44 [i.e. 34]-38.

American Mechanics' Magazine; Containing Useful Original Matter, on Subjects Connected with Manufactures, the Arts and Sciences (New York, 1825-1826)
1076 1825-1826, [Brief notes on ores and mining]: v. 1, 62, 92, 192, 208 (1825); v. 2, 18, 161-162, 255 (1826).
1077 1825, Plan for purifying coal mines from choke-damps: v. 1, no. 19, 295-297.

The American Medical and Philosophical Register; of Annals of Medicine, Natural History, Agriculture, and the Arts (New York, 1810-1814)
1078 1810, Coal discovered in Pennsylvania: v. 1, no. 1, 117-118.
1079 1810-1814, [Brief notices of recent geological publications]: v. 1, 138, 294-295 (1810); v. 3, no. 2, 221 (1812); v. 4, no. 4, 648 (1814).
    Notices and reviews of 2107, 4424 and 9602.
1080 1812, Memoirs of the Connecticut Academy of Arts and Sciences, vol. I. part I. New Haven. O. Steele & Co. 8vo. pp. 216. 1810 [review with excerpts]: v. 3, no. 2, 204-226.

The American Medical Recorder (Philadelphia, 1818-1829)
1081 1819, Sulphate of strontian [from Schoharie, New York]: v. 2, 151.
1082 1825, Lyceum of Natural History of the Berkshire Medical Institution: v. 8, no. 437. Notice of organization and officers.

The American Medical Review and Journal (Philadelphia, 1824-1826)
1083 1824, Medicinal properties of the waters of the Mississippi - mineral impregnation of the well of Henderson, Ohio: v. 1, 300-301.

The American Mineralogical Journal (New York, 1810-1814)
    The short-lived American Mineralogical Journal was published in four numbers, totaling one volume, under the editorship of Archibald Bruce. The following unsigned notes were probably selected and abstracted by Bruce. See also A. Bruce (2106 et seq.).
1084 1810, Mr. Greville: v. 1, no. 1, 55-56.
    Obituary of Charles F. Greville.
1085 1811, Eruption of Vesuvuis: vo. 1, no. 2, 120-121.
    Abstracted from J. du Commerce.
1086 1811, Mass of malleable iron: v. 1, no. 2, 124, fig.
    Notice of a meteorite from the Red River.

The American Mining Journal (see Mining Journal and American Railroad Gazette)

The American Monthly Magazine (New York, 1833-1838)
1087 1833, [Review of A new theory of terrestrial magnetism by S. Metcalf (see 7138)]: v. 2, 67-68.
1088 1835, Volcanoes and volcanic action: v. 5, 50-61; 102-107.
1089 1835, Vesuvius: v. 6, 184-189.
    Signed by "Oceanus."
1090 1837, [Review of Geology and mineralogy..., by W. Buckland (see 2160)]: v. 9, 508 (9 lines).

The American Monthly Magazine and Critical Review (New York, 1817-1819)
1091 1817, [Review of An elementary treatise..., by P. Cleaveland (see 2420)]: v. 1, no. 3, 183-187.
    Signed by "K."
1092 1817, Mr. Werner [obituary]: v. 2, no. 2, 155 (7 lines).
1093 1818, [Review of A geological essay on the imperfect evidence in support of a theory of the earth by J. Kidd]: v. 2, no. 5, 357 [i.e. 353]-356.
    Signed by "K."
1094 1818, [Review with excerpts of Essay on the theory of the earth by G. Cuvier (see 2754)]: v. 3, no. 1, 51-59; to be continued but no more found.
    Signed by "K."
1095 1818, Animal remains. - mammoth - crocodile [from the Isle of Wight, England]: v. 3, no. 6, 478.
1096 1819, Letter to editor, on ancient lake in central New York: v. 4, 385-387.
    Signed by "W. G."

The American Monthly Review (Cambridge, 1832; Boston, 1833)
1097 1832, [Review of "Remarks on the mineralogy and geology of Nova Scotia" by C. T. Jackson and F. Alger (see 5538)]: v. 1, no. 5, 398-407.
1098 1832, [Review of Monthly American journal of geology and natural science by G. W. Featherstonhaugh (see 3746)]: v. 1, no. 5, 407.
1099 1832, [Review of A geological manual by H. T. De la Beche (see 3036)]: v. 2, no. 1, 19-22.
1100 1832, [Review of Treatise on mineralogy by C. U. Shepard (see 9489)]: v. 2, no. 2, 89-95.
1101 1832, [Review of Report of a geological survey of Massachusetts. Part I. Economical geology by E. Hitchcock (see 5128)]: v. 2, no. 2, 95-101.
1102 1832, [Review with excerpts of Familiar lessons in mineralogy and geology by J. K. Welsh (see 10798)]: v. 2, no. 6, 482-502.
    Review of volume 1 only.

American Monthly Review (cont.)
1103   1833, [Review of Geology: comprising the elements of the science in its advanced state by D. J. Browne (see 2083)]: v. 3, no. 1, 46-50.
1104   1833, [Review of Lectures in natural history, geology, chemistry, etc. by T. Flint (see 3888)]: v. 3, no. 4, 261-267.
1105   1833, [Review of Scripture natural history..., by W. Carpenter (see 2258)]: v. 4, no. 1, 80-86.

American Moral and Sentimental Magazine (New York, 1797-1798)
1106   1797, An account of the ancient city of Herculaneum, destroyed by an eruption of Mount Vesuvius, with descriptions of some antiquities found there: v. 1, 70-76.
1107   1797, Account of a fire-damp, in a tin-mine in Cornwall: v. 1, 117-120.
1108   1797, Description of the Grotto of Antiparas: v. 1, 210-214; 249-253.
1109   1798, On the immensity of the works of Creation: v. 2, 690-693; 716-719. Includes remarks on mineralogy.

American Museum, New York
1110   1823, A companion to the American Museum; being a catalogue of upwards of fifty thousand natural and foreign curiosities... : NY, G. F. Hopkins, xii, 103 p. 1st edition, with notes by John Scudder.

The American Museum; Or, Annual Register of Fugitive Pieces, Ancient and Modern (Philadelphia, 1799)
1111   1799, Description of a newly-discovered cavern, on the north-east end of Mount Anthony, in Bennington: v. 1, 123-124.

The American Museum, or Universal Magazine (Philadelphia, 1787-1792)
1112   1787, Curious subterranean discovery [at Rutsens, New York]: v. 1, 205-206.
1113   1789, Description of the mineral springs of Saratoga [New York]: v. 4, 40-41.
1114   1789, Experiments on the mineral waters of Saratoga [New York]: v. 4, 41-42.
1115   1790, Of the enormous bones found in America: v. 8. Describes Kentucky fossils at Big Bone Lick.
1116   1792, Some particulars relative to the soil, situation, productions, &c. of Kentucky: v. 11, 11-51 [i.e. 15].

American Philosophical Society
1117   1786, Presents made to the American Philosophical Society, since its revival and incorporation in 1780, with the names of the donors: Am. Philos. Soc. Trans., v. 2, 10 p.
1118   1789, An account of communications and donations made to the American Philosophical Society, at Philadelphia, since the publication of their second volume of Transactions: Colum. Mag., v. 3, 360-361; 412-413; 484-485; 602-603; 673-674; to be continued, but no more found.
1119   1789, Some account of the American Philosophical Society, held at Philadelphia: Colum. Mag., v. 3, 703-706.
1120   1793, Presents received by the American Philosophical Society, since the publication of their 2d. vol. of Transactions, with the names of the donors: Am. Philos. Soc. Trans., v. 3, 351-365.
1121   1799, Donations received by the American Philosophical Society since the publication of their third volume of Transactions, with the names of the donors: Am. Philos. Soc. Trans., v. 4, xvii-xxxvi.
1122   1802, Presents received by the American Philosophical Society since the publication of their 4th vol. of Transactions, with the names of the donors: Am. Philos. Soc. Trans., v. 5, xiv-xix.
1123   1805, American Philosophical Society: Phila. Med. Mus., v. 1, 87-90; 234; 327-330; 445-446. Officers, communications, and donations.
1124   1809, Donations received by the American Philosophical Society, since the publication of the fifth vol. of their Transactions: Am. Philos. Soc. Trans., v. 6, ix-xx.
1125   1819, Donations received by the American Philosophical Society, since the publication of vol. VI - old series: Am. Philos. Soc. Trans., v. 1 (ns), 435-453.
1126   1824, Donations received by the American Philosophical Society, since the publication of vol. I - new series: Am. Philos. Soc. Trans., v. 2 (ns), 481-503.
1127   1834, Donations received by the American Philosophical Society, since the publication of vol. II - new series: Am. Philos. Soc. Trans., v. 3 (ns), 487-511.
1128   1834, Donations received by the American Philosophical Society, since the publication of vol. III - new series: Am. Philos. Soc. Trans., v. 4 (ns), 471-493.
1129   1837, Donations received by the American Philosophical Society, since the publication of vol. IV - new series: Am. Philos. Soc. Trans., v. 5 (ns), 459-482.
1130   1838-1850, Donations for the library: Am. Philos. Soc. Proc., v. 1-5, numerous entries.
1131   1838-1850, Donations for the cabinet: Am. Philos. Soc. Proc., v. 1-5, numerous entries.
1132   1839, Donations received by the American Philosophical Society, since the publication of vol. V - new series: Am. Philos. Soc. Trans., v. 6 (ns), 399-427.
1133   1840, Proceedings of the American Philosophical Society: Am. J. Sci., v. 38, no. 1, 153-193; no. 2, 396-405; v. 39, no. 2, 361-373.
1134   1841, American Philosophical Society: Am. J. Sci., v. 40, no. 2, 374-386.
1135   1841, Donations received by the American Philosophical Society, since the publication of vol. V [i.e. VI] - new series: Am. Philos. Soc. Trans, v. 7 (ns), 467-502.
1136   1841, Extracts from the proceedings of the American Philosophical Society: Am. J. Sci., v. 40, no. 1, 27-59.
1137   1843, Donations received by the American Philosophical Society, since the publication of vol. VII - new series: Am. Philos. Soc. Trans., v. 8 (ns), 327-357.
1138   1843, Hundredth anniversary of the American Philosophical Society: Am. J. Sci., v. 45, no. 1, 231-232.
1139   1846, Donations received by the American Philosophical Society since the publication of vol. VIII - new series: Am. Philos. Soc. Trans., v. 9 (ns), 381-413.

American Philosophical Society. Transactions of the (Philadelphia, 1769-1850)
1140   1793, An account of a hill, on the borders of N. Carolina, supposed to have been a volcano. In a letter from a continental officer, residing in that neighborhood, to Dr. J. Greenway, near Petersburg in Virginia: v. 3, 231-233.
Signed by "T. D." With a letter by Dr. Greenway (see 4564).

The American Quarterly Journal of Agriculture and Science (Albany, 1845-1846; New York, 1847-1848)
1141   1845, Greatest iron mines in the world [Newcomb, New York]: v. 2, 129-130.
1142   1845, Remarks ... suggested by a lecture of Mr. Murchison on the geology of Russia: v. 2, 320-323.
1143   1846, Recent discoveries: v. 3, 100.
Includes remarks on the Geological Society of London and on the activities of C. Lyell.
1144   1846, Visit of a distinguished foreigner: v. 3, no. 2, 328 (8 lines).
Visit of L. Agassiz announced.
1145   1846, Reward of merit: v. 3, no. 2, 328 (8 lines).
R. Murchison knighted in England and Russia.
1146   1846, Gradual rise of Newfoundland: v. 3, no. 2, 328 (5 lines).

American Quarterly Observer (Boston, 1833-1834)
1147   1834, Mineralogy and geology: v. 3, 377-378.
Goals of mineralogy and geology noted.

The American Quarterly Register and Magazine (Philadelphia, 1848-1850)
1148   1848, The coal of Pennsylvania: v. 1, no. 1, 97-101.
Production statistics for 1821 to 1846.
1149   1848, The manufacture of iron [in Pennsylvania]: v. 1, no. 1, 101-107.
1150   1848, Gold and silver: v. 1, no. 1, 148-149.
Describes world production to 1846.
1151   1848, Specie: v. 1, no. 2, 414-415.
Concerns gold imports to the United States from 1821.
1152   1849, California gold: v. 3, no. 1, 123.
1153   1849, California: v. 3, no. 2, 377-386.
Provides statistics on gold.
1154   1850, The late eruption of Vesuvius: v. 4, no. 1, 139.
1155   1850, The precious metals: v. 4, no. 1, 265-269.

The American Quarterly Review (Philadelphia, 1827-1837)
1156   1827, [Review of Historical researches on the wars and sports of the Mongols and Romans; in which elephants and wild beasts were slain, by John Ranking]: v. 1, no. 1, 78-105.
Review of theory that Tertiary fossil mammal remains are due to animals slain in wars and games.
1157   1829, [Review of the third edition of Introduction to geology by R. Bakewell, with "Appendix," by B. Silliman (see 1460 and 9759)]: v. 6, 73-104.
1158   1830, [Reviews of System of geology, by Andrew Ure and Outlines of geology, by W. T. Brande]: v. 7, 361-409.

American Rail-Road Journal (New York, 1831-1850)
This weekly periodical contains numerous brief notes on mines and mining. Consult annual subject indices for details. Weekly summaries of coal and iron trade statistics from Pennsylvania and England appear from volume 18 (1845) onward. The American Mining Journal was largely excerpted by the American Rail-Road Journal before their merger in 1848 as The Mining Journal and American Railroad Gazette.
1159   1834, Map of the gold region of Virginia: v. 3, 437, map woodcut.
1160   1837, The interior of a coal mine: v. 6, 638-639; 645-646.
1161   1838, The gold region in the state of Georgia: v. 8, 97-100, woodcut.
1162   1840, Earthquake in Connecticut, &c.: v. 11, 248-253.
Includes excerpt from description of earlier earthquake by H. Chapman (see 2313).
1163   1844, Allan's mineralogy [review of Phillips' Mineralogy, United States edition (see 8325)]: v. 17, 333-334.
1164   1846, The Maryland Mining Company and their works: v. 19, 361-363.
1165   1846, Mineral lands of Lake Superior: v. 19, 693-695.
1166   1846, Lead mines of the West: v. 19, 794-795.
1167   1847, Pennsylvania coal trade: v. 20, 117-119.
1168   1847, Improvement in treating metallic ores: v. 20, 455-456.
1169   1847, Smelting copper ores by electricity, again: v. 20, 489-490.
1170   1848, Mineral resources of the United States: v. 21, 23-25.
1171   1848, Upper Canada Mining Company: v. 21, 50-52.
1172   1848, Copper mining at Lake Superior: v. 21, 105, 579-580, 582.
1173   1848, Gold mines in Canada: v. 21, 617-618.
1174   1849, Gold in California, and other countries: v. 22, 165-166, 246-247, 310-311.
1175   1849, Terrestrial magnetism; and its effects on the semi-fluid surface of the earth: v. 22, 345-346.
1176   1849, The California gold region: v. 22, 640-642.
1177   1849, Lake Superior copper region: v. 22, 687-689, woodcut; 703-704; 751-752, plate.
1178   1849, Coal mine in Rhode Island: v. 22, 738-739.
1179   1850, Lake Superior copper region: v. 23, 325-326, 408
1180   1850, Alabama minerals: v. 23, 612-613.

The American Register, or, General Repository of History, Politics, & Science (Philadelphia, 1807-1811)
1181   1807, [Leopold von Buch's studies in Iceland]: v. 1, pt. 2, 43 (6 lines).
1182   1807, [New mineral from Cornwall]: v. 1, pt. 2, 49.
1183   1807, [Eruption of Vesuvius]: v. 1, pt. 2, 53.
1184   1807, [Fossil bones from Lincolnshire, England]: v. 1, pt. 2, 56.
1185   1808, [Coal in France]: v. 1, 377 (11 lines).
1186   1808, [Hot springs and lead mines of Portugal]: v. 2, 380 (6 lines).

American Register (cont.)
1187   1808, [Natural occurrence of rare metallic elements Cr, Pd, Rh, Ir, Os]: v. 2, 401-402.
1188   1808, [Fluorine found in fossil bones]: v. 2, 405.
1189   1810, A geographical and statistical sketch of the district of Mobile [Alabama]: v. 6, 332-341.
1190   1810-1811, Eruption of Mount Aetna: v. 6, 343-345; v. 7, 216-220.
1191   1811, [Earthquakes]: v. 7, 220-222.

The American Register; or, Summary Review of History, Politics, and Literature (Philadelphia, 1817)
1192   1817, [Mineral collection of Werner]: v. 1 (ns), 367.
1193   1817, [On optical properties of minerals]: v. 2 (ns), 422-423.
       Abstracted from Brande's J. In T. Cooper's review article (see 2621).

The American Repertory of Arts, Sciences, and Manufactures (New York, 1840-1842)
       Short notices on mining, fossils, earthquakes and volcanoes appear in all four volumes.
1194   1840, Geological reports: v. 1, 4-10; 81-87.
       Reviews of progress of all state surveys.
1195   1840, Supply of coal: v. 2, 57-58.
       Concerns world energy supplies.
1196   1840, Windows of mica: v. 2, 215-216.
       On uses of mica.
1197   1841, Lead: v. 3, 67-68.
       On the size and production of Missouri mines.
1198   1841, The remains of the gigantic mastodon [in the Genesee Valley, New York]: v. 3, 218-219.
1199   1841, Geological remains in the West: v. 3, 219-221.
       On fossil bones discovered in Missouri and Wisconsin.
1200   1841, New mode of reducing silver ores: v. 3, 223-224.
1201   1841, Lime and its compounds: v. 4, 1-5; 89-96; 189-194.
1202   1841, Terrible volcanic eruption at the Sandwich Islands: v. 4, 139-143.
1203   1841, The Siberian mammoth: v. 4, 302-303.
1204   1842, Force exerted by water in freezing - its application in splitting rocks: v. 4, 401-402.
       Signed by "P."
1205   1842, Bituminous shale [from Belgium]: v. 4, 452-453.

The American Repertory of Arts, Sciences, and Useful Literature (Philadelphia, 1830-1832)
1206   1831, Mineral kingdom, called mineralogy: v. 2, 13-18.
       A brief outline of the science of mineralogy.

The American Review, and Literary Journal (New York, 1801-1802)
1207   1801, Mineral substances from Pompeia: v. 1, 123.
1208   1801, Discoveries in mineralogy: v. 1, 364.
       Abstracted from Med. Repos. On minerals from New York and New Jersey.
1209   1801, Gypsum discovered in New-Jersey: v. 1, 500 (4 lines).
1210   1801-1802, [Review of the Medical repository edited by S. L. Mitchill]: v. 1, 137-146, 279-286, 381-390; v. 2, 1-10, 283-290.
1211   1801, Skeleton of the unknown quadruped: v. 1, 368-369.
       Abstracted from Med. Repos. Notice of the Peale Museum specimen.
1212   1802, Slate quarry [in Clinton, New York]: v. 2, 233 (6 lines).
1213   1802, [Review with excerpts of Memoir on the supply and application of the blowpipe by R. Hare (see 4762)]: v. 2, 432-439.

The American Weekly Messenger; or, Register of State Papers, History, and Politics (Philadelphia, 1813-1814)
1214   1813, Coal mine in Berks County [Pennsylvania]: v. 1, no. 2, 25.
1215   1813, Coal: v. 1, no. 9, 147-149.
       Coal localities in Pennsylvania. The need for a geological survey of Pennsylvania is noted.

The American Whig Review (New York, 1845-1850)
1216   1845, [Review with excerpts of Tour..., by C. Lyell (see 6557)]: v. 2, 403-411.
1217   1846, The author of Vestiges of the natural history of Creation [review with excerpts of Vestiges..., by R. Chambers (see 2292 et seq.)]: v. 3, 168-179.
       Isaac Taylor is suggested as the author of this unsigned work.
1218   1846, [Review of Elementary treatise on mineralogy by W. Phillips (see 8325)]: v. 3, 671-672.
1219   1846, [Review of Mineral springs of western Virginia by W. Burke (see 2196)]: v. 3, 674.
1220   1850, [Review of Aspects of nature by A. von Humboldt (see 5394)]: v. 11, 143-154.
1221   1850, [Review of Footprints of the Creator by H. Miller (see 7186)]: v. 12, 657.
1222   1850, [Review of Pre-Adamite earth by J. Harris (see 4830)]: v. 12, 657.

The Analectic Magazine (Philadelphia, 1813-1820)
1223   1813, Account of the late earthquake at Caracas: v. 2, 163-167.
       Abstracted from Philos. Mag. Signed by "J. H. S."
1224   1813, On the salt mines of Wielicska, in Poland: v. 2, 255-260.
       Abstracted from the Royal Acad. Sci. Mem.
1225   1814, Meteoric stones: v. 3, 410-413.
       Signed by "M." On the origin of meteorites.
1226   1814, [On meteoric stones]: v. 3, 488-492.
       Signed by "S." On the nature of meteorites.
1227   1814, Variation of the compass: v. 3, 526.
       Change since 1657 noted.
1228   1814, [Review with excerpts of Essay on the theory of the earth by G. Cuvier (British edition, see also 2754)]: v. 4, 206-218.
       From British Rev.
1229   1814, Earthquake at Venezuela: v. 4, 301-307.
       Signed by "W. D. R."

Analectic Magazine
1230 1816, Description of a volcano in Java: v. 8, 379-380.
1231 1817, [Review with excerpts of An elementary treatise on mineralogy and geology by P. Cleaveland (see 2420)]: v. 9, 301-314.
1232 1817, Transactions of the Geological Society [of London; notice of volume 2]: v. 10, 293.
1233 1817, Earthquakes: v. 10, 343.
       List of world earthquakes from January to April, 1817.
1234 1817, Cuvier's Theory of the earth [notice of Essay on the theory of the earth by G. Cuvier (see 2754)]: v. 10, 352.
1235 1817, [Review of Outlines of geology by T. Brandel: v. 10, 362-373.
       From British Critic.
1236 1817, Volcanic eruptions [in Batavia]: v. 10, 515-516.
1237 1818, Instrument to distinguish minerals: v. 12, 80.
       On the use of optical properties to identify minerals.
1238 1819, On the introduction of the steam engine to the Peruvian mines: v. 13, 198-203.
1239 1819, [Review of Outlines of the mineralogy and geology of Boston and its vicinity by J. F. Dana and S. L. Dana (see 2871)]: v. 13, 204-207.
1240 1819, [Reviews of Observations on the geology of the United States by W. Maclure, and An elementary treatise on mineralogy and geology by P. Cleaveland (see 6619, 6620, and 2420)]: v. 13, 322-326.
       From Edinburgh Rev.
1241 1819, York Springs, Adams County, Pennsylvania: v. 13, 353-359, plate.
1242 1820, The Natural Bridge [in Virginia]: v. 15, 75-76, plate.
1243 1820, [Review of A view of the soil and climate of the United States by Volney (see 10653)]: v. 16, 40-46.
       Signed by "G."
1244 1820, [Review with excerpts of Remarks made on a short tour between Hartford and Quebec by B. Silliman (see 9631)]: v. 16, 366-390.

Anderson, Paul
1245 1842, [Letter to Mr. Baker on fossil human foot-prints]: Am. J. Sci., v. 43, no. 1, 16-17.
       In "Regarding human foot-prints in solid limestone" by D. D. Owen (see 8051).

Andrews, Elnathan
1246 1811, New and copious salt-well at Montezuma, on the edge of the Seneca River: Med. Repos., v. 14, 182-183.

Andrews, George W.
1247 1842, To Charles Berg, Esq. [Footnote to "Geological and statistical notice of the coal mines in the vicinity of Richmond, Va." by A. S. Wooldridge (see 11050)]: Am. J. Sci., v. 43, no. 1, 10-11.

Anglada, J.
1248 1822, Mineral waters: Am. J. Sci., v. 5, no. 1, 187-188.
       On the origin of mineral water. Abstracted by J. Griscom.

Anglo-American, a Journal of Literature, News, Politics, the Drama, Fine Arts, etc. (New York, 1843-1847)
1249 1843, Temperature of the earth: v. 1, 538-539.
1250 1845, How rocks are formed: v. 5, 206-207.
1251 1845, Shooting stars and aerolites: v. 5, 537-539.
1252 1845, The gold mines of Siberia: v. 5, 563-564.
1253 1845, Account of a visit to the volcano of Kirauea, in Owhyhee, Sandwich Islands, in September 1844: v. 6, 178-180, fig.
       Includes figure map of the crater.
1254 1845, Diamond mine of Sincura: v. 6, 226-227.
1255 1846, Marble quarries at Carrara [Italy]: v. 7, 260.
1256 1846, M. Arago: v. 7, 540-542.
       Biographical sketch.
1257 1847, The silver mines of Peru: v. 9, 246-247.
1258 1847, Rising and sinking of land in northern Europe: v. 9, 486-488.
1259 1847, Scriptural geology: v. 10, 84-85.
       Proof of the deluge cited.

Annan, Robert (1742-1819)
1260 1793, Account of a skeleton of a large animal found near Hudson's River [New York]: Am. Acad. Arts Sci's Mem., v. 2, pt. 1, 160-164.

The Annual of Scientific Discovery; or, Year-book of Facts in Science and Art (Boston, 1850)
1261 1850, Analysis of waters of the Dead Sea: v. 1, 200 (11 lines).
       Abstracted from Poggendorff's Annalen.
1262 1850, Mode of coloring stones artificially: v. 1, 211-212.
       Abstracted from London J. Arts.
1263 1850, Geological surveys of the United States: v. 1, 222 (12 lines).
       Abstracted from Am. J. Sci.
1264 1850, Geology of the gold regions of California: v. 1, 222-225.
1265 1850, Eruption of a volcano in Java: v. 1, 234-235.
1266 1850, Draining the mines of Sierra-morena: v. 1, 252 (17 lines).
1267 1850, [Gold mines in Maryland, Africa and Siberia]: v. 1, 256-258.
1268 1850, [Mining in California]: v. 1, 259-260.
1269 1850, [Copper mines of Lake Superior]: v. 1, 260-261.
1270 1850, Argentiferous galena [from Arkansas]: v. 1, 263 (13 lines).
1271 1850, Zinc mines of New Jersey: v. 1, 263-265.
1272 1850, [Mineral resources of Russia and Japan]: v. 1, 265.
1273 1850, [Descriptions of American coal deposits]: v. 1, 267-272.
1274 1850, [Short notices on fossil vertebrates]: v. 1, 280, 282, 286.
1275 1850, The Dodo: v. 1, 328-329.
       Abstracted from Jameson's J.
1276 1850, American Association for the Promotion of Science: v. 1, 367.

Annual of Scientific Discovery (cont.)
1277   1850, Scientific reports: v. 1, 367-368.
        Abstracted from Am. J. Sci.
1278   1850, List of books: v. 1, 372-373.
        Describes American geological literature published in 1849.
1279   1850, Index to articles in scientific journals: v. 1, 374-379.

Ansted, David Thomas (1814-1880)
1280   1847, The ancient world, or picturesque sketches of creation: Philadelphia, Lee and Blanchard, xi, 382 p., illus.
        For review with excerpts, see 10527 and 10831. For review see 904, 2768, and 5467. For excerpts see Lit. World, v. 2, 65.
1281   1848, On mining, and the practical applications of geological science: Mining J., v. 1, 174, 190.
1282   1848, On the application of geology to engineering and architecture, and the supply of water to towns and cities: Franklin Inst. J., v. 15 (3s), 226-229.
1283   1849, The gold seeker's manual, a practical and instructive guide to all persons emigrating to the newly discovered gold regions of California: NY, D. Appleton and Co., 176 p.
        Also Philadelphia edition? For review see 6362 and 9179.

Anthon, M.
1284   1846, Solubility of sulphate of lime: Am. J. Sci., v. 2 (ns), no. 1, 114 (9 lines).
        Abstracted from Chemical Gaz.

Anthony, J. B. (see 1285 with J. G. Anthony)

Anthony, John Gould (1804-1877)
        See also 4468 with G. Graham.
1285   1825, and J. B. Anthony, [On mineral localities] in Rhode Island: Am. J. Sci., v. 9, no. 1, 46-47.
1286   1838, New trilobites (2) Ceratocephala ceralepta: Am. J. Sci., v. 34, no. 2, 379-380, 2 figs.
        On fossils from Springfield, Ohio.
1287   1839, Description of a new fossil (Calymene bucklandii) [from Cincinnati, Ohio]: Am. J. Sci., v. 36, no. 1, 106-107, 2 figs.
1288   1839, Fossil encrinite [from Cincinnati, Ohio]: Am. J. Sci., v. 35, no. 2, 359-360, fig.
1289   1843, [Fossils from Cincinnati, Ohio]: Bos. Soc. Nat. History Proc., v. 1, 76 (10 lines).
1290   1848, On an impression of the soft parts of an Orthoceras: Am. J. Sci., v. 6 (ns), no. 1, 132-133, fig.
        Abstracted from Q. J. Geol. Soc.

The Antiquarian and General Review (Schenectady, New York, 1845-1848)
1291   1845, Wiclif's council of the earthquake: v. 1, no. 7, 157-158.
        Provides a religious interpretation of the earthquake of May, 1382.
1292   1846, Geology: v. 2, no. 6, 147.
        Notice of miscellaneous geological facts.
1293   1847, Geological discovery at Gibralter: v. 3, no. 4, 96.
        From Bos. J. Phil. Arts. Describes fossil bones found in St. Michael Cave.
1294   1847, The Mammouth Cave [in Kentucky]: v. 3, no. 5, 111-117.
        Signed by "E. M." From J. Commerce.

Antisell, Thomas (1817-1893)
1295   1849, Analysis of Florida muck: Am. Agriculturist, v. 8, 363.
1296   1849, Analysis of New Jersey marl: Am. Agriculturist, v. 8, 376-377.

Arago, Dominique Francois Jean (1786-1853)
1297   1822, Magnetism affected by earthquakes: Mus. For. Liter. Sci., v. 1, 189.
        Abstracted from Philos. Mag.
1298   1836, Declination and inclination of the magnetic needle at Paris: Am. J. Sci., v. 31, no. 1, 190 (6 lines).
        Abstracted from Annuaire.
1299   1836, French scientific voyage: Army and Navy Chron., v. 3, 153-155; 230-232.
1300   1838, On the increases of temperature in the interior of the earth: Franklin Inst. J., v. 21 (ns), 427-428.

Archer, Robert
1301   1836, Cold weather, Florida soil and limestone. Sea ore and other manure, &c.: Farm's Reg., v. 4, 185-187.

Archives of Useful Knowledge (Philadelphia, 1810-1813)
1302   1811, Coal mines and iron forges in Scotland: v. 2, no. 2, 131-141.
        From Tradesman, or Com. Mag.

Arcturus, a Journal of Books and Opinions (New York, 1841-1842)
1303   1841, [Review of The elements of geology for popular use by C. A. Lee (see 6125)]: v. 1, no. 3, 194-195.

Arfvedson, Johann August (1792-1841)
1304   1818, An account of the new alkali lately discovered in Sweden: J. Sci. Arts, v. 5, no. 2, 337-340.
        On the discovery of lithium, in spodumene from Utoen, Sweden.

Army and Navy Chronicle (Washington, 1835-1842)
1305   1835-1843, [Brief notices of earthquakes and volcanoes]: v. 1, 183, 191, 210, 369; v. 3, 290; v. 6, 237-238; v. 7, 133-134; v. 9, 343.
1306   1835, The geology of the West: v. 1, 279.
        Notice of western United States geological studies.

Army and Navy Chronicle (cont.)
1307   1836, Geological Society of Fred'Bg.: v. 2, 333.
         Notice of formation of new society.
1308   1837, Filling up of Lake Superior: v. 4, 372.
1309   1838, [Review with excerpts of Exploring tour beyond the Rocky Mountains by S. Parker (see 8121)]: v. 7, 65-66.
1310   1839, Aerolites: v. 8, 279.
         Notice of meteorite falls at the Cape of Good Hope in 1838.
1311   1841, New scientific discovery: v. 12, 291.
         On the figure of the earth.
1312   1842, Lead mines: v. 13, 52.
         On the conditions of United States mines in 1842.

Army and Navy Chronicle, and Scientific Repository (Washington, 1843-1844)
1313   1843, Eruption of Mount Aetna: v. 1, 177-180.
1314   1843-1844, Earthquakes: v. 1, 248; v. 3, 797.
1315   1843, Depot of charts and instruments: v. 1, 643-646, plate.
         Notice of the United States Observatory's magnetic instruments.
1316   1843, Quarrying stones: v. 1, 553-554.

Arnell, David R.
1317   1809, A geological and topographical history of Orange County, New-York: Med. Repos., v. 12, 313-318.
1318   1812, A description of the Cheechunk Spring, in the town Goshen, Orange County, (N. Y.): Am. Min. J., v. 1, no. 3, 152-153.

The art of making common salt. Particularly adapted to the American Colonies. With an extract from Dr. Brownrigg's Treatise on the art of making bay-salt. Detached from the Pennsylvania Magazine for March 1776:
1319   1776, Phila., R. Aitken, 7 p., plate; also Bos., J. Gill, 15 p., plate.

Ashe, Thomas (1770-1835)
1320   1808, Travels in America performed in 1806, for the purpose of exploring the rivers Alleghany, Monongahela, Ohio, and Mississippi, and ascertaining the produce and condition of their banks, and vicinity: Newburyport, Mass., by E. M. Blunt for W. Sawyer and Co., 9, [11]-366 p.
         Also NY edition, 1811.

Ashmead
1321   1848, [Calcite from Rossie, New York]: Phila. Acad. Nat. Sci's Proc., v. 4, 6 (19 lines).

Association of American Geologists and Naturalists
1322   1841, Abstract of the proceedings of the Association of American Geologists, at their 1st meeting held in Philadelphia, April, 1840 [and] Abstract of the proceedings of the Association of American Geologists, at their 2nd meeting held in Philadelphia, April, 1841: NHav., B. L. Hamlen, 34 p.
         Usually bound with "Address" by E. Hitchcock, 1841 (see 5139). Reprinted from the Am. J. Sci.
1323   1842, Constitution and by-laws of the Association of American Geologists and Naturalists: Assoc. Am. Geol's Nat's Proc., v. 3, 4-6.
1324   1842, Proceedings of the third annual meeting of the Association of American Geologists and Naturalists: New York, Wiley and Putnam, 39 p.
         Usually bound with "Address" by B. Silliman, 1842 (see 9767). Reprinted from the Am. J. Sci.
1325   1843, Proceedings of the fourth annual meeting of the Association of American Geologists and Naturalists: Am. J. Sci., v. 45, no. 1, 135-165; no. 2, 310-353.
1326   1843, Reports of the first, second, and third meetings of the Association of American Geologists and Naturalists, at Philadelphia, in 1840 and 1841, and at Boston in 1842. Embracing its Proceedings and Transactions: Boston, Gould, Kendall, and Lincoln, i-viii, 9-544 p., 21 plates.
         For review with excerpts see 1604. For review see 6377.
1327   1843, Transactions of the Association of American Geologists and Naturalists, 1840-1842: Am. J. Sci., v. 45, no. 1, 220.
1328   1844, Abstract of the proceedings of the fifth session of the Association of American Geologists and Naturalists: NY, Wiley and Putnam, 43 p.
         Usually bound with "Address" by B. Silliman, 1844 (see 9809). Reprinted from Am. J. Sci.
1329   1845, Abstract of the proceedings of the sixth meeting of the Association of American Geologists and Naturalists, held at New Haven, Conn., April, 1845: NHav., B. L. Hamlen, 87 p.
         For review see 9950.
1330   1845, Proceedings of the American Association of Geologists and Naturalists [sixth meeting at New Haven]: Am. Q. J. Ag. Sci., v. 2, 132-170.
         Contains remarks by E. Emmons (see 3533).
1331   1845, The Association of American Geologists and Naturalists: Am. J. Sci., v. 48, no. 2, 404 (6 lines).
         Notice of the sixth meeting.
1332   1845, Sixth annual meeting of the Association of American Geologists: Am. J. Sci., v. 49, no. 1, 219.
1333   1846, Seventh meeting of the Association of American Geologists and Naturalists held in New York City, Sept. 2-8, 1846, and proceedings: New York Herald, Sept. 3, 4, 5, 7, 8, and 9.
1334   1846, Association of American Geologists and Naturalists: Am. J. Sci., v. 1 (ns), Appendix, [16].
         List of New York meeting.
1335   1846, Association of Geologists and Naturalists: Am. J. Sci., v. 2 (ns), no. 1, 144 (5 lines).
         Notice of New York meeting.

Association of American Geologists (cont.)
1336   1846, The Association of American Geologists and Naturalists: Am. J. Sci., v. 2 (ns), no. 3, 441 (6 lines).
       Notice of seventh meeting in New York.
1337   1847, American Association of Geologists and Naturalists: Am. Q. J. Ag. Sci., v. 6, 208-219; and [i.e. 246-247].
       Proceedings of the seventh meeting at Boston.
1338   1847, Association of American Geologists and Naturalists: Am. J. Sci., v. 4 (ns), no. 1, 146.
       List of officers and notices.
1339   1847, Eighth annual meeting of the Association of American Geologists and Naturalists: Am. J. Sci., v. 4 (ns), no. 3, 427-429.
       Abstract of the proceedings.
1340   1848, Proceedings of the meeting of the Association of American Geologists and Naturalists, held at Boston, September, 1847: Am. J. Sci., v. 5 (ns), no. 1, 102-116; no. 2, 243-250.
1341   1848, Association of Geologists and Naturalists: Am. J. Sci., v. 5 (ns), no. 2, 285 (6 lines).
1342   1848, [Proceedings of the 8th annual meeting in Boston of the Association of American Geologists and Naturalists]: Lit. World, v. 2, 227-231; 255-257.
       Abstracted from the Bos. Evening J. and Bos. Atlas. Notices of officers, meeting proceedings, and name change to American Association for the Promotion of Science.

At a meeting of salt manufacturers, convened at New-Bedford, on the 18th of 12th mo. (Dec.) 1827, to take into consideration the subject of the reduction of the present duty on imported salt...
1343   1827, New Bedford, Mass., B. J. Congdon (printer), 7, 1 p.

Atkins, Layton Y. (Or Lytton Y., from Stafford County, Virginia)
1344   1840, Blue marl and green sand: Farm's Reg., v. 8, 320.
1345   1847, The use of marl near Fredericksburg [Virginia]: Mon. J. Ag., v. 3, no. 2, 62-64.

The Atlantic Magazine (New York, 1824-1825)
1346   1824, Geology: v. 1, 159 (11 lines).
       On granite boulders.
1347   1824, Mineralogy: v. 1, 159.
       On Iron Mountain, Missouri, on sillimanite from Saybrook, Connecticut, and on the fusion of charcoal and diamond.
1348   1824, Zoology: v. 1, 160.
       On fossil mastodon bones from New Jersey.
1349   1825, Mineralogy: v. 2, 79.
       On jeffersonite, on New Jersey zinc ore, and on North Carolina gold.
1350   1825, Geology: v. 2, 79-80.
       On western New Jersey marls, and on fossil shells from Florida.
1351   1825, Chemistry: v. 2, 80.
       On the combustion of diamonds.
1352   1825, Mineralogy: v. 2, 160-161.
       Notice of native iron from Bogota, Columbia.
1353   1825, Zoology: v. 2, 162-163.
       On the fossil Megatherium from Savannah, Georgia.
1354   1825, Mineralogy: v. 2, 163.
       On native silver from Lake Ontario, and on I. Cozzen's collection of western United States minerals.

Atlee, Washington Lemuel (1808-1878)
1355   1838, On certain cavities in quartz, &c., in a letter to the editor: Am. J. Sci., v. 35, no. 1, 139-144.
       Description of calcite cavities found in Lancaster, Pennsylvania.

Atwater, Caleb (1778-1867)
1356   1818, On the prairies and barrens of the West: Am. J. Sci., v. 1, no. 1, 116-125.
       See also 1357.
1357   1819, Geology of the West: Niles' Wk. Reg., v. 17, 109.
       From the [Mansfield, Ohio] Olive. Records facts additional to those in the American Journal of Science article (see 1356).
1358   1819, Notice of the scenery, geology, mineralogy, botany, &c. of Belmont County, Ohio: Am. J. Sci., v. 1, no. 3, 226-230.
1359   1820, Geology: Saturday Mag., v. 4, 112.
       On paleo-climates.
1360   1820, On some ancient human bones, etc. with a notice of the bones of the mastodon or mammoth, and of various shells found in Ohio and the West: Am. J. Sci., v. 2, no. 2, 242-246, 11 figs. on 2 plates.
1361   1821, Alum-stone and oher minerals in Ohio, with remarks on vegetable remains: Am. J. Sci., v. 3, no. 2, 245-246.
1362   1826, Facts and remarks relating to the climate, diseases, geology and organized remains of parts of the state of Ohio, etc.: Am. J. Sci., v. 11, no. 2, 224-231.
1363   1831, Remarks made on a tour to Prairie du Chien, thence to Washington City, in 1829: Columbus, Ohio, I. N. Whiting, vii, 296 p.
1364   1838, A history of the state of Ohio, natural and civil: Cin., Glezen and Shepard, 407 p.

Atwater, William
1365   1821, Hill of serpentine: Am. J. Sci., v. 3, no. 2, 238.
       Serpentine from Westfield and Russell, Massachusetts.

Audouard, Mathieu Francois Maxime (1776-1856)
1366   1848, also M. Filhol, On the presence of arsenic in certain chalybeate waters: Am. J. Sci., v. 6 (ns), no. 3, 422 (10 lines).
       From Comptes Rendus. Translated and abstracted by G. C. Schaeffer.

Audubon, John James (1785-1851)
1367   1831, The earthquake: in Ornithological Biography, v. 1, 239-241, Phila., J. Dobson, 5 v.

Auldjo, John (d. 1857)
1368   1833, Spouting fountain of mineral water, discovered in 1832, near Cape Uncino, Kingdom of Naples: Am. J. Sci., v. 25, no. 1, 194-196.
       Abstracted from Bibliotheque Universelle by J. Griscom.

Austin, Moses (1761-1821)
1369   1804, A summary description of the lead mines in upper Louisiana. Also, an estimate of their produce for three years past: Wash., A. and G. Way, 22 p; Also in American State Papers, v. 28, 206-209 (1832).
1370   1838, Lead mine in Davidson County, N. C.: Am. Rr. J., v. 7, 119-120.

Ayres, Stephen
1371   1807, A description of the region of North Carolina where gold has been found: Med. Repos., v. 10, 148-150.

B., A. M. (see 7692)

B., J. H. (see 11124)

Babbage, Charles (1792-1871)
1372   1833, [On the mines of Cornwall, England] : Mech's Mag. and Reg. Inventions, v. 2, Appendix, 43-44.
1373   1835, Abstract of a theory of the elevation and depression of the earth's crust, by variations of temperature, as illustrated by the Temple of Serapis: Am. J. Sci., v. 27, no. 2, 408-411.
       Abstracted from Geol. Soc. London Proc.
1374   1841, The ninth Bridgewater treatise, a fragment: Phila., Lea and Blanchard, 250 p., illus.
       For excerpts see: Franklin Inst. J., v. 20 (ns), 356-360.

Bache, Alexander Dallas (1806-1867)
       See also 8842 with H. D. Rogers and 10530 by A. Ure.
1375   1832, [Annotations to "An essay on chemical nomenclature," by J. J. Berzelius (see 1662)] : Am. J. Sci., v. 22, no. 2, 248-276.
1376   1834, Observations on the disturbance in the direction of the horizontal needle. 1. During the Aurora Borealis, visible at Philadelphia, on the 17th of May, 1833; 2. During that of July 10, 1833: Am. J. Sci., v. 27, no. 1, 113-126.
       Also in: Franklin Inst. J., v. 13 (ns), 1-8, plate; also in: Hazard's Reg. Penn., v. 12, 81-82 (1834)
1377   1835, On the variation of the magnetic needle: Am. J. Sci., v. 27, no. 2, 385-386.
1378   1837, Observations to determine the magnetic dip at Baltimore, Philadelphia, New York, Providence, Springfield and Albany: Am. Philos. Soc. Trans., v. 5 (ns), 209-216.
1379   1837, On the diurnal variation of the horizontal needle: Am. Philos. Soc. Trans., v. 5 (ns), 1-21.
1380   1837, and E. H. Courtenay, On the relative horizontal intensities of terrestrial magnetism at several places in the United States, with the investigation of corrections for temperature, and comparisons of the methods of oscillation in full and in rarefied air: Am. Philos. Soc. Trans., v. 5 (ns), 427-457.
1381   1839, [Glass crystal-form models] : Am. Philos. Soc. Proc., v. 1, no. 7, 97 (9 lines).
1382   1839-1843, [Short notices on magnetic observations] : Am. Philos. Soc. Proc., v. 1, no. 9, 146-148, 151-155; no. 11, 185-186; no. 14, 294-295, 310-312, 320; v. 2, no. 21, 150-151, v. 3, no. 27, 90-92, 175-179; no. 28, 11-12.
1383   1841, Observations of the magnetic intensity at twenty-one stations in Europe: Am. Philos. Soc. Trans., v. 7 (ns), 75-100.
       Also reprint edition. Abstracted in Am. J. Sci., v. 40, no. 1, 30-31.
1384   1842, [On coral reefs] : Am. Philos. Soc. Proc., v. 2, no. 21, 150 (9 lines).
1385   1845, [Map of Sandy Hook, exhibiting the increase of that headland from the earliest surveys] : Am. Philos. Soc. Proc., v. 4, no. 33, 168-169.
1386   1847, Observations at the magnetic and meteorological observatory, at the Girard College, Philadelphia, made under the direction of A. D. Bache, L. L. D.... : Wash., Gales and Seaton (printers), 3 v., plates.
       For review see 974.

Bache, Richard (1794-1836)
1387   1827, Notes on Colombia, taken in the years 1822-3. With an itinerary of the route from Caracas to Bogata: Phila., Carey and Lea, viii, 9-303 p., maps.
       Signed by "an officer in the U. S. Army." For review see 2139.

Bachman, John (1790-1874)
1388   1843, An inquiry into the nature and benefits of an agricultural survey of the state of South Carolina: Char., Miller and Browne, 42 p.

Backer, Benjamin
1389   1823, Earthquakes at Allepo: Niles' Wk. Reg., v. 24, 172-173.

Bacon, George
1390   1845, Copper mines and the copper trade: Fisher's Nat'l Mag., v. 1, no. 2, 116-128.
       Mines noted from Norway, Lake Superior, England, and Wisconsin.

Bacon, John, jr.
1391   1846, [Microscopic examination of Sahara sand] : Bos. Soc. Nat. History Proc., v. 2, 126 (9 lines); 164.
1392   1848, [Mineral specimens] : Bos. Soc. Nat. History Proc., v. 3, 88-89.

Bacon, J. (cont.)
1393   1849, [Quartz crystals with dendrites]: Bos. Soc. Nat. History Proc., v. 3, 110 (7 lines).

Bacon, William
1394   1837, Medical topographical report of the county of Tompkins [New York]: Med. Soc. NY Trans., v. 3, 25-39.
        Notice of the geological structure of Tompkins County.
1395   1837, Medical topography of the county of Tioga [New York]: Med. Soc. NY Trans., v. 3, 151-166.

Baddeley, Frederick H. (Lieutenant)
1396   1830, Mineralogical examination of the sulphate of strontian from Kingston (U. C.), with miscellaneous notices of the geology of the vicinity: Am. J. Sci., v. 18, no. 1, 104-109.
1397   1830, On the red color of flame as produced by strontian and as characteristic of minerals of that genus: Am. J. Sci., v. 18, no. 2, 261-263.
1398   1834, A tabular view of metallic minerals, compiled from the best and latest authorities ... : Am. J. Sci., v. 25, Appendix, 1-16.
1399   1835, Adhesive power of the cement of the Castle Rock [Quebec]: Am. J. Sci., v. 28, no. 2, 367-368.

Baddeley, J. W. (Lieut.; i.e. F. H.?)
1400   1830, A new instrument for taking specific gravities: Am. J. Sci., v. 18, no. 2, 263-266, fig.

Baddeley, W. W. (Lieut.; i.e. F. H.?)
1401   1835, Discovery of gold in Lower Canada: Am. J. Sci., v. 28, no. 1, 112-113.
1402   1835, On the conjectured buoyancy of boulders at great depths in the ocean: Am. J. Sci., v. 28, no. 1, 111-112.
1403   1835, Water lime made from the rock of Quebec: Am. J. Sci., v. 28, no. 1, 113-114.

Baer, William
1404   1843, [Maryland soil analyses]: Am. Farm., v. 5 (3s), 326.
1405   1845, Analysis of soils [from Virginia]: Am. Farm., v. 1 (4s), 20.

Bagnold, Thomas (Captain)
1406   1830, Extraordinary effects of an earthquake at Lima, in 1828: Franklin Inst. J., v. 5 (ns), 71-72.

Bailey, Ebenezer
1407   1827, On the use of soapstone to diminish the friction of machinery: Am. J. Sci., v. 13, no. 1, 192-193.

Bailey, J. T.
1408   1844, Smelting copper in the United States: Hunt's Merch. Mag., v. 11, 290.

Bailey, Jacob Whitman (1811-1857)
1409   1837, Account of an excursion to Mount Katahdin, in Maine: Am. J. Sci., v. 32, no. 1, 20-34; also in: Me. Mon. Mag., v. 1, no. 12, 544-547.
1410   1837, Locality of hyalite at West Point: Am. J. Sci., v. 32, no. 1, 87.
1411   1837, On the common blowpipe: Am. J. Sci., v. 32, no. 2, 319-325, 3 figs.
1412   1838, On fossil infusoria, discovered in peat-earth at West Point, N. Y., with some notices of American species of Diatomae: Am. J. Sci., v. 35, no. 1, 118-124, plate with 16 figs.
        Abstract in: Am. Rep. Arts Sci's Manu's, v. 1, 256-258.
1413   1841, American Polythalmia from the Upper Mississippi, and also from the Cretaceous formation on the Upper Missouri: Am. J. Sci., v. 41, no. 2, 400-401, 4 figs.
1414   1841, Calcareous marl from the Cretaceous formation: Phila. Acad. Nat. Sci's Proc., v. 1, no. 5, 75.
1415   1841, Fossil foraminifera in the green sand of New Jersey: Am. J. Sci., v. 41, no. 1, 213-214.
1416   1841, Fossil infusoria: in Final report..., by E. Hitchcock (see 5137), 311-315, plate 20.
        On species of Massachusetts fossil infusoria.
1417   1841-1842, A sketch of the infusoria, of the family Bacillaria, with some account of the most interesting species which have been found in a recent or fossil state in the United States: Am. J. Sci., v. 41, no. 2, 284-305, 2 plates with 39 figs; v. 42, no. 1, 88-105, 2 plates with 42 figs; also in: Assoc. Am. Geol's Nat's Rept., v. 1, 112-164, 3 plates with 117 figs.
1418   1842, Ehrenberg's notices of American infusoria: Am. J. Sci., v. 43, no. 2, 393-395.
1419   1842, Sketch of the infusoria of the family Bacillaria: Am. J. Sci., v. 43, no. 2, 321-332, 12 figs.
1420   1843, [Description of fossil foraminifera from Egypt]: Assoc. Am. Geol's Nat's Rept., v. 1, 356-358, plate 15 (7 figs.).
        In "Notes on the geology of several parts of Western Asia" by E. Hitchcock (see 5150).
1421   1843, [On American fossil Bacillaria]: in Geology of New York, first district, by W. W. Mather (see 6907), 48-79; 238-245; plates 21, 22, and 42.
1422   1843, [On infusorial deposits at Petersburg, Virginia]: Am. J. Sci., v. 45, no. 2, 313.
1423   1844, Account of some new infusorial forms discovered in the fossil infusoria from Petersburg, Va., and Piscataway, Md.: Am. J. Sci., v. 46, no. 1, 137-141, plate with 32 figs.
1424   1844, Ehrenberg's Researches on the distribution of microscopic life [review of]: Am. J. Sci., v. 47, no. 1, 208-211.
1425   1844, [Fossil infusoria from the southern United States Tertiary formations]: Am. J. Sci., v. 47, no. 1, 117 (5 + 8 lines).
        Also in Assoc. Am. Geol's Nat's Proc., v. 5, 24 (5 + 8 lines).
1426   1844, Marl underlying Charleston: Farm's Cab., v. 9, 98-99.

Bailey, J. W. (cont.)
1427   1844, Notice of a memoir by C. G. Ehrenberg, "On the extent and influence of microscopic life in North and South America": Am. J. Sci., v. 46, no. 2, 297-313.
        See also 3456.
1428   1845, Ehrenberg's observations on the fossil infusoria of Virginia and Maryland, and comparison of the same with those found in the chalk formations of Europe and Africa: Am. J. Sci., v. 48, no. 1, 201-204.
        See also 3458.
1429   1845, Extract of a letter from Prof. Bailey to J. L. Smith: Am. J. Sci., v. 48, no. 1, 102.
        On fossil infusoria from Charleston, South Carolina.
1430   1845, New locality of fossil fluviatile infusoria in Oregon: Assoc. Am. Geol's Nat's Proc., v. 6, 64-66.
        Also in Am. Q. J. Ag. Sci., v. 2, 163-164. Includes discussion by H. D. Rogers.
1431   1845, Notice of some new localities of infusoria, fossil and recent: Am. J. Sci., v. 48, no. 2, 321-343, plate with 23 figs.
        Includes remarks by Ehrenberg (see 3459). Notices of infusorial deposits in Oregon, Bermuda, Virginia, Maryland, and Nova Scotia.
        Also, notice of Mastodon bones from Nova Scotia.
1432   1845, On some of the fossil coniferous trees of the United States: Assoc. Am. Geol's Nat's Proc., v. 6, 81-82.
1433   1845, [Report on fresh water infusoria from Oregon] : in Report of the exploring expedition to the Rocky Mountains by J. C. Fremont (see 4113), 302, plate.
1434   1846, On some new species of American Desmidiaceae, from the Catskill Mountains: Am. J. Sci., v. 1 (ns), no. 1, 126-127, 4 figs.
1435   1846, On the detection of spirally dotted, or sclariform ducts, and other vegetable tissues in anthracite coal: Am. J. Sci., v. 1 (ns), no. 3, 407-410, 4 figs.
1436   1846, [Report on infusorial deposits from Peacham and Maidstone, Vermont] : in second annual Vermont report by C. B. Adams (see 26), 150-152, 15 figs.
1437   1847, Foraminiferes fossiles du bassin Tertaire de Vienna, decrits par Alcide D'Orbigny [review of] : Am. J. Sci., v. 4 (ns), no. 3, 452-454.
1438   1847, [Structure of anthracite coal] : Am. Q. J. Ag. Sci., v. 6, 218 (7 lines).
        From Assoc. Am. Geol's Nat's Proc.
1439   1848, Notes concerning the minerals and fossils, collected by Lieutenant J. W. Abert, while engaged in the geographical examination of New Mexico: US SED 23, 30-1, v. 4 (506), 131-132, 3 plates.
        In Examination of New Mexico..., by J. W. Abert (see 7). Also in US HED 41, 30-1, v. 4 (517), 547-548, 3 plates.
1440   1849, New localities of infusoria in the Tertiary of Maryland: Am. J. Sci., v. 7 (ns), no. 3, 437.
1441   1849, Plumbic ochre from Mexico: Am. J. Sci., v. 8 (ns), no. 3, 420-421; also in: Am. Scien. Disc., v. 1, 274.
1442   1850, Discovery of an infusorial stratum in Florida: Am. J. Sci., v. 10 (ns), no. 2, 282.
1443   1850, On the process for detecting the remains of infusoria, &c., in sedimentary deposits: Am. Assoc. Adv. Sci. Proc., v. 2, 409.

Bailey, S. B.
1444   1843, [On the Zygodon] : Am. J. Sci., v. 45, no. 2, 320 (10 lines).
        From Assoc. Am. Geol's Nat's Proc.

Bain, Sir William (d. 1853)
1445   1818, An essay on the variation of the compass, showing how far it is influenced by a change in the direction of the ship's head, &c. &c.: J. Sci. Arts, v. 4, 102-112.

Baird, John
1446   1822, Geological remarks on the Rock of Gibralter and the adjacent country: Mus. For. Liter. Sci., v. 1, 49-51.

Baird, Robert (1798-1863)
1447   1832, View of the Valley of the Mississippi; or, the emigrant's and traveller's guide to the West: Phila., H. S. Tanner, 341 p., colored maps.
        Minerals and soils noted on 28-30. For review see 1676.
1448   1834, View of the Valley of the Mississippi; or, the emigrant's and traveller's guide to the West [2nd edition] : Phila., H. S. Tanner, 372 p., colored maps.
        Minerals and soils noted on 40-42.

Baird, Spencer Fullerton (1823-1887)
1449   1846, Hints for preserving objects of natural history prepared for Dickinson College, Carlisle, Pa.: Carlisle, Penn., Gitt and Hinckley (printers), 12 p.
1450   1850, On the bone caves of Pennsylvania: Am. Assoc. Adv. Sci. Proc., v. 2, 352-355; Also in: Ann. Scien. Disc., v. 1, 276-278.

Baker, Henry (1698-1774)
1451   1746, The following account of an extraordinary large fossil tooth and bones of an elephant, with reflections occasioned thereby: Am. Mag. and Hist. Chron., v. 3, number for December 1746, 541.
        On the discovery of fossil bones at Norwich, England, and on evidence for past changes in the earth's climate and ocean level.

Baker, J. S.
1452   1835, Dagger's Sulphur Springs: N. Am. Arch. Med. Surgical Sci., v. 2, no. 4, 252-254; also in: US Med. Surgical J., v. 2, 24-25.
        Notice of mineral springs at Pattonsburg, Botetourt County, Virginia.

Bakewell, Robert (1768-1843)
1453   1811, On the application of mineralogical and chemical sciences to the selection of stone, for the purposes of durable architecture: Arch. Use. Know., v. 2, no. 2, 157-169.
        Abstracted from Mon. Mag. (London).
1454   1825, Manufacture of salt by evaporation on faggots: Am. J. Sci., v. 10, no. 1, 180-181.
        Abstracted from London Philos. Mag. and J..
1455   1829, An introduction to geology; comprising the elements of the science in its present advanced state and all the recent discoveries, with an outline of the geology of England and Wales. First American edition from the 3rd London edition, edited by Benjamin Silliman: NHav., H. Howe, 400 p., illus.
        Contains 9681 and 9687. For review see 7340 and 11099.
1456   1831, A notice of the Salt Springs of Moutiers in the Tarentaise, (Alps) and of a peculiar method of evaporation: Am. J. Sci., v. 20, no. 2, 219-227, plate.
1457   1833, An introduction to geology (intended to convey a practical knowledge of the science and comprising the most important recent discoveries; with explanations of the facts and phenomena which serve to confirm or invalidate various geological theories): NHav., Hezekiah Howe and Co., xxiv, 479 p., illus., 6 plates, map.
        Contains 9681 and 9712. For review see 585. This is the "second American from the fourth London edition."
1458   1835, Visit of Prof. Agassiz to Mr. Mantell's Museum at Brighton: Am. J. Sci., v. 28, no. 1, 194-197.
        Abstracted from London Ath.
1459   1838, Probable duration of English coal beds: Franklin Inst. J., v. 22 (ns), 357-358; also in: Am. Rr. J., v. 7, 390-391.
1460   1839, An introduction to geology (intended to convey a practical knowledge of the science and comprising the most important recent discoveries; with explanations of the facts and phenomena which serve to confirm or invalidate various geological theories): NHav., B. and W. Noyes, printed by B. L. Hamlen, front., xxxvi, 596 p., illus., 7 plates, map.
        Contains 9681 and 9759. For notice see 702 and 715. For review see 727, 1157, 7818, 8569, and 9969. This is the "Third American from the fifth London edition."
1461   1847, Observations on the whirlpool and on the rapids, below the Falls of Niagara; designed by illustrations to account for the origin of both: Am. J. Sci., v. 4 (ns), no. 1, 25-36, 3 figs., plate.

Balard, Antoine Jérome (1802-1876)
1462   1827, Discovery of a new elementary substance, called brome or bromine: N. Am. Med. Surgical J., v. 3, no. 1, 205-206.
        Abstracted from Revue Medicale.

Baldwin, Ebenezer (d. 1837)
1463   1831, Annals of Yale College, in New Haven: N. Hav.
        Includes sketch of the geology of New Haven.

Baldwin, Joseph
1464   1818, New locality of fluor spar, or fluate of lime, and of galena, or sulfuret of lead [Shawnee Town, Illinois]: Am. J. Sci., v. 1, no. 1, 52-53.

Baldwin, Theron (1801-1870)
1465   1829, Letter of Mr. Theron Baldwin: Am. J. Sci., v. 15, no. 2, 228-232.
        In B. Silliman 9684. Description of an avalanche in the White Mountains, New Hampshire.

Ball, John (1794-1884)
1466   1832, Diluvial scratches on rocks [at Hebron, New Hampshire]: Am. J. Sci., v. 22, no. 1, 166 (13 lines).
1467   1835, Remarks upon the geology, and physical features of the country west of the Rocky Mountains, with miscellaneous facts: Am. J. Sci., v. 28, no. 1, 1-16.
        Includes notices of volcanic rocks from Hawaii, Tahiti, and Washington.

Ball, William
1468   1850, Improved gold washer: Franklin Inst. J., v. 19 (3s), 100.

Ballingall, Captain James
1469   1832, The new volcano of Hotham Island: Mon. Am. J. Geology: v. 1, no. 7, 314-317, plate.

Baltimore and Cuba Smelting and Mining Company
1470   1845, Exposition of the Baltimore and Cuba Smelting & Mining Company, ... : Baltimore, R. Neilson (printer), 15 p.

Baltimore and Ohio Railroad Company
1471   1835, Ninth annual report of the President and Directors to the stockholders: n.p., 174 p.
        Appendix A contains the Sixth Annual Report of the Chief Engineer.
1472   1837, Report upon the surveys for the extension of the Baltimore and Ohio Railroad from its present termination near Harper's Ferry on the Potomac to Wheeling and Pittsburgh on the Ohio River: n.p., 138 p.

Baltimore Medical and Philosophical Lyceum (Baltimore, 1811)
1473   1811, [Review of Notices concerning Cincinnati by D. Drake (see 3216]: v. 1, 97-100.
1474   1811, [Discovery of molybdenite in Baltimore]: v. 1, 414 (4 lines).

Baltimore Medical and Surgical Journal (Baltimore, 1833-1834)
1475   1833, [Review of "Memoir on the bones of the Mastodon and Tetracaulodon" by I. Hays (see 4925)]: v. 1, 252.

The Baltimore Monument; Devoted To Polite Literature, Science, and the Arts (Baltimore, 1836-1838)
1476   1837, The Giant's Causeway: v. 1, 173-174.
1477   1837, Age of the world: v. 1, 221, 310-311.

The Baltimore Repertory, on Papers on Literary and Other Topics (Baltimore, 1811)
1478   1811, Observations upon a chromat of iron, found on the estate of Thomas Rutter, Esq.; in the vicinity of Baltimore: v. 1, no. 2, 98-99.
        Signed by "O."
1479   1811, Perkiomen mines [Montgomery County, Pennsylvania]: v. 1, no. 6, 321-324.

The Baltimore Weekly Magazine (Baltimore, 1800-1801)
1480   1800, The Lake of Alba [Italy]: v. 1, 14-15.
        Description of changes in water level due to volcanic activity.

Baltzell, John
1481   1802, An essay on the mineral properties of the Sweet Springs of Virginia, and conjectures respecting the processes of their production by nature ... : Baltimore, Warner and Hanna, 30 p.

Bankart, Frederick
1482   1846, Improvements in treating certain metallic ores, and refining the products therefrom: Franklin Inst. J., v. 12 (3s), 329-332.

Barba, Alvaro Alonso (b. 1569)
1483   1763, Das zweyte Buch Von den Metallen, Darin gelehrt wird der gemeine Weg, wie man das Silber durch Quecksilber reiniget: Ephrata, Penn., Durch J. George Zeisiger, [119]-198, [4], 14 p. incl. plate (diagrs.).
        Translated from The Second Book of the Art of Mettals (London, 1670), which was translated from Arte De Los Metales (Madrid, 1640).
1484   1763, Grundlicher unterricht von den Mettalen, darinnen beschrieben wird, wie sie werden in der Erden generirt; und was man insgemein dabey findet. In zwey Buchen: Ephrata, Penn., durch J. George Zeisinger, 2 pts. in 1 volume; pt. 1, 118 p.

Barber, John Warner (1798-1885)
1485   1844, and H. Howe, Historical collections of the state of New Jersey; containing a general collection of the most interesting facts, traditional biographical sketches, anecdotes, etc. relating to its history and antiquities, with geographical descriptions of every township in the state: NY, S. Tuttle, 507 p.

Barker, Benjamin
1486   1822, Dreadful earthquake at Aleppo: Meth. Mag., v. 5, 185-189.

Barlow, William Henry (1812-1902)
1487   1850, On the cause of the diurnal variations of the magnetic needle: Am. J. Sci., v. 9 (ns), no. 3, 445-447.
        From Philos. Mag.

Barnard, Daniel Dewey (1796-1861)
1488   1840, Stephen Van Rensselaer: Bos. Wk. Mag., v. 2, 150-151.
        Description of Van Rensselaer's sponsorship of geological surveys.

Barnard, John (1681-1770)
1489   1727, Two discourses addressed to young persons, to excite them to seek the Lord in their youth. To which is added, a sermon occasioned by the earthquake, ... October 29. 1727: Bos., S. Gerrish (printer), iv, 99 p.
        Evans number 2840. Second title: "Earthquakes under the divine government. A sermon preach'd November 2, 1727, at the lecture in Marblehead after the terrible earthquake," 71-99.
1490   1728, Sin testify'd against by heaven and earth. A sermon preached on the Friday after the great and terrible earthquake, which occur'd on the Lord's-day-evening, between the 29th and 30th of October, 1727 ... : Bos., John Phillips, 132 p.
        Evans number 2989.

Barnes, Daniel Henry (1785-1828)
1491   1822, A geological section of the Canaan Mountain, with observations on the soil productions of the neighboring region: Am. J. Sci., v. 5, no. 1, 8-21, plate (with explanation of plate on p. [204]).
        Describes stratigraphy and mineral productions of Canaan Mountain, New York.

Barnes, George Orville
1492   1849, [Field notes, analyses, and reports on the Keweenaw Point District, Lake Superior]: in Michigan report by C. T. Jackson (see 5655), 494, 509-514, 627-638.

Barney, Joshua (d. 1867)
1493   1847, Iowa meteorite: Am. J. Sci., v. 4 (ns), no. 3, 429.
        Abstracted from a letter to J. J. Abert.

Barrande, Joachim (1799-1883)
1494   1850, The trilobites of Bohemia: Ann. Scien. Disc., v. 1, 284 (15 lines).
        Abstracted from London Ath.

Barratt, Joseph (1796-1882)
1495   1825, Notices of miscellaneous localities of minerals. 1. by Dr. Joseph Barratt: Am. J. Sci., v. 9, no. 1, 39-42.
        New York mineral localities, primarily in Putnam County.
1496   1845, On fossil footmarks in the red sandstone of the Connecticut Valley: Assoc. Am. Geol's Nat's Proc., v. 6, 23.
        Also in Am. Q. J. Ag. Sci., v. 2, 146-147.
1497   1845, On the evidence of congelation in the New Red Sandstone: Assoc. Am. Geol's Nat's Proc., v. 6, 26.
        Also in Am. Q. J. Ag. Sci., v. 2, 148-150.

Barratt, J. (cont.)
1498   1845, [On New England drift]: Am. Q. J. Ag. Sci., v. 2, 141.
1499   1846, Geology of Middletown and vicinity ... On the tracks of large birds, found at Middletown, Conn.: Middletown, Ct., "A Sentinel and Witness extra," broadside.

Barrelville Mining Company
1500   1844, Report by Professors Silliman, Renwick, Schoolcraft, and others, upon the mineral resources of the lands of the Barrelville Mining Company: Balt.?, 10 p.
1501   1844, Supplementary report by Professors Silliman, Renwick, Schoolcraft, and others, upon the mineral resources of the lands of the Barrelville Mining Company: Balt.?, 19 p.
         Signed by B. Silliman, J. Renwick, H. R. Schoolcraft, J. Pickell, W. H. Morell, and E. S. Renwick. On the geology and mineral resources of the coal basin of Alleghany County, Maryland.
1502   1847, Extracts from reports made by a committee of the Maryland Legislature: Phila., 64 p.
         Includes extracts from geological reports by J. T. Ducatel (see 3253), G. W. Featherstonhaugh (see 3785), Major Douglass (see 3205), G. W. Hughes (see 5363), A. Ure (see 10535), and J. F. Daniell (see 2897). Also contains "Supplementary reports" by B. Silliman, J. Renwick, and H. R. Schoolcraft (see 9281). Also contains numerous testimonial letters on the quality of Maryland coal.

Barreswil, Charles Louis (1817-1870)
1503   1846, Separation of cobalt from manganese: Am. J. Sci., v. 2 (ns), no. 2, 260 (11 lines).
         Abstracted from J. de Pharmacie by J. L. Smith.

Barrington, A.
1504   1850, A treatise on physical geography, comprising hydrology, geognosy, geology, meteorology, botany, zoology, and anthropology, by A. Barrington; edited by Charles Burdett: NY, M. H. Newman and Co., xii, 420 p.
         For review see 6351.

Barrow, Sir John, bart. (1764-1848)
1505   1802, An account of travels into the interior of southern Africa in the years 1797 and 1798: including cursory observations on the geology and geography ... the natural history of such objects as occurred in the animal, vegetable, and mineral kingdoms; ... : NY, G. F. Hopkins, [4], 386 p., front., folded map; and, Phila., G. F. Hopkins for John Conrad and Co., [4], 386 p., front., folded map.
         "First American edition."
1506   1846, Voyages of discovery and research within the Arctic Regions, from the year 1818, to the present time, under the command of the several Naval officers employed by sea and land in search of a North-west passage from the Atlantic to the Pacific, with two attempts to reach the North Pole: NY, Harper and Brothers, xii, 359 p., map.
         Also 1849 edition, NY, Harper, 433 p.

Bartels (Prof.)
1507   1796, Letter from Professor Bartels, describing his journey to Mount Aetna: NY Mag., v. 1 (ns), 561-568; 632-637, plate.

Bartenbach, Mr. (see 159 with Mr. Allain)

Bartlett, John (1820-1905)
1508   1846, [On Zeuglodon near Natchez, Mississippi]: Bos. Soc. Nat. History Proc., v. 2, 96.

Bartlett, Levi
1509   1807, Reasons for supposing clay or argillaceous earth to be of vegetable origin: Med. Repos., v. 10, 403-404.

Bartlett, William
1510   1846, and D. Tod, [On the mineral regions of Lake Superior]: US SED 160, 29-1, v. 4, 20-27.

Bartlett, William Holmes Chambers (1804-1893)
1511   1832, Experiments on the expansion and contraction of building stones by variation of temperature: Am. J. Sci., v. 22, no. 1, 136-140; also in: Franklin Inst. J., v. 10 (ns), 58-62, fig., table.

Barton, Benjamin Smith (1766-1815)
1512   1794, Fragments of natural history of Pennsylvania. Part I: Phila.
         Evans number 26625. Part I only published.
1513   1798, Collections for an essay toward a materia medica of the United States: Phila., Printed for the author by Way and Groff, vii, 49 p.
         First edition. Evans number 33377.
1514   1799, Fragments of natural history of Pennsylvania. Part I: Phila., Way and Groff (printers), xviii, 24, 14 p., plate, 18 figs.
         Evans number 35159.
1515   1801/1804, Collections for an essay toward a materia medica of the United States: Phila., Part I printed for the author by Robert Carr, 1801, 2 v. in 1, xii, 64 p.; Part 2 printed for the author by A. and G. Way, 1804.
1516   1804, Memorandums concerning the earthquakes of North America: Phila. Med. Phys. J., v. 1, pt. 1, 60-67.
1517   1804, [On a bone cave in Virginia]: Phila. Med. Phys. J., v. 1, pt. 1, 152-154.
1518   1804, [History of the mammoth]: Phila. Med. Phys. J., v. 1, pt. 1, 154-159.
1519   1804, [Oilstone from Buffalo Creek, near Lake Erie, in New York]: Phila. Med. Phys. J., v. 1, pt. 1, 161-162.
1520   1804, [Fossil shells in slate]: Phila. Med. Phys. J., v. 1, pt. 1, 162-163.
1521   1804, [Porcelain earth from Maryland]: Phila. Med. Phys. J., v. 1, pt. 1, 163.
1522   1804, [American coal]: Phila. Med. Phys. J., v. 1, pt. 1, 164.
         On known American localities.

Barton, B. S. (cont.)
1523   1805, Notice of the sulphur springs, in the county of Ontario, and state of New-York: Phila. Med. Phys. J., v. 1, pt. 2, 166-168.
1524   1805, [Notices of vertebrate fossil discoveries]: Phila. Med. Phys. J., v. 2, pt. 1, 60, 155-159.
1525   1805, [Manganese mine in Shenandoah County, Virginia]: Phila. Med. Phys. J., v. 2, pt. 1, 178 (6 lines).
1526   1805, [Zeolite in basalt from Reading, Pennsylvania]: Phila. Med. Phys. J., v. 2, pt. 1, 178-179 (6 lines).
1527   1806, Facts, observations, and conjectures, relative to the elephantine bones (of different species), that are found in various parts of North-America: Phila. Med. Phys. J., First Supplement, 22-35.
1528   1806, [Native sal-ammoniac from Williamsport, Maryland]: Phila. Med. Phys. J., First Supplement, 73 (7 lines).
1529   1807, A discourse on some of the principal desiderata in natural history, and on the best means of promoting the study of this science, in the United States: Phila., Denham and Town (printers), 90 p.
        For review see 1531.
1530   1807, Additional facts and observations, relative to the extinct species of American elephants: Phila. Med. Phys. J., Second Supplement, 166-173.
1531   1808, [Review of (his own) A discourse on some of the principal desiderata in natural history ..., by B. S. Barton (see 1529)]: Phila. Med. Phys. J., v. 3, pt. 1, 165-174.
1532   1809, [Obituary of Matthias Barton]: Phila. Med. Phys. J., Third Supplement, 309-311.
1533   1810, Collections for an essay toward a materia medica of the United States: Phila., Edward Earle and Co., xvi, 67, xv, 53, 1 p.
1534   1810, [On the American mammoth]: Port Folio, v. 4 (3s), 340-344.
1535   1811, Mineralogical notice respecting fluate of lime from Virginia: Am. Min. J., v. 1, no. 2, 79-80.
1536   1814, Archaeologiae americanae telluris collectanea et specimina. Or, collection, with specimens, for a series of memoirs on certain extinct animals and vegetables of North-America: Phila., For the author, vii, [9]-64 p.
1537   1848, Prof. Barton's opinion of the Virginia Springs: S. Med. Surgical J., v. 4 (ns), 310.

Barton, David Walker (d. 1863)
        See also 1542 with R. P. Barton
1538   1822, Mr. D. W. Barton on the Virginia fluorspar: Am. J. Sci., v. 4, no. 2, 277-278.
1539*   1822, Notice of the geology of the Catskills: Am. J. Sci., v. 4, no. 2, 249-251, plate with map.

Barton, Richard P.
1540   1804, On the use of plaster of Paris (different varieties of gypsum, or sulphate of lime) as a manure, in Virginia: Phila. Med. Phys. J., v. 1, pt. 1, 86-87.
1541   1821, Manganese and sulphat of lime in Virginia [Jefferson County]: Am. J. Sci., v. 3, no. 2, 245.
1542   1821, and D. W. Barton, Virginia and Illinois fluor spar [from Shawnee Town, Illinois]: Am. J. Sci., v. 3, no. 2, 243-245.

Bartram, John (1699-1777)
1543   1833, [Excerpts from Observations on the inhabitants, climate, soil, rivers, productions, animals, and other matters worthy of notice (i.e. Bartram's Travels, London, 1751), in "Geological observations upon Alabama, Georgia and Florida" by C. U. Shepard (see 9490)]: Am. J. Sci., v. 25, no. 1, 166-173.

Bartram, William (1739-1823)
1544   1791, Travels through North and South Carolina, Georgia, East and West Florida, the Cherokee country, the extensive territories of the Muscagulges or Creek Confederacy, and the country of the Chactaws; containing an account of the soil and natural productions of those regions ... : Phila., James and Johnson (printers), xxxiv, 522 p., plates, maps.
        Two impressions, Evans numbers 23159 and 23160. For review with excerpts see 10517.

Baruel, M.
1545   1838, Analysis of a double phosphate of lead and lime [from Nussiere, France]: Franklin Inst. J., v. 21 (ns), 343.
        Translated by J. Griscom from J. des Mines.

Batapolis Mining Company
1546   1826, Articles of association of the Batapolis Mining Company: Phila.?, n.p., 8 p.

Baudrimont, Alexandre Edouard (1806-1880)
1547   1848, On the structure and teratology of crystallized bodies: Am. J. Sci., v. 5 (ns), no. 3, 419-420.
        From Comptes Rendus. On the cleavage of calcite.

Baylies, William (1743-1826)
1548   1793, Description of Gay Head: Am. Acad. Arts Sci's Mem., v. 2, pt. 1, 150-155.

Baynes
1549   1845, Fossil remains from Algoa Bay, near the Cape of Good Hope: Am. J. Sci., v. 49, no. 1, 213 (14 lines).

Beall, Elias
1550   1830, Georgia meteor and aerolite [from Forsyeth, Georgia]: Am. J. Sci., v. 18, no. 2, 388-389.
        With notices by Dr. Boykin (see 1936) and B. Silliman (see 9699).

Beamish, L. (Major)
1551   1844, On the apparent fall or diminution of the water in the Baltic, and elevation of the Scandinavian coast: Am. J. Sci., v. 47, no. 1, 184-185.
        From British Assoc. Proc.

Beaumont, Jean B. Elie de (1798-1874)
1552   1836, From a memoir on the origin of Mt. Etna: Am. J. Sci., v. 31, no. 1, 168-170; also in: Am. Rr. J., v. 5, 824-825; Mech's Mag. and Reg. Inventions, v. 9, 38-39.
        From Edinburgh New Philos. J.
1553   1848, Extract from a discourse pronounced by M. Elie de Beaumont at the funeral of M. Alexandre Brongniart, October 9, 1847: Am. J. Sci., v. 5 (ns), no. 2, 155-159.
1554   1848, Tribute to American geologists: Am. J. Sci., v. 5 (ns), no. 1, 137-138.
        Translated by C. T. Jackson from Leçons de géologie practique.

Beaumont, John Thomas Barber (1774-1841)
1555   1840, On the origin of the vegetation of our coal fields and wealdens: Franklin Inst. J., v. 25 (ns), 205-207.
        From Mining J.

Beaver Meadow Railroad and Coal Company (Pennsylvania)
1556   1850, Annual report of the President and Managers of the Beaver Meadow Railroad and Coal Company to the Stockholders, Jan. 21,1850:
        Not seen. For excerpts see Franklin Inst. J., v. 19 (3s), 220-221.

Bechtler, Christopher
1557   1832, Machine for washing gold out of sand and pounded ore: Franklin Inst. J., v. 9 (ns), 226-227.

Beck, Lewis Caleb (1798-1853)
1558   1823, A gazetteer of the states of Illinois and Missouri; containing a general view of each state - a general view of their counties and a particular description of their towns, villages, rivers, etc., etc.: Alb., C. R. and G. Webster, vii, 352 p., map, 5 plates.
1559   1826, Account of the salt springs at Salina, in Onondaga County, state of New York; with a chemical examination of the water, and of several varieties of salt manufactured at Salina and Syracuse: NY, J. Seymour, 36 p.
        Also in: NY Med. Phys. J., v. 5, 176-199.
1560   1827, Notice and chemical examination of the mineral water recently discovered in the city of Albany: NY Med. Phys. J., v. 6, 92-97.
1561   1829, Chloro-hydrogen blowpipe: Am. J. Sci., v. 17, no. 1, 211.
1562   1829, Note on the presence of iron in the salt springs of Salina, N. Y.: Am. J. Sci., v. 16, no. 1, 187-188.
1563   1830, Lead mines: Alb. Inst. Trans., v. 1, pt. 2, 36-40.
        History of lead mining in the United States.
1564   1836, Researches on the commercial potash of the state of New York: Am. J. Sci., v. 29, no. 2, 260-273; also in: Am. J. Pharmacy, v. 2 (ns), 28-44.
1565   1837, Report - on the mineralogical and chemical department of the survey: NY Geol. Survey Ann. Rept., v. 1, 15-60.
1566   1838, Report - on the mineralogical and chemical department of the survey: NY Geol. Survey Ann. Rept., v. 2, 7-73, folding table.
1567   1839, Notices of the native copper, ores of copper, and other minerals found in the vicinity of New Brunswick, New Jersey: Am. J. Sci., v. 36, no. 1, 107-114.
1568   1839, Report - on the mineralogical and chemical department of the survey: NY Geol. Survey Ann. Rept., v. 3, 9-56.
1569   1840, Association of American Geologists: Am. J. Sci., v. 39, no. 1, 189-191; also in: Franklin Inst. J., v. 25 (ns), 219-220.
        Notice of the Association, by the secretary.
1570   1840, Communication - to the Governor, relative to the geological survey of the state: NY Geol. Survey Ann. Rept., v. 4, 37-43.
1571   1840, Report - on the mineralogical and chemical department of the survey: NY Geol. Survey Ann. Rept., v. 4, 45-111.
        Mineral localities listed by county.
1572   1841, Association of American Geologists: Am. J. Sci., v. 41, no. 1, 158-189.
        Abstract of the second meeting proceedings, by the secretary.
1573   1841, Letter - to the Governor, relative to the geological survey of the state: NY Geol. Survey Ann. Rept., v. 5, 3-4.
1574   1841, On the sulphur springs of the state of New York: Am. J. Sci., v. 41, no. 1, 162-163.
        Also in Assoc. Am. Geol's Nat's Proc., v. 2, 7-8. Also in Assoc. Am. Geol's Nat's Rept., v. 1, 15-16, 1843.
1575   1841, Report - on the mineralogical and chemical department of the survey: NY Geol. Survey Ann. Rept., v. 5, 5-23.
1576   1841, Proceedings of the Association of American Geologists and Naturalists. First Session. 1840: Assoc. Am. Geol's Nat's Proc., v. 1, 1-3.
        Also in Assoc. Am. Geol's Nat's Rept., v. 1, 9-11, 1843.
1577   1842, Mineralogy of New-York; comprising detailed descriptions of the minerals hitherto found in the state of New-York, and notices of their uses in the arts and agriculture: Alb., W. and A. White, and J. Visscher (printers), xxiv, 536, 4 p., 533 figs, 8 plates.
        This is part 3 of "Natural History of New York." For review with excerpts see 2807, 10029. Reprinted in part in Blake (see 1734).
1578   1843, Antediluvian climate: Am. J. Sci., v. 45, no. 1, 144.
        In Assoc. Am. Geol's Nat's Proc., with discussion by J. D. Dana and H. D. Rogers.
1579   1843, [Native copper at Somerville, New Jersey]: Am. J. Sci., v. 45, no. 2, 331 (6 lines), 332 (1 line).
1580   1843, Notices of some trappean minerals found in New Jersey and New York: Am. J. Sci., v. 44, no. 1, 54-60, 5 figs.

Beck, L. C. (cont.)
1581   1843, Occurrence of bituminous matter in several of the New York limestones and sandstones: Am. J. Sci., v. 45, no. 2, 335-336.
　　　In Assoc. Am. Geol's Nat's Proc., with discussion by E. Emmons (3516).
1582   1843, On certain phenomena of igneous action, chiefly as observed in the state of New York: Am. J. Sci., v. 45, no. 1, 143-145.
　　　In Assoc. Am. Geol's Nat's Proc., with discussion by J. D. Dana and J. Johnston.
1583   1843, On some pseudomorphous minerals of the state of New York: Assoc. Am. Geol's Nat's Rept., v. 1, 241-253.
1584   1844, [On New York minerals]: Am. J. Sci., v. 46, no. 1, 25-37, 5 figs.
　　　In "Mineralogy of New York" by J. D. Dana (see 2807).
1585   1844, Note on graptolites: Am. J. Sci., v. 47, no. 2, 372-374.
　　　On the characteristics of New York species.
1586   1844, Views concerning igneous action, chiefly as deduced from the phenomena presented by some of the minerals and rocks of the state of New York: Am. J. Sci., v. 46, no. 2, 333-343.
　　　Effects of igneous and metamorphic action on minerals. Contact metamorphism and origin of granite discussed.
1587   1848, Catalogue of the specimens in the mineralogical department of the geological survey: NY State Mus. Ann. Rept., v. 1, 21-33.
1588   1850, [Analysis of hypersthene]: Am. J. Sci., v. 10 (ns), no. 1, 147-148.
　　　In review of NY State Mus. Ann Rept., v. 3.
1589   1850, Report on the mineralogy of New York; comprising notices of the additions which have been made since 1842: NY State Mus. Ann. Rept., v. 3, 107-151, 21 figs.

Beck, Theodoric Romeyn (1791-1855)
　　　See also 3328 with A. Eaton
1590   1813, Annual address, delivered by appointment, before the Society for the Promotion of Useful Arts, at the capitol, in the city of Albany, on the 3rd of February, 1813: Alb., Websters and Skinners (printers), 44 p.; also in: NY Soc. Prom. Use. Arts Trans., v. 3, 1-41.
　　　For review see 6239 and 7101.
1591   1814, Circular [asking for mineral specimens]: NY Soc. Prom. Use. Arts Trans., v. 3, 258-259.
1592   1816, and J. Law and J. Green, Report on the marble quarry, in the town of Bennington, (Vermont): NY Soc. Prom. Use. Arts Trans., v. 4, Appendix, 19-23.
1593   1819, Memoir on alum: NY Soc. Prom. Use. Arts Trans., v. 4, pt. 2, 50-92.
1594   1822, [Review with excerpts of Elementary treatise ... , by P. Cleaveland (see 2422)]: NY Med. Phys. J., v. 1, 327-335.
1595   1828, Address delivered before the Lyceum of Natural History, (now the second department of the Institute,) at its first anniversary, March 1, 1824: Alb. Inst. Trans., v. 1, 137-147.
1596   1828, Tioga coal: Am. J. Sci., v. 13, no. 2, 381.
　　　Notice of coal from Covington, Wayne County, Pennsylvania.
1597   1830, Analysis of the impure limestone (hydraulic lime) used in the construction of the locks of the Erie Canal: Alb. Inst. Trans., v. 1, pt. 2, 52-53.
1598   1830, Bituminous coal of Tioga, (Pennsylvania.): Alb. Inst. Trans., v. 1, pt. 2, 34.
1599   1841, History and practical uses of minerals (A lecture delivered before the Young Men's Association of the city of Albany, January 15, 1841):
　　　This work is noted in No. Light, v. 1, 27, but was not seen. For excerpt see 1602.
1600   1841, [Review with excerpts of first New Hampshire report by C. T. Jackson (see 5576)]: No. Light, v. 1, 72-73.
1601   1841, Salt: No. Light, v. 1, 9-10.
　　　On the geological origin and uses of salt.
1602   1841, Sulphur: No. Light, v. 1, 27.
　　　From History and practical uses of minerals (see 1599). On geological origin and uses of sulphur. Also table of United States' imports from 1824 to 1839.
1603   1843, Notice of the fourth meeting of the Association of American Geologists and Naturalists - held at Albany, in April and May, 1843: No. Light, v. 3, 45.
1604   1843, [Review with excerpts from Reports ... of the Association of American Geologists and Naturalists (see 1326)]: No. Light, v. 3, 19-20.
　　　Includes extract from "Geology of western Asia" by E. Hitchcock (see 5150).

Beck, Wilhelm von (b. 1822)
1605   1849, Chemical analysis of glinkite [from Perm, Russia]: Am. J. Sci., v. 8 (ns), no. 1, 121 (12 lines).
　　　Abstracted from "Verhandl. Min. Gesel. zu St. Petersburg, 1847."

Beckett, H.
1606   1846, Fossil forest in the Parkfield Colliery, near Wolverhampton: Am. J. Sci., v. 2 (ns), no. 3, 422.

Beckett, Sylvester Breakmore (1812-1882)
1607   1849, Land disruption at Stroudwater, Me.: Bos. Daily Mail, June 11, 1849.
　　　Not seen.

Beckwith, John
1608   1822, A memoir on the natural walls, or solid dikes, in the state of North-Carolina; about which there have been debates, whether they were basaltic, or of some other formation: Am. J. Sci., v. 5, no. 1, 1-7.

Becquerel, Antoine César (1788-1878)
1609   1829, Electricity of the tourmaline: Am. J. Sci., v. 16, no. 2, 390-391.
　　　On thermoelectricity, abstracted from Ferussac's Bulletin.
1610   1830, Combinations and crystallizations, effected by the action of weak electrical forces: Am. J. Sci., v. 17, no. 2, 383-384.
　　　Abstracted from Bibliothèque Universelle. On the synthesis of CuO (cuprite), and unsuccessful attempts to synthesize diamond.

Becquerel, A. C. (cont.)
1611   1833, Carbonate of lime and its compound: Am. J. Sci., v. 23, no. 2, 387-388.
       From Revue Encyclopédique. On calcite and arragonite crystal forms.
1612   1836, Electricity of peroxyd of manganese: Am. J. Sci., v. 30, no. 1, 179 (9 lines).
       Abstracted by "D." from l'Institut. On electrical properties of pyrolusite.
1613   1836, Reduction of metals: Am. J. Sci., v. 31, no. 1, 164 (8 lines).
       On the reduction of silver ores by galvanism.

Beeman, N. S. S.
1614   1816, Marine petrifactions: Lit. and Philos. Rep., v. 2, 399-408.
       On the origins of fossils from southeast United States.

Belfast Natural History Society
1615   1825, Belfast Natural History Society: Am. J. Sci., v. 9, no. 2, 381-382.
       Activities of the Society. Communications with W. Maclure noted.
1616   1835, Belfast Natural History Society: Am. J. Sci., v. 28, no. 2, 369 (5 lines).
       Notice of activities of the Society.

Belknap, Jeremy (1744-1798)
1617   1784/1791/1792, The history of New Hampshire: v. 1, Phila., R. Aitken (printer), viii, 361, lxxxiv p.; v. 2, Bos., I. Thomas and E. T. Andrews, 493 p., map; v. 3, Bos., Belknap and Young, 480, 7 p.
       Numerous other editions. First edition is Evans number 18344. White Mountains natural history, v. 3, 39-54; minerals, stones, and fossils noted, v. 3, 188-196.
1618   1785, An account of large quantities of a fossil substance, containing vitriol and sulphur, found at Lebanon, in the state of New-Hampshire, accompanying a specimen: Am. Acad. Arts Sci's Mem., v. 1, 377.
1619   1786, Description of the White Mountains in New-Hampshire: Am. Philos. Soc. Trans., v. 2, 42-49.

Bell, Benjamin (1752-1836)
1620   1829, Strictures on the hypothesis of Mr. Joseph du Commun, on volcanoes and earthquakes: Am. J. Sci., v. 16, no. 1, 51-53.
       See also 3255.

Bell, John (1796-1872)
1621   1824, On baths and mineral waters: Phila. J. Med. Phys. Sci's, v. 8, 311-348; v. 9, 69-97; 309-349.
       Includes list of mineral springs of the world, their chemical and physical properties, and their medical value.
1622   1831, On baths and mineral water. In two parts. Part I. A full account of the hygienic and curative powers of cold, tepid, warm, hot, and vapour baths, and of sea bathing. Part II. A history of the chemical composition and medicinal properties of the chief mineral springs of the United States: Phila., H. H. Porter, xviii, 532 p.
       For review with excerpts see 7925 and 10309. For review see 1023 and 10837. Also 1832 Phila., Carey and Lea edition.
1623   1834, Observations on the mineral waters in the south western part of Virginia. In a series of lectures: Phila., 30 p.

Bell, William H.
1624   1844, Report on ... mineral lands of the Upper Mississippi: US HED 43, 28-1, v. 3 (441), 2-52.
       Includes table of mines in Illinois, Iowa, Missouri and Wisconsin. Deposits compared to those of England.

Bellevue, Fleuriau de
1625   1822, On meteorolites: Am. J. Sci., v. 5, no. 1, 170-171.
       Abstracted from J. de Physique by Brande's J.

Bender, George
1626   1832, Specific gravities of the rocks used in the construction of the Delaware breakwater: Hazard's Reg. Penn., v. 9, 192; also in: Mon. Am. J. Geology, v. 1, no. 7, 312-314.

Benedict, George W.
1627   1824, Notice of new localities of sahlite - augite - ceylanite, etc. [from Munroe, near Newburgh, New York]: Am. J. Sci., v. 8, no. 1, 88-92, plate with 5 figs.
1628   1825, By George W. Benedict: Am. J. Sci., v. 9, no. 2, 250-252.
       On New York mineral localities.

Bennet, John
1629   1791, Letters to a young lady: Am. Mus. or Univ. Mag., v. 10, 228-229, 307-308.
       On the pleasures of studying natural history.

Benton, Thomas Hart (1782-1858)
1630   1822, Col. Benton in reply [to H. R. Schoolcraft's "Remarks on the prints of human feet, observed in the Secondary limestone of the Mississippi Valley" (see 9264)]: Am. J. Sci., v. 5, no. 2, 230-231.

Berendt, George Carl (1790-1850)
1631   1839, Dr. Berendt's investigations on amber: Am. J. Sci., v. 37, no. 2, 365-366.
       Abstracted from Jameson's J.

Bergemann, C.
1632   1850, On the meteoric iron of Zacatecas [Mexico]: Am. J. Sci., v. 10 (ns), no. 2, 255.
       Abstracted from Poggendorff's Annalen.

Bernard, Simon (1779-1839)
1633   1823, and J. G. Totten, Report of the Board of Engineers on the Ohio and Mississippi Rivers made in the year 1821: US HED 35, 17-2, v. 3 (78), 7-22.
1634   1827, and W. T. Poussin, Letter from the Postmaster General transmitting the report of General Bernard on surveys of routes for a post road from Baltimore to Philadelphia: Wash.
       Not seen. Includes notes on the soils and geological formations along the route.
1635   1828, and W. T. Poussin, Report upon the reconnoissance of a route across the Cumberland Mountains, of the national road contemplated from Washington City to New Orleans: US HED 125, 20-1, 5-27.
       Not seen.
1636   1832, and W. T. Poussin, Report of the Board of Internal Improvements on the contemplated canal between the Atlantic and the Gulf of Mexico: US HED 185, 22-1, 9-58.
       Not seen.

Berthier, Pierre (1782-1861)
1637   1820, Analysis of two zinc ores from the United States of America [Franklin, New Jersey]: Am. J. Sci., v. 2, no. 2, 319-326.
1638   1824, Ores of manganese: Am. J. Sci., v. 7, no. 2, 366-367.
1639   1824, Roman cement: Bos. J. Phil. Arts., v. 1, no. 6, 593 (19 lines).
       Abstracted from Philos. Mag.
1640   1833, Analyses of fer titané of Baltimore, Maryland: Am. J. Sci., v. 24, no. 2, 375-376.
       Abstracted from Annales des Mines by J. Griscom.
1641   1838, Analysis of a vegetable soil from Cuba: Franklin Inst. J., v. 21 (ns), 341.
1642   1838, Analyses of two vegetable soils in the vicinity of Puiseaux, Department of Loiret: Franklin Inst. J., v. 21 (ns), 340-341.
1643   1838, Analysis of the sulphate of iron of Nordhausen: Franklin Inst. J., v. 21 (ns), 341-342.
1644   1839, Analysis of several bituminous minerals: Franklin Inst. J., v. 23 (ns), 345-347.
       From Annales des Mines.
1645   1839, Copper in Cuba: Franklin Inst. J., v. 23 (ns), 347 (6 lines).
       Abstracted from Annales des Mines.

Berthollet, Claude Louis (1748-1822)
1646   1808, [Chemical classification of some minerals]: Am. Reg. or Gen. Repos., v. 2, 387-398 [i.e. 389].

Berzelius, Jons Jakob (1779-1848)
1647   1811, Tantalum: Am. Min. J., v. 1, no. 2, 126 (10 lines).
       Abstracted from Annales de Chimie. See also article by W. H. Wollaston (11028).
1648   1817, [Excerpts from An attempt to establish a pure scientific system of mineralogy, by the application of electrochymical theory and the chymical proportions: J. Sci. Arts, v. 1, no. 2, 226-241.
       In review (see 5860).
1649   1817, Supplément pour l'éclaircissement de plusieurs objets dans la dissertation: J. Sci. Arts, v. 2, no. 2, 445.
       Abstract in English. Includes chemical analysis of gadolinite.
1650   1823, Achmite [from Egen, Norway]: Bos. J. Phil. Arts, v. 1, no. 3, 302 (16 lines).
1651   1823, Carlsbad water and uranium: Am. J. Sci., v. 7, no. 1, 185.
1652   1823, Turquoise and lazulite: Am. J. Sci., v. 6, no. 2, 382 (6 lines).
       Abstracted by J. Van Rensselaer.
1653   1824, Silicon: Bos. J. Phil. Arts, v. 2, no. 1, 102.
       Abstracted from Annals Phil.
1654   1825, Extract of a letter to the editor: Am. J. Sci., v. 9, no. 2, 376-378.
       On orthite and feldspar from Stockholm, Sweden.
1655   1825, Zircon, a phosphate of Yttria: Bos. J. Phil. Arts, v. 3, no. 1, 94 (16 lines).
       Abstracted from "Trommsdorff's N. Journal."
1656   1826, Discovery of lithia in the mineral waters of Bohemia: N. Am. Med. Surgical J., v. 1, no. 2, 468 (4 lines); also in: Bos. J. Phil. Arts, v. 3, no. 5, 461-462.
       Abstracted from Brewster's J.
1657   1830, Analysis of the Russian platina: Am. J. Sci., v. 18, no. 1, 162-164.
       Abstracted from Annales de Chimie.
1658   1830, Discovery of a new metal named thorium [at Brevig, Norway]: Am. J. Sci., v. 17, no. 2, 381; also in: N. Am. Med. Surgical J., v. 9, 233-234.
       Abstracted from Bibliothèque Universelle.
1659   1830, On iridium and osmium: N. Am. Med. Surgical J., v. 9, 445-446.
1660   1831, A new metal discovered: Am. J. Sci., v. 20, no. 2, 386.
       Abstracted from Edinburgh Philos. J. On the discovery of vanadium from Fahlun, Sweden; also in: N. Am. Med. Surgical J., v. 12, no. 1, 266-267.
1661   1831, Iron pyrites: Am. J. Sci., v. 19, no. 2, 387-388.
       Abstracted from Berzelius Arsberat.
1662   1832, An essay on chemical nomenclature, prefixed to the treatise on chemistry: Am. J. Sci., v. 22, no. 2, 248-276.
       Translated and annotated by A. D. Bache (see 1375).
1663   1834, Platinum in France: Am. J. Sci., v. 26, no. 2, 389 (7 lines).
       Translated by L. Feuchtwanger.
1664   1835, Memoir on tellurium: Am. J. Sci., v. 28, no. 1, 137-140.
       Translated and abstracted by C. U. Shepard from Annales des Mines.
1665   1839, Latanium - a new metal: Franklin Inst. J., v. 24 (ns), 113-114; also in: Am. J. Sci., v. 37, no. 1, 192-193.
       From London and Edinburgh Philos. Mag.
1666   1839/1840, Notice of advances made in the science of mineralogy during the year 1837: Franklin Inst. J., v. 24 (ns), 218-221, 298-301, 364-367; v. 25 (ns), 20-21.
1667   1839, On meteoric stones: Am. J. Sci., v. 37, no. 1, 93-99.
       Translated by W. A. Larned from Berzelius' Jahresbericht.

Berzelius, J. J. (cont.)
1668 1845, The use of the blowpipe in chemistry and mineralogy: Bos., W. D. Ticknor, xv, 237 p., plates.
Translated from the fourth enlarged and corrected edition by J. D. Whitney. For notice of the German edition see 502. For review see 818 and 4903.

Beswick, Samuel
1669 1847, [Excerpts from How worlds are formed (see 2768)]: The Daguerreotype, v. 1, 417-426.

Bethlehem, Northampton County, Pennsylvania
1670 1849, [Geological notes, and a request for development of coal and iron works]: n. p., circular letter, 1 p.

Beudant, François Sulpice (1787-1850)
See also 9135 by W. S. W. Ruschenberger.
1671 1818, [Use of chemistry and crystallography in mineralogy]: J. Sci. Arts, v. 3, no. 1, 178 (7 lines).
From Royal Inst. France Proc.
1672 1823, Account of the opals of Hungary: Bos. J. Phil. Arts, v. 1, no. 2, 158-164.
Excerpted by J. W. Webster from Travels in Hungary.
1673 1825, [Excerpts from An elementary treatise on mineralogy]: Bos. J. Phil. Arts, v. 3, no. 1, 71-81.
1674 1829, Specific gravity considered as a mineralogical character: Am. J. Sci., v. 16, no. 2, 260-261.

Beusted, W. H.
1675 1843, [Letter to G. A. Mantell]: Am. J. Sci., v. 45, no. 2, 244-246.
In "Notice of 'Molluskite,'" by G. A. Mantell (see 6742).

Biblical Repertory (Princeton, New Jersey, 1825-1850)
1676 1832, [Review with excerpts of Valley of the Mississippi by Rev. Robert Baird (see 1447)]: v. 4 (ns), 552-568.
Contains excerpts describing the minerals of Ohio, Kentucky, Alabama, and Tennessee.
1677 1832, [Review of Natural history of the Bible by F. A. Ewing]: v. 7 (ns), 559-573.
1678 1841, Report of the tenth meeting of the British Association for the Advancement of Science, held at Glasgow, Sept. 1840: v. 13 (ns), 132-149.
1679 1841, [Review of On the relation between the Holy Scriptures and some parts of geological sciences by J. P. Smith (see 9895)]: v. 13 (ns), 368-394.
1680 1845, [Review with excerpts of Vestiges of the natural history of Cretion by R. Chambers (see 2292 et seq.)]: v. 17 (ns), 505-557.
1681 1849, [Review of The earth and man by A. Guyot (see 4609)]: v. 21 (ns), 457-458.
1682 1850, [Review of Creation..., by J. Murphy (see 7531)]: v. 22 (ns), 337.
1683 1850, [Review of The testimony of science to the truth of the Bible by B. M. Smith (see 9855)]: v. 22 (ns), 678.

The Biblical Repository (Andover, Massachusetts, 1831-1850)
1684 1835, [Notice of geological lectures by C. U. Shepard]: v. 6, 254.
1685 1837, [Review of Geology and mineralogy..., by W. Buckland (see 2160)]: v. 9, 515-516.
1686 1840, [Review of Elements of geology by C. A. Lee (see 6125)]: v. 3 (2s), 512.
1687 1840, [Review of Relations between the Holy Scripture and some parts of geological sciences by J. P. Smith (see 9895)]: v. 4 (2s), 258-259.
1688 1840, New work on geology [notice of Elementary geology by E. Hitchcock (see 5132)]: v. 4 (2s), 264 (11 lines).
1689 1840, [Review of Elementary geology by E. Hitchcock (see 5132)]: v. 4 (ns), 497-498.
1690 1841, [Review of Lectures on geology by K. C. von Leonhard (see 6192)]: v. 5 (ns), 498-499.
1691 1845, [Review of Travels in North America by C. Lyell (see 6557)]: v. 1 (3s), 776-777 (9 lines).

Biddle, John (1789-1859)
1692 1834, Discourse delivered before the Historical Society of Michigan: Hist. Soc. Mich. Hist. Scien. Sketches, v. 1, [149]-175.

Bigarre, Pierre de la
1693 1794, Excursions in our Blue Mountains: NY Soc. Prom. Use. Arts Trans., v. 1, 128-139.

Bigelow, Artemas
1694 1846, Observations upon some sandstone rocks in Baldwin County, Ala.: Am. J. Sci., v. 2 (ns), no. 3, 419-422, fig.

Bigelow, James (i.e. Jacob, 1787-1879)
1695 1816, Some account of the White Mountains of New Hampshire: NE J. Medicine Surg., v. 5, 321-338.
1696 1817, An abstract of an account of the White Mountains: J. Sci. Arts, v. 2, no. 2, 392-399.

Bigsby, John Jeremiah (1792-1881)
1697 1820, Remarks on the environs of Carthage Bridge, near the mouth of the Genesee River: Am. J. Sci., v. 2, no. 2, 250-254, fig.
1698 1821, Geological and mineralogical observations on the northwest portion of Lake Huron: Am. J. Sci., v. 3, no. 2, 254-272, 7 figs. on plate (1 fig. is colored).
1699 1822, Extract of a letter: Am. J. Sci., v. 4, no. 2, 280-282.
On celestite from limestone at Put-in-Bay, Ohio, in Lake Erie.
1700 1822, Outline of the mineralogy, geology, etc. of Malbay, in Lower Canada: Am. J. Sci., v. 5, no. 2, 205-222, plate, 2 figs.

Bigsby, J. J. (cont.)
1701 1823, Geology of Lake Huron: Bos. J. Phil. Arts, v. 1, no. 2, 189 (5 lines).
          Abstracted from Annals of Phil.
1702 1824, A list of minerals and organic remains, occurring in the Canadas: Am. J. Sci., v. 8, no. 1, 60-88, fig.
1703 1824, Petalite [from Ontario]: Bos. J. Phil. Arts, v. 1, no. 6, 598 (8 lines).
          Abstracted from Phila. Acad. Nat. Sci's J.
1704 1825, Description of a new species of trilobite: Phila. Acad. Nat. Sci's J., v. 4, pt. 2, 365-368, plate.
          On Paradoxides boltoni from Lockport, New York.
1705 1825, Notice of a cave containing bones in Lanark, Upper Canada: Am. J. Sci., v. 9, no. 2, 354-355.
1706 1825, A sketch of the geology of the Island of Montreal: NY Lyc. Nat. History Annals, v. 1, 198-219, fig., map.

Billups, A.
1707 1834, Calcareous deposits on Plankitank River: Farm's Reg., v. 2, 348.

Bingham, Hiram (1789-1869)
1708 1845, Particulars of the fall of meteorites in the Sandwich Islands: Am. J. Sci., v. 49, no. 2, 407-408.

Binney, Amos (1803-1847)
1709 1845, [Review of engraved plates for Paleontology of New York by J. Hall (see 4712)]: Assoc. Am. Geol's Nat's Proc., v. 6, 82-83 (11 lines).
1710 1845, Remarks made at the annual meeting of the Boston Society of Natural History, June 2, 1845: Bos. Freeman and Bolles, 16 p.
          On the history and progress of the Society.
1711 1845-1846, Sigillaria and Stigmariae: Am. J. Sci., v. 49, no. 1, 227 (9 lines); v. 2 (ns), no. 2, 279.
          On fossil plants from England.
1712 1846, [The bluff formation at Natchez, Mississippi - notice of its Tertiary fossils]: Bos. Soc. Nat. History Proc., v. 2, 126-130.

Biot, Jean Baptiste (1774-1862)
1713 1818, Sur la réunion de la lepidolithe avec l'espece des mica, prouvée par la comparison des forces polarisantes: J. Sci. Arts, v. 3, no. 1, 184.
          From Annales de Chimie.
1714 1818, Observations relating to the operations undertaken to determine the figure of the earth: J. Sci. Arts, v. 5, no. 2, 340-351.
          Partly in French.
1715 1819, Notice of the operations undertaken to determine the figure of the earth: Analectic Mag., v. 13, 26-41.
          From Paris Acad. Sci. Proc.
1716 1822, Theory of earthquakes: Mus. For. Liter. Sci., v. 1, 287-288.
1717 1828, Figure of the earth: Am. J. Sci., v. 15, no. 1, 172-173.
          Abstracted from Revue Encyclopèdique by J. Griscom.
1718 1840, Note on the stony substance used in times of famine, under the name of flour of stone: Am. J. Pharmacy, v. 6 (ns), 161-164.
          From Annales de Chimie.

Bird, Isaac (1793-1876)
1719 1825, Notice of minerals, &c. from Palestine, Egypt, &c.: Am. J. Sci., v. 10, no. 1, 21-29.
1720 1829, Notices of Palestine: Am. J. Sci., v. 15, no. 2, 374-378.

Bird, Robert Montgomery (1806-1854)
1721 1837, The Mammoth Cave of Kentucky: Am. Mon. Mag., v. 9, 417-438, plate; 525-546; also in: New Yorker, v. 3, 101-104, 163-164, 181-183.

Birkmyre, William
1722 1849, On gold and gold mines: Am. Rr. J., v. 22, 470-471; 506; 514-515.

Bischof, Carl Gustav Christoph (1792-1870)
1723 1839, On the natural history of volcanoes and earthquakes: Am. J. Sci., v. 36, no. 2, 230-281, fig.; v. 37, no. 1, 41-75, 3 figs.
          See also "Reply" by C. Daubeny (2931).
1724 1841, Thermal springs: Am. Eclec., v. 1, no. 1, 181.
          Abstracted from Year-Book of Facts.
1725 1845, On the origin of quartz and metaliferous veins: Am. J. Sci., v. 49, no. 2, 396.
          Abstracted from Jameson's J.

Black, [James (1788?-1867)]
1726 1838, On a fossil stem of a tree recently discovered near Bolton-le-Moor: Franklin Inst. J., v. 22 (ns), 375.
          Abstracted from London and Edinburgh Philos. Mag.

Black, Joseph (1728-1799)
1727 1798, and J. T. Stanley, Account of the geyzer, a surprising spring in Iceland: Wk. Mag., v. 2, 152-154.

Blackburn, George (1765-1823)
1728 1809, Variation of the magnetic needle in Virginia: Mon. Anth., v. 7, 354-356.
          From NY Spectator.

Blake, John Harrison (1808-1899)
1729 1842, Coal mines in Cuba: Am. J. Sci., v. 42, no. 2, 388-390.
          With notes by M. Castales (see 2282).
1730 1843, Geological and miscellaneous notice of the Province of Tarapaca [Peru]: Am. J. Sci., v. 44, no. 1, 1-12, map; also in: The Friend, v. 16, 161-162, 169-170.

Blake, J. H. (cont.)
1731   1845, [Curious globular masses of quartz from Arequipa, Peru]: Bos. Soc. Nat. History Proc., v. 2, 26.
1732   1846, Gold at Dedham, Mass.: Am. J. Sci., v. 2 (ns), no. 3, 419 (11 lines).
1733   1848, Report to the directors and stockholders of the Isle Royale and Ohio Mining Company, in relation to their mineral locations on Isle Royale (Lake Superior) for 1847: Mining J., v. 1, 146.

Blake, William J.
1734   1849, The history of Putnam County; with an enumeration of its towns, villages, rivers, creeks, lakes, ponds, mountains, hills, and geological features ... : NY, Baker and Scribner, iv, 13-368 p.
        Includes extracts from New York reports (see 1577 and 6907).

Blake, William Phipps (1825-1910)
1735   1850, Occurrence of crystallized oxyd of chromium in furnaces for the manufacture of chromate of potash: Am. J. Sci., v. 10, no. 3, 352-354, fig.

Bliss, George, jr. (1830-1897; see 10789 and 10792 with D. A. Wells)

Bloomfield, J. E.
1736   1844/1845, Anthracite coal trade, by railways and canals: Hunt's Merch. Mag., v. 11, 541-544; v. 12, 103.

Blum, Johann Reinhard (1802-1883)
1737   1845, [Excerpts from Die Pseudomorphosen des Mineralreiches in J. D. Dana's review (see 2816)]: Am. J. Sci., v. 48, no. 1, 66-80.
1738   1848, Pseudomorphism: Am. J. Sci., v. 6 (ns), no. 2, 267-268.
        Abstracted from Die Pseudomorphosen des Mineralreiches.
1739   1848, and M. Delfess, Stilbite, a new mineral [from Zamara, Spain]: Am. J. Sci., v. 5 (ns), no. 2, 268 (6 lines).
        Abstracted from l'Insitut.

Blunt, J.
1740   1841, The coal business of the United States: Hunt's Merch. Mag., v. 4, 62-72.

Boblaye, Emile le Paillon de (1792-1843)
1741   1831, Elevation of the Morea: Mon. Am. J. Geology, v. 1, no. 4, 192 (5 lines).
        Abstracted from Paris Acad. Sci. Proc.

Bogert, J. G.
1742   1817, [New York Historical Society mineral collection catalogue]: Am. Mon. Mag. and Crit. Rev., v. 1, no. 4, 287.
1743   1818, [Classification of American fossil shells]: in Essay on the theory of the earth by G. Cuvier (see 2754), 419-424.

Bollman, Erick (1769-1821)
1744   1813, [Manufacture of platinum]: Emp. Arts Sci's, v. 1 (ns), 181.

Bolton, Richard
1745   1849, On the physical geography and geology of the northern portion of the state of Mississippi: Am. Assoc. Adv. Sci. Proc., v. 1, 71-74.

Bolton, Samuel
1746   1842, [Letter on fossil footprints]: Am. J. Sci., v. 43, no. 1, 22.
        In "Regarding human foot-prints in solid limestone" by D. D. Owen (see 8051).

Bomar, W. W.
1747   1827, Observations on the medical topography and diseases of the western district of Tennessee: Ohio Med. Repos., v. 1, no. 23, 89-90.

Bomford, George (1782-1848)
1748   1822, Report ... respecting the lead mines of the United States: US SED 94, 17-1, v. 2 (60), 14-17.

Bond, Josiah
1749   1844, Wisconsin and its resources: Hunt's Merch. Mag., v. 10, 541-557.

Bond, William Cranch (1789-1859)
        See also 6504 and 6506 with J. Lovering.
1750   1849, Description of the observatory at Cambridge, Massachusetts: Am. Acad. Arts Sci's Mem., v. 4 (ns), 177-188.

Bonnard, Auguste Henri de (1781-1857)
1751   1817, [Geognostic description of the Erzegebirge]: Am. Reg. or Summary Rev., v. 1 (ns), 441-444.

Bonner, William G.
1752   1849, Bonner's map of the state of Georgia with the addition of its geological features: Savannah, W. T. Williams, map, 46 x 52 cm.
        In G. White Statistics of the state of Georgia (see 10897).

Bonny, John S.
1753   1835, Schoharie [New York] minerals: Am. J. Sci., v. 28, no. 2, 381.
1754   1836, Crinoidea, or lily shaped animal: Am. J. Sci., v. 31, no. 1, 165-167, fig.
        From Schenectady Reflector.

Bonnycastle, Richard Henry (1791-1848)
1755   1830-1836, On the transition rocks of the Cataraqui [Ontario]: Am. J. Sci., v. 18, no. 1, 85-104, fig. (1830); v. 20, no. 1, 74-82, plate with 6 figs. (1831); v. 24, no. 1, 97-104, 2 figs. (1833); v. 30, no. 2, 233-248, fig., 2 plates (1836).

Bonomi, Joseph (1796-1878)
1756   1845, On a gigantic bird sculptured on the tomb of an officer of the household of Pharaoh: Am. J. Sci., v. 49, no. 2, 403-405.
       Abstracted from London Ath.

Bonsdorf, Pehr Adolph von (1791-1838)
1757   1822, Analysis of an ore of silver: Am. J. Sci., v. 5, no. 2, 377-378.
       Chemical analysis of miargyrite.
1758   1822, [Analysis of idocrase]: Am. J. Sci., v. 4, no. 2, 243-244.
       Abstracted from Thompson's Annals of Phil.
1759   1824, New analysis of red silver ore: Port Folio, v. 18 (5s), 384.

Booker (Mr., of Cardiff, England)
1760   1846, Production of bar-iron direct from the ore: Franklin Inst. J., v. 11 (3s), 286-287.
       Abstracted from Mining J.

Booth, James Curtis (1810-1888)
       See also 1929 with M. H. Boye
1761   1836, On the deutarseniuret of nickel, from Riechelsdorf in Hessia: Am. J. Sci., v. 29, no. 2, 241-243.
1762   1836, Report of the examination and survey of the coal lands, etc., belonging to the Boston Purchase, near Cumberland, in the state of Maryland: NY, D. Fanshaw, 8 p.
1763   1836, The subscriber has taken a laboratory in which he proposes to perform analyses of the various ores ... : Phila., broadside, 18 x 12.5 cm.
1764   1839, First and second report of the geological survey of Delaware: Dover, Del., 25 p; also in Del. House J., January, 1839, 63-82.
1765   1840, and M. C. Lea, Analysis of a chromic iron ore, first observed by R. C. Taylor, Esq., at Mahobal, near Gibara, Island of Cuba: Am. J. Sci., v. 38, no. 2, 243-245.
1766   1840, Notice of a few of the subjects contained in Poggendorff's Annals for 1840: Franklin Inst. J., v. 25 (ns), 391.
1767   1841, Analysis of various ores of lead, silver, copper, zinc, iron, &c., from King's Mine, Davidson County, North Carolina: Am. J. Sci., v. 41, no. 2, 348-352.
1768   1841, [Fossil shells for fertilizer, in Delaware]: Am. J. Sci., v. 41, no. 1, 170 (7 lines).
1769   1841, Memoir of the geological survey of the state of Delaware: including the application of the geological observations to agriculture: Dover, S. Kimmey (printer), xii, 9-188 p., illus.; also in: Del. Senate J., January, 1841, 41-170.
       For review see 3505. For excerpts see 9955.
1770   1841, Remarks relative to the fossiliferous ore of Pennsylvania, and its employment in the manufacture of iron: Franklin Inst. J., v. 2 (3s), 187-188.
1771   1841, [Giant oyster deposits on the Atlantic coast of the United States]: Assoc. Am. Geol's Nat's Proc., v. 2, 15 (12 lines).
       Also in Assoc. Am. Geol's Nat's Rept., v. 1, 23 (12 lines), 1843.
1772   1845, [Zinc ore from Pennsylvania]: Assoc. Am. Geol's Nat's Proc., v. 6, 82 (3 lines).
1773   1849, Metallurgical treatment of gold ores: Hunt's Merch. Mag., v. 20, 109-110.
       From Encyclopedia of Chemistry.
1774   1850, Booth's patent for the reduction of gold: Scien. Am., v. 6, no. 8, 59; no. 9, 67.
1775   1850, The encyclopedia of chemistry, practical and theoretical: embracing its application to the arts, metallurgy, mineralogy, geology, medicine and pharmacy: Phila., H. C. Baird, 4, 974 p., illus., 9 plates.
       Assisted by C. Morfit. Also a second printing in 1850 called "second edition."
       For review see 4096.

Borden, Simeon (1798-1856)
1776   1843, Comparison of the dimensions of the earth, obtained from measurements made in the survey of the state of Massachusetts, with accredited mean determinations: Am. Philos. Soc. Proc., v. 3, no. 27, 130-132.

Borré (see Boué, Ami)

Boston and California Mining and Trading Joint-Stock Company
1777   1848, By-laws of the Boston and California Mining and Trading Joint-Stock Company: n. p., 1 p.

Boston and New York Coal Company
1778   1837, Charter and by-laws of the Boston and New York Coal Company: NY.
       Not seen.

Boston Company (see Lake Superior Copper Mining Company)

The Boston Cultivator (Boston, 1839-1850)
1779   1839, Earths and soils: v. 1, no. 9, [1].
1780   1841, Shock of an earthquake [at New York City]: v. 3, no. 5, [2].
1781   1841, Salt mines [of Norwich, England]: v. 3, no. 15, [1].
1782   1841, A gold mine: v. 3, no. 38, [2] (4 lines).
       Gold discovered at Clear Creek, North Carolina.
1783   1841, Salt: v. 3, no. 38, [2].
       Production statistics for Grand River Valley, Michigan.
1784   1841, Mammoth Cave of Kentucky: v. 3, no. 44, [4].
1785   1841, Extraordinary shower of stones in Hungary: v. 3, no. 50, [4].
1786   1843, Decomposition of rocks: v. 5, 153.
1787   1845, Kanawha salt works: v. 7, 144.
1788   1845, A mammoth lead cave [in Jefferson County, Missouri]: v. 7, 184.
1789   1845, Manganese ores [from New York]: v. 7, 296.
1790   1845, The great copper region of America [Lake Superior District]: v. 7, 320.
1791   1846, New process of reducing silver ores in Mexico: v. 8, 21 [i.e. 24].
1792   1848, [On the origin of coal]: v. 10, 385, 394, 410.

Boston Cultivator (cont.)
1793   1849, Bruce's gold washer: v. 11, 106 (11 lines).

Boston Journal of Philosophy and the Arts ... (Boston, 1823-1826)
1794   1823, Primitive boulders [in Finland and Massachusetts]: v. 1, no. 1, 90-92.
        With notes on boulders of Roxbury, Massachusetts by J. W. Webster (see 10745).
        From Geol. Soc. London Trans.
1795   1823, New scientific work [notice of Travels in Brazil by Martius and Spix]: v. 1, no. 2, 188 (20 lines).
1796   1823, Method of obtaining iron from slags and cinders: v. 1, no. 3, 300 (12 lines).
1797   1823, Land slip [at Hyotte, Lower Canada]: v. 1, no. 3, 301 (18 lines).
1798   1824, [Review of Outlines of Oryctology. An introduction to the study of fossil organic remains; especially those found in the British strata..., by J. Parkinson]: v. 1, no. 6, 559-570.
1799   1824, Iodine in mineral waters: v. 2, no. 1, 103 (11 lines).
1800   1824, Fall of meteoric stones [at Arenazzo, Italy]: v. 2, no. 3, 298 (6 lines).
        Abstracted from Bulletin Universel.
1801   1825, A geognostical essay on the superposition of rocks in both hemispheres [review of Essai géognostique by A. Humboldt]: v. 3, no. 1, 49-71.
        From Royal Inst. J.
1802   1825, Nature of colour in mineral productions: v. 3, no. 1, 97-98.
        From Annals of Phil.
1803   1825, M. Bonpland: v. 2, no. 4, 386 (8 lines).
        From Ferussac's Bulletin. On his current work in Paraguay.
1804   1825, Discovery of a fossil bat [at Montmartre]: v. 2, no. 4, 392.
        From Brewster's J.
1805   1825, Biography of Professor Playfair: v. 2, no. 5, 466-474.
        From New Edinburgh Encyclopaedia.
1806   1825, Mr. Heuland's mineralogical collection: v. 2, no. 5, 500-501.
        From Edinburgh Philos. J.
1807   1825, Albany Institute: v. 2, no. 6, 603.
        On the progress of the Society.
1808   1825, Fossil remains [from Placquemine, Louisiana]: v. 2, no. 6, 604-607.
        From the Louisiana Gaz. On the discovery of fossil bones.
1809   1826, Notice regarding steatite or soap-stone, and its principal uses: v. 3, no. 4, 405-407.
        From Jameson's Edinburgh Philos. J.
1810   1826, Selenium from Lukawitz in Bohemia: v. 3, no. 5, 495.
        From Brewster's J.
1811   1826, Sketch of the mineralogy of Gay Head, and of Bird Island, with a description of a large sun fish, caught in Vineyard Sound: v. 3, no. 6, 588-592.
1812   1826, Structure of the Alps: v. 3, no. 6, 600-601.
        From Edinburgh New Philos. J.

The Boston Literary Magazine (Boston, 1832-1833)
1813   1832, [Review with excerpts of Familiar lessons ..., by J. K. Welsh (see 10798)]: v. 1, no. 7, 343-344.

The Boston Lyceum (Boston, 1827)
1814   1827, [Review of Elements of mineralogy by J. L. Comstock (see 2506)]: v. 1, no. 3, 159-160.
1815   1827, To the editor of the Lyceum [review of Elements of mineralogy by J. L. Comstock (see 2506)]: v. 1, no. 3, 137-143.
        On the popularity of mineral textbooks by J. L. Comstock, P. Cleaveland, and J. W. Webster (see also 2422).

The Boston Magazine (Boston, 1783-1786)
1816   1785, Description of the famous salt mines, at Williska, in Poland: v. 2, 100-102.
1817   1786, Extract of a letter from a gentleman in the Western Country: v. 3, 214-216.
        On the discovery of fossil bones at Big Bone Lick, Kentucky.

The Boston Mechanic and Journal of the Useful Arts and Sciences (Boston, 1832-1836)
1818   1835, New hydraulic blow-pipe: v. 4, 21-22, fig.
1819   1835, Description of coal: v. 4, 35-38.
1820   1835, Soap stone [from Middlefield, Massachusetts]: v. 4, 70-71.
1821   1835, Explanation of specific gravity: v. 4, 71-73.
1822   1835, Platina [from the Urals]: v. 4, 75 (8 lines).
1823   1835, Potash from stones: v. 4, 172 (5 lines).
        On the use of mica and feldspar as a source of potash.
1824   1835, Anthracite ware: v. 4, 172 (6 lines).
        On the use of jet.
1825   1835, American porcelain clay [from Monkton, Vermont]: v. 4, 180-181.
1826   1835, Coal [from Mansfield, Massachusetts]: v. 4, 194 (5 lines).

Boston Medical Intelligencer (Boston, 1823-1828)
1827   1824, Mineral springs: v. 2, 27-28.
1828   1824, Observations on the use and abuse of the mineral waters of the Saratoga Springs: v. 2, 53-54.
1829   1824, Mineralogy: v. 2, 100 (4 lines).
        On the presentation of Siberian minerals to the New York Lyceum of Natural History by G. Gibbs.
1830   1824, Human fossil: v. 2, 101-102.
1831   1824, Mineralogy [of the Labrador Coast]: v. 2, 112.
1832   1825, Quartz: v. 2, 104 [i.e. 204].
1833   1825, Mineralogy: v. 3, 28.
        Notice of the Portland Mineral Society.
1834   1825, Animal remains [from the Mississippi River Valley]: v. 3, 40.
1835   1825, Fossil bones: v. 3, 56.
        Discoveries of Cuvier near Lyons.
1836   1825, [On deposits of American ores and minerals]: v. 3, 112, 116, 201 [i.e. 120], 124, 152, 160, 208.
1837   1825, Petrified wood [from the Mississippi River Valley]: v. 3, 128.

Boston Medical Intelligencer (cont.)
1838   1826, Shell limestone [from Eutaw Springs, South Carolina]: v. 3, 148.
1839   1826, Aereolites: v. 3, 160 (7 lines).
        Notice of augite in a meteorite.
1840   1826, Temperature of the earth: v. 4, 15.

The Boston Monthly Magazine (Boston, 1825-1826)
1841   1825, [Lehigh Coal Company production]: v. 1, no. 2, 111 (11 lines).

The Boston Quarterly Review (Boston, 1838-1842)
1842   1842, [Review of Lectures on agricultural chemistry and geology by J. F. W. Johnston (see 5819)]: v. 5, no. 28, 255.
1843   1842, [Review of A muck manual for farmers by S. L. Dana (see 2885)]: v. 5, no. 29, 371-383.

Boston Society of Natural History
        For lists of officers and members, and notices, see also The Boston Journal of Natural History, and the Society's Proceedings.
1844   1837, Catalogue of the library of the Boston Society of Natural History: Bos. J. Nat. History, v. 1, no. 4, 497-512.
1845   1841, Additions to the library of the Boston Society of Natural History since 1837: Bos. J. Nat. History: v. 3, no. 4, 513-522.
1846   1841, Proceedings of the Boston Society of Natural History: Am. J. Sci., v. 40, no. 2, 386-390.
        Includes notice of a donation of minerals by Monticelli.

Boston Society of Natural History. Proceedings of the (Boston, 1841-1850)
1847   1841-1850, Donations to the library: numerous entries in all volumes.
1848   1841-1850, [Catalogue of, and donations to, the mineral and geological cabinets]: numerous entries in all volumes.

The Boston Spectator, Devoted to Politicks and Belles-Lettres (Boston, 1814-1815)
1849   1814, Stones from the clouds: v. 1, no. 1, 4.
        On early accounts of meteorites.
1850   1814, A chymical curiosity: v. 1, no. 10, 39.
        On the weathering of calcareous rocks at Malta, Goza, and Cumin Island.

The Boston Weekly Magazine (Boston, 1802-1806)
1851   1802, Chemistry: v. 1, 3.
        On the discovery of columbium in ore.
1852   1802-1804, [Brief notes on earthquakes]: v. 1, 7, 19 (1802); v. 1, 71, 110 (1803); v. 2, 30 (1804).
1853   1803, [Quicksilver discovered in Cottah, Columbo, India]: v. 1, 70 (9 lines).
1854   1804, [Mineral springs of Stafford, Connecticut]: v. 2, 170.

The Boston Weekly Magazine (Boston, 1838-1841)
1855   1838-1840, [Brief notices on American mining]: v. 1, 38, 47, 71, 79 (1838); v. 1, 294; v. 2, 68 (1839); v. 2, 310 (1840).
1856   1838, Gaseous state of the earth: v. 1, 91.
        On the origin of the earth.
1857   1838-1840, [Brief notices on fossils]: v. 1, 120, 123 (1838); v. 1, 211, 270, 290-291; v. 2, 27, 54, 67 (1839); v. 2, 210 (1840).
1858   1839, Extraordinary caverns in Moravia: v. 1, 150.
1859   1839-1840, [Brief notices on earthquakes and volcanoes]: v. 1, 198, 223, 277-278; v. 2, 51 (1839); v. 2, 147, 155, 264, 410; v. 3, 11, 66-67, 199 (1840).
1860   1839, Burning coal mines [in England]: v. 1, 223 (16 lines).
1861   1839, The earth like an egg: v. 1, 237 (8 lines).
1862   1839, Rocks: v. 1, 244.
        On the decomposition of rocks to form soil.
1863   1839, The wonders of the mines: v. 1, 263.
1864   1839, [Prof. Rogers, State Geologist of Virginia]: v. 2, 42.
        Notice of his lectures on the Blue Ridge Mountains.
1865   1840, Burning lakes of South America: v. 2, 263-264.
1866   1840, Mining: v. 2, 314.
        A popular article on the nature of mining.
1867   1840, Meteorolites, or stones which have fallen from the sky: v. 2, 331.
1868   1840, Theories of the earth: v. 2, 381.
1869   1840, Sulphur mines of Naples: v. 2, 382.
1870   1840, Burning mines of Commentry, in France: v. 2, 403.
1871   1840, Mammoth Cave of Kentucky: v. 3, no. 51, 402.

Bosworth, Nathaniel
1872   1836, Extracting gold from its ores: Franklin Inst. J., v. 17 (ns), 39.
        Patent description.

Botfield, Thomas
1873   1831, Invention of certain improvements in making iron, or in the methods, of smelting and making iron: Franklin Inst. J., v. 8 (ns), 196.

Boué, Ami (1794-1881)
1874   1823, Dr. Borré's [i.e., Boué] notices of European continental geology, with remarks on the prevailing geological arrangements in a letter to Dr. J. W. Webster of Boston: Am. J. Sci., v. 6, no. 1, 188-192.
1875   1825, Geological and miscellaneous observations: Am. J. Sci., v. 9, no. 1, 23-26.
        Describes observations in southern Europe.

Bouis, M.
1876   1828, Presence of ammonia in argillaceous minerals: Am. J. Sci., v. 15, no. 1, 182.
        Abstracted from Annales de Chimie by J. Griscom.

Bouisson, M.
1877   1846, Artificial marble: Am. J. Sci., v. 2 (ns), no. 2, 266 (13 lines).
        Translated and abstracted by J. L. Smith from "Jour. de Chem. Med."

Boulanger, Charles Louis (1810-1849)
1878   1836, On a double sulphuret of antimony and lead [from Molieres, France]: Am. J. Sci.,
        v. 30, no. 1, 177-178.
        Translated and abstracted by "D." from Annales des Mines.

Bourne, A.
1879   1820, On the prairies and barrens of the West: Am. J. Sci., v. 2, no. 1, 30-34.

Bourne, Henry
1880   1849, Improvement in the machine for washing gold: Franklin Inst. J., v. 18 (3s), 488
        (11 lines).

Bourne, William Oland
1881   1841, Notice of a locality of zeolites, &c. at Bergen, Bergen County, N. J.: Am. J.
        Sci., v. 40, no. 1, 69-73.

Boussingault, John Baptiste Joseph Dieudonné (1802-1887)
1882   1824, and M. Reveno, South America: Bos. J. Phil. Arts, v. 2, no. 1, 103 (12 lines).
        Abstracted from Philos. Mag. Discussion of meteorites, gold and mineral waters
        found in South America.
1883   1825, Meteoric iron [from Bogota, Colombia]: Am. J. Sci., v. 9, no. 1, 194 (9 lines).
        Abstracted from Annales de Chimie by J. Griscom.
1884   1827, Gay-lussite [from Merida, Colombia]: Am. J. Sci., v. 12, no. 1, 187 (11 lines).
        Abstracted from Annales de Chimie by J. Griscom.
1885   1833, Analysis of the water of Rio Vinagre: Am. J. Sci., v. 24, no. 1, 149-150.
        Translated by Oliver P. Hubbard.

Bouve, Thomas Tracy
1886   1843, [Conolites from Meadville, Pennsylvania]: Bos. Soc. Nat. History Proc., v. 1, 97
        (10 lines).
1887   1843, [Fossil shells from the Tertiary of Europe]: Bos. Soc. Nat. History Proc., v. 1,
        156-157.
1888   1844, [Fossils from the green sand of New Jersey]: Bos. Soc. Nat. History Proc., v. 1,
        171.
1889   1845, Review of Dr. C. T. Jackson's Final report on the geology and mineralogy of the
        state of New Hampshire [see 5611]: Am. J. Sci., v. 49, no. 1, 27-37.
1890   1845, [Collection of Tertiary fossils]: Bos. Soc. Nat. History Proc., v. 2, 80-81.
1891   1846, [Fossil shells from Cincinnati, Ohio]: Bos. Soc. Nat. History Proc., v. 2, 184
        (10 lines).
1892   1846, [Pygorhynchus gouldii, a new Echinus from the Millstone grit of Georgia]: Bos.
        Soc. Nat. History Proc., v. 2, 192; also in: Am. J. Sci., v. 3 (ns), no. 3, 437.
1893   1849, Review of M. Tuomey's Final report on the geological survey of South Carolina
        [see 10438]: Am. J. Sci., v. 8 (ns), no. 1, 61-74, figs.
1894   1849, [Flexible sandstone]: Bos. Soc. Nat. History Proc., v. 3, 149 (10 lines).

Bowditch, Nathaniel (1773-1838)
1895   1814, On the variation of the magnetical needle: Am. Acad. Arts Sci's Mem., v. 3, pt.
        2, 337-343.
1896   1818, On the calculation of the oblateness of the earth, by means of the observed
        lengths of a pendulum in different latitudes, according to the method given by La
        Place in the second volume of his "Mécanique Céleste," with remarks on other parts
        of the same work, relating to the figure of the earth: Am. Acad. Arts Sci's Mem.,
        v. 4, pt. 1, 30-49.
1897   1818, On the method of computing the dip of the magnetic needle in different latitudes,
        according to the theory of Mr. Biot: Am. Acad. Arts Sci's Mem., v. 4, pt. 1, 57-61,
        fig.
1898   1829, On the variation of the magnetic needle: Am. J. Sci., v. 16, no. 1, 64-69.
        In "Variation of the magnetic needle," by B. Silliman (see 9688).

Bowen, Eli (b. 1824)
1899   1845, The coal regions of Pennsylvania being a general, geological, historical and
        statistical review of the anthracite coal districts: Pottsville, Penn., B. Carvalic
        & Co., 72 p., illus., map.
        Also 1848 printing. Includes poetry about mining.

Bowen, Francis (1811-1890)
1900   1843, [Reviews of A muck manual for farmers by S. L. Dana (see 2885) and An essay on
        calcareous manures by E. Ruffin (see 9088)]: N. Am. Rev., v. 57, 243-244.
1901   1845, [Review with excerpts of Vestiges of the natural history of Creation by R.
        Chambers (see 2292 et seq.)]: N. Am. Rev., v. 60, 426-478.
1902   1849, [Review with excerpts of Ancient sea margins by R. Chambers (see 2299) and
        "Geological action of the tidal and other currents of the ocean" by C. H. Davis
        (see 2947)]: N. Am. Rev., v. 69, 256-269.
        Includes discussion of the possible author of Vestiges of the natural history of
        Creation (see 2292 et seq.), and correctly ascribes this unsigned work to Robert
        Chambers.
1903   1849, [Review with excerpts of A second visit ... by C. Lyell (see 6571)]: N. Am. Rev.,
        v. 69, 325-353.

Bowen, George Thomas (1803-1828)
1904   1822, Analysis of a variety of nephrite, from Smithfield, R. I.: Am. J. Sci., v. 5, no.
        2, 346-348.

Bowen, G. T. (cont.)
1905   1822, Analysis of the calcareous oxide of tungsten, from Huntington, Con.: Am. J. Sci., v. 5, no. 1, 118-121.
1906   1822, Analysis of the pyroxene sahlite, from the vicinity of New-Haven, Conn.: Am. J. Sci., v. 5, no. 2, 344-346.
1907   1822, Analysis of the sulphat of strontian from Lake Erie, and some sulphats of barytes: Am. J. Sci., v. 4, no. 2, 324-327.
1908   1824, Analysis of a siliceous hydrate of copper from New-Jersey, with a notice of the discovery of two localities of spodumene in the United States: Am. J. Sci., v. 8, no. 1, 118-121.
       On chrysocolla from Somerville, New Jersey, and spodumene from Conway and Sterling, Massachusetts.
1909   1824, Analysis of an ore of copper from New Jersey: Phila. Acad. Nat. Sci´s J., v. 3, pt. 2, 285-286; 295-297.
1910   1824, Description and analysis of the sillimanite, a new mineral: Am. J. Sci., v. 8, no. 1, 113-118; also in: Phila. Acad. Nat. Sci´s J., v. 3, 375-381.
       With introductory remarks by B. Silliman (see 9655), on the name "sillimanite." The mineral described is from Saybrook, Ct.
1911   1824, Notices of American spodumene [from Deerfield, Massachusetts]: Phila. Acad. Nat. Sci´s J., v. 3, pt. 2, 284-286.

Bowen´s Boston News-letter and City Record (Boston, 1825-1827)
1912   1825, Lead: v. 1, 8, 11.
       On the production of Missouri lead mines.
1913   1826, [Mr. Shepard´s lectures on mineralogy]: v. 1, 144 (6 lines).
1914   1826, Gold [from North Carolina]: v. 2, 43, 130.
1915   1826, [Earthquake in Bogota, Colombia]: v. 2, 83 (13 lines).
1916   1826, Lehigh coal [from Pennsylvania]: v. 2, 121 (12 lines).
       On the quality and production rate of coal.
1917   1826, [Silver mine in Adams County, Pennsylvania]: v. 2, 129 (4 lines).

Bowles, Don Guillermo
1918   1829, On the Spanish iron ore of Mondragon, in Guypsucoa; with some account of the famous Toledo sword-blades, so greatly valued formerly in England: Franklin Inst. J., v. 4 (ns), 307-310.

Bowman, John Eddowes (1819-1854)
1919   1837, [Bone cave in Denbighshire, England]: Am. J. Sci., v. 31, no. 2, 341-342.
       In British Assoc. Proc. By "J. B. [i.e. J. E.] Bowman."
1920   1841, Vegetable skeletons: Am. Eclec., v. 2, no. 2, 392.
       Abstracted from London Ath.
1921   1841, Fossil infusoria in England: Am. J. Sci., v. 40, no. 1, 174-176.
       Abstracted from J. of Botany.
1922   1842, On the Upper Silurian rocks of Denbighshire: Am. J. Sci., v. 42, no. 2, 325-326 (5 lines).
       In British Assoc. Proc.

Bowring, Sir John (1792-1872)
1923   1839, On the boric acid lagoons of Tuscany: Am. J. Sci., v. 37, no. 2, 270-274; also in: Am. J. Pharmacy, v. 6 (ns), 40-45 (1840).
       From Statistics of Tuscany, abstracted by Edingurgh New Philos. J.

Boyd, George W. (d. 1840)
1924   1832, [Report on George´s Mine]: in An act to incorporate the North American Mining Company (see 7926).
1925   1834, Observations on the mineralogy and geology of the gold region of the southern United States: Am. J. Sci., v. 27, no. 1, 151 (6 lines).
       Abstract in NY Lyc. Nat. History Proc.
1926   1838, Local and economic geology, Wayne County [New York]: NY Geol. Survey Ann. Rept., v. 2, 312-326.
1927   1838, Local and economical geology. Orleans County [New York]: NY Geol. Survey Ann. Rept., v. 2, 347-359.

Boyd, John
1928   1821, Fluor spar on the Genesee River [Ontario County, New York]: Am. J. Sci., v. 3, no. 2, 235.

Boye, Martin Hans (1812-1909)
       See also 8950 and 8951 with R. E. Rogers.
1929   1841, and J. C. Booth, Results of the analysis of three different varieties of feldspar from the primary rocks of the state of Delaware: Am. Philos. Soc. Proc., v. 2, 53-56.
1930   1842, [Analysis of a new variety of feldspar from Westchester, Pennsylvania]: Am. Philos. Soc. Proc., v. 2, no. 22, 190-191.
1931   1846, [Brown hematite ore from Chester Furnace, Huntingdon County, Pennsylvania]: Am. Philos. Soc. Proc., v. 4, no. 35, 238-239.
1932   1846, [Mineral specimens from the coal measures of Slazburg, Pennsylvania]: Am. Philos. Soc. Proc., v. 4, no. 35, 247 (15 lines).
1933   1847, Oxyd of cobalt with the brown hematite ore of Chester Ridge, Pa.: Am. J. Sci., v. 4 (ns), no. 2, 281 (12 lines).
       Abstracted from Am. Philos. Soc. Proc.
1934   1849, Analysis of the bittern of a saline on the Kiskiminetas River, near Freeport, Armstrong County, Pennsylvania: Am. J. Sci., v. 7 (ns), no. 1, 74-76.
1935   1849, On the composition of the Schuylkill water [Pennsylvania]: Am. Assoc. Adv. Sci. Proc., v. 1, 123-132; also in: Am. J. Sci., v. 9 (ns), no. 1, 123 (1850).

Boykin (Dr.)
1936   1830, Notice: Am. J. Sci., v. 18, no. 2, 388-389.
       In "Georgia meteor and aerolite" by E. Beall (see 1550).

Boyle
1937   1835, Sept. 9 - slaty structure: Am. J. Sci., v. 28, no. 1, 72 (11 lines).
       In British Assoc. Proc.

Boynton (Dr.)
1938   1849, Geology: Scien. Am., v. 4, no. 18, 139.
       Review of lecture on elementary geology.

Brace, John Pierce (1793-1872)
1939   1819, Observations on the minerals connected with the gneiss range of Litchfield County
       [Connecticut]: Am. J. Sci., v. 1, no. 4, 351-355.
1940   1820, [Geology and mineralogy of Litchfield, Connecticut]: Am. J. Sci., v. 2, no. 2,
       370.
       Abstracted in NY Lyc. Nat. History Proc.
1941   1823, By J. P. Brace: Am. J. Sci., v. 6, no. 2, 250-251.
       On the minerals of Connecticut.

Brackenridge, Henry Marie (1786-1871)
1942   1811, An account of the minerals of Louisiana: Am. Med. Philos. Reg., v. 2, no. 1,
       33-40.
1943   1814, Views of Louisiana (including what is now known as Missouri), together with a
       journal of a voyage up the Missouri River in 1811: Pittsburgh, Cramer, Spear and
       Eichbaum, 304 p.
       For review see 10806.
1944   1817, Views of Louisiana; containing geographical, statistical and historical notices
       of that vast and important portion of America: Balt., Schaeffer and Maund, 323 p.

Braconnot, Henri (1780-1855)
1945   1838, Indication of organic remains in the oldest rocks of the globe; means of
       distinguishing trap from basalt: Franklin Inst. J., v. 22 (ns), 341-342.
       Abstracted from Annales de Chimie.

Bradbury, Charles (1798-1864)
1946   1837, History of Kennebunk port, from its first discovery by Bartholomew Gosnold, May
       14, 1602, to A. D. 1837: Kennebunk, J. K. Remick (printer), 301 p.
       Includes account of granite quarries, with opinion of P. Cleaveland (p. 219).

Bradbury, John (fl. 1809)
1947   1817, Description of the minerals and plants found at the lead mines in the Missouri
       Territory: Med. Repos., v. 18, 135-138.
1948   1818, Alum formed by the agency of the atmosphere: Med. Repos., v. 19, 101.
       On native alum from Shawangunk Mountain, New York.

Bradford, William John Alden (1791-1858)
1949   1846, Notes on the Northwest or Valley of the Upper Mississippi: NY, Wiley and Putnam,
       vi, 302 p.
       Includes an appendix with geological observations.

Bradley, Abraham (1731-1824)
1950   1801, A new theory of the earth ... : Wilkesbarre, Pa., Joseph Wright, 63 p.
1951   1808, A philosophical retrospect on the general outlines of Creation and Providence:
       Wilkesbarre, Pa., Charles Miner, 194, [4] p.

Braga, Antonio
1952   1808, [Mining in Portugal]: Am. Reg. or Gen. Repos., v. 2, 380.

Braid, James
1953   1818, Account of the fatal accident which happened in the Leadhills Company's mines,
       the 1st March, 1817: Eclec. Rep. and Analytic Rev., v. 8, 125-128.

Brande, William Thomas (1788-1866)
1954   1816, Observations on the application of coal gas to the purposes of illumination:
       Eclec. Rep. and Analytic Rev., v. 6, 273-289; also in J. Sci. Arts, v. 1, no. 1,
       71-80.
1955   1817, Plan of an extended and practical course of lectures and demonstrations on
       chymistry, to be delivered in the laboratory of the Royal Institution: J. Sci.
       Arts, v. 2, no. 1, 213-216.
1956   1818, A descriptive account of Mr. Thompson's laboratory at Cheltenham, for the
       preparation of the Cheltenham salts; with a chymical analysis of the waters where
       they are produced: J. Sci. Arts, v. 3, no. 1, 54-71, plate.
1957   1818, Lectures on mineralogical chymistry, delivered in the theatre of the Royal
       Institution: J. Sci. Arts, v. 3, no. 2, 358-368; v. 4, no. 1, 66-74; no. 2,
       233-247; v. 5, no. 1, 64-73; no. 2, 291-300.
1958   1843, A dictionary of science, literature and art: comprising the history, description,
       and scientific principles of every branch of human knowledge; with the derivation
       and definition of all the terms in general use: NY, Harper and Brothers, iv, 1352
       p., illus.
       "Assisted by Joseph Cauvin." Other printings by Harper and Brothers in 1844,
       1845, 1846, 1847, and 1848.

Brandes, Rudolph (1795-1842)
1959   1828, Incompatible salts: Am. J. Sci., v. 15, no. 1, 185 (7 lines).
       Abstracted from Ferussac's Bulletin by J. Griscom. On the salt content of mineral
       waters.
1960   1829, Examination of a substance called shooting star, which was found in a wet meadow:
       Am. J. Sci., v. 16, no. 1, 20-27.

Brandt, Johann Friedrich von (1802-1879)
1961   1850, The fossil rhinoceros of Siberia: Ann. Scien. Disc., v. 1, 286 (14 lines).
       Abstracted from Jameson's J.

Bravais, Auguste (1791-1873)
1962   1846, Lines of ancient level of the sea in Finmark: Am. J. Sci., v. 1 (ns), no. 2, 273-274.

Braxton, Corbin
1963   1836, Different effects, compared, of the green sand, calcareous marl, and the mixture of both: Farm's Reg., v. 4, 276-278.
1964   1841, Comments on the articles of the editor on the effects of greensand as manure: Farm's Reg., v. 9, 233-237.

Brayley, Edward William (1802-1870)
1965   1824, Igneous meteors: Bos. J. Phil Arts, v. 2, no. 1, 101-102.
        Meteorites compared with igneous rocks.

Breithaupt, Johann Friedrich Auguste (1791-1873)
1966   1831, Pinguite, a new argillaceous mineral [from Erzgeberg]: Am. J. Sci., v. 20, no. 1, 197 (8 lines).
        Abstracted from Jahrbuch für mineralogie by C. U. Shepard.
1967   1834, Characteristic of the classes and orders of minerals: Geol. Soc. Penn. Trans., v. 1, 120-136.
        Translated by A. Del Rio (see 3086).
1968   1837, Breithaupt's new specific gravities of minerals: Am. J. Sci., v. 31, no. 2, 268-271.
        Abstracted by L. Feuchtwanger.
1969   1846, Bodenite [from Boden, Saxony]: Am. J. Sci., v. 2 (ns), no. 3, 415 (3 lines).
        Abstracted from Poggendorff's Annalen.
1970   1846, Chloanthite, a new binarseniet of nickel [from Schneeberg, Saxony]: Am. J. Sci., v. 1 (ns), no. 2, 266 (6 lines).
1971   1846, Digenite and cuproplumbite [from Chile]: Am. J. Sci., v. 2 (ns), no. 3, 414 (11 lines).
        Abstracted from Poggendorff's Annalen.
1972   1846, Dysclasite [from Copper Harbor, Lake Superior]: Am. J. Sci., v. 1 (ns), no. 2, 267 (4 lines).
1973   1846, Loxoclase [from Hammond, New York]: Am. J. Sci., v. 2 (ns), no. 3, 414 (13 lines).
        Abstracted from Philos. Mag.
1974   1847, and Plattner, Castor and pollux, two new minerals [from the Island of Elba]: Am. J. Sci., v. 3 (ns), no. 3, 430.
        Abstracted from Poggendorff's Annalen.
1975   1849, New minerals: Am. J. Sci., v. 8 (ns), no. 1, 127 (10 lines).
        Abstracted from Poggendorff's Annalen. On pliniane and stannine from Cornwall, England.
1976   1849, On pistomesite and mesitine: Am. J. Sci., v. 8 (ns), no. 1, 121 (8 lines).
1977   1849, On zygadite: Am. J. Sci., v. 8 (ns), no. 1, 124 (5 lines).

Brereton, M. John (fl. 1602)
1978   1843, A brief and true relation of the discovery of the north part of Virginia; being a most pleasant, fruitful and commodious soil, made this present year 1602: Mass. Hist. Soc. Coll., v. 8 (3s), [83]-123.
        Includes notice of gold, silver, and copper mines (121-123).

Breton, Peter
1979   1828, Account of the diamond workings and diamonds of Sumbhulpore [India]: Franklin Inst. J., v. 6, 46-50.
        Abstracted from Med. Phys. Soc. Calcutta Trans.

Breunner, Count Augustus
1980   1839, On the use of wire ropes in deep mines: Am. J. Sci., v. 35, no. 2, 319.
        In British Assoc. Proc.

Brewster, Sir David (1781-1868)
        See also 3440 to 3442 under Edinburgh Encyclopedia.
1981   1817, On the effects of mechanical pressure in communicating double refraction to regularly crystallized bodies: J. Sci. Arts, v. 2, no. 2, 460-461.
        On the effect of pressure on calcite optical properties.
1982   1817, [Optical properties of calcite]: J. Sci. Arts, v. 1, no. 2, 293.
        In Royal Soc. Edinburgh Proc.
1983   1818, On the difference between the optical properties of arragonite and calcareous spar: J. Sci. Arts, v. 4, no. 1, 112-114.
1984   1818, Optical structure of ice: J. Sci. Arts, v. 4, no. 1, 155.
1985   1823, Dr. Brewster: Am. J. Sci., v. 6, no. 2, 400 (13 lines).
        Abstracted by J. Griscom from Royal Soc. Edinburgh Trans. On optical properties of Brazilian topaz, and on brucite.
1986   1824, Account of the native hydrate of magnesia, discovered by Dr. Hibbert of Shetland: Am. J. Sci., v. 7, no. 2, 365 (11 lines).
        On the physical properties of brucite from Unst, Shetland.
1987   1824, Additional observations on the connection between the primitive forms of minerals and the number of their axes of double refraction: Am. J. Sci., v. 7, no. 2, 363-364.
1988   1824, Description of hopeite, a new mineral from Altenberg, near Aix-la-Chapelle: Bos. J. Phil. Arts, v. 2, no. 2, 119 (9 lines).
        Abstracted from Edinburgh Philos. J.
1989   1824, On the distribution of the colouring matter and on certain peculiarities in the structure and optical properties of the Brazilian topaz: Am. J. Sci., v. 7, no. 2, 364-365.
1990   1824, On the existence of two new fluids in the cavities of minerals, which are immiscible, and possess remarkable physical properties: Bos. J. Phil. Arts, v. 2, no. 2, 118-119 (14 lines).
        Abstracted from Edinburgh Philos. Trans.
1991   1825, Optical structure of minerals: Am. J. Sci., v. 9, no. 2, 384.
        On the optical properties of mica.

Brewster, D. (cont.)
1992 1827, Fluids in the cavities of minerals: Am. J. Sci., v. 12, no. 2, 214-227, plate with 10 figs.
1993 1832, [On the crystallographic system of F. Mohs]: Mon. Am. J. Geology v. 1, no. 10, 470 (6 lines).
1994 1832, On the law of the partial polarization of light by reflexion: Am. J. Sci., v. 22, no. 2, 277-292, 2 figs.
1995 1832, On the production of regular double refraction in the molecules of bodies by simple pressure, with observations on the origin of the doubly refracting structure: Am. J. Sci., v. 21, no. 2, 296-303.
1996 1833, Fluid in the cavities of minerals: Franklin Inst. J., v. 12 (ns), 213-216.
From "Letters on natural magic."
1997 1833, On the laws of the polarization of light by refraction: Am. J. Sci., v. 23, no. 2, 225-236, 2 figs.
1998 1835, Amber from Ava: Am. J. Sci., v. 28, no. 1, 71 (3 lines).
In British Assoc. Proc.
1999 1836, Diamond, matrix of, &c.: Am. J. Sci., v. 29, no. 2, 364.
On the matrix of diamonds from southern India.
2000 1838, Diamond: Am. J. Sci., v. 34, no. 1, 37-39, fig.
In British Assoc. Proc. On optical properties of the diamond.
2001 1845, An account of the cause of the colors in precious opals: Franklin Inst. J., v. 10 (3s), 195.
Abstracted from London Ath.
2002 1845, On crystals in the cavities of topaz, which are dissolved by heat, and re-crystallized on cooling: Franklin Inst. J., v. 9 (3s), 142.
Abstracted from London Ath.
2003 1846/1848, Fluids and cavities in topaz: Am. J. Sci., v. 1 (ns), no. 1, 121 (13 lines) (1846); v. 5 (ns), no. 3, 420-421.
2004 1850, Address of Sir David Brewster before the twentieth meeting of the British Association at Edinburgh, July 31, 1850: Am. J. Sci., v. 10 (ns), no. 3, 305-320.

Brickell, John (1749?-1810)
2005 1809, On the subject of the falling of stones from the atmosphere: Balt. Med. Phys. Rec'r, v. 2, 55-58; also in: Am. Mag. of Wonders, v. 1, 124-125.
From the Charleston City Gaz.

Briggs, Charles, jr.
2006 1838, and J. W. Foster, Geological section to illustrate the superposition of the rocks of the south part of Ohio between the great limestone deposits and the upper part of the coal series: in Ohio Geol. Survey first annual report by W. W. Mather (see 6891), folding plate.
2007 1838, Report of C. Briggs, jr., fourth assistant geologist: in Ohio Geol. Survey first annual report by W. W. Mather (see 6891), 71-98.
On the geology of Scioto and Hocking Counties.
2008 1838 [i.e. 1839], Report of Mr. Briggs: in Ohio Geol. Survey second annual report by W. W. Mather (see 6896), 109-154, plate, fig.
On the geology and stratigraphy of Wood, Crawford, Athens and Hocking Counties.
2009 1839, The mastodon of the Ohio Valley: Hesperian, v. 2, 333-335.

Bringler, L.
2010 1821, Notices of the geology, mineralogy, topography, productions, and aboriginal inhabitants of the regions around the Mississippi and its confluent waters: Am. J. Sci., v. 3, no. 1, 15-46.

Brisbane, Sir Thomas M. (1773-1860)
2011 1846, Observations in magnetism and meteorology, made at Makerstown, Scotland, in the observatory of Sir T. M. Brisbane, 1841 and 1842: Am. J. Sci., v. 1 (ns), no. 1, 138-139.

Bristed, John (1778-1855)
2012 1818, The resources of the United States of America ... : NY, J. Eastburn, xvi, 505 p.

Bristol, M. (Dr.)
2013 1831, Buffalo mineral springs [New York]: Am. J. Sci., v. 20, no. 1, 156-158.
With "Chemical examination" by C. U. Shepard (see 9484), and "Remarks" by B. Silliman (see 9703).

Bristol, T. W. (see 5322 with J. Houghton)

British Association for the Advancement of Science
2014 1832, British Association for the Promotion of Science: Am. J. Sci., v. 21, no. 2, 373-374.
Notice of officers and proceedings. Abstracted from Philos. Mag.
2015 1832, British Association for the Advancement of Science: Am. J. Sci., v. 23, no. 1, 179-182.
Notice of the 1832 meeting. Abstracted from London Med. Gaz.
2016 1834, Notice of the British Association for the Advancement of Science: Am. J. Sci., v. 25, no. 2, 411-412.
Notice of the third meeting.
2017 1835, Notice of the meetings of the British Association for the Advancement of Science, in 1833, at Cambridge, and in 1834, at Edinburgh; in two parts: Am. J. Sci., v. 28, no. 1, 55-84.
2018 1836, Fifth meeting of the British Association for the Advancement of Science at Dublin, Ireland, July, 1835: Am. J. Sci., v. 29, no. 2, 347-355.
2019 1836, Report of the fourth meeting of the British Association for the Advancement of Science: Am. J. Sci., v. 29, no. 2, 355-358.
2020 1836, Proceedings of the fifth meeting of the British Association: Am. J. Sci., v. 29, no. 2, 364 (7 lines).
Abstracted from London Ath.

British Association (cont.)
2021 1837, Proceedings of the British Association at Bristol in August 1836: Am. J. Sci., v. 31, no. 2, 332-381.
From Edinburgh New Philos. J.
2022 1838, Seventh meeting of the British Association for the Advancement of Science - Liverpool, Saturday, Sept. 9: Am. J. Sci., v. 33, no. 2, 265-296; v. 34, no. 1, 1-56.
2023 1838, Eighth meeting of the British Association for the Advancement of Science: Am. J. Sci., v. 35, no. 1, 200 (8 lines).
2024 1838, The salt mines of Northwich: Am. J. Sci., v. 34, no. 1, 55-56.
In British Assoc. Proc., seventh meeting. On a field trip by members of the Association.
2025 1839, An account of the proceedings of the eighth meeting of the British Association for the Advancement of Science: Am. J. Sci., v. 35, no. 2, 275-321.
From the London Ath.
2026 1839, Geological excursion: Am. J. Sci., v. 35, no. 2, 309.
In British Assoc. Proc., eighth meeting. On a field trip led by J. Hutton and A. Sedgwick to Tynemouth and Cullercoats.
2027 1840, Abstract of the proceedings of the ninth meeting of British Association for the Advancement of Science: Am. J. Sci., v. 38, no. 1, 93-138.
2028 1840, Magnetic observatories: Am. J. Sci., v. 38, no. 1, 108 (12 lines).
In British Assoc. Proc., ninth meeting. On magnetic measurements made in various British colonies.
2029 1840, Meeting of the British Association for the Advancement of Science: Am. J. Sci., v. 38, no. 2, 406-408.
2030 1841, Abstract of the proceedings of the tenth meeting of the British Association for the Advancement of Science: Am. J. Sci., v. 40, no. 2, 308-345; v. 41, no. 1, 40-68.
From London Ath.
2031 1842, Abstract of the proceedings of the eleventh meeting of the British Association for the Advancement of Science, held at Plymouth, September, 1841: Am. J. Sci., v. 42, no. 1, 147-164; no. 2, 317-346.
2032 1842, Abstract of the proceedings of the twelfth meeting of the British Association for the Advancement of Science: Am. J. Sci., v. 43, no. 2, 367-376; v. 44, no. 1, 158-172; v. 44, no. 2, 351-372.
From London Ath.
2033 1843, Meeting of the British Association at Dublin: Am. J. Sci., v. 45, no. 2, 403 (10 lines).
On sigillaria in coal from Liverpool, England.
2034 1844, Abstract of the proceedings of the thirteenth meeting of the British Association for the Advancement of Science: Am. J. Sci., v. 46, no. 2, 388-401; v. 47, no. 1, 182-187.
From London Ath.
2035 1846, The British Association: Am. J. Sci., v. 2 (ns), no. 3, 441 (6 lines).
Notice of meeting.
2036 1848, British Association: Am. J. Sci., v. 6 (ns), no. 3, 445-446.
From London Ath. Notice of 1848 meeting at Swansea.
2037 1849, British Association: Am. J. Sci., v. 8 (ns), no. 3, 444-448.
Notice of 1849 meeting at Birmingham.
2038 1850, [British Association meeting of 1850 at Edinburgh]: Am. J. Sci., v. 10 (ns), no. 1, 137 (2 lines).
2039 1850, British Association: Am. J. Sci., v. 10 (ns), no. 2, 287 (5 lines).
Notice of twentieth meeting at Edinburgh.
2040 1850, Extracts from the proceedings of the twentieth meeting of the British Association, held at Edinburgh, July, 1850: Am. J. Sci., v. 10 (ns), no. 3, 386-414.
2041 1850, The British Association: Ann. Scien. Disc., v. 1, 368 (16 lines).
On history and progress of the Association.

Broadmeadow, Simeon
2042 1844, Patent for manufacturing malleable iron directly from the ore, in a puddling furnace: Franklin Inst. J., v. 8 (3s), 119-122, fig.
2043 1845, Process for obtaining malleable iron directly from the ore, by treating in a puddling furnace: Franklin Inst. J., v. 9 (ns), 59.

Brocchi, Giovani Battista (1772-1826)
2044 1818, On the prehnite found in Tuscany: J. Sci. Arts, v. 3, no. 2, 398-399.

Brockelbank, M. T.
2045 1848, Gold mines of Virginia: Am. Q. J. Ag. Sci., v. 7, no. 10, 443-445.

Broderip, William John (1789-1850)
2046 1837, Table of the situations and depths at which recent genera of marine and estuary shell have been observed: in Researches in theoretical geology by H. T. De la Beche (see 3039), 335-342.
2047 1849, Zoological recreations: Phila., Lea and Blanchard, vi, [13]-376 p.
"From the second enlarged London ed." For review see 2771.

Brodie, Peter Bellinger (b. 1815)
2048 1842, Notice of the occurrence of fossil plants in the plastic clay at Bournemouth, Hants.: Franklin Inst. J., v. 4 (3s), 178-179.
Abstracted from London Ath.

Brongniart, Adolphe Théodore (1801-1876)
2049 1838, Considerations upon the nature of vegetables that have covered the surface of the earth, at different epoques of its formation: Am. J. Sci., v. 34, no. 2, 315-329.
Translated by R. W. Haskins.
2050 1846, The relations of Noggerathia with living plants: Am. J. Sci., v. 2 (ns), no. 2, 279-280.
Abstracted from "Ann. Sci. Nat."

Brongniart, A. T. (cont.)
2051   1848, On the changes of the vegetable kingdom in the different geological epochs: Am. J. Sci., v. 6 (ns), no. 1, 120-123.
            Abstracted from Edinburgh New Philos. J.

Brongniart, Alexandre (1770-1847)
        See also 2758 with G. Cuvier
2052   1818, [Geognostic situation of a calcareous rock at Montpelier, France]: J. Sci. Arts, v. 3, no. 1, 149.
            Abstracted from Royal Inst. France Proc.
2053   1818, Notice of M. Brongniart on organized remains: Am. J. Sci., v. 1, no. 1, 71-74.
            On techniques for collecting and preserving fossils.
2054   1818, [On sodalite from Mount Vesuvius]: J. Sci. Arts, v. 3, no. 1, 165-166.
            From Royal Inst. France Proc.
2055   1821, Miscellaneous observations relating to geology, mineralogy and some connected topics: Am. J. Sci., v. 3, no. 2, 216-227.
            With "remarks by the editor" by B. Silliman (see 9643). On coal and its fossils from Connecticut and Rhode Island.
2056   1822, Circular, relative to the manner of collecting, labelling, and conveying the specimens of fossil organick remains and their accompanying rocks: West. Q. Reporter, v. 1, no. 3, 284-286.
2057   1822, Notice of the vegetable fossils, which traverse the layers of coal formations: Am. J. Sci., v. 4, no. 2, 266-274, plate.
            From Annales des Mines, translated by I. Doolittle.
2058   1824, [Extracts from Memoir on the lignite in review by J. W. Webster (see 10751)]: Bos. J. Phil. Arts, v. 2, no. 1, 88-96.
2059   1828, Geological question [and] Reply: Am. J. Sci., v. 15, no. 1, 96-99.
            In "Observations concerning fossil organic remains" by J. E. Doornik (see 3202).
2060   1832, Extracts from the report on the memoir of M. Dufresnoy, &c.: Am. J. Sci., v. 22, no. 1, 91-95.
            Definition of the chalk formation, and on the use of Belemnites as a Cretaceous index fossil.

Bronn, Heinrich Georg (1800-1862)
2061   1831, Exchanges of organized remains: Am. J. Sci., v. 20, no. 2, 397-398.
2062   1839, Lethaea geognostica of Prof. Bronn. Exchanges of objects of natural history: Am. J. Sci., v. 37, no. 2, 369.
2063   1847, Organic remains: Am. J. Sci., v. 3 (ns), no. 2, 311 (7 lines).
            Notice of catalogue of fossil specimens.

Bronson, Abel
2064   1802, [Use of gypsum for fertilizer]: Ct. Soc. Promoting Ag. Trans., v. 1, 8-9.

Bronson, Eli
2065   1802, [Use of gypsum for fertilizer]: Ct. Soc. Promoting Ag. Trans., v. 1, 7, 9 (9 lines).

Brooke, Henry James (1771-1857)
2066   1821, Account of three new species of lead ore found at Lead Hills [Lanark, Scotland]: Am. J. Sci., v. 4, no. 1, 28-31.
2067   1823, Cleavelandite [from Chesterfield, Massachusetts]: Bos. J. Phil. Arts, v. 1, no. 2, 187.
2068   1824, New minerals: Bos. J. Phil. Arts, v. 2, no. 1, 102-103.
            On childrenite from Devonshire, England, and somervillite from Mount Vesuvius.

Brooks, Jared
2069   1819, A list of the shocks of the earthquakes of 1811 and 1812, giving their relative degrees of intensity, etc.: Louisville, Kentucky.
            Not seen.

Broun, John Allan (1817-1879)
2070   1847, On some results of the magnetic observations made at General Sir T. M. Brisbane's observatory, Makerstown: Am. J. Sci., v. 3 (ns), no. 1, 115-116.
            From London Ath.

Brown, Andrew
        See also 3179 with M. W. Dickeson.
2071   1849, and M. W. Dickeson, The sediment of the Mississippi River: Am. Assoc. Adv. Sci. Proc., v. 1, 42-55.

Brown, Charles Brockden (1771-1810; see 10653 by C. F. Volney)

Brown, Clark
2072   1810, A topographical description of Newtown, in the state of New York: Mass. Hist. Soc. Coll., v. 9, 120-126.

Brown, E.
2073   1836, [Geology and mineralogy of Nova Scotia]: NE Farm., v. 15, 65-66.

Brown, Edward
2074   1843, A new principle, applied to the roasting and refining of copper: Franklin Inst. J., v. 5 (3s), 143-144.
            From London J. Arts and Sci's.

Brown, J. B.
2075   1831, Singular impression in marble: Am. J. Sci., v. 19, no. 2, 361, fig.
            On a strange surface marking in marble from Philadelphia.

Brown, John Newton (1803-1868)
2076   1849, Weyer's Cave. In Augusta County, Virginia: S. Lit. Mess., v. 15, 516-517.
       A poem descriptive of the cave.

Brown, Richard (F. G. S.)
2077   1846, Salt lakes and coal beds of Cape Breton [Nova Scotia]: Am. J. Sci., v. 1 (ns), no. 2, 278-279.
2078   1847, [Fossil plants from Nova Scotia]: Phila. Acad. Nat. Sci's Proc., v. 3, no. 10, 229; no. 12, 317-318.
2079   1847, [Fossil trees from Cranberry Head, Nova Scotia]: Phila. Acad. Nat. Sci's Proc., v. 3, no. 7, 143-144.

Brown, Samuel R. (1775-1817)
2080   1809, A description of a cave on Crooked Creek, with remarks and observations on nitre and gun-powder [from Kentucky]: Am. Philos. Soc. Trans., v. 4, pt. 2, 235-247; also in: Am. Min. J., v. 1, no. 2, 100-113 (1811).
2081   1817, The western gazetteer or emigrants' directory, containing a geographical description of the western states and territories: Auburn, NY, H. C. Southwick, 360 p.
2082   1818, On a curious substance which accompanies the native nitre of Kentucky and Africa: Am. J. Sci., v. 1, no. 2, 146-148.

Browne, Daniel Jay (b. 1804)
2083   1832, Geology: comprising the elements of the science in its present advanced state. Designed for the use of schools and private learners: Bos., William Hyde and Co., 108 p.
       For review see 1103.

Browne, Peter A. (1782-1860)
2084   1826, An address, intended to promote a geological and mineralogical survey of Pennsylvania, the publication of a series of geological maps, and the formation of a state and county geological and mineralogical collection: Phila., P. M. Lafourcade (printer), 8 p.
2085   1827, Some account of the hot springs, in Bath County, Virginia: Franklin Inst. J., v. 4, 134-135.
2086   183?, A profile of the rocks between Phila. and Norristown: Geol. Soc. Penn., n. p., n. d., 1 p.
2087   1830, Geological Society [of Pennsylvania]: Hazard's Reg. Penn., v. 6, 271.
2088   1832, On the geological character of the beds upon which the city of Philadelphia stands: Mon. Am. J. Geology, v. 1, no. 8, 363-367.
2089   1832, On the rocks found in the vicinity of Philadelphia: Mon. Am. J. Geology, v. 1, no. 11, 517-519.
2090   1832, Geological survey [of Massachusetts]: Hazard's Reg. Penn., v. 10, 174-175.
       Notice of E. Hitchcock's work in Massachusetts.
2091   1833, On the geological character of the beds on which Philadephia stands: Phila., n. p.
       Not seen.
2092   1833, Essays on the physical history of the globe: Phila., n. p.
       Not seen.
2093   1833, The geology of Virginia: Phila., n. p., 7 p.
2094   1834, The geology of the Alleghanies: n. p., 7 p.
2095   1834, Geology and mineralogy of Virginia: Farm's Reg., v. 1, 504-506.
2096   1834, Mineral wealth of Virginia: S. Lit. Mess., v. 1, 91-93.
2097   1834/1835, Hints to students of geology: S. Lit. Mess., v. 1, 162-163; 300-304.
2098   1841, [First geological section of Pennsylvania]: Assoc. Am. Geol's Nat's Proc., v. 2, 28 (4 lines).
       Also in Assoc. Am. Geol's Nat's Rept., v. 1, 35 (4 lines), 1843. On the first geological section of Pennsylvania produced in 1825.
2099   1844, An essay on solid meteors: Bos. Soc. Nat. History Proc., v. 1, 208-209.
2100   1848, Mineralogy: Scien. Am., v. 4, no. 2, 11.
       On mullicite from Gloucester County, New Jersey.
2101   1849, [In support of a geological survey]: Am. Assoc. Adv. Sci. Proc., v. 1, 99-100.
       Article in favor of a national geological survey.
2102   1849, Meteorites: Am. Assoc. Adv. Sci. Proc., v. 1, 80-82.
       A summary of theories on the origin of meteorites.
2103   1849, Some notice of the fossil Cephalopoda Belemnosepia, long known by the name of "belemnite," and of the diphospate of iron called "mullicite," found together at Mullica Hill [New Jersey]: Am. Assoc. Adv. Sci. Proc., v. 1, 13-16.

Brownell, Thomas Church (1779-1865)
2104   1814, Fossil coal: NY Soc. Prom. Use. Arts Trans., v. 3, 253-255.
       Call for the search for coal in New York state.

Brownson's Quarterly Review (Boston, 1844-1850)
2105   1844, [Review of An elementary treatise on mineralogy by W. Phillips (see 8325)]: v. 1, no. 3, 414.

Bruce, Archibald (1777-1818)
2106   1808, Notices concerning fluate of lime and oxide of manganese, discovered in the state of New-York: Med. Repos., v. 11, 441-442.
2107   1810-1814, The American Mineralogical Journal: Being a collection of facts and observations tending to elucidate the mineralogy and geology of the United States of America: NY, Collins and Co., vi, 270 p., index 2 p., published in four parts.
       For reviews and notices see 1079, 3428, 7089, 7092, 7103, 7801.
2108   1810, Mineralogical notice respecting American fluates of lime: Am. Min. J., v. 1, no. 1, 32-33.
2109   1810, On native magnesia from New Jersey: Am. Min. J., v. 1, no. 1, 26-30.
       See also Am. Med. and Phys. Reg., v. 1, 117.
2110   1811, Benjamin D. Perkins [obituary]: Am. Min. J., v. 1, no. 2, 126 (8 lines).
2111   1811, Cabinet of minerals at Yale College: Am. Min. J., v. 1, no. 2, 126 (7 lines).

Bruce, A. (cont.)
2112 1811, Description and chemical examination of an ore of zinc from New Jersey: Am. Min. J., v. 1, no. 2, 96-100.
       On zincite from Sparta.
2113 1811, Indicolite [from Goshen, Massachusetts]: Am. Min. J., v. 1, no. 2, 122-123.
2114 1811, Mr. Greville's cabinet: Am. Min. J., v. 1, no. 2, 122 (8 lines).
2115 1811, Sulphuret of antimony [from Louisiana]: Am. Min. J., v. 1, no. 2, 125 (10 lines).
2116 1812, Meteoric stones lately fallen in France and Russia: Am. Min. J., v. 1, no. 3, 187-189.
       Abstracted with remarks by A. Bruce from J. de Physique and Philos. Mag.
2117 1814, Description of some of the combinations of titanium occurring within the United States: Am. Min. J., v. 1, no. 4, 233-243, plate.
2118 1814, Emerald: Am. Min. J., v. 1, no. 4, 263-265.
       On American localities of emerald and beryl.
2119 1814, Kirwan: Am. Min. J., v. 1, no. 4, 270 (9 lines).
       Obituary of Richard Kirwan.
2120 1814, Mineralogical premiums: Am. Min. J., v. 1, no. 4, 269-270.
       Notice of mineralogy prizes at Yale College.
2121 1814, White pyroxene [from Manhattan, New York]: Am. Min. J., v. 1, no. 4, 266-267 (16 lines).

Bruce, James (1730-1794)
2122 1791, An interesting narrative of the travels of James Bruce, esq., into Abyssinia to discover the source of the Nile; abridged from the original work by Samuel Shaw: NY, Berry and Rogers, ix, [11]-380 p.
       Also 1798 Boston edition by Thomas and Merriam, vii, xvi, [17]-388 p.
2123 1797, An account of the marble mountains in Egypt: Univ. Mag., v. 3, 443-445.

Bruckmann, M. de
2124 1832, Advantages of bored wells in communicating heat: Am. J. Sci., v. 22, no. 2, 373-374.
       Abstracted from "Bull. de la Soc. d'Encour." On geothermal heat.

Brumby, Richard Trapier (1804-1875)
2125 1839, Mineral resources of Alabama; mineral waters, etc.: in Alabama almanac by F. A. P. Barnard, Tuscaloosa, Alabama, 65-80.
2126 1840, A brief analysis of the Blount, Shelby, and Talladega Springs; also a few observations on the coalfields, iron ores, etc. of the state of Alabama: n. p..
       Not seen.
2127 1842, Address on the importance of a geological survey of the state of Alabama: Tuscaloosa, Alabama, 18 p.
2128 1845, Letter of Professor R. T. Brumby, on the importance of a geological survey of Alabama: Tuscaloosa, Alabama, Wm. H. Fowler (printer), 25 p.
       Also 1846 edition.
2129 1846, Analysis of Bladon Springs: Tuscaloosa, Alabama, M. D. J. Slade (printer), 27 p.
2130 1847, Notes on the geology and mineralogy of Alabama. Part first: De Bow's Rev., v. 3, 316-324.
2131 1849, An address on the sphere, interest and importance of geology: Columbia, SC, A. S. Johnston, 37 p.
       For review see 9965.

Brumley (i.e. Brumby, R. T.?)
2132 1850, The Charleston artesian well: Ann. Scien. Disc., v. 1, 251-252.

Brush, George Jarvis (1831-1912)
2133 1850, On American spodumene: Am. J. Sci., v. 10 (ns), no. 3, 370-371.

Bry, H. (Judge)
2134 1834, [Letter in "Notice of fossil bones found in the Tertiary formation of the state of Louisiana" by R. Harlan (see 4798)]: Am. Philos. Soc. Trans., v. 4 (ns), 397-401.

Bryan, James W.
2135 1839, The marl beds near Newbern [North Carolina]: Farm's Reg., v. 7, 687.
2136 1840, The marl and limestone of the borders of the Neuse and Trent: Farm's Reg., v. 8, 257-259.

Bryant, Edwin (1805-1869)
2137 1848, What I saw in California; being the journal of a tour by the emigrant route and south pass of the Rocky Mountains across the continent of North America, in the years 1846, 1847 ... : NY, D. Appleton and Co.; Phila., G. S. Appleton, 455 p.
       Other printings include "third edition" in 1849. See below for second edition.
       For review see 5469.
2138 1849, What I saw in California; ... [as above]: NY, D. Appleton and Co., and Phila., G. S. Appleton, 480 p.
       This is called "second edition". Other printings of this edition include "4th edition" and "5th edition" (1849), and "6th edition" and "7th edition" (1850).

Bryant, F. (see New England Coal Mining Company)

Bryant, William Cullen (1794-1878)
2139 1827, [Review of Notes on Columbia by R. D. Bache (see 1387)]: US Rev. and Lit. Gaz., v. 1, 418-432.

Bryce, James (1806-1877)
2140 1849, On the altered dolomites of the Island of Bute: Am. J. Sci., v. 8 (ns), no. 3, 420.
       From Philos. Mag.

Bryce, J. (cont.)
2141 1850, Curious effects of trap dikes on magnesian limestone: Ann. Scien. Disc., v. 1, 235-236 (14 lines).
Abstracted from Philos. Mag.

Brydone, Patrick (1743-1818)
2142 1775, Farther extracts from Brydone's travels through Sicily and Malta: Penn. Mag., v. 1, 80-84.
Includes a description of Mount Etna.

Bryson, Alex
2143 1841, Refractive powers of minute bodies: Am. J. Sci., v. 41, no. 1, 54 (17 lines).
In British Assoc. Proc. On the use of optical refraction by powders in mineral identification.

Buch, Leopold von, freiherr (1774-1853)
2144 1850, On the geographical limits of the chalk formation: Am. J. Sci., v. 10, no. 2, 268-272.
From Q. J. Geol. Soc. London.

Buchanan, Archibald H.
2145 1836, An essay on the medical topography and diseases of Middle Tennessee: Transylvania J. Medicine and Associated Sci's, v. 9, 459-496.

Bucholtz, Christian Friedrich (1770-1818)
2146 1807, [Ores of platinum]: Am. Reg. or Gen. Repos., v. 1, pt. 2, 52.
2147 1817, and Meissner, Expériences pour déterminer la quantité de strontiane contenue dans plusiers espèces d'arragonite: J. Sci. Arts, v. 2, no. 1, 164-165.
Abstract in English

Buck Mountain Coal Company
2148 1839, An act to incorporate the Buck Montain Coal Company: Phila., J. C. Clark, 5 p.
2149 1846, Report of the Board of Directors of the Buck Mountain Coal Company: Phila., 15 p.

Buckland, William (1784-1856)
2150 1821, Instructions for conducting geological investigations, and collecting specimens: Am. J. Sci., v. 3, no. 2, 249-251.
2151 1821, Opinion of Professor Buckland of the University of Oxford, respecting certain features of American geology: Am. J. Sci., v. 4, no. 1, 186.
2152 1823, Some account of an assemblage of fossil teeth and bones of the elephant, rhinoceros, hippopotamus, bear, tiger, hyena, and sixteen other animals; discovered in a cave at Kirkdale, Yorkshire, in the year 1821: Bos. J. Phil. Arts, v. 1, no. 1, 74-86.
2153 1824, [Excerpts and abstract of Reliquiae diluvianae]: Am. J. Sci., v. 8, no. 1, 150-168; no. 2, 317-338.
2154 1825, Professor Buckland's reply to some observations in Dr. Fleming's remarks on the distribution of British animals [see 3880]: Bos. J. Phil. Arts, v. 2, no. 6, 543-557.
From Edinburgh Philos. J.
2155 1828, Recent discovery of fossil bones in the eastern part of France: Am. J. Sci., v. 14, no. 1, 203-204.
Abstracted from "Bull. Univ."
2156 1830, Antediluvian human remains [from Bize, France]: Am. J. Sci., v. 18, no. 2, 393-394.
2157 1830, Opinion of Prof. Buckland as to the Heidelberg collections of geological specimens, noticed in the last vol. of this journal: Am. J. Sci., v. 18, no. 2, 394 (8 lines).
Notice of the quality of Prof. K. C. von Leonhard's mineral collection, which was for sale.
2158 1830, Reliquiae diluvianae: Am. J. Sci., v. 18, no. 2, 393 (5 lines):
Notice of the delay of the second volume.
2159 1831, Bones in caves, &c.: Mon. Am. J. Geology, v. 1, no. 6, 278-280.
2160 1837, Geology and mineralogy considered with reference to natural theology: Phila., Carey, Lea and Blanchard, 2 v. (xv, 13-443 p; vii, 1-131 p., 69 plates).
This work is also known as Buckland's Bridgewater Treatise. For review with excerpts see 2354, 7576, and 9949. For review see 674, 683, 1090, 1685, 5986, 7845, 7998, and 10773. For excerpts see 4553. For notice see 677.
2161 1838, Fossil remains: Bos. Wk. Mag., v. 1, 59-60.
2162 1839, An account of footsteps on sandstone near Liverpool: Am. J. Sci., v. 35, no. 2, 307-308.
In British Assoc. Proc.
2163 1839, On the application of small coal to economical purposes: Am. J. Sci., v. 35, no. 2, 308 (6 lines).
In British Assoc. Proc.
2164 1840, Geology and geography: Am. J. Sci., v. 38, no. 1, 121-127.
In British Assoc. Proc. On current geological work by C. Lyell, R. I. Murchison, L. Agassiz, and H. T. De la Beche in England.
2165 1841, Geology and mineralogy considered with reference to natural theology: Phila., Lea and Blanchard, 2 v. (468 p; vii, 1-131 p., 69 plates)
"A new edition with supplementary notes."
2166 1844, On artesian wells: Living Age, v. 3, 68-72.
From Edinburgh Philos. J.
2167 1844, Origin and history of coal: Parley's Mag., v. 12, 370-371.
2168 1846, Megatherium: Naturalist, v. 1, no. 12, 534-535.
2169 1848, Geological distribution of the elements of the mineral kingdom, in relation to the uses of mankind: Am. Rr. J., v. 21, 266.
From London Mining J.
2170 1848, Petrifaction of animals: Antiquarian and Gen. Rev., v. 3, no. 11, 256-257.
From Bridgewater Treatise (see 2160).

Buckley, Samuel Botsford (1809-1884)
2171   1841, Anti-diluvial remains [from Clarke County, Alabama]: Hazard's US Reg., v. 5, 191.
2172   1841, Some account of very large fossil bones found during the present months, in Clarke County, Alabama: Mon. Chron., v. 2, 423-425.
2173   1843, Notice of the discovery of a nearly complete skeleton of the Zygodon of Owen (Basilosaurus of Harlan) in Alabama: Am. J. Sci., v. 44, no. 2, 409-412.
2174   1846, On the Zeuglodon remains of Alabama: Am. J. Sci., v. 2 (ns), n. 1, 125-131, 2 figs.

Buddle, John
2175   1839, On the Newcastle coal field: Am. J. Sci., v. 35, no. 2, 305 (5 lines).
       In British Assoc. Proc.

Buehner (Dr.)
2176   1839, Analysis of sand stone. - on the correct method of ascertaining the resistability of stone to frost: Am. Rr. J., v. 9, 88-89.

Buel, David, jr. (1784-1860)
2177   1819, Observations on the geology of the counties of Montgomery and Schenectady, in the state of New-York: Plough Boy, v. 1, no. 30, 232.
       In Troy Lyceum Proc.
2178   1820, and I. M. Wells, On the geology of the region near Lake George, and a new locality of septarium: Plough Boy, v. 2, 65.
       In Troy Lyceum Proc. See also 2179.
2179   1820, Geology: Plough Boy, v. 2, 90.
       Signed "B". Error in 2178 corrected.

Buffum, Edward Gould (1820-1867)
2180   1850, Six months in the gold mines. From a journal of three years' residence in Upper and Lower California in 1847-8-9: Phila., Lea and Blanchard, 172 p.
       For review with excerpts see 6369. For review see 4820, 5229, 5486, and 9184.

Buist, George (1805-1860)
2181   1846, Petrified forest near Cairo [Egypt]: Am. J. Sci., v. 1 (ns), no. 3, 433-434.
       From London Ath.

Bulfinch, Thomas (1796-1867)
2182   1841, Boston Society of Natural History: Am. J. Sci., v. 41, no. 1, 189-190.
       Notice of proceedings of 1841 meeting.
2183   1843, [Fossil footprints from Cheshire, England]: Bos. Soc. Nat. History Proc., v. 1, 45.

Bull, James A.
2184   1849, Concentric centrifugal gold washer: Franklin Inst. J., v. 18 (3s), 380 (5 lines).

Bull, Marcus
2185   1827, Experiments to determine the comparative value of the principal varieties of fuel used in the United States, and also in Europe; and on the ordinary apparatus used for their combustion: Phila., J. Dobson; NY, G. and C. Carvill, x, 103 p., tables, front.
2186   1828, A defence of the experiments to determine the comparative value of the principal varieties of fuel used in the United States, and also in Europe: Phila., J. Dobson; NY, G. and C. Carvill, 51 p.
2187   1828, An answer to "A short reply to" A defence of the experiments to determine the comparative value of the principal varieties of fuel: Phila., J. Dobson, 16 p.
2188   1830, Experiments to determine the comparative quantities of heat evolved in the combustion of the principal varieties of wood and coal used in the United States, for fuel; and, also, to determine the comparative quantities of heat lost by the ordinary apparatus made use of for their combustion: Am. Philos. Soc. Trans., v. 3 (ns), 1-63, tables.
2189   1833, Statements of the origin, and subsequent history of the mining operations of the North American Coal Company: Pottsville, Penn., 10 p.

Bunburry, Sir Charles James Fox (1809-1886)
2190   1846, Description of Alabama coal plants: Am. J. Sci., v. 2 (ns), no. 2, 230-233, 3 figs.
2191   1846, Fossil ferns from Frostburg, Maryland, collected by Mr. Lyell: Am. J. Sci., v. 2 (ns), no. 3, 427-428.
       Abstracted from Q. J. Geol. Soc. London.
2192   1847, Descriptions of fossil plants from the coal field near Richmond, Virginia: Am. J. Sci., v. 4 (ns), no. 1, 114-115.
       Abstracted from London Geol. Soc. Proc.

Bunker, James Madison
2193   1833, Vegetable origin of anthracite: Am. J. Sci., v. 24, no. 1, 172-173.
       With a note by B. Silliman (see 9713).

Bunsen, Robert Wilhelm Eberhard (1811-1899)
2194   1846, Parisite [from Musso, New Grenada]: Am. J. Sci., v. 2 (ns), no. 3, 415 (11 lines).
       Abstracted from "Ann. der Chem. und Pharm."

Burat, Amedee (1809-1883)
2195   1850, Discovery of metalliferous deposits: Ann. Scien. Disc., v. 1, 253 (18 lines).
       Abstracted from Annales des Mines.

Burdett, Charles (b. 1815, see 1504)

Burke, William (M.D.)
2196   1842, The mineral springs of Western Virginia, with remarks on their use and the diseases to which they are applicable: NY, Wiley and Putnam, 391 [i.e. 291] p., illus., map.
        First edition. For review see 1219 and 5426.
2197   1843, Brief notice of "The mineral springs of Western Virginia": Phila., 12 mo.
        Not seen.
2198   1846, The mineral springs of Western Virginia, with remarks on their use and the diseases to which they are applicable: NY, Wiley and Putnam, 394 p., front., folded map.
        Second edition.

Burmeister, Hermann (1807-1892)
2199   1846, Trilobites: Am. J. Sci., v. 1 (ns), no. 2, 285 (10 lines).
        From Q. J. Geology. On the affinities of different genera of trilobites.

Burnett, Waldo Irving (1828-1853)
2200   1850, [Microscopic examination of vermiculite]: Bos. Soc. Nat. History Proc., v. 3, 288 (8 lines).

Burns (Lieutenant)
2201   1838, Magnetism: Am. J. Sci., v. 34, no. 1, 37.
        In British Assoc. Proc. On world magnetic observations.

Burr, Frederick
2202   1838, South Staffordshire coal fields: Franklin Inst. J., v. 22 (ns), 375-383, fig.
        From Mining Rev.

Burrall, William Porter (1806-1874)
2203   1827, Notice of native iron from Canaan, Conn.: Am. J. Sci., v. 12, no. 1, 154-156.
        With notice by C. A. Lee (see 6124) and report on properties by C. U. Shepard (see 9473).

Burrowes, F. S.
2204   1846, Topography and geology of the survey of a district of township lines south of Lake Superior, 1845: in Reports ..., by J. Houghton and T. W. Bristol (see 5322), 1-20.
2205   1849, Topography and geology of the survey of a district of township lines south of Lake Superior, 1845: in Lake Superior report by C. T. Jackson (see 5655), 811-832.
2206   1849, Geological report of the survey, with reference to mines and minerals of a district of township lines in the state of Michigan in the year 1846, and a tabular statement of minerals collected: in Lake Superior report by C. T. Jackson (see 5655), 842-875; 933-935.

Burt, John (d. 1775)
2207   1755, Earthquakes, the effects of God's wrath. A sermon preached at Bristol, the Lord's Day after a very terrible earthquake, which was on Tuesday, November 18, 1755, a few minutes after four o'clock in the morning ... : Newport, Rhode Island, J. Franklin (printer), 17 p., 4to.
        Evans number 7374.

Burt, William Austin (1792-1858)
2208   1846, Topography and geology of the survey of a district of township lines, south of Lake Superior, 1845: US SED 357, 29-1, v. 7 (476), 2-19.
        See also 4302.
2209   1849, Geological map of a district of township lines in the northern peninsula of Michigan surveyed by Wm. A. Burt in the year 1846: Balt., W. Weber and Co., black and white geological map, 36.5 x 25 cm.
        In Lake Superior report by C. T. Jackson (see 5655).
2210   1849, Geological map of township lines in the northern peninsula of Michigan surveyed by Wm. A. Burt D. S. in the year 1845 for D. Houghton: Balt., E. Weber and Co., black and white geological map, 26 x 17.5 cm.
        In Lake Superior report by C. T. Jackson (see 5655).
2211   1849, Geological report of survey [of the Lake Superior district]: in Lake Superior report by C. T. Jackson (see 5655), 811-832; 842-875; 933-935, 7 plates, 5 maps.
2212   1849, and B. Hubbard, Reports on the linear surveys with reference to the mines and minerals in the northern peninsula of Michigan: in Lake Superior report by C. T. Jackson (see 5655), 802-935, 7 plates, 5 maps.
        Contains reports and maps by W. A. Burt (see 2209 to 2211) and B. Hubbard (see 5341).

Burton's Gentleman's Magazine (Phildelphia, 1837-1840)
2213   1839, An account of all the volcanoes now burning: v. 4, 320-326.
        Translated from "the Annuaire."
2214   1839, [Review of Popular lectures ..., by von Leonhard (see 6192)]: v. 5, 115.

Busholz (see Bucholtz)

Butler, Mann (1784-1852)
2215   1834, A history of the Commonwealth of Kentucky: Louisville, Kentucky, Wilcox, Dickerman and Co., 396 p., illus.

Byles, Mather (1706-1788)
2216   1755, Divine power and anger displayed in earthquakes. A sermon occasioned by the late earthquake, in New-England, November 18, 1755. And preached the next Lord's-Day, at Point-Shirley ... : Bos., S. Kneeland, 31 p.
        Evans number 7375.

Byrd, William (1674-1744)
2217   1841, A progress to the mines: in Westover Papers, Petersburg, Va., v. 2, 41-82.
        Original manuscript written in 1732.

Byrnes, Daniel
2218   1830, Suggestions as to a union effort to obtain a correct account of the variation of
        the magnetic needle: Am. J. Sci., v. 18, no. 1, 380-381.

C., G. F. (see 7147)

C., T. (see 2620, 8471; i.e. Thomas Cooper?).

The Cabinet of Religion, Education, Literature, Science, and Intelligence (New York, 1830-1831)
2219   1830, The Kentucky cavern: v. 4, 105-107.
2220   1830, The natural resources of Algiers: v. 4, 155-156.
2221   1831, Earthquake at Caracas, in Colombia, South America, in 1812: v. 5, 323-327.
2222   1831, Moving mountains in Calabria: v. 5, 329.
        On mud slides in the ravine of Terranuora.

Cabot, Edward C.
        See also 8236 by M. H. Perley.
2223   1848, [Glacial scratches in Brookline, Massachusetts]: Bos. Soc. Nat. History Proc., v.
        3, 28 (3 lines).
2224   1849, [Glacial origin of Californian gold]: Bos. Soc. Nat. History Proc., v. 3, 95 (8
        lines).
2225   1850, Formation of fresh-water ponds on the coast: Ann. Scien. Disc., v. 1, 249 (19
        lines).
2226   1850, Occurrence of fresh water on beaches and sandy islands: Ann. Scien. Disc., v. 1,
        247-248.
2227   1850, [Ripple marks on a rock from Brookline, Massachusetts]: Bos. Soc. Nat. History
        Proc., v. 3, 208.

Cabot, Samuel
2228   1849, [Migration of the mastodon]: Bos. Soc. Nat. History Proc., v. 3, 104 (8 lines).

Caldwell, Joseph (1773-1835; see A tour through part of Virginia, 10298, ascribed by some
    to J. Caldwell)

Caldwell, T. (see A tour through part of Virginia, 10298, ascribed by some to T. Caldwell)

Callcott (Mrs.; see Graham, Marie)

Campbell, Charles
2229   1834, Marl on Nottoway River: Farm's Reg., v. 1, 701.

Campbell, H. U. (see 8505 with J. Porter)

Campbell, J. B.
2230   1845, [Lake Superior region observations]: US SED 175, 28-2, v. 11 (461), 4-8.

Campbell's Foreign Semi-Monthly Magazine; or, Select Miscellany of the Periodical
    Literature of Great Britain (Philadelphia, 1842-1844)
2231   1842, First report of the children's employment commissioners: mines and collieries: v.
        1, 159-184, 5 figs.
        On the use of child labor, and their working conditions, in English mines.
2232   1842, Lancashire coal mines: v. 1, 396 (7 lines).
2233   1842, Aden: native quicksilver: v. 1, 427 (16 lines).
2234   1843, Russian gold: v. 2, 346-347.
2235   1844, Artesian well on volcanic soil: v. 6, 141.
2236   1844, Sandstone pillars and caves of North-western Australia: v. 6, 142.
2237   1844, Microscope in geological research: v. 6, 428.

Candolle, Augustin-Pyramus de (1778-1841)
2238   1833, Memoir of G. Cuvier: Am. J. Sci., v. 23, no. 2, 303-308.
        Abstracted by J. Griscom from Bibliothèque Universelle.

Cantu, M.
2239   1826, Iodine in mineral waters [of Castel Nova d'Asti, Turin]: Bos. J. Philos. Arts, v.
        3, no. 3, 298 (8 lines).
        Abstracted from "Mem. de Torino."

Carey, Henry Charles (1793-1874)
2240   1850, The harmony of interests: agricultural, manufacturing, and commercial: Plough
        Loom Anvil, v. 2, no. 7, 389-419, 44 graphs.
        On United States production of many items including lead, iron and coal.

Carlisle, Sir Anthony (1768-1840)
2241   1839, A series of facts and observations respecting the natural causes of arborescent
        or dendritic figures in two divisions of animal and vegetable substances, and in
        mineral formations: Franklin Inst. J., v. 23 (ns), 415-416.
2242   1840, Geological speculations. - Origin of metals: Am. Rep. Arts Sci's Manu's, v. 2,
        121-122.

Carmichael, William
2243   1836, Remarks on Professor Ducatel's geological survey of the eastern shore of
        Maryland: Farm's Reg., v. 4, 300.

Carmichael, W. (cont.)
2244   1841, Remarks on the geological survey of the eastern shore of Maryland: Farm's Reg., v. 9, 347.
        Signed by "Agricola."

Carnegie, Captain
2245   1844, An account of the earthquake at the Islands of Antigua and Guadaloupe on the 8th of February, 1843: Am. J. Sci., v. 47, no. 1, 182-183.
        In British Assoc. Proc.

The Carolina Journal of Medicine, Science and Agriculture (Charleston, 1825)
2246   1825, Erlanite, a new mineral [from Saxony]: v. 1, 200-201.

Carpenter, Caleb
2247   1847, The Mississippi River in the olden time: De Bow's Rev., v. 3, 115-123.

Carpenter, F. W.
2248   1838, Interesting fossils found in Louisiana: Am. J. Sci., v. 34, no. 1, 201-203, 3 figs.
        Describes fossil teeth found in West Feliciana, Louisiana.

Carpenter, George Washington (1802-1860)
2249   1824-1825, By George W. Carpenter: Am. J. Sci., v. 8, no. 2, 236; v. 9, no. 1, 45.
        On mineral localities near Philadelphia.
2250   1825, On the mineralogy of Chester County; with an account of some of the minerals of Delaware, Maryland, and other localities: Phila., 19 p.
2251   1825-1826, and G. Spackman, Miscellaneous localities of minerals [in Delaware, Pennsylvania, and New Jersey]. By Messrs. Carpenter and Spackman: Am. J. Sci., v. 9, no. 2, 245-246; v. 10, no. 2, 218-224.
2252   1826, and G. Spackman, Miscellaneous localities of minerals [in Delaware and Pennsylvania]: Am. J. Sci., v. 10, no. 2, 218-224.
2253   1827, Exchange of minerals: Am. J. Sci., v. 13, no. 1, 199 (4 lines).
2254   1828, Mineralogical notes [on Pennsylvania]: Hazard's Reg. Penn., v. 2, 84-88.
2255   1828, On mineralogy of Chester County; with an account of some of the minerals of Delaware, Maryland, and other localities: Phila., by the author, 16 p.; also in: Am. J. Sci., v. 14, no. 1, 1-14.
2256   1828, On the muriate of soda, or common salt, with an account of the salt springs in the United States: Am. J. Sci., v. 15, no. 1, 1-6; also reprint, 8 p. (1829).
2257   1831, On the mineralogy of Chester County; with an account of some of the minerals of Delaware, Maryland, and other localities: Phila., G. W. Carpenter, 19 p.

Carpenter, William (1797-1874)
2258   1833, Scripture natural history; containing a descriptive account of the quadrupeds, birds, fishes, insects, reptiles, serpents, plants, trees, minerals, gems, and precious stones, mentioned in the Bible: Bos., Lincoln, Edmands and Co., 408 p.
        "First American, from the latest London edition with improvements, by Rev. Gorham D. Abbott." For review see 1104.
2259   1838, A guide to the reading and study of the Bible; being a comprehensive digest of the principles and details of Biblical criticism, interpretation, theology, history, natural science, usages, etc. ... : Brattleboro, Vt., Brattleboro Typographic Company, 198 p., illus.
        Also another printing: Phila., Lippincott and Co., 1846, 198 p.

Carpenter, William Benjamin (1813-1885)
2260   1844, On the application of the microscope to geological research: Eclec. Mag., v. 2, 428-429.
        From London Ath.
2261   1850, On the structure of Nummulina: Am. J. Sci., v. 10 (ns), no. 2, 275.
        Abstracted from Q. J. Geol. Soc. London.

Carpenter, William M.
2262   1839, Account of the bituminization of wood in the human era [at Port Hudson, Louisiana]: Am. J. Sci., v. 36, no. 1, 118-124, 2 figs.
2263   1839, Miscellaneous notices in Opelousas, Attakapas, etc.: Am. J. Sci., v. 35, no. 2, 344-346.
        On the formation of prairies in the American Midwest.
2264   1842, Geological survey of Louisiana: Am. J. Sci., v. 42, no. 2, 390-391, fig.
        On the discovery of a fossil tooth. For review see 9959.
2265   1846, Remarks on some fossil bones recently brought to New Orleans from Tennessee and from Texas: Am. J. Sci., v. 1 (ns), no. 2, 244-250, 4 figs.

Carpentier, M. de (i.e. Charpentier?)
2266   1833, Thermal spring in the bed of the Rhone [in Switzerland]: Am. J. Sci., v. 24, no. 1, 201-203.
        Abstracted from Bibliothèque Universelle by J. Griscom.

Carpmael, William (1804-1867; see 8316 with R. Phillips)

Carr, Ezra Slocum (1819-1894)
        See also 10587 with L. Vanuxem.
2267   1840, Appendix to the fourth annual report of the geological survey of the third district: NY Geol. Survey Ann. Rept., v. 4, 385-388.

Carter, S. (of Barren Hill, Nottoway)
2268   1836, Inquiry as to green sand: Farm's Reg., v. 4, 473-474.

Carver, Jonathan (1732-1780)
2269   1784, Three years of travels, through the interior parts of North America, for more than five thousand miles, containing, an account of the Great Lakes, and all the lakes, islands, and rivers, cataracts, mountains, minerals, soils, and vegetable productions of the north west regions of that vast continent: Phila., Joseph Crukshank, xxi, 23-217 p.
         Evans number 18391.
2270   1849, The copper and gold regions: Scien. Am., v. 4, no. 24, 186.
         Reprinted from Travels ... (see 2269).

Carver, Samuel D. (Dr.)
2271   1825, Aerolite: Bos. J. Phil. Arts, v. 2, no. 6, 604 (9 lines).
         Abstracted from Am. J. Sci.
2272   1825, Notice of a meteoric stone which fell at Nanjemoy, Maryland, February 10th, 1825: Am. J. Sci., v. 9, no. 2, 351-353.
         With "Statement of W. D. Harrison, Esq." (see 4838).

Casey, William R.
2273   1843, Advantages of Ralston, Lycoming Co., Penn. for the manufacture of iron; with a description of its mines of bituminous coal and peculiar iron ore: NY, 18 p., folded map.
2274   1845, Report: in Reports on the Bear Mountain Railroad by Edwin F. Johnson and William R. Casey (see 5753), 9-23, plate.

The Casket. Devoted to Literature, Science, the Arts, News, &c. (Cincinnati, 1846)
2275   1846, Greece. The late earthquakes: v. 1, no. 19, 152.
2276   1846, Earthquake at Boston: v. 1, no. 20, 168.

The Casket, or Flowers of Literature (see Graham's American Monthly Magazine)

Cass, Lewis (1782-1866, Governor of Michigan)
2277   1825, Letter ... on the advantage of purchasing the country upon Lake Superior where copper has been found: US SED 19, 18-2, v. 2, (no. 109), 3-4.
2278   1833, [Letter describing G. W. Featherstonhaugh]: US SED 35, 22-2, v. 1 (no. 230), 1-4.
         In report by G. W. Featherstonhaugh (see 3773).
2279   1849, Rocks of Lake Superior: Lit. Un., v. 2, 87.

Cassan, J. (fl. 1789)
2280   1800, Description of the volcano in the Island of St. Lucia: Mon. Mag. and Am. Rev., v. 2, 73-75.

Casselberry, Issac (1821-1873)
2281   1845, A description of certain fossil bones found near Evansville, Ia.: Evansville, Indiana, Printed at the Courier Office, 8 p.
         Also another printing, 1847.

Castales, M.
2282   1842, Coal mines in Cuba: Am. J. Sci., v. 42, no. 2, 388-390.
         In work by J. H. Blake (see 1729).

Catlin, George (1796-1872)
2283   1840, Account of a journey to the Côteau des Prairies, with a description of the red pipe stone quarry and granite boulders found there: Am. J. Sci., v. 38, no. 1, 138-146; also in: The Friend, v. 13, 129-131.

Cautley, Sir Proby Thomas (1802-1871)
2284   1837, [New fossil animal from the sub-Himalayas]: Am. J. Sci., v. 33, no. 1, 211 (13 lines).
         From London Geol. Soc. Proc.

Cavendish, Henry (1731-1810)
2285   1799, Experiments to determine the density of the earth: Med. Repos., v. 2, 448-450.

Celis, Don Michael Rubin de
2286   1790, An account of a mass of native iron found in South America: Colum. Mag., v. 4, 111-112.

The Central New York Farmer (Rome, New York, 1842-1844)
2287   1842, Bituminous coal [from Augusta, New York]: v. 1, 120.
2288   1843, Acknowledgments [review of Lectures on the application of chemistry and geology to agriculture, part III by J. F. W. Johnston (see 5819)]: v. 2, 152.

Chalmers, Lionel (1715?-1777)
2289   1788, A view of the climate, water, and soil in South Carolina ... : Am. Mus. or Univ. Mag., v. 3, 316-334.

Chaloner, A. D.
2290   1843, Verbal communication [on some fossil bones from Missouri]: Phila. Acad. Nat. Sci's Proc., v. 1, no. 33, 321-322.

Chambers, Robert (1802-1871)
2291   1845, Explanations: a sequel to "Vestiges of the natural history of Creation." By the author of that work: Phila., Carey, Hart, 142 p.
         This work was unsigned. For review with excerpts see 4503. For review see 836, 5452, and 7150. Also 1848 printing in NY.
2292   1845, Vestiges of the natural history of Creation: NY, Wiley and Putnam, vi, 7-291 p.
         This work was unsigned. For review with excerpts see 1217, 1680, 1901, 1902, 3410, 6381, 6386, 10049, 10823. For review see 188, 810, 3538, 3569, 5441, 6218, 7150, 9875, 10100. For notice see 802, 6356, 6391, 7767, 9943, 10825.

Chambers, R. (cont.)
2293   1845, Vestiges of the natural history of Creation: NY, Wiley and Putnam, xxviii, 280 p.
       "2nd ed. from 3rd London, greatly amended by the author, and an introduction by
       the Rev. George Cheever" (see 2332). Unsigned.
2294   1845, Vestiges of the natural history of Creation: NY, Wiley and Putnam, vi, 353 p.
       Unsigned. "3rd edition".
2295   1846, Explanations: A sequel to "Vestiges of the natural history of Creation". By the
       author of that work: NY, Wiley and Putnam, vii, 142 p.
       Second edition. Unsigned.
2296   1846, Vestiges of the natural history of Creation: NY, W. H. Colyer, 303 p.
       "2nd edition with a sequel." Unsigned.
2297   1846, Vestiges of the natural history of Creation: NY, W. H. Colyer, 303 p.
       "4th edition." Unsigned. Also 1847 and 1848 printings.
2298   1846, Vestiges of the natural history of Creation: NY, Wiley and Putnam, 353, vii,
       142 p.
       "4th edition with a sequel." Unsigned. Also 1847 and 1848 printings.
2299   1847, Ancient sea margins: Am. J. Sci., v. 4 (ns), no. 3, 323-325.
       See also 2301. For review see 1902, 2840, 6401, 6405.
2300   1848, [Request for information on river terraces]: Lit. World, v. 2, 228.
       Abstracted from Assoc. Am. Geol's Nat's Proc.
2301   1849, On ancient sea margins: Am. J. Sci., v. 8 (ns), no. 1, 33-35.
2302   1850, The glacial theory: Eclec. Mag., v. 21, 485-489.

Chambers, Thomas
2303   1845, The iron trade: Fisher's Nat'l Mag., v. 1, no. 2, 135-170.

Champion, George (Reverend)
2304   1836, Remarks on the topography, scenery, geology, &c., of the vicinity of the Cape of
       Good Hope: Am. J. Sci., v. 29, no. 2, 230-236.

Chandler, J.
2305   1848, Science - its advantages and pleasures: New Orleans Miscellany, v. 1, no. 3,
       115-118.
       Includes notes on the origin of the earth.

Chandler, R. W.
2306   1829, Map of the United States' lead mines on the Upper Mississippi River. Drawn &
       published by R. W. Chandler of Galena. Cin., Ebr. Martin engraver, 37.7 x 44.4 cm.
       For review see 506.

Channing, Walter (1786-1876)
2307   1817, [Review of An elementary treatise on mineralogy and geology by P. Cleaveland (see
       2420)]: N. Am. Rev., v. 5, 409-429.
       Includes many excerpts, and opinions of R. Jameson, R. Hauy, and A. Werner, on
       the use of crystallography in mineral classification.

Channing, William Francis (b. 1820)
2308   1847, [Report of explorations on the St. Mary's River in Upper Michigan]: in Lake
       Superior report by C. T. Jackson (see 5640), 199-208.
2309   1847, Synopsis of a survey of St. Mary's River: in Lake Superior report by C. T.
       Jackson (see 5640), 209.

Chapin, Alonzo Bowen (1808-1858)
2310   1834, Junction of trap and sandstone; Wallingford, Conn.: Am. J. Sci., v. 27, no. 1,
       104-112, 8 figs.
2311   1836, Ornithichnology [review of "Description of the footmarks of birds ... " by E.
       Hitchcock (see 5115 and 5116)]: Knickerbocker, v. 7, 578-582.
2312   1836, Ornithichnology reconsidered: Knickerbocker, v. 8, 456-458.

Chapman, Henry (Reverend)
2313   1840, [Description of New England earthquakes, 1791 to 1813]: Am. Rr. J., v. 11,
       248-253.
       In "Earthquakes in Connecticut", see 1162.

Chapman, Isaac
2314   1814, An account of some experiments with sulphuret of barytes as a manure: Phila. Soc.
       Promoting Ag. Mem., v. 3, 120-125.
2315   1826, On the sulphuret of lime as a manure: Phila. Soc. Promoting Ag. Mem., v. 5,
       80-83.

Chapman, W.
2316   1817, [Formation of the coal districts in England]: J. Sci. Arts, v. 2, no. 1, 205 (7
       lines).

Chaptal de Chanteloup, Jean Antoine Claude (1756-1832)
2317   1796, Elements of chemistry: Phila., Lang and Ustrich for M. Carey, 3 v. in 1.
       Evans number 30183. Includes instructions for chemical analysis.

The Charleston Medical Journal and Review (Charleston, 1846-1850)
2318   1847, The artesian well of Charleston: v. 2, 486.
2319   1848, [Review of "Memoir on the fossil genus Basilosaurus" by R. W. Gibbes (see 4350
       and 4351)]: v. 3, 95 (12 lines).
2320   1848, [Review of "Monograph on the fossil squalidae ... " by R. W. Gibbes (see 4352)]:
       v. 3, 700-701 (9 lines).

Charlesworth, Edward
2321   1837, [Age of Tertiary deposits]: Am. J. Sci., v. 31, no. 2, 380 (4 lines).
       In British Assoc. Proc.

Charlesworth, E. (cont.)
2322   1837, [On fossil bones from Norfolk and Suffolk, England]: Am. J. Sci., v. 31, no. 2, 339-341.
       In British Assoc. Proc.

Charpentier, Jean de (1786-1855)
2323   1829, Couzeranite: Am. J. Sci., v. 17, no. 1, 183.
       Abstracted from Annales des Mines by J. Griscom.

Chase, George
2324   1819, Scintillating limestone: Am. J. Sci., v. 1, no. 3, 241.
       "Extracted from a letter", on limestone from Randolph, Vermont.
2325   1821, Notice of a dolomite, and description of a soft green rock: Am. J. Sci., v. 3, no. 2, 246.
       On dolomite and "chlorite slate" from Vermont.

Chase, George J. (Professor at Brown University)
2326   1850, Fossil bones of the large birds of New Zealand: Ann. Scien. Disc., v. 1, 278-280.
2327   1850, On some bones on [sic] the Dinornis Novae Zealandiae: Am. Assoc. Adv. Sci. Proc., v. 2, 267-271.
       Discussion by L. Agassiz and others.

Chase, Stephen (1813-1851)
2328   1847, Consistency of scientific and religious truth: Bib. Repos., v. 3 (3s), 656-688.

Chateaubriand, M. de
2329   1807, An excursion to Vesuvius, 1804: Lit. Mag. and Am. Reg., v. 8, 301-305.

Chauncy, Charles (1705-1787)
2330   1755, Earthquakes a token of the righteous anger of God. A sermon preached at the old-brick-meeting-house in Boston, the Lord's-day after the terrible earthquake, which suddenly awoke us out of our sleep in the morning of the 18th of November, 1755 ... : Bos., Edes and Gill, 32 p.
       Evans number 7380.
2331   1756, The earth delivered from the curse to which it is, at present, subjected. A sermon occasioned by the late earthquakes in Spain and Portugal, as well as New-England; ... : Bos., Edes and Gill, 28 p.
       Evans number 7634.

Cheever, George Barrell (1807-1890)
2332   1845, Introduction: in Vestiges of the natural history of Creation by R. Chambers (see 2293), vii-xviii.

Chester County Cabinet of Natural Science (West-Chester, Pennsylvania)
2333   1826-1849, [Notices of officers, members, and additions to the geological collections]: Chester Co. Cab. Nat. Sci. Papers, in all 14 numbers.
2334   18??, Constitution of the Chester County Cabinet of Natural Sciencee: West-Chester, Pa., C. Hannum and J. A. Hemphill (printers), 11 p.

Chevallier, Jean-Baptiste-Alphonse (1793-1879)
2335   1831, Ammonia in native oxide of iron: Am. J. Sci., v. 21, no. 1, 155.
       Abstracted from Royal Inst. J. by J. Griscom. On the chemical analysis of iron ore from Cumba Mine, Marmalo.

Child, J.
2336   1849, [Section of Copper Falls Mine, Lake Superior District, Michigan]: folded plate in Lake Superior report by C. T. Jackson (see 5655).

Child, Maria (i.e. Lydia Maria Francis Child?)
2337   1843, Mammoth cave: Anglo-American, v. 2, 7-9.

Children, John George (1777-1852)
2338   1826, Mr. Children's summary view of the atomic theory according to the hypothesis adopted by M. Berzelius: Bos. J. Phil. Arts, v. 3, no. 3, 260-371 [i.e. 271]; no. 4, 350-363.
       Abstracted from Annals of Philos.

Childs, Cephas Greer (1793-1871)
2339   1847, The coal and iron trade, embracing statistics of Pennsylvania. A series of articles published in the Philadelphia Commercial List, in 1847: Phila., C. G. Childs, 24 p.
       For review see 5465. Includes list of Pennsylvania mines, and coal production statistics.
2340   1847, The iron trade of Europe and the United States: with special reference to the iron trade of Pennsylvania: Hunt's Merch. Mag., v. 16, 574-593.
2341   1849, The coal trade of the United States: Hunt's Merch. Mag., v. 21, 266-279.

Chilton
2342   1847, Analysis of coal: Am. Inst. NY Ann. Rept., v. 5, 498.

Chilton, George
2343   1810, Chemical examination of heavy spar from New Jersey: Am. Min. J., v. 1, no. 1, 16-19.
       Chemical analysis of barite from Newton, Sussex County, NJ
2344   1810, Manufacture of chemical tests and reagents: Am. Min. J., v. 1, no. 1, 56 (10 lines).
2345   1812, On the deoxidation of potash: Am. Min. J., v. 1, no. 3, 153-156, plate with 2 figs.
2346   1824, Results of the analysis of the principal brine springs of the state of New-Nork [i.e. New-York]: Am. J. Sci., v. 7, no. 2, 344-347.

Chilton, G. (cont.)
2347   1825, Analysis of the Maryland aerolite: Am. J. Sci., v. 10, no. 1, 131-137.
       See also Am. J. Sci., v. 9, no. 2, 351.
2348   1826, Appendix to the above letter [follows "Notice of rocks and minerals in Westfield, Mass." by E. Davis (see 2950)]: Am. J. Sci., v. 10, no. 2, 215-217.
2349   1830, Analysis of the Clinton [New York] mineral water: Am. J. Sci., v. 18, no. 2, 346-349.

Chilton, James R.
2350   1839, Staten Island [New York] granite: Am. Inst. NY J., v. 4, 312.
2351   1850, Report [on the Eagle Mine]: in Enterprise Mining Company Reports (see 3594).

Chipman, Nathaniel
2352   1828, On moving stones in lakes, ponds, &c.: Am. J. Sci., v. 14, no. 2, 303-305.

Chipman, Samuel
2353   1813, An account of some singular springs in the town of Honeoye, Ontario County, New-York: Lit. and Philos. Rep., v. 1, 310-312.

The Christian Review (Boston, 1836-1850)
2354   1837, [Review of Geology considered with respect to natural theology by W. Buckland (see 2160)]: v. 2, no. 4, 552-579.
2355   1840, [Review of The wonders of geology by G. A. Mantell (see 6732)]: v. 5, no. 1, 143-144.
2356   1843, [Review with excerpts of Geological cosmogony by E. Lord (see 6500)]: v. 8, no. 4, 627-628.
2357   1844, [Review of Elementary treatise on mineralogy by W. Phillip, edited by F. Alger (see 8325)]: v. 9, no. 3, 470-471.
2358   1850, Geology and Revelation: v. 15, no. 3, 380-399.
       Signed by "H. P."
2359   1850, [Review of Lake Superior by L. Agassiz (see 83)]: v. 15, no. 3, 474-476.
2360   1850, The territories on the Pacific: v. 15, no. 4, 573-595.
       On the California gold region.

The Christian's, Scholar's, and Farmer's Magazine (Elizabethtown, New Jersey, 1789-1791)
2361   1790, Account of a burning river [at Tremoulac, Florida]: v. 1, 628-629.
2362   1791, Account of a burning island that arose out of the sea: v. 2, 703-704.

Christy, David (b. 1802)
2363   1847, Letter on geology: erratic rocks of North America. To M. de Verneuil: Oxford, Ohio, 11 p.
2364   1847, Some views relating to North American geology, communicated in a letter from David Christy, Oxford, Ohio, to M. de Verneuil ... : Oxford, Ohio, 12 p.
       Not seen.
2365   1848, Letters on geology: being a series of communications originally addressed to Dr. John Locke, of Cincinnati, giving an outline of the geology of the West and Southwest, together with an essay on the erratic rocks of North America, addressed to M. de Verneuil, illustrated by geological sections and engravings of some rare fossils: Oxford, Ohio, D. Christy; Rossville, Ohio, J. M. Christy, 68, [2], 11 p., 5 plates.
       Not seen. For review, see 948.

Chubbuk, Bissel (see 10794 with others)

Church, William
2366   1823, An account of the Frankfort mineral springs, &c. &c. [in Pennsylvania?]: Phila. J. Med. Phys. Sci's, v. 6, 50-59.
2367   1824, Observations and experiments on the water of the Summerville mineral spring [in St. Clair, Pennsylvania]: Phila. J. Med. Phys. Sci's, v. 9, 286-294.
2368   1825?, An analysis of the waters of the Bedford mineral springs, &c... : Bedford, Penn., N. T. Chapman, jr., 8 p.
       Also in Phila. J. Med. Phys. Sci's, v. 11, 218-229.
2369   1830, P. S. Bedford mineral springs: Am. J. Sci., v. 19, no. 1, 204.

Churchman, John (1753-1805)
2370   1790, An explanation of the magnetic atlas, or variation chart, hereunto annexed: Phila., James and Johnson, x, 11-46 p., chart
       For review see 10514.
2371   1800, The magnetic atlas; or, variation charts of the whole terraqueous globe ... : NY, Gaine and Ten Eyck, viii, 82 p., plates.
       Third edition. Evans number 37183. For review see 10514. For notice see 7021.

The Cincinnati Literary Gazette (Cincinnati, 1824-1825)
       Uncaptioned weekly "Literary and Scientific Notices" not included.
2372   1824, Finch's geology [Notice of Introduction to the study of geology by J. Finch (see 3833)]: v. 1, 7 (5 lines).
2373   1824, [Mammoth from New Jersey]: v. 1, 191.
2374   1824, North Carolina: v. 1, 207.
       On value of gold mine.
2375   1824, Virgin silver [at Sault St. Marie]: v. 2, 159.
2376   1824, [Review of Ancient history ..., by C. S. Rafinesque (see 8635)]: v. 2, 202-204.
2377   1825, On the mines of Mexico: v. 3, 117-118.
2378   1825, Mammoth Cave: v. 3, 195-196.
2379   1825, White Cave: v. 3, 196.
2380   1825, Yellow Springs [in Green County, Ohio]: v. 4, 243.
2381   1825, Organic remains: v. 4, 268-269.
       Notice of fossil mastodon bones found at Plaque Mine.

The Cincinnati Mirror (Cincinnati, 1831-1836)
2382   1835, [Review of Virginia springs by P. H. Nicklin (see 7865)]: v. 4, 378.
2383   1835, Caves in Cuba: v. 4, 398.

Cincinnati Mirror (cont.)
2384   1836, [Brief notes on volcanoes and earthquakes]: v. 5, 8, 32, 64.
2385   1836, Bituminous coal [from Mercer County]: v. 5, 75.
2386   1836, Genesis and geology: v. 5, 236-237.

The Cincinnati Miscellany, or Antiquities of the West (Cincinnati, 1845-1846)
2387   1845, The Mammoth Cave: v. 2, no. 1, 34-35; 44-47.

Cist, Charles (1792-1868)
2388   1844, Fuel.--Coal: Cin. Miscellany, v. 1, no. 1, 31 (20 lines).
2389   1845, The anthracite coal of Pennsylvania: Cin. Miscellany, v. 1, no. 4, 170-171.
2390   1845, Dayton limestone: Cin. Miscellany, v. 2, no. 3, 109.
2391   1846, Coal: Cin. Miscellany, v. 2, no. 7, 257-258.
        On the use and supply of coal in Cincinnati.
2392   1846, Coal mines of Pennsylvania: Cin. Miscellany, v. 2, no. 7, 288.
        Abstracted from Miner's J., of Pottsville, Pennsylvania.
2393   1846, The iron manufacture of the United States: Cin. Miscellany, v. 2, no. 7, 287-288.

Cist, Jacob (1782-1825)
2394   1814, Gypsum [from western New York]: Am. Wk. Mess., v. 1, no. 23, 357-359.
2395   1814, Account of the beds of gypsum found in the western part of the state of New York: Phila. Soc. Promoting Ag. Mem., v. 3, 138-141.

Cist, Zachariah
2396   1815?, [On coal mining in Pennsylvania]: pamphlet.
        For notice see 2397. Not seen.
2397   1821, Account of the mines of anthracite in the region about Wilkesbarre, Pennsylvania: Am. J. Sci., v. 4, no. 1, 1-16, plate, map.
        With letters by "J. C.", "White & Hazard", and others on the economic uses of coal.
2398   1821, New locality of manganese [from Wilkesbarre, Pennsylvania]: Am. J. Sci., v. 4, no. 1, 38-39.
2399   1825, Notice of impressions of plants accompanying the anthracite of Wilkesbarre: Am. J. Sci., v. 9, no. 1, 165-166.

Civil Engineer and Herald of Internal Improvement (Columbus, Ohio, 1828)
2400   1828, Geauga County: v. 1, no. 11, 162.
        On the iron production of Geauga County, Ohio.

Clapp, A.
2401   1832, Greenstone dyke [in Montgomery, Vermont]: Am. J. Sci., v. 22, no. 1, 189 (5 lines).
2402   1841, Written communications: Phila. Acad. Nat. Sci's Proc., v. 1, no. 2, 18-19.
        On the geological equivalence of New Albany, Indiana formations with those described by R. I. Murchison in his Silurian System.
2403   1842, Geological equivalents of the rocks of the falls of the Ohio and other strata in the western states: Phila. Acad. Nat. Sci's Proc., v. 1, no. 14, 177-178.

Clark, Joseph
2404   1849, Indiana oolite: West. Q. Rev., v. 1, 184-185.

Clark, Thomas
2405   1836, On the application of the hot blast, in the manufacture of cast iron: Am. J. Sci., v. 31, no. 1, 180-183.
        Abstracted from "Rec. of Gen. Sci."

Clark, William (see 6212)

Clarke (Reverend)
2406   1838, Impressions in sandstone: Am. J. Sci., v. 33, no. 2, 271 (8 lines).
        In British Assoc. Proc. On fossil footprints from the Connecticut River Valley, Massachusetts.

Clarke, C. C. (Reverend; pseudonym for Sir Richard Phillips. See 8312)

Clarke, Daniel (Dr. of Edinburgh)
2407   1820, On volcanoes: Saturday Mag., v. 3, 220-222.
        Extract from a work published in 1793.
2408   1827, Vesuvius and its environment: N. Am. or Wk. J., v. 1, 91.

Clarke, Edward Daniel (1769-1822)
2409   1811, Blue iron earth [from Turkey]: Am. Min. J., v. 1, no. 2, 123 (8 lines).
        On vivianite.
2410   1817, Account of some experiments made with Newman's blow-pipe, by inflaming a highly condensed mixture of the gaseous constituents of water: J. Sci. Arts, v. 2, no. 1, 104-123, fig.; also in: Eclec. Rep. and Analytic Rev., v. 7, 137-156, fig.
2411   1820, Gehlenite, needle-stone, and datolite: Plough Boy, v. 2, 8 (10 lines).
2412   1825, Dr. Clarke's descent into the mines of Persberg: Am. Mech's Mag., v. 1, no. 22, 349-350.

Clary, Lyman (see 9842 with others)

Claubry, Henri François Gaultier de (1792-1878)
        See also 3016 and 3017 with M. Dechaud.
2413   1835, Origin of nitre: Am. J. Sci., v. 28, no. 2, 292-293.
        In C. U. Shepard's review of Boué's notice of the Transactions of the Geological Society of France (see 9497).

Claus, Carl Ernst (1796-1864)
2414   1846, New metals: Am. J. Sci., v. 1 (ns), no. 1, 103-104.
        Abstracted from Comptes Rendus. On the discovery of ruthenium.

Claussen, M.
2415   1846, Mexican fossils: Am. J. Sci., v. 1 (ns), no. 1, 122-123.
        Abstracted from "Ann. Mag. Nat. Hist." On South American fossil bones.

Clay (Mr.)
2416   1843, Making malleable iron direct from the ore: Franklin Inst. J., v. 5 (3s), 287-288.
        Abstracted from London Ath.

Clay, Joseph
2417   1802, Observations on the figure of the earth: Am. Philos. Soc. Trans., v. 5, 312-319.

Clay, Joseph A.
2418   1841, Verbal communications: Phila. Acad. Nat. Sci's Proc., v. 1, no. 3, 39 (11 lines).
        On the magnesian minerals agalmatolite, kerolite, picrosmine, picrolite, and metaxite, from Europe.

Cleaveland, Parker (1780-1858)
2419   1809, Account of fossil shells, with the author's reasons for attending to the same: Am. Acad. Arts Sci's Mem., v. 3, pt. 1, 155-158.
        On fossil shells from the Pleistocene deposits of Maine.
2420   1816, An elementary treatise on mineralogy and geology being an introduction to the study of those sciences, and designed for the use of pupils, - for persons, attending lectures on these subjects, - and as a companion for travellers in the United States of America: Bos., Cummings and Hilliard, xvi, 668 p., 6 plates, plate 6 is a colored geological map of the United States (see 2421).
        For review with excerpts see 1231, 2307, 3437, 7680, and 8542. For review see 332, 1091, 1240, 9624. For notice see 3434, 7934, and 7938.
2421   1816, The United States of America published by Cummings & Hilliard, No. 1 Cornhill, Boston: Colored geological map, 27.7 x 21.5 cm., published in Elementary treatise by P. Cleaveland (see 2420).
        This map is based on the field work of W. Maclure (see 6614).
2422   1822, An elementary treatise on mineralogy and geology ... [as in first edition]: Bos., Cummings and Hilliard, xii, 818 p., 6 plates, plate 6 is a colored geological map (see 2423).
        "Second edition, - in two volumes." Contains article by B. Silliman (see 9609). For review of second edition see 1594 and 1815. For notice see 356 and 392.
2423   1822, The United States, published by Cummings and Hilliard, No. 1 Cornhill, Boston: Colored geological map 42.7 x 27.9 cm., published in Elementary treatise (second edition) by P. Cleaveland (see 2422).
        This map is based on the field work of W. Maclure (see 6619).
2424   1823, Notice of the late meteor in Maine: Am. J. Sci., v. 7, no. 1, 170-171.
2425   1826, Vocabulary, containing an explanation of certain chemical terms; more particularly those, which relate to the chemical nomenclature: Brunswick, Me., J. Griffen (printer), 27 p.
2426   1829, An elementary treatise on mineralogy and geology: Third edition. Not seen and probably never published.
        For notice see 461 and 479.

Clemandat (see 4120 with E. Fremy)

Clemens, James W.
2427   1833, Notices of Wheeling, Virginia: Am. J. Sci., v. 24, no. 1, 186-187.
        Notice of geological production from the area.

Clemson, Thomas Green (1807-1888)
        See also 10124 and 10127 with R. C. Taylor.
2428   1829, Assay and analysis of an iron ore, (fer titané,) from the environs of Baltimore: Am. J. Sci., v. 17, no. 1, 42-43.
        On ilmenite from Baltimore, Maryland.
2429   1830, The Hartz. Physical geography, state of industry, etc.: Am. J. Sci., v. 19, no. 1, 105-130, plate with 4 figs.
        On the geology of the Hartz Mountains in Germany.
2430   1833, Analysis of American spathic iron and bronzite: Am. J. Sci., v. 24, no. 1, 170-171.
        Chemical analyses of bronzite from Amity, New Jersey and siderite from Plymouth, Vermont.
2431   1833, Magnetic oxide of iron: Am. J. Sci., v. 25, no. 1, 212-213.
        On magnetite from Franklin County, New York.
2432   1834, Analysis and observations on divers mineral substances: Franklin Inst. J., v. 13 (ns), 78-80.
2433   1834, Analysis of American Cologne earth: Am. J. Pharmacy, v. 6, 70-71.
2434   1834, Flemington copper ore [from Hunterdon County, New Jersey]: Geol. Soc. Penn. Trans., v. 1, 167.
2435   1834, Observations on the geology of York County, Pennsylvania: Geol. Soc. Penn. Trans., v. 1, Appendix 1-13.
2436   1834, Report of a committee appointed by the Geological Society of Pennsylvania on a geological survey of York County: Advocate of Sci. and Annals Nat. History, v. 1, 163-175.
2437   1835, Analysis of some of the coal from the Richmond Mines: Geol. Soc. Penn. Trans., v. 1, 295-297.
2438   1835, Analysis of the minerals accompanying Mr. E. Miller's donation: Geol. Soc. Penn. Trans., v. 1, 271-274.
        On coal and siderite from the Pennsylvania coal measures.
2439   1835, Examination and analysis of several coals and iron ores, accompanying Mr. R. C. Taylor's account of the coal field of Blossburg [in Tioga County, Pennsylvania]: Geol. Soc. Penn. Trans., v. 1, 220-223.
        See also article by R. C. Taylor (10121).
2440   1835, Notes on the character of mines: Advocate of Sci. and Annals of Nat. History, v. 1, 298-308.

Clemson, T. G. (cont.)
2441 1835, Notice of a geological examination of the country between Fredericksburg and Winchester in Virginia, including the gold region: Geol. Soc. Penn. Trans., v. 1, 298-313, plate (colored and folded).
2442 1835, Notice of native iron from Pen Yan, Yates County, New York: Geol. Soc. Penn. Trans., v. 1, 358-359.
2443 1835, Observations on the geology of York County, Pennsylvania: Hazard's Reg. Penn., v. 15, 184-188.
       Abstracted from the Geol. Soc. Penn. Trans.
2444 1836, Hints on the mineral wealth of the United States, suggested by the recent discovery of diamonds: Naval Mag., v. 1, 64-66.
2445 1842, [Specimens from Havana, Cuba]: Nat'l Inst. Prom. Sci. Bulletin, v. 2, 156.

Clifford, John D. (1778?-1820)
2446 1820, On the geology of the Valley of the Mississippi: West. Rev. and Misc. Mag., v. 2, 257-265.
       This article was unsigned, but an obituary of J. D. Clifford (West. Rev. and Misc. Mag., v. 2, 309-310) cites him as author. The article was continued by C. S. Rafinesque (see 8629).

Clingman, T. L.
2447 1847, [Corundum from North Carolina]: Am. Q. J. Ag. Sci., v. 6, 213 (4 lines).
2448 1849, [On North Carolina earthquakes]: in Letters ..., by C. Lanman (see 6034), 173-182.

Clinton, De Witt (1769-1828)
2449 1821, Notice on the hydraulic limestone: West. Minerva, v. 1, 38.
       With "Note" by C. S. Rafinesque (see 8632).
2450 1822, Letters on the natural history and internal resources of the state of New York: NY, E. Bliss and E. White, 224 p.
       Signed by "Hibernicus," i.e. De Witt Clinton.
2451 1825, A memoir on the antiquities of the western parts of the state of New York: Lit. and Philos. Soc. NY Trans., v. 2, [71]-84.
2452 1825, On certain phenomena of the Great Lakes of North America: Lit. and Philos. Soc. NY Trans., v. 2, pt. 1, 1-33.

Clinton, George William (1807-1885)
2453 1828, Notice of the graphite of Ticonderoga: Alb. Inst. Trans., v. 1, 233-235.
2454 1846, Sketches of Niagara Falls and River: Buffalo, W. B. & C. E. Peck, viii, [9]-142 p., front, plates.
       Pseudonym "Cousin George."

Clissold, Frederick
2455 1823, Account of the recent successful ascent of Mont Blanc: Bos. J. Phil. Arts, v. 1, no. 2, 102-107.
       Excerpted from Edinburgh Philos. J., with a note by Prof. Pictet (see 8332).

Cloud, Joseph
2456 1809, An account of experiments made on palladium, found in combination with pure gold: Am. Philos. Soc. Trans., v. 6, pt. 2, 407.
       For review see 7090; also in: Am. Min. J., v. 1, no. 1, 40-43.
2457 1811, Carolina gold: Am. Min. J., v. 1, no. 2, 125 (9 lines).
2458 1818, An account of some experiments made on crude platinum, and a new process for separating palladium and rhodium from that metal: Am. Philos. Soc. Trans., v. 1 (ns), 161-165.
2459 1839, Earths and soils: Farm's Cab., v. 3, 202-203.
2460 1839, Lime: Farm's Cab., v. 3, 233-234.
2461 1839, Magnesian earth: Farm's Cab., v. 3, 325.
2462 1839, [Calcium in soils]: Farm's Cab., v. 3, 376.

Cloz, François Stanislas (b. 1817)
2463 1846, Separation of oxide of cobalt from the oxide of manganese: Am. J. Sci., v. 1 (ns), no. 1, 106 (7 lines).
       Abstracted from J. de Pharmacie by J. L. Smith.

Cluny, or Clunie, Alexander
2464 1770, The American traveller: containing observations on the present state, culture and commerce of the British colonies in America, and the further improvements of which they are capable; ... : Phila., Printed by Crukshank and Collins, 89, [1] p.
       Evans number 11603.

Coan, Titus (1801-1882)
2465 1840, Eruption of Kilauea: Bos. Wk. Mag., v. 3, no. 48, 377-378.
2466 1841, Great eruption of the volcano of Kilauea in Hawaii, Sandwich Islands: New-Yorker, v. 11, 247; also in: West. Miscellany, v. 1, 316-319, (1849).

Coates, Benjamin Horner (1797-1881)
2467 1818, Description of a hydrostatic balance, by which the specific gravities of minerals may be ascertained without calculation: Phila. Acad. Nat. Sci's J., v. 1, pt. 2, 368-370, plate.

Coatsworth, J.
2468 1848, Description of mines in St. Francis County, Mo.: West. J. Ag., v. 1, no. 1, 610-613.

Cobbett, William (1763-1835)
2469 1818, A year's residence in the United States of America, treating of the face of the country, the climate, the soil, the products, the mode of cultivating the land ... : NY, Printed for the author by Clayton and Kingsland, 432 p. (in 3 pts.).
2470 1818, A year's residence in America: Bos., Small, Maynard and Co., xx, 275 p.
       Also 1819 printing.

Cocke, John Hartwell (1780-1866)
2471   1828, Virginia aerolite: Am. J. Sci., v. 15, no. 1, 195-196.

Cogswell, James (1720-1807)
2472   1755, The danger of disregarding the works of God: A sermon, delivered at Canterbury, November 23, 1755, being the next Sabbath after the late surprising earthquake: NHav., Printed by James Parker and Company, 24 p.
       Evans number 7391.

Cogswell, Joseph Green (1720-1807)
2473   1822, [Review of Narrative journal of travels and Memoirs on the geological position of a fossil tree, by H. Schoolcraft (see 9261)]: N. Am. Rev., v. 15, 224-250.

Cohan, C. C. C.
2474   1833, Essay on the analysis of mineral waters, together with a new analysis of Saratoga water, ... : Am. J. Pharmacy, v. 5, 186-194.
       See also article by H. D. Rogers (8841).

Colby
2475   1848, [The lead mine of Eaton, New Hampshire]: NY J. Medicine and Collateral Sci's, v. 10, 340-341.
       In NY Lyc. Nat. History Minutes.

Colden, Cadwallader David (1769-1834)
2476   1814, Observations on the intended application of the North-American coal and mining company, to the legislature of the state of New York: NY, n. p., 23 p.

Coleman, Lyman (1796-1882)
2477   1835, Soap stone [from Middlefield, Massachusetts]: Am. J. Sci., v. 28, no. 2, 370-371.

Collections, Topographical, Historical, and Bibliographical, Relating Principally to New Hampshire (Concord, New Hampshire, 1822-1824)
2478   1823, [Crystal of quartz from Conway, New Hampshire]: v. 2, 29 (5 lines).
2479   1823, Curiosities - natural or artificial: v. 2, 48-49.
       On foot prints in limestone.
2480   1824, Great earthquake of 1638: v. 3, 101.
       On the Newbury, Massachusetts earthquake.
2481   1824, Burning springs [Calf Killer River, Tennessee]: v. 3, Appendix 26-27.

Collier, Henry Watkins (1801-1855)
2482   1850, First biennial report of the geology of Alabama. By M. Tuomey [review of 10440]: Plough Loom Anvil, v. 3, no. 1, 59-60.

Collins, Lewis (1797-1870)
2483   1847, Historical sketches of Kentucky ... : Maysville, Kentucky, J. A. and V. P. James, xvi, 560 p., illus.
       Also 1848 and 1850 reprintings.

Collinson, Peter
2484   1789, An account of some very large fossil teeth found in North America: Am. Mus. or Univ. Mag., v. 5, 155-156.
       Fossil bones of the Ohio River in Kentucky compared to those of Siberia.
2485   1789, Sequel to the foregoing account of the large fossil teeth: Am. Mus. or Univ. Mag., v. 5, 156-157.

Colman, Benjamin (1673-1747)
2486   1727, The judgements of Providence in the hand of Christ: His voice to us in the terrible earthquake. And, the earth devoured by the curse: Bos., J. Phillips and T. Hancock, viii, 86 p.
       Evans number 2853.

Colman, Henry (1785-1849)
2487   1839, [Review with excerpts of third annual Maine report, by C. T. Jackson (see 5566)]: NE Farm., v. 18, 58; 66; 82-82; 98.
2488   1840, [Review of Rhode Island geology report, by C. T. Jackson (see 5571)]: NE Farm., v. 19, 70.

Colthurst (Mr.)
2489   1842, An account of the contortions and faults produced in the strata underneath and adjacent to the Great Embankment across the Valley of the Brent, on the line of the Great Western Railway: Franklin Inst. J., v. 4 (3s), 98-99.
       From London Ath.

Colton, Simeon (1783/84-1868)
2490   1825, By Simeon Colton: Am. J. Sci., v. 10, no. 1, 12-14.
       On Connecticut mineral localities.

Columbian Chemical Society of Philadelphia. Memoirs of the (Philadelphia, 1813)
2491   1813, Report of the committee to whom was referred the analysis of certain ores: v. 1, 208-210.

The Columbian Magazine, or, Monthly Miscellany (Philadelphia, 1786-1790)
2492   1786, Some observations on the structure of the surface of the earth in Pennsylvania and the adjoining countries: v. 1, no. 2, 49-53.
2493   1786, Description of bones, &c. found near the River Ohio: v. 1, no. 3, 103-107, plate.
2494   1787, Considerations on the alterations, made on the face of the earth by atterations: v. 1, no. 6, 267-271; no. 7, 303-306.
       Signed by "A. Z."
2495   1787, Account of remarkable cascades and caverns in the state of Virginia: v. 1, no. 7, 335-337.

Columbian Magazine (cont.)
2496 1787, A description of the natural bridge, called in Virginia, Rocky Bridge, with an elegant engraving annexed: v. 1, 617-618, plate.
2497 1787, Description of a remarkable tooth, in the possession of Mr. Peale: v. 1, 655, plate.
 Probably by C. W. Peale.
2498 1788, Remarks on the ores and fossils of America: v. 2, 153-156, plate; 208-211, plate; 282-283, plate; 333-334, plate; 389-390; 454, plate; 509-510.
 Signed by "B."
2499 1788, Account of a very extraordinary eruption of fire in Iceland, in 1783: v. 2, 323-325.
2500 1789, Specimens of fossils found in the United States: v. 3, 228, plate with 6 figs.
2501 1789, Chymical and economical essays.---Earths: v. 3, 525-531; 577-582.

The Commercial Review (see De Bow's Review of the Southern and Western States)

Commun, Joseph (see Du Commun, Joseph)

A compendius system of mineralogy & metallurgy; extracted from the American edition of the Encyclopaedia, now publishing.
2502 1794, Phila., Thomas Dobson, iv, 505 p., 12mo.
 Evans number 26801. Also 1798 title page, "Published by A. Bartram," on same printing.

Comstock, John Lee (1798-1858)
2503 1819, Extract of a letter: Am. J. Sci., v. 1, no. 4, 433-434.
 On the discovery of cinnabar near Lake Erie, and native lead at Fort Wayne, Indiana.
2504 1820, Sulphate of barytes: Am. J. Sci., v. 2, no. 2, 379-380.
 On mineral specimens from Hartford, Connecticut.
2505 1826, Manual of mineralogy:
 For review see 10501, 10761, 440. Not seen.
2506 1827, Elements of mineralogy, adapted to the use of seminars and private students: Bos., S. G. Goodrich, 76, 338 p., illus.
 First edition. For review see 1815, 1814, 7693 and 10764. For notice see 447.
2507 1829, Natural history of quadrupeds; with engravings, on a new plan, exhibiting their comparative size, adapted to the capacities of youth; with authentic anecdotes, illustrating the habits and characters of the animals, together with reflections moral and religious ... : Hart., Robinson and Co., 201 p., illus.
2508 1832, Elements of mineralogy, adapted to the use of seminars and private students: Hart., 343 p., illus.
 Second edition.
2509 1832, An introduction to mineralogy: NY, Robinson, Pratt and Co., 343 p, illus.
 First edition.
2510 1832, An introduction to mineralogy: Hart., B. B. Barber, 343 p., illus.
 Second edition. Also 1833 printing.
2511 1834, Outlines of geology; intended as a popular treatise on the most interesting parts of the science, together with the examination of the question, whether the days of Creation were indefinite periods ... : Hart., D. F. Robinson and Co., xii, 336 p. illus.
 First edition. For review see 583 and 6259. For notice see 581 and 667.
2512 1836, An introduction to mineralogy: NY, Robinson, Pratt and Co., 369 p., illus.
 Third edition. 4th edition through 15th edition are equivalent to yearly printings of this edition, and are not listed separately.
2513 1836, Outlines of geology; ...[as in first edition, see 2511]:NY, Robinson, Pratt, and Co., xii, 372, illus.
 Second edition.
2514 1838, Earthquake of Lisbon: Friend's Intell., v. 1, no. 2, 20-21.
 Description of the 1755 earthquake from Outlines of Geology.
2515 1838, Outlines of Geology; ...[as in first edition, see 2511]: NY,Robinson, Pratt and Co., 384 p., illus.
 Third edition. Also yearly printings of this edition through 1845; all are called "third edition."
2516 1841, Elements of mineralogy, adapted to the use of seminars and private students: NY, 369 p., illus.
 Third edition.
2517 1846, Outlines of geology; ... [as in first edition, see 2511] : NY, Robinson, Pratt and Co., 396, 12 p., illus.
 "16th edition." This edition followed the fifteenth printing of the "third edition."
2518 1847, Elements of geology, including fossil botany and paleontology, a popular treatise on the most interesting parts of the science ... : NY, Pratt, Woodford and Co., viii, [9]-432 p., illus.
 First edition. Also 1849 printing. For review see 6355.
2519 1848, Natural history of quadrupeds; ...[as in first edition, see 2507] : NY, Pratt, Woodford, and Co., iv, [5]-73, front., illus., folding plate.
2520 1849, A history of the precious metals, from the earliest periods to the present time, with directions for testing their purity and use: Hart., Belknap and Hamersley, viii, 222 p., illus, tables.
 For review see 6363.

Concord Natural History Society (Concord, Massachusetts)
2521 1846, Concord Natural History Society: Am. J. Sci., v. 2 (ns), no. 1, 144 (1 lines).
 Notice of officers.

Connecticut. Colony of
2522 1800, The heads of inquiry relative to the present state snd condition of his Majesty's Colony of Connecticut: Mass. Hist. Soc. Coll., v. 7, 231-239.
 From a manuscript dated 1774. Includes motice of mines.

Connecticut. State of
2523 1835, [Call for a geological survey] : Governor's Message, 1835.

Connecticut (cont.)
2524 1835, [Resolution to form a geological survey] : Ct. Gen. Assembly, 1835.
2525 1837, Report of the Committee on the geological survey: Ct. Gen. Assembly, 1837.
2526 1837, [Resolution to publish the state geology reports] : Ct. Gen. Assembly, 1837.
2527 1842, [Resolution to publish Percival's geological report] : Ct. Gen. Assembly, 1842.

The Connecticut Magazine (Bridgeport, Connecticut, 1801)
2528 1801, [Rich mine of quicksilver in Columbo, Ceylon] : v. 1, no. 1, 57 (3 lines).
2529 1801, [Fossil bone from Wales] : v. 1, no. 1, 59-60.
2530 1801, [Brief notes on earthquakes] : v. 1, no. 1, 62; no. 2, 112.
2531 1801, [Origin of Bath Hot Springs, England] : v. 1, no. 2, 114.
2532 1801, [Notice of minerlalogy classes taught by W. Watson of Bakewell, Derbyshire, England] : v. 1, no. 2, 115.

Connell, Arthur (1794-1863)
2533 1831, New analysis of brewsterite: Am. J. Sci., v. 20, no. 1, 198.
    Abstracted from Edinburgh New Philos. J. by C. U. Shepard.
2534 1831, A remarkable chalybeate water: Am. J. Sci., v. 20, no. 2, 384-385.
    Abstracted from Jameson's J. by J. Griscom. On mineral water from Vicar's Bridge, Scotland.
2535 1838, Analysis of gmelinite or hydrolite: Am. J. Sci., v. 35, no. 1, 195-196.
    Abstracted from Jameson's J.
2536 1838, Analysis of the scales of the fossil gavial of Caen, in Normandy: Am. J. Sci., v. 34, no. 1, 201.
    Abstracted from Edinburgh New Philos. J.
2537 1847, Analysis of the American mineral nemalite: Am. J. Sci., v. 3 (ns), no. 2, 265-266.
    Abstracted from London Ath.
2538 1847, On sulphato-chlorid of copper, a new mineral [from Cornwall, England] : Am. J. Sci., v. 4 (ns), no. 3, 415-416.
    Abstracted from London Ath.
2539 1848, Carbonate of copper and zinc: Am. J. Sci., v. 6 (ns), no. 3, 426 (11 lines).
    Abstracted from Jameson's J. On aurichalcite from Matlock.

Conrad, Solomon W.
2540 1811, Zircon [from Trenton, New Jersey] : Am. Min. J., v. 1, no. 2, 124 (6 lines).
2541 1812, Mineralogical notice respecting zircon from Trenton, N. J.: Am. Min. J., v. 1, no. 3, 127-128, plate with 2 figs.
2542 1829, Notice of a mineral which approaches to the bildstein of Werner; with a few remarks on the connexion of bildstein with feldspar: Phila. Acad. Nat. Sci's J., v. 6, pt. 1, 102-104.
    On feldspar from Wilmington, Delaware.

Conrad, Timothy Abbott (1803-1877)
    See also 6889 with W. W. Mather and 8875 with H. D. Rogers and others.
2543 1830, Description of fifteen new species of recent, and three of fossil shells, chiefly from the coast of the United States: Phila. Acad. Nat. Sci's J., v. 6, pt. 2, 256-268, plate with 21 figs.
2544 1830, On the geology and organic remains of a part of the Peninsula of Maryland: Phila. Acad. Nat. Sci's J., v. 6, pt. 2, 205-230, 2 plates with 30 figs.
    Includes list of known Maryland Tertiary fossils.
2545 1832/1833, Fossils shells of the Tertiary formations of North America; illustrated by figures drawn on stone from nature: Phila., Judah Dobson, viii, [9] -56 p., plates, folded map.
    Issued in four parts: 1 and 2 in 1832, and 3 and 4 in 1833. Part 3 was republished in 1835. For review see 556 and 559.
2546 1833, On some new fossil and recent shells of the United States: Am. J. Sci., v. 23, no. 2, 339-346.
    Includes notice of Tertiary fossils from Virginia, Maryland, and Alabama.
2547 1834, Clairborne, Alabama: Advocate of Sci. and Annals Nat. History, v. 1, no. 1, 26-31.
    On the stratigraphy and fossils of the region near Clairborne.
2548 1834, Mobile, Alabama: Advocate of Sci. and Annals Nat. History, v. 1, no. 2, 57-61.
    On the geology of the region near Mobile.
2549 1834, Sketches from the notebook of a traveller: Advocate of Sci. and Annals Nat. History, v. 1, no. 4, 153-162.
    Geological notes on Wilmington, North Carolina.
2550 1834, Descriptions of new Tertiary fossils from the Southern states: Phila. Acad. Nat. Sci's J., v. 7, pt. 1, 130-157.
2551 1834, Observations on the Tertiary and more recent formations of a portion of the Southern states: Phila. Acad. Nat. Sci's J., v. 7, pt. 1, 116-129, table.
2552 1835, Chronological analysis of the various papers hitherto published on the geology of the upper Secondary and Tertiary formation of the United States: Advocate of Sci. and Annals Nat. History, v. 1, no. 9, 393-400. To be continued, but no more found.
    This survey paper includes reviews of 22 articles and books by S. G. Morton, G. W. Featherstonhaugh, R. Harlan, J. Dekay, W. Maclure, and S. L. Mitchill.
2553 1835, Description of five new species of fossil shells in the collection presented by Mr. Edward Miller to the Geological Society: Geol. Soc. Penn. Trans., v. 1, 267-270, plate with 5 figs.
2554 1835, Notes on the geology of West Florida: Advocate of Sci. and Annals Nat. History, v. 1, no. 8, 351-352.
2555 1835, Observations on a portion of the Atlantic Tertiary region: Geol. Soc. Penn. Trans., v. 1, 335-341, plate with 4 figs.
    On Eocene fossils from Prince Georges County, Maryland.

Conrad, T. A. (cont.)

2556 1835, Observations on the Tertiary strata of the Atlantic coast: Am. J. Sci., v. 28, no. 1, 104-111, fig.; no. 2, 280-282.
Includes definitions of Pliocene, Miocene, and Eocene on the basis of percentage of species still living. Also includes list of Pliocene species from Benner's Plantation, North Carolina.

2557 1837, First annual report on the geological survey of the third district of the state of New-York: NY Geol. Survey Ann. Rept., v. 1, 155-186.
Includes plate by L. Vanuxem and J. Eights (see 10579).

2558 1838, Fossils of the Tertiary formations of the United States. Illustrated by figures, drawn from nature: Phila., J. Dobson, xvi, 80 p., 49 plates.

2559 1838, Report on the paleontological department of the survey: NY Geol. Survey Ann. Rept., v. 2, 107-119.

2560 1839, Fossils of the medial Tertiary of the United States: Phila.
Not seen. For notice of part 1, with price and printing schedule, see Am. J. Sci., v. 35, unnumbered page of errata (7 lines).

2561 1839, New species of fossil shells [from Virginia]: Phila., 1 p.

2562 1839, Notes on American geology: Am. J. Sci., v. 35, no. 2, 237-251.
With "remarks by the editors" (see 9758). Includes discussion of the use of fossils as indicators of paleoclimates, and review of known American Lower Paleozoic fossil localities.

2563 1839, Second annual report - on the paleontological department of the survey: NY Geol. Survey Ann. Rept., v. 3, 57-66.

2564 1840, New fossil shells from N. Carolina: Am. J. Sci., v. 39, no. 2, 387-388.

2565 1840, On the geognostic position of the Zeuglodon, or Basilosaurus of Harlan: Am. J. Sci., v. 38, no. 2, 381-382.

2566 1840, On the Silurian system, with a table of the strata and characteristic fossils: Am. J. Sci., v. 38, no. 1, 86-93.
On the correlation of Silurian rocks in New York and Wales.

2567 1840, Third annual report on the paleontological department of the survey: NY Geol. Survey Ann. Rept., v. 4, 199-207.

2568 1841, Appendix to Mr. Hodge's paper, describing the new shells, &c.,: Am. J. Sci., v. 41, no. 2, 344-348, plate with 18 figs.
Appendix to 5192. On new species of Tertiary fossil shells. Also in Assoc. Am. Geol's Nat's Rept., v. 1, 108-11, plate 6 with 18 figs., 1843.

2569 1841, Fifth annual report on the paleontology of the state of New-York: NY Geol. Survey Ann. Rept., v. 5, 25-57.

2570 1841, Twenty-six new species of fossil shells, discovered (by him) in the medial Tertiary deposits of Calvert Cliffs, Maryland: Phila. Acad. Nat. Sci's Proc., v. 1, no. 3, 28-33.

2571 1842, Description of twenty-four new species of fossil shells, chiefly from the Tertiary deposits of Calvert Cliffs, Maryland: Phila. Acad. Nat. Sci's J., v. 8, pt. 2, 183-190.

2572 1842, Observations on a portion of the Atlantic Tertiary region, with a description of a new species of organic remains: Nat'l Ist. Prom. Sci. Bulletin, v. 2, 171-194, 2 plates with 10 figs., woodcut.

2573 1842, Observations on the Silurian and Devonian systems of the United States, with descriptions of new organic remains: Phila. Acad. Nat. Sci's J., v. 8, pt. 2, 228-280, 6 plates with 91 figs.

2574 1842, On the identity of the Middle Cretaceous formation of the United States, with the Faxoe limestone of Europe: Phila. Acad. Nat. Sci's Proc., v. 1, no. 10, 143-144.

2575 1843, Descriptions of a new genus, and of twenty-nine new Miocene, and one Eocene, fossil shells of the United States: Phila. Acad. Nat. Sci's Proc., v. 1, no. 31, 305-311.

2576 1843, [Description of fossils from the Helderberg Division of New York]: in Geology of New York, by W. W. Mather (see 6907), 338, 339, 345, 349.

2577 1843, Descriptions of nineteen species of Tertiary fossils of Virginia and North Carolina: Phila. Acad. Nat. Sci's Proc., v. 1, no. 33, 323-329.

2578 1843, Observations on the lead bearing limestone of Wisconsin, and descriptions of a new genus of trilobites and fifteen new Silurian fossils: Phila. Acad. Nat. Sci's Proc., v. 1, no. 33, 329-335.

2579 1844, Descriptions of eight new fossil shells of the United States: Phila. Acad. Nat. Sci's Proc., v. 2, no. 6, 173-175.
On Tertiary fossils from Virginia, Alabama, and South Carolina.

2580 1844, [On the nature of the country rock in the Missouri lead region]: Am. J. Sci., v. 47, no. 1, 106 (4 lines).
Also in Assoc. Am. Geol's Nat's Proc., v. 5, 13 (4 lines).

2581 1845, Fossils of the Medial Tertiary of Miocene formations of the United States: Phila., J. Dobson, 80 p.

2582 1846, Descriptions of new species of fossil and recent shells and corals: Phila. Acad. Nat. Sci's Proc., v. 3, no. 1, 19-27, illus.
Includes notice of Paleozoic fossils from Ohio and Pennsylvania, and Tertiary fossils from Pennsylvania and Virginia.

2583 1846, Descriptions of new species of organic remains from the upper Eocene limestone of Tampa Bay [Florida]: Am. J. Sci., v. 2 (ns), no. 3, 399-400, 9 figs.

2584 1846, Eocene formation of the Walnut Hills, etc., Mississippi: Am. J. Sci., v. 2 (ns), no. 2, 210-215.

2585 1846, Observations on the Eocene formation of the United States, with descriptions of species of shells, etc. occurring in it: Am. J. Sci., v. 1 (ns), no. 2, 209-220; no. 3, 395-405, 3 plates with 23 figs.
With additional remarks by J. D. Dana (see 2819). Two plates with 17 figs., scheduled for publication with no. 2, 209-220, did not appear due to a printer's delay. These plates were not found in subsequent numbers.

2586 1846, Observations on the geology of a part of east Florida, with a catalogue of recent shells of the coast: Am. J. Sci., v. 2 (ns), no. 1, 36-48.

2587 1846, Tertiary of Warren Co., Mississippi: Am. J. Sci., v. 2 (ns), no. 1, 124-125.

2588 1847, Observations on the Eocene formation, and description of one hundred and five new fossils of that period, from the vicinity of Vicksburg, Mississippi, with an appendix: Phila. Acad. Nat. Sci's Proc., v. 3 (ns), no. 11, 280-299.

Conrad, T. A. (cont.)
2589   1848, Fossil shells from Tertiary deposits on Columbia River, near Astoria: Am. J. Sci., v. 5 (ns), no. 3, 432-433, 14 figs.
2590   1848, Observations on the Eocene formation, and descriptions of one hundred and five new fossils of that period, from the vicinity of Vicksburg, Mississippi; with an appendix: Phila. Acad. Nat. Sci's J., v. 1 (ns), no. 2, 111-134, 4 plates.
2591   1849, Descriptions of new fossil and recent shells of the United States: Phila. Acad. Nat. Sci's J., v. 1 (ns), no. 3, 207-209.
2592   1849, Fossils from northwestern America: in Geology of the United States exploring expedition ..., by J. D. Dana (see 2841).
2593   1850, Descriptions of one new Cretaceous, and seven new Eocene fossils: Phila. Acad. Nat. Sci's J., v. 2 (ns), pt. 1, 39-41, plate with 15 figs.
        Also reprint edition, 39 p.

Considerations on the practicability and utility of immediately constructing a central railway, from Pottsville to Sunbury and Danville, through the coal region of Mahanoy and Shamokin
2594   1830, Phila., n. p., 1, 37 p.

A controversy between the four elements, viz. fire, water, earth and air. Wherein each of them claim the superiority, and extol their own goodness and worth to mankind. With their various arguments why they ought to be esteemed superior:
2595   1775, Norwich, 20 p.
        Evans number 13890.

Conybeare, John Josias (1779-1824)
2596   1824, Plumbago in coal-gas retorts: Am. J. Sci., v. 7, no. 2, 380.
        Abstracted from Annals Philos. by J. Griscom.

Conybeare, William Daniel (1787-1857)
2597   1823, and W. Phillips, [Excerpts from Outlines of the Geology of England and Wales]: Bos. J. Phil. Arts, v. 1, no. 3, 239-250.
        In review by J. W. Webster (see 10743).
        For other reviews of Outlines of Geology of England and Wales, see 401, 417.
2598   1831, Communication from the president of the Geological Society of London, and other naturalists: Mon. Am. J. Geology, v. 1, no. 4, 175-176.

Cook, Captain James (1728-1779)
2599   1844, Description by Captains Cook and Flinders of bird's nest of enormous size on the coast of New Holland: Am. J. Sci., v. 47, no. 1, 217-218.
        In article by E. Hitchcock (see 5153).

Cook, Ransom
2600   1849, Improved electro-magnetic ore separator: Franklin Inst. J., v. 18 (3s), 308.

Cooke, Henry David (1825-1881)
2601   1845, Mammoth Cave: Arthur's Mag., v. 4, 25-28.
        Signed by "H. D. C." but index indicates H. D. Cooke.

Cooke, William S.
2602   1848, An improvement in the reduction of iron ores: Franklin Inst. J., v. 16 (3s), 112-113.

Cooley, Issac (see 10794 with others)

Cooper, James F.
2603   1845, [Diamond from Hall County, Georgia]: Am. Philos. Soc. Proc., v. 4, no. 34, 211.

Cooper, Paul
2604   1826, On the New Jersey marls: Phila. Soc. Promoting Ag. Mem., v. 5, 17.

Cooper, Susan Fenimore (1813-1894)
2605   1850, Rural hours, by a lady ... : NY, Putnam, 521 p.

Cooper, Thomas (1759-1839)
        See also 14 by F. Accum.
2606   1812, Account of the decomposition of potash and production of potassium, by heat: Am. Min. J., v. 1, no. 3, 134-139, plate with 1 fig.
2607   1813, Geology: Emp. Arts Sci's, v. 3 (ns), 412-444.
        With extract from Mineralogy, v. 3, 256-276, by R. Jameson (see 5702). Includes discussions of geological theories of A. Werner and J. Hutton.
2608   1813, Iron: Emp. Arts Sci's, v. 1 (ns), 15-155; 185-258.
        Survey of manufacture of iron, with reference to ores and smelting.
2609   1813, Magnesian limestone: Emp. Arts Sci's, v. 1 (ns), 318-323.
2610   1813, On mineral waters, and watering places: particularly the Carlisle and York Springs in Pennsylvania, with a method of making artificial mineral water: Emp. Arts Sci's, v. 1 (ns), 474-481.
2611   1813, On sulphat of lime, plaster of Paris, or gypsum, and the method of ascertaining it: Emp. Arts Sci's, v. 1 (ns), 323-326.
2612   1814, Lead: Emp. Arts Sci's, v. 3 (ns), no. 2, 167-266; no. 3, 339-344.
2613   1814, Notes to aid in the analysis of mineral waters: Emp. Arts Sci's, v. 3 (ns), no. 3, 491-494.
2614   1814, Of copper and the manufactures dependent on copper: Emp. Arts Sci's, v. 3 (ns), no. 1, 1-144.
2615   1814, Remarks by the editor [on iron assaying]: Emp. Arts Sci's, v. 3 (ns), no. 3, 452-454.
        See also 3581.
2616   1814, Tin: Emp. Arts Sci's, v. 3 (ns), no. 3, 345-443.
2617   1817, [Advertisement for mineralogy lectures]: Analectic Mag., v. 10, 352.
        Notice of lectures at the University of Pennsylvania.

Cooper, T. (cont.)

2618 1817, [Geological classification and nomenclature]: Am. Reg. or Summary Rev., v. 1 (ns), 438-441.
   On theories of mineral classification by R. Haüy and J. Berzelius.
2619 1817, Introductory lecture on mineralogy: Port Folio, v. 4 (5s), 482-498.
2620 1817, The new blowpipe: Port Folio, v. 3 (5s), 417-423.
   Signed by "T. C." On R. Hare's blowpipe.
2621 1817, [Optical properties of minerals]: Am. Reg. or Summary Rev., v. 1 (ns), 422-425.
2622 1818, Analysis of the blue iron earth of New Jersey: Am. Philos. Soc. Trans., v. 1 (ns), 193-199; also in: Eclec. Rep. and Analytic Rev., v. 8, 315-321.
2623 1818, Syllabus of the lectures of Thomas Cooper: Port Folio, v. 6 (5s), 117-120.
2624 1821, Syllabus of a course of lectures on the elements of geological mineralogy: Columbia, SC, D. Faust, 8 p.
2625 1822, On volcanoes and volcanic substances, with a particular reference to the origin of the rocks of the floetz trap formation: Am. J. Sci., v. 4, no. 2, 205-243.
   With an "Appendix" by T. D. Porter (see 8523).
2626 1828, Analysis of an "Essai sur la temperature de l'interieur de la terre, par M. Cordier" [see 2645]: Am. J. Sci., v. 15, no. 1, 109-131.
2627 1833, On the connection between geology and the Pentateuch, in a letter to Professor Silliman ...: Bos., J. Hall, 72 p.
   For review see 5984, 4107. Other editions include: Bos., A. Kneeland, 83 p. (1837) and 72 p. (1845); Coumbia, SC, Times and Gazette Office, 64 p. (1833) and 2, 64, v, 17 p. (1836).
2628 1838, Professor Cooper's analysis of various specimens of Pennsylvania limestone: Del. Reg. and Farm's Mag., v. 1, 281-285.
   From Emp. Arts Sci's. See also 2629.

Cooper, William (1798?-1864)
   See also 3021 with J. E. Dekay, and 7317 with J. A. Smith.
2629 1814, Analysis of various specimens of Pennsylvania limestone: Phila. Soc. Promoting Ag. Mem., v. 3, Appendix, 106-113.
2630 1814, Mode of analyzing and testing gypsum: Phila. Soc. Promoting Ag. Mem., v. 3, Appendix, 104-105.
2631 1824, On the remains of the Megatherium recently discovered in Georgia: NY Lyc. Nat. History Annals, v. 1, pt. 1, no. 4, 114-124. plate with 4 figs.
2632 1824, On the remains of the Megatherium recently discovered in Georgia: in American Natural History by J. D. Godman (see 4417), v. 2, 187-196.
2633 1827, Further discovery of fossil bones in Georgia; and remarks on their identity with those of the Megatheruim of Paraguay: NY Lyc. Nat. History Annals, v. 2, no. 9, 267-270.
2634 1831, Notices of Big Bone lick: Mon. Am. J. Geology, v. 1, no. 4, 158-174; 205-217, map.
   On the history of exploration, and types of fossil bones, of Big Bone Lick, Kentucky.
2635 1831, and J. A. Smith and J. Dekay, Scientific memoranda: Mon. Am. J. Geology, v. 1, 42-45.
   On Big Bone Lick, Kentucky.
2636 1831, and J. A. Smith and J. Dekay, Report of Messrs. Cooper, J. A. Smith, and De Kay, to the Lyceum of Natural History, on a collection of fossil bones, disinterred at Big Bone Lick, Kentucky, in September, 1830, and recently brought to New York: Am. J. Sci., v. 20, no. 2, 370-372.
   With remarks by B. Silliman (see 9704).
2637 1831, Mr. Cooper's disclaimer: Am. J. Sci., v. 20, no. 2, 413.
   On debate on nature of a fossil specimen. See also article by A. Eaton (see 3365).
2638 1834, [Big Bone Lick, Kentucky]: Am. J. Sci., v. 27, no. 1, 149 (12 lines).
   In NY Lyc. Nat. History Proc. On recent discoveries of fossils.
2639 1836, A report on some fossil bones of the Megalonyx from Virginia; with a notice of such parts of the skeleton of this animal as have been hitherto discovered, and remarks on the affinities which they indicate: NY Lyc. Nat. History Annals, v. 3, 166-173.

Copper Falls Mining Company
2640 1845, Articles of association and agreement of the Copper Falls Company: Bos., S. N. Dickinson and Co. (printer), 10 p.

Corbin, William
2641 1703, A sermon preached at Kingstown in Jamaica, upon the 7th June, being the anniversary fast for that dreadful earthquake which happened there in the year 1692: NY, W. Bradford, 4, 16 p., 4to.
   Evans number 1105.

Cordier, Peirre Louis Antoine (1777-1861)
2642 1817, [Analysis of volcanic rocks]: Am. Reg. or Summary Rev., v. 1 (ns), 445.
   Abstracted by T. Cooper.
2643 1817/1818, Memoire sur les substances minerales dites en masse qui entrent dans la composition des roches vocaniques de tous les ages: J. Sci. Arts, v. 2, no. 2, 434-438; v. 3, no. 1, 201-207.
   Abstract in English. From J. de Physique.
2644 1817, [Salt mountain at Cordonna, Catalonia]: Am. Reg. or Summary Rev., v. 1 (ns), 445-446.
   Abstracted by T. Cooper.
2645 1828, An essay on the temperature of the interior of the earth. Amherst, Mass., John S. and Charles Adams, 94 p.
   "Translated by the Junior Class in Amherst College", with "Explanatory" note by E. Hitchcock (see 5093). For review with excerpts see 2626 and 5096. For excerpt see: Am. J. Improvements Use. Arts, v. 1, 217 (11 lines). For notice see 9249.

Corelli (see Covelli, Niccolle)

Corliss, Hiram
2646   1850, Brief notices of the medical topography and diseases of Washington Co.: Med. Soc. NY Trans., v. 8, 225-229.

Cormack, William Epps (1796-1868)
2647   1824, Account of a journey across the Island of Newfoundland: Bos. J. Phil. Arts, v. 1, no. 6, 529-534.
       Abstracted from Edinburgh Philos. J.

Cornelius, Elias (1794-1832)
2648   1819, On the geology, mineralogy, scenery, and curiosities of parts of Virginia, Tennessee, and the Alabama and Mississippi Territories, &c. with miscellaneous remarks: Am. J. Sci., v. 1, no. 3, 214-226; no. 4, 317-331, fig.
2649   1820, Account of a singular position of a granite rock: Am. J. Sci., v. 2, no. 2, 200-201, fig.
       On a rocking stone from North Salem, New York.
2650   1821, Notice of the Salem sienite, jasper, amygdaloid, etc.: Am. J. Sci., v. 3, no. 2, 232-233.

The Correspondent (New York, 1827-1829)
2651   1828, Geology: v. 2, 339-340.
       On the concept of changes in water level over geological time.

The Corsair (New York, 1839-1840)
2652   1840, Extraordinary earthquakes in England: v. 1, no. 51, 813.

Cotta, Bernard von (1808-1879)
2653   1840, Notice of tracks of animals in variegated sandstone at Polzig, between Ronneburg and Weissenfels [Germany]: Am. J. Sci., v. 38, no. 2, 255-259, 8 figs.
       Abstracted from Neues Jahrbuch fur Mineralogie.

Cotte, P. Louis (1740-1815)
2654   1788, Variation of the magnetic needle, observed at Laon: Colum. Mag., v. 2, 682-683.
2655   1789, Variation of the magnetic needle; observed at Laon ... in the course of the year 1787: Colum. Mag., v. 3, 110, table.

Cotting, John Ruggles (1783-1867)
2656   1835, A synopsis of lectures on geology comprising the principles of the science designed as a text book: Taunton, Mass., published for the author, printed by E. Anthony, 120 p.
2657   1836, Report of a geological and agricultural survey of Burke and Richmond counties, Georgia, performed under the patronage of the citizens of the two counties: Augusta, Georgia, Guieu and Thompson, 198 p.
2658   1843, An essay on the soil and available manures of the state of Georgia, with the mode of application and management, founded on a geological and agricultural survey: Milledgeville, Georgia, Park and Rogers (printers), vi, 7-121 p.

Cotton, John (1691-1757)
2659   1727, A holy fear of God and his judgements exhorted to: in a sermon preach'd at Newtown November 3, 1727. On a day of fasting and prayer occasion'd by the late terrible earthquake, that shook New England, on the Lord's day night before ... with an appendix containing a remarkable account of the extraordinary impressions made on the inhabitants of Haverill: Bos., Printed by B. Green. jr., for S. Gerrish, xvi, 24, 7 p.
       Evans number 2861.
2660   1728, God's awful determination against a people, that will not obey his voice. ... A sermon preach'd ... in Boston, February 8, 1728. After repeated shocks of the earthquake: Bos., printed by G. Rogers, for Samuel Gerrish and Thomas Hancock, (4), viii, 42 p.
       Evans number 3015.

Coultas, Harland (d. 1877)
2661   1849, On the origin and diversity of soils: NE Farm., v. 1, no. 17, 266-267.

Couper, James Hamilton (1794-1866)
2662   1842, Collection of fossil bones and shells from the Brunswick Canal, Georgia: Phila. Acad. Nat. Sci's Proc., v. 1, no. 16, 189-190.
2663   1842, Description of the strata, in which the fossil bones and shells from the Brunswick Canal, lately presented by him to the Academy, were found: Phila. Acad. Nat. Sci's Proc., v. 1, no. 20, 216-217.
2664   1846, Observations on the geology of a part of the seacoast of the State of Georgia; with a description of the fossil remains of the Megatherium, mastodon, and other contemporaneous mammalia and fossil marine shells found in the Brunswick Canal at Skiddaway Island: in Memoir on the Megatherium by W. B. Hodgson (see 5213), 31-47.
2665   1846, [On the age of the Burr mill stone, from Bainbridge, Georgia]: Bos. Soc. Nat. History Proc., v. 2, 123-124.

Courtenay, Edward Henry (1803-1853; see 1380 with A. D. Bache)

Couthouy, Joseph Pitty (1808-1864)
2666   1841, Vocanic phenomena in Hawaii: Am. J. Sci., v. 41, no. 1, 200.
2667   1842, [On icebergs and diluvial phenomena]: Am. J. Sci., v. 43, no. 1, 154-165.
       Also in Assoc. Am. Geol's Proc., v. 3, 9-20. Also in Assoc. Am. Geol's Nat's Rept., v. 1, 49-59, 1843.
2668   1842, Remarks upon coral formations in the Pacific; with suggestions as to the causes of their absence in the same parellels of latitude on the coast of South America: Bos. J. Nat. History, v. 4, no. 1, 66-105, 4 figs; no. 2, 137-162, 4 figs.
       Also reprint edition, 68 p.

Couthouy, J. P. (cont.)
2669   1842, [Rising of land and origin of certain hollows]: Am. J. Sci., 43, no. 1, 153-154 (17 lines).
          Also in Assoc. Am. Geol's Nat's Proc., v. 3, 8-9. Also in Assoc. Am. Geol's Nat's Rept., v. 1, 48, 1843.
2670   1843, [On formation of coral islands]: Bos. Soc. Nat. History Proc., v. 1, 48 (9 lines).
2671   1843, Reply of J. P. Couthouy, to the accusations of J. D. Dana, geologist to the exploring expedition: Am. J. Sci., v. 45, no. 2, 378-389.
          See also 2810, 2811, 2813, 9777.
2672   1844, Influence of temperature on the development of corals: Am. J. Sci., v. 47, no. 1, 123-126.
          Also in Assoc. Am. Geol's Nat's Proc., v. 5, 31-35.
2673   1844, [Resolution of J. D. Dana's plagarism charge]: Am. J. Sci., v. 47, no. 1, 122-123.
          Also in Assoc. Am. Geol's Nat's Proc., v. 5, 30-31. See also 2810, 2811, 2813, 9777.
2674   1844, Review of and strictures on Mr. Dana's reply to Mr. Couthouy's vindication against his charge of plagarism: Am. J. Sci., v. 46, no. 2, Appendix, 1-9.
          See also 2810, 2811, 2813, 9777.

Covelli, Niccolo (1790-1829; see 7411 and 7412 with W. Monticelli)

Cox, Robert
2675   1838, Meteoric iron: Am. J. Sci., v. 33, no. 2, 257-258.
          Abstracted, with a letter, by W. C. Woodbridge (see 11039). On meteorites from Texas and Auvergne, France.

Cox, Ross (1793-1853)
2676   1832, Adventures on the Columbia River, including ... six years on the western side of the Rocky Mountains ... : NY, J. and J. Harper, 335 p.

Coxe, John Redman (1773-1864)
2677   1822, [Endorsement for Conversations on Mineralogy by D. Lowry (see 6514)]: in 6514, p. ii.

Coxe, Tench (1755-1824)
2678   1789, Observations on the agriculture, manufactures and commerce of the United States. In a letter to a member of Congress. By a citizen of the United States: NY, F. Childs and J. Swaine (printers), 102 p.
          Evans number 21774.

Cozzens, Issachar (1780-1865)
2679   1825, Examination of iron ores from the northern part of the state of New-York: NY Lyc. Nat. History Annals, v. 1, pt. 2, no. 12, 378-383.
2680   1843, A geological history of Manhattan or New York Island, together with a map of the island, and a suite of sections, tables and columns, for the study of geology, particularly adapted for the American student: NY, W. E. Dean, 114 p., illus., 9 colored plates, map.
          Contains map 2681.
2681   1843, A geological map of New York or Manhattan Island by I. Cozzens, Jr.: colored geological map, 31.6 x 20.8 cm.
2682   1846, Description of three new fossils from the Falls of the Ohio: NY Lyc. Nat. History Annals, v. 4, 157-159, plate with 3 figs.

Craft, George
2683   1826, On the New Jersey marls: Phila. Soc. Promoting Ag. Mem., v. 5, 16.

Craig, Hector (see 2476 by C. D. Colden)

Cram, Thomas Jefferson (1807-1883)
2684   1837, On the length of a degree of the terrestrial meridian - oblateness and axes of the earth - comparative oblateness of the planets - reduction of latitude - radius of the earth - and length of a degree of parallels of latitude; with appropriate tables: Am. J. Sci., v. 31, no. 2, 222-235, fig.

Crane (Mr.)
2685   1838, Iron. - On the use of anthracite coal by the combination of heated air to the purpose of smelting iron ore: Am. J. Sci., v. 33, no. 2, 266-267.
          In British Assoc. Proc.

Craw, William J.
2686   1850, Analyses of phlogopite from St. Lawrence County, NY: Am. J. Sci., v. 10 (ns), no. 3, 383-385.

Crawe, J. B.
2687   1834, and A. Gray, A sketch of the mineralogy of a portion of Jefferson and St. Lawrence Counties, (N.Y.): Am. J. Sci., v. 25, no. 2, 346-350.

Croghan, George (d. 1782)
2688   1789, A list of the teeth and bones sent over: Am. Mus. or Univ. Mag., v. 5, 156.

Croghan, John
2689   1845, Rambles in the Mammoth Cave during the year 1844, by a visitor: Louisville, Kentucky, Morton and Griswold, 101 p., 6 plates, folded map.

Cronstedt, Axel Frederik (1722-1765; see 2502)

Croom, Hardy Bryan (1798-1837)
2690   1834, Some account of the agricultural soil and products of middle Florida: Farm's Reg., v. 2, 1-3.

Croom, H. B. (cont.)
2691   1834, Some account of the organic remains found in the marl pits of Lucas Benners, Esq. in Craven County, N. C.: Am. J. Sci., v. 27, no. 1, 168-171.

Croom, Issac
2692   1834, Calcareous deposits in North Carolina: Farm's Reg., v. 1, 614-615.

Cropley, R.
2693   1849, [Chemical analysis of Lake Superior minerals]: in Lake Superior report by C. T. Jackson (see 5655), 482; 496-497.

Cross, Osborne (1803-1876)
2694   1850, Journal of the march of the regiment of mounted riflemen to Oregon, from May 10 to October 5, 1849: US SED 1, 31-2, v. 1, pt. 2 (587), 127-244, 35 plates.

Crosse, Andrew (1784-1855)
2695   1837, Artificial crystals and minerals: Am. J. Sci., v. 31, no. 1, 374-375.
       In British Assoc. Proc. On the synthesis of minerals including quartz, azurite, malachite, galena, etc.

Crossley, Richard
2696   1850, Analysis of algerite: Am. J. Sci., v. 10 (ns), no. 1, 77-78.

Crossman, G. H.
2697   1830, Petrified forest of Missouri: Cab. Nat. History, v. 1, 71-72.
2698   1831, Petrified forest [at Yellowstone]: Mon. Am. J. Geology, v. 1, no. 5, 233-234; also in: The Friend, v. 5, 197 (1832) and NE Mag., v. 2, 89 (1832).

Crum Walter
2699   1846, Test for manganese: Am. J. Sci., v. 1 (ns), no. 2, 262 (7 lines).
       Abstracted from "Annalen der Chem. und Pharm."

The Cultivator (Albany, 1834-1850)
       In addition to the articles listed below, most volumes contain brief notes on the analysis and improvement of soils
2700   1834, Soils: v. 1, 110-113; 126-128.
       On the types and origin of soils.
2701   1835, Calcareous manures [review with excerpts of text by E. Ruffin (see 9088)]: v. 2, 64-66; 99-101.
2702   1838, [Review of Elements of geology by W. W. Mather (see 6875)]: v. 4, 177 (4 lines).
2703   1838, Geological science applied to agriculture [notice of Re-examination ..., by E. Hitchcock (see 5128)]: v. 5, 70.
2704   1838, Geological survey of Ohio [review with excerpts of survey by W. W. Mather (see 6891)]: v. 5, 72-73.
2705   1838, Dr. Dana's new mode of analyzing soils [notice of 2888]: v. 5, 114-115.
2706   1838, [Poland's salt mine at Wilitska]: v. 5, 147.
2707   1839, New York geological reports [review with excerpts of 7740]: v. 6, 67-68; 82; 99-100.
2708   1839, Agricultural geology [review with excerpts of Maine geology report (third) by C. T. Jackson (see 5566)]: v. 6, 131; 137-139; 152-154.
2709   1840, Geological survey of the state [of New York]: v. 7, 88, 108.
2710   1840, [Review of Rhode Island geology report by C. T. Jackson (see 5571)]: v. 7, 159; 172.
       Signed by "N. N. D."
2711   1841, Geological survey of New Hampshire: v. 8, 13 (11 lines).
2712   1841, Feldspar: v. 8, 31.
       On the composition and importance of feldspar.
2713   1841, [Review of Elementary geology by E. Hitchcock (see 5132)]: v. 8, 186.
2714   1842, [Silliman's Journal [notice of "Geology of the western states" by J. Hall (see 4676)]: v. 9, 28.
2715   1842, [Review of Muck manual ..., by S. L. Dana (see 2885), with excerpts]: v. 9, 90-92.
2716   1843, [Notice of Natural History of New York with excerpts on New York mineralogy and geology]: v. 10, 11; 28-29; 46; 58; 68; 124.
2717   1843, [Review of Essay on calcareous manures by E. Ruffin (see 9088)]: v. 10, 61.
2718   1843, Prof. Johnton's lectures: v. 10, 124; 140.
       Includes notice of 5819.
2719   1844, [Review of Geology for beginners by Richardson (see 8771)]: v. 1 (ns).
2720   1844, Geological survey of New-York: v. 1 (ns), 266.
       On sales difficulties of Natural History of New York.
2721   1845, [Review of Lectures ..., by J. F. W. Johnston (see 5819)]: v. 2 (ns), 33-35.
2722   1845, [Review of South Carolina geology report by M. Tuomey (see 10438) with excerpts]: v. 2 (ns), 155-156.
2723   1847, Chalybeate springs: v. 4 (ns), 273 (10 lines).
       On the origin of iron-bearing mineral water.
2724   1848, Visit of Prof. Johnston to the United States: v. 5 (ns), 193 (11 lines).
2725   1849, Digging gold at home and abroad: v. 6 (ns), 124.
       Signed by "D. M." On danger of depletion of East Coast workforce due to gold rush.
2726   1849, The address by Prof. Johnston: v. 6 (ns), 306.
2727   1850, Prof. Johnston's lectures: v. 7 (ns), 91.
       Notice of lectures given at Albany, New York.

Cumberland Coal and Iron Company
2728   1841?, An act to incorporate the Cumberland Coal and Iron Company: Balt.?, 5 p.

Cuming, Fortescue (1762-1828)
2729   1810, Sketches of a tour to the western country, through the states of Ohio and Kentucky; a voyage down the Ohio and Mississippi Rivers, and a trip through the Mississippi Territory, and part of West Florida, commenced at Philadelphia in the winter of 1807, and concluded in 1809: Pittsburgh, Crammer, Spear, and Eichbaum, 504 p.

Cunningham, John
2730   1839, Solid impressions and casts of drops of rain [from Storeton Hill, Cheshire, England]: Am. J. Sci., v. 37, no. 2, 371.
        Abstracted from London and Edinburgh Philos. Mag.
2731   1847, Footmarks in the New Red Sandstone of Storeton, near Liverpool: Am. J. Sci., v. 3 (ns), no. 1, 142 (5 lines).
        Abstracted from Geol. Soc. J.

Currie, William (1754-1828)
2732   1792, An historical account of the climates and diseases of the United States of America, ... : Phila., T. Dobson, 409 p.
        Evans number 27239. Includes topographical descriptions.

Curson, Samuel
2733   1823, Narrative of an ascent to the supposed volcano of Arequipa, or Peak of Miste in Peru: Bos. J. Phil. Arts, v. 1, no. 4, 352-370.

A cursory review of the Schuylkill coal, in reference to its introduction into New York, and the other Atlantic cities
2734   1823, NY, G. F. Hopkins, 24 p.

Cushing, Caleb (1800-1879)
2735   1836, Anthracite coal trade of Pennsylvania: Fam. Mag., v. 3, 433-438.
        From N. Am. Rev.
2736   1836, Anthracite coal trade of Pennsylvania [review of five coal mining reports, including 6153, 8206, and 9289]: N. Am. Rev., v. 42, 241-256.
2737   1840, [Reviews with excerpts of Journal of an exploring tour ..., by S. Parker (see 8121), and Narrative of a journey ..., by J. K. Townsend (see 10301)]: N. Am. Rev., v. 50, 75-144.

Cushing, Jonathan Peter (1793-1835)
2738   1833, Importance of investigating the mineral resources of Virginia, for promoting agricultural improvement and national wealth: Farm's Reg., v. 1, 118-119.
2739   1833, President Cushing's address: Va. Hist. Philos. Soc. Coll., v. 1, 9-33.
        On the importance of science to society.

Cutbush, Edward (1772-1843)
2740   1805, [Condition of Mount Etna, January, 1805]: Phila. Med. Phys. J., v. 2, pt. 1, 179-180.

Cutbush, James (1788-1823)
2741   1810, Experiments on the sulphate of barytes, found at Perkiomen [Pennsylvania]: Eclec. Rep. and Analytic Rev., v. 1, 226-229.
2742   1810, Experiments on the mineral waters of Bath, (Bristol, Penn.): Eclec. Rep. and Analytic Rev., v. 1, 366-369.
2743   1811, Analysis of American limestone: Phila. Soc. Promoting Ag. Mem., v. 5, 305-307; also in: Arch. Use. Know., v. 2, no. 1, 28-30.
2744   1811, An analysis of the water of the sulphur springs, situated in Adams County, Pennsylvania: Eclec. Rep. and Analytic Rev., v. 2, 102-104.
2745   1811, On the blue earth of New Jersey: Am. Min. J., v. 1, no. 2, 86-88.
        On vivianite from Allentown, New Jersey.
2746   1812, Hydrostatics, or a treatise on specific gravity: Phila., printed for the author, 49 p.
2747   1823, Localities of minerals near West Point: Am. J. Sci., v. 7, no. 1, 57-58.

Cutler, Jervis (1768-1844)
2748   1812, A topographical description of the state of Ohio, Indiana Territory, and Louisiana, comprehending the Ohio and Mississippi Rivers and their principal tributary streams; the face of the country, soils, waters, natural productions, animal, vegetable and mineral ... : Bos., Charles Williams, 219 p., plate, map.

Cuvier, Georges, baron (1769-1832)
2749   1797, [Large fossil mammals from South America]: Univ. Mag., v. 1, 331.
2750   1797, [Classification of beings]: Univ. Mag., v. 2, 318.
        From Nat'l Inst. Paris Proc. Signed "Citizen Cuvier."
2751   1797, Notice concerning the skeleton of a very large species of quadruped, found in Paraguay, and deposited in the cabinet of natural history at Madrid: Univ. Mag., v. 1, 337-339.
        On the fossil bones of a giant sloth.
2752   1804, [Fossil bones from Paris, France]: Am. Reg. or Gen. Repos., v. 1, pt. 2, 52-53.
        On fossil bones of a Paleotherium from the Paris Basin.
2753   1808, [Fossil bones of Hungary and Germany]: Am. Reg. or Gen. Repos., v. 2, 386-387.
        On extinct species of bear, rhinoceros, elephant, etc.
2754   1818, Essay on the theory of the earth ... with mineralogical notes, and an account of Cuvier's geological discoveries, by Professor Jameson. To which are added, observations on the geology of North America; ... by Samuel L. Mitchill: NY, Kirk and Mercein, 431 p., 7 plates.
        Contains sections by J. G. Bogert (1743), R. Jameson (5705 and 5706), S. L. Mitchill (7307), and W. Marsden (6783).
        For review with excerpts see 1094, 1228, 4509, and 8545. For review see 7678. For notice see 322, 1234, and 4578.
2755   1818, [Fossil crab claws from Paris Basin]: J. Sci. Arts, v. 3, no. 1, 158.
        In Royal Inst. France Proc.

Cuvier, G. (cont.)
2756   1823, [Excerpts from Researches on the fossil bones of quadrupeds]: Bos. J. Phil. Arts, v. 1, no. 3, 260-262.
       In article on the American mammoth by J. Ware (see 10694).
2757   1831, A discourse on the revolutions of the surface of the globe and the changes thereby produced in the animal kingdom: Phila., Carey and Lea, iv, 252 p., illus., 9 plates.
       "Translated from the French with illustrations and a glossary." For review see 9970.
2758   1832, and A. Brongniart, Cuvier and Brongniart's report on M. Deshayes' "Tableau comparatif des coquilles vivantes avec les fossiles des terrains tertiares de l'Europe: Am. J. Sci., v. 23, no. 1, 196-198.
       Abstracted from Revue Encyclopèdique.
2759   1833, Remarks on the connection between the Mosaic history of the Creation and the discoveries of geology, occasioned by the lectures of Baron Cuvier on the history of the natural sciences, and published in Prof. Jamesons's Edinburgh New Philosophical Journal in 1832: Am. J. Sci., v. 25, no. 1, 26-41, table.
2760   1834, Organic remains [on the origin of fossils]: Lit. J. and Wk. Reg. Sci. Arts, v. 1, no. 47, 373.
2761   1834, [Excerpts from Ossemens Fossiles]: Geol. Soc. Penn. Trans., v. 1, 53-54.
       In "Critical notices of various organic remains hitherto discovered in North America" by R. Harlan (see 4796).

D., C. (see 421, 903, 3161, 3719, and 4280; i.e. Chester Dewey?)

D., N. N. (see 2710)

D., S. E. (see 436)

D., T. R. (see 712)

D'Abbadie, A.
2762   1849, Mineral coal on the Nile: Am. J. Sci., v. 7 (ns), no. 3, 436 (5 lines).
       Abstracted from London Ath.

Dabney, John B.
2763   1808, [Account of a volcano]: Balt. Med. Phys. Rec'r, v. 1, 341-344.
2764   1809, Account of a volcanic eruption at St. George's, one of the Western Islands: Med. Repos., v. 12, 254-256; also in: Mon. Anth., v. 6, 64-65 (1809).

Da Costa, H. M.
2765   1818, Extract from a paper on the north of Ireland: J. Sci. Arts, v. 4, no. 1, 160-163.
       On the lithologies and stratigraphy of Northern Ireland.

Da Costa, Hippolyto I.
2766   1800, Description of the city of Lisbon; shewing the utility of constructing the houses, and paving the streets of cities, with marble, limestone, or other calcareous materials, in preference to siliceous materials, or bricks of clay: Med. Repos., v. 3, 1-5.

The Daguerreotype (Boston, 1847-1849)
2767   1847, [Review of A canoe voyage ..., by G. W. Featherstonhaugh (see 3786)]: v. 1, 36-38.
       From Westminster Rev.
2768   1847, [Review of How are worlds formed by S. Beswick (see 1669), and The ancient world ..., by D. T. Ansted (see 1280)]: v. 1, 417-426.
       From Tait's Edinburgh Mag.
2769   1847, The copper mines of South Australia: v. 1, 526.
       From London Ath.
2770   1848, [Marble quarry in Maremma, Tuscany]: v. 2, 143.
2771   1848, [Review with excerpts of Zoological recreations, by W. J. Broderip (see 2047)]: v. 2, 225-236.
       From Q. Rev.
2772   1849, Platina [from the Alps]: v. 3, 190 (8 lines).
2773   1849, Norwegian silver: v. 3, 524.
2774   1849, A parachute for coalpits: v. 3, 524.
       On the use of parachutes to descend into coal pits.

Dakin, Moses (1775-1865)
2775   1847, Monterey, or the mountain city; containing a desciption of the boundaries, situation, mountains, lakes, rivers, soil and climate, geological features, natural scenery, religious, moral and literary institutions, number of inhabitants and of professional men, and manufacturing resources of the southern part of the county of Waldo, Maine; to which is added an appendix giving a brief account of some of the islands of Penobscot Bay: Bos., Mead's Press, 26 p.

Dakin, Samuel
2776   1811, A peculiar earth found in Vermont, promising to be useful in the manufacture of porcelain: Med. Repos., v. 14, 404-406.

Dale (Mr.)
2777   1847, On elliptic polarization: Am. J. Sci., v. 3 (ns), no. 2, 262-264.
       From London Ath.

Damour, Alfred (1817-1897)
2778   1844, Anatase and dioptase: Am. J. Sci., v. 47, no. 1, 215-216, fig.
       Abstracted from "Ann. de Ch. et de Phys."
2779   1846, Diaspore: Am. J. Sci., v. 1 (ns), no. 1, 120 (4 lines).
       Abstracted from l'Institut.

Damour, A. (cont.)
2780   1846, and A. Descloizeaux, Mowenite and harmotome: Am. J. Sci., v. 2 (ns), no. 3, 417.
        Abstracted from l'Institut.
2781   1846, Oriental jade and tremolite: Am. J. Sci., v. 2 (ns), no. 2, 267.
        Abstracted from "Ann. de Ch. et de Phys."
2782   1847, Herschelite: Am. J. Sci., v. 3 (ns), no. 3, 429 (4 lines).
        Abstracted from "Ann. de Ch. et de Phys."
2783   1848, Columbite from near Limoges, France: Am. J. Sci., v. 5 (ns), no. 3, 422 (6 lines).
        Abstracted from Comptes Rendus.
2784   1848, Dufrenoysite: Am. J. Sci., v. 5 (ns), no. 2, 268 (9 lines).
        Abstracted from "Ann. de Ch. et de Phys."
2785   1848, Hydrosilicate of alumina, allied to halloysite: Am. J. Sci., v. 5 (ns), no. 2, 268 (11 lines).
        Abstracted from Geol. Soc. France Bulletin.
2786   1849, and M. Salvetat, On a new hydrosilicate of alumina: Am. J. Sci., v. 8 (ns), no. 1, 122 (7 lines).
        Abstracted from "Ann. de Ch. et de Phys."
2787   1849, On the composition of heulandite: Am. J. Sci., v. 8 (ns), no. 1, 122-123 (8 lines).
        Abstracted from Comptes Rendus.
2788   1849, Telluric bismuth from Brazil: Am. J. Sci., v. 8 (ns), no. 1, 127 (12 lines).
        Abstracted from "Annales de Ch. et de Phys."

Damsel, Hu. L.
2789   1850, On the manufacture of iron in South Wales: Franklin Inst. J., v. 19 (3s), 339-344.

Dana, Edmund
2790   1819, Geographical sketches of the western country: designed for emigrants and settlers: being the result of extensive researches and remarks ... : Cin., Looker, Reynolds and Co., iv, 5-312 p.

Dana, James Dwight (1813-1895)
        See also American Journal of Science, new series, volumes 1 to 10 (1846-1850).
        Dana served as co-editor with B. Silliman and B. Silliman, jr. See also 9509 with C. U. Shepard.
2791   1835, On the condition of Vesuvius in July, 1834: Am. J. Sci., v. 27, no. 2, 281-288.
2792   1835, A new system of crystallographic symbols: Am. J. Sci., v. 28, no. 2, 250-262, plate.
2793   1836, On the formation of compound or twin crystals: Am. J. Sci., v. 30, no 2, 275-300, 2 plates with 28 figs.
2794   1837, Crystallographic examination of eremite [from Watertown, Connecticut]: Am. J. Sci., v. 33, no. 1, 70-75, 3 figs.
2795   1837, A new mineralogical nomenclature: NY Lyc. Nat. History Annals, v. 4, no. 1 and 2, 9-34.
2796   1837, On the drawing of figures of crystals: Am. J. Sci., v. 33, no. 1, 30-50, 14 figs.
2797   1837, On the identity of the torrelite of Thomson with columbite: Am. J. Sci., v. 32, no. 1, 149-153, 3 figs.
2798   1837, A system of mineralogy: including an extended treatise on crystallography: with an appendix containing the application of mathematics to crystallographic investigation, and a mineralogical bibliography: NHav., Durrie, Peck, Herrick, and Noyes, xiv, 452, 119 p., illus., 4 plates.
        For notice, see 672. For review, see 681.
2799   1838, Supposed new mineral at Bolton, Mass.: Am. J. Sci., v. 35, no. 1, 178-179.
2800   1842, [Recognition of J. D. Schöpf's Beyträge zur mineralogischen Kenntnis]: Assoc. Am. Geol's Nat's Proc., v. 3, 32 (6 lines).
        Also in Assoc. Am. Geol's Nat's Rept., v. 1, 69 (6 lines), 1843.
2801   1843, Eremite: Am. J. Sci., v. 45, no. 2, 402, 2 figs.
2802   1843, [On the origin of coral islands]: Am. J. Sci., v. 45, no. 1, 141-142 (13 lines).
        In Assoc. Am. Geol's Nat's Proc.
2803   1843, On the analogies between the modern igneous rocks and the so called primary formations, and the metamorphic changes produced by heat in the associated sedimentary deposits: Am. J. Sci., v. 45, no. 1, 104-129.
        In Assoc. Am. Geol's Nat's Proc.
2804   1843, On the areas of subsidence in the Pacific, as indicated by the distribution of coral islands: Am. J. Sci., v. 45, no. 1, 131-135, folding map.
        In Assoc. Am. Geol's Nat's Proc.
2805   1843, On the temperature limiting the distribution of corals: Am. J. Sci., v. 45, no. 1, 130-131; no. 2, 310-311.
        In Assoc. Am. Geol's Nat's Proc., with discussion by H. D. Rogers, W. C. Redfield (see 8712), and J. L. Hayes (see 4911).
2806   1843, [Reply to the reasoning of Dr. Beck]: Am. J. Sci., v. 45, no. 1, 145.
        In Assoc. Am. Geol's Nat's Proc. See also 1578 and 1582.
2807   1844, [Review with excerpts of Mineralogy of New York, by L. C. Beck (see 1577)]: Am. J. Sci., v. 46, no. 1, 25-37, 5 figs.
2808   1844, On the composition of corals and the production of the phosphates, aluminates, silicates, and other minerals by the metamorphic action of hot water: Am. J. Sci., v. 47, no. 1, 135-136.
2809   1844, On the occurrence of magnesia and phosphoric acid in recent corals, as discovered by Mr. B. Silliman, Jr. and its bearing on many important geological and mineralogical questions: Assoc. Am. Geol's Nat's Proc., v. 5, 28-30.
2810   1844, Reply of J. D. Dana to the foregiong article by Mr. Couthouy: Am. J. Sci., v. 46, no. 2, Appendix, 10-12.
        See also 2671, 2673, 2674, 9777.

Dana, J. D. (cont.)
2811   1844, Reply to Mr. Couthouy's vindication against the charge of plagiarism: Am. J. Sci., v. 46, no. 1, 129-136.
       See also 2671, 2673, 2674, 9777.
2812   1844, A system of mineralogy, comprising the most recent discoveries: with numerous wood cuts and four copper plates: NY and London, Wiley and Putnam, 633, [1], 4 plates.
       For review see 6377. For review with excerpts see 9808.
2813   1844, [Withdrawal of plagiarism charge against J. P. Couthouy]: Am. J. Sci., v. 47, no. 1, 122 (10 lines).
       Also in Assoc. Am. Geol's Nat's Proc., v. 5, 30 (10 lines). See also 2671, 2673, 2674, 9777.
2814   1845, [Nomenclature in geology]: Assoc. Am. Geol's Nat's Proc., v. 6, 69-74.
       Also in Am. Q. J. Ag. Sci., v. 2, 167.
2815   1845, Observations on pseudomorphism: Am. J. Sci., v. 48, no. 1, 81-92; no. 2, 397-398.
2816   1845, On pseudomorphous minerals (Die Pseudomorphosen des Mineral riechs), by Dr. J. Reinhard Blum [review with excerpts of 2168)]: Am. J. Sci., v. 48, no. 1, 66-80.
2817   1845, On the minerals of trap and the allied rocks: Assoc. Am. Geol's Nat's Proc., v. 6, 26-28; 31-32.
       With discussion by C. U. Shepard and C. T. Jackson. Also in Am. Q. J. Ag. Sci., v. 2, 151-152.
2818   1845, Origin of the constituent and adventitious minerals of trap and the allied rocks: Am. J. Sci., v. 49, no. 1, 49-64.
2819   1846, Additional remarks: Am. J. Sci., v. 1 (ns), no. 2, 220-221.
       Follows "Observations on the Eocene formation of the United States" by T. A. Conrad (see 2585).
2820   1846, Genera of fossil corals of the family Cyathophyllidae: Am. J. Sci., v. 1 (ns), no. 2, 178-186, 5 figs.
2821   1846, On the occurrence of fluorspar, apatite and chondrodite in limestone: Am. J. Sci., v. 2 (ns), no. 1, 88-89.
2822   1846, On the volcanoes of the moon: Am. J. Sci., v. 2 (ns), no. 3, 335-355, 3 figs.
2823   1847, Count Keyserling's geology of the northeastern extremity of Russia in Europe; by Sir R. I. Murchison: Am. J. Sci., v. 4 (ns), no. 3, 419-420.
       Abstracted from London Ath. Discussion by Dana of Murchison's review of Keyserling.
2824   1847, Descriptions of fossil shells of the collections of the exploring expedition under the command of Charles Wilkes, U. S. N., obtained in Australia, from the lower layers of the coal formation in Illawarra, and from a deposit probably of nearly the same age at Harper's Hill, valley of the hunter: Am. J. Sci., v. 4 (ns), no. 1, 151-160.
2825   1847, A general review of the geological effects of the earth's cooling from a state of igneous fusion: Am. J. Sci., v. 4 (ns), no. 1, 88-92.
2826   1847, Geological results of the earth's contraction in consequence of cooling: Am. J. Sci., v. 3 (ns), no. 2, 176-188, 7 figs.
2827   1847, Observations in reply to Mr. Lonsdale's "Remarks": Am. J. Sci., v. 4 (ns), no. 3, 359-362.
       See also 6481. On the Tertiary corals of the United States.
2828   1847, On certain laws of cohesive attraction: Am. J. Sci., v. 4 (ns), no. 3, 364-385, 20 figs; also in Assoc. Am. Geol's Nat's Proc. in Am. Q. J. Ag. Sci., v. 6, 201-202.
       On the origin of crystals, and the relation of crystal forms to molecular arrangements.
2829   1847, On the origin of continents: Am. J. Sci., v. 3 (ns), no. 1, 94-100.
       For review see 10031.
2830   1847, Origin of the grand outline features of the earth: Am. J. Sci., v. 3 (ns), no. 3, 381-398, 10 figs.
2831   1848, Fossils of the exploring expedition under the command of Charles Wilkes, U. S. N.: a fossil fish from Australia, and a Belemnite from Tierra del Fuego: Am. J. Sci., v. 5 (ns), no. 3, 433-435.
2832   1848, Manual of mineralogy, including observations on mines, rocks, reduction of ores, and the applications of the science to the arts ... designed for the use of schools and colleges: NHav., Durrie and Peck; Phila., H. C. Peck, xii, [13]-432 p., 260 figs., diagram.
       Also other printings in 1849 and 1850, as "2nd edition" and "3rd edition." For review see 946, 6340, 5515, and 9177.
2833   1848, On a law of cohesive attraction, as exemplified in a crystal of snow: Am. J. Sci., v. 5 (ns), no. 1, 100-102, fig.
2834   1848, [On the laws of cohesive attraction]: Lit. World, v. 2, 230.
       Abstracted from Assoc. Am. Geol's Nat's Proc.
2835   1849, Gold in California: Am. J. Sci., v. 7 (ns), no. 1, 125-126.
2836   1848, [i.e. 1849], Map of New South Wales [and] District of Illawarra: in Geology of the United States exploring expedition, colored geological map 20 x 30 cm., facing p. 449 (see 2841 & 2842).
2837   1849, Notes on Upper California: Am. J. Sci., v. 7 (ns), no. 2, 247-264, fig., plate.
2838   1849, Observations on some points in the physical geography of Oregon and Upper California: Am. J. Sci., v. 7 (ns), no. 3, 376-394, 5 figs.
2839   1849, Observations on terraces: Am. J. Sci., v. 8 (ns), no. 1, 86-89.
2840   1849, Review of Chambers's Ancient sea margins, with observations on the study of terraces: Am. J. Sci., v. 7 (ns), no. 1, 1-14, fig.
       Review of 2299, with notes on 2839.
2841   1849, United States exploring expedition during 1838, 1839, 1840, 1841, and 1842, under Charles Wilkes, U. S. N. Vol. X. Geology. By James D. Dana, A. M., Geologist of the expedition: Phila., C. Sherman (printer), xii, 10-756 p., 4 maps, illus.
       This is the "official issue." 100 copies were printed, of which 30 were destroyed by fire. For review see 6366, 6864. For notice see 992. Includes report by T. A. Conrad 2592 and map 2836. See also Atlas 2843.

Dana, J. D. (cont.)

2842 1849, United States exploring expedition ...[as above]: NY, Geo. P. Putnam, xii, 10-756 p., 4 maps, illus.
This is an "unofficial issue." 100 copies printed.

2843 1849, United States exploring expedition during 1838, 1839, 1840, 1841, and 1842, under Charles Wilkes, U. S. N. Atlas. Geology. By James D. Dana, A. M., Geologist of the expedition: Phila., C. Sherman (printer), 6 p., 21 plates, folio.
This is the "official issue" to accompany 2841.

2844 1849, United States exploring expedition ...[as above]: NY, Geo. P. Putnam, 6 p., 21 plates, folio.
This is the "unofficial issue" to accompany 2842.

2845 1850, The crater of Kalauea [Hawaii]: Fam. Vis., v. 1, no. 13, 98.

2846 1850, Described species: Am. J. Sci., v. 10 (ns), no. 2, 248-252.
Abstracted from "J. für Prakt. Ch." and Poggendorff's Annalen. On recently described minerals from Europe.

2847 1850, Geology of the Pacific: Ann. Scien. Disc., v. 1, 238-242.
Abstracted from Geology of the United States Exploring Expedition (see 2841 to 2844).

2848 1850, Historical account of the eruptions on Hawaii: Am. J. Sci., v. 9 (ns), no. 2, 347-364, 6 figs.

2849 1850, Isomorphism of miargyrite and augite: Am. J. Sci., v. 9 (ns), no. 3, 429.

2850 1850, Note on heteronomic isomorphism: Am. J. Sci., v. 9 (ns), no. 3, 407.

2851 1850, New species [of minerals]: Am. J. Sci., v. 10 (ns), no. 2, 245-248.
Compiled from Poggendorff's Annalen, and Annales des Mines.

2852 1850, Observations on the mica family: Am. J. Sci., v. 10 (ns), no. 1, 114-119.

2853 1850, On a new crystalline form of staurotide and isomorphism of staurotide with andalusite and topaz: Am. J. Sci., v. 10 (ns), no. 1, 121, fig.

2854 1850, On danburite [from Danbury, Connecticut]: Am. J. Sci., v. 9 (ns), no. 2, 286-287, fig.

2855 1850, On denudation in the Pacific: Am. J. Sci., v. 9 (ns), no. 1, 48-62, fig.

2856 1850, On isomorphism and atomic volume of some minerals: Am. J. Sci., v. 9 (ns), no. 2, 220-245, tables.

2857 1850, On some minerals recently investigated by M. Hermann: Am. J. Sci., v. 9 (ns), no. 3, 408-412.

2858 1850, On the degradation of the rocks of New South Wales and formation of valleys: Am. J. Sci., v. 9 (ns), no. 2, 289-294, fig.
Abstracted from Geology of the United States exploring expedition (see 2841 to 2844).

2859 1850, On the isolation of volcanic action in Hawaii, or volcanoes no safety valves: Assoc. Am. Geol's Nat's Proc., v. 2, 95-100.

2860 1850, On the ozarkite of Shepard: Am. J. Sci., v. 9 (ns), no. 3, 430-431.

2861 1850, On the trend of islands and axis of subsidence in the Pacific: Assoc. Am. Geol's Nat's Proc., v. 2, 321-325.

2862 1850, On the volcanic eruptions of Hawaii: Am. J. Sci., v. 10 (ns), no. 2, 235-244, fig.
See also 2848.

2863 1850, Spodumene [from Norwich, Massachusetts]: Am. J. Sci., v. 10 (ns), no. 1, 119-120, 2 figs.

2864 1850, Supposed staurotide of Norwich, Mass.: Am. J. Sci., v. 10 (ns), no. 3, 414 (7 lines).

2865 1850, A system of mineralogy, comprising the most recent discoveries: including full descriptions of species and their localities, chemical analyses and formulas, tables for the determination of minerals, and a treatise on mathematical crystallography and the drawing of figures of crystals: NY and London, George P. Putnam, 711, [1] p.
For review see 1014. "Third edition."

2866 1850, Table of atomic weights: Am. J. Sci., v. 9 (ns), no. 2, 217-219, table.

2867 1850, Volcanoes no safety-valves: Ann. Scien. Disc., v. 1, 232-234.

Dana, James Freeman (1793-1827)

2868 1817, Chemical examination of the water of the Congress Spring, Saratoga: NE J. Medicine Surg., v. 6, 19-23.
For review with excerpts see 7681.

2869 1817, [Review with excerpts of A manual of mineralogy by A. Aikin (see 101)]: N. Am. Rev., v. 5, 74-81.

2870 1818, and S. L. Dana, A geological map of Boston and its vicinity engraved for Danas' Outlines of the mineralogy and geology of Boston: Am. Acad. Arts Sci's Mem., v. 4, pt. 1, colored geological map 35.2 x 31.1 cm.
Published with 2871. Also private printing (see 2872).

2871 1818, and S. L. Dana, Outlines of the mineralogy and geology of Boston and its vicinity, with a geological map: Am. Acad. Arts Sci's Mem., v. 4, pt. 1, 129-223.
For review see 4509 and 1239. For notice see 7937.

2872 1818, and S. L. Dana, Outlines of the mineralogy and geology of Boston and its vicinity with a geological map: Bos., Cummings and Hilliard, 108 p., map.

2873 1820, Localities of minerals [in New Hampshire]: Am. J. Sci., v. 2, no. 2, 241.

2874 1822, A fragment - Putnam's rock: Am. J. Sci., v. 5, no. 1, 37-38.
Description of a large glacial erratic.

2875 1823, Miscellaneous localities of minerals, communicated by various persons: Am. J. Sci., v. 6, no. 2, 245.
On mineral localities in New Hampshire and Vermont.

2876 1824, By Prof. J. F. Dana: Am. J. Sci., v. 8, no. 2, 234-235 (14 lines).
On New Hampshire mineral localities.

2877 1824, New locality of cobalt: Am. J. Sci., v. 10, no. 1, 198 (5 lines).

2878 1824, Remarks on the common methods of detecting cobalt: Am. J. Sci., v. 8, no. 2, 301-304.

2879 1827, Analysis of the copper ore of Franconia, New-Hampshire, with remarks on pyritous copper: NY Lyc. Nat. History Annals, v. 2, no. 7 and 8, 253-258.

2880 1827, Analysis of the water of the Congress Spring, Saratoga, N. Y.: NY Med. Phys. J., v. 6, 66-73.

Dana, Samuel Luther (1795-1868)
See also 2870, 2871 and 2872 with J. F. Dana.
2881 1838, Rules of analysis: in Report of a reexamination ..., by E. Hitchcock (see 5128).
New method for soil analysis described. For review see 2705.
2882 1839, Letters ... on ashes, lime, &c.: Am. Farm., v. 6 (2s), 2-3; 12.
On agricultural geology.
2883 1839, [On soil analysis]: Am. J. Sci., v. 36, no. 2, 363-378.
In C. U. Shepard's review of Report of a reexamination ..., by E. Hitchcock (see 9523). See also 5128 and 2881.
2884 1840, Rules of analysis: Am. Rep. Arts Sci's Manu's, v. 1, 251-253.
In "Analysis of soils" by C. A. Lee (see 6126).
2885 1842, A muck manual for farmers: Lowell, Mass., D. Bixby, 242 p.
For review with excerpts see 2715. For review see 770, 1843, 1900, 9940, 9961.
Includes a chapter on soil geology.
2886 1842, Geology of soils: Am. Rr. J., v. 15, 312-319.
Extracted from Muck manual for farmers (see 2885).
2887 1843, A muck manual for farmers: Lowell, Mass., Bixby and Whiting, 2, iii-x, [11]-232 p.
"2nd ed. with additions."
2888 1845, Analysis of coprolites from the New Red Sandstone formation of New England: Am. J. Sci., v. 48, no. 1, 46-60.
With remarks by E. Hitchcock (see 5167).

Danforth, John (1660-1730)
2889 1728, A sermon occasioned by the late great earthquake, and the terrors that attended it. ... : Bos., printed by G. Rogers for John Eliot, [4], 46, 5 p.
Evans number 3016.

Danger, T. P.
2890 1829, Blowpipe simplified: Am. J. Sci., v. 17, no. 1, 163-164; also in: Mech's Mag., v. 1, 170 (1830); also in: Am. J. Sci., v. 22, no. 2, 376 (1832).
Abstracted from "Bulletin d'Encour."

Daniell, John Frederic (1790-1845)
2891 1817, On some phenomena attending the process of solution, and on their application to the laws of crystallization: J. Sci. Arts, v. 1, no. 1, 24-49, 16 figs on 3 plates.
2892 1818, On the strata of a remarkable chalk formation in the vicinity of Brighton and Rottingdean: J. Sci. Arts, v. 4, no. 2, 227-232, 2 folding plates with 4 figs.
2893 1818, On the theory of spherical atoms, and on the relation which it bears to the specific gravity of certain minerals: J. Sci. Arts, v. 4, no. 1, 30-42, 2 plates with 10 figs.
2894 1833, On a new oxy-hydrogen jet: Franklin Inst. J., v. 11 (ns), 217-221, 4 figs.
2895 1838, Blowpipe mouth for oxygen and hydrogen: Am. J. Sci., v. 35, no. 1, 187-188.
Abstracted from Philos. Mag.
2896 1839, [Extracts from geological report on the Maryland Mining Company]: see 6812.
Signed by J. W. Daniell [i.e. J. F.]
2897 1847, [Extract from geological report on the Barrelville Mining Company]: see 1409.

Danker, Albert
2898 1829, Magnetic variation [at Troy, New York]: Am. J. Sci., v. 17, no. 1, 198.

Dann, D.
2899 1818, Method of analyzing spring water: Mon. Scien. J., v. 1, 32-33.
Also in Scien. J., v. 1, 32-33.

Darby, William (1775-1854)
2900 1816, A geographical description of the state of Louisiana: presenting a view of the soil, climate, animal, vegetable and mineral productions; illustrative of its natural physiognomy, its geographical configuration, and relative situation; with an account of the character and manners of the inhabitants; being an accompaniment to the map of Louisiana: Phila., Printed for the author by J. Melish, ix, [11]-270, [xiv], 11, [2], xvii p., map.
2901 1817, A geographical description of the state of Louisiana ... [as above]: NY, James Olmstead, xi, [1], [13]-356, [3] p., 3 maps.
Second edition, enlarged and improved.
2902 1818, The emigrant's guide to the western and southwestern states and territories; comprising a geographical and statistical description of the states of Louisiana, Mississippi, Tennessee, Kentucky, and Ohio; the territories of Alabama, Missouri, Illinois and Michigan; and the western parts of Virginia, Pennsylvania, and New York ... : NY, Kirk and Mercein, 3, 1, 311, xiii p., 2 maps, tables.
2903 1819, [Syllabus for "Darby's geographical lectures"]: The Academician, v. 1, no. 22, 350-352.
2904 1819, A tour from the city of New-York to Detroit, in the Michigan territory, made between the 2d of May and 22d of September, 1818 ... [with] notices of the ... natural history and geography ... : NY, Kirk and Mercein, viii, [9]-228, lxiii, [7] p., 3 folded maps.
2905 1821, Memoir on the geography and natural and civil history of Florida attended by a map of that country, connected with the adjacent places: Phila., T. H. Palmer, vii, [5]-92 p., folded map.
2906 1824, Geographical view of Pennsylvania: Geog. Hist. Statistical Repos., v. 1, 9-35.
2907 1828, View of the United States, historical, geographical; and statistical; exhibiting in a convenient form the natural and artificial features of the several states, and embracing ... : Phila., H. S. Tanner, 654 p., folded maps.
2908 1841, Organic remains [from Clarke County, Alabama]: Hazard's US Reg., v. 5, 203.
2909 1841, Organic remains discovered in Alabama: Mon. Chron., v. 2, 421-423.
On the discovery of fossil saurian bones in Clarke County, Alabama.

Darlington (Mr.)
2910   1849, Mode of extinguishing fires in coal mines: Franklin Inst. J., v. 18 (3s), 152-154.
       Abstracted from London J. Arts.

Darlington, William (1782-1863)
2911   1826, On plaster of Paris: Phila. Soc. Promoting Ag. Mem., v. 5, 178-182.

Daru, Pierre Antoine Noël Bruno, comte (1767-1829)
2912   1821, Volney. Discourse pronounced in the Chamber of Peers, Jan. 14, 1820: Lit. Gaz., or J. Criticism, v. 1, no. 2, 20-23.
       From Revue Encyclopédique. Obituary of C. F. C. Volney.

Darwin, Charles Robert (1809-1882)
2913   1837, [Geology of the Rio de la Plata, Brazil]: Am. J. Sci., v. 33, no. 1, 210.
       Abstracted in London Geol. Soc. Proc.
2914   1841, Uprising of the earth: Am. Eclec., v. 1, no. 1, 186.
       Abstracted from Lit. Gaz.
2915   1846, Falkland Islands: Am. J. Sci., v. 2 (ns), no. 1, 124 (9 lines).
       Abstracted from London Ath.
2916   1846, Voyage of a naturalist, or journal of researches into the natural history and geology of the countries visited during the voyage of H. M. S. Beagle round the world, under the command of Capt. Fitz Roy, R. N.: NY, Harper and Brothers, 2 v. (324 p.).
       For notice see 3310, 10827. For review see 3546, 4487, 5455, 8175.
2917   1847, Cause of the absence of ancient marine formations from certain regions: Am. J. Sci., v. 3 (ns), no. 1, 120-121.
       From his "Geol. Obs."

Daubeny, Charles Giles Briddle (1795-1867)
2918   1820, Extract of a letter: Am. J. Sci., v. 2, no. 2, 351-352.
       Abstracted by J. Griscom.
2919   1825, Geonostical map of Sicily by Dr. Daubeny: Am. J. Sci., v. 10. no. 2, colored geological map, 25.5 x 19.7 cm., facing 256.
       Published with 2920.
2920   1826, Sketch of the geology of Sicily: Am. J. Sci., v. 10, no. 2, 230-256, map.
       Contains map (see 2919).
2921   1828, [Excerpts from Description of active and extinct volcanoes]: Am. J. Sci., v. 13, no. 2, 235-310; v. 14, no. 1, 70-91.
       In review by B. Silliman (see 9677). See also 9673.
2922   1830, Information desired respecting mineral waters: Am. J. Sci., v. 17, no. 2, 407-408.
2923   1831, On the development of Azotic gas in warm springs: Am. J. Sci., v. 20, no. 2, 383.
       Abstracted from Bibliothèque Universelle.
2924   1832, [Hot springs and their relationships to volcanic action]: Mon. Am. J. Geology, v. 1, no. 10, 474 (9 lines).
       In British Assoc. Proc.
2925   1832, Iodine and bromine in mineral waters: Am. J. Sci., v. 21, no. 2, 366.
       Abstracted from Royal Inst. J., by J. Griscom.
2926   1836, Carbonate of magnesia discovered in lava [from Vesuvius]: Am. J. Sci., v. 29, no. 2, 348 (3 lines).
       In British Assoc. Proc.
2927   1838, [Excerpts from "Excursion to the Lake Amsanctus and to Mount Vultur in Apulia" and from "Report on the present state of our knowledge with respect to mineral and thermal waters"]: NY Rev., v. 3, no. 1, 20-43.
       In review (see 7815).
2928   1839, Notice of the thermal springs of North America, being an extract from an unpublished memoir on the geology of North America: Am. J. Sci., v. 36, no. 1, 88-93.
2929   1839, On the geology and thermal springs of North America: Am. J. Sci., v. 35, no. 2, 307 (14 lines).
       In British Assoc. Proc.
2930   1839, Quantity of salt in sea water: Am. J. Sci., v. 36, no. 1, 188.
2931   1839, Reply of Dr. Daubeny to Professor Bischof's objections to the chemical theory of volcanos: Am. J. Sci., v. 37, no. 1, 78-84.
       See also article by G. Bischof (1723).
2932   1841, Sketch of the geology of North America, being the substance of a memoir read before the Ashmolean Society, November 26, 1838: Am. J. Sci., v. 41, no. 1, 195-199.
2933   1842, On the disintegration of the dolomitic rocks of the Tyrol: Am. J. Sci., v. 42, no. 2, 321-322.
       In British Assoc. Proc.
2934   1843, On the agricultural importance of ascertaining the minute portions of matter derived from organic sources that may be preserved in the surface soil, and on the chemical means by which its presence may be detected: Am. J. Sci., v. 44, no. 2, 352-355.
       In British Assoc. Proc.
2935   1846, Phosphorite rock of Estremadura, Spain, a good material for manure: Am. J. Sci., v. 1 (ns), no. 2, 277-278.
       From British Assoc. Rept.
2936   1848, Remarks on some of the applications of chemistry to geological research: Franklin Inst. J., v. 16 (3s), 211-212.
       From London Ath.

Daubrée, Auguste (1814-1896)
2937   1844, Description and theory of tin veins: In New Hampshire geology report, by C. T. Jackson (see 5611), 143-145.
2938   1846, Axinite and other minerals in a fossiliferous rock: Am. J. Sci., v. 2 (ns), no. 1, 123 (12 lines).
       Abstracted from Geol. Soc. France Bulletin.

Daubree, A. (cont.)
2939 1846, Gold washings of the Rhine: Am. J. Sci., v. 2 (ns), no. 3, 419.
      Abstracted from l'Institut.
2940 1850, On the artificial production of certain crystallized minerals, particularly oxyd of tin, oxyd of titanium, and quartz: Am. J. Sci., v. 9 (ns), no. 1, 120-121.
      Abstracted from Comptes Rendus.
2941 1850, On the origin of the titaniferous veins of the Alps: Am. J. Sci., v. 9 (ns)9, no. 1, 122.
      Abstracted from Comptes Rendus.

Dauphin and Susquehanna Coal Company
2942 1832, Act to incorporate. Apl. 4, 1826: Phila., 8vo.
      Not seen.
2943 1848, Report to the stockholders of the Dauphin and Susquehanna Coal Co.: Phila., T. K. and P. G. Collins (printers), 16 p., front., folded map.

Dauvergne, Dexter L.
2944 1848, Improvement in washing ores: Franklin Inst. J., v. 16 (3s), 179 (11 lines).

Davis, Charles Henry (1807-1877)
2945 1849, [Review of The earth and man by A. Guyot (see 4609)]: N. Am. Rev., v. 69, 250-255.
2946 1849, The theory of the geological action of the tides: Am. Assoc. Adv. Sci. Proc., v. 1, 27-28; also in: Ann. Scien. Disc., v. 1, 244-245 (1850).
2947 1849, A memoir upon the geological action of the tidal and other currents of the ocean: Am. Acad. Arts Sci's Mem., v. 4 (ns), 117-156, 2 plates; also in: Ann. Scien. Disc., v. 1, 242-244.
      For review see 986, 1902, 6863.

Davis, D. D. (see 9842 with others)

Davis, E. H.
2948 1847, Footprints and Indian sculpture: Am. J. Sci., v. 3 (ns), no. 2, 286-288, fig.
      On human footprints in limestone.

Davis, Emerson
2949 1825, By Emerson Davis: Am. J. Sci., v. 9, no. 2, 252.
      On mineral localities near West Springfield, Mass.
2950 1826, Notice of rocks and minerals in Westfield, Mass.: Am. J. Sci., v. 10, 213-215.
      With "appendix" by Mr. Chilton (see 2348).

Davis, Jacob P. (or Jacob S.?)
2951 1832, Geology of Wayne County, Pennsylvania: Mon. Am. J. Geology, v. 1, 520-523.

Davis, Jacob S. (or Jacob P.?)
2952 1829, Wayne County [Pennsylvania]: Hazard's Reg. Penn., v. 3, 135-139.

Davis, John (Judge)
2953 1809, An attempt to explain the inscription on the Dighton rocks: Am. Acad. Arts Sci's Mem., v. 3, pt. 1, 197-205.

Davis, N. S.
2954 1843, Medical and topographical sketches of Binghamton and the surrounding country: Med. Soc. NY Trans., v. 5, 459-465.

Davis, William
      See also 4839 with C. Harsleban.
2955 1833, A machine for washing gold and other ores: Franklin Inst. J., v. 11 (ns), 314 (4 lines); 322-323, fig.

Davy, C.
2956 1795, and F. Davy, A relation of a journey to the glaciers, in the Duchy of Savoy: Penn. Mag., v. 1, 520-523; 750 [i.e. 570]-573; 608-611.

Davy, F. (see 2956 with C. Davy)

Davy, Sir Humphrey (1778-1829)
2957 1808, Decomposition of fixed alkalis: Phila. Med. Mus., v. 5, 40-43.
      Abstracted from "Med. & Chir. Rev."
2958 1808, Earths, ascertained to be metallic oxides: Phila. Med. Mus., v. 5, 188-189.
      Abstracted from Med. and Phys. J.
2959 1809, Account of further discoveries in extracting metals from the earths and alkalis: Phila. Med. Mus., v. 6, 33-35.
2960 1809, A sketch of Mr. Davy's recent discoveries of the chemical agencies of electricity, and of the decomposition and composition of fixed alkalis, of ammonia, the alkaline earths, &c.: Phila. Med. Mus., v. 6, 89-115.
2961 1811, On the analysis of soils, as connected with their improvement: Agri. Mus., v. 1, 285-289; 252-256; 264-269; 273-279.
2962 1815, Elements of agricultural chemistry: Hart., Hudson and Co., 304 p.
      "Second edition", but first American-published edition. For excerpts see 3717.
2963 1815, Elements of agricultural chemistry: Balt., R. Gray and W. F. Gray, 332 p.
      Also Phila., NY and Bos. editions.
2964 1817, Observations on the preceeding paper: J. Sci. Arts, v. 1, no. 2, 262-264.
      On M. Faraday's analysis of caustic lime of Tuscany (see 3676).
2965 1817, On the wire-guage safe-lamps for preventing explosions from fire-damps, and for giving light in explosive atmospheres of coal mines: J. Sci. Arts, v. 1, no. 1, 1-5, plate.
2966 1818, Combustion of the diamond: J. Sci. Arts, v. 4, no. 1, 155.

Davy, H. (cont.)
2967   1819, Elements of agricultural chemistry: NY, Eastburn and Co.; Bos., Ward and Lilly, 332 p., plates.
2968   1820, Agricultural chemistry, No. XXV: Plough Boy, v. 1, no. 36, 283-284.
       On soil analysis.
2969   1821, Elements of agricultural chemistry: Phila., B. Warner, 304, 92 p.
2970   1824, On manures: Penn. Agri. Soc. Mem., v. 1, 258-281.
2971   1825, On soils: their constituent parts. On the analysis of soils. Of the uses of the soil. Of the rocks and strata found beneath soils. Of the improvement of soil: Carolina J. Medicine, Sci., Ag., v. 1, 72-94.
       From Elements of agricultural chemistry.
2972   1828, Volcanoes: Am. J. Improvements Use. Arts, v. 1, 217-218.
       On the nature of volcanoes.
2973   1829, Color of the sea: Am. J. Sci., v. 17, no. 1, 170 (6 lines).
       On iodine and bromine in sea water.
2974   1829, On the phenomena of volcanoes: Franklin Inst. J., v. 4 (ns), 154-161.
       From Philos. Trans.
2975   1838/1839, [Excerpts from Elements of agricultural chemistry (see 2962)]: Obs. and Rec. Ag. Sci. Art, v. 1, no. 1, 1-3; no. 2, 30-32; no. 3, 44-48; no. 4, 61-64; no. 5, 75-78; no. 6, 94-96; no. 7, 104-112; no. 8, 122-126; no. 9, 132-141; no. 10, 149-153; no. 11, 162-164; no. 12, 179-182; to be continued but no more published.

Davy, J.
2976   1850, Carbonate of lime as an ingredient of sea-water: Ann. Scien. Disc., v. 1, 201-202 (15 lines).
       Abstracted from Brewster's Mag.

Davy, John (1790-1868)
2977   1818, A description of Adam's Peak [in Ceylon]: J. Sci. Arts, v. 5, no. 1, 25-30.
2978   1818, Extract of a letter: J. Sci. Arts, v. 5, no. 2, 233-235.
       Geological notes on Ceylon.
2979   1823, Geology and mineralogy of Ceylon: Bos. J. Phil. Arts, v. 1, no. 1, 53-59.
       Abstracted by J. W. Webster from Geol. Soc. London Trans.

Dawson, John William (1820-1899)
2980   1846, Stigmaria [from Pictou, Nova Scotia]: Am. J. Sci., v. 1 (ns), no. 3, 435 (7 lines).
       Abstracted from Geol. Soc. London Proc.
2981   1847, [The gypsum of Antigonish, Nova Scotia]: Phila. Acad. Nat. Sci's Proc., v. 3, no. 11, 271-274.

Day, Jeremiah (1773-1867)
2982   1810, A view of the theories which have been proposed, to explain the origin of meteoric stones: Ct. Acad. Arts Sci's Mem., v. 1, pt. 1, 163-174; also in: Am. Med. Philos. Reg., v. 3, no. 2, 221-225 (1812).
       For review see 4242.

Dean, Philotus (Reverend)
2983   1849, Animalcules: Oberlin Q. Rev., v. 4, 36-58.
       On the geological significance of animalcules.

Deane, James (1801-1858)
2984   1843, Ornithichnites of the Connecticut River sandstone, and the Dinorthis of New Zealand: Am. J. Sci., v. 45, no. 1, 177-183.
2985   1844, Answer to the "Rejoinder" of Prof. Hitchcock: Am. J. Sci., v. 47, no. 2, 399-401.
       See article by E. Hitchcock (5162). On fossil footsteps in the red sandstone of the Connecticut River Valley.
2986   1844, On the discovery of fossil footmarks: Am. J. Sci., v. 47, no. 2, 381-390.
2987   1844, On the fossil footmarks of Turner's Falls, Massachusetts: Am. J. Sci., v. 46, no. 1, 73-77, 2 plates.
2988   1845, Description of fossil footprints in the New Red Sandstone of the Connecticut Valley: Am. J. Sci., v. 48, no. 1, 158-167, folding plate.
2989   1845, Fossil footmarks and rain-drops: Am. J. Sci., v. 49, no. 1, 213-214.
2990   1845, Illustrations of fossil footmarks [from the Connecticut River Valley]: Bos. J. Nat. History, v. 5, no. 2, 277-284, 3 figs., plate with 4 figs.
2991   1845, Notice of a new species of batrachian footmarks: Am. J. Sci., v. 49, no. 1, 79-81, 2 figs.
2992   1845, Notice of new fossil footmarks in the New Red Sandstone: Assoc. Am. Geol's Nat's Proc., v. 6, 25 (9 lines).
       Also in Am. Q. J. Ag. Sci., v. 2, 147-148.
2993   1847, Fossil footprints [from the Connecticut River Valley]: Am. J. Sci., v. 4 (ns), no. 3, 448-449.
2994   1847, Notice of new fossil footprints: Am. J. Sci., v. 3 (ns), no. 1, 74-79, 4 figs.
       On Triassic fossil footprints from Turner's Falls, Massachusetts.
2995   1848, Fossil footprints of a new species of quadruped [from Turner's Falls, Massachusetts]: Am. J. Sci., v. 5 (ns), 40-41, fig.
2996   1849, Illustrations of fossil footprints of the valley of the Connecticut: Am. Acad. Arts Sci's Mem., v. 4 (ns), 209-220, 9 plates with 27 figs.
2997   1850, Fossil footprints of Connecticut River: Phila. Acad. Nat. Sci's J., v. 2 (ns), pt. 1, 71-74, 2 plates with 5 figs.

Deane, Samuel (1733-1814)
2998   1785, An account of yellow and red pigment, found at Norton, with the process for preparing the yellow for use; accompanied with specimens: Am. Acad. Arts Sci's Mem., v. 1, 378-379.
       On the use of minerals from Bristol County, Massachusetts as dyes.

"Death of Denison Olmsted, jr."
2999   1846, in second Vermont report by C. B. Adams (see 29), 252-253.
        Unsigned article extracted from the New Haven Paladium, August 17.

De Bow, James Dunwoody Brownson (1820-1867)
3000   1846, Coins, weights, and measures: De Bow's Rev., v. 2, no. 5, 283-302.
        On the precious metal production of Mexican mines.
3001   1849, Kentucky: De Bow's Rev., v. 7, 191-205.
        Includes notice of Mammoth Cave, Big Bone Lick, mineral springs, and other geological features.

De Bow's Review of the Southern and Western States (New Orleans, 1846-1850)
3002   1846, American coal trade: v. 2, 210-211.
        Pennsylvania and foreign production statistics for 1820 to 1845.
3003   1847-1848, [Brief notes on American mineral resources]: v. 3, 273, 275, 445; v. 4, 92 (1847); v. 5, 93, 382 (1848).
3004   1848, Mineral lands [notice of Lake Superior report by C. T. Jackson (see 5640) and Wisconsin and Iowa report by D. D. Owen (see 8065)]: v. 6, 99.
3005   1848, [Review of Report of a geological exploration of part of Iowa, Wisconsin and Illinois by D. D. Owen (see 8065)]: v. 5, 107-108.
3006   1849, Commerce of Philadelphia: v. 7, 365-368.
        Pennsylvania coal trade statistics for 1820 to 1848.
3007   1849, Mineral resources of Arkansas: v. 7, 376-377.
3008   1849, Lead ore in Marion County, Arkansas: v. 7, 549-550.
3009   1850, Georgia and her resources. Her population; internal improvements; productions; enterprise; minerals; manufactures; mineral springs: v. 8, 39-45.
3010   1850, Coal trade of Ohio: v. 8, 163-164.
3011   1850, Pennsylvania and her resources: v. 8, 168-169.
3012   1850, Southern granite [from Columbia, South Carolina]: v. 8, 169-171; v. 9, 115-118; 434-436.
        Signed "by a Charleston working-man."
3013   1850, The Baron Humboldt's "Cosmos." The physical history of the universe examined and displayed [review with excerpts of 5395]: v. 9, 150-158; 271-275.
3014   1850, The gold of California: v. 9, 246-248.
3015   1850, Iron manufacture in East Mississippi: v. 9, 331-332.

Dechaud, M.
3016   1846, and M. Gualtier de Claubry, Extraction of copper from its ores by electricity: Franklin Inst. J., v. 11 (3s), 128-131.
3017   1847, and M. Gualtier de Claubry, Improvement in smelting copper ores by electricity: Am. Q. J. Ag. Sci., v. 6, 106-108.

Dechen, Heinrich von (1800-1889)
3018   1848, On the occurrence of ores of mercury in the coal formation of Saarbrück: Am. J. Sci., v. 6 (ns), no. 3, 426-427.
        Abstracted from "Geol. J."

Deck, Isaiah
3019   1847, General rules for the quantitative analysis of soils: Franklin Inst. J., v. 14 (3s), 417-419.
        Abstracted from Farm's Mag.

Dekay, James Ellsworth (1792-1851)
        See also 2635 and 2636 with T. Cooper.
3020   1823, Note on the organic remains, termed Bilobites, from the Catskill Mountains [New York]: NY Lyc. Nat. History Annals, v. 1, pt. 1, no. 2, 45-49, plate with 4 figs.
3021   1824, and J. Van Rensselaer and W. Cooper, Account of the discovery of a skeleton of the Mastodon giganteum: NY Lyc. Nat. History Annals, v. 1, pt. 1, no. 5, 143-147.
        Fossil bones discovered at Long Branch, Monmouth County, New Jersey.
3022   1824, Observations on the structure of trilobites, and description of an apparently new genus [from Trenton Falls, New York]: NY Lyc. Nat. History Annals, v. 1, 174-189, 2 plates with 5 figs.
        See also "Geology of Trenton Falls" by J. Renwick (8745).
3023   1825, Observations on a fossil crustaceous animal of the order Branchiopoda: NY Lyc. Nat. History Annals, v. 1, pt. 2, no. 12, 375-377, plate.
        On Eurypterus from Westmoreland, Oneida County, New York.
3024   1827, on a fossil skull in the cabinet of the Lyceum, of the genus Bos from the banks of the Mississippi; with observations on the American species of that genus: NY Lyc. Nat. History Annals, v. 2, 280-291.
3025   1827, Report on several multiocular shells from the state of Delaware: with observations of a second specimen of the new fossil genus Eurypterus: NY Lyc. Nat. History Annals, v. 2, 273-279, plate with 4 figs.
3026   1828, On the supposed transportation of rocks: Am. J. Sci., v. 13, no. 2, 348-350.
3027   1829, Remarks on certain phenomena exhibited upon the surface of the Primitive rocks in the vicinity of this city [New York City]: Am. J. Sci., v. 16, no. 2, 357 (11 lines).
        In NY Lyc. Nat. History Proc.
3028   1830, On the remains of extinct reptiles of the genera Mosasaurus and Geosaurus found in the Secondary formation of New Jersey; and on the occurrence of the substance recently named coprolite: NY Lyc. Nat. History Annals, v. 3, 134-141, plate with 6 figs.
3029   1834, Examination of the facts and arguments by which it is attempted to be proved that lava has not been subjected to great elevations of temperature: Am. J. Sci., v. 27, no. 1, 148 (9 lines).
        In NY Lyc. Nat. History Proc.
3030   1836, Observations on a fossil jaw of a species of gavial from west Jersey: NY Lyc. Nat. History Annals, v. 3, 156-165, plate with 4 figs.

Dekay, J. E. (cont.)
3031 1840, [Analysis of coal and iron ore from Maryland]: Am. Rep. Arts Sci's Manu's, v. 1, 178.
   In NY Lyc. Nat. History Trans.
3032 1842, [List of the fossil fishes of New York]: in Zoology of New York: Alb., Van Benthuysen (printer), v. 4, 385-389.
3033 1842, [Notes on fossil mammals of New York]: in Zoology of New York: Alb., Van Benthuysen (printer), v. 1, several notes, especially 98-106, plate 32.

De la Beche, Henry Thomas (1796-1855)
3034 1830, Sketch of a classification of the European rocks: Am. J. Sci., v. 18, no. 1, 26-37.
3035 1830, [Tabular view of rocks]: S. Rev., v. 6, no. 12, 304-305, table.
   In review of four geology texts.
3036 1832, A geological manual: Phila., Carey and Lea, viii, 535 p., 104 figs.
   For review see 1099 and 3756.
3037 1835, A geological manual: Phila.
   Not seen. "3rd ed." For notice see 609.
3038 1836, Application of geology to the useful purposes of life: NY, Scott, 22 p.
3039 1837, Researches in theoretical geology: NY, F. J. Huntington and Co.; Phila., Desilver, Thomas, and Co., xvi, 17-342 p., illus., front.
   With a preface and notes by E. Hitchcock (see 5124), and an appendix by W. J. Broderip (see 2046). For review, see 610. For review with excerpts see 3471.
3040 1846, and also L. McLane, Correspondence between Mr. McLane and Sir H. T. De la Beche, director of the Geological Survey of Great Britain and Ireland: Nat'l Inst. Prom. Sci. Bulletin, v. 1, 504-505.
3041 1846, [Excerpts from his works]: Am. Q. J. Ag. Sci., v. 3, 219-223.
   In "Notes on natural history," by J. Eights (see 3469).
3042 1848, Artificial colors in agate: Franklin Inst. J., v. 16 (3s), 285-286.
   Abstracted from Edinburgh New Philos. J.
3043 1849, Phosphate of lime in greensand and marl [in England]: Am. J. Sci., v. 8 (ns), no. 3, 422-424.
   Abstracted from Geol. Soc. London Proc.

De la Bigarre, Peter
3044 1797, Excursions on our Blue Mountains: NY Soc. Prom. Use. Arts Trans., v. 1, pt. 2, 128-139.

De Lacépède, Count Bernard-Germain-Etienne (1756-1825)
3045 1809, Concerning the fossil bones, presented to the National Institute by the President of the United States: Mon. Anth., v. 7, 69.

Delacoste, I. C.
3046 1804, Catalogue of the natural productions and curiosities which compose the collections of the cabinet of natural history opened for public exhibition at No. 38 William St., New York: NY, Printed by Isaac Collins and son, 87 p.
3047 1804, New York. Sir, I take the liberty of presenting you with a catalogue of the contents of the cabinet of natural history ... : NY, Broadside.

Delafield, James (Major)
3048 1822, Geological remarks on the Lake regions: Am. J. Sci., v. 4, no. 2, 282.
   On fossil trilobites from the Lake Huron region.
3049 1822, Notices of the sulphate of strontian of Lake Erie and Detroit, River: Am. J. Sci., v. 4, no. 2, 279-280.
3050 1823, American andalusite [from Litchfield, Connecticut]: Am. J. Sci., v. 6, no. 1, 176.

Delafield, John (1786-1853)
3051 1850, [On agricultural survey of Seneca County]: NY State Agri. Soc. J., v. 1, 63.
3052 1850, A general view and agricultural survey of the county of Seneca: NY State Agri. Soc. Trans., v. 1, 356-616, map.

Delafield, Joseph (1790-1875)
3053 1824, Notice of new localities of simple minerals, along the north coast of Lake Superior, and in the Indian territory NW. from Lake Superior to the river Winnepec: NY Lyc. Nat. History Annals, v. 1, 79-81.
3054 1829, Memorial: Am. J. Sci., v. 16, no. 2, 358-360.
   On the use of coal in New York City.
3055 1840, Report upon the minerals, geological specimens and fossils, from the Island of St. Lorenzo, presented to the New York Lyceum of Natural History, by D. Brinckerhoff: Am. J. Sci., v. 38, no. 1, 201-202.
   With report by J. H. Redfield (8697).

Delafield, Richard (1798-1873)
3056 1829, Report of the survey of the passes at the mouth of the Mississippi, La.: US HED 7, 21-1, v. 1 (195), 7-11.
   On the channel morphology at the mouth of the Mississippi River and techniques of sand bar removal.

Delamater, Jacob J.
3057 1844, Thesis on the use and abuse of the Saratoga waters: NY J. Medicine and Collateral Sci's, v. 3, 53-65.

De la Métherie, J. C.
3058 1817, Le cours de géologie donne au Collège de France: J. Sci. Arts, v. 2, no. 2, 429.
   Abstract in English.

Delaware. State of
3059 1836, Propriety of ordering a geological survey ... : Del. Governor's message, June, 1836.

Delaware (cont.)
3060 1837, An act to provide for a geological and mineralogical survey of this state: Del. Gen. Assembly, Dover, February 13, 1837.
3061 1837, Report on so much of Governor's message as relates to agricultural interests ... : Del. Senate J., January, 1837, 104-106.
3062 1839, Commission on the geological and mineral survey: Del. House J., January, 1839, 62-63.
3063 1841, Sec. State. Communication: Del. Senate J., 1841, 37-38.
   On the Delaware geological survey.

Delaware Chemical and Geological Society (Delaware County, New York)
3064 1822, Delaware Chemical and Geological Society: Am. J. Sci., v. 5, no. 1, 198.

Delaware Coal Company
3065 1830, Act of incorporation and by-laws: Wilmington, Del., R. Porter and Son, 13 p.
3066 1834, Charter and by-laws of the Delaware Coal Company: Phila., Printed by J. C. Clark, 18 p.
   Includes report by J. Wilde (see 10949).

The Delaware Register and Farmer's Magazine (Dover, Delaware, 1838-1839)
3067 1838, Lime: v. 1, no. 1, 48-58.

The Delaware Register; or, Farmer's, Manufacturers' and Mechanics' Advocate, Containing a Variety of Original and Selected Articles (Wilmington, Delaware, 1828-1829)
3068 1828-1829, [Brief notes on American mineral resources]: v. 1, no. 4, 27 (1828); no. 22, 177; no. 34, 273; no. 46, 369, 371 (1829).
3069 1829, The late earthquakes in Spain: v. 1, no. 30, 238.
   Abstracted from Gibralter Chron.
3070 1829, Meteoric explosion [in Forseyth, Georgia]: v. 1, no. 36, 287.
   Abstracted from Georgia Statesman.
3071 1829, Extraordinary animal remains [from the Mississippi Valley]: v. 1, no. 36, 290-291; no. 37, 294-295.
   Abstracted from NY Com. Advertiser.
3072 1829, Remains of a mammoth [from Pennsylvania]: v. 1, no. 46, 367 (20 lines).
   Abstracted from Chambersburg Penn. Repos.

Delesse, Achille Ernst Oscar Joseph (1817-1881)
3073 1844, Beaumontite [from Baltimore, Maryland]: Am. J. Sci., v. 47, no. 1, 216-217.
   Abstracted from Annales de Chimie.
3074 1844, Dipyre: Am. J. Sci., v. 47, no. 2, 417-418.
   Abstracted from Comptes Rendus. On the equivalence of dipyre and scapolite.
3075 1844, Sismondine - a new mineral: Am. J. Sci., v. 47, no. 1, 217.
   Abstracted by J. D. Dana.
3076 1846, Damourite, a new mineral: Am. J. Sci., v. 1 (ns), no. 1, 120 (7 lines).
   Abstracted from l'Institut.
3077 1847, Buratite, a new mineral: Am. J. Sci., v. 3 (ns), no. 3, 429 (11 lines).
   Abstracted from Comptes Rendus.
3078 1848, Effect of fusion on the density of rocks: Am. J. Sci., v. 6 (ns), no. 1, 133.
   Abstracted from "Jour. de Pharm. et de Chim."
3079 1848, On the fusion of rocks: Am. J. Sci., v. 6 (ns), no. 3, 423 (12 lines).
   Abstracted by G. C. Schaeffer from "Jour. de Pharm. et de Chim."
3080 1849, Analysis of talc of Rhode Island and steatite of Hungary: Am. J. Sci., v. 8 (ns), no. 1, 122 (13 lines).
   Abstracted from "Rev. Sci. et Indust."
3081 1849, Feldspar in the "orbicular diorite" of Corsica: Am. J. Sci., v. 7 (ns), no. 1, 113 (13 lines).
3082 1850, On the mineralogical and chemical composition of some rocks: Am. J. Sci., v. 10 (ns), no. 2, 252-255.
   Abstracted from Annales des Mines.

Delfess, M. (see 1739 with M. Blum)

Del Rio, Andres Manuel (1765-1849)
3083 1826, Analysis of a specimen of gold, found to be alloyed with rhodium: Am. J. Sci., v. 9, no. 2, 298-304.
   Translated by Wm. Smith.
3084 1832, The brown lead ore of Zimapan [Mexico]: Mon. Am. J. Geology, v. 1, no. 2, 69-70; no. 10, 438-444.
   With remarks by G. W. Featherstonhaugh (see 3769).
3085 1832, Elementos de orictognosia, ó del conocimiento de los fósiles, dispuestos, segunlos princípos de A. G. Werner, para el uso del real seminaro de minería de Mexico: Phila., 683 p.
   Second edition.
3086 1834, Observations on the Treatise of mineralogy of Mr. C. U. Shepard, with the translation of "the characteristic of the classes and orders of Breithaupt": Geol. Soc. Penn. Trans., v. 1, 113-136.
   Includes review with excerpts of C. U. Shepard, 9489. See also responses by C. U. Shepard, 9501 and 9505.
3087 1834, On the conversion of sulphuret of silver into native silver, after the method of Becquerel: Geol. Soc. Penn. Trans., v. 1, 137-138.
3088 1834, and J. Millington, Report of the committee appointed by the Geological Society of Pennsylvania, to investigate the Rappahannock gold mines in Virginia: Geol. Soc. Penn. Trans., v. 1, 147-166, 4 figs., map.
3089 1834, Silver ores by the method of Bequerel: Am. Philos. Soc. Trans., v. 4 (ns), 60-62.
   On methods of reducing ores.
3090 1836, A few observations on the reply of Professor Shepard: Am. J. Sci., v. 30, no. 2, 384-387.
   See also notes by C. U. Shepard (9501, 9505) and A. Del Rio (3086). Includes a defense of Del Rio's review of Treatise on mineralogy by Shepard (see 9502).

Del Rio, A. M. (cont.)
3091   1837, On the crystals developed in vermiculite by heat: Am. Philos. Soc. Trans., v. 5 (ns), 137-138.
3092   1849, Catalogue of the geological collection formed in the Isthmus of Tehuantepec ... : US House Rept. 145, 30-2, 160-163.
       Not seen.

De Luc, Jean André (1727-1817)
3093   1817, On the primitive matter of lavas: J. Sci. Arts, v. 2, no. 1, 158-159.
3094   1817, [Origin of volcanic minerals]: Am. Reg. or Summary Rev., v. 1 (ns), 444-445.
       Abstracted by T. Cooper.

Demaree, Samuel D.
3095   1808, Mineralogical notices from Kentucky: Med. Repos., v. 11, 306-307; also in: Mon. Anth., v. 5, 232.
       Includes notice of mineral spring at Harrodsburg, Kentucky.

Denham, Captain
3096   1838, Sediment: Am. J. Sci., v. 33, no. 2, 269-270.
       In British Assoc. Proc. On sediment transport in the Mersey River, Ireland, and on sediment filling of Liverpool Harbor.

De Normandie, John
3097   1771, An analysis of the chalybeate waters of Bristol, in Pennsylvania; in two letters from Dr. John De Normandie, of Bristol, addressed to Dr. Thomas Bond: Am. Philos. Soc. Trans., v. 1, 1st edition, 303-313; also in 2nd edition, 368-379 (1789).

Derby, George Horatio (1823-1861)
3098   1850, Geological reconnoissance in California: black and white geological map, 35 x 30 cm., in "Memoir on the geology and topography of California" (see 3099).
3099   1850, Memoir on the geology and topography of California: US SED 47, 31-1, v. 10, pt. 2 (558), 3-16, 2 maps.
       Includes geological map (3098).

Des Cloiseaux, Alfred Louis Olivier Legrand (1817-1897)
       See also 2780 with A. Damour and 6775 and 6776 with J. C. G. Marignac.
3100   1846, and M. Dumas, Baryto-calcite: Am. J. Sci., v. 1 (ns), no. 1, 121 (6 lines).
       Abstracted from Annales de Chimie.
3101   1846, Crystallization of sulphuret of cadmium and perowskite: Am. J. Sci., v. 1 (ns), no. 1, 120-121 (13 lines).
       Abstracted from Annales de Chimie.
3102   1847, Hecla: Am. J. Sci., v. 3 (ns), no. 2, 288.
       Abstracted from Comptes Rendus. Description of Mt. Hecla, in Iceland.
3103   1848, Temperature of the geysers of Iceland: Am. J. Sci., v. 5 (ns), no. 2, 269-273.
       Abstracted from Philos. Mag.

Descostils, M.
3104   1808, and J. H. Hassenfratz, [Fusibility of "spathic iron"]: Am. Reg. or Gen. Repos., v. 2, 376-377.

Descriptive catalogue of a collection of rare minerals, proper for the study of crystallography recently arranged after Phillips; with localities and references for the figures of crystals to his third edition.
3105   1829, NY, G. & C. & H. Carvill, 49 p.

Desor, Edouard (1811-1882)
3106   1847, [Fossils in drift of Brooklyn and Westport, New York]: Bos. Soc. Nat. History Proc., v. 2, 247.
3107   1847, [Observations on drift]: Am. Q. J. Ag. Sci., v. 6, 214, 215, 216-217.
       In Assoc. Am. Geol's Nat's Proc. See also 8719.
3108   1847, [On parallel trains of boulders in Berkshire County, Massachusetts]: Bos. Soc. Nat. History Proc., v. 2, 260-261 (12 lines).
3109   1847, On the phenomena of drift and glacial action in New England: Am. Q. J. Ag. Sci., v. 6, 213-214 [i.e. 261-262].
       In Assoc. Am. Geol's Nat's Proc.
3110   1847, On the relations which exist between the phenomena of erratic blocks in northern Europe and the elevations of Scandinavia: Am. J. Sci., v. 3 (ns), no. 3, 313-318.
3111   1848, [Belemnites from the mica slates of St. Gothard]: Bos. Soc. Nat. History Proc., v. 3, 19 (5 lines).
3112   1848, [Drift fossils from Nantucket, Massachusetts]: Bos. Soc. Nat. History Proc., v. 3, 79-80.
3113   1848, [Iceberg theory of glacial drift]: Lit. World, v. 2, 229 (12 lines), 256 (8 lines).
       Abstracted from Assoc. Am. Geol's Nat's Proc.
3114   1848, [Peculiarities in scratchings on puddingstone of Brookline, Massachusetts]: Bos. Soc. Nat. History Proc., v. 3, 28 (7 lines).
3115   1849, Deposit of drift shells in the cliffs of Sancati Island of Nantucket [Massachusetts]: Am. Assoc. Adv. Sci. Proc., v. 1, 100-101.
3116   1849, [Geological position of the mastodon]: Bos. Soc. Nat. History Proc., v. 3, 115-117.
3117   1849, [Glacial action on rocks]: Bos. Soc. Nat. History Proc., v. 3, 95 (10 lines).
3118   1849, [On the ribbon structure of the ice in glaciers]: Bos. Soc. Nat. History Proc., v. 3, 125-127.
3119   1850, [Comparison of English and American Tertiary formations]: Bos. Soc. Nat. History Proc., v. 3, 247.
       Includes discussion of article by J. Wyman (see 11084).
3120   1850, [Drift fossils from Nantucket]: Ann. Scien. Disc., v. 1, 281-282.

Desor, E. (cont.)
3121 1850, Interesting fossils [from the Potsdam Sandstone in New York]: Ann. Scien. Disc., v. 1, 285 (10 lines).
3122 1850, [Notes on terraces of Lake Erie]: Bos. Soc. Nat. History Proc., v. 3, 291-292.
       Contains letters by I. A. Lapham and C. Whittlesey.
3123 1850, [On a shark's tooth from Keokuk, Iowa]: Bos. Soc. Nat. History Proc., v. 3, 257-258.
3124 1850, [On clay and drift deposits in the vicinity of Lake Superior]: Bos. Soc. Nat. History Proc., v. 3, 207, 235-236.
3125 1850, [On deposits of marine shells in Maine, on Lake Champlain, and the St. Lawrence, and their probable origin]: Bos. Soc. Nat. History Proc., v. 3, 357-358.
       With discussion by H. D. Rogers.
3126 1850, [On mastodon remains from Galena, Missouri]: Bos. Soc. Nat. History Proc., v. 3, 207.
3127 1850, [On swamps bordering western rivers]: Bos. Soc. Nat. History Proc., v. 3, 376.
3128 1850, [On the parallelism of mountain chains in America]: Bos. Soc. Nat. History Proc., v. 3, 380-382.
3129 1850, [On the probable origin of the so-called fossil raindrops from the Connecticut Valley]: Bos. Soc. Nat. History Proc., v. 3, 200-202; also in: Am. J. Sci., v. 10 (ns), no. 1, 135.
3130 1850, [On the relation of the alluvium to the drift of the Mississippi]: Bos. Soc. Nat. History Proc., v. 3, 242-243.
3131 1850, [On the "Ridge Road" from Rochester to Lewiston, New York, and other terraces]: Bos. Soc. Nat. History Proc., v. 3, 358-359.
       With discussion by C. Stodder.
3132 1850, [On the extent of the Potsdam Sandstone]: Bos. Soc. Nat. History Proc., v. 3, 202, 212 (4 lines).
3133 1850, [Recent ornithichnites from Lake Superior]: Bos. Soc. Nat. History Proc., v. 3, 202.
3134 1850, [Subsidence of Newfoundland's shore]: Bos. Soc. Nat. History Proc., v. 3, 375.
3135 1850, [Variation of inclination in sandstone]: Bos. Soc. Nat. History Proc., v. 3, 341 (6 lines).

Despretz, César Mansuète (1789-1863)
3136 1849, Volatilization of carbon: Am. J. Sci., v. 8 (ns), no. 3, 413-414.
       Abstracted from Comptes Rendus. On the fusion of minerals by electricity.

Detmold, Christian Edward (1810-1887)
3137 1849, George's Creek Coal and Iron Company: Balt., 22 p.
       On the finances of the company.

Deuchar, J.
3138 1822, [On the origin of fluid inclusions]: Am. J. Sci., v. 4, no. 2, 244.
       Abstracted by J. W. Webster.
3139 1823, On the porosity of glass and siliceous bodies: Am. J. Sci., v. 7, no. 1, 192.
       Abstracted by J. Griscom from Edinburgh Philos. J.

Deville, Charles Joseph Sainte-Claire (1814-1876)
       See also 8158 with M. S. Pasteur.
3140 1847, Volcanic peak of the Island of Fogo, Cape Verds: Am. J. Sci., v. 3 (ns), no. 3, 432-433.
       Abstracted from Geol. Soc. France Bulletin.

Dewey, Chester (1784-1867)
       See also articles signed by "C.D.": 421, 903, 3161, 3719, and 4280. These works may be authored by Chester Dewey.
3141 1819, Description of two ranges of mountains in the state of Massachusetts: Am. Mon. Mag. and Crit. Rev., v. 4, no. 4, 284-285.
3142 1819, A geological map of the north-west part of Massachusetts. 1819: Am. J. Sci., v. 1, no. 4, colored map, 11.7 x 19.7 cm., facing p. 344.
       In 3143.
3143 1819, Sketch of the mineralogy and geology of the vicinity of Williams College, Williamstown, Mass.: Am. J. Sci., v. 1, no. 4, 337-346, map.
       Includes map 3142.
3144 1820, Geological section from Taconick Range, in Williamstown, to the city of Troy, on the Hudson: Am. J. Sci., v. 2, no. 2, 246-249.
       Includes a note on wavellite from Richmond, New York.
3145 1820, Localities of minerals [in Massachusetts, New Jersey, and New York]: Am. J. Sci., v. 2, no. 2, 236-238, 4 figs.
3146 1821, American wavellite [from Richmond]: Am. J. Sci., v. 3, no. 2, 239.
3147 1821, Crystals of snow: Am. J. Sci., v. 3, no. 2, 367.
3148 1821, Fetid dolomite: Am. J. Sci., v. 3, no. 2, 239 (9 lines).
3149 1822, From Prof. Dewey of Williams College: Am. J. Sci., v. 4, no. 2, 274-277.
       On minerals of Vermont.
3150 1822, Notice of crystallized steatite - ores of iron and manganese, &c. [from Bennington, Vermont]: Am. J. Sci., v. 5, no. 2, 249-251.
3151 1822, Prismatic mica [from Hinsdale, Vermont]: Am. J. Sci., v. 5, no. 2, 399 (4 lines).
3152 1823, Analysis of argentine and crystallized steatite [from New England]: Am. J. Sci., v. 6, no. 2, 333-336.
3153 1824, Additional notice of argentine [from Southampton, Massachusetts]: Am. J. Sci., v. 7, no. 2, 248-249.
3154 1824, Additional remarks on the geology of a part of Massachusetts, &c.: Am. J. Sci., v. 8, no. 2, 240-244, fig.
       See also 3156.

Dewey, C. (cont.)
3155   1824, A geological map of the county of Berkshire, Mass. and of a small part of the adjoining states 1824: Am. J. Sci., v. 8, no. 1, colored geological map, 20.2 x 29.8 cm.
        In 3156. Illustrates portions of Vermont, Connecticut, and New York, as well as Massachusetts.
3156   1824, A sketch of the geology and mineralogy of the western part of Massachusetts and a small part of the adjoining states: Am. J. Sci., v. 8, no. 1, 1-60; no. 2, 240-244; map, fig.
        See also additional notes, 3154. Contains map, 3155.
3157   1825, Notice of a singular conformation of limestone: Am. J. Sci., v. 9, no. 1, 19-20, fig.
        Description of folded limestone near Williams College.
3158   1825, Notice of the flexible or elastic marble of Berkshire County: Am. J. Sci., v. 9, no. 2, 241-242.
3159   1827, Porcelain clay [from Pownal, Vermont]: Am. J. Sci., v. 12, no. 2, 298-299.
3160   1829, Natural productions of Berkshire County, Mass. [and notes on the geology of the county]: in A history of the county of Berkshire, Massachusetts, in two parts by D. D. Field: Pittsfield, Mass., S. W. Bush (printer), iv, 468 p., front., folded map, plates, 36-86.
        Part one, "General view of the county," written by Dewey.
3161   1837, Bones of the Mammoth: Am. J. Sci., v. 33, no. 1, 201.
        Signed "C. D." On mammoth bones discovered at Rochester, New York.
3162   1837, Remarks on the rocks of New York: Am. J. Sci., v. 33, no. 1, 121-123.
        Cites evidence for the Transition age of most New York sediments, on the basis of fossils.
3163   1839, On the polished limestone of Rochester [New York]: Am. J. Sci., v. 37, no. 2, 240-242; also in: Assoc. Am. Geol's Nat's Rept., v. 1, 264-266 (1843).
3164   1843, Striae and furrows of the polished rocks of western New York: Am. J. Sci., v. 44, no. 1, 146-150.
3165   1845, On the gypsum beds of New York: Assoc. Am. Geol's Nat's Proc., v. 6, 38-39.

De Witt, Benjamin (1774-1819)
3166   1798, A memoir on the Onondaga salt springs and salt manufactories in the western part of New York ...: Alb., L. Andrews, 28 p.
3167   1801, A memoir on the Onondaga salt springs and salt manufactories in the state of New York: NY Soc. Prom. Use. Arts Trans., v. 1, pt. 3, 268-286.
3168   1804, An account of some of the mineral productions in the state of New York, (accompanying specimens transmitted for the cabinet of the American Academy of Arts and Sciences,) in a letter: Am. Acad. Arts Sci's Mem., v. 2, pt. 2, 73-81.
3169   1820, A catalogue of minerals contained in the cabinet of the late Benjamin De Witt, Professor of Mineralogy in the College of Physicians and Surgeons, New-York, consisting of more than 11,000 specimens, collected in Europe and America: Alb., G. J. Loomis and Co., 108 p.

De Witt, Simeon (1756-1834)
3170   1807, Description of a petrified horn, from Helderberg, a mountain westward of Albany: Med. Repos., v. 10, 350-352, woodcut.
3171   1817, [Letter to S. L. Mitchill "describing certain fossils" from Cayuga County, New York]: Am. Mon. Mag. and Crit. Rev., v. 1, 289.

The Dial (Boston, 1840-1844)
3172   1842, Association of state geologists: v. 3, no. 1, 133.
        Notice of Boston meeting.

Dick, Thomas Lander
3173   1823, Account of the travelled stone near Castle Stuart Invernesshire: Am. J. Sci., v. 6, no. 1, 158-162.

Dickenson, George J.
3174   1849, [Chemical analysis of Lake Superior minerals]: in Lake Superior report by C. T. Jackson (see 5655), 480.
3175   1849, Mr. Dickenson's report [on the minerals and geology of Isle Royale]: in Lake Superior report by C. T. Jackson (see 5655), 503-506.

Dickeson, Montroville Wilson
        See also 2071 with A. Brown.
3176   1845, On the geology of the Natchez bluffs [in Mississippi]: Assoc. Am. Geol's Nat's Proc., v. 6, 77-79.
        Also in Am. Q. J. Ag. Sci., v. 2, 168.
3177   1846, [On fossil bones from the vicinity of Natchez, Mississippi]: Phila. Acad. Nat. Sci's Proc., v. 3, no. 5, 106-107.
3178   1847, Tracks of alligators: Am. J. Sci., v. 3 (ns), no. 1, 125 (4 lines).
        From Phila. Acad. Nat. Sci's Proc. On the comparison of recent and fossil tracks.
3179   1849, and A. Brown, Sediment of the Mississippi River: West. Q. Rev., v. 1, 175-178.

Dickinson, James T.
3180   1839, Geological specimens from the East Indian Archipelago: Am. J. Sci., v. 35, no. 2, 381.

Dickson, James
3181   1831, On the silver, gold, and platina, of Russia: Mon. Am. J. Geology, v. 1, no. 3, 118-124.
3182   1834, An essay on the gold region of the United States: Geol. Soc. Penn. Trans., v. 1, 16-32; also in: Eclec. J. Sci., v. 2, 617-619 (1834); also in: Am. J. Sci., v. 27, no. 2, 348-351 (1835).
        On geology and mining procedures in Virginia, North Carolina, and Georgia.
3183   1835, On the science and practice of mining: Geol. Soc. Penn. Trans., v. 1, 360-408, 2 plates.
        Describes general mining procedures and the mines of Cornwall, England.

Dickson, John (d. 1847)
3184 1821, Confirmation of the genuineness of the locality of American corundum, mentioned p. 7 of this volume: Am. J. Sci., v. 3, no. 2, 229-230.
      On corundum found in the Carolinas.
3185 1821, Notices of the mineralogy and geology of parts of South and North Carolina: Am. J. Sci., v. 3, no. 1, 1-5.
      With "Postscript by the editor," B. Silliman (see 9640).

Dietz, R.
3186 1824, Description of a testaceous formation at Anastasia Island [Florida], extracted from notes made on a journey to the southern part of the United States, during the winter of 1822 and 1823: Phila. Acad. Nat. Sci's J., v. 4, pt. 1, 73-80.
      With note by T. Say (see 9208).

Dille, Israel
3187 1845, Mineral resources of southern Missouri: Hunt's Merch. Mag., v. 13, 222-227.
3188 1846, Mineral region and resources of Missouri: Hunt's Merch. Mag., v. 15, 28-34.

Dille, J.
3189 1845, The great metaliferous geological belt: which traverses the continent of North and South America: The Investigator, v. 1, 168-170.

Disbrow (Mr.)
3190 1827, Notice of some recent experiments in boring for fresh water, and of a pamphlet on that subject: Am. J. Sci., v. 12, no. 1, 136-144.
      Introduction, with remarks, by B. Silliman (see 9672). Notices by S. Hepburn (see 4994) and the New York Times.

Ditson, George
3191 1845, Copper smelting in the United States: Hunt's Merch. Mag., v. 12, 551-554; v. 13, 256-259.
3192 1845, The copper trade - England and America: Am. Rr. J., v. 18, 462.

Dix, John Adams (1798-1879)
3193 1841, Temperature of the earth: No. Light, v. 1, 73-74.
      With extracts from reports by Mr. Fox (see 3958) and E. Hodgkinson (see 5210).

Dize (Mr.)
3194 1826, Improved mode of separating gold from silver: Am. Mech's Mag., v. 2, no. 33, 113-114.
      Abstracted from London Mech. J.

Dobson, John
3195 1835, Machine for separating gold from the soil: Franklin Inst. J., v. 15 (ns), 200.

Dobson, Peter
3196 1826, Remarks on bowlders: Am. J. Sci., v. 10, no. 2, 217-218.
      On scratches and polished surfaces of boulders from Connecticut.
3197 1844, Hints on the iceberg theory of drift: Am. J. Sci., v. 46, no. 1, 169-172.

Domeyko, Ignacio (1802-1889)
3198 1846, Bismuth silver: Am. J. Sci., v. 2 (ns), no. 3, 418 (4 lines).
      Abstracted from Annales des Mines. Chemical analysis of silver bismuth from Copiapo, South America.
3199 1849, On a native antimonite of mercury: Am. J. Sci., v. 8 (ns), no. 1, 127-128 (5 lines).
      From Annuaire de Chimie. On chemical analysis of a mineral from Chile.

Donovan
3200 1836, Composition of organized structures; and similarity of the diamond and charcoal: Am. Mag. Use. Ent. Know., v. 3, no. 2, 55.

Doolittle, Thomas (1632?-1707)
3201 1693, Earthquakes explained and practically improved: occasioned by the late earthquake on Sept. 8, 1692, in London, many other parts of England, and beyond the sea: Bos., Reprinted by Benjamin Harris, 56 p.
      Evans number 634.

Doornik, Jacob Elisa (1777-1837)
3202 1828, Observations concerning fossil organic remains: Am. J. Sci., v. 15, no. 1, 90-109.
      Abstracted and translated by C. U. Shepard. With article by A. Brongniart (see 2059).

D'Orbigny, Alcide Charles Victor Dessalines (1802-1857)
3203 1846, On the paleontology of South America: Am. J. Sci., v. 1 (ns), no. 2, 279-280.
      Abstracted from his memoir on South America.
3204 1847, Note by M. D'Orbigny on the Orbitolina [from North America]: Am. J. Sci., v. 4 (ns), no. 2, 282.

Douglas, Major
3205 1847, [Extract from a report on the geology of the Barrellville Mining Company lands]: see 1502.
3206 1847, Extracts from Major Douglas' report [on the Phoenix Mining Company lands]: see 8328, p. 26-29.

Douglass Houghton Mining Company
3207   1847, Report of the trustees of the Douglass Houghton Mining Company of Lake Superior: Detroit, C. Wilcox (printer), 38 p., plate.
  Contains reports by J. R. Grout and C. C. Douglass (see 4598). Includes list of British mining companies.

Douglass (Senator)
3208   1850, Geological survey [of the United States]: West. J. Ag., v. 3, no. 4, 275.

Douglass, Columbus C.
  See also 4598 with J. R. Grout.
3209   1839, Report [on Ingham County and parts of Eaton and Jackson Counties]: Mich. State Geologist Ann. Rept., v. 2, 66-77.
  In 5299.
3210   1840, Report [on Allegan, Calhoun, Eaton, Ionia, Jackson, Kalamazoo, Kent, Ottawa, and Van Buren Counties]: Mich. State Geologist Ann. Rept., v. 3, 53-75.
  In 5301.
3211   1841, Report [on the northern portion of the southern peninsula of Michigan]: Mich. State Geologist Ann. Rept., v. 4, 97-111.
  In 5303.

Douglass, David Bates (1790-1849)
3212   1820, New locality of crystallized sulphat of barytes, &c.: Am. J. Sci., v. 2, no. 2, 241 (12 lines).
3213   1821, Sulphat of strontian: Am. J. Sci., v. 3, no. 2, 363-364.
  On celestite from Mouse Island, Ohio.
3214   1838, Report on the coal and iron formation of Frostburg and the Upper Potomac in the state of Maryland and Virginia: Brooklyn?, 29 p., map.

Doveri, M.
3215   1847, Observations upon silica: Franklin Inst. J., v. 14 (3s), 209-210.
  Abstracted from Comptes Rendus.

Drake, Daniel (1785-1852)
3216   1810, Notices concerning Cincinnati: Cin., Printed for the author by John W. Browne and Co., 60, 4 p.
  For review see 1473.
3217   1810/1811, Strictures on Volney's "View of the soil and climate of the United States" [see 10653]: Port Folio, v. 4 (3s), 587-591; v. 5 (3s), 320-324; v. 6 (3s), 203-209.
  Signed by "D."
3218   1815, Natural and statistical view; or, picture of Cincinnati and the Miami Coutry ...: Cin., Looker and Wallace, 251 p., 2 maps, front., folded plate.
  For review with excerpts see 8540. For review see 8467.
3219   1816, Medical topography [of Cincinnati]: Eclec. Rep. and Analytic Rev., v. 6, 137-149.
3220   1818, Natural and statistical view of Cincinnati and the Miami country: J. Sci. Arts, v. 3, no. 1, 81-88.
3221   1819, An introductory lecture on the utility and pleasures of the study of mineralogy and geology, delivered in the Western Museum, December 18, 1819 ...: "in Cincinnati newspapers of the day" (see Meisel, 1926, v. 2, 391).
3222   1820, On the utility and pleasures of the study of mineralogy: Port Folio, v. 9 (5s), 86-93.
3223   1825, Geological account of the Valley of the Ohio: Am. Philos. Soc. Trans., v. 2 (ns), 124-139, plate.
3224   1828, Notices of the mineral springs of Kentucky and Ohio: West. J. Med. Phys. Sci's, v. 2, no. 2, 142-167.
3225   1829, Bath springs [at Oxford, Ohio]: Lit. Reg., v. 2, 195.

Draper, John William (1811-1882)
3226   1834, Chemical analysis of the native chloride of carbon, a singular mineral [from Kent, England]: Franklin Inst. J., v. 14 (ns), 295-298.
3227   1835, Coins and medals: Am. J. Sci, v. 29, no. 1, 157-160.
  Evidence of ancient gold and silver mines from coins.

Drummond
3228   1841, Infusoria in Ireland: Am. Eclec., v. 1, no. 2, 390-391.
  Abstracted from Mag. Nat. History.

Drury, E. H.
3229   1845, Letter from Dr. E. H. Drury: in First annual report on the geology of the state of Vermont by C. B. Adams (see 23), 80-81.
  On black lead from Brandon, Vermont.

Dublin Geological Society (Dublin, Ireland)
3230   1835, Journal of the Geological Society of Dublin, vol. 1: Am. J. Sci., v. 28, no. 2, 368-369.
  Table of contents of the journal, with short review.

Ducatel, Julius Timoleon (1796-1849)
3231   1834, and J. H. Alexander, Report on the projected survey of the state of Maryland, pursuent to a resolution of the General Assembly: Annapolis, 39 p., 8vo., map; also 43 p. and 58 p. editions.
  This edition was authorized by the Maryland House of Delegates, December Session.
3232   1834, and J. H. Alexander, Report on a projected geological and topographical survey of the state of Maryland: Am. J. Sci., v. 27, no. 1, 1-38. See also "Remarks" by B. Silliman (9717).
3233   1834, To the public: Am. Farm., v. 16, 22-23.
  On the activities of the Maryland state geologist, and request for information.
3234   1835, Report of the geologist to the legislature of Maryland, 1834: n. p., 50 p., 2 maps, folded table.
  Also another edition, not seen. For review see 9064.

Ducatel, J. T. (cont.)
3235 1835, and J. H. Alexander, Report on the new map of Maryland, 1834: Annapolis?, 59, [1] p., 2 maps, folding table; also another edition.
For review see 627.
3236 1836, Extracts from the report of Professor Ducatel's survey of Maryland: Farm's Reg., v. 4, 292-300, 409-411; also in: Am. Rr. J., v. 5, 791-792.
3237 1836, Geological report: Am. Farm., v. 2 (2s), 357-359, 364-366, 371-374, 380-382, 388-390, 398.
Extensive excerpts from 3240.
3238 1836, and J. H. Alexander, Report on the engineer and geologist in relation to the new map to the executive of Maryland: Annapolis, W. M'Neir (printer), 84, 1 p., 6 maps and plates.
3239 1836, and J. H. Alexander, Report on the new map of Maryland, 1835: Annapolis?, 96, [1] p., maps, plates.
3240 1836, Report of the geologist: Annapolis, 35-84, plate.
Also in 3238 and 3239. Largely republished in 3237. For review see 9740. For notice see 259, 665.
3241 1837, [Excerpts from the geological report on Maryland (see 3245)]: Am. Farm., v. 3 (2s), 365-367, 373-375, 380-381, 389-391, 398-399.
3242 1837, Extract from Professor Ducatel's last report on the geological survey of Maryland. Condition of agriculture in Calvert County: Farm's Reg., v. 5, 49-52.
3243 1837, [Letter on lead ore of the Osage Mining Company]: in 8042, p. 17.
3244 1837, Outline of the physical geography of Maryland, embracing its prominent geological features: Md. Acad. Sci. Liter. Trans., v. 1, 24-54, map.
3245 1837, and J. H. Alexander, Report on the new map of Maryland, 1836: Annapolis, 104 p., 5 maps, illus.; also 117 p. edition.
Authorized by Maryland House of Delegates, December Session. Geologist's report pp. 1-60. For review see 263.
3246 1838, Annual report of the geologist of Maryland, 1837: Annapolis, 39, 1 p., 2 maps.
3247 1838, [Maryland geological report for 1837]: Am. Farm., v. 5 (2s), 29-31, 38-39, 44-46, 53-54, 62-63, 69-70, 77-79, 86-89.
Extensive excerpts from 3246.
3248 1839, Annual report of the geologist of Maryland, 1838: Annapolis, 33, [1] p., 3 plates.
3249 1840, Annual report of the geologist of Maryland, 1839: Annapolis, 45, [1] p., 4 plates; also 59 p. and 43 p. editions.
3250 1840, Geological survey of Maryland: Am. Farm., v. 2 (3s), 62-63.
Excerpted from 3249.
3251 1841, Annual report of the geologist of Maryland, 1840: Annapolis, 46 p., map, plates; also 59 p. and 43 p. editions.
3252 1843, A general view of the physical geography and geology of the state of Maryland: Am. Philos. Soc. Proc., v. 3, no. 27, 157-158.
3253 1847, [Extract from report on the Barrelville Mining Company lands]: see 1502.
3254 1847, Extract from the annual report of J. T. Ducatel: in Phoenix Mining Company "Documents," p. 29, see 8328.

Du Commun, Joseph
3255 1828, Hypothesis on volcanoes and earthquakes: Am. J. Sci., v. 15, no. 1, 12-27, fig.; also in: Scholar's Q. J., v. 1, no. 4, 15-16.
With objections by B. Silliman (see 9676). See also "Strictures ... ", by B. Bell (1620).

Dudley, Benjamin Winslow (1785-1870)
3256 1806, A sketch of the medical topography of Lexington and its vicinity ... : Phila., T. & G. Palmer, 21 p.

Dudley, J. (Governor of Massachusetts)
3257 1814, Letter from Governour [sic] Dudley to the Reverend Cotton Mather; Roxbury, 10 July, 1706: Mass. Hist. Soc. Coll., v. 2 (2s), 263-264.
On the discovery of a fossil tooth near Albany, in 1706. This may be the earliest recorded fossil vertebrate discovered in North America.

Dudley, Paul (1675-1751)
3258 1789, An account of earthquakes which have happened in New England, since the first settlement of the English in that country, especially of that, which happened on October 27, 1727: Am. Mus. or Univ. Mag., v. 5, 363-365, 595-597.

Duflos, Adolf (1802-1889)
3259 1848, and N. W. Fischer, Analysis of the meteoric iron that fell near Braunau in Bohemia, on the 14th of July, 1847: Am. J. Sci., v. 5 (ns), no. 3, 338-342.
See also 3841 by N. W. Fischer.
3260 1848, Meteoric iron of Seelasgen in Brandenburg: Am. J. Sci., v. 6 (ns), no. 3, 426.
Abstracted from Annalen der Physik und Chemie, by C. U. Shepard.

Dufrenoy, Ours Pierre Armand Petit (1792-1857)
3261 1831, Huraulite and hetepozite: Am. J. Sci., v. 19, no. 2, 371.
Abstracted by J. Griscom from "Ann. de Chim. et de Phys."
3262 1835, Remarks upon the nature of the coal employed in the furnaces using crude coal: Am. Rr. J., v. 4, 597-598.
Abstracted from his "Report to the Board of Directors of Bridges, Public Roads, and Mines, upon the use of heated air in the iron works of Scotland and England."
3263 1836, Dreelite; a new mineral species: Am. J. Sci., v. 30, no. 1, 380-381.
Abstracted from "Ann. de Chim. et Phys."
3264 1849, Compact diamond from Brazil: Am. J. Sci., v. 7 (ns), no. 3, 433.
Abstracted from l'Institut.
3265 1850, New adamantine mineral: Ann. Scien. Disc., v. 1, 273-274.
Abstracted from Jameson's Philos. J. On diamond from Brazil.

Dufresnoy, M.
3266 1842, [Letter acknowledging receipt of a specimen of Missouri iron ore, donated to the Paris School of Mines]: Nat'l Inst. Prom. Sci. Bulletin, v. 2, 137.

Dufresnoy, M. (cont.)
3267 1850, Analysis of California gold: Ann. Scien. Disc., v. 1, 255-256.

Dugard, T.
3268 1809, English improvements in the working mines of iron ore and coal: Med. Repos., v. 12, 400-401.

Dumas, Jean Baptiste-André (1800-1884)
  See also 3100 with A. L. O. L. Descloiseaux.
3269 1846, Extracting copper from its ores by electricity: Franklin Inst. J., v. 11 (3s), 69-70.
  Abstracted from Paris Soc. Encouragement Nat'l Industry Proc.

Dunbar, William (1749-1810)
3270 1804, Description of the river Mississippi and its delta, with that of the adjacent parts of Louisiana: Am. Philos. Soc. Trans., v. 6, pt. 1, 165-187.
3271 1804, Extracts from a letter, from William Dunbar Esq. of the Natchez, to Thomas Jefferson, President of the Society: Am. Philos. Soc. Trans., v. 6, pt. 1, 40-42.
  On fossil bones from the "Country of the Apelousas."
3272 1806, and W. Hunter, Observations: in 10471, pp. 116-171.
3273 1809, Appendix to a memoir on the Mississippi, No. XXX. of the 1st part of this volume: Am. Philos. Soc. Trans., v. 6, pt. 2, 191-201.
  See also 3270.
3274 1809, and W. Hunter, Observations made in a voyage commencing at St. Catherine's landing, on the east bank of the Mississippi, proceeding downwards to the mouth of Red River, and from thence ascending that river, the Black River, and the Washita River, as high as the hot springs in the proximity of the last mentioned river: Am. Reg. or Gen. Repos., v. 5, 311-345.
  For review see 7059.

Duncan, Henry (1774-1846)
3275 1828, Account of the tracks of foot-marks of animals found impressed in sandstone in the quarry of Corncockle Muir, Dumfries-shire: Am. J. Sci., v. 15, no. 1, 84-90.

Dundonald, Lord Archibald Cochrane (1749-1831)
3276 1808, On the analysis of soils: Phila. Soc. Promoting Ag. Mem., v. 1, Appendix, 27-32.
  From "Treatise on the connection of agriculture with chemistry."

Dupaty, President
3277 1791, Account of the first known eruption of Vesuvius, and of the death of the elder Pliny in visiting it: NY Mag., v. 2, 719-722.
3278 1791, Description of Mount Vesuvius: NY Mag., v. 2, 717-719.

Duperrey, Louis-Isidore (1786-1865)
3279 1825, No diurnal variation of the needle at the equator: Am. J. Sci., v. 9, no. 2, 387 (4 lines).

Du Ponceau, M.
3280 1838, Magnetic dip in Ohio: Franklin Inst. J., v. 22 (ns), 269-270.
  Abstracted from Am. Philos. Soc. Proc.

Duralde, Martin
3281 1804, Abstract of a communication from Mr. Martin Duralde, relative to fossil bones, &c. of the country of Apelousas west of the Mississippi to Mr. William Dunbar of the Natchez, and by him transmitted to the Society: Am. Philos. Soc. Trans., v. 6, pt. 1, 55-58.

Durand, Elie Magloire (1794-1873)
3282 1833, On the alum and copperas manufactory of Cape Sable, Maryland: Am. J. Pharmacy, v. 5, 12-16.
3283 1833, On the preparation of magnesia and its salts from magnesite: Am. J. Pharmacy, v. 5, 1-10.

D'Urville, Commodore
3284 1840, French exploring expedition to the antarctic regions: Am. J. Sci., v. 39, no. 1, 201-203.
  Translated by the Singapore Free Press. On the progress of the expedition.

Du Simitière, Pierre Eugene (c. 1736-1784)
3285 1782, American Museum: Phila., Printed by John Dunlap, broadside 4to.
  Evans number 17523. Notice of the museum.

Dutrochet, Rene-Joachim-Henri (1776-1847)
3286 1842, Observations on the diurnal variation of the magnetic needle: Cambridge Miscellany, v. 1, 81-82.
  Abstracted from Comptes Rendus.

Dutton, T. R.
3287 1847, Observations on the basaltic formation on the northern shore of Lake Superior: Am. J. Sci., v. 4 (ns), no. 1, 118-119, fig.

Dwight, Henry Edwin (1797-1832)
3288 1820, Account of the Kaatskill Mountains: Am. J. Sci., v. 2, 11-29; also in: Plough Boy, v. 3, 121-123, 132-134 (1821).

Dwight, Sereno Edwards (1786-1850)
3289 1813, A dissertation on the origin of springs: Ct. Acad. Arts Sci's Mem., v. 1, pt. 3, 311-328.

Dwight, S. E. (cont.)
3290   1826, Description of the eruption of Long Lake and Mud Lake, in Vermont, and of the desolation effected by the rush of the waters through Barton River, and the lower country, towards Lake Memphremagog, in the summer of 1810: Am. J. Sci., v. 9, no. 1, 38-54, map.
Also reprint edition, 18 p.

Dwight, Theodore (1796-1866)
3291   1825, The Northern traveller; containing the routes to Niagara, Quebec, and the Springs; with descriptions of the principal scenes, and useful hints to strangers: NY, Wilder and Campbell, 222 p., maps and plates.
3292   1826, The Northern traveller; containing the routes to Niagara, Quebec, and the Springs; with the tour of New England, and the route to the coal mines of Pennsylvania: NY, A. T. Goodrich, iv, 382 p., maps and plates.
Second edition.
3293   1828, The Northern traveller; (combined with the Northern tour.) ... With the tour of New-England, and the route to the coal mines of Pennsylvania: NY, C. and G. Carvill, 403 p., maps and plates.
Third edition.
3294   1830, The Northern traveller, and Northern tour; with the routes to the Springs, Niagara, and Quebec, and the coal mines of Pennsylvania ... : NY, J. and J. Harper, viii, 444 p., maps and plates.
Fourth edition. Also 1831 printing as "new edition."
3295   1834, The Northern traveller, and Northern tour; ... [as above] : NY, Goodrich and Wiley, 432 p., maps and plates.
Fifth edition.
3296   1841, The Northern traveller; containing the routes to the Springs, Niagara, Quebec, and the coal mines; with the tour of New-England, and a brief guide to the Virginia Springs, and southern and western routes: NY, J. P. Haven, 50 [i.e. 250] p.
Sixth edition.

Dwight, Timothy (1752-1817)
3297   1821/1822, Travels in New England and New York: New Haven, T. Dwight, 4 volumes; 524, 527, 534, 527 p.
For correction see 432. For notice see 40.

Dwight's American Magazine, and Family Newspaper (New York, 1845-1850)
Each of the first three volumes (1845-1847) contains numerous brief notes on American mines and mining. Consult annual indices for specific articles.
3298   1845, A geological theory undermined by a favorite mollusca: v. 1, 23.
3299   1845-1846, [Brief notes on volcanoes and earthquakes]: v. 1, 123, 631, 670 (1845); v. 2, 282 (1846).
3300   1845-1847, [Brief notes on caves]: v. 1, 135, 269, 285-286, 328-woodcut, 335 (1845); v. 2, 554-555 (1846); v. 3, 524-525 (1847).
3301   1845, Minerals: v. 1, 139, 157, 172, 189, 204-205, 221, 236, 253-254, 268, 285.
Elementary descriptions of common minerals and rocks.
3302   1845-1846, [Brief notes on mineral springs]: v. 1, 219, 270, 356-357 (1845); v. 2, 572 (1846).
3303   1845-1847, [Brief notes on fossils]: v. 1, 308-309, 564-565, 602, 635-636, 716-717, 776, 790, 815 (1845); v. 2, 531, 707 (1846); v. 3, 73-74, inc. 6 figs. (1847).
3304   1845, Metals: v. 1, 317-318, 367, 383, 398-399, 413-414, 463.
Description of the metals and their ores.
3305   1845, A volcano at sea. Formation of Hotham Island: v. 1, 457-459, 2 woodcuts.
3306   1845, The skeleton of the great sea-serpent; or, Hydrargos sillimanii: v. 1, full page woodcut on pp. 513.
3307   1845, The great fossil sea-serpent, or Hydrargos: v. 1, 514-515, fig.
3308   1845, Meteoric stone: v. 1, 603-604.
3309   1845, On the known thickness of the crust of the earth: v. 1, 655.
3310   1845, Darwin's researches in geology [notice of 2916]: v. 1, 699.
3311   1845, Dr. Houghton [obituary]: v. 1, 771 (15 lines).
3312   1845, Rocks and mountains: v. 1, 797, fig.
On the geological structure of mountains.
3313   1846, Bluffs on the Mississippi: v. 2, 545-546, fig.
3314   1846, Volcanic wonders of the Azores: v. 2, 676-677, 692-694.
3315   1846, Stromboli: v. 2, 697-698, fig.
3316   1847, Great eruption of the volcano Kilauea in 1840: v. 3, 6-7, 19-20.
3317   1847, Scientific survey of the state of New York: v. 3, 38-39, 52-53.
3318   1847, Pompeii: v. 3, 137-138, fig.
3319   1847, Professor Agassiz's lectures on glaciers: v. 3, 251-253.
Abstracted from the Boston Daily Advertiser.

Earl, Samuel F.
3320   1813, Analysis of the Bordentown (N. J.) spring: Colum. Chemical Soc. Phila. Mem., v. 1, 205-207.

The earthquake Naples; September, 21. 1694:
3321   1694, Bos., Reprinted by B. Green, broadside 8vo.
Evans number 715.

Earthquakes, tokens of God's power and wrath....Being a warning to sinners and comfort to the children of God... :
3322   1744, Bos., broadside folio.
Evans number 5383. Also "second edition," 1744, Evans 5384.

Eaton, Amos (1776-1842)
3323   1818, Account of the strata perforated by, and of the minerals found in, the great adit to the Southampton lead mine [in Massachusetts]: Am. J. Sci., v. 1, no. 1, 136-139.

Eaton, A. (cont.)

3324 1818, Index to the geology of the northern states, with a transverse section from Catskill Mountain to the Atlantic; prepared for the geological classes of Williams College: Leicester, Mass., Hari Brown (printer), 52 p., 1 folded plate.
For review with excerpts see 10729. For review see 7941 and 8625. For notice see 323. There were 800 copies printed.

3325 1820, An index to the geology of the northern states, with transverse sections, extending from the Susquehanna River to the Atlantic, crossing Catskill Mountains. To which is prefixed a geological grammar ... Second ed., wholly written over anew ... : Troy, NY, W. S. Parker; Alb., Websters and Skinners, xi, [1], [13]-286 p., 2 plates.
For notice see 93.

3326 1820, Localities [of minerals in Massachusetts and Connecticut]: Am. J. Sci., v. 2, no. 2, 238-239.

3327 1820, Observations on the geology of the district of country lying between the rivers Hudson and Susquehanna in the state of New York: Plough boy, v. 1, no. 36, 282.
From Troy Lyc. Nat. History Trans.

3328 1820, and T. R. Beck, A geological survey of the county of Albany: Alb., 56 p., 1 fig., 1 table.
Also in NY State Board Ag. Mem., v. 1, 1821. For review see 373. For notice see 363, 7805, 8419, and 8422.

3329 1821, The globe had a beginning: Am. J. Sci., v. 3, no. 2, 238.

3330 1822, An outline of the geology of the Highlands on the River Hudson: Am. J. Sci., v. 5, no. 2, 231-235.

3331 1822, A geological and agricultural survey of Rensselaer County in the state of New-York; to which is annexed, a geological profile, extending from Onondaga Salt Springs, across said country, to Williams College in Massachusetts: Alb., E. and E. Hosford (printers), 70 p., folding table, plate.
For review see 386 and 7942.

3332 1822, On a singular deposit of gravel: Am. J. Sci., v. 5, no. 1, 22-23.
Description of gravel deposits at Troy, New York.

3333 1822, To gentlemen residing in the vicinity of the Erie Canal: Troy (?), NY, 10 p.
Notice of the proposed survey for the Erie Canal.

3334 1823, Explanation of the plate representing a profile view, or transverse section of the rock strata, from Onondaga Salt Springs, in the state of New-York, to Williams College, in Massachusetts: NY State Board Ag. Mem., v. 2, 41-43, plate.

3335 1823, A geological and agricultural survey of Rensselaer County: NY State Board Ag. Mem., v. 2, 3-18, table.

3336 1823, On the probable origin of certain salt springs [at Onondaga, New York]: Am. J. Sci., v. 6, no. 2, 242-243.

3337 1823, Proposed geological nomenclature: Troy, NY.
Not seen. See also 3348.

3338 1824, A geological and agricultural survey of the district adjoining the Erie Canal in the state of New York. Taken under the direction of the Hon. Stephen Van Rensselaer. Part I containing a description of the rock formations together with a geological profile extending from the Atlantic to Lake Erie: Alb., Packard and van Benthuysen, 163 p., plate.
Plate entitled "Geological profile extending from the Atlantic to Lake Erie, running near the 43° N. L. and embracing 9 degrees of Longitude. Taken 1822 & 3 under the direction of Amos Eaton. Engraved by Rawdon, Clark & Co. Albany," 141 x 17 cm.
With remarks by E. Hitchcock (see 5086). For review see 464. For notice see 409, 421, 443.

3339 1824, A geological and agricultural survey of Rensselaer County, state of New-York: Mass. Agri. Repos. J., v. 8, no. 2, 146-177.
Reprint of 3331.

3340 1824, Ought American geologists to adopt the changes in the science proposed by Phillips and Conybeare?: Am. J. Sci., v. 8, no. 2, 261-263.

3341 1824, Progress of the geological survey on the grand canal: Am. J. Sci., v. 8, no. 1, 195-198.

3342 1827, Analysis of soils: Am. J. Sci., v. 12, no. 2, 370-372.

3343 1827, Improvement in the manufacture of magnetic needles: Am. J. Sci., v. 12, no. 1, 14-16.
See also "Remarks" (455).

3344 1827, Notices respecting ... diluvial deposits in the state of New York and elsewhere: Am. J. Sci., v. 12, no. 1, 17-20.

3345 1828, General geological strata: Am. J. Sci., v. 14, no. 2, 359-368; also in: Am. J. Improvements Use. Arts, v. 1, 103 [i.e. 203]-208.

3346 1828, Geological nomenclature, classes of rocks, etc.: Am. J. Sci., v. 14, no. 1, 145-159.

3347 1828, Geological nomenclature, exhibited in a synopsis of North American rocks and detritus: Am. J. Sci., v. 14, no. 1, 4 p. of which 2-4 are colored.

3348 1828, A geological nomenclature for North America; founded upon geological surveys, taken under the direction of the Hon. Stephen Van Rensselaer: Alb., Packard and Van Benthuysen, 31 p., map, 3 colored tables.
For notice see 466.

3349 1828, Geological profile extending from the Atlantic to Lake Erie. Running near the 43o N. L. and embracing 9 degrees of Longitude. Taken 1822 & 3 under the direction of the Hon. Stephen Van Rensselaer, by Amos Eaton. Corrected by a re-survey Feb. 1st, 1828 A. E.: Alb., Engraved by Rawdon, Clark and Co., plate, 142.8 x 18.0 cm, published in Am. J. Sci., v. 14, no. 2.
See also 3338.

3350 1828, Notice: Am. J. Sci., v. 14, no. 2, 400 (5 lines).
Announcement of American geology (i.e. Geological Prodromus?, see 3356).

3351 1828, Observations on the coal formations in the state of New-York; in connexion with the great coal beds in Pennsylvania: Alb. Inst. Trans., v. 1, 126-130.
See also remarks by D. Thomas (10224).

3352 1828, Tabular view of North American rocks: Am. J. Sci., v. 13, 384-385.

3353 1829, Anthracite coal, and liquids, in quartz crystals: Am. J. Sci., v. 15, no. 2, 362 (8 lines).

3354 1829, Argillite embracing anthracite coal: Am. J. Sci., v. 16, 299-301.
On the extent of New York argillite formations.

Eaton, A. (cont.)
3355 1829, Gases, acids, and salts, of recent origin and now forming, on and near the Erie Canal, in the state of New-York; also living antediluvial animals: Am. J. Sci., v. 15, no. 2, 233-249.
3356 1829, Geological Prodromus: Alb., 7 p.
For notice see 3350.
3357 1829, Geological prodromus: Am. J. Sci., v. 17, no. 1, 63-69.
On the classification of United States rocks.
3358 1830, All primitive strata, below granular quartz, are co-temperaneous and schistose: Am. J. Sci., v. 17, no. 2, 334-335.
3359 1830, Colored map featuring a general view of the economical geology of New York and parts of adjoining states: colored map, 31 x 38 cm., in A geological text-book (see 3361).
3360 1830, Direction and extent of Primitive ranges: Am. J. Sci., v. 18, no. 2, 376.
With article by D. Thomas (see 10223).
3361 1830, Geological text-book, prepared for popular lectures on North American geology; with applications to agriculture and the arts: Alb., Websters and Skinners, vii, [9]-63, [1] p., 3 figs., folding map ("map sold separately," see 3359).
For reviews see 3744, 3750, and 4250.
3362 1830, The gold of the Carolinas in talcose slate: Am. J. Sci., v. 18, no. 1, 50-52.
3363 1830, Observations on the coal formations in the state of New York; in connexion with the great coal beds of Pennsylvania: Am. J. Sci., v. 19, no. 1, 21-26; also in: Hazard's Reg. Penn., v. 6, 289-290.
For review see 10224.
3364 1830, Travelling term of Rensselaer School, for 1830, with a notice of the nature of the institution: Am. J. Sci., v. 19, no. 1, 151-159.
3365 1831, Crotalus? reliquus, or Arundo? crotaloides: Am. J. Sci., v. 20, no. 1, 122-123, fig.; no. 2, 204 (8 lines).
See also article by W. Cooper (2637). On a fossil fish from Montrose, Susquehanna County, Pennsylvania.
3366 1831, Four cardinal points in stratigraphical geology, established by organic remains: Am. J. Sci., v. 21, no. 1, 199-200.
3367 1831, Geological equivalents: Am. J. Sci., v. 21, no. 1, 132-138.
3368 1831, The gold of Mexico in a rock, equivalent to that which contains the gold of the Carolinas: Am. J. Sci., v. 20, no. 1, 124.
3369 1831, Improvement in the reflecting goniometer: Am. J. Sci., v. 20, no. 1, 158-159.
3370 1831, [Letter in "Scratches on elevated strata of horizontal graywacke in the Alleghany range; probably diluvial" by W. A. Thompson (see 10234)]: Am. J. Sci. v. 20, no. 1, 124.
3371 1832, Application of geology; as the basis of the science of agriculture: Am. Farm., v. 14, 210-211.
3372 1832, Correction of an error in Prof. Green's Monograph of North American trilobites; with additional explanations: Am. J. Sci., v. 23, no. 2, 400-401 (10 lines).
See 4535.
3373 1832, Geological nomenclature, exhibited in a synopsis of North American rocks and detritus: in The history and topography of the United States by Hinton (see 5064), v. 2, 53-56, tables.
Also in the 1834 Boston edition, v. 2, 67-70.
3374 1832, Geological text-book, for aiding the study of North American geology: being a systematic arrangement of facts, collected by the author and his pupils ... : Alb., Webster and Skinners, 134 p., 4 figs, 5 plates, map (see 3359).
For review see 700 and 719. For notice see 545. This is the second edition.
3375 1832, Trilobites: Am. J. Sci., v. 22, no. 1, 165-166.
On possible living species.
3376 1833, Application of geology, as the basis of the science of agriculture: Farm's Reg., v. 1, 246-249.
3377 1833, The coal beds of Pennsylvania equivalent to the great secondary coal measures of Europe: Am. J. Sci., v. 23, no. 2, 399-400.
3378 1834, Geology and meteorology west of the Rocky Mountains: Am. J. Sci., v. 25, no. 2, 351-353; also in: Eclec. J. Sci., v. 2, 222-224.
Notice of James Hall's geological studies in the West. See also 6680.
3379 1835, Strontian in Marcellus, Onondaga Co., New York: Am. J. Sci., v. 28, no. 2, 380-381.
From Geological text-book.
3380 1839, Cherty lime-rock, or corniferous lime-rock, proposed as the lime of reference, for state geologists of New York and Pennsylvania: Am. J. Sci., v. 36, no. 1, 61-71; 198 (17 lines).
3381 1840, References to North American localities, to be applied in illustration of the equivalency of geological deposits on the eastern and western sides of the Atlantic: Am. J. Sci., v. 39, no. 1, 149-156.
3382 1841?, Eaton's geological note book, for the Troy class of 1841: Bos.?, 13 p.

Eaton, Eben
3383 1827, On the use of soapstone to diminish the friction of machinery: New Harmony Gaz., v. 3, 90.

Ebelman (i.e. Jacques Joseph Ebelmen?, 1814-1852)
3384 1838, New method of analysing the ores of manganese: Franklin Inst. J., v. 22 (ns), 332-333.
Abstracted from Annales des Mines.
3385 1839, Description of a new mode of chemical analysis: Am. J. Pharmacy, v. 5 (ns), 39-44.
Abstracted from Annales des Mines.
3386 1846, On the artificial production of diaphanous quartz: Franklin Inst. J., v. 12 (3s), 64-65.
Abstracted from London, Edinburgh and Dublin Philos. Mag.

Ebelman (cont.)
3387 1846, Production of diaphanous quartz and hydrophane: Am. J. Sci., v. 1 (ns), no. 1, 261-262.
   Abstracted by J. L. Smith from Comptes Rendus.
3388 1847, Artificial minerals: Franklin Inst. J., v. 14 (3s), 408-410.
   Abstracted from Civil Engineer and Architect J.
3389 1848, M. Ebelman on artificial hyalite and hydrophane: Am. J. Sci., v. 5 (ns), no. 3, 412-413.
   Abstracted from l'Institut.
3390 1849, Method of obtaining crystalline combinations by heat, and of reproducing thereby various mineral species: Am. J. Sci., v. 7 (ns), no. 3, 427.
   Abstracted from Comptes Rendus.
3391 1849, On the decomposition of rocks: Am. J. Sci., v. 8 (ns), no. 3, 421-422; also in: Ann. Scien. Disc., v. 1, 236.
   Abstracted from l'Institute.

Eccentricities of Literature and Life; or the Recreative Magazine (Boston, 1822)
3392 1822, Stone showers - stone candles - stone petticoats - stone animals - stone eaters: v. 1, no. 2, 123-127.
   Includes description of meteorites.
3393 1822, Petrifaction: v. 1, no. 3, 205-206.

The Eclectic, and Medical Botanist (Columbus, Ohio, 1832-1833)
3394 1833, Petrifaction of water [from Persia]: v. 1, 317-318.
3395 1833, The diamond [of Brazil]: v. 1, 365-366.
3396 1833, Meteoric explosion [at Weston, Connecticut]: v. 1, 282-283.

The Eclectic Jounal of Science (Columbus, Ohio, 1834-1835)
3397 1834, Gold mine in Virginia: v. 2, 48.
3398 1834, Iron mine in Sweden: v. 2, 187-189.
3399 1834, The pitch lake [of Trinidad]: v. 2, 350-351.
3400 1834, Geology: v. 2, 616-617.
   On the origin of meteorites.
3401 1834, Geological surveys [in the United States]: v. 2, 672.
3402 1834, Geological treat: v. 2, 749-750.
   On geological lectures by B. Silliman, by "a female correspondent of the Portsmouth Journal, at Lowell, Massachusetts".
3403 1834, Remarkable caverns [at Erpfingen, Hollenberg]: v. 2, 766.
3404 1834, Lake Superior: v. 2, 783.
   Notice of geological surveys by Schoolcraft and Houghton.

The Eclectic Magazine of Foreign Literature, Science and Art (New York, 1844-1850)
3405 1844-1846, [Brief notes on volcanoes and earthquakes]: v. 1, 427; v. 2, 106 (1844); v. 4, 471, 568; v. 5, 508; v. 6, 429, 566 (1845); v. 7, 287.
3406 1844-1848, [Brief notes on fossils]: v. 2, 141; v. 3, 569 (1844); v. 6, 141 (1845); v. 14, 432 (1848).
3407 1844-1846, [Brief notes on mining]: v. 2, 141 (1844); v. 6, 569 (1845); v. 7, 430-431 (1846).
3408 1844, The scriptural difficulties of geology: v. 3, 185-200.
   Includes reviews of Twelve lectures ..., by N. Wiseman (see 11007), Geology and Scripture by J. P. Smith (see 9895), and Recreations in geology by R. M. Zornlin.
3409 1845, Geology of Gibralter: v. 4, 284.
3410 1845, [Review with excerpts of Vestiges of the natural history of Creation by R. Chambers (see 2292 et seq.)]: v. 6, 43-91.
3411 1845, [Review of Nimshi. The adventures of a man to obtain a solution of Scriptural geology ... (see 7910)]: v. 6, 342-345.
   From Lit. Gaz.
3412 1846, [Review with excerpts of Cosmos by A. Humboldt (see 5395)]: v. 7, 353-375.
3413 1846, The microscope and its revelations [review of Thoughts on Aminalcules by G. A. Mantell]: v. 9, 452-470.
   From For. Q. and Westminster Rev.
3414 1848, [Review with excerpts of Cosmos by A. Humboldt (see 5395)]: v. 13, 296-329.
   From Edinburgh Rev.
3415 1848, Central fires in the earth: v. 14, 432.
   On the conflicting views of D. Arago and C. Morton.
3416 1848, Lapis Lazuli [from Lake Baikal, Russia]: v. 15, 142.
3417 1849, Lyell's second visit to the United States [review with excerpts of 6571]: v. 18, 27-50.
   From Q. Rev.
3418 1850, Origin of the Giant's Causeway: v. 20, 43.
3419 1850, [Review of Aspects of nature, in differnet lands and different climates, with scientific elucidation by A. Humboldt (see 5394)]: v. 19, 374-396.
   From North British Rev.
3420 1850, California: its past progress, present condition, and future prospects: v. 19, 548-557.
   From Sharpe's Mag.
3421 1850, California - the gold hunters [reviews with excerpts of Eldorado by B. Taylor (see 10098) and Personal adventures ..., by W. R. Ryan (see 9147)]: v. 21, 289-301.
   From British Q. Rev.
3422 1850, A visit to Mammoth Cave: v. 21, 474-484.
   Abstracted from Fraser's Mag.

The Eclectic Museum of Foreign Literature, Science and Art (New York, 1843-1849)
3423 1843, [Review of On the glacial theory by R. I. Murchison]: v. 1, 383-393.
   From Edinburgh New Philos. J.
3424 1843, [Brief notes on earthquakes and volcanoes]: v. 1, 570, 572, 574.
3425 1843, Animal skeletons [from Chelsea, England]: v. 3, 284 (6 lines).
3426 1843, Mr. Nott "On terrestrial magnetism" [notice of]: v. 3, 426.
3427 1843, Egyptian gold mine: v. 3, 426 (4 lines).

The Eclectic Repertory and Analytic Review, Medical and Philosophical (Philadelphia, 1810-1819)
3428 1811, [Review of American mineralogical journal (see 2107)]: v. 2, 210-218.
       From Edinburgh Rev.
3429 1815, The volcano of Albay [Philippines]: v. 5, 219-220.
3430 1815, [Shower of meteorites near Toulouse]: v. 5, 405-406.
3431 1815, [New bitumen found by Bucholz in Halle, Saxony]: v. 5, 542-543.
3432 1816, Native epsom salt [from Louisville, Kentucky]: v. 6, 253.
3433 1816, Earthquakes [in western Massachusetts]: v. 6, 254.
3434 1816, Mineralogy and geology [notice of Elementary treatise ..., by P. Cleaveland (see 2420)]: v. 6, 270.
3435 1817, [Review with excerpts of Analysis of the mineral waters of Dublane and Pitcaithly by J. Murray (see 7532)]: v. 7, 64-76.
       From Edinburgh Med. Surgical J.
3436 1818, Memoir of Abraham Gottlob Werner: v. 8, 102-110.
       From Philos. Mag. and J.
3437 1819, [Reviews with excerpts of Observations ..., by W. Maclure (see 6619, 6620 and 2420)]: v. 9, 241-258.
       From Edinburgh Rev.
3438 1819, Geological Society of Connecticut: v. 9, 422.
       From NY Daily Advertiser. On the organization of the society.

Edes, Oliver
3439 1850, Submerged rocker for separating ores: Franklin Inst. J., v. 20 (3s), 49.

Edinburgh Encyclopaedia
3440 1808-1824, Phila., 30 v.
       "Improved American edition." Contains articles on geology, mineralogy, earthquakes, volcanoes, fossils, etc.
3441 1813, Phila., Edward Parker and Joseph Delaplaine.
       "The American edition of the New Edinburgh Encyclopaedia. Conducted by David Brewster ... assisted by upwards of one hundred gentlemen in Europe ... "
3442 1813-1831, NY, Whiting and others, 18 v.
       "Second American edition of the New Edinburgh Encyclopaedia, conducted by David Brewster".

Edmiston, Joseph W.
3443 1821, Marbles of Kentucky: Am. J. Sci., v. 3, no. 2, 234 (8 lines).

Edrington, E. G.
3444 1845, An intermittent spring [at Pittsburgh, Pennsylvania]: Am. J. Sci., v. 48, no. 2, 400 (18 lines).

Edwards, Timothy
3445 1789, Description of a horn or bone lately found in the River Chemung or Tyoga, a western branch of the Susquehanna, about twelve miles from Tyoga Point: Am. Mus. or Univ. Mag., v. 4, 42; also in: Am. Acad. Arts Sci's Mem., v. 2, pt. 1, 164-165 (1793); also in: Hazard's Reg. Penn., v. 1, 432.

Egede, Hans Poulsen (1686-1758)
3446 1744, A new account of Greenland from a Danish book in quarto: Am. Mag. and Hist. Chron., April and May, 1744, 326-329; 359-365.

Egerton, Sir Philip
3447 1838, [Gravel near Avon, England]: Am. J. Sci., v. 33, no. 2, 287 (7 lines).
       In British Assoc. Proc.

Ehrenberg, Christian Gottfried (1795-1876)
3448 1837, Fossil infusoria: Franklin Inst. J., v. 20 (ns), 198-200.
       From Edinburgh New Philos. J.
3449 1838, and J. F. L. Hausmann, Discoveries regarding two varieties of siliceous earth found near Oberohe in the Hanoverian Province of Luneberg: Franklin Inst. J., v. 22 (ns), 396-398.
       From Edinburgh Philos. J.
3450 1839, Communication respecting fossil and recent infusoria made to the British Association at Newcastle: Am. J. Sci., v. 35, no. 2, 371-374, 3 figs.
3451 1840, Fossil infusoria of West Point, New York: Am. J. Sci., v. 39, no. 1, 191-193.
       Abstracted from Royal Prussian Acad. Sci. Berlin Proc.
3452 1840, On the remarkable diffusion of coralline animalcules from the use of chalk in the arts of life, as observed by Ehrenberg: Am. J. Sci., v. 39, no. 1, 205-206.
       Abstracted from Annals of Nat. History.
3453 1841, Amicular constitution of chalk: Am. Eclec., v. 2, no. 2, 389-390.
       Abstracted from London Ath.
3454 1841, Professor Ehrenberg's microscopical discoveries: Franklin Inst. J., v. 1 (3s), 61-62.
       Abstracted from Edinburgh Philos. J. See also: Am. Mag. and Repos. of Useful Liter., v. 1, 93.
3455 1844, [Excerpts from "Researches on the distribution of microscopic life"]: Am. J. Sci., v. 47, no. 1, 208-211.
       In review by J. W. Bailey (see 1424).
3456 1844, [Excerpts from his works]: Am. J. Sci., v. 46, no. 2, 297-307, fig.
       In "Notice of a memoir by C. G. Ehrenberg ...," by J. W. Bailey (see 1427).
3457 1845, Infusoria [in volcanic tuff]: Am. J. Sci., v. 49, no. 2, 397.
       Abstracted from Berlin Acad. Proc.
3458 1845, [Observations on fossil infusoria of Maryland and Virginia]: Am. J. Sci., v. 48, no. 1, 201-204.
       In article by J. W. Bailey (see 1428).
3459 1845, [On fossil infusoria]: Am. J. Sci., v. 48, no. 2, 321-343.
       In "Notice of some new localities of infusoria, fossil and recent" by J. W. Bailey (see 1431).

Ehrenberg, C. G. (cont.)
3460 1846, Infusoria: Am. J. Sci., v. 1 (ns), no. 1, 123-124.
       Notice of their discovery at Ascension Island, Patagonia, Pompeii, and in the Rhine volcanics.
3461 1846, On the microscopic constituents of the ash of fossil coal: Am. J. Sci., v. 1 (ns), no. 1, 124-126.
       Abstracted from "Ann. Mag. Nat. Hist."
3462 1850, On infusorial deposits on the River Chutes in Oregon: Am. J. Sci., v. 9 (ns), no. 1, 140; also in: Ann. Scien. Disc., v. 1, 288 (17 lines).
       Abstracted from "Monatsh. Acad. Berlin."

Eights, James
       See also 10579 with L. Vanuxem
3463 1835, Synopsis of the rocks of the state of New-York: Zodiac, v. 1, no. 2, 27-28, fig.; to be continued but no more found.
3464 1835/1836, Naturalist's every day book: Zodiac, v. 1, no. 1, 4-8; no. 2, 23-25; no. 3, 33-35; no. 4, 60-63; no. 9, 129-132.
       Notes on the natural history of New York
3465 1836, Marl [of New York]: The Cultivator, v. 3, 73-74.
3466 1836, Notes of a pedestrian: Zodiac, v. 1, no. 7, 111-112; no. 8, 113-116, fig.; no. 9, 141-143; no. 10, 146-147; no. 12, 177-178; v. 2, no. 1, 10-11, fig., no. 2, 28-29; to be continued but no more found.
       Geological notes on New York and the coal regions of Pennsylvania.
3467 1837, [Section of strata at Montezuma, New York]: NY Geol. Survey Ann. Rept., v. 1, 173-175.
       In "First annual report on the geological survey of the third district of the state of New York" (see 2557).
3468 1842, Description furnished by Dr. James Eights: in Geology of New York, second district by W. W. Mather (see 6907), 433-434, fig.
       Description of Sphaeroma bumastiformis.
3469 1846, Notes on natural history: Am. Q. J. Ag. Sci., v. 3, 219-223.
       With excerpts by H. T. De la Beche (see 3041).
3470 1846, Outlines of the geological structure of Lake Superior mineral region belonging to the New York and Lake Superior Mining Company: Alb., 21 p.
       Published as appendix to New York and Lake Superior Mining Company first annual report (see 7760). For review see 3545.
3471 1847, Notes on natural history, etc.: Am. Q. J. Ag. Sci., v. 5, 56-57.
       With excerpts from Geological researches ..., by H. T. De la Beche (see 3039).
3472 1848, Notes of a geological examination and survey of Mitchill's Cave, town of Root, county of Montgomery, N. Y.: Am. Q. J. Ag. Sci., v. 7, no. 1, 21-27.

Ekeberg, Anders Gustaf (1767-1813; see 10245 with T. Thompson)

Elder, Thomas, jr.
3473 1838, Notice of three specimens of garnet: Franklin Inst. J., v. 21 (ns), 253.
       On garnets from Vesuvius; Dauphin County, Pennsylvania; Ceylon; and Northern New Jersey.

Eldredge, N. T.
3474 1837, Report of the special agent sent to examine the mines of the company: NY, 13 p.
       Geological notes on the lands of the Boston and New York Coal Company.

Eliot, Jared (1685-1763)
3475 1762, An essay on the invention, or art of making very good, if not the best iron, from black sea sand: NY, John Holt, 34 p.
       Evans number 9109.

Ellet, W. H. (Dr.)
3476 1839, On the value of marls and calcareous deposits: S. Agriculturist, v. 12, 25-26.

Ellicott, Andrew (1754-1820)
3477 1790, Description of the Falls of Niagara: Univ. Asylum and Colum. Mag., v. 4, 331-332, plate.
       Also in Am. Mus. or Univ. Mag., v. 8, 215-216.
3478 1799, Miscellaneous observations relative to the Western parts of Pennsylvania, particularly those in the neighbourhood of Lake Erie: Am. Philos. Soc. Trans., v. 4, 224-229.
3479 1803, The journal of ... late commissions on behalf of the United States during part of the year 1796, the years 1797, 1798, 1799, and part of the year 1800: For determining the boundary between the United States and the possessions of his Catholic Majesty in America, containing occasional remarks on the situation, soil, rivers, natural productions, and diseases of the different countries of the Ohio, Mississippi and the Gulph of Mexico: Phila., Budd and Bartram (printers), 7, 299, 151 p., illus., 6 maps.
3480 1814, The journal of ... [as above]: Phila., W. Fry (printer), 7, 299, 151 p., illus., 8 plates and maps.

Elliott, Stephen (1771-1830)
3481 1814, An address to the Literary and Philosophical Society of South Carolina; delivered in Charleston, on Wednesday, the 10th of August, 1814: Charleston, SC, W. P. Young (printer), 20 p.

Ellis, William (1794-1872)
3482 1825, A journal of a tour around Hawaii, the largest of the Sandwich Islands: Bos., Crocker and Brewster, 264 p.
       For excerpts see 4438.

Ells, B. F.
3483 1848, Rock Bridge in Virginia: West. Miscellany, v. 1, 36-37, fig.

Elmore, Franklin Harper (1799-1850)
3484 1835, Statements of the constituent parts of soils of the prairies of Alabama: Farm's Reg., v. 2, 715-717.
3485 1838, Marl of South Carolina: Farm's Reg., v. 5, 693.

Emmet, John Patton (1796-1842)
3486 1830, Bromine and iodine in Kenawha waters, Virginia: Am. J. Sci., v. 18, no. 2, 260 (16 lines).
3487 1832, Experiments upon the solidification of raw gypsum: Am. J. Sci., v. 23, no. 2, 209-212.

Emmons, Ebenezer (1799-1863)
See also 8875 with others.
See also discussion of article by C. B. Adams (31).
3488 1824, Miscellaneous localities: Am. J. Sci., v. 7, no. 2, 254-256.
      On mineral localities in Massachusetts.
3489 1824, Notice of the granitic veins and beds in Chester, Mass.: Am. J. Sci., v. 8, no. 2, 250-252, 5 figs.
3490 1825, By Dr. E. Emmons: Am. J. Sci., v. 9, no. 2, 249-250; v. 10, no. 1, 11.
      On mineral localities in Massachusetts.
3491 1826, Manual of mineralogy and geology: designed for the use of schools; and for persons attending lectures on these subjects, as also a convenient pocket companion for travellers, in the United States of America: Alb., Websters and Skinners, xxiii, 230 p.
      For review see 10502 and 10760. For notice see 447.
3492 1832, Manual of mineralogy and geology: Alb., Webster and Skinners, xii, 299 p., illus.
      For review see 570.
3493 1834, Strontianite discovered in the United States: Am. J. Sci., v. 27, no. 1, 182-183.
      With note by C. U. Shepard (see 9491).
3494 1836, Notice of a scientific expedition [to Nova Scotia]: Am. J. Sci., v. 30, no. 2, 330-354, 15 figs.
3495 1837, First annual report of the second geological district of the state of New York: NY Geol. Survey Ann. Rept., v. 1, 97-153.
      With "Ores of iron" by J. Hall (see 4670).
3496 1838, Map of the Tertiary of Essex Co. [New York]: colored geological map, Alb., G. W. Merchant (engraver), 22 x 55 cm.
      In NY Geol. Survey Ann. Rept., v. 2, (see 3497).
3497 1838, Report - of the 2nd geological district of the state of New-York: NY Geol. Survey Ann. Rept., v. 2, 185-252, 17 figs., 9 plates, 2 folding maps.
      Contains map (see 3496).
3498 1839, and J. Hall, Communication - relative to a place of deposite for the different specimens collected by the geologists: NY Geol. Survey Ann. Rept., v. 3, 517.
3499 1839, Third annual report - of the survey of the second geological district: NY Geol. Survey Ann. Rept., v. 3, 201-239.
3500 1840, Fourth annual report - of the survey of the second geological district: NY Geol. Survey Ann. Rept., v. 4, 259-353.
3501 1841, Fifth annual report - of the survey of the second geological district: NY Geol. Survey Ann. Rept., v. 5, 113-136.
3502 1841, Geology of the Montmorenci [River in Quebec]: Am. Mag. and Repos. Use. Liter., v. 1, 146-150, fig.
3503 1841, Iron ore at Duane, New York: Mon. Chron., v. 2, 81-83.
3504 1841, Report of Prof. E. Emmons, in answer to a resolution of the Assembly calling for information in relation to the steel ore at Duane: Alb., 4 p.
      NY State Documents, Assembly Document 182 of the 64th State Legislature.
3505 1841, [Review of Memoir of the geological survey of the State of Delaware by J. C. Booth (see 1769)]: Am. Mag. and Repos. Use. Liter., v. 1, 77-79.
3506 1841, Utility of natural history: Am. Mag. and Repos. Use. Liter., v. 1, 163-165.
3507 1842, Geological map of Clinton County: NY, Endicott, geological, map 20 x 23 cm.
      In Geology of New York (see 3510).
3508 1842, and J. Hall, W. W. Mather and L. Vanuxem, Geological map of the state of New York: colored geological map, 92 x 99 cm.
      Published separately, with Geology of New York, 4 v.
3509 1842, Geological observations [on Montmorenci River, Quebec]: Am. Mag. and Repos. Use. Liter., v. 2, no. 1, 5-9, 2 figs.
3510 1842, Geology of New York. Part II, comprising the survey of the second geological district: Alb., W. and A. White and J. Visscher, x, 437 p., 17 plates, illus.
      For review see 8066. Contains maps 3507 and 3511.
3511 1842, Map of the county of Jefferson [New York]: NY, Endicott (lithographer), colored geological map, 25 x 28 cm.
      In Geology of New York, second district (see 3510), plate 16.
3512 1842, [Notes on the classification of drift and diluvial action]: Am. J. Sci., v. 43, no. 1, 166-167.
      Also in Assoc. Am. Geol's Nat's Proc., v. 3, 21-22. Also in Assoc. Am. Geol's Nat's Rept., v. 1, 60, 1843.
3513 1842, Topography, geology, and mineral resources of the state of New York: in Gazetteer of the state of New York, Alb., J. Disturnell, 5-25.
3514 1843, [Artificially heated rocks showing columnar structure]: Am. J. Sci., v. 45, no. 1, 146 (14 lines).
      In Assoc. Am. Geol's Nat's Proc.
3515 1843, and J. Hall, Communication from Messrs. Emmons and Hall, State geologists: Alb., 9 p.
      NY State Legislature, SD v. 2, no. 60, 66th session. On the geological survey.
3516 1843, [On bituminous matter in New York rocks]: Am. J. Sci., v. 45, no. 2, 336, 10 lines).
      In Assoc. Am. Geol's Nat's Proc. See also 1581.
3517 1843, [Remarks on metamorphism]: Am. J. Sci., v. 45, no. 1, 142; 146 (15 lines).
      In Assoc. Am. Geol's Nat's Proc.

Emmons, E. (cont.)

3518 1844, Agricultural and geological map of the state of New York by legislative authority 1844: NY, Sherman and Smith (engravers), colored geological map, 110 x 93 cm.
3519 1844, Analysis of soils from Alabama: The Cultivator, v. 1 (ns), 145-146.
3520 1844, The Taconic system; based on observations in New York, Massachusetts, Maine, Vermont and Rhode Island: Alb., 65, 3 p., illus.
    For review see 4629. For notice see 800.
3521 1845, Agricultural geology: Am. Q. J. Ag. Sci., v. 2, 1-14; 179-198, fig., 2 plates.
3522 1845, [Analyses of soil from Lyonsdale, Lewis County, New York]: Am. Q. J. Ag. Sci., v. 2, 126-127.
3523 1845, Conglomerate of the granular quartz: Am. Q. J. Ag. Sci., v. 2, 368 (7 lines).
3524 1845, Fertilizers in the rocks: Am. Q. J. Ag. Sci., v. 1, 62-64.
3525 1845, Geological and agricultural report of New-Hampshire, by C. T. Jackson [review of 5611]: Am. Q. J. Ag. Sci., v. 1, 232-240.
3526 1845, Native lamellar iron [from New York]: Am. Q. J. Ag. Sci., v. 2, 367 (15 lines).
3527 1845, On drift and the changes which have been effected in the position of soils, etc.: Am. Q. J. Ag. Sci., v. 2, 26-33.
3528 1845, On the supposed Zeuglodon cetoides of Prof. Owen: Am. Q. J. Ag. Sci., v. 2, 58-63; 366, plate with 3 figs.
3529 1845, Oxide of copper [from Lake Superior]: Am. Q. J. Ag. Sci., v. 2, 367 (4 lines).
3530 1845, Phosphate of lime [from New York] Am. Q. J. Ag. Sci., v. 1, 60-61.
3531 1845, Phosphate of lime and other fertilizers in the older rocks: Am. Q. J. Ag. Sci., v. 1, 219-221.
3532 1845, The relation of clay to sandy soils: Am. Q. J. Ag. Sci., v. 1, 118.
3533 1845, Remarks [on the sixth meeting "Proceedings of the American Association of Geologists and Naturalists"]: Am. Q. J. Ag. Sci., v. 2, 132-170.
    In 1330.
3534 1845, First annual report of the geology of the state of Vermont. By C. B. Adams [review of 23]: Am. Q. J. Sci., v. 2, 315.
3535 1845, Specimens of soils from Wisconsin: The Cultivator, v. 2 (ns), 215.
3536 1845, Travels in North America, in the years 1841-42; with geological observations on the United States, Canada and Nova Scotia: by Charles Lyell [review of 6557]: Am. Q. J. Ag. Sci., v. 2, 265-271.
3537 1845, Veins of hematite [from Adams, Massachusetts]: Am. Q. J. Ag. Sci., v. 2, 367 (11 lines).
3538 1845, Vestiges of the natural history of Creation [review of text by R. Chambers (see 2292 et seq.)]: Am. Q. J. Ag. Sci., v. 1, 240-250.
3539 1846, Agricultural geology of Onondaga County [New York]: Am. Q. J. Ag. Sci., v. 3, 161-193, 3 figs.
    For review see 4295.
3540 1846, Agricultural map of the state of New York: NY, G. and W. Endicott (lithographer), colored geological map, 60 x 45 cm.
3541 1846, Agriculture of New York, vol. 1: Alb., C. van Benthuysen and Co., 11, 371 p., 21 plates, map.
    See also volume II (3573). For review see 7769.
3542 1846, Analysis of mineral waters [of Saratoga, New York]: Am. Q. J. Ag. Sci., v. 4, 332.
3543 1846, [Review of Catechism of agricultural chemistry and geology, by J. F. W. Johnston (see 5825)]: Am. Q. J. Ag. Sci., v. 3, 277-278.
3544 1846, Conularia vernuelia n. s. [from Des Moines River, Iowa]: Am. Q. J. Ag. Sci., v. 4, 330, 2 figs.
3545 1846, Copper mines [reviews of Lake Superior surveys by C. T. Jackson (see 5621), J. Eights (see 3470), and Capt. Bayfield]: Am. Q. J. Ag. Sci., v. 3, 59-67, plate with 4 maps.
3546 1846, Darwin's voyage of a naturalist [review of 2916]: Am. Q. J. Ag. Sci., v. 3, 265-271.
3547 1846, Description of some of the bones of the Zeuglodon cetoides of Prof. Owen [from Alabama]: Am. Q. J. Ag. Sci., v. 3, 223-231, 2 plates with 13 figs.
3548 1846, A new locality of pyroxene [from St. Lawrence County, New York]: Am. Q. J. Ag. Sci., v. 3, 158 (4 lines).
3549 1846, The New York system: Am. Q. J. Ag. Sci., v. 4, 199-202.
3550 1846, The quarterly journal of the Geological Society of London [review of]: Am. Q. J. Ag. Sci., v. 3, 159 (15 lines).
3551 1846, Remarks on the Taconic system: Am. Q. J. Ag. Sci., v. 4, 202-209.
3552 1846, Reports on the geological survey of the Province of Canada [review of report by W. E. Logan]: Am. Q. J. Ag. Sci., v. 3, 52-54.
3553 1846, Some of the mineral resources of New York: Am. Q. J. Ag. Sci., v. 4, 27-44.
3554 1846, Structure of granitic mountains: Am. Q. J. Ag. Sci., v. 3, 207-210.
3555 1847, Agriculture of New York: Am. Q. J. Ag. Sci., v. 6, 132-143.
3556 1847, Analysis of argillaceous slate [from Steuben County, New York]: Am. Q. J. Ag. Sci., v. 6, 41-42.
3557 1847, Analysis of soils [from Mississippi River valley]: Am. Q. J. Ag. Sci., v. 5, 50-52.
3558 1847, [Discussion of glacial theory of drift]: Am. Q. J. Ag. Sci., v. 6, 214 [i.e., 262].
    In Assoc. Am. Geol's Nat's Proc.
3559 1847, Empire Spring, Saratoga [New York]: NY J. Medicine and Collateral Sci's, v. 9, 253.
3560 1847, Geology of Vermont [review of second annual Vermont report, by C. B. Adams (see 26)]: Am. Q. J. Ag. Sci., v. 5, 60-61.
3561 1847, Harrowgate Springs of Massena, St. Lawrence County [New York]: Am. Q. J. Ag. Sci., v. 5, 102.
3562 1847, The limestones and lime: Am. Q. J. Ag. Sci., v. 5, 65-82; 113-126.
3563 1847, Mining report, no. 1: Am. Q. J. Ag. Sci., v. 6, 192-198 [i.e., 240-246], 2 figs.
    On the geology and structure of the iron ore bed at Clintonville, New York.

Emmons, E. (cont.)
3564 1847, No coal in the New York rocks: Am. Q. J. Ag. Sci., v. 6, 125-129.
3565 1847, [Review of "Memoir on the fossil genus Basilosaurus ..., by R. W. Gibbes (see 4350 and 4351)]: Am. Q. J. Ag. Sci., v. 6, 285 (5 lines).
3566 1847, [On drift phenomena]: Am. Q. J. Ag. Sci., v. 6, 218 [i.e. 266].
       In Assoc. Am. Geol's Nat's Proc. See also 8922.
3567 1848, Elements of soil: Am. Q. J. Ag. Sci., v. 7, no. 4, 171-174.
3568 1848, Geology - Taconic range of moutains: Am. Q. J. Ag. Sci., v. 7, 401, plate.
3569 1848, [Review of Vestiges of the natural history of Creation by R. Chambers (see 2292 et seq.)]: Am. Q. J. Ag. Sci., v. 7, no. 5, 200.
3570 1848, View of the head of the gorge at Summit [New York]: Am. Q. J. Ag. Sci., v. 7, no. 4, 165-167, plate.
3571 1849, Address: NY State Agri. Soc. Trans., v. 8, 138-151.
       On agricultural geology.
3572 1849, Agricultural geology. Niagara Falls - its past, present and prospective condition: Genesee Farm. and Gardener's J., v. 10, 116-117, fig.
3573 1849, Agriculture of New York, Vol. II: Alb., C. Van Benthuysen and Co., 8, 343, 46 p., 42 plates.
       See also v. 1 (3541).
3574 1849, The Empire Spring, its composition and medical uses, together with a notice of the mineral waters of Saratoga, and those of other parts of New York: Alb., Van Benthuysen, 36 p.
3575 1849, [On gold in Montgomery County, Maryland]: Am. Philos. Soc. Proc., v. 5, no. 43, 84-86.
3576 1849, On the identity of the Atops trilineatus and the Triarthus beckii (Green), with remarks upon the Eliptocephalus asaphoides: Am. Assoc. Adv. Sci. Proc., v. 1, 16-19.
3577 1850, Foster's geological chart [review of 3920]: Ohio Cultivator, v. 6, 120.

Emory, William Hemsley (1811-1887)
3578 1848, Notes of a military reconnaissance from Fort Leavenworth in Missouri, to San Diego in California, including part of the Arkansas, Del Norte, and Gila Rivers: US SED 7, 30-1, v. 3 (505), 416 p., 43 plates, map.
       For review with excerpts see 949. See also reports by J. W. Abert (7) and J. W. Bailey (1439).
3579 1848, Notes of a military reconnaissance ... [as above]: US HED 41, 30-1, v. 4 (517), 614 p., 67 plates, 3 maps.

The Emporium of Arts and Sciences (Philadelphia, 1812-1814)
3580 1812, Account of a descent into the crater of Mount Vesuvius by eight Frenchmen on the night between the 18th and 19th of July 1801: v. 2, 12-20.
       From Tilloch's J.
3581 1814, Preliminary essay upon the theory of assaying as applied to the ores of iron: v. 3 (ns), no. 3, 444-454.
       With remarks by T. Cooper (see 2615). Signed "J. H. H."
3582 1814, Variation of the compass: v. 3 (ns), no. 3, 495 (10 lines).
       From "Anal. Rev."
3583 1814, [Frozen mammoth from Siberia]: v. 3 (ns), no. 3, 495 (9 lines).
       From "Anal. Rev."

Encyclopaedia; or, a dictionary of arts, sciences, and miscellaneous literature; constructed on a plan, by which the different sciences and arts are digested into the form of distinct treatises or systems, comprehending the history, theory, and practice, of each, according to the latest discoveries and improvements; ... :
3584 1790-1798, Phila., Thomas Dobson (printer), 18 v., 542 plates.
       Based on the third edition of the Encyclopaedia Britanica Major articles on mineralogy, chemistry, volcanoes, etc. For notice see 530.

Encyclopaedia, or dictionary of arts, science and miscellaneous literature. ... :
3585 1818, Phila., Thomas Dobson, 3 v., plates.
       Includes articles on mineralogy (v. 2, 485-553), magnetism, etc.

Encyclopedia Americana. A popular dictionary of arts, science, literature, history, politics and biography, ... :
3586 1829-1833, Phila., Carey, Lea, and Carey, 13 v.
       Francis Lieber, editor. Includes articles on earthquakes, geology, mines and mining, mineralogy, volcanoes, etc. For review of geology article see 530.
       Other editions include: 1836, Phila., 13 v.; 1843-1844, Phila., Lea and Blanchard, 13 v., 1850, Phila., Lea and Blanchard, 13 v.
3587 1847, Phila., Lea and Blanchard, 663 p. = v. 14.
       Supplementary volume, editied by H. Vethake. With additions to articles on mineralogy, geology, etc.

Encyclopedia Perthensis, or universal dictionary of knowledge:
3588 1805-1811, NY, 7 v.
       Not seen.

Endicott, G. (Engraver)
3589 1845, and W. Endicott, Hacket's Town Warren Co. N. J. 5 miles from Schooley's Mountain - The Delaware water gap seen in the distance - showing the spot where the antediluvial monster the Mastodon was found: NY, G. and W. Endicott, colored lithograph, 45.5 x 31.0 cm.
       "Drawn by J. W. Hill", "Pub. by W. A. Colman".

Endicott, W. (see 3589 with G. Endicott)

Engelmann, George (1809-1884)
3590 1842, Separation of silver or gold from lead: Am. J. Sci., v. 42, no. 2, 394-395.
3591 1847, Remarks on the Melonites multipora: Am. J. Sci., v. 3 (ns), no. 1, 124-125.
       On a fossil crinoid? from St. Louis, Missouri.

Engelmann, G. (cont.)
3592   1847, Remarks on the St. Louis limestone: Am. J. Sci., v. 3 (ns), no. 1, 119-120.

English, Michael
3593   1850, A gold washer: Franklin Inst. J., v. 19 (3s), 173 (4 lines).

Enterprise Mining and Manufacturing Company
3594   1850, The Enterprise Mining and Manufacturing Company. Chartered by the state of Virginia: NY, Wm. Osborn (printer), 15 p., map.
       Includes reports on the Eagle Mine by R. C. Taylor (see 10150) and J. R. Chilton (see 2351).

Erdmann, Otto Linné (1804-1869)
3595   1846, Bucholzite of Chester Pennsylvania: Am. J. Sci., v. 2 (ns), no. 3, 418 (5 lines).
       Abstracted from Berzelius Jahresbericht.
3596   1846, Keilhauite [from Arendel, Norway]: Am. J. Sci., v. 2 (ns), no. 3, 415-416.
       Abstracted from "Kongl. Vet. Akad. Handl."
3597   1849, and M. Gerathewohl, On chloritoid, Am. J. Sci., v. 8 (ns), no. 1, 123 (6 lines).
       Abstracted from "Jour. für Prakt. Ch."

Erickson, Captain [i.e. John Ericsson, 1803-1889?]
3598   1839, Report of Captain Erickson, civil engineer, London, showing the cost of the coal of the Maryland Mining Company per ton, delivered at the several cities of Washington, Baltimore, Philadelphia and New York: n.p.
       May be a London publication.

Erman, George Adolf (1806-1877)
3599   1839, Extract of a letter from M. Erman, junior, to M. Arago, upon the temperature of the ground in Siberia: Am. J. Sci., v. 36, no. 1, 205-206.
       Abstracted from Comptes Rendus.
3600   1847, Gauss's magnetic constants: Am. J. Sci., v. 3 (ns), no. 1, 140 (9 lines).
       Abstracted from London Ath.
3601   1850, Travels in Siberia. Including excursions northwards, down the Obi, to the Polar Circle, and southwards, to the Chinese frontier: Phila., Lea and Blanchard, 2 v., 371, 400 p.
       Translated by W. D. Cooley. Includes notes on Siberian mines, fossils, and magnetic observations.

Esmark, Jens (1763-1839)
3602   1807, [Description of "datholite" from Arendel, Norway]: Am. Reg. or Gen. Repos., v. 1, pt. 2, 39.
3603   1831, Prunnerite [from Hestoe Island, Faroes]: Am. J. Sci., v. 20, no. 1, 197 (7 lines).
       Abstracted from Edinburgh New Philos. J. by C. U. Shepard.

Espy, Josiah Murdoch (1771-1847)
3604   1822, Bedford Spring: Am. Med. Rec'r, v. 5, 762-763.

An essay on the agitation of the sea, and some other remarkables attending the earthquakes of the year m,dcc,l,v. To which are added, some thoughts on the causes of earthquakes, written in the year 1756:
3605   1761, Bos., B. Mecom (printer), 40 p.
       Evans number 8851.

An essay on the art of boring the earth for the obtainment of a spontaneous flow of water
3606   1826, New Brunswick, Rutgers Press, 46, 1 p.

Essex County Natural History Society (Essex County, Massachusetts)
3607   1839, Journal of the Essex County (Mass.) Natural History Society, 8vo., Salem [review of contents]: Am. J. Sci., v. 37, no. 1, 187-188.

Eustis, William (1753-1825)
3608   1818, American copper [from Lake Superior]: Niles' Wk. Reg., v. 13, 323.
3609   1818, Experiments made by the assay-master of the King of the Netherlands, at the mint of Utrecht, on the native copper existing in huge blocks on the South side of Lake Superior: Am. Mon. Mag. and Crit. Rev., v. 2, no. 5, 366-367.

Evans, J. (see 8082 with D. D. Owen and J. G. Norwood)

Evans, Lewis (1700-1756)
3610   1749, A map of Pennsylvania, New-York, and the three lower counties on Delaware: NY, James Parker (printer), map.
       Evans number 6316.
3611   1755, A general map of the middle British colonies; ... : Phila., B. Franklin and D. Hall, map, 68 x 51 cm.
       Evans number 7413.
3612   1755, Geographical, historical, political and mechanical essays. ... : Phila., Franklin and Hall, iv, 32 p., 4to.
       Evans number 7411. Analysis of map (see 3611).
       Also "second edition", Evans numbers 7412 and 7413 (two impressions).

The Evening Fireside (Philadelphia, 1804-1806)
3613   1805, Important fact in zoology, by which it appears that the mammoth was an herbiverous animal: v. 1, no. 43, 340-341.
       From Amoenitates Graphicae.

Everard (Mr.)
3614   1794, Two letters from Mr. Everard, F. S. M. containing an adventure, of which he was
        an eye witness, at the quicksilver mine of Idria: Royal Am. Mag., v. 1, 243-244;
        302.

Evesham
3615   1838, On lime and marl: Del. Reg. and Farm's Mag., v. 1, 438-442.
3616   1839, Application of marl: Del. Reg. and Farm's Mag., v. 2, 49-50.
            Abstracted from Farm's Cab.

The Examiner, and Journal of Political Economy (Philadelphia, 1834-1835)
3617   1835, The gold coinage: v. 2, 91-96.
            Gold production from Southern United States mines.

The Expositor; a Weekly Journal of Foreign and Domestic Intelligence, Literature, Science,
    and the Fine Arts (New York, 1838-1839)
3618   1839, Vesuvius: v. 1, 155.
3619   1839, The deluge: v. 1, 167.
3620   1839, [Review of Wonders of geology by G. A. Mantell (see 6732)]: v. 1, 145-147.
3621   1839, Coal formation in France: v. 1, 239.

The Expositor and Universalist Review (Boston, 1831-1840)
3622   1840, The Mosaic account of creation: v. 4 (ns), 5-21.
            Signed by "G. W. M."
3623   1840, The order of creation: v. 4 (ns), 349-360.
            Signed by "W. E. M."

F., A. T. (see 4134)

Fahnestock, George Wolff (b. 1823)
3624   1847, Coal mining at Pittsburgh: Lit. Rec. and J. Linnaean Assoc. Penn. College, v. 3,
        279-280.

Fairbairn, Sir William, bart. (1789-1874)
3625   1844, On the reduction of the magnetic ores of Samakoff, (Turkey): Franklin Inst. J.,
        v. 8 (3s), 103-104.
            Abstracted from London Ath.

Fairburn
3626   1838, Iron: Am. J. Sci., v. 33, no. 2, 292-296.
            In British Assoc. Proc.

Fairburn, H.
3627   1850, On smelting magnetic iron ores: Franklin Inst. J., v. 19 (3s), 125-131.

Fairholme, George
3628   1833, General view of the geology of Scripture, in which the unerring truth of the
        inspired narrative of the early events in the world ... is exhibited ... : Phila.,
        Key and Biddle, 281 p.
            For review see 7149. For notice see 581.
3629   1843, General view of the geology of Scripture, ... [as above]: Phila., H. Hooker, 282
        p.

The Family Lyceum (Boston, 1832-1833)
3630   1832/1833, Geology: v. 1, 2; 6; 13; 18; 22; 24; 26; 30; 34; 42-43; 46-47; 51; 143; 148;
        164; 187; 202; 205-206; Extra no. 2, 3.
            Elementary introduction to geology.
3631   1832/1833, [Notices on the activities of scientific and natural history lyceums in the
        United States]: v. 1, numerous short notices in all numbers.
3632   1832/1833, [Brief notices on geological surveys]: v. 1, 7; 21; 31; 115; 121,
3633   1832/1833, [Brief notices on volcanoes and earthquakes]: v. 1, 8; 12; 62; 115; 117,
        129, 137, 141, 145-146, 149.
3634   1832/1833, [Brief notices on mining]: v. 1, 26; 36; 40; 48; 58; 91; 115; 121; 190.
3635   1832, Mineralogy and geology [notice and review of elementary text by J. K. Welsh (see
        10798)]: v. 1, 27, 87.
3636   1832/1833, Caves: v. 1, 66 (Wier's Cave); 71 (Indiana epsom salt cave); 87 (Fingals
        Cave).
3637   1832, The internal structure of the earth: v. 1, 72.
3638   1832, The form of the earth: v. 1, 72.
3639   1832/1833, [On mineral springs]: v. 1, 72; 207.
3640   1833, Magnetism: v. 1, 110-111.
            Description of magnetite, variety loadstone.
3641   1833, Meteoric stones: v. 1, 136.
3642   1833, Geological course of cholera: v. 1, 142.
            On the effects of local lithologies on the prevalence of cholera.
3643   1833, Goniometer: v. 1, 199.

The Family Magazine, or General Abstract of Useful Knowledge (New York, 1833-1841)
3644   1833, Earthquakes: v. 1, 113; 121-122; 129-130; 140-141, fig.; 148-149, fig.; 156-157,
        fig.; 161-162, fig.; 172-174, 2 figs.; 178-181.
3645   1833-1839, [Brief notes on earthquakes and volcanoes]: v. 1, 163; v. 2, 342-345, 399;
        v. 4, 141-143; v. 6, 415-416.
3646   1835-1841, [Brief notes on mining]: v. 3, 39-40; v. 7, 299; v. 8, 157-158; 255; 285;
        316; 338-340.
3647   1835, Interior structure of the solid parts of the earth: v. 3, 50-52; 85-86; 140-143,
        168-171.
3648   1838, White sulphur springs, Va.: v. 5, 123-124.
3649   1839, The science of geology: v. 6, 25-29, 2 figs.; 57-61.
            From the Glasgow Treatises.
3650   1839/1841, [Brief notes on fossils]: v. 6, 189-190; v. 8, 60 [i.e., 460]-461, fig.

Family Magazine (cont.)
3651   1840/1841, The mineral kingdom: v. 7, 116-117, fig.; 127-128, fig.; 165, fig.; 226-227, fig.; 245, 3 figs.; 332-333, fig.; 426-427, fig.; v. 8, 96, fig.; 164-166, 2 figs.
3652   1840, [Brief notes on caves]: v. 7, 300, 324.
3653   1840, Geology of South America: v. 7, 324.
       Notice of C. Darwin's research.
3654   1841, [Review with excerpts of Mineral lands ... , by D. D. Owen (see 8046)]: v. 8, 130-137.
3655   1841, Terrestrial magnetism: v. 8, 42 [i.e., 421]-423.

The Family Visitor (Cleveland, Ohio, 1850)
3656   1850, [Brief notes on fossils, earthquakes, volcanoes, mining]: numerous entries throughout v. 1. See annual index.
3657   1850, The Giant's Causeway: v. 1, no. 1, 5, fig.
3658   1850, Fingal's Cave: v. 1, no. 2, 12, fig.
3659   1850, Trap rocks: v. 1, no. 4, 29, 2 figs.
       On the origin of columnar basalt.
3660   1850, Mounted bowlder: v. 1, no. 6, 44, fig.
       Description of rocking stone in North Salem, Westchester County, New York.

Fanning, John C. (see 11127 with J. B. Zabriskie and N. A. Garrison)

Fanning, Tolbert (1810-1874)
3661   1833, Geological Museum: Agriculturist, v. 4, 2.
3662   1843, Geology: Agriculturist, v. 4, 19-21, 34-36.
3663   1843, Acknowledgments [for fossil donations]: Agriculturist, v. 4, 127, 139, 180.
3664   1845, Donations to Franklin College [from W. Maclure and D. D. Owen]: Agriculturist, v. 6, 171, 187.
3665   1847, Observations on a tour to the South: Christian Rev. (Tenn.), v. 4, 2-7, 37-45.
3666   1847, Geological observations on the South: Christian Rev. (Tenn.), v. 4, 14-20.
3667   1850, Geology for practical man: Naturalist, v. 1, 13-14.
3668   1850, Notes on South Carolina: Naturalist, v. 1, 15-17.
3669   1850, Geology for the people: Naturalist, v. 1, 26-30, 73-78.
3670   1850, The present state of geological knowledge: Naturalist, v. 1, 54-57.
3671   1850, Natural science in Mississippi: Naturalist, v. 1, 58-59.
3672   1850, Do the phenomena of nature contravene the facts of revelation?: Naturalist, v. 1, 87-103.
3673   1850, The manner in which animals and plants become distributed over the earth: Naturalist, v. 1, 169-170.
3674   1850, Analysis of soil as practiced at Franklin College: Naturalist, v. 1, 193-196.

Fansher, Sylvanus
3675   1840, The burning of Monkton Pond, Vt.: Am. J. Sci., v. 39, no. 2, 399 (10 lines).

Faraday, Michael (1791-1867)
3676   1817, Analysis of the native caustic lime: J. Sci. Arts, v. 1, no. 2, 261-262.
       With observations by H. Davy (see 2964).
3677   1839, Cold Bokkeveld meteorites: Am. J. Sci., v. 37, no. 1, 190.
       Abstracted from London and Edinburgh Philos. Mag.
3678   1847, Diamond converted to coke: Am. J. Sci., v. 4 (ns), no. 3, 409 (12 lines).
       Abstracted from London Ath.

Farmer
3679   1756, Two very circumstantial accounts of the late dreadful earthquake at Lisbon. Giving a more particular relation of that event than any hitherto publish'd. ... : Bos., D. Fowle and Z. Fowle, 16 p.
       Evans number 7653.
3680   1756, Two very circumstantial accounts ... [as above]. The second edition. To which is added an account of the late earthquake of Boston: Bos., D. Fowle and Z. Fowle, 32 p.
       Evans number 7654.

The Farmer and Mechanic, Devoted to Agriculture, Mechanics, Manufactures, Science and the Arts (New York, 1847-1850)
3681   1847-1850, [Notices on mining, earthquakes, volcanoes, fossils, geology]: All volumes, v. 1 (ns) - v. 4 (ns), have numerous brief entries. Consult annual indices for specific entries.
3682   1848-1850, California gold (the gold region): v. 2 (ns), 597-598; 610; 620; v. 3 (ns), 8; 20; 45; 63; 80; 117; 224; 345; 453-454; v. 4 (ns), 199; 519-520.
3683   1849, [Notices of gold washers]: v. 3 (ns), 34, 63-64, 2 figs., 111, fig.
3684   1849-1850, [Notices of Lake Superior copper mines]: v. 3 (ns), 102-103; 294; 318; 342-343; 414; 450-451; v. 4 (ns), 4-5; 102-103; 258; 413; 462.
3685   1849, Hot Springs of Washitaw [Arkansas]: v. 3 (ns), 66-67, map.
3686   1849, Mineralogical society: v. 3 (ns), 176-177.
       Proposal to start United States Society.
3687   1849, River terraces of the Connecticut Valley; and on erosion of the earth's surface: v. 3 (ns), 414-415.
       Review of E. Hitchcock's views on river terraces.
3688   1850, Lead ores in the United States: v. 4 (ns), 29-30; 42; 53; 65, fig.; 77-78; 159-160; 172; 184-185; 197-198.
3689   1850, Statistics of coal [review of book by R. C. Taylor (see 10147)]: v. 4 (ns), 117 (12 lines); 199.
3690   1850, Louis Agassiz: v. 4 (ns), 366-367.
       Brief biographical sketch.
3691   1850, Facts illustrative of the interior heat of the globe: v. 4 (ns), 401-402.

The Farmers' Cabinet; Devoted to Agriculture, Horticulture, and Rural Economy
(Philadelphia, 1836-1848)
3692   1836-1845, [Brief notices on the nature and improvement of soils]: v. 1, 81-82; 94; v. 5, 34; 69; 98-99; v. 7, 358; v. 10, 21.
3693   1841-1842, [Notes on fossils and footprints in limestone]: v. 6, 62-63; v. 7, 44-45; 102; 117.
3694   1844, Report of the committee on agriculture, relative to the application of lime to the defficient qualities of soil: v. 8, 306-309; 338-341.
3695   1847, Anthracite coal: v. 11, 320, table; 325.
From Pottsville Miner's J. Table of Pennsylvania production.

The Farmer's, Mechanic's Manufacturer's, and Sportsman's Magazine (New York, 1826-1827)
3696   1827, Nature of soils: v. 1, 163-165.
3697   1827, Water-limes: v. 1, 224.
On the use of lime in cement.
3698   1827, Polishing granite: v. 1, 231-232.
3699   1827, Emery: v. 1, 346-347.
3700   1827, Mining: v. 1, 472.
On methods of blasting granite.

The Farmer's Monthly Visitor (Concord, New Hampshire, 1839-1849)
3701   1839-1849, [Brief notes on mining and mineral resources]: v. 1, 4; 32; v. 2, 22; 32; v. 3, 32; 98-99; v. 9, 79; 89; 112; 142; 187-188; v. 10, 47; 62; 173; 184; v. 11, 11; 76; 121-122.
3702   1839, Geology [notice of the second Maine report by C. T. Jackson (see 5560)]: v. 1, 4 (9 lines).
3703   1839, [Evidence for the flood in New Hampshire]: v. 1, 146.
Signed by "Geologist."
3704   1840-1844, [Brief notices of volcanoes and earthquakes]: v. 2, 137-138; v. 3, 46; v. 4, 101-102; v. 5, 47; 173-174; v. 6, 19; 31.
3705   1840, Geological survey of New Hampshire: v. 2, 173.
3706   1841, The Mammoth Cave: v. 3, 20-21.
3707   1841, Dr. Jackson's survey of Rhode Island [(notice of 5571)]: v. 3, 40.
3708   1841, The greatest natural curiosity: v. 3, 89.
On a fossil mammoth from Missouri in A. Koch's collection.
3709   1842, The geological survey of the state [review of 2nd New Hampshire report by C. T. Jackson (see 5592)]: v. 4, 132.
3710   1846, Dr. Jackson's geological report [review of 5611]: v. 8, 123.
3711   1847, Quincy granite in New Orleans: v. 9, 182-183.
On the use of granite from Massachusetts as a building stone.
3712   1848, The progress of the glaciers: v. 10, 75.
On glacial movement and growth in Switzerland.
3713   1849, Interior of the earth: v. 11, 15.

The Farmers' Register (Shellbanks, Virginia, 1833-1835; Petersburg, Virginia, 1835-1843)
3714   1833-1841, [Brief notes on the nature and improvement of soils]: v. 1, 9; 337; 338, 403-404; 534; 581, 631; v. 2, 444; v. 3, 269; 319; v. 5, 545; 570; 579; v. 6, 265; v. 8, 176-177; 614-616; v. 9, 127-128; 469-470.
3715   1833-1841, [Notes on mining in Virginia]: v. 1, 244; 501-503; v. 2, 242-243; v. 3, 143; v. 4, 343-344; v. 6, 117-118; v. 9, 458-459.
3716   1833, Organic remains found in the marl pits of Lucas Benners, esq., in Craven County, N. C., by H. B. Croom: v. 1, 298-299.
3717   1833, Extracts from several authors, showing their opinions of the effects of magnesia, in different forms, on soils: v. 1, 426-432.
With excerpts by J. Headrick (see 4971) and H. Davy (see 2962).
3718   1833/1834, Geological: v. 1, 473-474; 529-530; 605-606.
Signed by "Galen." On the deluge.
3719   1834, Remarks, topographical, geological and general, respecting Preston's and King's saltworks, and the surrounding district: v. 1, 497-501.
Signed by "C. D." (i.e. Chester Dewey?)
3720   1835, Importance of geological surveys to Virginia: v. 2, 517-518.
3721   1835, [Review of Essay on calcareous manures by E. Ruffin (see 9088)]: v. 2, 632-634.
3722   1836, [Review of Geological Society of Pennsylvania Transactions (see 4314)]: v. 3, 613-614.
3723   1836, Geological survey of the state of New York: v. 4, 29-30; 3 352-353.
3724   1836, Fine white marble discovered near Gaston [North Carolina]: v. 4, 30.
3725   1836, [Review of Essay on calcareous manures by E. Ruffin (see 9088)]: v. 4, 104-105.
From London's Gardener's Mag.
3726   1836, Practical uses of geology: v. 4, 302.
From Advocate of Sci.
3727   1836, Geological wonder: v. 4, 557-558.
On a granite quarry at Kennebunkport, Maine.
3728   1837, Geology and geography of New York: v. 5, 402-404.
3729   1841, English mines and mining [at Newcastle]: v. 9, 45.
3730   1841, The Missourium, or leviathan skeleton: v. 9, 654-656.
From Farm's Cab.

Farnham, John Hay (1791?-1833)
3731   1820, Extract of a letter describing Mammoth Cave, in Kentucky: Am. Antiquarian Soc. Trans. and Coll., v. 1, 355-361.

Farrand, William Powell (d. 1839)
3732   1833/1834, The Virginia mines near Richmond and James River: Penn. Senate J., v. 2, 567-568.

Farrar, S. C.
3733   1826, A description of the White Sulphur Springs of Virginia, with some description of their medicinal properties: NY Med. Phys. J., v. 5, 199-202.

Fauvelle, M.
3734   1847, On boring for artesian wells: Am. J. Sci., v. 3 (ns), no. 1, 135-136.
            Abstracted from London Ath.

Favre, Pierre, Antoine (1813-1880)
3735   1846, and Silberman, Arragonite and calc spar: Am. J. Sci., v. 2 (ns), no. 3, 417 (5 lines).
            Abstracted from l'Institut.

Faxar, Palacio
3736   1816, Account of a soda lake in South America [at Maracaybo]: Eclec. Rep. and Analytic Rev., v. 6, 467-471; also in J. Sci. Arts, v. 1, no. 2, 188-192 (1817).
3737   1817, Account of the earthquake at Caracas [Venezuela]: J. Sci. Arts, v. 2, no. 2, 400-402.

Featherstonhaugh, George William (1780-1866)
3738   1821, On soils: NY State Board Ag. Mem., v. 1, 51-97.
3739   1831, Adjudication of the Wollaston Medal to William Smith: Mon. Am. J. Geology, v. 1, no. 1, 29-33.
3740   1831, Anthracite coal applied to generate steam power: Mon. Am. J. Geology, v. 1, no. 2, 72-74.
3741   1831, Bone caves in New Holland: Mon. Am. J. Geology, v. 1, no. 1, 47.
3742   1831, Discovery of rionium (vanadium) in Scotland: Mon. Am. J. Geology, v. 1, no. 5, 232-233.
            Notice of J. T. W. Johnson's (?) discovery in Wanlockhead.
3743   1831, The Earl of Bridgewater's bequest: Mon. Am. J. Geology, v. 1, no. 1, 33-35.
3744   1831, Eaton's geology: Mon. Am. J. Geology, v. 1, no. 2, 82-91.
            Review of geological theories of A. Eaton, H. T. De la Beche, and J. Van Rensselaer. See also 3361.
3745   1831, An epitome of the progress of natural science: Mon. Am. J. Geology, v. 1, no. 2, 49-58; no. 3, 97-104; no. 4, 145-158; no. 5, 193-205; no. 6, 241-257.
            On the history of science.
3746   1831/1832, Monthly American Journal of Geology and Natural Science: Phila., 576 p., 14 plates.
            G. W. Featherstonhaugh was founder and editor of this journal, as well as its principal contributor.
            For review see 539, 1098, 3988, 4951, 7687 and 8661.
3747   1831, On nomenclature: Mon. Am. J. Geology, v. 1, no. 1, 28-29.
3748   1831, On the ancient drainage of North America and the origin of the cataract of Niagara: Mon. Am. J. Geology, v. 1, no. 1, 13-21, plate.
3749   1831, On the value of geological information to engineers, and on the inequalities of the earth's surface, and their true levels above tide water: Mon. Am. J. Geology, v. 1, no. 3, 128-135, plate with 2 figs.
3750   1831, [Review of A geological text-book by A. Eaton (see 3361)]: N. Am. Rev., v. 31, 471-490.
3751   1831, Rhinoceroides alleghaniensis [from Somerset County, Pennsylvania]: Mon. Am. J. Geology, v. 1, no. 1, 10-12, plate.
3752   1832, British Association for the Advancement of Science: Mon. Am. J. Geology, v. 1, no. 10, 468-480.
            Proceedings of the first meeting, 1831, with remarks by G. W. Featherstonhaugh.
3753   1832, Continuation of Conybeare and Phillip's outlines of the geology of England and Wales: Mon. Am. J. Geology, v. 1, no. 7, 327-328.
            Announcement of second volume.
3754   1832, General remarks on the constituents of primary rocks: Hazard's Reg. Penn., v. 9, 191-192.
3755   1832, General remarks on the constituents of primary rocks: Mon. Am. J. Geology, v. 1, no. 7, 308-312.
3756   1832, A geological manual, by Henry T. De la Beche [review of 3036]: Mon. Am. J. Geology, v. 1, no. 11, 527 (5 lines).
3757   1832, Geological Society of London, and Megatherium from Buenos Ayres: Mon. Am. J. Geology, v. 1, no. 12, 570-571.
3758   1832, Geological Society of Pennsylvania: Mon. Am. J. Geology, v. 1, no. 9, 425-430.
            On the objectives and officers of the society.
3759   1832, Geology; No. 1, on the crust of the earth: Mon. Am. J. Geology, v. 1, no. 7, 289-296.
3760   1832, Geology; No. 2, on the order of succession of the rocks composing the crust of the earth: Mon. Am. J. Geology, v. 1, no. 8, 337-347.
3761   1832, Geology; No. 3, on the constituent minerals, and the structure of the primary rocks: Mon. Am. J. Geology, v. 1, no. 9, 385-391.
3762   1832, Interesting discovery of fossil animals [at Argenton]: Mon. Am. J. Geology, v. 1, no. 7, 331-332 (11 lines).
3763   1832, Limestone caves in Schoharie, State of New York: Mon. Am. J. Geology, v. 1, no. 8, 381-382.
3764   1832, On mineral and metallic veins: Mon. Am. J. Geology, v. 1, no. 11, 481-490, plate with 9 figs.
3765   1832, Mineralogy and geology of Nova Scotia [review of survey by C. T. Jackson and F. Alger (see 5538)]: Mon. Am J. Geology, v. 1, no. 9, 431 (8 lines).
3766   1832, Natural bridge in Rockbridge County, Virginia: Mon. Am. J. Geology, v. 1, no. 9, 414-416, plate.
3767   1832, Petrefacta musei bonensis, by Professor Goldfuss [review of]: Mon. Am. J. Geology, v. 1, no. 11, 527 (8 lines).
3768   1832, Rafinesque's Atlantic journal [review of]: Mon. Am. J. Geology, v. 1, no. 11, 508-515.
3769   1832, Remarks: Mon. Am. J. Geology, v. 1, no. 10, 440-444.
            On Prof. Del Rio's discovery of vanadium in Mexico (see 3084).
3770   1832, Remarks by the editor: Mon. Am. J. Geology, v. 1, no. 8, 352-353.
            On the Blue Ridge Mountains, the Great Valley, and drainage of Virginia.
            Following Lang's description of a natural tunnel (see 6478).
3771   1832, Teleo-saurus: Mon. Am. J. Geology, v. 1, no. 7, 332 (9 lines).

Featherstonhaugh, G. W. (cont.)
3772   1832, Volcanic island of Pantellaria: Mon. Am. J. Geology, v. 1, no. 9, 431-432 (11 lines).
3773   1833, [Letter requesting aid, and giving reasons for a geological survey]: US SED 35, 22-2, v. 1 (230), 5-7.
       With letter by L. Cass (see 2278).
3774   1833, On the importance of geological knowledge to agriculturists: Farm's Reg., v. 1, 153-155.
3775   1834, Mineral resources of Virginia: Farm's Reg., v. 1, 520-523.
       A letter to James Madison on the proper manner of opening the state's mineral resources to the public.
3776   1834, [Warm Springs, Bath County, Virginia]: Geol. Soc. Penn. Trans., v. 1, 169-170.
3777   1835, Account of the travertine deposited by the waters of the Sweet Springs, in Alleghany County, in the state of Virginia, and of an ancient travertin [sic] discovered in the adjacent hills: Geol. Soc. Penn. Trans., v. 1, 328-334, 3 figs; abstracted in: Farm's Reg., v. 3, 557-559 (1836).
3778   1835, Geological report of an examination made in 1834 of the elevated country between the Missouri and Red Rivers: Wash., Gales and Seaton (printers), US HED 151, 23-2, v. 4 (274), 97 p., plate.
       For review see 626, 3997, and 4327.
3779   1835, A report of the mineralogical and geological investigations made by G. W. Featherstonhaugh: US SED 153, 23-2, v. 4 (269), 2-43.
3780   1835, Importance of geological principles to mining operations: Farm's Reg., v. 3, 196.
3781   1836, Report of a geological reconnoissance made in 1835, from the seat of government, by the way of Green Bay and the Wisconsin Territory to the Coteau de Prairie, an elevated ridge dividing the Missouri from the St. Peter's River: Wash., Gales and Seaton (printers), US SED 333, 24-1, v. 4 (282), 168 p., 3 plates.
       For review with excerpts see 7244, 7629. For review see 9740.
3782   1836, Iron mountain in Missouri: NE Farm., v. 14, 340; also in: The Friend, v. 9, 218.
3783   1837, [Letter in response to the criticisms of W. W. Mather]: Naval Mag., v. 2, 569-580.
       In "Reply of G. W. Featherstonhaugh, Esq. U. S. Geologist to Professor W. W. Mather", with appended letters by W. W. Mather (see 6886), J. J. Abert (see 8), W. Hood (see 5249), and A. Macomb (see 6651). See also 7630.
3784   1844, Excursion through the slave states, from Washington on the Potomac to the frontier of Mexico; with sketches of popular manners and geological notices: NY, Harper and Brothers, x, 11-168.
       For review with excerpts see 6375. For review see 4486 and 7965.
3785   1847, [Extract from report on the geology of the Barrelville Mining Company lands]: In 1409.
3786   1847, A canoe voyage [Extracts from Account of the lead and copper deposits in Wisconsin; of the gold region in the Cherokee Country]: Daguerreotype, v. 1, 36-38.
       In review (see 2767).

Fell, Jesse (Judge)
3787   1827, On the discovery and first use of anthracite in the Valley of Wyoming [Pennsylvania]: Penn. Hist. Soc. Mem., v. 2, pt. 1, 162-164.
3788   1830, Notice of the first anthracite coal on the Susquehanna: Hazard's Reg. Penn., v. 6, 83.

Fenn, H. R.
3789   1821, Fluor spar of Genesee: Am. J. Sci., v. 3, no. 2, 367.
       On fluorite from Rochester, New York.

Fenn, Horatio N.
3790   1823, Coal, gypsum and barytes [from Rochester, New York]: Am. J. Sci., v. 7, no. 1, 56.

Fergus, T. H.
3791   1848, Mica originating from hornblende [at Boston]: Am. J. Sci., v. 6 (ns), no. 3, 425-426.

Ferrara, Sig. Abate
3792   1824, An account of the earthquake which occurred in Sicily, in March, 1823: Bos. J. Phil. Arts, v. 2, no. 2, 138-159.
       Translated by W. S. Emerson. Also in Am. J. Sci., v. 9, no. 2, 216-239.
3793   1824, Notices of the geology and mineralogy of Sicily, from a work entitled Storia naturale della Sicilia. Cat. 1813; del Ab. F. Ferrara. Translated and condensed by James G. Percival: Am. J. Sci., v. 8, no. 2, 201-213.
       See also 8228.

Fessenden's Silk Manual and Practical Farmer (Boston, 1835-1837)
3794   1835, Coal trade of Pennsylvania: v. 1, no. 2, 21-22.
3795   1835, Bituminous coal in Massachusetts: v. 1, no. 7, 110 (11 lines).
3796   1836, Rossie lead mines [in New York]: v. 2, no. 4, 64 (8 lines).
3797   1837, Mansfield coal [in Massachusetts]: v. 3, no. 9, 136 (12 lines).

Feuchtwanger, Lewis (1805-1876)
3798   1830, Minerals not yet described in the common systems of mineralogy: Am. J. Sci., v. 18, no. 2, 391-392.
       Descriptions of thirteen new minerals.
3799   1831, Amber: Mag. Use. Ent. Know., v. 2, no. 2, 95-98.
3800   1831, Mercury: Mag. Use. Ent. Know., v. 2, no. 4, 248-251.
3801   1831, On gold: Mag. Use. Ent. Know., v. 2, no. 2, 98-101.
3802   1831, Ore of platinum: Mag. Use. Ent. Know., v. 2, no. 1, 40-42.
3803   1831, Seleniuret of palladium, etc.: Am. J. Sci., v. 19, no. 2, 369 (4 lines).
3804   1833, Epistilbite from Elba: Am. J. Sci., v. 24, no. 1, 194.

Feuchtwanger, L. (cont.)
3805   1838, A treatise on gems, in reference to their practical and scientific value; a useful guide for the jeweller, amateur, artist, lapidary, mineralogist, and chemist; accompanied by a description of the most interesting American gems, and ornamental and architectural materials: NY, Printed by A. Hanford, i-vi, 7-178 p., illus.
        For notice see 709.
3806   1842, Mineralogical notices: Am. J. Sci., v. 42, no. 2, 386-388.
        Notice of ten new minerals from Europe.
3807   1847, The agates from Oberstein: Hunt's Merch. Mag., v. 16, 533.
3808   1847, Iron: a short sketch of its production: Hunt's Merch. Mag., v. 17, 536-537.
3809   1847, The iron mountains of Missouri: Hunt's Merch. Mag., v. 16, 94-95.
3810   1847, The mineral resources of Missouri: Hunt's Merch. Mag., v. 16, 177-181.
3811   1847, Real chalk in the United States: Hunt's Merch. Mag., v. 17, 114.
3812   1848, Mineral wealth of Missouri: De Bow's Rev., v. 5, 93-94.
3813   1848, Minerals and mines in Missouri and Illinois: Hunt's Merch. Mag., v. 18, 225.
3814   1849, Mineral wealth of the South-western Territory of the U. States: Hunt's Merch. Mag., v. 20, 319-320.
3815   1850, The discovery of a new cave in Kentucky: Am. Assoc. Adv. Sci. Proc., v. 2, 355-356.

Field (Dr.)
3816   1846, An analysis of barren and improved soils and the muck used: NY State Agri. Soc. Trans., v. 5, 508-510.

Field, David Dudley (see 3160 on Berkshire County, Massachusetts)

Field, M. F.
3817   1850, Native copper containing silver from Chile: Am. J. Sci., v. 10 (ns), no. 2, 255 (6 lines).
        Abstracted from Q. J. Chemical Soc.

Field, Martin (General)
3818   1824, Chrysoprase and pimelite [from New Fane, Vermont]: Am. J. Sci., v. 8, no. 2, 234.
3819   1826, Correction by Gen. Martin Field: Am. J. Sci., v. 11, no. 2, 384-385.
        On fibrolite from Bellows Falls, Vermont.
3820   1826, Notice of the discovery and geological situation of the native gold of Vermont: Bos. J. Phil. Arts, v. 3, no. 6, 592-594.
3821   1826, Vermont gold: Niles' Wk. Reg., v. 31, 2.

Filhol (see 1366 by M. Audouard)

Filson, John (1747-1788)
3822   1784, The discovery, settlement and present state of Kentucke: and an essay towards the topography, and natural history of that important country ... : Wilmington, Del., James Adams (printer), 118 p., map.
        Evans number 18467.
3823   1787, History of the discovery, settlement, and present state of Kentucke, with an account of the soil, and produce of that country: NY Mag. and Mon. Advertiser, v. 1, no. 3, 102-110.

The Financial Register of the United States (Philadelphia, 1838)
3824   1838, [Brief notes on mining, especially gold mining]: v. 1, 112-114; 142; 236; 319; v. 2, 159-160; 206; 239; 252.

Finch, Heneage (Second Earl of Winchilsea, d. 1689)
3825   1669, A true and exact relation of the late prodigious earthquake & eruption of Mount Aetna, or, Monte-Gibello; as it came in a letter written to his majesty from Naples by the right honorable the Earl of Winchilsea, his majesties late ambassador at Constantinople, who in his return from thence, visiting Catania in the island of Sicily, was an ey-witness of that dreadful spectacle. Together with a more particular narrative of the same, as it is collected out of several relations sent from Catania. Published by authority: Cambridge, Mass., Printed by S. G[reen] and M. J[ohnson], 19 p.
        Evans number 139.

Finch, John
3826   1823, Geological essay on the Tertiary formations in America: Am. J. Sci., v. 7, no. 1, 31-43.
        Also reprint edition, 13 p.
3827   1823, Observations on what are termed alluvial formations in the United States: NY Med. Phys. J., v. 2, 116-120.
3828   1823, On the Celtic antiquities of America: Am. J. Sci., v. 7, no. 1, 149-161.
        On the locations and origins of American rocking stones.
3829   1824, A sketch of the geology of the country near Easton, Penn.; with a catalogue of the minerals, and a map: Am. J. Sci., v. 8, no. 2, 236-240, map.
        Contains map 3830.
3830   1824, Geology of Easton &c: Am. J. Sci., v. 8, no. 2, geological map, 11.4 x 18.1 cm.
        In 3829.
3831   1826, Memoir on the new or variegated sandstone of the United States: Am. J. Sci., v. 10, no. 2, 209-212, fig.
3832   1826, On the Tertiary formations on the borders of the Hudson River: Am. J. Sci., v. 10, no. 2, 227-229, colored plate.
3833   1827, Introduction to the study of geology, containing some account of the coal mines of Pennsylvania, with a geological profile of the country between Philadelphia and Sunbury on the Susquehanna: c. 150 p.
        Not seen. For notice see 397, 457, and 2372.
3834   1828, On the geology and mineralogy of the country near West Chester, Penn.: Am. J. Sci., v. 14, no. 1, 15-18.

Finch, J. (cont.)
3835 1829, Notice of the locality of the bronzite, Jameson; or diallage metaloid, Haüy and Brongniart; at Amity, Orange County, state of New York: Am. J. Sci., v. 16, no. 1, 185-186.
3836 1830, Notice of a locality of arragonite, near New Brunswick, (N.J.): Am. J. Sci., v. 18, no. 1, 197-198.
3837 1831, Essay on the mineralogy and geology of St. Lawrence County, State of New York: Am. J. Sci., v. 19, no. 2, 220-228.
3838 1834, [Notes on the geology of Lake Erie]: Am. J. Sci., v. 27, no. 1, 151 (6 lines).
       In NY Lyc. Nat. History Proc.

Fischer, Ernst Gottfried (1754-1831)
3839 1822, Life of Klaproth, the celebrated chemist: Mus. For. Liter. Sci., v. 1, 548-553.
3840 1824, Memoir on the life of Martin Henry Klaproth: Bos. J. Phil. Arts, v. 2, no. 1, 1-15.

Fischer, Nicholas Wolfgang (1782-1850; see also 3259 with A. Duflos)
3841 1849, Analysis of the Braunau meteoric iron: Am. J. Sci., v. 7 (ns), no. 2, 171-175.
       Continued from 3259.

Fish, Samuel
3842 1835, Outlines of geology: Scien. Tracts and Fam. Lyc., v. 4 (ns), 5-22.
3843 1836, Granite rock: Scien. Tracts, v. 1, 265-280.
3844 1836, Theory of the earth: Scien. Tracts, v. 1, 281-306.

Fisher
3845 1837, [Mineral spring in Baltimore]: Am. J. Sci., v. 31, no. 2, 398 (4 lines).
       In Md. Acad. Sci. Liter. Sci.

Fisher, Lieutenant Colonel
3846 1825, A discourse, delivered at the opening of the session of 1824, of the Helvetic Society of Natural Sciences, held at Schaffhausen the 26th, 27th, and 28th of July: Am. J. Sci., v. 9, no. 2, 368-373.

Fisher, Coleman, jr.
3847 1849, Examination of the telluret of bismuth, from Virginia: Am. J. Sci., v. 7 (ns), no. 2, 282-283.

Fisher, Redwood S. (1782-1856)
3848 1847, The precious metals: Am. Whig Rev., v. 5, 418-424.
       On the political effects of increased production.

Fisher, Samuel B.
3849 1836, Map of the first and second anthracite coal fields. Scale 200 perches to inch: Pottsville, Penn.; Phila., Watson's lithography.
       Also second edition, 1849, not seen.

Fisher, William
3850 1850, Analyses of several minerals: Am. J. Sci., v. 9, no. 1, 83-85.

Fisher's National Magazine and Industrial Record (New York, 1845-1846)
3851 1845, The manufacture of iron: v. 1, 35-50.
3852 1845, Anthracite coal: v. 1, 72-73.
3853 1845, Iron trade of England: v. 1, 276-278.
3854 1845-1846, [Brief notes on mining]: v. 1, 278; 376; 383; v. 3, 204.
3855 1845, Mining district of Brazil: v. 1, 305-321.
3856 1845, Coal and iron mines of France and Belgium: v. 1, 376-377.
3857 1845, Coal field of Alleghany County, Maryland: v. 1, 431-443.
3858 1845, Iron trade of Great Britain: v. 1, 567-568.
3859 1845, [Review of Travels in North America by C. Lyell (See 6557)]: v. 1, 385-386.
3860 1846, Present state of the iron trade: v. 2, 657-662.
3861 1846, The iron trade of Europe, &c.: v. 2, 753-757.
3862 1846, Mining and other intelligence: v. 2, 859-864.
3863 1846, Iron and copper trade of Europe: v. 2, 954-961.
3864 1846, English trade in metals: v. 2, 1064-1065.
3865 1846, Russia, its industry and commerce: v. 2, 1077-1099.
3866 1846, Foreign trade in metals: v. 2, 1164-1167.
3867 1846, Commerce of Sweden and Norway: v. 3, 1-19.
3868 1846, Mining and other intelligence: v. 3, 95-96.
3869 1846, Report of the Coal and Iron Association of Pennsylvania: v. 3, 133-153.
3870 1846, Trip to the coal mines of Virginia: v. 3, 224-229.
3871 1846, Rock salt: v. 3, 248-249.
3872 1846, Facts connected with iron and coal: v. 3, 250-251.
3873 1846, Mining and other intelligence: v. 3, 281-285; 385-387; 487-489; 586-588.
       On the use, production, and prices of metals in Europe.

Fisk, Wilber (1792-1839)
3874 1838, Visit to Vesuvius: Richmond Lyc. J., v. 1, 22-23.
       From Fisk's Travels in Europe.

Fitch, Asa (1809-1879)
3875 1850, Geological map of Washinton County: NY State Agri. Soc. Trans., v. 9, map facing 907.
       In 3876.
3876 1850, Rocks and soils of Washington County: NY State Agri. Soc Trans., v. 9, 816-909, 14 figs., folding plate, map.
       Contains map (see 3875)

Fitch, Jabez (1672-1746)
3877   1728, The work of the Lord in the earthquake to be duly regarded by us. A discourse at Portsmouth after the earthquake which happened on the night of the 29th October 1727: Bos., 17 p.
        Evans number 3024.

Fitton, William Henry (1780-1861)
3878   1826, Instructions for collecting geological specimens: Bos. J. Phil. Arts, v. 3, no. 6, 584-588.
3879   1836, Geological notice on the new country passed over by Captain Back during his late expedition: in Narrative of the Arctic land expedition to the mouth of the Great Fish River and along the shores of the Arctic Ocean in the years 1833, 1834, and 1835 by Captain George Back, Phila., Carey and Hart, 399-411.
        Also in Second edition, 1837.

Fleming, John (1785-1857)
3880   1825, Remarks illustrative of the influence of society on the distribution of British animals: Bos. J. Phil. Arts, v. 2, no. 4, 315-332.
        From Edinburgh Philos. J. See also W. Buckland 2154.
3881   1826, The geological deluge, as interpreted by Baron Cuvier and Professor Buckland, inconsistent with the testimony of Moses and the phenomena of nature: Bos. J. Phil. Arts, v. 3, no. 5, 470-485; no. 6, 553-568.
        On the deluge.

Flint, James
3882   1822, Geology of the Falls of the Ohio: West. Q. Reporter, v. 1, no. 2, 153-164, 2 figs.

Flint, Timothy (1780-1840)
3883   1828, A condensed geography and history of the Western states, or the Mississippi Valley: Cin., E. H. Flint, 2 v., 596, 520 p.
        Also Second edition, 1832, as above.
3884   1828, Recollections of the last ten years passed in occasional residence and journeyings in the Valley of the Mississippi from Pittsburgh and the Missouri to the Gulf of Mexico; and from Florida to the Spanish Frontier: in a series of letters to the Rev. James Flint, of Salem, Massachusetts: S. Rev., v. 2, no. 3, 192-216.
3885   1829, Earthquakes on the Mississippi; extracted from the travels of Mr. Flint: Am. J. Sci., v. 15, no. 2, 366-368.
3886   1830, Notice of earthquakes on the Mississippi: Lit. Port Folio, v. 1, no. 1, 4.
3887   1831, The history and geography of the Mississippi Valley to which is appended a condensed physical geography of the Atlantic United States and the whole American continent: Cin., E. H. Flint and L. R. Lincoln, 2 v. (469, 310 p.)
        Also second edition, 1832; for review see 563. Also third edition, 1833 and 1838 printings.
3888   1833, Lectures upon natural history, geology, chemistry, the application of steam, and interesting discoveries in the arts: Bos., Lilly, Wait, Colmen and Holdan, xii, [13]-408.
        For review see 1104, 4665, and 5979.

Floyd, John (Governor of Virginia)
3889   1834, General view of public works in Virginia. Geological researches recommended: Farm's Reg., v. 1, 472-473.

Folger, Walter (1765-1849)
3890   1826, Memoir on the subject of aerolites; read before the Nantucket Philosophical Institute, Nov. 2, 1826: Nantucket, Mass., n.p., 12 p.

Folsom, Joseph Libbey (1816/17-1885)
3891   1849, The gold region: Mining J., v. 2, 114-115.
        California gold region described.

Foord, Alvin
3892   1835, Medical topographical report of the county of Madison: Med. Soc. NY Trans., v. 2, 36-61.

Foot, Lyman (Dr.)
3893   1821, Notices in geology and mineralogy [of New York]: Am. J. Sci., v. 4, no. 1, 35-37.

Foot, William
3894   1834, Soils and farming of Fairfax County [Virginia]: Farm's Reg., v. 1, 552-553.

Forbes (Mr. of Windsor, Connecticut)
3895   1848, [Stalactitic formation of Sharon, Vermont]: Bos. Soc. Nat. History Proc., v. 3, 56.

Forbes, Edward (1815-1854)
3896   1826, Distance to which minutely-divided matter may be carried by the wind: Bos. J. Phil. Arts, v. 3, no. 3, 299 (14 lines).
        Abstracted from New Mon. Mag.
3897   1846, Fresh water formation of the Smyrna Harbor: Am. J. Sci., v. 1 (ns), no. 2, 278.
        Abstracted from Q. J. Geol. Soc.

Forbes, Eli (1726-1804)
3898   1785, An account of the effects of lightning on a large rock in Gloucester [Massachusetts]: Am. Acad. Arts Sci's Mem., v. 1, 253-256, plate.

Forbes, James David (1809-1868)
3899   1822, Heat of the earth: Am. J. Sci., v. 4, no. 2, 372 (7 lines).
        Abstracted by J. Griscom.

Forbes, J. D. (cont.)
3900   1839, On subterranean temperature, and notice of a brine spring emiting carbonic acid gas: Am. J. Sci., v. 35, no. 2, 293-294.
        From British Assoc. Proc.
3901   1840, Temperature of the earth: Am. J. Sci., v. 38, no. 1, 108-109.
        In British Assoc. Proc.
3902   1840, On the use of mica in polarizing light: Am. J. Sci., v. 38, no. 1, 101-102.
        In British Assoc. Proc.
3903   1844, Notice of travels in the Alps of Savoy, and other parts of the Pennine Chain, with observations on the phenomena of glaciers: Am. J. Sci., v. 46, no. 1, 172-192.
3904   1846, Cuchullin Hills in Skye: Am. J. Sci., v. 1 (ns), no. 3, 434.
        Abstracted from Jameson's J.

Forbes, James Grant
3905   1821, Sketches, historical and topographical, of the Floridas, more particularly of East Florida: NY, C. S. Van Winkle, 226 p., map.

Forchhammer, Johan Georg (1794-1865)
3906   1831, Recent formation of zeolite: Am. J. Sci., v. 20, no. 2, 382 (12 lines).
        Abstracted from Edinburgh Philos. J. by J. Griscom.
3907   1836, Oerstedite: Am. J. Sci., v. 30, no. 1, 179 (7 lines).
        Abstracted from "Karsten Archiv für Min." by "D."

Fordyce, George
3908   1811, A new method of assaying copper ores: Agri. Mus., v. 1, 110-118.

Forest River Lead Company
3909   1849, By-laws of the Forest River Lead Company, in Salem: Bos., C. C. Mead (printer), 8 p.

Forman, Joshua (1777-1848)
3910   1830, Remarks upon the salt formation of Salina, N. Y. and other places: Am. J. Sci., v. 19, no. 1, 141-143.

Forry, Samuel (1811-1844)
3911   1843, Meteorology: comprising a description of the atmosphere and its phenomena, the laws of climate in general, and especially the climatic features peculiar to the region of the United States, with some remarks upon the climate of the ancient world as based on fossil geology: NY, 48 p.
        Contains article by C. A. Lee (see 6131).

Forshey, Caleb Goldsmith (1812-1881)
3912   1846, Geology of Western states. Facts in relation to the Mississippi: De Bow's Rev., v. 2, 213.
3913   184?, An address to the members of the Mississippi Legislature on the subject of the geological survey of the state:
        Not seen.
3914   1850, Memoir on the physics of the Mississippi River: New Orleans, Office of "The Daily Crescent", 7 p; another edition: New Orleans, Office of "The Bee", 8 p.; also in: West. J. Ag., v. 5, no. 1, 6-16.
3915   1850, Louisiana geology and hydrography: De Bow's Rev., v. 8, 495-496.

Forsyth, Gideon C.
3916   1809, Geological, topographical and medical information concerning the eastern part of the state of Ohio: Med. Repos., v. 12, 350-358.

Foster
3917   1831, Enormous quantity of iron manufactured, and of coal consumed in Wales: Am. J. Sci., v. 20, no. 1, 182-183.
        Abstracted from Jameson's J.

Foster, A.
3918   1828, Notice of the Tockoa and Tallulah Falls in Georgia: Am. J. Sci., v. 14, no. 2, 209-215.

Foster, George G. (d. 1850)
3919   1848, The gold mines of California, and also a geographical, topographical and historical view of that country, from official documents and authentic sources, with a map of the country, and particularly of the gold region: NY, Dewitt and Davenport, 80 p., map.
        Extracts from many reports and letters, including R. B. Mason (see 6818). Three printings in 1848, as "first," "second," and third editions.
        Alternate title is "The gold regions of California; being a succinct description of the geography, history, topography, and general features of California: including ... the gold regions of that fortunate country."
        For review see 6360.

Foster, James T.
3920   1850, Foster's geological chart: Alb.?
        Not seen. For review see 996, 1001, and 3577. For notice see 6918.
3921   1850, Introduction to the study of geology together with a key to Foster's geological chart: Alb., J. Munsell, 137 p.

Foster, John S.
3922   1825, Notice of firestones used in the manufacture of glass: Am. J. Sci., v. 10, no. 1, 19-21.

Foster, John Wells (1815-1873)
See also 2006 with C. Briggs, Jr.

3923   1838, New locality of iolite, with other minerals associated: Am. J. Sci. v. 33, no. 2, 399-400.
       On minerals from Brimfield, Ohio.
3924   1838 [i.e. 1839], Report [on Muskingham County]: Ohio Geol. Survey Ann. Rept, v. 2, 9-10; 73-107, fig., colored plate.
3925   1839, Head of the Mastodon giganteum: Am. J. Sci., v. 36, no. 1, 189-191, 2 figs.
3926   1839, On the vegetable origin of coal: Hesperian, v. 2, 204-210.
3927   1847, [Report of field work in the Lake Superior land district]: US SED 2, 30-2, v. 2 (530), 159-163.
3928   1849, and J. D. Whitney, Geological map of the district between Portage Lake and Montreal River, Lake Superior, Michigan: Phila., P. S. Dural's steam lithography, colored geological map, 75.5 x 39 cm.
       In Lake Superior report by C. T. Jackson (see 5655).
3929   1849, and J. D. Whitney, Geological map of the district between Keweenaw Bay and Chocolate River, Lake Superior, Michigan: Phila., Ackerman's (lithographer), colored geological map, 60 x 49 cm.
       In Lake Superior report by C. T. Jackson (see 5655).
3930   1849, and J. D. Whitney, Geological map of Isle Royale/Lake Superior, Michigan: NY, Ackerman (lithographer), colored map, 60 x 49 cm.
       In Lake Superior report by C. T. Jackson (see 5655).
3931   1849, Geological notes submitted to C. T. Jackson [on the Northern Peninsula of Michigan and Isle Royale]: in report by C. T. Jackson (see 5655), 766-771.
3932   1849, and J. D. Whitney, Geological map of Keweenaw Point, Lake Superior, Michigan: Phila., P. S. Duval's steam lith. press, colored geological map, 52 x 27 cm.
       In Lake Superior report by C. T. Jackson (see 5655).
3933   1849, Notes on the geology and topography of portions of the country adjacent to Lake Superior and Michigan: in Lake Superior report by C. T. Jackson (see 5655), 773-785.
3934   1849, [On the geological position of Mastodon giganteus]: Bos. Soc. Nat. History Proc., v. 3, 111-116.
       Discussion of 10698.
3935   1849, and S. W. Hill, Statistics of the mines of Keweenaw Point:
       In Lake Superior report by C. T. Jackson (see 5655).
3936   1849, and J. D. Whitney, Synopsis of the explorations of the geological corps in the Lake Superior land district in the northern peninsula of Michigan: in report (see 5655), 605-626.
3937   1850, and J. D. Whitney, Extract from the report submitted to Congress...: in Piscataqua Mining Company Charter (see 8380), 15-17.
       From "Synopsis ... " (see 3936), 609-610.
3938   1850, and J. D. Whitney, Geological map of the Lake Superior land district in the state of Michigan: NY, Ackerman (lithographer), colored geological map, 110 x 63 cm.
       In 3940.
3939   1850, and J. D. Whitney, Mineral reports [on Lake Superior land district]: US SED 2, 31-2, v. 2 (588), 147-152.
       Also in US HED 9, 31-2, 147-152, not seen.
3940   1850, and J. D. Whitney, Report on the geology and topography of a portion of the Lake Superior land district in the state of Michigan; Part I, copper lands: US HED 69, 31-1, v. 9 (578), 3-224, 12 plates, 3 maps, 55 figs.
       Contains maps 3929, 3938, and 3941. With "Tabular statement of the mines in Lake Superior land district," 145-151, and a geological glossary.
3941   1850, and J. D. Whitney, Sketch and diagram illustrating the geology of the region between the Northern shores of Lake Superior and Michigan: NY, Ackerman (lithographer), colored geological map, 116 x 15 cm.
       In 3940.

Foudrinier
3942   1850, Prevention of mine accidents: Ann. Scien. Disc., v. 1, 66-67.

Fourcroy, Antoine François, Compte (1755-1809)
3943   1797, and L. N. Vauquelin, [On strontian earth and barytes]: Univ. Mag., v. 1, 328-329; v. 2, 317.
       In Nat'l Inst. Paris Proc.

Fourier, Jean Baptiste (1768-1830)
3944   1837, General remarks on the temperature of the terrestrial globe and the planetary spaces: Am. J. Sci., v. 32, no. 1, 1-20.
       Translated by E. Burgess. Communicated by E. Hitchcock.

Fournel, Henri Jérôme Marie (1799-1876)
3945   1847, On the salt and salt lakes of Algeria: Am. J. Sci., v. 3 (ns), no. 3, 432.
       Abstracted from Annales des Mines.
3946   1848, Argentiferous galena and iron ore in Algeria: Am. J. Sci., v. 6 (ns), no. 2, 271-272.
       Abstracted from l'Institut.

Fowler, Samuel (MD)
3947   1825, An account of some new and extraordinary minerals discovered in Warwick, Orange Co., N. Y.: Am. J. Sci., v. 9, no. 2, 242-245.
3948   1825, New American minerals [from Warwick, New York]: Bos. J. Phil. Arts, v. 2, no. 6, 603 (15 lines).
       Abstracted from Am. J. Sci.
3949   1832, An account of the sapphire and other minerals in Newton township, Sussex County, New Jersey: Am. J. Sci., v. 21, no. 2, 319-320.

Fox, George (see 4386 with J. B. Griscom)

Fox, Robert
3950   1831, On the electro-magnetic properties of metalliferous veins in the mines of
       Cornwall: Am. J. Sci., v. 20, no. 1, 136-143.
3951   1837, [Effect of galvanism on minerals]: Am. J. Sci., v. 31, no. 1, 373-374.
       In British Assoc. Proc.
3952   1837, On the formation of mineral veins: Franklin Inst. J., v. 19 (ns), 228-229.
3953   1837, Questions relative to mineral veins, submitted to practical miners: Am. J. Sci.,
       v. 33, no. 1, 135-139.
3954   1838, Heat in mines: Am. J. Sci., v. 34, no. 1, 46-47.
       In British Assoc. Proc.
3955   1838, Origin of mineral veins: Franklin Inst. J., v. 21 (ns), 251-252, fig.
3956   1839, [Artificial formation of a mineral vein by voltaic action]: Am. J. Sci., v. 35,
       no. 2, 308 (5 lines).
       In British Assoc. Proc.
3957   1839, Formation of metallic veins by galvanic agency: Am. J. Sci., v. 37, no. 1,
       199-200.
       Abstracted from l'Institut.
3958   1841, Report on 'subterranean temperature': No. Light, v. 1, 74.
       In article by J. A. Dix (see 3193).
3959   1846, On certain pseudomorphous crystals of quartz: Am. J. Sci., v. 1 (ns), no. 3,
       430-432.
       Abstracted from Jameson's J.
3960   1847, High temperature of mines: Franklin Inst. J., v. 14 (3s), 138-139.
       Abstracted from London Mech's Mag.

Foxcroft, Thomas
3961   1727, The voice of the lord, from the deep places of the earth. A sermon preached in
       Boston, in the audience of the general court, at the opening of the sessions,
       November 23, 1727: Bos., Printed for S. Gerrish, (4), 52 p.
       Evans number 2874.
3962   1756, The earthquake, a divine visitation. A sermon preached ... in Boston, January 8,
       1756, being a day of public humiliation and prayer ...; upon the occasion of the
       repeated shock of an earthquake on this continent, and the very destructive
       earthquake and inundations in divers parts of Europe, all in the month of November
       last ... : Bos., S. Kneeland and T. Rand, (2), 51 p.
       Evans number 7665.

France. New Academy of Arts and Sciences.
3963   1797, Account of the New Academy of Arts and Sciences of France: Univ. Mag., v. 1,
       324-325.
3964   1797, Account of the proceedings of the first and second sittings of the National
       Institution at Paris: Univ. Mag., v. 1, 325-334.
3965   1797, Proceedings of the third sitting of the National Institution, held at Paris ... :
       Univ. Mag., v. 2, 316-319.

Francis, John Wakefield (1789-1861)
3966   1834, Observation on the Avon mineral waters: Am. Rr. J., v. 3, 518-519; also in: US
       Med. Surgical J., v. 1, 14-18.
       From the NY Mirror.
3967   1835, Mineral waters of the United States: Am. Inst. NY J., v. 1, 95-100.

Franklin
3968   1816, Thoughts, on philosophical science, on creation, and on the order and
       constitution of nature: Am. Mag., a Mon. Miscellany, v. 1, no. 11, 388-401; no. 12,
       411-421.
       Dated "Feb. 18, 1816."

Franklin, Benjamin (1706-1790)
3969   1792, Letter to Abbe Soulavie, occasioned by his sending me some notes he had taken of
       what I had said to him in conversation on the theory of the earth: Univ. Asylum and
       Colum. Mag., v. 9, 106-108.
3970   1792, On the nature of sea coal: Mass. Mag., v. 4, 152.
3971   1793, Conjectures concerning the formation of the earth, &c. in a letter from Dr. B.
       Franklin, to the Abbé Soulavie: Am. Philos. Soc. Trans., v. 3, 1-5.
3972   1793, Queries and conjectures relating to magnetism, and the theory of the earth, in a
       letter from Dr. B. Franklin, to Mr. Bodoin: Am. Philos. Soc. Trans., v. 3, 10-13.
3973   1822, Original letters of Dr. Franklin and others addressed to the late Rev. Jared
       Eliot of Killingworth, Con.: Am. J. Sci., v. 4, no. 2, 357-370 [especially
       365-366].

Franklin Institute (Philadelphia)
3974   1825, Franklin Institute: Am. J. Sci., v. 9, no. 2, 391-392.
       Signed by "C. H." Formation of organization and officers announced.

Franklin Institute Journal (Philadelphia, 1826-1850)
3975   1826, Meteoric iron [from Bogota, Colombia]: v. 1, 185.
3976   1826, Vermont gold [from New Fane]: v. 2, 189-190.
3977   1826, Proposed geological survey of the state of Pennsylvania: v. 2, 296-297.
       Notice of speech by P. A. Browne.
3978   1826, Observations on the hydraulic cement, used on the Pennsylvania Union Canal: v. 2,
       287-288.
3979   1827, Conflagration in the quicksilver mines of Idria, in 1803: v. 4, 40-41.
       Abstracted from Brewster's J.
3980   1827, Notice regarding steatite or soap-stone, and its principal uses: v. 4, 52-53.
       Abstracted from Edinburgh J.
3981   1829, On the cloth and paper made from amianthus, a species of asbestos: v. 4 (ns),
       78-82.
       From Va. Lit. Mus. Signed "P."
3982   1830, Mills for grinding, washing, and separating gold and silver from ores, earth, or
       in whatever state these metals may be found: v. 5 (ns), 132-133.
       Unsigned patent description.

Franklin Institute Journal (cont.)
- 3983   1830, Blowpipe simplified: v. 6 (ns), 144.
- 3984   1830, Cloth of amianthus: v. 6, 419 (8 lines).
  Abstracted from Am. J. Sci., on techniques of making asbestos.
- 3985   1830, On the uses of steatite; and particularly in the lubrication of machinery to reduce friction: v. 5 (ns), 274-277.
  Abstracted from Recuil Industriel by Technological Repos.
- 3986   1831, Account of the processes followed in Mexico in the reduction of the precious metals. Extracts from a review of a work entitled "Commentaries on the mining ordinances of Spain": v. 7 (ns), 112-121.
  Abstracted from Royal Inst. Great Britain J.
- 3987   1831, Proposed plan for smelting iron ore with anthracite: v. 8 (ns), 123 [i.e. 231]-232.
- 3988   1831, [Review of Monthly American Journal of Geology by G. W. Featherstonhaugh (see 3746)]: v. 8 (ns), 143-144.
- 3989   1831, [Review of Scientific Tracts, v. 1, no. 2]: v. 8 (ns), 419.
  Review of article on geology.
- 3990   1832, On the use of heated air and uncoked coal in the smelting of iron ores: v. 10 (ns), 130-131.
  From Brewster's J.
- 3991   1832, On the employment of heated air in the smelting of iron: v. 9 (ns), 339-340.
- 3992   1832, Expansion and contraction of coping stones: v. 10 (ns), 365.
  Signed by "W.T."
- 3993   1833, On the accidents which have occurred in the coal mines of the Department of the Loire from 1817 to 1831: v. 12 (ns), 268-269.
  Abstracted from Annales des Mines.
- 3994   1835, Expansion and contraction of coping stones: v. 15 (ns), 92.
  Signed by "R.L."
- 3995   1835, On the practice of the blow pipe: v. 16 (ns), 202-204, 6 figs.
  From London Mech's Mag.
- 3996   1835, The Russian platina mines: v. 15 (ns), 293.
  Abstracted from London Mech's Mag.
- 3997   1836, [Review of The elevated country ..., by G. W. Featherstonhaugh (see 3778, 3779)]: v. 17 (ns), 184-190.
- 3998   1836, Flint and chalcedony not simple minerals: v. 17 (ns), 286.
- 3999   1836, Temperature of the earth, as shown by that of the waters of artesian wells: v. 17 (ns), 286 (4 lines).
- 4000   1836, Vapour cave at Pyrmont [Italy]: v. 17 (ns), 366 (4 lines).
- 4001   1836, Notice of coal mines in Illinois: v. 17 (ns), 375.
- 4002   1836, On the mineral waters of Nevis: v. 17 (ns), 435 (10 lines).
- 4003   1836, Analysis of the waters of the Grey Sulphur Springs [Monroe and Giles Counties, Virginia]: v. 17 (ns), 435-436 (11 lines).
- 4004   1836, Mineral pitch lake of Trinidad: v. 17 (ns),437 (8 lines).
- 4005   1836, Lines of equal magnetic dip in Great Britain: v. 18 (ns), 126-127 (11 lines).
- 4006   1836, Change in the chemical character of minerals induced by galvanism: v. 18 (ns), 336.
  Notice of the research of R. W. Fox.
- 4007   1836, Artificial crystals and minerals: v. 18 (ns), 336-337.
  Notice of the research of A. Cross.
- 4008   1837, Mining in Cornwall: v. 19 (ns), 153-154.
- 4009   1837, On the calcination of ores: v. 19 (ns), 215-216.
- 4010   1837, Analysis of iron ores: v. 19 (ns), 218 (4 lines).
- 4011   1837, Mining product in England and Ireland: v. 19 (ns), 226-227.
- 4012   1837, Central heat of the earth not alarming: v. 19 (ns), 238.
- 4013   1837, Temperature of the globe: v. 19 (ns), 305-306.
- 4014   1837, Earthquake at anchor [in Callao Roads Harbor]: v. 19 (ns), 308.
- 4015   1837, Deepest mine in Great Britain: v. 19 (ns), 405-406.
- 4016   1837, Amber: v. 20 (ns), 71 (5 lines).
  Abstracted from Mining J. On localities.
- 4017   1837, Mineral resources of Belgium: v. 20 (ns), 151-153.
  Abstracted from Mining J.
- 4018   1837, [Review of Principles of Geology by C. Lyell (see 6532)]: v. 20 (ns), 263-265.
- 4019   1837, Museum of economic geology: v. 20 (ns), 289-290.
  Abstracted from Mining J.
- 4020   1837, Preservation of mineral specimens from decomposition: v. 20 (ns), 369-370.
  Abstracted from Mag. Nat. History.
- 4021   1837, Simultaneous magnetic observations: v. 20 (ns), 371.
- 4022   1838, Mica as a substitute for glass: v. 21 (ns), 133-134.
  Abstracted from Mining J.
- 4023   1838, Artificial production of rubies: v. 21 (ns), 138.
  Abstracted from Comptes Rendus.
- 4024   1838, On the various uses of steatite: v. 21 (ns), 197-198.
  Abstracted from "J. des Conn. Usuelles".
- 4025   1838, Estimated quantity of coal in the Derbyshire and Yorkshire coal fields: v. 21 (ns), 213-214.
  Signed "Alpha," from the Mining J.
- 4026   1838, Coal in France: v. 21 (ns), 214 (7 lines).
  Abstracted from Mining J.
- 4027   1838, Vegetable origin of the diamond: v. 21 (ns), 215.
  Abstracted from Mag. Popular Sci.
- 4028   1838, Minerals in Jamaica: v. 21 (ns), 287 (11 lines).
  Abstracted from Mining J.
- 4029   1838, Origin of coal: v. 21 (ns), 287.
  Abstracted from Mining J.
- 4030   1838, Fossil bones: v. 21 (ns), 339-340.
  On the chemical composition of fossil bones.
- 4031   1838, Description of the borax lagoons of Tuscany: v. 21 (ns), 346-347.
  Abstracted from Mining J.
- 4032   1838, Geology of Bretagne: v. 21 (ns), 352-353.
  Abstracted from Mining J.

Franklin Institute Journal (cont.)

4033 1838, Singular discovery of a subterranean river [in Flintshire, England]: v. 21 (ns), 353.
4034 1838, Russian gold mines. - important discovery: v. 21 (ns), 428-429.
       Abstracted from Mining J. On the productivity of the mines.
4035 1838, European lead mines: v. 22 (ns), 137.
       Abstracted from Mining J.
4036 1838, Asphaltic mine in Pyrimont: v. 22 (ns), 276-279.
       Abstracted from Mining J.
4037 1839, New mines in Egypt: v. 23 (ns), 213-214.
       Abstracted from London Mech's Mag.
4038 1839, Aerolites: v. 23 (ns), 275-276.
       From Nautical Mag.
4039 1839, [Review of Popular lectures ..., by K. C. Von Leonhard (see 6192)]: v. 23 (ns), 378-380.
4040 1839, Remains of the mammoth dredged up in the English Channel and German Ocean: v. 24 (ns), 275-276.
       Abstracted from Mining Rev.
4041 1839, Five fossil trees found in the excavations for the Manchester and Bolton Railway: v. 24 (ns), 276.
       Abstracted from Mining Rev.
4042 1839, Coal-fields of Nova Scotia and Cape Breton: v. 24 (ns), 276-277.
       Abstracted from "Edin. Cab. Lib."
4043 1839, Fossil bones: v. 24 (ns), 287.
       Abstracted from London Ath. On prices of specimens in the British Museum.
4044 1839, Natural history of coal: v. 24 (ns), 355-357.
4045 1839, Establishment of magnetic observatories, and naval expedition, for magnetic purposes: v. 24 (ns), 404-408.
4046 1839, Submarine volcano [at Bahama Banks]: v. 23 (ns), 284-285 (11 lines).
4047 1840, Connexion of geology with the arts: v. 25 (ns), 128.
       Abstracted from Mining J.
4048 1840, Microscopic geology: v. 25 (ns), 209.
       Abstracted from Mining J. On Ehrenberg's research.
4049 1840, [Review of Silurian System by R. I. Murchison]: v. 25 (ns), 353-355.
       Abstracted from Mining Rev.
4050 1840, Iron pyrites and calamine: v. 25 (ns), 358.
       Abstracted from Mining J.
4051 1840, Burning coal mines [of Allier, France]: v. 26 (ns), 68-69.
       Abstracted from Mining J.
4052 1841, Southern magnetic expedition: v. 1 (3s), 70-71.
       Abstracted from Nautical Mag.
4053 1841, Report of the committee on combined magnetic observations: v. 1 (3s), 133-134.
4054 1841, Magnetic observatory at Munich: v. 1 (3s), 134.
4055 1841, Geological specimens found on the Great Western Railway: v. 1 (3s), 132.
       Abstracted from Mining J.
4056 1841, On the best method of ventilating coal mines: v. 1 (3s), 139-140.
       Abstracted from Mining J.
4057 1841, Comparative longevity of miners: v. 1 (3s), 142.
       Abstracted from Mining J.
4058 1841, [Review of "Remarks on the mineralogy and geology of Nova Scotia" by A. Gesner]: v. 1 (3s), 206-210.
4059 1841, Salt mine in Switzerland: v. 1 (3s), 215 (5 lines).
       Abstracted from Mining J.
4060 1843, Phillip's Mineralogy [notice of 8325]: v. 5 (3s), 72.
4061 1843, Geological survey of Great Britain: v. 6 (3s), 422 (7 lines).
4062 1843, Earthquakes in Scotland: v. 6 (3s), 423 (3 lines).
4063 1844, Manufacture of oil of vitriol from iron pyrites: v. 7 (3s), 141.
       Abstracted from London Mining J. On production of sulfuric acid.
4064 1844, Extraction of palladium from Brazil: v. 7 (3s), 255.
       Abstracted from Philos. Mag.
4065 1844, Statistics of the coal trade: v. 7 (3s), 295-296.
       Abstracted from London Mining J.
4066 1844, The diamonds of Brazil: v. 8 (3s), 72.
       Abstracted from Ann. de Chimie.
4067 1844, Platina sands of the Rhine: v. 8 (3s), 72 (3 lines).
4068 1844, Mineral and metallic statistics: v. 8 (3s), 221-223.
       Abstracted from London Mining J. On world production.
4069 1844, Copper mining in Cornwall: v. 8 (3s), 223-226.
       Abstracted from London Mining J.
4070 1844, Coal mines of France: v. 8 (3s), 431-432.
       Abstracted from London Mining J.
4071 1845, [Review with excerpts of Elementary treatise on mineralogy by W. Phillips (see 8325)]: v. 9 (3s), 259-265.
4072 1845, Mineral produce of South Wales: v. 10 (3s), 341.
       Abstracted from London Mining J.
4073 1846, Anthracite coal trade of the United States: v. 11 (3s), 100.
4074 1846, Iron and coal statistics: v. 12 (3s), 124-141, tables.
4075 1846, Reduction of silver ores without quicksilver: v. 12 (3s), 204-205.
       Abstracted from Mech's Mag.
4076 1846, Statistics of the coal trade. - The coal measures of Britain: v. 12 (3s), 266-271.
       Abstracted from Mining J.
4077 1846, Production of coal in the different states of Europe: v. 12 (3s), 284-286.
       Abstracted from Mining J.
4078 1846, On the proper coal for smelting sulphuretted lead and silver ores: v. 12 (3s), 353-357.
       Signed by "J.M." from Mining J.
4079 1846, Paris Royal School of Mining: v. 12 (3s), 359.
       From Mining J. List of admissions requirements.
4080 1847, Review of the mines and mining industry of Belgium: v. 14 (3s), 144.
       Abstracted from London Mining J.
4081 1848, Production of silver in Spain: v. 15 (3s), 351-352.

Franklin Institute Journal (cont.)
4082   1848, New method of treating the ore of platinum: v. 15 (3s), 388.
         Abstracted from "Civ. Eng. & Arch. J."
4083   1848, English and foreign mining glossary. To which is added, the smelting terms used in France, Spain, and Germany: v. 15 (3s), 293-320; 461-464.
         Abstracted from London Mining J.
4084   1848, Analysis of moulding sand: v. 16 (3s), 65 (9 lines).
         Abstracted from London Builder.
4085   1848, [Review of Statistics of coal by R. C. Taylor (see 10147)]: v. 16 (3s), 68-69.
4086   1848, Produce of gold in Russia: v. 16 (3s), 215.
         Abstracted from Hunt's Merch. Mag.
4087   1848, The coal fields of England: v. 16 (3s), 215-216.
         Abstracted from Hunt's Merch. Mag.
4088   1848, The Lackawanna and Wyoming coal regions: v. 16 (3s), 280-285.
         Abstracted from Hunt's Merch. Mag. Signed by "J. H. L."
4089   1848, Death of Berzelius: v. 16 (3s), 343-348.
         Abstracted from London Ath.
4090   1848, Copper smelting at the Royal Saxon works, at Freyberg: v. 16 (3s), 407-408.
         Abstracted from London Mining J. Signed by "C."
4091   1849, Magnetic and meteorological observations made at the Royal Observatory, Greenwich, in 1846: v. 17 (3s), 128-129.
         Abstracted from London Ath.
4092   1849, [Review of The miner's guide ..., by J. W. Orton (see 8041)]: v. 17 (3s), 407-408.
4093   1849, Description of some minerals from Brazil: v. 18 (3s), 47-49.
         Abstracted from l'Institut.
4094   1849, Quantity of gold from the Ural and Siberia: v. 18 (3s), 234-235.
         Abstracted from "Geneva Arch. Phys. & Nat. Sci."
4095   1850, Coal for the steam Navy [review of report by H. T. De la Beche and L. Playfair]: v. 19 (3s), 52-54.
         Abstracted from London, Edinburgh, and Dublin Philos. Mag.
4096   1850, [Review of Encyclopaedia of chemistry by J. C. Booth (see 1775)]: v. 19 (3s), 360.

Franklin Institute of New Haven (Connecticut)
4097   1831, Franklin Institute of New Haven: Am. J. Sci., v. 21, no. 1, 185-186.

Franklin Institute of Providence (Rhode Island)
4098   1837, General directions for collecting and preserving articles in the various departments of natural history: Naval Mag., v. 2, 91-97.

Frary, Robert H. (see 10542 with P. Van Beuren)

Fraser, Donald
4099   1791, The young gentleman and lady's assistant; containing I. geography, II. natural history, III. rhetoric, IV. miscellany: NY, Tho's Greenleaf (printer), xii, 273, 22 p.
         Evans number 23387. Also second edition, Danbury, Ct., N. Douglas (1794) and third edition, NY, T. and J. Sword, 216 p., 3 plates (1796).

Frazer, John Fries (1812-1872)
4100   1850, [Gold from Bloomington, Indiana]: Am. Philos. Soc. Proc., v. 5, no. 45, 150 (7 lines).
4101   1850, Report on minerals [collected in Oregon and California]: US SED 47, 31-1, v. 10, pt. 1 (558), 116-117.
         In report by P. T. Tyson (see 10457).

Frederick Mining Company
4102   1847, Articles of association and agreement of the Frederick Mining Company: Bos., Allen and Co., 12 p.

Fredly, Frederick
4103   1838, Machine for washing iron and other ores: Franklin Inst. J., v. 21 (ns), 46 (11 lines).

The Free Enquirer (New Harmony, Indiana, 1830-1835)
         See also New Harmony Gazette.
4104   1830, The great earthquake in Calabria in 1783: v. 1 (2s), 131.
4105   1832, Relics of a mammoth [from Rochester, New York]: v. 5 (2s), 215.
4106   1834, Shape of the earth: v. 1 (3s), 330.
4107   1835, Dr. Cooper and Prof. Silliman: v. 1 (3s), 127-128; 130-131.
         On the Pentateuch debate. See also 2627 et seq., and 9712.

Freeman, Charles (1794-1853)
4108   1831, An account of Limerick [Maine]: Maine Hist. Soc. Coll., v. 1, 245-253.

Freeman, William
4109   1837, Experiments on the force required to fracture and crush stone: Mech's Mag. and Reg. Inventions, v. 9, 277-278.
         With article by F. C. Lukis (see 6518). Abstracted from "Trans. Inst. Civil Eng."

Free-Masons Magazine and General Miscellany (Philadelphia, 1811-1812)
4110   1811, Natural history of the earth: v. 2, 133-137.
         General geological principles.

Fremont, John Charles (1813-1890)
4111   1843, A report on an exploration of the country lying between the Missouri River and the Rocky Mountains: US SED 243, 27-3, v. 4 (416), 207 p., 6 plates, map.

Fremont, J. C. (cont.)
4112 1845, Narrative of the exploring expedition to the Rocky Mountains in the year 1842 and to Oregon and North California in the years 1843-44: Wash., H. Polkinhorn, viii, 9-278 p.
Also second edition: Wash., Taylor, Wilde and Co., 278 p. Other editions include Cooperstown, NY, H. and E. Phinney, 186 p. (1846); Syracuse, NY, L. W. Hall, 305 p. (1846); Syracuse, NY, Hall and Dickson; NY, A. S. Barnes and Co., 427 p. (1847); NY, D. Appleton and Co.; Phila., G. S. Appleton, 186 p. (1849).

4113 1845, Report of the exploring expedition to the Rocky Mountains in the year 1842, and to Oregon and California in the years 1843-44: US SED 174, 28-2, v. 11 (461), 693 p., 22 plates, 5 maps.
Includes reports by J. Hall (4700, 4703) and J. Bailey (1433). Printed in Washington by Gales and Seaton. For notice see 832.

4114 1845, Report of the exploring expedition ... [as above]: US HED 166, 28-2, v. 4 (467), 583 p., 22 plates, 5 maps.
Printed in Washington by Blair and Rivers.

4115 1847, Bear River Springs, Rocky Mountains: Am. J. Sci., v. 3 (ns), no. 3, 450-451.
Abstracted from the expedition report (see 4113).

4116 1847, Observations on the Rocky Mountains and Oregon: Am. J. Sci., v. 3 (ns), no. 1, 192-202, 2 figs.

4117 1848, Geographical memoir upon Upper California in illustration of his map of Oregon and California: Wash., Wendell and Van Benthuysen (printers), US SMD 148, 30-1, v. 1, (511), 67 p., map.
Also 1849 editions: Wash., Tippin and Streepen (printers), US HMD 5, 30-2, v. 1 (544), 40 p., map. Phila., W. McCarty, 80 p.

4118 1849, Notes of travel in California, comprising the prominent geographical, agricultural, geological, and mineralogical features of the country: NY, Appleton, 29, 83 p., maps.

4119 1849, Oregon and California. The exploring expedition to the Rocky Mountains, Oregon, and California. ... with recent notices of the gold region ... : Buffalo, G. H. Derby and Co., 456 p., front., 2 plates.
Also an 1850 printing.

Frémy, Edmond (1814-1894)
4120 1846, and Clemandat, Artificial aventurine: Am. J. Sci., v. 1 (ns), no. 3, 430 (11 lines).
Abstracted from l'Institut.

Fresenius, Carl Remigius (1818-1897)
4121 1846, On the solubility of the basic phosphate of magnesia and ammonia, and on the quantitative determination of phosphoric acid and magnesia by means of this salt: Am. J. Sci., v. 1 (ns), no. 2, 258.
Abstracted by J. L. Smith from "Annalen der Chem. und Pharm."

The Friend, a Periodical Work (Albany, 1815-1816)
4122 1815, Natural curiosity: v. 1, 73-75.
Description of two caves in Bethlehem, NY.

The Friend. A Religious and Literary Journal (Philadelphia, 1827-1850)
4123 1827-1847, [Notes on Pennsylvania coal mining]: v. 1, 72; 80; 96; 104; v. 5, 348; v. 11, 384; v. 15, 74-85; v. 18, 167; v. 20, 183;
4124 1828/1831, [Notes on North Carolina gold mining]: v. 1, 91-92; v. 4, 244.
4125 1828-1845, [Notes on vertebrate fossils]: v. 1, 121; v. 3, 96; v. 4, 103; v. 13, 176; 399-400; v. 15, 341; 357; v. 16, 259; 264; v. 17, 266; v. 18, 155; 360;
4126 1829, Avalanches of the White and Green Mountains: v. 2, 108-109; 115-116; 122.
4127 1829-1850, [Notes on earthquakes and volcanoes]: Numerous entries in all volumes, v. 2 - v. 23. Consult annual indices.
4128 1829, Geological phenomena: v. 2, 380.
On an oil well found in Cumberland, Kentucky.
4129 1829, Geology: v. 3, 16.
On the beauty of the science.
4130 1831, Temperature of the earth: v. 4, 295.
4131 1832-1844, [Notes on fossil plants]: v. 5, 189; v. 9, 136; v. 14, 103; v. 18, 12.
4132 1833-1849, [Notes on salt mining]: v. 6, 242; v. 8, 203-204; 210-211; 282; v. 19, 370-371; v. 22, 63;
4133 1833-1848, [Notes on foreign mining]: v. 7, 28-29; v. 11, 329-311; v. 13, 332-333; 340; v. 18, 161-162; v. 20, 384; 415; v. 21, 133.
4134 1834, Geology of Scripture: v. 7, 249-250; 257-258; 265-266.
Signed by "A. T. F."
4135 1834, Meteorolites, or stones which have fallen from the sky: v. 7, 409-410.
4136 1835-1849, [Notes on caves]: v. 8, 324-325; v. 11, 337-338; v. 15, 337-339; v. 17, 99; 164; v. 18, 247; 303; v. 22, 42.
4137 1836, Geological facts - formation of deltas, &c.: v. 9, 192-193.
Signed by "B."
4138 1836, The marl district of New Jersey: v. 9, 401-402; 409-410; v. 10, 1-2.
4139 1836, The Giant's Causeway: v. 10, 108.
4140 1837, Geology: v. 10, 225-226; 235-236; 253-254; 297-298; 305-306; 313-314; 322-324; 330-331; 338-339; 346-347; 353.
4141 1838, Bog earth fuel [from Finland]: v. 12, 89 (10 lines).
4142 1839, Geological wonder: v. 12, 164 (10 lines).
On a frog in rock, Oldham, England.
4143 1839, Quicksilver mine [at Blue Mountain, Pennsylvania]: v. 12, 312 (7 lines).
4144 1840, How rocks are formed: v. 13, 298-300.
4145 1840, Grotto of Adelsberg: v. 13, 386-387.
Signed by "W. J. M."
4146 1840, History of the Lehigh Coal and Navigation Company: v. 13, 362-363; 370-371; 379-380; 387-388.
4147 1840, [Geology lectures by Dr. Joseph Thomas]: v. 14, 36 (6 lines); 48 (4 lines).
4148 1841, Colouring marble [Italian techniques]: v. 14, 116.
4149 1841, Steel ore [from Duane, Franklin County, Pennsylvania]: v. 14, 151-152.

The Friend (cont.)
4150 1841, [Notes on the Missourium from Benton County, Missouri]: v. 14, 155-156; 204; 404; v. 15, 52.
4151 1841, Minerals in Canada: v. 14, 240.
4152 1841, Geology of Pennsylvania: v. 14, 267-268.
       Notice of H. D. Rodgers' survey.
4153 1841, Melting of rock crystal: v. 14, 355.
4154 1841, Insects in chalk: v. 14, 360 (11 lines).
       Notice of Ehrenberg's studies.
4155 1841, Hot Springs of St. Michael's, one of the Azores: v. 14, 387-388.
4156 1841, An essay on the geological evidence of the existence and divine attributes of the creator: v. 15, 25-26; 33-34; 41-42; 50-51; 58-59; 66-67.
       Signed by "L. L. N."
4157 1842, Silver mine in North Carolina [at Conrad Hill]: v. 16, 35-36.
4158 1842, Terra di sienna [from Lancaster, Pennsylvania]: v. 16, 80 (10 lines).
4159 1843, Fountain of fire [at Newbridge, Glamorganshire]: v. 16, 170.
       On a natural gas well.
4160 1843, Iron ore [from Reading, Pennsylvania]: v. 16, 214 (3 lines).
4161 1843, Stalactites and stalagmites: v. 16, 234-235.
4162 1843, Siliceous stone - Tripoli: v. 16, 236.
4163 1843, The asphaltum or pitch lake of Trinidad: v. 16, 345-346.
4164 1843-1848, [Notes on the copper region of Lake Superior]: v. 17, 52; v. 19, 245; v. 22, 111.
4165 1845, Geology of new Hampshire: v. 18, 159.
4166 1845, The paint rock, Tennessee: v. 18, 263.
4167 1845, Iron Mountain [Missouri]: v. 18, 320.
4168 1845, More important discoveries of iron ore: v. 18, 339.
4169 1845, Silver mines in North Carolina [Davidson County]: v. 18, 410.
4170 1846, Cannel coal [from Culloway County, Missouri]: v. 19, 230 (10 lines).
4171 1846, Visit of a naturalist: v. 19, 332.
       On L. Agassiz's visit.
4172 1847, Burning wells of Kanawha: v. 20, 412.
4173 1847, The Mammoth Cave, Edmondson County, Kentucky: v. 20, 389-390; 397-398; 404-405; 414-415; v. 21, 7-8; 15.
4174 1848, Resources of Virginia: v. 21, 171.
4175 1848, Nitre lake in Egypt: v. 21, 250.
4176 1848, A new mineral useful in the arts: v. 21, 266.
       On the discovery of a wax-like mineral.
4177 1848, Mineral waters: v. 21, 307.
       On the arsenic content of water.
4178 1849, Glaciers: v. 22, 310-312.
4179 1849, Coal [from the Mississippi Valley]: v. 22, 328 (8 lines).
4180 1849, Cumberland coal region: v. 22, 397-398.
4181 1850, A large diamond [from India]: v. 23, 276.
4182 1850, Mineral wealth of New Jersey: v. 23, 296.
4183 1850, Movement of glaciers: v. 23, 376.

The Friends' Intelligencer (Philadelphia, 1838-1839)
4184 1838, Wonderful subterranean forest in Lincolnshire [England]: v. 1, 171-173.

Frost, James
4185 1824, Improved process for preparing hydraulic cement: Bos. J. Phil. Arts, v. 1, no. 6, 593-594.
4186 1835/1836, On calcareous cements: Franklin Inst. J., v. 16 (ns), 217-219; 376-377; v. 17 (ns), 234-239; v. 18 (ns), 17-19; v. 19 (ns), 193-195; 277-280. Also in: Mech's Mag. and Reg. Inventions, v. 6, 337.

Frost, John (1800-1859)
4187 1850, History of the state of California: Auburn, NY, Derby and Miller, 508 p., plates.

Fuchs, Johann Nepomuk von (1774-1856)
4188 1831, Potash obtained from feldspar: Am. J. Sci., v. 31, no. 1, 157 (7 lines).
       Abstracted from Royal Inst. J.
4189 1835, Triphylme, a new mineral: Am. J. Sci., v. 28, no. 2, 394 (4 lines).
       Abstracted from l'Institut.

Fulton, Hamilton (d. 1834)
4190 1819, Report of sundry surveys, made by Hamilton Fulton, esq. ... : Raleigh, NC, Thomas Henderson, 70 p.
4191 1825, Anthracite [from the Lehigh Coal Company]: Am. Mech's Mag., v. 1, no. 6, 96.

Furguson, James
4192 1838, [On Dr. Sherwood's magnetic discoveries]: New-Yorker, v. 6, 8.

Fyfe, Andrew (1792-1861)
4193 1822, [Hydrate of magnesia from Hoboken, New Jersey]: Am. J. Sci., v. 4, no. 2, 245 (6 lines).
4194 1849, On the comparative value of different kinds of coal for the purpose of illumination; and on methods not hitherto practised for ascertaining the value of the gases they afford: Am. J. Sci., v. 7 (ns), no. 1, 77-86; no. 2, 157-167.

Gadolin, Johan (1760-1852; see Garolin, 4213)

Gairdner, Meredith
4195 1836, Geology: Mus. For. Liter. Sci., v. 28, 431.
       On volcanoes of California and British Columbia.
4196 1838, Physico-geognostic sketch of the Island of Oahu, one of the Sandwich Island group: Hawaiian Spectator, v. 1, no. 2, 1-18.
       With appended notes by G. P. Judd (see 5888).

Gale, Leonard Dunnell (1800-1883)
4197   1838, Fossil fishes in the red sandstone of New Jersey: Am. J. Sci., v. 35, no. 1, 192 (8 lines).
4198   1839, Report - on the geology of New York County: NY Geol. Survey Ann. Rept., v. 3, 177-199.
4199   1840, [Lignite beds of Amboy, New Jersey]: Am. Rep. Arts Sci's Manu's, v. 1, 177 (6 lines).
       In NY Lyc. Nat. History Proc.
4200   1840, [On New York Island garnet crystals and boulders]: Am. Rep. Arts Sci's Manu's, v. 1, 346.
       In NY Lyc. Nat. History Proc.
4201   1843, Diary of a geological survey of the Island of New York: in Geology of New York first district by W. W. Mather (see 6907), 581-604.
4202   1843, Sketch illustrating the extent of the primary limestone on the N. end of New-York Island: in Geology of New York, first district by W. W. Mather (see 6907), colored geological map 9 x 9 cm., fig. 7 on plate 30.

Gale, N. D.
4203   1847, On the Natchez Bluff formation: Am. Q. J. Ag. Sci., v. 6, 208-209 [i.e. 256-257].
       In Assoc. Am. Geol's Nat's Proc., with discussion by W. B. Rogers and L. Agassiz.
       Also in: Am. J. Sci., v. 5 (ns), no. 2, 249-250 (1848).
4204   1848, [On the Natchez Bluff formation, Mississippi]: Lit. World, v. 2, 255 (6 lines).
       In Assoc. Am. Geol's Nat's Proc.

Galen (pseudonym, see 3718)

Galindo, Juan
4205   1835, Eruption of the volcano of Cosiguina [in Nicaragua]: Am. J. Sci., v. 28, no. 2, 332-336.

Gallagher, William Davis (1808-1894)
4206   1838, Ohio in eighteen hundred thirty-eight: Hesperian, v. 1, no. 2, 95-103.

Gallatin, Albert (1761-1849)
4207   1811, American manufactures in 1810: Am. Min. J., v. 1, no. 2, 113-120.

Galt, John (1779-1893)
4208   1821, The earthquake; a tale ... : NY, C. S. Van Winkle and Clayton and Kingsland, 2 v.

Gamma (pseudonym, see 9945)

Gannal, Jean Nicolas (1791-1852)
4209   1829, Manufacture of diamonds: Franklin Inst. J., v. 3 (ns), 140-141.
       Abstracted from Lit. Gaz.
4210   1829, Artificial production of diamonds: Am. J. Pharmacy, v. 1, 76.
       Abstracted from "J. de Chim. Med."

Gardiner, George A.
4211   1820, A brief and correct account of an earthquake which happened in South America ... : Poughkeepsie, NY, P. Potter (printer), 24 p.

Gardner, Daniel Pereira (d. 1853)
4212   1846, The chemical principles of the rotation of crops: Am. Agri. Assoc. Trans., v. 1, 13-29.
       Also reprint edition.

Garolin (i.e. Johan Gadolin?)
4213   1822, A new mineral substance: Am. J. Sci., v. 4, no. 2, 377 (7 lines).
       Abstracted by J. Griscom.

Garratt, William
4214   1828, Artificial emery: Am. J. Improvements Use. Arts, v. 1, 450 (16 lines).

Garrison, Nelson A. (see 11127 with J. B. Zabriskie)

Gass, Patrick (1771-1870)
4215   1807, A journal of the voyage and travels of a corps of discovery, under the command of Capt. Lewis and Capt. Clarke of the Army of the United States ... : Pittsburgh, David M'Keehan, viii, 9-262 p., illus.
       Other editions: Phila., M. Carey, 262 p. (1810); Dayton, Ohio, Ellis, Claflin and Co., 238 p. (1847).

Gassiot, John Peter (1797-1877)
4216   1850, A fused diamond: Scien. Am., v. 6, no. 3, 18 (15 lines).

Gatewood, Bennett P. (Captain)
4217   1811, Another meteoric stone, possessing, apparently, new qualities: Med. Repos., v. 14, 178-179.

Gaudin, Marc Antoine Augustin (1804-1880)
4218   1841, Rock crystal spun: Am. Eclec., v. 1, no. 2, 393-394.
       Abstracted from "J. Pract. Chem."

Gaumhaner (Mr.)
4219   1822, Eruption of the volcano of Goonong-Api: Am. J. Sci., v. 4, no. 2, 375-376.
       Abstracted by J. Griscom from Annales de Chimie.

Gaylord, Reuben (Reverend)
4220   1847, Fall of meteoric stones in Iowa: Am. J. Sci., v. 4 (ns), no. 2, 288-289; also in: The Friend, v. 21, 50-51.

Gaylord, Willis (1792-1844)
4221   1836, Nature of plaster, and its mode of action: Genesee Farm. and Gardener's J., v. 6, 45-46.
4222   1840, Bone cavern [at Harrisburg, Pennsylvania]: Am. J. Sci., v. 39, no. 2, 399-400.
4223   1843, Figure of the earth: No. Light, v. 3, 65-66.
4224   1843, Foot prints in rocks: No. Light, v. 3, 56.
        On human footprints in limestone from St. Louis.
4225   1843, Geology as connected with argiculture: in NY Tribune Extra, "Useful works for the people", no. 2, 71-80.
4226   1845, Analysis of soils; and the difference in the several parts appropriated by different crops grown upon them: NY State Agri. Soc. Trans., v. 4, 61-75.
4227   1845, Soils: S. Planter, v. 5, 177-178.

Gay-Lussac, Joseph Louis (1778-1850)
4228   1818, and P. Pinel, [Baths of Mont d'Or, France]: J. Sci. Arts, v. 3, no. 1, 174-175.
        Abstracted from Royal Inst. France Proc.
4229   1824, Reflections on volcanos: Bos. J. Phil. Arts, v. 1, no. 5, 416-426; also in: Meth. Mag., v. 7, 25-27; 56-59; 98-100.
        Abstracted from London Philos. Mag. and Annales de Chimie.
4230   1837, Observations on the assay of silver in the moist way: Franklin Inst. J., v. 20 (ns), 62-64.
        Abstracted from Annales de Chimie.
4231   1837, On the decomposition of lime by heat: Franklin Inst. J., v. 20 (ns), 62-64.
        Abstracted from Annales de Chimie.

Gazlay, Sayrs (Reverend)
4232   1830, Notice of the osseous remains at Big Bone Lick, Kentucky: Am. J. Sci., v. 18, no. 1, 139-141.
4233   1830, Origin of bituminous coal: Am. J. Sci., v. 17, no. 2, 397-398.
4234   1833, Notices of fossil wood in Ohio: Am. J. Sci., v. 25, no. 1, 104-107.

Gebhard, John
4235   1835, On the geology and mineralogy of Schoharie, (N. Y.): Am. J. Sci., v. 28, no. 1, 172-177, fig.

Geddes, James (1763-1838)
4236   1826, Observations on the geological features of the south side of the Ontario Valley: Am. J. Sci., v. 11, no. 2, 213-218, 4 figs.
        In a letter to "F. Romeyn Beck" [i.e. T. R. Beck].
4237   1828, Observations on the geological features of the south side of the Ontario Valley: Alb. Inst. Trans., v. 1, 55-59, 4 figs.

Gemmellaro, Carlo (1787-1866)
4238   1825, An account of the eruption of Mount Etna, on the 27th May, 1819: Carolina J. Medicine Sci. Ag., v. 1, 344-348.
4239   1835, Origin of sulphur: Am. J. Sci., v. 28, no. 2, 293-294 (8 lines).
        In C. U. Shepard's review of A. Boue's notice of the Geol. Soc. France Trans. (see 9497)
4240   1836, Sopra i vulcani estinti del val di noto, del Professore Carlo Gemmellaro [notice of]: Am. J. Sci., v. 30, no. 2, 382 (6 lines).

The General Repository and Review (Cambridge, 1812-1813)
4241   1812, Geological and mineralogical papers: v. 2, no. 2, 327-347.
        Reviews with excerpts of papers by S. Godon (see 4422 and 4423), B. H. Latrobe (see 6067), W. Maclure (see 6614 and 6615), and B. Silliman (see 9605).
4242   1813, [Review of A view of the theories which have been proposed to explain the origin of meteoric stones by Jeremiah Day (see 2982)]: v. 3, no. 1, 140-164.

The Genesee Farmer and Gardener's Journal (Rochester, New York, 1831-1839)
4243   1831, Geology: v. 1, 7.
        On Vermont Governor Craft's support of geological survey.
4244   1831-1838, [Brief notes on the nature and improvement of soils]: v. 1, 26-27; 38; v. 4, 398; v. 5, 129-130; 274; v. 6, 146-147; 342-343; 364-365; 404-407; v. 7, 73; 209; 316-317; 403; v. 8, 158; 302;
4245   1831, Geology. The Valley of the Genesee: v. 1, 52-53; 60; 77-78.
4246   1831, Zinc: v. 1, 297-298.
4247   1832, Agricultural geographical geology: v. 2, 17.
        Notice of research by J. H. Steele (see 10002).
4248   1832, Ball's Cave [in Schoharie, New York]: v. 2, 39-40.
4249   1832, Geology of Mass. [notice of E. Hitchcock's surveys]: v. 2, 155.
4250   1832, Eaton's geological text book [review with excerpts of 3361]: v. 2, 265-267.
4251   1833, Oil from the earth [at Marietta, Ohio]: v. 3, 216.
4252   1833, Extraordinary earthquake [at St. Leon, central United States]: v. 3, 296.
4253   1833, Geological course of cholera: v. 3, 312.
4254   1833, Speculations on the Asiatic cholera: v. 3, 340-341.
        Signed by "----K." On the geological course of cholera.
4255   1834, Awful earthquake in S. America. The city of Pasto destroyed: v. 4, 152.
4256   1834, Coal vs. gold: v. 4, 168.
4257   1834, American mammoth: v. 4, 200.
4258   1834, [Review of Essay on calcareous manures by E. Ruffin (see 9088)]: v. 4, 385-386; v. 6, 145-146; 163.
4259   1835, The primitive earths: v. 5, 23 (11 lines).
4260   1835, Geology in schools: v. 5, 34-35.
4261   1835, Geology: v. 5, 155.
        Review of B. Silliman's lectures.
4262   1835, Prof. Silliman on geology: v. 5, 256.

Genesee Farmer (cont.)
4263 1835, Land rising above the level of the sea: v. 5, 265-266.
       Notice of C. Lyell's work.
4264 1835, Ball's Cave, Schoharie [New York]: v. 5, 304.
4265 1835, Geological changes in Great Britain: v. 5, 383.
4266 1836, Geological survey of the state [of New York]: v. 6, 49.
4267 1836, Geological: v. 6, 216 (8 lines).
       Notice of the Connecticut geological survey.
4268 1836, [Review of On the use of lime as a manure by H. Puvis (see 8612)]: v. 6, 241; 249-250; 258.
4269 1836, The geological survey of the state [of New York]: v. 6, 256.
4270 1836, Central heat of the earth: v. 6, 468.
4271 1836, Plumbago or black lead: v. 6, 377.
4272 1837, Gold mines of Virginia: v. 7, 10.
4273 1837, Mineralogical and geological school: v. 7, 31.
       Development of a school in Pennsylvania proposed.
4274 1837, Alum mine discovered in New York [in Chatauqua County]: v. 7, 45 (5 lines).
4275 1837, Rock salt [from Louisiana Territory]: v. 7, 66.
4276 1837, Artificial crystals: v. 7, 105.
       On experiments by Mr. Cross in Somerset.
4277 1837, Report of the state geologists [of New York]: v. 7, 113-114.
4278 1837, South American salt works [at New Granada, Zipaguera]: v. 7, 153-154.
4279 1837, Geological survey of Ohio: v. 7, 218.
4280 1837, Polished rocks [from Rochester, New York]: v. 7, 250.
       Signed by "C. D." [i.e. Chester Dewey].
4281 1837, Burning springs [in Allegheny Mountains, Pennsylvania]: v. 7, 330-331.
4282 1837, [Mammoth bones from Jackson County, Ohio]: v. 7, 355 (9 lines).
4283 1837, [Geological survey of Maryland]: v. 7, 360 (8 lines).
4284 1838, [Review of Pennsylvania geology report, by H. D. Rogers (see 8855)]: v. 8, 155-156.
4285 1838, The geological reports [review of New York report (see 7737)]: v. 8, 170; 180; 194-195; 212; 219-220; 226.
       Signed by "A." With extensive excerpts.
4286 1838, Notes on western New York - geological structure: v. 8, 324; 332; 340; 356; 372.
       Signed by "Viator."
4287 1839, Geological survey of the state [of New York]: v. 9, 137.
4288 1839, The geological reports [review with excerpts of New York report (see 7740)]: v. 9, 171; 211; 219; 242-243; 259; 281-282; 402.
4289 1839, [Review of the third Maine report by C. T. Jackson (see 5566)]: v. 9, 314-315.

The Genesee Farmer and Gardener's Journal (Rochester, New York, 1840-1850)
4290 1841-1849, [Brief notes on the nature and improvement of soils]: v. 2, 6 [i.e. 63]; v. 3, 9; 88-89; v. 7, 57; v. 9, 68-69; v. 10, 154-155; 157, fig.
4291 1841, [Notice with excerpts of third Pennsylvania report, by H. D. Rogers (see 8860 and 8861)]: v. 2, 71.
4292 1841, Michigan coal [in Corunna, Shiawasse County]: v. 2, 71.
4293 1841, Geology of North Sherbrooke, U. C.: v. 2, 176.
       Notice of work by E. Wilson.
4294 1845, Greatest iron mines in the world [at Newcomb, Essex County, New York]: v. 6, 123.
4295 1846, Agricultural geology of Onondaga County [notice of article by E. Emmons (see 3539)]: v. 7, 130-131.
4296 1849, [Review of Scientific agriculture, by M. M. Rodgers (see 8829)]: v. 10, 21, 5 figs.
4297 1849, Extensive coal field [in the Mississippi River valley]: v. 10, 167 (6 lines).

Genet, Edmond Charles (1763-1834)
4298 1818, Memorial on the alluvions or obstructions at the head of the navigation of the River Hudson: Alb., I. W. Clark, 46 p.
4299 1825, Address on the several subjects of science, useful knowledge ... in reference to rivers, canals, navigation, and commerce ... : Alb., Packard and Van Benthuysen, 43 p.

Genth, Friedrich August Ludwig Karl Wilhelm (1820-1893)
4300 1849, On the products of eruption at Hecla: Am. J. Sci., v. 7 (ns), no. 1, 114.

The Gentleman and Lady's Town and Country Magazine (Boston, 1784)
4301 1784, Description of the Island of Elephanta, in the East-Indies: v. 1, no. 5, 188-189.

Geological diagram of the field notes of the surveys of township and subdivision lines in the Northern Peninsula of Michigan, in the years 1844 and 1845:
4302 1846, Wash., C. B. Graham's lith., geological map, 85 x 64 cm.
       After W. A. Burt (see 2208) and B. Hubbard (see 5337). In report of J. H. Relfe (see 8739).

Geological Society of France
4303 1833, Geological Society of France: Am. J. Sci., v. 24, no. 1, 192-193.
       Circular listing officers and notices.
4304 1846-1847, The Geological Society of France: Am. J. Sci., v. 1 (ns), no. 1, 146 (4 lines); v. 2 (ns), no. 3, 440 (6 lines); v. 3 (ns), no. 2, 271-272; v. 4 (ns), no. 3, 449 (11 lines).
       On activities of 1846 and 1847. Abstracted from l'Institut.

Geological Society of London
4305 1810, Geological inquiries: Am. Min. J., v. 1, no. 1, 43-52.
       List of questions to be answered in preparing a survey. Also in: Portico, v. 1, 384-385 (1816).
4306 1820, Geological Society of London: Am. J. Sci., v. 2, no. 2, 353-355.
       The Society's proceedings, abstracted by J. Griscom.

Geological Society of London (cont.)
4307 1831, Biennial election of president of the Geological Society of London: Mon. Am. J. Geology, v. 1, no. 1, 45-46 (6 lines).
Notice of the election of R. I. Murchison.
4308 1836, The Geological Society of London: Am. J. Sci., v. 30, no. 2, 382 (4 lines).
Notice of Wollaston Medal given to L. Agassiz.
4309 1837, Geological Society - April 19: Am. J. Sci., v. 33, no. 1, 208-211.
Abstracted from London Ath. On their meeting proceedings.
4310 1846, Wollaston Medal: Am. J. Sci., v. 2 (ns), no. 3, 440 (4 lines).
Awarded to W. Lonsdale.
4311 1849, The anniversary of the Geological Society for 1849, held in London: Am. J. Sci., v. 7 (ns), no. 3, 451.

Geological Society of Pennsylvania
4312 1834, The constitution and bye-laws of the Geological Society of Pennsylvania to which is added a list of the officers and members of the Society: Phila., J. and W. Kite (printers), 24 p.
4313 1834, [Officers, members, and activities of the Society]: Geol. Soc. Penn. Trans., v. 1, many notices.
4314 1834/1835, Transactions of the Geological Society of Pennsylvania: v. 1, pt. 1, Phila., W. P. Gibbons, 180, 13 p., 6 colored plates, 1834; v. 1, pt. 2, Phila., James Lay, jun. and Co., x, 177-428 p., 20 colored plates.
For review see 650 and 3722. For notice see 598.

Geological Society of Pennsylvania. Transactions of the (Philadelphia, 1834-1835)
4315 1834, [Dr. G. Troost, state geologist of Tennessee]: v. 1, 168-169.
4316 1834, [Mr. T. G. Clemson's activities in 1834]: v. 1, 171.
4317 1834, [Progress of the Maryland geological survey]: v. 1, 171 (9 lines).
4318 1834, [Progress of the Massachusetts geological survey]: v. 1, 172.
4319 1834, [Survey of Schuylkill County, Pennsylvania]: v. 1, 172-173.
4320 1834, [Mercury as an amalgam]: v. 1, 173 (6 lines).
4321 1834, [Dr. Douglass Houghton]: v. 1, 173 (9 lines).
4322 1834, [New Jersey geological survey]: v. 1, 173 (3 lines).
4323 1834, [Geological surveys of the Carolinas]: v. 1, 173-174 (7 lines).
4324 1834, [Recent geological investigations]: v. 1, 174-175.
4325 1834, [Mr. Taylor's memoir on the fossil (see 5538) fucoides]: v. 1, 175.
Notice of the donation of R. C. Taylor's fossil fucoid collection to the Society. See also 10114.
4326 1835, Geological and topographical survey of the state of Maryland: v. 1, 412-413.
4327 1835, Mr. Featherstonhaugh's geological report: v. 1, 413.
Review of 3778.
4328 1835, Mr. Lyell's geology [notice of 6532]: v. 1, 416-417.
4329 1835, [Diamond found in Carolina]: v. 1, 417 (6 lines).

Georges Creek Coal and Iron Company
4330 1836, An act to incorporate the George's Creek Coal Mining Company: n.p., 34 p., 6 folding plates.
Includes report by J. H. Alexander and P. T. Tyson (see 126).
4331 1836, Charter, &c., of the Georges Creek Coal and Iron Company, containing a detailed account of the geology, &c., of this locality: Balt.?, 36 p., 6 plates.

Georgia. State of
4332 1836, [Resolution supporting a state geological survey]: Georgia Senate, December 7, 1836.
4333 1840, [Resolution abolishing the Georgia geological survey]: Georgia House of Representatives, November 27, 1840.
4334 1840, [Resolution re-instating the Georgia geological survey]: Georgia Senate, December 18, 1840.

Gerathewohl, M. (see 3597 with Erdmand)

Gere
4335 1850, Salt in New York: Ann. Scien. Disc., v. 1, 93 (9 lines).

Gerhardt, Charles Frederic (1816-1856)
4336 1847, On the atomic volume of some isomorphous oxyds of the regular system: Am. J. Sci., v. 4 (ns), no. 3, 405-408.
On isomorphism in spinels.
4337 1849, On epidote: Am. J. Sci., v. 8 (ns), no. 1, 123-124.
Abstracted from "Jour. de Pharm. et de Chim."

Germain, L. J.
4338 1849, On the forces in nature which rupture, contort, upheave, and depress the superficial strata of the earth: Am. Assoc. Adv. Sci. Proc., v. 1, 40-41.

Gerry, James T. (lieutenant, United States Navy)
4339 1842, Tubular concretions of iron and sand from Florida: Am. J. Sci., v. 42, no. 1, 207-209.
In article by B. Silliman (see 9771).

Gervais, Paul (1816-1879)
4340 1850, Fossil ape: Ann. Scien. Disc., v. 1, 284-285 (12 lines).
4341 1850, Fossil elephant and mastodon from Africa: Ann. Scien. Disc., v. 1, 287 (15 lines).

Gesner, Abraham (1797-1864)
4342 1841, [Excerpts from "Remarks on the mineralogy and geology of Nova Scotia"]: Franklin Inst. J., v. 1 (3s), 206-210.
4343 1844, Coal formation of Nova Scotia: Franklin Inst. J., v. 7 (3s), 399-402.

Gettysburg College Linnaean Association (see Literary Record and Journal of the Linnaean Association of Pennsylvania College)

Gibbes, Lewis Reeve (1810-1894)
4344 1850, [Experiment to determine the density of the earth] : Am. Assoc. Adv. Sci. Proc., v. 3, 57.

Gibbes, Robert Wilson (1809-1866)
       See also discussion of articles by G. J. Chase (2327) and J. C. Warren (10701).
4345 1845, Description of the teeth of a new fossil animal found in the green sand of South Carolina: Phila. Acad. Nat. Sci's Proc., v. 2, no. 9, 254-256, plate with 5 figs.
       Also reprint edition, 3 p. On the Zeuglodon, or Dorudon.
4346 1845, Dorudon [from South Carolina] : Am. J. Sci., v. 49, no. 1, 216 (7 lines).
4347 1845, Gypsum in South Carolina: Mon. J. Ag., v. 1, no. 1, 59 (5 lines).
4348 1846, On the fossil Squalidae of the United States: Phila. Acad. Nat. Sci's Proc., v. 3, no. 2, 41-43.
4349 1847, Description of new species of squalides from the Tertiary beds of South Carolina: Phila. Acad. Nat. Sci's Proc., v. 3, no. 11, 266-268.
4350 1847, On the fossil genus Basilosaurus, Harlan, (Zeuglodon, Owen,) with a notice of species from the Eocene green sand of South Carolina: Phila. Acad. Nat. Sci's J., v. 1 (ns), 5-15, plate.
       Also reprint edition (see 4351).
4351 1847, Memoir on the fossil genus Basilosaurus, with a notice of species from the Eocene green sand of South Carolina: Phila., Merrihew and Thompson (printers), 13 p., 5 plates.
       For review see 2319 and 3565.
4352 1848, Monograph of the fossil Squalidae of the United States: Phila. Acad. Nat. Sci's J., v. 1 (ns), no. 2, 139-147, 4 plates; no. 3, 191-206, 3 plates.
       Also reprint edition. For review see 2320 and 9962.
4353 1848, On the fossil genus Basilosaurus, Harlan, (Zeuglodon, Owen,) with a notice of specimens from the Eocene green sand of South Carolina: Am. J. Sci., v. 5 (ns), no. 2, 303 (11 lines).
       Abstracted from Phila. Acad. Nat. Sci's J.
4354 1848, [On the validity of the name Dorudon] : Phila. Acad. Nat. Sci's Proc., v. 4, no. 3, 57.
4355 1849, [Fossil fish from Western United States] : Am. Assoc. Adv. Sci. Proc., v. 1, 71 (8 lines).
4356 1849, Fossil Squalidae of the United States: Am. J. Sci., v. 7 (ns), no. 3, 441.
       Abstracted from Phila. Acad. Nat. Sci's J.
4357 1849, The present world the remains of a former world; a lecture delivered before the South Carolina Institute, Sept. 6, 1849: Columbia, SC, A. S. Johnston, 31 p.
4358 1850, Fossils common to several formations: Am. Assoc. Adv. Sci. Proc., v. 3, 70-71.
       With discussion by L. Agassiz and M. Tuomey. On fossil Squalidae.
4359 1850, A memoir on the Mosasaurus and three allied new genera, Holcodus, Conosaurus, and Amphrosteus: Wash., Smithsonian Inst., Contributions to Know., v. 2, article 5.
4360 1850, New species of fossil Myliobates, from the Eocene of South Carolina, and new fossils from the Cretaceous, Eocene, and Pliocene of South Carolina, Alabama, and Mississippi: Am. Assoc. Adv. Sci. Proc., v. 2, 193-194.
       With discussion by L. Agassiz.
4361 1850, New species of Myliobates from the Eocene of South Carolina, with other genera not heretofore observed in the United States: Phila. Acad. Nat. Sci's J., v. 1 (ns), no. 4, 299-300, plate.
4362 1850, On Mosasaurus and other allied genera in the United States: Am. Assoc. Adv. Sci. Proc., v. 2, 77.
4363 1850, On the genera of Mosasaurus: Ann. Scien. Disc., v. 1, 288 (10 lines).
4364 1850, Remarks on the Mastodon angustidens: Am. Assoc. Adv. Sci. Proc., v. 3, 69-70.
4365 1850, Remarks on the fossil Equus: Am. Assoc. Adv. Sci. Proc., v. 3, 66-68.
4366 1850, Remarks on the northern Elephas of Prof. Agassiz: Am. Assoc. Adv. Sci. Proc., v. 3, 69.

Gibbon, J. H.
4367 1837, A visit to the salt works of Zipaquera, near Bogota, in New Granada: Am. J. Sci., v. 32, no. 1, 89-95.
4368 1845, Gold of North Carolina: Am. J. Sci., v. 48, no. 2, 398-399.
4369 1850, Meteorite in North Carolina: Am. J. Sci., v. 9 (ns), 143-146. Also in: Fam. Vis., v. 1, no. 10, 74; and Ann. Scien. Disc., v. 1, 275-276.

Gibbons, Henry (1808-1884)
4370 1834, A sketch of the history of geology: Advocate of Sci. and Annals Nat. History, v. 1, no. 3, 112-120.
4371 1834/1835, Geology: Advocate of Sci. and Annals Nat. History, v. 1, no. 5, 201-207, 2 figs.; no. 6, 261-266; no. 7, 297-297 [i.e. 307]; no. 8, 360-367; no. 9, 401-406.
4372 1835, Geology: Scien. Tracts and Fam. Lyc., v. 4 (ns), 309-313.
4373 1836, Geology: Genesee Farm. and Gardener's J., v. 6, 246.

Gibbs, George (1776-1833)
4374 1810, Mineralogical notice respecting the West River Mountain, Connecticut River: Am. Min. J., v. 1, no. 1, 19-20.
4375 1810, Observations on the Franconia Iron Works: Am. Min. J., v. 1, no. 1, 5-7.
4376 1811, On the iron works at Vergennes, Vermont: Am. Min. J., v. 1, no. 2, 81 [i.e. 80]-83.
4377 1812, Crystallized bodies discovered in meteoric stone. Extract from a letter from Col. Gibbs, to the editor: Am. Min. J., v. 1, no. 3, 190 (17 lines).
4378 1814, Observations on the mass of iron from Louisiana: Am. Min. J., v. 1, no. 4, 218-221.

Gibbs, G. (cont.)
4379   1814, Proposed classification of some of the ores of iron: Am. Min. J., v. 1, no. 4, 268.
4380   1817, Circular: Am. Mon. Mag. and Crit. Rev., v. 1, no. 1, 125.
        On the mineral collection of the NY Hist. Soc.
4381   1817, Report on mineralogy [for the New York Historical Society]: Am. Mon. Mag. and Crit. Rev., v. 1, no. 1, 48.
4382   1819, On the tourmalines and other minerals found at Chesterfield and Goshen, Massachusetts: Am. J. Sci., v. 1, no. 4, 346-351.

Gibbs, William P.
4383   1849, Field notes [on the Lake Superior region]: in Lake Superior report by C. T. Jackson (see 5655), 702-711.

Gibbs, Wolcott (1822-1908)
4384   1845, An inaugural dissertation on a natural system of chemical classification: Princeton, NJ, J. T. Robinson, 59 p.
        For review see 821.

Gibson, John Bannister (1780-1853)
4385   1825, Observations on the trap rocks of the Connewago Hill near Middletown, Dauphin County, and of the stony ridge near Carlisle, Cumberland County, Pennsylvania: Am. Philos. Soc. Trans., v. 2 (ns), 156-166.
4386   1832, and G. Fox, Geological Society of Pennsylvania. Circular: Hazard's Reg. Penn., v. 10, 306-307.
4387   1833, and R. Harlan and H. S. Tanner, [Letter supporting a geological survey of Pennsylvania]: Hazard's Reg. Penn., v. 11, 226-228.
        See also 8207.
4388   1836, Remarks on the geology of the lakes and the valley of the Mississippi, suggested by an excursion to the Niagara and Detroit Rivers, in July, 1833: Am. J. Sci., v. 29, no. 2, 201-213.

Gillespie, M. (see 174 with M. Allis)

Gillespie, William Mitchell (1816-1868)
4389   1842, Magnetic variation: Am. Rr. J., v. 14, 52-53.

Gillet, George
4390   1832, Declination of the magnetic needle: Am. J. Sci., v. 23, no. 1, 205-206.

Gilliam, Albert M. (d. 1859)
4391   1846, Travels over the table lands and Cordilleras of Mexico, during the years 1843 and 1844; including a description of California, the principal cities and mining districts of that republic, and biographies of Iturbide and Santa Anna: Phila., J. M. Moore, xv, 455 p., map, plates.
        For review see 5453.

Gilliss, James Melville (1811-1865)
4392   1846, Magnetical and meteorological observations, made at the Naval Observatory, Washington, under the orders of the honorable Secretary of the Navy, dated August 13th, 1838: Wash., Gales and Seaton, xxv, 671 p., folded plate.
4393   1846, Magnetical and meteorological observations made at Washington: Am. J. Sci., v. 1 (ns), no. 1, 143-145.
4394   1846, Report on the erection of a depot of charts and instruments at Washington City: Am. J. Sci., v. 1 (ns), no. 2, 294-297.

Gilmer, Francis William
4395   1818, On the geological formation of the Natural Bridge of Virginia: Am. Philos. Soc. Trans., v. 1 (ns), 187-192.
        Also reprint edition.

Gilmor, Robert, jr. (1774-1848)
4396   1814, A descriptive catalogue of minerals occurring in the vicinity of Baltimore, arranged according to the distribution methodique of Hauy: Am. Min. J., v. 1, no. 4, 221-233.
4397   1821, Microscopic crystals of iron pyrites: Am. J. Sci., v. 3, no. 2, 233.

Gilpin, Thomas (1776-1853)
4398   1843, An essay on organic remains, as connected with an ancient tropical region of the earth: Phila., E. H. Butler, 39 p.
4399   1843, Essay on the position of the organic remains, as connected with a former tropical region of the earth: Am. Philos. Soc. Proc., v. 4, no. 28, 27-29.

Gipperich, Frederick
4400   1847, Report on the gold mines of the Philadelphia and North Carolina Mining and Smelting Company: Phila., J. H. Schwacke, 17 p., map.

Girdwood, Jonathan
4401   1845, Rotation of crops: The Cultivator, v. 2 (ns), 172.
        Table of soil analyses.

Giroud (Mr.)
4402   1802, Gold and platina found among the mountains of St. Domingo: Am. Rev. and Lit. J., v. 2, 104-106; also in: Med. Repos., v. 5, 340-342.

Glibbon, J. H. (see Gibbon, J. H.)

The Globe (New York, 1819)
4403   1819, Valuable discovery: v. 1, 19.
        On the discovery of a cobalt mine in Chatham County, Connecticut.

Glocker, Ernst Friedrich (1793-1858)
4404   1846, Turquoise in Silesia: Am. J. Sci., v. 1 (ns), no. 2, 266-267 (5 lines).
        Abstracted from Poggendorff's Annalen.
4405   1848, Smaelite [from Tekibang, Hungary]: Am. J. Sci., v. 5 (ns), no. 2, 268 (6 lines).
        Abstracted from "Jour. für Prakt. Chem."

Glooker, (i.e. Glocker, E. F.?)
4406   1834, Ozokerite, a new combustible mineral [from Slauik, Moldavia]: Am. J. Sci., v. 26, no. 2, 388-389 (12 lines).
        Translated by L. Feuchtwanger from Berzelius' Jahresbericht.

Gmelin, Christian Gotlob (1792-1860)
4407   1822, Lava [from Tubingen]: Am. J. Sci., v. 4, no. 2, 271 (6 lines).
        Abstracted by J. Griscom.
4408   1825, Analyses of several minerals: Am. J. Sci., v. 9, no. 2, 329-330.
4409   1828-1829, Analysis of tourmaline: Am. J. Sci., v. 14, no. 2, 384; v. 15, no. 2, 389-390.
        Abstracted from Annales de Chimie.
4410   1828, The Dead Sea: Am. J. Sci., v. 13, no. 2, 395.
        Abstracted from Naturwissenschaftliche Abhandlungen.
4411   1831, Geology: Am. J. Sci., v. 19, no. 2, 380.
        Abstracted from Edinburgh J. Sci. by J. Griscom.

Goddard, Paul Beck (1811-1866)
4412   1841, New mastodon fossil [Missourium kochii]: Med. Exam., v. 4, 731.
        Abstracted from Phila. Acad. Nat. Sci's Proc.
4413   1841, The so-called "Missourium kochii": Phila. Acad. Nat. Sci's Proc., v. 1, no. 7, 115-116.
4414   1842, and R. C. Taylor and H. D. Rogers, The supposed human footprints found near St. Louis, in the Carboniferous limestone: Phila. Acad. Nat. Sci's Proc., v. 1, no. 21, 225-226.

Godding, Miss D. W.
4415   1847, First lessons in geology; comprising its most important and interesting facts, simplified to the understanding of children: Hart., H. S. Parsons, 142 p.

Godman, John Davidson (1794-1830)
4416   1824, Description of the os hyeides of the mastodon: Phila. Acad. Nat. Sci's J., v. 4, pt. 1, 67-72, plate with 3 figs.
4417   1826/1828, American natural history ... pt. 1, Mastology: v. 1 and 2, Phila., H. C. Carey and I. Lea, xvi, 17-362 p., plates; v. 3, Phila., Carey, Lea and Carey, 331 p., plates.
        First edition. contains section by W. Cooper (see 2632).
        Volume 3 "Appendix" contains notes on fossil mammals.
        Also "second edtion", Phila., Key and Mielkie, 3v (1831); and "third edition", Phila., Hogan and Thompson, 2v (1836). Other printings include: Phila., R. W. Pomeroy, 2v., 1842; Phila., U. Hunt and Sons, 2v. in 1, 1846.
4418   1830, Description of a new genus and new species of extinct mammiferous quadurped [Tetraculodon mastodontordeum]: Am. Philos. soc. Trans., v. 3 (ns), 478-485, 2 plates.
        Also reprint edition.
4419   1830, also I. Hays, New fossil animals [from Orange County, New York]: Va. Lit. Mus., v. 1, no. 49, 772-774.
        Abstracted from Am. Philos. Soc. Trans.

Godon, Sylvain (1774?-1840)
4420   1807, Mineralogy: Mon. Anth., v. 4, 658-660.
        Notice of Godon's introductory lectures.
4421   1808, On the manganese found in Vermont: Mon. Anth., v. 5, 344.
4422   1809, Mineralogical observations made in the environs of Boston, in the years 1807 and 1808: Am. Acad. Arts Sci's Mem., v. 3, 127-154, table.
        Also reprint edition, 28 p. For review see 4241.
4423   1809, Observations to serve for the mineralogical map of the state of Maryland: Am. Philos. Soc. Trans., v. 6, pt. 2, 319-323.
        For review see 4241 and 7090.
4424   1810, Elementary treatise on mineralogy. Prospectus of a treatise on mineralogy: Adopted to the present state of science; including important applications to the arts and manufactures. By S. Godon. To be published by Birch and Small, Philadelphia: Phila., Birch and Small, Broadside, April, 1810.
        For notice of the Treatise see also 1079, 4427, 7804. No proof that this work was actually sold has been found.
4425   1810, Mineralogical note respecting phosphated lime, and phosphated lead, from Pennsylvania [Germantown]: Am. Min. J., v. 1, no. 1, 30.
4426   1811, Chromic yellow [from Baltimore, Maryland]: Am. Min. J., v. 1, no. 2, 125 (9 lines).
4427   1811, New elementary work on mineralogy [notice of Treatise of mineralogy (see 4424)]: Am. Min. J., v. 1, no. 2, 126 (5 lines).

Goeppert (see Göppert)

Gold, T. S.
4428   1838, New locality of crichtonite: Am. J. Sci., v. 35, no. 1, 179 (14 lines).

Goldfuss, Georg August (1782-1848)
4429   1832, Pterodactylus crassirostris [from Daiting, Bavaria]: Mon. Am. J. Geology, v. 1, no. 7, 335 (18 lines).

Goldsmith, Oliver (1728-1774)
4430   1795, An history of the earth, and animated nature: Phila., Mathew Carey, 4 v., 55 plates.
        Also at least 19 later edtions and printings, many as Goldsmith's natural history. Includes discussion of the origin of fossils.
4431   1805, Theory of the earth: Lit. Tablet, v. 3, 7.
        Brief summary of Burnet's theory.
4432   1819, The Grotto of Antiparos: Meth. Mag., v. 2, 24-26.

Good, John Mason (1764-1827)
4433   1826, The book of nature: Bos., Wells and Lilly, 2 v.
        Also many other editions, including: NY, E. Duyckinck, 530 p., 1827; NY, J. and J. Harper, vii, 530 p., 1827, 1828, 1830, and 1831; Hart., Belknap, 467 p., 1829; Hart., Belknap and Hamersley, xx, 25-467 p., "To which is now prefixed, a sketch of the author's life", 1840.
        Chapters 6 and 7 are on geology.
4434   1830, Alternate advances and recessions of the sea: Young Ladies' J. Liter. Sci., v. 1, 58-65.
4435   1830, Formation of coral islands: Young Ladies J. Liter. Sci., v. 1, 24-27.
        From Tour through creation and science.
4436   1830, Quicksilver [from Idria, Spain]: Young Ladies J. Liter. Sci., v. 1, 432-434.
        From Tour through creation and science.

Goode, Thomas (1789-1858)
4437   1839, The invalid's guide to the Virginia Hot Springs: Richmond, Va., P. D. Bernard, 44 p; also 1846 (95 p.) and 1848 (106 p.) editions.

Goodrich, Joseph
4438   1826, Notice of the volcanic character of the Island of Hawaii, in a letter to the editor, and of various facts connected with late observations of the Christian missionaries in that country, abstracted from a journal of a tour around Hawaii, the largest of the Sandwich Islands: Am. J. Sci., v. 11, no. 1, 1-36.
        With introduction by B. Silliman (see 9668), and excerpts from Journal of a tour ..., by W. Ellis (see 3482).
4439   1829, Letters from the Sandwich Islands: Am. J. Sci., v. 16, no. 2, 345-347.
        With notes by B. Silliman (9685).
4440   1831, Notices: Am. J. Sci., v. 20, no. 2, 228-229.
        In "Hawaii ... " by C. Stewart (see 10019).
4441   1833, Notices of some of the volcanoes and volcanic phenomena of Hawaii, (Owyhee,) and other islands in that group: Am. J. Sci., v. 25, no. 1, 199-203.

Goodrich, Samuel Griswold (1793-1860)
4442   1832, The child's geology: Brattleboro, Vt., G. H. Peck and Co., 132 p., front., map, illus.
4443   1842, A pictorial natural history: embracing a view of the mineral, vegetable and animal kingdoms, for the use of schools: Bos., J. Munroe and Co., 415 p., illus.
4444   1844, A glance at the physical sciences; or, the wonders of nature, in earth, air, and sky: Bos., Bradbury, Sodon and Co., 352 p., front., illus.
        Also several other printings and editions including: Bos., J. E. Hickman, 352 p., 1844; NY, J. Allen, 352 p., 1844; Phila., Thomas, Cowperthwait and Co., 352 p., 1844; Bos., Peirce and Rand, 352 p., 1848; Bos., 1849.
4445   1844, The wonders of geology, by the author of Peter Parley's tales: Bos., J. E. Hickman, v, 6-291 p., front.
        Also several other printings including: NY, 291 p., 1844; Phila., 291 p., 1844; Bos., Bradbury, Sodon and Co., 291 p., 1845; Phila., Thomas, Cowperthwait and Co., 291 p., 1846; Bos., C. H. Peirce and G. C. Rand, 291 p., 1848; Bos., Rand and Mann, 291 p., 1849.

Goodsell, N.
4446   1845, Marl: Genesee Farm. and Gardener's J., v. 6, 68.

Gookin, Nathaniel (1687-1735)
4447   1728, The day of trouble near, the tokens of it, and a due preparation for it ... And an appendix, giving some account of the earthquake as it was in Hampton: Bos., D. Henchman, 75 p.
        Evans number 3033.
4448   1834, An appendix, giving some account of the earthquake as it was at Hampton. To which is added something remarkable of thunder and lightning in the same town in the year 1727: NH Hist. Soc. Trans., v. 4, 92-97.
        From The day of trouble near ... , see 4447.

Göppert, Johann Heinrich Robert (1800-1884)
4449   1838, Origin of amber: Franklin Inst. J., v. 21 (ns), 70 (12 lines).
        Abstracted from Edinburgh New Philos. J.
4450   1847, [Distribution of fossil plants by species per period]: Am. Q. J. Ag. Sci., v. 5, 167 (12 lines).
4451   1847, On the origin of the coal of Silesia: Franklin Inst. J., v. 13 (3s), 215-216.
        Abstracted from London Ath.

Gordon, Alexander
4452   1833, Analysis of soils: Am. Farm., v. 14, 348-349.

Gordon, Thomas Francis (1787-1860)
4453   1832, A gazetteer of the state of Pennsylvania. Part first ... general description of the state ... geological construction, canals and rail-roads, bridges, revenue, expenditures, public debt &c. Part second ... counties, towns, cities, villages, mountains, lakes, rivers, creeks, &c.: Phila., T. Belknap, 63, 508 p., map.
4454   1834, A gazetteer of the state of New Jersey, comprehending a general view of its physical and moral condition together with a topographical and statistical account of its counties, towns, villages, canals, railroads, etc., accompanied by a map: Trenton, NJ, Daniel Felton, 266 p., map.

Gordon, T. F. (cont.)
4455   1836, Gazetteer of the state of New York: comprehending its colonial history, general geography, geology, and internal improvements, its political state, a minute description of its several counties, towns, and villages ... : Phila., Printed for the author, xii, 102, 801 p., maps.
4456   1836, [Extracts from Gazetteer of New Jersey ...] : in Report on the geological survey of the state by H. D. Rogers (see 8850 to 8852), Freehold, NJ edition, 107-109; Phila. edition, 118-120.

Gorham, John (1783-1829)
4457   1814, Analysis of sulphate of barytes, from Hatfield, Massachusetts: Am. Acad. Arts Sci's Mem., v. 3, pt. 2, 237-240.

Gould, Augustus Addison (1805-1866)
    See also 75 with L. Agassiz.
4458   1837, Report of special committee on the geological survey of the state proposing an extension of Hitchcock's first survey, with special attention to the minerals of the state: Mass. House Document 26, 1837, 16 p.
4459   1839, Proceedings of the Boston Society of Natural History, from September 19th, 1838, to March 21st, 1839: Am. J. Sci., v. 37, no. 2, 391-397.
4460   1839, Scientific proceedings of the Boston Society of Natural History in the months of June, July, and August, 1838; drawn up from the records of the Society: Am. J. Sci., v. 36, no. 2, 379-393.
4461   1843, On corallines: Bos. Soc. Nat. History Proc., v. 1, 16.
    On the nature and origin of corals.
4462   1844, [Limestone from Lake Erie]: Bos. Soc. Nat. History Proc., v. 1, 212-213 (8 lines).
4463   1848, [Fossil fish from Maine]: Bos. Soc. Nat. History Proc., v. 3, 64 (5 lines).

Gould, Hannah Flagg (1789-1865)
4464   1846, The mastodon: Sartain's Union Mag., v. 1, 8.

Goulding, Joseph
4465   1833, An apparatus for separating iron ore from any extraneous matter: Franklin Inst. J., v. 11 (ns), 98-99.

Graham, Edward
4466   1838, Limestones of Rockingham County: Farm's Reg., v. 5, 669.

Graham, George
4467   1829, Inundated lands on the Mississippi: US HED 99, 20-2, v. 3 (186), 1-10.
    On fluvial features of the Mississippi River.
4468   1846, and J. G. Anthony and U. P. James, Two species of fossil Asterias in the blue limestone of Cincinnati: Am. J. Sci., v. 1 (ns), no. 3, 441-442, fig.

Graham, James Duncan (1799-1865)
4469   1843, Earthquake of Feb. 8, 1843 in Washington: Am. Philos. Soc. Proc., v. 2, no. 25, 259-260.
4470   1845-1846, Observations for the magnetic dip, made at several positions, chiefly on the south-western and north-eastern frontiers of the United States; and of the magnetic declination at two positions on the River Sabine, in 1840: Am. Philos. Soc. Proc., v. 4, no. 34, 205-207; also in Am. Philos. Soc. Trans., v. 9 (ns), 329-380.

Graham, James G.
4471   1801, Further account of the fossil bones in Orange and Ulster Counties: Med. Repos., v. 4, 213-214; also in: Mon. Mag. and Am. Rev., v. 3, 393-395.

Graham, Maria (later Mrs. Callcott; 1785-1842)
4472   1835, On the reality of the rise of the coast of Chili [sic], in 1822: Am. J. Sci., v. 28, no. 2, 236-247.
    With discussion by President Greenough (see 4562).

Graham, R. (see 6474 with S. H. Long and J. Philips)

Graham, Thomas (1805-1869)
4473   1846, Report on the composition of the fire-damp of the newcastle coal mines, and the means of preventing accidents from its explosion: Franklin Inst. J., v. 11 (3s), 64-66.
    From London Ath.

Graham's American Monthly Magazine (Philadelphia, 1826-1850)
4474   1826, The natural bridge; or a scene in Virginia: v. 1, 242-243.
4475   1826-1835, [Notes on earthquakes and volcanoes]: v. 1, 262; 352-353; v. 9, 42-43; v. 10, 286, fig.
4476   1827, Remarks upon the use of anthracite, and its application to the various purposes of domestic economy: v. 2, 59-61.
4477   1827, [Cavern at Soli, Java]: v. 2, 382.
4478   1828, The natural bridge [in Virgina]: v. 3, 361, fig.
4479   1828, Nature of peculiar diamonds: v. 3, 373.
    On the largest diamonds in the world.
4480   1835, Pennsylvania: v. 10, 1-15, colored map.
    Includes description of the coal region.
4481   1836, Sketch of Mexico and Texas: v. 11, 109-114, colored map.
4482   1837, Kentucky: v. 12, 193-194, colored map.
4483   1837, State of New York: v. 12, 289-291, colored map.
4484   1838, The state of Virginia: v. 13, 49-54, colored map.
4485   1843, [Review of Dictionary ..., by A. Ure (see 10532)]: v. 22, 260 (12 lines).
4486   1844, [Review of Slave states ..., by G. W. Featherstonhaugh (see 3784)]: v. 26, 192 (5 lines).
4487   1846, [Review of Voyage ..., by C. Darwin (see 2916)]: v. 28, 284 (11 lines).
4488   1849, [Review of Earth and man by A. Guyot (see 4609)]: v. 35, 131.

Graham's American Monthly Magazine (cont.)
4489   1849, [Review of A second visit ..., by C. Lyell (see 6571)]: v. 35, 251.
4490   1850, [Review of Cosmos by A. von Humboldt (see 5395)]: v. 36, 347.

Grammer, John, jr.
4491   1818, Account of the coal mines in the vicinity of Richmond, Virginia: Am. J. Sci., v. 1, no. 2, 125-130.

Granger, Ebenezer
4492   1821, Notice of vegetable impressions on the rocks connected with the coal formation of Zanesville, Ohio: Am. J. Sci., v. 3, no. 1, 5-7, 13 figs on 2 plates.
        With "Remarks by the editor" by B. Silliman (see 9641).
4493   1823, Notice of a curious fluted rock at Sandusky Bay, Ohio: Am. J. Sci., v. 6, no. 1, 179-180.

Grant, Henry Allen
4494   1844, A week among the glaciers: Am. J. Sci., v. 46, no. 2, 281-294. Also in: Eclec. Mag., v. 3, 109-113.
        From London Ath., on a field trip to the French Alps.

Granville, Augustus Bozzi (1783-1872)
4495   1817, A report on a memoir of Signor Monticelli, ..., entitled "Descrizione dell' Eruzione del Vesuvio avvennuta ne' giorni 25 e 26 Dicembre, 1813": J. Sci. Arts, v. 2, no. 1, 25-34.

Graves, R., jr.
4496   1832, A simplification of Dr. Wollaston's reflective goniometer: Am. J. Sci., v. 23, no. 1, 75-78, 5 figs.

Graves, Rufus
4497   1820, Account of a gelatinous meteor: Am. J. Sci., v. 2, no. 2, 335-337.

Gray, A. B.
4498   1845, [Mineral lands of Lake Superior region]: US SED 175, 28-2, v. 11 (461), 14-22.
        For notice see 915.
4499   1846, [Report] on mineral lands of Lake Superior: US HED 211, 29-1, v. 7 (486), 2-23, map.
        Includes map by G. Talcott (see 10086).
4500   1847, Map of that part of the mineral lands adjacent to Lake Superior, ceded to the United States by the Treaty of 1842 with the Chippewas. Comprising that district lying between Chocolate River and Fond du Lac. Projected and drawn by A. B. Gray: Wash., geological map, 94 x 116 cm.

Gray, Asa (1810-1888)
        See also 2687 with J. B. Crawe
4501   1842, [Moraine in Andover, Massachusetts]: Am. J. Sci., v. 43, no. 1, 151 (2 lines).
        Also in Assoc. Am. Geol's Nat's Proc., v. 3, 6 (5 lines). Also in Assoc. Am. Geol's Nat's Rept., v. 1, 45-46 (5 lines), 1843.
4502   1843, [Review with excerpts of Travels in North America by C. Lyell (see 6557)]: N. Am. Rev., v. 61, 498-518.
4503   1846, [Review with excerpts of Explanations: a sequel to the vestiges of the natural history of Creation by R. Chambers (see 2291 et seq.)]: N. Am. Rev., v, 62, 465-506.
4504   1846, [Contents of a Mastodon's stomach]: Bos. Soc. Nat. History Proc., v. 2, 92-93.
4505   1846, Martius, genera et species Palmarum [review of Synopsis plantarum fossilium by Prof. Unger]: Am. J. Sci., v. 2 (ns), no. 1, 152 (12 lines).
4506   1847, Food of the Mastodon: Am. J. Sci., v. 3 (ns), no. 3, 436.
        Abstracted from Bos. Soc. Nat. History Proc.
4507   1847, Nomenclator zoologicus, continens nomina systematica generum animalium, tam viventium quam fossilium, etc.; auctore L. Agassiz [review of]: Am. J. Sci., v. 3 (ns), no. 2, 302-309.
4508   1849, Notice of Dr. Hooker's flora antarctica: Am. J. Sci., v. 8 (ns), no. 2, 161-180.
        With many excerpts (see 5252).

Gray, Francis Calley (1790-1856)
4509   1819, [Review with excerpts of Essay on the theory of the earth by G. Cuvier (see 2754), and review of Outlines of the mineralogy and geology of Boston by J. F. Dana and S. L. Dana (see 2871)]: N. Am. Rev., v. 8, 396-414.

Great Western Iron Company
4510   1839, Charter and by-laws of the Great Western Iron Company; also an act to encourage the manufacture of iron in the state of Pennsylvania: n.p., 16 p.
        Contains "An act to encourage the manufacture of iron in the state of Pennsylvania" (see 8214).

Greeley, Horace (1811-1872)
4511   1847, Copper region: Niles' Wk. Reg., v. 72, 323.
4512   1847, [Report on mines of the North-west Mining Company]: in 10055.
4513   1848, Process of working a Lake Superior copper mine: Hunt's Merch. Mag., v. 19, 559-560.

Green Mountain Gem (Bradford, Vermont, 1843-1849)
4514   1843, Appalling particulars of the late awful earthquake: v. 1, 54-55.
4515   1843, Wonderful organic remains [from Warsaw, Osage County, Missouri]: v. 1, 144.
4516   1844/1847, The natural bridge [in Virginia]: v. 2, 17-18, plate; v. 5, 85, plate.
4517   1844, The petrified forest [in Texas]: v. 2, 256-257.
4518   1846, Geology: v. 4, 147.
        On the definition and scope of geology.
4519   1849, More gold marvels [from California]: v. 7, 15; 46.

Green Mountain Gem (cont.)
4520   1849, [Review with excerpts of Cosmos by A. von Humboldt (see 5395)]: v. 7, 105-110.

The Green Mountain Repository, For the Year 1832 (Burlington, Vermont, 1832)
4521   1832, Mud volcanoes [of Java]: v. 1, no. 2, 41-42.
        From Library of Use. Know.
4522   1832, Remarks upon the direction and use of the magnetic needle: v. 1, no. 11, 254-259; no. 12, 270-272.

Green, Ashbel (1762-1848)
4523   1804, On the medical virtues of the warm and hot springs, in the county of Bath, in Virginia: Phila. Med. Phys. J., v. 1, pt. 1, 24-30.
4524   1805, Notices of the Sulphur, Sweet, and other mineral springs, in the western parts of Virginia: Phila. Med. Phys. J., v. 1, pt. 2, 148-159.

Green, Duff (1791-1875)
4525   1836, A letter addressed to the General Assembly of Maryland, by Duff Green, on the bill incorporating the Union Company: Annapolis, Md., 8 p.

Green, Jacob (1790-1841)
        See also 1592 with t. R. Beck and J. Law.
4526   1816, Description of a cave and intermitting lake, in Saratoga County, New-York: NY Soc. Prom. Use. Arts Trans., v. 4, 79-82.
4527   1816, Description of a cavern near Bennington, Vermont: NY Soc. Prom. Use. Arts Trans., v. 4, 109-111.
4528   1820, On the crystallization of snow: Am. J. Sci., v. 2, no. 2, 337-339, 7 figs. on plate.
4529   1821, A new blow-pipe: Am. J. Sci., v. 4, no. 1, 164-166, 2 figs.
        For notice see 7806.
4530   1821, Some curious facts respecting the bones of the rattle snake: Am. J. Sci., v. 3, no. 1, 85-86.
4531   1822, Dr. Hosack's donation of minerals: Am. J. Sci., v. 4, no. 2, 396-397.
4532   1822, Notice of a mineralized tree - rocking stone, etc.: Am. J. Sci., v. 5, no. 2, 251-254, fig.
        On a fossil tree from Chitteningo, Sullivan County, New York; a rocking stone (with figure) from Durham, New Hampshire; and minerals from Princeton, New Jersey.
4533   1830, Monograph of the cones [Conus] of North America, including three new species: Alb. Inst. Trans., v. 1, 121-125.
4534   1830, Notes of a traveller, during a tour through England, France, and Switzerland, in 1828: NY, G. and C. and H. Carvill, 3 v.
4535   1832, A monograph of the trilobites of North America, with coloured models of the species: Phila., J. Brano, 94, [1] p., plate.
        Published with a set of plaster casts of trilobites. See also Supplement..., (4543). For review with excerpts see 557. For correction see 3372.
4536   1832, Notes of a naturalist: Cab. Nat. History, v. 2, 155-156.
        On a burning well at Canonsburg, Washington County, Pennsylvania.
4537   1832, Synopsis of the trilobites of North America: Mon. Am. J. Geology, v. 1, no. 12, 558-560, plate with 10 figs.
4538   1833, Nature of the trilobite [and] Asaphus myrmecoides: Am. J. Sci., v. 23, no. 2, 396-398.
        In review of trilobite monograph (see 557).
4539   1834, Description of a new trilobite from Nova Scotia: Geol. Soc. Penn. Trans., v. 1, 37-39, plate with 1 fig.
4540   1834, Descriptions of some new North American trilobites: Am. J. Sci., v. 25, no. 2, 334-337.
        See also letter by R. Harlan (4800).
4541   1834, Some experiments on a sulphated ferruginous earth from Kent County in the state of Delaware, with a view to ascertain its commercial value: Geol. Soc. Penn. Trans., v. 1, 33-36.
4542   1835, New trilobite, &c.: Am. J. Sci., v. 27, no. 2, 351 (11 lines).
        In "Notice of the transactions of the Geological Society of Pennsylvania". On a new species of Asaphus.
4543   1835, A supplement to the "Monograph of the trilobites of North America": Phila., J. Brano, 24 p.
        See also 4535.
4544   1837, Description of a new trilobite: Am. J. Sci., v. 32, no. 1, 167-169.
        On trilobites from Springfield, Ohio and Huntington County, Pennsylvania.
4545   1837, Description of several new trilobites: Am. J. Sci., v. 32, no. 2, 343, 2 figs.
        On Cryphaeus (i.e. Greenops) from Huntington County, Pennsylvania.
4546   1837, Description of two new species of trilobites: Phila. Acad. Nat. Sci's J., v. 7, pt. 2, 217-226, 2 figs.
        On Asaphus and Greenops from Huntington County, Pennsylvania.
4547   1838, Description of a new trilobite: Am. J. Sci., v. 33, no. 2, 406-407.
        On Calymene (i.e. Protus) rowii from Cooperstown, New York.
4548   1838, New trilobites, Asaphus polypleurus: Am. J. Sci., v. 34, no. 2, 380.
4549   1838, Some remarks on the genus Paradoxides of Brongniart, and on the necessity of preserving the genus Triarthus, proposed in the monograph of the trilobites of North America: Am. J. Sci., v. 33, no. 2, 341-344.
4550   1839, Description of a new trilobite: Am. J. Sci., v. 37, no. 1, 40.
        On Asaphus diurus.
4551   1839, Interior surface of the trilobite discovered: Phila.
        Not seen.
4552   1839, Remarks on the trilobite: The Friend, March 16, 1839.
        Not seen.
4553   1839, Remarks on the trilobite: Am. J. Sci., v. 37, no. 1, 25-39.
        With extensive quotations by W. Buckland (see 2160).
4554   1840, An additional fact illustrating the interior surface of the Calymene bufo: Am. J. Sci., v. 38, no. 2, 410.

Green, Thomas J.
4555   1850, Improvements in the rockers of gold washers: Franklin Inst. J., v. 19 (3s), 251 (7 lines).

Greenhow, Robert (1800-1854)
4556   1840, Memoir, historical and political, on the northwest coast of North America and the adjacent territories; illustrated by a map and a geographical view of those countries: NY, Wiley and Putnam, 228 p., map.
       Also Wash., Blair and Rivers, 228 p., map. See also 4557.
4557   1844, The history of Oregon and California, and the other territories on the northwest coast of North America, accompanied by a geographical view and map of those countries ... : Bos., Little and Brown, 18, 482 p., plate, maps.
       This is an enlarged version of Memoir, ..., (see 4556).
       Other editions and printings include Bos., Little and Brown, xviii, 492 p., 1845; NY, Appleton, xviii, 492 p.; NY and Phila., Appleton, xviii, 492 p., 1846; Bos., Freeman and Bolles, xviii, 491 p., 1847.
4558   1845, The geography of Oregon and California, ... illustrated by a new and beautiful map: Bos., Freeman and Bolles (printers), 42 p., map.
       Another printing, NY, M. H. Newman, 42 p., map, 1845.

Greenleaf, Moses (1777-1834)
4559   1816, A statistical view of the district of Maine; more especially with reference to the value and importance of its interior ... : Bos., Cummings and Hilliard, 154 p.
4560   1829, A survey of the state of Maine in reference to its geographical features, statistics, and political economy: Portland, Me., Shirley Hyde, vii, [9]-468 p., map, table, diagram.
       This is a much-enlarged version of A statistical view ... (see 4559).

Greenock, Lord
4561   1835, Coal of Fyfeshire and Edinburgh: Am. J. Sci., v. 28, no. 1, 73-74.
       In British Assoc. Proc.

Greenough, George Bellas (1778-1855)
4562   1835, Extract from Mr. President Greenough's address to the Geological Society, delivered on the 4th of June, 1834: Am. J. Sci., v. 28, no. 2, 239-242.
       In "Rise of the coast of Chili" by M. Graham (see 4472).
4563   1835, Idocrase from the Isle of Skye: Am. J. Sci., v. 28, no. 2, 395 (6 lines).
       Abstracted from Jameson's Edinburgh New Philos. J.

Greenway, Joseph
4564   1793, Farther remarks: extracted from a letter from Dr. Greenway to Dr. Barton: Am. Philos. Soc. Trans., v. 3, 233 (10 lines).
       Following "An account of a hill ... supposed to have been a volcano" (see 1140).

Greenwood, Francis William Pitt (1797-1843)
4565   1832, Falls of the Niagara: Naturalist, v. 2, no. 1, 1-10, plate.

Gregg, Amos
4566   1804, A topographical and medical sketch of Bristol, in Pennsylvania: Phila. Med. Phys. J., v. 1, pt. 1, 15-22.

Gregory, George (1754-1808)
4567   1815-1816, A dictionary of arts and sciences ... 1st American from 2d London ed: Charlestown, S. Etheridge, Jun., and I. Peirce, 3v.
       Also 2nd American edition, NY, W. T. Robinson and Collins and Co., 3 v. (1821-22).

Griscom, J. H. (i.e. John?)
4568   1835, Professor Griscom's cabinet for sale: Am. J. Sci., v. 29, no. 1, Prospectus, 5 (5 lines).

Griscom, John (1774-1852)
       John Griscom was responsible for many of the translations and abstracts of foreign geological and mineralogical literature which appeared in the American Journal of Science from c. 1820 to 1845.
4569   1810, Chemical examination of mineral water from Litchfield, state of New York: Am. Min. J., v. 1, no. 1, 20-26.
4570   1812, Observations and experiments on several mineral waters in the state of New York: Am. Min. J., v. 1, no. 3, 156-163.
       For review with excerpts see 7681.
4571   1814, A vocabulary of scientific terms: NY, c. 35 p.
       Not seen. Also in The New York Expositor by R. Wiggins, 250-285, c. 1825; also NY, S. S. and W. Wood, 1842.
4572   1820, Foreign literature and science: Am. J. Sci., v. 2, no. 2, 340-357.
4573   1821, [Pompeii]: Am. J. Sci., v. 3, no. 2, 374 (9 lines).
4574   1821, Salt [from European mines]: Am. J. Sci., v. 3, no. 2, 371 (3 lines).
4575   1822, Dolcoath mine: Am. J. Sci., v. 4, no. 2, 372 (6 lines).
4576   1822, Extract from a French work on lime, mortar, and artificial puzzolana: Am. J. Sci., v. 4, no. 2, 373-375.
4577   1822, Extract of a letter: Am. J. Sci., v. 5, no. 2, 364-366.
       On the fusion of charcoal by the blowpipe.
4578   1822, Organic remains [notice of a new edition of Essay on the theory of the earth by G. Cuvier (see 2754)]: Am. J. Sci., v. 4, no. 2, 386 (9 lines).
4579   1823, Abbeville department of Somme France. - fossil remains: Am. J. Sci., v. 6, no. 1, 199 (9 lines).
4580   1823?, Appendix [to third American edition of Mineralogy ..., by W. Phillips (see 8322)].
       Not seen.
4581   1823, Geology: Am. J. Sci., v. 6, no. 2, 398-399.
       On fossil wood from Russia.
4582   1823, New work on fossil shells: Am. J. Sci., v. 6, no. 1, 197-198 (9 lines).
       Notice of work on fossil trilobites by Desmarest and Brongniart.

Griscom, J. (cont.)
4583   1823, Sal-ammoniac [from St. Etienne, France]: Am. J. Sci., v. 6, no. 2, 395-396.
4584   1823, Skeletons of the mammoth and elephant: Am. J. Sci., v. 6, no. 2, 386 (2 lines).
        On the discovery of fossil bones in Hautes, Hungary.
4585   1823, A year in Europe. Comprising a journal of observations in England, Scotland, Iceland, France, Switzerland, the North of Italy, and Holland. In 1818 and 1819: NY, Collins and Co.; Phila., H. C. Carey and J. Lea, 2 v.
        Also "second edition", NY, Collins and Hannay, 2 v.
        For notice see 8227.
4586   1824, A mineral spring [in Sales, Piedmont, Italy]: Am. J. Sci., v. 7, no. 2, 387.
4587   1825, Coal formation within the United States: US Lit. Gaz., v. 1, 317.
4588   1825, Diamond [from Brazil]: Am. J. Sci., v. 9, no. 1, 195 (3 lines).
4589   1825, Remarks upon the coal formation of the Susquehanna and Lackawanna, and on the advantages of a water communication between that region and the Hudson: NE Farm., v. 3, 220-221.
4590   1828, Obituary [of William Phillips]: Am. J. Sci., v. 15, no. 1, 160.
4591   1829, Dr. Wollaston [Obituary]: Am. J. Sci., v. 16, no. 1, 216 (7 lines).
4592   1829, Iodine in Saratoga mineral waters: Am. J. Sci., v. 16, no. 1, unnumbered page following 216.
4593   1831, Analysis of the mineral water of Saratoga, and Ballston, with an account of their medicinal properties, &c. by John Steel, M. D. [review with excerpts of 9998]: Am. J. Sci., v. 21, no. 1, 182-184.
4594   1831, Journal of the Philadelphia College of Pharmacy. - Edited by Benjamin Ellis, M. D. [review and extracts of the first volume]: Am. J. Sci., v. 21, no. 1, 173-179.
4595   1833, Documents in commemoration of Baron Cuvier: Am. J. Sci., v. 23, no. 2, 303-311.

Griswold, Samuel
4596   1813, Information concerning the earthquakes which have prevailed in the United States since December, 1811; particularly in the states and territories adjoining to the Mississippi: Med. Repos., v. 16, 304-309.

Gros, Jean Baptist Louis, Baron (1793-1870)
4597   1835, Ascent to the summit of the Popocatepetl, the highest point of the Mexican Andes, 18,000 feet above the level of the sea: Am. J. Sci., v. 28, no. 2, 220-231; Also in: Mus. For. Liter. Sci., v. 26, 104-108.
        From London Ath.

Grout, John R.
4598   1847, and C. C. Douglass, Report [on the Douglass Houghton Mining Company]: in 3765.

Groye, Menard de la
4599   1817, [Geognostic appearances at Beaulieu, Rhone]: Am. Reg. or Summary Rev., v. 1 (ns), 444.
        Abstracted by T. Cooper.

The Guardian, a Family Magazine (Columbia, Tennessee, 1841-1850)
4600   1841, Niagara Falls: v. 1, 59.

Gueniveau
4601   1814, Memoir upon the de-sulphuration of metals: Emp. Arts Sci's, v. 3 (ns), no. 3, 478-491.

Guernsey, J. A.
4602   1831, Mastodon, near Rochester, N. Y.: Am. J. Sci., v. 19, no. 2, 358-359, fig.

Guerrant, Daniel, jr.
4603   1833, Gold in Buckingham [Virginia]: Farm's Reg., v. 1, 244.

Guillemin, M.
4604   1828, On a gelatinous quartz: Am. J. Sci., v. 14, no. 2, 391-392.
        Abstracted from "Bull. Univ."

Gulich
4605   1846, Scolezite and natrolite: Am. J. Sci., v. 2 (ns), no. 3, 418 (8 lines).
        Abstracted from Poggendorff's Annalen.

Gurney
4606   1850, Carbonic acid as a means of extinguishing fires in coal-mines: Ann. Scien. Disc., v. 1, 194-196.

Guthrie, William (1708-1770)
4607   1822, Natural curiosities of Italy: Ohio Misc. Mus., v. 1, no. 3, 122-125.
        From Guthrie's Geography, on Vesuvius and Etna.

Gutmuths, M.
4608   1832, Delta of Oroonoko and Maragnon: Mon. Am. J. Geology, v. 1, no. 8, 384.
        Abstracted from Mag. Nat. History. On growth of a delta in Guiana.

Guyot, Arnold (1807-1884)
4609   1849, The earth and man: lectures on comparative physical geography, in its relation to the history of mankind: Bos., Gould, Kendall, and Lincoln, xviii, 310, 6 plates, illus.
        Translated from the French by C. C. Felton.
        For review see 980, 1681, 2945, 4488, 5471, 5517, 7152, 7694, and 9180. For notice see 8595.

Guyot, A. (cont.)
4610   1850, The earth and man ... [as above]: Bos., Gould, Kendall and Lincoln, 334 p., illus.
"Second edition".
4611   1850, On the erratic phenomena of the central Alps: Am. Assoc. Adv. Sci. Proc., v. 2, 311-320; 321.
With discussion by L. Agassiz, J. Hall, P. Leslie, and H. D. Rogers.
4612   1850, On the erratic phenomena of the White Mountains: Am. Assoc. Adv. Sci. Proc., v. 2, 308-311.
With discussion by L. Agassiz, J. Hall, C. T. Jackson (5669), and W. C. Redfield.
4613   1850, [Remarks on the elevation and structure of the Jura Mountains]: Am. Assoc. Adv. Sci. Proc., v. 2, 115-117; 118.
Discussion of article by H. D. Rogers (see 8941).

Gwyne, James S.
4614   1848, Raritan Mining and Manufacturing Company: Hunt's Merch. Mag., v. 19, 111.

H., D. L. (see 696)

H., E. N. (see 1033)

H., F. B. (see 912)

H., J. E. (see 5526; i.e. James Ewell Heath?)

H., O. P. (see 658; i.e. Oliver Payson Hubbard?)

H., W. E. (see 1024)

Habersham, Joseph Clay (d. 1855)
4615   1846, Memorandum by Dr. Joseph Habersham of the most important fossil bones and shells, now in his possession, which were discovered in the year 1842, on the Island of Skiddaway on the sea coast of Georgia: in Memoir on the Megatherium ..., by W. B. Hodgson (see 5213), 24-30.

Hadley, G.
4616   1847, Crystallized carbonate of lead, at Rossie, New York: Am. J. Sci., v. 3 (ns), no. 1, 117 (10 lines).

Hadley, James
4617   1823, Notice of alum slate, sulphuret of zinc and limpid quartz: NY Med. Phys. J., v. 2, 132-133.

Haidinger, Wilhelm Karl, ritter von (1795-1871)
4618   1824, Remarks concerning the natural historical determination of diallage: Bos. J. Phil. Arts, v. 2, no. 2, 120 (7 lines).
Abstracted from Edinburgh Philos. J.
4619   1846, On iolite [and its pseudomorphs]: Am. J. Sci., v. 2 (ns), no. 3, 418-419.
Abstracted from Poggendorff's Annalen.
4620   1846, Piauzite, a new mineral [from Neustadt]: Am. J. Sci., v. 1 (ns), no. 2, 267 (7 lines).
Abstracted from l'Institut.
4621   1846, Transparent anandalusites [sic] from Brazil: Am. J. Sci., v. 2 (ns), no. 1, 119 (4 lines).
Abstracted from Geol. Soc. France Bulletin.
4622   1846, Trichroism of crystals: Am. J. Sci., v. 1 (ns), no. 2, 267 (8 lines).
Abstracted from Poggendorff's Annalen.
4623   1847, Hauerite, a new mineral species: Am. J. Sci., v. 4 (ns), no. 1, 108-109.
Abstracted from Poggendorff's Annalen.
4624   1847, Pleochroism: Am. J. Sci., v. 3 (ns), no. 3, 430 (5 lines).
4625   1848, Brandesite, a new mineral [from Fassatol]: Am. J. Sci., v. 5 (ns), no. 2, 267 (8 lines).
Abstracted from Neues Jahrbuch fur Mineralogie.

Haines, Charles Glidden (1792-1825)
4626   1819, [Letter to H. R. Schoolcraft]: Saturday Mag., v. 2, 345.

Haldeman, Samuel Stehman (1812-1880)
See also 5790 with W. R. Johnson.
4627   1839, An analysis of marl from New Jersey: Phila. Acad. Nat. Sci's J., v. 8, pt. 1, 150 (11 lines).
4628   1840, [Supplement to number one of "A monograph of the Limniadae, or freshwater univalve shells of North America ... "]: Phila., 3 p.
On fossil fucoides from Berks and Columbia Counties, Pennsylvania.
4629   1845, [On the Taconic system of Emmons (see 3520)]: Assoc. Am. Geol's Nat's Proc., v. 6, 66-68.
Contains note by J. Hall (4706), and discussion by H. D. Rogers.
Also in Am. Q. J. Ag. Sci., v. 2, 164-165.
4630   1847, Description of several new and interesting animals: Am. Q. J. Ag. Sci., v. 6, 191-192.
On a new trilobite from Bedford County, Pennsylvania.
4631   1847, [Discussion of the formation of concretions]: Am. Q. J. Ag. Sci., v. 6, 207-208 [i.e. 255-256]. ?In Assoc. Am. Geol's Nat's Proc.
4632   1847, Report on the supposed identity of Atops trilineatus, (Emmons), with Triarthus beckii: Am. Q. J. Ag. Sci., v. 6, 194-195.
See also remarks by J. Hall (4722). Also in Am. J. Sci., v. 5 (ns), no. 1, 107-108.
4633   1848, Chikiswalungo iron furnace, near Columbia, Pa.: Am. J. Sci., v. 5 (ns), no. 2, 296.
4634   1848, [Fibrous lava of Hawaii]: Phila. Acad. Nat. Sci's Proc., v. 4, no. 1, 5 (8 lines).

Haldeman, S. S. (cont.)
4635 1848, [Migration of animals deduced from the fossil record]: Lit. World, v. 2, 228.
       From Assoc. Am. Geol´s Nat´s Proc.
4636 1848, On the construction of blast-furnaces for the smelting of iron with anthracite:
       Am. J. Sci., v. 6 (ns), no. 1, 74-80, 3 figs.
4637 1849, Bibliographia zoologiae et geologiae, &c.; by Prof. L. Agassiz. Corrected and
       enlarged by H. E. Strickland [review of]: Am. J. Sci., v. 7 (ns), no. 3, 454-455.

Hale, C. S. (Professor)
4638 1847, The delta of the Alabama: De Bow´s Rev., v. 3, 469-475.
4639 1848, Geology of South Alabama: Am. J. Sci., v. 6 (ns), no. 3, 354-363.

Hale, E. (see 5070 by E. Hitchcock)

Hale, Moses
4640 1821, Geological notice of Troy: Am. J. Sci., v. 3, no. 1, 72-73.
       On the rocks and minerals of Troy, New York.

Hall, Basil (1788-1844)
4641 1828, Voyage to the Eastern seas - in 1816: Am. J. Sci., v. 14, no. 1, 206-208.
       On the formation of coral reefs.

Hall, Frederick (1780-1843)
4642 1808, Note: Phila. Med. Phys. J., v. 3, pt. 1, vii-viii.
       On H. Davy´s discovery of oxygen as a principal component of earthy minerals.
4643 1812, Vermont marble: Am. Min. J., v. 1, no. 3, 189-190.
4644 1821, Agaric mineral: Am. J. Sci., v. 3, no. 2, 234.
       On decomposed limestone from Lyndon, Caledonia County, Vermont.
4645 1821, Notice of iron mines and manufactures in Vermont and of some localities of earthy
       minerals: Am. J. Sci., v. 4, no. 1, 23-25.
4646 1821, Notice of ores of iron and manganese, and of yellow ochre in Vermont: Am. J.
       Sci., v. 3, no. 1, 57-58.
4647 1823, Localities of minerals: Am. J. Sci., v. 7, no. 1, 58-59.
       On localities of fossils in New York and minerals in Vermont.
4648 1823, Notice of a curious water fall and of excavations in the rocks: Am. J. Sci., v.
       6, no. 2, 252-254.
       On river erosion in the Green Mountains of Vermont.
4649 1823, Notice of the plumbago of Ticonderoga: Am. J. Sci., v. 6, no. 1, 178.
4650 1824, Catalogue of minerals, found in the state of Vermont, and in the adjoining
       states; together with their localities: including a number of the most interesting
       minerals, which have been discovered in other parts of the United States; arranged
       alphabetically: Hart., P. B. Goodsell (printer), 44 p.
       For notice see 405, 7637 and 8495. For correction see 4652.
4651 1825, Description of minerals from Palestine: Am. J. Sci., v. 9, no. 2, 337-351.
4652 1828, Errors corrected in Dr. Robinson´s catalogue of minerals: Am. J. Sci., v. 15, no.
       1, 197-199.
       See also 4650. On localities of coccolite.
4653 1828, Miscellaneous notices among the White Mountains and other places: Am. J. Sci., v.
       13, no. 2, 373-376.
       Also reprint edition, 4 p.
4654 1832, Statistical account of the town of Middlebury, in the state of Vermont: Mass.
       Hist. Soc. Coll., v. 9 (2s), 123-158.
4655 1836, A synopsis of a course of lectures on mineralogy; delivered at the Medical
       College, Washington, in the winter of 1835-6: Wash., J. Gideon, jr. (printer),
       24 p.
4656 1837, A trip from Boston to Littleton, through the notch of the White Mountains ... :
       Wash., J. Gideon, jr., 30 p.
4657 1837, Notes on a tour in France, Italy, and Elba, with a notice of its mines of iron:
       Am. J. Sci., v. 32, no. 1, 74-84.
       Also reprint edition, 11 p.
4658 1838, Mammoth Cave: Balt. Monument, v. 2, 338-339.
4659 1838, Notice of oriental minerals: Am. J. Sci., v. 33, no. 2, 249-255.
4660 1840, Iron in Missouri: Bos. Wk. Mag., v. 2, 264; also in: Hazard´s US Reg., v. 2, 207.
4661 1840, Letters from the East and from the West: Wash., F. Taylor and W. M. Morrison, xi,
       168 p.
       First published in Nat´l Intell. Includes notes on the minerals of Nashville,
       149-160.

Hall, G. E.
4662 1833, The calcareous soils of Alabama, and their effect in preserving health: Farm´s
       Reg., v. 1, 276-277.

Hall, James (1744-1826)
4663 1799, An account of a supposed artificial wall discovered under the surface of the
       earth in North Carolina, in a letter to James Woodhouse: Med. Repos., v. 2,
       272-278.
       See also reply by J. Woodhouse (see 11040).
4664 1801, A brief history of the Mississippi Territory, to which is prefixed, a summary
       view of the country between the settlements on the Cumberland-River & the
       territory: Salisbury, NC, F. Coupee, 1, 70 p.

Hall, James (1793-1868)
4665 1833, Flint on the natural sciences [review of 3888]: West. Mon. Mag., v. 1, pt. 1,
       262-273.
4666 1837, Statistics of the West, at the close of the year 1836: Cin., J. A. James, 284 p.
4667 1838, Notes on the Western States; containing descriptive sketches of their soil,
       climate, resources, and scenery: Phila., H. Hall, xxiii, 13-304 p.
4668 1848, The West: its soil, surface, and production: Cin: Derby, Bradley & Co.,
       [5]-260 p.

Hall, James (1811-1898)
See also 3498 and 3515 with E. Emmons; see also discussion of articles 4611 and 4612 by A. Guyot, and 10409 by G. Troost.

4669 1837, Descriptions of two species of trilobites, belonging to the genus Paradoxides: Am. J. Sci., v. 33, no. 1, 139-142, 2 figs.

4670 1837, Ores of iron [of the second geological district in New York]: NY Geol. Survey Ann. Rept., v. 1, 127-149.
In report by E. Emmons (see 3495).

4671 1838, Geology of the Genesee River: NY Geol. Survey Ann. Rept., v. 2, geological map, E. N. Horsford, del., 32 x 40 cm.

4672 1838, Second annual report of the fourth geological district of New York: NY Geol. Survey Ann. Rept., v. 2, 287-374, 3 figs., 2 plates, map.

4673 1839, Third annual report of the fourth geological district of the state of New-York: NY Geol. Survey Ann. Rept., v. 3, 287-339.

4674 1840, Fourth annual report of the survey of the fourth geological district: NY Geol. Survey Ann. Rept., v. 4, 389-456.

4675 1841, Fifth annual report of the fourth geological district: NY Geol. Survey Ann. Rept., v. 5, 149-179.

4676 1841, Notes upon the geology of the western states: Am. J. Sci., v. 42, no. 1, 51-62.
Also reprint edition, 12 p. For review see 2714.

4677 1842, Niagara Falls - their physical changes, and the geology and topography of the surrounding country: Bos. J. Nat. History, v. 4, no. 1, 106-134, 4 figs.
Also reprint edition.

4678 1842, Observations on the resources of the Western States: No. Light, v. 1, 184-185.

4679 1842, [Review of final Massachusetts report by E. Hitchcock (see 5135)]: No. Light, v. 1, 187-188.

4680 1843, Communication from Mr. James Hall, one of the state geologists: NY SD 59, v. 2, 66th session, 9 p.
On the progress of the survey.

4681 1843, The crinoidea of the rocks of New York, their geological and geographical distribution: Am. J. Sci., v. 45, no. 2, 349-351.
In Assoc. Am. Geol's Nat's Proc.

4682 1843, and others, Geological map of the middle and western states: in Geology of New York, fourth district (see 4683), colored geological map, NY, Endicott (lithographer), 80 x 57 cm.

4683 1843, Geology of New York. Part IV, comprising the survey of the fourth geological district: Alb., Carroll and Cook (printers), xxii, 525 p., 34 plates and sheets, 663-683, 25 plates, map, 192 figs. in text.
Contains map 4682. For review see 8066.

4684 1843, [Geology of the region of Niagara Falls]: Bos. Soc. Nat. History Proc., v. 1, 52.

4685 1843, [Glaciated cherty limestone from near Niagara, New York]: Am. J. Sci., v. 45, no. 2, 332.
In Assoc. Am. Geol's Nat's Proc. With discussion by W. C. Redfield, H. D. Rogers, and B. Silliman, jr.

4686 1843, Notes explanatory of a section from Cleveland, Ohio, to the Mississippi River, in a southwest direction; with remarks upon the identity of the western formations with those of New York: Assoc. Am. Geol's Nat's Rept., v. 1, 267-293, 3 figs, plate.

4687 1843, On the geographical distribution of fossils in the older rocks of the United States: Am. J. Sci., v. 45, no. 1, 157-160; 162-163.
In Assoc. Am. Geol's Nat's Proc.

4688 1843, On wave lines and casts of mud furrows: Am. J. Sci., v. 45, no. 1, 148-149.
In Assoc. Am. Geol's Nat's Proc.

4689 1843, Remarks upon casts of mud furrows, wave lines, and other markings upon rocks of the New York system: Assoc. Am. Geol's Nat's Rept., v. 1, 422-432, plate with 4 figs.

4690 1843, [On the shore of Lake Erie and Portage, New York]: Am. J. Sci., v. 45, no. 2, 327-330.
In Assoc. Am. Geol's Nat's Proc.

4691 1844, An address delivered before the Society of Natural History of the Auburn Theological Seminary ... August 15, 1843: Auburn, Oliphant, 20 p.

4692 1844, The geological survey of New-York: its influence upon the productive pursuits of the community: NY State Agri. Soc. Trans., v. 3, 241-278, 30 plates with 100's of figs.

4693 1844, The mountains of Northern New-York: No. Light, v. 4, no. 5, 77-80, 2 plates.

4694 1844, Observations on brachiopoda and orthocerata: Am. J. Sci., v. 47, no. 1, 109.

4695 1844, [On the nomenclature of Paleozoic rocks, based on the New York system]: Am. J. Sci., v. 47, no. 1, 112.
Also in Assoc. Am. Geol's Nat's Proc., v. 5, 19.

4696 1844, On the geographical distribution of fossils in the Paleozoic strata of the United States: Am. J. Sci., v. 47, no. 1, 117-118.
Also in Assoc. Am. Geol's Nat's Proc., v. 5, 24-25.

4697 1844, Report of a reconnoisance [sic] of a route for a railroad from Portland to Montreal: Portland, Me., A. Shirley and Son (printers), 16 p.

4698 1844, Trenton Falls [New York]: No. Light, v. 4, no. 3, 33-36, 15 figs.

4699 1845, Description of some microscopic shells from the decomposing marl slate of Cincinnati: Am. J. Sci., v. 48, no. 2, 292-295.

4700 1845, Description of organic remains collected by Captain J. C. Fremont, in the geographical survey of Oregon and North California: in A report of the exploring expedition ..., by J. C. Fremont (see 4113), pp. 304-310.

4701 1845, Fossil vegetables and shells from Oregon: Assoc. Am. Geol's Nat's Proc., v. 6, 66.
Also in Am. Q. J. Ag. Sci., v. 2, 164.

4702 1845, [Map of the coal regions of Pennsylvania]: in Reports on the Bear Mountain Railroad by E. F. Johnson and W. R. Casey (see 5753), geological map, 32 x 20.5 cm.

Hall, J. (cont.)
4703 1845, Nature of the geological formations occupying the portion of Oregon and North California included in a geographical survey under the direction of Captain Fremont: in A report of the exploring expedition ..., by J. C. Fremont (see 4113), 295-303, plate.
4704 1845, Nature of the strata, and geographical distribution of the organic remains in the older formations of the United States: Bos. J. Nat. History, v. 5, no. 1, 1-20.
4705 1845, [New York fossils, including a trilobite from New Bedford]: Assoc. Am. Geol's Nat's Proc., v. 6, 77 (8 lines).
Also in Am. Q. J. Ag. Sci., v. 2, 168.
4706 1845, [On the Taconic system of E. Emmons]: Assoc. Am. Geol's Nat's Proc., v. 6, 68 (16 lines).
In article (see 4629). Also in Am. Q. J. Ag. Sci., v. 2, 166.
4707 1845, The coal and iron ores of the Bear Valley coal basin: in Reports on the Bear Mountain Railroad by E. F. Johnson and W. R. Casey (see 5753), 37-80, map.
4708 1846, Notice of the geological position of the cranium of the Castoroides ohioensis: Bos. J. Nat. History, v. 5, no. 3, 385-391.
4709 1847, Coal on the Rocky Mountains, discovered by Capt. Fremont: Am. J. Sci., v. 3 (ns), no. 2, 273-274.
4710 1847, Letter from Professor James Hall, on certain fossils in the red sand-rock of Highgate [Vermont]: in Third annual report on the geology of the state of Vermont by C. B. Adams (see 32), 31.
4711 1847, On the general results of investigations in the paleontology of the lower strata of New York: Am. Q. J. Ag. Sci., v. 6, 210 [i.e. 258] (13 lines).
In Assoc. Am. Geol's Nat's Proc.
4712 1847, Paleontology of New York, volume 1, containing descriptions of the organic remains of the lower division of the New York system (equivalent of the lower Silurian rocks of Europe): Alb., C. van Benthuysen (printers), 23, 338 p., 98 plates of fossils.
For notice see 894. For review see 914 and 1709.
4713 1848, An address delivered at the anniversary meeting of the Harvard Natural History Society, May 24, 1848: Cambridge, Mass., Metcalf and Co., 39 p.
4714 1848, Agricultural geology: Am. Q. J. Ag. Sci., v. 7, no. 3, 91-94, fig.
On the agricultural geology of the Niagara shale in New York.
4715 1848, Agricultural geology: Genesee Farm. and Gardener's J., v. 9, 103-105, fig.
4716 1848, Catalogue of specimens in the Geological Department of the Geological Survey: NY State Mus. Ann. Rept., v. 1, 39 p.
4717 1848, Catalogue of the specimens in the paleontological department of the geological survey: NY State Mus. Ann. Rept., v. 1, 15 p.
4718 1848, Communication from Mr. James Hall, one of the state geologists. In senate, March 6, 1843: Alb., NY SD 59, 9 p.
4719 1848, Crinoidea of the inferior strata New-York system: Alb., C. van Benthuysen (printer), 1, 1, 20 p., 6 plates.
From Paleontology of New York, volume 1, (see 4712).
4720 1848/1849, [Translation of and annotations in] On the parallelism of the Paleozoic deposits of North America with those of Europe, by E. Verneuil (see 10620): Am. J. Sci., v. 5, no. 1, 176-183; no. 2, 359-370; v. 7, no. 1, 45-51; no. 2, 218-231.
4721 1848, [Paleontology of the lower Paleozoic strata of New York]: Lit. World, v. 2, 255 (8 lines).
4722 1848, Remarks on the observations of S. S. Haldeman "on the supposed identity of Atops trilineatus with Triarthrus beckii [see 4632]: Am. J. Sci., v. 5 (ns), no. 3, 322-327, 10 figs.
4723 1848, [Terraces from the Lake Ontario region]: Lit. World, v. 2, 228 (5 lines).
From Assoc. Am. Geol's Nat's Proc.
4724 1848, Upon some of the results of the paleontological investigations in the state of New York: Am. J. Sci., v. 5 (ns), no. 2, 243-249.
In Assoc. Am. Geol's Nat's Proc. On stratigraphy, strata, and correlations of fossils in New York.
4725 1849, [Obituary of Lardner Vanuxem]: Am. Assoc. Adv. Sci. Proc., v. 1, 91-92.
4726 1849, List of minerals, geological specimens and fossils, added to the collection during 1847 and 1848; with a memorandum of Prof. Hall, as to the additions made to the paleontological cabinet since April 11, 1848: NY State Mus. Ann. Rept., v. 2, 65-70.
4727 1850, Description of new species of fossils, and observations upon some other species previously not well known, from the Trenton limestone: NY State Mus. Ann. Rept., v. 3, 167-175, 5 plates with 23 figs.
4728 1850, List of minerals, geological specimens and fossils, added to the state cabinet of natural history, (including the collection of the late Mr. De Rham): NY State Mus. Ann. Rept., v. 3, 27-46.
4729 1850, On graptolites, their duration in geological periods, and their value in the identification of strata: Am. Assoc. Adv. Sci. Proc., v. 2, 351-352.
4730 1850, On the brachiopoda of the Silurian period; particularly the Leptaenidae: Am. Assoc. Adv. Sci. Proc., v. 2, 347-351.
With discussion by L. Agassiz.
4731 1850, On the trails and tracks in the sandstones on the Clinton group of New York, their probable origin, etc.; and a comparison of some of them with Nereites and Myrianites: Am. Assoc. Adv. Sci. Proc., v. 2, 256-260.
With discussion by L. Agassiz and H. D. Rogers.

Hall, John
4732 1821, Coal found in Somers and in Ellington, Connecticut: Am. J. Sci., v. 3, no. 2, 248.
4733 1821, Fossil bones found in East-Windsor, Connecticut: Am. J. Sci., v. 3, no. 2, 247.

Hall, Samuel Read (1795-1877)
4734 1845, and Z. Thompson, Report of Messrs. Hall and Thompson: in First annual report on the geology of the state of Vermont by C. B. Adams (see 23), 68-76.

Hall, S. R. (cont.)
4735   1846, Report of Mr. Hall: in Second annual report on the geology of the state of Vermont by C. B. Adams (see 26), 174-214.
4736   1847, Report of S. R. Hall: in Third annual report on the geology of the state of Vermont by C. B. Adams (see 32), 27-31.

Hall, W. E.
4737   1841, Popular views and economical applications of geological science: Am. Rep. Arts Sci's Manu's, v. 3, 209-216, 441-442.

Haller, Baron
4738   1797, The topography and natural history of the Swiss Alps: NY Mag., v. 2 (ns), 308-313; 345-349.

Hallowell, Edward
4739   1846, [On the fossil bones of a young mastodon found near Plattsburg, New Jersey]: Phila. Acad. Nat. Sci's Proc., v. 3, no. 6, 130.

Hamilton, J. (i.e. James Hamilton, 1796-1849?)
4740   1839, Terrestrial magnetism: Am. J. Sci., v. 37, no. 1, 100-103.

Hamilton, Mathie
4741   1841, Observations on great earthquakes on the west coast of South America: Am. J. Sci., v. 41, no. 1, 57-59.
        In British Assoc. Proc.

Hamilton, William James (1805-1867)
4742   1837, and H. E. Strickland, [Tertiary formation on the island of Cephalonia]: Am. J. Sci., v. 33, no. 1, 211 (14 lines).
4743   1839, Hot springs [at Singerli, Asia Minor]: Am. J. Sci., v. 37, no. 2, 372 (7 lines). Abstracted from London and Edinburgh Philos. Mag.
4744   1840, The Katakekaumene [near Smyrna]: Am. J. Sci., v. 38, no. 1, 206-208. Abstracted from London and Edinburgh Philos. Mag. On volcanic terrains.

Hamilton, Sir William (1730-1803)
4745   1783, Account of the late earthquakes in Calabria and Sicily: Bos. Mag., v. 1, 95-98.

Hamilton, William (Actuary)
4746   1835, Report on Mr. Thomas S. Ridgway's smelting furnace: Franklin Inst. J., v. 16 (ns), 78-79.
        See also description by T. S. Ridgway (8791).
4747   1837, Report on the stone of Leiper's quarry [Crum Creek, Delaware County, near Philadelphia]: Franklin Inst. J., v. 19 (ns), 367-369.

Hamlin, Elijah L.
4748   1825, By Elijah L. Hamlin [on mineral localities in Maine]: Am. J. Sci., v. 10, no. 1, 14-18.

Hamlin, Hannibal (1809-1891)
4749   1837, Geology: Me. Farm., v. 5, 325-326; 333-334; 340-342.
        From Bangor J. An introduction to geology.

Hammond, James Henry (1807-1864)
4750   1844, [On the South Carolina geological survey]: Niles' Wk. Reg., v. 67, 227.
4751   1846, Marl: Mon. J. Ag., v. 1, no. 11, 559-564; no. 12, 559-605.

Hanau, Menge de (Professor)
4752   1822, Iceland: Am. J. Sci., v. 4, no. 2, 370-371.
        Abstracted by J. Griscom. On a geyser.

Hance, William
4753   1834, Geology: Eclec. J. Sci., v. 2, 705-707.
        On the importance of geology.
4754   1834, Common salt: Eclec. J. Sci., v. 2, 769-770.

Hanchett, I. W. (see 9842 with others)

Hancock, James
4755   1850, [Report on the mineral lands of the Siskowit Mining Company]: in Charter ..., (see 9843), 16-18.

Hand, George
4756   1850, Analysis of the Byron Acid Spring, near Batavia, with a short account of its remedial properties: NY J. Medicine and Collateral Sci's, v. 5 (ns), 141-142.
        From Med. Soc. NY Trans.

Hannum, Cheyney
4757   1834, Chester County cabinet: Hazard's Reg. Penn., v. 13, 327-329.
        Notice of the officers and collections of the Society.

Hanstein (Prof.)
4758   1822, Intensity of the magnetic force in different parts of the world: Mus. For. Liter. Sci., v. 1, 90.
        From J. Sci.
4759   1330, Professor Hanstein. - terrestrial magnetism: Am. J. Sci., v. 17, no. 2 392.

Harcourt, W. W. (Reverend)
4760   1836, Long continued effect of heat on mineral substances: Am. J. Sci., v. 29, no. 2, 357-258.
        In British Assoc. Proc.

Harden, John M. B.
4761   1845, Observations on the soil, climate and diseases of Liberty County, Georgia: S. Med. Surgical J., v. 1 (ns), 545-569.

Hare, Robert (1781-1858)
        See also 10530 by A. Ure.
4762   1802, Memoir on the supply and application of the blow-pipe. Containing an account of a new method of supplying the blowpipe either with common air or oxygen gas; and also the effects of the intense heat produced by the combustion of the hydrogen and oxygen gases: Phila., Chemical Society of Philadeophia, H. Maxwell (printer), 34 p., plate.
        For review with excerpts see 1213.
4763   1804, Account of the fusion of strontites, and volatilization of platinum, and also of a new arrangement of apparatus: Am. Philos. Soc. Trans., v. 6, pt. 1, 99-105, plate.
4764   1820, Strictures on a publication, entitled Clark's gas blowpipe: Am. J. Sci., v. 2, no. 2, 281-302, plate.
        Also reprint editions, Phila., 24 and 28 p.
4765   1823, Description of an improved blowpipe by alcohol, in which the inflammation is sustained by opposing jets of vapour, without a lamp: Also, of the means of rendering the flame of alcohol competent for the purpose of illumination: Am. J. Sci., v. 7, no. 1, 110-111, fig.
4766   1823, Description of an improved blowpipe, by alcohol, in which the inflammation is sustained by opposing jets of vapour, without a lamp: Phila. J. Med. Phys. Sci's, v. 6, 106-107.
4767   1824, Remarks respecting Mr. Vanuxem's memoir on a fused product, erroneously identified with the fused carbon of Professor Silliman; with some additional facts and observations: Am. J. Sci., v. 8, no. 2, 288-292.
        See also memoir by L. Vanuxem (10569). Also reprint edition, 8 p.
4768   1824, Remarks respecting Mr. Vanuxem's memoir ... [as above]: Phila. J. Med. Phys. Sci's, v. 8, 348-352.
4769   1825, Examination of the projections which arise upon charcoal intensely ignited between the poles of a galvanic deflagrator: Am. J. Sci., v. 10, no. 1, 118-119.
4770   1825, Letter from Dr. Hare to the editor: Am. J. Sci., v. 10, no. 1, 114-117.
        On the fusion of carbon.
4771   1825, Strictures by Robert Hare, M. D. Professor of Chemistry, &c. &c. upon Professor Vanuxem's memoir on plumbago, anthracite, fused carbon &c.: Am. J. Sci., v. 10, no. 1, 111-114.
        See also articles by L. Vanuxem (10569 et seq).
4772   1826, An account of the hydrostatic blowpipe, now used in laboratory of the University of Pennsylvania: Franklin Inst. J., v. 1, 160-163, fig.; 193-197, 3 figs.
4773   1826, On specific gravity: Am. J. Sci., v. 11, no. 1, 121-144, 11 figs; also in: Franklin Inst. J., v. 1, 42-47, 4 figs.; 99-103, 2 figs; 157-160, fig.
4774   1831, On blasting rocks: Franklin Inst. J., v. 8 (ns), 232-234.
4775   1833, Description of a process and an apparatus for blasting rocks, by means of galvanic ignition: Franklin Inst. J., v. 11 (ns), 221-226, 2 figs.
4776   1838, [On the compound blowpipe]: Am. Philos. Soc. Proc., v. 1, no. 5, 59-60.
4777   1844, [Mica used for light filter]: Am. Philos. Soc. Proc., v. 4, no. 31, 114.
4778   1846, Improvements in the compound hydro-oxygen or hydrostatic blowpipe: Franklin Inst. J., v. 11 (3s), 388-389.
4779   1847, On certain improvements in the construction and supply of the hydro-oxygen blowpipe, ... : Franklin Inst. J., v. 13 (3s), 196-206, 4 figs.
4780   1847, On certain improvements in the construction and supply of the hydro-oxygen blowpipe, by which rhodium, iridium, or the osmiuret of iridium, also platinum in the large way, have been fused: Am. J. Sci., v. 4 (ns), no. 1, 37-45, 4 plates.

Harlan, Richard (1796-1843)
        See also 4387 with J. B. Gibson.
4781   1821, Description of the fossil bones of Megalonyx, discovered in White Cave, Kentucky: 8vo. pamphlet.
        Not seen.
4782   1823, Observations on fossil elephant teeth of North America: Phila. Acad. Nat. Sci's J., v. 3, pt. 1, 65-67, plate with 3 figs.
4783   1824, On a new extinct fossil species of the genus Ichthyosaurus: Phila. Acad. Nat. Sci's J., v. 3, pt. 2, 338-340, plate with 3 figs.
4784   1824, On a new fossil genus, (of the order Enalio Sauri of Conybeare): Phila. Acad. Nat. Sci's J., v. 3, pt. 2, 331-337, plate with 5 figs.
4785   1824, On an extinct species of crocodile not before described; and some observations on the geology of West Jersey: Phila. Acad. Nat. Sci's J., v. 4, pt. 1, 15-24, plate with 8 figs.
4786   1825, Fauna Americana, being a description of the mammiferous animals inhabiting North America: Phila., A. Finley, x, 11-318 p.
4787   1825, Notice of the Plesiosaurus, and other fossil reliquiae, from the state of New Jersey: Phila. Acad. Nat Sci's J., v. 4, pt. 2, 232-236, plate with 12 figs.
4788   1828, Note, from R. Harlan, M. D. on the examination of the large bones disinterred at the mouth of the Mississippi River, and exhibited in the city of Baltimore, January 22nd, 1828: Am. J. Sci., v. 14, no. 1, 186-187.
4789   1831, Description of an extinct species of fossil vegetable, of the family Fucoides: Phila. Acad. Nat. Sci's J., v. 6, pt. 2, 289-295, plate with 3 figs.
4790   1831, Description of the fossil bones of the Megalonyx, discovered in "White Cave," Kentucky: Phila. Acad. Nat. Sci's J., v. 6, pt. 2, 269-288, 3 plates with 26 figs.
        Abstracted in Am. J. Sci., v. 20, no. 2, 414-415.
4791   1831, Description of the jaws, teeth, and clavicle of the Megalonyx laqueatus: Mon. Am. J. Geology, v. 1, 74-76, plate with 7 figs.

Harlan, R. (cont.)

4792 1831, Megalonyx laqueatus [from White Cave, Kentucky]: Mon. Am. J. Geology, v. 1, no. 1, 45.
4793 1831, Tour to the caves in Virginia: Mon. Am. J. Geology, v. 1, 58-67.
4794 1832, [Letter on Atlantic Journal by C. S. Rafinesque]: Mon. Am. J. Geology, v. 1, no. 11, 514.
4795 1832, On a new extinct fossil vegetable of the family Fucoides [from Lockport, New York]: Mon. Am. J. Geology, v. 1, no. 7, 307-308.
4796 1834, Critical notices of various organic remains hitherto discovered in North America: Geol. Soc. Penn. Trans., v. 1, 46-112, plate.
    With excerpts by Cuvier (see 2761). Includes lists of vertebrate and invertebrate fossils found in America.
4797 1834, [Footnote to "On the localities in Tennessee in which bones of the gigantic Mastodon and Megalonyx are found" by G. Troost (see 10364)]: Geol. Soc. Penn. Trans., v. 1, 146.
4798 1834, Notice of fossil bones found in the Tertiary formation of the state of Louisiana: Am. Philos. Soc. Trans., v. 4 (ns), 397-403.
    With a letter by Judge H. Bry (see 2134).
4799 1834, Notice of the discovery of the remains of the Ichthyosaurus in Missouri, N. A.: Am. Philos. Soc. Trans., v. 4 (ns), 405-408, [2], plate.
4800 1834, [Letter to Jacob Green on new North American trilobites]: Am. J. Sci., v. 25, no. 2, 337 (7 lines).
    See article by J. Green (4540).
4801 1834, On the structure of the teeth in the "Edentata", fossil and recent: Geol. Soc. Penn. Trans., v. 1, 40-45.
4802 1835, Description of a new fossil plant from Pennsylvania of the genus Equisetum: Geol. Soc. Penn. Trans., v. 1, 260-262, plate.
4803 1835, Description of the remains of the "Basilosaurus", a large fossil marine animal recently discovered in the horizontal limestone of Alabama: Geol. Soc. Penn. Trans., v. 1, 348-357, plate with 4 figs., folding plate with 6 figs.
4804 1835, Fossil zoology and comparative anatomy: Am. J. Sci., v. 27, no. 2, 351-354.
    From Geol. Soc. Penn. Trans. On fossil vertebrates of North America.
4805 1835, Medical and physical researches; or, original memoirs in medicine, surgery, physiology, geology, zoology and comparative anatomy ... : Phila., Lydia R. Bailey (printer), xxxix, [9]-653 p., 39 plates (4 colored, 12 folded), folded table.
    For review see 661.
4806 1835, Notice of fossil vegetable remains from the bituminous coal measures of Pennsylvania, being a portion of the illustrative specimens accompanying Mr. Miller's essay or geological section of the Alleghany Mountain, near the Portage Railway: Geol. Soc. Penn. Trans., v. 1, 256-259, plate with 3 figs.
4807 1835, Notice of nondescript trilobites, from the state of New York, with some observations on the genus Triarthus, &c.: Geol. Soc. Penn. Trans., v. 1, 263-266, plate with 7 figs.
4808 1835, Notice of the os illium of the Megalonyx laqueatus from Big Bone Cave, White County Tennessee: Geol. Soc. Penn. Trans., v. 1, 347, plate.
4809 1835, On the affiliation of the natural and physical sciences: being the introduction to "Medical and physical researches" ... : Phila., 31 p.
4810 1841, Fossil bones and models of Dinortherium and Basilosaurus: Am. J. Sci., v. 41, no. 1, 178-180.
    Also in Assoc. Am. Geol's Nat's Proc., v. 2, 23-25. Also in Assoc. Am. Geol's Nat's Rept., v. 1, 31-32, 1843.
4811 1842, Bones of the Orycterotherium: Am. J. Sci., v. 42, no. 2, 392 (12 lines).
    Opinions of Dr. Perkins refuted.
4812 1842, Description of a new extinct species of dolphin; from Maryland: Nat'l Inst. Prom. Sci. Bulletin, v. 2, 195-196, 3 folding plates.
4813 1842-1843, Description of the bones of a fossil animal of the order Edentata: Am. Philos. Soc. Proc., v. 2, 109-111; Am. J. Sci., v. 44, no. 1, 69-80, 3 plates with 32 figs.
    On fossil bones from the collection of A. Koch, from Benton County, Missouri, identified as the genus Orycterotherium.
4814 1842, Notice of two new fossil mammals from Brunswick Canal, Georgia; with observations on some of the fossil quadrupeds of the United States: Am. J. Sci., v. 43, no. 1, 141-144, plate with 3 figs.
    See also article by R. Owen (8092). On Sus and Chelonia from Georgia, and on the nomenclature of Megalonyx vs. Mylodon.
4815 1843, Remarks on Mr. Owen's letter to the editors on Dr. Harlan's new fossil mammalia: Am. J. Sci., v. 45, no. 1, 208-211.
    See also article by R. Owen (8092). On Megalonyx vs. Mylodon.

Harlem Royal Society of Sciences. (Netherlands)
4816 1812, Prize question concerning the new metals: Am. Min. J., v. 1, no. 3, 189 (10 lines).
    Abstracted from Nicholson's J. On the manufacture of sodium and potassium.

Harmon, Daniel Williams (1778-1845)
4817 1820, A journal of voyages and travels in the interior of North America, between the 47th and 58th degrees of latitude ... illustrated by a map of the country: Andover, Mass., Flagg and Gould, xxiii, 25-432 p., folded map, plate.

Harper's New Monthly Magazine (New York, 1850)
4818 1850, Eruption of Mount Etna in 1669: v. 1, no. 1, 35.
4819 1850, [Pelorosaurus of Sussex, England]: v. 1, no. 1, 125.
4820 1850, [Review of Six months in the gold mines by E. G. Buffum (see 2180)]: v. 1, no. 2, 286.
4821 1850, [Gold in California and Oregon]: v. 1, no. 3, 419; no. 4, 566; no. 5, 706.
4822 1850, Terrestrial magnetism: v. 1, no. 5, 651-656.
4823 1850, Early history of the use of coal: v. 1, no. 5, 656-657.
4824 1850, Salt mines of Europe: v. 1, no. 5, 759-761.
4825 1850, [Earthquake at Cleveland Ohio]: v. 1, no. 6, 851 (7 lines).
4826 1850, [Notice of Footprints of the Creator by H. Miller (see 7186)]: v. 1, no. 6, 857.

Harris, Edward
4827 1845, (On the geology of the Upper Missouri): Phila. Acad. Nat. Sci's Proc., v. 2, no. 9, 235-238.

Harris, Frederick (of Frederickshall, Louisiana)
4828 1833, Tar from pit coal, a cheap substitute for paint, for the roofs of houses: Farm's Reg., v. 1, 289 (14 lines).

Harris, James (c. 1666-1719)
4829 1809, An account of the production of a large island out of the sea, between the Canary Islands, in the year 1707: Visitor, v. 1, 35-36.

Harris, John (1802-1856)
4830 1847, The Pre-Adamite earth; contributions to theological science ... : Bos., Gould, Kendall and Lincoln, 294 p.
Also an 1849 printing. For review see 1222, 5231, 8173, 9339.
4831 1847, Antiquity of the earth: Living Age, v. 14, 516-518.
Extracted from Pre-Adamite earth.
4832 1849, Man primeval: or, the constitution and primitive condition of the human being. A contribution to theological science: Bos., Gould, Kendall and Lincoln, xvi, 480 p., plate.
Also 1850 printing as "second thousand." For review see 1009, 5472, 5518, 8173.
4833 1850, The Pre-Adamite earth; contributions to theological science... : Bos., Gould, Kendall and Lincoln, 300 p.
"3rd thousand, revised and enlarged."

Harris, Thaddeus Mason (1768-1842)
4834 1793, The natural history of the Bible: or a description of all the beasts, birds, fishes, insects, reptiles, trees, precious stones, &c. mentioned in the sacred Scriptures: Bos., I. Thomas and E. T. Andrews, 272 p.
Evans number 25586. Also 1820 edition: Bos., Wells and Lilly, 476 p.
4835 1803, The minor encyclopedia, or cabinet of general knowledge: being a dictionary of arts, sciences, and polite literature: Bos., West and Greenleaf, 4 v.
4836 1805, The journal of a tour into the territory Northwest of the Alleghany Mountains, made in the spring of the year 1803. With a geographical and historical account of the state of Ohio: Bos., Manning and Losing, viii, [11]-271 p., folded plate, 3 folding maps.

Harrison, Randolph (1769-1839)
4837 1840, Plaster of Paris: Farm's Reg., v. 8, 582.
From Va. Soc. Promoting Ag. Mem.

Harrison, W. D.
4838 1825, Statement of W. D. Harrison, Esq.: Am. J. Sci., v. 9, no. 2, 353.
On a meteorite fall in Maryland. See also article by S. D. Carver (2272).

Harsleban, Charles
4839 1831, and W. Davis, Machine for facilitating the washing of ores and alluvial soils: Franklin Inst. J., v. 7 (ns), 83.

Hart (Mr., of Berlin, Connecticut)
4840 1802, [Use of gypsum for fertilizer]: Ct. Soc. Promoting Ag. Trans., v. 1, 6-7.

Hart, Reuban
4841 1811, Topographical sketch of the county of Ontario, in the state of New York: Am. Med. Philos. Reg., v. 2, no. 2, 150-154.

Hartwell, Charles
4842 1850, and E. Hitchcock, Description of certain mineral localities, chiefly in the northern part of Worcester and Franklin Counties in Massachusetts: Am. Assoc. Adv. Sci. Proc., v. 2, 159-160.
4843 1850, On a new spodumene locality at Norwich, Mass.: Am. J. Sci., v. 10, no. 2, 264-265, fig.

Harvard Natural History Society (Cambridge, Massachusetts)
4844 1837, A catalogue of the officers and members of the Harvard Natural History Society, founded 1837: Cambridge, Mass., Metcalf and Co., 19 p.
4845 1848, A catalogue ... [as above]: Cambridge, Mass., 22 p.

Haskell, Charles T. (of Abbeville District, South Carolina)
4846 1841, Calcareous earth discovered in a new form and a new locality: Farm's Reg., v. 9, 217-218.

Hassenfratz, Jean Henri (1755-1827; see 3104 with M. Descostils)

Hastings, G. H.
4847 1842, A visit to the caves of Virginia: Bos. Miscellany Liter. and Fashion, v. 2, 272-277.

Hastings, Lansford Warren (1819-c.1870)
4848 1845, The emigrants' guide to Oregon and California, containing scenes and incidents of a party of Oregon emigrants; and a description of California; with a description of the different routes to those countries; and all necessary information relative to the equipment, supplies, and the method of travelling: Cin., G. Conclin, 152 p.
4849 1847, A new history of Oregon and California: containing complete descriptions of those countries, together with the Oregon treaty and correspondence, and a vast amount of information relating to the soil, climate, productions, rivers and lakes, and the various routes over the Rocky Mountains: Cin., G. Conclin.
Also other printings in 1848 (160 p.) and 1849 (168 p.).

Hatchett, Charles (1765-1847)
4850   1803, Analysis of the American mineral substance containing a metal hitherto unknown: Med. Repos., v. 6, 322-324.
       On the discovery of columbium. Also an obituary of T. P. Smith.

Hausmann, Johann Friedrich Ludwig (1782-1859)
       See also 10045 with F. Stromeyer, and 3449 with C. G. Ehrenberg.
4851   1823, Prof. Hausmann on the geology of the Apennines: Bos. J. Phil. Arts, v. 1, no. 4, 377-383.
       Abstract of "De Apenninorum constitutione geognostica commentatio", translated by J. W. Webster.
4852   1835, Antimonial nickel: Am. J. Sci., v. 28, no. 2, 395-396.
       Abstracted from Jameson's Edinburgh New Philos. J.
4853   1848, Cause of irised colors on minerals: Am. J. Sci., v. 6 (ns), no. 2, 254-255.
       Abstracted from l'Institut.
4854   1849, On glaucophane [from Cyclades Island]: Am. J. Sci., v. 8 (ns), no. 1, 123 (9 lines).
       Abstracted from "Jour. fur Prakt. Ch."

Haüy, René-Just (1743-1822)
4855   1797, [Crystal structure of zeolites]: Univ. Mag., v. 1, 330.
       In Nat'l Inst. Paris Proc.
4856   1811, New edition of Haüy's mineralogy: Am. Min. J., v. 1, no. 2, 121 (4 lines).
4857   1815, Haüy on American mineralogy: Med. Repos., v. 17, 86-87.
4858   1817, Observations on tourmalines, and particularly on those which are found in the United States: Am. Mon. Mag. and Crit. Rev., v. 1, no. 2, 128-130.
       Translated by P. S. Townsend.
4859   1818, Magnetism applied as a test for iron: J. Sci. Arts, v. 5, no. 1, 136-137.
       Abstracted from Annales des Mines.

Hawaiian Spectator (Honolulu, 1838-1839)
4860   1838, Great crater on the summit of Mauna Loa, Hawaii: v. 1, no. 2, 98-103.
4861   1839, Natural curiosity on Kauai: v. 2, no. 1, 120.
4862   1839, Sandwich Islands. - earthquake at Hilo: v. 2, no. 2, 231-132 [i.e. 232].

Hawkes, Pitty
4863   1827, American companion, or a brief sketch of geography ... : Phila., R. Desilver and R. H. Small, viii, 237 p.

Hawkins (Mr.)
4864   1825, Hawkins mode of preparing emery: Bos. J. Phil. Arts, v. 3, no. 2, 204-205.
       Abstracted from Technical Reports.

Hawkshaw, Sir John (1811-1891)
4865   1841, Fossil trees [from Manchester and Chesterfield, England]: Am. Eclec., v. 1, no. 2, 392-393.
       Abstracted from London and Edinburgh Philos. Mag. and from Annals Nat. History.
4866   1843, Notice of fossil footsteps in the New Red Sandstone at Lymm, in Cheshire: Am. J. Sci., v. 44, no. 2, 370 (10 lines).
       In British Assoc. Proc.

Hay, Charles Augustus (1821-1893)
4867   1844, How salt is procured: Lit. Rec. and J. Linnaean Assoc. Penn. College, v. 1, 27-29.
4868   1844, The silver mines of Andreasberg, in the Harz: Lit. Rec. and J. Linnaean Assoc. Penn. College, v. 1, 141-145.

Hayden, C. B. (of Smithfield, Virginia?)
4869   1841, Capacity of the clays and rock-marl of Virginia, to form hydraulic cement, and the applicability of rock-marl to burning lime: Farm's Reg., v. 9, 270-272.
4870   1841, A chemical and geological account of the Shocco Springs: Farm's Reg., v. 9, 145-146.
4871   1843, Analysis of the Scott Spring, Scott County, Virginia: Am. J. Sci., v. 44, no. 2, 409 (15 lines).
4872   1843, On the ice mountain of Hampshire County, Virginia, with a proposed explanation of its low temperature: Am. J. Sci., v. 45, no. 1, 78-83, fig.
4873   1843, On the rock salt and salines of the Holsten [River, Virginia]: Am. J. Sci., v. 44, no. 1, 173-179; also in: Farm's Reg., v. 11, 51-54.

Hayden, Horace Handel (1769-1844)
4874   1811, Geological sketch of Baltimore, &c.: Balt. Med. Philos. Lyc., v. 1, 255-271; Also in: Am. Min. J., v. 1, no. 4, 243-248 (1814).
4875   1819, Extract of a letter from Dr. H. H. Hayden of Baltimore, to the editor, dated January 5, 1819: Am. J. Sci., v. 1, no. 3, 306-307.
       See also 331. On the mineral "necronite" from Maryland.
4876   1819, Red pyroxene augite. Extract of a letter to the editor: Am. J. Sci., v. 1, no. 3, 244 (13 lines).
4877   1820, Geological essays; or, an enquiry into some of the geological phenomena to be found in various parts of America, and elsewhere: Balt., J. Robinson (printer), for the author, viii, 412 p.
       For review with excerpts see 10708. For review see 443. For notice see 358, 4883.
4878   1821, Fluor spar in Tennessee: Am. J. Sci., v. 4, no. 1, 51.
4879   1822, Notice of a singular ore of cobalt and manganese [from Baltimore]: Am. J. Sci., v. 4, no. 2, 283-284.
4880   1824, An inquiry into some of the geological phenomena to be found in various parts of America, and elsewhere: Meth. Rec'r, v. 1, no. 1, 26-32; no. 2, 58-68; no. 3, 115-118; no. 5, 189-197.

Hayden, H. H. (cont.)
4881 1830, Notices of the geology of the country near Bedford Springs in Pennsylvania, and the Bath or Berkeley Spring in Virginia, with remarks upon those waters: Am. J. Sci., v. 19, no. 1, 97-104.
With "Remark" by B. Silliman (see 9697).
4882 1830, Notices of the geology of the country near Bedford Springs in Pennsylvania: Hazard's Reg. Penn., v. 6, 290-292.
4883 1831, To mineralogists, geologists, &c.: Am. J. Sci., v. 20, no. 1, 164.
Notice of Geological essays (see 4877) for sale.
4884 1833, Description of the Bare Hills near Baltimore: Am. J. Sci., v. 24, no. 2, 349-360, fig.

Hayes, Augustus Allen (1806-1882)
See also 10172 with J. E. Teschemacher.
4885 1827, Localities of minerals in Vermont: Am. J. Sci., v. 13, no. 1, 195-196.
4886 1829, Eaton [New Hampshire] galena: Am. J. Sci., v. 17, no. 1, 196 (11 lines).
4887 1830, Hydro-bromic acid and potash in the Saratoga Springs: Am. J. Sci., v. 18, no. 1, 142-143.
4888 1831, Bromine [in the Hingham, Massachusetts mineral springs]: Am. J. Sci., v. 20, no. 1, 161 (10 lines).
4889 1831, Floating pumice. - Extract of a letter: Am. J. Sci., v. 20, no. 1, 161 (13 lines).
4890 1831, Notice of cobalt, nickel, &c. of the Chatham Mine, Connecticut: Am. J. Sci., v. 21, no. 1, 195-196.
4891 1833, Analysis of ledererite: Am. J. Sci., v. 25, no. 1, 80-84.
In "A description of a new mineral from Nova Scotia" by C. T. Jackson (see 5540).
4892 1833, Details of a chemical analysis of danaite, a new ore of iron and cobalt [from Franconia, New Hampshire]: Am. J. Sci., v. 24, no. 2, 386-388.
4893 1840, Notice of native nitrate of soda, containing sulphate of soda, chloride of sodium, iodate of soda, and chloriodide of sodium, from the Province of Tarapaca, Pampa of Tamarugal, in South Peru; and of Algoroba Wood, from the buried forests beneath the Pampa of Tamarugal: Bos. J. Nat. History, v. 3, no. 1-2, 279-280.
Abstracted in Am. J. Sci., v. 38, no. 2, 410 (1840).
4894 1840, On the native nitrate of soda found in South Peru: Am. J. Sci., v. 39, no. 2, 375-378.
In Bos. Soc. Nat. History Proc.
4895 1842, [Qualitative analysis of a uranium mineral from Chesterfield, Massachusetts]: Bos. J. Nat. History, v. 4, no. 1, 36.
In article by J. E. Teschemacher (see 10173).
4896 1844, [Analysis of meteoric iron]: Bos. Soc. Nat. History Proc., v. 1, 207 (9 lines).
4897 1844, Description and analysis of pickeringite, a native magnesium alum: Am. J. Sci., v. 46, no. 2, 360-362.
On a mineral from the Pampa of Iquique, South Peru.
4898 1844, Hydrous borate of lime: Am. J. Sci., v. 47, no. 1, 215 (9 lines).
4899 1844, Re-examination of microlite and pyrochlore: Am. J. Sci., v. 46, no. 1, 158-166.
4900 1845, Analysis of red oxide of zinc: Am. J. Sci., v. 48, no. 2, 261-264.
In article on Franklin, New Jersey, by F. Alger (see 139).
4901 1845, Letter from Mr. A. A. Hayes on the same subject, with remarks on the origin of the chlorine found in the Alabama iron, and a description of new methods employed in the analysis of meteoric iron: Am. J. Sci., v. 48, no. 1, 147-156.
4902 1846, Acadiolite of Nova Scotia, (chabasite): Am. J. Sci., v. 1 (ns), no. 1, 122 (7 lines).
4903 1846, [Review of The use of the blowpipe in chemistry and mineralogy by J. J. Berzelius (see 1668)]: N. Am. Rev., v. 62, 260-262.
4904 1847, Report [on the Pittsburgh and Boston Copper Harbor Mining Company]: in 8381.
4905 1850, On the blowpipe characters of the mineral from the Azores identified with pyrrhite by J. E. Teschemacher: Am. J. Sci., v. 9 (ns), no. 3, 423-424.
4906 1850, On the red zinc ore of New Jersey: Am. J. Sci., v. 9 (ns), 424 (11 lines).
On zincite from Franklin, New Jersey.

Hayes, George Edward (1804-1882)
4907 1837-1838, Remarks on the geology of western New York: Am. J. Sci., v. 31, no. 2, 241-247; v. 35, no. 1, 86-105, fig.
4908 1838, Evidences of diluvial currents - petrifactions - metallic models of shells: Am. J. Sci., v. 35, no. 1, 191.
4909 1839, The geology of the United States north and west of the Alleghany Mountains: Am. Inst. NY J., v. 4, 489-490.
Abstracted from Am. J. Sci. and Buffalo J.

Hayes, John Lord (1812-1887)
4910 1842, Explanation of the fossil footprints in the sandstone of Connecticut River valley: Am. J. Sci., v. 43, no. 1, 172 (4 lines).
Also in Assoc. Am. Geol's Nat's Proc., v. 3, 27 (4 lines). Also in Assoc. Am. Geol's Nat's Rept., v. 1, 65-66 (4 lines), 1843. Am. J. Sci. note is ascribed to "John S. Hayes."
4911 1843, [Dana's coral reef theories]: Am. J. Sci., v. 45, no. 1, 312 (5 lines).
In Assoc. Am. Geol's Nat's Proc. See also 2805.
4912 1843, [Origin of ornithichnites]: Am. J. Sci., v. 45, no. 2, 316 (6 lines).
In Assoc. Am. Geol's Nat's Proc.
4913 1843, Probable influence of icebergs upon drift: Am. J. Sci., v. 45, no. 2, 316-319.
In Assoc. Am. Geol's Nat's Proc.
4914 1844, Letter [on the geology of Portsmouth, New Hampshire]: in Final report on the geology and mineralogy of the state of New Hampshire by C. T. Jackson (see 5611), 279-281.
4915 1844, Probable influence of icebergs upon drift: Bos. J. Nat. History, v. 4, no. 4, 426-452.

Hayes, J. L. (cont.)
4916 1844, Report on the geographical distribution and phenomena of volcanoes: Am. J. Sci., v. 47, no. 1, 127-128.
Also in Assoc. Am. Geol's Nat's Proc., v. 5, 36-37.
4917 1846, Notice of Baron Wolfgang Sartorius von Waltershausen's work on Mount Etna: Am. J. Sci., v. 2 (ns), no. 2, 157-162.
4918 1849, The gold deposits of Siberia: Am. Rr. J., v. 22, 213-214.
4919 1850, Gold dust in California: Ann. Scien. Disc., v. 1, 256 (11 lines).
4920 1850, Memorial of the iron manufacturers of New England, asking for a modification of the tariff of 1846: Phila., C. Sherman (printer), 39 p.
4921 1850, [Recent ornithichnites from Nova Scotia]: Bos. Soc. Nat. History Proc., v. 3, 227-228.

Hayes, P.
4922 1824, An account of the inflammable springs in Ontario county state of New York: NY Med. Phys. J., v. 3, 49-54.

Haymond, Rufus
4923 1844, Notice of remains of Megatherium, Mastodon, and Silurian fossils [from Ohio and Indiana]: Am. J. Sci., v. 46, no. 2, 294-296.

Hays, Isaac (1796-1879)
See also 4419 with J. D. Godman, and 5267 with W. E. Horner
4924 1830, Description of a fragment of the head of a new fossil animal, discovered in a marl pit, near Moorestown, New Jersey: Am. Philos. Soc. Trans., v. 3 (ns), 471-477, plate.
On the Saurodon leanus. also reprint edition, 7 p.
4925 1834, Descriptions of the specimens of inferior maxillary bones of mastodons in the cabinet of the American Philosophical Society, with remarks on the genus Tetracaulodon (Godman), &c.: Am. Philos. Soc. Trans., v. 4 (ns), 317-339, plate.
Also reprint edition, 23 p., plate. For review see 1475.
4926 1841, Fossil bones, chiefly of the Mastodon [from the collection of A. Koch]: Am. Philos. Soc. Proc., v. 2, no. 19, 102-103, 105-106.
4927 1842, [Remarks on Professor Owen's paper on Missouri fossils]: Am. Philos. Soc. Proc., v. 2, no. 22, 183-184.
On the genus Tetracaulodon.
4928 1843, On the family Proboscidae, their general character and relations, their mode of dentition and geological distribution: Am. Philos. Soc. Proc., v. 3, no. 27, 44-48.
4929 1843, [On three papers relative to the mastodontoid fossil bones in the collections of A. Koch]: Am. Philos. Soc. Proc., v. 2, no. 26, 264-266.

Haywood, John (1762-1826)
4930 1823, The natural and aboriginal history of Tennessee, up to the first settlements therein by the white people in the year 1768: Nashville, Tenn., George Wilson (printer), viii, 390, liv p.
See especially chapters 1 and 2, pp. 1-61.

Haywood, John
4931 1843, Gazetteer of Maine: Bos. and Portland, Me.
Not seen. Pages 30-36 are supposedly on rocks and minerals of the state.

Hazard, Erskine (1789-1865)
4932 1827, History of the introduction of anthracite coal into Philadelphia: Penn. Hist. Soc. Mem., v. 2, pt. 1, 155-162.

Hazard, Samuel (1784-1870)
4933 1834, Our coal formation [in Pennsylvania]: Hazard's Reg. Penn., v. 14, 408-410.

Hazard's Register of Pennsylvania. Devoted to the preservation of facts and documents and every other kind of useful information respecting the state of Pennsylvania. Edited by Samuel Hazard. (Philadelphia, 1828-1835)
4934 1828, Pennsylvania salt: v. 1, 29-30, 61-62.
4935 1828-1833, [Notes on Pennsylvania caves]: v. 1, 132-133; v. 3, 265-267; v. 9, 222-223; v. 12, 22.
4936 1828-1834, [Notes on Pennsylvania earthquakes]: v. 1, 192; v. 2, 25; 379; 382; 386; v. 13, 112, 128.
4937 1828, Chester County Cabinet of Natural Sciences: v. 1, 302-304.
Description of the mineral collection.
4938 1828-1833, [Notes on Pennsylvania mineral springs]: v. 2, 110-112, 115-117; v. 12, 96, 112.
4939 1828-1829, [Notes on Pennsylvania vertebrate fossils]: v. 2, 142; v. 4, 160.
4940 1829, A sketch of Crawford County, Pa.: v. 3, 10-11.
4941 1829, Portraiture of Bedford County [Pennsylvania]: v. 3, 145-146.
4942 1829, Franklin County [Pennsylvania]: v. 3, 233-235.
From the Franklin Repub.
4943 1829-1835, [Notes on Pennsylvania coal]: v. 3, 240, 256; v. 3, 336; 399; 410; v. 4, 64; 72-73; 92; 144; 380; v. 5, 240; 403-405; v. 6, 222-223; 314; 413; v. 7, 313; 400; v. 8, 176; 366; 383-384; 416; v. 9, 303-304; v. 10, 14-15; 319-320; 415-416; v. 12, 57-58; 62; 80; 111; 253; v. 14, 136; 318; v. 16, 64; 84-88; 157.
4944 1829, Aereoliths: v. 4, 250-251.
4945 1829, Franklin Institute: v. 4, 284.
On the mineral collection of the Franklin Institute.
4946 1830-1833, [Notes on Pennsylvania marble]: v. 5, 256; v. 10, 157-158; v. 12, 31.
4947 1830, Rocking stone [from Mine Hill, Pennsylvania]: v. 5, 256.
4948 1830-1832, [Notes on Geological Society of Pennsylvania]: v. 6, 272; v. 9, 208; v. 10, 335.
4949 1831-1835, [Notes on Pennsylvania lead ore]: v. 7, 31; v. 8, 164-165; v. 16, 112, 192.
4950 1831, [Geological society of Bradford County]: v. 7, 96.

Hazard's Register of Pennsylvania (cont.)
4951 1831, Monthly American Journal of Geology [review of journal by G. W. Featherstonhaugh (see 3746)] : v. 8, 16.
4952 1831-1835, [Notes on soils, limestone and lime] : v. 8, 24, 48; v. 14, 141; 159; 175; v. 15, 288; v. 16, 16; 22; 286.
4953 1831, A description of Bald Eagle Valley [Bedford County, Pennsylvania] : v. 8, 36-38.
4954 1831-1833, [Notes on gold in Pennsylvania]: v. 8, 106; v. 12, 65-68; v. 14, 34.
4955 1832, Geology of Bradford County: v. 10, 201.
       From Geol. Soc. Penn. Minutes.
4956 1832-1833, [Notes on Pennsylvania iron ore] : v. 10, 208; v. 11, 128; v. 12, 143.
4957 1833, A silver mine [from Sugarcreek Township, Armstrong County, Pennsylvania] : v. 11, 287 (12 lines).
4958 1833, Mineralogy [of Lancaster, Pennsylvania] : v 12, 390.
4959 1834, Franklin Institute: v. 13, 76-78.
       On their support of geology.

Hazard's United States Commercial and Statistical Register, Containing Documents, Facts, and Other Useful Information Illustrative of the History and Resources of the American Union, and of Each State. Edited by Samuel Hazard. (Philadelphia, 1839-1842)
4960 1839-1842, [Hundreds of short notes on the United States' production, import, export, and use of coal, lead, iron, copper, salt, tin, salt, slate, gold, silver, diamonds, marble, mercury and granite appear in this periodical] : Consult annual indices for specific notes.
4961 1839-1841, [Notes on American caves] : v. 1, 95; v. 5, 126-127.
4962 1839-1840, New mineral fountain [at Saratoga, New York] : v. 1, 102;
4963 1839-1842, [Notes on American vertebrate fossils] : v. 1, 175; v. 6, 181; 292; 368.
4964 1839, Geological exploration of Iowa, &c.: v. 1, 229.
       Notice of survey by D. D. Owen.
4965 1839, [C. T. Jackson to do New Hampshire survey]: v. 1, 232 (3 lines).
4966 1839, [Vermont geological survey appropriations]: v. 1, 339 (3 lines).
4967 1839-1840, [On North American earthquakes]: v. 1, 420; v. 3, 88; 352.
4968 1840, Catlinite: v. 2, 349 (8 lines).
4969 1840, Connecticut meteor [at Stratford]: v. 3, 172.
4970 1842, Remarkable geological changes [of the Merrimack River] : v. 6, 325.

Headrick, James (d. 1841)
4971 1833, [Effects of magnesia in different forms on soils] : Farm's Reg., v. 1, 426-433.
       In 3717.

Heart, Jonathan (1748-1791)
4972 1793, A letter from Major Jonathan Heart, to Benjamin Smith Barton, M. D. corresponding member of the Society of the Antiquaries of Scotland, Member of the American Philosophical Society, and Professor of Natural History and Botany in the University of Pennsylvania, - containing observations on the ancient works of art, the native inhabitants, &c. of the Western-country: Am. Philos. Soc. Trans., v. 3, 214-222.
       On Big Bone Lick, Kentucky.

Heath, James Ewell (1792-1862)
4973 1833, General description of Virginia from the Encyclopedia Americana: Farm's Reg., v. 1, 1-5.
       Signed by "J. E. H."
4974 1834, A description of the valley of the Kanawha, by a low-lander, in a letter to a friend: Farm's Reg., v. 1, 525-528.

Hebert (see L. Herbert)

Heine, M.
4975 1846, Quantitative estimation of bromine in mineral waters: Am. J. Sci., v. 2 (ns), no. 1, 113.
       Abstracted from "Jour. fur Prakt. Chem."

Heintz, M.
4976 1846, Colors of quartz: Am. J. Sci., v. 2 (ns), no. 3, 413 (5 lines).
       Abstracted from Poggendorff's Annalen.

Helmersen, M. S. de (i.e. Grigorii Petrovich, 1803-1885)
4977 1846, The Oust-Urt, and shores of Lake Aral: Am. J. Sci., v. 1 (ns), no. 1, 123 (10 lines).
       Abstracted from l'Institut. On the historical geology of this Mediterranean region.

Helvetic Society of Natural Sciences
4978 1830, Helvetic Society of Natural Sciences: Am. J. Sci., v. 18, no. 1, 168-170.

Hemans, Felicia Dorothea (1793-1835)
4979 1836, Epitaph on a mineralogist: Am. Mag. Use. Ent. Know., v. 3, no. 2, 72.
       A poem.

Hemming
4980 1834, A new tenantite [from Cornwall, England] : Am. J. Sci., v. 26, no. 2, 386 (6 lines).
       Abstracted from Berzelius' Jahresbericht by L. Feuchtwanger.

Henderson, Ebenezer (1784-1858)
4981 1831, Iceland: or the journal of a residence in that island, during the years 1814 and 1815; containing observations on the natural phenomena, history, literature, and antiquities of the island; and the religion, character, manners and customs of its inhabitants: Bos., Perkins and Marvin, 252 p., map, plates.
       For review see 9984. "Abridged from the second Edinburgh edition."

Henderson, Hugh
4982 1811, A topographical description of Jefferson County in the state of New-York: Med. Repos., v. 14, 21-27.

Henfrey, Benjamin
4983 1795, [Broadside on the promotion of mining in America]: Phila., Snowden and M'Corkle, 4to.
Evans number 28822.
4984 1797, A plan with proposals for forming a company to work mines in the United States; and to smelt and refine ores whether of copper, lead, tin, silver, or gold: Phila., Snowden and M'Corkle (printer), 34 p., folded chart.
Evans number 32245.

Henmann
4985 1825, Changes in the contents of brine springs: Am. J. Sci., v. 10, no. 1, 193.
Abstracted by J. Griscom from "Schweigger."

Henry, Joseph (1797-1878)
4986 1828, Topographical sketch of the state of New-York, designed chiefly to show the elevations and depressions of its surface: Alb. Inst. Trans., v. 1, 87-112.
4987 1832, On a disturbance of the earth's magnetism, in connexion with the appearance of an Aurora Borealis, as observed at Albany, April 19th, 1831: Am. J. Sci., v. 22, no. 1, 143-155.
4988 1841, Glossary of geological terms, selected from Lyell, Mantell and others: Princeton, NJ, R. E. Horner, 16 p.
4989 1843, On phosphorogenic emanation: Am. Philos. Soc. Proc., v. 3, no. 27, 38-44.

Henry, M.
4990 1828, Comparative analysis of the elastic bitumen of England and France: Am. J. Sci., v. 14, no. 2, 371-372; also in: Am. J. Improvements Use. Arts, v. 1, 489-490.
Abstracted from Bulletin des Sciences.

Henry, William (1774-1836)
4991 1808, Analysis of mineral waters: Phila. Med. Mus., v. 5, pt. 2, 111-141.
From Henry's Chemistry.

Henslow, John Stevens (1796-1861)
4992 1845, Coprolitic nodules: Am. Q. J. Ag. Sci., v. 2, 319 (7 lines).
On phosphatic nodules from England.

Henwood, William Jory (1805-1875)
4993 1847, Notices of the superposition of certain minerals in some of the metalliferous deposits of Cornwall and Devon: Am. J. Sci., v. 3 (ns), no. 2, 269-271.
Abstracted from London, Edinburgh and Dublin Philos. Mag.

Hepburn, Samuel
4994 1827, Extract of a letter [on rock drilling techniques at Milton, Pennsylvania]: Am. J. Sci., v. 12, no. 1, 143-144.
Following Mr. Disbrow's "Notice of some recent experiments in boring for fresh water" (see 3190).
4995 1831, Geology of Lycoming County [Pennsylvania]: Hazard's Reg. Penn., v. 7, 38-40; 70-72; 82-83.

Hepites, P. C.
4996 1831, The Black Sea: Am. J. Sci., v. 20, no. 1, 188-189.
Abstracted from Bibliothèque Universelle by J. Griscom.

Herapath, Thornton J.
4997 1849, On some newly discovered substances from the African guano deposits: Am. J. Sci., v. 8 (ns), no. 1, 129-131.
Abstacted from Q. J. Chemical Soc.

Herbert, L.
4998 1830, On the substitution of plumbago for oil, in chronometers: Franklin Inst. J., v. 5 (ns), 86-89.
4999 1830, Plumbago, instead of oil, in watches and chronometers: Mech's Mag., v. 1, 191.
Abstracted from Soc. Arts Trans.

Hermann, M. R.
5000 1846, Antimoniate of lead: Am. J. Sci., v. 2 (ns), no. 3, 414 (6 lines).
Abstracted from J. d'Erdmann. On mineral from Nertschinsk, Russia.
5001 1846, Fischerite [from Nische Tagilsk, Urals]: Am. J. Sci., v. 2 (ns), no. 3, 415 (5 lines).
Abstracted from Berzelius' Jahresbericht.
5002 1846, Native tin [from the Urals]: Am. J. Sci., v. 2 (ns), no. 3, 415 (3 lines).
Abstracted from "Jour. für Prakt. Chem."
5003 1846, Stroganowite [from Sludanka, Russia]: Am. J. Sci., v. 2 (ns), no. 3, 413 (8 lines).
Abstracted from J. d'Erdmann.
5004 1846, Turgite [from the Urals]: Am. J. Sci., v. 2 (ns), no. 3, 415 (5 lines).
Abstracted from "Jour. für Prakt. Chem."
5005 1846, Turquoise: Am. J. Sci., v. 2 (ns), no. 3, 415 (2 lines).
5006 1846, Xylite [from the Urals]: Am. J. Sci., v. 2 (ns), no. 3, 413-414 (6 lines).
Abstracted from J. d'Erdmann.
5007 1849, Analysis of phosphates of copper from Nische Tagilsk [Russia]: Am. J. Sci., v. 8 (ns), no. 1, 127.
Abstracted from "Jour. für Prakt. Chem."
5008 1849, On volknerite, a new mineral: Am. J. Sci., v. 7 (ns), no. 1, 113 (10 lines); v. 8 (ns), no. 1, 122 (5 lines).
Abstracted from "Leonh. und Bronn, Neues Jahrb." On a new mineral from Slatoust, Siberia.

Hermann, M. R. (cont.)
5009   1849, Monazitoid, a new mineral from near Lake Ilmen: Am. J. Sci., v. 8 (ns), no. 1, 125 (9 lines).
       Abstracted from "J. für Prakt. Ch."

Heron, J. (see 11113 and 11116 with J. P. Young)

Herrick, Edward Claudius (1811-1862)
5010   1839, Fall of a meteorite in Missouri, February 13, 1839: Am. J. Sci., v. 37, no. 2, 385-386.

Herschel, John Frederick William (1792-1871)
5011   1837, On the theory of volcanic phenomena: Franklin Inst. J., v. 20 (ns), 353-356.
5012   1838, On subterraneous and ominous sounds: Bos. Wk. Mag., v. 1, 75-76.
5013   1839, Notice of a chemical examination of a specimen of native iron, from the east bank of the Great Fish River, in South Africa: Am. J. Sci., v. 36, no. 1, 213-214.
       Abstracted from London and Edinburgh Philos. Mag.
5014   1844, Magnetic observations by the antartic [sic] expedition: Franklin Inst. J., v. 7 (3s), 68.

Herschel, William (1732-1822)
5015   1787, An account of volcanos in the moon: Am. Mus. or Univ. Mag., v. 2, 488-489.
5016   1803, Dr. Herschel's account of volcanos in the moon: The Medley, v. 1, no. 9, 171-173.

The Hesperian (Columbus, Ohio, 1838-1839)
5017   1838, Geological investigations [reviews with excerpts of first and second Maine reports by C. T. Jackson (see 5550 and 5558), and of second Pennsylvania report by H. D. Rogers (see 8855)]: v. 1, 281-286, 360-365.
       Signed by "S. T." See also 5018.
5018   1838, Geological investigation: v. 1, 455-456.
       Signed by "Q." C. T. Jackson defended, and review (see 5017) criticized.
5019   1838, Hints on geology: v. 1, 465-466.
       From Colburn's New Monthly. On the variety of fossil life.
5020   1838, The Mammoth Cave: v. 2, 156-157.
       From Louisville J.
5021   1839, [Review of third annual New York report (see 7740)]: v. 3, 375-376.
       Signed "by a westerner" in the index. See also 5022.
5022   1839, [Review of second annual Ohio report by W. W. Mather (see 6896)]: v. 3, 421-428.
       Signed "S. Y." after article, and "By a westerner" in the index. See also 5021.

Hess, Germain Henri (1802-1850)
5023   1834, Worthite [from Petersburg, Russia]: Am. J. Sci., v. 26, no. 2, 387 (11 lines).
       Translated by L. Feuchtwanger from Berzelius' Jahresbericht.
5024   1835, Hydroboracite, a new mineral: Am. J. Sci., v. 28, no. 2, 394 (7 lines).
       Abstracted from Jameson's Edinburgh New Philos. J.

Hess, J.
5025   1848, New method of treating platinum ore: Hunt's Merch. Mag., v. 18, 324.

Heuland, Henry (1777-1856)
5026   1823, Matrix of the diamond [from Brazil]: Bos. J. Phil. Arts, v. 1, no. 3, 300 (18 lines).
       Abstracted from Philos. Mag.

Heustis, Jabez Wiggins (1784-1841)
5027   1817, Observations on the topography of Louisiana: Eclec. Rep. and Analytic Rev., v. 7, 292-325.
5028   1817, Physical observations, and medical tracts and researches, on the topography and diseases of Louisiana: NY, T. and J. Swords (printers), 165 p.
5029   1831, Topographical remarks on the climate, soil, &c. of the middle section of Alabama, more especially in relation to the county of Dallas: Am. J. Med. Sci's, v. 8, 75-94.
5030   1836, Topographical and medical sketches of Mobile for the year 1835: Am. J. Med. Sci's, v. 19, 65-85.

Hibbert-Ware, Samuel (1782-1848)
5031   1833, Some essential views of geology: Atl. J. and Friend Know., v. 1, no. 7, 191-195.
       With "Notes" by C. S. Rafinesque (see 8667).

Hibernicus (pseudonym; see Clinton, H. E. de Witt)

Hickling, Thomas
5032   1812, Submarine volcano in the Azores: Med. Repos., v. 15, 96-99.

Hickok, Milo Judson (1809-1873)
5033   1838, Earthquakes and volcanoes. Abstract of a lecture, delivered before the Delaware Academy of Natural Science, April 6, 1838: Del. Reg. and Farm's Mag., v. 1, 366-380.

Hicks, Henry
5034   1848, Fall of an aerolite at Forest Hill, Arkansaw: Am. J. Sci., v. 5 (ns), no. 2, 293-294.

Higgason, Josiah
5035   1836, An essay on the medical topography, and diseases of the Western district of Tennessee: Transylvania J. Medicine and Associated Sci´s, v. 8, 34-56.

Higgins, James (Maryland State Agricultural Chemist)
5036   1850, Analysis of soils: Am. Farm., v. 5 (4s), 225-226.
5037   1850, Report of James Higgins, M. D., State Agricultural Chemist, to the House of Delegates: Annapolis, Md., 92 p.
        Also in Am. Farm., v. 5 (4s), 310; 361-365; 392-397; 439-440; v. 6 (4s), 57-62.

Higgins, S. W.
5038   1839, Report [on topography and drainage in Michigan] : in Michigan second annual report (see 5299).
5039   1840, Report [on diurnal magnetic variations] : in Michigan third annual report (see 5301).
5040   1841, Report [on the variation of the magnetic needle] : in fourth Michigan report (see 5303).
5041   1846, Report from the geological department: Mich. JD 12.
        Not seen.
5042   1849, and B. Hubbard, Geological map of a district, E. & W. of the Ontonagon subdivided by Messrs. Higgins and Hubbard under a contract bearing the date April 23rd, 1846: Balt., E. Weber and Co. (engravers), geological map 46 x 25 cm., in Lake Superior report by C. T. Jackson (see 5655).

Higgins, William (1732-1822)
5043   1818, Account of a shower of meteoric stones which fell in the county of Limerick: Am. Mon. Mag. and Crit. Rev., v. 3, no. 4, 312-314.

Higgins, William Mullinger
5044   1836, The earth; its physical condition and most remarkable phenomena: NY, Harper and Brothers, 408 p., illus.
        Other printings dated 1838, 1839, 1840, 1842, 1844, and 1846.

Hildreth, Samuel Prescott (1783-1863)
5045   1809, A concise description of Marietta, in the state of Ohio; with an enumeration of some vegetable and mineral productions in its neighbourhood: Med. Repos., v. 12, 358-363.
5046   1825, Facts relating to certain parts of the state of Ohio: Am. J. Sci., v. 10, no. 1, 1-8.
5047   1826, Notes on certain parts of the state of Ohio: Am. J. Sci., v. 11, no. 2, 231-238.
5048   1827, Miscellaneous observations on the coal, diluvial, and other strata of certain portions of the state of Ohio: Am. J. Sci., v. 13, no. 1, 38-40.
5049   1827, Notice of fossil trees, near Gallipolis, Ohio: Am. J. Sci., v. 12, no. 2, 205-206.
5050   1829, Bowlder stones of primitive rocks [in Ohio] : Am. J. Sci., v. 16, no. 1, 154-159.
5051   1833, Observations on the saliferous rock formation, in the valley of the Ohio: Am. J. Sci., v. 24, no. 1, 46-68.
5052   1834, Ten days in Ohio; from the diary of a naturalist: Am. J. Sci., v. 25, no. 1, 217-257.
5053   1835, Observations on the bituminous coal deposits of the valley of the Ohio, and the accompanying rock strata; with notices of the fossil organic remains and the relics of vegetables and animal bodies, illustrated by a geological map, by numerous drawings of plants and shells, and by views of interesting scenery: Am. J. Sci., v. 29, no. 1, 1-154, 36 plates, 19 figs, front., map.
        Contains map (5054), and "Appendix" by S. G. Morton (see 7471). Also reprint edition, 153 p., plates, map, illus. For review see 666. Abstracted in Mech´s Mag. and Reg. Inventions, v. 6, 358-360.
5054   1835, A topographical & geological map of the coal measures of the muriatiferous & ferruginous deposits, in the Secondary region of the Valley of the Ohio: Am. J. Sci., v. 29, no. 1, geological map, 41.6 x 38.0 cm.
        In 5053.
5055   1836, Miscellaneous observations made during a tour in May, 1835, to the Falls of the Cuyahoga, near Lake Erie: Am. J. Sci., v. 31, no. 1, 1-84, 20 figs.
5056   1836, Note by Dr. S. P. Hildreth on the Lias of the West: Am. J. Sci., v. 30, no. 2, 395.
        On sediments from Lawrence and Scioto Counties, Ohio.
5057   1836, [On the best method of obtaining an Ohio geological survey] : in Ohio General Assembly "Report", 3-16, 2 plates.
        For review see 9740.
5058   1838, Report of Dr. S. P. Hildreth, first assistant geologist [on the coal measures] : in first annual report of the Ohio geological survey by W. W. Mather (see 6891), 25-63.
5059   1842, History of an early voyage on the Ohio and Mississippi Rivers, with historical sketches of the different points along them, &c. &c.: Am. Pioneer, v. 1, 89-145.

Hill, Ira (c. 1783-1838)
5060   1823, An abstract of a new theory of the formation of the earth: Balt., N. G. Maxwell, 211 p.
        For review with excerpts see 9980. For review see 10490.

Hill, Samuel Worth (b. 1815)
        See also 3935 with J. W. Foster.
5061   1849, [Sections of Lac Labelle, North West, and Cliff Mines, in the Lake Superior District] : 3 folded plates in report by C. T. Jackson (see 5655).

Hinde, Thomas S.
5062   1822, Remarks upon the early settlements of the Western Country, with some account of its soil, climate, and production: Meth. Mag., v. 5, 223-226; 260-264.

Hinton, John Howard (1791-1873)
5063   1832, A geological map of the United States: in The history and topography of the United States ..., (see 5064), Fenner, Sears and Co. (engravers and printers), colored geological map, 39 x 25 cm, in v. 2.
5064   1830/1832, The history and topography of the United States of North America, from the earliest period to the present time, comprising political and biographical history, geography, geology, mineralogy, zoology and botany; agriculture, manufactures, and commerce; ... : London and Phila., T. Wardle and I. T. Hinton, 2 v. (xvi, 476 p.; viii, 580 p.), numerous plates and maps.
Contains sections by A. Eaton (see 3373) and E. Hitchcock (see 5102), and map (see 5063). For review see 7568.
5065   1834, The history and topography of the United States ... [as above]: Bos., S. Walker, 2 v.
Other printings in 1843, 1844, 1845 and 1846. "A new and improved edition, with additions and corrections, by Samuel L. Knapp."
5066   1850, The history and topography of the United States ... [as above]: London and NY, John Tallis and Co., 2 v.

Historical and Philosophical Society of Ohio
5067   1834, Circular to all such as feel disposed to promote the geological and mineralogical knowledge of Ohio: Eclec. J. Sci., v. 2, 757-758.
5068   1834, Scientific: Eclec. J. Sci., v. 2, 818-822.
Circular on how to collect scientific specimens.

A history of the county of Berkshire, Mass. in two parts, the first being a general view of the whole country, and the second an account of the several towns; by gentlemen of the county, clergymen and laymen:
5069   1829, Pittsfield, Mass., S. W. Bush (printer), iv [7]-468 p., front., plates, folded colored chart.
For review see 518.

Hitchcock, Edward (1793-1864)
See also 4842 with C. Hartwell
5070   1815, Southampton lead mine. Basaltick columns [in Massachusetts]: N. Am. Rev., v. 1, 334-338.
Signed by "E. H.", i.e. Edward Hitchcock or E. Hale?
5071   1818, A geological map of a part of Massachusetts on Connecticut River 1817.[with] Transverse section of rock strata from Hoosack Mountain to eleven miles east of Connecticut River: Am. J. Sci., v. 1, no. 2, geological map, 42 x 18 cm.
In 5072.
5072   1818, Remarks on the geology and mineralogy of a section of Massachusetts on Connecticut River, with a part of New Hampshire and Vermont: Am. J. Sci., v. 1, no. 2, 105-116; 436-439.
Contains map 5071. See also 5074.
5073   1819, On a singular disruption of the ground, apparently by frost: Am. J. Sci., v. 1, no. 3, 286-292, plate.
On frost action at Deerfield, Massachusetts.
5074   1819, Supplement to the "Remarks on the geology and mineralogy of a section of Massachusetts, on Connecticut River, &c" contained in no. 2, art. I, of this journal: Am. J. Sci., v. 1, no. 4, 436-439.
See also 5071 and 5072.
5075   1820, Oxid of manganese, and chromat of iron: Am. J. Sci., v. 2, no. 2, 374 (5 lines).
5076   1821, List of organic remains and accompanying rocks contained in a box forwarded to Professor Silliman: Am. J. Sci., v. 3, no. 2, 366.
On Triassic fossils from Sunderland, Masschusetts.
5077   1822, Fluor spar and oxide of titanium [from Conway, New Hampshire]: Am. J. Sci., v. 5, no. 2, 405-406 (10 lines).
5078   1822, Fluate of lime and noble agates in Deerfield, Mass.: Am. J. Sci., v. 5, no. 2, 407 (5 lines).
5079   1822 (i.e. 1823), A geological map of the Connecticut. 1822: Am. J. Sci., v. 6, no. 1, colored geological map, 56.5 x 19.2 cm.
In 5081.
5080   1823, New mineralogical hammer: Am. J. Sci., v. 7, no. 1, 715 [i.e. 175], plate with 2 figs.
5081   1823, A sketch of the geology, mineralogy, and scenery of the regions contiguous to the River Connecticut; with a geological map and drawings of organic remains; and occasional botanical notices: Am. J. Sci., v. 6, no. 1, 1-86, 2 figs, map; no. 2, 201-236, 2 plates with 7 figs.; v. 7, no. 1, 1-30, plate with 3 figs.
Contains map (see 5079), and report by J. W. Webster (see 10738). For notice see 418. Also reprint edition, N. Hav., Converse, iv, 154, iv p.
5082   1823, Utility of natural history. A discourse delivered before the Berkshire Medical Institution, at the organization of the Lyceum of Natural History, in Pittsfield, Sept. 10, 1823: Pittsfield, Mass., P. Allen, 32 p.
5083   1824, Notice of a singular conglomerate, and of an interesting locality of trap tuff or tufa [at Deerfield, Massachusetts]: Am. J. Sci., v. 8, no. 2, 244-247.
5084   1824, Notices of the geology of Martha's Vineyard, and the Elizabeth Islands: Am. J. Sci., v. 7, no. 2, 240-248, colored map.
Contains map 5085.
5085   1824, Outlines of the geology of Marthas Vineyard &c.: Am. J. Sci., v. 7, no. 2, colored geological map on plate 4, 10 x 7.4 cm.
In 5084.
5086   1824, [Remarks on the geology of the district adjoining the Erie Canal]: in Erie Canal survey by A. Eaton (see 3338), 158-163.
5087   1825, Notice of several localities of minerals in Massachusetts: Am. J. Sci., v. 9, no. 1, 20-23.
5088   1825, Topaz? [from Goshen, Connecticut]: Am. J. Sci., v. 9, no. 1, 180.
With "Remark" by B. Silliman (see 9664).
5089   1826, Andalusite [from Westford, Massachusetts]: Am. J. Sci., v. 10, no. 2, 384 (9 lines); also in: Bos. J. Phil. Arts, v. 3, no. 5, 492.
5090   1826, Chlorophoeite [from Turner's Falls, Massachusetts]: Am. J. Sci., v. 10, no. 2, 393-394.

Hitchcock, E. (cont.)

5091 1826, and B. D. Silliman, jr., Topaz [from various American localities]: Am. J. Sci., v. 10, no. 2, 352-358.

5092 1827, Scientific agriculture. An address delivered before the Hampshire, Franklin, and Hampden Agricultural Society ... : Amherst, Mass., The Society, 24 p.

5093 1828, Explanatory: introductory note in Essay on the temperature of the interior of the earth by L. Cordier (see 2645), p. [2].

5094 1828, Miscellaneous notices of mineral localities, with geological remarks: Am. J. Sci., v. 14, no. 2, 215-230.

5095 1829, Tin in [Goshen] Massachusetts: Am. J. Sci., v. 16, no. 1, 188-191.

5096 1829, [Reviews with excerpts of Essai sur la temperature de l´interieur de la terre by L. Cordier; the Amherst English translation of same (see 2645); and Considerations on volcanoes by P. Scrope]: N. Am. Rev., v. 28, 265-294.

5097 1830, German collection of rocks, minerals, &c. communicated by Prof. Hitchcock: Am. J. Sci., v. 17, no. 2, 400-405.
Describes minerals for sale by David and Adolph Zimmern.

5098 1831, topaz in the White Mountains of New Hampshire: Am. J. Sci., v. 20, no. 2, 410 (12 lines).

5099 1832, Geological map of Massachusetts, by Edwd. Hitchcock. 1832. Executed under the direction of the government of the state: Am. J. Sci., v. 22, no. 1, colored geological map, 70.5 x 45.5 cm.
In 5100.

5100 1832, Report on the geology of Massachusetts; examined under the direction of the government of that state, during the years 1830 and 1831: Am. J. Sci., v. 22, no. 1, 1-70, map.
Contains geological map 5099.

5101 1832, Report on the geology of Massachusetts. Part I. Economical geology [half-title]. Report of a geological survey of Massachusetts; made under an appointment by the governor, and pursuant to a resolve of the legislature of the state: Part I. Economical geology: Amherst, Mass., J. S. and C. Adams (printers) 4, 70 p.
600 copies printed. For notice see 558. Contains map 5099.

5102 1832, Tabular arrangement of the rock formations along the Connecticut, after the method of Conybeare and Phillips: in History and topography of the United States ... , by J. H. Hinton (see 5064), v. 2.

5103 1833, A geological map of Massachusetts, by Edwd. Hitchcock. 1833. Executed under the direction of the government of the state: Bos., Pendleton´s Lithography, colored geological map, 72.2 x 45.7 cm.
In 5105. For notice see 7649.

5104 1833, Ores: Me. Farm., v. 1, 269-270.
From E. Hitchcock´s report of 1833 (see 5105).

5105 1833, Report on the geology mineralogy, botany, and zoology of Massachusetts: Amherst, Mass., J. S. and C. Adams, v. 1, xii, 692 p, illus.; v. 2, "Plates illustrating the geology and scenery of Massachusetts," 19 plates.
1200 copies printed. for review see 584, 603, 6297. For notice see 569. Contains map 5103.

5106 1834, A geological map of Massachusetts, by Edwd. Hitchcock. 1834. Second edition. Executed under the direction of the government of the state: Bos., Pendleton´s Lithography, colored geological map, 72.1 x 45.4 cm.
Included in 5110.

5107 1835, The connection between geology and natural religion: Bib. Repos., v. 5, 113-138; 439-450; v. 6, 261-332.
For review see 5984.

5108 1835, On certain causes of geological change now in operation in Massachusetts: Bos. J. Nat. History, v. 1, no. 2, 69-82.

5109 1835, Organic remains [of Massachusetts]: Am. Mag. Use. Ent. Know., v. 1, no. 10, 452.
From Massachusetts geology report (see 5110).

5110 1835, Report on the geology, mineralogy, and zoology of Massachusetts made and published by order of the government of that state, in four parts ... with a descriptive list of the specimens of rocks and minerals collected for the government. Illustrated by numerous wood cuts and an atlas of plates: Amherst, Mass., J. S. and C. Adams, v. 1, xii, 702 p., illus.; v. 2, 19 plates.
500 copies printed. Contains map 5106. "Second edition, corrected and enlarged."
For review see 5548.

5111 1835, Supposed evidence of volcanic agency at Gay Head: Am. Mag. Use. Ent. Know., v. 1, no. 5, 270.

5112 1836, Geology of Portland, & its vicinity: Bos. J. Nat. History, v. 1, no. 3, colored map and section, 15 x 13.5 cm.
In 5118.

5113 1836, Marl [from Springfield, Massachusetts]: Am. Mag. Use. Ent. Know., v. 3, no. 3, 97 (9 lines).

5114 1836, Ornithichnites in Connecticut: Am. J. Sci., v. 31, no. 1, 174-175.

5115 1836, Ornithichnology defended: Knickerbocker, v. 8, 289-295.
On controversy with Rev. Mr. Chapin (see 2311 and 2312).

5116 1836, Ornithichnology. - Description of the foot marks of birds, (Ornithichnites) on New Red Sandstone in Massachusetts: Am. J. Sci., v. 29, no. 2, 307-340, 3 plates with 24 figs.
See also 2217 and 2218.

5117 1836, Remarks on Professor Stuart´s examination of Gen. 1. In reference to geology: Bib. Repos., v. 7, 448-487.
See article by also M. Stuart 10057.

5118 1836, Sketch of the geology of Portland and its vicinity: Bos. J. Nat. History, v. 1, no. 3, 306-347, map, 11 figs.
Contains map 5112.

5119 1837, Circular containing questions about the economical geology of Massachusetts: Amherst, Mass., 7 p.

5120 1837, Fossil footsteps in sandstone and graywacke [from the Connecticut River valley]: Am. J. Sci., v. 32, no. 1, 174-176.

5121 1837/1838, The historical and geological deluges compared: Bib. Repos., v. 9, 78-139; v. 10, 328-374; v. 11, 1-27.

Hitchcock, E. (cont.)

5122 1837, Idle search after gold and silver: Am. Mag. Use. Ent. Know., v. 3, no. 7, 305-306.
From Massachusetts geological report (see 5110).

5123 1837, Letter on the geological survey of Massachusetts: Mass. SD 9, 7 p.

5124 1837, Preface by the American editor [and notes throughout]: in Researches in theoretical geology by H. T. De la Beche (see 3039), iii-viii.

5125 1838, Fossil animalculae: The Friend, v. 12, 59.

5126 1838, Geology of Massachusetts: NE Farm., v. 17, 58-59; 69; 84-85; 92-93; 100-101.

5127 1838, [Letter on age of coal beds]: in Report of the hearing on the memorial of the New England coal mining company (see 7631), 54-55.

5128 1838, Report on a re-examination of the economical geology of Massachusetts: Bos., Dutton and Wentworth, 139 p.
Contains "Rules of analysis" by S. L. Dana (see 2881). For review see 1101, 2703, 7662, 7947. See also articles by S. L. Dana (2883 and 2884). For excerpts see 5130, 8760, and 9523.

5129 1839, Analysis of marl from Farmington, Conn.: Am. J. Sci., v. 36, no. 1, 176 (13 lines).

5130 1839, Basaltic rocks at Mount Holyoke, Mass.: Parley's Mag., v. 7, 267-268.
From Massachusetts geology report (see 5128).

5131 1839, [Excerpts from Massachusetts geology report (5128)]: Bos. Cult., v. 1, no. 11, [4]; no. 15, [1]; no. 18, [1]; no. 19, [1].

5132 1840, Elementary Geology: Amherst, Mass., J. S. and C. Adams, 320 p., illus., plate, diagram.
Contains "Introduction" by J. P. Smith (see 9896). This is the first edition. For review see 744, 1033, 1689, 2713. For notice see 21, 1688.

5133 1840, Specimens of minerals and rocks: Am. J. Sci., v. 39, no. 1, 199-200.
Specimens for sale by L. W. Zimmern of Heidelberg noted.

5134 1841, Elementary geology: Amherst, Mass., J. S. and C. Adams; NY, Dayton and Saxton; and London, xii, 13-246 p., illus., 118 figs., 2 colored plates.
Contains "Introduction" by J. P. Smith (see 9896). This is the second edition. For review see 229 and 7950.

5135 1841, Final report on the geology of Massachusetts: In four parts: I. Economical geology. II. Scenographical geology. III. Scientific geology. IV. Elementary geology. With an appended catalogue of the specimens of rocks and minerals in the state collection: Amherst, Mass., J. S. and C. Adams; Northampton, Mass., J. H. Butler, 2 v. in 1, xii, 13-831 p., 54 plates (some colored), colored map.
Contains notes by J. W. Bailey (see 1416) and map (see 5140). 1500 copies printed.
For review with excerpts see 21. For review see 185, 753, and 4679.

5136 1841, Final report on the geology of Massachusetts: Vol. I. containing I. Economical geology II. Scenographical geology: Amherst, Mass., J. S. and C. Adams; Northampton, Mass., J. H. Butler, xii, 12-299 p., 14 plates, colored map, illus.

5137 1841, Final report on the geology of Massachusetts: Vol. II. containing III. Scientific geology. IV. Elementary geology. With an appended catalogue of the specimens of rocks and minerals in the state collection: Amherst, Mass., J. S. and C. Adams; Northampton, Mass., J. H. Butler, 301-831 p., plates 15-51, 53-55, illus.
This two volume edition is a variant of 5135. A third variant lists only J. H. Butler, Northampton, Mass. as the publisher.
Contains article by J. W. Bailey (see 1416) and map (see 5140).

5138 1841, First anniversary address before the Association of American Geologists, at their second annual meeting in Philadelphia, April 5 [i.e. 8], 1841: Am. J. Sci., v. 41, no. 2, 232-275.

5139 1841, First anniversary address ... [as above] . With an abstract of the proceedings of the Association, at their sessions in 1840 and 1841: NHav., B. L. Hamlin (printer), 48, 34 p.
Reprinted from Am. J. Sci. (Usually bound with 1322.)

5140 1841, A geological map of Massachusetts by Edwd. Hitchcock 1841 fourth edition executed under the direction of the government of the state: Bos., B. W. Thayer's lithography, colored geological map, 71.3 x 45.5 cm.
Published separately, and as plate 52 in the Final report ..., (see 5135).

5141 1841, [On joints in Triassic rocks of the Connecticut River Valley]: Am. J. Sci., v. 41, no. 1, 173 (16 lines).
Also in Assoc. Am. Geol's Nat's Proc., v. 2, 18. Also in Assoc. Am. Geol's Nat's Rept., v. 1, 25-26, 1843.

5142 1842, Elementary geology: NY, Dayton and Newman, xii, [13]-348 p., illus., folded plate, diagram.
"Third revised edition." Also 1843 printing by M. H. Newman, NY. contains J. P. Smith's "Introduction" (see 9896).

5143 1842, On the phenomena of drift in this country: Am. J. Sci., v. 43, no. 1, 151-154.
Also in Assoc. Am. Geol's Nat's Proc., v. 3, 6-9.
Also in Assoc. Am. Geol's Nat's Rept., v. 1, 45-49, 1843.

5144 1842, Papers re. coal mines of the state: Mass. HD 19, 128 p.
Not seen.

5145 1842, Remarks upon Mr. Murchison's anniversary address before the London Geological Society: Am. J. Sci., v. 43, no. 2, 396-398.

5146 1843, Description of five new species of fossil footmarks, from the red sandstone of the valley of Connecticut River: Assoc. Am. Geol's Nat's Rept., v. 1, 254-264, plate with 9 figs.

5147 1843, Description of several species of fossil plants, from the New Red Sandstone formation of Connecticut and Massachusetts: Assoc. Am. Geol's Nat's Rept., v. 1, 294-296, plate with 5 figs.

5148 1843, [Exhibition of fossil footprints from the Connecticut River Valley]: Am. J. Sci., v. 45, no. 2, 316 (15 lines).
In Assoc. Am. Geol's Nat's Proc. See also article by W. C. Redfield (8717).

5149 1843, [Native copper in Massachusetts]: Am. J. Sci., v. 45, no. 2, 331 (7 lines).
In Assoc. Am. Geol's Nat's Proc.

Hitchcock, E. (cont.)

5150   1843, Notes on the geology of several parts of Western Asia: founded chiefly on specimens and descriptions from American missionaries: Assoc. Am. Geol's Nat's Rept., 348-421, plate with 7 figs.
Contains 1420. For review with excerpts see 1604. See also 5173.

5151   1843, [On drift phenomena]: Am. J. Sci., v. 45, no. 2, 324-325.
In Assoc. Am. Geol's Nat's Proc. See also 5602.

5152   1843, The phenomena of drift, of glacioaqueous action in North America, between the Tertiary and Alluvial periods: Assoc. Am. Geol's Nat's Rept., v. 1, 164-221, 3 plates with 17 figs, 2 maps.

5153   1844, Description by Captains Cook and Flanders of bird's nest of enormous size on the coast of New Holland: Am. J. Sci., v. 47, no. 1, 217-218.
See article by J. Cook (see 2599).

5154   1844, Discovery of more native copper in the town of Whately in Massachusetts, in the valley of the Connecticut River, with remarks upon its origin: Am. J. Sci., v. 47, no. 2, 322-323.

5155   1844, Discovery of the yttro-cerite in Massachusetts: Am. J. Sci., v. 47, no. 2, 351-353.

5156   1844, Dispersion of blocks of stone at the drift period in Berkshire County, Massachusetts: Am. J. Sci., v. 47, no. 1, 132-133.
Also in Assoc. Am. Geol's Nat's Proc., v. 5, 41-42.

5157   1844, Elementary geology: NY, Mark H. Newman, xii, 13-348 p., illus. 122 figs, 2 colored plates.
"Third edition revised and improved." Contains "Introduction" by J. P. Smith (see 9896).

5158   1844, Explanation of the geological map attached to the topographical map of Massachusetts: Bos., C. Hickling, 22 p.
See also map (5160).

5159   1844, Extract of a letter from Prof. Hitchcock, respecting the lincolnite: Am. J. Sci., v. 47, no. 2, 416.

5160   1844, Geological map of Massachusetts. Scale, 5 miles to 1 inch or 1:316800: inset on Topographical map of Massachusetts, Bos., Simeon Borden, colored geological map, 57.5 x 93.5 cm.
See also 5158. This is the fifth geological map of Massachusetts by Hitchcock.

5161   1844, [On the importance of geological surveys]: Am. J. Sci., v. 47, no. 1, 116 (8 lines).
Also in Assoc. Am. Geol's Nat's Proc., v. 5, 23 (8 lines).

5162   1844, Rejoinder to the preceding article of Dr. Deane: Am. J. Sci., v. 47, no. 2, 390-399.
See also article by J. Deane (2985). On the priority of discovery of fossil footmarks in the Connecticut River Valley.

5163   1844, Report on ichnolithology or fossil footmarks, with a description of the coprolites of birds, discovered recently in the Connecticut Valley: Am. J. Sci., v. 47, no. 1, 113-114.
Also in Assoc. Am. Geol's Nat's Proc., v. 5, 20-21.

5164   1844, Report on ichnolithology, or fossil footmarks, with a description of several new species, and the coprolites of birds, from the valley of Connecticut River, and of a supposed footmark from the valley of the Hudson River: Am. J. Sci., v. 47, no. 2, 292-322, 2 plates.

5165   1844, The trap tufa, or volcanic grit of the valley of the Connecticut River: Am. J. Sci., v. 47, no. 1, 103-104.
Also in Assoc. Am. Geol's Nat's Proc., v. 5, 10-11.

5166   1845, An attempt to name, classify, and describe, the animals that made the fossil footmarks of New England: Assoc. Am. Geol's Nat's Proc., v. 6, 23-25.
Also in Am. Q. J. Ag. Sci., v. 2, 147.

5167   1845, [On coprolites]: Am. J. Sci., v. 48, no. 1, 46-60.
In "Analysis of coprolites from the New Red Sandstone formation of New England" by S. L. Dana (see 2888).

5168   1845, Description of a singular case of the dispersion of blocks of stone connected with drift, in Berkshire County, Massachusetts: Am. J. Sci., v. 49, no. 2, 258-265, 2 figs.

5169   1845, Extract of a letter from Prof. E. Hitchcock, embracing miscellaneous remarks upon fossil footmarks, the lincolnite, &c.: Am. J. Sci., v. 48, no. 1, 61-65.
With letter by R. Owen (see 8094).

5170   1845, Letter from President Hitchcock: in First annual report on the geology of the state of Vermont by C. B. Adams (see 23), 64-68.
On the useful stones of Vermont.

5171   1845, [On New England drift]: Am. Q. J. Ag. Sci., v. 2, 141-142.
In Assoc. Am. Geol's Nat's Proc.

5172   1845, Remarkable facts respecting the magnetic polarity of trap rocks in New England: Assoc. Am. Geol's Nat's Proc., v. 6, 32.

5173   1845, Supplement to notes on the geology of western Asia: Assoc. Am. Geol's Nat's Proc., v. 6, 22-23.
Also in Am. Q. J. Ag. Sci., v. 2, 144-145. See also 5150.

5174   1846, Letter from President Hitchcock: in Second annual report on the geology of the state of Vermont by C. B. Adams (see 26), 247-252.
On drift phenomena in New England.

5175   1846, The mutual defense between agriculture and other pursuits: Mass. Agri. Societies Trans., v. 1, 151-182.

5176   1847, Description of two new species of fossil footmarks found in Massachusetts and Connecticut, or, of the animals that made them: Am. J. Sci., v. 4 (ns), no. 1, 46-57, 3 figs.

5177   1847, Elementary geology: NY and Cin., M. H. Newman and Co., xii, [13]-361 p., illus.
"7th edition." Another printing in 1847, as "8th edition." For review of "8th edition" see 897.

5178   1847, [Fossil footprints in the Connecticut River Valley]: Am. Q. J. Ag. Sci., v. 6, 218-219.
In Assoc. Am. Geol's Nat's Proc.

5179   1847, [Nature of Mounts Monadnock and Holyoke]: Am. Q. J. Ag. Sci., v. 6, 216.
In Assoc. Am. Geol's Nat's Proc. See also article by W. C. Redfield (see 8719).
On glacial polishing.

Hitchcock, E. (cont.)
5180 1847, [On terraces of North America]: Am. Q. J. Ag. Sci., v. 6, 212-213.
         In Assoc. Am. Geol's Nat's Proc.
5181 1847, On the trap tuff, or volcanic grit, of the Connecticut Valley, with the bearings of its history upon the age of the trap rock and sandstone generally in that valley: Am. J. Sci., v. 4 (ns), no. 2, 199-207, 2 figs.
5182 1847, Ornithichnites: Am. J. Sci., v. 3 (ns), no. 2, 276.
         On the different species of fossil footprints from the Connecticut River Valley.
5183 1848, An attempt to discriminate and describe the animals that made the fossil footmarks of the United States and especially of New England: Am. Acad. Arts Sci's Mem., v. 3 (ns), 129-256, 24 plates with 132 figs.
         For review see 976.
5184 1848, [Fossil footprints and glacial phenomena of New England]: Lit. World, v. 2, 229.
5185 1850, [On glacial potholes in Massachusetts]: Bos. Soc. Nat. History Proc., v. 3, 324 (4 lines).
5186 1850, On the river terraces of the Connecticut Valley, and on the erosions of the earth's surface: Am. Assoc. Adv. Sci. Proc., v. 2, 148-156.
         Abstracted in Ann. Scien. Disc., v. 1, 229.
5187 1850, [On the origin of cleavage]: Bos. Soc. Nat. History Proc., v. 3, 226 (5 lines).

Hobhouse (Mr.)
5188 1834, Great cave [at Mount Parre]: Genesee Farm. and Gardener's J., v. 4, 312.

Hoblyn, Richard Dennis (1803-1886)
5189 1850, A dictionary of scientific terms: NY, D. Appleton, ii, 386, 1 p.

Hodge, James Thacher (1816-1871)
5190 1838, On the Allagash section, from the Penobscot to the St. Lawrence River: in Second annual report on the geology of the public lands ..., by C. T. Jackson (see 5558), 49-73.
5191 1841, [Observations on the Secondary and Tertiary deposits of the Carolinas]: Am. J. Sci., v. 41, no. 1, 182-183.
         Also in Assoc. Am. Geol's Nat's Proc., v. 2, 27-28. Also in Assoc. Am. Geol's Nat's Rept., v. 1, 34-35, 1843.
5192 1841, Observations on the Secondary and Tertiary formations of the southern Atlantic states: Am. J. Sci., v. 41, no. 2, 332-344.
         With an appendix by T. A. Conrad (see 2568). Also in Assoc. Am. Geol's Nat's Rept., v. 1, 94-111, 1843.
5193 1841, and C. B. Trego, [On the supposed unconformity in Carboniferous rocks in Savage Mountain, Pennsylvania]: Am. J. Sci., v. 41, no. 1, 186 (6 lines).
         Also in Assoc. Am. Geol's Nat's Proc., v. 2, 31 (6 lines). Also in Assoc. Am. Geol's Nat's Rept., v. 1, 38 (6 lines), 1843.
5194 1842, [Mineral resources of Georgia]: Am. Rr. J., v. 15, 158-160.
5195 1842, On the Wisconsin and Missouri lead region: Am. J. Sci., v. 43, no. 1, 35-72, 4 figs.
5196 1847, Report of a tour of exploration through the mineral locations of Montreal River, Lake Superior: n.p., 19 p., map.
5197 1847, Report. To James Phalen, esq., Samuel Ward esq., Auguste Belmont, esq., Thomas Dixon, esq., C. G. Hamilton, esq., trustees of the Montreal River Mining Company: N. Y.?, 19 p., front.
         Not seen. Same as 5196?
5198 1848, [Economic geology of the Berkshire Valley, Massachusetts]: Lit. World, v. 2, 256 (8 lines).
         In Assoc. Am. Geol's Nat's Proc.
5199 1848, [Remarks on alluvial fossils from Maine]: NY J. Medicine and Collateral Sci's, v. 10, 338 (5 lines).
         In NY Lyc. Nat. History Minutes.
5200 1848, [Remarks on the lead mines of the United States]: NY J. Medicine and Colateral Sci's, v. 10, 343-344.
         In NY Lyc. Nat. History Minutes.
5201 1849, Iron ores and the iron manufacture of the United States: Am. Rr. J., v. 22, 273-275; 290-291; 305-306; 323-324; 340-341; 353-354; 369-370; 385-386; 465-467, woodcut; 481-481 [i.e. 483], table; 494-498, woodcut; 511-513, woodcut; 527-529; 543-544; 558-562, 2 woodcuts; 575-577, 2 woodcuts; 591-595, tables; 607-609, table; 623-624; 639-640; 655-657, woodcut; 670-672, woodcut; 735-737; 767-768; 783-785; 799-801, woodcut; 815-817.
         Includes descriptions of major American iron ore deposits, and their mode of mining and reduction. For notice see 5203.
5202 1849, The manufacture of iron in Georgia: Hunt's Merch. Mag., v. 20, 507-511.
5203 1849, Mining [notice of 5201 and 5204]: Am. Rr. J., v. 22, 273.
5204 1849, Copper ores of Lake Superior: Am. Rr. J., v. 22, 401-402; 417-418; 433-434; 449-452, woodcut; 467-468; 481-482; 498-513; 529; 544-545.
         For notice see 5203.
5205 1850, Ancient mining operations on Lake Superior: Ann. Scien. Disc., v. 1, 360.
5206 1850, Lake Superior copper mines: Am. Rr. J., v. 23, 305-306, plate; 433-435.
5207 1850, Lead ores of the United States: Am. Rr. J., v. 23, 1-2; 33-35; 49-50; 177-178, woodcut; 193-194, woodcut; 209-211.
5208 1850, On the mineral region of Lake Superior: Am. Assoc. Adv. Sci. Proc., v. 2, 301-308.
         With remarks by C. T. Jackson and H. D. Rogers.

Hodge, S. H.
5209 1847, The boiling well of Green County, Kentucky: The Friend, v. 20, 179.

Hodgkinson, Eaton (1789-1861)
5210 1841, On the temperature of the earth in the deep mines in the neighborhood of Manchester: No. Light, v. 1, 74.
         In article by J. A. Dix (see 3193).

Hodgkinson, E. (cont.)
5211 1843, Experimental inquiries on the strength of stones and other materials: Am. J. Sci., v. 44, no. 1, 168-169.
In British Assoc. Proc.

Hodgson, Brian Houghton (1800-1894)
5212 1849, The Himalayan alpine land: Am. J. Sci., v. 8 (ns), no. 1, 133-135.
Abstracted from Jameson's J.

Hodgson, William Brown (1800-1871)
5213 1846, Memoir on the Megatherium, and other extinct gigantic quadrupeds of the coast of Georgia, with observations on its geologic features: NY, Bartlett and Welford, 47 p., illus.
Contains sections by J. Couper and J. Habersham (see 2664 and 4615).

Hoffman, E.
5214 1836, Analysis of chabasie: Am. J. Sci., v. 30, no. 1, 366-368.
Translated by C. Cramer from Poggendorff's Annalen. On the physical properties of chabsite.

Hoklham
5215 1835, Formation of calcareous earths: Genesee Farm. and Gardener's J., v. 5, 164-165.

Holbrook (Mr., of Derby, Connecticut)
5216 1802, [Use of gypsum for fertilizer]: Ct. Soc. Promoting Ag. Trans., v. 1, 7-8.

Holbrook, Josiah (1788-1854)
5217 1828, Agricultural and geological surveys: NE Farm., v. 6, 337-338.
5218 1833, First lessons in geology for the use of families, schools, and lyceums: Bos., Brown and Peirce, 64 p.
5219 1835, First lessons in geology ... [as above]: Phila., 35 p.
5220 1839, Geology and agriculture of Wheatland [Monroe, County, New York]: NE Farm., v. 18, 12.
5221 1841-1842, Agricultural geology: The Cultivator, v. 8, 100; also in: Me. Farm., v. 9, 252-253; Youth's Medallion, v. 1, no. 12, 90; Am. Farm., v. 3 (3s), 278 (1842);
On the geological alphabet, i.e. 10 simple and common minerals.
5222 1848, Geological cabinet, an introduction and aid to books ..: NY, J. J. Reed, 16 p.
5223 1849, Geology: Green Mtn. Gem, v. 7, 237-238; 241-242; 255-256; 259-260; 270-271.
5224 1849, Geology for schools and families: Ohio Sch. J., v. 4, no. 8, 121-122; no. 9, 137-139; no. 10, 149-152; no. 11, 168-170; no. 12, 184-186.
Also in part in Farm's Mon. Vis., v. 11, 111, 127, 144.

Holcombe, George
5225 1826, On the New Jersey marls: Phila. Soc. Promoting Ag. Mem., v. 5, 8-15.

Holden's Dollar Magazine (New York, 1848-1850)
5226 1849, [Review with excerpts of Oregon and California in 1849 by J. Q. Thornton (see 10259)]: v. 3, 243.
5227 1849, [Review with excerpts of A tour of duty in California, including a description of the gold region by J. R. Revere (see 8756)]: v. 3, 243-244.
5228 1849, [Review with excerpts of Second visit to the United States by C. Lyell (see 6571)]: v. 4, 569-570.
5229 1850, [Review of Six months in the gold mines by E. Gould Buffum (see 2180)]: v. 6, 634 (4 lines).
5230 1850, [Review of Foot-prints of the Creator by H. Miller (see 7186)]: v. 6, 759 (6 lines).
5231 1850, [Review of The pre-Adamite earth by J. Harris (see 4830)]: v. 6, 759-760 (13 lines).

Holley, Mary Austin (1784-1846)
5232 1833, Texas observations, historical, geographical, and descriptive: Balt., Armstrong and Plaskitt, 167 p., map.
5233 1836, Texas observations, historical, geographical, and descriptive: Lexington, Kentucky, J. Clarke and Co., viii, 410 p., map.

Holley, Myron (1779-1841)
5234 1821, [On hydraulic limestone used for the Erie Canal]: Am. J. Sci., v. 3, no. 2, 231-232.
In article by B. Wright (see 11066).

Holly, Horace (i.e. Horace Holley, 1781-1821,?)
5235 1808, An investigation of the facts relative to the descent of stones from the atmosphere to the earth, on the 14th of December, 1807, in the towns of Fairfield, Weston, and Huntington, Connecticut, and to the meteor whence these earthy bodies proceded: Med. Repos., v. 11, 418-421.

Holmes, A. F.
5236 1831, Note: Am. J. Sci., v. 19, no. 2, 360 (5 lines).
In "Analysis of the supposed anthophyllite of New York," by T. Thompson (see 10250).

Holmes, Ezekiel (1801-1865)
5237 1839, Report of an exploration and survey of the territory on the Aroostook River, during the spring and autumn of 1838: Augusta, Me., Smith and Robinson (printers), 78 p.
Also in Maine Public Documents 1, 1839.

Holmes, Francis Simmons (1815-1882)
5238 1848, Notes on the geology of Charleston: Char. Med. J. and Rev., v. 3, 655-671.

Holmes, F. S. (cont.)
5239 1849, Geology of Charleston: Scien. Am., v. 4, no. 45, 355.
5240 1849, Notes on the geology of Charleston, S. C.: Am. J. Sci., v. 7 (ns), no. 2, 187-201.
Also reprint edition, 17 p.
5241 1850, Geology of Charleston, S. C.: Ann. Scien. Disc., v. 1, 230-231.
Abstracted from Am. J. Sci.
5242 1850, [No fossil mammalian remains except cetacean, in the Eocene marl of South Carolina]: Am. Assoc. Adv. Sci. Proc., v. 3, 68-69.
5243 1850, Observations on the geology of Ashoey River, South-Carolina: Am. Assoc. Adv. Sci. Proc., v. 3, 201-204.

Holms, Charles (see 6074 with A. Laurent)

Home, Sir Everard (1756-1832)
5244 1817, [Fossil rhinoceros bones from England]: Am. Mon. Mag. and Crit. Rev., v. 1, no. 3, 196-197.
5245 1818, [On fossil remains]: J. Sci. Arts, v. 5, no. 1, 168.
From Royal Soc. London Proc.
5246 1823, On a new species of rhinoceros found in the interior of Africa, the skull of which bears a close resemblance to that found in a fossil state in Siberia, and other countries: Bos. J. Phil. Arts, v. 1, no. 2, 142-146.
Excerpted from Philos. Trans.

Honestas (pseudonym, see 6157)

Hood, Thomas
5247 1841, Geological discoveries: Me. Farm., v. 9, 152.
5248 1848, Hood on geology: Eclec. Mag., v. 14, 286-287.

Hood, Washington (1808-1840)
5249 1837, [Letter on controversy between W. W. Mather and G. W. Featherstonhaugh]: Naval Mag., v. 2, 578-579.
In "Reply of G. W. Featherstonhaugh" (see 7630).

Hooker (Mr., of Farmington, Connecticut)
5250 1802, [Use of Gypsum for fertilizer]: Ct. Soc. Promoting Ag. Trans., v. 1, 7 (8 lines).

Hooker, Joseph Dalton (1817-1911)
5251 1848, The fossil forest: Lit. World, v. 3, 591.
Extracted from Voyage to India.
5252 1849, [Excerpts from Flora Antarctica]: Am. J. Sci., v. 8 (ns), no. 2, 161-180.
In notice by A. Gray (see 4508).
5253 1849, On the probable extent of the flora of the coal-formation in Britain: Am. J. Sci., v. 8 (ns), no. 1, 131-133.
Abstracted from Jameson's J.
5254 1850, The table land of Tibet: Am. J. Sci., v. 9 (ns), no. 2, 298-300.
Abstracted from London Ath.

Hooker, Sir William Jackson (1785-1865)
5255 1812, Account of the geysers [of Iceland]: Select Review, v. 7, 408-422.
Extracted from Journal of a tour in Iceland.

Hopkins, Evan (d. 1867)
5256 1837, [Geological phenomena of elevation]: Am. J. Sci., v. 31, no. 2, 365-366.
In British Assoc. Proc.
5257 1843, Motion of glaciers: Franklin Inst. J., v. 6 (3s), 422-423.
From British Assoc. Proc.
5258 1848, Geology and topography of the Isthmus of Panama: Am. J. Sci., v. 6 (ns), no. 1, 123-129.
Abstracted from Mining J.

Hornblower, Josiah (1729-1809)
5259 1800, A letter from Mr. Hornblower to Mr. Kitchell, on the subject of Schuyler's Copper Mine, in New Jersey: Phila., 4 p.
Evans number 37648.

Horner, Leonard (1785-1864)
5260 1846, Museum of economic geology in Great Britain: Am. J. Sci., v. 2 (ns), no. 3, 441-442.
Abstracted from Q. J. Geology.
5261 1846, Theories on the formation of coal: Illus. Fam. Mag., v. 4, 123-126.

Horner, T. (i.e. Leonard Horner?)
5262 1847, T. Monticelli [obituary]: Am. J. Sci., v. 3 (ns), no. 1, 143.
Abstracted from Q. J. Geol. Soc.

Horner, William Edmonds (1793-1853)
5263 1834, Observations on the mineral waters in the south western part of Virginia: Phila., J. Thompson, 30 p.
5264 1840, Note of the remains of the mastodon, and some other extinct animals, collected together in St. Louis, Mo.: Am. Philos. Soc. Proc., v. 1, no. 13, 279-282.
Abstracted in Am. J. Sci., v. 40, no. 1, 56-59 (1841).
5265 1840, [Supposed earthquakes in Pennsylvania]: Am. Philos. Soc. Proc., v. 1, no. 14, 301.

Horner, W. E. (cont.)

5266   1840-1843, Remarks on the dental system of the Mastodon, with an account of some lower jaws in Mr. Koch's collection, St. Louis, Missouri, where there is a solitary tusk on the right side: Am. Philos. Soc. Proc., v. 1, no. 14, 306-308 (1840); v. 2, no. 15, 6-7 (8 lines); Am. Philos. Soc. Trans., v. 8 (ns), 53-59 (1843).
Abstracted in Am. J. Sci., v. 40, no. 2, 377-378 (1841).

5267   1843, and I. Hays, Description of an entire head and various other bones of the Mastodon: Am. Philos. Soc. Trans., v. 8 (ns), 37-47, fig., 4 plates.
Also reprint edition, 48 p., fig., plates.

Horrebow, Niels (1712-1760)

5268   1758, Some account of Iceland, an island belonging to the King of Denmark: Am. Mag. and Mon. Chron., v. 1, no. 8, 369-374; no. 9, 417-421; to be continued but no more published.

5269   1795, Account of extraordinary springs in Iceland: NY Mag., v. 6, 269-272.
From Horrebow's Natural History.

Horsfield, Thomas (1773-1859)

5270   1849, Mineralogical description of the Island of Bánká: Am. J. Sci., v. 7 (ns), no. 1, 86-101.

Horsford, Eben Norton (1818-1893)

5271   1840, Report to James Hall, on the geology of Cattaraugus County: NY Geol. Survey Ann. Rept., v. 4, 457-472.

5272   1846, Chemical essays relating to agriculture: Bos., 12 mo.
Not seen.

5273   1848, Discovery of a new cave in Kentucky: West. J. Medicine Surg., v. 4 (3s), 364-365.

5274   1850, Connection between the atomic weights and the physical and chemical properties of barium, strontium, calcium and magnesium, and some of their compounds: Am. J. Sci., v. 9 (ns), no. 2, 176-184, 4 tables.

5275   1850, Note on soda in the ashes of anthracite coal: Am. Assoc. Adv. Sci. Proc., v. 2, 233-234.

Hort, William P.

5276   1847, Analysis of Texas sugar soils, etc.: De Bow's Rev., v. 3, 553-555.

Horton, William P.

5277   1839, Report - on the geology of Orange County: NY Geol. Survey Ann. Rept., v. 3, 135-175.

5278   1842, Outlines of an essay on the structure of the earth: Am. Farm., v. 3 (3s), 333; 341.
For notice see 271. See also 5279.

5279   1842, Reply to strictures on his theory of the earth: Am. Farm., v. 3 (3s), 388.
See also 5278.

5280   1843, List of minerals observed in making the examination of the county of Orange: in Geology of New York, first district, by W. W. Mather (see 6907), 577-579.

5281   1843, Opinions of Dr. Horton on the superposition, &c. of the rocks of Orange County, N. Y.: in Geology of New York, first district, by W. W. Mather (see 6907), 580.

Hosack, David (1769-1835)

5282   1810, Answer to Dr. Seaman's "Examination" of a review of his dissertation on the Saratoga and Ballston waters, in the second number of this journal: NY Med. Philos. J. Rev., v. 2, 145-175.
See also 9384, 9385, and 7803.

5283   1810, Observations on the use of Ballston mineral waters in various diseases: Am. Med. Philos. Reg., v. 1, no. 1, 42-47.
Review with excerpts of 7681. Also reprint edition.

5284   1810?, Analysis of Ballston water. Observations on the use of the Ballston mineral waters, in various diseases: Alb., n.p., 8 p.
Edited by John Cook of Albany.

5285   1822, Extract of a letter from Dr. Hosack to Prof. Green: Am. J. Sci., v. 4, no. 2, 397-398.
On D. Hosack's career and mineral collection.

Hough, Franklin Benjamin (1822-1885)

5286   1845, Burning well [of Trumbull County, Ohio]: Am. J. Sci., v. 49, no. 2, 406-407.

5287   1847, Observations on the geology of Lewis County [New York]: Am. Q. J. Ag. Sci., v. 5, 267-274; 314-327.

5288   1850, New mineral localities in New York: Am. J. Sci., v. 9 (ns), no. 3, 288-289.

5289   1850, On the discovery of sulphuret of nickel in northern New York: Am. J. Sci., v. 9 (ns), no. 2, 287-288.
With "additional observations" by S. W. Johnson (see 5758).

5290   1850, On the existing mineral localities of Lewis, Jefferson, and St. Lawrence Counties, New York: Am. J. Sci., v. 9 (ns), no. 3, 424-429.

Houghton, Douglass (1809-1845)
See also 9805 with B. Silliman, Jr.

5291   1832, Report ... on the copper of Lake Superior: US HED 152, 22-1, v. 4 (219), 17-20.

5292   1834, Report on the copper of Lake Superior: In Narrative of an expedition through the Upper Mississippi to Itasca Lake ..., by H. R. Schoolcraft (see 9273), 287-292.

5293   1838, Communication from the state geologist: Mich. HD 46, 457-460.

5294   1838, Report of the state geologist of the state of Michigan: Mich. HD 24, 276-317.
Also in Mich. SD 16. Also published separately as Document 14, 39 p. Also in History of Michigan by Lanman (see 6035), 347-366, 1839.

5295   1838, Statement of the expenditures on account of the state geological survey for the year 1837: Mich. HD 8, 115-118; Mich. SD 21, 315-318.

Houghton, D. (cont.)
5296 1839, Communication ... relative to the geological survey ... March 7, 1839: Mich. SD 25, 463-466.
5297 1839, Report of the state geologist in relation to the iron ore, etc., on the school section in town five south, range seven west in Branch County ... : Mich. HD 21, 342-343.
5298 1839, Report of the state geologist in relation to the improvement of state salt springs: Mich. HD 2, 39-45; Mich. SD 1, 1-7.
  Published in Detroit.
5299 1839, Second annual report of the state geologist of the state of Michigan: Mich. SD 12, 264-391; Mich. HD 23, 380-507.
  Contains reports by S. W. Higgins (see 5038), C. C. Douglass (see 3209), and B. Hubbard (see 5332). For review see 5346.
5300 1840, Report of the state geologist relative to the improvement of the salt spring: Mich. SD 8, v. 2, 153-158; Mich. HD 2, v. 1, 18-23.
5301 1840, [Third annual report of the state geologist of the state of Michigan]: Mich. SD 7, v. 2, 66-153; Mich. HD 27, v. 2, 206-293.
  Also published separately as Document 8, 120 p.
  Includes reports by S. W. Higgins (see 5039), C. C. Douglass (see 3210), and B. Hubbard (see 5333).
5302 1841, [Abstracts and excerpts of the geological reports of Michigan]: Mich. Farm., v. 1, 28-29; 36-37; 84-85.
5303 1841, [Fourth] annual report of the state geologist: Mich. HD, SD, and JD 11, 472-607.
  Also published separately as Document 27, 184 p., Detroit. In part, with title "General geology of the Upper Peninsula" in US HED 591, 29-2, 6-38, map.
  Contains reports by C. C. Douglass (see 3211), S. W. Higgins (see 5040), B. Hubbard (see 5335) and F. Hubbard (see 5342).
5304 1841, Geology of Michigan: Hazard's US Reg., v. 4, 59-60.
5305 1841, [Magnetic anomalies near the Great Lakes]: Am. J. Sci., v. 41, no. 1, 171 (4 lines).
  Also in Assoc. Am. Geol's Nat's Proc., v. 2, 16 (4 lines). Also in Assoc. Am. Geol's Nat's Rept., v. 1, 24 (4 lines), 1843.
5306 1841, Metalliferous veins of the Northern Peninsula of Michigan: Am. J. Sci., v. 41, no. 1, 183-186.
  Also in Assoc. Am. Geol's Nat's Proc., v. 2, 28-31. Also in Assoc. Am. Geol's Nat's Rept., v. 1, 35-38, 1843.
5307 1841, Report of the state geologist relative to county and state maps: Mich. HD 35, 94-98.
5308 1841, Special massage concerning state salt springs: Mich. HD, SD, and JD 5, 235-254.
5309 1842, [Fifth] annual report of the state geologist: Mich. JD 2, 436-441.
5310 1842, Report of the state geologist relative to the state salt springs: Mich. HD 2, 15-21; Mich. SD 1, 1-9.
5311 1843, [Age of Lake Superior sandstone and limestone]: Am. J. Sci., v. 45, no. 1, 160-161.
  In Assoc. Am. Geol's Nat's Proc.
5312 1843, [Age of Michigan copper]: Am. J. Sci., v. 45, no. 2, 332 (4 + 7 lines).
  In Assoc. Am. Geol's Nat's Proc.
5313 1843, [Rocks about the Great Salt Lake]: Am. J. Sci., v. 45, no. 1, 155 (3 lines).
  In Assoc. Am. Geol's Nat's Proc. See also 7873.
5314 1843, [Sixth] annual report of the state geologist: Mich. JD 8, 398-402.
5315 1843, Report of the state geologist relative to the state salt springs: Mich. SD 9, 402-408.
5316 1844, [Nature of the country rock in the Missouri lead region]: Am. J. Sci., v. 47, no. 1, 106 (8 lines).
  In Assoc. Am. Geol's Nat's Proc.
5317 1844, Remarks on the importance and practicability of connecting geological surveys with the linear United States surveys: Am. J. Sci. v. 47, no. 1, 115-116.
  In Assoc. Am. Geol's Nat's Proc.
5318 1844, [Seventh] annual report of the state geologist: Mich. JD 11, 3 p.
5319 1846, [Geology of the copper region of Lake Superior] extracted from his report: in "Mineral lands ...," by J. Relfe (see 8739).
5320 1849, Geological map of townships in the Northern Peninsula of Michigan subdivided by D. Houghton D. S. in the year 1845: Balt., E. Weber and Co. (lithographers), geological map, 26 x 23.5 cm.
  In Michigan report by C. T. Jackson (see 5655).

Houghton, Jacob, jr.
5321 1846, The mineral regions of Lake Superior: ... : Buffalo, NY, O. G. Steele, 191 p., front., folding map.
5322 1846, and T. W. Bristol, Reports of Wm. A. Burt, and Bela Hubbard on the geography, topography, and geology of the U. S. surveys of the mineral region of the south shore of Lake Superior: Detroit, Mich., C. Wilcox (printer), 109 p., folded map, folded chart.
  Published with report by F. S. Burrowes (see 2204).

Hovey, Sylvester
5323 1838, Geology of Antigua: Am. J. Sci., v. 35, no. 1, 75-85.
5324 1838, Geology of St. Croix: Am. J. Sci., v. 35, no. 1, 64-74.

Howard, William (1793-1834)
5325 1821, Narrative of a journey to the summit of Mount Blanc, made in July 1819: Balt., F. Lucas, jr., 49 p., front.
5326 1828, Report [on] a route for a national road from the city of Washington to the northwestern frontier of the state of New York: US HED 38, 20-2, v. 2 (185), 2-22, 2 maps.

Howe, Henry (1816-1893; see 1485 with J. W. Barber)

Howe, Timothy
5327   1804, History of the medicinal springs at Saratoga and Ballston: Brattleboro, n.p., 22 p.

Howson, John
5328   1848, On the separation of sulphur from ore: Franklin Inst. J., v. 16 (3s), 139-140. Abstracted from London Mining J.

Hoyt, H. (see 9842 with others)

Hoyt, Thomas
5329   1850, Improved ore washer: Franklin Inst. J., v. 19 (3s), 378 (4 lines).

Hubbard, Austin Osgood
5330   1822, Oilstone of Lake Memphremagog. --- Notice of two quarries of stone, lately discovered in Lake Memphremagog, Lower Canada: Am. J. Sci., v. 5, no. 2, 406.
5331   1825, Remarks on lead veins of Massachusetts: Am. J. Sci., v. 9, no. 1, 166-167.

Hubbard, Bela (1814-1896)
       See also 2212 with W. A. Burt and 5042 with S. W. Higgins.
5332   1839, Report [on Wayne and Monroe counties]: in "Second annual report of the state geologist" by D. Houghton (see 5299), 79-114.
5333   1840, Report [on nine counties]: in third "Annual report" by D. Houghton (see 5301), 77-111.
5334   1841, Agricultural geology: Mich. Farm., v. 1, 33; 41; 54; 73; 83; 90-91; 106.
5335   1841, Report [on the geology of the organized counties of Michigan]: in fourth "Annual report" by D. Houghton (see 5303), 113-146.
5336   1845, [Answers to queries on the soil of Michigan]: Mich. Farm., v. 3, 17-18.
5337   1846, General observations upon the geology and topography of the district south of Lake Superior: US SED 357, 29-1, v. 7 (476), 20-29.
       See also 4302.
5338   1849, Analysis of the soils of our state: Mich. Farm., v. 7, 197.
5339   1849, General observations upon the geology and topography of the district south of Lake Superior, subdivided in 1845 under Houghton: in Lake Superior report by C. T. Jackson (see 5655).
5340   1849, and Ives, Geological map of the district subdivided by Messrs Hubbard & Ives under contract bearing date Sept. 7th 1846: Balt., E. Weber and Co. (lithographers), geological map, 38.5 x 20 cm., in report by C. T. Jackson (see 5655).
5341   1849, Report upon the geology and topography of the district on Lake Superior, subdivided in 1846 by William Ives, with tabular statement of specimens collected: in Lake Superior report by C. T. Jackson (see 5655), 899-932.

Hubbard, Frederick
5342   1841, Report [on magnetic variation]: in fourth Michigan "Annual report" by D. Houghton (see 5303).

Hubbard, Oliver Payson (1809-1900)
       See also articles signed by O. P. H. (658) and 9768 and 9793 with B. Silliman.
5343   1837, Geological and mineralogical notices: Am. J. Sci., v. 32, no. 2, 230-235, fig. On jointing, glacial boulders, and minerals from Boonville, Oneida County, New York.
5344   1838, Observations made during an excursion to the White Mountains, in July, 1837: Am. J. Sci., v. 34, no. 1, 105-124, 6 figs.
5345   1840, Notice of "Third annual reports on the geological survey of the state of New York, to the General Assembly, Document 275, Feb. 27, 1839": Am. J. Sci., v. 39, no. 1, 95-108.
       Review with excerpts of 7740.
5346   1841, Notice of geological surveys [reviews of second Michigan report by D. Houghton (see 5299), second Ohio report by W. W. Mather (see 6896), and Indiana survey report by D. D. Owen (see 8048)]: Am. J. Sci., v. 40, no. 1, 126-137.
5347   1841, Notice of the geological survey of the state of New York, presented to the legislature, Jan. 24, 1840 [review of 7743]: Am. J. Sci., v. 40, no. 1, 73-85.
5348   1841, [Waterville, Maine slates]: Am. J. Sci., v. 41, no. 1, 163 (13 lines).
       Also in Assoc. Am. Geol's Nat's Proc., v. 2, 8 (13 lines). Also in Assoc. Am. Geol's Nat's Rept., v. 1, 16 (14 lines), 1843.
5349   1845, Gray antimony [from Hanover, New Hampshire]: Am. J. Sci., v. 49, no. 1, 228 (5 lines).
5350   1850, The condition of trap dikes in New Hampshire an evidence and measure of erosion: Am. J. Sci., v. 9 (ns), no. 2, 158-171, fig.
5351   1850, On rutile and chlorite in quartz: Am. J. Sci., v. 10 (ns), no. 3, 350-352, 2 figs.
       From Am. Assoc. Adv. Sci. Proc.

Hughes, George Wurtz (1806-1870)
       See also 6812 by the Maryland Mining Company.
5352   1828, Tioga coal beds [in Pennsylvania]: Hazard's Reg. Penn., v. 1, 143-144.
5353   1829, Report made to the committee of the citizens of Elmira, of the Tioga Coal and Iron Mines situated on the Tioga River: Alb., John B. Van Steenbergh, 8 p.
5354   1836, Report of an examination of the coal measures, including the iron ore deposites, belonging to the Maryland Mining Company, in Alleghany County; ... : Wash., 59 p., plate.
       Includes report by T. P. Jones (see 5848).
5355   1837, Extracts from reports of an examination of the coal measures belonging to the Maryland Mining Company in Alleghany County; ... : Wash., Gales and Seaton (printer), 33 p.
       With extracts of reports by T. P. Jones (see 5849) and J. Renwick (see 8750).

Hughes, G. W. (cont.)
5356 1837, Petition ... praying congress to establish a mineralogical cabinet: US SED 167, 24-2, v. 2 (298), 1-2.
Also signed by 11 others.
5357 1837, A report of the survey of the harbor of Harve-de-Grace: US HED 134, 24-2 (303), v. 3, 10 p.
On the variation of the coast in Chesapeake Bay.
5358 1841, [Description of the mines and geology of Wales, Devon and Cornwall]: Nat´l Inst. Prom. Sci. Bulletin, v. 1, 49-65.
5359 1841, [Geological notes on the Isle of Arran, and proceedings of the Geological Section of the British Association]: Nat´l Inst. Prom. Sci. Bulletin, v. 1, 33-42.
5360 1843, Survey of the Ohio River: Wash., 47 p.
Conducted by the United States Engineering Department.
5361 1844, Report ... relative to the working of copper ore: US SED 291, 28-1, v. 5 (435), 2-58, folding table.
5362 1844, Extracts from the report ... relative to the working of copper ore: Franklin Inst. J., v. 8 (3s), 31-43; 124-133; 194-202; 261-267.
5363 1847, [Extract from report on the mineral lands of the Barrelville Mining Company]: see 1502.

Hughes, Jeremiah
5364 1845, A brief sketch of Maryland, its geography, boundaries, history, government, legislation, internal improvements, &c.: Annapolis, Md., printed for the publishers, 41, 156 p.

Huguent Coal and Iron Manufacturing Company
5365 1838, Proposals for taking sales of stock in the Huguent Coal and Iron Manufacturing Company: Richmond, Va., 12 p.

Hulett, Thomas G.
5366 1843, Every man his own guide to the falls of Niagara, or, the whole story in a few words: Buffalo, NY, Faxon and Co.
Several edition. Third through seventh editions contain "Recession of the falls from Prof. Lyell´s lectures on geology" (see 6549).

Humboldt, Alexander Friedrich Heinrich von (1769-1859)
See also 8801 with M. de Rivero.
5367 1804, [On volcanoes]: Lit. Mag. and Am. Reg., v. 2, 207-212.
5368 1806, Volcanic fish: Lit. Mag. and Am. Reg., v. 5, 366-367.
On mud volcanoes which discharge fish.
5369 1806, Volcanoes [of the Andes]: Evening Fireside, v. 2, no. 14, 110; no. 51, 403-404.
5370 1810, [Abstract of "Tableau physique des regions equitoriale"]: Select Reviews, v. 4, 217-235.
In review (see 9409).
5371 1810, On the volcanoes of Jorullo: Select Reviews, v. 4, 343-346.
5372 1818, On caverns of rocks and on their relation to the strata in which they are found: J. Sci. Arts. v. 4, no. 1, 85 (8 lines).
In France Acad. Sci. Proc.
5373 1819, [Excerpts from Personal narrative of travels to the equinoctial regions of the new continent, volume 4]: NY Lit. J., v. 1, 65-75; 138-142; 209-212; 266-271; 371-380.
See also review (8539).
5374 1820, An account of the earthquake in South America, on the 26th March, 1812: Saturday Mag., v. 3, 42-43.
5375 1821, Description of the volcano of Cotopaxi: Meth. Mag., v. 4, 100-102.
5376 1823, On the constitution and mode of action of volcanoes, in different parts of the earth: The Minerva, v. 2, 229; 237; 245; 253; 261.
5377 1824, On rock formation: Bos. J. Phil. Arts, v. 2, no. 1, 15-27; no. 2, 105-118.
Translated and extracted from Essai Géognostique.
5378 1825, [Excerpts from A geognostical essay on the superposition of rocks in both hemispheres]: Bos. J. Phil. Arts, v. 3, no. 1, 49-71.
Extracted from J. Royal Inst.
5379 1826, Aerolite [at Santa Fe de Bogota, Colombia]: Am. Mech´s Mag., v. 2, no. 45, 293-294.
5380 1827, Mines of platinum in Russia: N. Am. Med. Surgical J., v. 3, no. 1, 206 (4 lines).
5381 1827, On the subterranean sounds heard at Nakous, on the Red Sea: Franklin Inst. J., v. 3, 257-258.
Edinburgh J. of Sci.
5382 1830, Peruvian geography and geognosy: Am. J. Sci., v. 18, no. 1, 182-184.
Abstracted from Bibliothèque Universelle.
5383 1831, Gold and silver of the Uralian Mountains: N. Am. Med. Surgical J., v. 12, no. 1, 268 (10 lines).
5384 1832, On the produce of gold and silver in the Russian Empire: Am. J. Sci., v. 21, no. 2, 372-373.
Abstracted by J. Griscom.
5385 1833, Air volcano, at Turbaco, South America: Fam. Lyc., v. 1, 153.
5386 1833, Cavern at Guachara: Mon. Repos., v. 4, 24-28.
5387 1833, Earthquake at Cumana: Fam. Lyc., v. 1, 153.
5388 1833, Sepulchral cave in South America: Fam. Lyc., v. 1, 153.
5389 1841, Gold: - Its history, fluctuations and present sources: Am. Eclec., v. 2, no. 3, 514-536.
Translated by Selah B. Treat.
5390 1844, [Note on the gold of the Urals]: Am. J. Sci., v. 46, no. 1, 212.
Abstracted from Annales des Mines. See also 6008.
5391 1845, Hot springs and volcanoes: Living Age, v. 7, 251-256.
Abstracted from Edinburgh Philos. J.

Humboldt, A. von (cont.)
5392 1845, Mean height of the continents above the surface of the sea: Am. J. Sci., v. 49, no. 2, 397.
    Abstracted from Jameson's J.
5393 1848, Earthquakes and terrestrial changes: Scien. Am., v. 3, no. 50, 398.
5394 1849, Aspects of Nature, in different lands and different climates; with scientific elucidations: Phila., Lea and Blanchard, [xvi], [25]-475.
    For review see 1220 and 3419.
5395 1850, Cosmos. A sketch of the physical description of the universe: NY, Harper and Brothers, 2 v.
    "Translated from the German by E. C. Otte".
    For review with excerpts see 3013, 3412, 3414, 4520, 6396 and 6399.
    For review see 4490, 5520 and 6350.

Hume, William
5396 1837, A bed of limestone recently discovered at Ashley River: S. Agriculturist, v. 10, 542-544.
5397 1838, Limestone discovered near Charleston, S. C.: Farm's Reg., v. 5, 597-599.

Humphrey, Heman (1779-1861)
5398 1837, The Giant's Causeway: Am. Mag. Use. Ent. Know., v. 3, no. 7, 294-295, fig.

Humphreys, Hector (1797-1857)
5399 1845, Remarks on the saltness of the ocean, and the effects of light on turbid waters: Am. J. Sci., v. 49, no. 1, 208-209.

Humphreys, Reuben
5400 1809, Mineralogical notices in the county of Onondaga, state of New York: Med. Repos., v. 12, 89-90.

Hunt, Edward Bissell (1822-1863)
5401 1850, Terrestrial thermonics: Buchanan's J. of Man, v. 1, no. 11, 542-546.

Hunt, Robert (1807-1887)
5402 1848, Electricity of mineral veins: Mining J., v. 2, 22.
5403 1848, History and practice of mining in the British Isles, in relation to metallurgy: Mining J., v. 1, 150; 160; 170-171.
5404 1848, Lecture on the electricity of mineral veins: Franklin Inst. J., v. 16 (3s), 129-133.
    Abstracted from London Mining J.
5405 1849, Total quantity of lead ore raised and lead smelted in the United Kingdom in 1848: Am. J. Sci., v. 8 (ns), no. 3, 448.
    Abstracted from Mining J.

Hunt, Thomas Sterry (1826-1892)
    See also 9820 with B. Silliman, Jr.
5406 1846, Description and analysis of a new mineral species; containing titanium; with some remarks on the constitution of titaniferous minerals: Am. J. Sci., v. 2 (ns), no. 1, 30-36.
5407 1846, On the artificial formation of specular iron: Am. J. Sci., v. 2 (ns), no. 3, 411-412.
5408 1847, Report of T. S. Hunt [on analyses of minerals]: in Third annual report on the geology of the state of Vermont, by C. B. Adams (see 32), 23-27.
5409 1848, Chemical analyses by Mr. Hunt: in Fourth annual report on the geological survey of Vermont, by C. B. Adams (see 34), 8.
5410 1848, On the analysis of chromic iron: Am. J. Sci., v. 5 (ns), no. 3, 418-419.
5411 1849, Chemical examination of algerite, a new mineral species, including a description of the mineral by F. Alger: Bos. J. Nat. History, v. 6, 118-123; also in: Am. J. Sci., v. 8 (ns), no. 1, 103-106.
    Abstracted in Bos. Soc. Nat. History Proc., v. 3, 150-151 (1850). Contains "A description of the mineral" by F. Alger (146).
5412 1849, Chemical examination of the water of the Tuscarora Sour Spring, and of some other mineral waters of western Canada: Am. J. Sci., v. 8 (ns), no. 2, 364-372.
5413 1849, On the acid springs and gypsum deposits of the Onondaga salt group: Am. J. Sci., v. 7 (ns), no. 2, 175-178.
5414 1850, Chemical examinations of the waters of some mineral springs of Canada: Am. J. Sci., v. 9 (ns), no. 2, 266-275.
5415 1850, Geology of Canada: Am. J. Sci., v. 9 (ns), no. 1, 12-19; also in: Am. Assoc. Adv. Sci. Proc., v. 2, 325-334.

Hunter, W. Perceval
5416 1835, Some account of the salt of the mountain of Gern, at Cardona, in Catalonia, Spain; with some facts indicative of the little esteem entertained by Spaniards for naturalists: The Friend, v. 8, 123-124.

Hunter, William
    See also 3272 and 3274 with W. Dunbar.
5417 1789, Observations on the bones, commonly supposed to [be] elephants' bones, which have been found near the River Ohio in America: Am. Mus. or Univ. Mag., v. 5, 152-155; also in: Rural Mag. or Vt. Repos., v. 2, 582-586 (1796).

The Huntingdon Literary Museum, and Monthly Miscellany (Huntingdon, Pennsylvania, 1810)
5418 1810, Huntingdon County: v. 1, 67-70.
    Account of warm springs and topography.
5419 1810, Description of the Grotto at Swatara [at Hummelstown, Dauphin County, Pennsylvania]: v. 1, 156-157.
5420 1810, Description of the iron mine at Dannemora, in Sweden: v. 1, 206-209.
5421 1810, The blowing cave [of Virginia]: v. 1, 221-222.
5422 1810, Description of the copper-mines at Fahlun, in Sweden: v. 1, 257-263.
5423 1810, The pulpit rocks [at Huntingdon, Pennsylvania]: v. 1, 471-472.

Hunt's Merchant's Magazine, and Commercial Review (New York, 1839-1850)

5424   1839-1850, [Brief notes on mines and mining in America]: all volumes (v. 1-v. 23) contain numerous notices of coal, iron, precious metals, etc. Consult annual indices.
5425   1842, [Review of A dictionary of arts, manufactures, and mines by A. Ure (see 10532)]: v. 6, 392 (7 lines); v. 7, 204 (5 lines).
5426   1842, [Review of Mineral springs of West Virginia by W. D. Burke (see 2196)]: v. 7, 201.
5427   1842, Agricultural and mineral riches of Spain: v. 7, 282.
5428   1842, [Review of Elements of agricultural chemistry and geology by J. F. W. Johnston (see 5818)]: v. 7, 299.
5429   1843, Trade and manufacture of salt in the United States: v. 8, 357-364.
5430   1843, Coal trade of Pennsylvania [review of the eleventh annual report of the Board of Trade to the Coal Mining Association of Schuylkill County (see 8206)]: v. 8, 544-549.
5431   1843, [Review of Lectures parts 2 and 3, by J. F. W. Johnston (see 5819)]: v. 9, 296.
5432   1843, [Review of Geological cosmogony by E. Lord (see 6500)]: v. 9, 492.
5433   1844, Product of precious metals since the discovery of America: v. 10, 146-152.
5434   1844, [Review of Lectures on agricultural chemistry and geology by J. F. W. Johnston (see 5819)]: v. 10, 199.
5435   1844, Manufacture of salt in New York: v. 10, 442-447.
5436   1844, The gold mines of North Carolina: v. 11, 62-65.
5437   1844, The copper mines of Cuba: v. 11, 143-146.
5438   1844, Commerce and resources of Louisiana: v. 11, 411-423.
5439   1845, Resources of the Lackawana Valley [Pennsylvania]: v. 12, 167-170.
5440   1845, Produce of the Russian and other gold mines: v. 12, 200-201.
5441   1845, [Review of Vestiges..., by R. Chambers (see 2292)]: v. 12, 205.
5442   1845, Resources of Pennsylvania: v. 12, 237-254.
5443   1845, The precious metals in Russia: v. 12, 347-351; 456-459.
5444   1845, [Review of Elementary treatise by W. Phillips (see 8325)]: v. 12, 493.
5445   1845, The gold sands of Siberia: v. 12, 554-556.
5446   1845, [Review of Statistics of coal by R. C. Taylor (see 10147)]: v. 12, 586.
5447   1845, The coal policy of Pennsylvania: v. 13, 242-245. Signed by "L."
5448   1845, [Review of Travels ..., by C. Lyell (see 6557)]: v. 13, 302.
5449   1845, Trade and commerce of Mobile, and the resources of Alabama: v. 13, 417-426.
5450   1846, The silver mines of Mexico: v. 14, 165-167.
5451   1846, Variation of the needle: v. 14, 190 (7 lines).
5452   1846, [Review of Explanations: a sequel to Vestiges ..., by R. Chambers (see 2291)]: v. 14, 300.
5453   1846, [Review of Travels over the table lands ..., by A. M. Gilliam (see 4391)]: v. 14, 303.
5454   1846, Mineral wealth and resources of Virginia: v. 14, 343-346. Signed by "D."
5455   1846, [Review of Voyage of a naturalist by C. Darwin (see 2916)]: v. 14, 494 (9 lines).
5456   1846, Coal region of the Schuylkill and Wyoming Valley: v. 14, 539-542.
5457   1846, [Review of Mineral springs ..., by J. Moormann (see 7422)]: v. 15, 124.
5458   1846, [Review of Wisconsin ..., by I. Lapham (see 6047)]: v. 15, 127.
5459   1846, The discovery of diamond mines in the Province of Bahia: v. 15, 600-601.
5460   1847, The coal trade of Pennsylvania: v. 16, 202-207.
5461   1847, [Notice of Mineral combustibles ..., by R. C. Taylor (see 10147]: v. 16, 327.
5462   1847, Coal mines and trade of Pennsylvania: v. 16, 327-329.
5463   1847, [Review of Incentives to the cultivation ..., by S. Randall (see 8684)]: v. 16, 645.
5464   1847, The American mining journal [review of]: v. 17, 112-113.
5465   1847, [Notice of The coal and iron trade by C. G. Childs (see 2339)]: v. 17, 115.
5466   1847, Commerce and resources of the state of New York: v. 17, 451-465.
5467   1847, [Review of Picturesque sketches ..., by D. T. Ansted (see 1280)]: v. 17, 542.
5468   1848, [Review of Statistics of coal by R. C. Taylor (see 10147)]: v. 19, 122.
5469   1848, [Review of What I saw ..., by E. Bryant (see 2137)]: v. 19, 125.
5470   1849, Commerce and resources of Cuba: v. 20, 34-40.
5471   1849, [Review of Earth and Man by A. Guyot (see 4609 and 4610)]: v. 20, 137.
5472   1849, [Review of Man Primeval by J. Harris (see 4832)]: v. 20, 141.
5473   1849, Virginia: her history and resources: v. 20, 181-191.
5474   1849, [Review of Sights in the gold region by T. T. Johnson (see 5760 and 5761)]: v. 20, 701.
5475   1849, The gold region of California: v. 21, 55-64.
5476   1849, [Review of Elements of geology by D. Page (see 8107)]: v. 21, 240 (5 lines).
5477   1849, The Cumberland semi-bituminous coal region: v. 21, 320-325.
5478   1849, The gold mines of Russia: v. 21, 326-328.
5479   1849, [Review of Oregon and California ..., by J. Q. Thornton (see 10259): v. 21, 351.
5480   1849, [Review of A tour of duty in .... by J. Revere (see 8756)]: v. 21, 352.
5481   1849, Silver mines and mining in Spain: v. 21, 452-453.
5482   1849, The coal fields and coal trade of Ohio: v. 21, 558-560.
5483   1849, The gold mines or deposits of Siberia: v. 21, 675-678.
5484   1849, [Review of Gold mines of the Gila by C. W. Webber (see 10718)]: v. 21, 693 (6 lines).
5485   1850, [Review of Manufacture of iron ..., by F. Overman (see 8045)]: v. 22, 253.
5486   1850, [Review of Six months in the gold region by E. G. Buffum (see 2180)]: v. 23, 709.
5487   1850, [Review of Footprints of the Creator by H. Miller (see 7186)]: v. 23, 591.
5488   1850, [Review of Lake Superior by L. Agassiz (see 83)]: v. 23, 253.
5489   1850, [Review of Eldorado by B. Taylor (see 10098)]: v. 23, 139.

Huntt, Henry (1792-1838)
5490   1838, A visit to the Red Sulphur Spring of Virginia during the summer of 1837 with observations on the water: Wash., Gideon, 27 p.
        Also 1839 editions: Bos., Dutton and Wentworth, 40 p., plate; and Phila., T. Cowperthwait, 44 p.

Hurd, Erastus
5491   1845, Report [on the geology of Mineral Creek, Michigan]: in Articles of Association of the Mineral Creek Mining Company (see 7198), 9-16.
5492   1845, Map of the Marshall & Boston Lake Superior Mining Company location mineral district: in Articles of Association of the Mineral Creek Mining Company (see 7198), colored geological map, 33 x 25 cm.
        Illustrates part of the south shore of Lake Superior.

Hutchins, Thomas (1730-1789)
5493   1784, An historical narrative and topographical description of Louisiana, and West-Florida, comprehending the River Mississippi with its principal branches and settlements, and the Rivers Pearl, Pascagoula, Mobile, Perdido, Escambia, Chacta-hatcha, &c.: The climate, soil, and produce whether animal, vegetable, or mineral; ... : Phila., R. Aitken, iv, 94 p.
        Evans number 18532.
5494   1786, Description of a remarkable rock and cascade, near the western side of the Youghiogeny river, a quarter of a mile from Crawford's Ferry, and about twelve miles from Union-Town, in Fayette County, in the state of Pennsylvania: Am. Philos. Soc. Trans., v. 2, 50-51. Also in: Univ. Mag., v. 3, 380-381.
5495   1787, A topographical description of Virginia, Pennsylvania, Maryland and North-Carolina, comprehending the rivers Ohio, Kenhawa, Sioto, Cherokee, Wasash, Illinois, Mississippi, &c. The climate, soil and produce, whether animal, vegetable, or mineral: Bos., John Norman, ii, 30, 2 p.
        Evans number 20424.
5496   1792, Description of the Upper Mississippi: Univ. Asylum and Colum. Mag., v. 9, 35-38.
        Also in: Lit. Mus. or Mon. Mag., v. 1, no. 4, 171-175.

Hutton, William (1798-1860)
5497   1832, Whin sill [in Northern England]: Mon. Am. J. Geology, v. 1, no. 10, 471 (12 lines).
        In British Assoc. Proc. See also 7507.

Hyatt, James
5498   1849, The agricultural uses of lime and marl: NY State Agri. Soc. Trans., v. 8, 386-402.

Ibbetson, B.
5499   1846, Fossil shark, (Hybodus): Am. J. Sci., v. 1 (ns), no. 2, 280 (8 lines).
        Abstracted from Q. J. Geol. Soc.

Ibbotson, Arvah J. (lithographer)
5500   184?, Vaucluse gold mine. Property of the Orange Grove Mining Company, Virginia. J. Stitz, del.: Sinclair's lith., colored lithograph, 54 x 86 cm.
        View of Vaucluse Mine site.

Ii, John
5501   1839, Fall of meteoric stones at Honolulu: Hawaiian Spectator, v. 1, no. 2, 132 [i.e. 232]-233.
        Translated from the Kuma Hawaii.

The Illinois Medical and Surgical Journal (Chicago, Illinois, 1844-1847)
5502   1847, Geological survey: v. 4, 188-189.
        On D. D. Owen's survey of Indiana.

The Illinois Monthly Magazine (Vandalia, Illinois, 1830-1832)
5503   1830, Geology of Illinois: v. 1, 43-46.
5504   1831, Earthquake at Caracas: v. 2, 85-93.

Illustrated Family Magazine (Boston, 1845-1846)
5505   1845, Wier's Cave, Augusta Co, Virginia: v. 1, 42-44.
5506   1845, Encroachments of land on the sea: v. 2, 37-40.
5507   1845, Encroachments of sea on the land: v. 2, 56-59.
5508   1845, The Giant's Causeway: v. 2, 71-72, fig.
5509   1845, The glaciers of the alps: v. 2, 84-87.
5510   1845, Rocking stones [of Great Britain]: v. 2, 190-191.

Imlay, Gilbert (1754?-1828?)
5511   1792, Some particulars relative to the soil, productions, and commerce of Kentucky, &c. with observations on the vast inland navigation of America: Univ. Asylum and Colum. Mag., v. 9, 307-311.
5512   1793, A topographical description of the Western Territory of North America; containing a succinct account of its climate, natural history, population, agriculture, manners and customs: NY, Samuel Campbell (printer), 2 v. (260 p., map, plan; 204 p., map).
        Evans number 25648. "Third edition" (editions 1, 2, and 4 were not American). Geology noted especially p. 110 et seq.

Imrie, Major
5513   1831, Bone caves in New Holland: Am. J. Sci., v. 20, no. 2, 380-381.
        Abstracted from Philos. J. by J. Griscom.

The Independent (New York, 1848-1850)
        Weekly intelligence and uncaptioned news items are not included.
5514   1849, Geological discovery: v. 1, no. 6, 21.
        On the discovery of fossil Megalosaurus bones from Monmouth County, New Jersey.

The Independent (cont.)
5515   1849, [Review of Manual of mineralogy by J. D. Dana (see 2832)]: v. 1, no. 14, 56 (10 lines).
5516   1849, Retrospect of geological mutations: v. 1, no. 19, 76.
5517   1849, [Review of Earth and man by A. Guyot (see 4609 and 4610)]: v. 1, no. 27, 108.
5518   1849, [Review of Man primeval by J. Harris (see 4832)]: v. 1, no. 31, 124.
5519   1849, [Review of A second visit ..., by C. Lyell (see 6571)]: v. 1, no. 48, 192.
5520   1850, [Review of Cosmos by A. von Humboldt (see 5395)]: v. 2, no. 70, 56.
5521   1850, The eruption of Vesuvius: v. 2, no. 71, 60.
5522   1850, [Review of Footprints of the Creator by H. Miller (see 7186)]: v. 2, no. 100, 180.

Indiana. State of
5523   1837, Act to provide for a geological survey of the state of Indiana: Indiana Gen. Assembly, February 6, 1837.
5524   1850, A joint resolution upon the subject of a grant of land for a geological survey of the state of Indiana: Indiana Gen. Assembly, January 21, 1850.

Indiana Historical Society
5525   1833, Indiana Historical Society: Am. J. Sci., v. 24, no. 1, 181-182.
       Notices of the Society's scientific activities.

The Indicator (Amherst, Massachusetts, 1848-1850)
5526   1849, Mineralogical reflections: v. 1, 200-202.
       Signed by "J. E. Hr." Thoughts on seeing a beautiful mineral; an allegory on Christian behavior.
5527   1849, The poetry of geology: v. 2, 109-111.
       Signed by "M*****S." On the beauty of the science of geology.

Indicus (pseudonym; see 7847).

Ingersoll, J.
5528   1846, Yazoo marl - Mississippi: Am. Q. J. Ag. Sci., v. 3, 295-296.

Inglis, Sir. Robert Harry (1786-1855)
5529   1847, Seventeenth meeting of the British Association for the Advancement of Science: Am. J. Sci., v. 4 (ns), no. 2, 238-257.
       Address on science.

Ingraham (Prof.)
5530   1839, Earthquake - New-Madrid: New-Yorker, v. 7, 389.

The International Monthly Magazine of Literature, Science, and Art (New York, 1850)
5531   1850, Prof. Agassiz: v. 1, 72.
       His views on the unity of the races of man debated.
5532   1850, Dr. Buckland: v. 1, 176 (5 lines).
       Notice of his entry into an Oxford insane asylum.

The Iris: Devoted to Science, Literature and the Arts (Richmond, Virginia, 1848)
5533   1848, Magnet cove - Arkansas: v. 1, no. 1, 23.
       Description abstracted from De Bow's Rev.

Irving, Washington (1783-1859)
5534   1836, Astoria; or, anecdotes of an enterprise beyond the Rocky Mountains ... : Phila., Carey and Lea, 2 v. (285 p.; 279 p., map).
       Other editions include Phila., Lea and Blanchard, 2 v., 1841; and NY, Putnam, viii, 519 p., 1849.
5535   1837, The Rocky Mountains; or, scenes, incidents and adventures, in the far West, digested from the journal of Captain B. L. E. Bonneville of the Army of the United States; and illustrated from various other sources: Phila., Carey, Lea and Blanchard, 2 v., 2 maps (248 p.; 248 p.).
       Other editions, some as The adventures of Captain Bonneville.

Isham, Warren (d. 1863)
5536   1849, Prof. Johnson's [i.e. Johnston's] address: Mich. Farm., v. 7, 292-293.

Ivanoff, M.
5537   1846, Kaliphite [from Hungary]: Am. J. Sci., v. 2 (ns), no. 3, 416 (12 lines).
       Abstracted from Berzelius' Jahresbericht.

Ives, William (see 5340 with B. Hubbard)

J., J. H. (see 6687 and 6697)

Jackson, Charles Thomas (1805-1880)
       See also discussion of articles by J. E. Teschemacher (10177), J. D. Dana (2817), J. T. Hodge (5208), A. Guyot (4612).
5538   1828/1829, and F. Alger, A description of the mineralogy and geology of a part of Nova Scotia: Am. J. Sci., v. 14, no. 2, 305-330, colored map; v. 15, no. 1, 132-160, 2 plates; no. 2, 201-217.
       Contains map 5539. For review see 549, 1097, 4324, 3765, 7689.
5539   1828, and F. Alger, A geological map of part of Nova Scotia with a section: Am. J. Sci., v. 14, no. 2, colored map, 44.2 x 20.0 cm.
       In 5538.
5540   1833, A description of a new mineral species from Nova Scotia: Am. J. Sci., v. 25, no. 1, 78-84, fig.
       With "chemical analysis" by A. A. Hayes (see 4891). On ledererite from Cape Blomidon, Nova Scotia.
5541   1833, and F. Alger, A new geological map of the peninsula of Nova Scotia with a section: Am. Acad. Arts Sci's Mem., v. 1 (ns), colored geological map, 43.5 x 23.1 cm.
       In 5542.

Jackson, C. T. (cont.)

5542 1833, and F. Alger, Remarks on the mineralogy and geology of Nova Scotia: Am. Acad. Arts Sci´s Mem., v. 1 (ns), 217-330, 4 colored plates, colored map.
Contains map 5541. Also reprint edition, 116 p.

5543 1834, An account of the chiastolite or macle of Lancaster [Massachusetts]: Bos. J. Nat. History, v. 1, no. 1, 55-62, colored plate with 25 figs.

5544 1835, Chemical analysis of chrysocolla from the Holquin copper mines, near Gibara, Cuba: Bos. J. Nat. History, v. 1, no. 2, 206-208.

5545 1836, Appendix. On the collection of geological specimens and on geological surveys: Am. J. Sci., v. 30, no. 1, 203-208.

5546 1836, Chemical analysis of mineral waters from the Azores: Am. J. Sci., v. 31, no. 1, 94-97.

5547 1836, Chemical analysis of three varieties of bituminous coal, and one of anthracite: Bos. J. Nat. History, v. 1, no. 3, 357-360.

5548 1836, [Review with excerpts of Report on the geology of Massachusetts, second edition, by E. Hitchcock (see 5110)]: N. Am. Rev., v. 42, 422-448.

5549 1837, The coast of Maine: Am. Farm., v. 3 (2s), 413.
On the geology of Quoody Head. From First Report by C. T. Jackson (see 5550).

5550 1837, First report on the geology of the state of Maine: Augusta, Me., Smith and Robinson (printers), viii, 9-127 p.
Also another edition, 190 p. For review see 7944, 9740. For review with excerpts see 5017. Published with Atlas (5551).

5551 1837, Atlas of plates illustrating the geology of the state of Maine, accompanying the first report on the geology of the state: Augusta, Me., Smith and Robinson (printers), 24 plates (21 colored).

5552 1837, First report on the geology of the public lands in the state of Maine: Mass. SD 89, 47 p.
For review see 7944.

5553 1837, List of mines and minerals belonging to the Maine Mining Company: Bos., Beals and Greene, 16 p.

5554 1838, Bituminization of peat and conversion into coal [and] The coal measures of Mansfield, Mass.: Am. J. Sci., v. 34, no. 2, 395.

5555 1838, Chemical analysis of meteoric iron, from Claiborne, Clarke Co., Alabama: Am. J. Sci., v. 34, no. 2, 332-337.

5556 1838, Final report on the geology of the public lands belonging to the states of Massachusetts and Maine: Mass. HD 70, 12, 8 p.
For review with excerpts see 723.

5557 1838, Miscellaneous remarks on certain portions of the geology of Maine: Am. J. Sci., v. 34, no. 1, 69-73.

5558 1838, Second annual report on the geology of the public lands belonging to the two states of Maine and Massachusetts: Augusta, Me., Luther Severance (printer), xi, 100 p., 6 plates, xvii p.
Contains report by J. T. Hodge (see 5190). For review with excerpts see 5017. For review see 6703. For notice see 718.

5559 1838, Second annual report on the geology of the public lands belonging to the two states of Maine and Massachusetts: Bos., Mass. HD 70, 12, 93 p.

5560 1838, Second report on the geology of the state of Maine: Augusta, Me., Luther Severance (printer), xiv, 17-168 p.
For review with excerpts see 7946. For review see 6702. For notice see 701, 718, 3702.

5561 1839, Extracts from Dr. Jackson´s second report on the geology of the state of Maine: NE Farm., v. 17, 258-259.

5562 1839, Catlinite or Indian pipe stone [from Coteau du Prairie]: Am. J. Sci., v. 35, no. 2, 388.

5563 1839, Dr. Jackson´s geological report [excerpts from third report on Maine geology (see 5566)]: Me. Farm., v. 7, 185-186.

5564 1839, Geology of Maine: Farm´s Mon. Vis., v. 1, 13-14; 26-27; 45-46; 57-58; 112; 134-135; 154-155; 166-167.
Extracted from 5566.

5565 1839, A new mineral: Am. J. Sci., v. 37, no. 2, 398 (10 lines).
On beaumontite from Chessy, France.

5566 1839, Third annual report on the geology of the state of Maine: Augusta, Me., Smith and Robinson (printers), xiv, 276, lxiv, 1 p., illus.
Contains report by S. L. Stephenson (see 10015).
For review with excerpts see 723, 2487, 2708, 6766. For review see 4289. For excerpts see 5563, 5564, 5567, and 10015.

5567 1840, Agricultural geology: S. Agriculturist, v. 1 (ns), 74-78.
From the third Maine report (see 5566).

5568 1840, [Extract from a speech on agricultural geology]: Farm´s Mon. Vis., v. 2, 19-21.
5569 1840, A geological map of the state of Rhode Island: Bos., C. Cook Del., colored geological map, 31.5 x 51 cm.
In 5571.

5570 1840, Iron in Maine: Hazard´s US Reg., v. 2, 414.

5571 1840, Report on the geological and agricultural survey of the state of Rhode Island: Providence, Rhode Island, B. Cranston and Co., 312 p., illus., colored map.
Contains map 5569. For review with excerpts see 6713. For review see 2488, 2710, 3707, 9795. See also "Act" (8762).

5572 1840, Speech of Dr. C. T. Jackson [on the geology of New England]: Farm´s Reg., v. 8, 152-157.

5573 1841, [American fossils and minerals]: Am. J. Sci., v. 41, no. 1, 161-162.
Also in Assoc. Am. Geol´s Nat´s Proc., v. 2, 6-7. Also in Assoc. Am. Geol´s Nat´s Rept., v. 1, 14-15, 1843.

5574 1841, [Columnar structure in dikes of Nova Scotia]: Am. J. Sci., v. 41, no. 1, 173 (8 lines).
Also in Assoc. Am. Geol´s Nat´s Proc., v. 2, 18 (8 lines). Also in Assoc. Am. Geol´s Nat´s Rept., v. 1, 26 (8 lines), 1843.

5575 1841, [Effects of magnesia in dolomite on vegetation]: Am. J. Sci., v. 41, no. 1, 159.
Also in Assoc. Am. Geol´s Nat´s Proc., v. 2, 4. Also in Assoc. Am. Geol´s Nat´s Rept., v. 1, 12, 1843.

Jackson, C. T. (cont.)

5576 1841, First annual report on the geology of New Hampshire: Concord, New Hampshire, Barton and Carroll (printers), 164 p., 7 figs. on wrappers.
Contains reports by J. D. Whitney and M. Williams (see 10904 and 10905).
For reviews see 754 and 1600. For excerpts see 5577, 5579, and 10171.

5577 1841, First annual report on the geology of the state of New Hampshire: Farm's Mon. Vis., v. 3, 113-117; 129-135; 145-150; 161-166, 6 figs.
Extracted from 5576.

5578 1841, and F. Alger, [Geological map of the peninsula of Nova Scotia]: Bos., folio.
New edition. Not seen.

5579 1841, Geological survey of New Hampshire: Hazard's US Reg., v. 5, 160.
From 5576. Notes on Coos County.

5580 1841, [Glaciation in New England]: Am. J. Sci., v. 41, no. 1, 176.
Also in Assoc. Am. Geol's Nat's Proc., v. 2, 21-22. Also in Assoc. Am. Geol's Nat's Rept., v. 1, 28-29, 1843.

5581 1841, [Infusorial deposit at Newfield, Maine]: Am. J. Sci., v. 41, no. 1, 174 (12 lines).
Also in Assoc. Am. Geol's Nat's Proc., v. 2, 19. Also in Assoc. Am. Geol's Nat's Rept., v. 1, 26, 1843.

5582 1841, [The nature of dolomite]: Am. J. Sci., v. 41, no. 1, 171 (8 lines).
Also in Assoc. Am. Geol's Nat's Proc., v. 2, 16 (8 lines). Also in Assoc. Am. Geol's Nat's Rept., v. 1, 24 (8 lines), 1843.

5583 1841, [On joints in rocks]: Am. J. Sci., v. 41, no. 1, 172.
Also in Assoc. Am. Geol's Nat's Proc., v. 2, 18, 19. Also in Assoc. Am. Geol's Nat's Rept., v. 1, 25, 26, 1843.

5584 1841, [On the construction of geological maps]: Am. J. Sci., v. 41, no. 1, 186 (7 lines).
Also in Assoc. Am. Geol's Nat's Proc., v. 2, 31 (7 lines). Also in Assoc. Am. Geol's Nat's Rept., v. 1, 38 (7 lines), 1843.

5585 1841, [On the origin of soils]: Farm's Mon. Vis., v. 3, 36-37.

5586 1841, [On the slates of Waterville, Maine]: Am. J. Sci., v. 41, no. 1, 163-164.
Also in Assoc. Am. Geol's Nat's Proc., v. 2, 8-9 (13 lines). Also in Assoc. Am. Geol's Nat's Rept., v. 1, 16-17 (13 lines), 1843.

5587 1841, The ore of Jackson [Coos County, New Hampshire]: Hazard's US Reg., v. 5, 63.
With chemical analysis by J. E. Teschemacher (see 10170).

5588 1841, [Potash in soils]: Am. J. Sci., v. 41, no. 1, 159-160 (9 lines).
Also in Assoc. Am. Geol's Nat's Proc., v. 2, 4-5. Also in Assoc. Am. Geol's Nat's Rept., v. 1, 12-13, 1843.

5589 1842, [Meteorite from Clayborne County, Alabama]: Am. J. Sci., v. 43, no. 1, 169 (4 lines).
Also in Assoc. Am. Geol's Nat's Proc., v. 3, 24 (4 lines). Also in Assoc. Am. Geol's Nat's Rept., v. 3, 62 (3 lines), 1843.

5590 1842, [On the application of the glacial theory in America]: Am. J. Sci., v. 43, no. 1, 151.
Also in Assoc. Am. Geol's Nat's Proc., v. 3, 6. Also in Assoc. Am. Geol's Nat's Rept., v. 1, 46, 1843.

5591 1842, [Potholes near Canaan, New Hampshire]: Am. J. Sci., v. 43, no. 1, 154 (12 lines).
Also in Assoc. Am. Geol's Nat's Proc., v. 3, 9. Also in Assoc. Am. Geol's Nat's Rept., v. 1, 48-49, 1843.

5592 1842, Second annual report on the geology of the state of New Hampshire: Concord, NH, Cyrus Barton, 8 p.
For review see 3709.

5593 1842, Report of the state geologist: NH House of Representatives Joint Session for 1842, 239-244.
Also reprinted in 5611.

5594 1842, Third annual meeting of the Association of American Geologists and Naturalists: Am. J. Sci., v. 43, no. 1, 146-184.
Jackson was secretary of the Association, and thus probably authored its proceedings.

5595 1843, [Agricultural chemistry and the analysis of soils]: Bos. Soc. Nat. History Proc., v. 1, 8-10.

5596 1843, Description of the tin veins of Jackson, N. H.: Assoc. Am. Geol's Nat's Rept., v. 1, 316-321, 2 figs.

5597 1843, An hypothesis to explain, &c. the changes of the surface of the earth: Bos. Soc. Nat. History Proc., v. 1, 123.

5598 1843, [Lava from Kilauea, Hawaii]: Bos. Soc. Nat. History Proc., v. 1, 22-23.

5599 1843, Letter of state geologist giving contents and make-up of proposed geological report: NH House J., June, 1843, 301-302.

5600 1843, [Metamorphic rocks of Pequawket Mountain, New Hampshire]: Am. J. Sci., v. 45, no. 1, 145-146.
In Assoc. Am. Geol's Nat's Proc.

5601 1843, [Minerals of New Hampshire]: Bos. Soc. Nat. History Proc., v. 1, 45-46.

5602 1843, [On glacial drift]: Am. J. Sci., v. 45, no. 2, 319-323.
In Assoc. Am. Geol's Nat's Proc. With discussion by E. Hitchcock (see 5151), J. N. Nicollet, and W. C. Redfield (see 8715).

5603 1843, [Ores of New Hampshire]: Bos. Soc. Nat. History Proc., v. 1, 90.

5604 1843, [Remarks on the origin of limestone and coral rock]: Am. J. Sci., v. 45, no. 1, 140-141.
In Assoc. Am. Geol's Nat's Proc.

5605 1843, Remarks on zinc, lead, and copper ores of New Hampshire: Assoc. Am. Geol's Nat's Rept., v. 1, 321-322.

5606 1843, [Yttro-cerite from Bolton, Massachusetts]: Am. J. Sci., v. 45, no. 2, 331 (1 line).
In Assoc. Am. Geol's Nat's Proc.

5607 1844, Analysis of pink scapolite, and of cerium ochre, from Bolton, Mass.: Bos. J. Nat. History, v. 4, no. 4, 504-506.

5608 1844, [Analysis of yttro-cerite from Bolton, Massachusetts]: Am. J. Sci., v. 47, no. 2, 353 (10 lines).

5609 1844, Chemical analyses of soils sent to me by Dr. Silas Meacham, Illinois: Prairie Farm., v. 4, 214-215.

Jackson, C. T. (cont.)

5610   1844, [Description and analysis of pink scapolite from Bolton, Massachusetts]: Bos. Soc. Nat. History Proc., v. 1, 167.

5611   1844, Final report on the geology and mineralogy of the state of New Hampshire; with contributions toward the improvement of agriculture and metallurgy: Concord, NH, Carroll and Baker (printers), viii, 376 p., illus., front., plates, 2 folded color plates, folded map.
    Contains reports or letters by A. Daubrée (see 2937), J. L. Hayes (see 4914), New Hampshire (see 7700 and 7701), E. Pierce (see 8345), and J. D. Whitney and M. B. Williams (see 10904 to 10907); map (see 5612); and first report (see 5593). For review see 1889, 3525, 3710, 10526. For notice see 799. 600 copies printed.

5612   1844, A geological map of New-Hampshire. By Charles T. Jackson. 1844: in Final report ... (see 5611), geological map, 39.2 x 69.6 cm.
    Also in 5624.

5613   1844, Formula of the pink scapolite of Bolton: Am. J. Sci., v. 47, no. 2, 418 (5 lines).
    Abstracted from Bos. J. Nat. History.

5614   1844, [Fremont's exploring expedition]: Bos. Soc. Nat. History Proc., v. 1, 161 (8 lines).

5615   1844, [On minerals from Keweenaw Point, Lake Superior]: Bos. Soc. Nat. History Proc., v. 1, 203.

5616   1844, [On yttro-cerite and pink scapolite from Bolton, Massachusetts]: Bos. Soc. Nat. History Proc., v. 1, 161, 165-167.

5617   1844, Remarks on the Alabama meteoric iron, with a chemical analysis of the drops of green liquid which exude from it: Bos. Soc. Nat. History Proc., v. 1, 207-208; also in: Am. J. Sci., v. 48, no. 1, 145-146 (1845).

5618   1845, [On copper ores of the Lake Superior region]: Bos. Soc. Nat. History Proc., v. 2, 57-58.

5619   1845, [On minerals from Litchfield, Maine]: Assoc. Am. Geol's Nat's Proc., v. 6, 44-49.
    Also in Am. Q. J. Ag. Sci., v. 2, 157.

5620   1845, [On New England drift and diluvial evidence]: Am. Q. J. Ag. Sci., v. 2, 141-142. In Assoc. Am. Geol's Nat's Proc.

5621   1845, On the copper and silver of Keweenaw Point, Lake Superior: Am. J. Sci., v. 49, no. 1, 81-93; also in: Assoc. Am. Geol's Nat's Proc., v. 6, 53-61; Am. Q. J. Ag. Sci., v. 2, 159-163.
    With discussion by C. U. Shepard. For review see 3545. Abstracted in Niles' Wk. Reg., v. 68, 409-410.

5622   1845, [Review of American coal report by W. R. Johnson (see 5796)]: Bos. Soc. Nat. History Proc., v. 2, 25-26 (8 lines).

5623   1845, Report to the trustees of Lake Superior Copper Company: Bos., Beals and Greene (printers), 19 p., plate, 4 diagrams.

5624   1845, Views and map illustrative of the scenery and geology of the state of New Hampshire: Bos., Thurston, Jorry, and Co., 20 p., plates, map.
    Contains map 5612.

5625   1846, Analysis of soils: Am. Q. J. Ag. Sci., v. 4, 220-238.

5626   1846, Cancrinite, nepheline, and zircon, from Litchfield, Maine: Am. J. Sci., v. 1 (ns), no. 1, 119-120.
    Abstracted from Assoc. Am. Geol's Nat's Proc.

5627   1846, [Chemical analysis of Sahara sand]: Bos. Soc. Nat. History Proc., v. 2, 170.

5628   1846, Chemical analyses of the ores [of copper from Lake Superior]: in Mineral lands ..., by J. H. Relfe (see 8739), 38-44.

5629   1846, Chemical and mineralogical fragments: Bos. J. Nat. History, v. 5, no. 3, 405-412.
    Abstracted in Am. Q. J. Ag. Sci., v. 5, 227-228 (1847)

5630   1846, [Fossils from the lead mines of Wisconsin]: Bos. Soc. Nat. History Proc., v. 2, 122 (8 lines).

5631   1846, Obituary [of Douglass Houghton]: Am. J. Sci., v. 1 (ns), no. 1, 150-152.

5632   1846, [On copper and zinc ores from Warren, New Hampshire]: Bos. Soc. Nat. History Proc., v. 2, 147.

5633   1846, [On the composition of lava from the crater of Kilauea in Hawaii]: Bos. Soc. Nat. History Proc., v. 2, 120-121.

5634   1846, [On the geology of the White Mountains, New Hampshire]: Bos. Soc. Nat. History Proc., v. 2, 147-148.

5635   1846, [On the Lake Superior mining district]: Bos. Soc. Nat. History Proc., v. 2, 110-114.
    For abstract see Am. J. Sci., v. 2 (ns), no. 1, 118-119.

5636   1847, [Cetacean vertebra fossil from Machias, Maine]: Bos. Soc. Nat. History Proc., v. 2, 255 (6 lines).

5637   1847, [Copper from Lake Superior]: Bos. Soc. Nat. History Proc., v. 2, 259-260 (12 lines).

5638   1847, [On three divisions in the diluvium of Maine]: Bos. Soc. Nat. History Proc., v. 2, 256 (7 lines).

5639   1847, [Ore from Cliff Mines, Lake Superior]: Bos. Soc. Nat. History Proc., v. 2, 256 (9 lines).

5640   1847, Synopsis (of a report) of the geological survey of the mineral lands of the United States in Michigan: US SED 2, 30-1, v. 2 (504), 175-230.
    Contains reports by W. F. Channing (2308, 2309), J. Locke (6461) and J. D. Whitney (10909).
    "Printing authorized late in 1847 and some bibliographies use this date" (Pestana, 1972, p. 71).
    For notice with excerpts see 3004.

5641   1847, Synopsis of a report on the progress of the geological survey of the mineral lands of the United States in Michigan: US SED 2, v. 2 (530), 30-2, 185-191.

5642   1847, [Tertiary of Maine]: Bos. Soc. Nat. History Proc., v. 2, 213 (5 lines).

5643   1848, [Direction of drift scratches and cleavage planes of the Roxbury, Massachusetts graywacke]: Bos. Soc. Nat. History Proc., v. 3, 28 (6 lines).

5644   1848, Discovery of tellurium in Virginia: Am. J. Sci., v. 6 (ns), no. 2, 188.

5645   1848, A new method of extracting pure gold from alloys and from ores: Am. J. Sci., v. 6 (ns), no. 2, 187.

Jackson, C. T. (cont.)
5646 1848, [New zeolite mineral from Lake Superior] : Bos. Soc. Nat. History Proc., v. 3, 76-77.
    See also "Correction," p. 228 (5 lines), 1850.
5647 1848, [On metamorphic rocks, particularly of Rhode Island] : Bos. Soc. Nat. History Proc., v. 3, 19-20.
5648 1848, [On the quality of work of J. W. Foster and J. D. Whitney] : US SED 2, 30-2, v. 2 (530), 153.
5649 1848, [Uplift and age of American continent] : Bos. Soc. Nat. History Proc., v. 3, 88 (8 lines).
5650 1849, Copper of the Lake Superior region: Am. J. Sci., v. 7 (ns), no. 2, 286-287.
5651 1849, Full exposure of the conduct of Dr. Charles T. Jackson, leading to his discharge from the government service, and justice to Messrs. Foster and Whitney, U. S. Geologists: n. p., 32 p.
    Contains letters and testimony by C. T. Jackson.
5652 1849, Geological map of Isle Royale Lake Superior: NY, Ackerman (lithographer), colored geological map, 54 x 38 cm., in 5655.
5653 1849, Geological map of Keweenaw Point, Lake Superior: NY, Ackerman (lithographer), colored geological map, 56 x 34 cm., in 5655.
5654 1849, [Medical geology] : Bos. Soc. Nat. History Proc., v. 3, 168-169.
    On the relationship between geological formations and prevalence of cholera.
5655 1849, Report on the geological and mineralogical survey of the mineral lands of the United States in the state of Michigan: US SED 1, 31-1, v. 3, pt. 3 (551), 371-935, 26 plates, 6 maps.
    Contains numerous reports, letters, and maps, including those of: G. O. Barnes (1492), F. S. Burrowes (2205 and 2206), W. A. Burt and B. Hubbard (2209 to 2212), J. Child (2336), R. Cropley (2693), G. J. Dickenson (3174 and 3175), J. T. Foster and J. D. Whitney (3928 to 3932; 3936), J. T. Foster (3933), J. T. Foster and S. Hill (3935), W. Gibbs (4383), D. S. Houghton (5320), B. Hubbard (5339 to 5341), B. Hubbard and W. Ives (5340), C. T. Jackson (5652 and 5653), J. Locke (6463), J. McIntyre (6606), J. D. Whitney (10917; 10918; 10919), S. Higgins (5042), S. Hill (5061). 15,000 copies printed. Also in US HED 5, 31-1, v. 5, pt. 3 (571), 371-935, 26 plates, 6 maps.
5656 1849, [Minerals from Virginia gold mines] : Bos. Soc. Nat. History Proc., v. 3, 122 (13 lines).
5657 1849, [On fissures in pudding stone of Roxbury, Massachusetts] : Bos. Soc. Nat. History Proc., v. 3, 127.
5658 1849, [On the structure of glaciers] : Bos. Soc. Nat. History Proc., v. 3, 123-124, 126.
5659 1850, [An analysis of the new mineral algerite] : Bos. Soc. Nat. History Proc., v. 3, 278-279.
5660 1850, [Analysis of manganese-bearing cast iron and ore] : Bos. Soc. Nat. History Proc., v. 3, 232-235.
5661 1850, Analysis of water from a hot spring in the region of the Great Salt Lake: Am. J. Sci., v. 10 (ns), no. 1, 134.
    Abstracted from Bos. Soc. Nat. History Proc.
5662 1850, Anhydrous prehnite: Am. J. Sci., v. 10, no. 1, 121 (9 lines).
    Abstracted from Bos. J. Nat. History.
5663 1850, [Artificial minerals from furnace slags] : Bos. Soc. Nat. History Proc., v. 3, 282 (10 lines).
5664 1850, Description of the vermiculite of Milbury, Mass.: Bos. Soc. Nat. History Proc., v. 3, 243-246; also in: Am. J. Sci., v. 9 (ns), no. 3, 422-423.
    With discussion by J. E. Teschemacher (10212).
5665 1850, Dr. Martin Gay [obituary] : Am. J. Sci., v. 9 (ns), no. 2, 305-306.
5666 1850, [Franklin, New Jersey minerals] : Bos. Soc. Nat. History Proc., v. 3, 326 (4 lines).
5667 1850, [Iron ore from Pennsylvania] : Bos. Soc. Nat. History Proc., v. 3, 319 (8 lines).
5668 1850, [Jacksonite, identity with prehnite] : Bos. Soc. Nat. History Proc., v. 3, 247-248.
5669 1850, [Nonglaciation of the White Mountain district] : Am. Assoc. Adv. Sci. Proc., v. 2, 309-310.
    With discussion of article by A. Guyot (see 4612).
5670 1850, [On asphaltum recently discovered in New Brunswick] : Bos. Soc. Nat. History Proc., v. 3, 279-280.
5671 1850, [On phosphate of iron in greensand] : Bos. Soc. Nat. History Proc., v. 3, 257.
5672 1850, [On potholes at Orange, New Hampshire, and elsewhere] : Bos. Soc. Nat. History Proc., v. 3, 324.
5673 1850, [On the age of the sandstones of the United States] : Bos. Soc. Nat. History Proc., v. 3, 335-339.
    With discussion by L. Agassiz.
5674 1850, On the existence of manganese in water: Ann. Scien. Disc., v. 1, 202 (9 lines).
5675 1850, On the geological structure of Keweenaw Point [Michigan] : Am. Assoc. Adv. Sci. Proc., v. 2, 288-301; also in: Am. J. Sci., v. 10 (ns), no. 1, 65-77.
5676 1850, On the telluric bismuth of Virginia: Am. J. Sci., v. 10 (ns), no. 1, 78-80; also in: Bos. Soc. Nat. History Proc., v. 3, 297-299.
5677 1850, [On Tertiary fossils from Marshfield, Massachusetts] : Bos. Soc. Nat. History Proc., v. 3, 323-324, 329.
5678 1850, [Origin of New Jersey greensand] : Bos. Soc. Nat. History Proc., v. 3, 249 (5 lines).
5679 1850, Remarks on the geology, mineralogy, and mines of Lake Superior: Am. Assoc. Adv. Sci. Proc., v. 2, 283-287.
5680 1850, [Salt from Great Salt Lake, California] : Bos. Soc. Nat. History Proc., v. 3, 223-224.
5681 1850, [Test of building stone strength] : Bos. Soc. Nat. History Proc., v. 3, 241 (7 lines).
5682 1850, Tin in plumbaginous slate: Am. J. Sci., v. 10 (ns), no. 1, 134 (5 lines).
    Abstracted from Bos. Soc. Nat. History Proc.

Jackson, C. T. (cont.)
5683   1850, [Water of the Jordan River] : Bos. Soc. Nat. History Proc., v. 3, 260 (4 lines).

Jackson, Julien R. (1790-1853)
5684   1849, [Excerpts from Minerals and their uses] : Living Age, v. 22, 155-157.
        Not seen. For review see 6403.

Jackson, John Barnard Swett (1806-1879)
5685   1845, [On Mastodon giganteus from Schooley's Mountain, New Jersey] : Bos. Soc. Nat. History Proc., v. 2, 60-62.
5686   1846, [Dentition of the Mastodon] : Bos. Soc. Nat. History Proc., v. 2, 140-141.
5687   1849, [Distribution of the Mastodon] : Bos. Soc. Nat. History Proc., v. 3, 104.
5688   1849, [Recent Ornithichnites] : Bos. Soc. Nat. History Proc., v. 3, 104.

Jackson, William (see 10508 with others)

Jacobson, M.
5689   1846, Staurotide: Am. J. Sci., v. 2 (ns), no. 3, 418 (6 lines).
        Abstracted from Poggendorff's Annalen.

Jamaica. Minister of
5690   1795, Account of the earthquake at Port-Royal, in Jamaica, 1692: Mass. Hist. Soc. Coll., v. 4, 223-230.

James, Edwin (1797-1861)
5691   1820, On the geology of a part of the state of Vermont, the shores of Lake Champlain, St. John's River, and Montreal: Plough Boy, v. 1, no. 32, 250.
        In Troy Lyc. Proc.
5692   1822, Geological sketches of the Mississippi Valley: Phila. Acad. Nat. Sci's J., v. 2, pt. 2, 326-329, plate.
5693   1823, On the identity of the supposed pumice of the Missouri, and a variety of amygdaloid found near the Rocky Mountains: NY Lyc. Nat. History Annals, v. 1, pt. 1, no. 1, 21-23.
        Abstracted in NY Med. Phys. J., v. 2, 505-507, and in Am. J. Sci., v. 7, no. 1, 173.
5694   1823, Account of an expedition from Pittsburgh to the Rocky Mountains performed in the years 1819 and '20 ... under ... command of Major Stephen H. Long: Phila., H. C. Carey and I. Lea, 2 v. (503 p.; 442, xcviii p.), atlas (2 maps, 9 plates).
        Contains notes by S. H. Long (see 6475). Extract in West. Q. Reporter, v. 2, no. 2, 131-159.
5695   1825, Remarks on the sandstone and floetz trap formations of the western part of the valley of the Mississippi: Am. Philos. Soc. Trans., v. 2 (ns), 191-215.
5696   1827, Remarks on the limestones of the Mississippi lead mines: Phila. Acad. Nat. Sci's J., v. 5, pt. 2, 376-380.

James, John (M. D.)
5697   1820, [On Mount Vesuvius] : Saturday Mag., v. 3, 393-395.
        Extracted from Sketches of travels in Sicily, Italy and France.

James, Lorenzo (of Monroe County, Alabama)
5698   1841, Inquiries and remarks upon the calcareous rocks and soils of South Alabama: Farm's Reg., v. 9, 59-60.
5699   1841, Marl in Alabama: Farm's Reg., v. 9, 423.

James, Thomas Chalkley (1766-1835)
5700   1826, A brief account of the discovery of anthracite coal on the Lehigh: Penn. Hist. Soc. Mem., v. 1, 313-320.
5701   1829, History of the discovery and use of Pennsylvania coal: Hazard's Reg. Penn., v. 3, 301-303.

James, Uriah Pierson (1811-1889; see 4468 with G. Graham and J. G. Anthony)

Jameson, Robert (1774-1854)
5702   1814, [Extract from Mineralogy, v. 3, 256-276] : Emp. Arts Sci's, v. 3 (ns), 426-444.
        In "Geology" by T. Cooper (see 2607). On the "relative age of metals."
5703   1814, On the topaz of Scotland: Am. Min. J., v. 1, no. 4, 253-258.
        Abstracted from Wernerian Nat. History Soc. Mem.
5704   1814, Zircon found in Scotland: Am. Min. J., v. 1, no. 4, 261 (3 lines).
5705   1818, Preface [to the English edition of Essay on the theory of the earth by G. Cuvier] : in 2754, v-xv.
5706   1818, Appendix, containing mineralogical notes, and an account of Cuvier's geological discoveries: in Essay on the theory of the earth by G. Cuvier (see 2754), 185-315.
5707   1818, Cheltenham [England] waters: J. Sci. Arts, v. 3, no. 2, 380.
5708   1825, Fossils and live shells of the same species differ, according to locality, distance, &c: Am. Ath., v. 1, 208.
5709   1826, Notes on the geology of the countries discovered during Captain Parry's second expedition, A. D. 1821-22-23: in Journal of a third voyage for the discovery of a northwest passage, from the Atlantic to the Pacific ..., by W. E. Parry, Phila., Carey and Lea, 232 p., map.
5710   1831, Arctic geology: in Narrative of discovery and adventure in the polar seas and regions with illustrations of their climate, geology, and natural history; and an account of the whale-fishery by J. Leslie, NY, J. and J. Harper, 373 p., illus., 352-373.
        Other printings by Harper in 1833, 1836, 1840, 1842, and 1844.
5711   1831, and H. Murray and J. Wilson, Narrative of discovery and adventure in Africa, from the earliest ages to the present time; with illustrations of the geology, mineralogy, and zoology: NY, J. and J. Harper, 359 p., illus.
        Other printings by Harper in 1832, 1833, 1836, 1839, 1842, 1844, and 1846.

Jameson, R. (cont.)
5712 1836, Fossil fishes [review with excerpts of Poissons fossiles by L. Agassiz (see 57)]: Am. J. Sci., v. 30, no. 1, 33-53.

Jamin, Jules Celestin (1818-1886)
5713 1850, On the elliptical double refraction of quartz: Franklin Inst. J., v. 19 (3s), 351-352.
Abstracted from Comptes Rendus.

Jamineau, Isaac (1710?-1789)
5714 1769, An account of the eruption of Vesuvius, in 1767, communicated in a letter from an English gentleman residing at Naples, to John Morgan: Am. Mag. or Gen. Repos., v. 1, 22-28. Also in Am. Philos. Soc. Trans., v. 1, 281-285 (1771); v. 1 (2nd edition), 345-349 (1789).

Janos, C.
5715 1805, Vesuvius and Aetna: Lit. Tablet, v. 2, 47.

Jarvis, Samuel Farmar (1786-1851)
5716 1836, Address: Nat. History Soc. Hartford Trans., v. 1, no. 1, 5-64.
On the history and progress of science.

Jaudon, Daniel (1767-1826)
5717 1797, A short system of polite learning: being a concise introduction to the arts and sciences. Adapted for schools: Litchfield, Ct., T. Collier (printer), 112 p.
Evans number 32316.

Jefferson, Thomas (1743-1826)
See also 10470 and 10471.
5718 1787, Comparative view of the animals of America and those of Europe, - being a refutation of Mr. Buffon's assertion, "that the animals, common to both the old and new world, are smaller in the latter: Colum Mag., v. 1, 366-369, 407-416.
From Notes on the state of Virginia (see 5719).
5719 1788, Notes on the state of Virginia: Phila., Prichard and Hall, ii, 244 p., table.
Evans number 21176. Numerous other editions including "2nd American ed.", Phila., M. Carey, 1794 (Evans number 27162); Balt., W. Pechin, 194, 53 p., 1800 (Evans numbers 37702 and 37703); "3rd edition", NY, Furman and Newark, NJ, Pennington and Gould, 1801; "4th ed.", NY, Jansen, 1801; Phila., R. T. Rowle, 1801; "8th American ed.", Bos., Thomas and Andrews; Walpole, NH, Thomas and Thomas, 1801; "9th ed.", Bos., H. Sprague, 1802; Trenton, NJ, Wilson and Blackwell, 1803; Phila., Carey and Lea, 1825; Bos., Well and Lilly, 1829; Bos., Lilly and Wait, 1832.
5720 1799, A memoir on the discovery of certain bones of a quadruped of the clawed kind in the western parts of Virginia: Am. Philos. Soc. Trans., v. 4, 246-260.
5721 1800, Description of the most remarkable cascades and caverns in the state of Virginia: Pol. Mag., v. 1, no. 2, 87-89.
From Notes on the State of Virginia (see 5719).
5722 1802, Description of the mammoth: Juv. Mag., v. 1, no. 2, 109-114.
From Notes on the State of Virginia (see 5719).
5723 1810, The natural bridge: Huntingdon Lit. Mus., v. 1, 220-221.
Also in Va. Hist. Reg., v. 1, no. 1, 39 (1848) and in West. Miscellany, v. 1, 36-37, fig. (1848).
From Notes on the State of Virginia (see 5719).
5724 1822, [Letter on a fossil tree]: Am. J. Sci., v. 5, no. 1, 25.
See also article by H. R. Schoolcraft (9263).

The Jeffersonian (Albany, 1838-1839)
5725 1838, Copper [from Guilford County, North Carolina]: v. 1, no. 4, 32 (4 lines).
5726 1838, Gold [United States production]: v. 1, no. 18, 144; no. 23, 182.
5727 1838, Sam Slick's idea of geology: v. 1, no. 31, 236 (5 lines).
A geological pun.
5728 1838, Bones of a monster [from Bucyrus]: v. 1, no. 31, 248.
5729 1838, A lake of quicksilver [at Marengo, Alabama]: v. 1, no. 42, 331 (6 lines); no. 43, 344 (6 lines).
5730 1839, Earthquake [at Pesaro, Italy]: v. 1, no. 49, 394 (7 lines).
5731 1839, [Gold in Georgia]: v. 1, no. 51, 408 (3 lines).
5732 1839, Mather's geology [review of 6890]: v. 1, no. 52, 416 (5 lines).

Jeffreys, Jules
5733 1841, On the solvent power exercised by water, at high temperature, on siliceous minerals: Am. J. Pharmacy, v. 7 (ns), 227-229.
Abstracted in Am. J. Sci., v. 41, no. 1, 60 (12 lines), from British Assoc. Proc.

Jeffries, William
5734 1828, Improvement in calcining, or roasting, and smelting or extracting, metals and semi-metals, from various kinds of ores, and matter containing metals and semi-metals: Franklin Inst. J., v. 5, 177-179.
5735 1842, Improvement in obtaining copper, spelter, and other metals, from ores: Franklin Inst. J., v. 3 (3s), 354-355, fig.

Jenkins, John P.
5736 1821, Notice of some facts at Hudson [New York]: Am. J. Sci., v. 4, no. 1, 33-35.

Jenkins, John Stilwell (1818-1852)
5737 1850, United States exploring expeditions: Voyage of the U. S. Exploring Squadron, commanded by Captain Charles Wilkes ... in 1838, 1839, 1840, 1841 and 1842 ... and an account of the expedition to the Dead Sea under Lieutenant Lynch ... : Auburn, NY, James Alden, 517 p., illus.
Also an 1850 edition, Hudson, NY, Wynkoop.

Jenkins, L. W.
5738   1833, Report of the select committee relative to the expediency of procuring a map of the state: Md. House of Delegates, December Session, 1832, 10 p.

Jenks, Lemuel P.
5739   1850, Improved arrangement of the conductors in centrifugal gold washers: Franklin Inst. J., v. 19 (3s), 243 (4 lines).

Jenners, Abriel
5740   1826, On plaster of Paris: Phila. Soc. Promoting Ag. Mem., v. 5, 183-188.

Jennings, Lewis
5741   1850, Improved gold washer: Franklin Inst. J., v. 19 (3s), 17.

Jennison, William H.
5742   1850, Improved gold washer: Franklin Inst. J., v. 18 (3s), 380 (3 lines).

Jersey, West
5743   1849, Coal ashes valuable as manure: Hort. and J. Rural Arts, v. 3, no. 10, 463-464.

Jessup, Augustus E.
5744   1820, Fetid fluor spar [from Shawnee, Illinois]: Am. J. Sci., v. 2, no. 1, 176.
5745   1821, Geological and mineralogical notice of a portion of the northeastern part of the state of New-York: Phila. Acad. Nat. Sci's J., v. 2, pt. 1, 185-191.
        Abstracted in Phila. J. Med. Phys. Sci's, v. 4, 389-390 (1822)

Jeter, J. B. [i.e. Jeremiah Bell Jeter?, 1802-1880]
5746   1843, Trip to the coal mines: Niles' Wk. Reg., v. 65, 108-109.

Jewell, William
5747   1820, An inaugural dissertation, on the medical topography of Shepherdsville, and the vicinity ... : Danville, Ky., James Armstrong, 47 p.

Johnson
5748   1834, Plumbacalcite and Vanadiate of lead [from Wanlockhead, Scotland]: Am. J. Sci., v. 26, no. 2, 386 (8 lines).
        Translated and abstracted by L. Feuchtwanger from Berzelius' Jahresbericht.
5749   1834, Vanadiate of lead [from Wanlockhead, Scotland]: Am. J. Sci., v. 26, no. 2, 386 (9 lines).
        Translated and abstracted by L. Feuchtwanger from Berzelius' Jahresbericht.

Johnson, Arthur L.
5750   1839, A plan of operations for the Osage Mining and Smelting Company: Balt., J. W. Woods (printer), 11 p.

Johnson, Benjamin Pierce (1793-1869)
5751   1850, Republication of Professor Johnston's Lectures: Am. Agriculturist, v. 9, 313-314, fig.
        Notice of 5823, with woodcut portrait of J. F. W. Johnston.

Johnson, Cuthbert William (1799-1878)
5752   1850, The formation of soils: Working Farm., v. 2, no. 10, 230-231.

Johnson, Edwin Ferry (1803-1872)
5753   1845, and W. R. Casey, Reports on the Bear Mountain Railroad: NY, 88 p.
        Contains reports by J. Hall (4702 and 4707) and W. R. Casey (2274). For notice see 847.

Johnson, George William (1802-1886)
5754   1830, Outlines of horticultural chemistry - analysis of soils: S. Agriculturist, v. 3, 608-612.

Johnson, J. W.
5755   1843, [On lead ore and fossils near Sunbury, Pennsylvania]: Bos. Soc. Nat. History Proc., v. 1, 43-45.

Johnson, Joseph
5756   1839, On the state and prospects of the iron trade in Scotland and South Wales, in May, 1839: Am. Rr. J., v. 9, 335-340, 369-372.
        Abstracted from "C. E. & A. Journal."

Johnson, Samuel William (1830-1909)
5757   1849, Analyses of limestone: The Cultivator, v. 6 (ns), 187.
5758   1850, On the discovery of sulphuret of nickel in northern New York: Am. J. Sci., v. 9, no. 2, 287-288.
        In article by F. B. Hough (see 5289).

Johnson, Sidney L.
5759   1834, Ascent of Mount Etna, February, 1832: Am. J. Sci., v. 26, no. 1, 1-10.

Johnson, Theodore Taylor (b. 1818)
5760   1849, Sights in the gold region, and scenes by the way: NY, Baker and Scribner, 278 p.
        For review see 5474, 6368, and 9183.
5761   1850, Sights in the gold region, and scenes by the way: NY, Baker and Scribner, xii, 324 p., front., plates.
        "Second edition," illustrated, revised and enlarged.

Johnson, Walter Rogers (1794-1852)
See also 8910 with H. D. Rogers and S. G. Morton.

5762   1834, On the specific heats of solids: Franklin Inst. J., v. 14 (ns), 306-317.
5763   1835, The scientific class book; or familiar introduction to the principles of physical science, for the use of schools and academies, on the basis of Mr. J. M. Moffat ... : Phila., Key and Biddle, 473 p., illus., diagram.
       Several other printings, some as A system of natural philosophy, in 1842, 1845, 1847, etc. Pages 236-298 contain "Mineralogy and Geology."
5764   1839, Analysis of some of the anthracites and iron ores found on the headwaters of Beaver Creek; in the counties of Luzerne, Northampton, and Schuylkill, Pennsylvania: Franklin Inst. J., v. 24 (ns), 289-298.
5765   1839, Analysis of some of the minerals found at Karthaus and Three Runs, on the west branch of the Susquehanna River, Clearfield County, Pennsylvania: Franklin Inst. J., v. 23 (ns), 73-80.
5766   1839, Examination of some of the anthracite found in Sugar Loaf Township, Luzerne County, Pennsylvania: Franklin Inst. J., v. 24 (ns) 73-77.
5767   1839, Experiments on two varieties of iron, manufactured from the magnetic ores at the Adirondack iron works, Essex County, N. Y.: Am. J. Sci., v. 36, no. 1, 94-105.
5768   1839, Report of an examination of the coal and iron ore lands known as the Wilson survey, lying on the south side of the Great Kanawha River, in the counties of Kanawha and Fayette, state of Virginia: Wash., Buell and Blanchard (printers), 8 p., folded map.
5769   1839, Report of an examination of the mines, iron works, and other property belonging to the Clearfield Coke and Iron Company, together with some examinations of the minerals employed in the manufacture of iron found at Karthaus and Three Runs, on the West Branch of the Susquehanna River ... : Phila., L. R. Bailey, 22 p., folded plate.
5770   1840, Extract from a report - of experiments on the iron, manufactured at the Village of McIntyre, Essex County, New-York: NY Geol. Survey Ann. Rept., v. 4, 305-311, 3 figs.
5771   1840, Professor Johnson's analyses of anthracite and iron ore: Am. J. Sci., v. 38, no. 2, 382-385.
5772   1840, Report of a geological, mineralogical, and topographical examination of the coal field of Carbon Creek, the property of the Towanda Rail Road and Coal Company, Bradford County, Pa. with an analysis of the minerals, accompanied by a map of the surveys, profile of the road, and sections of the mineral ground: Phila., 47 p.
       For review with excerpts see 740.
5773   1841, Anthracite of Rhode Island: Phila. Acad. Nat. Sci's Proc., v. 1, no. 8, 118-119.
5774   1841, Application of anthracite, to the smelting of the magnetic iron ores of New Jersey: Phila. Acad. Nat Sci's Proc., v. 1, no. 3, 42-43.
5775   1841, Examination and analysis of coal found in the Province of Arauco, coast of Chile, 30 miles south of Bio Bio River: Phila. Acad. Nat. Sci's Proc., v. 1, no. 2, 21-23.
5776   1841, [The Frostburg, Maryland coal basin]: Am. J. Sci., v. 41, no. 1, 186 (5 lines).
       In Assoc. Am. Geol's Nat's Proc.
5777   1841, Notes on the use of anthracite in the manufacture of iron, with some remarks on its evaporating power: Bos., C. C. Little and J. Brown, vi, 156 p., illus.
       For review see 762.
5778   1841, On the relation between the coal of South Wales and that of some Pennsylvania anthracites: Phila. Acad. Nat. Sci's Proc., v. 1, no. 3, 40-42. Also in Phila. Acad. Nat. Sci's J., v. 8, pt. 2, 195-199, table (1842).
5779   1841, Report of a survey and exploration of the coal and ore lands belonging to the Alleghany Coal Company, in Somerset County, Pennsylvania, accompanied by maps, profiles, and sections: Phila., Joseph and William Kite (printers), 64 p., 3 folding plates.
5780   1841, Report of an examination of the Bear Valley coal district in Dauphin County, Pennsylvania: Phila., J. and W. Kite (printers), 36 p., illus., tables.
5781   1841, Remarks on the Bear Valley coal district in Dauphin County, Pennsylvania: Franklin Inst. J., v. 2 (3s), 318-327, fig.
5782   1841, Some observations on the mechanical structure of coal, with evidences of the contemporaneous origin of its various kinds: Phila. Acad. Nat. Sci's Proc., v. 1, 9-12. Also in Phila. Acad. Nat. Sci's J., v. 8, pt. 2, 173-178 (1842).
5783   1841, Verbal communications [on crystalline forms of coal] : Phila. Acad. Nat. Sci's Proc., v. 1, no. 5, 73-75.
5784   1842, Determination of copper in analysis: Phila. Acad. Nat. Sci's Proc., v. 1, no. 16, 187 (11 lines).
5785   1842, Examination and analysis of coal found in the Province of Arauco, coast of Chili, thirty miles south of Bio Bio River: Phila. Acad. Nat. Sci's J., v. 8, pt. 2, 180-182.
5786   1842, On the practical determination of the heating power of fuel: Nat'l Inst. Prom. Sci. Bulletin, v. 2, 165-168.
5787   1842, Residues from a number of anthracite and bituminous coals of Europe and this country: Phila. Acad. Nat. Sci's Proc., v. 1, no. 12, 155-156.
5788   1842, The so-called natural coke from Virginia: Phila. Acad. Nat. Sci's Proc., v. 1, no. 21, 223-224.
5789   1842, Spontaneous combustion of bituminous coal: Phila. Acad. Nat. Sci's Proc., v. 1, no. 10, 140.
5790   1843, and S. S. Haldeman and Peale, An earthy matter from Lancaster County, Pennsylvania, supposed to be identical with terra di Sienna: Phila. Acad. Nat. Sci's Proc., v. 1, no. 25, 258.
5791   1844, Evaporative power and other properties of coal from different coal formations of the United States: Am. J. Sci., v. 47, no. 1, 126-127.
       Also in Assoc. Am. Geol's Nat's Proc., v. 5, 33-36.
5792   1844, [On some specimens of rocks from the White Mountains] : Phila. Acad. Nat. Sci's Proc. v. 2, no. 5, 89-90.
5793   1844, [Plumbago mica] : Phila. Acad. Nat. Sci's Proc., v. 2, no. 4, 74-75.
5794   1844, Preliminary report of experiments on the evaporative power and other properties of American coals: Wash., Gales and Seaton, [v] -xii.

Johnson, W. R. (cont.)
5795   1844, [Properties of different varieties of coal]: Phila. Acad. Nat. Sci's Proc., v. 2, no. 1, 8-10.
5796   1844, A report to the Navy Department of the United States, on American coals applicable to steam navigation, and to other purposes: Wash., Gales and Seaton (printers), US SED 386, 28-1, v. 6 (436), xii, 607 p., 3 folded plates.
        Also Wash., Blair and Rivers (printers), US HED 276, 28-1, v. 6 (444), iv, 5-607, 3 folded plates.
        For review see 807, 815, and 5622. For notice see 801.
5797   1844, [Spontaneous combustion of coal]: Phila. Acad. Nat. Sci's Proc., v. 2, 163-164, 165-166.
5798   1845, Abstract of Professor Johnson's report to the secretary of the Navy, of the United States, respecting forest improvement coal: NY, G. F. Nesbitt, 8 p., table.
5799   1845, Examination and analyses of samples of the alluvial soil of the Nile, from Korosco, in Nubia [Egypt]: Phila. Acad. Nat. Sci's Proc., v. 2, no. 12, 318-324.
5800   1845, [On the relative heating power of fuels]: Phila. Acad. Nat. Sci's Proc., v. 2, no. 7, 202-206.
5801   1846, Alluvial soil of the Nile: Am. Q. J. Ag. Sci., v. 3, 141-148.
5802   1846, Buck Mountain Coal Company [extract from report]: Am. Rr. J., v. 19, 90.
5803   1846, [Diluvial evidence in St. John's, New Brunswick]: Phila. Acad. Nat. Sci's Proc., v. 3, no. 5, 109 (10 lines).
5804   1846, An elementary treatise on chemistry, together with treatises on metallurgy, mineralogy, chrystallography, geology, oryctology, and meteorology. Designed for the use of schools and academies, on the basis of the book of science by Mr. J. M. Moffat. With additions ... 8th edition: Phila., E. C. and J. Biddle, 478 p., illus.
5805   1846, [Publications of the United States exploring expedition]: Phila. Acad. Nat. Sci's Proc., v. 3, no. 1, 18-19.
5806   1848, A section of the coal seams and accompanying measures of the Hazleton coal basin in Luzerne Co., Penn., as an illustration of the construction and coloring of geological sections: Am. J. Sci., v. 5 (ns), no. 1, 111-113.
        In Assoc. Am. Geol's Nat's Proc.
5807   1850, The coal trade of British America, with researches on the characters and practical values of American and foreign coals: Wash., Taylor and Maury, 179 p., tables, diagram.
5808   1850 Durability of stone: Ann. Scien. Disc., v. 1, 274-275.
5809   1850, On some experimental determinations of the economic values of British and American coals: Am. Assoc. Adv. Sci. Proc., v. 2, 221-231.

Johnston, A. R.
5810   1845, Remarks on the geology of the vicinity of Fort Washita: Assoc. Am. Geol's Nat's Proc., v. 6, 75-77.
        Also in Am. Q. J. Ag. Sci., v. 2, 167-168.

Johnston, Alexander Keith (1804-1871)
5811   1850, The physical atlas of natural phenomena for the use of colleges, academies, and families: Phila., Lea and Blanchard, 122 p., 26 maps.
        For review see 1005.

Johnston, J. C.
5812   1835, Fossil corn? (Zea maize) [from Wheeling, West Virginia]: Geol. Soc. Penn. Trans., v. 1, 414-415.

Johnston, James Finley Weir (1796-1855)
5813   1832, An account of the Society of German Naturalists and Physicians: Mon. J. Med. Liter., v. 1, no. 3, 65-86.
5814   1838, On the composition of certain mineral substances of organic origin: Franklin Inst. J., v. 22 (ns), 383-389.
        Abstracted from London and Edinburgh Philos. Mag.
5815   1839, Exceptions to the law of isomorphism: Am. J. Sci., v. 35, no. 2, 302 (3 lines).
        In British Assoc. Proc.
5816   1841, Chemical geology: Am. Eclec., v. 2, no. 2, 395-396.
        Abstracted from Lit. Gaz.
5817   1841, Report on chemical geology: Am. J. Sci., v. 41, no. 1, 56-57.
        In British Assoc. Proc.
5818   1842, Elements of agricultural chemistry and geology: NY, Wiley and Putnam, xi, [13]-249 p.
        For review see 184, 797, 5428.
5819   1842/1843, Lectures on agricultural chemistry and geology: NY, Wiley and Putnam, 2 v. (pt. 1, "Organic elements of plants," 1842; pt. 2 and 3, "Inorganic elements of plants," 1843), 175 p.
        For review with excerpts see 9956. For review see 273, 760, 779, 1842, 2721, 5431, 5434, 8577, 9961. For notice see 2288, 2718.
5820   1842, Extract from ... lectures on geology as applied to agriculture: Am. Agriculturist, v. 1, 7-8.
5821   1842, Geology as applied to agriculture: Farm's Mon. Vis., v. 4, 58.
5822   1842, Magnesian limestone: Am. Agriculturist, v. 1, 122-123.
5823   1844, Lectures on the application of chemistry and geology to agriculture: NY, Wiley and Putnam, vi, iv, [11]-619, 89 p.
        "New edition with an appendix." Other printings by Wiley and Putnam in 1847 and 1849, and by NY, C. M. Saxton in 1850.
        For notice see 5751.
5824   1845/1846, Agricultural chemistry and geology: Am. Agriculturist, v. 4, 203; 268-269; 315-316; v. 5, 18; 82-83; 190-191.
        Extracted from Catechism ..., (see 5825).

Johnston, J. F. W. (cont.)
5825 1845, Catechism of agricultural chemistry and geology with an introduction by John Pitkin Norton from the eighth English ed.: Alb., Erastus H. Pease, 74 p.
Contains "Introduction" by J. P. Norton (see 7967) and letter of endorsement by B. Silliman, jr. (see 9813).
For review see 3543. For notice see 797.
5826 1849, The agricultural school book. Section I. - Catechism of agricultural chemistry and geology: Plough Loom Anvil, v. 1, no. 9, 547-554; no. 10, 623-631; no. 11, 690-702.
5827 1850, [Greensands of England]: Bos. Soc. Nat. History Proc., v. 3, 256-567.
With discussion by C. T. Jackson, J. Pickering, and H. D. Rogers.
5828 1850, Lectures ... on the general relations of science to practical agriculture: NY State Agri. Soc. Trans., v. 9, 162-268.
5829 1850, Professor Johnston's address: The Cultivator, v. 7 (ns), 71-77.
5830 1850, [Syllabus of lectures presented] before the N. Y. State Ag. Society: The Cultivator, v. 7 (ns), 62.

Johnston, John (1791-1880)
5831 1836, Notice of a large crystal of columbite [from Middletown, Connecticut]: Am. J. Sci., v. 30, no. 2, 387-388, fig.
5832 1841, Notice of a new variety of beryl, recently discovered at Haddam, Conn.: Am. J. Sci., v. 40, no. 2, 401-402, 3 figs.
5833 1843, [Beryl at Haddam, Connecticut]: Am. J. Sci., v. 45, no. 1, 145 (5 lines).
In Assoc. Am. Geol's Nat's Proc. See also 1582.

Johnston, Joseph Eggleston (1807-1891)
5834 1850, and others, Reconnaissance of routes from San Antonio to El Paso: US SED 64, 31-1, v. 4 (562), 3-54, map.

Jones
5835 1829, Sketches of Naval life, with notices of men, manners and scenery, on the shores of the Mediterranean, in a series of letters from the Brandywine and Constitution frigates, in two volumes; by a civilian: Am. J. Sci., v. 16, no. 2, 320-341.
Contains notice of minerals from Milo Island, in the Mediterranean.

Jones, Alexander (1802-1863)
5836 1834, Miscellaneous facts: Am. J. Sci., v. 26, no. 1, 191 (12 lines).
On ferruginous sandstone and petrified wood from Alabama.
5837 1834, Bituminous coal [from Shelby and Bibb Counties, Alabama]: Am. J. Sci., v. 26, no. 1, 190-191.

Jones, Calvin (1775-1846)
5838 1815, A description of Wier's Cave, in Augusta County, Virginia ... : Alb., Henry C. Southwick (printer), 8 p., map.
Reprint editions in Port Folio, v. 12 (5s), 325-332, fig., plate (1821); Am. Farm., v. 3, 273-274, fig. (1821); Balt., The American Farmer, Broadside, 2 p., map (1821); Plough Boy, v. 4, no. 15, fig.; 113-115 (1822).

Jones, Daniel
5839 1785, An account of West River Mountain, and the appearance of there having been a volcano in it: Am. Acad. Arts Sci's Mem., v. 1, 312-315. Also in Am. Mus. or Univ. Mag., v. 1, 204-205 (1787); Mass. Mag., v. 3, 745-746 (1789); Rural Mag. or Vt. Repos., v. 1, 615.
Evidence for the volcanic origin of West River Mountain, Cheshire County, New Hampshire.

Jones, Daniel (of North Carolina)
5840 1831, Machine for washing gold: Franklin Inst. J., v. 7 (ns), 159 (5 lines).

Jones, George Wallace (1804-1896)
5841 1838, Some observations in Holland, connected with our prairie region: Am. J. Sci., v. 33, no. 2, 226-230.
On the topography and origin of mid-American prairie.

Jones, John
5842 1841, Human foot-prints in solid limestone: Farm's Cab., v. 7, 117-118.

Jones, Joseph
5843 1831, Invention of an improvement in certain parts of the process of smelting, or obtaining malleable copper, from copper ore: Franklin Inst. J., v. 7 (ns), 33.

Jones, Joseph Seawell (1811-1877)
5844 1838, Memorials of North Carolina: NY, Scatcherd and Adams, 87 p.
Section on gold mines, p. 82-85.

Jones, Thomas P. (1774-1848)
5845 1826/1827, Observations on specific gravity: Franklin Inst. J., v. 2, 353-354; v. 3, 41-44.
5846 1827, Remarks additional on anthracite: Franklin Inst. J., v. 3, 123-124.
5847 1828, On azure stone, or lapis lazuli, ultramarine, and the manufacture of artificial ultramarine: Franklin Inst. J., v. 6, 270-273.
5848 1836, Report [on the mineral lands of the Maryland Mining Company]: in Report ..., by G. W. Hughes (see 5354).
5849 1837, [Extract from a report on the coal measures of the Maryland Mining Company]: in Report ..., by G. W. Hughes (see 5355).
Also in 6812 (1839).
5850 1847, Report [on the Pittsburgh and Boston Copper Harbor Mining Company]: see 8381.

Jonnes, Alexandre Moreau de (1778-1870)
5851   1818, [Geological notes on the Antilles] : J. Sci. Arts, v. 3, no. 1, 151-152.
       In France Royal Soc. Proc.

Jordan, Henry
5852   1830, Art of digging and procuring gold from the mine: Franklin Inst. J., v. 6 (ns), 372.

Jordan, T. B.
5853   1842, On copying fossils by a galvanic deposit: Am. J. Sci., v. 42, no. 2, 327-328.
       In British Assoc. Proc.

Joslin, Benjamin Franklin (1796-1861)
5854   1831, Hints and conjectures respecting some of the causes of the earth's magnetism: Am. J. Sci., v. 19, no. 2, 398-399.

The Journal of Foreign Medical Sciences and Literature (Philadelphia, 1821-1824)
5855   1821, Burning spring [at Licking River, Kentucky] : v. 1 (ns), 128.
5856   1822, Analysis of an aerolite: v. 2 (ns), 711.
       Abstracted from Annals of Phil.
5857   1823, Abbé Haüy [obituary] : v. 3, 159-160.
       Abstracted from London Med. Phys. J.

The Journal of Health and Monthly Miscellany (Boston, 1846-1850)
5858   1846, Arkansas oil stone: v. 1, 342-343.

The Journal of Science and the Arts (New York, 1817-1818)
5859   1817, [Catalogue of British mineral specimens at the Royal Institution noted] : v. 1, no. 1, 138.
5860   1817, [Review with excerpts of System of Mineralogy by J. J. Berzelius] : v. 1, no. 2, 226-241.
       See also 1648.
5861   1817, [Review of Mineralogical nomenclature by T. Allan (see 101)] : v. 1, no. 2, 242-244.
5862   1817, Miscellaneous observations on the volcanic eruptions at the islands of Java and Sumbawa, with a particular account of the mud volcano at Grobogar: v. 1, no. 2, 245-258.
5863   1817, Royal institution, January 1, 1817: v. 2, no. 1, 465-468.
       Abstracted from Royal Inst. Proc. Lectures in chemistry and geology noted.
5864   1818, [Fossil elephant bones discovered at Lyons, France] : v. 3, no. 1, 164-165.
       Abstracted from France Royal Inst. Proc.
5865   1818, [Review of Esquisse mineralogique des environs de la Chausée des gêans by "B."] : v. 3, no. 1, 197 (7 lines).
       Abstracted from Biblio des Sciences et des Arts (Geneva).
5866   1818, Building materials [of England] : v. 3, no. 2, 381-382.
5867   1818, Royal Institution [of Great Britain] : v. 4, no. 1, 131-150.
       Notices of lectures, buildings, and objectives.
5868   1818, Temperature on and beneath the surface of the earth: v. 5, no. 1, 123 (7 lines).
5869   1818, Increase of a glacier [Tyrol] : v. 5, no. 1, 134.
5870   1818, [Earthquakes in Europe] : v. 5, no. 1, 134-135; no. 2, 372.
5871   1818, Pargasite, a new mineral [from Ersby, Finland] : v. 5, no. 1, 138-139.
5872   1818, Instrument to distinguish minerals [by optical properties] : v. 5, no. 1, 139 (9 lines).
5873   1818, Silicated hydrate of alumine: v. 5, no. 1, 139-140.
5874   1818, Native copper [from Lake Superior] : v. 5, no. 1, 140 (9 lines).
5875   1818, [European fossil bones] : v. 5, no. 1, 140-141; no. 2, 377-378.
5876   1818, On the figure of the earth, and on the length of the seconds pendulum in different latitudes: v. 5, no. 2, 235-249.
5877   1818, Meteoric iron: v. 5, no. 2, 372.
       On etch patterns in polished meteoric iron.
5878   1818, Description of a lake which has been formed in the Valley of Bagne, in the Valaise, May 19: v. 5, no. 2, 372-374.
5879   1818, New mineral. - Hydrate of silica and alumina [from Saint Sever, France] : v. 5, no. 2, 376.
5880   1818, Siliciferous sub-sulphate of alumine [from an English coal mine] : v. 5, no. 2, 376-377.
5881   1818, Sliding mountain [at Soncebos, Switzerland] : v. 5, no. 2, 377 (7 lines).
5882   1818, Embedded diamonds [from Brazil] : v. 5, no. 2, 378 (7 lines).
5883   1818, Zircon [from Sutherland, England] : v. 5, no. 2, 378 (8 lines).

Journal of the Times (Baltimore, 1818-1819)
5884   1819, Milford "mineral spring" [Massachusetts] : v. 1, no. 16, 243.
5885   1819, An essay on asbestus [from Newburyport, Maine] : v. 1, no. 21, 330-332.

Joy, Captain
5886   1836, An extract from the letter: Am. J. Sci., v. 30, no. 1, 113.
       In 654. On the effects of an earthquake on the coast of Chile.

Juben, H. M.
5887   1839, Meteoric iron from Potosi [Bolivia] : Am. J. Sci., v. 37, no. 1, 190-191.
       Abstracted from London and Edinburgh Philos. Mag.

Judd, Gerrit Parmalee (1803-1873)
5888   1838, [Notes appended to "Physico-geognostic sketch of the island of Oahu," by M. Gairdner (see 4196)] : Hawaiian Spectator, v. 1, no. 2, 1-18.

Jukes, Fred
5889   1832, Trilobite [from Great Barr, Staffordshire, England] : Am. J. Sci., v. 23, no. 1, 203, fig.

Jurasky, J.
5890   1849, Ceramohalite [from Königsberg, Hungary]: Am. J. Sci., v. 7 (ns), no. 1, 113 (11 lines).
         Abstracted from "Leon. und Bronn, Neues Jahrbuch."

Justice, George M.
5891   1846, [On itacolumbite from Stokes County, North Carolina]: Am. Philos. Soc. Proc., v. 4, no. 35, 244 (9 lines).
5892   1849, Gold from Montgomery County, Maryland: Am. Philos. Soc. Proc., v. 5, 84-85.

The Juvenile Magazine, or Miscellaneous Repository of Useful Information (Philadelphia, 1802-1803)
5893   1802, Interesting description of gems and precious stones: v. 1, 28-34.

The Juvenile Miscellany (Boston, 1826-1834)
5894   1831, [Tabnez marble from Persia]: v. 1, (3s), 159.
5895   1831, Diamonds: v. 1, (3s), 170-177.
5896   1833, Aerolites, or meteoric stones: v. 5 (3s), 99-105.

Juvenile Rambler, or Family and School Journal (Boston, 1832-1833)
5897   1832, The great diamond [from Brazil]: v. 1, no. 6, 24; no. 7, 28; no. 23, 91; no. 27, 106.
5898   1832, Gold found in Germany: v. 1, no. 6, 24 (6 lines).
5899   1832-1833, [Notes on earthquakes and volcanoes]: v. 1, no. 8, 30; no. 23, 90; no. 33, 131; no. 37, 146; no. 41, 163; v. 2, no. 6, 24; no. 9, 36; no. 10, 40; no. 11, 44; no. 12, 48; no. 13, 52; no. 16, 64; no. 29, 114; no. 32, 128; no. 34, 136; no. 36, 144; no. 37, 148; no. 44, 175; 175-176.
5900   1832, Petrified forest [in Rome, Italy]: v. 1, no. 19, 74-75.
5901   1832, A salt mine [of Cracow, Poland]: v. 1, no. 34, 135 (11 lines).
5902   1832, Showers of stones: v. 1, no. 34, 136.
5903   1832, The burning spring [in Floyd County, Kentucky]: v. 1, no. 39, 154-155.
5904   1833, Lake of vitriol [at Mount Idienne, Java]: v. 2, no. 8, 31.
5905   1833, Delany's Cave [at Union, Fayette County, Pennsylvania]: v. 2, no. 22, 87.
5906   1833, Air volcanoes [at Carthagena, South America]: v. 2, no. 29, 114.
5907   1833, Weyer's Cave: v. 2, no. 29, 114 (7 lines).
5908   1833, Dialogue on geology: v. 2, no. 36, 143; no. 40, 158; no. 42, 166; no. 47, 187.
5909   1833, Mammoth caves in Ireland: v. 2, no. 41, 163.
5910   1833, Petrifactions: v. 2, no. 43, 171.
         Various fossil plants noted.

Kain, John H. (1759-1831)
5911   1818, Remarks on the mineralogy and geology of the northwestern part of the state of Virginia and eastern part of the state of Tennessee: Am. J. Sci., v. 1, no. 1, 60-67.
5912   1819, Account of several ancient mounds, and of two caves, in east Tennessee: Am. J. Sci., v. 1, no. 4, 428-430.

Kain, John Henry (d. 1849)
5913   1845, On the prairies of Alabama: Assoc. Am. Geol's Nat's Proc., v. 6, 68.
         Also in Am. Q. J. Ag. Sci., v. 2, 166.
5914   1845, [Oolitic limestone from Cumberland Mountain, Alabama]: Assoc. Am. Geol's Nat's Proc., v. 6, 68-69.

Kames, Lord Henry Home (1696-1782)
5915   1797, On the theory of agriculture, and the means of fertilizing soils: SC Wk. Mus., v. 1, no. 3, 69-72.

Kammerer
5916   1834, Wolchouskoit [from Siberia]: Am. J. Sci., v. 26, no. 2, 387 (5 lines).
         Translated by L. Feuchtwanger from Berzelius' Jahresbericht.

Kanawha Company, Virgina Manufacturers of Salt
5917   1830, Memoirial of the manufacturers of salt in Kanawha County, Virginia: praying for a restoration of the duty on imported salt: Kanawha Court House, Virginia, office of the Kanawha Banner, 19, 7 p.

Karkeek, William Floyd (1802-1858)
5918   1842, The geological history of the horse: Farm's Reg., v. 10, 431-436.
         From the Veterinarian.

Karsten, Karl Johann Bernhard (1782-1853)
5919   1837, On the extraction of copper from poor ores, as practised successfully in Germany: Franklin Inst. J., v. 20 (ns), 183.
         Abstracted from Mining J.
5920   1846, Martinsite, a new mineral [from Stassfurth, Germany]: Am. J. Sci., v. 2 (ns), no. 1, 119 (6 lines).
         Abstracted from l'Institut.
5921   1827, Bismuth cobalt ore [from Schneeberg, Saxony]: Am. J. Sci., v. 13, no. 1, 187.
         Abstracted from Jameson's J. by C. U. Shepard.
5922   1832, On the chemical composition of the brown lead ore: Am. J. Sci., v. 22, no. 2, 307-320.
         Abstracted from Neues Jahrbuch der Chemie und Physik by C. U. Shepard.
5923   1835, Account of artificial felspar: Am. J. Sci., v. 28, no. 2, 396.
         Abstracted from Jameson's J.
5924   1839, Upon the mode of determining the quantity of copper in ores and smelting products, by the humid way: Franklin Inst. J., v. 24 (ns), 261-263.
         Abstracted from Mining Rev.
5925   1847, On an amorphous boracite: Am. J. Sci., v. 4 (ns), no. 3, 415.
         Abstracted from London Ath.

Karsten, K. J. B. (cont.)
5926 1849, Analysis of lardite from near Voigtsberg, in Saxony: Am. J. Sci., v. 8, no. 1, 121 (11 lines).
Abstracted from "Jour. für Prakt. Chem."

Kastner
5927 1831, Discovery of bromine in the Baltic: Am. J. Sci., v. 19, no. 2, 382 (3 lines).
Abstracted by J. Griscom from "Archiv für die Ges. Naturlehre."

Keating, William Hypolitus (1799-1840)
See also 10563, 10565, and 10571 with L. Vanuxem.
5928 1821, Considerations upon the art of mining, to which are added, reflections on its actual state in Europe, and the advantages which would result from an introduction of this art into the United States: Phila., M. Carey and Sons, 87 p.
5929 1822, Observations upon the cadmia found at the Ancram iron works in Columbia County, New York, erroneously supposed to be a new mineral: Phila. Acad. Nat. Sci's J., v. 2, pt. 2, 289-296.
Also in Am. J. Sci., v. 6, no. 1, 180-185 (1823).
5930 1822, Syllabus of lectures on mineralogy and chemistry: Phila.
Not seen.
5931 1824, Geology and organic remains: Phila. Acad. Nat. Sci's Rept. Trans. for 1824, 11-14.
5932 1824, Mineralogy: Phila. Acad. Nat. Sci's Rept. Trans. for 1824, 14-16.
5933 1824, Narrative of an expedition to the source of St. Peter's River Lake Winnepeck, Lake of the Woods, &c., &c., performed in the year 1823: Phila., H. C. Carey and I. Lea, 2 v. (439; 459 p.), illus., 15 plates.
Contains notes by S. H. Long (see 6476).

Keeney, J. C.
5934 1829, Novaculite in Georgia: Am. J. Sci., v. 16, no. 1, 185.

Keferstein, Christian (1784-1866)
5935 1819, A new variety of serpentine: Am. Mon. Mag. and Crit. Rev., v. 4, no. 4, 294.
Attributed to "Mr. Kelferstein."

Keith, Thomas (1759-1824)
5936 1811, A new treatise on the use of the globes; or, A philosophical view of the earth and heavens: NY, Samuel Whiting & Co., xviii, 346 p., 5 plates.
Other editions: NY, Whiting & Watson, 344 p., 5 plates (1815); NY, S. Wood, xvi, 352 p., plates (1819); NY, S. Wood, 334 p., plates (1826); NY, S. Wood & Sons, xxiii, [25]-334, 6 plates (1832). The 1826 and 1832 editions were revised and corrected by Robert Adrain.

Kelley, Edward G.
5937 1841, Remarks on the geological features of the Island of Owyhee or Hawaii, the largest of the group called the Sandwich Islands, with an account of the condition of the volcano of Kirauea, situated in the southern part of the Island near the foot of Mauna Roa. Drawn up from statements made by Captain Chase, of the ship Charles Carroll, and Captain Parker, of the ship Ocean, who visited it in 1838: Am. J. Sci., v. 40, no. 1, 117-122, plate.
Abstract in The Friend, v. 14, 162-163.

Kellogg, Orson
5938 1849, A remarkable geological development in Elizabethtown, Essex County, N. Y.: Am. Assoc. Adv. Sci. Proc., v. 1, 135-138.

Kemp, M.
5939 1842, Process for separating gold from platina: Am. J. Pharmacy, v. 8 (ns), 135 (8 lines).
Abstracted from Am. Philos. Soc. Proc.

Kenall, J. L.
5940 1846, Washingtonite of Shepard: Am. J. Sci., v. 1 (ns), no. 1, 122 (6 lines).
On the equivalence of ilmenite and washingtonite.

Kendall, Edward Augustus (1776?-1842)
5941 1809, Account of the writing-rock in Taunton River [on Dighton Rock]: Am. Acad. Arts Sci's Mem., v. 3, pt. 1, 165-191, plate.

Kennedy, William (1799-1871)
5942 1844, Texas: its geography, natural history and topography: NY, Benjamin and Young; Bos., Redding and Co., 118 p.

Kenhawa Company (see Kanawha Company)

Kentucky. State of
5943 1838, [Resolution for establishing a Kentucky geological survey]: Kentucky General Assembly, February 16, 1838.

Kentucky. Committee on the Coal Trade and Iron Interests
5944 1843?, Report: n.p., 20 p.

Kercheval, Samuel (1786-1845?)
5945 1833, A history of the Valley of Virginia: Winchester, Va., S. H. Davis, 486 p.
Also second edition, revised: Woodstock, Va., J. Gatewood, 347 p. (1850).
5946 1840, The ice mountain of Hampshire, Va. from Kercheval's History of the Valley: Farm's Reg., v. 8, 667-668.

Kerndt, Thomas
5947 1846, Crystalline form of geochronite [from Val di Costello]: Am. J. Sci., v. 1 (ns), no. 2, 267 (6 lines).
Abstracted from Poggendorff's Annalen.
5948 1849, Bodenite: Am. J. Sci., v. 8 (ns), no. 1, 124 (7 lines).
Abstracted from "Jour. für Prakt. Chem."
5949 1849, Muramontite, a new mineral [from Marienberg, Erzgebirge]: Am. J. Sci., v. 8 (ns), no. 1, 125 (8 lines).
Abstracted from "jour. für Prakt. Chem."

Kernick, Richard
5950 1846, [Report on Lake Shore Mining Company holdings]: in Articles of Association (see 6021).

Kerr, John Bozman (1809-1878)
5951 1838, Report of the Select Committeee appointed to enquire into the expediency of repealing the act to provide for completing a new map and geological survey of this state: Md. General Assembly, December session, 1837, Public Document, Annapolis, Md., 8 p.

Kersh, William D. (Dr.)
5952 1848, On a remarkable slide of a rock in Fairfield District, S. C.: Am. J. Sci., v. 6 (ns), no. 3, 443-444.
Abstracted from Bos. Soc. Nat. History Proc.

Kersten, Charles (see Karsten, Karl)

The Key (Fredericktown, Maryland, 1798)
5953 1798, Account of some remarkable springs: v. 1, no. 3, 21-22.
On an oil creek in Pennsylvania.

Kincaid, Eld. E.
5954 1840, Great earthquakes in Burmah: Am. J. Sci., v. 38, no. 2, 385-387.
From the Utica Reg.

King, Alexander
5955 1795, Mineral springs. The natural history of the mineral spring in Suffield, Connecticut: Rural Mag. or Vt. Repos., v. 1, 400-403.

King, Alfred T.
5956 1844-1845, Description of fossil foot marks, supposed to be referable to the classes birds, Reptilia and Mammalia, found in the Carboniferous series, in Westmoreland County, Pennsylvania: Phila. Acad. Nat. Sci's Proc., v. 2, no. 6, 175-180, 6 figs; no. 12, 299-300.
Also in Am. J. Sci., v. 48, no. 2, 343-352, 12 figs. (1845).
5957 1845, Footprints [from the Alleghany Mountains, Pennsylvania]: Am. J. Sci., v. 49, no. 1, 215; 216-217, 2 figs.
5958 1845, Fossil footmarks found in strata of the Carboniferous series in Westmoreland County, Pennsylvania: Am. J. Sci., v. 48, no. 1, 217-218.
5959 1846, Footprints in the coal rocks of Westmoreland Co., Pennsylvania: Am. J. Sci., v. 1 (ns), no. 1, 268, 2 figs.

King, Henry
5960 1840, Report of a geological reconnaissance of that part of the Missouri River adjacent to the Osage River, 1839: n.p., 19 p.
5961 1842, [Position of lead-bearing limestone of the upper Mississippi River]: Am. J. Sci., v. 43, no. 1, 173 (11 lines).
Also in Assoc. Am. Geol's Nat's Proc., v. 3, 28 (11 lines). Also in Assoc. Am. Geol's Nat's Rept., v. 1, 66 (11 lines), 1843.
5962 1844, [Effect of igneous activity on the Missouri lead formation]: Am. J. Sci., v. 47, no. 1, 107 (9 lines).
Also in Assoc. Am. Geol's Nat's Proc., v. 5, 14 (9 lines).
5963 1844, Geology of the Valley of the Mississippi, from the southern part of the state of Missouri to Wisconsin River, in the Territory of Iowa: Am. J. Sci., v. 47, no. 1, 128-130.
Also in Assoc. Am. Geol's Nat's Proc., v. 5, 37-39.
5964 1849, A geological survey of the state of Missouri: West. J. Ag., v. 3, no. 1, 12-29; no. 2, 76-83.
5965 1849, Report of an examination of the Town of Birmingham, Perry County, Missouri; and of an exploration of the lands owned by the St. Louis and Birmingham Iron Mining Company: St. Louis, T. W. Ustick, 12 p.

King, John
5966 1820, Variation of the needle: Rural Mag. and Lit. Evening Fireside, v. 1, 351.

King, Thomas Butler (1800-1864)
5967 1850, California: the wonder of the age; A book for every one going to, or having an interest in, that golden region: NY, W. Gowans, 34 p.
For review see 6368.
5968 1850, Report on California: US HED 59, 31-1, v. 8 (577), 2-32.
For excerpts see 10098. Contains notes on soils (12-17) and mineral wealth (23-32).
5969 1850, Report on the gold mines of California: West. J. Ag., v. 4, no. 1, 38-50.

Kingsley, James Luce (1778-1852; see 9599 to 9602 with B. Silliman).

Kip, Leonard (1826-1906)
5970 1850, California sketches, with recollections of the gold mines: Alb., E. H. Pease and Co., 57 p.
For review see 6368.

Kippart, John H.
5971   1850, Discourse upon paleontology: Fam. Vis., v. 1, no. 5, 37; no. 6, 41; no. 7, 49; no. 8, 57; no. 9, 65; no. 10, 73; no. 11, 81, 2 figs; no. 14, 109.

Kirwan, Richard (1733-1812)
5972   1800, Geological facts, corroborative of the Mosaic account of the deluge: Mon. Mag. and Am. Rev., v. 3, 297-301; 382-387.
5973   1810, Geological facts, corroborative of the Mosaic account of the deluge, with an inquiry into the origin, progress, and still permanent consequences of that catastrophe: Christian's Mag., v. 3, 74-82; 140-151.

Klaproth, Martin Heinrich (1743-1817)
5974   1807, [Analysis of cinnabar]: Am. Reg. or Gen. Repos., v. 1, pt. 2, 44.

Klarick (Mr.)
5975   1824, [On the use of loadstone as a panacea]: Aesculapian Reg., v. 1, 82.

Klöden, Karl Friedrich (1786-1856)
5976   1835, On the origin of the erratic blocks of the North of Germany: Am. J. Sci., v. 28, no. 2, 389-390.
       Abstracted from Jameson's J.

Knapp, John Leonard (1767-1845)
5977   1831, Journal of a naturalist: Phila., Carey and Lea, 286 p.
       Other printings include Phila., G. W. Donahue, 1837; Phila., Gihon and Smith, 1846; Phila., J. and J. L. Gihon, 1850.
       For review see 547.
       Contains notes on soil analysis, p. 20-22.

Knapp, Samuel Lorenzo (1784-1838)
5978   1835, Reminiscences of Ballston and Saratoga Springs: Knickerbocker, v. 6, 96-106.

The Knickerbocker; or New-York Monthly Magazine (New York, 1833-1850)
5979   1833, [Review of Lectures ..., by T. Flint (see 3888)]: v. 1, 193-194.
5980   1834, Life and labors of Baron Cuvier: v. 3, 17-28.
5981   1834, [Review of Contributions to geology by I. Lea (see 6093)]: v. 3, 314.
5982   1834, Mountain of salt [in Egypt]: v. 4, 144 (4 lines).
5983   1835, [Review of Ancient mineralogy by Moore (see 7419)]: v. 5, 79-80.
5984   1836, Geology and revealed religion [reviews of Connection ..., by T. Cooper (see 2627), Consistancy ..., by B. Silliman (see 9712), "Connection ...," by E. Hitchcock (see 5107), and "Critical examination ...," by M. Stuart (see 10057)]: v. 7, 441-452.
5985   1836, Organic remains: v. 8, 125-132; 377-388.
5986   1837, [Review of Mineralogy and geology by W. Buckland (see 2160)]: v. 9, 416-418.
5987   1839, [Review of Popular lectures on geology by K. C. von Leonhard (see 6192)]: v. 14, 83-84.
5988   1840, A visit to the Lackawana Mines [Pennsylvania]: v. 15, 102-106.
5989   1841, Sensations and reflections. Caused by the earthquake in January last: v. 18, 27-28.
       A poem, signed by "Flaccus".
5990   1845, [Review of Travels ..., by C. Lyell (see 6557)]: v. 26, 261-262.

Knight, J. E.
5991   1850, The mines near Little Rock, Arkansas: De Bow's Rev., v. 8, 296-298.

Knight, Jonathan (1787-1858; see 6650 with A. Macomb)

Knox, Sir George
5992   1827, Bitumen, and other volatile ingredients in stones: Am. J. Sci., v. 12, no. 1, 147-149.

Kobell, Franz, ritter von (1803-1882)
5993   1830, On the silicate of iron, of Badenmais [Germany]: Am. J. Sci., v. 18, no. 1, 164.
       Abstracted from Annales de Chimie.
5994   1845, Spadaite, a new mineral: Am. J. Sci., v. 49, no. 2, 394 (6 lines).
       Abstracted from Jameson's J.
5995   1846, Amoibite [from Lichtenberg]: Am. J. Sci., v. 2 (ns), no. 3,
       Abstracted from "Jour. für Prakt. Chem."
5996   1849, Analysis of brandesite: Am. J. Sci., v. 7 (ns), no. 1, 113
       Abstracted from Leonhard und Bronn Neues Jahrbuch.
5997   1849, On disterrite, from the valley of Fassa in Tyrol: Am. J. Sci., v. 8 (ns), no. 1, 123 (7 lines).
       Abstracted from "Jour. für Prakt. Ch."

Koch, Albrecht Karl
       See also 812, 3306, 3307, 4412, and 7901.
5998   1839, Remains of the mastodon in Missouri: Am. J. Sci., v. 37, no. 1, 191-192.
5999   1840, A short description of fossil remains found in the state of Missouri by the author: St. Louis, Missouri, Churchill and Stewart (printers), 8 p., illus.
6000   1841, A description of the Missourium or Missouri leviathan, together with its supposed habits; Indian traditions concerning the location from whence it was exhumed; also comparisons of the whole, crocodile and Missourium with the leviathan as described in the 41st Chapter of the Book of Job: St. Louis, Missouri, Charles Keemle (printer), 16 p.
6001   1841, A description of the Missourium ... [as above]: Louisville, Kentucky, Prentice and Weissinger (printers), 20 p.
       "2nd edition, enlarged".

Koch, A. (cont.)
6002 1845, Description of the Hydrarchos sillimanii: (Koch), a gigantic fossil reptile, or sea serpent: lately discovered by the author in the state of Alabama, March, 1845; together with some geological observations made on different formations of the rocks, during a geological tour through the eastern, western and southern parts of the United States, in the years 1844-1845: NY, 16 p., illus.
6003 1845, Description of the Hydrarchos harlani Koch (the name sillimani is changed to harlani by the particular desire of Professor Silliman), a gigantic fossil reptile, lately discovered by the author in the state of Alabama; ... [as above]: NY, B. Owen (printer), 24 p.
Second edition.

Koehler, H.
6004 1835, On the anthracite deposit at Tamaqua, Schuylkill County, Pennsylvania, with a map and a section: Geol. Soc. Penn. Trans., v. 1, 326-327, map.
Contains map 6005.
6005 1835, Petrographical map of the coal-region of Tamaqua: Geol. Soc. Penn. Trans., v. 1, plate 20, geological map, 42 x 16 cm.
In 6004.

Kohler, Friedrich Wilhelm (1805-1871)
6006 1831, Affinity of the diallage family, in chemical constitution, with augite: Am. J. Sci., v. 20, no. 1, 168 (7 lines).
Abstracted from Zeitschrift für Mineralogie.
6007 1831, [Specific gravity of common pyrite]: Am. J. Sci., v. 19, no. 2, 388 (3 lines).
Abstracted from Poggendorff's Annalen.
6008 1844, Upon the deposit of gold recently discovered in the Ural: Am. J. Sci., v. 46, no. 1, 211-212. Also in Franklin Inst. J., v. 8 (3s), 66-67.
Abstracted from Annales des Mines. With note by A. Humboldt. (see 5390).
6009 1848, Bagrationite, a new mineral from the Urals: Am. J. Sci., v. 6 (ns), no. 2, 267 (14 lines).
Abstracted from Poggendorff's Annalen.
6010 1849, Crystallization of uralorthite: Am. J. Sci., v. 8 (ns), no. 1, 125-126, 4 figs.
Abstracted from "Verhandl. der Russ.-Kais. Min. Gesellsch. zu St. Peters."

Kolenati, Friedrich August (1813-1864)
6011 1846, Ascension of the Snowy Peak, Kasbek, Persia, in 1844: Am. J. Sci., v. 1 (ns), no. 2, 308 (9 lines).
Abstracted from l'Institut.

Köllner, Augustus (b. 1813)
6012 1848?, Highrock-iodine and Empire Springs [Saratoga, New York]: NY, Goupil, Vibert and Co., colored lithograph by Deroy after a drawing by A. Köllner.
6013 1848?, Pavillion Fountain [Saratoga, New York]: NY, Goupil, Vibert and Co., colored lithograpgh by Deroy after a drawing by A. Köllner.
6014 1848?, Congress Spring [Saratoga, New York]: NY, Goupil, Vibert and Co., colored lithograph by Deroy after a drawing by A. Köllner.

Kopp, Hermann (1817-1892)
6015 1847, Silica [formula of]: Am. J. Sci., v. 3 (ns), no. 2, 261 (12 lines).
Abstracted from l'Institut.

Kropff, Frederick C. (see 9792 with B. Silliman)

Kuhlman, M.
6016 1827, Action of anhydrous sulphuric acid on fluor spar: Am. J. Sci., v. 13, no. 1, 174 (5 lines).
Abstracted from Annales de Chimie.

Kuhlmann, Frédéric (1803-1881)
6017 1849, On the formation of hydraulic limestones, cements, and other minerals, in the moist way: Franklin Inst. J., v. 17 (3s), 201-202.
Abstracted from London Chemical Gaz.

Kupffer, Adolphe Theodor von (1790-1865)
6018 1846, Annuaire magnetique et meteorologique de Russie, par A. T. Kupffer, 1842. 4to.: Am. J. Sci., v. 1 (ns), no. 1, 139-140.

L., J. H. (see 4088)

Lacharme, Louis
6019 1850, Improvement in gold washers: Franklin Inst. J., v. 19 (3s), 245.

The Lady's Magazine and Repository of Entertaining Knowledge (Philadelphia, 1792-1793)
6020 1793, An account of the interior parts of Sumatra, and of a neighboring island never known to have been visited by any European: v. 1 [i.e. 2], no. 1, 9-16.

Lake Shore Mining Company
6021 1846, Articles of Association of the Lake Shore Mining Company: NY, Israel Sackett (printer), 11 p., map.
Contains report by R. Kernick, jr. (see 5950).

Lamming, R.
6022 1850, On the purification of coal-gas: Am. J. Sci., v. 10, no. 1, 133-134.
Abstracted from London Ath.

Lamont, Johann von (1805-1879)
6023 1846, Ueber das magnetische observatorium der sternwarte bei Munchen: Am. J. Sci., v. 1 (ns), no. 1, 141-142.

Lancaster Hive (Lancaster, Pennsylvania, 1803-1805)
6024   1804, Gold: v. 2, 32.
6025   1804, Silver: v. 2, 34-35.
6026   1804, Lead: v. 2, 38.

Landgrebe, Georg (1802-1873)
6027   1835, Chiastolite: Am. J. Sci., v. 28, no. 2, 395 (8 lines).
        Abstracted from Jameson's J. On the equivalence of chiastolite and andalusite.

Lane, Ephram
6028   1838, Substitute for emery: Am. J. Sci., v. 34, no. 2, 381 (11 lines).
        On the use of topaz as an abrasive from Mr. Lane's mine in Monroe, Connecticut.

Lang, John
6029   1814, On lime and marls: Phila. Soc. Promoting Ag. Mem., v. 3, 204-209.
6030   1825, On lime and marls: NJ and Penn. Agri. Mon. Intell., v. 1, 65-67.

Langstaff, W.
6031   1823, By Dr. W. Langstaff [on New Jersey mineral localities]: Am. J. Sci., v. 6, no. 2, 250 (13 lines).

Lanman, Charles (1819-1895)
6032   1847, The lead region [of Illinois]: Hunt's Merch. Mag., v. 16, 181-182.
6033   1847, A summer in the wilderness; embracing a canoe voyage up the Mississippi and around Lake Superior: NY, D. Appleton and Co., 208 p.
        Also 1848 printing.
6034   1849, Letters from the Alleghany Mountains: NY, Putnam, 198 p.
        Contains articles by C. U. Shepard (see 9563), E. Mitchill (see 7259), and T. L. Clingman (see 2448).

Lanman, James Henry (1812-1887)
6035   1839, History of Michigan, civil and topographical ... : NY, E. French, 397 p., map.
        Contains D. Houghton's first annual report (see 5294).
6036   1840, History of Michigan, from its earliest colonization to the present time: NY, Harper, 269 p.
        Other printings in 1841, 1842, 1843, and 1845.
6037   1842, The iron trade of the United States: Hunt's Merch. Mag., v. 6, 511-530.
6038   1842, Resources of the United States: Hunt's Merch. Mag., v. 7, 430-442.
6039   1844, Maine and its resources: Hunt's Merch. Mag., v. 11, 313-327.
6040   1846, The copper mines of Lake Superior: Hunt's Merch. Mag., v. 14, 439-443.
6041   1848, The Lackawana and Wyoming coal region: Hunt's Merch. Mag., v. 19, 290-294.

Lapham, Darius
6042   1832, and I. A. Lapham, Observations on the Primitive and other boulders of Ohio: Am. J. Sci., v. 22, no. 2, 300-303.
6043   1842, An essay on the importance of lime in soils: Un. Agriculturist and West. Prairie Farm., v. 2, 38; 46; 54; 62; 70; 78; 86; 94.
        Abstracted from West. Farm. and Gardener.

Lapham, Increase Allen (1811-1875)
        See also 6042 with D. Lapham, and discussion of 3122 by E. Desor.
6044   1828, Notice of the Louisville and Shippingsport Canal and of the geology of the vicinity [Ohio River]: Am. J. Sci., v. 14, no. 1, 65-69, 2 figs. on plate.
6045   1837, Miscellaneous observations on the geology of Ohio: in Report ... , by J. L. Riddell (see 8785), 31-34.
6046   1844, A geographical and topographical description of Wisconsin, with brief sketches of its history, geology, mineralogy, natural history, population: Milwaukee, Wisconsin, P. C. Hale, 255 p., map.
        For review see 280.
6047   1846, Wisconsin, its geography and topography, history, geology, and mineralogy, together with brief sketches of its natural history, soil, productions, population and government: NY, Paine and Burgess, viii, 208 p.
        For review see 5458.
6048   1846, Wisconsin, sectional map with the most recent surveys: Milwaukee, Wisconsin, P. C. Hale, map, 57 x 56 cm.
        Not seen.
6049   1847, On the existence of certain lacustrine deposits, in the vicinity of the Great Lakes, usually confounded with the "drift": Am. J. Sci., v. 3 (ns), no. 1, 90-94, 3 figs.
6050   1850, [Letter on medical geology]: Am. Assoc. Adv. Sci. Proc., v. 2, 406-407.

Lardner, Dionysius (1793-1859)
6051   1845-1846, Popular lectures on science and art; delivered in the principal cities and towns of the United States: NY, Greeley and McElrath, 2 v., plates.
        Published in several parts. Other printings in 1848, 1849, 1850.
6052   1846, Shape of the earth: Scien. Am., v. 1, no. 52, [3].
6053   1846, Temperature of the interior of the earth: Scien. Am., v. 1, no. 52, 11.

La Rive, see Rive, A. A. de la

Lartet, Edouard Armand Isidore Hippolyte (1801-1871)
6054   1841, Fossil organic remains: Am. Eclec., v. 1, no. 2, 386-387.
        Abstracted from London Ath.

Lassaigne, Jean Louis (1800-1859)
6055   1849, Mud of the Nile: Am. J. Sci., v. 8 (ns), no. 2, 275 (11 lines).
        Abstracted from "Jour. de Pharm."
6056   1850, Arsenic in chalybeate springs [from Wattviller, Haut Rhine]: Ann. Scien. Disc., v. 1, 200-201 (17 lines); also in Am. J. Sci., v. 9 (ns), no. 3, 418.
        Abstracted from "Jour. de Chim. Med."

Lathrop, John (1772-1820)
6057 1809, An account of the springs and wells on the Peninsula of Boston, with an attempt to explain the manner in which they are supplied: Am. Acad. Arts Sci. Mem., v. 3, pt. 1, 57-68, plate.
For extract see 10753.

Lathrop, John Hiram (1799-1866)
6058 1840, On the connexion between the theory of the earth and the secular variations of the magnetic needle: Am. J. Sci., v. 38, no. 1, 68-72.
6059 1840, Applications of the igneous theory of the earth: Am. J. Sci., v. 39, no. 1, 90-95.

Lathrop, Leonard E.
6060 1825, The farmer's library, or, essays designed to encourage the pursuits and promote the science of agriculture: Rutland, Vt., W. Fay, 215 p.
Other editions include Windsor, Vt., W. Spooner, 300 p. (1826); and Rochester, NY, Marshall and Dean, 344 p. (1828).

Lathrop, S. Pearl
6061 1844, Notice of an ice mountain in Wallingford, Rutland County, Vermont: Am. J. Sci., v. 46, no. 2, 331-332.

Latrobe, Benjamin Henry (1764-1820)
6062 1798, American copper-mines. To the chairman of the committee of commerce and manufactures, to whom has been referred the petition of N. I. Roosevelt and his associates, praying for an act of incorporation of a mine and metal company: Phila., 8 p.
Evans number 33987. See also 10509.
6063 1799, On the sand-hills of Cape Henry in Virginia: Am. Philos. Soc. Trans., v. 4, 439-444, plate. Also in Am. Min. J., v. 1, no. 4, 248-252 (1814).
With a supplement by W. Tatham (see 10096).
6064 1800, American copper-mines: Wash., 8 p.
Evans number 37785. See also 6062.
6065 1803, Description of the Schuyler Copper Mine in New-Jersey: Med. Repos., v. 6, 319-321.
6066 1807, (Deposit of silt in the Potomac Estuary), in [Papers relative to the contemplated bridge across the Potomac]: Wash., p. 6-11.
6067 1809, An account of the freestone quarries on the Potomac and Rappahannock rivers: Am. Philos. Soc. Trans., v. 6, pt. 2, 283-293.
For review with excerpts see 4241 and 7090.
6068 1812, Opinion on a project for removing the obstructions to a ship navigation to Georgetown, Col.: Wash., 25 p.
6069 1838, [Experiments on burning coal]: in "Report made to the Maryland Coal Company" by B. Silliman (see 9748), p. 21-22.

Latrobe, Charles Joseph (1801-1875)
6070 1835, The rambler in North America ... : NY and London, Seeky, 2 v. (11, 321 p.; 8, 336 p.), maps.

Laugier, M.
6071 1807, [Chromium in meteorites]: Am. Reg. or Gen. Repos., v. 1, pt. 2, 42.
6072 1823, Composition of meteoric stones: Am. J. Sci., v. 6, no. 2, 397.
Abstracted by J. Griscom.
6073 1828, Analysis of the massive cinnamon stone [from Ceylon]: Am. J. Sci., v. 14, no. 1, 204 (7 lines).
Abstracted from Bulletin Universelle.

Laurent, August (1807-1853)
6074 1836, and C. Holms, Albite of Chesterfield [Massachusetts]: Am. J. Sci., v. 30, no. 2, 381 (14 lines).
Abstracted from Annales de Chimie.
6075 1848, General formulas for the silicates and borates: Am. J. Sci., v. 5 (ns), no. 3, 405-407.
Abstracted from Comptes Rendus.
6076 1849, Composition of the tungstates: Am. J. Sci., v. 7 (ns), no. 2, 281-282.
Abstracted from "J. de Pharm. et de Chim."

Laurie, James
6077 1848, Paper on the coal and iron trade of Great Britain and the United States, read before the Boston Society of Civil Engineers, September 4, 1848: Mining J., v. 2, 50-51.

Law, James (see 1592 with T. R. Beck and J. Green)

Law, Thomas (1756-1834)
6078 1806, Ballston springs ... : NY, S. Gould, 48, 46 p.

Lawrence, Abbott (1792-1855)
6079 1847, Arts and sciences at Harvard: Am. J. Sci., v. 4 (ns), no. 2, 294-297.

Lawrence, Byrem
6080 1843, A concise description of the geological formations and mineral localities of the western states; designed as a key to the geological map of the same: Bos., S. N. Dickinson, 48 p.
Primarily on the Ohio Valley. See also 6081.
6081 1843, A geological map of the western states: Bos., colored geological map.
Not seen. Copied from D. D. Owen's geological chart of the Ohio Valley. See also 6080.

Lea, Albert Miller (1807-1890)
6082   1836, Notes on the Wisconsin Territory, (particularly with reference to the Iowa District or Black Hawk purchase ...): Phila., H. S. Tanner, 53 p., map.
6083   1837, Report of the chief engineer of the state of Tennessee, on the surveys and examinations for the Central Railroad and for the Central Turnpike. Under an act of Assembly, passed October 25th, 1836: Nashville, Tenn., S. Nye and Co., 80 p.

Lea, Henry Charles (1825-1909)
6084   1841, Description of some new species of fossil shells, from the Eocene at Claiborne, Alabama: Am. J. Sci., v. 40, no. 1, 92-103, plate with 24 figs.
6085   1842, Remarks upon an examination of the peroxide of manganese: Am. J. Sci., v. 42, no. 1, 81-87.
6086   1843, Description of some new fossil shells from the Tertiary of Virginia: Am. Philos. Soc. Proc., v. 3, no. 27, 162-165.
6087   1846, Description of some new fossil shells, from the Tertiary of Petersburg, Virginia: Am. Philos. Soc. Trans., v. 9 (ns), 229-274, 4 plates.
6088   1848, Catalogue of the Tertiary Testacea of the United States: Phila. Acad. Nat. Sci's Proc., v. 4, no. 5, 95-107.

Lea, Isaac (1792-1886)
       See also 6621 with W. Maclure and R. M. Patterson.
6089   1818, An account of the minerals at present known to exist in the vicinity of Philadelphia: Phila. Acad. Nat. Sci's J., v. 1, pt. 2, 462-482.
       Abstracted in Hazard's Reg. Penn., v. 2, 17-21 (1828).
6090   1822, Notice of a singular impression in sandstone: Am. J. Sci., v. 5, no. 1, 155.
       On fossils from the Monongahelia River, Pennsylvania.
6091   1823, A sketch of the history of mineralogy: Phila. J. Med. Phys. Sci's, v. 7, 270-289.
       Abstracted in Mon. J. Medicine, v. 3, 318-319 (1824).
6092   1825, On earthquakes - their causes and effects: Am. J. Sci., v. 9, 209-215.
6093   1833, Contributions to geology: Phila., Carey, Lea and Blanchard, viii, 227 p.
       Contains "Tertiary formation of Alabama", 1-208; "New Tertiary fossil shells, from Maryland and New Jersey," 209-216; "New genus of fossil shell, from New Jersey," 217-220; and "Tufaceous lacustrine formation of Syracuse, Onondaga County, N. Y.", 218-227.
       For review with excerpts see 576. For review see 5981.
6094   1840, Notice of the Oolitic formation in America, with descriptions of some of its organic remains: Am. Philos. Soc. Proc., v. 1, no. 12, 225-227.
       Abstracted in Am. J. Sci., v. 40, no. 1, 41-42 (1841).
6095   1841, Notice of the Oolitic formation in America, with descriptions of some of its organic remains: Am. Philos. Soc. Trans., v. 7 (ns), 251-260, 3 plates.
       Includes "The Oolite of Cuba", and descriptions of Mezozoic fossils from New Granada.
6096   1842, [On specimens of anthracite coal from Pine Grove, Pennsylvania]: Am. Philos. Soc. Proc., v. 2, no. 23, 229-230.
6097   1843, On coprolites: Am. Philos. Soc. Proc., v. 3, no. 27, 143.
6098   1844, [On variation of form in shells with growth]: Am. J. Sci., v. 47, no. 1, 109-110.
       Also in Assoc. Am. Geol's Nat's Proc., v. 5, 16-17.
6099   1846, [On a specimen of flexible quartz, called itacolumbite, from Spartanburg district, South Carolina]: Am. Philos. Soc. Proc., v. 4, no. 35, 244 (8 lines).
6100   1849, Appendix [on fossil footprints in coal]: Am. J. Sci., v. 8 (ns), no. 1, 160.
6101   1849, [On reptilian footmarks in the gorge of the Sharp Mountain near Pottsville, Pennsylvania]: Am. Philos. Soc. Proc., v. 5, 91-94, fig. Also in Am. J. Sci., v. 9 (ns), 124-126, fig. (1850).
6102   1849, Report to the directors of the Pequa Railroad and Improvement Company: Phila., T. K. and P. G. Collins (printers), 24 p., map, 3 folded plates.
       Contains excerpts from geology reports by H. D. Rogers and R. C. Taylor.
6103   1850, Footprints in the Old Red Sandstone: Ann. Scien. Disc., v. 1, 281.

Lea, Mathew Carey (1823-1897)
       See also 1765 with J. C. Booth.
6104   1841, On the first, or Southern coal field of Pennsylvania: Am. J. Sci., v. 40, no. 2, 370-374. Also in Am. Rr. J., v. 12, 342-345.
6105   1845, Coal of Pennsylvania and other states: Hunt's Merch. Mag., v. 13, 67-72.
6106   1845, The first coal region of Pennsylvania: Hunt's Merch. Mag., v. 13, 426-434.

Leavitt, Benjamin
6107   1836, Slate in Foxcroft [Maine]: Me. Farm., v. 4, 145.
6108   1841, Foxcroft slate quarry: Me. Farm., v. 9, 164-165.

Leavitt, Dudley (1772-1851)
6109   1823, Principles of geology, or the history of opinions concerning the origin and formation of the world. Collected from various authors: Coll. Topo. Hist. Bibliographical NH, v. 2, 257-259.

Leblanc, M.
6110   1846, Air of mines: Am. J. Sci., v. 1 (ns), no. 1, 118-119.
       Abstracted from l'Institute.
6111   1848, Talus slopes: Am. J. Sci., v. 6 (ns), no. 1, 133 (7 lines).
       On the angle of talus slopes.

Le Conte, John Lawrence (1825-1883)
       See also discussion of article 8941 by H. D. Rogers.
6112   1846, New metals: S. Med. Surgical J., v. 2 (ns), 117 (9 lines).
       On Rhoduim, Niobium, and Pelopium from Platinum.
6113   1847, On coracite, a new ore of uranium: Am. J. Sci., v. 3 (ns), no. 1, 117 (9 lines); no. 3, 173-175.
       From the North Shore of Lake Superior.

Le Conte, J. L. (cont.)
6114   1847, Plumbo-resinite and cuprous sulphato-carbonate of lead in Missouri: Am. J. Sci., v. 3 (ns), no. 1, 117.
6115   1848, [Fossil horse teeth in the collection of the New York Lyceum of Natural History]: NY J. Medicine and Collateral Sci´s, v. 10, 341-342.
       Abstracted from NY Lyc. Nat. History Minutes.
6116   1848, Notice of five new species of fossil mammalia from Illinois: Am. J. Sci., v. 5 (ns), no. 1, 102-106, 4 figs.
       In Assoc. Am. Geol´s Nat´s Proc. Abstracted in Lit. World, v. 2, 228 (5 lines).
6117   1848, On Platygonus compressus; a new fossil pachyderm: Am. Acad. Arts Sci´s Mem., v. 3 (ns), 257-274, 4 plates with 29 figs.

Lee, Charles Alfred (1801-1872)
6118   1822, On certain rocks supposed to move without any apparent cause [at Salisbury, Connecticut]: Am. J. Sci., v. 5, 34-37, plate.
       Signed by "Petros". See also 40 and 6122.
6119   1824, Notice of the Ancram lead mine [Columbia County, North Carolina]: Am. J. Sci., v. 8, no. 2, 247-250.
6120   1824, Sketch of the geology and mineralogy of Salisbury, Con.: Am. J. Sci., v. 8, no. 2, 252-261.
6121   1825, By Mr. Charles A. Lee: Am. J. Sci., v. 9, no. 1, 42-44.
       On mineral localities in New York, Massachusetts and Connecticut.
6122   1825, Remarks on the moving rocks of Salisbury [Connecticut]: Am. J. Sci., v. 9, no. 2, 239-241.
       "Petros" revealed to be C. A. Lee. See also 6118.
6123   1827, Localities of minerals [in Connecticut]: Am. J. Sci., v. 12, no. 1, 169-170.
6124   1827, Native iron: Am. J. Sci., v. 12, no. 1, 154-155.
       In "Notice of native iron from Canaan, Conn." by W. Burrall (see 2203).
6125   1840 [or 1839?], The elements of geology, for popular use; containing a description of the geological formations and mineral resources of the United States: NY, Harper and Brothers, viii, [9]-375 p.
       The Preface is dated Oct. 14, 1839. Other printings by Harper and Brothers appeared in 1844, 1846, and 1848.
       For review see 1686 and 1303.
6126   1840, Analysis of soils: Am. Rep. Arts Sci´s Manu´s, v. 1, 251-253.
       Contains "Rules of analysis" by S. L. Dana (see 2884).
6127   1840, [On United States mineral deposits]: Am. Rep. Arts Sci´s Manu´s, v. 2, 143-149.
6128   1840, Geology of the state of New York [review of 7740]: Am. Rep. Arts Sci´s Manu´s, v. 1, 328-335; 404-412.
6129   1840, On the geology of Palestine, and the destruction of Sodom and Gomorrah: Bib. Repos., v. 3 (2s), 324-352.
6130   1841, Minerals in the United States: Fam. Mag., v. 8, 220-223.
       From Elements of geology (see 6125).
6131   1843, On ancient climate as viewed in the light of fossil geology: in Meteorology ..., by S. Forry (see 3911), 45-48.

Lee, Daniel (1806-1895)
       As editor of The Genesee Farmer and Gardener´s Journal, D. Lee wrote many unsigned editorial notes on soils and their analyses.
6132   1846, Agricultural geology: Genesee Farm. and Gardener´s J., v. 7, 177.
6133   1846, Geological excursion: Genesee Farm. and Gardener´s J., v. 7, 199-201.
6134   1846, Study the soil: S. Planter, v. 7, 229-231.
       Abstracted from Genesee Farm. and Gardener´s J.
6135   1847, The analysis of soils: Genesee Farm. and Gardener´s J., v. 8, 12-13.
6136   1847, Analysis of soils: Genesee Farm. and Gardener´s J., v. 8, 154-155; 203-204.
6137   1847, Study the soil: Am. Farm., v. 3 (4s), 41-42.
       Abstracted from Genesee Farm. and Gardener´s J.

Lee, Hamlin
6138   1846, Abundant occurrence of rare infusoria in the Scallop [from the Miocene of Virginia]: Am. J. Sci., v. 1 (ns), no. 1, 124.
       Abstracted from Annals Nat. History.

Lee, Richard
6139   1829, Machine for separating gold dust, grains and particles from the ore: Franklin Inst. J., v. 4 (ns), 252-253.

Legaré, John D.
6140   1835, Account of mineral springs: S. Agriculturist, v. 8, 63-72; 120-128.
6141   1836, An account of the medical properties of the Grey Sulphur Springs, Virginia: Char., A. E. Miller, 18 p.
6142   1840, Geological survey: S. Agriculturist, v. 1 (ns), 667-670.
       Support given for geological survey of South Carolina.

Lehigh Coal and Navigation Company
6143   1821, An address of the Lehigh Coal and Navigation Company to their fellow citizens: Phila., William Brown (printer), 8 p.
6144   1824, Facts illustrative of the character of the anthracite, or Lehigh coal, found in the great mines at Mauch Chunk: Phila., S. W. Conrad (printer), 12 p.
       Includes many testimonial letters. Also 1825 editions, including NY, Gray and Bunce (printers), 18 p.; Bos., T. R. Marvin, 22 p. Also 1827 edition, Phila., S. W. Conrad (printer), 20 p.
6145   1826, Report of the engineers of the Lehigh Coal and Navigation Company, who were appointed, on the nineteenth instant, a committee on the subject of an expose of the property of the company: Phila., S. W. Conrad (printer), 12 p., map.
6146   1828, Report of the board of managers of the Lehigh Coal and Navigation Company, presented to the stockholders, January 14th, 1828: Phila., S. W. Conrad (printer).
       This is the first annual report. Not seen.

Lehigh Coal and Navigation Company (cont.)

6147 1829, Report ... [as above] ... stockholders, January 12th, 1829: Phila., S. W. Conrad (printer), 15 p.

6148 1830, Report ... [as above] ... stockholders, January 11th, 1830: Phila., T. A. Conrad (printer), 16 p.

6149 1831, Report ... [as above] ... stockholders, January 17th, 1831: Phila., Wm. F. Geddes (printer), 16 p.

6150 1832, Report ... [as above] ... stockholders, January 9, 1832: Phila., Wm. F. Geddes (printer), 16 p.

6151 1833, Report ... [as above] ... stockholders, January 14, 1833: Phila., Wm. F. Geddes (printer), 8 p.

6152 1834, Report ... [as above] ... stockholders, January 13, 1834: Phila., James Kay, jr. and Co. (printers), 11 p.

6153 1835, Report ... [as above] ... stockholders, January 12, 1835: Phila., James Kay, jr. and Co. (printers), 12 p.
For review see 2736.

6154 1836, Report ... [as above] ... stockholders, January 11, 1836: Phila., James Kay, jr. and Co. (printers), 16 p.

6155 1837, Report ... [as above] ... stockholders, January 9, 1837: Phila., James Kay, jr. and Co. (printer), 27 p.

6156 1839, Report ... [as above] ... stockholders, January 14, 1839: Phila., James Kay, jr. and Co. (printer), 60 p.

6157 1840, A defense of the Lehigh Coal and Navigation Company, from the assaults made upon its interests by X: Phila., J. Harding (printer), 66 p., map.
Signed by "Honestas".

6158 1840, History of the Lehigh Coal and Navigation Company: Phila., W. S. Young (printer), 69 p., 2 maps.

6159 1840, Report ...[as in 6146] ... stockholders, January 13, 1840: Phila., W. S. Young (printer), 48 p.

6160 1841, Report ... [as above] ... stockholders, January 11, 1841: Phila., W. S. Young (printer), 19 p.

6161 1842, Report ... [as above] ... stockholders, January 10, 1842: Phila., Brown, Bicking and Guilbert (printers), 24 p.

6162 1843, Report ... [as above] ... stockholders, January 9, 1843: Phila., W. S. Young (printer), 30 p.

6163 1844, Report ... [as above] ... stockholders, January 8, 1844: Phila., W. S. Young (printer), 35 p.

6164 1845, Report ... [as above] ... stockholders, January 13, 1845: Phila., W. S. Young (printer), 16 p.

6165 1846, Report ... [as above] ... stockholders, May 5, 1846: Phila., W. S. Young (printer), 24 p.

6166 1847, Report ... [as above] ... stockholders, May 4, 1847: Phila., J. C. Clark (printer), 16, [1] p.

6167 1847, Report of the board of managers of the Lehigh Coal and Navigation Company, to the stockholders: Am. Rr. J., v. 20, 486-488.
Extract from 6166.

6168 1848, Report ... [as in 6146] ... stockholders. May 2, 1848: Phila., J. C. Clark (printer), 24 p.

6169 1849, Report ... [as above] ... stockholders. May 1, 1849: Phila., J. C. Clark (printer), 22 p.

6170 1850, Report ... [as above] ... stockholders. May 7, 1850: Phila., J. C. Clark (printer), 23 p.

Leidy, Joseph (1823-1891)

6171 1845, Notes taken on a visit to White Pond, in Warren Co., New Jersey: Phila. Acad. Nat. Sci´s Proc., v. 2, no. 11, 279-281.

6172 1847, On a new genus and species of fossil ruminantia: Poebrotherium wilsoni: Phila. Acad. Nat. Sci´s Proc., v. 3, no. 12, 322-326, 6 figs.
Also reprint edition. Abstracted in Am. J. Sci., v. 5 (ns), no. 2, 276-279 (1848).

6173 1847, On the fossil horse of America: Phila. Acad. Nat. Sci´s Proc., v. 3, no. 11, 262-266; no. 12, 328, plate with 6 figs.

6174 1847, Report of the curators for the year 1847 [on the mineral cabinet] : Phila. Acad. Nat. Sci´s Proc., v. 3, no. 12, 359-362.

6175 1848, On a new fossil genus and species of ruminantoid Pachydermata: Merycoidodon culbertsonii: Phila. Acad. Nat. Sci´s Proc., v. 4, no. 2, 47-50, plate.
Also reprint edition.

6176 1848, Report of the curators for 1848 [on the mineral cabinet] : Phila. Acad. Nat. Sci´s Proc., v. 4, no. 6, 132-136.

6177 1849, Report of the curators for 1849 [on the mineral cabinet] : Phila. Acad. Nat. Sci´s Proc., v. 4, no. 12, 254-256.

6178 1849, Tapirus americanus fossilis: Phila. Acad. Nat. Sci´s Proc., v. 4, no. 9, 180-182.
On fossil bones from Opelousas, Louisiana. Abstracted in Am. J. Sci., v. 9 (ns), no. 1, 140 (1850).

6179 1850, [Descriptions of mammalian remains from Missouri Territory] : Phila. Acad. Nat. Sci´s Proc., v. 5, no. 6, 121-122.

6180 1850, [On Eucrotaphus jacksoni and Archacotherium mortoni from the Badlands of South Dakota] : Phila. Acad. Nat. Sci´s Proc., v. 5, no. 5, 90-93.

6181 1850, [On Rhinoceros occidentalis from the Missouri Territory] : Phila. Acad. Nat. Sci´s Proc., v. 5, no. 6, 119 (5 lines).

6182 1850, Report of the curators for 1850 [on the mineral cabinet] : Phila. Acad. Nat. Sci´s Proc., v. 5, no. 6, 129-132.

Leighton, T. H.

6183 1849, Commercial importance of the metallic sulphurets: Hunt´s Merch. Mag., v. 21, 237-239.

Lenz, Emil Khristianovich (1804-1865)

6184 1832, On the temperature and saltness of the waters of the ocean at different depths: Am. J. Sci., v. 23, no. 1, 10-14.
Abstracted from Edinburgh J. Sci.

Leonhard, Gustav (Prof.; i.e. Karl Casar Leonhard?)
6185   1846, Notice of Mr. Alger's Phillips' mineralogy [review of 8325]: Am. J. Sci., v. 1 (ns), no. 1, 148.
        Abstracted from Annals of the University of Heidelberg.

Leonhard, Karl Casar, ritter von (1779-1862)
6186   1820, Biographical account of M. Werner, late Professor of Mineralogy Freiberg [obituary and bibliography]: Saturday Mag., v. 4, 248-251.
6187   1831, Agenda geognostica [notice and outline of Help book for travelling geologists, by Leonhard]: Am. J. Sci., v. 17, no. 1, 200.
        Translated by Fiske.
6188   1831, Arrangement of rocks: Am. J. Sci., v. 20, no. 1, 182.
        Abstracted from Jameson's J. by J. Griscom.
6189   1831, Heidelberg collection of minerals, petrifactions, and models of crystals. - I. Oryctognostic collection: Am. J. Sci., v. 20, no. 2, 398-399.
6190   1831, Trap, and rocks altered by it: Am. J. Sci., v. 20, no. 1, 170.
        On contact metamorphism by basalt on sediments.
6191   1834, Chalk and chalk fossils in granite [from Bohemia and Saxony]: Am. J. Sci., v. 26, no. 1, 218 (5 lines).
6192   1839-1841, Popular lectures on geology, treated in a very comprehensive manner, translated [from the German] by Rev. J. G. Morris, and edited by Prof. F. Hall: Balt., N. Hickman, 400 p., illus.
        Issued in 4 parts. For review see 757, 1690, 2214, 5987, 7371, 4039, 9277.
6193   1842, [On fossil human footprints]: Am. J. Sci., v. 42, no. 1, 26 (16 lines).
        In "Regarding human footprints in solid limestone," by D. D. Owen (see 8051).

Le Page du Pratz, Antoine Simon (1718-1758)
6194   1804, An account of Louisiana, exhibiting a compendious sketch of its political and natural history and topography ... : Newbern, Franklin and Garrow, 360 p.

Le Row, George L.
6195   1837, Trilobite: Poughkeepsie Telegraph, v. 12, no. 31, fig.

Leschenault de la Tour, Jean Baptist Claude Théodore (1773-1826)
6196   1814, Lake of sulphuric acid on the Isle of Java: Analectic Mag., v. 3, 161-164.
        Abstracted from Report to the Imperial Institute.

Leslie, John (1766-1832; see 5710)

Leslie, P. (i.e. Peter Lesley, Jr.?; see discussion of article 4611 by A. Guyot)

Lester, William, jr.
6197   1832, Geological map of New London and Windham counties, Connecticut: Hart.
        See also 6876 and 6879.

Le Sueur, Charles Alexandre (1778-1846)
        See also 10349 to 10354 with G. Troost.
6198   1818, Observations on a new genus of fossil shells: Phila. Acad. Nat. Sci's J., v. 1, pt. 2, 310-313, plate with 3 figs.
        On Maclurites from New York and Kentucky.

Lettson, John Coakley (1744-1815)
6199   1795, In collecting minerals and fossil substances, the following particulars are to be attended to: Mass. Hist. Soc. Coll, v. 4, 14.
        Reprinted in v. 1 (2s), 25-26.

Levadiefs, Nicholaïdes
6200   1837, Greece: Am. J. Sci., v. 33, no. 1, 207-208.
        Abstracted from London Ath. On gold and silver mines in ancient Greece and Macedonia, and coal deposits near Negroport and Argos.

Levol. A.
6201   1850, Native gold [from California]: Am. J. Sci., v. 10 (ns), no. 2, 255 (8 lines).
        Abstracted from Annales de Chimie.

Levy, Armand (1794-1841)
6202   1824, Forsterite [from Mt. Vesuvius]: Bos. J. Phil. Arts, v. 1, no. 6, 599 (7 lines).
6203   1824, New mineral: Bos. J. Phil. Arts, v. 1, no. 6, 598.
        On bucklandite from Norway.
6204   1827, Identity of epistilbite and heulandite: Am. J. Sci., v. 13, no. 1, 185-186 (10 lines).
        Abstracted from Philos. Mag. and Annals Phil. by C. U. Shepard.

Levy, M.
6205   1840, On haydenite and beaumontite, two minerals from the United States [Baltimore]: Franklin Inst. J., v. 26 (ns), 154-156.

Lewis, Alonzo (1794-1861)
6206   1844, The history of Lynn, including Nahant: Bos., Samuel N. Dickinson (printer), 278 p., plates, illus.
        Includes notes on geology and earthquakes of Massachusetts.

Lewis, Ellis (see 10794 with H. Wells and others)

Lewis, James A.
6207   1845, Kenawha gas [Kenawha Court House, Virginia]: Am. J. Sci., v. 49, no. 1, 209-211; also in Dwight's Am. Mag., v. 1, 654-655.

Lewis, John
6208   1830, Virginia gold [from Spottsylvania County, Virginia]: Am. Farm., v. 12, 230.

Lewis, John (of Franklin County)
6209   1840, Different soils in Kentucky, and the conjectured formation of the rich lands around Lexington: Farm's Reg., v. 8, 350-352.
        From the Franklin Farmer.

Lewis, Meriwether (1774-1809)
6210   1806, [Report on the expedition]: in Message from the President of the United States ..., (see 10471).
6211   1810, Extract of a letter ... to the President of the United States: Omnium Gatherum, v. 1, no. 1, 10-11.
6212   1814, History of the expedition under the command of Captains Lewis and Clark, to the sources of the Missouri River, thence across the Rocky Mountains and down the Columbia River, performed during the years 1804, 1805, and 1806: Phila., Bradford and Inskeep; NY, A. H. Inskeep and J. Maxwell, 2 v., 6 maps.
        Edited by Paul Allen. Contains numerous geological notes. Also an 1817 printing.
6213   1824, The history of the expedition ... [as above]: Phila., 2 v., 470 p.; 522 p.).
6214   1842, History of the expedition ... [as above]: NY, Harper and Brothers, 2 v., maps, plates.
        Edited by Archibald McVicker. Other printings in 1843, 1844, 1845, and 1847.

Lewis, Samuel
6215   1842, Lime: Farm's Cab., v. 6, 201; 233-234; 265-266.
6216   1842, Marl: Farm's Cab., v. 6, 274-275.
6217   1844, On the geology of soils: Farm's Cab., v. 9, 122-123; 137-139.

Lewis, Samuel G. (see 6614 by W. Maclure)

Lewis, Tayler (1802-1877)
6218   1845, [Review of Vestiges of the natural history of Creation by R. Chambers (see 2292 et seq.)]: Am. Whig Rev., v. 1, 525-543.

Lewis, Zechariah
6219   1801, Remarks on a subterranean wall in North Carolina: Med. Repos., v. 4, 227-234.
6220   1802, Letter on subterranean wall on the Yadkin in North Carolina [at Salisbury]: Med. Repos., v. 5, 397-407.

Lhotsky, John
6221   1841, Mineral springs in New South Wales: Am. Eclec., v. 1, no. 1, 181.
        Abstracted from Lit. Gaz.

Liebig, Justus, freiherr von (1803-1873)
6222   1845, Mineral manures: Am. Q. J. Ag. Sci., v. 2, 319 (14 lines).

Lightner, Joel
6223   1828, Fossil bones: Hazard's Reg. Penn., v. 1, 98.
        On fossil bones from Lancaster County, Pennsylvania.

Limber, John
6224   1841, Fossil remains in Lenoir County, N. C.: Am. J. Sci., v. 40, no. 2, 405; also in Am. Rr. J., v. 13, 21-22.
6225   1846, Petrified wood in Texas: Am. J. Sci., v. 2 (ns), no. 1, 124 (11 lines).

Lincklaen, Ledyard
6226   1845, The geology of Madison County [New York]: Madison Co. Agri. Soc. Trans., 1842-1845, 30-46, fig., map, colored plate.
        Signed by "L." Contains map 6227.
6227   1845, Geological map of the county of Madison, N. Y. from the state survey: Hamilton, NY, "engraved, printed & colored for the Madison County Agricultural Society", colored geological map, 12.5 x 20 cm.
        In 6226.

Lincoln, Almira Hart (see 10602 by L. N. Vauquelin; see also 8256 and 8257).

Lincoln, Benjamin (1733-1810)
6228   1785, An account of several strata of earth and shells on the branches of the York River, in Virginia; of a subterraneous passage, and the sudden descent of a very large current of water from a mountain, near Carlisle; of a remarkably large spring near Reading, in Pennsylvania; and also of several remarkable springs in the states of Pennsylvania and Virginia: Am. Acad. Arts Sci's Mem., v. 1, 372-376.
        Extracts published in NHav. Gaz., v. 1, 124 (1786); Am. Mus. or Univ. Mag., v. 1, 208-209 (1787); NY Mag., v. 3, 7-8 (1792); Univ. Asylum and Colum. Mag., v. 9, 39-40 (1792); and Rural Mag. or Vt. Repos., v. 1, 403-404 (1795).

Lindsay, Colonel
6229   1831, Volcano in New Zealand: Am. J. Sci., v. 20, no. 2, 381-382.
        Abstracted from Philos. J. by J. Griscom.

Lines made after the great earthquake, in 1755:
6230   1755, Bos., broadside folio.
        Evans number 7450. A poem of 36 verses in three columns.

Linn, Lewis Fields (1795-1843)
6231   1836, Letters ... relative to obstructions ... of the White, Big Black, and St. Francis Rivers [in Missouri and Arkansas]: US SED 113, 24-1, v. 2 (280), 1-7.
     Notes on the effects of the earthquake of 1811, with description of the natural damming of rivers by logs.
6232   1839, Observations on the La Motte mines and domain, in the state of Missouri, with some account of the advantages and inducements there promised to capitalists and individuals desirous of engaging in mining, manufacturing, or farming operations: n.p., Royston and Brown (printers).
     Dated "April 10th, 1839".

Linnard, T. B.
6233   1844, Report on improvement of navigation of Red River, Louisiana: US SED 1, 28-2, v. 1 (449), 283-293.

Linsley (Mr.)
6234   1843, [On the connection between hurricanes and earthquakes]: Bos. Soc. Nat. History Proc., v. 1, 15.

Lister, George
6235   1846, [On the fossil bones from Washington County, Alabama, exhibited with the name Hydrarchos]: Bos. Soc. Nat. History Proc., v. 2, 94-96.

Literary and Historical Society of Quebec
6236   1829, Catalogue of the mineralogical collection, belonging to the Literary and Historical Society of Quebec: N.Hav., H. Howe, 23 p.

The Literary and Philosophical Repertory (Middlebury, Vermont, 1812-1816)
6237   1812, Earthquakes [in South America]: v. 1, 87-88.
6238   1812, An account of the earthquakes, which have occurred in the years 1811-12: v. 1, 137-148.
6239   1813, [Abstract and review of Annual address before the Society for the Promotion of Useful Arts by T. R. Beck (see 1590)]: v. 1, 294-306.
6240   1813, Georgia porcelain earth: v. 1, 306-307.
6241   1813, Coccolite [from Lake Champlain, Vermont]: v. 1, 379-382.
6242   1813, Volcanick eruption [at Souffriere, St. Vincent]: v. 1, 391-392.
6243   1814, Clay near Pittsburgh [at Wilkinsburgh, Pennsylvania]: v. 1, 451.
6244   1814, Oxyd of manganese [from Lancaster, Pennsylvania]: v. 1, 451-452.
6245   1814, Emerald [from Chesterfield, Massachusetts; Haddam, Connecticut; Philadelphia; etc.]: v. 1, 452-453.
6246   1815, Late fall of stones in France: v. 2, 153-154.
6247   1815, Earthquake [in New England]: v. 2, 158-160.
6248   1815, Lead mine found in Ancram, Columbia County, N. Y.: v. 2, 238-239.
6249   1815, Natural curiosity: v. 2, 287-288.
     On a cave on Lake Ontario's North Shore.
6250   1815, Onondaga salt spring: v. 2, 290 (5 lines).
6251   1815, Diamonds: v. 2, 297-298.
6252   1816, Burning springs [Vernon, New York]: v. 2, 367-368.
6253   1816, Variation of the magnetic needle [in Vermont]: v. 2, 388-389.
6254   1816, American flint [localities given]: v. 2, 482.

Literary and Philosophical Society of New York (see New York Literary and Philosophical Society)

Literary and Scientific Repository, and Critical Review (New York, 1820-1822)
6255   1820, [Reviews of A critical examination of the first principles of geology by G. B. Greenough]: v. 1, no. 1, 37-51.
     From Mon. Rev., London and Edinburgh Rev.
6256   1820-1822, [Lists of recent geological and mineralogical works]: v. 1, no. 1, 257-258; v. 2, no. 3, 255; v. 4, no. 7, 248.
6257   1821, [Review of King Coal's Levee, or geological etiquette (3rd ed.); and A geological primer in verse; and Court news; or, the peers of King coal; popular works on geology by John Scafe]: v. 3, no. 5, 159-159 [i.e. 167].
     Includes poetry excerpts from all three books.
6258   1821, Memoirs of Frederick Accum: v. 2, no. 3, 221-226.
     Abstracted from European Mag.

The Literary and Theological Review (New York, 1834-1839)
6259   1834, [Review of Outlines of geology by J. L. Comstock (see 2511)]: v. 1, no. 2, 335.

The Literary Casket: Devoted to Literature, the Arts, and Science (Hartford, 1826-1827)
6260   1826, Rich shell marl [from Maryland and Virginia]: v. 1, 99.
6261   1826, The magnetic needle [its variation in Russia]: v. 1, 99.
6262   1826, [Coal fires may cause "volcanoes"]: v. 1, 121.
6263   1826, Topaz [from Connecticut]: v. 1, 130.
6264   1826, Fossil animals: v. 1, 130.
6265   1826, Causes of volcanoes: v. 1, 147.
6266   1826, Diurnal variation of the compass: v. 1, 155.

The Literary Focus, a Monthly Periodical (Oxford, Ohio, 1827-1828)
6267   1828, Mount Hecla: v. 1, 214-215.

The Literary Gazette (Concord, New Hampshire, 1834-1835)
6268   1834, Sub-marine mines [Cornwall]: v. 1, 39.
6269   1834, Hot springs at Arkansas: v. 1, 96.
6270   1834, [Review of Lecture on Earthquakes by J. Winthrop (see 11002)]: v. 1, 151.

The Literary Gazette and American Athenaeum (New York, 1825-1827)
6271   1825, [Cuvier's discovery of fossil bones at Lyon, France] : v. 1, 14.
6272   1825, Falls of West Canada Creek, NY: v. 1, 46-47.
6273   1825, Natural bombs of Maryland: v. 1, 79.
        On geodes from Bladensburg, Maryland.
6274   1826, Music of the rocks: v. 1, 409.
        On resonating rocks in wind, at Oronoco River, South America.
6275   1826, Woodbridge spa spring [New Jersey] : v. 2, 142.

The Literary Gazette; or, Journal of Criticism, Science, and the Arts (Philadelphia, 1821)
6276   1821, Earthquakes: v. 1, no. 8, 128.
        Opinions on the origin of earthquakes from Percy's Anecdotes.

The Literary Inquirer; a Semi-Monthly Journal Devoted to Literature and Science (Buffalo, New York, 1833-1834)
6277   1833, Coal mines [of England] : v. 1, 22.
6278   1833, New cave of bones [in Mallet, France] : v. 1, 34.
6279   1833, Geology: v. 1, 72.
        On fossils in graywacke inclusions in granite, from the Hartz.
6280   1833, Silver mine [from Coquimbo, Chile] : v. 1, 84 (6 lines).
6281   1833, Lake of vitriol [in Java] : v. 1, 84 (13 lines).
6282   1833, Salt spring [in Upper Canada] : v. 1, 104 (4 lines).
6283   1833, Geological Society: v. 1, 112 (5 lines).
        On activities of researchers in Madagascar.
6284   1833, A new mineral [from Corsica] : v. 1, 126.
6285   1833, A loadstone mine [from Cintra, Portugal] : v. 1, 140.
6286   1834, [Notes on gold mines of southern United States] : v. 2, 46; 157; v. 3, 66.
6287   1834, Valuable copper mine [from Flemington, New Jersey] : v. 1, 142.
6288   1834, Awful earthquake in South America [at Pasto, New Granada] : v. 1, 149.
6289   1834, American mammoth [from Berlin, Connecticut] : v. 2, 182.
6290   1834, Earthquakes in South America [at Santa Martha] : v. 2, 198 (7 lines); v. 3, 18 (9 lines).
6291   1834, Geological theories: v. 3, 12 (13 lines).
6292   1834, Mountain of salt: v. 3, 35 (5 lines).
6293   1834, Geology: v. 3, 74 (12 lines).
        Notice of S. G. Clemson's activities; on the discovery of gold at York, Pennsylvania; and appointment of G. Troost as geologist of Tennessee.

The Literary Journal, and Weekly Register of Science and the Arts (Providence, Rhode Island, 1833-1834)
6294   1833, Formation of soil: v. 1, no. 1, 8.
6295   1833, Quicksilver mine in Idria [Spain] : v. 1, no. 8, 64.
6296   1833, The Pitch Lake [in Trinidad] : v. 1, no. 9, 68.
6297   1834, Hitchcock's geological report [review of 1833 Massachusetts report (see 5105)] : v. 1, no. 40, 316.

Literary Magazine and American Register (Philadelphia, 1803-1807)
6298   1804, A description of a species of coal found near Woodstock [near Hudson River, New York] : v. 1, 206-207.
6299   1804-1806, [Notes on meteorites] : v. 1, 379; v. 2, 257-258; v. 6, 412.
6300   1804, Produce of gold and silver mines: v. 2, 104-105.
6301   1804, A preservative against earthquakes and thunder: v. 2, 183.
        On the use of sacred relicts to prevent natural disasters.
6302   1804, A grotto described: v. 2, 214-215.
6303   1804, Curious facts relating to stones and other substances, said to have fallen at different periods, and at different places, from the clouds: v. 2, 223-231; 318; 385-389.
        Includes list of falls, chemical analyses, and review of theories of meteorite origins.
6304   1804, A cavern newly discovered [at Nice, France] : v. 2, 312.
6305   1804, On coal as a fuel in America: v. 2, 422-424.
        Signed by "A." Support given for using coal.
6306   1804, Natural bridge [of Virginia] : v. 2, 441-442.
6307   1804, On volcanic and Neptunian mountains: v. 2, 487.
6308   1804, On the manufacture and quality of Cheshire salt: v. 2, 504-513.
6309   1804, Origin of pebbles: v. 2, 560-561.
6310   1804, Account of the Philadelphia Museum: v. 2, 576-579.
6311   1804, Cavern near Nice: v. 2, 669.
6312   1805, Volcanoes: v. 3, 290-291.
6313   1805, Situation of coal: v. 3, 405-406.
        Signed by "Q."
6314   1805, Why are diamonds valuable?: v. 3, 423-424.
6315   1805, Petrifaction: v. 4, 212-214.
6316   1805, New metals: v. 4, 252-253.
        On the noble metals.
6317   1806, An account of the earthquake in the Kingdom of Naples, on the 26th of July, 1805, and of the eruption of Mount Vesuvius, on the 12th of August: v. 5, 138-140.
6318   1806, Meteoric stones: v. 6, 412.
        On the volcanic origin of meteorites.
6319   1807, Volcanic islands [Tristan de Chuna] : v. 7, 202-203.
6320   1807, [Antimony found at Sagherties, New York] : v. 7, 465 (11 lines).
6321   1807, [New cave near Madison Cave in Virginia] : v. 7, 465.
6322   1807, [Fossil bones near Paris] : v. 8, 100; 153 (7 lines).
6323   1807, Alum works [at Hurlett, Scotland] : v. 8, 254.
6324   1807, [Marble quarry of Sing-Sing, New York] : v. 8, 325 (8 lines).
6325   1807, [Earthquake at Nieuwied] : v. 8, 326.

The Literary Miscellany (Cambridge, Massachusetts, 1805-1806)
6326   1805, [Review of Tableau du climat, by Volney (see 10653)] : v. 1, 167-178; 200; 273-281.
6327   1805, Natural history, with an account of the professorship for this science recently founded in the University in Cambridge: v. 1, 395-400.

The Literary Miscellany, or Monthly Review (New York, 1811)
6328   1811, Description of a romantic grotto in St. Anne's Parish, Jamaica: v. 1, 14-19.
6329   1811, A petrifying lake [in Iceland]: v. 1, 20.

Literary Port Folio (Philadelphia, 1830)
6330   1830, Fishes in the London clay: v. 1, no. 2, 16.
6331   1830, Mine-hunting mania: v. 1, no. 5, 40.
       On the search for coal in Sunbury, Northumberland County, Pennsylvania.
6332   1830, History and geology mutually confirmatory: v. 1, no. 19, 152.
6333   1830, [Humboldt's Russian excursion]: v. 1, no. 20, 159-160.

The Literary Record and Journal of the Linnaean Association of Pennsylvania College
   (Gettysburg, Pennsylvania, 1844-1848)
6334   1846, Geology: v. 2, 116-120; 123-126, fig.; 147-150; 169-172.
6335   1846, Fossil forest near Cairo, in Egypt: v. 2, 192.
6336   1846, Dinotherium [found near Orthes, Eppleshiem]: v. 2, 222-225.
6337   1847, Visit to the Schuylkill coal region: v. 4, 1-6.
6338   1847, The great American coal field: v. 4, 40-44.
6339   1847, The iron man - petrifaction: v. 4, 70-72.
6340   1847, [Review of Manual of mineralogy by J. D. Dana (see 2832)]: v. 4, 240 (8 lines).

The Literary Register, a Weekly Paper (Oxford, Ohio, 1828-1829)
6341   1828, Subterranean walls: v. 1, 205.
       Discussion as to whether dikes are natural or man-made.
6342   1828, Singular discovery: v. 1, 330-331.
       Organic material found after 35 feet of boring.
6343   1829, Ohio oil stone: v. 2, 74.
6344   1829, Gold mines [of North Carolina]: v. 2, 142.

The Literary Union (Syracuse, New York, 1849-1850)
6345   1849, Geological changes of our own time: v. 1, 70.
6346   1849, A brown stone [loadstone deposit]: v. 1, 87.
6347   1849, Geological fact: v. 1, 103.
       Agassiz notes no fossil flowers.
6348   1849, Curiosities of the earth: v. 1, 103.
       On plant remains in the strata at Modina, Italy.
6349   1849, Salt springs of New York: v. 1, 310-311; 323-325.
6350   1850, [Notice and review of Cosmos, by A. von Humboldt (see 5395)]: v. 1 (ns), 227 (5 lines); 245-254.
6351   1850, Review of A treatise on physical geography, by A. Barrington (see 1504)]: v. 2 (ns), 81-82.

The Literary World (New York, 1847-1850)
6352   1847, [Copper mines in Australia]: v. 1, 421 (4 lines).
       Abstracted from London Ath.
6353   1848, Extraordinary features in the geology of Kerguelin Island: v. 2, 40.
6354   1848, News for the geologists: v. 2, 65.
       On fossil bones found in Russia.
6355   1848, [Review of Elements of geology by J. L. Comstock (see 2518)]: v. 2, 139.
6356   1848, Authorship of the Vestiges of Creation [notice of 2292 et seq.]: v. 2, 566.
6357   1848, Boiling spring [at Broseley, Shropshire, England]: v. 3, 50.
6358   1848, Survey of the copper mines: v. 3, 432.
       Notice of survey by C. T. Jackson and D. D. Owen.
6359   1848, [Notice of Elements of geology third American edition, by D. Page (see 8108)]: v. 3, 898.
6360   1848, [Review of Gold regions of California edited by G. G. Foster (see 3919)]: v. 3, 947.
6361   1849, Siftings from the gold region: v. 4, 61-62.
6362   1849, [Review of Gold seeker's manual by D. T. Ansted (see 1283); and Miner's guide and metallurgists' directory by J. W. Orton (see 8041)]: v. 4, 267-268.
6363   1849, [Review of History of the precious metals by J. L. Comstock (see 2520)]: v. 4, 337.
6364   1849, [Review of A second visit to the United States by C. Lyell (see 6571)]: v. 5, 85-86.
6365   1850, Geology of Australia: v. 6, 129.
       Abstracted from N. Am. Rev.
6366   1850, [Review with excerpts of Geology of the United States exploring expedition, by J. D. Dana (see 2841 to 2844)]: v. 6, 29-32; 55-57; 148-149.
6367   1850, [Review with excerpts of Lake Superior by L. Agassiz (see 83)]: v. 6, 370-371; 390-391.
6368   1850, [Review of California as I saw it by W. M'Cullem (see 6590); California sketches by L. Kip (see 5970); Notes on California and the placers (see 7985); California: the wonder of the age by T. B. King (see 5967); and Sights in the gold region by T. F. Johnson (see 5760 and 5761)]: v. 6, 444.
6369   1850, [Review with excerpts of Six months in the gold mines by E. G. Buffum (see 2180)]: v. 6, 609-610.
6370   1850, The great Indian diamond [Koh-i-noor diamond]: v. 7, 96-97.

Little, Daniel
6371   1785, Observations upon the art of making steel: Am. Acad. Arts Sci's Mem., v. 1, pt. 2, 525-528.

Little, George
6372   1850, Preventing explosions in coal mines: Franklin Inst. J., v. 19 (3s), 8-9, fig.

Little, Norman (see 7160 under Michigan, State of)

Little Schuykill Navigation, Rail Road and Coal Company
6373   1830, The act, and supplement therto, authorising the incorporation of the ... : Reading, Pa., Douglass W. Hyde, 33 p.
       Another edition: Phila., J. & W. Kite, 48 p. (1835).

Liverpool Natural History Society
6374   1839, Footsteps and impressions of the Chirotherium, and of various animals, in sandstone: Am. J. Sci., v. 36, no. 2, 394-398.

The Living Age (Boston, 1844-1850)
6375   1844, [Review with excerpts of Slave states ... , by G. W. Featherstonhaugh (see 3784)]: v. 1, 674-676.
        From London Ath.
6376   1844-1850, [Notes on foreign mining]: v. 2, 61; 367; 439-440; v. 3, 75; v. 7, 229; v. 19, 415-416; v. 20, 606; v. 26, 345-346.
6377   1844, [Reviews of Association of American Geologists and Naturalists Transactions (see 1326) and System of mineralogy by J. D. Dana (see 2812)]: v. 2, 549-550.
6378   1844, Cave in Ireland: v. 2, 618.
        Abstracted from London Ath.
6379   1844, The colouring matter of flint, carnelian, and amethyst: v. 3, 75 (9 lines).
6380   1844, Magnetic observations: v. 3, 468 (12 lines).
        On plans for measurements in the Southern hemisphere.
6381   1845, [Review with excerpts of Vestiges of the natural history of Creation by R. Chambers (see 2292 et seq.)]: v. 4, 60-64.
        From Examiner.
6382   1845, [Gigantic fossil bones found in Benton County, Missouri]: v. 4, 204.
6383   1845, [Review of Geology by D. Ansted]: v. 4, 250-251.
        Abstracted from Spectator.
6384   1845, [Review with excerpts of Travels ..., by C. Lyell (see 6557)]: v. 6, 340-343
        Abstracted from Spectator.
6385   1845-1848, [Notes on Lake Superior copper mining]: v. 6, 488; v. 7, 136; v. 10, 536; v. 18, 375.
6386   1845, [Review with excerpts of Vestiges of the natural history of Creation by R. Chambers (see 2292 et seq.)]: v. 6, 564-582.
        From British N. Am. Rev.
6387   1845, Artificial quartz: v. 7, 152.
        On Tibelmen's experiments.
6388   1845, Kenawha gas: v. 7, 258.
        Abstracted from Am. J. Sci.
6389   1845, On the known thickness of the crust of the earth: v. 7, 258-259.
        Abstracted from Edinburgh Philos. J.
6390   1845, [Review of Memoirs of William Smith by J. Phillips]: v. 7, 505-520.
6391   1845, [On the authorship of Vestiges of the natural history of Creation by R. Chambers (see 2292 et seq.)]: v. 7, 551 (8 lines).
        Statement that it is not by Sir Richard Vyvyan.
6392   1846, Earthquake at Caracas: v. 8, 162-163.
6393   1846, [Review with excerpts of Travels ..., by C. Lyell (see 6557)]: v. 8, 605-613.
6394   1846, [Review of "Description of a skeleton of an extinct gigantic sloth" by R. Owen]: v. 11, 109-118.
6395   1847, The geologist's wife: v. 12, 231.
        A humorous poem.
6396   1847, [Review with excerpts of Cosmos by A. von Humboldt (see 5395)]: v. 12, 327-330.
6397   1847, Assaying metals: v. 14, 153.
6398   1847, Terrestrial magnetism: v. 15, 225-228.
        From Chamber's Mag.
6399   1848, [Review with excerpts of Cosmos by A. von Humboldt (see 5395)]: v. 16, 385-411; v. 17, 559-560.
        From Edinburgh Rev. and the Examiner.
6400   1848, Peak of Popocatepetl: v. 18, 49-51.
        From J. of Commerce.
6401   1848, [Review with excerpts of Ancient sea margins by R. Chambers (see 2299)]: v. 18, 147-149.
        From the Spectator.
6402   1849, [Notes on California gold mining]: v. 20, 305-308; v. 22, 44-45.
6403   1849, Gossip on minerals [review with excerpts of Minerals and their uses by J. R. Jackson (see 5684)]: v. 22, 155-157.
        From Chamber's J.
6404   1849, [Review with excerpts of A second visit ..., by C. Lyell (see 6571)]: v. 22, 170-175; 337-356.
        From Spectator, and Quarterly Review.
6405   1850, [Reviews with excerpts of Footprints of the Creator by H. Miller (see 7186); and Old Red Sandstone by H. Miller (see 7187); and reviews of Ancient sea margins by R. Chambers (see 2299); La science et la foi sur l'oeuvre de la Creation by H. B. Waterkeyn; La deluge by F. Klee; and Passages in the history of geology by A. C. Ramsay (both 1848 and 1849 editions)]: v. 25, 145-162.
        From N. British Rev.
6406   1850, The terrors of Vesuvius: v. 25, 466-467.
        Abstracted from Tribune.

Livingston, Chancellor Robert R. (1746-1813)
6407   1794, Experiments and observations on calcareous and gypsious earths: NY Soc. Prom. Use. Arts Trans., v. 1, pt. 1, 25-54.
        Also in v. 1 (2nd edition), 34-56 (1801).
6408   1801, Thoughts on lime and gypsum: NY Soc. Prom. Use. Arts Trans., v. 1 (second edition), pt. 4, 330-335.
6409   1810, Analysis of the Ballston water: Am. Med. Philos. Reg., v. 1, 40-42.

Lloyd, Humphrey (1800-1881)
6410   1838, Magnetical observatory at Dublin: Am. J. Sci., v. 34, no. 1, 3-8; 8.
        In British Assoc. Proc.
6411   1839, [Improvement in instruments for magnetic observation]: Am. Philos. Soc. Proc., v. 1, no. 6, 77-78.
        Reported by A. D. Bache.
6412   1839, Recalculation of the observations of the magnetic dip and intensity in Ireland, with additional elements: Am. J. Sci., v. 35, no. 2, 296-297.
        In British Assoc. Proc.

Lloyd, H. (cont.)
6413 1841, Corresponding magnetic observation, by Prof. A. D. Bache of Philadelphia, and Prof. Lloyd of Dublin: Am. J. Sci., v. 41, no. 1, 210-213.
6414 1842, History of the present magnetic crusade: Cambridge Miscellany, v. 1, 130-136.
6415 1843, Diurnal changes of the magnetic elements: Franklin Inst. J., v. 6 (3s), 423-424.
6416 1844, On the regular variations of the direction and intensity of the earth's magnetic force: Am. J. Sci., v. 46, no. 2, 391-392.
In British Assoc. Proc.

Lloyd, J. A.
6417 1849, Notes respecting the Isthmus of Panama: US House Report 145, 30-2, 455-472.
Not seen.

Lloyd, John W.
6418 1847, Trap rock as manure: Mon. J. Ag., v. 2, no. 7, 303-304.

Locke, John (1796-1856)
6419 1821, Some account of the copperas mines and manufactory in Strafford, Vt.: Am. J. Sci., v. 3, no. 2, 326-330.
Abstract in Am. J. Pharmacy, v. 3, 55-59 (1831).
6420 1836, Analysis of the limestone of Cincinnati and Dayton: in Report of the committee on a geological survey of the state (of Ohio) (see 8003), 77-78.
Also in separate report by Ohio Gen. Assembly, 16-17.
6421 1836, Geology no. III: Cin. Mirror, v. 5, 181.
On erosion and alluvial deposits.
6422 1837, Dr. Buckland, the geologist: New-Yorker, v. 3, 339-340.
6423 1838, [Magnetic observation in Ohio]: Am. Philos. Soc. Proc., v. 1, no. 3, 24-25.
6424 1838 [i.e. 1839], Geological map of Adams County, Ohio. With a section of the same applicable to every part of it: in second Ohio report by W. W. Mather (6896), geological map, 33 x 40 cm., facing p. 238.
6425 1838 [i.e. 1839], Prof. Locke's geological report, communicated by the governor to the general assembly of Ohio. December, 1838: in second Ohio report by W. W. Mather (see 6896), 201-274, 12 plates, map.
6426 1839, An introductory lecture on chemistry and geology, delivered November 6th, 1838, before the class of the Medical College of Ohio: Cin., 18, [1] p.
"Published at the request of the class."
6427 1839, On the magnetic dip at several places in the state of Ohio, and on the relative horizontal magnetic intensities of Cincinnati and London: Am. Philos. Soc. Trans., v. 6 (ns), 267-273.
6428 1840, [Magnetic observations at Louisville, Kentucky and Cincinnati, Ohio]: Am. Philos. Soc. Proc., v. 1, no. 13, 271.
6429 1840, On terrestrial magnetism: Am. J. Sci., v. 39, no. 2, 319-328.
6430 1840, Report of John Locke, M. D.: in Report of a geological exploration of part of Iowa, Wisconsin, and Illinois by D. D. Owen (see 8050 and 8070), 116-159.
See also 6454. On magnetic observations.
6431 1841, [Glacial planing and boulders in Ohio]: Am. J. Sci. v. 41, no. 1, 175-176 (17 + 4 lines).
Also in Assoc. Am. Geol's Nat's Proc., v. 2, 20-21. Also in Assoc. Am. Geol's Nat's Rept., v. 1, 28, 1843.
6432 1841, Observations to determine the horizontal magnetic intensity and dip at Louisville, Kentucky, and at Cincinnati, Ohio: Am. Philos. Soc. Trans., v. 7 (ns), 261-264.
6433 1841, On a new species of trilobite found at Cincinnati, Ohio [Isotelus maximus]: Am. J. Sci., v. 41, no. 1, 161 (9 lines).
Also in Assoc. Am. Geol's Nat's Proc., v. 2, 6 (9 lines). Also in Assoc. Am. Geol's Nat's Rept., v. 1, 14, (11 lines), 1843.
6434 1841, On the geology of some parts of the United States west of the Alleghany Mountains: Am. J. Sci., v. 41, no. 1, 160-161.
Also in Assoc. Am. Geol's Nat's Proc., v. 2, 5-6. Also in Assoc. Am. Geol's Nat's Rept., v. 1, 13-14, 1843.
6435 1841, On the magnetical dip in the United States: Am. J. Sci., v. 41, no. 1, 15-21.
6436 1841, On terrestrial magnetism: Am. J. Sci., v. 40, no. 1, 149-156.
6437 1841, [Remarks on the connection between magnetism and geology]: Am. J. Sci., v. 41, no. 1, 171 (8 lines).
Also in Assoc. Am. Geol's Nat's Proc., v. 2, 16 (8 lines). Also in Assoc. Am. Geol's Nat's Rept., v. 1, 23-24 (8 lines), 1843.
6438 1842, Alabaster in the Mammoth Cave of Kentucky: Am. J. Sci., v. 42, no. 1, 206-207, fig.
6439 1842, Cryptolithus tesselatus: Phila. Acad. Nat. Sci's Proc., v. 1, no. 16, 186 (8 lines); 196-197, fig.
Complete trilobite specimen described.
6440 1842, On a new species of trilobite of very large size: Am. J. Sci., v. 42, no. 2, 336-368, folding plate.
On Isotelus megistos from Cincinnati, Ohio.
6441 1842, On the manipulation of the dipping compass: Am. J. Sci., v. 42, no. 2, 235-238.
6442 1842, Sections of rocks of the lead regions of the Upper Mississippi: Am. J. Sci., v. 43, no. 1, 147-149.
Also in Assoc. Am. Geol's Nat's Proc., v. 3, 2-4. Also in Assoc. Am. Geol's Nat's Rept., v. 1, 43-45, 1843.
6443 1843, Fossil Cryptolithus tesselatus: Phila. Acad. Nat. Sci's Proc., v. 1, no. 22, 236, fig.
6444 1843, A new reflecting level and goniometer: Assoc. Am. Geol's Nat's Rept., v. 1, 238-239, 2 figs.
6445 1843, Notice of a new trilobite, Ceraurus crosotus: Am. J. Sci., v. 44, no. 2, 346, fig.
6446 1843, Notice of a prostrate forest under the diluvium of Ohio: Assoc. Am. Geol's Nat's Rept., v. 1, 240-241.

Locke, J. (cont.)
6447 1843, On a new species of trilobite of a very large size: Assoc. Am. Geol's Nat's Rept., v. 1, 221-224, plate.
   On Isotelus from Cincinnati, Ohio.
6448 1843, Supplementary notice of the Ceraurus crosotus [from Cincinnati, Ohio]: Am. J. Sci., v. 45, no. 1, 222-224, 3 figs.
6449 1844, [Additional magnetic observations in the United States]: Am. Philos. Soc. Proc., v. 4, no. 31, 109.
6450 1844, Connection between geology and magnetism: Am. J. Sci., v. 47, no. 1, 101-103.
   Also in Assoc. Am. Geol's Nat's Proc., v. 5, 8-10.
6451 1839 [i.e. 1844], Magnetical chart exhibiting the observations made by J. Locke during the survey of the mines of the United States by D. D. Owen: in Report of a geological exploration ... by D. D. Owen (see 8065), magnetic map, 27 x 27 cm., plate 21.
6452 1844, [Magnetic observations in the United States]: Am. Philos. Soc., v. 4, no. 30, 63-65.
6453 1844, [Nature of the country rock in the Missouri lead region]: Am. J. Sci., v. 47, no. 1, 106 (5 lines).
   Also in Assoc. Am. Geol's Nat's Proc., v. 5, 13 (5 lines).
6454 1844, Report of John Locke, M. D.: in Report of a geological exploration by D. D. Owen (see 8065), 147-189, 7 plates, map.
   See also 6430. On magnetic observations in the Midwest.
6455 1845, Geological: Niles' Wk. Reg., v. 69, 138.
   On calcareous conglomerate from Ohio.
6456 1845, [List of fossils from Cincinnati, Ohio]: Nat'l Inst. Prom Sci. Bulletin, v. 3, 237.
6457 1846, [Description of an Asterias from the Blue Limestone of Cincinnati]: Phila. Acad. Nat Sci's Proc., v. 3, no. 2, 32-34, fig.
6458 1846, Geological and magnetical chart of Copper Harbor [at Lake Superior, Michigan]: Am. Philos. Soc. Trans., v. 9 (ns), magnetic map, plate 4.
   In 6460.
6459 1846, Magnetic chart of the United States: Am. Philos. Soc. Trans., v. 9 (ns), map, 37.8 x 12.1.
   In 6460.
6460 1846, Observations made in the years 1838, '39, '40, '41, '42, and '43 to determine the magnetical dip and intensity of magnetical force, in several parts of the United States: Am. Philos. Soc. Trans., v. 9 (ns), 283-328, 3 figs., plate, map, magnetical chart.
   Also reprint edition. Contains map 6458 and chart 6459.
6461 1847, [Geological observations in the Upper Peninsula of Michigan]: in Geological survey of the mineral lands ... of Michigan by C. T. Jackson (see 5640), 183-199, fig.
6462 1847, The pictured rocks [of Lake Superior]: Dwight's Am. Mag., v. 3, 797-798.
6463 1849, Catalogue of specimens forwarded to Dr. Jackson [and report of field work in the Lake Superior region]: in Michigan survey by C. T. Jackson (see 5655), 563-572.
6464 1850, Fossil remains in the West [Cincinnati and Xenia, Ohio]: Fam. Vis., v. 1, no. 26, 201.

Lockwood, M. B.
6465 1843, Variation of the magnetic needle at Providence, R. I., from A. D. 1717 to 1843: Am. J. Sci., v. 44, no. 2, 314.

Logan, George (1753-1821)
6466 1797, Agricultural experiments on gypsum, or plaister of Paris; with some observations on the fertilizing quality and natural history of that fossil: Phila., Baileys, 18 p.
   For review see 7008.
6467 1807, [Gypsum from America and Europe]: Am. Reg. or Gen. Repos., v. 1, pt. 2, 38.

Lohmeyer, M.
6468 1846, Mica: Am. J. Sci., v. 2 (ns), no. 3, 417 (9 lines).
   On biotite from Vesuvius, and chemical analysis of a lithium mica by Zinnwald.

London Geological Society (see Geological Society of London)

London Microscopical Society (see Microscopical Society of London)

London Mining Journal (for extensive excerpts from this journal, see The Mining Journal and American Rail-road Gazette)

London Paleontological Society (see Paleontological Society of London)

Long
6469 1839, A description of a bone cavern in the Mendip Hills: Am. J. Sci., v. 35, no. 2, 304-305.
   In British Assoc. Proc.

Long, George W.
6470 1830, On the origin of springs and fountains: Am. J. Sci., v. 17, no. 2, 336-338.

Long, Silas
6471 1843, Analysis of prairie soil: Prairie Farm., v. 3, 35-36.
6472 1844, Lime and lime-stone: Prairie Farm., v. 4, 195.

Long, Stephen Harriman (1784-1864)
6473 1818, A description of the Hot Spring, near the River Washitaw, and of the physical geography of the adjacent country: Am. Mon. Mag. and Crit. Rev., v. 3, no. 2, 85-87, map.

Long, S. H. (cont.)
6474   1819, and R. Graham and J. Philips, Topographical reports, made with a view to ascertain the practicability of uniting the waters of Illinois River with those of Lake Michigan: US HED 17, 16-1, v. 2 (32), 5-10.
6475   1823, [Geological notes]: in Account of an expedition ..., by E. James (see 5694).
6476   1824, [Geological notes]: in Narrative of an expedition ..., by W. H. Keating (see 5933).
6477   1827, Surveys of proposed routes of a national road from the city of Washington to Buffalo in the state of New York: US HED 105, 19-2, v. 5 (152), 7-39, 3 tables.
6478   1832, Description of a natural tunnel, in Scott County, Virginia: Mon. Am. J. Geology, v. 1, no. 8, 347-352, plate.
   With remarks by G. W. Featherstonhaugh (see 3770).
6479   1832, Fossil remains of a mastodon found in Tennessee [at Walker's Mountain]: Mon. Am. J. Geology, v. 1, no. 12, 565-566.
6480   1841, Report on the improvement of Red River: US SED 64, 27-1, v. 1 (390), 22 p., map.

Lonsdale, William (1794-1849)
6481   1847, Remarks on the characters of several species of Tertiary corals from the United States, in reply to Mr. Dana: Am. J. Sci., v. 4 (ns), no. 3, 357-359.
   See also 2827.

Loomis, Elias (1811-1899)
6482   1836, Observations on the variation of the magnetic needle, made at Yale College in 1834 and 1835: Am. J. Sci., v. 30, no. 2, 221-233.
6483   1838, On the variation and dip of the magnetic needle in different parts of the United States: Am. J. Sci., v. 34, no. 2, 290-309, chart (see 6484).
6484   1838, Magnetic chart of the United States: Am. J. Sci., v. 34, no. 2, chart, 19.5 x 19.2 cm.
   In 6483.
6485   1839, Observations to determine the magnetic dip at various places in Ohio and Michigan: Am. Philos. Soc. Proc., v. 1, no. 8, 116-117; no. 9, 144-145.
   Abstract in Am. J. Sci., v. 40, no. 2, 378-380 (1841).
6486   1840, On the variation and dip of the magnetic needle in the United States: Am. J. Sci., v. 39, no. 1, 41-50, 2 charts.
6487   1841, Additional observations of the magnetic dip in the United States: Am. Philos. Soc. Trans., v. 7 (ns), 101-111.
6488   1841, Observations to determine magnetic dip at various places in Ohio and Michigan: Am. Philos. Soc. Trans., v. 7 (ns), 1-6.
6489   1841, On the magnetic dip in the United States: Am. J. Sci., v. 40, no. 1, 85-92.
6490   1842, On the dip and variation of the magnetic needle in the United States: Am. J. Sci., v. 43, no. 1, 93-116.
6491   1843, Observations to determine the magnetic intensity at several places in the United States, with some additional observations of the magnetic dip: Am. Philos. Soc. Trans., v. 8 (ns), 61-72.
6492   1843, Observations of the magnetic dip in the United States. Fourth series: Am. Philos. Soc. Trans., v. 8 (ns), 285-304.
6493   1844, Comparison of Gauss's theory of terrestrial magnetism with observation: Am. J. Sci., v. 47, no. 2, 278-281.
6494   1847, Notice of some recent additions to our knowledge of the magnetism of the United States and its vicinity: Am. J. Sci., v. 4 (ns), no. 2, 192-198.
6495   1850, Experiments on the electricity of a plate buried in the earth: Am. J. Sci., v. 9 (ns), no. 1, 1-11.

Loomis, I. Newton
6496   1846, An account of the geology of the Harpeth Ridge, Davidson Co., Tenn.: Am. J. Sci., v. 1 (ns), no. 2, 222-224.
6497   1846, Fossils of Tennessee: Natur. and J., v. 1, 57-60, fig.
6498   1846, Geologic encampment: Natur. and J., v. 1, 337-343, 385-387.
   On a field trip to Mammoth Cave, Kentucky.
6499   1846, Geology: Natur. and J., v. 1, 5-11; 145-148; 198-200.

Lord, Eleazar (1788-1871)
6500   1843, Geological cosmogony; or an examination of the geological theory of the origin and antiquity of the earth, and of the causes and object of the changes it has undergone: NY, R. Carter, 167 p.
   Signed "by a Layman." For review with excerpts see 2356. For review see 5432 and 7148.

Lorimer, John (1732-1795)
6501   1771, Extracts of a letter from Dr. Lorimer of West-Florida, to Hugh Williamson, M. D.: Am. Philos. Soc. Trans., v. 1, 250-255.
   Also in Am. Philos. Soc. Trans., v. 1 (2nd edition), 320-325. On the need for United States magnetic observatories.

Loughborough, John
6502   1850, Western geography, comprising all the territory west of the Mississippi, north of New Mexico and east of the main ridge of the Rocky Mountains: West. J. Ag., v. 4, no. 6, 356-379.

Louyet, M.
6503   1849, Process of extracting nickel and cobalt followed in a manufactory at Birmingham: Am. J. Sci., v. 8 (ns), no. 1, 112-113.
   Abstracted from Chemical Gaz.

Lovering, Joseph (1813-1892)
6504   1841, and W. C. Bond, Account of magnetic observations made at the observatory of Harvard University, Cambridge: Am. Philos. Soc. Proc., v. 2, no. 19, 101-102.
6505   1841, Terrestrial magnetism: US Mag. and Dem. Rev., v. 9, 251-260.
   Abstracted from Am. Acad. Arts and Sci's Mem.

Lovering, J. (cont.)
6506   1846, and W. C. Bond, An account of the magnetic observations made at the observatory of Harvard University, Cambridge: Am. Acad. Arts Sci's Mem., v. 2 (ns), 1-84.
       Also reprint edition, 84 p. For review see 7949.
6507   1846, An account of the magnetic observations made at the observatory of Harvard University: Am. Acad. Arts Sci's Mem., v. 2 (ns), 85-160.

Low
6508   1840, The properties of soils, as determined by chemical analysis: Farm's Mon. Vis., v. 2, 118.

Lowe, L. G.
6509   1849, Analysis of soils: NE Farm., v. 1, no. 25, 389.
6510   1849, Soils: Mich. Farm., v. 7, 356.

The Lowell Offering (Lowell, Massachusetts, 1840-1844)
6511   1840, The pleasures of science: v. 1, no. 1, 7-8.
       Signed by "Ella."
6512   1842, Chapters on the sciences. Geology and mineralogy: v. 2, 264-273, 300-306.
       Signed by "D."

Lowrey, Thomas
6513   1828, Water cement of Southington, Conn.: Am. J. Sci., v. 13, no. 2, 382-383.

Lowry, Delvalle
6514   1822, Conversations on mineralogy: Phila., Uriah Hunt, 332 p., 11 plates (1 colored).
       "First American from the last London ed." Contains letter of endorsement by J. R. Coxe (see 2677). For excerpts see The Minerva, v. 2, 6 (1823).

Lowthorp, T.
6515   1831, Emery cloth: Franklin Inst. J., v. 8 (ns), 66.

Lozano, Pedro (1697-1752)
6516   1749, A true and particular relation of the dreadful earthquake, which happen'd at Lima, the capital of Peru, and the neighbouring part of Callao, on the 28th of October, 1746 ... : Phila., B. Franklin and D. Hall, 52 p.
       Evans number 6348. Also second edition, Bos., D. Fowle, 8 p. (1750), Evans number 6531; third edition, Bos., D. and Z. Fowle, 8 p. (1755), Evans number 7453.

Ludwig, M. (see 10658 with M. Wackenroder)

Lukens, I.
6517   1827, On improving the colours of agates, and on an improved manner of using Florence oil flasks: Franklin Inst. J., v. 4, 332-333.

Lukis, Frederick Corbin
6518   1837, and Walker, Remarks on herm granite, ... in reply to enquiries from the President; with some experiments ... on the wear of different granites: Mech's Mag. and Reg. Inventions, v. 9, 276-277.
       Abstracted from "Trans. Inst. Civil Eng." With an article on fracture of stone by W. Freeman (see 4109).

Lund, Peter Wilhelm (1801-1880)
6519   1841, Brazil [geological notes]: Am. Eclec., v. 1, no. 2, 391-392.
       Abstracted from London Ath.
6520   1843, On the occurrence of fossil human bones of the prehistorical world: Am. J. Sci., v. 44, no. 2, 277-280.
       Translated by E. E. Salisbury from a letter.
6521   1846, Geology of Brazil: Am. J. Sci., v. 1 (ns), no. 2, 308 (5 lines).
       Abstracted from l'Institut.

Lyceum of Natural History of New York (see New York Lyceum of Natural History)

Lycoming Navigation, Rail Road and Coal Company
6522   1828, A brief description of the property belonging to the Lycoming Coal Company, with some general remarks on the subject of the coal and iron business: Poughkeepsie, NY, P. Potter, 32 p.

Lycoming Valley Iron Company
6523   1838, Articles of association and agreement of the Lycoming Valley Iron Company: NY, Coolidge and Lambert, 33 p.

Lyell, Charles (1797-1875)
6524   1830, [Excerpts from Principles of Geology]: Cab. Nat. History, v. 1, 9-12, 178-179, 224-226.
6525   1835, Gradual elevation of parts of Sweden, &c.: Am. J. Sci., v. 28, no. 1, 72-73.
       In British Assoc. Proc.
6526   1835, Wollaston medal [awarded to G. Mantell]: Am. J. Sci., v. 28, no. 2, 391-393.
       Abstracted from Jameson's Edinburgh New Philos. J.
6527   1836, Corals and Entomostraca in chalk: Am. J. Sci., v. 30, no. 2, 382 (10 lines).
       Abstracted from "Address before the Geological Society of London, 1836."
6528   1836, Glossary of geological and other scientific terms: in Report on the geological survey of the state of New Jersey (second edition), by H. D. Rogers (see 8851), 175-188.
       Also in Fifth Pennsylvania Report (see 8868 and 8869).
6529   1836, Gradual rising of parts of Sweden, and of other countries: Am. J. Sci., v. 29, no. 2, 363-364.

Lyell, C. (cont.)

6530 1837, Address delivered at the anniversary meeting of the Geological Society of London, on the 17th of February, 1837: Am. J. Sci., v. 33, no. 1, 76-117.

6531 1837, Phenomena of petrefaction: Franklin Inst. J., v. 20 (ns), 141-143.

6532 1837, Principles of geology, being an inquiry how far the former changes of the earth's surface are referable to causes now in operation. In two volumes. First American edition from the fifth and last London edition: Phila., James Kay, jr. and brother; Pittsburgh, John I. Kay and Co., 2 v. (546, 553 p.), 226 figs., 15 plates (several colored).
For review see 611, 641, 686, 4018, 7559 (English ed.), 7564, 7725, 10851. For excerpts of British first edition see 6524. For notice see 678 and 4328. For excerpts of American editions see Naval Mag., v. 2, 185-186 (1837); Natur. and J., v. 1, no. 12, 533-534.

6533 1839, Elements of geology: Phila., J. Kay, jr. and brothers; Pittsburgh, C. H. Kay and Co., xi, 12-316 p., colored front., illus.
First American from the first London edition. Also 1845 printing as "second American edition." Also 1846 and 1849 printings by Kay and Troutman of Philadelphia and C. H. Kay of Pittsburgh as "second American edition."
For review with excerpts see 9950. For review see 716 and 761.

6534 184?, Principles of geology: NHav., Hezekiah Howe and Co.
Not seen.

6535 1840, Tubular cavities filled with gravel and sand in the chalk: Am. J. Sci., v. 38, no. 1, 122-123.
In British Assoc. Proc.

6536 1841, Elements of geology: Bos., Hilliard, Gray and Co., 2 v., 7 plates, map.
Second American from the second London edition. See also 6533.

6537 1841, Relative ages of the crag of Norfolk and Suffolk (Eng.): Am. Eclec., v. 1, no. 1, 182.
Abstracted from Mag. Nat. History.

6538 1842, [Cause of dip of New Red Sandstone in the eastern United States]: Am. J. Sci., v. 43, no. 1, 170-171 (15 lines).
Also in Assoc. Am. Geol's Nat's Proc., v. 3, 25 (3 lines); 25-26 (15 lines). Also in Assoc. Am. Geol's Nat's Rept., v. 1, 63 (4 lines); 64 (15 lines), 1843.

6539 1842, [Distribution of glacial evidence in Europe]: Am. J. Sci., v. 43, no. 1, 151 (3 lines); 152 (4 lines); 153 (4 lines).
Also in Assoc. Am. Geol's Nat's Proc., v. 3, 6 (3 lines); 7 (4 lines); 8 (4 lines). Also in Assoc. Am. Geol's Nat's Rept., v. 1, 46 (3 lines), 47 (4 lines), 48 (4 lines), 1843.

6540 1842, Eight lectures on geology, delivered at the Broadway Tabernacle in the city of New York ... reported for the New York Tribune: NY, Greeley and McElraeth, 56 p., illus.

6541 1842, Mr. Lyell's fifth lecture on geology - origin of coal: Am. Rr. J., v. 14, 284-288, 305-311.

6542 1842, On the fossil footprints of birds and impressions of raindrops in the valley of the Connecticut: Am. J. Sci., v. 45, no. 2, 394-397.

6543 1842, Principles of geology: or, the modern changes of the earth and its inhabitants, considered as illustration of geology. In three volumes reprinted from the sixth English edition from the original plates and wood cuts, under the direction of the author: Bos., Hilliard, Gray and Co. (Isaac Butts, printer), 3 v. (v. 1, xxiv, 442 p., 16 figs., 4 plates; v. 2, xii, 479 p., figs. 17-73, plates 5-11; v. 3, xii, 476 p., figs. 74-101).
For review with excerpts see 9950. For review see 761 and 899.

6544 1843, Eight lectures on geology, delivered at the Broadway Tabernacle in the city of New York ... reported for the New York Tribune: NY, Greeley and McElraeth, viii, [9]-55 p., front., illus.
"Second edition, revised and corrected." Also another printing as "third edition" in 1843.

6545 1843, On the coal formation of Nova Scotia, and on the age and relative position of the gypsum and accompanying marine limestones: Am. J. Sci., v. 45, no. 2, 356-359.

6546 1843, On the ridges, elevated beaches, inland cliffs and boulder formations of the Canadian lakes and valley of St. Lawrence: Am. J. Sci., v. 46, no. 2, 314-317.

6547 1843, On the Tertiary strata of the Island of Martha's Vineyard in Massachusetts: Am. J. Sci., v. 46, no. 2, 318-320.

6548 1843, On the upright fossil trees found at different levels in the coal strata of Cumberland, Nova Scotia: Am. J. Sci., v. 45, no. 2, 353-356.

6549 1843, Recession of the falls from Prof. Lyell's lectures on geology: in Every man his own guide to the Falls of Niagara, by T. G. Hulett (see 5366).

6550 1844, Mineral riches of the U. States: Niles' Wk. Reg., v. 67, 196.

6551 1844, Notes on the Cretaceous strata of New Jersey and parts of the United States bordering the Atlantic: Am. J. Sci., v. 47, no. 1, 213-214.

6552 1844, On the geological position of the Mastodon giganteum and associated fossil remains at Bigbone Lick, Kentucky, and other localities in the United States and Canada: Am. J. Sci., v. 46, no. 2, 320-323.

6553 1844, On the probable age and origin of a bed of plumbago and anthracite, occurring in mica schist, near Worcester, Mass.: Am. J. Sci., v. 47, no. 1, 214-215.

6554 1845, Birdseye view of the Falls of Niagara and adjacent country colored geologically: in Travels in North America (see 6557), front. to v. 1.

6555 1845, Geological map of the United States, Canada &c. compiled from the state surveys of the U. S. and other sources: in Travels in North America (see 6557), colored geological map, 50 x 38.5 cm., front. to v. 2.

6556 1845, Map of the Niagara district: in Travels in North America (see 6557), geological map facing p. 30.

6557 1845, Travels in North America, in the years 1841-42; with geological observations on the United States, Canada, and Nova Scotia: NY, Wiley and Putnam, 2 v. (v. 1, viii, 251 p., 6 figs., plates; v. 2, vi, 231 p., figs. 7-18, plates).
Contains 6554, 6555, 6556. For review with excerpts see 4502, 6384, 6393, 9950. For review see 276, 816, 1216, 1691, 3536, 3859, 5448, 5990.
For excerpts see Niles' Wk. Reg., v. 69, 126-127 (1845); US Mag. and Dem. Rev., v. 17, 199-202 (1846).

6558 1846, Coal field of Tuscaloosa, Alabama: Am. J. Sci., v. 1 (ns), no. 3, 371-376, fig.

Lyell, C. (cont.)
6559   1846, Mr. Lyell on the Mississippi delta: Anglo-Am., v. 8, 32-33.
         Abstracted from British Assoc. Proc.
6560   1846, Observations on the fossil plants of the coal field of Tuscaloosa, Alabama: Am. J. Sci., v. 2 (ns), no. 2, 228-230.
6561   1846, On the coal fields of Alabama: Franklin Inst. J., v. 13 (3s), 288.
         Abstracted from London Ath.
6562   1846, On the evidence of fossil footprints of quadruped allied to the Cheirotherium, in the coal strata of Pennsylvania: Am. J. Sci., v. 2 (ns), no. 1, 25-29.
6563   1846, [On the valley and delta of the Mississippi River]: De Bow's Rev., v. 2, 439.
6564   1846, Postscript [on the Zeuglodon and nummulitic limestone of Clarke County, Alabama]: Am. J. Sci., v. 1 (ns), no. 2, 313-315.
6565   1847, [Geological notes on Alabama]: De Bow's Rev., v. 3, 275 (8 lines).
         In "Mineral resources and trade of Alabama."
6566   1847, On the alleged coexistence of man and the Megatherium [at Natchez, Mississippi]: Am. J. Sci., v. 3 (ns), no. 2, 267-269.
         Abstracted from The Times.
6567   1847, On the delta and alluvial deposits of the Mississippi, and other points in the geology of North America, observed in the years 1845, 1846: Am. J. Sci., v. 3 (ns), no. 1, 34-39, 2 figs.
6568   1847, On the Mississippi delta: Am. J. Sci., v. 3 (ns), no. 1, 118-119.
6569   1847, On the relative age and position of the so-called nummulite limestone of Alabama: Am. J. Sci., v. 4 (ns), no. 2, 186-191, 2 figs.
6570   1847, On the structure and probable age of the coal field of the James River, near Richmond, Virginia: Am. J. Sci., v. 4 (ns), no. 1, 113-114.
         Abstracted from Geol. Soc. London Proc.
6571   1849, A second visit to the United States of North America: NY, Harper and Brothers, 2 v. (273; 287 p.), illus.
         Also an 1850 printing.
         For review with excerpts see 1903, 3417, 5228, 6404, 7153, 9966. For review see 4489, 5519, 6364, 9182. For excerpts see Lit. World, v. 5, 52-53, 70-71 (1849); Ann. Scien. Disc., v. 1, 226-228 (1850).
6572   1850, Age of the nummulitic formation of the Alps: Am. J. Sci., v. 10 (ns), no. 2, 265-268.
         Abstracted from Q. J. Geol. Soc.

Lyford, William Gilman (1784-1852)
6573   1841, Maryland, and its resources: Hunt's Merch. Mag., v. 5, 50-58.

Lyman and Ralston
6574   1820, Notice. The subscribers, agents of the Lehigh Coal Company of Pennsylvania, take this method of informing the public that they are prepared to execute orders for coal ... : Phila., Broadside.

Lyman, Chester Smith (1814-1890)
6575   1848, Mines of cinnabar in Upper California: Am. J. Sci., v. 6 (ns), no. 2, 270-271.
         Abstracted in Am. J. Pharmacy, v. 15 (ns), 44-46 (1849).
6576   1849, Appendix - observations on California: Am. J. Sci., v. 7 (ns), no. 2, 305-309.
6577   1849, Observations on California: Am. J. Sci., v. 7 (ns), no. 2, 290-292; no. 3, 305-309.
         See also 6576.
6578   1849, Observations on the "old crater" adjoining Kilauea on the east: Am. J. Sci., v. 7 (ns), no. 2, 287.
         Abstracted in Ann. Scien. Disc., v. 1, 235 (1850).
6579   1849, Notes on the California gold region: Am. J. Sci., v. 8 (ns), no. 3, 415-419.
6580   1850, Gold of California: Am. J. Sci., v. 9 (ns), no. 1, 126-127.

Lynch, William Francis (1801-1865)
6581   1848, Report on an examination of the Dead Sea: US SED 34, 30-2, v. 4 (532), 88 p., plate.
         For review see 7154.
6582   1849, Notice of the narrative of the U. S. expedition to the River Jordan and the Dead Sea: Am. J. Sci., v. 8 (ns), no. 3, 317-333.
6583   1850, The Dead Sea: Ann. Scien. Disc., v. 1, 236-237.
6584   1850, Pillar of salt ("Lot's Wife"): Ann. Scien. Disc., v. 1, 237-238.

Lyon, George Francis (1795-1832)
6585   1821, Geological notices in Northern Africa, from Quarterly Review, No. 49: Am. J. Sci., v. 4, no. 1, 32-33.

Lyon, Lucius (1800-1851)
6586   1827, Pyrites investing quartz [from Mackinac, Michigan]: Am. J. Sci., v. 12, no. 1, 162-163.

M., E. D. (see 10856)

M., G. W. (see 3622 and 10524)

M., J. C. (see 905)

M., W. E. (see 3623)

M., W. J. (see 4145)

Macaulay, Patrick (1792-1849)
6587   1836, Maryland Academy of Science and Literature: Am. J. Sci., v. 30, no. 1, 192-194.

Macauley, James
6588 1829, The natural, statistical, and civil history of the state of New York, in three volumes: NY, Gould and Banks; Alb., William Gould and Co., 3 v. (xxiv, 539; xi, 459; xvi, 451 p.).
For review see 513. Contains, "A sketch of the geology and mineralogy of the state of New York," p. 281-362.

M´Causlin
6589 1793, An account of an earthy substance found near the Falls of Niagara and vulgarly called the spray of the falls: together with some remarks on the Falls: Am. Philos. Soc. Trans., v. 3, 17-24.

M´Collum, William S.
6590 1850, California as I saw it. Pencillings by the way of its gold and gold diggers, and incidents of travel by land and water: Buffalo, NY, G. H. Derby and Co., 72 p.
For review see 6368.

M´Conihe, Isaac
6591 1820, On the geology of the southern part of Indiana: Plough Boy, v. 1, no. 43, 336.
In Troy Lyc. Proc.

M´Coy, Sir Frederick (1823-1899)
6592 1848, On the fossil botany and geology of the rocks associated with the coal of Australia: Am. J. Sci., v. 5 (ns), no. 2, 273-276.
Abstracted from Annals and Mag. Nat. History.

MacCulloch, John (1773-1835)
6593 1818, [On chlorophacite and conite from Scotland]: Am. Mon. Mag. and Crit. Rev., v. 3, no. 6, 461.
6594 1822, [Chlorophoeite from Scotland]: Am. J. Sci., v. 4, no. 2, 245.
Abstracted by J. W. Webster from Edinburgh Philos. J.
6595 1823, On certain elevations of land, connected with the actions of volcanoes: Bos. J. Phil. Arts, v. 1, no. 1, 31-48; no. 2, 108.
From Royal Inst. J.
6596 1824, On animals preserved in amber, with remarks on the nature and origin of that substance: Bos. J. Phil. Arts, v. 2, no. 1, 55-61.
From Royal Inst. J.
6597 1825, On a method of splitting rocks by fire: Bos. J. Phil. Arts, v. 2, no. 4, 380-384.
Abstracted from Brewster´s J.

M´Donald, Archibald
6598 1850, Calcined granite as a material for fictile purposes: Ann. Scien. Disc., v. 1, 96 (14 lines).
Abstracted from Practical Mech´s J.

Mace, Jean
6599 1797, Natural history of the Isle of France: Univ. Mag., v. 3, 75-77.

McElwee, Thomas B.
6600 1829, Bedford County [Pennsylvania]: Hazard´s Reg. Penn., v. 4, 36-41, 49-51.
On the springs and ores of Bedford County.

Maceroni, Francis (1788-1846)
6601 1839, Effects of an earthquake in Calabria: Poughkeepsie Casket, v. 2, 196.
From his Memoirs.

McEuen, Thomas (see 10606 with R. Vaux)

Macgillivray, William (1796-1852)
6602 1822, The travels and researches of Alexander von Humboldt... : NY, J. & J. Harper, 367, 4 p.

McGoffern, J. (of Southern Alabama)
6603 1839, Rotten limestone soils of Alabama: Farm´s Reg., v. 7, 617.
Signed by "Agricola."

McGuire, W. W.
6604 1834, On the prairies of Alabama: Am. J. Sci., v. 26, no. 1, 93-98.
Also in Farm´s Reg., v. 2, 182-184.

Macintosh, A. F.
6605 1846, Grooves or scratches in North Wales: Am. J. Sci., v. 1 (ns), no. 2, 277.
Abstracted from Q. J. Geol. Soc.

McIntyre, James
6606 1849, Mr. McIntyre´s report [on the geology of Isle Royale, Lake Superior]: in Michigan survey by C. T. Jackson (see 5655), 506-509.

Mackenzie, Sir George Steuart, 7th bart. (1780-1848)
6607 1818, Description of a remarkable alternating boiling spring: Mon. Scien. J., v. 1, 133-135.
Extracted from Travels in Iceland. See also review 9413.
6608 1824, On the formation of chalcedony: Bos. J. Phil. Arts, v. 2, no. 2, 119 (9 lines).
Abstracted from Edinburgh Philos. J.

Mackubin, George
6609   1838, Report of the treasurer of the Western Shore to the House of Delegates, respecting the expenses incurred in making the geographical and geological surveys of the state [of Maryland] : Annapolis, Md., Md. Public Documents, December Session 1837, 3 p.
         See also 6800.

McLane, Louis
6610   1846, also H. T. De la Beche, Correspondence between Mr. McLane and Sir H. T. De La Beche, Director of the Geological Survey of Great Britain and Ireland: Nat'l Inst. Prom. Sci. Bulletin, v. 4, 504-505.
         See also 3040.

Maclaren, Charles (1782-1866)
6611   1842, The glacial theory of Prof. Agassiz: Am. J. Sci., v. 42, no. 2, 346-365, 9 figs.

Maclay, W. B. (i.e. William Brown?, 1812-1882)
6612   1845, Bear Mountain coal and iron regions: Fisher's Nat'l Mag., v. 1, no. 2, 171-174.

Mcleod, Donald (1779-1879)
6613   1846, History of Wiskonsan [sic.], from its first discovery to the present period. Including a geological and topographical description of the territory, with a correct catalogue of all its plants: Buffalo, NY, Steele's Press, xii, 310 p., map.

Maclure, William (1763-1840)
6614   1809, A map of the United States of America. By Samuel G. Lewis: Am. Philos. Soc. Trans., v. 6, pt. 2, colored geological map, 54.2 x 42.1 cm.
         In 6615. Maclure's geology was printed and colored on S. G. Lewis's earlier map, which was cut in half for Maclure's Observations. Only the right-hand half of the original map was used. Scale 1" = c. 75 miles.
6615   1809, Observations on the geology of the United States, explanatory of a geological map: Am. Philos. Soc. Trans., v. 6, pt. 2, 411-428, colored map; notice on [430].
         Contains map 6614.
         For review with excerpts see 3437, 4241, and 7090.
6616   1817, Observations on the geology of the United States of America; with some remarks on the effect produced on the nature and fertility of soils, by the decomposition of the different classes of rocks; and an application to the fertility of every state in the Union, in reference to the accompanying geological map. With two plates: Phila., printed for the author by Abraham Small, and sold by him and J. Melish, x, 11-127, 2 p., 2 colored plates, colored map.
         Contains map 6614, which exists in several variants in the 1817 Observations. Plates are frequently uncolored.
         For review with excerpts see 8543. For review see 7939.
6617   1817. Observations on the geology of the West India Islands, from Barbados to Santa Cruz, inclusive: Phila. Acad. Nat. Sci's J., v. 1, pt. 1, no. 6, 134-149.
         Abstracted in J. Sci. Arts, v. 5, no. 2, 311-323 (1818).
6618   1818, Essay on the formation of rocks, or an inquiry into the probable origin of their present form and structure: Phila. Acad. Nat. Sci's J., v. 1, pt. 2, 261-276; 327-345; 385-310.
         Also reprint edition.
6619   1818, Map of the United States of America. Designed to illustrate the geological memoir of Wm. Maclure Esqr.: Phila., John Melish, colored geological map, 44.3 x 33.3 cm.
         Accompanies Observations (see 6620). Scale 1" = c. 95 miles.
         See also 11038.
6620   1818, Observations on the geology of the United States of North America; with remarks on the probable effects that may be produced by the decomposition of the different classes of rocks on the nature and fertility of soils: applied to the different states of the Union, agreeably to the accompanying geological map: Am. Philos. Soc. Trans., v. 1 (ns), 1-91, colored plate, colored map.
         Contains map 6619.
         For review see 1240, 3437, 8626.
6621   1818, and R. M. Patterson and I. Lea, Report of a committee on a new hydrostatic balance, invented by Isaiah Lukens, and submitted to the Society: Phila. Acad. Nat. Sci's J., v. 1, no. 2, 260-261, plate.
6622   1819, Hints on some of the outlines of geological arrangement, with particular reference to the system of Werner: Am. J. Sci., v. 1, no. 3, 209-213.
         With introductory remarks (see 324).
6623   1821, Remarks on the study of geology: Am. J. Sci., v. 3, no. 2, 363.
6624   1822, Extracts of a letter from Wm. M'Clure, esq. to the editor, dated, Madrid, Dec. 4, 1821: Am. J. Sci., v. 5, no. 1, 197-198.
         On the comparative geology of Europe and the United States.
6625   1823, Some speculative conjectures on the probable changes that may have taken place in the geology of the continent of North America east of the Stoney Mountains: Am. J. Sci., v. 6, no. 1, 98-102.
6626   1824, Difficulty of mineralogical excursions in Spain: Am. J. Sci., v. 8, no. 1, 187 (15 lines).
6627   1824, Gift to the American Geological Society: Am. J. Sci., v. 8, no. 1, 187 (11 lines).
         Donation by Maclure of volcanic rocks from Humilla, Spain.
6628   1824, Miscellaneous remarks on the systematic arrangement of rocks, and on their probable origin, especially of the Secondary: Am. J. Sci., v. 7, no. 2, 261-264.
6629   1824, Observations on Mr. Beudant's geological travels in Hungary, &c. with miscellaneous remarks on coal, &c.: Am. J. Sci., v. 7, no. 2, 256-261.
6630   1824, Opinions as to the principal rock formations: Am. J. Sci., v. 8, no. 1, 187-188.

Maclure, W. (cont.)
6631 1824, Remarks on the rocks accompanying anthracite at Wilkesbarre and elsewhere: Am. J. Sci., v. 7, no. 2, 260-261.
Abstracted in Bos. J. Phil. Arts, v. 3, no. 1, 100-101.
6632 1825, Extracts of letters, addressed to the editor: Am. J. Sci., v. 9, no. 1, 157-164.
On Mosaic geology, the cupidity of European mineral dealers, European systems of rock classification, and the Giant's Causeway.
6633 1825, Geological systems. - Geological maps. - Chatoyant feldspar: Am. J. Sci., v. 9, no. 2, 253-256.
On the Wernerian system of rock classification, progress in European geological mapping, and the origin of schiller in feldspar.
6634 1825, Notice of the anthracite region of Pennsylvania: Am. J. Sci., v. 10, no. 1, 205.
6635 1826, Notice of Mr. Owen's establishment, in Indiana - in a letter from William Maclure, esq. to the editor: Am. J. Sci., v. 11, no. 1, 189-192.
6636 1829, Extract of a letter to the editor, from an American resident in Mexico: Am. J. Sci., v. 16, no. 1, 159-163, fig.
On the volcanic geology of Halcotal, Mexico, and on the classificatin of volcanic rocks.
6637 1829, Remarks on the igneous theory of the earth: Am. J. Sci., v. 16, no. 2, 351-452 [i.e. 352].
6638 1829, Remarks on the theory of a central heat in the earth, and on other geological theories: Am. J. Sci., v. 15, no. 2, 384-386.
Comments on the internal heat of the earth, and on the controversy between Neptunists and Vulcanists.
6639 1831, Geological remarks relating to Mexico, &c. in a letter dated Mexico, May 30th, 1830: Am. J. Sci., v. 20, no. 2, 406-408.
6640 1832, Essay on the formation of rocks, or an inquiry into the probable origin of their present form and structure: New Harmony, Indiana, for the author, 53 p.
6641 1832, Observations on the geology of the West India Islands from Barbados to Santa Cruz, inclusive: New Harmony, Indiana, for the author, 17 p.
6642 1838, Essay on the formation of rocks, or an inquiry into the probable origin of their present form and structure: Phila., printed by John Wilbank, for the author, 32 p.
6643 1840, Catalogue of mineralogical and geological specimens, at New-Harmony, Indiana. Collected in various parts of Europe and America: New Harmony, Indiana, 15, [1] p.
6644 1843, Catalogue of geological specimens illustrating the formation of the Ohio Valley: New Harmony, Indiana.
Not seen.

McMurtrie, Henry (1793-1865)
6645 1819, Sketches of Louisville and its environs; including among a great variety of miscellaneous matter, a Florula Louisvillensis: or, A catalogue of nearly 400 genera and 600 species of plants that grow in the vicinity of the town, exhibiting their generic, specific and vulgar English names ... : Louisville, Kentucky, S. Penn, jr. (printer), viii, 255 p., map.
6646 1847, Lexicon scientiarium - a dictionary of terms used in the various branches of anatomy, astronomy, botany, chemistry, geology, geometry, hygiene, mineralogy, natural philosophy, physiology, zoology, &c., (for the use of all who read or study in college, school, or private life): Phila., Biddle, 264 p.
Also second edition, 1849. For review see 911.

McNair, Alexander (1775-1826)
6647 1805, Lead mines and saline: report of Alexander McNair, Register for the district of St. Louis, Missouri, United States: US SED 248, 14-2, v. 4, 235-238.

McNeill, William Gibbs (1801-1853)
6648 1838, and G. W. Whistler and W. H. Swift, Reports of the engineers of the Western Rail Road Corporation, made to the directors, in 1836-7: Springfield, Merriam, Wood, and Co. (printers), 20 p., 2 folded maps.

McNeven, William James (1763-1841)
See also 7297 with S. L. Mitchill.
6649 1815, Chemical analysis of the mineral water of Schooley's Mountain, in New Jersey: Lit. and Philos. Soc. NY Trans., v. 1, 539-557.

Macomb, Alexander (1782-1841)
6650 1826, and J. Knight, Report of the chief engineer in relation to the survey ... of the road from the .. Ohio ... to the state of Missouri: US HED 51, 19-1, v. 4 (134), 32 p.
6651 1837, [Letter supporting G. W. Featherstonhaugh]: Naval Mag., v. 2, 579.
In "Reply of G. W. Featherstonhaugh" (see 7630).

Macpherson, John
6652 1782, An introduction to the study of natural philosophy: Phila., 2 p.
Evans number 17580.

Macquart, Louis Charles Henri (1745-1818)
6653 1810, Description of the celebrated salt mine near Wieliczk in Poland: Huntingdon Lit. Mus., v. 1, 359-366.

Macrae, F.
6654 1835/1836, On the soils and agricultural advantages of Florida: Farm's Reg., v. 3, 179-181; 228-229; 372-374; 515-516.

Macrery, Joseph
6655 1806, A description of the hot springs and volcanic appearances in the county adjoining the River Ouachitta, in Louisiana: Med. Repos., v. 9, 47-50.

McVickar, Archibald (see 6214 by M. Lewis)

Madden
6656   1830, The Dead Sea: Mag. Use. Ent. Know., v. 1, no. 6, 377-379.

Madison, James (1749-1812)
6657   1805, Notices of the warm spring in the county of Bath, Virginia: Phila. Med. Phys. J., v. 2, pt. 1, 62-65.
6658   1805, Observations on the mammoth, or American elephant: Phila. Med. Phys. J., v. 2, pt. 1, 58-60.
6659   1812, Abstract of the late Bishop Madison's memoir on elephantine bones, discovered in Virginia, during 1811: Med. Repos., v. 15, 388-390.
6660   1812, A description of the Yellow Spring in Virginia: Med. Repos., v. 15, 17-23.
6661   1812, Meteoric stone that fell in North Carolina in Jan. 1810: Med. Repos., v. 14, 390.

Madison, James (1751-1836)
6662   1786, A letter from J. Madison, Esq. to D. Rittenhouse, Esq. containing experiments and observations upon what are commonly called the Sweet Springs [at Botetourt County, Virginia]: Am. Philos. Soc. Trans., v. 2, 197-199.
6663   1822, [Letter on a fossil tree discovered by H. R. Schoolcraft]: Am. J. Sci., v. 5, no. 1, 25.
        See also H. R. Schoolcraft 9263.

The Magazine of Useful and Entertaining Knowledge (New York, 1830-1831)
6664   1830, The American mammoth: v. 1, no. 3, 91-95, fig.

Magill, John (1759-1842)
6665   1832, The pioneer to the Kentucky emigrant, a brief topographical and historical description of the state of Kentucky: Frankfort, Kentucky, J. B Marshall (printer), 83 p.

Magnus (Dr.)
6666   1839, Temperature of the earth: Am. J. Sci., v. 36, no. 1, 203-204.
        Abstracted from Edinburgh New Philos. J.

Maillet, Benoit de (1656-1738)
6667   1797, Telliamed; or, the world explained: containing discourses between an Indian philosopher and a missionary, on the diminution of the sea - the formation of the earth, - the origin of men and animals and other singular subjects, relating to natural history & philosophy: Balt., W. Pechin for D. Porter, 268 p.
        Evans number 32414. "A very curious work."

Maine, State of
6668   1832, Recommendation to employ scientific persons to make geological exploration of uninhabited part of state: in Ann. Report of the Land Agent, 1-10.
6669   1836, [Geological survey recommended]: in Governor's Message 1836, 6-7.
6670   1836, Report on so much of the Governor's message as relates to geological survey; containing resumé of known mineral resources of Maine: In Me. SD 53, 20 p.
        Also in Resolves of Me., 53-68.
6671   1836, Report [on the geological survey]: Me. Farm., v. 4, 57-58; 68-69.
6672   1836, Report of a committee to whom was referred an order relating to a geological survey in Maine: Me. Legislative Document 31, 8 p., Allen Putnam, Chairman.
6673   1836, Resolve authorizing the board of internal improvements to commence a geological survey of the state: Resolve of the 16th legislature of the state of Maine, Chapter 66, 67-68.
6674   1836, Resolve relating to a geological survey of the Commonwealth's lands in Maine: Mass., House of Representatives, December 31, 1836, 8 p.
6675   1837, [Resolution continuing the geological survey through a second year]:
        Not seen.
6676   1838, [Resolution continuing the geological survey through a third year]:
        Not seen.
6677   1839, Account of progress of geological survey together with expenditures for prosecuting same: in Me. HD 1, 5 p.
        Also in Me. Public Documents for 1839, v. 2.

The Maine Farmer and Journal of the Useful Arts (Winthrop, Maine, 1833-1850)
6678   1833, Meteorlites, or stones which have fallen from the sky: v. 1, 248.
6679   1833-1839, [Brief notes on nature and improvement of soils]: v. 1, 331, 364; v. 2, 177-178; 194; 226; 285-286; v. 3, 89-90; v. 4, 240-241; v. 5, 193; v. 7, 66, 106.
6680   1834, The Rocky Mountains [notice of article by A. Eaton (see 3378)]: v. 2, 30-31.
6681   1834, Geology: v. 2, 220.
6682   1834, Petrifaction: v. 2, 225-226.
6683   1834, Formation of amber: v. 2, 226.
6684   1834, Geology: v. 2, 228.
        On granite.
6685   1834-1839, [Notes on Maine mining]: v. 2, 234; v. 3, 121, 257; v. 4, 389; v. 5, 91; v. 6, 329; v. 7, 201.
6686   1834, Geology: v. 2, 236.
        On rock types.
6687   1834, Geology and mineralogy: v. 2, 284.
        Signed by "J. H. J." On the importance of geology to agriculture.
6688   1834, Mineralogy of Hancock County: v. 2, 350.
6689   1835, Geological survey recommended: v. 2, 403.
        Signed by "Upandbedoing."
6690   1835, Geology: v. 3, 50.
        On the effect of corals on geological formations.
6691   1835, Geology in schools: v. 3, 61.
        Abstracted from Scien. Tracts. On reasons for teaching geology in school.
6692   1835, The geological lectures: v. 3, 76-77.
        Notice of B. Silliman's series of lectures.
6693   1835, Geological surveys: v. 3, 107.

The Maine Farmer (cont.)
6694 1835, Geology: v. 3, 107.
  Notice of B. Silliman's lectures.
6695 1835, Geology for schools: v. 3, 211.
6696 1835, Geology of Portland and vicinity: v. 3, 236-237.
  Abstracted from Portland Mag.
6697 1835, Geological survey of Maine: v. 3, 273-274.
  Letter in support of a geological survey signed by "J. H. J."
6698 1836, Lectures on geology: v. 4, 115.
  On C. T. Jackson's third public lecture.
6699 1836, Geological survey [of Maine]: v. 4, 172.
6700 1837, Opening of a vast lead mine in Wisconsin Territory: v. 5, 110.
6701 1838, Dr. Jackson's lectures: v. 6, 37-38, 45-46, 53-54, 61-62, 69-70, 84-85.
6702 1838, Geological survey of Maine [review of second report by C. T. Jackson (see 5560)]: v. 6, 73.
6703 1838, Geological survey of the public lands [review of the second report by C. T. Jackson (see 5558 and 5559)]: v. 6, 89.
6704 1840, [On anthracite coal]: v. 8, 51.
6705 1840, Ararat - extinct volcano near Mount Ararat: v. 8, 56.
6706 1840, An internal volcano [in Asia]: v. 8, 56.
6707 1840, Earthquake at Tabriz [in Asia]: v. 8, 56.
6708 1840, Iron ores [from Beaver Creek, Pennsylvania]: v. 8, 187.
6709 1840, Skeleton of mastodon [from St. Louis]: v. 8, 235.
6710 1840, Geology: v. 8, 277-278.
  On the common rocks and minerals.
6711 1840, Geological survey of the state [of Maine]: v. 8, 341.
6712 1840, Geological statistics: v. 8, 376.
6713 1841, Dr. Jackson's geological and agricultural survey of Rhode Island [notice and excerpts of 5571]: v. 9, 49-50; 57-58; 68.
6714 1841, Discovery of platina [in New South Wales, near Melborne]: v. 9, 210.
6715 1841, Resources of Maine: v. 9, 347; 354; 365-366.

The Maine Monthly Magazine (Bangor, Maine, 1836-1837)
6716 1836, Outline of geology: v. 1, no. 5, 193-200.
  Signed by "X." Review of the major geological theories including Wernerian and Plutonian. Neptunian views are favored as more consistent with the Scriptures.

Mair, Robert
6717 1823, Extract of a letter: Am. J. Sci., v. 7, no. 1, 56-57.
  On pyrite and sulphur from Connecticut.

Malden, Massachusetts. Church of
6718 1727, Articles drawn up by the members of the church of Malden on a day of public fasting and prayer (December, 21, 1727). Occasioned by a terrible earthquake, on the Lord's day-night, October 29th, 1727: Bos., 4 p.
  Evans number 2896.

Mallet, Robert (1810-1881)
6719 1838, Mechanism of the motion of glaciers: Franklin Inst. J., v. 21 (ns), 353-354.
  From British Assoc. Proc.
6720 1846, On the vorticose movement assumed to accompany earthquakes: Am. J. Sci., v. 2 (ns), no. 2, 270-272.
  Abstracted from Philos. Mag.

Manchester Geological Society
6721 1842, Transactions of the Manchester Geological Society, vol. I: Am. J. Sci., v. 43, no. 1, 201-202.
  Notice of officers, activities, and contents of the Trans.

Mansfield, Jared (1759-1830)
6722 1810, On the figure of the earth: Ct. Acad. Arts Sci's Mem., v. 1, pt. 1, 111-118.
  Abstracted in Am. Med. Philos. Reg., v. 3, no. 2, 214-217 (1812).

Mantell, Gideon Algernon (1790-1852)
6723 1832, The geological age of reptiles: Am. J. Sci., v. 21, no. 1, 359-363. Also in Free Enquirer, v. 5 (2s), 70-71.
  Abstracted from Edinburgh New Philos. J.
6724 1834, Fossil jaws of the tapir [from Darmstadt]: Am. J. Sci., v. 26, no. 1, 218 (7 lines).
6725 1835, The Maidstone Iguanodon: Am. J. Sci., v. 27, no. 2, 420.
6726 1835, Notice of the discovery of the remains of the Iguanodon in the Lower Green Sand formation of the South-east of England: Am. J. Sci., v. 27, no. 2, 355-360, 6 figs.
6727 1835, Notice of the work of T. Hawkins: Am. J. Sci., v. 27, no. 2, 413-415.
  With comments by B. Silliman (see 9724). Includes review by Mantell of T. Hawkin's Memoirs of Ichthyosauri and Plesiosauri.
6728 1835, Opossum in the Stonesfield slate, near Oxford, England: Am. J. Sci., v. 27, no. 2, 412 (11 lines).
6729 1836, Impressions of the feet of Mammalia in sandstone: Am. J. Sci., v. 30, no. 1, 191-192.
  On fossil footprints from Hildburghausen, Germany.
6730 1836, Remains of birds in the strata of Tilgate Forest, Sussex, England: Am. J. Sci., v. 29, no. 2, 362.
6731 1838, Lectures and remarks of Dr. Gideon Mantell: Am. J. Sci., v. 33, no. 2, 328-334.
  With remarks by B. Silliman (see 9752). On geological philosophy, and the nature of zoophytes.

Mantell, G. A. (cont.)

6732 1839, The wonders of geology; or, a familiar exposition of geological phenomena, being the substance of a course of lectures delivered at Brighton: NHav., A. H. Maltby, 2 v., illus.
For review see 705, 739, 933 (6th ed., 1848), 2355, 3620, 7128, 7817.
For notice see 713, 927 (6th ed., 1848), 9757, 9761.

6733 1841, Fossil saurians: Am. J. Sci., v. 41, no. 1, 205 (4 lines).
On bones of Iguanodon and Hylaesaurus from Tilgate Forest.

6734 1841, Fossil turtle: Am. J. Sci., v. 41, no. 1, 205 (3 lines).
On fossil bones from the Kent Chalk.

6735 1841, Geological drawings [for sale]: Am. J. Sci., v. 41, no. 1, 206-207 (7 lines).

6736 1841, Glaciers, moraines, &c.: Am. J. Sci., v. 41, no. 1, 207 (13 lines).
On British evidence for L. Agassiz' glacial theory.

6737 1841, Microscopical observations and microscopes: Am. J. Sci., v. 41, no. 1, 205-206.

6738 1842, [On a human footprint in limestone]: Am. J. Sci., v. 42, no. 1, 25-26.
In "Regarding human foot-prints in solid limestone" by D. D. Owen (see 8051).

6739 1843, Animal of the belemnite: Am. J. Sci., v. 45, no. 2, 403 (10 lines).
Description of a Belemnite preserved with soft parts in the Oolite of Chippenham, England.

6740 1843, Fossil birds [from New Zealand]: Am. J. Sci., v. 44, no. 2, 417-418.

6741 1843, Fossil fruits described by Dr. Gideon Algernon Mantell: Am. J. Sci., v. 45, no. 2, 401-402.

6742 1843, Notice of "molluskite," or the fossilized remains of the soft parts of mollusca: Am. J. Sci., v. 45, no. 2, 243-247.
With notes by W. H. Beusted (see 1675).

6743 1843, Reply of Dr. G. A. Mantell to Dr. Deane: Am. J. Sci., v. 45, no. 1, 184-185.
On the possible relationship between the fossil footmarks in the Connecticut River Valley, and the giant fossil birds of New Zealand.

6744 1845, [Excerpts from Medals of Creation (London ed., 1845)]: Am. J. Sci., v. 48, no. 1, 105-137.
In review by B. Silliman (see 9783). No American editions of this work have been found.

6745 1845, Silicification [of fossil infusoria in Southeast England]: Am. J. Sci., v. 49, no. 1, 227 (9 lines).

6746 1846, Bones of the Iguanodon and other colossal reptiles recently found in the Isle of Wight, England: Am. J. Sci., v. 1 (ns), no. 2, 275-276.

6747 1846, The soft bodies of Polythalmia found in a fossil state: Am. J. Sci., v. 2 (ns), no. 2, 275-276.
Abstracted from Annals of Nat. History.

6748 1846, Supposed birds' bones of the Wealden: Am. J. Sci., v. 1 (ns), no. 2, 274-275.

6749 1847, Notice of Dr. Mantell's Isle of Wight: Am. J. Sci., v. 4 (ns), no. 2, 230.

6750 1848, [Fossil Iguanodon jaw and Foraminifera from England]: Bos. Soc. Nat. History Proc., v. 3, 46-47.

6751 1848, Notices of the fossil bones of the ancient birds of New Zealand: Am. J. Sci., v. 5 (ns), no. 3, 431.

6752 1848, On the fossil remains of the soft parts of Foraminifera, discovered in the Chalk and Flint of the Southeast of England: Am. J. Sci., v. 5 (ns), no. 1, 70-74.

6753 1848, On the structure of the jaws and teeth of the Iguanodon: Am. J. Sci., v. 6 (ns), no. 3, 429-431.
Abstracted from Lit. Gaz., London.

6754 1849, Additional observations on the osteology of the Iguanodon and Hylaesaurus: Am. J. Sci., v. 7 (ns), no. 3, 439-441.

6755 1849, New discoveries of bones of the Iguanodon: Am. J. Sci., v. 7 (ns), no. 3, 438, fig.

6756 1849, On the fossil remains of birds collected in various parts of New Zealand by Mr. Walter Mantell, of Wellington: Am. J. Sci., v. 7 (ns), no. 1, 28-44.

6757 1850, Discovery of another huge reptile: Am. J. Sci., v. 9 (ns), no. 1, 147 (6 lines).
Also in Ann. Scien. Disc., v. 1, 288 (6 lines).

6758 1850, Notice of the remains of the Dinorthis and other birds, and of fossil and rock specimens recently collected by Walter Mantell, Esq., from the Middle Island of New Zealand: Am. J. Sci., v. 9 (ns), no. 3, 437-438.
Abstracted from Geol. Soc. London Proc.

6759 1850, On the Pelorosaurus; an undescribed gigantic terrestrial reptile, whose remains are associated with those of the Iguanodon and other saurians, in the strata of Tilgate Forest: Am. J. Sci., v. 9 (ns), no. 3, 439-444.
Abstracted from Royal Soc. Proc.

6760 1850, Supplementary observations on the structure of the Belemnite and Belemnoteuthis: Am. J. Sci., v. 9 (ns), no. 3, 438-439.
Abstracted from Royal Soc. Proc.

Mantell, Reginald Neville

6761 1850, An account of the strata and organic remains exposed in the cuttings of the railway from the Great Western line near Corsham, through Trowbridge to Westbury in Wiltshire: Am. J. Sci., v. 9 (ns), no. 3, 436-437.
Abstracted from Geol. Soc. London Proc.

6762 1850, [On the existence of man with extinct species of vertebrates]: Am. Assoc. Adv. Sci. Proc., v. 2, 271 (14 lines).
Discussion of article by G. J. Chase (see 2327).

6763 1850, [Recent and fossil sponge spicules from England]: Am. Assoc. Adv. Sci. Proc., v. 2, 91 (5 lines).

Mantell, Walter

6764 1850, Colossal birds of New Zealand: Am. J. Sci., v. 9 (ns), no. 1, 147.
Abstracted from Lit. Gaz., London.

Manual of the naturalist; or, directions practical for collecting, preparing and preserving subjects of natural history:
6765   1831, Bos.
       Not Seen.

Mapes, James Jay (1806-1866)
6766   1840, Agricultural geology [review of third annual Maine report by C. T. Jackson (see 5566)]: Am. Rep. Arts Sci's Manu's, v. 1, 99-100.
6767   1849, The chemical analysis of soils: Working Farm., v. 1, no. 3, 36-37.
6768   1849, Composition of soils: Working Farm., v. 1, no. 9, 130.

Marcet, Alexander John Gaspard (1770-1822)
6769   1809, Analysis of the waters of the Dead Sea, and of the River Jordan: Phila. Med. Mus., v. 6, 168-174.
       From Philos. Trans.
6770   1823, Some experiments and researches on the saline contents of sea-water, undertaken with a view to correct and improve its chemical analysis: Bos. J. Phil. Arts, v. 1, no. 2, 146-152.
       From Philos. Trans.

Marchand, Richard Felix (1813-1850?)
6771   1849, Analysis of the water of the Dead Sea: Am. J. Sci., v. 8 (ns), no. 3, 444 (14 lines).
       Abstracted from Poggendorff's Annalen.

Marcy, Randolph Barnes (1812-1887)
       See also 9841 with J. H. Simpson.
6772   1850, [Report on the expedition from Fort Smith to Santa Fe, New Mexico]: US SED 64, 31-1, v. 14 (562), 169-233.
       Includes notes on the landforms and lithologies of parts of Arkansas, Texas, and New Mexico.

Marget, Alexander (see Marcet, Alexander)

Marguerite, M.
6773   1846, Analysis of tungstates: Am. J. Sci., v. 1 (ns), no. 1, 108 (9 lines).
       Abstracted by J. L. Smith from Comptes Rendus.
6774   1846, A new method for the quantitative determination of iron: Am. J. Sci., v. 2 (ns), no. 2, 257-258.
       Abstracted by J. L. Smith from Comptes Rendus.

Marignac, Jean Charles Gallisard de (1817-1894)
6775   1844, and A. L. O. L. Descloiseaux, Penine [from Zermatt]: Am. J. Sci., v. 47, no. 1, 216 (12 lines).
       Abstracted by "D." (i.e. J. D. Dana?) from "Ann. de Ch. et de Phys."
6776   1844, and A. L. O. L. Descloiseaux, Talc [from Chamouni]: Am. J. Sci., v. 47, no. 1, 216 (4 lines).
       Abstracted by "D." (i.e. J. D. Dana?) from "Ann. de Ch. et de Phys."
6777   1848, Liebenerite - a new mineral: Am. J. Sci., v. 6 (ns), no. 2, 275.
       Abstracted from Philos. Mag. From the Fassa Valley.
6778   1849, On humite: Am. J. Sci., v. 8 (ns), no. 1, 123 (2 lines).
6779   1849, Philippsite and gismondine: Am. J. Sci., v. 8 (ns), no. 1, 122 (8 lines).
       Abstracted from "Ann. de Ch. et de Phys."

Mariti, Abbe
6780   1795, Description of the Dead Sea: NY Mag., v. 6, 451-455.

Mark, Jacob (see 10509 with N. I. Roosevelt)

Markoe, Francis
6781   1842, [List of fossils from Calvert and St. Mary's, Maryland]: Nat'l Inst. Prom. Sci. Bulletin, v. 2, 132.

Marquerite, M. (see Marguerite, M.)

Marr, Robert A.
6782   1850, Report of observations made at the Navy Yard at Memphis, during the months of April, May, June, and a part of July, 1849: Am. Assoc. Adv. Sci. Proc., v. 2, 335-346, fig., tables.
       With introduction by M. F. Maury (see 6928). On sedimentation in the Mississippi River.

Marsden, William (1754-1836)
6783   1818, [Extract of a letter]: in Essay on the theory of the earth by G. Cuvier (see 2754), 316-317.

Marsh, Charles (of Woodstock, Vermont)
6784   1841, Shell marl under peat, in Vermont: Farm's Reg., v. 9, 453.

Marsh, Dexter
6785   1848, Fossil footprints [from the Connecticut River Valley]: Am. J. Sci., v. 6 (ns), no. 2, 272-275, fig.

Marshall
6786   1823, Manures: Plough Boy, v. 4, no. 47, 397-398.
       From Rural Economy of Norfolk, England.

Marshall, Humphrey (1760-1841)
6787 1812, The history of Kentucky. Including an account of the discovery - settlement - progressive improvement - political and military events - and present state of the country: Frankfort, Kentucky, Henry Gore (printer), 407 p.
Intended to be 2 volumes, but only 1 published.
6788 1824, The history of Kentucky: Frankfort, Kentucky, G. S. Robinson, 2 v.

Martin, Anthony B.
6789 1822, Remarks on agriculture and a method of improving soils, by creating artificial manures, founded on chemical and scientific principles: Balt., J. D. Toy (printer), 19 p.

Martin, Falkland H.
6790 1849, and others, Memorial to the fifteenth General Assembly in Missouri, concerning a geological survey of the state: Missouri Hist. and Philos. Soc., 18 p.

Martin, Joseph
6791 1835, A new and comprehensive gazetteer of Virginia, and the District of Columbia: Charlottesville, Va., Moseley and Tompkins (printers), 529 p., map.

Martin, Samuel Davies (b. 1791)
6792 1845, Analysis of Kentucky soil: The Cultivator, v. 2 (ns), 88-89.
Also in Agriculturist, v. 6, 107-108.

Martinet, Joannes Florentius (1729-1795)
6793 1793, The catechism of nature. For the use of children: Bos., Young and Etheridge, for D. West, 108 p.
Evans number 25759.

Martins, Charles Frédéric (1806-1889)
6794 1848, Memoire sur les temperatures de la Mer Glaciale a la surface, a des grandes profondeurs, et dans le voisinage des Glaciers du Spitzberg: Am. J. Sci., v. 6 (ns), no. 1, 143-144.
Abstract of glacial study.

Martius, K. F. P. von
6795 1824, Letter from the Chevalier de Martius to the editor [on paleobotany of Brazil]: Am. J. Sci., v. 8, no. 2, 382-384.
6796 1825, Antediluvian plants: Am. J. Sci., v. 9, no. 2, 375 (6 lines).
Abstracted by "C. H."

Marye, John Lawrence (see 10508 under United States Mining Company)

Maryland. State of
6797 1833, Resolution relative to a geological survey: Md. Gen. Assembly, March 18, 1833.
6798 1834, An act to provide for making a new and complete map and a geological survey of this state: Md. Gen. Assembly, February 25, 1834.
6799 1839, Report of the Select Committee appointed by the House of Delegates to report a bill to abolish the office of State Geologist: Md. Public Documents, December Session 1838, 3 p.
6800 1839, Report of the Treasurer of the Western Shore to the House of Delegates of Maryland. In obedience to their order of the 28th ultimo stating the expenses incurred in making the geographical and geological surveys of the state: Annapolis, Md., Md. Public Documents, December Session 1838, 2 p.
See also 6609.
6801 1842, [An act to abolish the geological survey of Maryland]: Md. Gen. Assembly, February 24, 1842.
6802 1844, Report of the Committee on Agriculture relative to the application of lime to the different qualities of soil and the use of calcareous matter for agricultural purposes: Md. House of Delegates, December Session 1843, Annapolis, Md., 15 p.
6803 1846, Report of the Committee on Agriculture in relation to the appointment of an agricultural chemist: Md. House of Delegates, December Session 1846, 8 p.
6804 1847, An act entitled an act to provide for the appointment of an agricultural chemist for the state: Md. House of Delegates, December, 1847.

Maryland Academy of Science and Literature
6805 1836, Transactions of the Maryland Academy of Science and Literature: Am. J. Sci., v. 30, no. 2, 395-398.
6806 1837, Directions for preparing specimens of natural history: Md. Acad. Sci. Liter. Trans., v. 1, 148-156.
6807 1837, [Donations to the library and cabinet]: Md. Acad. Sci. Liter. Trans., v. 1, 159-174, 187-198.
6808 1837, Proceedings of the Maryland Academy of Science and Literature, 1836: Am. J. Sci., v. 31, no. 2, 395-399.
6809 1837, Transactions of the Maryland Academy of Science and Literature, 1836: Am. J. Sci., v. 32, no. 1, 204-207.

Maryland and New York Iron and Coal Company
6810 1839, Charter and by-laws of the Maryland and New York Iron and Coal Company: NY, Stationer's Hall Press, 20 p.
See also 10787.

The Maryland Medical and Surgical Journal (Baltimore, 1840-1843)
6811 1840, Red Sulphur Springs: v. 1, no. 4, 534-536.
Abstracted from Med. Exam.

Maryland Mining Company
6812 1839, Loan of the Maryland Mining Company: Balt.?, 46 p.
With extracts from geological reports by G. W. Hughes, J. Renwick, T. P. Jones, B. Silliman, B. Silliman, jr., D. Mushet, and J. W. Daniell.

Mascarene, John (1722-1779)
6813   1757, The manufacture of potash in the British North-American plantations recommended: Bos., Z. Fowle, 4, (11), p.
        Evans number 7940.

Mason, Francis (1799-1874)
6814   1843, [On some geologic features of British Burma] : Bos. Soc. Nat. History Proc., v. 1, 118-119.

Mason, O.
6815   1825, Miscellaneous localities of minerals [in Rhode Island] 1. By O. Mason: Am. J. Sci., v. 10, no. 1, 10-11.
6816   1825, Notice of a rocking stone [in North Providence, Rhode Island] : Am. J. Sci., v. 10, no. 1, 9-10, 2 figs. on plate.

Mason, Richard Barnes (1797-1850)
6817   1848, California gold mines: Farm's Mon. Vis., v. 10, 186-187.
6818   1850, Letter from Col. Richard B. Mason: US HED 17, 31-1, 528-536.
        Also in The gold mines of California, by G. G. Foster (see 3919).
        First official report of California gold discovery.

Massachusetts. Commonwealth of
6819   1830, Resolve authorizing further appropriations for a survey of the Commonwealth: Mass. Gen. Assembly, June 5, 1830.
6820   1830, Utility of connecting geographical surveys and examination of geological features of the state ... : in Governor's Message, May, 1830, 16.
6821   1831, Progress of explorations ... : in Governor's Message, January, 1831.
6822   1831, Resolve in relation to the geological survey of the Commonwealth: Mass. Gen. Assembly, February 2, 1831.
6823   1831, Resolve making further appropriations for a survey and geological examination of the Commonwealth: Mass. Gen. Assembly, June 22, 1831.
6824   1832, Resolve for the distribution of the first part of the report on the geological survey of the Commonwealth: Mass. Gen. Assembly, March 24, 1832.
6825   1833, [Report of the Committee on Education on the geological survey] : Mass. HD 36, 4 p.
6826   1833, Resolve for the publication and distribution of the report on the geological survey of the Commonwealth: Mass. Gen. Assembly, March 2, 1833.
6827   1833, Resolve making a further appropriation for the survey of the Commonwealth: Mass. Gen. Assemlby, March 25, 1833.
6828   1834, Report on the expediency of publishing ... [the results of the Massachusetts geological survey] : Mass. HD 23, 6 p.
6829   1834, [Resolve for new edition of Hitchcock's geology] : Mass. Gen. Assembly, February 19, 1834.
6830   1835, [Report of the Library Committee of the geological survey] : Mass. SD 63, 8 p.
6831   1835, [Brief statement of the progress of the geological survey] : in Governor's Message, 1835, 39-42.
6832   1836, [Report on the expediency of a geological survey] : Mass. HD 31, 8 p.
6833   1836, Resolve of the legislature of Massachusetts [on the joint survey of the public lands of Maine and Massachusetts] : Mass. Gen. Assembly, March 21, 1836.
6834   1837, [Comment of the Governor on the geological survey] : in Governor's Message, 1837, p. 9.
6835   1837, Report &c., relating to the geological survey of the state: Mass. HD 26, 16 p.
6836   1837, Resolution providing for an agricultural survey of the state: Mass. Gen. Assembly, April 12, 1837.
6837   1837, Statement of appropriations [for the geological survey] : Mass. HD 39, 3 p.
6838   1839, [Resolution providing for publication of the Massachusetts natural history survey] : Mass. Gen. Assembly, April 9, 1839.
6839   1839, Treasurer's statement [on the geological survey] : Mass. HD 11, 3 p.
6840   1840, Tables showing expenditures [of the geological survey] : Mass. HD 22, 31-32.
6841   1841, Detailed financial statement from Board of Treasury [on the geological survey] : Mass. HD 9, 20 p.
6842   1841, [A resolution concerning the agricultural survey of the state] : Mass. Gen. Assembly, February 15, 1841.
6843   1849, Statement of the cost of several scientific surveys ordered by the state since 1830; also the aggregate amount paid to agricultural societies ... : Mass. HD 18, 5 p.

Massachusetts Agricultural Repository and Journal (Boston, 1815-1832)
6844   1819, Remarks on soil: v. 5, no. 4, 379-389.
        Signed by "Agricola."

Massachusetts Agricultural Society. Papers on Agriculture (Boston, 1793-1811)
6845   1804, On the anlaysis of lime and marl: for 1804, 32-35.
6846   1805, Particulars respecting the history and the use of the species of gypsum, called plaster of Paris, especially as it concerns agriculture: for 1804 [i.e. 1805] , 9-24.

Massachusetts. Agricultural Survey
6847   1838-1841, Report of the agriculture of Massachusetts: Bos., Dutton and Wentworth, 4 v. Edited by Henry Colman.
6848   1840, Report of the agricultural meeting, held in Boston January 13, 1840, containing the remarks on that occasion of the Hon. Daniel Webster ... and of Professor Silliman ... with notes by Henry Colman: Salem. Mass., Gazette Office, 36, 7 p.

Massachusetts Bay
6849   1780, An act to incorporate and establish a society for the cultivation and promotion of arts and sciences: Bos., Benjamin Edes and Sons, 2 p.
        Evans number 16841.

Massachusetts Historical Society. Collections of the (Boston, 1792-1850)
6850 1792-1850, Topographical descriptions of New England towns include brief mineral notices.
6851 1798, An account of the present state and government of Virginia: v. 5, 124.
       Written c. 1697, with some notes on geological features.

The Massachusetts Magazine (Boston, 1789-1796)
6852 1789, The following curiosities have lately been presented to Mr. Peale's American Museum, in Philadelphia: v. 1, 54.
6853 1789, On mines, &c. in Massachusetts: v. 1, 172.
       Signed by "M. P."
6854 1789, American natural curiosities: v. 1, 335-336; 416-417; 580-581; 645-646.
6855 1790 Method for preparing a liquor that will sink into and penetrate marble; so that a picture drawn on its surface will appear also in its inmost parts: v. 2, 50.
6856 1790, Description of the grotto of Antiparos: v. 2, 198-199.
6857 1791, Account of some remarkable springs [in Virginia and Pennsylvania] : v. 3, 407-408.
6858 1792, Thought on the origin of coal mines: v. 4, 221-222.
       Signed by "H."
6859 1792, Reflections on earthquakes: v. 4, 241-242.
6860 1792, Description of the Saratoga Springs: v. 4, 621-623.
6861 1792, Description of Sepascot Cave [at Ryhnbeck] : v. 4, 656.

The Massachusetts Quarterly Review (Boston, 1847-1850)
6862 1847, The life and writings of Agassiz: v. 1, no. 1, 96-119.
6863 1848, A new theory of the effect of tides: v. 2, no. 1, 77-82.
       Review of theory of C. H. Davis on the geological effects of tides and the origin of coastal morphology (see 2947).
6864 1850, [Review of Geology of the United States exploring expedition, by J. D. Dana (see 2841)] : v. 3, no. 4, 459-483, fig.

Mather (Reverend)
6865 1819, Beryl [from Chatham, Connecticut] : Am. J. Sci., v. 1, no. 3, 242.

Mather, Cotton (1663-1728)
6866 1727, Boanerges. A short essay to preserve and strengthen the good impressions produced by earthquakes on the minds of people that have been awakened with them ... With an historical appendix, giving an account of all the observable occurrences of the present year; more especially the earthquakes that have been in Europe and the West Indies: Bos., S. Kneeland, (2), 53 p.
       Evans number 2908.
6867 1727, The terror of the Lord. Some account of the earthquake that shook New-England, in the night, between the 29 and the 30 of October. 1727 ... : Bos., Printed by T. Fleet, for S. Kneeland, (4), 37, 6 p.
       Evans number 2919.

Mather, Increase (1639-1723)
6868 1706, A discourse concerning earthquakes. Occasioned by the earthquakes which were in New-England, in the Province of Massachusetts-Bay, June 16. and in Conecticot-Colony, June 22, 1705. Also, two sermons shewing, that sin is the greatest evil; and that to redeem is the greatest wisdom ... : Bos., Printed by Timothy Green, for Benjamin Eliot at his shop, 131 p.
       Evans number 1268.

Mather, John C.
6869 1842, Lime: Un. Agriculturist and West. Prairie Farm., v. 2, 28.
6870 1845, Agriculture and mines of N. Carolina: The Cultivator, v. 2 (ns), 242.

Mather, Samuel (1651-1728)
6871 1727, Essay on the good impressions produced by earthquakes ... : Bos., 53 p.
       Evans number 2923.

Mather, William Williams (1804-1859)
6872 1830, On xanthite and its crystalline form [from Amity, New York], with a notice of mineral localities [in Rhode Island and Connecticut] : Am. J. Sci., v. 18, no. 2, 359-361, fig.
6873 1831, Geological notices [on New York and Connecticut] : Am. J. Sci., v. 21, no. 1, 94-99, plate.
6874 1831, On the principles involved in the reduction of iron and silver ores, with a supplementary notice of the principal silver mines in Mexico and South America: n.p., 27 p.
6875 1833, Elements of geology, for the use of schools: Norwich, Ct., William Lester, jr., vi, [5]-139 p., illus.
       This is the first edition. See also 6890.
       For review see 592 and 2702.
6876 1833, Geological map: Am. J. Sci., v. 23, no. 2, 404.
       Notice of preparation of a map of New London and Windham Counties, Connecticut (see 6879). See also 6197.
6877 1833, On the principles involved in the reduction of iron and silver ores, with a supplementary notice of some of the principal silver mines of Mexico and South America: Am. J. Sci., v. 24, no. 2, 213-237.
       See also 6874.
6878 1833, Sulphurets of bismuth: Am. J. Sci., v. 24, no. 1, 189-190.
6879 1834, A geological map of Windham and New London counties [Connecticut] : colored geological map, 1" = 5 miles.
       In 6880. For notice see 6876. See also 6197.
6880 1834, Sketch of the geology and mineralogy of New London and Windham counties in Connecticut: Norwich, Ct., W. Lester, jr., 36 p., colored map.
       Contains map 6879.
6881 1835, Contributions to chemical science: Am. J. Sci., v. 27, no. 2, 241-266.
       On gold from Georgia and Lane's silver mine, Munroe, Connecticut.

Mather, W. W. (cont.)

6882 1835, New work by Prof. Brown of Heidelberg - sharks teeth - Conrad on shells, &c.: Am. J. Sci., v. 28, no. 2, 378, 2 figs.
Notice of Lethaea geognostica by Bronn; shark's teeth from New Jersey correlated to the Chalk formation of England; T. A. Conrad's studies on Tertiary shells of Maryland and Virginia noted.

6883 1837, First annual report of the first geological district of New-York: NY Geol. Survey Ann. Rept., v. 1, 61-95.

6884 1837, [Geological report on the lands of the Norwich and Worcester Rail-road Company] : in Annual report ..., (see 7979).

6885 1837, Geology and agriculture: New-Yorker, v. 3, 46.

6886 1837, [Letter on the conduct of G. W. Featherstonhaugh] : Naval Mag., v. 2, 487-488, 573.
Appended to review of Featherstonhaugh's 1835 report (see 7629), and appended to "Reply of G. W. Featherstonhaugh" (see 7630).

6887 1837, Minerals, ores, mines, &c. examined: Am. J. Sci., v. 31, no. 2, 418.
Announcement by Mather that he will assay ores for a $5.00 fee at his Albany office.

6888 1837, Protest of Lt. Mather: Am. J. Sci., v. 33, no. 1, 205-206.
G. W. Featherstonhaugh accused of plagiarism.

6889 1837, and T. A. Conrad, Queries proposed by the geologists of the new survey of the state of New York: Am. J. Sci., v. 33, no. 1, 124-133.

6890 1838, Elements of geology, for the use of schools: NY, American Common School Union, 286 p., illus.
This is the second edition. Other printings include: "third edition," as above, 1839; "fourth edition," NY, Clement and Packard, 1841; "fifth edition," NY, Turner, Hughes and Hayden, xii, 286 p., illus., 1844, 1845, and 1846.
For review see 7821 and 5732.

6891 1838, and others, First annual report on the geological survey of the state of Ohio: Columbus, Ohio, Samuel Medary (printer), 134 p., plate.
Contains letters and reports by C. Briggs (see 2007), C. Briggs and J. W. Foster (see 2006), and S. Hildreth (see 5058).
6000 copies were printed.
For review with excerpts see 703, 2704, and 9749.

6892 1838, A series of geological queries contained in the first annual report on the geological survey of Ohio: Columbus, Ohio, Samuel Medary (printer), 38 p.
Also in First annual report ..., (see 6891), 111-121.

6893 1838, Geological survey of Ohio: Hesperian, v. 1, no. 6, 495-498.
Extracted from First annual report ..., (see 6891) on soil analyses.

6894 1838, Remarks in addition to and explanation of the review of the report of the geological survey of Ohio: Am. J. Sci., v. 34, no. 2, 362-364, 3 figs.
Includes notice of Mastodon bones from Jackson, Ohio.

6895 1838, Report - of the 1st geological district of the state of New-York: NY Geol. Survey Ann. Rept., v. 2, 121-184.

6896 1838 [i.e. 1839], and others, Second annual report on the geological survey of the state of Ohio: Columbus, Ohio, S. Medary (printer), 286 p., 18 plates (several colored or folded), map.
Contains reports by C. Briggs (see 2008), J. Locke (see 6424, 6425), and C. Whittlesey (see 10929).
For review see 5022 and 5346.

6897 1839, On cupellation, an easy, an accurate, and new method: Am. J. Sci.,v. 35, no. 2, 321-323.

6898 1839, Report on a geological reconnoissance of Kentucky made in 1838: Frankfort, Kentucky, 40 p.
Also in Kentucky House of Representatives J. for 1838-1839, Appendix, 239-278.

6899 1839, Third annual report - of the first geological district of the state of New York: NY Geol. Survey Ann. Rept., v. 3, 67-134.

6900 1840, Fourth annual report - of the first geological district of the state of New-York: NY Geol. Survey Ann. Rept., v. 4, 209-258.

6901 1841, [Analogy between fossils of Hudson slates in Rensselaer and Saratoga counties and in western New York] : Am. J. Sci., v. 41, no. 1, 164 (3 lines).
Also in Assoc. Am. Geol's Nat's Proc., v. 2, 9 (3 lines). Also in Assoc. Am. Geol's Nat's Rept., v. 1, 17 (3 lines), 1843.

6902 1841, Fifth annual report on the geological survey of the first geological district of New-York: NY Geol. Survey Ann. Rept., v. 5, 59-112.

6903 1841, [On glacial boulders and diluvial scratches in the United States] : Am. J. Sci., v. 41, no. 1, 174-176.
Also in Assoc. Am. Geol's Nat's Proc., v. 2, 19-22. Also in Assoc. Am. Geol's Nat's Rept., v. 1, 26-29, 1843.

6904 1841, [On joints in rocks] : Am. J. Sci., v. 41, no. 1, 172 (9 lines).
Also in Assoc. Am. Geol's Nat's Proc., v. 2, 17 (9 lines). Also in Assoc. Am. Geol's Nat's Rept., v. 1, 24-25 (9 lines), 1843.

6905 1842, Catalogue of the geological specimens, collected, on the late survey of the state of Ohio: Columbus, Ohio, 7 p., 11 tables.

6906 1843, Geological map of Long & Staten Islands with the environs of New York: NY, Endicott (lithographer), colored geological map, 124 x 55 cm.
Published as plate I of Geology of New York, first district (see 6907).

6907 1843, Geology of New York; part I: comprising the geology of the first geological district: Alb., Carroll and Cook (printers), xxxvi, 653, [1], [15], 42 plates, 4 maps.
Contains reports by T. A. Conrad (see 2576), L. D. Gale (see 4201 and 4202), W. Horton (see 5280 and 5281), H. D. Rogers (see 8890), W. H. Sidell (see 9598), J. Eights (see 3468), J. P. Young and J. Heron (see 11116), H. D. Rogers (see 8900), J. W. Bailey (see 1421), and maps (see 6906, 6908 to 6910).
For excerpts see 1734. For review see 8066.

6908 1843, [Map of the Hudson River district near Rhinebeck, Germantown, etc.] : NY, Endicott (lithographer), colored geological map, 17 x 20 cm.
Published as plate 29 of Geology of New York, first district (see 6907).

6909 1843, Sketch of the left bank of the Hudson from Barnegat to Fishkill: NY, Endicott (lithographer), colored geological map, 21 x 17 cm.
Published as plate 28 of Geology of New York, first district (see 6907).

Mather, W. W. (cont.)

6910   1843, Topographical sketch of the Sterling iron-mines and the vicinity: NY, Endicott (lithographer), colored geological map, 8 x 10 cm.
    Published on plate 30 of Geology of New York, first district (see 6907).

6911   1844, On the origin of the sedimentary rocks of the United States, and on the causes that have led to their elevation above the level of the sea: Am. J. Sci., v. 47, no. 1, 95-98.
    Also in Assoc. Am. Geol's Nat's Proc., v. 5, 2-5.

6912   1844, Report of the Coal Grove Company, Lawrence County, Ohio: Cin., Daily Atlas Office, 10 p.

6913   1845, Bromine and iodine [in salt springs at Athens, Ohio]: Am. J. Sci., v. 49, no. 1, 211 (11 lines).

6914   1845, On the physical geology of the United States east of the Rocky Mountains, and on some of the causes affecting the sedimentary formations of the earth: Am. J. Sci., v. 49, no. 1, 1-20; no. 2, 284-301.

6915   1847, On cupellation with the blowpipe: Am. J. Sci., v. 3 (ns), no. 3, 409-414.

6916   1848, Agricultural geology: Am. Q. J. Ag. Sci., v. 7, no. 2, 36-37.
    Extracted from Geology of New York, first district (see 6907).

6917   1848, The mineral kingdom, being the source of the matters in the animal and vegetable kingdom ... : Ohio Board Ag. Ann. Rept., v. 3, 389-401.

6918   1850, Foster's geological chart [notice of 3920]: Am. J. Sci., v. 9 (ns), no. 3, 444.
    On the forgery of W. W. Mather's signature.

Matteucci, Carlo (1811-1868)

6919   1850, New researches on the conductability of the earth: Am. J. Sci., v. 10 (ns), no. 3, 406-409.
    In British Assoc. Proc. Also in Franklin Inst. J., v. 20 (3s), 389-392.

Matthews, Oliver

6920   1847, Orange Grove Mining Company, (Vaucluse Gold Mine): Phila., Thomas H. Town (printer), 19 p.

Matthews, Thomas Johnston (1787-1852)

6921   1824, A lecture on Symmes' theory of concentric spheres, read at the Western Museum: Cin., A. N. Deming (printer), 14 p.
    See also 10079.

Maury, Matthew Fontaine (1806-1873)
    See also 10508 under United States Mining Company.

6922   1837, Description of an alembic for distilling amalgam of gold: Am. J. Sci., v. 33, no. 1, 66-70, 2 figs.

6923   1837, Notice of the gold veins of the United States' mine near Fredericksburg, Va.: Am. J. Sci., v. 32, no. 2, 325-330.

6924   1841, [Letter to the National Institute on the Tertiary formations of the United States]: Nat'l Inst. Prom. Sci. Bulletin, v. 1, 17-18.

6925   1844, The currents of the sea as connected with geology: Army and Navy Chron. and Scien. Repos., v. 3, 661-667.

6926   1848, Geological map from soundings [of the United States east coast]: Am. J. Sci., v. 6 (ns), no. 1, 149 (8 lines).

6927   1850, [Importance of Alabama coal fields]: Am. Assoc. Adv. Sci. Proc., v. 3, 73 (9 lines).

6928   1850, Sediment of the Mississippi River: Am. Assoc. Adv. Sci. Proc., v. 2, 334-335.
    As an introduction to report of R. A. Marr (see 6782).

Mawe, John (1764-1829)

6929   1810, Mineralogy of the Brazils: Am. Min. J., v. 1, no. 1, 52-53 (15 lines).

6930   1816, Travels in the interior of Brazil, particularly in the gold and diamond districts of that country, by authority of the Prince regent of Portugal: including a voyage to the Rio de la Plata, and an historical sketch of the revolution of Buenos Ayres. Illustrated with five engravings: Phila., M. Carey; Bos., Wells and Lilly, viii, 9-373 p., front., 5 plates, folded map.

6931   1818, On the tourmalin and apatite of Devonshire: J. Sci. Arts, v. 4, no. 2, 369-372.

Maxwell, Hugh (1787-1873)

6932   1831, Dauphin County coal region: Hazard's Reg. Penn., v. 7, 312-313.

Maxwell, J. B.

6933   1845, [On the discovery of Mastodon bones near Hackettstown, New Jersey]: Am. Philos. Soc. Proc., v. 4, no. 32, 118-121, 126-127.

Mayall, J. E.

6934   1846, Charles Lyell Esqr. F.R.S. F.G.S.: P.S. Duval, lithographer; portrait engraving b/w, 39 x 54 cm.
    Mayall was the Daguerreotypist and publisher.

Mayer, Brantz (1809-1870)

6935   1844, Mexico as it was, and is: NY, J. Winchester, xii, 390 p., plates.
    Other printings include Balt., W. Taylor, 1846, xv, 390 p.; NY, J. Winchester, 1847, 390 p.; and Phila., G. B. Zieber and Co., 1847, xii, 390 p. For extract see Hunt's Merch. Mag., v. 10, 118-131.

6936   1845, Thomaite; a new mineral species [from Siebengebirge, Germany]: Am. J. Sci., v. 49, no. 2, 393 (9 lines).
    Abstracted from Jameson's J.

6937   1850, Mexican mines and mineral resources in 1850: De Bow's Rev., v. 9, 31-43.

Mayerbach, Augustus

6938   1822, New theory of the deluge: Saturday Mag., v. 2 (ns), 114-117.

Mayhew, Jonathan (1720-1766)
6939 1755, A discourse on Rev. XV. 3d, 4th, occasioned by the earthquakes in November, 1755. Delivered in the West-Meeting-house, Boston, Thursday December 18, following. In five parts, with an introduction. Part I. Of the greatness of God's works. Part II. Of their marvellous and unsearchable nature. Part III. Of the moral perfections and government of God. Part IV. Of our obligation to fear, glorify and worship Him. Part V. Practical reflections upon the subject, relative to this occasion ... : Bos., Edes and Gill (printers), 72, (2) p.
Evans number 7486.
6940 1755, The expected dissolution of all things, a motive to universal holiness. Two sermons preached in Boston, N. E. on the Lord's Day, Nov. 23, 1755; Occasioned by the earthquakes, which happened on the Tuesday morning and Saturday evening preceding ...: Bos., Edes and Gill (printers), 76, 5 p.
Evans number 7487.
6941 1760, Practical discourses delivered on occasion of the earthquakes in November, 1755 ...: Bos., R. Draper, Edes and Gill, and T. and J. Fleet, 69 p.
Evans number 8667.

Meade, William
6942 1808, An inquiry into the chymical character and properties of that species of coal lately discovered in Rhode Island; together with observations on the useful application of it to the arts and manufacture of the eastern states: Bos., 21 p.
For review see 7347. Extracts in: Am. Min. J., v. 1, 34-40.
6943 1810, Description and analysis of an ore of lead from Louisiana: Am. Min. J., v. 1, no. 1, 7-10.
On lead ore from Ste. Genevieve, Missouri.
6944 1811, Mineralogical notice respecting elastic marble, from Massachusetts: Am. Min. J., v. 1, no. 2, 93-95.
6945 1812, A description of several combinations of lead, lately discovered at Northampton [Massachusetts]: Am. Min. J., v. 1, no. 3, 149-151.
6946 1814, Elastic marble: Am. Min. J., v. 1, no. 4, 267-268.
6947 1817, An experimental enquiry into the chemical properties and medicinal qualities of the principal mineral waters of Ballston and Saratoga, in the state of New-York ... To which is added an appendix containing a chemical analysis of the Lebanon spring in the state of New-York: Phila., H. Hall, 195 p., illus., plates.
For review with excerpts see 7681. For review see 8468 and 8473.
6948 1818, A chemical analysis of the waters of New Lebanon in the state of New-York: Burlington, NJ, David Allinson, 45 p.
6949 1822, Letter from Dr. William Meade, communicating an account of a travelled stone, &c.: Am. J. Sci., v. 6, no. 1, 158.
6950 1823, Localities of minerals [in Massachusetts and Connecticut]: Am. J. Sci., v. 7, no. 1, 49-54.
6951 1827, Account of the new mineral spring at Albany, with an analysis and remarks: Am. J. Sci., v. 13, no. 1, 145-158.
6952 1827, Chemical analysis and description of the coal lately discovered near Tioga River, in the state of Pennsylvania: Am. J. Sci., v. 13, no. 1, 22-35.
6953 1827, Observations on the analogy between the minerals of the north of Europe and of America, more particularly as connected with the uniformity of their geological situation in both countries: Am. J. Sci., v. 12, no. 2, 303-309.
6954 1827, Remarks on the anthracites of Europe and America: Am. J. Sci., v. 12, no. 1, 75-83.
6955 1828, A chemical analysis of the Pittsburgh mineral spring: Am. J. Sci., v. 14, no. 1, 124-135.
6956 1828, Death of the Hon. George Knox: Am. J. Sci., v. 15, no. 1, 189.
6957 1828, Pittsburgh mineral spring: Hazard's Reg. Penn., v. 2, 81-84.
6958 1830, Description of a new locality of zircon, particularly referring to its geological character [at Orange, New York]: Am. J. Sci., v. 17, no. 1, 196-197.
6959 1830, On the use of black mica, as a substitute for colored glasses in spectacles: Am. J. Sci., v. 18, no. 2, 374-375.

Mease, James (1771-1846)
6960 1807, A geological account of the United States; comprehending a short description of their animal, vegetable, and mineral productions, antiquities and curiosities: Phila., Birch and Small, [8], 496, xiv p., plates.
For review see 7065 and 7344.
6961 1811, Iron and lead [in the United States]: Arch. Use. Know., v. 1, no. 3, 247-249.
6962 1811, On iron: Arch. Use. Know., v. 2, no. 2, 142-151.
6963 1811, The picture of Philadelphia, giving an account of its origin, increase and improvements in arts, sciences, manufactures, commerce and revenue ... : Phila., B. and T. Kite, xii, 376 p., folded front.

Meaux, Thomas (of Amelia, Virginia)
6964 1833, Marl discovered in the granite and coal region of Virginia: Farm's Reg., v. 1, 424-426.

Mechanic Apprentice (Boston, 1845-1846)
6965 1845, Saratoga Springs, as they were and are: v. 1, 29.

The Mechanics' Magazine, and Journal of Public Internal Improvement (Boston, 1830)
6966 1830, [Anthracite coal used in brick making]: v. 1, 188-189.
Abstracted from Am. J. Sci.
6967 1830, Gold and platina: v. 1, 208-209.
Abstracted from Annals of Phil.
6968 1830, Franconia iron works [in New Hampshire]: v. 1, 253-254.
Abstracted from Am. J. Sci.
6969 1830, Volcano [Mount Etna's eruption]: v. 1, 286 (11 lines).
6970 1830, Density of the earth: v. 1, 310-311.
6971 1830, Temperature of the interior of the earth: v. 1, 311.
6972 1830, Descent from Mauch Chunk coal mine: v. 1, 334-335.
Abstracted from Am. J. Sci.

The Mechanics' Magazine, and Journal (cont.)
6973　1830, Lead, produced at the United States lead mines, annually from 1823 to 1829: v. 1, 364 (9 lines).

Mechanics' Magazine, and Register of Inventions and Improvements (New York, 1833-1837)
6974　1833-1837, [Notes on American mines and mining]: v. 1, 20-21; v. 2, 71; 129-130; v. 3, 118, 256; v. 4, 46, 128; v. 5, 126-127; 127-128; 224; v. 6, 108-114; v. 7, 128, 148-152; v. 8, 251-252; v. 9, 72; 137-139; 350.
6975　1833-1837, [Notes on foreign mines and mining]: v. 1, 58-59; v. 4, 171, 352; v. 7, 251, 284-286; v. 9, 272.
6976　1833, Earthquakes [at Swansea, England]: v. 1, 113-114.
6977　1833, Cuvier [on his contributions to science]: v. 2, 25.
6978　1833, Inflammable spring [at Wales, New York]: v. 2, 104 (9 lines).
　　　　On natural gas from the vicinity of Buffalo, New York.
6979　1834, On the dip and declination of the needle: v. 3, 368; v. 4, 16.
6980　1834, Fossil horns [from Lancaster Canal, Pennsylvania]: v. 4, 16 (7 lines).
6981　1834, Geological surveys: v. 4, 224.
　　　　On the progress of the several state surveys.
6982　1835, Curiosities of caves: v. 5, 96 (14 lines).
　　　　On bone caves of the Hartz Mountains.
6983　1835, Fossil wax of Moldavia: v. 5, 96 (14 lines).
　　　　On amber with fossil insects.
6984　1835, Great mass of meteoritic iron from Louisiana: v. 5, 127.
6985　1835, Primary geology [review of text by Dr. Boase]: v. 5, 230.
6986　1835, The volcano of Popocatepetl: v. 6, 229-230.
6987　1836, Geological survey of the state (NY): v. 7, 191-192.
6988　1836, Fossil tree [at Sunderland]: v. 7, 192 (17 lines).
6989　1836, Marble cement: v. 7, 323-324.
6990　1836, Ornamental slate manufacture: v. 8, 174.
6991　1837, Natural history of Missouri earthquakes: v. 9, 21-22.
6992　1837, Declination and inclination of the magnetic needle at Paris: v. 9, 48 (8 lines).
6993　1837, Pennsylvania College of Mines: v. 9, 90-91.
6994　1837, Precious stones: v. 9, 341-343, 269-270 [i.e. 369-370].
　　　　Abstracted from Scien. and Lit. J.

The Mechanics' Mirror (Albany, 1846)
6995　1846, Coal mines. Fuel: v. 1, 71.
　　　　On need for both fuel and mine safety
6996　1846, Mineral structure: v. 1, 71.
　　　　On the grain size of minerals and texture in rocks.
6997　1846, On the art of assaying gold and silver: v. 1, 95-96, 106-107.
6998　1846, Mount Hecla: v. 1, 217 (14 lines).
6999　1846, [On fossil mammalia of the British Isles]: v. 1, 283.
　　　　Abstracted from British Assoc. Proc.

Mecklenburg Gold Mining Company, North Carolina
7000　1830, Report on incorporating the Mecklenburg Gold Mining Company: Raleigh, NC, Lawrence and Lemay (printers), 8 p.
7001　1833, Act of incorporation, and report of secretary of the company ... with accompanying documents: NY, W. Tolefree, 31 p.

The Medical and Agricultural Register (Boston, 1806-1807)
7002　1806, Gypsum, better known by the name "plaster of Paris" - signs by which to judge its purity: v. 1, 24-25, 40-43.
　　　　"By a member of the Kennebeck Ag. Soc."
7003　1806, Gold discovered in North Carolina: v. 1, 44-45.
　　　　Abstracted from Med. Repos.

The Medical Examiner (Philadelphia, 1838-1850)
7004　1840, Letters from the Virginia Springs: v. 3, no. 31, 485-486; no. 33, 518-521.
7005　1844, The artesian wells at Naples: v. 7, 300 (6 lines).
　　　　Abstracted from Medical Times.
7006　1847, [Review of The Virginia springs by J. Moormann (see 7422)]: v. 3 (ns), 408-409.

The Medical Magazine (Boston, 1832-1835)
7007　1834, Mineral and thermal springs: v. 2, no. 21, 518-521.
　　　　Abstracted from Med. and Chirurgical Rev.

The Medical Repository (New York, 1798-1818)
　　　　Samuel Latham Mitchill was editor of The Medical Repository, and he probably authored or excerpted many of the following unsigned notes and reviews.
7008　1798, [Review of "Agricultural experiments on gypsum ..." by G. Logan (see 6466)]: v. 1, 350-351.
7009　1798, [Review of "Agricultural inquiries on plaister of Paris" by R. Peters (see 8247)]: v. 1, 351-353.
7010　1798, [On combustion of the diamond]: v. 1, 259 (9 lines).
7011　1798, [An earthquake in Lancaster, Pennsylvania]: v. 1, 572 (7 lines).
7012　1799, [Analysis of a Peruvian emerald]: v. 2, 222 (6 lines).
　　　　Notice of Vauquelin's research.
7013　1799, [A new species of fluorite]: v. 2, 225 (9 lines).
7014　1799, Magnetism: v. 2, 443.
　　　　On the magnetic properties of basalt and serpentine.
7015　1799, [Formation of American Mineralogical Society]: v. 2, 114-116.
7016　1799, Mineralogy: v. 2, 215-217.
　　　　A circular asking for mineral specimens for the cabinet of the American Mineralogical Society. S. L. Mitchill is listed as the President.
7017　1800, [Review of "Situation du volcan de la Guadaloupe" by Victor Hughes]: v. 3, 51-54.
7018　1800, The Chemical Society of Philadelphia: v. 3, 68-69.
　　　　On mineral specimens received by the Society.

Medical Repository (cont.)
7019 1800, Combustion of diamonds: v. 3, 313-314.
       Notice of experiments by Guiton.
7020 1801, Fossil coal on the River Lehigh: v. 4, 71 (7 lines).
7021 1801, Churchman's magnetic atlas [notice of third edition (see 2371)]: v. 4, 73.
7022 1801, Mineralogical Society of Jena [Saxony]: v. 4, 78.
       Charles Loss elected a corresponding member from New York.
7023 1801, Grand specimen of Mexican gold [described]: v. 4, 201-202.
7024 1801, Chemical Society of Philadelphia [notice of geological essay by George Lee]: v. 4, 303 (8 lines).
7025 1801, Tusk of the mammoth [from Hudson River, New York]: v. 4, 308.
7026 1801, Meteor and earthquake [at Pittsburgh, Pennsylvania]: v. 4, 324-325.
7027 1801, Discoveries in mineralogy: v. 4, 419.
       On American mineral localities.
7028 1801, Extinct species of animals: v. 4, 419-421.
       Notice of Cuvier's research on fossil quadrupeds.
7029 1802, [Review of Introduction to a course of lectures on natural history and Discourses introductory to a course of lectures by C. W. Peale (see 8179 and 8180)]: v. 5, 55-58.
7030 1802, Skeleton of the unknown quadruped: v. 5, 83.
       On a specimen in the Peale Museum.
7031 1802, Premium of the Chemical Society [of Philadelphia]: v. 5, 349.
       For the discovery of pottery clay in America.
7032 1803, New American metal [on Columbium]: v. 6, 212 (6 lines).
7033 1803, Platina from the mines of Chaco, in Terra Firma [Jamaica]: v. 6, 213-214.
7034 1803, Mineral spring near Passaick Falls [New Jersey]: v. 6, 214-215.
7035 1803, Fossil shells of Long-Island: v. 6, 217-218.
7036 1803, Coal trade between the United States of America and foreign parts: v. 6, 437-438.
7037 1804, Curious mineralogical appearance in the City of Washington: v. 7, 199-200.
       Evidence for an old forest found in excavation.
7038 1804, Remarkable facts touching the geology of the Atlantic Territory of Virginia [whale skeleton found]: v. 7, 201.
7039 1804, Slate-quarry on the west bank of the Hudson [at New Paltz]: v. 7, 295 (9 lines).
7040 1804, Meteoric stones [from Ensisheim, Germany]: v. 7, 296 (12 lines).
7041 1804, Description of a cavern in Ulster County, in the state of New-York: v. 7, 303-304.
7042 1804, Lead, iron and tin mines in the Western country: v. 7, 306.
7043 1804, Native gold discovered in North-Carolina: v. 7, 307.
7044 1804, Earthquake at New-York: v. 7, 416 (13 lines).
7045 1804, Barytes discovered in New-Jersey: v. 7, 427 (11 lines).
7046 1805, Cabinet of French minerals at Cambridge (Mas.): v. 8, 209 (12 lines).
7047 1805/1806, [Review with excerpts of A view of the soil ..., by Volney (see 10653)]: v. 8, no. 2, 172-196; no. 4, 410-420; v. 9, no. 3, 276-286.
7048 1805, New mineral spring in Virginia [at Harrisonburgh]: v. 8, 355.
7049 1805, Other discoveries of slate for roofing [in Pennsylvania and Maryland]: v. 8, 433 (9 lines).
7050 1805, Elegant and instructive collections of minerals: v. 8, 433-434.
       On the collections of A. Bruce and B. D. Perkins.
7051 1805, Place where the ore of columbium was found [New London, Connecticut]: v. 8, 437.
7052 1805, More gold picked up in North-Carolina [in Cabarrus County]: v. 8, 439-440.
7053 1806, Caverns in Virginia, Kentucky, and Tennessee, which afford an inexhaustable supply of salt-petre: v. 9, 86-87.
7054 1806, The greatest lead mines in the world existing in Upper Louisiana: v. 9, 87-88.
7055 1806, Domestic supply of sulphur in New-York [in Ontario County]: v. 9, 88-89.
7056 1806, Surprising extrication of inflammable air at Licking River, Kentucky]: v. 9, 89-90.
7057 1806, Native sulphate of magnesia and soda in Virginia [in caverns of Greenbriar County]: v. 9, 112.
7058 1806, Lead mine of Pennsylvania [at Perkiomen Creek]: v. 9, 112-113.
7059 1806, [Review of Journey up the Washita by W. Dunbar and W. Hunter (see 3274)]: v. 9, 305-308.
7060 1806, Sulphate of barytes in New-Jersey and Maryland: v. 9, 332.
       On deposits in Sussex County, New Jersey and Sugar Loaf Mt., Maryland.
7061 1806, Extensive layers of marine shells found in Georgia and the Mississippi Territory: v. 9, 436.
7062 1807, A splendid cavern discovered in the lime-stone country of Virginia, in 1806: v. 10, 298-300.
7063 1807, Antimony discovered in New-York [at Sagherties]: v. 10, 304 (6 lines).
7064 1807, English native rock-salt, or Sal Gem: v. 10, 413-416.
7065 1808, [Review of A geological account of the United States by J. Mease (see 6960)]: v. 11, 42-44.
7066 1808, Mexican mineralogy [notice of text by M. Del Rio]: v. 11, 68-69.
7067 1808, American ore of titanium [in New Jersey]: v. 11, 69-70.
7068 1808, Mineralogical discoveries: v. 11, 96.
       On molybdenite from Chester County, Pennsylvania.
7069 1808, Petrified wood found in Maryland: v. 11, 199-200.
7070 1808, Gibb's grand collection of minerals: v. 11, 213-214.
7071 1808, Analysis of Balltown waters: v. 11, 214-215.
7072 1808, American tourmaline: v. 11, 307.
7073 1808, Farther discoveries in fossil zoology: v. 11, 318-319.
       On fossil bones discovered in New York and Kentucky.
7074 1808, The mammoth really a northern elephant: v. 11, 319-320.
       On Cuvier's opinions of the mammoth.
7075 1808, Uncommon petrifactions, from Georgia and Kentucky: v. 11, 415-416, 5 figs.
       On fossil echinoids.
7076 1808, Ferruginous oxide of manganese [from Ancram, New York]: v. 11, 442-443.
7077 1808, Specimens of minerals lately discovered in Europe: v. 11, 443 (9 lines).
7078 1809, Arrangement for a museum of minerals: v. 12, 91-92.
       On Prof. Leske's organization of minerals in his museum.
7079 1809, Progress of finding gold in North-Carolina: v. 12, 192-193.
7080 1809, American ochres [from near Philadelphia]: v. 12, 193-194.

Medical Repository (cont.)
7081   1809, Discovery of valuable minerals [in New Jersey for making paint] : v. 12 194-195.
7082   1809, Sulphuric acid ascertained to exist in a free or uncombined state [at Clifton Springs, Connecticut] : v. 12, 200.
7083   1809, [Review of The geographical, natural, and civil history of Chile by J. I. Molina (see 7332)] : v. 12, 257-272.
7084   1809, Maclure's geological enquires: v. 12, 295-296.
         On the progress of W. Maclure's United States field work.
7085   1809, A natural saltpetrous earth, discovered near the South Branch of Potomac, in Virginia: v. 12, 296-297.
7086   1809, Encouragement for discovering antimony in the United States: v. 12, 299-300.
         Notice of failure of the mine at Sagherties, New York.
7087   1810, Another species of atmospheric stone, descended incrusted with ice [from France] : v. 13, 189-190.
7088   1810, Mineralogical and metallurgical institution: v. 13, 194.
         Notice of lectures by Gibbs.
7089   1810, Mineralogical journal: v. 13, 202.
         Prospectus of Bruce's Am. Min. J. (see 2107).
7090   1811, [Review with excerpts of Transactions of the American Philosophical Society, Vol. IV, pt. 2] : v. 14, 59-73.
         Includes review with excerpts of articles by B. H. Latrobe (see 6067); J. Cloud (see 2456); W. Maclure (see 6615); and reviews of articles by S. Godon (see 4423); and B. Silliman and J. L. Kingsley (see 9601).
7091   1811, Scudder's Museum of Natural History in New-York: v. 14, 88.
7092   1811, Journal of mineralogy: v. 14, 99-100.
         Notice of the progress of Bruce's journal (see 2107).
7093   1811, Plaster of Paris in Madison County, N. Y.: v. 14, 182.
7094   1811, Earthquake in Georgia: v. 14, 393-394.
7095   1812, Quartz crystallized around lead [from Shawangunk Mt., Orange County, New York] : v. 15, 85-86.
7096   1812, Large mass of malleable iron [from Red River, Louisiana] : v. 15, 88.
7097   1812, Extraordinary crystal of quartz: v. 15, 88-89.
7098   1812, Surprising extrication of inflammable air [at Licking River, Kentucky] : v. 15, 94-95.
7099   1813, Native malleable iron: v. 16, 424.
         From Bruce's Am. Min. J.
7100   1813, Earthquakes in Venezuela: v. 16, 425-428.
7101   1815, [Review of Annual address ..., by T. R. Beck (see 1590)] : v. 17, 151-153.
7102   1815, Green River, or Mammoth Cave, Henderson County, Kentucky: v. 17, 391-393, plate.
7103   1815, [On Bruce's American Mineralogical Journal (see 2107)] : v. 17, 415 (12 lines).
7104   1817, [On the progress of geology] : v. 18, preface, p. v.
7105   1818, Native copper of North America [from Lake Superior] : v. 19, 101-102.
7106   1820, [Notice of An essay on the geology of the Hudson River by S. Akerly (see 115)] : v. 20, 442 (6 lines).
7107   1821, Petrified elephant tooth [from Tuscany] : v. 21, 366.
7108   1822, Remarks on the origin of amber: v. 22, 98-101.

Medical Society of Orange County, New York
7109   1820, Report of the committee appointed ... to analyze the waters of Chechunk Spring ... : Goshen, NY, T. B. Crowell (printer), 24 p.

The Medley or Monthly Miscellany (Louisville, Ky, 1803)
7110   1803, v. 1, contains short notes on Lake Superior, lunar volcanoes, and Virginia Mountains.

Meeker, John
7111   1815, An inaugural dissertation on the principal mineral waters of the states of New-York and New-Jersey ... : NY, n.p., 37 p.

Meissner (see 2147 with Bucholtz)

Mellen, John
7112   1795, An account of some effects of the great earthquake, in the year 1755, ... : Mass. Hist. Soc. Coll., v. 4, 231-232.

Melloni, Macedonio (1798-1854)
7113   1835, Transmission of radiant heat through different solid and liquid bodies: Am. J. Sci., v. 27, no. 2, 228-236.
         Abstracted by J. Griscom from Bibliothèque Universelle.

Melograni, M. L. Abbé
7114   1807, [A new blowpipe] : Am. Reg. or Gen. Repos., v. 1, pt. 2, 68.

Melville, Alexander Garden (see 10038 with H. E. Strickland)

Memorial of the salt manufacturers in the town of Salina, Onandago County, New York against the repeal of duty on imported salt.
7115   1828, Wash., Duff Green, 4 p. (SED 48).

Mercer, Charles Fenton (1778-1858)
7116   1834, Report of the Hon. Charles Fenton Mercer [on the Chesapeake and Ohio Canal] : US HMD 414, 23-1, 378 p.
         Appendix "Z" is on the coal and iron trade.

Meredith, Thomas
7117   1827, Notice of the Belmont anthracite mines in Pennsylvania: Am. J. Sci., v. 12, no. 2, 301-302.

Meriam, Ebenezer (1794-1864)
7118   1841, Salt rock [from Syracuse, New York] : Hazard's US Reg., v. 5, 211.
7119   1842, Virginia salt mine: Farm's Reg., v. 10, 21.
7120   1845, The arkansite [from Hot Springs, Arkansas] : S. Platner, v. 5, 221.

Meriam, E. (cont.)
7121   1847, Earthquakes, lightning storms, hurricanes, etc.: Scien. Am., v. 2, no. 26, 203.
7122   1847, Volcanoes and earthquakes: Scien. Am., v. 2, no. 37, 291.
7123   1849, Geological tour in the state of New-York: Am. Inst. NY Ann. Rept., v. 7, 156-164.
7124   1850, Manufacture of salt at state works - marl deposits, &c.: NY State Agri. Soc. Trans., v. 9, 412-414.

Meriwether, David (General)
7125   1803, Particulars of a remarkable body of sea-shells now existing in the interior part of the state of Georgia: Med. Repos., v. 6, 329.

Merrimack Magazine and Ladies Literary Cabinet (Newburyport, Massachusetts, 1805-1806)
7126   1805, Great earthquake [at Naples, Italy]: v. 1, 43.

Merry's Museum (New York, 1841-1850)
7127   1841, The mammoth: v. 2, 152.
7128   1842, The wonders of geology [review of G. A. Mantell's English edition (see 6732)]: v. 3, 3-7, 3 figs.
7129   1842, Discovery of the mines of Potosi: v. 4, 165-166, fig.
7130   1846, Mineral coal: v. 11, 66-67, 4 figs.
7131   1847, Precious stones: v. 13, 172.
7132   1847, Chalk: v. 14, 63.
7133   1848, Changes of the earth's surface: v. 15, 141-142.
7134   1848, Wonders of geology: v. 16, 183, fig.
       On Buckland's opinions on extinct quadrupeds.
7135   1849, Wonders of the West: v. 17, 69-70.
       On the Arkansas Hot Springs.
7136   1849, Lumps of gold: v. 17, 134.
       List of world's largest gold nuggets.
7137   1850, The coal and the diamond: v. 20, 78.
       A poem on the chemical relationships between coal and the diamond.

Metcalf, John (see 10508 under United States Mining Company)

Metcalf, Samuel Lyther (1798-1856)
7138   1833, A new theory of terrestrial magnetism: NY, G. and C. Carvill, v, 158 p., tables.
       "Read before the New-York Lyceum of Natural History."
       For review see 1087 and 10855. Abstracted in Transylvania J. Medicine and Associated Sci's, v. 6, 432-435 (1833) and Am. J. Sci., v. 27, no. 1, 153-154.
7139   1834, The interest and importance of scientific geology as a subject for study: Knickerbocker, v. 3, 225-235.

The Methodist Magazine (Philadelphia, 1797-1798)
7140   1797, A description of the famous copper mine, belonging to his Grace the Duke of Devonshire, at Ecton-Hill, in the County of Stafford: v. 1, 457-460.
7141   1798, A description of the mines of salt at Wiliska, in Poland: v. 2, 131-133, 163-164.
7142   1798, An account of the ancient City of Herculaneum, destroyed by an eruption of Mount Vesuvius, with descriptions of some antiquities found there: v. 2, 176-179, 223-226.

The Methodist Magazine (New York, 1818-1850)
7143   1819, Remarks on the Giant's Causeway, in Ireland: v. 2, 294-295.
       Abstracted from Armenian Mag.
7144   1821, Account of a volcano in the sea: v. 4, 457-458.
       Abstracted from London Meth. Mag.
7145   1821, Account of the geysers, or boiling springs, in Iceland: v. 5, 19-21.
       Abstracted from London Meth. Mag.
7146   1833, Mount Etna: v. 15, 118-120.
       Abstracted from NY Mess.
7147   1837, Geology: v. 8 (ns), 100-103.
       Signed by "G. F. C." On the miracle of Creation, and the futility of physical explanations of this event.
7148   1843, [Review of Geological cosmogony by E. Lord (see 6500)]: v. 3 (3s), 639 (3 lines).
7149   1845, [Review of Geology of Scriptures by G. Fairholme (see 3628) and Scripture and geology by J. P. Smith (see 9895)]: v. 5 (3s), 198-220.
7150   1846, [Review with excerpts of Vestiges of the natural history of Creation and Explanations: a sequel to Vestiges ..., by R. Chambers (see 2291 et seq.)]: v. 6 (ns), 292-327.
7151   1847, [Review of Incentives to the cultivation of the science of geology by S. S. Randall (see 8684)]: v. 7 (3s), 633 (5 lines).
7152   1849, [Review of Earth and man by A. Guyot (see 4609 and 4610)]: v. 9 (3s), 501-502.
7153   1849, [Review with excerpts of A second visit ..., by C. Lyell (see 6571)]: v. 9 (3s), 667-669.
7154   1849, The Jordon and the Dead Sea [review of report by the Lynch expedition (see 6581)]: v. 9 (3s), 633-653.

The Methodist Quarterly Review (see The Methodist Magazine; New York, third series, 1841-1850)

Methuon
7155   1817, Découverte de la manière dont se forment les cristaux terreux et metalliques non solins, &c.: J. Sci. Arts, v. 1, no. 1, 123-130.
       Translated by A. B. Granville.

Michener, Ezra (1794-1887)
7156   1847, Fossil corn [from Wheeling, Pennsylvania]: Farm's Cab., v. 11, 286-287.

Michigan. State of
7157 1837, An act to provide for a geological survey of the state: Mich. Gen. Assembly, February 23, 1837.
7158 1838, An act to provide for the improvement of certain state salt springs: Mich. Gen. Assembly, March 24, 1838.
7159 1838, Report of a select committee of the board of regents on a collection of the state geologist: Mich. HD 1, 1-2.
　　Also in Mich. HD 55, 1. Also in Mich. SD 1, 1.
　　Zina Pitcher, chairman of the committee.
7160 1839, Report of the committee on the state geologist's report in relation to the improvement of the state salt springs: Mich. HD 4, p. 123.
　　Norman Little, chairman of the committee.
7161 1839, Report of the committee of the Senate on manufactures, to whom was referred the communication of the state geologist relative to salt springs and the salines of the state: Mich. SD 3, 85-86.
7162 1840, Report of the select committee to whom was referred the several reports of the state geologist: Mich. HD 46, v. 2, 455-461.
7163 1840, Report of the majority of the committee on finance on the communication and accounts of the state geologist for 1839; report of the minority of the committee of finance on the same subject; report of the select committee on the state geologist's report and account relative to improvement of salt springs, etc.; State geologist's account for the year 1839 ... : Mich. SD 15, 16, 17, and 18, pp. 209-224.
7164 1842, An act making an appropriation for the improvement of the state salt springs: Mich. Gen. Assembly, February 1, 1842.
7165 1842, An act making appropriations for the current expenses of the government for the year 1842: Mich. Gen. Assembly, February 17, 1842.
　　Includes salary for a state geologist.
7166 1842, Report of the select committee in relation to the report of the state geologist: Mich. HD 19, 77-79.
　　F. C. Annable, chairman of the committee.
7167 1846, [Remarks on the geological survey of Michigan] : in Governor's Message, January, 1846.
7168 1846, Report of the joint committee relative to the geological survey: Mich. JD 15, 8 p.
7169 1846, [A resolution for the appointment of an individual to organize the geological notes of the late Douglass Houghton] : Mich. Gen. Assembly, May 15, 1846.
7170 1846, [A resolution transferring the state geological collection to the University] : Mich. Gen. Assembly, April 7, 1846 and May 11, 1846.

The Michigan Farmer, and Western Horticulturist (Jackson, Michigan, 1843-1850)
7171 1844, Copper mines on Lake Superior: v. 2, 3 (14 lines).
7172 1844, Michigan salt works [at Grand Rapids, Michigan]: v. 2, 18.
7173 1845/1848, Analysis of soils: v. 3, 143; v. 6, 225-226.
7174 1849, The Michigan gold region: v. 7, 107.
　　A poem on the relative value of gold mines and agriculture.
7175 1850, The nature of soils: v. 8, 94.

Michler, Nathaniel (1827-1881)
7176 1850, [Report of a reconnaissance of a route from the upper valley of the south branch of Red River to the Rio Pecos] : US SED 64, 31-1, v. 14 (562), 30-39.
　　Also in US HED 67, 31-1, v. 8 (577), 3-12.

The Microscope and General Advertiser (Louisville, Kentucky and New Albany, Indiana, 1824-1825)
7177 1825, Gold finders [in North Carolina] : v. 2, no. 15, [3].

Microscopical Society of London
7178 1840, Proceedings of the Microscopical Society of London: Am. J. Sci., v. 39, no. 1, 203-205.

Middendorff, Alexsandr Fedorovich (1815-1894)
7179 1846, Expedition to Siberia: Am. J. Sci., v. 1 (ns), no. 1, 146.
　　Abstracted from l'Institut.

Middleton, J.
7180 1844, Fluorine in bones: Am. J. Sci., v. 47, no. 2, 419 (7 lines).
　　Abstracted from Philos. Mag. On the chemical analysis of fossil bones.
7181 1845, On the comparative composition of recent and fossil bones: Am. J. Sci., v. 48, no. 1, 186.
　　Abstracted by J. L. Smith from London and Edinburgh Philos. Mag.

Mifflin, John Houston (1807-1888)
7182 1837, Vesuvius: Burton's Gentleman's Mag., v. 1, 421.
　　A poem about an ascent of Vesuvius in 1837.

Mighels, Jesse W.
7183 1842, and C. B. Adams [Description of fossil shells Nucula and Bulla occurring at Westbrook, Maine] : Bos. J. Nat. History, v. 4, 53-54.

Miller, A. E.
7184 1836, Account of the medical properties of the Grey Sulphur Springs, Virginia: Char., 18 p.
　　Contains analyses by C. U. Shepard. For review with excerpts see 9951.

Miller, Edward
7185 1835, Geological description of a portion of the Alleghany Mountain, illustrated by drawings and specimens: Geol. Soc. Penn. Trans., v. 1, 251-255, plate.

Miller, Hugh (1802-1856)
7186   1850, Footprints of the Creator; or, the Asterolepis of Stromness: Bos., Gould, Kendall and Lincoln, 337 p., illus.
        With a memoir of the author by L. Agassiz (see 87).
        For review see 1221, 4826, 5230, 5487, 5522, 6405, 9344 and 10529.
7187   1850, [Excerpts from The Old Red Sandstone]: Living Age, v. 25, 145-162.
        In review (see 6405).

Miller, John Peter (1709-1796)
7188   1786, Description of the Grotto at Swatara [in Pennsylvania]: Am. Philos. Soc. Trans., v. 2, 177-178.
        Also in Colum. Mag., v. 1, 525-526 (1787); Am. Mus. or Univ. Mag., v. 3, 141-142 (1788); Mass. Mag., v. 3, 101-102 (1790); Lancaster Hive, v. 1, 26-27 (1803).

Miller, Samuel (1769-1850)
7189   1803, A brief retrospect of the eighteenth century. Part first; in two volumes; containing a sketch of the revolutions in science, art, and literature, during that period: NY, T. and J. Swords, 3 v.

Miller, Sylvanus
7190   1801, Account of large bones dug up in Orange and Ulster Counties (State of New-York): Med. Repos., v. 4, 211-213.
7191   1815, A letter to De Witt Clinton on the fossil bones of the mammoth, discovered in the state of New York, with some observations on the adjacent country: NY, Nicholas van Riper (printer), 15 p.
7192   1836, Retrospective notice of the discovery of fossil mastodon bones in Orange County, (N. Y.): Am. J. Sci., v. 31, no. 1, 171-172.
        Abstracted in Mech's Mag. and Reg. Invention, v. 9, 28-29.

Millington, John (see 3088 with A. Del Rio)

Millon, Auguste Nicolas Eugène (1812-1867)
7193   1846, On the quantitative determination of mercury: Am. J. Sci., v. 2 (ns), no. 2, 258-259.
        Abstracted by J. L. Smith from Annales de Chimie.

Mills, Robert (1781-1855)
7194   1826, Statistics of South Carolina, including a view of its natural, civil, and military history, general and particular: Char., Hurlburt and Lloyd, vii, [17]-782, [48] p., map.
        Contains section by L. Vanuxem (see 10574).

Milne, David, later David Milne Home (1805-1890)
7195   1839, On the Berwick and North Durham coal-field [in England]: Am. J. Sci., v. 35, no. 2, 308 (3 lines).
        In British Assoc. Proc.

Miner, Charles (1780-1865)
7196   1815?, Certificates from a number of persons, shewing the use and value of the Lehigh Stone Coal. With some prefatory remarks: Wilkes-Barre, Pa., n.p., 8 p.

Miner, Thomas (1777-1841)
7197   1829, Information concerning the digging, preparation, and use of peat; from a memoir of Ribaucourt, published by the Council of Mines: Am. J. Sci., v. 15, no. 2, 250-260.

Mineral Creek Copper Mining Company
7198   1845, Articles of Association: Bos., 16 p., colored geological map.
        Contains "Report" by E. Hurd (see 5491 and 5492).

Mineralogical Society of Virginia
7199   1836, Proceedings of the Mineralogical Society of Virginia: Farm's Reg., v. 4, 315-316.
        Organized at Prince Edward court house, July 23, 1836.

The Minerva; or Literary, Entertaining, and Scientific Journal (New York, 1822-1825)
7200   1822-1824, [Notes on fossils]: v. 1, 12; 46-47; 94; 151; 222; 246; 335; v. 2, 15; 54; 166; 358; 374; v. 1 (ns), 186; v. 2 (ns), 68, 155.
7201   1822, African mineralogy: v. 1, 54.
7202   1822-1824, [Notes on mines and mining]: v. 1, 63; 214; 270; v. 2, 53; 166; 319; 333-334; 366; 373; 382; 408; v. 1 (ns), 235; 266; 266-267.
7203   1822, A crystal of quartz [from Greenville, South Carolina]: v. 1, 70.
7204   1822, Scientific institutions in Paris: v. 1, 86-87.
7205   1822-1825, [Notes on volcanoes and earthquakes]: v. 1, 134; 150; 269-270; v. 2, 326; v. 1 (ns), 75, 123; v. 2 (ns), 155, 298.
7206   1822, On geology: v. 1, 246.
7207   1822, The ruby: v. 1, 246.
        On its color changes with changes in temperature.
7208   1822, An aerolite [which fell near La Baffe, France]: v. 1, 302.
7209   1823, Remarkable glacier [in the Behring Strait]: v. 1, 318.
7210   1823, Structure of the earth: v. 2, 30.
        On W. Buckland's opinions.
7211   1823, Blowpipe: v. 2, 31.
        On new model made by Gurney.
7212   1823, Encroachments of the sea: v. 2, 38.
7213   1823, Asphaltum of the Gulf of Mexico: v. 2, 53.
7214   1823, Geology [notice of Geognostic essay, by A. Humboldt]: v. 2, 70.
7215   1823, Figure of the earth: v. 2, 134.
7216   1823, Bitumen in minerals: v. 2, 214.
        Found in many minerals by G. Knox.
7217   1823, On the variation of the needle: v. 2, 254.

The Minerva (cont.)
7218   1823, Pumice stone: v. 2, 270 (9 lines).
7219   1823, On the phenomena of the deluge: v. 2, 277.
7220   1823, New application of Bath stone: v. 2, 278.
            Oolite use for jars and vessels.
7221   1824, Anatomy of the earth: v. 2, 357.
            Notice of unsigned work.
7222   1824, Volcanic eruption in Iceland: v. 2, 398.
7223   1824, Figure of the earth: v. 1 (ns), 104-105.
7224   1824, To geologists: v. 1 (ns), 121.
            On the strata at a well in Owego, Tioga County, New York.
7225   1824, The grotto of Antiparos: v. 1 (ns), 261-262.
7226   1824, Aerolites, or meteoric stones: v. 1 (ns), 280-281.
7227   1824, [On Cuvier's visit to Lyme, England]: v. 2 (ns), 74 (4 lines).
7228   1824, [Hare's blowpipe]: v. 2 (ns), 91.
7229   1825, Mineralogy in Ireland: v. 2 (ns), 299 (9 lines).

The Mining Journal and American Rail-Road Gazette. Devoted to Improvements in Mining Manufactures, Inter-Communications and the Arts: (New York, 1847-1849)
7230   1847-1849, [Notices of new mines, mineral deposits, and advertisements for mining equipment and supplies]: numerous articles and advertisements in all numbers.
7231   1847-1849, Coal intelligence: a weekly feature on Pennsylvania and foreign coal production.
7232   1847-1879, Foreign mining intelligence: a weekly feature on foreign mining techniques and production.
7233   1847-1849, Iron mining and manufacturing: a weekly feature.
7234   1847-1849, Mining correspondence: a weekly feature with letters on mining.
7235   1847-1849, Lake Superior copper mines: a weekly feature, with many descriptions of individual mines.
7236   1847, New Jersey copper mines: v. 1, 22, 26.
7237   1848, A glossary of Cornwall mining terms: v. 1, 147.

The Miscellaneous Magazine (Trenton, New Jersey, 1824)
7238   1824, Account of the boiling springs in Iceland: v. 1, no. 2, 34-35.
            Abstracted from London Methodist Mag.
7239   1824, Number of known species of organized animals: v. 1, no. 2, 45.
7240   1824, Natural Bridge in Virginia: v. 1, no. 3, 65-68.
            Abstracted from Christian Herald.
7241   1824, Great volcano of the Lipara Isles [Stromboli]: v. 1, no. 7, 148-149.
7242   1824, Account of the earthquake in Chile, in November, 1822: v. 1, no. 10, 229-237.

Mississippi. State of
7243   1850, An act to further endow the University of Mississippi: Mississippi Gen. Assembly, March 5, 1850.
            This act established the Mississippi Geological Survey.

Missouri Iron Company (Missouri City, Missouri)
7244   1837?, Prospectus of the Missouri Iron Company. With acts of incorporation: n.p., 36 p., plate, map.
            Includes extracts from reports by H. R. Schoolcraft (see 9255), G. W. Featherstonhaugh (see 3781), and many endorsements.
7245   1837, Prospectus of the Missouri Iron Company and Missouri and Iron Mountain Cities: Bos., Marden and Kimbell (printers), 40 p., map.
            Also printing by P. Canfield, Hart.

Mitchell (of London)
7246   1841, Gases in well-digging: Am. Eclec., v. 1, no. 1, 178-179.
            Abstracted from London Ath.

Mitchell, Elisha (1793-1857)
7247   1827, Report on the geology of North Carolina, conducted under the Board of Agriculture. Part III: Raleigh, NC, J. Gales and Sons, 43 p.
            Contains report by C. E. Roth (see 9038). See also 8029 and 8030.
7248   1827, Minerals from North Carolina: The Parthenon, v. 1, no. 10, 154-155.
7249   1828, Geological report on North Carolina ... conducted under the direction of the Board of Agriculture; part 3: NC Board Ag. Papers, 101-108.
7250   1828, On the character and origin of the low country of North Carolina: Am. J. Sci., v. 13, no. 2, 336-347.
7251   1829, Geology of the gold region of North Carolina: Am. J. Sci., v. 16, no. 1, colored geological map, 27.5 x 18.5 cm.
            In 7252.
7252   1829, On the geology of the gold region of North Carolina: Am. J. Sci., v. 16, no. 1, 1-19, colored map.
            Contains map 7251. See also 7253.
7253   1830, Geology of the gold region of North Carolina: Am. J. Sco., v. 17, no. 2, 400.
7254   1831, Analysis of the Protogae of Leibnitz: Am. J. Sci., v. 20, no. 1, 56-64.
7255   1839, Notice of the height of mountains in North Carolina: Am. J. Sci., v. 35, no. 2, 377-380.
7256   1842, Elements of geology, with an outline of the geology of North Carolina; for the use of students of the University [Yale]: n.p., [9]-141 p., folding map.
7257   1842, [Geological map of North Carolina]: colored geological map, in 7256.
            Not seen.
7258   1846, Report on the turnpike from Raleigh, West: Raleigh, NC, W. R. Gales, 18 p.
            In Governor's Message, commemorating Mitchell's survey.
7259   1849, [On North Carolina soils and topography]: in Letters from the Alleghany Mountains by C. Lanman (see 6034), 192-198.

Mitchell, I. (i.e. Israel?)
7260   1849, Mitchell's guide to California: Phila., Thomas, Copperthwait and Co.
            Not seen.

Mitchell, Samuel Augustus (1792-1868)
7261   1837, Illinois in 1837; a sketch descriptive of the situation, boundaries, face of the country, prominent districts, prairies, rivers, minerals ... : Phila., S. A. Mitchell.
       Not seen.
7262   1849, Description of Oregon and California, embracing an account of the gold region; to which is added, an appendix, containing descriptions of various kinds of gold, and methods of testing its genuineness; ... : Phila., T. Copperthwait and Co., 76 p., illus., folded map.

Mitchell, William (1791-1868)
7263   1844, The variation and dip of the magnetic needle at Nantucket, Mass.: Am. J. Sci., v. 46, no. 1, 157-158.

Mitchill, Samuel Latham (1764-1831)
       As editor of The Medical Repository, S. L. Mitchill probably authored or excerpted many of the unsigned notes and reviews in that journal (see 7008 to 7108).
7264   1787, Observations, anatomical, physiological, and pathological on the absorbent tubes of animal bodies. To which are added geological remarks on the maritime parts of the state of New York: NY, J. M'Lean and Co., 16 p.
       Evans number 20527.
7265   1787, On the nature and origin of peat or turf: Colum. Mag., v. 1, 581-584.
7266   1788, On the petrifactions near Claverak [Columbia County, New York]: Am. Mag., v. 1, no. 7, 493-494.
7267   1789, Geological remarks on certain maritime parts of the state of New York: Am. Mus. or Univ. Mag., v. 5, 123-126.
7268   1792, Outline of the doctrines of natural history, chemistry, and economics ... now delivering in the College of New-York: NY, 8vo.
       Not seen. Evans number 24549.
7269   1795, Description of the Blue Mountains in the state of New-York: NY Mag., v. 6, 465-471. Also in: Lit. Mus. or Mon. Mag., v. 1, no. 2, 65-72 (1797).
7270   1798, American Mineralogical Society: Phila. Mon. Mag., v. 2, 109.
       On the objectives of the Society, and a call for mineral specimens.
7271   1798, Gun flints [from Lake Erie, New York]: Phila. Mon. Mag., v. 2, 104.
7272   1798, A short account of the Blue Mountains, and the Falls of the Kaat's Kill - from a tour through the state of New-York, in the vicinity of Hudson's River: Phila. Mon. Mag., v. 1, no. 3, 127-133.
7273   1798/1800, A sketch of the mineralogical history of the state of New York: Med. Repos., v. 1, no. 3, 293-314; no. 4, 445-452; v. 3, no. 4, 325-335. Also in NY Soc. Prom. Use. Arts Trans., v. 1, pt. 4, 124-152.
       See also 7278.
7274   1799, Address of the American Mineralogical Society to the public: Wk. Mag., v. 3, 83-84.
       On the organization of the American Mineralogical Society, and on the importance of minerals to America's growth.
7275   1799, A new variety of iron-ore of the argillaceous kind, and resembling basaltes: Med. Repos., v. 2, 219-220.
7276   1799, Outlines of medical geography: being an inquiry how far calcareous soils and strata counteract the septic exhalations which occasion distempers of a febrile or pestilential type: Med. Repos., v. 2, 39-47.
7277   1801, Excellence of calcareous materials for building and paving cities, particularly as respects their power to overcome the exciting cause of fevers: Med. Repos., v. 4, 91-93.
7278   1802, Additional articles of my report to the Agricultural Society on the mineralogy of New York: Med. Repos., v. 5, 212-215.
7279   1802, Observations on the soda, magnesia, and lime, contained in the water of the ocean; shewing that they operate advantageously there by neutralizing acids, and among others the septic acid, and that sea-water may be rendered fit for washing clothes without the aid of soap: Am. Philos. Soc. Trans., v. 5, 139-147. Also in: Bos. Wk. Mag. Devoted to Morality, v. 1, 70 (1803).
7280   1804, Remarks on some parts of New-York, made September, 1802: Med. Repos., v. 7, 285-289.
       On the Hudson River valley.
7281   1805, Disclosures in mineralogy [on mineral localities in Stockbridge, Massachusetts; Newport, Rhode Island; Woodbridge, New Jersey; Louisiana; and Cuba]: Med. Repos., v. 8, 81-83.
       1809, A tour through part of Virginia: see 10298.
       Ascribed by some to S. L. Mitchill.
7282   1810, A concise description of Schooley's Mountain, in New Jersey ... : NJ, J. Seymour, 14 p.
7283   1810, Descriptive catalogue accompanying a suite of mineral specimens presented to the editor by his colleague: Am. Min. J., v. 1, no. 1, 1-5.
7284   1810, Outlines of Professor Mitchill's lectures on natural history, in the College at New-York, delivered in 1809-1810: Med. Repos., v. 13, 257-267.
7285   1810, Reasons for supposing that the Great Lakes of North-America were originally composed of salt water: Med. Repos., v. 13, 404-406. Also in: Am. Reg. or Gen. Repos., v. 6, 341-343.
       Extracted from Tour to Niagara.
7286   1811, Account of the remains of marine animals in a fossil state, in New Jersey: Am. Min. J., v. 1, no. 2, 95-96.
7287   1811, A concise description of Schooley's Mountain, in New Jersey, with some experiments on the water of its chalybeate spring: Eclec. Rep. and Analytic Rev., v. 2, 317-324.
7288   1811, Copious sources of inflammable air, in western New-York and in Upper Canada, near Niagara: Med. Repos., v. 14, 412-415.
7289   1811, Description of the sulphureous springs in the county of Ontario: Med. Repos., v. 14, 412-415.
       Extracted from Tour to Niagara.
7290   1811, The physical geography of the first range of mountains extending across New Jersey, from the Hudson to the Delaware; with some experiments on the chalybeate spring at Schooley's Mountain: Am. Min. J., v. 1, no. 2, 70-79.

Mitchill, S. L. (cont.)

7291 1812, An amendment proposed to the geological chart of the United States, as respects the character of the north side of Long Island, which is shown to be Alluvial and not Primitive, as therein states: Am. Min. J., v. 1, no. 3, 129-133.

7292 1812, Geological observations on the United States: Am. Med. Philos. Reg., v. 2, no. 3, 253-256.

7293 1812, and G. E. Pendergrast and W. Ross, Report of the commissioners, stating the existence of flint-stones on the banks of the Musconetcong River in the state of New Jersey: Niles' Wk. Reg., v. 2, 390-392.

7294 1814, Geology of Long Island: Am. Min. J., v. 1, no. 4, 261-263.

7295 1814, A sketch of the scenery in the region around Harpers Ferry, where the ridge of Blue Mountains is penetrated by the joint waters of the Potomac and Shenandoah Rivers: Am. Min. J., v. 1, no. 4, 211-218.

7296 1814, [Testimonial letter]: in North-American Coal and Mining Company Report (see 7921).

7297 1815, and W. J. McNeven, A chymical examination of the mineral water of Schooley's Mountain springs, together with a physical geography of the first range of mountains extending across New Jersey, from the Hudson to the Delaware ... : NY, Van Winkle, 21 p.
Other editions in 1828 (NY, Sickles, 24 p.; Morristown, New Jersey, J. Mann, 23 p.); 1841 (Morristown, NJ, S. P. Hull, 24 p.); 1845 (20 p.).

7298 1815, Description of the volcano and earthquake which happened in the Island of St. Vincents, on the 30th day of April, 1812: Lit. and Philos. Soc. NY Trans., v. 1, 315-323.

7299 1815, A detailed narrative of the earthquakes which occurred on the 16th day of December, 1811, and agitated the parts of North America that lie between the Atlantic Ocean and Louisiana; and also a particular account of the other quakings of the earth occasionally felt from that time to the 23rd and 30th of January, and the 7th and 16th of February, 1812, and subsequently to the 18th of December, 1813, and which shook the country from Detroit and the Lakes to New-Orleans and the Gulf of Mexico: Lit. and Philos. Soc. NY Trans., v. 1, 281-307.

7300 1815, History of the earthquakes and volcanoes in the Azores, particularly in the Islands of St. George, Pico, and St. Michael, and in the adjoining Ocean, during the years 1808 and 1811: Lit. and Philos. Soc. NY Trans., v. 1, 324-331.

7301 1815, The leading facts relative to the earthquakes which desolated Venezuela, in South America, in the months of March and April, 1812: Lit. and Philos. Soc. NY Trans., v. 1, 308-315.

7302 1817, [Fossil remains of New York]: Am. Mon. Mag. and Crit. Rev., v. 1, no. 2, 126-127.

7303 1817, [Notes on the tract between the Highlands and Catskill Mountains, New York]: Am. Mon. Mag. and Crit. Rev., v. 1, no. 3, 195-196.

7304 1818, Account of recent scientific transactions at the New-York Institution: Am. Mon. Mag. and Crit. Rev., v. 4, no. 2, 133-137.
On the discovery of fossils in New York, Georgia, and Tennessee.

7305 1818, An account of the impression of fish in the rock of Oneida County, New-York: Am. Mon. Mag. and Crit. Rev., v. 3, no. 4, 291.
On Eurypterus remipes De Kay from Westmoreland, Oneida County, New York.

7306 1818, Notes and additions on America articles: in An elementary introduction to the knowledge of mineralogy by W. Phillips (see 8321).
Also in 8323.

7307 1818, Observations on the geology of North America; illustrated by the description of various organic remains found in that part of the world: in Essay on the theory of the earth by G. Cuvier (see 2754), 319-431, 3 plates with 17 figs.
See also note by J. Stranger (10036). For review see 8545.

7308 1820, [Fossil tooth from Belleville, New York]: Agri. Intell., 1, 147.

7309 1821, A discourse on the character and services of Thomas Jefferson, more especially as a promoter of natural and physical science, pronounced by request before the N. Y. Lyceum of Natural History, on the 11th day of October, 1821: NY, D. Fanshaw (printer), 40 p.

7310 1822-1825, Notices of a conversatione at Dr. Mitchill's: The Minerva, v. 1, 198; 398; v. 1 (ns), 58; 184; 217; 233-234; 248; 263-265; 279-280; 361; v. 2 (ns), 41; 56-58; 73; 89-90; 106-107; 185; 250-251; 265; 281; 345; 363.
These are brief notes on fossils, minerals, and mining.

7311 1823, Brief notice of a new work on crustaceous animals found in a fossil state: NY Med. Phys. J., v. 2, 216-222.

7312 1823, Observations on the teeth of the Megatherium recently discovered in the United States [at Skidaway Island, Georgia]: NY Lyc. Nat. History Annals, v. 1, pt. 1, no. 2, 58-61, 2 figs.

7313 1823, Plumbago, or graphite [from Bristol and Francestown, New Hampshire]: Coll. Topo. Hist. Bibliographical NH, v. 2, 30-31.

7314 1826, Catalogue of the organic remains, which, with other geological and some mineral articles, were presented to the New York Lyceum of Natural History, in August, 1826, by their associate, Samuel L. Mitchill: NY, New York Lyceum of Natural History, printed by J. Seymour, 40 p.

7315 1827, Another locality for a trilobite [York, Pennsylvania]: The Parthenon, v. 1, no. 12, 185.

7316 1827, Converzationes at Dr. Mitchill's. Organic remains from the chalk formation along the River Thames: The Parthenon, v. 1, no. 6, 90-91.

7317 1827, and J. A. Smith and W. Cooper, Discovery of a fossil walrus in Virginia: NY Lyc. Nat. History Annals, v. 2, no. 9/11, 271-272.

7318 1827, Geology. Mackey's theory: The Parthenon, v. 1, no. 1, 3-6.

7319 1827, Mammoth bones: Tha Parthenon, v. 1, no. 11, 171.
On bones found at Schooley's Mountain, New Jersey.

7320 1827, More bones of the Mastodon found [at Schooley's Mountain, New Jersey]: The Parthenon, v. 1, no. 12, 185-186.

7321 1828, A lecture on some parts of the natural history of New Jersey, delivered before the Newark Mechanic Association for Mutual Improvement in the Arts and Sciences, on Tuesday, June 3, 1828: NY, Elliott and Palmer (printers), 34 p.

7322 1831, Magnetism: Young Ladies' J. Liter. Sci., v. 1, 372-374.
From First lines of science.

Mitchill, S. L. (cont.)
7323   1843, On the origin of peat: in Geology of New York, first district by W. W. Mather (see 6907), 229-232.
        From the Med. Repos.

Mitchill, Thomas Duché (1779-1865)
7324   1813, Analysis of a mineral spring at the Willow Grove, Montgomery County, Pennsylvania: Colum. Chemical Soc. Mem., v. 1, 93-95.
7325   1813, Analysis of malachite, or green carbonate of copper of Perkioming, Pennsylvania: Colum. Chemical Soc. Phila. Mem., v. 1, 125-126.
7326   1833, Sulphuric acid spring [near the Erie Canal in New York]: West. Med. Gaz., v. 1, 110.
        Signed by "T. D. M."

Mitscherlich, Eilhard (1794-1863)
7327   1824, Difference of crystalline forms of the same substance: Am. J. Sci., v. 8, no. 2, 376 (14 lines).
        Abstracted by J. Griscom. On the polymorphism of sulphur.
7328   1825, Effect of heat on the form and double refraction of calcareous spar: Bos. J. Phil. Arts, v. 2, no. 4, 387-388.
        Abstracted from Ferussac's Bulletin.
7329   1825, Minerals produced by Heat: Bos. J. Phil. Arts, v. 2, no. 4, 393; Am. J. Sci., v. 10, no. 1, 190.
        Abstracted from Brewster's J. On the synthesis of pyroxene.

Mix, Stephen (1672-1738)
7330   1728, Extraordinary displays of the divine majesty & power, are to try men, and impress the fear of God on their hearts, that they sin not. Being the substance of two sermons occasioned by a terrible earthquake in New-England, and other parts of North America; in the night immediately following the Sabbath-day, October 29, 1727 ... : New-London, Ct., T. Green (printer), 2, 36 p.

Moffat, John M. (see 5763 and 5804 by W. R. Johnson)

Moldenhauer, Mr.
7331   1827, Notice of the Heidelberg collections of rocks and petrifactions: Am. J. Sci., v. 12, no. 1, 399-400.

Molina, Juan Ignacio (1740-1829)
7332   1808, The geographical, natural and civil history of Chile: Middletown, Ct., I. Riley, 2 v., folded map.
        Translated by R. Alsop. Also 1809 printing.
        For review see 7083.

Monell, Gilbert Chester (1816-1881)
7333   1846, Human fossils: Am. Q. J. Ag. Sci., v. 3, 311-312.

Monette, John Wesley (1803-1851)
7334   1846, History of the Discovery and Settlement of the Valley of the Mississippi: NY, Harper, 2 v., front., plates.
        Also 1848 edition.
7335   1847, Geology of the Mississippi Valley: De Bow's Rev., v. 3, 124-125; 215-225.

Monison, Isaac
7336   1789, A brief account of Kentucky: Am. Mus. or Univ. Mag., v. 5, 57-59.

Monthly American Journal of Geology and Natural Science (Philadelphia, 1831-1832)
7337   1831, Herds of frozen elephants, rhinoceros, &c. &c.: v. 1, no. 4, 191 (16 lines).
        Abstracted from Mag. Nat. History.
7338   1831, Geology of India: v. 1, no. 4, 192 (6 lines).
        Abstracted from Jameson's J. On survey by T. Christie.
7339   1831, A new skeleton of the Megatherium [found near Buenos Ayres]: v. 1, no. 4, 192 (8 lines).
        Abstracted from Jameson's J.
7340   1832, On the causes which retard geological knowledge [reviews of Introductin to Geology by R. Bakewell (see 1455) and "Appendix" by B. Silliman (see 9681)]: v. 1, no. 7, 296-301.
        From a "London correspondent".
7341   1832, Description of a crystal of native copper: v. 1, no. 7, 318-319.
        Signed by "A."

The Monthly Anthology and Boston Review (Boston, 1803-1811)
7342   1805, An account of the earthquake, which took place at Naples, July 26th, 1805: v. 2, 628-630.
7343   1806, Salt and sulphur springs [at Onondaga, New York]: v. 3, 393-395.
7344   1807, [Review of A geological account of the United States by J. Mease (see 6960)]: v. 4, 666-670.
7345   1808, Col. Gibbs's grand collection of minerals: v. 5, 57-58.
7346   1808, [S. Godon's mineralogical lectures]: v. 5, 209.
7347   1808, [Review of An inquiry into the chemical character of Rhode Island coal by W. Meade (see 6942)]: v. 5, 677-679.
7348   1809, Progress of finding gold in North Carolina: v. 6, 208.
        Abstracted from Med. Repos.
7349   1809, American ochres [from near Philadelphia]: v. 6, 208-209.
        Abstracted from Med. Repos.
7350   1809, Discovery of valuable minerals [in New Jersey]: v. 6, 209.
        Abstracted from Med. Repos.
7351   1809, Mineralogical notices in the county of Onondaga, state of New York: v. 6, 209-210.
        Abstracted from Med. Repos.
7352   1809, Arrangement for a museum of minerals: v. 6, 212-213.
        Abstracted from Med. Repos.

Monthly Anthology (cont.)
7353   1809, American mineralogy: v. 6, 213 (4 lines).
       Notice of work by A. Seybert (see 9433).
7354   1809, Interesting discovery in Virginia: v. 6, 285-286.
       On a copper mine at Stanardsville, Orange County, Virginia.
7355   1809, Maclure's geological enquiries: v. 6, 352.
       Abstracted from Med. Repos.
7356   1809, A natural saltpetrous earth, discovered near the south branch of Potomack, in Virginia: v. 6, 353-354.
       Abstracted from Med. Repos.
7357   1809, Encouragement for discovering antimony in the United States: v. 6, 354-355.
       Abstracted from Med. Repos.
7358   1809, [Eruption of Mount Vesuvius in 1809]: v. 7, 139.
7359   1810, [Review of Travels ..., by B. Silliman (see 9603)]: v. 9, 184-190.
       With excerpts on the mines of Great Britain.
7360   1810, American porcelain [from Monkton, Vermont]: v. 9, 355-356.
       From The Washintonian (Windsor, Vermont).
7361   1811, Eruption of Mount Vesuvius: v. 10, 278-280.

The Monthly Chronicle of Events, Discoveries, Improvements and Opinions (Boston, 1840-1842)
7362   1840-1842, [Notes on coal mining]: v. 1, 511; v. 3, 468-469.
7363   1841-1842, [Notes on earthquakes and volcanoes]: v. 2, 94; 185; 320-323; 466; v. 3, 280-281; 363-367.
7364   1842, The mines of Almaden, in Spain: v. 3, 500-503.

The Monthly Journal of Agriculture (New York, 1845-1848)
7365   1845, Composition of soils: v. 1, no. 3, 126.
       Abstracted from Edinburgh Rev.

The Monthly Magazine and American Review (New York, 1800-1802)
7366   1801, Encroachments of the ocean [in Asia]: v. 2, 423.
       Abstracted from Asiatic Researches.
7367   1802, Specimen of sulphureous minerals from the Solfatarra [Italy]: v. 3, 75-76.
7368   1802, [Limestone from Westchester County, New York]: v. 3, 154-155.
7369   1802, [Lehigh Pennsylvania coal mine]: v. 3, 155 (12 lines).
7370   1802, Grand specimen of Mexican gold: v. 3, 393.
       Abstracted from Med. Repos.

The Monthly Magazine of Religion and Literature (Gettysburg, Pennsylvania, 1840-1841)
7371   1840, [Review of Popular lectures on geology by K. C. von Leonhard (see 6192)]: v. 1, no. 1, 23-27.
7372   1841, Coloring marble: v. 1, no. 11, 354.

The Monthly Miscellany of Religion and Letters (Boston, 1839-1843)
7373   1840, Lectures before the Lowell Institute [by B. Silliman]: v. 2, 171-172.
7374   1840, Earthquake in Burmah: v. 2, 177.

The Monthly Repository and Library of Entertaining Knowledge (New York, 1830-1834)
7375   1831, Rock bridge in Virginia: v. 1, 267-270, plate.
7376   1831, Earthquakes: v. 1, 11-13.
7377   1831, General features of the earth's surface: v. 1, 69-72.
7378   1831, Geology: v. 1, 203-207; 232-237.
       An introduction to the science.
7379   1831, A petrified forest [on the banks of the Missouri River]: v. 1, 320-321.
7380   1831, Glaciers of Switzerland: v. 2, 75-77, plate.
7381   1831, Crater of Kirauea, in Hawaii: v. 2, 183-187, plate.
7382   1832, Two scenes in Virginia - Wier's Cave: v. 2, 263-267.
7383   1832, Basaltic formation [at Buanaauia?]: v. 2, 370.
7384   1832, Eruption of Mount Galoungoun [at Java]: v. 2, 413-415.
7385   1832, The great Kentucky caverns: v. 3, 9-12; 96-99.
7386   1832, Fingal's Cave: v. 3, 206-209, plate.
       Includes a descriptive poem.
7387   1833, Mount Hecla, in Iceland: v. 3, 262-268, plate.
       Includes a poem on an eruption.
7388   1833, Mount Vesuvius - near Naples: v. 3, 304-311, plate.
7389   1833, The geysers [of Iceland]: v. 3, 370-374, woodcut.
7390   1833, Naples [and Vesuvius]: v. 4, 1-6, plate.
7391   1833, Crater of Mount Vesuvius: v. 4, 53-56, plate.
7392   1833, Intermitting springs [on their origin]: v. 4, 121-126, 2 woodcuts.
7393   1833, Earthquakes: v. 4, 166-170, plate.
7394   1833, Cornwall rocking stone: v. 4, 197-200, plate.
7395   1833, A great cavern discovered in Ireland [in Cather]: v. 4, 200-202.
7396   1833, The mineral kingdom - gold: v. 4, 220-222.
7397   1833, Buckstone [rocking stone in England]: v. 4, 228-229.
7398   1834, Earthquakes in Sicily and the two Calabrias: v. 4, 260-269, 2 plates; 306-310, plate.
7399   1834, The mineral kingdom - silver: v. 4, 275-277.
7400   1834, Needle rocks, Isle of Wight: v. 4, 286-287, plate.
7401   1834, The mineral kingdom - iron: v. 4, 322-324.
7402   1834, The mineral kingdom - copper: v. 4, 358-360.
7403   1834, The mineral kingdom - mercury: v. 4, 386-388.
7404   1834, Mud and air volcanoes: v. 4, 409-412, plate.
7405   1834, The mineral kingdom - lead: v. 4, 416-419.

The Monthly Review and Miscellany of the United States (Charleston, South Carolina, 1805-1806)
7406   1805, Account of a shower of stones [in France]: v. 1, 148-149.

The Monthly Scientific Journal (New York, 1818)
7407   1818, Death of De Luc, the geologist: v. 1, 14 (11 lines).
7408   1818, Tests for different metals: v. 1, 56-57.

Monticelli, Teodoro (1759-1845)
7409   1817, A description of the eruption of Vesuvius, which took place on the 25th and 26th of December, 1813: Analectic Mag., v. 9, 117-183.
         Abstracted from J. Sci. Arts.
7410   1818, Report to the Royal Academy of Sciences at Naples, upon the eruption of Vesuvius in December 1817: J. Sci. Arts, v. 5, no. 2, 199-201.
7411   1824, and Covelli, New Vesuvian minerals: Bos. J. Phil. Arts, v. 2, no. 2, 207 (16 lines).
         Abstracted from Bibliotheque Universelle.
7412   1826, and Covelli, Account of some new Vesuvian minerals: Am. J. Sci., v. 11, no. 2, 250-267, 34 figs.
         Translated by Dr. J. Van Rensselaer.

Montin, M.
7413   1822, Lithography: Am. J. Sci., v. 5, no. 2, 386 (6 lines).
         On lithographic stone from Cervesa, Spain.

Montlosier, François Dominique, comte de Reynaud de (1755-1838)
7414   1850, What becomes of the skeletons of wild animals after death? Ann. Scien. Disc., v. 1, 287 (13 lines).
         Abstracted from Jameson's J.

Moore (Dr.)
7415   1842, On fossil bones found on the surface of a raised beach, at the Hoe, near Plymouth: Franklin Inst. J., v. 4 (3s), 97-98.
         From the London Ath.

Moore, Francis (1803-1864)
7416   1840, Map and description of Texas, containing sketches of its history, geology, geography, and statistics ... and some brief remarks upon the character and customs of its inhabitants: Phila., H. Tanner; NY, Tanner and Disturnell, 143 p., 8 plates, folded map. Also second edition, NY, T. R. Tanner, 143 p. (1844).
7417   1841, [Geology of Texas]: Am. Rep. Arts, Sci's Manu's, v. 3, 9-10.
         In NY Lyc. Nat. History Proc.

Moore, Jacob Bailey (1797-1853)
7418   1823, On a rocking stone in Durham, New Hampshire: Am. J. Sci., v. 6, no. 2, 243-244, plate with 4 figs.

Moore, Nathaniel Fish (1782-1872)
7419   1834, Ancient mineralogy, or an inquiry respecting the mineral substances mentioned by the ancients; with occasional remarks on the uses to which they were applied: NY, G. and C. Carvill, viii, 192 p.
         For review see 616, 5983 and 7691. For extract on "Minerals mentioned in the Bible" see Am. Mag. Use. Ent. Know., v. 2, no. 5, 186-187 (1835).

Moormann, John Jennings (1802-1885)
7420   1839, A directory for the use of the White Sulphur waters with practical remarks on their medical properties, and applicability to particular diseases: Phila., T. K. and P. G. Collins, 35 p.
7421   1840?, Water from the White Sulphur Springs, Greenbrier County, Virginia. With practical remarks on its medical properties, and applicability to particular diseases: n.p., 8 p.
7422   1847, The Virginia Springs, with their analysis; and some remarks on their character, together with a directory for the use of the White Sulphur water, and an account of the diseases to which it is applicable, etc., and an account of the different routes to the springs: Phila., Lindsay and Blakiston, 219 p.
         For review see 5457, 7006, 7768, 9952 and 10828.

Morand, M.
7423   1789, Account of a strange phenomenon, which happened at the village of Bonnevallée near Vintimille, in France: Gentlemen and Ladies Town and Country Mag., v. 1, no. 1, 7.
         Death of a woman ascribed to volcanic eruption.

Moreau de St.-Méry, Médéric Louis Elie (1750-1819)
7424   1797, A general view of the arts and sciences, adapted to the capacity of youth: Phila., the editor, xii, 363, (3) p.
         Evans number 32505. Translated by Michael Fortune.

Morey, Samuel (1762-1843)
7425   1821, Fetid crystallized limestone: Am. J. Sci., v. 3, no. 2, 234-235.
         On limestone from Oxford, New Hampshire.
7426   1821, On artificial mineral waters, with some remarks on artificial light: Am. J. Sci., v. 3, no. 1, 94-100.

Morfit, Campell (1820-1897; see 1775 by J. C. Booth)

Morgenstern (Prof.)
7427   1819, Visit to Mount Vesuvius: Robinson's Mag., v. 2, 394-397.
         From New Mon. Mag.

Morlot, Adolphe (1820-1867)
7428   1848, On dolomisation: Am. J. Sci., v. 6 (ns), no. 2, 268-269.
         Abstracted from "Naturwill. Abhandl. von W. Haidinger."

Mornley (Mr.)
7429   1817, [Meteorite from Bahia, Brazil]: Am. Reg. or Summary Rev., v. 1 (ns), 446 (9 lines).
         Abstracted by T. Cooper.

Morrell, William H. (see 1500 and 1501 with others)

Morrill, Nathaniel (1698-1730)
7430   1728, The Lord's voice in the earthquake crieth to careless & secure sinners, shewed in a sermon preach'd in the Parish of Rye in New-Castle, in New Hampshire, in New-England, Novemb. 16, 1727 ... : Bos., Richard Jenness and Joseph Lock, (2), iv, 32 p.
        Evans number 3068.

Morris, Eastin
7431   1834, The Tennessee Gazetteer or topographical dictionary ... : Nashville, Tenn., W. H. Hunt and Co., [4], 178, [18] p.

Morris, James (1752-1820)
7432   1802, [On the use of gypsum as fertilizer]: Ct. Soc. Promoting Ag. Trans., v. 1, 9 (7 lines).

Morris, John Gottlieb (1803-1895)
7433   1838, An excursion to the Bare Hills [near Baltimore]: Balt. Lit. Monument, v. 1, 123-125.
7434   1841, An address on the study of natural history ... : Balt., Publication Rooms, 23 p.
7435   1846, Glaciers: Lit. Rec. and J. Linnaean Assoc. Penn. College, v. 3, 202-205.
        Signed by "J. G. M."
7436   1846, The Hydrarchos, or fossil sea serpent: Lit. Rec. and J. Linnaean Assoc. Penn. College, v. 2, 49-51, plate.
        Signed by "J. G. M." On Koch's display.
7437   1846, Paleontology, or fossil remains: Lit. Rec. and J. Linnaean Assoc. Penn. College, v. 2, 99-102; 121-123.
        Signed by "J. G. M."

Morris, M.
7438   1850, Geology of the state of Mississippi: Am. J. Sci., v. 10, no. 1, 133 (13 lines).

Morris, R. (Reverend)
7439   1847, To mineralogists: Ohio Sch. J., v. 2, no. 7, 112 (7 lines).
        On the mineral collection of the Mount Sylvan Academy, Lafayette County, Mississippi. Offer to trade for specimens.
7440   1847, Notes on the study of geology: Ohio Sch. J., v. 3, no. 4, 51-52; no. 5, 66-67.

Morrison,
7441   1838, Religious history of man:
        Not seen. For excerpt see Hesperian, v. 1, 244-245.

Morrison, M. F.
7442   1840, A lecture on geology: Farm's Mon. Vis., v. 2, 184-187.

Morse, Ebenezer
7443   1832, Manual of mineralogy and geology: Alb., Webster and Skinner, 297 p.

Morse, James O.
7444   1817, Mineralogical sketch of Cherry-Valley, in the county of Montgomery [New York]: Med. Repos., v. 18, 420.
7445   1828, Observations on the great greywacke region of the state of New-York: Alb. Inst. Trans., v. 1, 84-85.

Morse, Jedidiah (1761-1826)
7446   1784, Geography made easy: being a short but comprehensive system of that useful and agreeable science: ... : NHav., Meigs, Bowen and Dana (printers), 214 p., 2 maps.
        Evans number 18615. This is the first edition.
7447   1789, The American geography; or, a view of the present situation of the United State of America. ... : Elizabethtown, NJ, Shepard Kollock (printer), xii, 534 p., 2 maps.
        Also many later printings and editions. Evans number 21978.
        Contains many notes on geological features and mineral deposits of the United States.
7448   1790, Geography made easy: Being an abridgement of the American Geography. ... : Bos., I. Thomas and E. T. Andrews (printers), 332 p., 8 maps.
        Evans number 22681. At least 22 printings by 1820.
7449   1793, The American universal geography, or, a view of the present state of all the empires, kingdoms, states, and republics of the known world, and of the United States of America in particular. In two parts: Bos., I. Thomas and E. T. Andrews, 2 v. (696 p., 8 maps; 552 p., 3 maps).
        Evans number 25847. Also many other editions and printings.
7450   1795, Elements of geography ... : Bos., I. Thomas and E. T. Andrews, 143 p., map.
        Evans number 29112. Many other editions and printing.
7451   1797, The American gazetteer, exhibiting, in alphabetical order, a much more full and accurate account, than has been given, of the states, provinces, counties, cities, towns, villages, rivers, bays, harbours, gulfs, sounds, capes, mountains, forts, indian tribes, & new discoveries, on the American continent ... : Bos., E. Larkin, viii, 619 p., 7 maps.
        Evans number 32509. Several other editions.
7452   1797, A description of the soil, productions, commercial, agricultural and local advantages of the Georgia Western Territory; ... extracted and published in this form, (by permission) from Rev. Dr. Morse's American Gazetteer, a new work: Bos., Thomas and Andrews (printer), 24 p., map.
        Evans number 32510.

Mortimer, George
7453 1791, Extraordinary account of certain hot springs in the island of Amsterdam: Univ. Asylum and Colum. Mag., v. 7, 174.
Extracted from Observations during a voyage.

Morton, Samuel George (1799-1851)
See also 8910 with H. D. Rogers and W. R. Johnson.
7454 1827, Geology, mineralogy, & organic remains: Phila. Acad. Nat. Sci's Rept. Trans., v. 2, 4-5.
7455 1829, Analysis of tabular spar from Bucks County, Pennsylvania; with a notice of various minerals found at the same locality: Phila. Acad. Nat. Sci's J., v. 6, pt. 1, 46-49.
7456 1829, Description of a new species of Ostrea; with some remarks on the O. convexa of Say: Phila. Acad. Nat. Sci's J., v. 6, pt. 1, 50-51, fig.
7457 1829, Description of the fossil shells which characterize the Atlantic Secondary formation of New Jersey and Delaware; including four new species: Phila. Acad. Nat. Sci's J., v. 6, pt. 1, 72-100, 4 plates with 22 figs.
7458 1829, Description of two new species of fossil shells of the genera Scaphites and Crepidula: with some observations on the ferruginous sand, plastic clay, and upper marine formations of the United States: Phila. Acad. Nat. Sci's J., v. 6, pt. 1, 107-119, plate with 3 figs.
Also reprint edition.
7459 1829, Essays on the geology and organic remains of a part of the Atlantic front: Phila., Mifflin, 67 p.
7460 1829, Geological observations on the Secondary, Tertiary, and Alluvial formations of the Atlantic coast of the United States of America. Arranged from the notes of Lardner Vanuxem: Phila. Acad. Nat. Sci's J., v. 6, pt. 1, 59-71; 107.
Also reprint edition.
7461 1829, Geology and organic remains: Phila. Acad. Nat. Sci's Rept. Trans., v. 3, 4-6.
7462 1829, Mineralogy: Phila. Acad. Nat. Sci's Rept. Trans., v. 3, 7-8.
7463 1829, Note: Containing a notice of some fossils recently discovered in New Jersey: Phila. Acad. Nat. Sci's J., v. 6, pt. 1, 120-128, plate.
7464 1830, Additional observations on the geology and organic remains of New Jersey and Delaware: Phila. Acad. Nat. Sci's J., v. 6, pt. 2, 189-204, plate with 12 figs.
7465 1830, Note [on doubtful occurrence of Belemnites in the alluvial beds of Cockspur Island, Georgia]: Phila. Acad. Nat. Sci's J., v. 6, pt. 2, 244 (9 lines).
7466 1830, Synopsis of the organic remains of the ferruginous sand formation of the United States, with geological remarks: Am. J. Sci., v. 17, no. 2, 274-295; v. 18, no. 2, 243-250, 3 plates with 36 figs.
See also 7469 and 7470. Also reprint edition, 31 p.
7467 1831, Notice of the Academy of Natural Sciences of Philadelphia: Phila., Mifflin and Parry (printers), 16 p.
On the objectives, members, and collections.
7468 1832, On the analogy which exists between the marl of New Jersey, &c. and the Chalk Formation of Europe: Am. J. Sci., v. 22, no. 1, 90-91.
7469 1833, Supplement to the "Synopsis of the organic remains of the ferruginous sand formation of the United States", contained in vols. XVII and XVIII of this journal: Am. J. Sci., v. 23, no. 2, 288-294, 2 plates with 15 figs.; v. 24, no. 1, 128-132, 2 plates with 24 figs.
See also 7466 and 7470.
7470 1834, Synopsis of the organic remains of the Cretaceous group of the United States, illustrated by nineteen plates, to which is added an appendix, containing a tabular view of the Tertiary fossils hitherto discovered in North America: Phila., Key and Biddle, vi, [7]-88, [4], 8, [8] p., 19 plates with 170 figs.
For review see 52 and 599.
For notice see 512. See also 7466 and 7470.
7471 1835, Appendix: Am. J. Sci., v. 29, no. 1, 149-154.
Follows "Observations on the bituminous coal deposits of the valley of the Ohio" by S. P. Hildreth (see 5053 and 5054).
7472 1835, Notice of the fossil teeth of fishes of the United States, the discovery of the Galt in Alabama, and a proposed division of the American Cretaceous group: Am. J. Sci., v. 28, no. 2, 276-278.
7473 1840, Obituary. - Death of Wm. Maclure: Am. J. Sci., v. 39, no. 1, 212.
7474 1841, Description of several new species of fossil shells from the Cretaceous deposits of the United States: Phila. Acad. Nat. Sci's Proc., v. 1, no. 7, 106-110.
7475 1841, A memoir of William Maclure: Phila., T. K. and P. G. Collins (printers), 37 p., front.
7476 1841, Two new species of fossils from the lower Cretaceous strata of New Jersey: Phila. Acad. Nat. Sci's Proc., v. 1, no. 9, 132-133.
7477 1842, Description of some new species of organic remains of the Cretaceous group of the United States: with a tabular view of the fossils hitherto discovered in this formation: Phila. Acad. Nat. Sci's J., v. 8, pt. 2, 207-227, 2 plates with 12 figs.
7478 1844, Description of the head of a fossil crocodile from the Cretaceous strata of New Jersey: Phila. Acad. Nat. Sci's Proc., v. 2, no. 4, 82-85, fig.
7479 1844, A memoir of William Maclure, Esq., late president of the Academy of Natural Sciences of Philadelphia: Am. J. Sci., v. 47, no. 1, 1-17, front.
7480 1844, A memoir of William Maclure: Phila., Merrihew and Thompson (printers), 33, [1] p., front.
7481 1844, [On some fossil bones of a Mosasaurus from Mount Holly, New Jersey]: Phila. Acad. Nat. Sci's Proc., v. 2, no. 6, 132-133.
7482 1844, [On the fossil crocodile]: Phila. Acad. Nat. Sci's Proc., v. 2, no. 4, 72-73.
7483 1845, Description of the head of a fossil crocodile from the Cretaceous strata of New Jersey: Am. J. Sci., v. 48, no. 1, 265-267.
7484 1846, Description of two new species of fossil Echinodermata, from the Eocene of the United States: Phila. Acad. Nat. Sci's Proc., v. 3, no. 3, 51. Also in: Am. J. Sci., v. 2 (ns), no. 2, 273.
7485 1846, [On Cretaceous fossils from Burlington, New Jersey]: Phila. Acad. Nat. Sci's Proc., v. 3, no. 2, 32; 39.

Morton, S. G. (cont.)
7486   1850, [On the antiquity of some races of dogs] : Phila. Acad. Nat. Sci's Proc., v. 5, no. 5, 85-89.
       Fossil evidence of dogs.

Morton, W. S. (of Virginia)
7487   1833, Calcareous manures of central Virginia: Farm's Reg., v. 1, 117-118.
7488   1833, Discovery of marl in a new district: Farm's Reg., v. 1, 59.
7489   1834, Calcareous rock in Prince Edward: Farm's Reg., v. 2, 154-155.
7490   1834, Coal in Prince Edward: Farm's Reg., v. 1, 506.
7491   1834, Gypseous deposits [in Prince Edward County] : Farm's Reg., v. 1, 700.
7492   1835, The connexion of agriculture with other sciences, and remarks on soils and manures: Farm's Reg., v. 2, 713-715.
7493   1836, Coal deposits near Farmville [Virginia] : Farm's Reg., v. 4, 473.
7494   1836, Buhr stone - Iron ore - Clay for bricks - Charcoal: Farm's Reg., v. 3, 677.
7495   1836, Calcareous rocks in Mecklenburg, Prince Edward and Chesterfield, and gypsum in Cumberland: Farm's Reg., v. 3, 516-517.
7496   1837, Coal deposits near Farmville: Mech's Mag. or Reg. Inventions, v. 9, 55-56.
       Ascribed to "W. S. Morotn" (i.e. Morton).

Mosander, Carl Gustaf (1797-1858)
7497   1839, New metal discovered in Sweden: Am. J. Pharmacy, v. 5 (ns), 259.
       Abstracted from "Acad. des Scien." On lanthanum from Bassnaes.

Moseley, William S.
7498   1850, [Missouri lead mines] : West J. Ag., v. 4, no. 6, 412-413.

Moss, Theodore F.
7499   1843, On the blowpipe: Franklin Inst. J., v. 5 (3s), 284-287; v. 6 (3s), 104-109; to be continued but no more found.
7500   1849, The school of mines at Freiberg in Saxony: Mining J., v. 2, 150-151.
7501   1850, Description of a new carpolite from Arkansas: Phila. Acad. Nat. Sci's Proc., v. 5, no. 3, 59 (16 lines), fig.

Moultrie, James (1793-1869)
7502   1837, Description of the skull of the Guadaloupe fossil human skeleton: Am. J. Sci., v. 32, no. 2, 361-364.
       With "Introductory remarks" by C. U. Shepard (see 9510).

Muller, Johannes Peter (1801-1858)
7503   1847, Basilosaurus [and its equivalence to Hydrarchus] : Am. J. Sci., v. 4 (ns), no. 3, 421-422.
7504   1847, [On the Basilosaurus or Zeuglodon of Alabama] : Am. Philos. Soc. Proc., v. 4, no. 38, 338-339 (12 lines).

Mulvany, Daniel H.
7505   1832, Cabinet of natural science. First annual report. [in Montgomery County, Pennsylvania] : Hazard's Reg. Penn., v. 9, 116-119.

Murchison, Roderick Impey (1792-1871)
       See also 9393 with A. Sedgwick
7506   1832, [Geological excursion in York] : Mon. Am. J. Geology, v. 1, no. 10, 473-474.
7507   1832, [On the origin of the Whin Sill] : Mon. Am. J. Geology, v. 1, no. 10, 471 (14 lines).
       See also 5497.
7508   1835, Geological position of fossil fishes in England and Scotland: Am. J. Sci., v. 28, no. 1, 75-76.
       In British Assoc. Proc. See also 55.
7509   1836, The ancient stratified deposits of the border counties of England and Wales: Am. J. Sci., v. 29, no. 2, 352-353.
       In British Assoc. Proc.
7510   1837, [A map of the geology of England] : Am. J. Sci., v. 31, no. 2, 375-376.
       In British Assoc. Proc.
7511   1838, Fossil fishes - Agassiz's great work going on: Am. J. Sci., v. 34, no. 1, 46.
       In British Assoc. Proc.
7512   1840, Classification of rocks: Am. J. Sci., v. 38, no. 2, 408 (12 lines).
7513   1840, Letter from Roderick Impey Murchison, Esq., to Prof. Silliman: Am. J. Sci., v. 38, no. 2, 406.
       Request that Silliman attend British Association meeting.
7514   1841, British Association - Mr. Murchison - His journey to the Ural Mountains - Opinions, &c: Am. J. Sci., v. 41, no. 1, 207-208.
7515   1841, Carboniferous and Devonian systems of Northwestern Germany: Am. Eclec., v. 1, no. 1, 183-184.
       Abstracted from Lit. Gaz.
7516   1842, Address delivered at the anniversary meeting of the Geological Society of London, Feb. 18, 1842; and the announcement of the award of the Wollaston Medal and donation fund for the same year: Am. J. Sci., v. 43, no. 1, 197-201.
7517   1843, [Comments on fossil footprints from the Connecticut River Valley] : Am. J. Sci., v. 45, no. 1, 187-188.
7518   1843, On the salt steppe south of Orenburg [Russia] , and on a remarkable freezing cavern: Am. J. Sci., v. 44, no. 1, 205-206.
       Abstracted from London, Edinburgh and Dublin Philos. Mag.
7519   1844, On the important additions recently made to the fossil contents of the Tertiary basin of the Middle Rhine: Am. J. Sci., v. 47, no. 1, 183-184.
       In British Assoc. Proc.
7520   1845, On Russia and the Ural Mountains: Eclec. Mag., v. 5, 569-570.
7521   1846, [Extracts from The Geology of Russia on mineral wealth of the Urals] : Am. J. Sci., v. 2 (ns), no. 1, 119-123.
7522   1846, Paleozoic deposits of Scandinavia: Am. J. Sci., v. 1 (ns), no. 2, 271-273.
       Abstracted from Q. J. Geol. Soc.
7523   1847, British Association: Am. J. Sci., v. 3 (ns), no. 1, 133-135.
       Abstracted from London Ath. On work of L. Agassiz and Prof. von Middendorf.

Murchison, R. I. (cont.)
7524   1847, A few remarks on Silurian classification: Am. J. Sci., v. 3 (ns), no. 3, 404-407.
7525   1850, Distribution of gold ore over the surface of the globe: Ann. Scien. Disc., v. 1, 253-255.
       Abstracted from London Ath.
7526   1850, Geology of Scinde, British India: Ann. Scien. Disc., v. 1, 238 (19 lines).
7527   1850, Gold mining: Am. Rr. J., v. 23, 246.
       Abstracted from Am. J. Sci.
7528   1850, Ice-cave in Russia: Ann. Scien. Disc., v. 1, 249-250.
7529   1850, Identity of the nummulitic formation with the Eocene Tertiary: Ann. Scien. Disc., v. 1, 230.

Murphey, Archibald Debow (1777-1832)
7530   18??, Memorial to the General Assembly of North Carolina [proposing a scientific survey]: Hillsborough, NC, D. Heartt (printer), 11 p.

Murphy, James (Reverend)
7531   1850, Creation; or the Bible and geology consistent; together with the moral design in Mosaic history: NY, Carters, 254 p.
       For review see 1682.

Murray, H. (see 5711 by R. Jameson)

Murray, John (d. 1820)
7532   1817, [Excerpts from "Analysis of the mineral waters of Dublane and Pitcaithly"]: Eclec. Rep. and Analytic Rev., v. 7, 64-76.
7533   1817, The new blow-pipe: Eclec. Rep. and Analytic Rev., v. 7, 264-265.
7534   1818, Superior free-stone from Callao, Fife, in Scotland: J. Sci. Arts, v. 4, no. 2, 381-382.
7535   1822, On a species of earthy matter spontaneously combustible: Mus. For. Liter. Sci., v. 1, 183-184.
       Abstracted from Edinburgh Philos. J.

Muse, Joseph Ennals
7536   1845, Address to the Agricultural Society of New-Castle County: New-Castle Co. Agri. Soc. Trans., 10th Meeting, 47-61.

The Museum of Foreign Literature and Science (Philadelphia and New York, 1822-1845)
7537   1822, [Review of "Account of an assemblage of fossil teeth and bones ...; discovered in a cave at Kirkdale, Yorkshire" by W. Buckland]: v. 1, 229-240.
       Abstracted from Annals of Phil.
7538   1822, [Review of "A description of the Shetland Islands ..., by S. Hibbert]: v. 1, 481-493.
       Abstracted from British Critic.
7539   1822, Mineralogy [notice of text by F. Mohs]: v. 1, 576 (12 lines).
7540   1823, Haüy's collection of minerals: v. 2, iv (3 lines).
       An advertisement.
7541   1823, Organic remains: v. 2, iv (5 lines).
       On fossil rhinoceros bones from Wirksworth, Derbyshire.
7542   1823, Aerolite [fall at Vosges]: v. 2, 525-527.
7543   1823, Haüy's collection of minerals: v. 3, 199 (3 lines).
       Notice of auction.
7544   1823, Insects in amber: v. 3, 576 (11 lines).
7545   1824, Mineralogy and geology: v. 4, 287 (8 lines).
       Obituary of Haüy.
7546   1824, Biographical account of M. Haüy, a celebrated mineralogist and botanist: v. 4, 375-378.
       From Mon. Mag.
7547   1825, On the mines of Mexico: v. 6, 268-272.
7548   1826, Scientific institutions [of Europe]: v. 9, 219-231.
7549   1827, Subterranean noises heard at Nakous: v. 11, 188-189.
7550   1827, Gold mines of the Urals: v. 11, 285-286.
7551   1827, Burning cliff near Weymouth [England]: v. 11, 287.
7552   1828, Temperature of the earth: v. 13, 433.
7553   1828, Traces of ancient currents: v. 13, 433.
       Bedding features in the Juras.
7554   1828, The Newcastle coal-field: v. 13, 437.
7555   1828, Growth of rocks: v. 13, 437 (10 lines).
       On the erroneous observations of Tournefork.
7556   1829, Descent into the crater of the vulcano, one of the Lipari Islands [Stromboli]: v. 15, 189-190.
7557   1830, Phenomena of the great earthquake of 1783 in Calabria and Sicily: v. 16, 97-108.
7558   1830, A uniformity of climate prevailed over the earth prior to the time of the deluge: v. 17, 72-73.
       Signed by "H-----n." From Edinburgh Philos. J.
7559   1830, [Review of Principles of geology by C. Lyell (see 6532)]: v. 18, 13-27, 107-120.
       From Q. Rev.
7560   1831, On the produce of gold and silver in the Russian Empire: v. 19, 119.
7561   1831, Diamonds [from Brazil]: v. 19, 358.
7562   1832, Volcano of St. Vincent: v. 20, 570-573.
7563   1832, Notice of an eruption of Vesuvius: v. 20, 609.
7564   1832, [Review with excerpts of Principles of geology by C. Lyell (see 6532)]: v. 20, 597-607.
       From Mon. Rev.
7565   1832, The diamond district of the Serro do Frio: v. 20, 625-629.
       From Mon. Rev.
7566   1832, An earthquake in Chile: v. 21, 63-170.
       From New Mon. Mag.
7567   1832, Cuvier and his cabinet: v. 21, 171-172.

Museum of Foreign Literature and Science (cont.)

7568 1832, [Review of History and topography of the United States by J. Hinton (see 5064)]: v. 21, 248-253.
7569 1835, [Review of Voyage dans le district des diamonds et sur Littard du Bresil by A. de Saint-Hillaire]: v. 26, 6-15.
　　　　From Westminster Rev.
7570 1835, Meeting of geologists: v. 26, 115 (9 lines).
　　　　Notice of the Association of American Geologists meeting.
7571 1835, The Russian platina mines: v. 26, 142.
7572 1835, Life and labors of Cuvier: v. 26, 209-225.
　　　　From For. Q. Rev.
7573 1835, The deepest mine in Great Britain [at Monkwearmouth Colliery, Sunderland]: v. 26, 228-229.
7574 1835, [Review of A visit to Iceland by way of Tronyem by J. Barrow, jr.]: v. 26, 648-655.
7575 1836, The life and works of Baron Cuvier: v. 28, 477-489.
　　　　From Edinburgh Rev.
7576 1836, [Review of Mineralogy and geology by W. Buckland (see 2160) with excerpts]: v. 29, 77-92.
　　　　From Q. Rev.
7577 1836, Uses of slate: v. 29, 93-94.
7578 1836, To a fossil fern: v. 29, 572.
　　　　A poem.
7579 1836, An ascent of Vesuvius: v. 30, 278-283.
　　　　Signed by "D." From Mon. Mag.
7580 1837, Geological Society [of London]: v. 31, 47-48.
7581 1841, Visit to the volcano Kirauea: v. 42, 374-376.
　　　　From London Ath.
7582 1841, Geological phenomena of the falls of Niagara: v. 43, 435-440.
　　　　From Christian Obs. Signed by "Vespucius."
7583 1842, Erratic blocks and moraines: v. 44, 146.
7584 1842, Fossil fucoides [from Cheshire, England]: v. 44, 147-148.
7585 1842, Modern fossil wood: v. 44, 148.
7586 1842, Glaciers of Switzerland: v. 44, 288.
　　　　Notice of L. Agassiz's work.
7587 1842, Geological map of France [by Dufresnoy and Beaumont]: v. 44, 436.
7588 1842, Fossil remains of a sloth [from South America]: v. 44, 438.
7589 1842, Erratic blocks [in Northern Europe]: v. 44, 559.
7590 1842, Separation of gold and platina: v. 45, 672.

Mushet, David (see 1502 under Barrelville Mining Company and 6812 under Maryland Mining Company.)

Muzzey, Artemas Bowers (1802-1892)
7591 1843, Ascent of Vesuvius: Mon. Miscellany of Religion and Letters, v. 9, 253-255.

Muzzy, J.
7592 1812, On the Monkton porcelain earth: Lit. and Philos. Rep., v. 1, no. 1, 64-75.

N., L. L. (see 4156)

Nance, C. W.
7593 1837, Report of examinations and surveys made at Randolph Fulton, mouth of Cool Creek. and Ashport: Tenn. State Engineer Rept., for 1837, 63-81.

Napier, James (1810-1847)
7594 1847-1848, Improvement in smelting copper ores: Franklin Inst. J., v. 14 (3s), 190-191; v. 15 (3s), 255-256; v. 16 (3s), 160 (4 lines).
　　　　Abstracted from Mining J. and from London J. Arts Sci's.
7595 1848, Improvements in smelting copper and other ores: Franklin Inst. J., v. 15 (3s), 256-258.
　　　　Abstracted from London J. Arts Sci's.
7596 1848, Patent granted to James Napier, Middlesex, for improvements in smelting copper and other ores: Am. J. Sci., v. 5 (ns), no. 2, 260-261.
　　　　Abstracted from Chemical Gaz.

Nash, Alanson
7597 1827, Geology of the lead mines and veins of Hampshire County, Mass.: Am. J. Sci., v. 12, no. 2, colored geological map, 30 x 20 cm.
　　　　In 7598.
7598 1827, Notice of the lead mines and veins of Hampshire County, Mass., and of the geology and mineralogy of that region: Am. J. Sci., v. 12, no. 2, 238-270, 14 figs., colored map.
　　　　Contains map 7597.

Nasmyth, James (1808-1890)
7599 1848, New property of coke: Am. J. Sci., v. 6 (ns), no. 3, 424.
　　　　Abstracted from Mining J.
7600 1848, On the slow transmission of heat through loosely coherent clay and sand: Am. J. Sci., v. 6 (ns), no. 1, 119-120.
　　　　Abstracted from London Q. J. Geol. Soc.

National Anthracite Coal Company of Pennsylvania
7601 1827, Maps of the lands of the National Anthracite Coal Co. and of the anthracite coal region of Pa.: NY, n.p.
　　　　Not seen.

National Institution at Washington
7602   1841, National Institution at Washington for the promotion of science: Am. J. Sci., v. 41, no. 1, 203-205.

National Magazine; or, Cabinet of the United States (Washington, 1801-1802)
7603   1801, The nature and formation of soils: v. 1, 6-8; 9-11.

Natural History of the Earth
7604   1803, Bos., n.p.

The Naturalist (Nashville, Tennessee, 1850)
       Unsigned scientific notes are by the editor, Rev. Tolbert Fanning (see 3667 et. seq.)

The Naturalist, and Journal of Natural History, Agriculture, Education, and Literature (Nashville, Tennessee, 1846)
7605   1846, Geology of Harpeth Ridge: v. 1, 107.
7606   1846, Geology: v. 1, 145; 196; 488.
7607   1846, Tennessee marble: v. 1, 256.
7608   1846, Fossils of Tennessee: v. 1, 59; 108.
7609   1846, The fossil giant: v. 1, 97.
7610   1846, Association of Geologists and Naturalists: v. 1, 490.
7611   1846, Gold in New England: v. 1, 456.
7612   1846, Geology of Tennessee: v. 1, 532-533.
7613   1846, Agricultural science: v. 1, 535-538.
7614   1846, Geological view of England: v. 1, 538-539.

The Naturalist, Containing Treatises on Natural History, Chemistry, Domestic and Rural Economy, Manufactures and the Arts (Boston, 1830-1832)
7615   1830, Platina: v. 1, no. 1, 32.
7616   1831, Gold: v. 1, no. 2, 62-64.
7617   1831, Mineralogy: v. 1, no. 4, 97-101.
7618   1831, Silver: v. 1, no. 4, 125-128.
7619   1831, Mercury: v. 1, no. 5, 159-160.
7620   1831, Iron: v. 1, no. 7, 221-224.
7621   1831, Copper: v. 1, no. 8, 254-256.
7622   1831, Lead: v. 1, no. 9, 286-288.
7623   1831, Tin: v. 1, no. 10, 316-319.
7624   1831, Zinc: v. 1, no. 10, 319-320.
7625   1831, Arsenic: v. 1, no. 11, 351-352.
7626   1832, On volcanoes: v. 2, no. 7, 219-221.
7627   1832, Structure of the earth: v. 2, no. 12, 376-379, fig., plate.
       From First Lessons in Natural History.

Naumann, Karl Friedrich (1797-1873)
7628   1849, On polymerous isomorphism: Am. J. Sci., v. 8 (ns), no. 1, 128 (13 lines).
       Abstracted from "J. für Prakt. Ch."

The Naval Magazine (New York, 1836-1837)
7629   1837, Review of recent geological reports [reviews with excerpts of Report of a geological reconnoissance made in 1835 ..., by G. W. Featherstonhaugh (see 3781) and the first annual report of the New York geological survey (see 7736)]: v. 2, 473-503, 2 figs.
       See also response to review of Featherstonhaugh (7630).
7630   1837, Reply of G. W. Featherstonhaugh, Esq. U. S. Geologist to Prof. W. W. Mather: v. 2, 569-580.
       Contains letters by G. W. Featherstonhaugh (see 3783), W. W. Mather (see 6886), J. J. Abert (see 8), W. Hood (see 5249), and A. Macomb (6651).

New England Coal Mining Company
7631   1838, A report of the important hearing on the memorial of the New England Coal Company for encouragement from the state; and on the numerous petitions from the freeholders in aid of the same; before the special select committee of the General Assembly of Rhode Island and Providence Plantations, together with the report of the committee ... : n.p., 148 p.
       Contains letter by E. Hitchcock (see 5127).

The New England Farmer (Boston, 1822-1845)
7632   1822, Putnam's rock [large glacial boulder in New York]: v. 1, 24.
7633   1822, Mineralogical: v. 1, 124-125; 132; 140; 153; 161.
       From National Aegis. Basic concepts in mineralogy.
7634   1822-1845, [Notes on soils and their analysis]: v. 1, 143; 166; 403; v. 2, 68-69; v. 3, 189; v. 6, 19; 229; v. 22, 145; 389; v. 24, 182; 212-213.
7635   1822-1845, [Notes on American mines and mining]: Short notes appear in all 24 volumes.
       Consult annual indices for details. Notes on coal mining, southern gold deposits, and New England mines are predominant.
7636   1823, Mineral spring [at Bedford, Massachusetts]: v. 1, 351 (7 lines).
7637   1824, Mineralogy [notice of Catalogue of minerals by F. Hall (see 4650)]: v. 2, 245.
7638   1824-1837, [Notes on the importance of geological surveys]: v. 2, 321-322; v. 6, 69; 322; v. 7, 222-223; v. 9, 141; v. 15, 276; 332.
7639   1824-1845, [Notes on the discovery of fossil bones]: v. 2, 360; v. 5, 200; v. 7, 189; v. 8, 6; 13; v. 9, 149-150; 235; v. 11, 259; v. 15, 69; 107; v. 16, 88; v. 24, 8.
7640   1827-1835, [Notes on earthquakes and volcanoes]: v. 5, 232; v. 6, 3; 17-18; v. 7, 52; v. 9, 96; v. 11, 216; v. 12, 264; v. 13, 315.
7641   1827, Subterranean forest [at Norfolk, England]: v. 6, 69.
7642   1828, Collection of minerals: v. 6, 375.
       Notice of specimens for sale by Josiah Holbrook.
7643   1828, State collection of minerals: v. 6, 393-394.
       On the need for a Massachusetts state rock and mineral cabinet.
7644   1828, Geology [on conglomerate from Roxbury, Massachusetts]: v. 7, 68.

The New England Farmer (cont.)
7645 1829, Geology for schools: v. 7, 213-214.
        Geology suggested as a subject for grade schools.
7646 1832, Geology of Massachusetts [notice of work by E. Hitchcock]: v. 11, 8.
7647 1832, Curious geological fact: v. 11, 16.
        On a piece of coal embedded in a rock, at Broad Mountain, Pennsylvania.
7648 1832, Formation of the mountains: v. 11, 191.
7649 1833, Geology of Massachusetts [notice of map by E. Hitchcock (see 5103)]: v. 11, 259.
7650 1833, Analysis of different sorts of salt, &c.: v. 11, 337; 345.
7651 1833, [Ornamental objects carved from jet coal]: v. 11, 349.
7652 1833, Burning springs [near Buffalo, New York at Aurora]: v. 12, 45.
7653 1834, Pyrites: v. 12, 419 (10 lines).
7654 1834, The copper springs [at Wicklow, Ireland]: v. 13, 24.
7655 1834, Hot springs of the Arkansas: v. 13, 34.
7656 1834, Mineralogy in Hancock Co., Maine: v. 13, 145.
7657 1834, Geological treat: v. 13, 168; 323.
        Notice of B. Silliman's field lectures.
7658 1836, Lectures on geology: v. 14, 334-335; 342; 353-354; v. 5, 150; 290.
        Notice and review of C. T. Jackson's public lectures.
7659 1836, Making diamonds: v. 15, 144.
        Presents B. Silliman's opinions.
7660 1837, Appointments by the governor: v. 15, 374.
        E. Hitchcock appointed as geologist.
7661 1837, Meteoric stone [in New Hampshire]: v. 16, 74.
7662 1838, Geology of Massachusetts and coal mines [notice of Re-examination ... , by E. Hitchcock (see 5128)]: v. 16, 318.
7663 1838, The salt mines at Norwich: v. 17, 2.
        On the British Association field trip.

The New England Farmer; A Semi-Monthly Journal, Devoted to Agriculture, Horticulture, and their Kindred Arts and Sciences (Boston, 1848-1850)
7664 1849, Platina metal [from Russia]: v. 1, no. 9, 141.
7665 1849-1850, [Notes on the nature and improvement of soils]: v. 1, no. 17, 257-258; 275; v. 2, no. 3, 47; no. 8, 133; no. 13, 231; no. 18, 293.
7666 1849, Mineral cements: v. 1, no. 21, 333-334.
7667 1850, Wonders of geology: v. 2, no. 6, 103.
        On the number of extinct fossil animals.
7668 1850, [Litchfield, Connecticut copper mines]: v. 2, no. 23, 362 (6 lines).
7669 1850, James F. W. Johnston: v. 2, no. 23, 369, fig.
        Biographical sketch.
7670 1850, A mine of paint [in Springfield, Massachusetts]: v. 2, no. 24, 383.
7671 1850, How coal was made: v. 2, no. 24, 407.
        Abstracted from Chamber's Miscellany.

The New-England Journal of Medicine and Surgery (Boston, 1812-1826)
7672 1812, [Amber from Köningsberg, Russia]: v. 1, 104 (5 lines).
7673 1812-1819, [Notes on meteorites]: v. 1, 198-199; v. 2, 203; v. 8, 89.
7674 1813, Description of the eruption of the Souffrier Mountain, on Thursday night, the 30th of April, 1812, in the Island of St. Vincent: v. 2, 95-101.
7675 1814, Effect of pressure of the air on crystallization: v. 3, 403.
        Notice of opinions of Gay-Lussac.
7676 1815, [Cuvier's examination of fossil bones from Aeningen, Switzerland]: v. 4, 400 (11 lines).
7677 1816, Some account of Harvard University, in Cambridge, Massachusetts: v. 5, 109-123.
7678 1816, [Review of Essay on the theory of the earth by G. Cuvier (see 2754)]: v. 5, 168-176.
7679 1816, Museum of natural history [at the Linnaen Association of New England]: v. 5, 189-191.
7680 1817, [Notice and review with excerpts of Elementary Treatise ... , by P. Cleaveland (see 2420)]: v. 6, 208 (3 lines); 283-297.
7681 1817, [Reviews with excerpts of "Dissertation ..." by V. Seaman (see 9385), "Analysis ... " by J. Steele (see 9998), "Experimental inquiry ..." by W. Meade (see 6947), "Observations ... " by D. Hosack (see 5283), "Observations and experiments ..." by J. Griscom (see 4570), and "Chemical analysis ... " by J. F. Dana (see 2868)]: v. 6, 363-385.
        Review of the chemistry of American mineral waters.
7682 1818, Observations of the progress of the physical sciences during the last year: v. 7, 1-18.
7683 1818, Siliceous spar: v. 7, 91-92.
        On the chemical analyses of feldspars.

The New-England Magazine (Boston, 1831-1835)
7684 1831, Fossil bones [from Kentucky]: v. 1, 92-93.
        Bones exhibited at New York.
7685 1831, Arkansas springs [Hot Springs]: v. 1, 276-277.
7686 1831, Kentucky cavern [Mammoth Cave]: v. 1, 275-276.
7687 1831, [Review of Monthly American Journal of Geology by G. W. Featherstonhaugh (see 3746)]: v. 1, 542.
7688 1831, Anthracite coal [from Pennsylvania]: v. 1, 544-545.
7689 1832, [Review of Nova Scotia study by C. T. Jackson and F. Alger (see 5538)]: v. 2, 264.
7690 1832, Newly discovered cave in Pennsylvania [in Peters Township, Washington County]: v. 3, 174.
7691 1835, [Review of Ancient mineralogy by N. F. Moore (see 7419)]: v. 8, 238.

The New-England Medical Eclectic, and Guide to Health (Worcester, Massachusetts, 1846)
7692 1846, Hot springs in Arkansas: v. 1, 350-351.
        Signed by "A. M. B." From Botanico-Med. Rec'r.

The New Englander (New Haven, 1843-1850)
7693 1847, [Review of Elements of geology by J. L. Comstock (see 2506)]: v. 5, no. 3, 479.
7694 1850, [Review of The earth and man by A. Guyot (see 4609)]: v. 8, no. 3, 365-377.

New Hampshire. State of
- 7695 1836, Advisability of the state geological survey: in Governor's Message, June, 1836.
- 7696 1836, Adoption of measures for commencing geological survey of the state ... : in Governor's Message, June, 1838.
- 7697 1839, An act to provide for the geological and mineralogical survey of the state: NH Gen. Assembly, June 24, 1839.
- 7698 1840, [On the New Hampshire geological survey]: Niles' Wk. Reg., v. 59, 242-243. From the Governor's Message, June, 1840.
- 7699 1840, Progress of the state geological survey: In Governor's Message, 1840.
- 7700 1844, An act to provide for a geological and mineralogical survey of the state: in Final report ... , by C. T. Jackson (see 5611), iii-iv.
- 7701 1844, Resolve of the legislature to continue the survey: in Final report ... , by C. T. Jackson (see 5611), v.
- 7702 1844, Report of Secretary of State on printing of the Report of the geological survey: NH House J., November, 1844, 401-402.
- 7703 1844, Statment of progress on geological report; printing and cost thereof: NH House J., June, 1844, 260.

New Hampshire Historical Society. Transactions of the (Concord, New Hampshire, 1824)
- 7704 1827, Memoir of James Freeman Dana, M. D.: v. 2, 290-300.

New Hampshire Iron Factory Company
- 7705 1808, An act to incorporate certain persons, by the name of the New Hampshire Iron Factory Company; together with the bye-laws of said company: Salem, NH, Pool and Palfrey (printers), 16 p.

The New-Harmony Gazette (New Harmony, Indiana, 1825-1827)
See also The Free Enquirer.
- 7706 1825, Pompeii and Vesuvius: v. 1, 46.
- 7707 1825, Mineralogy. Dr. Gerard Troost's introductory lecture: v. 1, 270-271.
- 7708 1827, Vesuvius and its environs: v. 2, 298-299.
- 7709 1827, Mineral spring in New-York [at Congress Spring, Saratoga]: v. 2, 382-383.
- 7710 1827, Cornish mines: v. 2, 411.
- 7711 1827, Singular cave [at Soli, Java]: v. 2, 414.
- 7712 1827, A volcano [Popocatepotl, Mexico]: v. 3, 48.
- 7713 1827, Quicksilver mines of Idria: v. 3, 88.

New Haven Gazette, and the Connecticut Magazine (New Haven, Connecticut, 1786-1789)
- 7714 1786, [Mineral spring at New Haven]: v. 1, 16.
- 7715 1786, [Mine discovered at Leverett, Hampshire County, Massachusetts]: v. 1, 23-24.
- 7716 1786, Account of an adventure at the quicksilver mine of Idria: v. 1, 22-23; 26-27.
- 7717 1786-1787, [Notes on earthquakes and volcanoes]: v. 1, 46; 62; 77; 183; v. 2, 108; 315-316; 340.
- 7718 1786, [Virginia's burning spring at Fincastle County]: v. 1, 72.
- 7719 1786, An account of the salt mines at Wiliska in Poland: v. 1, 229-230.
- 7720 1786, [New cave found at Asturia]: v. 1, 340.

New Jersey. State of
- 7721 1834, [Advantages of a geological survey]: in Governor's Address, 1834.
- 7722 1835, To provide for a geological and mineralogical survey of the state of New Jersey: NJ Gen. Assembly, February 26, 1835.

New Monthly Magazine and Literary Journal (Boston and Philadelphia, 1821-1834)
- 7723 1833, Curious cave [in Tipperary County, Ireland]: v. 2 (ns), 293.
- 7724 1833, Increasing productiveness of the gold and platina mines in the Ural Mountains: v. 2 (ns), 452.
- 7725 1834, [Review of Principles of geology by C. Lyell (see 6532)]: v. 3 (ns), 221-222.
- 7726 1834, Slate: v. 3 (ns), 390.
- 7727 1834, New fact in mineralogy: v. 3 (ns), 464. On galena and ledererite from Nova Scotia.

New York. State of
- 7728 1825, Report of the commissioners appointed to perform certain duties relative to the salt springs, in the county of Onondaga: together with the bill introduced by said commissioners, to regulate the manufacture of salt, in the town of Salina: Alb., Croswell and Van Benthuysen (printers), 82 p.
- 7729 1827, [Act to encourage discoveries of mines, etc.]: NY Gen. Assembly, 1827.
- 7730 1835, [Resolution by the House of Assembly to determine the most expedient method of obtaining a geological survey of the state]: NY Gen. Assembly, 1835.
- 7731 1836, An act to provide for a geological survey of the state: NY Gen. Assembly, 1836.
- 7732 1836, The geological survey of the state [of New York]: Am. Rr. J., v. 5, 500-501.
- 7733 1836, Report of the Secretary of State in relation to a geological survey of the state of New York: Alb., Crosswell, Van Benthuysen and Burt, 60 p. This is New York Assembly document number 9, for 1836.
- 7734 1837, Geological survey: Am. Inst. NY J., v. 2, 531-542. Extracted from the first annual report (see 7736).
- 7735 1837, Geological survey of the state: Genesee Farm. and Gardener's J., v. 7, 178-178 [i.e. 179]; 188-189. A list of geological queries proposed by the state geologists.
- 7736 1837, State of New York No. 161. In assembly, February 11, 1837. Communication from the Governor, relative to the geological survey of the state: Alb., 212 p. This is the first annual report of the state geologists. For review with excerpts see 7629 and 7843.
- 7737 1838, State of New-York. No. 200. In assembly, February 20, 1838. Communication from the Governor, relative to the geological survey of the state: Alb., 384 p., table. This is the second annual report of the state geologists. For review with excerpts see 722 and 4285. For excerpts see 302.
- 7738 1839, A glossary of technical terms: NY Geol. Survey Ann. Rept., v. 3, 341-347.
- 7739 1839, Message from the Governor, transmitting several reports in relation to the geological survey of the state ... : NY State Assembly Document 406.

New York (cont.)

7740   1839, State of New-York. No. 275. In assembly, February 27, 1839. Communication from the Governor, relative to the geological survey of the state: Alb., 351 p.
This is the third annual report of the state geologist.
For review with excerpts see 2707, 4288 and 5345. For review see 6128, 7816 and 7819. For excerpts see 305 and 5021.

7741   1840, An act to continue the geological survey of the state: NY Gen. Assembly, May 8, 1840.

7742   1840, Glossary of technical terms: NY Geol. Survey Ann. Rept., v. 4, 473-480.

7743   1840, State of New York. No. 50. In assembly, January 24, 1840. Communication from the Governor, transmitting several reports relative to the geological survey of the state: Alb., 484 p.
This is the fourth annual report of the state geologist.
For review see 7852, 5347.

7744   1840, [Report of an investigating committee on the progress of the state geological survey]: NY Gen. Assembly, 1840.

7745   1841, [Resolution to distribute geological reports]: NY Senate, May 26, 1841.

7746   1841, State of New-York. No. 150. In assembly, February 17, 1841. Communication from the Governor, transmitting several reports relative to the geological survey of the state: Alb., 184 p.
This is the fifth annual report of the state geologist.
For notice see 1037.

7747   1842, An act relating to the geological survey of the state: NY Gen. Assembly, April 9, 1842.
On appropriations for the survey.

7748   1842, Geological reports of the state of New York for 1840, communicated by the Governor to the Assembly, Feb. 17, 1841: Am. J. Sci., v. 42, no. 2, 227-235.

7749   1842, Geological map of the state of New York by legislative authority: NY, Sherman and Smith (engravers), colored geological map, 94 x 89 cm.
Compiled from the surveys of the New York state geologists, and H. D. Rogers.

7750   1843, An act in relation to the natural history of New York: NY Gen. Assembly, April 8, 1843.
Appropriation for the continuation of the survey.

7751   1844, An act in relation to the natural history of New York: NY Gen. Assembly, May 3, 1844.

7752   1846, An act concerning the natural history of the state of New York: NY Gen. Assembly, May 5, 1846.

7753   1847, An act in relation to the natural history of the state of New York: NY Gen. Assembly, May 5, 1847.
On the distribution of the reports.

7754   1847, An act for completing the publication of the natural history of New York: NY Gen. Assembly, May 7, 1847.

7755   1848, Annual report of the regents of the University, on the condition of the state cabinet of natural history. With catalogues of the same. Vol. I: Alb., C. van Benthuysen, 33, 39, 15 p.
This is NY Senate Document 72 for 1848.

7756   1849, Annual report of the regents of the University ... [as above] ... Vol. II: Alb., Weed, Parsons and Co., 91 p., 5 plates.
This is NY Senate Document 20 for 1849.

7757   1850, Annual report of the regents of the University ... [as above] ... Vol. III: Alb., Weed, Parsons and Co., 175 p., 27 plates.
This is NY Senate Document 75 for 1850.
For notice see 1018.

7758   1850, An act to provide for the completion of the geological survey of the state: NY Gen. Assembly, April 10, 1850.

7759   1850, Report of the select committee of the legislature of 1849 on the publication of the natural history of the state of New York: Alb., Weed, Parson and Co., 178, 1 p., 28 tables.
This is NY Assembly Document 9, 73rd session, v. 1.

New York and Lake Superior Mining Company

7760   1846, [First annual report of the New York and Lake Superior Mining Company]: Alb.
Not seen. See 3470.

New York and Schuylkill Coal Company

7761   1823, A cursory review of the Schuylkill coal, in reference to its introduction into New-York, and other Atlantic cities: NY, G. F. Hopkins (printer), 24 p.

7762   1826, History of the coal lands and other real estate owned by the New-York and Schuylkill Coal Company: NY, G. F. Hopkins (printer), 21 p.
Contains several testimonial letters.

New York City. Mechanics' Institute.

7763   1837, History and proceedings of the Mechanic' Institute of the City of New York, from the corresponding secretary: Am. J. Sci., v. 31, no. 2, 415-417.

New York Historical Society

7764   1817, New-York Historical Society: Am. Mon. Mag. and Crit. Rev., v. 1, no. 1, 44-48; no. 2, 124-126; no. 3, 193-194; no. 4, 286-288; no. 5, 374-376; no. 6, 452; v. 2, no. 1, 55-56; no. 2, 121-122.
Transactions of the society reported.

The New York Journal of Medicine and Surgery (New York, 1839-1841)

7765   1840, [Review of Saratoga waters by M. L. North (see 7955)]: v. 3, 434 (5 lines).

The New York Journal of Medicine, and the Collateral Sciences (New York, 1843-1850)

7766   1844, Artesian wells: v. 2, 280 (12 lines).
Abstracted from Bulletin of Med. Sci.

7767   1845, [Notice of Vestiges of the natural history of Creation by R. Chambers (see 2292)]: v. 4, 269.

7768   1847, [Review with excerpts of The Virginia springs by J. J. Moormann (see 7422)]: v. 9, 82-85.

7769   1847, [Review of Agriculture of New York by E. Emmons (see 3541)]: v. 9, 235-237.

New York Literary and Philosophical Society
7770 1815, Donations for the library and cabinet of the Literary and Philosophical Society of New York: Lit. Philos. Soc. NY Trans., v. 1, 567-570.
7771 1817, Literary and Philosophical Society of New-York: Am. Mon. Mag. and Crit. Rev., v. 1, no. 2, 126-127; no. 3, 194-195; no. 4, 288; no. 5, 376; no. 6, 451-452; v. 2, no. 1, 57-58; no. 2, 122-123. Transactions noted.

New-York Literary Journal, and Belles-Lettres Repository (New York, 1819-1821)
7772 1819, Distillation of coal: v. 1, 37.
7773 1819, Catskill Mountains [geology of]: v. 1, 202-207; 251-255.
7774 1819, Great Cave of Ellora: v. 2, 73-85.
  From Fitzclarence's Journey through Egypt. On a cave in India.
7775 1819, [Review of Tour through Sicily by G. Russel (see 9143)]: v. 2, 170-180.
7776 1819, Hot springs of Ouachitta, (Washitaw): v. 2, 217-219.
7777 1819, Precious stone of Missouri: v. 2, 219-220.
7778 1820, [Review with excerpts of View of the lead mines by H. R. Schoolcraft (see 9255)]: v. 3, 2-9.
7779 1820, Silver ore [from Zanesville, Ohio]: v. 3, 88.

New York Lyceum of Natural History
7780 1817, Lyceum of Natural History: Am. Mon. Mag. and Crit. Rev., v. 1, no. 2, 127-133; no. 3, 195-196; no. 4, 288-290; no. 5, 376-379; no. 6, 452-454. v. 2, no. 1, 56-57; no. 2, 123.
7781 1820-1846, [Proceedings, officers and notices of the Lyceum]: Am. J. Sci., v. 2, no. 2, 366-372; v. 6, no. 2, 361-366; v. 7, no. 1, 171-174; no. 2, 359-360; v. 8, no. 1, 192-193; v. 9, no. 2, 387-391; v. 10, no. 1, 198-201; no. 2, 397; v. 13, no. 2, 378-381; v. 14, no. 1, 190-194; v. 15, no. 1, 191-194; no. 2, 357-360; v. 16, no. 1, 205-209; no. 2, 354-357; v. 18, no. 1, 193-195; v. 19, no. 1, 159-160; no. 2, 353-355; v. 27, no. 1, 148-163; v. 28, no. 1, 189-192; v. 31, no. 1, 204; v. 36, no. 1, 195; v. 48, no. 2, 404; v. 1 (ns), no. 3, 453.
7782 1825, Catalogue of books in the library of the Lyceum of Natural History: NY Lyc. Nat. History Annals, v. 1, pt. 2, 392-400; Addenda, 1-12.
7783 1827, Catalogue of books in the Lyceum of Natural History: NY Lyc. Nat. History Annals, v. 2, 453-366.
7784 1827, Donations to the cabinet of the Lyceum of Natural History: NY Lyc. Nat. History Annals, v. 2, 467-472.
7785 1830, Proceedings: Mag. Use. Ent. Know., v. 1, no. 2, 63; no. 3, 129; no. 4, 185-186; no. 5, 234.
7786 1840, Transactions of the New York Lyceum of Natural History: Am. Rep. Arts Sci's Manu's, v. 1, 177-178.
7787 1840-1841, Lyceum of Natural History proceedings: Am. Rep. Arts Sci's Manu's, v. 1, 256-258; 343-346; 418-421; v. 2, 8-12; 249-253; v. 3, 9-13; 197-199; 260-263; 413-416; v. 4, 5-8; 242-245.
7788 1848, Extracts from the minutes of the lyceum of natural history: NY J. Medicine and Collateral Sci's, v. 10, 338-344.

New York Magazine; or, Literary Repository (New York, 1790-1797)
7789 1790, A description of Mount Aetna: v. 1, 623-628, plate.
7790 1791, Some particulars relative to the soil, situation, production, &c. of Kentucky: v. 2, 702-706.
  From the Nat'l Gaz.
7791 1792, An enquiry into the formation of islands: v. 3, 210-212.
7792 1792, Account of an earthquake at St. Paul's Bay, Eboulement's, &c. in Canada: v. 3, 263-264.
7793 1794, Eruption of Mount Vesuvius: v. 5, 528-529.
7794 1794, Description of the tin and copper mines in Cornwall: v. 5, 675-680.
  From A tour through England in 1791.
7795 1797, Account of the eruption of a singular species of volcano in an island of the Crimea: v. 2 (ns), 192.
7796 1797, Account of the dreadful effects of the earthquake of Quito, and its neighbourhood, which happened on the 4th of February, 1797: v. 2 (ns), 482-483.

New York Medical and Philosophical Journal and Review (New York, 1809-1811)
7797 1809, American mineralogy: v. 1, 150 (4 lines).
  Notice of A. Seybert's catalogue of United States minerals.
7798 1809, Jameson's Elements of geognosy [review of]: v. 1, 160.
7799 1809, [Review of Dissertation on the mineral waters of Saratoga by V. Seaman (see 9385)]: v. 1, 286-290.
7800 1809, Salt spring [at Butler, Pennsylvania]: v. 1, 300 (8 lines).
7801 1810, [Notice of American mineralogical journal by A. Bruce (see 2107)]: v. 2, 107-108.
7802 1810, Meteoric stones [from Russia]: v. 2, 115-116 (6 lines).
7803 1810, [Review of "An examination of "A review of a Dissertation on the mineral waters of Saratoga" by Valentine Seaman", by Valentine Seaman (see 9384)]: v. 2, 139-175.
  With a letter by David Hosack (see 5282). Abstracted from Med. Repos.
7804 1810, Prospectus of a new mineralogical work [notice of Treatise on mineralogy by S. Godon (see 4424)]: v. 2, 285 (7 lines).

New York Medical and Physical Journal (New York, 1822-1830)
7805 1822, Notices of some mineral springs in the county of Albany [notice with excerpts of A geological survey ..., by A. Eaton and T. R. Beck (see 3328)]: v. 1, 237-239.
7806 1822, New blowpipe [notice of article by J. Green (see 4529)]: v. 1, 515 (6 lines).
7807 1822, Mineralogy [notice of textbook by Mohs]: v. 1, 517 (12 lines).
7808 1823, New locality of cerium: v. 2, no. 2, 250.
7809 1823, [Review of Annals of the New York Lyceum of Natural History]: v. 2, 502-513.
7810 1823, [Review of Essay on salt by J. Van Rensselaer (see 10548)]: v. 2, 514-515.

New York Medical and Physical Journal (cont.)
7811   1823, Onondaga salt works: v. 2, 515.
7812   1828, [Review of Annals of the New York Lyceum of Natural History, volumes 1 and 2]: v. 7, 115-126.

The New-York Monthly Chronicle of Medicine and Surgery (New York, 1824-1825)
7813   1825, On the connexion between natural history and medicine: v. 1, no. 9, 257-263.
       Signed by "L."

The New York Monthly Magazine (New York, 1824)
7814   1824, The mines of Idria: v. 1, 19.

The New York Review (New York, 1837-1842)
7815   1838, [Review with excerpts of Narrative of an excursion to the Lake Amsanctus and to Mount Vultur in Apulia and of Report on ... mineral and thermal waters by C. Daubeny]: v. 3, no. 1, 20-43.
7816   1839, [Review of reports on the geological survey of New York (see 7740)]: v. 4, no. 1, 71-108.
7817   1839, [Review of Wonders of geology by G. A. Mantell (see 6732)]: v. 4, no. 1, 255-256.
7818   1839, [Review with excerpts of An introduction to geology by R. Bakewell (see 1460) and "Appendix" by B. Silliman (see 9759)]: v. 5, no. 2, 457-477.
7819   1839, [Review of geological survey of New York (see 7740)]: v. 5, no. 2, 477-490.
7820   1840, Terrestrial magnetism [reviews of theories by C. F. Gauss, W. Weber, and A. Kircher]: v. 7, no. 2, 475-501.
7821   1841, [Review of Elements of geology by W. W. Mather (see 6890)]: v. 8, no. 1, 272-273.

The New York Review or Atheneum Magazine (New York, 1825-1826)
7822   1825, [Review of Travels in the central portions of the Mississippi Valley by H. R. Schoolcraft (see 9268)]: v. 1, no. 2, 85-103.
7823   1825, [Review of Lectures on geology by J. van Rensselaer (see 10551)]: v. 1, no. 6, 429-441.
7824   1826, A meditation on Rhode Island coal: v. 2, no. 5, 386-388.
       Signed by "B." A poem.

New York State Board of Agriculture. Memoirs of the (Albany, 1821-1826)
7825   1823, Remarks on soils: v. 2, 285-292.
       Signed by "Agricola."

The New York State Mechanic, a Journal of the Manual Arts, Trades, and Manufactures (Albany, 1841-1843)
7826   1841-1843, [Notes on mines and mining]: v. 1, 3; 11; 67; 70; 91; 107; v. 1, pt. 2, 19; 43; 56; 111; 183; 199; 202; v. 2, 86; 101, 102; 145.
7827   1841-1842, [Notes on earthquakes and volcanoes]: v. 1, 3; v. 1, pt. 2, 19; 56; 142; 184.
7828   1841-1842, [Notes on fossil bones]: v. 1, 3; 35; 139; v. 1, pt. 2, 103; 167; v. 2, 20.
7829   1841-1843, [Notes on hot springs and mineral springs]: v. 1, 11; v. 1, pt. 2, 9; 186; 187.
7830   1842, The metals: v. 1, pt. 1, 152-154; 161-162; 169-170; 177-178; 209-210; pt. 2, 65-66.
       Signed by "E. G. S."
7831   1842, The Yankee geologist: v. 1, 190.
       On a giant excavating machine.
7832   1842, Prof. Amos Eaton [obituary]: v. 1, pt. 2, 6.
7833   1842, Levels of the Mediterranean Sea: v. 1, pt. 2, 14 (4 lines).
7834   1842-1843, [Notes on caves]: v. 1, pt. 2, 87; 105; v. 2, 152.
7835   1842, Terra di Sienna [from Lancaster, Pennsylvania]: v. 2, 14.
7836   1842, The enchanted rock: v. 2, 40.
       On an odd rock formation with mica in Texas.
7837   1843, Highland granite quarry [in New York]: v. 2, 76.
7838   1843, National Association of Geologists: v. 2, 181.
       Notice of meeting of the Association of American Geologists.

The New York Weekly Magazine (New York, 1795-1797)
7839   1795, Account of a burning river [at Tremoulac, France]: v. 1, no. 10, 73.
7840   1796, Description of the famous salt-mines of Williska in Poland: v. 2, no. 53, 1; no. 54, 9.

The New-Yorker (New York, 1836-1841)
7841   1836-1839, [Notes on earthquakes and volcanoes]: v. 1, 355; v. 4, 582-584; v. 7, 52; 184; v. 8, 51-52.
7842   1836, Plumbago, or black lead: v. 2, 231.
7843   1837, [Review of first annual New York report (see 7736)]: v. 3, 78.
       From Am. J. Sci.
7844   1837, The science of geology: v. 3, 258-259.
       Abstracted from Edinburgh Rev.
7845   1837, Geology - Dr. Buckland's new work [review of 2160]: v. 3, 294-295.
       Abstracted from Edinburgh Rev.
7846   1837, Visit to the salt mines of Salzburg: v. 4, 467-468.
       Abstracted from New Mon. Mag.
7847   1838, Caves of Ellora [in India]: v. 5, 94-95.
       Signed by "Indicus".
7848   1838, The hot springs [of Virginia]: v. 5, 344.
7849   1838, The natural bridge [of Virginia]: v. 5, 344.
7850   1838, Hints on geology: v. 5, 404.
7851   1839, Geology in Russia: v. 7, 200.
7852   1840, Geological reports of 1840 [review of New York survey report (see 7743)]: v. 9, 242-244.
       Signed by "W. G."

The New-Yorker (cont.)
7853   1840, Meteorolites: v. 9, 388.
7854   1841, Geological remains of the West: v. 11, 40.
        On mammoth, Missourium, and petrified wood fossils from Missouri and Wisconsin.

Newbold, Captain
7855   1845, On the Kunker, a tufaceous deposit in India: Am. J. Sci., v. 49, no. 2, 398.
        Abstracted from Philos. Mag.

Newhall, John B.
7856   1841, Sketches of Iowa; or, the emigrant's guide, containing a correct description of the agricultural and mineral resources, geological features and statistics of the territory of Iowa: NY, J. H. Colton, 252 p., map.
        For extract see New-Yorker, v. 11, 54-55.
7857   1846, A glimpse of Iowa in 1846; or, the emigrant's guide, and state directory; with a description of the New Purchase embracing much practical advice and useful information to intending emigrants: Burlington, Iowa, W. D. Skillman, 106 p.
        Second edition. Mineral notes, especially on 55-58.

Newland, Jeremiah
7858   1755, Earthquakes improved; or a solemn warning to the world; by the tremendous earthquakes which happened on Tuesday morning, the 18th of November, 1755, between four and five oclock: Bos., sold by J. Green, Broadside folio, woodcut.
        Evans number 7518. A poem of 40 verses.

Newman, John
7859   1816, Account of a new blowpipe: Eclec. Rep. and Analytic Rev., v. 6, 295-297; also in J. Sci. Arts, v. 1, no. 1, 65-66; no. 2, 379-382, fig. (1817).
        Abstracted from London J. Sci. Arts.

Newman, John B.
7860   1846, Texas and Mexico, in 1846; comprising the history of both countries, with an account of the soils climate, and productions of each: NY, J. K. Wellman, 32 p.

Newton, Willoughby (1802-1874)
7861   1835, Speculations on the nature and fertilizing properties of the earth called "Jersey marl," or "green sand": Farm's Reg., v. 3, 419-422.

Nichols, Andrew (1785-1853)
7862   1841, Project for procuring extensive analyses of soils: NE Farm., v. 19, 290-291; 294; 318.

Nicholson, William (1753-1815)
7863   1816/1817, American edition of the British encyclopedia, or dictionary of arts and sciences, comprising an accurate and popular view of the present improved state of human knowledge: Phila., S. A. Mitchell and H. Ames, 7 v., plates.
        Other editions include: "second edition", Phila., Mitchell, Ames, and White, 12 volumes, 1818; and "third edition", as in second edition, 1819-1821; and Nashville, Tenn., Ingram and Lloyd, 12 v.

Nicklès, François Joseph Jérôme (1821-1869)
7864   1848, On the crystallized hydrated oxyd of zinc: Am. J. Sci., v. 6 (ns), no. 2, 257.
        Abstracted by "G. C. S." from Annales de Chimie.

Nicklin, Philip Holbrook (1786-1842)
7865   1835, Letters descriptive of the Virginia springs: Phila., H. S. Tanner (printer), 99 p.
        Written under the pseudonym "Peregrine Prolix".
        For review with excerpts see 9948. For review see 2382 and 7945.
7866   1836, Letters descriptive of the Bedford spring: Phila., H. S. Tanner, 148 p.
7867   1837, Letters descriptive of the Virginia springs: Phila., H. S. Tanner (printer), 248 p.
        Second edition. Signed by "Peregrine Prolix". This edition includes "Eight more letters", p. [117]-224.

Nicol, Andrew
7868   1836, The peat soils of Scotland, compared with the juniper soil of the Dismal Swamp: Farm's Reg., v. 4, 528-529.

Nicol, William (1768-1851)
7869   1835, Sept. 11. - Structure of recent and fossil wood: Am. J. Sci., v. 28, no. 1, 74 (6 lines).
        In British Assoc. Proc.

Nicollet, Joseph Nicolas (1786-1843)
        See also discussion of article by C. T. Jackson (5602).
7870   1841, Observations of magnetic dip, made at Baltimore: Am. Philos. Soc. Proc., v. 2, no. 19, 83-85.
        Abstracted by A. D. Bache.
7871   1841, Remarks on the geology of the region on the Upper Mississippi, and the Cretaceous formation of the Upper Missouri: Am. J. Sci., v. 41, no. 1, 180-182. Also in Assoc. Am. Geol's Nat's Proc., v. 2, 25-27; also in Assoc. Am. Geol's Nat's Rept., v. 1, 32-34 (1843).
7872   1843, Observations on the magnetic dip, made in the United States in 1841: Am. Philos. Soc. Trans., v. 8 (ns), 315-326.
7873   1843, On the Cretaceous formation of the Missouri River: Am. J. Sci., v. 45, no. 1, 153-155.
        In Assoc. Am. Geol's Nat's Proc. With discussion by H. D. Rogers, D. Houghton, and W. C. Redfield (8711).

Nicollet, J. N. (cont.)
7874   1843, On the mineral region of the state of Missouri: Am. J. Sci., v. 45, no. 2, 340-341.
        In Assoc. Am. Geol's Nat's Proc.
7875   1843, Report intended to illustrate a map of the hydrographical basin of the Upper Mississippi River: US SED 237, 26-2, v. 5, pt. 2 (380), 170 p., map.
        Also in US HED 52, 28-2, v. 2 (464), 170 p., map, 1845.
        Translated by J. T. Ducatel (part 1) and J. H. Alexander (part 2). Appendix C lists fossils found, by formation and locality. For abstract see Am. J. Sci., v. 1 (ns), no. 2, 270-271.

The Nightingale, or, a Melange de Literature (Boston, 1796)
7876   1796, Natural curiosity. on a newly discovered cavern, on the north-east end of Mount Anthony, in Bennington, Vermont: v 1, no. 36, 424-426.

Niles, Hezekiah (1777-1839)
        1812-1850, As editor of Niles' Weekly Register, H. Niles was probably the author of many of the short notes and news items listed below.

Niles' Weekly Register (Baltimore, 1812-1850)
7877   1812-1850, [Notes on earthquakes and volcanoes]: Most volumes contain short reports of recent occurrences from around the world. Consult annual indices for details.
7878   1812-1850, [Notes on mines and mining]: Numerous short notes appear in most volumes. Progress and statistics of American mines including special emphasis on salt, coal, lead, copper, gold, silver and zinc, are included.
7879   1812, Plaister of Paris [from Cayuga Lake, New York]: v. 2, 10.
7880   1814, Meteoric stones [fall at Malpas]: v. 5, supplement, 185-186.
        Signed by "Edward H----d."
7881   1814-1849, [Notes on vertebrate fossils]: v. 5, supplement, 189, 189-190; v. 16, supplement, 104, 190-191; v. 26, 220; v. 29, 6; v. 31, 68; v. 32, 304; v. 45, 132; v. 54, 405-406; v. 58, 240; v. 63, 224; v. 68, 249; v. 71, 198; v. 72, 281; v. 74, 251-252; v. 75, 64.
7882   1815-1845, [Notes on American caves]: v. 9, supplement, 176; v. 10, 420-421; v. 22, 270; v. 60, 256; v. 66, 336; v. 68, 216.
7883   1818, Mineralogy [notice of textbook by Brongniart]: v. 15, 196.
7884   1822-1841, [Notes on mineral and hot springs]: v. 23, 2; v. 31, 286; v. 32, 259; 292; 354; v. 43, 43; v. 61, 103.
7885   1828, The interior [of the earth]: v. 35, 67.
7886   1829, Petroleum [from Cumberland County, Kentucky]: v. 36, 117.
7887   1829, Geological phenomena: v. 37, 4.
        Salt water well struck oil at Cumberland County, Kentucky.
7888   1831, Dr. Mitchill [obituary]: v. 41, 37.
7889   1832, General convention of the friends of domestic industry, assembled at New York, October 26, 1831: v. 41, Addendum, 64 p.
        On iron manufacturing (1-32) and salt manufacturing (36-40).
7890   1835, Hydraulic cement [from Onondaga County, New York]: v. 49, 203.
7891   1836, Geological changes in Great Britain: v. 51, 83.
7892   1836, Geological wonder: v. 51, 114.
        On granite at the Kennebunkport, Maine quarry.
7893   1837, Geology of Virginia: v. 52, 405.
        On the progress of a geological survey.
7894   1841, Virginia. Geology: v. 60, 188-189.
        Contains extracts and notice of report of 1840, by W. B. Rogers (see 8980).
7895   1841, New Hampshire. Geological survey: v. 61, 20.
        On the progress of C. T. Jackson.
7896   1842, Geologic survey of New York: v. 62, 150-151.
7897   1843, The geological Association: v. 64, 166.
        Meeting of the Association of American Geologists and Naturalists noted.
7898   1843, Fossil copal [from the Oregon Territory]: v. 64, 272.
7899   1844, American Association of Geologists and Naturalists: v. 66, 112, (10 lines).
        Notice of meeting.
7900   1845, Pennsylvania. Geological survey: v. 67, 386 (9 lines).
        Notice of H. D. Roger's field work.
7901   1845, Geological wonders: v. 69, 70.
        Notice of Koch's museum.
7902   1846, Geology: v. 71, 61.
        Notice of the British Mining Museum.
7903   1847, Geologists of the Lake Superior region: v. 72, 145 (12 lines).
7904   1847, Survey of the United States mineral lands [by D. D. Owen]: v. 72, 274.
7905   1847, Notes taken during a trip to Fond du Lac: v. 72, 378-383.
7906   1847, The geological explorers [of Lake Superior]: v. 23 (6s) [i.e. 73], 48.
7907   1847, The close of the mineral season for 1847 [at Lake Superior]: v. 23 (6s) [i.e. 73], 150.
7908   1847, Geological survey of Iowa, Wisconsin and Minnesota: v. 23 (6s) [i.e. 73], 220.

Nilsson, Prof.
7909   1850, The fossil bovine animals of Scandinavia: Ann. Scien. Disc., v. 1, 283-284.
        Abstracted from Skandin's Daggdjur.

Nimshi. The adventure of a man to obtain a solution of Scriptural geology.
7910   1845, [Excerpts in review of London edition (see 3411)]: Eclec. Mag., v. 6, 342-345.
        No American editions were found.

Noehden, George Heinrich (1770-1826)
7911   1817, Some account of the meteoric stones, in the Imperial Museum at Vienna: J. Sci. Arts, v. 2, no. 2, 314-320.

Nöggerath, Johann Jakob (1788-1877)
7912   1829, Compression of air. - An account of a remarkable accident which occurred in a mine of Bovey coal, in consequence of the compression of air; by the inspector: Am. J. Sci., v. 17, no. 1, 38-41.

Non, M. de
7913   1795, Description of a visit to Mount Etna: Mass. Mag., v. 7, 107-108.

Nordenskiold, M. G.
7914   1823, Sordawalite [from Finland]: Bos. J. Phil. Arts, v. 1, no. 3, 302; also in Am. J. Sci., v. 10, no. 1, 186 (1825).
7915   1825, Meteoric stones: Bos. J. Phil. Arts, v. 2, no. 4, 386 (9 lines).
       Abstracted from Ferussac's Bulletin.
7916   1834, Amphodellite [from Lozo, Finland]: Am. J. Sci., v. 26, no. 2, 387-388.
       Includes notice of other zeolites from Finland, and Xanthite from Orange County, New York.
       Translated by L. Feuchtwanger from Berzelius' Jahresbericht.
7917   1834, Pyrargillite [from Finland]: Am. J. Sci., v. 26, no. 2, 387 (18 lines).
       Translated by L. Feuchtwanger from Berzelius' Jahresbericht.
7918   1847, M. Nordenskiold upon diphanite, a new mineral species from the emerald mines of the Ural in the neighborhood of Catherinenburg: Am. J. Sci., v. 4 (ns), no. 2, 277-278.
       Abstracted by W. C. Lettsom from Poggendorff's Annalen.

Norlin, E. C.
7919   1846, Iberite, a new mineral from Montalvum, province of Toledo, Spain: Am. J. Sci., v. 1 (ns), no. 1, 120 (12 lines).
       Abstracted from "Comptes Rend. Acad. Stockholm."

Normandie, John de (see De Normandie, John)

Norris, William
7920   1838, Stratification of minerals by voltaic electricity: Franklin Inst. J., v. 22 (ns), 69.
       Abstracted from Edinburgh New Philos. J.

North-American Coal and Mining Company
7921   1814, Observations on the intended application of the North-American Coal & Mining Company, to the legislature of the state of New York: NY, 23 p.
       Includes testimonial letters by S. L. Mitchill (see 7296) and B. Silliman (see 9615); and extracts from J. Williams' The Mineral Kingdom (see 10967), p. 13-14.
7922   1814?, Memorial of the North-American Coal and Mining Company, to the legislature of the state of New York, and the report of the select committee: NY, by order of the NY House, 16 p.

North American Coal Company
7923   1827, A brief sketch of the property belonging to the North American Coal Company; with some general remarks on the subject of coal and coal mines: NY, G. F. Hopkins (printer), 23 p., 2 maps.
       On coal mines in Carbon Mountain and Centerville, Pennsylvania.
7924   1838, An act to incorporate the North American Coal Company: NY, J. Navine (printer), 18 p.
       Includes report on the mines by J. Penman (see 8205). Describes holdings in Ulster and Sullivan Counties, New York.

The North American Medical and Surgical Journal (Philadelphia, 1826-1831)
7925   1831, [Review with excerpts of On baths and mineral waters by J. Bell (see 1622)]: v. 12, no. 2, 302-343.

North American Mining Company
7926   1832, An act to incorporate the North American Mining Company: n.p., 21 p.
       Includes geology report by G. W. Byrd (see 1924). Report contains descriptions of the Keith Mine and New Potosi Mine in Hall County, Georgia.

The North American, or, Weekly Journal of Politics, Science, and Literature (Baltimore, 1827)
7927   1827, Figure of the earth: v. 1, 21.
7928   1827, Important discovery: v. 1, 22.
       On zinc mines in Jefferson County, Missouri.
7929   1827, Meteoric stones [fall at Charles County, Louisiana]: v. 1, 39.
7930   1827, [On a display of mammoth bones in New Orleans]: v. 1, 43.
7931   1827, [Volcano in Sumner County, Kentucky]: v. 1, 88.
7932   1827, Earthquake [in Illinois]: v. 1, 141.
7933   1827, Account of the conflagration in the quicksilver mines of Idria, in 1803: v. 1, 147.

North American Review (Boston, 1815-1850)
7934   1816, New work on mineralogy [notice of An elementary treatise on mineralogy ... , by P. Cleaveland (see 2420)]: v. 4, 143.
7935   1816, Mr. Maclure [and his activities]: v. 4, 144.
7936   1816, On geological systems: v. 3, 209-212.
       Review of Neptunist and Vulcanist theories.
7937   1817, Mineralogy of Boston and its vicinity [notice of study by J. F. and S. L. Dana (see 2872)]: v. 6, 142-143 (11 lines).
7938   1817, Cleaveland's mineralogy [notice of an English edition of 2420]: v. 6, 145 (7 lines).
7939   1817, Maclure's geology of the United States: v. 6, 146.
       Notice of different editions, and brief review of 6616.
7940   1818, Professor Silliman's Journal of the sciences, &c.: v. 7, 143.
       Prospectus of the American Journal of Science.
7941   1818, [Review of Geology of the northern states by A. Eaton (see 3324)]: v. 6, 416.
7942   1822, [Review of A geological and agricultural survey of Rensselaer County by A. Eaton (see 3331)]: v. 14, 378-381.
7943   1822, [Review with excerpts of A description of the island of St. Michael by J. W. Webster (see 10730)]: v. 14, 34-51.

North American Review (cont.)
7944 1837, [Review of First report ... on Maine and First report ... on the public lands by C. T. Jackson (see 5550 and 5552)]: v. 45, 240-243.
7945 1837, [Review of Letters descriptive of the Virginia springs by P. Nicklin (see 7865)]: v. 45, 257-258.
7946 1838, [Review with excerpts of Second report ... on Maine by C. T. Jackson (see 5560)]: v. 47, 241-244.
7947 1838, [Review of Report on a reexamination of the economical geology of Massachusetts by E. Hitchcock (see 5128)]: v. 47, 250-253.
7948 1838, [Review of Inquiries in the Province of Kemaon relative to geology by John McClelland]: v. 46, 533-536.
7949 1841, [Review of An account of magnetic observations made at the observatory of Harvard University, Cambridge by J. Lovering and W. C. Bond (see 6506)]: v. 53, 520-521.
7950 1842, [Review of Elementary geology by E. Hitchcock, second edition (see 5134)]: v. 54, 238-239.

North Carolina. State of
7951 1823, An act directing a geological and mineralogical survey to be made of the state of North Carolina: NC Gen. Assembly, 1823.

North Carolina Gold-Mine Company
7952 1806, North Carolina gold-mine company: Wash., n.p., 20 p.

North, Elisha (1771-1843)
7953 1826, On fuel: Am. J. Sci., v. 11, no. 1, 66-78.

North, Milo Linus
7954 1839, Saratoga waters: Med. Exam., v. 2, 252-253.
    From Bos. Med. Surgical J. Ascribed to "N. L. North (i.e. Milo Linus North).
7955 1840, Saratoga waters, or the invalid at Saratoga: NY, M. W. Dodd, 70 p.
    For review see 7765. Another printing as "second edition" by Saxton and Miles, NY, 72 p., 1843.
    Abstract in NY J. Medicine and Collateral Sci's, v. 9, 326-336; also in West. J. Medicine Surg., v. 8 (ns), 66-70.
7956 1846, Analysis of Saratoga waters: also of Sharon, Avon, Virginia and other mineral waters of the United States: NY, Saxton & Miles, 72 p.
    This is "3rd edition" of 7955. Also 4th ed. (1846).

North-West Mining Company
7957 1847, Report to the trustees and stockholders, of a visit to the locations and on examination of the works: NY, J. A. Fraetus and Co. (printers), 16 p.
    Contains report on mines by H. Greeley (see 4512).

Northampton Slate Quarry Company
7958 1828, Memorial of the Northampton Slate Quarry Company in Pennsylvania, praying that further protecting duty be laid on imported slate for roofing: Wash., Duff Green, 3 p. (SED 146).

Northern Academy of Arts and Sciences
7959 1842, Constitution and by-laws of the Northern Academy of Arts and Sciences and first annual report of the curators: Hanover, NH, W. A. Patten (printer), 28 p.
    Also 1843 edition, 18 p.

The Northern Light (Albany, 1841-1844)
7960 1841, Earthquakes in 1840 [world summary]: v. 1, 7-8.
7961 1841, Temperature of the earth: v. 1, 10.
7962 1841, Geological distribution of coal: v. 1, 39.
    Abstracted from London Q. Rev.
7963 1841, Statistics of mines [in France]: v. 1, 46-47.
7964 1844, [Review of First lessons on geology by G. F. Richardson (see 8771)]: v. 3, 172-173.
7965 1844, [Review of Excursion through the slave states by G. W. Featherstonhaugh (see 3784)]: v. 4, 48.
7966 1844, Geology patronized by the United States government: v. 4, 63.

Norton, John Pitkin (1822-1852)
7967 1845, Introduction: in Catechism of agricultural chemistry and geology by J. F. W. Johnston (see 5825).
7968 1845, Mr. Norton's letters: The Cultivator, v. 2 (ns), 171-172, 233.
7969 1849, Analysis of soils: Prairie Farm., v. 9, 650.
7970 1849, Origin of the soil - scientific agriculture: Farm's Mon. Vis., v. 11, 58-59.
7971 1849, The structure, the physical properties, and the chemical composition of the soil: Mass. Agri. Societies Trans., for 1848, 221-243.
7972 1850, Address delivered before the Ontario Co., Agricultural Society (Oct. 2nd, 1850): Canandaigua, NY, J. J. Mattison (printer), 25 p.
7973 1850, Elements of scientific agriculture; or, the connection between science and the art of practical farming ... : Alb., E. H. Pease & Co., x, 208 p.
7974 1850, Elements of scientific agriculture: NY State Agri. Soc. Trans., v. 9, 602-735, 11 figs.
7975 1850, The necessity for the improvement of the soil, and the means of effecting it: Mass. Agri. Societies Trans., for 1849, 379-396.

Norton, William Augustus (1810-1853)
7976 1847, On terrestrial magnetism: Am. J. Sci., v. 4 (ns), no. 1, 1-12, 8 figs.; no. 2, 207-230, 3 figs.
7977 1849, On the diurnal variations in the declination of the magnetic needle, and in the intensities of the horizontal and vertical magnetic forces: Am. J. Sci., v. 8 (ns), no. 1, 35-55, 4 figs.; no. 2, 216-226; no. 3, 350-364, 10 figs.

Norton, W. A. (cont.)
7978   1850, On the diurnal and annual variations in the declination of the magnetic needle, and in the horizontal and vertical magnetic intensities: Am. J. Sci., v. 10 (ns), no. 3, 330-341.
    Abstracted in Ann. Scien. Disc., v. 1, 135-138 (1850).

Norwich and Worcester Rail-road Company
7979   1837, [First] Annual report of the Norwich and Worcester Rail-road Company: Bos., 58 p.
    Contains geology report by W. W. Mather (see 6884).

Norwood, Joseph Granville (1807-1895)
    See also 8075, 8076, 8080, and 8082 with D. D. Owen.
7980   1846, and D. D. Owen, Description of a new fossil fish from the Paleozoic rocks of Indiana [from Madison, Jefferson County]: Am. J. Sci., v. 1 (ns), no. 3, 367-371, 2 figs.
7981   1846, and D. D. Owen, Description of a remarkable fossil echinoderm, from the limestone formation of St. Louis, Missouri: Am. J. Sci., v. 2 (ns), no. 2, 225-228, 3 figs.
7982   1846, and D. D. Owen, [On a Devonian fossil fish from southern Indiana]: Phila. Acad. Nat. Sci´s Proc., v. 3, no. 2, 30 (15 lines).
7983   1846, and D. D. Owen, [On a Devonian fossil fish, Macropetalichthys rapheidolabris, from southern Indiana]: Bos. Soc. Nat. History Proc., v. 2, 102, 116.
7984   1848, Report [on the geology of part of Wisconsin and the Lake Superior region]: in A geological reconnaissance of the Chippewa land district ... , by D. D. Owen (see 8079), 73-129.

Notes on California and the placers; how to get there, and what to do afterwards. By one who has been there:
7985   1850, NY, Long snd Brother, 128 p., 2 plates.
    Copyright by James Delavan, as proprietor. For review see 6368.

Notes on Cuba: Containing an account of its discovery and early history; a description of the face of the country, its population, resources, and wealth ... :
7986   1845, Bos., James Munroe and Co.
    Signed "By a physician." Not seen.

Nott, Josiah Clark (1804-1873)
7987   1849, Two lectures on the connection between the Biblical and physical history of man, delivered by invitation, from the chair of political economy, etc., of the Louisiana University in December, 1848: NY, Bartlett and Welford, 146 p., map.
7988   1850, Chronology, ancient and scriptural; being a reply to an article contained in the Southern Presbyterian Review: Charleston, SC, Walker and James, 44 p.

Nova Scotia. Surveyor General
7989   1794, Miscellaneous remarks and observations on Nova Scotia, New Brunswick, and Cape Breton. Supposed to be written by the Surveyor General of Nova Scotia: Mass. Hist. Soc. Coll., v. 3, 94-101.

Nugent, Nicholas
7990   1818, Notices of geology in the West Indies: Am. J. Sci., v. 1, no. 1, 140-142.

Nutt, Rush
7991   1832, Miscellaneous geological topics relating to the lower part of the vale of the Mississippi; alluvion by rain; up filling and extension of valleys; subsidence of the sea, original course of the river with its wings and present channel - from unpublished mss. on the theory of the earth: Am. J. Sci., v. 23, no. 1, 49-65.
7992   1832, On the origin, extension, and continuance of prairies, extracted and abridged from unpublished mss. on a theory of the earth: Am. J. Sci., v. 23, no. 1, 40-45.

Nuttall, Thomas (1786-1859)
7993   1821, A journal of travels into the Arkansas territory, during the year 1819. With occasional observations on the manners of the aborigines ... : Phila., T. H. Palmer, xii, 296 p., 5 plates, map.
7994   1821, Observations on the geological structure of the Valley of the Mississippi: Phila. Acad. Nat. Sci´s J., v. 2, no. 1, 14-52.
    Abstracted in Phila. J. Med. Phys. Sci´s, v. 4, 382-384 (1822).
7995   1821, Observations on the serpentine rocks of Hoboken in New Jersey and on the minerals which they contain: Am. J. Sci., v. 4, no. 1, 16-23.
7996   1822, Observation and remarks on the minerals of Paterson and the valley of Sparta, in New Jersey: Am. J. Sci., v. 5, no. 2, 239-248. Also in NY Med. Phys. J., v. 1, 194-204.
    See also "Protest" of H. Seybert (9446).
7997   1823, Letter of Mr. Nuttall: Am. J. Sci., v. 6, no. 1, 171-173.
    In reply to H. Seybert (see 9441).

The Oasis, a Monthly Magazine, Devoted to Literature, Science and the Arts (Oswego, New York, 1837-1838)
7998   1837, Geology - Dr. Buckland´s new work [review of 2160]: v. 1, 6-7.

The Observer and Record of Agriculture, Science, and Art (Philadelphia, 1838-1839)
7999   1838/1839, Definition of terms: v. 1, no. 3, 40-42; no. 4, 58-61; no. 5, 70-74; no. 6, 89-91; no. 7, 102-104; no. 8, 121-122; no. 9, 139-141; no. 10, 155-158; no. 11, 164-175; no. 12, 182-186.
    A glossary of scientific terms.

Ohio. State of
8000   1832, [Importance of a geological survey]: in Governor´s Message, December 22, 1832.
8001   1835, [General scientific survey of the state urged]: in Governor´s Message, December, 1835.

Ohio (cont.)
8002 1835, Report [made by Mr. Creed to the House of Representatives] of the select committee on a mineralogical and geological survey ... : Ohio Gen. Assembly, Executive Documents, 18 p.
8003 1836, Report of the special committee, appointed by the legislature to report ... on the best method of obtaining a complete geological survey of the state of Ohio: Columbus, Ohio, James B. Gardiner (printer), 18 p.
  Also in Ohio State Legislature Rept. 1-D, p. 65-80, 1837. Contains note by J. Locke (6420).
8004 1836, [Resolution] appointing a committee to make certain geological observations and estimates of this state: Ohio Gen. Assembly, March 14, 1836.
8005 1837, An act providing for a geological survey of the state of Ohio, and other purposes: Ohio Gen. Assembly, March 27, 1837.
8006 1838, [On the progress of the geological survey] : in Governor's Message, December, 1838.
8007 1838, Report from the geological board in reply to a resolution .. : Ohio House J., 1837-1838, 814-816.
8008 1838, [Resolution on the disposition of geological survey property] : Ohio Gen. Assembly, January 10, 1838.
8009 1839, [Order for publication of the geological report] : Ohio Gen. Assembly, March 18, 1839.
8010 1839, [Resolution on the disposition of geological survey property] : Ohio Gen. Assembly, March 18, 1839.
8011 1841, [Appropriation for organizing the geological collections] : Ohio Gen. Assembly, March, 1841.
8012 1841, [On the progress of the geological survey] : in Governor's Message, December, 1841.
8013 1841, Specimens procured during the geological survey ... : in Governor's Message, December, 1841, 11.

The Ohio Cultivator (Columbus, Ohio, 1845-1850)
8014 1845, Analysis of soils: v. 1, 166 (7 lines).
  An advertisement by Dr. C. A. Raymond who offers to analyze soils, and advise farmers.
8015 1846, Study the soil: v. 2, 142-143.
8016 1847, Analysis of soils: v. 3, 157.

The Ohio Medical Repository (Cincinnati, 1826-1827)
8017 1826, Fossil bones: v. 1, no. 9, 36.
  On an exhibit of fossil bones (the Koch collection?).

Ohio Miscellaneous Museum (Lebanon, Ohio, 1822)
8018 1822, Stupendous cavern [in Jefferson County, New York] : v. 1, no. 5, 222-223.

Olbers, Heinrich Wilhelm Matthias (1758-1840)
8019 1822, Supposed volcanoes in the moon: Mus. For. Liter. Sci., v. 1, 94 (12 lines).
  Abstracted from Edinburgh Philos. J.

The Olio, a Literary and Miscellaneous Paper (New York, 1813-1814)
8020 1813, Upper Canada: v. 1, 140-141.
  Notes on minerals and fossils.
8021 1813, Antimony: v. 1, 253 (11 lines).
  On its origins and uses.

Oliver
8022 1808, [On the paleogeography of Turkey, Greece and Persia] : Am. Reg. or Gen. Repos., v. 2, 398 [i.e. 389]-390.

Oliver (Dr.)
8023 1830, Rock crystals substituted for crown glass in making telescopes: Am. J. Sci., v. 18, no. 1, 191 (8 lines).

Olmsted, Denison (1791-1858)
8024 1819, Outlines of the Lectures on chemistry, mineralogy, and geology, delivered at the University of North-Carolina: Raleigh, J. Gales, 44 p.
8025 1820, Red sandstone formation of North Carolina: Am. J. Sci., v. 2, no. 1, 175-176.
8026 1821, Letter to the Board of Internal Improvements on the geology of North Carolina: in Report to the General Assembly by the Board of Internal Improvement, Raleigh, NC, J. Gales (printer), 65-68 (of xxviii, 68 p.)
8027 1822, Descriptive catalogue of rocks and minerals collected in North Carolina, and forwarded to the American Geological Society: Am. J. Sci., v. 5, no. 2, 257-264.
  ?Includes notice of 87 mineral localties in North Carolina.
8028 1822, Useful minerals in North Carolina: Am. J. Sci., v. 5, no. 2, 407. Also in West. Q. Reporter, v. 1, no. 4, 400-401.
8029 1824, Report on the geology of North Carolina, conducted under the direction of the Board of Agriculture. Part I.: Raleigh, NC, J. Gales and Son, 44 p.
8030 1825, Report on the geology of North Carolina, conducted under the direction of the Board of Agriculture. Part II: Raleigh, NC, J. Gales and Son, 58 p.
  For review with excerpts see 470 and 9968. For notice see 385. Also as NC Board Ag. Rept., no. 2, 85-142. See also part III by E. Mitchell (7247). For extract see Del. Reg., vol. 1, no. 52, 407 (1829).
8031 1825, On the gold mines of North Carolina: Am. J. Sci., v. 9, no. 1, 5-15.
  Abstracted in Bos. J. Phil. Arts, v. 2, no. 4, 388-389.
8032 1836, Observations on the use of anthracite coal: Am. Almanac and Repos., for 1837, 61-69.
  Also reprint edition: Cambridge, Folson, Wells and Thurston, 11 p.
8033 1846, Report of Mr. Denison Olmsted, jr.: in Second annual report on the geology of the state of Vermont by C. B. Adams (see 26), 253-259, also chemical analyses of minerals, 215-243.

Omnium Gatherum (Boston, 1810)
8034 1810, A remarkable subterranean wall [at Salisbury, Rowen County, North Carolina]: v. 1, 23-24.
8035 1810, [On coal from Richmond, Virginia]: v. 1, 29-31.

Onderdonk, Henry M.
8036 1850, To the president and directors of the Winifrede Mining and Manufacturing Company of Virginia: n.p., 4 p.
    On the value of holdings in Kanawha, Virginia.

Ontonagon Copper Company
8037 1846, By-laws of the Ontonagon Copper Company, ... with Forrest Shepherd's report: Bos., Dutton and Wentworth (printers), 12 p., plate.
    See also report by F. Shepherd (see 9574).

Ontonagon Mining Company
8038 1845, Articles of agreement of the Ontonagon Mining Company: n.p., 15 p.

Orange Grove Mining Company (see Vaucluse Mine)

Orbigny, Alcide d' (see D'Orbigny, Alcide)

Origin of the material universe; with a description of the manner of formation of the Earth, and events connected therewith, from its existence in a fluid state to the time of the Mosaical narrative.
8039 1850, Bos., n.p., 83 p.

Orr, Isaac (1793-1844)
8040 1823, An essay on the formation of the universe: Am. J. Sci., v. 6, no. 1, 128-149.

Orton, J. W.
8041 1849, The miner's guide and metallurgist's directory: NY, A. S. Barnes and Co.; Cin., H. W. Derby and Co., 86 p., plates.
    For review see 4092 and 6362.

Osage Mining and Smelting Company
8042 1837, Osage Mining and Smelting Company, of Missouri: Balt., 20 p.
    Contains testimonial letters on the value of the mine's lead ore, including J. T. Ducatel (see 3243).

Osborn, A.
8043 1846-1847, Progressive changes of matter: Am. Q. J. Ag. Sci., v. 4, 212-220; v. 5, 89-90; 127-136.
    On lithification, erosion, and the formation of continents.

Oswald, F.
8044 1850, Analysis of California gold: Am. J. Sci., v. 16 (ns), 178-179.

Overman, Frederick (1803?-1852)
8045 1850, The manufacture of iron in all its various branches ... : Phila., Henry C. Baird, 492 p., 150 woodcuts.
    For review see 5479.

Owen, David Dale (1807-1860)
    See also 7980 to 7983 with J. G. Norwood.
8046 1838, Report of a geological reconnaissance of the state of Indiana made in the year 1837, in conformity to an order of the legislature: Indianapolis, Indiana, J. W. Osborn and J. S. Willets (printers), 34 p.
    Another printing by Bolton and Livingston, Indianapolis, 34 p. Also another edition in Indianapolis, 38 p., not seen. Also in 8048.
    For review with excerpts see 3654.
8047 1839, Geological survey of the state of Indiana made in the year 1838 ... : Indianapolis, Indiana, 46 p.
8048 1839, Geological survey of the state of Indiana made in the year 1838 ... : Indianapolis, Indiana, 54 p.
    Second edition. Published with first report (see 8046). For review see 5346.
8049 1840, Catalogue of mineralogical and geological specimens at New Harmony, Indiana, collected in various parts of Europe and America, by William Maclure, late president of the Academy of Natural Sciences of Philadelphia; arranged for distribution at the request of Miss Maclure and Alexander Maclure, his executors: New Harmony, Indiana, 16 p.
    List of 290 species.
8050 1840, Report of a geological exploration of part of Iowa, Wisconsin, and Illinois: US HED 239, 26-1, v. 6 (368), 9-161, 25 plates.
    Contains report by J. Locke (see 6430). See also 8065 and 8070. There were two different editions of this work, the first with spelling "Wiskonsin." For extract see US Mag. and Dem. Rev., v. 8, 30-42.
8051 1842, Regarding human foot-prints in solid limestone: Am. J. Sci., v. 43, no. 1, 14-32, plate with 11 figs.
    Contains opinions by P. Anderson (see 1245), G. A. Mantell (see 6738), H. R. Schoolcraft (see 9278), K. C. Leonhard (see 6193), and S. Bolton (see 1746).
8052 1843, Fossil palm trees found in Posey County, Indiana: Am. J. Sci., v. 45, no. 2, 336-337.
    In Assoc. Am. Geol's Nat's Proc.
8053 1843, On a universal system of geological coloring and symbols [for maps]: Am. J. Sci., v. 45, no. 2, 351-353.
    In Assoc. Am. Geol's Nat's Proc.
8054 1843, On geological paintings and illustrations: Am. J. Sci., v. 45, no. 1, 136-137.
    In Assoc. Am. Geol's Nat's Proc.
8055 1843, Verbal communication [on some fossil trees from New Harmony, Indiana]: Phila. Acad. Nat. Sci's Proc., v. 1, no. 26, 270-271.

Owen, D. D. (cont.)

8056 1843, On the geology of the western states: Am. J. Sci., v. 45, no. 1, 151-152; 163-165.
  In Assoc. Am. Geol's Nat's Proc.

8057 1843, On the western states of North America: Am. J. Sci., v. 44, no. 2, 365-368.
  In British Assoc. Proc.

8058 1843, Verbal communication [on geology of the western country]: Phila. Acad. Nat. Sci's Proc., v. 1, no. 26, 272 (8 lines).

8059 1844, Chart of the great Illinois coal field: in Report ... , (see 8065), colored geological map, 16 x 11 cm., plate 4.

8060 1843 [i.e. 1844], Chart of the great Illinois coal field: in letter on Indiana coal (see 8063), folded map, 25 x 21 cm.

8061 1844, Geological chart of part of Iowa, Wisconsin, and Illinois: Balt., E. Weber and Co. (lithographer), colored geological map, 40 x 55 cm.
  In Report ... (see 8065), plate 2.

8062 1844, Geological section from the mouth of Rock River through the blue mounds to the Wisconsin River in connection with a geological chart of part of Iowa, Wisconsin and Illinois N. W. of the section line: in Report ... (see 8065), colored geological map, 27 x 40 cm., plate 3.

8063 1844, [Letter on the coal formations of Indiana and adjoining states]: US SED 78, 28-1, v. 2 (432), 8-9, map.
  Contains map 8060.

8064 1844, [Nature of country rock in Missouri lead district]: Am. J. Sci., v. 47, no. 1, 106 (3 lines).
  Also in Assoc. Am. Geol's Nat's Proc., v. 5, 13 (3 lines).

8065 1844, Report of a geological exploration of part of Iowa, Wisconsin, and Illinois, made under instructions from the Secretary of the Treasury of the United States, in the autumn of the year 1839: US SED 407, 28-1, v. 7 (437), 191 p., 25 maps and plates.
  Contains "Report," by J. Locke (see 6451 and 6454); and maps (see 8059, 8061, and 8062).
  For review see 3005. For notice see 3004.

8066 1844-1847, Review of the New York geological reports [reviews with excerpts of reports by E. Emmons (see 3510), J. Hall (see 4683), W. W. Mather (see 6907), and L. Vanuxem (see 10589)]: Am. J. Sci., v. 46, no. 1, 143-157; v. 47, no. 2, 354-380, 76 figs. (with "Note on graptolites" by L. C. Beck, see 1585); v. 48, no. 2, 296-316, 76 figs.; v. 1 (ns), no. 1, 43-70, 61 figs.; v. 3 (ns), no. 1, 57-74, 61 figs.; v. 3 (ns), no. 2, 164-171, 17 figs.

8067 1845, Chart of the great Illinois coal field: in Report ... (see 8070), colored geological map, 19 x 16.5 cm.

8068 1845, Geological chart of part of Iowa, Wisconsin and Illinois: in Report ... (see 8070), colored geological map, 55 x 70 cm.
  C. B. Graham, lithographer.

8069 1845, Geological section ... in connection with a geological chart of part of Iowa, Wiconsin and Illinois: in Report ... (see 8070), colored geological map, 26 x 32 cm.
  Smaller version of 8068 with cross-section.

8070 1845, Report of a geological exploration of part of Iowa, Wisconsin, and Illinois: US HED 168, 28-2, v. 4, pt. 2 (467), 161 p., 27 plates and maps.
  Contains report by J. Locke (see 6430) and maps (see 8067 to 8069).
  For review with excerpts see 8613.

8071 1846, Geology: Q. J. and Rev., v. 1, 46-70, 2 figs.; 132-148, 33 figs., plate; 262-270; 345-362.
  These are "Introductory lectures on geology."

8072 1846, Scientific pursuits: Q. J. and Rev., v. 1, 40-46.
  This is in "Introductory lectures on geology."

8073 1846, Dr. Owen's report on the mineral lands of the United States: Am. J. Sci., v. 2 (ns), no. 2, 294-296.

8074 1847, Preliminary report, containing outlines of the progress of the geological survey of Wisconsin and Iowa up to October 11, 1847: US SED 2, 30-1, v. 2, (504 p.), 160-173.

8075 1847, and J. G. Norwood, Researches among the Protozoic and Carboniferous rocks of central Kentucky made during the summer of 1846: St. Louis, Keemle and Field (printers), 12 p., plates (1 folded).

8076 1847, and J. G. Norwood, Abstract of Researches among the Protozoic and Carboniferous rocks of central Kentucky, made during the summer of 1846: West. J. Medicine Surg., v. 7 (ns), 149-152.

8077 1847, Termination of the Palaeozoic period and commencement of the Mesozoic: Am. J. Sci., v. 3 (ns), no. 3, 365-368.

8078 1848, Provisional geological map of part of the Chippeway land district of Wisconsin, and incidently of a portion of the Kickapoo country, and of a part of Iowa & of Minnesota Territory: in A report ... (see 8079), colored geological map, 62 x 85 cm.

8079 1848, A report of a geological reconnaisance of the Chippewa land district of Wisconsin, and incidently of a portion of the Kickapoo country, and a part of Iowa and of Minnesota Territory: US SED 57, 30-1, v. 7 (509), 5-134, 37 plates, 1 map.
  Contains report by J. G. Norwood (see 7984) and map (see 8078). For review see 1000 and 1010.

8080 1848, and J. G. Norwood, Researches among the Protozoic and Carboniferous rocks of central Kentucky, made during the summer of 1846; by D. D. Owen, M. D., and J. G. Norwood, M. D.: Am. J. Sci., v. 5 (ns), no. 2, 268-269.
  Abstract of 8075 contained in review.

8081 1849, [Letter on the progress of the geological survey of the Chippewa Land District]: US SED 1, 31-1, v. 3, pt. 2 (551), 241; also in US HED 5, 31-1, v. 5, pt. 2 (571), 241.

8082 1850, and J. G. Norwood and J. Evans, Notice of fossil remains brought by Mr. J. Evans from the "Mauvais Terres," or bad lands of White River, 150 miles west of the Missouri: Phila. Acad. Nat. Sci's Proc., v. 5, no. 4, 66-67.

8083 1850, and B. F. Shumard, Descriptions of fifteen new species of Crinoidea from the Subcarboniferous limestone of Iowa, collected during the U. S. geological survey of Iowa, Wisconsin, and Minnesota, in the years 1848-9: Phila. Acad. Nat. Sci's J., v. 2 (ns), pt. 1, 57-70, plate with 16 figs.

Owen, J. S.
8084   1831, Fossil remains, found in Anne Arundel County, Maryland: Mon. Am. J. Geology, v. 1, no. 3, 114-118.

Owen, John (British)
8085   1797, Visit to Mount Vesuvius and the ruins of Pompeia: SC Wk. Mus., v. 1, 240-241.
        Extracted from Owen's Travels.

Owen, Richard (1804-1892)
8086   1837, On the cranium of the Toxodon, a new extinct gigantic animal, referable by its dentition to the Rodentia, but with affinities to Pachydermata and herbivorous Cetacea: Am. J. Sci., v. 33, no. 1, 208-210.
        Abstracted from Geol. Soc. London Proc.
8087   1838, Prof. Owen on the fossil animals collected by Mr. Charles Darwin: Am. J. Sci., v. 35, no. 1, 196-197.
        Abstracted from Edinburgh New Philos. J.
8088   1839, Megatherium: Am. J. Sci., v. 37, no. 2, 371-372 (7 lines).
        Abstracted from London and Edinburgh Philos. Mag.
8089   1839, On the structure of fossil teeth: Am. J. Sci., v. 35, no. 2, 307.
        In British Assoc. Proc.
8090   1842, Report on British fossil reptiles: Am. J. Sci., v. 42, no. 2, 328-333.
        In British Assoc. Proc. Abstract in Am. Eclec., v. 2, no. 3, 587-588 (1841).
8091   1843, Letter of Prof. Owen to Prof. Silliman on the Ornithichnites and Dinorthis: Am. J. Sci., v. 45, no. 1, 185-187.
8092   1843, On Dr. Harlan's notice of new fossil mammalia: Am. J. Sci., v. 44, no. 2, 341-345.
        See also 4814 and 4815. on Mylodon, Megalonyx and Orycterotherium.
8093   1844, Report on the fossil mammalia of Great Britain: Am. J. Sci., v. 47, no. 1, 186-187.
        In British Assoc. Proc.
8094   1845, A letter from Professor Richard Owen, on the great birds' nests of New Holland: Am. J. Sci., v. 48, no. 1, 61-62.
        See also article by E. Hitchcock (5169).
8095   1845, On Dinorthis, an extinct genus of tridactyle struthious birds, with descriptions of portions of the skeletons of six species, which formerly existed in New Zealand: Am. J. Sci., v. 48, no. 1, 194-201.
        Abstracted by "H." (i.e., E. Hitchcock?)
8096   1846, Mastodon giganteus: Am. J. Sci., v. 2 (ns), no. 1, 131-133, fig.
        Abstracted from British fossil mammalia.
8097   1846, Observations on certain fossils from the collection of the Academy of Natural Sciences of Philadelphia: Phila. Acad. Nat. Sci's Proc., v. 3, no. 4, 93-96.
8098   1846, On Belemnites: Am. J. Sci., v. 1 (ns), no. 2, 285.
        Abstracted from Q. J. Geology.
8099   1846, On the extinct mammals of Australia, with additional observations on the genus Dinorthis of New Zealand: Am. J. Sci., v. 1 (ns), no. 1, 129-130.
        Abstracted from Mag. Nat. History.
8100   1846, [On the equivalence of Zeuglodon and Dorudon]: Phila. Acad. Nat. Sci's Proc., v. 3, no. 1, 15 (7 lines).
8101   1847, Food and climate of the mammoth: Hort. and J. Rural Arts, v. 2, no. 4, 192.
        Abstracted in part from British fossil mammals.
8102   1847, General geological distribution and probable food and climate of the mammoth: Am. J. Sci., v. 4 (ns), no. 1, 13-19.
        Extracted from British fossil mammalia.
8103   1847, Harlanus, a new genus of fossil Pachyderms: Am. J. Sci., v. 3 (ns), no. 1, 125 (5 lines).
        Abstracted from Phila. Acad. Nat. Sci's Proc.
8104   1847, Observations on certain fossil bones from the collection of the Academy of Natural Sciences of Philadelphia: Phila. Acad. Nat. Sci's J., v. 1 (ns), no. 1, 18-20, plate.

Owen, Robert Dale (1801-1877)
8105   1849, Hints on public architecture, containing among other illustrations, views and plans of the Smithsonian Institution; together with an appendix relative to building materials: NY, G. P. Putnam, 119 p., illus.

P., J. G. (see Percival, James Gates)

Paddock, James A.
8106   1827, Kellyvale serpentine [from Vermont]: Am. J. Sci., v. 13, no. 1, 200 (9 lines).

Page, David (1814-1879)
8107   1846, Rudiments of geology: Phila., Sorin and Ball, 288 p., illus.
        For review see 5476.
8108   1849, Elements of geology: Phila., J. Ball; NY, A. S. Barnes, 332 p.
        With "Introductory observations" by D. M. Reese (see 8728). For review see 9181. For notice see 6359.
        Also "third edition," NY, 288 p., 1849. Second edition not found.

Paideaux, John
8109   1837, Suggestions for the use of the blowpipe by working miners: Franklin Inst. J., v. 20 (ns), 161-166, 3 figs; 266-270.
        Abstracted from Mining Rev.

Paine, Thomas (1697-1757)
8110   1728, The doctrine of earthquakes. Two sermons preached at a particular fast in Weymouth, November 3, 1727. The Friday after the earthquake. Wherein this terrible work appears not from natural second causes, in any orderly way of their producing: but from the mighty power of God immediately interposing; and is to the world, a token of God's anger, &c. and presage of terrible changes. With examples of many earthquakes in history, - illustrating this doctrine: Bos., printed for D. Henchman, 87 p.
Evans number 3079.

Paine, Timothy
8111   1792, and W. Young and S. Stearns, Particulars relating to Worcester, in the state of Massachusetts: Mass. Hist. Soc. Coll., v. 1, 112-116.

Paleontological Society of London
8112   1847, Paleontological Society of London: Am. J. Sci., v. 4 (ns), no. 2, 299.
Abstracted from their Prospectus.

Palisot de Beauvois, A. M. F. J. (see 8178 with C. W. Peale)

Palmer, Aaron Haight
8113   1848, Memoir, geographical, political, and commercial, on ... Siberia, Manchuria, and Asiatic Islands of the Northern Pacific Ocean: US SMD 80, 30-1, v. 1 (511), 77 p., map.

Palmer, Green B.
8114   1831, Machine for separating gold from earth: Franklin Inst. J., v. 8 (ns), 169.

Palsey, C. W. (Major General)
8115   1843, On the recent great mining operations near Dover: Franklin Inst. J., v. 6 (3s), 28-40.
From "Civ. Eng. & Arch. J."

Paris, John Ayrton (1785-1856)
8116   1819, Extract from a paper on a recent formation of sandstone, occurring in various parts of the northern coast of Cornwall: Am. J. Sci., v. 1, no. 3, 234-235.

Park, Mungo (1771-1806)
8117   1800, Travels in the interior districts of Africa in the years 1795, 1796, and 1797 ... with an appendix, containing geographical illustrations of Africa: by Major [James] Russell ... : Phila., James Humphreys, 484 p., map. Other editions: Phila., J. Tiebout, 354, [2], 86 p., map (1800) and NY, Evert Duychinck, 261 p., map (1813).
8118   1815, The journal of a mission to the interior of Africa, in the year 1805: Phila., Edward Earle, xlii, 302 p., illus.
8119   1842, The life and travels of Mungo Park; with the account of his death from the journal of Isaaco, the substance of later discoveries relative to his lamented fate, and the termination of the Niger: NY, Harper, 248 p., map.

Parker, Peter (1804-1888)
8120   1841, Volcanic ashes [in the South Pacific]: Am. J. Sci., v. 40, no. 1, 198.

Parker, Samuel (1779-1866)
8121   1838, Journal of an exploring tour beyond the Rocky Mountains under the direction of the American board of commissioners for foreign missions, performed in the years 1835, '36 and '37; containing a description of the geography, geology, climate, and productions; and the number, manners, and customs of the natives ... : Ithaca, NY, The author, 371 p., tables, map.
For review with excerpts see 1309 and 2737. Other editions: Ithaca, NY, The author, 400 p., map (1840); Ithaca, NY; Mack, Andrews, and Woodruff; Bos., Crocker and Brewster, 408 p., map (1842); Ithaca, NY; Andrews, Woodruff, and Gauntlett; Bos., Crocker and Brewster, 416 p., map (1844); Auburn, NY, J. C. Derby; NY, M. H. Newman, 422 p., map, front. (1846).

Parkes, Alexander
8122   1850, Improvement in the reduction of ores: Franklin Inst. J., v. 19 (3s), 288.

Parkes, Samuel (1779-1866)
8123   1807, A chymical catechism ...: Phila., J. Humphreys, 334 p., plate.
"With a front is piece of the economical laboratory of James Woodhouse, M. D."
Also later editions.
8124   1825, On plumbago, or carburet of iron, with an account of the mine at Borrowdale: Bos. J. Phil. Arts, v. 2, no. 4, 332-340.
Abstracted from Chemical essays.
8125   1826, Essay on earthenware and porcelain: Bos. J. Phil. Arts, v. 3, no. 5, 439-461; no. 6, 525-548.
Abstracted from Chemical essays.

Parkinson, James (1755-1824)
8126   1817, The mammoth, elephant, and hippopotamus, formerly natives of England: Am. Mon. Mag. and Crit. Rev., v. 1, no. 2, 131.
8127   1824, [Excerpts from Outlines of Oryctology]: Bos. J. Phil. Arts, v. 1, no. 6, 559-570.
In review, excerpted from Eclec. Rev.

Parley's Magazine (New York and Boston, 1833-1844)
8128   1833, Volcanoes: v. 1, 17-19, woodcut.
8129   1833, Eruption of Mount Vesuvius: v. 1, 161-162, woodcut.
8130   1833, The asbestos stone: v. 1, 206.
8131   1834, Volcano in the ocean [in the Azores]: v. 2, 208 (7 lines).
8132   1834, Mount Vesuvius: v. 2, 343.
8133   1835, The Natural Bridge [of Virginia]: v. 3, 14, 2oodcut.

Parley's Magazine (cont.)
8134   1835, Great earthquake in Chile: v. 3, 183.
8135   1835, The petrified buffalo: v. 3, 310.
8136   1835, The great salt mines of Wielicska [in Poland]: v. 3, 314-317.
8137   1836, Singular cavern, in Illinois [at Quincy]: v. 4, 236-237.
8138   1836, A coal fish [fossil from Mansfield, Massachusetts]: v. 4, 248 (6 lines).
8139   1836, Visit to Wier's Cave [in Rockingham County, Virginia]: v. 4, 259-262.
8140   1836, Visit to the great geyser in Iceland: v. 4, 299-303.
8141   1836, Earthquake at Lisbon [in 1755]: v. 4, 332.
8142   1837, Cities buried by volcanoes: v. 5, 43-46, woodcut.
8143   1839, Martha's Vineyard [fossils from Gay Head, Massachusetts]: v. 7, 36-39.
8144   1839, The fireside: v. 7, 95-99.
        On the origin, sources, and fossils of coal.
8145   1839, Basaltic rocks: v. 7, 256-259.
8146   1840, [On garnet and graphite from Mount Monadnock]: v. 8, 140.
        Footnote to "Wolves in the western states."
8147   1843, Remarkble spring at Niagara: v. 11, 341-342.
        On a burning spring.
8148   1844, A town in the heart of the earth. Austrian silver mine: v. 12, 100-101.
8149   1844, About cloth that will not burn [asbestos]: v. 12, 102-103.
8150   1844, Descent into a Swedish copper mine [at Coperbright]: v. 12, 227-228.
8151   1844, A beautiful crystallized cavern [at Bradford, Derbyshire, England]: v. 12, 371-373.

The Parlour Companion (Philadelphia, 1817-1819)
8152   1817, Description of the tin mines of Cornwall: v. 1, 18-19; 22-23; 26.

Parrot, Georg Friedrich (1767-1852)
8153   1834, Considerations upon the temperatures of the terrestrial globe: Am. J. Sci., v. 26, no. 1, 10-23.
        Abstracted by J. Griscom.

Parry, Harrison
8154   1850, Improved rotary gold washer: Franklin Inst. J., v. 18 (3s), 412 (4 lines).

Parry, William Edward (1790-1855; see 5709)

Parsons, General Samuel Holden (1737-1789)
8155   1793, Discoveries in the western country: Am. Acad. Arts Sci's Mem., v. 2, pt. 1, 119-127.
        Includes notice of Big Bone Lick and soils in Kentucky.

Partsch, Paul Maria (1791-1856)
8156   1846, Die meteoriten, or meteorites in the Imperial Cabinet at Vienna: Am. J. Sci., v. 1 (ns), no. 1, 148-149.

Pasch, M.
8157   1829, On preparing hydraulic cement: Franklin Inst. J., v. 4 (ns), 16-18.
        Abstracted from Technical Repos.

Pasteur, M. S.
8158   1849, and M. Ch. Deville, On the crystallization of sulphur: Am. J. Sci., v. 7 (ns), no. 2, 282.
        Abstracted from Comptes Rendus.

Patera, Ad.
8159   1850, A new method of determining uranium in its ores: Franklin Inst. J., v. 19 (3s), 415 (9 lines).
        Communicated by Theo. F. Moss.

Patrick, William
8160   1840, The age of the earth: The Friend, v. 13, 185-187.

Patterson (Dr.)
8161   1838, [Discussion of H. H. Sherwood's study of a dipping magnetic needle (see 9585)]: Franklin Inst. J., v. 22 (ns), 270-273.
8162   1848, [Minerals from the diamond mines of Brazil]: Am. Philos. Soc. Proc., v. 5, no. 41, 33.

Patterson, John
8163   1818, On gypsum: Phila. Soc. Promoting Ag. Mem., v. 4, 115-117.

Patterson, Robert
8164   1837, On the organic remains in the coal formation at Wardie, Scotland: Franklin Inst. J., v. 20 (ns), 297.
        Abstracted from Edinburgh New Philos. J.
8165   1839, Heat of the interior of the earth: Am. J. Sci., v. 37, no. 2, 357-358.
        Abstracted from Jameson's J.
8166   1840, Experiments and observations on the temperature of artesian springs or wells, in Mid-Lothian, Stirlingshire, and Clackmanshire: Franklin Inst. J., v. 25 (ns), 274-278.
        Abstracted from Edinburgh New Philos. J.
8167   1841, Temperature of artesian wells: Am. Eclec., v. 1, no. 1, 179 (8 lines).
        Abstracted from Year-book of Facts.

Patterson, Robert Maskell (1787-1854; see 6621 with W. Maclure and I. Lea)

Paulding, N.
8168   1820, [Red tourmaline]: Am. J. Sci., v. 2, no. 2, 366-367.
        In NY Lyc. Nat. History Proc.
8169   1820, [Marine fossils in Virginia]: Am. J. Sci., v. 2, no. 2, 371 (5 lines).
        In NY Lyc. Nat. History Proc.

Payne, John (fl. 1800)
8170 1798-1800, A new and complete system of universal geography; describing ... curiosities, of nature and art, also giving a general account of fossil and vegetable productions of the earth: NY, John Law, 4 v. (xlviii, 518, ii p., 1798; 578, [12] p., 1799; 710, [8] p., plates, maps, 1800; 525, [1], 36, [4], 17, [11] p., 1800.
   Evans numbers 34316, 36047, 38199.

Payson, Edward (1657-1732)
8171 1728, Pious heart-elations: being the substance of a sermon in publick on November 29th. In consideration of present awful providences amongst us; and on the Sabbath following in the forenoon, December 3d. 1727. From those words of Jeremiah, in Lamentations III. 41. Let us lift up our heart with our hands unto God in the Heavens: Bos., printed by B. Green for J. Phillips, [6], 23 p.
   Evans number 3080.

Peabody, Andrew Preston (1811-1893)
8172 1845, The connection between science and religion: Bos., C. C. Little & J. Brown, 29 p.
8173 1850, [Reviews of The pre-Adamite earth and Man primeval by J. Harris (see 4830 and 4832)]: N. Am. Rev., v. 70, 391-405.
   Unsigned.

Peabody, J.
8174 1846, Analysis of the glassy scoria of Kilauea, Hawaii: Am. J. Sci., v. 2 (ns), no. 2, 273 (6 lines).
   Abstracted from Bos. Soc. Nat. History Proc.

Peabody, Oliver William Bourn (1799-1848)
8175 1845, [Review of Journals of researches ..., by C. Darwin (see 2916)]: N. Am. Rev., v. 61, 181-199.

Peach, C. W.
8176 1842, An account of the fossil organic remains of the southeast coast of Cornwall and of Bodmin and Menheniott: Am. J. Sci., v. 42, no. 2, 327.
   Abstracted in British Assoc. Proc.

Peale (see 5790 with W. R. Johnson and S. S. Haldeman)

Peale, Charles Willson (1741-1827)
8177 1792, An address ... to the visitors and directors of Peale's Museum: Phila.?, n.p., broadside.
8178 1796, and A. M. F. J. Palisot de Beauvois, A scientific and descriptive catalogue of Peale's Museum ... : Phila., S. H. Smith (printer), 12, 44 p.
   Evans number 30967.
8179 1800, Discourse introductory to a course of lectures on the science of nature ... delivered in the Hall of the University of Pennsylvania, Nov. 8, 1800: Phila., Z. Poulson, jr., 50 p., 5 folded plates.
   Evans number 38203.
   For review see 7029.
8180 1800, Introduction to a course of lectures on natural history, delivered in the University of Pennsylvania, Nov. 16, 1799: Phila., Francis and Robert Bailey (printers), 28 p.
   Evans number 38204. For review see 7029.
8181 1805, Guide to the Philadelphia Museum: Phila., Museum Press, 8 p.
8182 1816, Address delivered ... to the corporation and citizens of Philadelphia on the 18th day of July, in 1816, in Academy Hall on Fourth Street: Phila., C. W. Peale, 23 p.
8183 1824, A sketch of the history of the museum: Phila. Mus., v. 1, 1-3.
8184 1824, On the polity of nature: Phila. Mus., v. 1, 3-6.
   On the dependence of organic life to inorganic nature.

Peale, Rembrandt (1778-1860)
8185 1802, A short account of the behemoth or mammoth: NY, Broadside, dated April, 1802.
8186 1804, A short account of the mammoth: Lit. Mag. and Am. Reg., v. 1, 292-297.
8187 1806, Dimensions of the skeleton of the extinct species of elephant dug up in about seventy miles north of the city of New-York, in 1801: Med. Repos., v. 9, 322-323.
8188 1823, [Extracts from "Historical disquisition on the mammoth"]: Bos. J. Phil. Arts, v. 1, no. 4, 400-402.
   In article on the American mammoth by J. Ware (see 10694).
8189 1824?, Prospectus of a museum of arts and sciences to be established in Baltimore ... : n.p., 2 p.
8190 1830, [Discovery and excavation of the New Jersey mastodon]: Cab. Nat. History, v. 1, iii-v.

Peale, Titian Ramsay (1799-1885)
8191 1831, Circular of the Philadelphia Museum: Containing directions for the preparation and preservation of objects of natural history: Phila., J. Kay, Jun. & Co., 29, [1] p., 5 plates.

Pearce, J. Chaning
8192 1842, On the mouths of ammonites, and on other fossils, found in the Oxford clay, near Christian Malford, on the line of the Great Western Railway: Franklin Inst. J., v. 4 (3s), 177-178.
   Abstracted from London Ath.
8193 1846, Notice of what appears to be the embryo of an Ichthyosaurus in the pelvic cavity of Ichthyosaurus (communis?): Am. J. Sci., v. 1 (ns), no. 2, 276-277.
   Abstracted from Annals and Mag. Nat. History.

Pease, G.
8194 1839, Brief statements respecting mineral deposits in the states of Missouri and Illinois: Bos., 12 p.

Pease, William H.
8195 1848, Observations on the geology and natural history of Mexico: Phila. Acad. Nat. Sci's Proc., v. 4, no. 5, 91-94.

Pease, William S.
8196 1847, [Fossils from Baltimore, Maryland]: Phila. Acad. Nat. Sci's Proc., v. 3, no. 10, 242-243.

Peck, Jacob (Judge)
8197 1832, Geological and mineralogical account of the mining districts in the state of Georgia - western part of North Carolina and of east Tennessee, with a map: Am. J. Sci., v. 23, no. 1, 1-10, map.
       Contains map 8198.
8198 1832, Geological map of the mining districts in the state of Georgia, western parts of N. Carolina, and in east Tennessee by Jacob Peck: Am. J. Sci., v. 23, no. 1, geological map, 36.6 x 29.6 cm.
       In 8197.

Peligot, M.
8199 1846, Analyses of glass: Am. J. Sci., v. 2 (ns), no. 1, 114 (11 lines).
       Abstracted from l'Institut.

Pell, Robert Livingston (1811-1880)
8200 1846, On the use of various manures: Am. Agri. Assoc. Trans., v. 1, 39-60.

Pelletier, Bertrand (1761-1797)
8201 1797, [On strontian earth from Scotland]: Univ. Mag., v. 1, 328-329.
       In Nat'l Inst. Paris Proc.

Pelouze, Theophile-Jules (1807-1867)
8202 1846, A new method of estimating copper: Am. J. Sci., v. 2 (ns), no. 2, 259.
       Abstracted by J. L. Smith from Comptes Rendus.

Pencil, Mark (pseudonym)
8203 1839, White Sulphur papers; or life at the springs of western Virginia: NY, S. Colman, 166 p.

Pendergrast, G. E. (see 7293 with S. L. Mitchill and W. Ross)

Penington, John (1768-1793)
8204 1790, Chemical and economical essays, designed to illustrate the connection between the theory and practice of chemistry, and the application of that science to some of the arts and manufactures of the United States of America: Phila., Joseph James (printer), viii, 200 p.
       Evans number 22757.

Penman, John
8205 1838, [Reports on North American Coal Company mines]: in An act to incorporate ... (see 7924).

Pennsylvania. State of
8206 1832, [First] annual report made by the Board of Trade to the Coal Mining Association of Schuylkill County: Pottsville, Penn., 12 p.
       Annual reports were probably published in successive years through at least 1843. Third annual report (1834) is reviewed in 2736 and 5430. Eleventh annual report (1843) is abstracted in 8216. Other reports not seen.
8207 1833, Geological survey of the state: Hazard's Reg. Penn., v. 11, 225-228.
       Legislation by the Penn. House of Representatives, with letter by J. B. Gibson, R. Harlan, and H. S. Tanner (see 4387).
8208 1833, Report of a committee of the House of Representatives recommending an appropriation by the Legislature to make a geological survey of the state, under the direction of the Geological Society of Pennsylvania: Harrisburg, Penn., H. Welsh, 10 p.
       B. Say, chairman.
8209 1834, Report of the committee of the Senate of Pennsylvania, upon the subject of the coal trade. S. J. Parker, chairman. Read in the Senate, March 4, 1834 ... : Harrisburg, Penn., H. Welsh, 126 p., folded map.
       Also in Penn. Senate J., v. 2, 449-572.
8210 1834, Report on the coal trade [by the state senate]: Hazard's Reg. Penn., v. 13, 185-189; 193-197; 209-218; 228-234; 246-252; 257-280; 289-295.
8211 1836, An act to provide for a geological and mineralogical survey of the state: Penn. Gen. Assembly, March 29, 1836.
8212 1837, A supplement to the act entitled "An act to provide for a geological and mineralogical survey of the state": Penn. Gen. Assembly, March 21, 1837.
8213 1838, [Act to provide for a geological survey]: Penn. Gen. Assembly, April 13, 1838.
8214 1839, An act to encourage the manufacture of iron in the state of Pennsylvania: in Charter and by-laws of the Great Western Iron Company (see 4510).
8215 1840, [Act to continue the geological survey]: Penn. Gen. Assembly, 1840.
8216 1843, Eleventh annual report, made by the Board of Trade to the coal mining association of Schuylkill County: Niles' Wk. Reg., v. 64, 23-25.
       See also 8206.
8217 1844, [Act for completion of the geological survey]: Penn. Gen. Assembly, 1844.
8218 1846, Report [of the Committee to the Iron and Coal Association]: Phila., J. Harding (printer), 30 p.
8219 1848, [Act to deposit mineral and geological specimens in museums]: Penn. Gen. Assembly, 1848.

Pennsylvania Coal Company
8220 1848, An act to incorporate the Pennsylvania Coal Company, by an act to incorporate the Washington Coal Company: NY, 16 p.

Pennsylvania Geological Society (see Geological Society of Pennsylvania)

The Pennsylvania Magazine; or, American Monthly Museum (Philadelphia, 1775-1776)
8221 1776, Practical chemistry: v. 2, 67-64 [i.e. 68].

Penot (Dr.)
8222 1841, Observations on the manufacture of coal gas: Am. Rep. Arts Sci's Manu's, v. 4, 267-269.

Percival, James Gates (1795-1856)
8223 1822, Notice of the locality of sulphate of barytes, from which a specimen was analysed by Mr. G. T. Bowen; (see p. 325, vol. IV of this Journal) and of various other minerals in Berlin, Conn.: Am. J. Sci., v. 5, no. 1, 42-45.
8224 1823, Analysis of a treatise, "Sur la classification et la distribution des vegetaux fossiles, (On the classification and distribution of fossil vegetables,) par M. Adolphe Brogniart": Am. J. Sci., v. 7, no. 1, 178-185.
8225 1824, Analysis of a memoir, 'Sur les caracteres zoologiques des formations, avec l'application de ces caracteres à la determination, de quelques terrains de Craie; par Alexandre Brogniart' (Ann. des Mines, 1821): Am. J. Sci., v. 8, no. 2, 213-218.
8226 1824, Dr. Van Rensselaer on salt: Am. J. Sci., v. 7, no. 2, 360-362.
       Signed by "J. G. P."
8227 1824, Griscom's journal [notice of A year in Europe, by J. Griscom (see 4585)]: Am. J. Sci., v. 7, no. 2, 360.
8228 1824, Notices of the geology and mineralogy of Sicily, from a work entitled Storia naturale della Sicilia. Cat. 1813. del Ab. F. Ferrara. Translated and condensed by James G. Percival: Am. J. Sci., v. 8, no. 2, 201-213.
8229 1842, A geological map of Connecticut: in Report on the geology of Connecticut (see 8230), geological map, 38 x 48 cm.
8230 1842, Report on the geology of the state of Connecticut: NHav., Osborn and Baldwin (printers), 495 p., map.
       Contains map 8229. Over 1000 copies printed. For review see 778.
8231 1846, On the hematite in Connecticut: Am. J. Sci., v. 2 (ns), no. 2, 268-269.
       Abstracted from his Geology of Connecticut (see 8230).

Percy, John (1817-188?)
8232 1848, Report on crystalline slags produced in the smelting and manufacture of iron: Am. J. Sci., v. 5 (ns), no. 1, 127-132.
       Abstracted from British Assoc. Repts.

Perkins, Henry Colt (d. 1873)
8233 1842, Note to the editors respecting fossil bones from Oregon: Bos. J. Nat. History, v. 4, no. 1, 134-136.
8234 1842, Notice of fossil bones from Oregon Territory: Am. J. Sci., v. 42, no. 1, 136-140, 4 figs.

Perkins, James Handasyd (1810-1849)
8235 1846, Annals of the West: Cin., J. R. Albach, 592 p.
       Also 1847 printing. Second edition: St. Louis, J. R. Albach, 808 p.

Perley, Moses Henry (1804-1862)
8236 1850, [On subsidence of the shore of Newfoundland]: Bos. Soc. Nat. History Proc., v. 3, 374-376.
       With discussion by J. H. Abbot, J. E. Cabot, E. Desor, and J. Wyman (11096).

Perrine, Henry (1797-1840)
8237 1839, The peculiar calcareous soil of southern Florida, and its effects on health: Farm's Reg., v. 7, 683-684.

Perry, John
8238 1848, Perry's lead mines [in St. Francis County, Missouri]: West. J. Ag., v. 1, no. 11, 608-610.

Perry, Thomas Hobart
8239 1839, Method of adjusting the dipping needle: Am. J. Sci., v. 37, no. 2, 277-278.
8240 1844, Observations upon the dip of the magnetic needle: Am. J. Sci., v. 47, no. 1, 84-88.
       Also reprint edition.

Person, Charles Cléopas (1801-1884)
8241 1848, Researches on the latent and specific heat of bodies: Am. J. Sci., v. 6 (ns), no. 1, 111-112.
       Abstracted from Comptes Rendus.

Peschier
8242 1827, Composition of feldspar and serpentine: Am. J. Sci., v. 12, no. 1, 187-188.
       Abstracted from Annales de Chimie by J. Griscom.

Peter, Robert (1805-1894)
8243 1849, Professor Peter's essay and analysis of lime: Am. Farm., v. 5 (4s), 129-132.
8244 1849, Remarks on the agricultural value of the Blue Limestone of Kentucky; with its analysis: The Cultivator, v. 6 (ns), 105-107.
8245 1850, Lower Blue Lick Spring; the quantitative chemical analysis of the water of the Lower Blue Lick Spring, in Nicholas County, Ky., with remarks on some other salt springs of the blue limestone formation: n.p., 4 p.

Petermann, Augustus (i.e. August Heinrich Petermann, 1822-1878)
8246 1850, On the fall of rivers: Ann. Scien. Disc., v. 1, 246-247.
       Abstracted from Jameson's J.

Peters, Richard (1744-1828)
8247   1797, Agricultural enquiries on plaister of Paris. Also facts, observations and conjectures on that subttance [sic.], when applied as manure: Phila., Charles Cist (printer), [2], 111, [1], [2] p.
        Evans number 32673. For review see 7009.
8248   1808, On gypsum: Phila. Soc. Promoting Ag. Mem., v. 1, 156-175.
8249   1808, Gypsum; whether it is found in the United States: Phila. Soc. Promoting Ag. Mem., v. 1, 310-315.
8250   1811, Inquiries, facts, observations and conjectures on plaister of Paris: Phila., J. Aitkin, xvi, [18]-129 p.
        Also published with the second volume of Phila. Soc. Promoting Ag. Mem.
8251   1814, On American gypsum: Phila. Soc. Promoting Ag. Mem., v. 3, 266-268.
8252   1814, On a fire stone [from Belmont, Pennsylvania]: Phila. Soc. Promoting Ag. Mem., v. 3, 389-390.

Petersburg Academy of Science (see St. Petersburg Academy of Science)

Petrarch (pseudonym; see 10521)

Petros (pseudonym; see C. A. Lee 6118)

Pettit, William
8253   1847, Remarks respecting the copper district of Lake Superior: Franklin Inst. J., v. 13 (3s), 338-345, map.

Peyssonnel, Jean Andre (1694-1759)
8254   1789, Observations ... made on the Brimstone-Hill, in the island of Guadeloupe: Gentlemen and Ladies Town and Country Mag., v. 1, no. 3, 136-138; no. 4, 180.

Peyton, William H.
8255   1836, Remarks on the supposed gypsum discovered in New York, and the probable existence of a similar substance in Western Virginia: Farm's Reg., v. 4, 314-315.

Phelps, Mrs. Almira Hart Lincoln (1793-1884)
        See also 10602.
8256   1832, The Child's geology: Brattleboro, Vt., Geo. H. Peck and Co.; Bos., Carter Hendee and Co.; Hart., F. J. Huntington, 132 p., illus., colored map (see 8257).
8257   1832, Geological map of the United States: in The Child's Geology (see 8256), colored geological map, 19.7 x 15.6 cm.

Phelps, Noah
8258   1802, [Use of gypsum for fertilizer]: Ct. Soc. Promoting Ag. Trans., v. 1, 6 (8 lines).

Phelps, Noah Amherst (1788-1872)
8259   1845, A history of the copper mines and Newgate Prison at Granby, Conn.: Hartford, Case, Tiffany & Burnham, 34 p.

Phenix Mining Company (see Phoenix Mining Company)

Philadelphia Academy of Natural Sciences
8260   1817-1825, Catalogue of the library of the Academy of Natural Sciences: Phila. Acad. Nat. Sci's J., v. 1, no. 1, 203-212; no. 2, 491-498; v. 2, no. 2, 383-393; v. 3, no. 2, 463-469; v. 4, no. 2, 391-398.
8261   1817-1834, List of donations to the museum of the Academy of Natural Sciences: Phila. Acad. Nat. Sci's J., v. 1, no. 1, 213-219; no. 2, 499-504; v. 2, no. 2, 394-403; v. 3, no. 2, 470-472; v. 4, no. 2, 399-402; v. 5, no. 2, 391-400; v. 6, no. 2, 315-324.
8262   1824, Journal of the Academy of Natural Sciences of Philadelphia: Am. J. Sci., v. 8, no. 2, 399-400.
        Officers and contents of v. 3 listed.
8263   1830, Notice of the Academy of Natural Sciences of Philadelphia: Am. J. Sci., v. 19, no. 1, 88-96.
        On the history, officers, and current progress of the Academy.
8264   1831, Academy of Natural Sciences of Philadelphia: Am. J. Sci., v. 19, no. 2, 355-356.
        Notices of the Academy.
8265   1831, Journal of the Academy of Natural Sciences of Philadelphia: Am. J. Sci., v. 20, no. 2, 414-415.
8266   1832, List of officers of the Academy of Natural Sciences of Philadelphia for the year 1832: Am. J. Sci., v. 22, no. 1, 183 (7 lines).
8267   1834/1835, Proceedings [of the Academy]: Advocate of Sci. and Annals Nat. History, v. 1, 46-49; 67-72; 121-122; 192-193; 231-234; 267-269; 313-319; 374-378; 421-423.
8268   1836, Academy of Natural Sciences of Philadelphia: Am. J. Sci., v. 30, no. 1, 187-188.
        Notices of the library, and of W. Maclure's retirement in Mexico.
8269   1837, Catalogue of the library of the Academy of Natural Sciences of Philadelphia. Phil. J. Dobson. 1837: Am. J. Sci., v. 33, no. 1, 181-182.
        Notice of large donation of books by W. Maclure.
8270   1839, New hall of the Academy of Natural Sciences at Philadelphia: Am. J. Sci., v. 37, no. 2, 399 (18 lines).
        Notice of W. Maclure's contributions to building fund.
8271   1841-1850, [Donations to the Academy's museum and library]: Phila. Acad. Nat. Sci's Proc., v. 1 through 5, numerous entries in all volumes.
8272   1841, Proceedings of the Academy of Natural Sciences of Philadelphia: Am. J. Sci., v. 41, no. 1, 215-216.
8273   1848, Academy of Natural Sciences of Philadelphia: Am. J. Sci., v. 5 (ns), no. 2, 294-295.
        Progress of the museum, abstracted from Ann. Rept.

Philadelphia and Lake Superior Mining Company
8274   1846, Deed of trust, and articles of association of the Philadelphia and Lake Superior Mining Company: Phila., 20 p.

Philadelphia and New Jersey Copper Company
8275   1847, Report of the trustees ... : Phila., office of the "Evening Bulletin", 20 p.

Philadelphia and North Carolina Mining and Smelting Company
8276   1847, Report on the gold mines of the ..., with maps and the act of incorporation: Phila; J. H. Schwacke, vi, 17 p., fold. map.

Philadelphia College and Academy
8277   1758, Account of the College and Academy of Philadelphia: Am. Mag. and Mon. Chron., v. 2, no. 13, 630-640.
    Lists courses and hours of instruction.

Philadelphia Linnaean Society
8278   1810, [List of members, and notices]: Am. Min. J., v. 1, no. 1, 53-55.

Philadelphia Magazine and Review (Philadelphia, 1799)
8279   1799, Some account of the earthquake in Calabria, in the year 1783: v. 1, no. 1, 11-14; no. 2, 61-64; no. 3, 129-133; no. 4, 199-201.
8280   1799, [Review of Elements of mineralogy by R. Kirwan]: v. 1, no. 1, 40-41.

The Philadelphia Magazine and Weekly Repository (Philadelphia, 1818)
8281   1818, Phenomenon in 1817: v. 1, 214-215.
    On an earthquake in Austria.
8282   1818, Natural curiosities of Poland: v. 1, 215.
    On inflammable oil springs at Cracow.

The Philadelphia Medical and Physical Journal (Philadelphia, 1804-1808)
8283   1809, Mineralogical and chemical account of the Yellow Springs, County of Chester, in Pennsylvania: Third Supplement, 207-216.
8284   1809, An enquiry into the chemical character and properties of that species of coal lately discovered at Rhode Island: together with observations on the useful application of it to the arts and manufactures of the eastern states: Third Supplement, 221-235.

Philadelphia Medical Museum (Philadelphia, 1805-1809; 1810-1811)
8285   1805, [Sulphate of magnesia from Munroe County, Virginia]: v. 1, 95 (5 lines).
8286   1805, [Mineral springs of Bedford, Pennsylvania]: v. 1, 95-97.
8287   1805, [Discovery of "ochroit" earth by Prof. Klaproth]: v. 1, 231 (5 lines).
8288   1805, [Discovery of osmium and iridium from platinum]: v. 1, 232 (7 lines).
8289   1808, Jenite, a new mineral [from Elba Island]: v. 5, 146.
    Abstracted from Med. Phys. J.

The Philadelphia Monthly Magazine (Philadelphia, 1798)
8290   1798, Original account of the remarkable salt spring at Onondaga: v. 1, no. 2, 90-92.
8291   1798, Description of the famous salt mines at Williska, Poland: v. 1, no. 6, 323-325.
8292   1798, Mineralogy: v. 2, no. 9, 160.
    On the importance of mineralogy. Signed by "A chemist."

The Philadelphia Repertory (Philadelphia, 1810-1812)
8293   1810, American millstone manufactory [at Allentown, Pennsylvania]: v. 1, 96.
8294   1810, [Eruption of Vesuvius]: v. 1, 239 (6 lines).
8295   1811, Earthquake [in the Azores]: v. 1, 277.
8296   1811, Mammoth tooth [from Scioto salt lick, Ohio]: v. 2, 8 (8 lines).
8297   1811, Surprising extrication of inflammable air [at Licking River, Kentucky]: v. 2, 24.
8298   1811, Curious discovery [elephant bones from York River, Williamsburg, Virginia]: v. 2, 87-88.
8299   1812, An earthquake [at Charleston, South Carolina]: v. 2, 255.

The Philadelphia Repository, and Weekly Register (Philadelphia, 1800-1806)
8300   1801, Earthquake [at Philadelphia]: v. 2, 7.
8301   1802, An Indian tradition respecting the mammoth: v. 2, 247.
8302   1802, Dimensions of the skeleton of the mammoth, lately exhibited in New York: v. 2, 275.
8303   1803, Petrifaction [from Schoharie County, New York]: v. 3, 222.
    Signed by "Puer."
8304   1803, Gold ore from Cabarrus [County, North Carolina]: v. 3, 407.

Phillips (see 8805 and 8806 with Rivot)

Phillips, J. (or Philips, J.?; see 6474 with S. H. Long and R. Graham)

Phillips, John (1800-1874)
8305   1831, Condensed view of the discoveries respecting the structure of the earth, which have produced the modern practical system of geology. Extracted from Phillip's Geology of Yorkshire (1829): Am. J. Sci., v. 21, no. 1, 2-26.
    See also 9702.
8306   1837, Experiments with a view to determine the interior temperature of the earth: Mech's Mag. and Reg. Inventions, v. 9, 120-121.
    Abstracted from Franklin Inst. J. Attributed to "Phillips," (i.e. John Phillips?).
8307   1838, [On gravel in England]: Am. J. Sci., v. 33, no. 2, 287-288.
    In British Assoc. proc.
8308   1842, Occurrence of some minute fossil crustaceans in Palaeozoic rocks: Am. J. Sci., v. 42, no. 2, 326.
    In British Assoc. Proc.
8309   1849, Thoughts on ancient metallurgy and mining in Brigantia and other parts of Britain, suggested by a page of Pliny's Natural History: Am. J. Sci., v. 8 (ns), no. 1, 96-102; no. 2, 258-262.

Phillips, Sir Richard (1778-1851)
8310 1810, A view of the earth, containing an account of its internal structure; its caves and subterranean passages; its mountains, its rivers and cataracts ... : Phila., n.p., 51 p.; Also Phila., J. Bouvier, 56 p.
8311 1818, A comparative analysis of the green and blue carbonates of copper: J. Sci. Arts, v. 4, no. 2, 273-281.
8312 1821, The hundred wonders of the world, and of the three kingdoms of nature: N. Hav., John Babcock & Sons; Char., S. & W. R. Babcock, i-xii, 1-660 p., 44 plates.
"First American from the tenth London edition." Contains lengthy text and plates on volcanoes, earthquakes, caves, mining, meteorites, etc.
8313 1823, Green ore of uranium: Am. J. Sci., v. 7, no. 1, 194-195 (4 lines).
Abstracted by J. Griscom.
8314 1824, Analysis of uranite: Am. J. Sci., v. 7, no. 1, 380 (12 lines).
Abstracted by J. Griscom from Revue Encyclopédique.
8315 1832, [Roasting of copper ore of Anglesea]: Mon. Am. J. Geology, v. 1, no. 10, 472 (4 lines).
In British Assoc. Proc. Ascribed to Robert Phillips (i.e. Richard Phillips?).
8316 1840, and W. Carpmael, Report on an improved process for the calcination of copper ore: Am. Rep. Arts Sci's Manu's, v. 1, 300-302.
8317 1845, On the state of iron in soils: Am. J. Sci., v. 49, no. 2, 394-396.
Abstracted from London, Edinburgh and Dublin Philos. Mag.
8318 1848, Process for separating silver from its ores without the use of mercury: Am. Rr. J., v. 21, 41.
See also note by Mr. Widder (10945).

Phillips, Samuel (1690-1771)
8319 1728, Three plain practical discourses, preach'd at Andover, October 29th. 1727. The day preceding the late earthquake. On Decemb. 21st. 1727. On Decemb. 24th. Prefac'd by the very Reverend Mr. Colman: Bos., Printed for J. Phillips, vi, 226 p.
Evans number 3091.

Phillips William (1775-1828)
See also 2597 by W. D. Conybeare.
8320 1816, An outline of mineralogy and geology, intended for the use of those who may desire to become aquainted with the elements of those sciences; especially of young persons. illustrated by four plates: NY, Collins and Co., xii, 192 p., illus., 4 plates.
8321 1818, An elementary introduction to the knowledge of mineralogy: including some account of mineral elements and constituents; explanations of terms in common use; brief accounts of minerals, and of the places and circumstances in which they are found. Designed for the use of the student. With notes and additions on American articles, by Samuel L. Mitchill (see 7306): NY, Collins and Co., x, xxxiv, [10], 246, [10] p.
8322 1823?, Elementary introduction to mineralogy: Third edition.
Not seen. Contains "Appendix" by J. Griscom (see 4580).
8323 1828, Elementary introduction to mineralogy. By William Phillips, with notes and additions on American articles, by Samuel L. Mitchill: NY, 8vo.
Not seen. Fourth edition?
8324 1833, Essay on the Georgia gold mines: Am. J. Sci., v. 24, no. 1, 1-18, 22 figs.
8325 1844, An elementary treatise on mineralogy: comprising an introduction to the science; by William Phillips. Fifth edition from the latest discoveries in American and foreign mineralogy; with numerous additions to the introduction by Francis Alger: Bos., William D. Ticknor and Co., cl, 662 p., illus.
Contains notes by F. Alger (see 136 and 137).
Some copies are bound with blank pages interleaved, for owner's notes.
For review with excerpts see 794 and 4071. For review see 791, 1163, 1218, 2105, 2357, 5444, 6185, 10744, and 10768. For notice see 781 and 4060.
This work is also known as "Allan's Mineralogy" and "Algers's Phillips's Mineralogy".

The Philosophical Medical Journal, or Family Physician (New York, 1844-1848)
8326 1847, Shower of ashes [in the Philippines]: v. 2, 30.
8327 1847, Geology: v. 2, 125-126.
Notice of C. Lyell's activities in England.

Phoenix Mining and Manufacturing Company
8328 1847, Documents, relating to the Phoenix Mining and Manufacturing Company: comprising extracts from various official reports, made under direction of government officers and others: NY, Sibell and Mott, 34 p., map.
Contains extracts from reports by F. Shepherd (see 9578), C. U. Shepard (see 9552), J. T. Ducatel (see 3254), W. B. Rogers (9001), A. Ure (10536), J. Renwick (see 8753), and Major Douglas (see 3206).
On the economic geology near Richmond, Virginia.

Phoenix Mining Company
8329 1831, Proposals of the Phenix [sic] Mining Company, with a statement of the history and character of their mines in Granby, Conn.: NY, 30 p.

Pickell, John (see 1500 and 1501 with others)

Pickering, Charles (1805-1878)
See also discussion of article by J. F. W. Johnston (5827).
8330 1849, [Geological position of the Mastodon]: Bos. Soc. Nat. History Proc., v. 3, 114-116.

Pickett, F. J. (Professor)
8331 1845, On the diluvial epoch: Dwight's Am. Mag., v. 1, 654.

Pictet, Marc-Auguste (1752-1825?)
8332   1823, Notes by Professor Pictet, on the specimens from Mont Blanc: Bos. J. Phil. Arts, v. 1, no. 2, 102-107.
        In article by F. Clissold (see 2455), extracted from Edinburgh Philos. J.

The Pictorial National Library, a Monthly Miscellany of the Useful and Entertaining in Science, Art and Literature (Boston, 1848-1849)
8333   1849, The romance of geology: v. 2, 25-30, 8 figs.
8334   1849, Precious minerals: v. 2, 30.
8335   1849, California gold: v. 2, 45.
8336   1849, The Mammoth Cave, Kentucky: v. 2, 137, plate.
8337   1849, Volcanic heat: v. 2, 172 (6 lines).
8338   1849, The geology of the ancients: v. 2, 298-299.
8339   1849, Curiosities of rock formations: v. 3, 28-31, 6 figs.
8340   1849, The diamond mines of Golconda [in India]: v. 3, 85-86.
8341   1849, Boulders and drift rocks: v. 3, 94-96, 2 figs.
8342   1849, New cave in Kentucky: v. 3, 228 (9 lines).
8343   1849, Destruction of Caracas by an earthquake: v. 3, 295.

Piddington, Henry
8344   1850, On the great diamond in the possession of the Nizam: Am. J. Sci., v. 9 (ns), no. 3, 434-436, fig.

Pierce, Elis
8345   1844, [On potholes in New Hampshire rocks]: in final New Hampshire report by C. T. Jackson (see 5611), 282.
        Also Warwick, Massachusetts examples cited.

Pierce, James
8346   1817, [On native magnesia from Hoboken, New Jersey]: Am. Mon. Mag. and Crit. Rev., v. 1, no. 2, 130.
8347   1818, Carbonate of magnesia, and very uncommon amianthus, discovered near New-York. - Extract of a letter from Mr. James Pierce to the editor: Am. J. Sci., v. 1, no. 1, 54-55.
8348   1818, Discovery of native crystallized carbonate of magnesia on Staten Island, with a notice of the geology and mineralogy of that island: Am. J. Sci., v. 1, no. 2, 142-146.
8349   1820, Account of the geology, mineralogy, scenery, &c. of the Secondary region of New-York and New-Jersey, and the adjacent regions: Am. J. Sci., v. 2, no. 1, 181-199.
8350   1820, [Kaolin from Weehawken, New Jersey]: Am. J. Sci., v. 2, no. 2, 368-369.
        Abstract in NY Lyc. Nat. History Proc.
8351   1820, [On mineralogy and geology of the Secondary region of New Jersey]: Am. J. Sci., v. 2, no. 2, 367-368.
        In NY Lyc. Nat. History Proc.
8352   1821, Catskill Lyceum, &c.: Am. J. Sci., v. 3, no. 2, 237-238.
        Includes notes on the mineralogy of the Catskill Mountains, New York.
8353   1821, Chalybeate spring at Litchfield [Connecticut]: Am. J. Sci., v. 3, no. 2, 235-236.
8354   1821, Chalybeate spring at Catskill - Marl and tufa, at the same place: Am. J. Sci., v. 3, no. 2, 236-237.
8355   1822, Geology, mineralogy, scenery, &c., of the Highlands of New-York and New-Jersey: Am. J. Sci., v. 5, no. 1, 26-33.
8356   1823, A memoir on the Catskill Mountains with notice of their topography, scenery, mineralogy, zoology, economical resources, &c.: Am. J. Sci., v. 6, no. 1, 86-97.
8357   1823, Notice of the alluvial district of New-Jersey, with remarks on the application of the rich marl of that region to agriculture: Am. J. Sci., v. 7, no. 2, 237-242.
8358   1824, Notice of an excursion among the White Mountains of New Hampshire, and to the summit of Mount Washington, in June, 1823, with miscellaneous remarks: Am. J. Sci., v. 8, no. 1, 172-181.
8359   1825, Notices of the agriculture, scenery, geology, and animal, vegetable and mineral productions of the Floridas, and of the Indian tribes, made during a recent tour in these countries: Am. J. Sci., v. 9, no. 1, 119-136.
8360   1826, Notice of the Peninsula of Michigan, in relation to its topography, scenery, agriculture, population, resources, &c.: Am. J. Sci., v. 10, no. 2, 304-319.
8361   1826, Practical remarks on the shell marl region of the eastern parts of Virginia and Maryland, and upon the bituminous coal formation in Virginia and the contiguous region: Am. J. Sci., v. 11, no. 1, 54-59.
8362   1827, Observations relative to some of the mountain districts of Pennsylvania and the mineral resources of that state, in its anthracite, bituminous coal, salt, and iron, with miscellaneous remarks: Am. J. Sci., v. 12, no. 1, 54-74.
8363   1828, Anthracite region of Pennsylvania: Hazard's Reg. Penn., v. 1, 310-316.

Pierce, William Leigh
8364   1812, An account of the great earthquakes, in the Western states ... December 16-23, 1811: Newburyport, Mass., the Herald Office, 16 p.
8365   1814, The earthquake of Caracas: West. Gleaner, v. 1, 197-198.
        A poem describing the earthquake.

Pierronet, Thomas
8366   1799, Remarks made during a residence at Stabroek Rio Demeray (Lat. 6.10.N) in the latter part of the year 1798: Mass. Hist. Soc. Coll., v. 6, 1-5.

Pike, Benjamin, jr.
8367   1848, Pike's illustrated catalogue of optical, mathematical, and philosophical instruments, manufatured, imported, and sold by the author; with the prices affixed at which they are offered in 1848: NY, by the author, 2 v (x, 10-346, 408 figs.; front., 282 p., figs. 409-782).

Pike, Zebulon Montgomery (1779-1813)
8368 1807, An account of a voyage up the Mississippi River, from St. Louis to its source; made under the orders of the War Department, by Lieut. Pike, of the United States Army, in the years, 1805 and 1806. Compiled from Mr. Pike's Journal: n.p., 68 p. Extracted in Am. Reg. and Gen. Repos., v. 5, pt. 2, 273-311.
8369 1810, An account of expeditions to the sources of the Mississippi, and the western parts of Louisiana, to the sources of the Arkansas, Kans, La Platte, and Pierre Jaun, Rivers; ... : Phila., C. and A. Conrad and Co., 5, 277, 65, [1], 53, 87 p., maps, tables, front.

The Pilot (New Haven, 1821-1924)
8370 1821, Volcano in the Isle of Bourbon: v. 1, no. 7, [4].
8371 1822, Pictured rocks [of Lake Superior]: v. 1, no. 22, [4].
8372 1822, Heat of the earth: v. 1, no. 31, [4].
8373 1822, Putnam's Rock: v. 1, no. 51, [4].
      On a giant glacial boulder near West Point, New York.
8374 1823, Gold mine [in Ansom County, North Carolina]: v. 2, no. 92, [2].
8375 1824, Heat of the earth: v. 3, no. 143, [2].
8376 1824, Gold [from Aruba]: v. 3, no. 154, [4].

Pinel, Philippe (1745-1820; see 4228 with J. L. Gay-Lussac)

Pingel, Christian (1793-1852)
8377 1836, Subsidence of the coast of Greenland: Am. J. Sci., v. 30, no. 1, 379-380. Abstracted from Geol. Soc. London Proc.

Piozzi, Mrs. Hester Lynch (1741-1821)
8378 1789, Earthquake at Messina: Gentlemen and Ladies Town and Country Mag, v. 1, no. 11, 526-527.
      From Journey through France, Italy and Germany. The story of a mother and child during the earthquake of 1783.
8379 1793, Affecting picture of an earthquake scene: Lady's Mag. and Repos. Ent. Know., v. 2, 187-188. Also in Mass. Mag., v. 5, 708-710; NY Mag., v. 4, 141-142.

Piscataqua Mining Company
8380 1850, Charter and by-laws of the Piscataqua Mining Co. of Michigan: Phila., Grattan and McLean, 19 p.
      With extracts from reports by J. W. Foster and J. D. Whitney (see 3937) and C. Whittlesey (see 10942).

Pitcher, Zina (see 7159)

Pittsburgh and Boston Copper Harbor Mining Company
8381 1847, Report of the committee of the stockholders of the Pittsburgh and Boston Copper Harbor Mining Company, with Appendices: Bos., S. N. Dickinson (printer), 48 p., plate.
      With reports by T. Jones (see 5850) and A. A. Hayes (see 4904).

Pittsfield Lyceum of Natural History (Massachusetts)
8382 1825, Officers of the Pittsfield Lyceum of Natural History of the Berkshire Medical Institution, elected September 9, 1824: Am. J. Sci., v. 9, no. 1, 177 (11 lines).

Place, F.
8383 1824, Account of the earthquake in Chili [sic], in November, 1822, from observations made by several Englishmen residing in that country: Bos. J. Phil. Arts, v. 2, no. 1, 27-33.
      Extracted from Royal Inst. J.

The Plaindealer (New York, 1836-1837)
8384 1836, Vesuvius: v. 1, no. 1, 12 (4 lines).
8385 1836, Coal for the next season [from Mauch Chunk, Pennsylvania]: v. 1, no. 3, 45 (6 lines).
8386 1836, Slaves in the diamond mines: v. 1, no. 3, 46.
8387 1837, A scrap for geologists: v. 1, no. 6, 94 (10 lines).
      On a millstone with metal inclusions from Salem, Ohio.
8388 1837, Geological survey [of New York, appropriations for]: v. 1, no. 20, 316.
8389 1837, Insects in flint: v. 1, no. 20, 317 (9 lines).
8390 1837, An extensive coal bed [in the Mississippi River Valley]: v. 1, no. 20, 318.
8391 1837, The mammoth cave of Kentucky: v. 1, no. 24, 379-381.
8392 1837, Moutains of ice [glaciers of Mont Blanc]: v. 1, no. 33, 524 (6 lines).
8393 1837, Virginia gold mines: v. 1, no. 33, 524 (4 lines).
8394 1837, Coal [from New York]: v. 1, no. 34, 542 (7 lines).

Plattes, Gabriel (fl. 1683)
8395 1784, A discovery of subterranean treasure: containing useful explorations, concerning all manner of mines and minerals, from the gold to the coal; with plain directions and rules for the finding of them in all kingdoms and countries. In which the art of melting, refining, and assaying them is plainly disclosed, ... : Phila., Robert Bell, 37 p.
      Evans number 18732. Another edition: Phila., 24 p. (1792) Evans number 24697.
8396 1796, Select pamphlets: viz. ... 8. Platte's discovery of subterranean treasures - containing useful explanations concerning all manner of mines and minerals: Phila., M. Carey, 76, 87, [103], [71], 44, 13, [116], 24.
      Evans number 31174. Pamphlet number 8 of 8.

Plattner, Karl Friedrich (1800-1858)
      See also 1974 with Breithaupt.
8397 1849, Analysis of copper blende: Am. J. Sci., v. 8 (ns), no. 1, 127 (12 lines).
      Abstracted from Poggendorff's Annalen.

Playfair, John (1748-1819)
8398   1817, Dissertation second: exhibiting a general view of the progress of mathematical and physical science, since the revival of letters in Europe: Bos., Wells and Lilly, 197 p.

Pleasants, Thomas S.
8399   1837, Account of the soils and agriculture of western New York: Farm's Reg., v. 5, 441-444, 547-552.

Pliny the Younger
8400   1841, Eruption of Vesuvius, A. D. 73: Iris, v. 1, no. 3, 116-118.
        Translated from Latin.

The Plough Boy (Albany, 1819-1823)
8401   1819, Geology applied to practical agriculture: v. 1, no. 6, 42; no. 8, 58-59.
        Signed by "A."
8402   1820, Scientific notice: v. 1, no. 43, 342.
        Notice of System of geology sponsored by the Troy Lyceum.
8403   1820, Fossil whale [from Carse of Falkirk, Scotland]: v. 1, 373 (11 lines).
8404   1820, Colouring of agate: v. 1, 373 (7 lines).
8405   1820, Discovery of human skulls in the same formation as that which contains remains of elephants, rhinoceri, &c. [in the West Indies]: v. 1, 313.
8406   1820, Feldspar and pitchstone [from the islands of Arran, Mull, Egg, and Skye]: v. 1, 313 (8 lines).
8407   1820, Subterranean sounds in granite rocks: v. 1, 313-314.
8408   1820, Coal not of vegetable origin: v. 1, 392.
8409   1820, Salt mines of Meurthe [France]: v. 2, 24.
8410   1820, Work on petrifactions [notice of text by Lamouroux]: v. 2, 24.
8411   1820, Jameson's geognosy [notice of]: v. 2, 37.
8412   1820, Discovery of ornamental emerald mines: v. 2, 37 (10 lines).
8413   1820, New set of rocks discovered in Iceland: v. 2, 37.
8414   1820, Cadium [discovery of]: v. 2, 37.
8415   1820, Earthquake at Copiapo: v. 2, 37-38.
8416   1820, Isle of Elba [and its magnetic anomaly]: v. 2, 38 (9 lines).
8417   1820, Mean temperature of the earth: v. 2, 38 (7 lines).
8418   1820, Dr. Davy's scientific tour in Ceylon: v. 2, 102.
8419   1820, Geology of Albany County [notice of survey by A. Eaton and T. R. Beck (see 3328)]: v. 2, 125.
8420   1820, Mineralogy. Extract of a letter from a scientific correspondent dated Canal Line, 65 miles W. of Genesee River, Oct. 11, 1820: v. 2, 161.
        On pisolite from Sandy Creek, New York.
8421   1821, Topography of the Northwest: v. 2, 279.
        Notice of H. R. Schoolcraft's expedition.
8422   1821, Geological survey of Albany: v. 2, 299.
        Notice of survey by A. Eaton and T. R. Beck (see 3328).
8423   1821, North American hot springs [at "Ouchitta", Kansas]: v. 2, 312.
8424   1821, Variation of the magnetic needle: v. 2, 312.
8425   1821, [Catskill Lyceum of Natural History]: v. 2, 342.
        Notice of the founding of the Lyceum.
8426   1821, Organic remains [from Lexington, Rockbridge County, Virginia]: v. 3, no. 3, 23.
8427   1821, Geology: v. 3, no. 17, 134.
        On 2 specimens of Ichthyosaurus found at Lyme Regis, England.
8428   1821, Discovery of petrified bones [at Bromyary]: v. 3, no. 37, 296.
8429   1821, Coal [from Tioga River, Pennsylvania]: v. 3, no. 38, 303.
8430   1822, Popular retrospect of the progress of philosophy and science: v. 4, no. 7, 49-51.
8431   1822, Virginia husbandry. Wier's Cave: v. 4, no. 15, 113.
8432   1822, Ruby: v. 4, no. 24, 192.
        Notice of changing color with temperature based on experiments by H. Davy.
8433   1822, Known species of organized beings: v. 4, no. 24, 192.
        On the number of known fossil species.

The Plough, the Loom, and the Anvil (Philadelphia, 1848-1850)
8434   1848, The anthracite coal trade of Pennsylvania: how it affects the farmer and the planter: v. 1, no. 2, 73-78.
8435   1848, The iron trade of the Union, and its influence upon the interests of the farmer and planter: v. 1, no. 3, 145-151.
8436   1848, Soils - constituent elements: v. 1, no. 6, 382-383.
        From the Pharmaceutical Times.
8437   1849, Pennsylvania anthracite coal trade: v. 1, no. 9, 564-568.
8438   1849, The Maryland state agricultural chemist: v. 1, no. 9, 575-576.
        On the appointment and activities of Dr. Higgins of Maryland.
8439   1849, Coal mines in the South [at Tuscaloosa, Alabama]: v. 1, no. 11, 704.
8440   1849, Zinc: v. 2, no. 1, 12.
8441   1849, Iron ore [from Reading, Pennsylvania]: v. 2, no. 1, 28.
8442   1849, New Jersey zinc [Sussex Zinc Company]: v. 2, no. 2, 84 (11 lines).
        Abstracted from Farm. and Mechanic (New York).
8443   1849, Alabama coal and iron trade, etc.: v. 2, no. 5, 281-283.
8444   1849, Virginia [on lead ore from Albemarle, and other Virginia ores]: v. 2, no. 5, 300.
8445   1850, A new and important resource of national industry and independence - American steel: v. 2, no. 12, 800-802.

Plucker, Julius (1801-1868)
8446   1849, On the magnetic relations of the positive and negative optic axes of crystals: Am. J. Sci., v. 8 (ns), no. 3, 430-431.
        Abstracted from Philos. Mag.

Plummer, John Thomas (1807-1865)
8447   1843, Suburban geology, or rocks, soil, and water, about Richmond, Wayne Co., Indiana: Am. J. Sci., v. 44, no. 2, 281-313, 15 figs.

Plummer, Jonathan (1761-1819)
8448   1812, The dreadful earthquake and the fatal spotted fever. A funeral sermon ... : Newburyport, Mass.?, the author, broadside.

Poinsett, Joel Roberts (1779-1851)
8449   1841, Discourse, on the objects and importance of the National Institution for the Promotion of Science: Wash., P. Force, 52 p.
        For extracts see Nat'l Inst. Prom. Sci. Bulletin, v. 2, 19-30.

Poisson, Simeon-Denis (1781-1840)
8450   1838, Memoir upon the temperature of the solid parts of the globe, of the atmosphere, and of those regions of space traversed by the earth: Am. J. Sci., v. 34, no. 1, 57-69.
        Translated from the French by R. W. Haskins.

Pomeroy, Samuel Wyllys (1802-1882)
8451   1832, Remarks on the coal region between Cumberland and Pittsburgh and on the topography, scenery, etc., of that portion of the Alleghany Mountains: Am. J. Sci., v. 21, no. 2, 342-347.

Pond, Enoch (1791-1882)
8452   1838, Geology and Revelation: Bib. Repos., v. 12, 1-21.

Pope, John (1822-1892)
8453   1850, Report of an exploration of the Territory of Minnesota: US SED 42, 31-1, v. 10 (558), 2-56, map.

The Port Folio (Philadelphia, 1801-1827)
8454   1807, Peale's Museum: v. 4 (2s), 293-297.
        A room-by-room description.
8455   1809, Mineral waters: v. 1 (3s), 311-316.
8456   1810, The mineralogist, no. 1: v. 3 (3s), 109-111.
        On the importance of minerals.
8457   1810, Discovery of a real and entire mammoth [in Siberia]: v. 3 (3s), 111-113.
8458   1810, On the origin of stones that have fallen from the atmosphere: v. 3 (3s), 475-479.
8459   1810, Description of the Yellow Springs, in Pennsylvania: v. 4 (3s), 44-47, plate.
        Signed by "A."
8460   1810, A ride to Niagara: v. 4 (3s), 50-53; 162-171; 220-238; 297-303, map.
8461   1810, Science for the Port Folio [on meteorites]: v. 4 (3s), 557-564.
        Signed by "F."
8462   1811, Bristol mineral waters [in Pennsylvania]: v. 6 (3s), 38-42.
        Signed by "J. D."
8463   1811, Eruption of the new volcano near St. Michaels [in the Azores]: v. 6 (3s), 135-137.
        Signed by "T. H." and "J. S."
8464   1812, Observations on earthquakes: v. 7 (3s), 421-436; 572-574.
        Signed by "C.," and by "M."
8465   1812, Mineralogy and zoology: v. 8 (3s), 281.
        Brief notes on European progress.
8466   1814, Description of natural walls: v. 3 (4s), 355-357.
        On dikes at the surface discovered by the Lewis and Clark expedition.
8467   1816, [Review with excerpts of View of Cincinnati by D. Drake (see 3218)]: v. 1 (5s), 25-38.
8468   1817, [Review of Mineral waters of Ballston and Saratoga by W. Meade (see 6947)]: v. 3 (5s), 178 (11 lines).
8469   1817, New blow-pipe: v. 3 (5s), 345.
        On J. Murray's experiments.
8470   1817, Burning springs [in Portland, New York]: v. 3 (5s), 354.
8471   1817, The new blow pipe: v. 3 (5s), 417-423.
        Signed by "T. C." (i.e., Thomas Cooper?). On Hare's blow-pipe.
8472   1817, Bedford mineral springs: v. 3 (5s), 507-509, plate.
8473   1817, [Review of Mineral waters of Ballston and Saratoga by W. Meade (see 6947)]: v. 4 (5s), 58-60, table.
8474   1817, Hare's blow-pipe: v. 4 (5s), 65.
8475   1817, Nashville - salt: v. 4 (5s), 261-262.
        On well borings for salt.
8476   1818, Salt springs [in Wayne County, Ohio]: v. 5 (5s), 87.
8477   1818, Copper [from Lake Superior]: v. 5 (5s), 248.
8478   1818, American marble [from New Haven, Connecticut]: v. 5 (5s), 247.
8479   1818, Mineralogy: v. 5 (5s), 400.
        Notice of T. Cooper's lecture.
8480   1818, The lead mines of Missouri: v. 6 (5s), 72.
8481   1818, Geology of England [reviews of six works by William Smith, including Delineation of the strata of England and Wales; Memoir of the map ... ; Geological section from London to Snowdon; Series of county maps; Strata identified by organized fossils; and Stratigraphical system]: v. 6 (5s), 91-117.
        From Edinburgh Rev.
8482   1818, American copper [from Lake Superior]: v. 6 (5s), 120 (8 lines).
8483   1818, Mammoth cave in Indiana: v. 6 (5s), 397-398.
8484   1818, Encroachment of the sea at Bridgetown, N. J.: v. 6 (5s), 399.
8485   1819, Lizard embedded in coal [found alive]: v. 8 (5s), 175-176.
8486   1819, Trenton, N. J. [mineral spring]: v. 8 (5s), 344.
8487   1819, Volcanic water [from Mount Vesuvius]: v. 8 (5s), 429-430.
8488   1820, On the manufacture of iron and the expense of a furnace in New Jersey: v. 9 (5s), 245-250.
        Signed by "J. B. Q."
8489   1823, Gold mine [at Ancram, North Carolina]: v. 16 (5s), 85.
8490   1824, The last eruption of Vesuvius: v. 17 (5s), 52-57.
8491   1824, Analysis of the principle varieties of coal of Great Britain: v. 17 (5s), 246-248.
8492   1824, Analysis of a meteorolite: v. 17 (5s), 325.

The Port Folio (cont.)
8493   1824, Illinois. Stone coal at the sea line [Saline, Illinois]: v. 17 (5s), 438.
8494   1824, Silver mines of Mexico: v. 18 (5s), 56-57.
8495   1824, [Notice of Catalogue of minerals by F. Hall (see 4650)]: v. 18 (5s), 71 (9 lines).
8496   1825, [History of alum manufacturing]: v. 20 (5s), 198-200.
8497   1826, The yellow springs [of Chester County, Pennsylvania]: v. 1 (Hall's second series), 408-409, plate.
8498   1827, The York Springs [in Pennsylvania]: v. 2 (Hall's second series), 177-180, plate. Signed by "M."

Porter, Elijah
8499   1833, Medical topographical account of the County of Saratoga: Med. Soc. NY Trans., v. 1, 342-347.

Porter, George Richardson (1792-1852)
8500   1832, A treatise on the origin, progressive improvement, and present state of the manufacture of porcelain and glass: Phila., Carey and Lea, xiv, [17]-252 p. Other printings in Philadelphia in 1834, 1837 and 1846.
8501   1839, A statistical view of the recent progress and present amount of mining industry in France; drawn from the official reports of the "Direction Generale des Ponts et Chaussees et des Mines": Franklin Inst. J., v. 24 (ns), 31-35; 106-111.
8502   1844, The commerce and resources of British America: Hunt's Merch. Mag., v. 10, 15-26.
8503   1845, The mining industry of France: Hunt's Merch. Mag., v. 12, 341-347.
8504   1847, On the iron manufacture of Great Britain: Am. J. Sci., v. 3 (ns), no. 2, 291-293. Abstracted from London Ath.

Porter, Jacob G.
8505   1823, and H. U. Cambridge and T. H. Webb, Account of the Roxbury rocking stone [in Massachusetts]: Am. J. Sci., v. 7, no. 1, 59-61, plate.
8506   1823, By Dr. Jacob Porter: Am. J. Sci., v. 6, no. 2, 246-249.
       On mineral localities in Connecticut, New York, and Ohio.
8507   1823, Localities of minerals [in New Jersey and Massachusetts]: Am. J. Sci., v. 7, no. 1, 58 (3 lines).
8508   1823, Note [on Connecticut minerals]: Am. J. Sci., v. 7, no. 1, 56-57.
8509   1824-1825, By Jacob Porter: Am. J. Sci., v. 8, no. 2, 233; v. 9, no. 1, 54-55; v. 10, no. 1, 18-19.
       On mineral localities in Massachusetts, Connecticut and Pennsylvania.
8510   1825, Localities of minerals, principally in Massachusetts: Am. J. Sci., v. 7, no. 2, 252-253.
8511   1825, Notice of a rocking stone in Savoy, Massachusetts, with a drawing: Am. J. Sci., v. 9, no. 1, 27-28, plate.
8512   1826, An account of Plainfield, in Hampshire County, Massachusetts: Mass. Hist. Soc. Coll., v. 8 (2s), 167-173.
8513   1827, Localities of minerals [in Massachusetts]: Am. J. Sci., v. 12, no. 2, 378 (11 lines).
8514   1829, Gifts to the Geological Society: Am. J. Sci., v. 17, no. 1, 202 (7 lines).
8515   1831, Localities of minerals [in Massachusetts]: Am. J. Sci., v. 20, no. 1, 170 (5 lines).
8516   1834, Topographical description and historical sketch of Plainfield, in Massachusetts: Greenfield, Mass., 44 p.
8517   1835, Chromate of iron [from Townsend, Vermont]: Am. J. Sci., v. 28, no. 2, 383 (3 lines).

Porter, James Madison (1793-1862)
8518   1845, [Large mass of native copper from Ontonagon, Lake Superior district]: Nat'l Inst. Prom. Sci. Bulletin, v. 3, 319.

Porter, Richard
8519   1822, Gypsum: Plough Boy, v. 4, no. 8, 61-62.

Porter, Sir Robert Ker (1777-1842)
8520   1809, Travelling sketches in Russia and Sweden: Phila., Hopkins and Earle, xii, 475 p. Not seen.
8521   1839, [Excerpts from Geological and other physical facts and of antiquities in Asia]: Am. J. Sci., v. 37, no. 2, 347-356.
       In review (see 731).

Porter, Timothy Dwight
8522   1821, Cursory notice of some parts of North and South Carolina: Am. J. Sci., v. 3, no. 2, 227-229.
8523   1822, Appendix: Am. J. Sci., v. 4, no. 2, 241-242.
       Follows "On volcanoes and volcanic substances, with a particular reference to the origin of the rocks of the floetz trap formation" by T. Cooper (see 2625).
8524   1824, Molybdena [from Saybrook, Connecticut]: Am. J. Sci., v. 8, no. 1, 194 (5 lines).
8525   1824, Sillimanite [from Saybrook, Connecticut]: Am. J. Sci., v. 8, no. 1, 195 (7 lines).
8526   1825, Localities of minerals on Connecticut River: Am. J. Sci., v. 9, no. 1, 177 (6 lines), map.

Porter, William S.
8527   1827, Sketches of the geology, &c., of Alabama: Am. J. Sci., v. 13, no. 1, 77-79.

The Portfolio, and Companion to the Select Circulating Library (Philadelphia, 1835-1836)
8528   1835, Curious discovery in France: v. 1, pt. 1, 63 (12 lines).
       On the discovery of a fossil elephant at Homaize.
8529   1835, Austrian salt mines: v. 1, pt. 1, 117.
8530   1835, The greatest volcano in the world [Kilauea, Hawaii]: v. 1, pt. 1, 184-185.
8531   1835, Vesuvius: v. 1, pt. 1, 198 (8 lines).

The Portfolio (cont.)
8532 1835, Meteorolites, or stones which have fallen from the sky: v. 1, pt. 1, 199-200.
8533 1835, Diamonds: v. 1, pt. 2, 89.
8534 1835, Petrifactions: v. 1, pt. 2, 157.
8535 1836, Eruption of Mount Etna: v. 2, pt. 1, 6-7; 28-29.
8536 1836, Fossil trees [at Balognie]: v. 2, pt. 1, 38.
8537 1836, Diamonds: v. 2, pt. 1, 203 (6 lines).
       Notice of Perrot's opinions on the volcanic origin of diamonds.
8538 1836, Cornish mines: v. 2, pt. 2, 165-166.

The Portico, a Repository of Science and Literature (Baltimore, 1816-1818)
8539 1816, [Review with excerpts of Personal narrative of travel by A. Humboldt]: v. 1, 169-184.
       Signed by "S." See also 5373.
8540 1816, [Review with excerpts of View of Cincinnati by D. Drake (see 3218)]: v. 1, 265-276.
8541 1816, Geology: v. 2, 336-340.
       On the scope of geology, and its recent advances.
8542 1817, [Review of Elementary treatise by P. Cleaveland (see 2420)]: v. 3, 345-352.
8543 1817, [Review with excerpts of Observations ... , by W. Maclure (see 6616)]: v. 4, 188-196.
8544 1817, [Review with excerpts of Essay on chemical reagents by F. C. Accum (see 17)]: v. 4, 11-15.
8545 1818, [Review with excerpts of Theory of the earth by G. Cuvier with notes by S. L. Mitchill (see 2754 and 7307)]: v. 5, 205-213.
       Pages 206-212 are misnumbered 306-312.

The Portland Magazine, Devoted to Literature (Portland, Maine, 1834-1835)
8546 1835, Geology of Portland and vicinity: v. 1, no. 10, 334-341.

Portlock, Lieutenant
8547 1824, Observations of the lakes of Canada, &c.: Bos. J. Phil. Arts, v. 2, no. 1, 104 (17 lines).
       Abstracted from Annals Phil.

Potomac and Alleghany Coal and Iron Manufacturing Company
8548 1841, Documents relating to the Potomac and Alleghany Coal and Iron Manufacturing Company: NY, D. Felt and Co. (printer), 39 p., folded map, folded plate.
       Contains reports by F. Shepherd (see 9573), C. U. Shepard (see 9535), and W. Wagstaff (see 10662).

Potter, Chandler Eastman (1807-1868)
8549 1833, Notice of a rocking stone [at Hanover, New Hampshire]: Am. J. Sci., v. 24, no. 1, 185-186, 4 figs.

The Poughkeepsie Casket (Poughkeepsie, New York, 1836-1841)
8550 1836, Precious stones: v. 1, 93.
8551 1839, Natural soda fountain: v. 3, 127.
8552 1840, Geology: v. 4, 43.
       Statistics of the earth's size and shape.
8553 1840, The last eruption of Vesuvius: v. 4, 130.

Poussin, William Tell (see 1634 to 1636 with S. Bernard)

Powell, Baden (1796-1860)
8554 1847, On certain cases of elliptic polarization of light by reflexion: Am. J. Sci., v. 3 (ns), no. 2, 264-265.
       Abstracted from London Ath.

Powell, Elisha
8555 1821, Great value of, and mode of using marls: Am. Farm., v. 3, 199.

Powell, John
8556 1831, Machine for washing gold ore: Franklin Inst. J., v. 8 (ns), 108.

Powell, William Byrd (1799-1866)
8557 1830, Formation of mineral coal: Hazard's Reg. Penn., v. 6, 346-347; 393-394. Also in Mech's Mag., v. 1, 304-305.
       Abstracted from Miner's J.
8558 1832, Coal mines of Nantico in the Alleghany Mountains: Atl. J. and Friend Know., v. 1, no. 3, 115-116.
8559 1835, Geology: Cin. Mirror, v. 4, 97-98.
       Debate with W. Wood on the origin of coal. See also 11035 and 11036.
8560 1835, Geology of mineral coal: Cin. Mirror, v. 4, 90-91.
       See also articles by W. Wood (11035 and 11036).
8561 1835, Mineral coal: Cin. Mirror, v. 4, 114-115; 129-130; 236 [i.e. 136]-137.
       See also articles by W. Wood (11035 and 11036). See also 8781
8562 1836, Geology of the Mississippi: West. Mess., v. 3, 552-557.
       Also in Army and Navy Chron,, v. 4, 257-258 (1837).
8563 1839, Mineral coal is not a vegetable: New-Yorker, v. 8, 2-3; 18-19; 33-34; 50-51.
8564 1842, A geological report upon the Fourch Cove and its immediate vicinity, with some remarks upon their importance to the science of geology, and upon the value of their productions to the arts of civilized society, accompanied with a suite of specimens and a catalogue: Little Rock, Arkansas, The Society, N. E. Woodruff (printer), 22 p., map.
       For review see 186. Map not seen (geological?).

The Prairie Farmer (Chicago, 1843-1850)
8565 1843, Petrifactions [in the Secondary formations of the West]: v. 3, 38.
8566 1843, Asbestos [from Pennsylvania]: v. 3, 253 (8 lines).
8567 1844, Electricity the cause of earthquakes: v. 4, 269 (10 lines).
8568 1845, [Analysis of soils]: v. 5, 196-197.

The Prairie Farmer (cont.)
8569   1846, [Review of Introduction to geology by R. Bakewell (see 1460)] : v. 6, 325.
8570   1847, Geological survey of Illinois [call for] : v. 7, 181.
8571   1847, Sand, clay, iron and lime: v. 7, 272.
        On the composition of soils.
8572   1848, Iron ore [from Schuyler County, Illinois] : v. 8, 38.
8573   1849, Lead and mining: v. 9, 27 [i.e. 37].
8574   1849, Concerning gold in California: v. 9, 69.
8575   1849, Wisconsin marble [from Richland County] : v. 9, 233.
8576   1850, Illinois coal [on the Kingston Coal Company] : v. 10, 37.
8577   1850, [Review of Lectures on the application of chemistry and geology to agriculture by J. F. W. Johnston (see 5819)] : v. 10, 369-371, woodcut.

Prendergrast, Garrett Elliot
8578   1803, A physical and topographical sketch of the Mississippi territory ... : Phila., Bronson and Chauncey (printers), 34 p.

Prentice, George Dennison (1802-1870)
8579   1845, Mammoth cave: Living Age, v. 7, 53.
        A descriptive poem.

Prentice, Thomas (1702-1782)
8580   1756, Observations moral and religious, on the late terrible night of the earthquake. A sermon preached at the Thursday lecture, in Boston, January 1st, 1756. ... : Bos., printed by S. Kneeland for D. Henchman, 24 p.
        Evans number 7768.

Prescott, Calvin
8581   1821, Black oxid of manganese [from Sheffield, Berkshire County, Massachusetts] : Am. J. Sci., v. 4, no. 1, 189.

Prescott, William (1788-1875)
8582   1822, Sketch of the history, geology, &c. of Gilmantown, in New-Hampshire: Coll. Topo. Hist. Bibliographical NH, v. 1, 72-79.
8583   1839, A sketch of the geology and mineralogy of the southern part of Essex County, in Massachusetts: Essex Co. Nat. History Soc. J., v. 1, no. 2, 78-91.

Preston, Daniel R.
8584   1807, The wonders of Creation, natural and artificial. Being an account of the most remarkable mountains, rivers, lakes, cataracts, mineral springs, miscellaneous curiosities, and antiquities in the world: Bos., John M. Dunham, 300, [16] p.
8585   1808, The wonders of Creation ... [as above] : Bos., John M. Dunham, 2 v. (336 p., plate).
        Second edition. Volume 2 not seen.

Preston, Samuel
8586   1829, Copper ore - internal improvements [at Lake Superior] : Niles' Wk. Reg., v. 36, 203.

Prevost, Louis Constant (1787-1856)
8587   1828, French Institute, July 9, 1827: Am. J. Sci., v. 13, no. 2, 397-398.
        Abstracted from Revue Encyclopédique.

Priest, Josiah (1788-1851)
8588   1825, The wonders of nature and providence, displayed. Compiled from authentic sources, both ancient and modern, giving an account of various and strange phenomena existing in nature, of travels, adventures, singular providences, &c. ... : Alb., J. Priest, viii, [9]-616 p., plates.
8589   1833, American antiquities, and discoveries in the West: being an exhibition of the evidence that an ancient population of partially civilized nations, differing entirely from those of the present Indians, peopled America, many centuries before its discovery by Columbus. And inquiries into their origin, with a copious description of many of their stupendous works, now in ruins. With conjectures concerning what may have become of them: Alb., Hoffman and White, viii, [9]-400, folding front., illus.
        This is the second revised edition of 8588. Other editions include 3rd revised edition, Alb., as above (1833); 4th edition, as above (1834); 5th edition, as above (1835 and 1838 printings).

Prime, A. J.
8590   1845, Great American mastodon [from Newburgh, Orange County, New York] : Am. Q. J. Ag. Sci., v. 2, 203-212.

Prince, Thomas (1687-1758)
8591   1727, Earthquakes the work of God, & tokens of his just displeasure ... Wherein among other things is offered a brief account of the natural causes of these operations in the hands of God: with a relation of some late terrible ones in other parts of the world, as well as those that have been perceived in New-England since it's [sic] settlement by English inhabitants ... : Bos., D. Henchman, [6], 45, [3] p.
        Evans number 2945.
8592   1755, An improvement of the doctrine of earthquakes, being the works of God, and tokens of his just displeasure. Containing an historical summary of the most remarkable earthquakes in New-England, from the first settlement of the English here, as also in other parts of the world since 1666. ... To which is added, a letter to a gentleman, giving an account of the dreadful earthquake felt in Boston, New-England Nov. 18, twenty five minutes past 4 in the morning: Bos., D. and Z. Fowle, 16 p.
        Evans number 7550.

Prince, T. (cont.)
8593   1755, Earthquakes the works of God, and tokens of his just displeasure; Being a discourse on that subject, wherein is given a particular description of this awful event of Providence. And among other things, is offer'd a brief account of the natural, instrumental, or secondary causes of these operations in the hands of God. After which, our thoughts are led up to Him as having the highest and principal agency in this stupendous work. ... Made public at this time on occasion of the late dreadful earthquake, which happen'd on the 18th of Nov. 1755: Bos., D. and Z. Fowle, 23, [1] p.
    Evans number 7549.

The Princeton Magazine (Princeton, New Jersey, 1850)
8594   1850, [Review of Lake Superior, by L. Agassiz (see 83)]: v. 1, 143-144.
8595   1850, [Notice of Earth and man, by A. Guyot (see 4609 and 4610)]: v. 1, 335.

Prinse, M.
8596   1838, On the carbonate of magnesia of southern India: Franklin Inst. J., v. 21 (ns), 340 (7 lines).
    Abstracted from J. des Mines. Translated by J. Griscom.

Prinsep, James (1799-1840)
8597   1836, Extract from a letter: Am. J. Sci., v. 31, no. 1, 170 (9 lines).
    Abstracted from l'Institut. On fossil bones from India.

Prinsep, Macaire
8598   1830, On a new mineral, hydro-carbon: Am. J. Sci., v. 18, no. 1, 164-165.
    Abstracted from Edinburgh Q. J.

Pritchett, Jacob
8599   1850, Improvements in ore washers: Franklin Inst. J., v. 19 (3s), 247 (6 lines).

Proctor, William, jr.
8600   1840, Description and analysis of a mineral, a saline incrustation, and a bark, brought from the country lying between Santa Fe, and the head waters of the Arkansas River: Am. J. Pharmacy, v. 6 (ns), 108-114.

Prolix, Peregrine (pseudonym; see Nicklin, P. H.)

Proposals for making a geological and mineralogical survey of Pennsylvania, for publishing a series of geological maps, and forming state and county geological and mineralogical collections.
8601   1827, Phila., n.p., 20 p.

Proust, Joseph Louis (1754-1826)
8602   1822, On the existence of mercury in the waters of the ocean: NY Med. Phys. J., v. 1, 250-251.
    Abstracted from J. Sci. Arts.

Prout, Hiram A. (d. 1862)
8603   1846, Gigantic Palaeotherium [from St. Louis, Missouri]: Am. J. Sci., v. 2, no. 2, 288-289, fig.
8604   1847, Description of a fossil maxillary bone of Palaeotherium, from near White River: Am. J. Sci., v. 3 (ns), no. 2, 248-250, 2 figs.
8605   1848, The geology and mineral resources of the state of Missouri: West. J. Ag., v. 1, no. 1, 6-9.
8606   1848, On the classification of the mineral masses composing the earth: West. J. Ag., v. 1, no. 3, 115-123, table.
    Includes stratigraphic correlation chart for New York, Europe, and the western states.
8607   1848, On the economical geology of the state of Missouri: West. J. Ag., v. 1, no. 8, 429-439.
8608   1848, On the geology of the Valley of the Mississippi: West. J. Ag., v. 1, no. 5, 242-252, plate.

Providence Franklin Society (Rhode Island)
8609   1830, Providence Franklin Society: Am. J. Sci., v. 18, no. 1, 195-196.
    Lists officers and notices.
8610   1831, Proposed exchanges by the Franklin Society of Providence, R. I.: Am. J. Sci., v. 20, no. 2, 415 (7 lines).
    Exchange of mineral specimens proposed.

Puer (pseudonym; see 8303)

Putney, James
8611   1844, The burning-wells of Kenawha: West. J. Medicine Surg., v. 1 (ns), 373-375.

Puvis, Marc Antoine (1776-1851)
8612   1835, Essay on the use of lime as a manure: NY, Scatcherd and Adams, 25 p. Also Shellbank, Va. edition from Farm's Reg. Office.
    For review see 656, 4268, 8749. Also extracted in Am. Farm., v. 2 (2s), 220-222; 229-231; 235-238; and 245-246.

Q., J. B. (see 8488)

The Quarterly Journal and Review (Cincinnati, 1846)
8613   1846, The lead region - geological exploration [review with excerpts of Geological exploration of part of Iowa, Wisconsin, and Illinois by D. D. Owen (see 8070)]: v. 1, 223-241.

Quetelet, Lambert Adolphe Jacques (1796-1874)
8614   1835, Magnetism: Am. J. Sci., v. 28, no. 1, 61-63.
       In British Assoc. Proc. On the earth's magnetic field.
8615   1848, Dip of the magnetic needle: Am. J. Sci., v. 6 (ns), no. 3, 446 (5 lines).
8616   1848, Magnetic perturbations: Am. J. Sci., v. 6 (ns), no. 2, 296 (10 lines).
       Abstracted from l'Institut.

Quillard, Claude S.
8617   1845, An improvement in the furnaces for the manufacture of malleable iron directly from the ore: Franklin Inst. J., v. 10 (3s), 107-108.

Quinby, J. B.
8618   1841, Mineral resources and physical geography of that portion of Peru which embraces the eastern ridges of the Andes: Phila. Acad. Nat. Sci's Proc., v. 1, no. 6, 82-84.
8619   1841, Spontaneous combustion of bituminous coal: Phila. Acad. Nat. Sci's Proc., v. 1, no. 8, 121.

Quincy, Josiah (1772-1864)
8620   1840, History of the Harvard College mineralogical cabinet: in History of Harvard College by J. Quincy, Cambridge, J. Owen, v. 2, 399-402; 544-545.

R., J. L. (see 279; i.e. J. L. Riddell?)

R., W. D. (see 1229)

Raffles, Sir Thomas Stafford Bingley (1781-1826)
8621   1817, Account of the Sunda Islands and Japan: J. Sci. Arts, v. 2, no. 1, 190-198.

Rafinesque, Constantine Samuel (1783-1840)
8622   1817, Description of Glomeris eurycephalus: Am. Mon. Mag. and Crit. Rev., v. 1, 452.
8623   1818, Discoveries in natural history, made during a journey through the western region of the United States: Am. Mon. Mag. and Crit. Rev., v. 3, no. 5, 354-356.
8624   1818, Farther account of discoveries in natural history, in the western states: Am. Mon. Mag. and Crit. Rev., v. 4, no. 1, 39-42.
8625   1818, [Review of An index to the geology of the northern states by A. Eaton (see 3324)]: Am. Mon. Mag. and Crit. Rev., v. 3, no. 3, 175-178.
8626   1818, [Review of Observations on the geology of the United States by W. Maclure (see 6619 and 6620)]: Am. Mon. Mag. and Crit. Rev., v. 3, no. 1, 41-44.
8627   1820, Annals of nature or annual synopsis of new genera and species of animals, plants, &c. discovered in North America: Lexington, Kentucky, C. S. Rafinesque, 16 p.
       Several new species of fossils from Kentucky are listed.
8628   1820, Carbonat of barytes [from Lexington, Kentucky]: Am. J. Sci., v. 2, no. 2, 374 (5 lines).
8629   1820, On the geology of the Valley of the Mississippi, No. II: West. Rev. and Misc. Mag., v. 2, 321-329.
       From the notes of the late J. D. Clifford (see 2446).
8630   1821, Description of a fossil medusa forming a new genus Trianisities cliffordi [from Lexington, Kentucky]: Am. J. Sci., v. 1, no. 2, 285-287, fig.
8631   1821, New mineral species discovered or ascertained in 1818 [from European sources]: West. Minerva, v. 1, 37-38.
8632   1821, [Note on hydraulic limestone]: West. Minerva, v. 1, 38 (15 lines).
       Following article by H. E. De Witt Clinton on the same subject (see 2449).
8633   1822, The cosmonist - no. 1. On a large fossil trilobite of Kentucky: Kentucky Gaz., v. 1, no. 5, 2-3, column 6 and 1.
8634   1822, The cosmonist - no. 7. On the geological meteoric formation: Kentucky Gaz., v. 1, no. 11, 2, column 6.
8635   1824, Ancient history, or annals of Kentucky; with a survey of the ancient monuments of North America, and a tabular view of the principal languages and primitive notions of the whole earth: Frankfort, Kentucky, for the author, iv, 39 p.
       For review see 2376.
8636   1824, Ancient annals of Kentucky; or, introduction to the history and antiquities of the state of Kentucky: in The history of Kentucky by H. Marshall, Frankfort, Kentucky, v. 1, ix-xii, 13-47.
8637   1831, Continuation of a monograph of the bivalve shells of the river Ohio, and other rivers of the western states, containing 46 species from 75 to 121, including an appendix on some bivalve shells of the rivers of Hindoostan, with a supplement on the fossil bivalve shells of the western states, and the Tulosities, a new genus of fossils: Phila.
       Not seen.
8638   1831, Enumeration and account of some remarkable natural objects in the cabinet of C. S. Rafinesque in Philadelphia, being animals, plants, shells, and fossils, collected by him in North America, between 1816-1831: Phila., William Sharpless (printer), 9 p., illus.
8639   1832, On the salses of Europe and America: Atl. J. and Friend Know., v. 1, no. 2, 73-74.
       On salses, or mud volcanoes.
8640   1832, Ancient volcanoes of North America: Atl. J. and Friend Know., v. 1, no. 4, 137-140.
8641   1832, Arcibites rhombifera, a new encrinite, from the cabinet of Dr. Cohen, of Baltimore [found at Lockport, New York]: Atl. J. and Friend Know., v. 1, no. 3, 116.
8642   1832, Atlantic review: Atl. J. and Friend Know., v. 1, 34-35.
       List of recent scientific literature.
8643   1832, The caves of Kentucky: Atl. J. and Friend Know., v. 1, 27-30, fig.
8644   1832, Description of some fossil teeth found in a cave in Pennsylvania: Atl. J. and Friend Know., v. 1, no. 3, 109-110.
8645   1832, Extracts of a series of geological letters to Prof. Al. Brongniart: Atl. J. and Friend Know., v. 1, no. 2, 65-67.
       Rafinesque enumerates four "schools of geology" used in the United States, including his own.

Rafinesque, C. S. (cont.)
8646   1832, Fossils of Sherman Creek [Pennsylvania]: Atl. J. and Friend Know., v. 1, no. 3, 122.
8647   1832, Geological strata of Ohio and Kentucky: Atl. J. and Friend Know., v. 1, no. 1, 30-31.
8648   1832, Geological survey of the Alleghany Mountains of Pennsylvania, in 1818, from west to east: Atl. J. and Friend Know., v. 1, no. 3, 105-109.
8649   1832, Geology and history. History of China before the flood: Atl. J. and Friend Know., v. 1, no. 1, 22-26.
8650   1832, Gold mines of North America: Atl. J. and Friend Know., v. 1, no. 1, 31.
8651   1832, Licks and sucks of Kentucky: Atl. J. and Friend Know., v. 1, no. 2, 74-77.
8652   1832, Lucilities nigra, a new univalve shell, from the Alleghany Mountains of Pennsylvania: Atl. J. and Friend Know., v. 1, no. 3, 116-117.
       From Bedford Springs.
8653   1832, New fossil shells of Pennsylvania [at Sherman Creek in the Alleghany Mountains]: Atl. J. and Friend Know., v. 1, no. 4, 142-143, 7 figs.
8654   1832, On Cavulites and Antrosites [from eastern Kentucky]: Atl. J. and Friend Know., v. 1, no. 2, 70-71.
8655   1832, On the false rhinoceros of Featherstonhaugh and Harlan: Atl. J. and Friend Know., v. 1, no. 3, 114-115.
       Rhinoceros alleghaniensis disputed.
8656   1832, On the genera of fossil trilobites or glomerites of North America: Atl. J. and Friend Know., v. 1, no. 2, 71-73.
8657   1832, On the Lamellites N. G. of American fossils [from Glen's Falls, New York]: Atl. J. and Friend Know., v. 1, no. 2, 74.
8658   1832, Oolites of North America: Atl. J. and Friend Know., v. 1, no. 4, 140-141.
8659   1832, Relative age of mountains: Atl. J. and Friend Know., v. 1, no. 3, 105.
8660   1832, Remarks on the geodes and geodites: Atl. J. and Friend Know., v. 1, no. 2, 69-70.
8661   1832, Remarks on the Monthly journal of geology and natural science of G. W. Featherstonhaugh, for May 1832, (but only published in July) [review of 3746]: Atl. J. and Friend Know., v. 1, no. 3, 110-114.
8662   1832, Remarks on the siliceous fossils of North America: Atl. J. and Friend Know., v. 1, no. 2, 67-69.
8663   1832, Stratipora and Flexulites: Atl. J. and Friend Know., v. 1, no. 4, 143.
       On new species of fossil corals from the Appalachian Mountains.
8664   1832, Visit to Big Bone Lick, in 1821: Mon. Am. J. Geology, v. 1, no. 8, 355-358.
8665   1832, Vulgar names of fossils and petrifactions in North America: Atl. J. and Friend Know., v. 1, no. 4, 137.
8666   1833, Huge water volcano [geyser in Guatemala]: Atl. J. and Friend Know., v. 1, no. 7, 201-202.
8667   1833, Notes [on "Some essential views of geology" by S. Hibbert (see 5031)]: Atl. J. and Friend Know., v. 1, no. 7, 194-195.
8668   1833, Of the Atlantic plains of North America: Atl. J. and Friend Know., v. 1, no. 8, 209-211.
8669   1836, A life of travels and researches in North America and South Europe; or outlines of the life, travels and researches of C. S. Rafinesque ... : Phila., printed for the author by F. Turner, 148 p.
8670   1836, The world, or instability. A poem: Phila., C. S. Rafinesque, 248 p. Also "second edition," Phila., J. Dobson, 248 p.
       "A scientific poem intended to teach the theory of evolution."
8671   1840, The good book, and amenities of nature, or annals of historical and natural sciences: Phila., Printed for the Eleutherium of Knowledge, 84 p., "with 1000 figures."
       Geology and paleontology especially on 5-12, 28-36, 66-67.

Raines, George W. (i.e. George Washington Rains?, 1817-1898)
8672   1848, Notes on the mines of a portion of the state of Mexico: Am. J. Sci., v. 6 (ns), no. 3, 427-429.
8673   1850, Geology of Florida: Scien. Am., v. 5, no. 21, 165, fig.

Rammelsberg, Karl Friedrich (1813-1840)
8674   1846, Arsenical antimony [from Allemont, Frankreich]: Am. J. Sci., v. 2 (ns), no. 3, 418 (4 lines).
       Abstracted from Berzelius' Jahresbericht.
8675   1846, Brochantite and krisuvigite: Am. J. Sci., v. 2 (ns), no. 3, 417.
       On the equivalence of these two minerals. Abstracted from Berzelius' Jahresbericht.
8676   1846, Phacolite [from Leipa, Bohemia]: Am. J. Sci., v. 2 (ns), no. 3, 417 (4 lines).
       Abstracted from Berzelius' Jahresbericht.
8677   1846, Quantitative determination of lithia: Am. J. Sci., v. 1 (ns), no. 2, 260-261.
       Abstracted from Chemical Gaz. by J. L. Smith.
8678   1849, Analysis of pyrophyllite of Spaa: Am. J. Sci., v. 8 (ns), no. 1, 122 (4 lines).
       Abstracted from Poggendorff's Annalen.
8679   1850, Analysis of the schorlomite of Shepard [from Arkansas]: Am. J. Sci., v. 9 (ns), no. 2, 429 (9 lines).
       Abstracted from Poggendorff's Annalen.

Ramsay, Sir Andrew Crombie (1814-1891)
8680   1848, On the submergence of ancient land in Wales; the accumulation of newer strata around and above it; and the re-appearance of the same land by elevation and denudation: Franklin Inst. J., v. 16 (3s), 355-357.
       Abstracted from London Ath.

Ramsay, David (1749-1815)
8681   1796, A sketch of the soil, climate, weather, and diseases of South Carolina: Charleston, SC, W. P. Young (printer), [4], 30 p., 3 tables.
       Evans number 31071.

Ramsay, J. G. M.
8682 1832, An essay on the medical topography of East Tennessee: Transylvania J. Medicine and Associated Sci's, v. 5, 363-375.

Randall, Henry Stephens (1811-1876)
8683 1846, Running notes, agricultural and geological, of a trip to Carbondale [Pennsylvania]: Am. Q. J. Ag. Sci., v. 4, 1-19.

Randall, Samuel Sidwell (1809-1881)
8684 1846, Incentives for the cultivation of the science of geology. Designed for the use of the young: NY, Greeley and McElrath, 189 p.
For review see 5463, 7151 and 9850.

Randolph, Jeremiah F. (see 2476 by C. D. Colden)

Raspail, Francois Vincent (1794-1878)
8685 1829, Belemnites [from the Alps]: Am. J. Sci., v. 17, no. 1, 184 (7 lines).
Abstracted by J. Griscom from Annales des Mines.

Rauch
8686 1839, Volborthite, a new mineral [from Russia]: Am. J. Sci., v. 36, no. 1, 187.
Abstracted from l'Institut.

Ravenel, Edmund (1797-1871)
8687 1841, Description of two species of fossil Scutella from South Carolina: Phila. Acad. Nat. Sci's Proc., v. 1, no. 6, 80-82.
8688 1842, Description of two new species of fossil Scutella from South Carolina: Phila. Acad. Nat. Sci's J., v. 8, pt. 2, 333-336, 2 figs.
8689 1844, Description of some new species of fossil organic remains, from the Eocene of South Carolina: Phila. Acad. Nat. Sci's Proc., v. 2, no. 5, 96-98.
8690 1848, Echinidae, recent and fossil, of South Carolina, January 1848: Char., Burges and James (printers), 4 p., 2 plates.
8691 1849, On the medical topography of St. John's, Berkley, S. C., and its relations to geology: Char. Med. J. and Rev., v. 4, 697-704.
8692 1850, On the recent squalidae of the coast of South Carolina, and a catalogue of the recent and fossil echinoderms of Carolina: Am. Assoc. Adv. Sci. Proc., v. 3, 159-161.

Raymond, Benjamin
8693 1823, On the soils, productions, &c. of the County of St. Lawrence: NY State Board Ag. Mem., v. 2, 84-88.
Also in Plough Boy, v. 4, no. 43, 366-367.

Read, Stephen
8694 1848, Canaan (Ct.) native iron: Am. J. Sci., v. 5 (ns), no. 2, 292-293.

Recluz, C.
8695 1840, Note upon the white and opaque variety of amber: Am. J. Pharmacy, v. 6 (ns), 148-149.

Redfield, Heman J.
8696 1815, Sulphur springs [from Canandaigua, New York]: Lit. and Philos. Rep., v. 2, 126-129.

Redfield, John Howard (1815-1895)
8697 1840, Reports on the shells and minerals presented by Dr. Brinckerhoff to the New York Lyceum of Natural History: Am. J. Sci., v. 38, no. 1, 198-202.
With report by J. Delafield (see 3055).
8698 1845, A catalogue of the fossil fish of the United States, at present known, with descriptions of those which occur in the New Red Sandstone: Assoc. Am. Geol's Nat's Proc., v. 6, 16 (6 lines).

Redfield, William Charles (1789-1857)
See also discussion of article by A. Guyot (4612).
8699 1836, The fossil fishes of Connecticut and Massachusetts: NY.
Not seen.
8700 1837, Fossil fishes of Connecticut and Massachusetts, with a notice of a undescribed genus: NY Lyc. Nat. History Annals, v. 4, no. 1 and 2, 35-40, 2 plates.
Also reprint edition.
8701 1838, Fossil fishes in Virginia [from Richmond]: Am. J. Sci., v. 34, no. 1, 201 (12 lines).
8702 1838, Some account of two visits to the mountains in Essex County, New York, in the years 1836 and 1837; with a sketch of the northern sources of the Hudson: Am. J. Sci., v. 33, no. 2, 301-323, map.
8703 1840/1841, [On fish fossils]: Am. Rep. Arts Sci's Manu., v. 1, 257; 343; 345; v. 3, 414-415.
In NY Lyc. Nat. History Proc.
8704 1840, [On the formation of bars in river mouths]: Am. Rep. Arts Sci's Manu's, v. 1, 420.
In NY Lyc. Nat. History Proc.
8705 1841, [Fossil fishes from the Connecticut River valley]: Am. J. Sci., v. 41, no. 1, 164-165.
Also in Assoc. Am. Geol's Proc., v. 2, 9-10. Also in Assoc. Am. Geol's Nat's Rept., v. 1, 17, 1843.
8706 1841, [Fossil shells from Tertiary marl beds at Washington, Beaufort County, North Carolina]: Am. J. Sci., v. 41, no. 1, 161 (8 lines).
Also in Assoc. Am. Geol's Nat's Proc., v. 2, 6 (9 lines). Also in Assoc. Am. Geol's Nat's Rept., v. 1, 14 (8 lines), 1843.
8707 1841, [New York City flagstone - with fossils]: Am. J. Sci., v. 41, no. 1, 164 (6 lines).
Also in Assoc. Am. Geol's Nat's Proc., v. 2, 9 (6 lines). Also in Assoc. Am. Geol's Nat's Rept., v. 1, 17 (6 lines), 1843.

Redfield, W. C. (cont.)
8708   1841, Short notices of American fossil fishes: Am. J. Sci., v. 41, no. 1, 24-28.
8709   1842, [Origin of drift near the city of New York]: Am. J. Sci., v. 43, no. 1, 152 (13 lines).
        Also in Assoc. Am. Geol's Nat's Proc., v. 3, 7. Also in Assoc. Am. Geol's Nat's Rept., v. 1, 46, 1843.
8710   1842, [Raindrops and fossil fish from the red sandstone of New Jersey]: Am. J. Sci., v. 43, no. 1, 172.
        Also in Assoc. Am. Geol's Nat's Proc., v. 3, 27. Also in Assoc. Am. Geol's Nat's Rept., v. 1, 65, 1843.
8711   1843, [Cretaceous fossil in deep well in Brooklyn, New York]: Am. J. Sci., v. 45, no. 1, 156 (7 lines).
        In Assoc. Am. Geol's Nat's Proc. See also note by J. N. Nicollet (7873).
8712   1843, [Dana's coral reef theories]: Am. J. Sci., v. 45, no. 2, 311-312.
        In Assoc. Am. Geol's Nat's Proc. See also 2805.
8713   1843, Notice of newly discovered fish beds and a fossil foot mark in the red sandstone formation of New Jersey: Am. J. Sci., v. 44, no. 1, 134-136, fig.
8714   1843, [On a double system of diluvial markings on American rocks]: Am. J. Sci., v. 45, no. 2, 333 (11 lines).
        In Assoc. Am. Geol's Nat's Proc. See also note by J. Hall (4685).
8715   1843, [On drift accumulation]: Am. J. Sci., v. 45, no. 2, 325-327.
        In Assoc. Am. Geol's Nat's Proc. See also note by C. T. Jackson (5602).
8716   1843, [On mobility of heated waters at bottom of the ocean]: Am. J. Sci., v. 45, no. 1, 138-140.
        In Assoc. Am. Geol's Nat's Proc.
8717   1843, Remarks on some new fishes and other fossil memorials from the New Red Sandstone of New Jersey: Am. J. Sci., v. 45, no. 2, 314-315.
        In Assoc. Am. Geol's Nat's Proc. With discussion by E. Hitchcock (5148) and by H. D. Rogers, 315-316.
8718   1844, [On drift phenomena in Portage County, Ohio]: Am. J. Sci., v. 47, no. 1, 120-121.
        Also in Assoc. Am. Geol's Nat's Proc., v. 5, 27-28.
8719   1847, On the remains of marine shells of existing species, found interspersed in deep portions of the hills of drift and boulders, in the heights of Brooklyn, on Long Island, near New York City: Am. Q. J. Ag. Sci., v. 6, 213-217.
        Also in: Am. J. Sci., v. 5 (ns), no. 1, 110-111 (1848).
        In Assoc. Am. Geol's Nat's Proc. With discussion by C. B. Adams, E. Desor, E. Hitchcock, H. D. Rogers, and B. Silliman.
8720   1848, [Fossils from drift at Brooklyn, New York]: Lit. World, v. 2, 228-229.
        In Assoc. Am. Geol's Nat's Proc.
8721   1848, [Glacial and iceberg theories of drift]: Lit. World, v. 2, 229.
        In Assoc. Am. Geol's Nat's Proc.
8722   1848, [On drift of western Long Island, New York]: NY J. Medicine and Collateral Sci's, v. 10, 338 (11 lines).
        In NY Lyc. Nat. History Minutes.
8723   1849, [Geological action of the tides]: Am. Assoc. Adv. Sci. Proc., v. 1, 28-30.
8724   1850, On some fossil remains from Broome County, N. Y.: Am. Assoc. Adv. Sci. Proc., v. 2, 255-256.
        Abstracted in Ann. Scien. Disc., v. 1, 287-288.

Reed, John (1757-1845)
8725   1847, The perils of mining: Scien. Am., v. 2, no. 45, 358.

Reed, Stephen
8726   1845, A chain of erratic serpentine rocks passing through the centre of Berkshire County, Mass.: Assoc. Am. Geol's Nat's Proc., v. 6, 12.
        Also in Am. Q. J. Ag. Sci., v. 2, 140-141, 142. With discussion by H. D. Rogers.

v (Assistant-Surgeon)
8727   1841, A late visit to the volcano of Kirauea, in Hawaii, one of the Sandwich Islands: Mus. For. Liter. Sci., v. 43, 236-241.

Reese, David Meredith (1800-1861)
8728   1849, Introductory observations: in Elements of geology by D. Page (see 8108).

Reeves, Mark
8729   1826, On the New Jersey marls: Phila. Soc. Promoting Ag. Mem., v. 5, 1-7.

Regnault, Henri Victor (1810-1878)
8730   1839, Analysis of two micas, with bases of potash and lithium: Franklin Inst. J., v. 23 (ns), 347-348.
        Abstracted from Annales des Mines.
8731   1839, Mica containing potash and lithia: Am. J. Sci., v. 37, no. 2, 356-357.
        Abstracted from London and Edinburgh Philos. Mag.

Reich (Prof.)
8732   1840, Electrical currents in metalliferous veins: Am. J. Sci., v. 38, no. 1, 120 (6 lines).
        In British Assoc. Proc.

Reid [i.e. David Boswell Reid?, 1805-1863]
8733   1844, Analyses of three sulphur springs at Sharon, Schoharie County, N. Y.: Phila. Acad. Nat. Sci's Proc., v. 2, no. 5, 120-122.

Reid, Stephen (see Reed, Stephen)

Reimarus, Johann Albert Heinrich (1729-1814)
8734   1808, Questions relative to the American elephantine bones, &c.: Phila. Med. Phys. J., v. 3, pt. 1, 137-138.

Reinhardt, D.
8735   1829, Gold mines of North Carolina: Am. J. Sci., v. 16, no. 2, 360-363.

Reinhardt, J. C.
8736   1846, Report of J. C. Reinhardt, naturalist: Nat'l Inst. Prom. Sci. Bulletin, v. 4, 533-567.
       On the voyage of the U. S. S. Constitution to China from 1844 to 1846.

Reiss, M.
8737   1849, Phosphorescence of the diamond: Am. J. Sci., v. 7 (ns), no. 3, 433 (5 lines).
       Abstracted from "Annuaire de Ch."

A relation of a remarkable providence, which fell out at the time of the great earthquake at Jamaica, very proper to be reflected on at this time of immenent [sic] danger, and after having lately had a warning from God, by a smaller shock of the like kind in this place. A paper proper to be given away:
8738   1755, Phila., James Chattin.
       Evans number 7553.

Relfe, James H.
8739   1846, Mineral lands, sale of ... May 4, 1846: US HED 591, 29-1, v. 3 (490), 51 p., folded map.
       Contains report by D. Houghton (see 5319) and C. T. Jackson (see 5628) and map (see 4302).

Relics from the wreck of a former world; or, splinters gathered on the shores of a turbulent planet. ... :
8740   1847, NY, W. H. Graham, illus.
       Not seen.

The Religious Magazine, or Spirit of Foreign Theological Journals and Reviews (Philadelphia, 1828-1830)
8741   1829, Scriptural geology [review with excerpts of Scriptural geology; or, geological phenomena consistent only with the literal interpretation of the sacred Scriptures (London edition)]: v. 3, 445-452.

Rennie, George
8742   1818, Account of the mineral springs of Caldas de Rainha, in the north of Portugal, with an analysis of the water: J. Sci. Arts, v. 5, no. 1, 60-63.

Rennie, Robert
8743   1815, On the natural history and origin of peat: Mass. Ag. Repos. J., v. 3, no. 4, 281-313.

Rensselaer, Jeremiah van (see Van Rensselaer, Jeremiah)

Renssalaer, Stephen van (see Van Rensselaer, Stephen)

Renwick, Edward Sabine (1823-1912; see 1500 to 1502 under Barrelville Mining Company)

Renwick, James (1790-1863)
       See also 1500 to 1502 under Barrelville Mining Company.
8744   1823, Examination of a new mineral from Andover Furnace, Sussex County, New-Jersey: NY Lyc. Nat. History Annals, v. 1, pt. 1, no. 2, 37-42.
       Abstracted in NY Med. Phys. J., v. 2, 510-511.
8745   1824, Notes on the geology of Trenton Falls [New York]: NY Lyc. Nat. History Annals, v. 1, pt. 1, no. 6, 185-189.
       See also note by J. E. Dekay (3022).
8746   1824, Torrelite, a new mineral [from Sussex County, New Jersey]: Am. J. Sci., v. 8, no. 1, 192-193.
8747   1826, [Review of "An account of experiments to determine the figure of the earth" by E. Sabine]: Franklin Inst. J., v. 1, 164-172, 203-213.
8748   1832, Adaptation of Stony Creek Valley as a site for a national foundry: NY, broadside.
       Dated "Columbia College, 1832."
       Reprinted in Am. Rr. J., v. 11, 39-41, 1840.
       On the Pennsylvania coal region.
8749   1836, Reviews of new works on calcareous manures [reviews of Essay on calcareous manures by E. Ruffin (see 9088) and On the use of lime as a manure by M. Puvis (see 8612)]: Farm's Reg., v. 4, 95-104.
8750   1837, [Extracts from a report on lands of the Maryland Mining Company]: in Extracts ..., by G. W. Hughes (see 5355).
8751   1838, Outlines of geology, prepared for the use of the Junior Class of Columbia College: NY, Henry Ludwig, 96 p.
       For review see 694.
8752   1839, [Extracts from a report on lands of the Maryland Mining Company]: in Loan of the Maryland Mining Company (see 6812).
8753   1847, [Extract from a report on the lands of the Phoenix Mining Company]: in Documents ... (see 8328), 29-30.

Repetti, Emanuel
8754   1822, A soft crystal of quartz [from Vossa del l'Angelo quarry, Italy]: Am. J. Sci., v. 5, no. 2, 394-395.
       Abstracted from Bibliothèque Universelle.

Retzius, Anders Adolf (1796-1860)
8755   1837, Fossil infusoria [used as food]: Franklin Inst. J., v. 20 (ns), 370.

Reveno (i.e. Mariano Rivero?; see 1882 with Boussingault)

Revere, Joseph Warren (1812-1880)
8756   1849, A tour of duty in California including a description of the gold region, and an account of the voyage around Cape Horn, with notices of Lower California, the Gulf and Pacific Coasts, etc.: Bos. and NY, J. H. and C. S. Francis and Co., 305 p., illus., maps.
    Edited by Joseph N. Balestier. For reviews with excerpts see 5227 and 5480.

Reynolds, Jeremiah N. (1799-1858)
8757   1827, Remarks on a review of Symmes' theory, which appeared in the American quarterly review, by a "citizen of the United States": Wash., Gales and Seaton, 75 p.
    See 10079.

Reynolds, William G. (M.D.)
8758   1819, Outline of a theory of meteors: Am. J. Sci., v. 1, no. 3, 266-276.

Reynoso, Bernard
8759   1850, Nature and office of earth and soil: NE Farm. A Semi-Mon. J., v. 2, no. 11, 173-175.

Rhode Island. State of
8760   1838, A report of the important hearing on the memorial of the New England Coal Mining Company, for encouragement from the state; ... Together with the report of the committee unanimously adopted by the assembly, in favor of the prayer of the memorialists and petitioners, and of a geological and agricultural survey of the state: NE, viii, 148 p.
    Contains excerpts from the Massachusetts geology report of E. Hitchcock (see 5128), and numerous testimonials and petitions in favor of mining operations in the state.
    F. Bryant, Committee Chairman.
8761   1839, [Act to incorporate the geological survey]: Rhode Island Gen. Assembly, 1839.
8762   1840, [Act to publish the geological report of C. T. Jackson (see 5571)]: Rhode Island Gen. Assembly, 1840.

Rhode Island Coal Company
8763   1809, An act to incorporate the Rhode Island Coal Company: Bos., J. Eliot (printer), 24 p.
8764   1814?, Observations on the Rhode Island coal and certificates with regard to its qualities, value, and various uses: Bos., 17 p.
8765   1825, Report of the Committee of the Rhode Island Coal Company: NY, J. Seymour, 16 p.
    See also Circular by H. D. Sedgwick (9389).

Rhode (Mr., of New Jersey)
8766   1797, Observations on erecting and working mines in this country ... : NY Mag., v. 2 (ns), 449-453.

Rhodius, R.
8767   1849, Analysis of ehlite: Am. J. Sci., v. 7 (ns), no. 1, 113 (7 lines).
    Abstracted from "Wöhl. & Liebig's Annal." On the equivalence of ehlite and libethenite.

Ribero, Mariano (see Rivero, Mariano)

Richards (Mr.)
8768   1850, [Plumbago from the Southampton Mine, Bucks County, Pennsylvania]: Am. Philos. Soc. Proc., v. 5, no. 45, 171 (9 lines).

Richards, Joseph
8769   1839, Specification of a patent for an improvement in the mode of smelting iron ores, and in the manufacturing of steel: Franklin Inst. J., v. 24 (ns), 385.

Richardson (Mr.)
8770   1832, Copper ore of Strafford, Vt. &c.: Am. J. Sci., v. 21, no. 2, 383-384.

Richardson, George Fleming (1796-1848)
8771   1844?, First lessons on geology: NY, Wiley and Putnam.
    Not seen. For review see 2719 and 7964.

Richardson, Sir John (1787-1865)
8772   1828, Appendix. Topographical and geological notices: in Narratives, a second expedition to the shores of the Polar Sea in the years 1825, 1826, and 1827 by J. Franklin, 263-318.

Richardson, S.
8773   1850, [Report on the mineral lands of the Siskowit Mining Company]: in Charter (see 9843), 19-23.

Richardson, Thomas (1816-1867)
8774   1836, On donium, a new substance discovered in davidsonite [from Aberdeen]: Am. J. Sci., v. 31, no. 1, 163.
    Abstracted from London Ath.
8775   1840, On the composition of idocrase: Am. J. Sci., v. 38, no. 1, 120.
    In British Assoc. Proc.

Richardson, William Harvie (1795-1876)
8776   1850, Report. Governor's message relative to the world's fair to be held in London in 1851: Va. House of Delegates, Document 25 for 1850-1851, 5-11.
    On plans to send a large geology map of Virginia to the London World's Fair.

Richter (of Freiburg)
8777   1832, Pelokonite, a new mineral [from Terra Amarillo, Chile]: Am. J. Sci., v. 22, no. 2, 387 (10 lines).
    Abstracted from Jameson's J. by C. U. Shepard.

Richter (cont.)
8778 1834, Pelokonite [from Chile]: Am. J. Sci., v. 26, no. 2, 386-387 (6 lines).

Riddell, J. R.
8779 1831, Alum [from Erie, Pennsylvania]: Hazard's Reg. Penn., v. 8, 271-272.

Riddell, John Leonard (1807-1867)
See also 279, signed "J. L. R."
8780 1833, Observations on the geology of the central parts of the state of Ohio: West. J. Med. Phys. Sci's, v. 7, no. 3, 356-368.
Abstracted in Eclec. J. Sci., v. 2, 469-473 (1834).
8781 1835, On the vegetable origin of coal: Cin. Mirror, v. 4, 98-99, 104.
See also debate by W. B. Powell (8559 to 8561) and W. Wood (see 11035 and 11036).
8782 1836, A geological ramble to Little Mountain, near Cleveland: Cin., 6 p.
8783 1836, Remarks on the geological features of Ohio, and some of the desiderata which might be supplied by a geological survey of the state: Cin., 12 p.
For abstract see Am. J. Sci., v. 30, no. 2, 394.
8784 1836, Remarks on the geological features of Ohio: West. Mon. Mag., v. 5, 160-172.
8785 1837, Report of John L. Riddell, one of the special committee ... on the method of obtaining a complete geological survey of this state, with miscellaneous observations on the geology of Ohio by Increase A. Lapham: Columbus, Ohio, Ohio Gen. Assembly Executive Document 60, 34 p.
Contains report by I. A. Lapham (see 6045).
8786 1839, Observations on the geology of the Trinity country, Texas, made during an excursion there in April and May, 1839: Am. J. Sci., v. 37, no. 2, 211-217.
8787 1840, Hog wallow prairies: Am. J. Sci., v. 39, no. 1, 211-212.
On large mud cracks and hog-wallows of the prairies.
8788 1846, Deposits of the Mississippi and changes at its mouth: De Bow's Rev., v. 2, no. 6, 433-439.
8789 1850, Remarks on the dynamics of the Mississippi River, and other matters pertaining thereto: New Orleans, office of the "Bee", 15 p.

Ridgway, Thomas S.
8790 1835, Abstract of a specification of a patent for a furnace for preparing and smelting iron ore, with anthracite coal: Am. Rr. J., v. 4, 508-509, woodcut.
Abstracted from Franklin Inst. J.
8791 1835, Furnace for preparing and smelting iron ore, with anthracite coal: Franklin Inst. J., v. 16 (ns), 24, 40-41, fig.
See also review of the patent by W. Hamilton (4746).
8792 1850, Anthracite coal in Massachusetts [at Mansfield]: Ann. Scien. Disc., v. 1, 271 (9 lines).

Ridley, William
8793 1850, Coal formation on the Pacific [at Costa Rica]: Ann. Scien. Disc., v. 1, 266-267.

Ridolfi, Marquis of
8794 1816, Method of separating platina from the other metallic substances which are found with it in the state of ore: Eclec. Rep. and Analytic Rev., v. 6, 539-540.
Translated by London J. Sci. Arts from Giornale di Scienze ed Arti. Also in J. Sci. Arts, v. 1, no. 2, 259-260 (1817).
8795 1817, On the native caustic lime of Tuscany: J. Sci. Arts, v. 1, no. 2, 260-261.

Riess, Peter Theophil (1804-1883)
8796 1846, Phosphorescence of the diamond: Am. J. Sci., v. 1 (ns), no. 2, 267 (6 lines).

Riley (see 10060 with S. Stutchbury)

Rio, Andres Manuel del (see Del Rio, Andres Manuel)

Risk, T. F.
8797 1849, Geological survey of Missouri: West. J. Ag., v. 2, no. 5, 333-337.

Rive, Auguste Arthur de la (1801-1873)
8798 1849, On the diurnal variation of the magnetic needle, and the Aurora Borealis: Franklin Inst. J., v. 17 (3s), 284 (6 lines); v. 18 (3s), 40-46.
Abstracted from Annales de Chimie.

Rivero, Mariano Eduardo de (1795-1857)
8799 1822, The muriate of copper and nitrat of soda of Peru [Tarapaca mines]: Am. J. Sci., v. 5, no. 2, 387-388.
8800 1822, A new mineral [on humboldtine from Berlin, Bohemia]: Am. J. Sci., v. 5, no. 1, 193.
Abstracted by J. Griscom. Ascribed to M. Riviero (i.e. Rivero).
8801 1825, and A. Humboldt, Analysis of the waters of the Rio Vinagre, in the Andes of Popayan; with geognostic and physical illustrations of some phenomena which are exhibited by sulphur, sulphuretted hydrogen, and water, in volcanoes: Bos. J. Phil. Arts, v. 2, no. 5, 460-466; no. 6, 513-521.
From Philos. Mag.
8802 1829, View, & topographical plan, of the new town of the Hill of Pasco taken from the Lake of Quiulacocha, 1827: Am. J. Sci., v. 17, no. 1, colored geological map, 27.3 x 18.6 cm.
In 8803.
8803 1829, Sketch of the mine of Pasco: Am. J. Sci., v. 17, 43-63.
Contains introduction by B. Silliman (see 9683) and maps (see 8802).

Rives, Thomas
8804 1832, Machine for washing and separating gold, from earthy matters: Franklin Inst. J., v. 10 (ns), 82-83.

Rivot, Louis Edouard (1820-1869)
8805   1848, and Phillips, New metallurgical treatment of the ores of copper: Franklin Inst. J., v. 16 (3s), 342-343.
         Abstracted from "Bull. Soc. Enc. Ind. Nat."
8806   1849, and Phillips, Description of the process ... for smelting copper ores: Franklin Inst. J., v. 17 (3s), 60-68.
         Abstracted from "Civ. Eng. & Arch. J."
8807   1849, Analysis of California gold: Am. J. Sci., v. 8 (ns), no. 1, 128 (14 lines).
         Abstracted from Annales des Mines.

Robb, James (d. 1861)
8808   1841, Geology of the country around the River St. John, in New Brunswick: Am. J. Sci., v. 41, no. 1, 55-56.
         In British Assoc. Proc.

Robbins, James (M. D.)
8809   1845, Letter from James Robbins, M. D.: in First annual report on the geology of the state of Vermont by C. B. Adams (see 23), 79-80.
         On the minerals of Chester, Vermont.

Robert, Louis Eugene (1806-1879)
8810   1836, Extracts from an account of a visit to Iceland: Am. J. Sci., v. 31, no. 1, 167-168.
         Abstracted from Geol. Soc. France Bulletin.
8811   1837, Visit to Iceland: Am. J. Sci., v. 32, no. 1, 196.
         Abstracted from Geol. Soc. France Bulletin.

Roberts, Algernon S.
8812   1849, On the advantages of geology to agriculture: NY State Agri. Soc. Trans., v. 8, 362-365.

Roberts, Martyn J.
8813   1841, On assaying copper by electro-chemical action: Am. Rep. Arts Sci´s Manu´s, v. 3, 118-120.

Roberts, William F.
8814   1846, The anthracite coal trade: Am. Rr. J., v. 19, 539-540; 571-572; 587-588.
8815   1846, Reports upon the West Hazelton and Cattawissa Falls, and the East Mahanoy coal and iron estates, situate in Luzerne and Schuylkill Counties, Pa.: Phila., J. C. Clark (printer), 22 p.
8816   1846, Abstract of a report on the coal and iron estate of the Little Schuylkill Navigation Rail Road and Coal Company, from actual survey made by authority of the managers of the company in October, 1845: Phila., J. C. Clark, 21 p.

Robertson, Felix
8817   1805, Additional observations on the Falls of Niagara and particularly on their (supposed) original position: Phila. Med. Phys. J., v. 1, pt. 2, 61-68.

Robertson, J. J.
8818   1836, Delos - Greece - titanium - iron, &c.: Am. J. Sci., v. 31, no. 1, 175-176.
         Also in Mech´s Mag. and Reg. Inventions, v. 9, 21.

Robinson (Dr., of Armagh)
8819   1840, Geology of the Moon: Franklin Inst. J., v. 25 (ns), 283-285, fig.

Robinson, Fayette (d. 1859)
8820   1849, California and its gold regions; with a geographical and topographical view of the country, its mineral and agricultural resources: NY, Stringer and Townsend, 137 p., map.
         For review see 9964.

Robinson, Samuel (Dr.)
8821   1824-1825, By Dr. Samuel Robinson: Am. J. Sci., v. 8, no. 2, 230-232; v. 9, no. 1, 49-53; v. 10, no. 2, 225-227.
         On mineral localities of New England.
8822   1825, A catalogue of American minerals, with their localities; including all which are known to exist in the United States and British Provinces, and having the town, counties, and districts in each state and province arranged alphabetically. With an appendix, containing additional localities and a tabular view: Bos., Cummings, Hilliard, and Co., vii, 316 p.
         For review see 425, 10758 and 10759. For notice see 406.

Robinson, Solon (1803-1880)
8823   1845, Notes of travel in the West: The Cultivator, v. 2 (ns), 92-94; 124-126; 142-143; 178-179; 239-240; 271-273; 303-304; 334-335; 365-366.
         Includes notes on the soil, mines and topography of Illinois, Indiana, Kentucky, Tennessee, Missouri, and Michigan.

Robinson´s Magazine, a Weekly Repository of Original Papers; and Selections from the English Magazines (Baltimore, 1818-1819)
8824   1819, Journey to Mount Etna: v. 2, 7-11.
         With notes by G Russell (see 9144). Abstracted from Lit. Gaz.

Robison, I.
8825   1836, Notice of the earthquake in Chili [sic] in November, 1822: Am. J. Sci., v. 30, no. 1, 110-113.
         In 654.

Robles, Manuel
8826   1849, Geology of the Isthmus of Tehuantepec: US House Rept. 145, 30-2, 111-117.
         Not seen.

Rochaz, Charles Andre Felix
8827   1849, Certain improvements in treating zinc ores, and in manufacturing oxide of zinc: Franklin Inst. J., v. 17 (3s), 184-185.
Abstracted from London J. Arts.

Rodgers, Miles M.
8828   1846, The farmer's agricultural chemistry: Geneva, NY, G. H. Derby and Auburn, NY, J. C. Derby, 106 p.
8829   1848, Scientific agriculture, or the elements of chemistry, geology, botany, and meteorology, applied to practical agriculture: Rochester, NY, E. Darrow, 280 p.
For review see 4296. Also second edition, 296 p., illus. (1850).

Rodman, W. W.
8830   1838, Crichtonite, in R. I. [at Westerly]: Am. J. Sci., v. 35, no. 1, 180 (7 lines).
8831   1838, Stilbite, chabasie, and other minerals, at Storington, Ct.: Am. J. Sci., v. 35, no. 1, 179-180.

Roe, E. R.
8832   1842, Fluor spar [from Shawneetown, Illinois]: West. J. Medicine Surg., v. 6, 313.

Roemer, Ferdinand (1818-1891)
8833   1846, A sketch of the geology of Texas: Am. J. Sci., v. 2 (ns), no. 3, 358-365.
8834   1847, [Report on the results of a geological tour recently made in Texas]: Am. Q. J. Ag. Sci., v. 6, 211 (7 lines).
In Assoc. Am. Geol's Nat's Proc.
8835   1848, Contributions to the geology of Texas: Am. J. Sci., v. 6 (ns), no. 1, 21-28.
8836   1848, [Notes on the geology of Texas]: Lit. World, v. 2, 228.
In Assoc. Am. Geol's Nat's Proc.
8837   1848, [Notes on the geology of Texas]: NY J. Medicine and Colateral Sci's, v. 10, 339.
In NY Lyc. Nat. History Minutes.
8838   1848, [Notes on the Cretaceous of New Jersey]: NY J. Medicine and Collateral Sci's, v. 10, 339 (7 lines).
In NY Lyc. Nat. History Minutes.

Rogers (Mr.)
8839   1846, Native titanium [from Merthyr-Tydrill, Cornwall, England]: Am. J. Sci., v. 2 (ns), no. 3, 414 (5 lines).

Rogers, H. E. (i.e. Henry Darwin?)
8840   1832, Geological memoir of the country between Massachusetts and Narragansett Bays. In illustration of reports on projected rail-roads from Boston to Providence, and also to Taunton: in Report of the Board of Directors to the stockholders of the Boston and Providence Railroad Company ..., Bos., J. E. Hinckley, 87 p., 2 folded maps, [49]-77.
Sections of the report include: "Report exhibiting the geological features of the country between Boston and the Narragansett waters" (51-65), "On the situation of building-stone" (66-68), "Boston and Providence tables" (69-77).

Rogers, Henry Darwin (1808-1866)
See also articles 4414 with P. B. Goddard and R. C. Taylor, and 8972, 8977, 8992, and 8997 with W. B. Rogers. See also discussions of articles by L. Agassiz (86), J. W. Bailey (1430), J. D. Dana (2805), E. Desor (3677), S. S. Haldeman (4629), J. Hall (4685 and 4731), C. T. Jackson (5208), J. N. Nicollet (7873), W. C. Redfield (8717 and 8719), S. Reed (8726), J. C. Warren (10698 and 10701), and J. Wyman (11084).
8841   1833, On the proposed method of analysing mineral waters by alcohol: Am. J. Pharmacy, v. 5, 279-284.
Discussion of errors in method of Cohen (see 2474).
8842   1834, and A. D. Bache, Analysis of some of the coals of Pennsylvania: Phila. Acad. Nat. Sci's J., v. 7, pt. 1, 158-177.
Also reprint edition.
8843   1834, British Association for the Advancement of Science: Am. J. Sci., v. 26, no. 2, 402 (9 lines).
Announcement of 1834 meeting in Edinburgh.
8844   1835, Description of the principal rocks and statements of their component parts: Farm's Reg., v. 3, 198-201, table.
8845   1835, Geology of North America: Am. J. Sci., v. 28, no. 1, 74-75.
In British Assoc. Proc., abstracted by R. I. Murchison. The Cretaceous of the United States is shown to be of similar age, but a different formation, from that of Europe.
8846   1835, A guide to a course of lectures on geology, delivered in the University of Pennsylvania: Phila., W. P. Gibbons (printer), 43 p.
8847   1835, On the Falls of Niagara and the reasonings of some authors respecting them: Am. J. Sci., v. 27, no. 2, 326-335, fig.
8848   1836, First annual report of the state geologist [of Pennsylvania]: Harrisburg, Penn., Samuel D. Patterson, 22 p.
For review see 9740. Also German edition (not seen).
8849   1836, Marl - its nature and effects: Am. Farm., v. 3 (2s), 82-83.
Abstracted from Franklin Mercury.
8850   1836, Report on the geological survey of the state of New Jersey: Phila., De Silver Thomas and Co., 174 p., plate.
First edition. Also German language edition (not seen).
8851   1836, Report on the geological survey of the state of New Jersey: Phila., De Silver Thomas and Co., 188 p., plate.
Second edition, with glossary (p. 175-188) by C. Lyell (6528). 1000 copies printed.
8852   1836, Report on the geological survey of the state of New Jersey: Freehold, NJ, Bernard Connolly, 157, [1] p.
Contains extracts from Gazetteer of New Jersey by T. F. Gordon (see 4456).

Rogers, H. D. (cont.)
8853 1837, A sketch of what has been accomplished by the geological survey during the year: in NJ Governor's Message for 1837, 130-131.
8854 1838, A sketch of what has been accomplished by the geological survey during the year: in NJ Governor's Message, October, 1838.
8855 1838, Second annual report on the geological exploration of the state of Pennsylvania: Harrisburg, Penn., Packer, Barrett and Parke, 91, [1] p., plate.
Also in Penn. Senate Documents. Also German language edition (Harrisburg, Penn., Gebrundt bei Baab und Hummel, 104 p.). For review with excerpts see 5017 and 4284.
8856 1838, Second annual report on the geological exploration of the state of Pennsylvania: Harrisburg, Penn., Thompson and Clark (printers), 93 p., plate.
8857 1838, Zweites Jahres-Bericht des Staats Geologen: Harrisburg, 104 p.
8858 1839, and W. B. Rogers, Contributions to the geology of the Tertiary formations of Virginia: second series: Am. Philos. Soc. Proc., v. 1, no. 7, 88-90.
8859 1839, Marl - its nature and effects: Bos. Cult., v. 1, no. 20, [1]. Also in Farm's Mon. Vis., v. 1, 74.
From New Jersey geological report (see 8850).
8860 1839, Third annual report of the geological survey of the state of Pennsylvania: Harrisburg, Penn., E. Guyer (printer), 119 p.
Also in Penn. Senate Documents. For notice with excerpts see 4291.
8861 1839, Third annual report of the geological survey of the state of Pennsylvania: Harrisburg, Penn., Boas and Caplan (printers), 118 p.
8862 1839, Dritter jahlicher bericht der geologischen untersuchung des Staats Pennsylvanien: Harrisburg, J. Ehrenfried, 142 p.
8863 1840, Description of the geology of the state of New Jersey, being a final report: Phila., C. Sherman and Co., 301 p., colored plate, colored map.
Contains map 8864. For excerpts see 9955.
8864 1839 (i.e. 1840), A geological map of New Jersey: Phila., P. S. Duval (lithographer), colored geological map, 39 x 67 cm.
In Description of the geology of the state of New Jersey (see 8863).
8865 1840, Fourth annual report on the geological survey of the state of Pennsylvania: Harrisburg, Penn., Holbrook, Henlock, and Bratton (printers), 252 p.
Also German language edition (not seen).
8866 1840, Fourth annual report on the geological survey of the state of Pennsylvania: Harrisburg, Penn., William D. Boas (printer), 215 p.
8867 1841, [Deposition of drift in Virginia and Pennsylvania]: Am. J. Sci., v. 41, no. 1, 175 (12 lines, 10 lines).
Also in Assoc. Am. Geol's Nat's Proc., v. 2, 20. Also in Assoc. Am. Geol's Nat's Rept., v. 1, 27, 1843.
8868 1841, Fifth annual report on the geological exploration of the Commonwealth of Pennsylvania: Harrisburg, Penn., James S. Wallace (printer), 179 p.
Contains geological glossary by C. Lyell (see 6528), 175-179. Also German language edition (not seen). For excerpts see 9456.
8869 1841, Fifth annual report on the geological exploration of the Commonwealth of Pennsylvania: Harrisburg, Penn., Elliott and McCurdy (printers), 156 p.
Contains geological glossary by C. Lyell (see 6528), 153-156.
8870 1841, [Fossils in flagstone from Philadelphia]: Am. J. Sci., v. 41, no. 1, 164 (3 lines).
Also in Assoc. Am. Geol's Nat's Proc., v. 2, 9 (3 lines). Also in Assoc. Am. Geol's Nat's Rept., v. 1, 17 (3 lines), 1843.
8871 1841, [Iron sulphide in New Jersey marls]: Am. J. Sci., v. 41, no. 1, 159 (6 lines).
Also in Assoc. Am. Geol's Nat's Proc., v. 2, 4 (7 lines). Also in Assoc. Am. Geol's Nat's Rept., v. 1, 12 (7 lines), 1843.
8872 1841, [New Jersey and Pennsylvania trap dikes with columnar structure]: Am. J. Sci., v. 41, no. 1, 173 (5 lines).
Also in Assoc. Am. Geol's Nat's Proc., v. 2, 18 (5 lines). Also in Assoc. Am. Geol's Nat's Rept., v. 1, 26 (5 lines), 1843.
8873 1841, Observations upon the geological structure of Berkshire, Mass., and the neighboring parts of New York: Am. Philos. Soc. Proc., v. 2, no. 15, 3-4.
8874 1841, [Origin of overturned folds in Pennsylvania]: Am. J. Sci., v. 41, no. 1, 177.
Also in Assoc. Am. Geol's Nat's Proc., v. 2, 22 (9 lines). Also in Assoc. Am. Geol's Nat's Rept., v. 1, 29-30 (9 lines), 1843.
8875 1841, and L. Vanuxem, R. C. Taylor, E. Emmons, T. A. Conrad, Report on the ornithichnites or footmarks of extinct birds in the New Red Sandstone of Massachusetts and Connecticut: Am. J. Sci., v. 41, no. 1, 165-168.
Also in Assoc. Am. Geol's Nat's Rept., v. 1, 18-21, 1843.
1842, with the New York state geologists, Geological map of New York: see 7749.
Note in Geology of New York by J. Hall (see 4683) states: "In some of the maps, a portion of the state of Pennsylvania has been colored under the direction of Prof. H. D. Rogers", (preface, xxiii).
8876 1842, An inquiry into the origin of the Appalachian coal strata, bituminous and anthracite: Am. J. Sci., v. 43, no. 1, 178-179.
Also in Assoc. Am. Geol's Nat's Proc., v. 3, 33-34. Also in Assoc. Am. Geol's Nat's Rept., v. 1, 71, 1843.
8877 1842, [Nature of the dip of the Triassic of the eastern United States]: Am. J. Sci., v. 43, no. 1, 170-171.
Also in Assoc. Am. Geol's Nat's Proc., v. 3, 25. Also in Assoc. Am. Geol's Nat's Rept., v. 1, 63-64, 1843.
8878 1842, and W. B. Rogers, Observations on the geology of the western peninsula of Upper Canada, and the western part of Ohio: Am. Philos. Soc. Proc., v. 2, no. 20, 120-125.
8879 1842, [On striated surfaces in northeastern Pennsylvania and on the origin of conglomerate]: Am. J. Sci., v. 43, no. 1, 180-181.
Also in Assoc. Am. Geol's Nat's Proc., v. 3, 35-36. Also in Assoc. Am. Geol's Nat's Rept., v. 1, 72-73, 1843.
8880 1842, and W. B. Rogers, On the structure of the Appalachian chain, as exemplifying the laws of great mountain chains generally: Am. J. Sci., v. 43, no. 1, 177-178.
Also in Assoc. Am. Geol's Nat's Proc., v. 3, 32-33. Also in Assoc. Am. Geol's Nat's Rept., v. 1, 70-71, 1843.

Rogers, H. D. (cont.)

8881 1842, The recent earthquake in St. Domingo: Phila. Acad. Nat. Sci's Proc., v. 1, no. 15, 181-182.

8882 1842, [Remarks on C. Lyell's views on the uplift of North American terraces]: Am. J. Sci., v. 43, no. 1, 153.
Also in Assoc. Am. Geol's Nat's Proc., v. 3, 8. Also in Assoc. Am. Geol's Nat's Rept., v. 1, 47-48, 1843.

8883 1842, Sixth annual report on the geological survey of Pennsylvania: Harrisburg, Penn., Henlock and Bratton, 28 p.
Also German language edition (not seen).

8884 1842, [Stranded icebergs and their effect on drift]: Am. J. Sci., v. 43, no. 1, 154 (4 lines).
Also in Assoc. Am. Geol's Nat's Proc., v. 3, 9 (4 lines). Also in Assoc. Am. Geol's Nat's Rept., v. 1, 48 (4 lines), 1843.

8885 1843, [Cause of the crescent-formed dikes of trap in New Jersey and Connecticut]: Am. J. Sci., v. 45, no. 2, 334.
In Assoc. Am. Geol's Nat's Proc.

8886 1843, [Earthquakes and their motions]: Am. Philos. Soc. Proc., v. 2, no. 26, 267 (7 lines).

8887 1843, General observations on the geology of the western states: Phila. Acad. Nat. Sci's Proc., v. 1, no. 26, 272-273 (7 lines).
On the similar trends of anticlinal axes.

8888 1843, Geological notices: Am. Philos. Soc. Proc., v. 3, no. 27, 181-183.
On coprolite and coal fossils, and on the relation of earthquakes to the formation of the Appalachian Mountains.

8889 1843, An inquiry into the origin of the Appalacian coal strata, bituminous and anthracite: Assoc. Am. Geol's Nat's Rept., v. 1, 433-474.

8890 1843, [Metamorphism of rocks and metamorphic minerals from New York]: in Geology of New York, first district by W. W. Mather (see 6907), 468-476.

8891 1843, [Objections to J. D. Dana's theory of metamorphism]: Am. J. Sci., v. 45, no. 1, 142 (5 lines).
In Assoc. Am. Geol's Nat's Proc.

8892 1843, Verbal communications [on the form of anticlines in Russia and America]: Phila. Acad. Nat. Sci's Proc., v. 1, no. 25, 256.

8893 1843, On hydrated minerals and antediluvian temperatures: Am. J. Sci., v. 45, no. 1, 147.
In Assoc. Am. Geol's Nat's Proc. Discussion of article by L. C. Beck (see 1578).

8894 1843, On the earthquake of January 4, 1843: Am. Philos. Soc. Proc., v. 2, 258-259.

8895 1843, [On Marcellus and Hamilton shales of the South and West]: Am. J. Sci., v. 45, no. 1, 161-162.
In Assoc. Am. Geol's Nat's Proc.

8896 1843, [On the origin of mud furrows]: Am. J. Sci., v. 45, no. 1, 149 (4 lines).
In Assoc. Am. Geol's Nat's Proc.

8897 1843, [On the absence of materials from the South in the drift]: Am. J. Sci., v. 45, no. 2, 329.
In Assoc. Am. Geol's Nat's Proc.

8898 1843, and W. B. Rogers, On the phenomena of the great earthquakes which occurred during the past winter, one in this country and the other in the West Indies, and on a general theory of earthquake motion, by which they propose to elucidate several points in geological dynamics: Am. Philos. Soc. Proc., v. 3, no. 27, 64-67.

8899 1843, and W. B. Rogers, On the physical structure of the Appalachian chain, as exemplifying the laws which have regulated the elevation of great mountain chains generally: Am. J. Sci., v. 44, no. 2, 359-362.
In British Assoc. Proc.

8900 1843, [Origins of the red sandstone formation]: in Geology of New York, first district, by W. W. Mather (see 6907), 289-292.
From Geological report of New Jersey, by H. D. Rogers (see 8863).

8901 1843, and W. B. Rogers, Theory of earthquake action: Am. J. Sci., v. 45, no. 2, 341-347.
In Assoc. Am. Geol's Nat's Proc.

8902 1844, Address delivered at the meeting of the Association of American Geologists and Naturalists, held in Washington, May, 1844: Am. J. Sci., v. 47, no. 1, 137-160; no. 2, 247-278.
On the present state of the science of geology in the United States.

8903 1844, Address delivered at the meeting of the Association of American geologists and naturalists, held in Washington, May, 1844. ... With an abstract of the proceedings at their meeting: NY and London, Wiley and Putnam, 58, 43 p.
For review see 9950.

8904 1844, [Age of beds containing fossil infusoria of South Carolina]: Am. J. Sci., v. 47, 117 (4 lines).
Also in Assoc. Am. Geol's Nat's Proc., v. 5, 24 (4 lines).

8905 1844, and W. B. Rogers, On the phenomena and theory of earthquakes, and the explanation they afford of certain facts in geological dynamics: Am. J. Sci., v. 47, no. 1, 182 (17 lines).
In British Assoc. Proc.

8906 1844, On the probable constitution of the atmosphere at the period of the formation of coal: Am. J. Sci., v. 47, no. 1, 105.
Also in Assoc. Am. Geol's Nat's Proc., v. 5, 12.

8907 1844, and W. B. Rogers, A system of classification and nomenclature of the Paleozoic rocks of the United States, with an account of their distribution more particularly in the Appalachian mountain chain: Am. J. Sci., v. 47, no. 1, 111-112.
Also in Assoc. Am. Geol's Nat's Proc., v. 5, 18-19.

8908 1845, and others, Changes in the earth's surface: The Investigator, v. 1, 170-171.
A joke on the pivoting of land. Extracted from Assoc. Am. Geol's Nat's Proc.

8909 1845, [Coal beds and strata of Pennsylvania]: Assoc. Am. Geol's Nat's Proc., v. 6, 25 (10 lines).
Also in Am. Q. J. Ag. Sci's, v. 2, 148.

8910 1845, and S. G. Morton and W. R. Johnson, Of the red pumice of the Missouri [and] Proofs of a freshwater formation, near the mouth of the Yellowstone River: Phila. Acad. Nat. Sci's Proc., v. 2, no. 9, 239-240.

Rogers, H. D. (cont.)
8911 1845, and W. B. Rogers, [On boulder trains in Berkshire County, , Massachusetts]: Bos. Soc. Nat. History Proc., v. 2, 79-80.
8912 1845, [On rhomboidal cleavage in New Red Sandstone]: Am. Q. J. Ag. Sci's, v. 2, 150.
   In Assoc. Am. Geol's Nat's Proc.
8913 1845, [On the bones of Zeuglodon, recently exhibited by A. Koch under the name of Hydrarchos]: Bos. Soc. Nat. History Proc., v. 2, 79.
8914 1845, On the direction of the slaty cleavages in strata of the southeastern belts of the Appalachian chain, and the parallelism of the cleavage dip with the planes of maximum temperature: Assoc. Am. Geol's Nat's Proc., v. 6, 49-50.
8915 1845, [Oolite in Florida]: Phila. Acad. Nat. Sci's Proc., v. 2, no. 8, 210-211 (7 lines).
8916 1846, and W. B. Rogers, An account of two remarkable trains of angular erratic blocks in Berkshire, Massachusetts, with an attempt at an explanation fo the phenomena: Bos. J. Nat. History, v. 5, no. 3, 310-330, map.
   Also reprint edition.
8917 1846, Composition of the marls of New Jersey: Am. Q. J. Ag. Sci's, v. 3, 293-294.
   On the chemical analysis of greensand from Mullica Hill, New Jersey.
8918 1846, [Crystals of fluorite in cannel coal from England]: Bos. Soc. Nat. History Proc., v. 2, 109-110.
   Abstracted in Am. J. Sci., v. 2 (ns), no. 1, 124 (10 lines).
8919 1846, and W. B. Rogers, On the geological age of the White Mountains: Am. J. Sci., v. 1 (ns), no. 3, 411-421.
8920 1846, [On the geology and mineralogy of the southern shore of Lake Superior]: Bos. Soc. Nat. History Proc., v. 2, 124-125.
8921 1846, Report to the Trustees of the Franklin Copper Company: Bos., Eastburn's Press, 23 p.
8922 1847, On the drift of New England and the River St. Lawrence: Am. Q. J. Ag. Sci., v. 6, 214 [i.e. 262].
   In Assoc. Am. Geol's Nat's Proc. With discussion by L. Agassiz and E. Emmons.
8923 1848, [Age of bituminous vs. anthracite coals]: Bos. Soc. Nat. History Proc., v. 3, 9 (9 lines).
8924 1848, [Altered shales and sandstones from New Hope, Pennsylvania]: Bos. Soc. Nat. History Proc., v. 3, 30.
8925 1848, [Anthracite coal of Rhode Island]: Bos. Soc. Nat. History Proc., v. 3, 44 (9 lines).
8926 1848, Fossils in the White Mountains [of New Hampshire]: Am. J. Sci., v. 5 (ns), no. 1, 116 (5 lines).
   In Assoc. Am. Geol's Nat's Proc.
8927 1848, [Infusorial deposits of the southern states]: Bos. Soc. Nat. History Proc., v. 3, 16.
8928 1848, [On the cause of metamorphism in rocks]: Bos. Soc. Nat. History Proc., v. 3, 20.
8929 1849, [On cleavage in Roxbury Conglomerate]: Bos. Soc. Nat. History Proc., v. 3, 127 (10 lines).
8930 1849, [On the structure of glaciers]: Bos. Soc. Nat. History Proc., v. 3, 122-123; 125-127.
8931 1850, [Centers of drift dispersion in New England and Europe]: Am. Assoc. Adv. Sci. Proc., v. 2, 309.
8932 1850, [Distribution of drift in the Alleghany Mountains]: Bos. Soc. Nat. History Proc., v. 3, 242.
8933 1850, [Nova Scotia sandstone and its qualities as a building stone]: Bos. Soc. Nat. History Proc., v. 3, 248.
8934 1850, [On fossil plants in coal]: Bos. Soc. Nat. History Proc., v. 3, 262 (10 lines).
8935 1850, On the analogy of the ribbon structure of glaciers to the slaty cleavage of rocks: Am. Assoc. Adv. Sci. Proc., v. 2, 181-182.
8936 1850, On the deposits of common salt and climate: Scien. Am., v. 5, no. 51, 402.
8937 1850, [On the origin of New York and Pennsylvania beds of limestone]: Bos. Soc. Nat. History Proc., v. 3, 258-259.
8938 1850, [On the origin of salt lakes]: Bos. Soc. Nat. History Proc., v. 3, 259-260; 266.
8939 1850, On the origin of the drift, and of the lake and river terraces of the United States and Europe, with an examination of the laws of aqueous action connected with the inquiry: Am. Assoc. Adv. Sci. Proc., v. 2, 239-255.
   With discussion by L. Agassiz.
8940 1850, [On the origin of the greensand of New Jersey]: Bos. Soc. Nat. History Proc, v. 3, 248-249.
8941 1850, On the structural features of the Appalachians, compared with those of the Alps and other disturbed districts of Europe: Am. Assoc. Adv. Sci. Proc., v. 2, 113-115; 118.
   With discussion by L. Agassiz, J. Le Conte, and A. Guyot (see 4613). Abstracted in Ann. Scien. Disc., v. 1, 225-226.
8942 1850, [On the origin of fossil raindrops]: Bos. Soc. Nat. History Proc., v. 3, 202 (7 lines).
8943 1850, Report on the coal lands of the Zerbe's Run and Shamokin improvement company, ... , with charter: Bos., Thurston, Torry & Co., 24 p., chart, map.
8944 1850, Report on the iron ores of the Pottsville coal basin: Phila., L. R. Bailey (printer), 16 p., plate.
8945 1850, [Test of the strength of building stone]: Bos. Soc. Nat. History Proc., v. 3, 241 (8 lines).

Rogers, James Blythe (1802-1852)
   See also 8955 and 8957 with R. E. Rogers.
8946 1841, [Chemical composition of dolomite]: Am. J. Sci., v. 41, no. 1, 172 (4 lines). Assoc. Am. Geol's Nat's Proc., v. 2, 17 (4 lines). Also in Assoc. Am. Geol's Nat's Rept., v. 1, 24 (4 lines), 1843.
8947 1842, [Analysis of meteoric iron]: Am. J. Sci., v. 43, no. 1, 169 (9 lines). Also in Assoc. Am. Geol's Nat's Proc., v. 3, 24 (9 lines). Also in Assoc. Am. Geol's Nat's Rept., v. 1, 62-63 (9 lines), 1843.

Rogers, John (1684-1755)
8948   1728, The nature and necessity of repentence, with the means and motives to it. A discourse occasion'd by the earthquake. Preach'd at Boxford, in part on the publick fast. December 21. 1727. ... : Bos., printed for S. Gerrish, 78 p.
        Evans number 3100.
8949   1756, Three sermons on different subjects and occasions. ... The last, on the terribleness, and the moral cause of earthquakes ... : Bos., Printed by Edes and Gill for S. Kneeland, 61 p.
        Evans number 7785.

Rogers, Robert Empie (1813-1884)
        Also articles with W. B. Rogers (8998, 9003, 9004, 9008 and 9009).
8950   1839, and M. H. Boye, An analysis of a specimen of the iron ore from the celebrated "Iron Mountain", Missouri: Franklin Inst. J., v. 23 (ns), 361-362.
8951   1840, and M. H. Boye, On the analysis of limestone, especially the magnesian kind, and a method of completely seaparating lime from magnesia, when both are present in large quantity: Franklin Inst. J., v. 25 (ns), 158-162.
8952   1841, [Magnesian limestone of Pennsylvania] : Am. J. Sci., v. 41, no. 1, 171 (12 lines).
        Also in Assoc. Am. Geol's Nat's Proc., v. 2, 16 (12 lines). Also in Assoc. Am. Geol's Nat's Rept., v. 1, 24 (12 lines), 1843.
8953   1844, [Origin of minerals in the Missouri lead region] : Am. J. Sci., v. 47, no. 1, 106 (5 lines).
        Also in Assoc. Am. Geol's Nat's Proc., v. 5, 13 (5 lines).
8954   1848, and W. B. Rogers, New method of determining the carbon in native and artificial graphites, &c.: Am. J. Sci., v. 5 (ns), no. 3, 352-359, fig.
        Also reprint edition.
8955   1848, and J. B. Rogers, On the alleged insolubility of copper in hydrochloric acid; with an examination of Fuch's method for analyzing iron ores, metallic iron, etc.: Am. J. Sci., v. 6 (ns), no. 3, 395-396.
        In Am. Assoc. Prom. Sci. Proc.
8956   1848, and W. B. Rogers, Oxidation of the diamond in the liquid way: Am. J. Sci., v. 6 (ns), no. 1, 110-111.
8957   1849, and J. B. Rogers, On the alleged insolubility of copper in hydrochloric acid; with an examination of Fuch's method for analyzing iron ores, metallic iron, &c.: Am. Assoc. Adv. Sci. Proc., v. 1, 39.

Rogers, William Barton (1804-1882)
        See also articles with H. D. Rogers (8858, 8878, 8880, 8898, 8899, 8901, 8905, 8907, 8911, 8916 and 8919) and with R. E. Rogers (8954 and 8956).
8958   1834, Analysis of shells: Am. J. Sci., v. 26, no. 2, 361-365.
8959   1834, Apparatus for analyzing marl, and the carbonates: Farm's Reg., v. 2, 364-365, fig.
8960   1834, Chemical analysis of shells: Farm's Reg., v. 1, 589-591.
8961   1834, Magnesian marl of Hanover [Virginia] : Farm's Reg., v. 1, 462-463.
8962   1834, On the discovery of green sand in the calcareous deposit of Eastern Virginia, and on the probable existence of this substance in extensive beds near the western limits of our ordinary marl: Farm's Reg., v. 2, 129-131.
8963   1835, Apparatus for analyzing calcareous marl and other carbonates: Am. J. Sci., v. 27, no. 2, 299-301, fig.
8964   1835, Further observations on the green sand and calcareous marl of Virginia: Farm's Reg., v. 2, 747-751.
8965   1835, Report of the select committee, to whom was referred the memorials from Morgan, Frederick, and Shenandoah counties, praying for a geological survey of the state: Farm's Reg., v 2, 688-692.
8966   1835, Self-filling syphon for chemical analysis: Am. J. Sci., v. 27, no. 2, 302-303, fig.
8967   1836, Extracts from the report of the geological reconnoissance of Virginia: Farm's Reg., v. 3, 627-634; 666-674.
8968   1836, Report of the geological reconnoissance of the state of Virginia ...: Phila., De Silver Thomas and Co., 143, [1] p., folded color plate.
        For review see 9740. For excerpts see 8967.
8969   1836, Report of the geological reconnoissance of the state of Virginia ...: Richmond, Va., Va. State Document 24, 52 p., plate.
8970   1836, Report of the Walton's Gold Mine in Louisa County, Virginia: in Report by B. Silliman and F. Shepherd (see 9736), 11-13.
8971   1836, Virginia springs: Am. Mag. Use. Ent. Know., v. 3, no. 3, 105.
8972   1837, and H. D. Rogers, Contributions to the geology of the Tertiary formations of Virginia: Am. Philos. Soc. Trans., v. 5 (ns), 319-341.
        Also reprint edition.
8973   1837, Report of the progress of the geological survey of Virginia for 1836: Farm's Reg., v. 4, 713-721.
8974   1837, First report of the progress of the geological survey of the state of Virginia for the year 1836: Phila., C. Sherman and Co., 30 p.
        Also in Va. State Documents 34, 14 p.
8975   1838, Second report of the progress of the geological survey of the state of Virginia for the year 1837: Phila., C. Sherman and Co. (printers), 31-87.
        Published separately and with 8974.
8976   1838, Second report of the progress of the geological survey of the state of Virginia for the year 1837: Richmond, Va., State Document 45, 24 p.
8977   1839, and H. D. Rogers, Contributions to the geology of the Tertiary formations of Virginia. - Second series: Am. Philos. Soc. Trans., v. 6 (ns), 347-377, 5 plates.
        Includes "Description of several species of Miocene and Eocene shells, not before discribed". Also reprint edition.
8978   1839, Report of the progress of the geological survey of the state of Virginia for the year 1838: Richmond, Va., Va. State Document 56, 32 p.
8979   1840, Report of the progress of the geological survey of the state of Virginia for the year 1839: Richmond, Va., Samuel Shepherd (printer), 161 p., 2 folded color plates.
8980   1841, Report of the progress of the geological survey of the state of Virginia for the year 1840: Richmond, Va., Samuel Shepherd (printer), 132 p.

Rogers, W. B. (cont.)
8981 1842, Geological age of the coal formation of the vicinity of Richmond: Phila. Acad. Nat. Sci's Proc., v. 1, no. 10, 142.
Communicated by H. D. Rogers.
8982 1842, [Meteorite from Roanoke County, Virginia]: Am. J. Sci., v. 43, no. 1, 169, (8 lines).
Also in Assoc. Am. Geol's Nat's Proc., v. 3, 24 (8 lines). Also in Assoc. Am. Geol's Nat's Rept., v. 1, 63 (8 lines), 1843.
8983 1842, Observations on subterranean temperature made in the mines of eastern Virginia: Am. J. Sci., v. 43, no. 1, 176 (11 lines).
Also in Assoc. Am. Geol's Nat's Proc., v. 3, 31 (11 lines). Also in Assoc. Am. Geol's Nat's Rept., v. 1, 69 (11 lines), 1843.
8984 1842, Occurrence of oxide of tin in Virginia: Am. J. Sci., v. 43, no. 1, 168-169.
Also in Assoc. Am. Geol's Nat's Proc., v. 3, 23-24. Also in Assoc. Am. Geol's Nat's Rept., v. 1, 62 (16 lines), 1843.
8985 1842, [On erosion of strata underlying the Oriskany Sandstone]: Am. J. Sci., v. 43, no. 1, 181-182.
Also in Assoc. Am. Geol's Nat's Proc., v. 3, 36-37. Also in Assoc. Am. Geol's Nat's Rept., v. 1, 73-74, 1843.
8986 1842, On the age of the coal rocks of eastern Virginia: Am. J. Sci., v. 43, no. 2, 175 (12 lines).
Also in Assoc. Am. Geol's Nat's Proc., v. 3, 30 (12 lines). Also in Assoc. Am. Geol's Nat's Rept., v. 1, 68 (12 lines), 1843.
8987 1842, [On the cause of the dip in the Triassic of the eastern United States]: Am. J. Sci., v. 43, no. 1, 171-172; 173.
Also in Assoc. Am. Geol's Nat's Proc., v. 3, 26-27; 28. Also in Assoc. Am. Geol's Nat's Rept., v. 1, 64-65; 66, 1843.
8988 1842, On the connection of thermal springs in Virginia with anticlinal axes and faults: Am. J. Sci., v. 43, no. 1, 176.
Also in Assoc. Am. Geol's Nat's Proc., v. 3, 31. Also in Assoc. Am. Geol's Nat's Rept., v. 1, 69 (14 lines), 1843.
8989 1842, On the porous anthracite or natural coke of eastern Virginia: Am. J. Sci., v. 43, no. 1, 175-176 (12 lines).
Also in Assoc. Am. Geol's Nat's Proc., v. 3, 30-31 (12 lines). Also in Assoc. Am. Geol's Nat's Rept., v. 1, 68 (12 lines), 1843.
8990 1842, Report of the progress of the geological survey of the state of Virginia for the year 1841: Richmond, Va., Samuel Shepherd (printer), 12 p.
8991 1843, Observations of subterranean temperature in the coal mines of eastern Virginia: Assoc. Am. Geol's Nat's Rept., v. 1, 532-538.
8992 1843, and H. D. Rogers, Observations on the geology of the western peninsula of Upper Canada, and the western part of Ohio: Am. Philos. Soc. Trans., v. 8 (ns), 273-284.
8993 1843, On the age of the coal rocks of eastern Virginia: Assoc. Am. Geol's Nat's Rept., v. 1, 298-316, plate.
8994 1843, Verbal communication [on the age of the Fredericksburg sandstone]: Phila. Acad. Nat. Sci. Proc., v. 1, no. 24, 250 (13 lines).
Communicated by H. D. Rogers.
8995 1843, On the connexion of thermal springs in Virginia with anticlinal axes and faults: Assoc. Am. Geol's Nat's Rept., v. 1, 323-347, 2 tables, plate with 10 figs.
8996 1843, On the limits of the infusorial stratum in Virginia: Am. J. Sci., v. 45, no. 2, 313-314.
In Assoc. Am. Geol's Nat's Proc.
8997 1843, and H. D. Rogers, On the physical structure of the Appalachian chain, as exemplifying the laws which have regulated the elevation of great mountain chains generally: Assoc. Am. Geol's Nat's Rept., v. 1, 474-531, 4 plates.
8998 1844, and R. E. Rogers, An account of some new instruments and processes for the analysis of the carbonates: Am. J. Sci., v. 46, no. 2, 346-359, 3 figs.
Also reprint edition.
8999 1844, Extract of a letter by Prof. Wm. B. Rogers, to the Junior Editor [on fossil infusoria of Maryland and Virginia]: Am. J. Sci., v. 46, no. 1, 141-142.
9000 1844, [Oolite structure of some lower Appalachian limestone and chert]: Am. J. Sci., v. 47, no. 1, 119 (7 lines).
In Assoc. Am. Geol's Nat's Proc.
9001 1847, Extracts from the official report of Professor William B. Rogers: in Documents of the Phoenix Mining Company (see 8328), p. 26.
9002 1848, [On the elevation of river terraces]: Lit. World, v. 2, 228 (6 lines).
In Assoc. Am. Geol's Nat's Proc.
9003 1848, and R. E. Rogers, On the decomposition and partial solution of minerals, rocks, &c., by pure water, and water charged with carbonic acid: Am. J. Sci., v. 5 (ns), no. 3, 401-405.
9004 1848, and R. E. Rogers, On the decomposition of rocks, &c., by meteoric waters, and on the action of the mineral acids upon feldspar, &c.: Am. J. Sci., v. 6 (ns), no. 3, 396-397.
In Assoc. Am. Geol's Nat's Proc.
9005 1848, On the transporting power of currents: Am. J. Sci., v. 5 (ns), no. 1, 115-116.
In Assoc. Am. Geol's Nat's Proc.
9006 1849, On acid and alkali springs: Am. Assoc. Adv. Sci. Proc., v. 1, 94-95.
9007 1849, [On the analogy of primary formations on the north shore of Lake Superior to those of the Blue Ridge of Virginia]: Am. Assoc. Adv. Sci. Proc., v. 1, 79-80.
9008 1849, and R. E. Rogers, On the comparative solubility of the carbonate of lime and the carbonate of magnesia: Am. Assoc. Adv. Sci. Proc., v. 1, 95-97.
9009 1849, and R. E. Rogers, On the decomposition of rocks by meteoric water: Am. Assoc. Adv. Sci. Proc., v. 1, 60-62.
9010 1850, On acid and alkaline springs: Am. J. Sci., v. 9, no. 1, 123-124.
Abstracted from Am. Assoc. Adv. Sci. Proc.
9011 1850, Gold formation: Scien. Am., v. 5, no. 31, 402.

Romans, Bernard (1720-1784)
9012 1775, A concise natural history of East and West Florida; containing an account of natural produce of all the southern part of British America, in three kingdoms of nature, particularly the animal and vegetable; likewise the artificial produce now raised and manufactured there ... : NY, Printed for the author, [4], 342 p., 6 plates, 3 maps.
     Evans number 14440.
9013 1776, A concise natural history of East and West Florida ... [as above]: NY, R. Aitken, [4], 4, 342 p., 6 plates, folded table.
     Evans number 15069.

Romayne, Nicholas (1756-1817)
9014 1809, Anniversary address to the Medical Society of the state [of New York]: Med. Soc. NY Trans., for 1808, 37-46.
9015 1810, Anniversary address to the Medical Society of the state [of New York]: Med. Soc. NY Trans., for 1809, 67-84.

Roosevelt, Nicholas I. (1764-1854; see 10509 under United States Papers)

Root, Erastus (1773-1846)
9016 1817, An inaugural dissertation on the chemical and medicinal properties of the mineral spring in Guilford ... : Brattleborough, Vt., Simeon Ide (printer), 15 p.

Rose, Gustav (1798-1873)
     See also Heinrich Rose, his brother, with whom his articles may be included.
9017 1820, Fluoric acid in mica: Am. J. Sci., v. 2, no. 2, 376 (5 lines).
9018 1840, On the identity of edwardsite and monazite: Franklin Inst. J., v. 25 (ns), 289-290.
     Extracted by M. H. Boye and J. C. Booth from Poggendorff's Annalen.
9019 1846, Anatase, brookite, and rutile: Am. J. Sci., v. 2 (ns), no. 3, 416.
     Abstracted from Jameson's J.
9020 1846, Columbite and wolfram: Am. J. Sci., v. 1 (ns), no. 2, 267 (10 lines).
     Abstracted from Poggendorff's Annalen.
9021 1849, Niobite [from Lake Ilmen]: Am. J. Sci., v. 8 (ns), no. 1, 126 (4 lines).

Rose, Heinrich (1795-1864)
     See also Gustav Rose, his brother, with whom his articles may be confused.
9022 1833, Characteristic of epistilbite according to Rose: Am. J. Sci., v. 24, no. 1, 194 (13 lines).
9023 1835, Phosphate of lime in the teeth and silica in the skin of the infusoria: Am. J. Sci., v. 28, no. 2, 386 (6 lines).
     From Jameson's Edinburgh New Philos. J.
9024 1845, Columbite: Am. J. Sci., v. 49, no. 1, 228 (14 lines).
9025 1845, New metals, pelopium and niobium [from Bavaria]: Am. J. Sci., v. 48, no. 2, 400-401.
     Abstracted from Comptes Rendus.
9026 1846, New metals: Am. J. Sci., v. 1 (ns), no. 1, 103-104.
     Abstracted from Comptes Rendus. On niobium and pelopium from Bavaria.
9027 1846, Perowskite: Am. J. Sci., v. 2 (ns), no. 3, 417 (3 lines).
     Abstracted from Berzelius Jahresbericht.
9028 1847, On the acid contained in the North American columbite: Am. J. Sci., v. 4 (ns), no. 3, 408-409.
     Abstracted from London, Edinburgh, and Dublin Philos. Mag.
9029 1847, On a new metal, pelopium, contained in the Bavarian tantalite: Am. J. Sci., v. 3 (ns), no. 3, 357-365.
9030 1848, On the composition of urano-tantalite and columbite: Am. J. Sci., v. 5 (ns), no. 3, 422.
     Abstracted from Comptes Rendus Mensuels (Berlin Academy).
9031 1848, Samarskite: Am. J. Sci., v. 6 (ns), no. 2, 266-267.
9032 1849, On the existence of mercury in the Tyrol: Am. J. Sci., v. 8 (ns), no. 2, 275 (14 lines).
     Abstracted from l'Institut.
9033 1849, On the yttrotantallite of Ytterby: Am. J. Sci., v. 8 (ns), no. 1, 126 (4 lines).
     Abstracted from "Jour. fur Prakt. Ch."

Rose, R. H.
9034 1823, Soil analysis: NE Farm., v. 2, 20.
9035 1834, Soil analysis: Me. Farm., v. 2, 165.

Ross, Sir James Clark (1800-1862)
9036 1849, Notice of, and citations from a voyage of discovery and research in the southern and Antarctic regions, during the years 1839-43: Am. J. Sci., v. 7 (ns), no. 3, 313-329; v. 8, no. 1, 14-33.

Ross, John
9037 1849, Nova Scotia mines. Cast steel by simple fusion direct from the ore: Scien. Am., v. 4, no. 35, 278.

Ross, William (see 7293 with S. L. Mitchill and G. E. Pendergrast)

Rothe, Charles Edward (Karl Eduard)
     See also 7247 with E. Mitchell.
9038 1827, Remarks on the gold mines of North Carolina: in Report on the geology of North Carolina, part III, by E. Mitchell and C. E. Rothe (see 7247), 29-43.
9039 1828, Remarks on the gold mines of North Carolina: Am. J. Sci., v. 13, no. 2, 201-217, fig.

Rouelle, John (i.e. Jean, b. 1751)
9040   1792, A complete treatise on the mineral waters of Virginia containing a description of their situation, their natural history, their analysis, contents, and their use in medicine: Phila., Charles Cist (printer), xix, 68 p.
        Evans number 24757. For review with excerpts see 10518.

Rousseau, J. A.
9041   1847, Analysis of soils: Prairie Farm., v. 7, 246-248.

Rowell, George Augustus (1804-1892)
9042   1847, On the cause of the Aurora, and the declination of the needle: Franklin Inst. J., v. 14 (3s), 343-345.
        Abstracted from London Ath.

Rowles, W. P. (d. ca. 1850)
9043   1841, Letters in the study of geology: Southron, v. 1, 298-301; 364-366; 398-400.

Rowley, S. A.
9044   1838, Encrinite, tufa, &c. [from Becraft's Mountain, New York]: Am. J. Sci., v. 33, no. 2, 405-406.

The Royal American Magazine (Boston, 1774-1775)
9045   1774, [On fossil bones from the Ohio River]: v. 1, 149-150.
        Signed by "D. M."
9046   1774, Very curious and accurate account of the climate, soil, and weather, at Port Egmont in Faulkland's Islands ... : v. 1, 424-426.
        Signed by "A mariner."

Rozet, Claude Antoine (1798-1858)
9047   1831, Geological notices of Barbary: Mon. Am. J. Geology, v. 1, no. 6, 273-277, plate.
        Abstracted from Geological notices of Barbary.
9048   1832, Geology of Africa: Mon. Am. J. Geology, v. 1, no. 7, 334 (10 lines).

Rozier, Francois (1734-1793)
9049   1821, On earths and soils: Am. Farm., v. 3, 348-349.

Ruffin, Edmund (1794-1865)
9050   1821, On the composition of soils, and their improvement by calcareous manures: Am. Farm., v. 3, 313-320.
9051   1824, Marl, its application and effects on various soils: Am. Farm., v. 6, 5.
9052   1828, Gypseous earth [from Petersburg, Virginia]: Am. Farm., v. 10, 3.
9053   1832, Calcareous manures. Effects of calcareous manures and directions for this most profitable application. From An essay on calcareous manures by E. Ruffin: Am. Farm., v. 14, 114-116.
9054   1832, An essay on calcareous manures: Petersburg, Va., J. W. Campbell, xii, 13-241 p.
        See 9088 for list of reviews.
9055   1832, Marl: Am. Farm., v. 14, 331-332.
9056   1833, Discovery of marl in a new district: Am. Farm., v. 15, 234-236.
        From Farm's Reg.
9057   1833, The gypseous earth of James River [in Virginia]: Farm's Reg., v. 1, 207-211.
9058   1833, Supplementary chapter to An essay on calcareous manures: Farm's Reg., v. 1, 76-79.
9059   1834, Plain directions for analyzing gypsum: Farm's Reg., v. 2, 33.
        Signed by "E. R."
9060   1834, Plain directions for analyzing marl, and other calcareous manures: Farm's Reg., v. 1, 609-610.
        Signed by "E. R."
9061   1835, An essay on calcareous manures: Shellbanks, Va., Published at the office of the Farmer's Register, viii, 116 p.
        Second edition. "Issued as supplement to Farmers' Register. v. 2." See 9088 for list of reviews and notices.
9062   1835, Notes of a hasty view of the soil and agriculture of part of the county of Northampton: Farm's Reg., v. 3, 233-240.
        Signed "A Gleaner."
9063   1835, Observations on marl and lime: Am. Farm., v. 2 (2s), 173-175.
        Abstracted from An essay on calcareous manures (see 9061).
9064   1835, [Review of Report of the geologist to the legislature of Maryland, 1834 by J. T. Ducatel (see 3234)]: Farm's Reg., v. 3, 36-43.
9065   1836, Description and statement of the ingredients of the earth improperly called "marl" in New Jersey: Farm's Reg., v. 4, 86-89.
9066   1836, Means to facilitate the analyzing of marl and other calcareous manures, and soils: Farm's Reg., v. 3, 575-576.
9067   1837, [Black lead in North Carolina and Virginia]: Farm's Reg., v. 5, 319.
9068   1837, British opinions on the "Essay on calcareous manures," and the original discovery of one of its positions: Farm's Reg., v. 5, 380-382.
9069   1837, The marl of Virginia now in use in Connecticut: Farm's Reg., v. 5, 506.
9070   1837, Notes of a three-day excursion into Goochland, Chesterfield, and Powhaton [Virginia]: Farm's Reg., v. 5, 315-319; 361-373.
9071   1838, Remarks on the soil and agriculture of Gloucester County: Farm's Reg., v. 6, 178-191; 193-194.
9072   1839, The almost purely calcareous soil of tropical Florida: Farm's Reg., v. 7, 684-685.
9073   1840, Remarks on calcareous earths and soils: Farm's Reg., v. 8, 406-408.
        From S. Cab.
9074   1840, Remarks on the soil and marling of the Pamunkey lands. Introductory to the queries and answers thereon: Farm's Reg., v. 8, 679 [i.e. 681]-683.
9075   1840, Remarks on the soil in general, and particularly the ridge lands, of eastern Virginia: Farm's Reg., v. 8, 168-171.
9076   1841, Experiments with and observations on green-sand earth, as manure, on Coggin's Point farm: Farm's Reg., v. 9, 118-124.

Ruffin, E. (cont.)
9077   1841, Green-sand of James River: Farm's Reg., v. 9, 645-646.
9078   1841, Lime in Prairie soils: Union Agriculturist and West. Prairie Farm., v. 1, 85.
9079   1841, On the soils, and marling improvements of King William County. Introductory to answers to the general queries on marling: Farm's Reg., v. 9, 21-24.
9080   1841, Remarks on the soils of part of Surry County: Farm's Reg., v. 9, 563-565.
9081   1842, Beneficial effects of the green-sand earth of James River, recently observed: Farm's Reg., v. 10, 252-253.
9082   1842, Description and account of the different kinds of marl, and of the gypseous earth, of the tide-water region of Virginia. Report to the state board of agriculture: in Report of board of agriculture, Va. House J. for 1842, Document 12, 75-103.
9083   1842, An essay on calcareous manures. Third edition: Petersburg, Va., for the author, vii, [13]-316 p.
       Printed as Farmer's Register, v. 10, no. 12, and Supplement.
9084   1842, Some account of the green-sand earth of Gloucester and Salem Counties, New Jersey, and the effects as manure: Farm's Reg., v. 10, 418-430.
9085   1842, Some account of the green-sand earth of Gloucester and Salem Counties, N. J., and the effects as manure: Farm's Cab., v. 7, 137-142.
       Abstracted from Farm's Reg.
9086   1843, Analyzing specimens of marls: S. Agriculturist, v. 3 (ns), 231-233; 273-274.
9087   1843, Report of the commencement and progress of the agricultural survey of South Carolina for 1843: Columbia, SC, A. H. Pemberton, 120, 55 p., illus.
       With "Analysis" by C. U. Shepard (see 9540). For review see 9941.
9088   1844, An essay on calcareous manures: Phila., Laurens Wallozz, [13]-316 p.
       "Fourth edition." See also 9054, 9061, and 9083.
       For review with excerpts see 656, 2701, and 2717.
       For review see 1902, 3721, 3725, 4258, 8749, 9940, and 249.
9089   1844, Errata in Ruffin's agricultural report: S. Agriculturist, v. 4 (ns), 80.
       See also 9088.
9090   1844, [Extracts from Report of the agricultural survey of South Carolina (see 9087)]: S. Agriculturist, v. 4 (ns), 169-176; 220-226; 246-251; 286-301.
9091   1844, General characters, extent and distribution of the calcareous formations of lower South Carolina: S. Agriculturist, v. 4 (ns), 136-143.
       Extracted from Report ... , (see 9087).
9092   1844, Secondary and Miocene marls on and near Lynch's Creek: in Report on the geological and agricultural survey of the state of South Carolina by M. Tuomey (see 10430), 59-63.
9093   1844, Supplementary report of the agricultural survey for 1843. Secondary and Miocene marls on and near Lynches Creek in Darlington, Sumter, Williamsburg, and Marion districts, S. C.: S. Agriculturist, v. 4 (ns), 121-127.
9094   1845, [List of Tertiary fossils from Virginia]: Nat'l Inst. Prom. Sci. Bulletin, v. 3, 253-254.
9095   1850, Description of a nut found in Eocene marl: Am. J. Sci., v. 9 (ns), no. 1, 127-129, 4 figs.

Ruffner, David
9096   1815, Kenawha salt-works: Niles' Wk. Reg., v. 8, 135.

Ruggles, Daniel (1810-1897)
9097   1836, Geological and miscellaneous notice of the region around Fort Winnebago, Michigan Territory: Am. J. Sci., v. 30, no. 1, 1-8, fig., map.
9098   1845, Considerations respecting the copper mines of Lake Superior: Am. J. Sci., v. 49, no. 1, 64-72, 2 maps.
9099   1845, Large trilobite; Iowa coralline marble: Am. J. Sci., v. 49, no. 1, 216 (11 lines).

Rupp, Israel Daniel (1803-1878)
9100   1844, History of Lancaster County [Pennsylvania]: Lancaster, Penn., G. Hills, 531 p., front., plate, table.
       Includes sections on "Geology" and "Natural history," 465-482.
9101   1844, History of the counties of Berks and Lebanon [Pa.]: Lancaster, G. Hills, vi, [13]-512 p., plate.
9102   1845, History of Lancaster and York Counties [Pa.]: Lancaster, Pa., G. Hills, 740 p. With geological description.
9103   1845, History of Northampton, Lehigh, Monroe, Carbon and Schuylkill Counties [Pa.]: Harrisburg, Hickok Canting, xiv, 568 p.
9104   1846, The History and Topography of Dauphin, Cumberland, Franklin, Bedford, Adams, and Perry Counties [Pa.]: Lancaster, G. Hills, xii, [25]-606 p., front, 4 plates.
9105   1847, History and Topography of Northumberland, Huntingdon, Mifflin, Centre, Union, Columbia, Juniata and Clinton Counties: Lancaster, G. Hills, 551, [557]-564, [4] p., front, plates.

The Rural Magazine (Newark, New Jersey, 1798-1799)
9106   1798, Account of the mines in Sweden: v. 1, no. 19, [2].
9107   1798, Earthquake at Sienna: v. 1, no. 32, [2-3].
9108   1798, Description of the salt mines at Williska, in Poland: v . 1, no. 33, [1].

The Rural Magazine and Literary Evening Fireside (Philadelphia, 1820)
9109   1820, Earthquake [at Kutch, India]: v. 1, 76 (6 lines).
9110   1820, A species of limestone: v. 1, 116.
       On hydraulic limestone used for the New York Canal.
9111   1820, [Oil spring at Morgan County, Ohio]: v. 1, 145.
9112   1820, Antediluvian oak: v. 1, 148.
9113   1820, Treatise on agriculture. sect. IV. On the analysis of soils, and of the agricultural relations between soils and plants: v. 1, 211-213.
9114   1820, Oil stones [from Easton, Pennsylvania]: v. 1, 276.
9115   1820, Density, weight, &c. of the earth: v. 1, 315.
9116   1820, Salt mines of Meurthe [France]: v. 1, 357.
9117   1820, More silver [from Salem, Indiana]: v. 1, 357.
9118   1820, Mineralogy: v. 1, 395-396.

The Rural Magazine (cont.)
9119 1820, Cave in Virginia [Wier's Cave]: v. 1, 396.
9120 1820, Staples of Missouri: v. 1, 418-420.
        Lead, iron, and salt production noted.
9121 1820, The diamond [optical properties of]: v. 1, 434.
9122 1820, Precious stones [from Brazil]: v. 1, 437.
9123 1820, The diamond: v. 1, 461.

The Rural Magazine: or, Vermont Repository (Rutland, Vermont, 1795-1796)
9124 1795, Medicinal springs at Saratoga: v. 1, 451-453.
9125 1795, Medicinal springs at Lebanon: v. 1, 453-454.
9126 1795, Natural history: v. 1, 510-512.
        On fossil bones from Biggin Swamp, South Carolina.
9127 1795, Curiosity at Adams, Massachusetts: v. 1, 512-513.
        On marble polished by a stream.
9128 1796, A description of two remarkable ponds in Vermont: v. 2, 244.
        On a mineral spring and pond at Brunswick.
9129 1796, An account of Antiparos [the Grotto]: v. 2, 261-266.
9130 1796, Memoir of the life, character, and writings of the late Professor Winthrop, of Cambridge: v. 2, 278-281.
        Obituary, with mention of his Lecture on Earthquakes (see 11002).
9131 1796, Cabinet of ores and other minerals in the University of Cambridge [Harvard College]: v. 2, 286-288.

Ruschenberger, William Samuel Waithman (1807-1895)
9132 1833, A letter from Dr. R. to the editor, dated Callao Roads, April 15, 1832: Am. J. Sci., v. 23, no. 2, 269-271.
        Geological notes on Chile.
9133 1834, Three years in the Pacific; including notices of Brazil, Chile, Bolivia, and Peru, by an officer of the United States Navy ... : Phila., Carey, Lea and Blanchard, xi, [1], [9]-441 p.
9134 1838, A voyage around the world; including an embassy to Muscat and Siam, in 1835, 1836, and 1837: Phila., Carey, Lea snd Blanchard, 559 p.
9135 1846, Elements of geology: Prepared for the use of schools and colleges ... from the text of Beudant, Milne-Edwards and Achille Comte: Phila., Grigg and Elliott, 235 p., illus., plate, front.
        Volume 1 of "First books of natural history". Based on text by François Sulpice Beudant.
9136 1850, A lexicon of terms used in natural history ... : Phila., Lippincott, Grambo and Co., 161 p.
9137 1850, Elements of natural history, embracing zoology, botany and geology: Phila., J. B. Lippincott, 2 v.

Rush, Benjamin (1745-1813)
9138 1783, A syllabus of a course of lectures on chemistry, for the use of the students of medicine in the College of Philadelphia: Phila., printed by Charles Cist, 39 p., 1 folded leaf.
        Evans number 18173.

Russager, M.
9139 1841, [Letter on the mineral wealth of Africa]: Fam. Mag., v. 8, 95.

Russel
9140 1830, Salt mines of Wieliezka [Poland]: The Friend, v. 3, 35-36.
        Extracted from Tour in Germany.
9141 1830, Stalactite cavern of Adelsberg: The Friend, v. 3, 65-66.
        Extracted from Tour in Germany.

Russell, George
9142 1819, Description of the eruption on Mount Macaluba: Am. Mon. Mag. and Crit. Rev., v. 4, no. 6, 471-472.
9143 1819, [Excerpts from A tour through Sicily]: NY Lit. J., v. 1, 437-441.
        In review (see 7775).
9144 1819, Observations made ... in ascending Mount Etna, on the 30th and 31st May 1815: Robinson's Mag., v. 2, 12.
        Extracted from Lit. Gaz. Follows "Journey to Mount Etna" (see 8824).

Russell, William (see 10794 with others)

Rutter, John Obadiah Newell (1799-1888)
9145 1833, Notice of a new oxy-hydrogen blow-pipe apparatus: Franklin Inst. J., v. 11 (ns), 149-152.
9146 1833, Reinvention of Hare's compound blow-pipe: West. Med. Gaz., v. 1, 165-166.
        Abstracted from Franklin Inst. J.

Ryan, William Redmond
9147 1850, [Excerpts from Personal adventure in Upper and Lower California, in 1848-49]: Eclec. Mag., v. 21, 289-301.
        In review of 1850 London edition (see 3421).

S., E. G. (see 7830)

S., J. H. (see 1223)

S., T. P. (see 10777)

Sabine, Sir Edward (1788-1883)
9148 1829, Observations on the magnetism of the earth, especially of the Arctic regions: Am. J. Sci., v. 17, no. 1, 145-156, chart.
9149 1836, The polarity and dip of the compass needle in different places: Am. J. Sci., v. 29, no. 2, 353-354.
        In British Assoc. Proc.

Sabine, E. (cont.)
9150   1839, Report on the variation in the magnetic intensity: Am. J. Sci., v. 35, no. 2, 296-297.
         In British Assoc. Proc.
9151   1840, [On the importance of combined magnetic observations]: Am. Philos. Soc. Proc., v. 1, no. 12, 242-243.
         Communicated by A. D. Bache.

Saemann, L.
9152   1850, Remarks on boltonite: Am. Assoc. Adv. Sci. Proc., v. 2, 105-109.

Sagra, Ramón de la (1798-1871)
9153   1828, [On the mines and minerals of Cuba]: S. Rev., v. 4, no. 8, 285-321.
         Abstracted from Anales de Ciencias, Agricultura, Comercia y Artes.

St. Croix and Lake Superior Mining Company
9154   1845, A few remarks on the operations of the companies at present organized, for digging & smelting of copper and other ores, on Lake Superior and the Saint Croix River: n.p., 16 p.
         Signed by "an explorer".

St. John, John R.
9155   1846, A true description of the Lake Superior country; it rivers, coasts, bays, harbours, islands, and commerce. With Bayfield's chart (showing the boundary line as established by joint commission.) Also a minute account of the copper mines and working companies, accompanied by a map of the mineral region; showing, by their no. and place, all the different locations: and containing a concise mode of assaying, treating, smelting, and refining copper ores: NY, William H. Graham, 118 p., maps.

Saint Petersburg Academy of Science
9156   1831, St. Petersburg Academy of Science: Am. J. Sci., v. 20, no. 2, 389-391.
         Abstracted by J. Griscom from Revue Encyclopédique. On the proceedings of the Academy.

Sainte-Claire Deville, Charles Joseph (1814-1876; see Deville, Charles Joseph)

Salisbury, James Henry (1823-1905)
9157   1850, [Analysis of soil from Vernon, New York]: NY State Agri. Soc. J., v. 1, 83-84.

Salisbury, Samuel, jr. (1806-1850)
9158   1835, Observations on the waters of the Avon New-Bath Spring and Long's Spring, at Avon, Livingston County, N.Y.: Rochester, William Alling, 7 p.
9159   1838, Analysis of the mineral waters of Avon: Rochester, W. Alling, 32 p.
9160   1838, Analysis of the mineral waters of Avon: Am. J. Sci., v. 35, no. 1, 188-189.
9161   1841, The mineral waters of Avon, Livingston County, N.Y., from a treatise published in 1833: Rochester, Wills and Hayes, 18 p.
9162   1843, Classification of sulphurous waters, with an analysis of those of Avon, Livingston County, and Sharon, Schoharie County, New York: NY J. Medicine and Collateral Sci's, v. 1, 348-352.
9163   1845, A descriptive, historical, chemical and therapeutical analysis of the Avon sulphur springs, Livingston County, N.Y.; with directions for their use: Rochester, D. M. Dewey, 1, [5]-95 p.

Salvetat, M.
       See also article 2786 with M. Damour.
9164   1849, Randanite, a native hydrated silica from Algiers: Am. J. Sci., v. 8 (ns), no. 1, 120-121.
         Abstracted from "Ann. de Ch. et de Phys."

Sample, James R.
9165   1822, Bituminous substances of Barbados: Am. J. Sci., v. 5, no. 2, 406.

Sanders, George Nicholas (1812-1873)
9166   1845, [Mineral region of Lake Superior]: US SED 174, 28-2, v. 11 (461), 8-14.
         Also in US SED 117, 28-2, v. 7 (457), 3-9.

Sandwich Island Gazette, & Journal of Commerce (Hawaii, 1836-1839)
9167   1837, Manufacture of the diamond: v. 1, no. 37, 4.
9168   1837, Wonderful curiosity: v. 1, no. 43, 1.
         On a fossil palm tree in coal from Anzin Mine.
9169   1837, More about the strange tide and the volcano at Hawaii: v. 2, no. 18, 2.
         Signed by "H."
9170   1838, Latest from California [on the earthquake of 1838]: v. 3, no. 16, 2.
9171   1839, Artificial crystals: v. 3, no. 43, 2.
         Notice of experiments by R. Fox in making artificial mineral veins with electricity.

Sanford, Edward (Professor, Polytechnic School of Chitteningo, New York)
9172   1830, An account of depositions of calcareous tufa at Chitteningo, Madison County, N. York: Am. J. Sci., v. 18, no. 2, 354-356.

Sanno, Frederick D.
9173   1831, Machine for separating gold from sand, earth, &c.: Franklin Inst. J., v. 7 (ns), 253-254.

Sanson, Joseph (1765/6-1826)
9174   1817, Sketches of Lower Canada, historical and descriptive; with the author's recollectins of the soil, and aspect; the morals, habits, and religious institutions of that isolated country ... : NY, Kirk and Mercein, 301, xvi p., front.

Sargent, Winthrop (1753-1820)
9175   1814, Account of several shocks of an earthquake in the southern and western parts of the United States: Am. Acad. Arts Sci´s Mem., v. 3, pt. 2, 350-358.

Sartain´s Union Magazine (New York and Philadelphia, 1847-1850)
9176   1847, Oil of stones [from France]: v. 1, 144.
9177   1848, [Review of Manual of mineralogy by J. D. Dana (see 2832)]: v. 3, 239.
9178   1849, [Review of Oregon and California in 1848 by J. Q. Thornton (see 10259)]: v. 4, 288.
9179   1849, [Review of The gold-seeker´s manual by D. T. Ansted (see 1283)]: v. 4, 351.
9180   1849, [Review of The earth and man by A. Guyot (see 4609)]: v. 5, 125-126.
9181   1849, [Review of Elements of geology by D. Page (see 8108)]: v. 5, 127 (13 lines).
9182   1849, [Review of A second visit to the United States by C. Lyell (see 6571)]: v. 5, 256.
9183   1850, [Review of Sights in the gold region by T. T. Johnson (see 5760 and 5761)]: v. 6, 102 (11 lines).
9184   1850, [Review of Six months in the gold mines by E. G. Buffum (see 2180)]: v. 7, 124-125.

The Saturday Magazine (Philadelphia, 1819-1822)
9185   1819, Valuable discovery [of cobalt at Chatham County, Connecticut]: v. 1, 90.
9186   1819-1821, [Notes on mines and mining]: v. 1, 188-189; v. 2, 188-189; 402; v. 3, 191; 240; 233-234; v. 1 (ns), 42.
9187   1819, Geological Society [of Connecticut, organized]: v. 1, 400.
9188   1819-1822, [Notes on volcanoes]: v. 1, 430; v. 3, 80; 190; v. 5, 800; v. 2 (ns), 139-140.
9189   1819, Geology of Munroe [Kansas]: v. 2, 60.
9190   1819, Institute of France: v. 2, 415.
       Includes abstract of La Places´s theory of the nature and history of the earth.
9191   1820, Anglo-Gallic operation, for determining the figure of the earth, &c.: v. 3, 63-64.
9192   1820, Analysis of sea water: v. 3, 64.
9193   1820, Measurement of the earth: v. 3, 80.
9194   1820, Fossil whales [from Falkirk, Ayrshire]: v. 3, 144.
9195   1820, Geology of the Cape of Good Hope: v. 3, 144 (8 lines).
       Notice of Jameson´s studies.
9196   1820, Colouring of agates: v. 3, 191.
       On staining techniques.
9197   1820, Subterranean sounds in granite rock: v. 3, 191.
9198   1820, Geological Society: v. 4, 316.
       Proceedings of the American Geological Society noted.
9199   1821, Method of staining marble: v. 5, 60-61.
       Signed by "J. B."
9200   1821, Remarkable petrifaction [fossil trees from Glasgow]: v. 5, 208.
9201   1821, [Notice of Hauy´s mineralogy text]: v. 5, 240 (6 lines).
9202   1821, Rock crystal containing globules of water formed, and forming, in decaying granite in Elba: v. 5, 255 (8 lines).
9203   1821, Aerolite [fall at Javinas, France]: v. 1 (ns), 597 (6 lines).
9204   1822, A new mineral substance: v. 2 (ns), 202 (7 lines).
       On steinheilite from Finland, discovered by Gadolin.

Saussure, Horace Benedicte de (1740-1799)
9205   1812, Agenda; or a collection of observations and researches the results of which may serve as the foundations for a theory of the earth: Emp. Arts Sci´s, v. 1, 122-130; 193-208; 262-267; 328-343; 456-473.
       From J. des Mines. This is a catalogue of field observations and queries to be made in geological research.

Say, Benjamin (see 8208 under Pennsylvania)

Say, Thomas (1787-1834)
9206   1819-1820, Observations on some species of zoophytes, shells, etc. principally fossil: Am. J. Sci., v. 1, no. 4, 381-387; v. 2, no. 1, 34-35.
       On fossil corals from the United States and Guadaloupe.
9207   1824, An account of some of the fossil shells of Maryland: Phila. Acad. Nat. Sci´s J., v. 4, pt. 1, 124-155, 7 plates with 38 figs.
9208   1824, Fossil shells found in a shell mass from Anastasia Island: Phila. Acad. Nat. Sci´s J., v. 4, pt. 1, 78-80.
       In "Description of a testaceous formation at Anastasia Island" (Florida) by R. Dietz (see 3186).
9209   1825, On two genera and several species of Crinoidea: Phila. Acad. Nat. Sci´s J., v. 4, pt. 2, 289-296.

Scacchi, Arcangelo (1810-1893)
9210   1844, Periclase, a new mineral [from Vesuvius]: Am. J. Sci., v. 46, no. 1, 212-213.
       Also in Living Age, v. 3, 75.
       Abstracted from Annales des Mines.

Scafe, John
9211   1821, [Excerpts from King Coal´s Levee, A Geological Primer in Verse, and Court News in review]: Lit. and Scien. Repos. and Crit. Rev., v. 3, no. 5, 159-159 [i.e. 167].
9212   1822, Geological poems: Am. J. Sci., v. 5, no. 2, 272-285.

Scenes in the Rocky Mountains, and in Oregon, California, New Mexico, Texas, and the Grand Prairies; ... :
9213   1846, Phila., Carey and Hart.

Schaeffer, Frederick Christian (1792-1831)
9214   1818, [Fluid inclusions in quartz]: Am. Mon. Mag. and Crit. Rev., v. 4, no. 1, 66.
9215   1818, On the peat of Dutchess County - read before the Lyceum of Natural History of New York: Am. J. Sci., v. 1, no. 2, 139-140.

Schaeffer, F. C. (cont.)
9216   1819-1820, Localities of minerals [in New York, New Jersey, and Pennsylvania]: Am. J. Sci., v. 1, no. 3, 236-237; v. 2, no. 2, 241.

Schaeffer, George Christian (1815-1873)
9217   1847, On the presence of fluorine in anthracite: Am. J. Sci., v. 3 (ns), no. 3, 422-423.
9218   1848, Action of sulphuretted hydrogen upon nitric acid, with a new process for obtaining sulphur from sulphate of lime by C. Lecomte: Am. J. Sci., v. 5 (ns), no. 2, 261-262.
        Abstracted from Annales de Chimie.
9219   1848, New and economical preparation of the chromates; by V. A. Jacquelin: Am. J. Sci., v. 5 (ns), no. 2, 262.
        Abstracted from Comptes Rendus.
9220   1848, On the artificial production of minerals and precious stones: Am. J. Sci., v. 5 (ns), no. 1, 125-126.
        Abstracted from Comptes Rendus.
9221   1850, Fossil coniferous wood from the lower Devonian strata, Lebanon, Marion Co., Ky.: Proc. Am. Assoc. Adv. Sci., v. 2, 193-194.

Schafhäutl, Karl Franz Emil von (1803-1890)
9222   1846, Margarodite [from Zillerthal]: Am. J. Sci., v. 2 (ns), no. 3, 417 (4 lines).

Scheerer, Thomas
9223   1845, Descriptions of polycrase and malacrone, two new minerals: Am. J. Sci., v. 49, no. 2, 394.
        Abstracted from Jameson's J. On minerals from Hitterōe, Sweden.
9224   1845, Scheerer on aventurine feldspar: Am. J. Sci., v. 49, no. 2, 394 (7 lines).
        Abstracted from Jameson's J.
9225   1846, Remarkable discoveries in isomorphism: Am. J. Sci., v. 2 (ns), no. 1, 115-116.
        Abstracted from Berzelius' Jahresbericht. On substitution of magnesium, iron, and nickel in minerals.
9226   1847, Aspasiolite, a new mineral [from Kragerö, Norway]: Am. J. Sci., v. 3 (ns), no. 3, 429-430 (7 lines).
        Abstracted from Berzelius' Jahresbericht.
9227   1848, Upon a peculiar kind of isomorphism that plays an important part in the mineral kingdom: Am. J. Sci., v. 5 (ns), no. 3, 381-389; v. 6, no. 1, 57-73; no. 2, 189-206.
        Translated by W. G. Lettsom from Poggendorff's Annalen.
9228   1849, Neolite, a new mineral [from Arendal, Norway]: Am. J. Sci., v. 8 (ns), no. 1, 121-122 (11 lines).
        Abstracted from "Oefvers. af K. Vet. Ak. Foerh."
9229   1849, On cystallized pitchblende: Am. J. Sci., v. 8 (ns), no. 1, 126 (6 lines).
        Abstracted from "Ann. der Ph. und Ch."
9230   1849, On eukolite, a new mineral: Am. J. Sci., v. 8 (ns), no. 1, 126, (7 lines).
        Abstracted from "Ann. der Ph. und Ch."
9231   1849, On euxenite from Tvedenstrand: Am. J. Sci., v. 8 (ns), no. 1, 126 (8 lines).
        Abstracted from "Ann. der Ph. und Ch."

Scheidauer, C. H.
9232   1846, Cuban of Breithaupt: Am. J. Sci., v. 1 (ns), no. 2, 266 (4 lines).
        Abstracted from Poggendorff's Annalen.
9233   1846, Kyrosite of Breithaupt: Am. J. Sci., v. 1 (ns), no. 2, 266 (4 lines).
        Abstracted from Poggendorff's Annalen.

A scheme for improving the mines the mineral and the battery works, in New-England:
9234   1712, n.p.
        Possibly a London publication.

Schmidt, M.
9235   1846, Saccharite [from Silesia]: Am. J. Sci., v. 2 (ns), no. 3, 415 (9 lines).
        Abstracted from l'Institut.

Schmidt, Peter von
9236   1850, Improvements in ore washers: Franklin Inst. J., v. 19 (3s), 249 (10 lines).

Schmitz, M.
9237   1829, Hydrate of silex [from Pfaffenreith Mine]: Am. J. Sci., v. 15, no. 2, 390 (12 lines).
        Abstracted from Bibliothèque Universelle.

Schnabel, M.
9238   1849, On mendipite: Am. J. Sci., v. 8 (ns), no. 1, 127 (2 lines).
        Abstracted from Poggendorff's Annalen.

Schneider, Robert
9239   1850, Wolfram [from Hartz Mines]: Am. J. Sci., v. 10 (ns), no. 2, 254-255.
        Abstracted from "J. fur Prakt. Ch."

Schoepf, Johann David (see Schöpf, Johann David)

The Scholar's Journal (Westfield, Massachusetts, 1829)
9240   1829, Meteoric stones: v. 2, no. 1, 1-2.
9241   1829, Petrifaction: v. 2, no. 2, 21.
9242   1829, Precious metals found in the U. States: v. 2, no. 3, 22-23.
9243   1829, Changes on the earth's surface: v. 2, no. 5, 42-43.
9244   1829, High rock spring at Saratoga: v. 2, no. 6, 46-47.
9245   1829, Lithium: v. 2, no. 7, 53-54.
        On a new element found in petalite.

The Scholar's Quarterly Journal (Westfield, Massachusetts, 1828)
9246   1828, Mineralogical: v. 1, 10-12.
On the variety of minerals, and on New England marble.
9247   1828, Geology: v. 1, 13-14.
On the types of rocks.
9248   1828, Earthquakes: v. 1, 17-18.
9249   1828, Temperature of the interior of the earth: v. 1, 12-13.
Notice of E. Hitchcock's translation of Cordier (see 2645).

Schomburgk, Sir Robert Hermann
9250   1847, On the geological structure of Barbados, and on the fossil infusoria, described by Prof. Ehrenberg, from the Tertiary marls of that island: Am. J. Sci., v. 4 (ns), no. 3, 416-417.
Abstracted from London Ath.

The School Journal, and Vermont Agriculturist (Windsor, Vermont, 1847-1850)
9251   1847, Marl: v. 1, 27.
Question on marl. See also answer by C. B. Adams (27).
9252   1847, Gold mines in Russia: v. 1, 48.

The school of wisdom; or repository of the most valuable curiosities of art and nature. Compiled from various authors:
9253   1787, New-Brunswick, NJ, Printed by Shelly Arnett (?).
Evans number 20695. Signed by "D. L."

Schoolcraft, Henry Rowe (1793-1864)
See also 1500 and 1501 under Barrelville Mining Company.
9254   1819, [Letter on the lead mines of Missouri]: Saturday Mag., v. 2, 345-347, 364-365.
9255   1819, A view of the lead mines of Missouri including some observations on the mineralogy, geology, geography, antiquities, soil, climate, population, and productions of Missouri and Arkansaw, and other sections of the western country: NY, Charles Wiley and Co., 299 p., 3 plates.
For review with excerpts see 7778. For review see 359. For excerpts see 7244.
9256   1819/1820, [Excerpts from Journal of a tour into the interior of Missouri and Arkansas]: NY Lit. J., v. 2, 256-265; 330-344; 393-408; v. 3, 100-111, 169-183.
9257   1820, From Schoolcraft's Lead mines of Missouri: NY Lit. J., v. 3, 345.
9258   1820, On the geology of Missouri, Arkansaw and Illinois: Plough Boy, v. 1, 45, 353.
In Troy Lyc. Proc.
9259   1821, Account of the native copper on the southern shore of Lake Superior, with historical citations and miscellaneous remarks, in a report to the Department of War: Am. J. Sci., v. 3, no. 2, 201-216, fig.
9260   1821, Geological character of the limestone of the Missouri lead mine region: Am. J. Sci., v. 3, no. 2, 248-249.
9261   1821, Narrative journal of travels through the northwestern regions of the United States, extending from Detroit through the great chain of American lakes to the sources of the Mississippi River, performed as a member of the expedition under Governor Cass, in the year 1820: Alb., E. and E. Horsford, 419, 4 p., 8 plates, map.
For review see 2473 and 10101.
9262   1822, A memoir on the geological position of a fossil tree discovered in the Secondary rocks of the River Des Plaines ... : Alb., NY, E. and E. Hosford, 18 p.
9263   1822, Remarkable fossil tree, found about fifty miles S. W. of Lake Michigan, by his Excellency Gov. Lewis Cass and Mr. Henry R. Schoolcraft, in August, 1821, on the River Des Plaines, in the N. E. angle of the state of Illinois: Am. J. Sci., v. 4, no. 2, 285-291.
See also letters by J. Adams (42), T. Jefferson (5724), and J. Madison (6663).
9264   1822, Remarks on the prints of human feet, observed in the Secondary limestone of the Mississippi valley: Am. J. Sci., v. 5, no. 2, 223-230, plate with 2 figs.
With a reply by Col. Thomas H. Benton (see 1630).
9265   1822, Report on the number, value, and position of the copper mines on the southern shore of Lake Superior: US SED 5, 17-2, v. 1 (73), 7-28.
9266   1823, Notice of a recently discovered copper mine on Lake Superior, with several other localities of minerals: Am. J. Sci., v. 7, no. 1, 43-49.
9267   1825, [Letter on Lake Superior copper region]: Niles' Wk. Reg., v. 28, 11.
9268   1825, Travels in the central portions of the Mississippi valley; comprising observations on its mineral geography, internal resources, and aboriginal population ... performed under the sanction of the government ... in 1821: NY, Collins and Hannay, iv, 459 p., 3 plates, 2 maps, front.
For review see 7822 and 10496.
9269   1825, Remarks on native silver from Michigan: NY Lyc. Nat. History Annals, v. 1, pt. 2, no. 8, 247-248.
9270   1831, Prints of human feet in rocks: Mon. Repos., v. 1, 242-245; plate.
9271   1833, Geology [letter to P. Cleaveland on Mississippi Valley geology]: Lit. Inquirer, v. 1, 7.
9272   1834, and J. Allen, Letter from the Secretary of War, transmitting a map and report of Lieutenant Allen, and H. B. [i.e. R.] Schoolcraft's visit to the Northwest Indians in 1832: US HED 323, 23-1, v. 4 (257), 68 p., map.
Printed in Washington by Gales and Seaton.
9273   1834, Narrative of an expedition through the Upper Mississippi to Itasca Lake, the actual source of this river, embracing an exploratory trip through the St. Croix and Burntwood (or Broule) Rivers, in 1832: NY, Harper and Brothers, 308 p.
Contains report by D. Houghton (see 5292).
9274   1834, Natural history: Hist. Soc. Mich. Hist. Scien. Sketches, v. 1, [177]-191.
9275   1834, Remarks on the prints of human feet in a rock: Am. Mag. Use. Ent. Know., v. 1, no. 3, 181-182, fig.
9276   1839, Sketches of a trip to Lake Superior: Knickerbocker, v. 13, 211-215, 428-432.
9277   1841, Leonhard's geology [review of Popular lectures on geology by K. C. von Leonhard (see 6192)]: No. Light, v. 1, 125.
9278   1842, [On fossil human footprints]: Am. J. Sci., v. 43, no. 1, 22-25.
In "Regarding human footprints in solid limestone," by D. D. Owen (see 8051).
9279   1843, The copper rock [of Lake Superior]: Farm's Mon. Vis., v. 5, 185.
Signed by "Colcraft" (i.e. H. R. Schoolcraft).

Schoolcraft, H. R. (cont.)
9280 1843, On the action of the North American lakes: Am. J. Sci., v. 44, no. 2, 368-370.
        In British Assoc. Proc.
9281 1847, [Extracts from report on the Barrelville Mining Company] : see 1502.
9282 1848, Mount Joliet, Illinois: West. Miscellany, v. 1, 60, woodcut.
9283 1849, Diluvial gold deposite of California: Mining J., v. 2, 124.

Schöpf, Johann David (1752-1800)
9284 1788, Travels in the Confederation: Phila., 2 v. (426, 344 p.).
9285 1791, Account of the mineral productions of New-Jersey and the neighboring states:
        Univ. Asylum and Colum Mag., v. 7, 3-8.
        Extracted from "Dr. Schoepf's Travels in America."

Schuyler, General
9286 1828, Table of variation of the magnetic needle: Alb. Inst. Trans., v. 1, 4-7.
9287 1829, Table of variations of the magnetic needle: Am. J. Sci., v. 16, no. 1, 60-63.
        In "Variation of the magnetic needle," by B. Silliman (see 9688).

Schuylkill County, Pennsylvania
9288 1842?, Public meeting of men engaged in the coal and iron trade: n.p., broadside.

Schuylkill Navigation Company
9289 1835, Annual report of the Schuylkill Navigation Company: n.p.
        Not seen. For review see 2736.

Schweigger-Seidel, Franz Wilhelm (1795-1836)
9290 1830, On the blue colouring matter of lapis-lazuli, and on artificial ultramarine: Am.
        J. Pharmacy, v. 2, 136-143.
        Abstracted from Jahrbuch der Chemie.

Schweitzer, G.
9291 1840, Analysis of sea water as it exists in the English Channel near Brighton: Am. J.
        Sci., v. 38, no. 1, 12-21.

The science of geology, from the Glasgow Treatises, with additions:
9292 1838, NHav., B. and W. Noyes.
        First American edition. Not seen. For review see 717.

Scientific American (New York, 1845-1850)
9293 1845-1850, [Brief notes on American and foreign mining] : numerous entries in all
        volumes; consult annual indices for details.
        Reports of discoveries, techniques, and production statistics for copper, lead,
        gold, diamonds, oil, granite, iron, coal, silver, mercury, talc, salt, zinc,
        marble, etc.
9294 1845-1850, [Brief notes on fossils] : numerous entries in all volumes; consult annual
        indices for details.
        Reports on discoveries of Mastodon, Mammoth.
9295 1845-1850, [Brief notes on caves] : v. 1, no. 3, [3]; no. 5, [3]; no. 6, [3]; no. 10,
        [1]; v. 2, 89; v. 3, 298; v. 4, 59; v. 5, 75; 248; 305; 386; v. 6, 106.
9296 1845-1850, [Brief notes on volcanoes, earthquakes, landslides] : numerous entries in all
        volumes; consult annual indices.
9297 1845-1850, [Brief notes on magnetic observations] : v. 1, no. 11, [1]; no. 23, [3]; v.
        2, 44; v. 3, 251, 270; v. 4, 400;
9298 1846, Subterranean heat: v. 1, no. 17, [3].
9299 1846, The "rock harmonium": v. 1, no. 19, [3].
9300 1846, Barytes vs. white lead [from Whately, Massachusetts] : v. 1, no. 29, [1].
9301 1846, Artificial coloring of marble: v. 1, no. 34, [2].
9302 1846, Heat of the earth: v. 1, no. 34, [3].
9303 1846, Geological gleanings in Mississippi [notice of a new work by B. L. C. Wailes (see
        10664)] : v. 1, no. 42, [3] (5 lines).
9304 1846, The Deerfield (N. H.) phenomena [subterranean sounds] : v. 2, 2.
9305 1847-1850, [Brief notes on mineral and hot springs] : v. 2, 218, 227; v. 3, 154; v. 4,
        373, 379.
9306 1847, Formation of rocks: v. 2, 291; 302; 310.
9307 1847, Geological wonders in Alabama: v. 2, 299.
        Notice of studies by M. W. Dickeson.
9308 1847, Association of geologists: v. 3, 18 (10 lines).
        Notice of meeting.
9309 1847, Gradual rise of Newfoundland from the sea: v. 3, 19.
9310 1847, Divisibility of matter: v. 3, 27.
9311 1847, Geology system makers: v. 3, 29.
        On the basic concepts introduced by Cuvier and others.
9312 1847, Agricultural chemistry: v. 3, 59.
9313 1847, Scientific expedition to the Dead Sea: v. 3, 74 (12 lines).
9314 1847, Rowe's universal pulverizing pressure ore mill: v. 3, 81, fig.
9315 1847, Natural bridge in Illinois [Jackson County] : v. 3, 106.
9316 1848, The Azores rising: v. 3, 130 (8 lines).
9317 1848, Diamonds converted to charcoal: v. 3, 131.
9318 1848, New blow pipe: v. 3, 140.
        On the blow pipe improved by M. A. Haughton.
9319 1848, Geology: v. 3, 171.
        Summary of C. Lyell's observations.
9320 1848, Figure of the earth: v. 3, 203.
9321 1848, Collections of specimens from soundings: v. 3, 318.
        Collected on the Bache survey of the Atlantic coast.
9322 1848, Two new minerals: v. 3, 378.
        On medjidite and liebigite from Turkey, discovered by J. L. Smith.
9323 1848, The salt lake of the Rocky Mountains: v. 3, 382.
9324 1848, Meteorites - their origin: v. 4, 94.
9325 1849, Stone cutting machine: v. 4, 129, fig.

Scientific American (cont.)
9326　1849, The mineralogist. - The description and locality of every important mineral in the United States: v. 4, 179; 187; 195; 203; 227; 235; 267; 275; 283; 291; 299; 307; 315; 323; to be continued, but no more found.
　　　　Minerals from A through S completed. For notice see 9327.
9327　1849, Mineralogy: v. 4, 181.
　　　　Notice of Scientific American series, see 9326.
9328　1849, Improved gold washer: v. 4, 233, woodcut.
9329　1849, Meteoric iron in South Carolina [at Chesterfield]: v. 4, 298.
9330　1849, Limestone rock: v. 4, 374.
9331　1849, Soils: v. 5, 46.
9332　1849, The blow-pipe and its uses: v. 5, 61.
9333　1850, The Arctic region: v. 5, 169.
9334　1850, An American mineralogist in Turkey: v. 5, 248 (10 lines).
　　　　On the explorations of J. L. Smith.
9335　1850, Curious discovery [in Brighton, Pennsylvania]: v. 5, 268 (15 lines).
　　　　On a curious rock.
9336　1850, Red oxide of zinc and franklinite [from Franklin, New Jersey]: v. 5, 291.
9337　1850, Charcoal melted and the Washington globe: v. 5, 301.
　　　　On R. Hare's blow pipe experiments with carbon.
9338　1850, Geological survey of Canada: v. 5, 378.
9339　1850, [Review of The pre-Adamite earth by J. Harris (see 4830)]: v. 6, 64.
9340　1850, Battin's coal breaker: v. 6, 17, 3 figs.
　　　　On coal mining equipment.
9341　1850, Gold in aerolite [from Marne, France]: v. 6, 73.
9342　1850, Geology: v. 6, 78.
　　　　A history of the progress of geology.
9343　1850, The physical aspects of geological systems: v. 6, 78.
　　　　On the relation between geology and topography.
9344　1850, [Review of The footprints of the Creator by H. Miller (see 7186)]: v. 6, 78.

Scientific and Literary Journal, for the Diffusion of Useful Knowledge (Boston, 1837)
9345　1837, Meteors: v. 1, 7-18.
9346　1837, [Notes on earthquakes and volcanoes]: v. 1, 45-46; 145; 192.
9347　1837, Diamond: v. 1, 68-70.
9348　1837, Remarkable avalanche [at Troy, NY]: v. 1, 71-72.
9349　1837, Fossil remains [of a Dinotherium, from Alzey, Germany]: v. 1, 72.
9350　1837, Earthquakes: v. 1, 77-86.
9351　1837, Artesian wells: v. 1, 91-93.
9352　1837, Mineral tallow [from Loch Tyne, Scotland]: v. 1, 144.
9353　1837, Preservation of animal matter in mines: v. 1, 167.
9354　1837, Precious stones: v. 1, 211-215, 229-233.
　　　　From Bingley's Useful Knowledge.

The Scientific Journal (Perth Amboy, New Jersey, 1818-1819)
9355　1818, Death of De Luc, the geologist: v. 1, no. 1, 14 (11 lines).
9356　1818, Tests for different metals: v. 1, no. 3, 56-57.

Scientific Mechanic, Inventors' Advocate, Patent Office Reporter, and Expositor of Arts and Trade (New York, 1847)
9357　1847, Precious stones: v. 1, no. 1, [1].
9358　1847, Gold mines [of the Urals and Siberia]: v. 1, no. 1, [1].
9359　1847, Gold and silver mines of America: v. 1, no. 3, [1].
9360　1847, Silver and copper in New Jersey [from Princeton]: v. 1, no. 4, [1].
9361　1847, Wonders of geology: v. 1, no. 12, [1].
　　　　On the number of species of fossil plants.
9362　1847, Coal mountain in New Hampshire [at Ossipee]: v. 1, no. 19, [1] (9 lines).
9363　1847, Pennsylvania anthracite: v. 1, no. 23, [3].

Scoresby, William (1789-1857)
9364　1824, Account of Captain Scoresby's magnetical discoveries, and of his magnetimeter and chronometrical compass: Bos. J. Phil. Arts, v. 1, no. 5, 446-458.
　　　　Abstracted from Edinburgh Philos. J.
9365　1845, Magnetical investigations: Am. J. Sci., v. 48, no. 1, 33-36.

Scott (Mr.)
9366　1799, Account of a natural curiosity at Abingdon, in Virginia: Wk. Mag., v. 3, 19.
　　　　On a cave at Abingdon.

Scott, J. W.
9367　1844, Ohio - its wealth and resources: Hunt's Merch. Mag., v. 11, 223-230.

Scott, Joseph
9368　1795, The United States gazetteer: containing an authentic description of the several states, their situation, extent, boundaries, soil, produce, climate, population, trade, and manufactures: Phila., Bailey, vi, 286 p., 12 maps.
　　　　Evans number 29476.
9369　1806, A geographical description of Pennsylvania ... : Phila., Robert Cochran (printer), 148 p., map.
9370　1807, A geographical description of the states of Maryland and Delaware ... : Phila., Kimber, Conrad and Co. (printer), 191 p., maps.

Scoville, S. S.
9371　1829, A large bowlder in southern Ohio: Cin. Soc. Nat. History J., v. 1, 56.
　　　　Not seen.

Scrivenor, Harry
9372　1842, Iron trade of Sweden and Norway: Hunt's Merch. Mag., v. 6, 425-433.
　　　　Extracted from Scrivenor's Comprehensive history of the iron trade.

Scrope, George Julius Duncombe Poulett (1797-1876)
9373   1823, An account of the eruption of Vesuvius, in October, 1822: Bos. J. Phil. Arts, v. 1, no. 4, 345.
       From Royal Inst. J.
9374   1826, Observations on the volcanic formations on the left bank of the Rhine: Bos. J. Phil. Arts, v. 3, no. 6, 569-583.
       Abstracted from Brewster's J.
9375   1827, [Excerpts from Considerations on volcanoes]: Am. J. Sci., v. 13, no. 1, 106-145; no. 2, 235-310.
       In review by B. Silliman (see 9670).
9376   1827, Volcanoes. - Have the elevating effects of volcanic power been perceived on the eastern side of the American continent?: Am. J. Sci., v. 13, no. 1, 190-192.
       With remarks by B. Silliman (see 9671).
9377   1828, Descriptive arrangement of volcanic rocks: Am. J. Sci., v. 15, no. 1, 28-40.

Seaborn, Adam (psuedonym: see Symmes, John Cleaves, 1780-1829)

Seabrook, Whitemarsh Benjamin (1795-1855)
9378   1840, Report on Professor Shepard's analysis of the soils of Edisto Island: Farm's Reg., v. 8, 519-528.
       W. B. Seabrook, chairman of the committee.
9379   1849, Agricultural and physical capabilities of South Carolina: De Bow's Rev., v. 7, 145-150.

Seal, Thomas
9380   1821, New locality of beryl [at Westchester, Pennsylvania]: Am. J. Sci., v. 4, no. 1, 39-40.

Seaman, Ezra Champion (1805-1880)
9381   1846, Essays on the progress of nations, in productive industry, civilization, population, and wealth; illustrated by statistics of mining ... : NY, Baker and Scribner, viii, 455, [1], 192 p.
9382   1850, The precious metals, coins and bank notes: Hunt's Merch. Mag., v. 23, 159-171, 270-283.

Seaman, Valentine (1770-1817)
9383   1793, A dissertation on the mineral waters of Saratoga; containing a topographical description of the country, also a method of making artificial mineral water, resembling that of Saratoga: NY, Campbell, 40 p.
       Evans number 26149. For extracts see NY Mag., v. 6, 426-427 (1795).
9384   1808, An examination of the account of an analysis of the Balltown Waters, as published in the last number of the Medical Repository: Med. Repos., v. 11, 253-256.
       For review see 7803. See also article by D. Hosack (5282).
9385   1809, A dissertation on the mineral waters of Saratoga, including an account of the waters of Ballston: NY, Collins and Perkins, 131 p., map.
       Second edition. For review with excerpts see 7681. For review see 7799.

Sears, Reuben
9386   1819, A poem on the mineral waters of Ballston and Saratoga ... : Ballston Spa, NY, J. Comstock, 108 p.

Seay, Thomas
9387   1842, Machine for separating gold from its ores: Franklin Inst. J., v. 4 (3s), 38 (9 lines).

Seckendorf, M. de
9388   1833, Geology: Am. J. Sci., v. 24, no. 1, 203 (5 lines).
       Abstracted from Revue Encyclopedique. On graywacke xenoliths in granite from the Hartz.

Sedgewick, Henry Dwight
9389   1827, Circular: n. p. (Bos.?), 20 p.
       A plea for the Rhode Island Coal Company to resume mining, with testimonial letters on the quality of the coal. See also "Report" (8765).

Sedgwick, Adam (1785-1873)
9390   1834, [On the present state of geology]: Am. Q. Obs., v. 3, 266-267.
       In review of Discourse on the studies of the University (of Cambridge).
9391   1834, [Excerpts from Discourse on the studies of the University]: Select J. For. Periodical Liter., v. 4, pt. 2, 160-164.
       In review (see 9401).
9392   1835, Geology: Am. Mag. Use. Ent. Know., v. 1, no. 10, 412.
       On the agreement between religion and geology.
9393   1837, and R. I. Murchison, A classification of the old slate of Devonshire, and on the true position of the Culm deposits of the central portion of that country: Am. J. Sci., v. 31, no. 2, 349-356, fig.
       In British Assoc. Proc.
9394   1838, Gravel, bowlders, &c.: Am. J. Sci., v. 33, no. 2, 287-288.
       In British Assoc. Proc. On glacial erratics in England.
9395   1838, Catastrophe in a mine [accident at Workington Colleries]: Am. J. Sci., v. 34, no. 1, 9-10.
       In British Assoc. Proc.
9396   1846, The older fossiliferous rocks of England: Am. Q. J. Ag. Sci., v. 3, 158 (14 lines).
9397   1847, Correspondence: Am. Q. J. Ag. Sci., v. 5, 105-107.
       On division of lower Paleozoic rocks of England.

Seeley, Lloyd
9398   1821, Garnet rock [from Reading, Connecticut]: Am. J. Sci., v. 3, no. 2, 241-242.

The Select Journal of Foreign Periodical Literature (Boston, 1833-1834)
9399 1833, Baron Cuvier [obituary]: v. 1, pt. 2, 88-91.
9400 1834, [Review of The analogy of revelation and science by F. Nolan]: v. 4, pt. 2, 23-27.
   From London Lit. Gaz.
9401 1834, [Review of A discourse on the studies of the University by A. Sedgwick (see 9391)]: v. 4, pt. 2, 160-164.
   From London Ath.

Select Reviews of Literature, and Spirit of the Foreign Magazines (Philadelphia, 1809-1812)
9402 1809, Discovery of a new mephitick grotto: v. 1, 284.
9403 1809, A brief account of the earliest discovery of diamonds in Brasil: v. 2, 133-135.
9404 1809, Account of works constructed for the manufacture of mineral tar, pitch, and varnish [from coal, in Staffordshire]: v. 2, 429-430.
9405 1810, An Icelandick tour: v. 3, 132.
   From Philos. Mag. Notice of Hooker's travels in Iceland.
9406 1810, Description of the Grotto of Zinzanusa, or the ancient Temple of Minerva, near the town of Castro in Apulia: v. 3, 207-208.
9407 1810, Singular local changes in the relative situations of France, England, and Holland; occasioned by the encroachments of the sea: v. 3, 387.
9408 1810, Description of the salt mines of Wieliczka: v. 3, 424-426.
9409 1810, [Review of Tableau physique des regions equitoriales by A. Humboldt]: v. 4, 217-239.
   From Edinburgh Rev.
9410 1811, [Review of Faroe Islands by G. Landt]: v. 5, 289-296.
9411 1811, [Review of Memoir sur les elephans vivans et fossiles by G. Cuvier]: v. 6, 304-315.
   From Edinburgh Rev.
9412 1812, Meteoric stones [from Orleans, France]: v. 7, 352.
9413 1812, [Review of Travels in the Island of Iceland by G. Mackenzie (see 6607)]: v. 8, 123-142.
   From Edinburgh Rev.
9414 1812, [The earthquake of Caracas]: v. 8, 143-146.
9415 1812, Earthquake [at Rome]: v. 8, 264.

Selligue, M. (of France)
9416 1840, A new process for making gas for illumination from bituminous schist: Franklin Inst. J., v. 26 (ns), 335-337.
   Abstracted from "Civ. Eng. & Arch. J."

Sementini, Luigi (1775-1847)
9417 1818, Rain of earthy matter [at Naples, Italy]: J. Sci. Arts, v. 5, no. 2, 370-374.

Semmola, M. S.
9418 1845, Tenorite: Am. J. Sci., v. 48, no. 1, 219 (10 lines).
   Abstracted from Geol. Soc. France Bulletin.

Senarmont, Henri Hureau de (1808-1862)
9419 1848, On the conductibility of heat by crystallized bodies: Am. J. Sci., v. 5 (ns), no. 3, 414-415.
   Abstracted from Comptes Rendus.
9420 1849, On the formation of minerals: Am. J. Sci., v. 8 (ns), no. 3, 421.
   Abstracted from Comptes Rendus.
9421 1850, Artificial formation of minerals in the dry way: Ann. Scien. Disc., v. 1, 212-213.
   Abstracted from Comptes Rendus.

Seram (Mr.)
9422 1848, [Observations on the Onondaga Salt Group in New York]: NY J. Medicine and Collateral Sci's, v. 10, 342-343.
   In NY Lyc. Nat. History Minutes.

Serra, J. Corres de
9423 1818, Observations and conjectures on the formation and nature of the soil of Kentucky: Am. Philos. Soc. Trans., v. 1 (ns), 174-180.

Serres, Pierre Marcel Toussaint de (1783-1862)
9424 1831, Letter addressed to M. Cordier, member of the Royal Academy of Sciences, on a certain new bone cave [at Bize, France]: Am. J. Sci., v. 21, no. 1, 56-59.
   Abstracted by J. Griscom from Annales des Mines.
9425 1833, New cave of bones [at Mialet, Department du Gand, France]: Am. J. Sci., v. 23, no. 2, 388-389.
   Abstracted from Revue Encyclopédique.
9426 1841, New observations on the infusoria of rock salt: Am. J. Sci., v. 41, no. 1, 193.
   Abstracted from Annals and Mag. Nat. History.
9427 1845, On the physical facts contained in the Bible compared with the discoveries of the modern science: Living Age, v. 5, 376-387.
   From Edinburgh Philos. J.

Sewall, Joseph (1688-1769)
9428 1727, The duty of a people to stand in aw [sic.] of God, and not sin, when under his terrible judgments. A sermon preach'd at the South Meeting House in Boston, the evening after the earthquake, which was the night between the 29th & 30th of October, 1727. ... : Bos., D. Henchman, 24 p.
   Evans number 2954. Also "second edition," Bos., D. Henchman, 28 p.

Sewall, J. (cont.)
9429 1727, Repentance the sure way to escape destruction. Two sermons on Jrr. [sic] 18.7.8. Preach'd December 21st, on a publick fast occasioned by the earthquake, the night after the Lord's-day Octob. 29th. And on the Lord's-day December 24th. 1727. ... Publish'd with some enlargement ... : Bos., D. Henchman, 55 p.
Evans number 2958.

Sewall, Stephen (1734-1804)
9430 1785, Magnetical observations, made at Cambridge: Am. Acad. Arts Sci's Mem., v. 1, 322-326.

Seward, William Henry (1801-1872)
9431 1839, [Bill for funding the New York geological survey]: Am. Inst. NY J., v. 4, 389-390.

Seybert, Adam (1773-1825)
9432 1800, Adamantine spar and basaltes [from Pennsylvania]: Med. Repos., v. 3, 202-203.
9433 1808, A catalogue of some American minerals, which are found in different parts of the United States: Phila. Med. Mus., v. 5, 152-159, 256-268.
Forty species are listed. For notice see 7353.
9434 1808, Facts to prove that blende, or sulphuret of zinc, may be worked with advantage in the United States: Phila. Med. Mus., v. 5, 209-216.
On Perkiomen Creek, Montgomery County, Pennsylvania mines. Includes discussion of articles by J. Woodhouse (see 11044 and 11048).

Seybert, Henry (1801-1883)
9435 1821, Analysis of American minerals: Phila. Acad. Nat. Sci's J., v. 2, pt. 1, 139-146.
On amphibole from Wilmington, Delaware, copper ore from Lebanon County, Pennsylvania, and apatite from Chester County, Pennsylvania.
9436 1821, An analysis of amorphous garnet, from New Jersey, and fetid sulphate of barytes, from Albemarle County, Virginia: Phila. J. Med. Phys. Sci's, v. 3, 247-250.
9437 1822, Additional facts respecting the condrodite and its identity with the Sparta mineral, (maclureite of Seybert, brucite of Cleaveland,) in a letter addressed to the editor: Am. J. Sci., v. 5, no. 2, 366-367.
9438 1822, Analysis of a sulphuret of molybdenum: Am. J. Sci., v. 4, no. 2, 320-321.
On molybdenite from Chester, Pennsylvania.
9439 1822, Analysis of American minerals: Phila. J. Med. Phys. Sci's, v. 4, 387-388.
Abstract in review of Phila. Acad. Nat. Sci's J.
9440 1822, Analysis of the American chromat of iron: Am. J. Sci., v. 4, no. 2, 321-323.
On chromite from Bare Hills, Baltimore, Maryland.
9441 1822, Analysis of the maclureite, or fluo-silicate of magnesia, a new mineral species, from New-Jersey: Am. J. Sci., v. 5, no. 2, 336-344.
See also article by T. Nuttall (see 7997). On chondrodite from Franklin, New Jersey.
9442 1822, Analysis of the tabular spar, from the vicinity of Willsborough, Lake Champlain, and of the pyroxene and colophanite, which accompany it: Am. J. Sci., v. 5, no. 1, 113-118.
9443 1823, Analysis of a manganesian garnet, from Haddam, Connecticut, with a notice of boric acid in tourmalines: Am. J. Sci., v. 6, no. 1, 155-157.
9444 1823, Analysis of the glassy actynolite from Concord Township, Delaware Co., Penn.: Am. J. Sci., v. 6, no. 2, 331-333.
9445 1823, Analysis of the pyroxene, found at the Franklin Iron Works, near Sparta, Sussex Co. New-Jersey: Am. J. Sci., v. 7, no. 1, 145-149.
On jeffersonite.
9446 1823, Protest of Mr. Henry Seybert, in vindication of his claim to the discovery of fluoric acid in the chondrodite (brucite of Col. Gibbs, maclureite of Mr. Seybert) in reference to a passage in the memoir on the minerals, &c. of Paterson, and the Valley of Sparta," &c. by Mr. Thos. Nuttall, with the reply of the latter: Am. J. Sci., v. 6, no. 1, 168-173.
See also article by T. Nuttall (7996).
9447 1823, Vindication of Mr. Henry Seybert's claim to the discovery of fluoric acid in the condrodite, (Maclureite of Mr. Seybert, yellow mineral of Sparta, N. J.): Am. J. Sci., v. 6, no. 2, 356-361.
9448 1824, Analysis of an hydrate of iron, (bog iron ore,) from Monmouth Co., New Jersey: Am. J. Sci., v. 8, no. 2, 298-299.
9449 1824, Analyses of the chrysoberyls from Haddam and Brazil: Am. J. Sci., v. 8, no. 1, 105-112.
9450 1824, Analysis of the melanite, from Franklin Furnace, Sussex Co., New Jersey: Am. J. Sci., v. 8, no. 2, 300-301.
9451 1825, Analyses of the chyrsoberyls from Haddam and Brazil: Am. Philos. Soc. Trans., v. 2 (ns), 116-123.
9452 1825, Analysis of the hydraulic lime used in constructing the Erie Canal in the state of New York: Am. Philos. Soc. Trans., v. 2 (ns), 229-231.
9453 1826, Analysis of the green earth from Rancocas Creek, New Jersey: Phila. Soc. Promoting Ag. Mem., v. 5, 18-21.
9454 1830, Tennessee meteorite: Am. J. Sci., v. 17, no. 2, 326-328.

Seymour, E. Sanford
9455 1849, Emigrants' guide to the gold mines: Chicago, Illinois, R. L. Wilson, 104 p., map.

Shamokin Coal and Iron Company
9456 1841, A brief sketch of the advantages of the Shamokin Coal and Iron Company, situate in Northumberland County, state of Pennsylvania; with the act of incorporation, &c.: Phila., Brown, Bicking and Guilbert (printers), 33 p.
With testimonial letters, an extract from report by H. D. Rogers (see 8868 and 8869).

Shamokin Coal Company, Northumberland, Pa.
9457   1839, Extract from "An act to incorporate the Queen's Run Rail Road and Coal Company: Phila., n.p.
9458   1840, A brief sketch of the peculiar advantages of the Shamokin Coal Company situate in Northumberland County, state of Pennsylvania: Phila., Brown, Bicking & Gilbert, 12 p. 8868 and 8869).

Sharpe, Daniel
9459   1832, Description of a new species of Ichthyosaurus [from Stratford-on-Avon, England]: Mon. Am. J. Geology, v. 1, no. 7, 330.
9460   1847, On slaty cleavage in North Wales: Am. J. Sci., v. 3 (ns), no. 3, 430-432.
       Abstracted from Q. J. Geol. Soc.
9461   1847, On slaty cleavage [at Tintagel, Cornwall, England]: Am. J. Sci., v. 4 (ns), no. 1, 110-113, 8 figs.
       Abstracted from Q. J. Geol. Soc.

Shattuck, George Cheyne (1783-1854)
9462   1808, Some account of an excursion to the White Hills of New-Hampshire, in the year 1807: Phila. Med. Phys. J., v. 3, pt. 1, 26-35.

Shepard, Charles Upham (1804-1886)
       See also discussions of articles by C. T. Jackson (5621 and J. D. Dana (2817).
9463   1824-1825, By Charles W. [i.e. U.] Shepard [on Massachusetts mineral localities]: Am. J. Sci., v. 8, no. 2, 235; v. 9, no. 1, 47-48; no. 2, 248-249.
9464   1824, Yenite and green feldspar [from New England]: Am. J. Sci., v. 7, no. 2, 251-252.
9465   1825, Notice of several new localities of American minerals: Bos. J. Phil. Arts, v. 2, no. 6, 607-610.
       Massachusetts mineral localities.
9466   1827, A comparison of the crystallographical characters of the cyanite and sillimanite: Am. J. Sci., v. 12, no. 1, 159-160, 2 figs.
9467   1827, Measurements of crystals of topaz from Huntington, Conn.: Am. J. Sci., v. 12, no. 1, 158-159, 3 figs.
9468   1827, Minerals from Antigua: Am. J. Sci., v. 12, no. 1, 378-379.
9469   1827, Minerals from New South Shetland: Am. J. Sci., v. 12, no. 1, 161-162 (10 lines).
9470   1827, Notice of minerals from Plymouth, Conn.: Am. J. Sci., v. 12, no. 1, 161.
9471   1827, Notice of sulphuret of antimony, automalite and pleonaste, at Haddam, Connecticut; with various other localities of minerals: Am. J. Sci., v. 12, no. 1, 156-158.
9472   1827, Phosphate of manganese in Connecticut, new locality of tabular spar, &c. [from Washington, Connecticut]: Am. J. Sci., v. 13, no. 1, 196-198.
9473   1827, Physical and chemical properties of the native iron of Canaan, ascertained in the laboratory of Yale College: Am. J. Sci., v. 12, no. 1, 155-156.
       In "Notice of native iron from Canaan, Connecticut," by W. Burrall (see 2203).
9474   1828, Chemical examination [of native iron from Bedford County, Pennsylvania]: Am. J. Sci., v. 14, no. 1, 183-186.
       In article by B. Silliman (see 9674).
9475   1828, Measurements of crystals of zircon, from Buncomb, North Carolina: Am. J. Sci., v. 13, no. 2, 392-393, 4 figs.
9476   1829, Analysis of the meteoric iron of Louisiana, and discovery of the stanniferous columbite in Massachusetts: Am. J. Sci., v. 16, no. 2, 217-224, fig.
9477   1829, A mineralogical and chemical description of the Virginia aerolite: Am. J. Sci., v. 16, no. 1, 191-205, fig.
9478   1829, Native alum, in Milo: Am. J. Sci., v. 16, no. 1, 203-205.
9479   1829, New minerals [from Europe]: Am. J. Sci., v. 15, no. 2, 386-389.
       Descriptions of eight new minerals discovered in Europe by various mineralogists.
9480   1829, On crystallized native terrestrial iron, ferro-silicate of manganese, and various other American minerals: Am. J. Sci., v. 17, no. 1, 140-145, 2 figs.
       See also 9481 for correction.
9481   1830, Correction [to article 9480]: Am. J. Sci., v. 18, no. 2, 390 (6 lines).
9482   1830, Mineralogical journey in the northern parts of New England: Am. J. Sci., v. 17, no. 2, 353-360, 3 figs.; v. 18, no. 1, 126-136, 3 figs; no. 2, 289-303.
       On New England mineral localities.
9483   1830, On the mineralogical and chemical characters of the deweylite, and the probable identity of the "magnesian hydrate of silica": Am. J. Sci., v. 18, no. 1, 81-84.
       Follows "Notice of some localities of minerals," by P. T. Tyson (see 10453).
9484   1831, Chemical examination [of Buffalo, New York, mineral waters]: Am. J. Sci., v. 20, no. 1, 157.
       In "Buffalo mineral spring," by Dr. M. Bristol (see 2013).
9485   1831, Notice of the mine of spathic iron (steel ore) of New Milford, and of the iron works of Salisbury, in the state of Connecticut: Am. J. Sci., v. 19, no. 2, 311-326, 2 figs.
9486   1832, Datholite and iolite in Connecticut: Am. J. Sci., v. 22,v no. 2, 389-391.
9487   1832, A sketch of the mineralogy and geology of the counties of Orange (N. Y.), and Sussex (N. J.): Am. J. Sci., v. 21, no. 2, 321-334, fig., plate with map.
9488   1832, Soda alum of Milo, a sulphate of alumine: Am. J. Sci., v. 22, no. 2, 387-389.
9489   1832, Treatise on mineralogy: NHav., Hezekiah Howe, xx, 256 p., 154 figs.
       For review with excerpts see 553, 3086. For review see 1100. For notice see 507, 546.
9490   1833, Geological observations upon Alabama, Georgia, and Florida: Am. J. Sci., v. 25, no. 1, 162-173.
       With excerpts from "Travels in North America" by J. Bartram (see 1543).
9491   1834, Note [on strontianite from Schoharie, New York]: Am. J. Sci., v. 27, no. 1, 183.
       See also article by E. Emmons (3493).
9492   1835, Aerolites: Am. J. Sci., v. 28, no. 2, 288-290.
       In his review of "Notice of the transactions of the Geological Society of France" by A. Boué (see 9497). On the origin of meteorites.

Shepard, C. U. (cont.)

9493 1835, Chemical mineralogy: Am. J. Sci., v. 28, no. 2, 290-292.
In his review of "Notice of the transactions of the Geological Society of France" by A. Boué (see 9497). On the origin of mineral veins by hydrothermal action.

9494 1835, General view of the progress of geology for 1833: Am. J. Sci., v. 28, no. 2, 294-296.
In his review of "Notice of the transactions of the Geological Society of France" by A. Boué (see 9497).

9495 1835, Microlite, a new mineral species [from Chesterfield, Massachusetts]: Am. J. Sci., v. 27, no. 2, 361-362.

9496 1835, New societies and publications: Am. J. Sci., v. 28, no. 2, 283-287.
In his review of "Notice of the transactions of the Geological Society of France" by A. Boué (see 9497).

9497 1835, Notice of the transactions of the Geological Society of France, for 1833, by M. Boue, 8vo. p. 506 [review of]: Am. J. Sci., v. 28, no. 2, 283-296.
Contains articles 2413, 4239, 9492 to 9494, 9496, 9500 and 121.

9498 1835, Notice of the "Travels of a naturalist in the Alps," read before the Society of Naturalists at Soleure, by their President, Fr. Jos. Hugi: Am. J. Sci., v. 28, no. 2, 296-303.

9499 1835, On the strontianite of Schoharie, (N. Y.) with a notice of the limestone cavern in the same place: Am. J. Sci., v. 27, no. 2, 363-370, 3 figs.

9500 1835, Origin of fossil pyrites: Am. J. Sci., v. 28, no. 2, 293 (8 lines).
In his review of "Notice of the transactions of the Geological Society of France" by A. Boué (see 9497).

9501 1835, Reply to "Observations on the Treatise of mineralogy of Mr. C. U. Shepard, by Andres Del Rio": Am. J. Sci., v. 27, no. 2, 312-325.
See also articles by A. Del Rio (3086 and 3090).

9502 1835, Treatise on mineralogy: Second part consisting of descriptions of the species, and tables, illustrative of their natural and chemical affinities. With five hundred wood cuts. In two volumes: NHav., Hezekiah Howe and Co., and Herrick and Noyes, 2 v. (v. 1, xviii, 300 p., figs. 1-271; v. 2, 331 p., figs. 272-500).
For review see 625 and 708.

9503 1835, Uranite at Chesterfield, Mass.: Am. J. Sci., v. 28, no. 2, 382 (6 lines).

9504 1836, Chemical examination of the water of the Gray Sulphur Springs of Virginia: Am. J. Sci., v. 30, no. 1, 100-109.

9505 1836, Rejoinder of Prof. Shepard to Prof. Del Rio: Am. J. Sci., v. 31, no. 1, 131-134.
See also articles by A. Del Rio (3086 and 3090).

9506 1837, Analysis of soil from the sugar plantation of Mr. John Renny, of Willow-Island, near Donaldsonville, Louisiana: S. Agriculturist, v. 10, 65-67.

9507 1837, Chemical examination of microlite [from Massachusetts]: Am. J. Sci., v. 32, no. 2, 338-341.

9508 1837, Description of edwardsite, a new mineral [from Norwich, Connecticut]: Am. J. Sci., v. 32, no. 1, 162-166.

9509 1837, and J. D. Dana, Edwardsite: Am. J. Sci., v. 33, no. 1, 202-203, 2 figs.

9510 1837, Introductory remarks [to "Description of the Guadaloupe fossil human skeleton," by J. Moultrie (see 7502)]: Am. J. Sci., v. 32, no. 2, 361.

9511 1837, Notice of eremite, a new mineral species [from Watertown, Connecticut]: Am. J. Sci., v. 32, no. 2, 341-342, fig.

9512 1837, A report on the geological survey of Connecticut: NHav., Printed by B. L. Hamlen, 188 p., figs.
2000 copies printed. For notice see 685 and 708.

9513 1838, Calstronbarite, a new mineral species [from Schoharie, New York]: Am. J. Sci., v. 34, no. 1, 161-163.

9514 1838, Columbite and tin-ore at Beverly, Mass.: Am. J. Sci., v. 34, no. 2, 402 (3 lines).

9515 1838, Geology of upper Illinois: Am. J. Sci., v. 34, no. 1, 134-161, 10 figs., 2 maps.
Also reprint edition.

9516 1838-1839, Missouri iron mountain: The Friend, v. 12, 83. Also in Niles' Wk. Reg., v. 55, 216 (1838); Bos. Wk. Mag., v. 1, 91 (1838); Am. Inst. NY J., v. 4, 220-221 (1839).
Also reprint edition.

9517 1838, Notice of a second locality of topaz in Connecticut [from Middletown], and of the phenakite in Massachusetts [from Goshen]: Am. J. Sci., v. 34, no. 2, 329-331.

9518 1838, Notice of danburite, a new mineral species [from Danbury, Connecticut]: Am. J. Sci., v. 35, no. 1, 137-139.

9519 1838, Notice of warwickite, a new mineral species [from Warwick, New York]: Am. J. Sci., v. 34, no. 2, 313-315.

9520 1838, Address ... Sept. 25, 1838: NHav., Babcock and Galpin, 24 p.
Delivered to the New Haven Horticultural Society.

9521 1838, [Chemical analysis of coal]: in Report made to the Maryland Mining Company by B. Silliman and B. Silliman, jr. (see 9748), 19-20.

9522 1839, Analysis of warwickite [from Warwick, New York]: Am. J. Sci., v. 36, no. 1, 85-87.

9523 1839, Notice of a report on a re-examination of the economical geology of Massachusetts; by Edward Hitchcock: Am. J. Sci., v. 36, no. 2, 363-378.
With excerpts by E. Hitchcock (see 5128) and S. L. Dana (see 2883).

9524 1839, On meteoric iron from Ashville, Buncombe County, N. C.: Am. J. Sci., v. 36, no. 1, 81-84.

9525 1839, [Footnotes to "On two new cobalt minerals, from Modum in Norway" by F. Wohler (see 11024)]: Am. J. Sci., v. 36, no. 2, 332-335.

9526 1840, Analysis of a meteorite: Am. J. Sci., v. 38, no. 1, 120 (3 lines).
In British Assoc. Proc.

9527 1840, Chemical notices: S. Agriculturist, v. 1 (ns), 153-154.
On the chemical analysis of limestones and marls of North Carolina.

9528 1840, Notice ... of certain mineral substances submitted to examination at the laboratory of the Medical College of South-Carolina: S. Agriculturist, v. 1 (ns), 22-25.

9529 1840, On a supposed new mineral species [from New York and Canada]: Am. J. Sci., v. 39, no. 2, 357-360, 2 figs.

Shepard, C. U. (cont.)

9530 1840, On the identity of edwardsite with monazite, (mengite,) and on the composition of the Missouri meteorite: Am. J. Sci., v. 39, no. 2, 249-255.

9531 1840, and F. Shepherd, Reports of Professor Charles U. Shepard and Forrest Shepherd esq. respecting mineral deposits in the states of Missouri and Illinois: Bos., 12 p., map.

9532 1841, Analysis of prairie soils, from Montgomery County, (Ala.): S. Agriculturist, v. 1 (ns), 133-134.

9533 1841, On native and meteoric iron [from New York]: Am. J. Sci., v. 40, no. 2, 366-370, fig.

9534 1841, On two decomposed varieties of iolite [from New England]: Am. J. Sci., v. 41, no. 1, 354-358.

9535 1841, Report [on geology of the Potomac and Alleghany Coal and Iron Company]: in 8548.

9536 1842, Analysis of meteoric iron from Cocke County, Tennessee, with some remarks upon chlorine in meteoric iron masses: Am. J. Sci., v. 43, no. 2, 354-363.

9537 1842, On the want of identity between microlite and pyrochlore: Am. J. Sci., v. 43, no. 1, 116-121.

9538 1842, On washingtonite (a new mineral), the discovery of euclase in Connecticut, and additional notices of the supposed phenakite of Goshen, and Calstron-baryte of Schoharie, N. Y.: Am. J. Sci., v. 43, no. 2, 364-366, 2 figs.

9539 1843, Analysis of soils [from Georgetown, South Carolina]: S. Agriculturist, v. 3 (ns), 253-254.

9540 1843, Analysis of soils from the tide-swamp plantation of Col. R. F. W. Allston: in Report of the commencement and progress of the agricultural survey of South Carolina, by E. Ruffin (see 9087), appendix.

9541 1843, On phosphate of lime (apatite), in the Virginia meteoritic stone: Am. J. Sci., v. 45, no. 1, 102-103.

9542 1844, A treatise on mineralogy: NHav., A. H. Maltby; Char., Babcock and Co.; Amherst, Mass., J. S. and C. Adams, viii, 168 p., 120 figs.
Second edition. For review see 794.

9543 1845, Analysis of marls from the vicinity of Charleston: S. Agriculturist, v. 5 (ns), 133-139.

9544 1845, Meteoric stones: West. J. Medicine Surg., v. 4 (ns), 455-457.

9545 1845, [On the occurrence of itacolumite and diamonds]: Assoc. Am. Geol's Nat's Proc., v. 6, 41-43.
Also in Am. Q. J. Ag. Sci., v. 2, 154-156.

9546 1845, Reply to a notice of Shepard's mineralogy, with various mineralogical observations: Am. J. Sci., v. 48, no. 1, 168-180.
Discussion of review by B. Silliman, jr. (see 9815). See also "Remarks on uranium and pyrochlore" by J. E. Teschemacher (see 10190).

9547 1845, [Spearhead of meteoric iron from Madagascar]: Assoc. Am. Geol's Nat's Proc., v. 6, 40-41.
Also in Am. Q. J. Ag. Sci., v. 2, 153.

9548 1846, Analysis of soils, - marl, clay and burnt-marsh: S. Agriculturist, v. 4 (ns), 145-151.

9549 1846, On three new mineral species from Arkansas and the discovery of the diamond in North Carolina: Am. J. Sci., v. 2 (ns), no. 2, 249-254, 2 figs.

9550 1846-1848, Report on meteorites: Am. J. Sci., v. 2 (ns), no. 3, 377-392; v. 4 (ns), no. 1, 74-87, 5 figs., 1847; v. 6 (ns), no. 3, 402-417, 2 figs.
Also reprint edition.

9551 1847, Corrigenda to vol. XXXII. East Bridgewater meteorite: Am. J. Sci., v. 50, 322 (11 lines).
Meteorite shown to be slag from a furnace.

9552 1847, Extracts from the report of Professor Charles U. Shepard: in Documents of the Phoenix Mining Company (see 8328), 19-25.

9553 1847, Mineralogical notices: Am. J. Sci., v. 4 (ns), no. 2, 278-281, fig.
See also article by J. D. Whitney (10913). On tautolite from Lake Superior, arkansite from Arkansas, and native platinum from North Carolina.

9554 1847, Reclamation respecting the identity of pinite, chlorophyllite, and other minerals, with the species iolite: Am. J. Sci., v. 3 (ns), no. 2, 266.
Abstracted by Poggendorff's Annalen. With notes by the editors.

9555 1848, An account of the meteorite of Castine, Maine, May 20, 1848: Am. J. Sci., v. 6 (ns), no. 2, 251-253.

9556 1848, Analysis of marls from the vicinity of Charleston: in Report on the geology of South Carolina by M. Tuomey (see 10438), xxxiv-xliii.

9557 1848, The deficiency of scientific information in the agricultural community, and the need of its being supplied: Mass. Agri. Societies Trans., for 1847, 248-274.

9558 1848, Observations on Rammelsberg's analysis of the Juvenas meteoric stone, and on the conclusion of Fischer's examination of the Braunau meteoric iron: Am. J. Sci., v. 6 (ns), no. 3, 346-349.

9559 1848, On new minerals from Texas. Lancaster Co., Penn.: Am. J. Sci., v. 6 (ns), no. 2, 249-250.
On williamsite.

9560 1849, Meteorites - their origin: Niles' Wk. Reg., v. 75, 28.
Abstracted from Am. J. Sci.

9561 1849, Notices of American minerals [from North Carolina, Georgia, and Massachusetts]: Am. J. Sci., v. 8 (ns), no. 2, 274-275, 2 figs.

9562 1849, On meteoric iron in South Carolina [Chesterville]: Am. J. Sci., v. 7 (ns), no. 3, 449-450.

9563 1849, [On the mineral resources of North Carolina]: in Letters from the Alleghany Mountains by C. Lanman (see 6034), 190-192.

9564 1850, Account of three new American meteorites, with observations upon the geographical distribution of such bodies generally: Am. Assoc. Adv. Sci. Proc., v. 3, 147-157, 2 figs.
Also reprint edition.

Shepard, Forrest (see Shepherd, Forrest)

Shepherd, Forrest
See also report with C. U. Shepard (9531).

9565 1834, Geological description of the gold region of Virginia: Farm's Reg., v. 2, 407.
9566 1834, Gold mines in Virginia [in Spotsylvania County]: Farm's Reg., v. 2, 406, 407.

Shepherd, F. (cont.)
9567 1834, The gold region of Virginia: Mech's Mag. and Reg. Inventions, v. 4, 33-34, map.
       Article ascribed to F. Shepard (i.e. Forrest Shepherd).
9568 1836, [Report]: in reports to the Richmond Mining Company by B. Silliman (see 9733), 15-16.
9569 1836, Virginia Exploring and Mining Company: in Report on the Walton Mining Company by B. Silliman (see 9736), 13-20.
9570 1838, [Report on the geology of the West Virginia Iron Mining and Manufacturing Company]: in 10800.
9571 1839, Report to the Potomac and Alleghany Coal and Iron Manufacturing Company: n.p.
       Not seen.
9572 1841, Analysis of prairie soils from Montgomery County, Ala.: Farm's Reg., v. 9, 220.
9573 1841, [Report on the geology of the lands of the Potomac and Alleghany Coal and Iron Company]: in 8548.
9574 1846, Report: in By-laws of the Ontonagon Copper Company (see 8037).
9575 1846, Report of Forrest Shepherd, esq., on the mineral lands of the Montreal mining company on the shores of Lake Superior: Montreal, 32 p.
9576 1846, Geological survey of the mineral lands on the southern shore of Lake Superior, belonging to the Pittsburgh and Boston Copper Harbor Mining Company: Pittsburgh, G. Parkin, 35, 8 l., maps.
       With ore analyses by B. Silliman, jr.
9577 1847, The copper mines of Lake Superior: De Bow's Rev., v. 3, 344-345.
       Ascribed to Prof. Sheppard (i.e. Forrest Shepherd).
9578 1847, Extracts from the reports of Professor Forrest Shepherd ... in relation to the character, location, topography, and value of the mineral property owned by the Phoenix Company: in Documents of the Phoenix Mining Company (see 8328), 7-19.
9579 1847, Observations on the drift furrows, grooves, scratches, and polished surfaces of the rocks of Lake Superior: Am. J. Sci., v. 4 (ns), no. 2, 282-283.
9580 1847, Remarks on a boulder mass of native copper from the southern shore of Lake Superior: Am. J. Sci., v. 4 (ns), no. 1, 115-116.

Sherlock, Thomas (1678-1761)
9581 1750, A letter from the lord Bishop of London, to the clergy and people of London and Westminster; on the occasion of the late earthquake: "London printed: Boston; N.E. re-printed and sold by John Draper," 15 p.
       Evans number 6707.

Sherman, John (1772-1828)
9582 1822, Trenton Falls: Oldenbarnveld [i.e. Trenton Falls], NY, 7 p.
9583 1827, A description of Trenton Falls, Oneida County, N. Y.: Utica, NY, W. Williams, 18 p.
       Also printings in 1828, 1830, 1835, 1838 and 1844. Another edition in 1847 by W. H. Colyer, NY, 23 p.

Sherman, Richard M.
9584 1849, Method of washing gold dust in California: Hunt's Merch. Mag., v. 20, 232.

Sherwood, Henry Hall
9585 1838, Dipping needle for finding longitude: Franklin Inst. J., v. 22 (ns), 270-273.
       In Am. Philos. Soc. Proc. With discussion by Dr. Patterson (8161) and Mr. Walker (10671).
9586 1838, Magnetic discoveries: New-Yorker, v. 6, 7-8.
9587 1841, Terrestrial magnetism: Am. Rep. Arts Sci's Manu., v. 4, 174-180.

Sherwood, John D. (1818-1891)
9588 1845, Some observations upon the valley of the Jordan and the Dead Sea: Am. J. Sci., v. 48, no. 1, 1-16.
       With analysis of Dead Sea water by B. Silliman, jr. (see 9817).

Shocking earthquakes. Charleston, S.C., Feb. 7, 1812. Yesterday morning, about half past three o'clock, the inhabitants of this place were very much alarmed by another shock ... :
9589 1812, Bos., broadside.

Shriver, Howard
9590 1824?, Catalogue of fossils found at Cumberland, Maryland: n.p., 4 p.
       Not seen.

Shriver, James
9591 1824, An account of surveys and examinations, with remarks and documents, relative to the projected Chesapeake and Ohio, and Ohio and Lake Erie Canals: Balt., F. Lucas, 116 p., illus., 2 maps.

Shumard, Benjamin Franklin (1820-1869; see 8083 with D. D. Owen, and 11094 and 11095 with L. P. Yandell.)

Shurtleff, Nathaniel Bradsteet (1810-1874)
9592 1846, [On the Mastodon giganteum from Newburg, Orange County, New York]: Bos. Soc. Nat. History Proc., v. 2, 96-98.

Sibley, John (1757-1837)
9593 1803, A topographical and statistical account of the province of Louisiana, containing a description of its soils, climate, trade and produce: Balt., Martin & Pratt.
       Not seen. Also Providence, RI and Washington, DC editions as "An account of Louisiana."
9594 1806, Some account of the country and productions near the Red River, in Louisiana: Med. Repos., v. 9, 425-427.
9595 1809, Louisiana. An account of the Red River and country adjacent: Am. Reg. or Gen. Repos., v. 4, 49-65.
9596 1845, Naptha [from Natchitoches]: The Investigator, v. 1, 171 (7 lines).

Sickels, T. E.
9597   1846, The Bear Valley coal basin and Bear Mountain Railroad: Hunt's Merch. Mag., v. 14, 141-145.

Sidell, William Henry (1810-1873)
9598   1843, Topographical sketch of Oyster Pond Point and the vicinity: in Geology of New York, first district, by W. W. Mather (see 6907), plate 2 (colored).
On short-term changes in alluvial deposits on Long Island.

Silberman (see 3735 with P. A. Favre)

Silliman, Benjamin (1779-1864)
Benjamin Silliman was founder, and editor from 1818 to 1850, of The American Journal of Science. Many of the unsigned notes, editorial comments, and reviews were probably authored by him (see 318 to 1021).
See also 1500 to 1502 under Barrelville Mining Company. See also discussion of article by W. C. Redfield (8719).
9599   1808, and J. L. Kingsley, Account of a remarkable fall of meteoric stones, in Connecticut: Phila. Med. Phys. J., v. 3, pt. 1, 39-57.
Originally published in Connecticut Herald.
9600   1808, and J. L. Kingsley, Fall of meteoric stones in Connecticut: Med. Repos., v. 11, 202-213.
9601   1809, and J. L. Kingsley, Memoir on the origin and composition of the meteoric stones which fell from the atmosphere, in the County of Fairfield, and state of Connecticut, on the 14th of December, 1807: Am. Philos. Soc. Trans., v. 6, pt. 2, 323-345.
For review see 7090.
9602   1810, and J. L. Kingsley, An account of the meteor which burst over Weston in Connecticut, in December 1807, and of the falling of stones on that occasion: Ct. Acad. Arts Sci's Mem., v. 1, pt. 1, 141-161.
For review see 1079. Also reprint edition.
9603   1810, A journal of travels in England, Holland and Scotland, and of two passages over the Atlantic, in the years 1805 and 1806: NY, Ezra Sargent, 2v.
Also second edition: Bos., Howe and Deforest, 2 v. For review with excerpts see 7359. See also 9628.
9604   1810, Logan rock: Rural Visiter [sic.], v. 1, 14 (11 lines).
From A journal of travels.
9605   1810, Sketch of the mineralogy of the town of New Haven: Ct. Acad. Arts Sci's Mem., v. 1, pt. 1, 83-96.
For review with excerpts see 4241.
9606   1811, Account of the production of the new metals, by the decomposition of soda and potash: Am. Min. J., v. 1, no. 2, 88-93.
9607   1811, Particulars relative to the lead mine near Northampton (Massachusetts): Am. Min. J., v. 1, no. 2, 63-69.
9608   1811, Mineralogy and geology of New Haven: in A statistical account of the city of New Haven by Timothy Dwight, N.Hav., Ct. Acad. Arts and Sci., 5-15.
9609   1812, Mineralogical and geological observations on New Haven and its vicinity: Am. Min. J., v. 1, no. 3, 139-149.
Also in Elementary treatise by P. Cleaveland (see 2422).
9610   1812, Particulars relative to a late accidental explosion of fulminating silver, in the chemical laboratory at Yale College. Communicated by Mr. Griscom: Am. Min. J., v. 1, no. 3, 163-166.
9611   1812, Sketch of the mineralogy of the town of New Haven: Am. Med. Philos. Reg., v. 3, no. 2, 210-211.
Abstracted from Ct. Acad. Arts Sci's Mem.
9612   1813, Experiments on the fusion of various refractory bodies, by the compound blow-pipe of Mr. Hare: Ct. Acad. Arts Sci's Mem., v. 1, pt. 3, 329-339, fig.
Also reprint edition.
9613   1814, Native bismuth [from New Stratford, Connecticut]: Am. Min. J., v. 1, 267, no. 4 (11 lines).
9614   1814, On the powers of the compound blowpipe: Am. Min. J., v. 1, no. 4, 199-210, fig.
9615   1814, [Testimonial letter]: in North-American Coal and Mining Company report (see 7921).
9616   1817, Mr. Hare's blowpipe: NE J. Medicine Surg., v. 6, 301-303.
9617   1817, Professor Silliman on the blow-pipe: Eclec. Rep. and Analytic Rev., v. 7, 289-292.
From NY Daily Advertiser.
9618   1818, Cautions regarding fulminating powders: Am. J. Sci., v. 1, no. 2, 168-170.
9619   1818, Introductory remarks [to readers of the American Journal of Science]: Am. J. Sci., v. 1, no. 1, [iii]-iv, 1-8.
9620   1818, Native copper [from Hartford, Connecticut]: Am. J. Sci., v. 1, no. 1, 55-56.
9621   1818, New localities of agate, chalcedony, chabasie, stilbite, analcime, titanium, prehnite, &c. [in Connecticut]: Am. J. Sci., v. 1, no. 2, 134-135.
9622   1818, The northwest passage, the North Pole, and the Greenland ice: Am. J. Sci., v. 1, no. 2, 101-104.
Contains quote from Q. J.
9623   1818, On the compound blowpipe. Extract from the Journal de Physique, of Paris, for January 1816: Am. J. Sci., v. 1, no. 2, 97-101.
9624   1818, Review of an Elementary treatise on mineralogy and geology, ... By Parker Cleaveland [review with excerpts of 2420]: Am. J. Sci., v. 1, no. 1, 35-52.
9625   1819, Additional notice of the tungsten and tellurium, mentioned in our last number [from Connecticut]: Am. J. Sci., v. 1, no. 2, 312, 405-410.
9626   1819, Conclusion [to volume 1 of the American Journal of Science]: Am. J. Sci., v. 1, no. 4, 440-442.
Evaluation of the contributions to the first volume, and an announcement of the second volume.
9627   1820, Fibrous sulphat of barytes from Carlisle, thirty-four miles west of Albany [New York]: Am. J. Sci., v. 2, no. 1, 172.
9628   1820, A journal of travels in England, Holland and Scotland, and of two passages over the Atlantic, in the years 1805 and 1806: NHav., S. Converse, 3 v.
Third edition. See also 9603.

Silliman, B. (cont.)

9629   1820, [Note on chrysoberyl from Haddam, Connecticut]: Am. J. Sci., v. 2, no. 2, 240 (11 lines).
       In "Localities of minerals" by J. W. Webster (see 10727).

9630   1820, [On geology of New Haven and Litchfield, Connecticut]: Am. J. Sci., v. 2, no. 1, 141-142.
       In Am. Geol. Soc. Proc.

9631   1820, Remarks made on a short tour between Hartford and Quebec in the autumn of 1819: NHav., Converse, [3]-407 p., front., 8 plates.
       For review with excerpts see 1244 and 10876. See also 9658.

9632   1820, Sketches of a tour in the counties of New-Haven and Litchfield in Connecticut, with notices of the geology, mineralogy and scenery, &c.: Am. J. Sci., v. 2, no. 2, 201-235.

9633   1821, Circumstances connected with the formation of ice on still waters, and with the continued action of cold upon the fluid beneath: Am. J. Sci., v. 3, no. 1, 179-182.

9634   1821, Map of mountains: Am. J. Sci., v. 3, no. 2, 364.

9635   1821, Fossil fish [from Sunderland, Massachusetts]: Am. J. Sci., v. 3, no. 2, 365.
       Preface to "List of organic remains and accompanying rocks, contained in a box forwarded to Professor Silliman by Edward Hitchcock."

9636   1821, Massive yellow oxide of tungsten [from Huntington, Connecticut]: Am. J. Sci., v. 4, no. 1, 187-188.

9637   1821, New locality of fluor spar [from Putney, Vermont]: Am. J. Sci., v. 4, no. 2, 188-189.

9638   1821, Notice of an argentiferous galena, from Huntington, and of another lead ore from Bethlem, [sic.] Conn. with miscellaneous observations on lead ores - the latter extracted chiefly from authors: Am. J. Sci., v. 3, no. 1, 173-179.

9639   1821, Notice of minerals and rocks chiefly in Berkshire, Mass. and contiguous to the waters of the upper Hudson and the Lakes George and Champlain, with occasional remarks on other subjects: Am. J. Sci., v. 4, no. 1, 40-55.

9640   1821, Postscript by the editor: Am. J. Sci., v. 3, no. 1, 1-5.
       In "Notices of the mineralogy and geology of parts of South and North Carolina," by J. Dickson (see 3185).

9641   1821, Remarks by the editor: Am. J. Sci., v. 3, no. 1, 5-7, fig.
       In "Notice of vegetable impressions on the rocks connected with the coal formations of Zanesville, Ohio" by E. Granger (see 4492).

9642   1821, Remarks by the editor: Am. J. Sci., v. 3, no. 1, 89-93.
       Discussion of R. Hare's blowpipe.

9643   1821, Remarks by the editor: Am. J. Sci, v. 3, no. 2, 216-227.
       In "Miscellaneous observations relating to geology ...," by A. Brongniart (see 2055).

9644   1822, Fusion and volatilization of charcoal, by the editor, with remarks on these experiments and on the galvanic instruments of Dr. Hare, by Professor John Griscom, of New York: Am. J. Sci., v. 5, no. 2, 361-364.

9645   1822, Miscellaneous notices in mineralogy, and geology: Am. J. Sci., v. 5, no. 1, 39-42.
       Contains 44 notices on United States mineral localities.

9646   1823, Additional notice on the fused carbonaceous bodies: Am. J. Sci., v. 6, no. 2, 378-379.

9647   1823, Expedition of Major Long and party, to the Rocky Mountains: Am. J. Sci., v. 6, no. 2, 374-375.

9648   1823, Experiments upon diamond, anthracite and plumbago with the compound blow pipe, in a letter addressed to Prof. Robert Hare: Am. J. Sci., v. 6, no. 2, 349-353.

9649   1823, Fossil vegetables [notice of Végétaux fossiles by A. Brongniart]: Am. J. Sci., v. 6, no. 2, 381 (6 lines).

9650   1823, Fusion of carbonaceous bodies: Bos. J. Phil. Arts, v. 1, no. 2, 191 (16 lines).

9651   1823, Fusion of plumbago: Am. J. Sci., v. 6, no. 2, 341-349.

9652   1823, Geological surveys on the Great Canal [Erie Canal, New York]: Am. J. Sci., v. 6, no. 2, 373-374 (11 lines).
       Notice of A. Eaton's field work.

9653   1823, Hudson marble [from New York]: Am. J. Sci., v. 6, no. 2, 371.

9654   1823, Mineral caoutchouc [from Southbury, Connecticut]: Am. J. Sci., v. 6, no. 2, 370.

9655   1824, Introductory remarks: Am. J. Sci., v. 8, no. 1, 113-114.
       In "Description and analysis of the sillimanite, a new mineral" by G. T. Bowen (see 1910).

9656   1824, Notice and review of the "Reliquiae diluvianae; observations on the organic remains contained in caves, fissures, and diluvial gravel, and on other geological phenomena, attesting the action of an universal deluge. By the Rev. William Buckland": Am. J. Sci., v. 8, no. 1, 150-168; no. 2, 317-338.
       Review, with abstract and excerpts.

9657   1824, Peculiar form of some of the beryls of Haddam, Conn.: Am. J. Sci., v. 8, no. 2, 395-396.

9658   1824, Remarks made on a short tour, between Hartford and Quebec in the autumn of 1819 ... : NHav., G. Converse, [3]-10, [9]-443 p., 9 plates.
       Second edition. See also 9631.

9659   1824, Remarks upon Prof. Vanuxem's paper of fused charcoal, published in vol. IV, p. 371, of the Journal of the Acad. of Nat. Sci. at Philadelphia: Am. J. Sci., v. 8, no. 1, 147-149.
       See also article by L. Vanuxem (10570).

9660   1825, Notice of a mineral supposed to be a phosphate of lime from Williamsburgh [sic.], Massachusetts, and of the localities of several other minerals: Am. J. Sci., v. 9, no. 1, 174-177.

9661   1825, Notices of minerals, etc., from Palestine, Egypt, etc., in a letter from the Rev. Issac Bird: Am. J. Sci., v. 10, no. 1, 21-29.

9662   1825, Notice of some recent experiments on charcoal, etc.: Am. J. Sci., v. 10, no. 1, 119-126.

9663   1825, Papers relating to the fusion of carbon. I. Remarks by the editor: Am. J. Sci., v. 10, no. 1, 109-110.

9664   1825, Remarks: Am. J. Sci., v. 9, no. 1, 180.
       In "Topaz?" by E. Hitchcock (see 5088).

Silliman, B. (cont.)

9665 1825, Remarks by the editor: Am. J. Sci., v. 10, no. 1, 206-207.
       In "New locality of rubellite, beryl, tourmaline, &c." by S. C. Williams (see 10978).

9666 1826, Anthracite coal of Pennsylvania, &c. Remarks upon its properties and economical uses: Am. J. Sci., v. 10, no. 2, 331-351.

9667 1826, Anthracite coal of Rhode-Island - Remarks upon its properties and economical uses: with an additional notice of the anthracites of Pennsylvania, &c.: Am. J. Sci., v. 11, no. 1, 78-100.

9668 1826, [Introduction to "Notice of the volcanic character of the island of Hawaii" by J. Goodrich (see 4438)]: Am. J. Sci., v. 11, no. 1, 1-36.

9669 1826, Topaz in Connecticut: Am. J. Sci., v. 11, no. 1, 192-194.

9670 1827, Volcanoes [review with excerpts of Considerations on Volcanoes by J. P. Scrope (see 9375)]: Am. J. Sci., v. 13, no. 1, 106-145; no. 2, 235-310.

9671 1827, Remarks: Am. J. Sci., v. 13, no. 1, 191-192.
       In query on Eastern United States volcanoes by G. P. Scrope (see 9376).

9672 1827, [Introduction and remarks to "Notice of some recent experiments in boring for fresh water ... " by Mr. Disbrow (see 3190)]: Am. J. Sci., v. 12, no. 1, 136-144.

9673 1828, Conclusion of the notice and analysis of Professor Daubeny's work on active and extinct volcanoes, from Vol. 13, page 310: Am. J. Sci., v. 14, no. 1, 70-91.
       See also 9677.

9674 1828, Native iron? slightly arseniuretted: Am. J. Sci., v. 14, no. 1, 183-186.
       On iron (meteoritic?) from Bedford County, Pennslyvania. With chemical analysis by C. U. Shepard (see 9474).

9675 1828, Note: Am. J. Sci., v. 14, no. 2, 397 (7 lines).
       Acknowledgement to S. Van Rensselaer for sponsoring geology plates in v. 14 of Am. J. Sci.

9676 1828, Objections to the above theory: Am. J. Sci., v. 15, no. 1, 12-27, fig.
       Follows "Hypotheses on volcanoes and earthquakes" by J. du Commun (see 3255).

9677 1828, Notice and analysis of "A description of active and extinct volcanos, with remarks on their origin, their chemical phenomena and the character of their products, as determined by the condition of the earth, during the period of their formation ; being the substance of some lectures delivered before the University of Oxford, with much additional matter; by Charles Daubeny": Am. J. Sci., v. 13, no. 2, 235-310, fig.
       Review with abstract and excerpts of 2921. See also 9673.

9678 1828, Vermont manganese [from Bennington]: Am. J. Sci., v. 14, no. 1, 190 (6 lines).

9679 1829, Dr. Wollaston [obituary]: Am. J. Sci., v. 17, no. 1, 159-160.

9680 1829, Igneous origin of some trap rocks: Am. J. Sci., v. 17, no. 1, 119-132, plate.
       Notice of ancient volcanoes of the Connecticut River Valley.

9681 1829, Introduction: in Introduction to Geology by R. Bakewell (see 1455).
       Also in second edition, 1833 (see 1457), and third edition, 1839 (see 1460).
       Silliman was the editor of the American editions of Bakewell's Geology.
       For review see 7340.

9682 1829?, Silver Mines in Mexico: NY?, 14 p.
       Not seen.

9683 1829, Introductory notice of specimens of silver from Peru and Chili [sic]: Am. J. Sci., v. 17, no. 1, 43-46.
       In "Sketch of the rich mine of Pasco" by M. de Rivero (see 8803).

9684 1829, Miscellaneous notices of mountain scenery, and of slides and avalanches in the White and Green Mountains: Am. J. Sci., v. 15, no. 2, 217-232.
       Contains letters by C. Wilcox (see 10948) and T. Baldwin (see 1465).

9685 1829, [Notes on "Letters from the Sandwich Island" by J. Goodrich (see 4439)]: Am. J. Sci., v. 16, no. 2, 345-347.

9686 1829, Obituary [of D. H. Barnes, G. T. Bowen, and T. D. Eaton]: Am. J. Sci., v. 15, no. 2, 401-404.

9687 1829, Outline of the course of geological lectures given in Yale College: NHav., H. Howe, 128 p.
       Published as an appendix to Bakewell's Geology (see 1455).
       For review see 9969.

9688 1829, Variation of the magnetic needle: Am. J. Sci., v. 16, no. 1, 60-69.
       Contains letters by Gen. Schuyler (see 9287) and N. Bowditch (see 1898).

9689 1830, Big Bone Lick, Kentucky: Cab. Nat. History, v. 2, 199-201.
       Abstracted from Am. J. Sci.

9690 1830, Bromine in the natural brine of Salina [New York]: Am. J. Sci., v. 18, no. 1, 143-144.

9691 1830, [Excerpts from Geological lectures (see 9687)]: S. Rev., v. 6, no. 12, 284-307.
       In review.

9692 1830, Notice of the anthracite region in the valley of the Lackawanna and of Wyoming on the Susquehanna: Am. J. Sci., v. 18, no. 2, 308-328, 3 figs, map.
       Abstracted in Hazard's Reg. Penn., v. 6, 70-77.

9693 1830, A fountain of petroleum, called the oil spring [in Alleghany County, New York]: Cab. Nat. History, v. 2, 196-199.
       Abstracted from Am. J. Sci.

9694 1830, Native silver from the mines of Pasco: Mech's Mag., v. 1, 191 (16 lines).
       Abstracted from Am. J. Sci.

9695 1830, Notes on a journey from New Haven, Conn., to Mauch Chunk and other anthracite regions of Pennsylvania: Am. J. Sci., v. 19, no. 1, 1-21, 3 plates with 9 figs.
       Abstract in Hazard's Reg. Penn., v. 6, 273-278.

9696 1830, Porcelain clay [from Granby, Connecticut]: Am. J. Sci., v. 18, no. 1, 199 (7 lines).

9697 1830, Remark by the editor: Am. J. Sci., v. 19, no. 1, 104 (7 lines).
       In "Notices of the country near Bedford Springs in Pennsylvania ... ," by H. H. Hayden (see 4881).

9698 1830, Remarks: Am. J. Sci., v. 18, no. 2, 210-211.
       In "Illustrations of a view taken from the Upper Falls of the Genesee River," by D. Wadsworth (see 10659).

9699 1830, Remarks: Am. J. Sci., v. 18, no. 2, 389.
       In "Georgia meteor and aerolite" by E. Beall (see 1550).

Silliman, B. (cont.)

9700   1830, Report on the coal formation of the valleys of Wyoming and Lackawanna, to which are added miscellaneous remarks and communications: NHav., H. Howe, 29 p., illus.

9701   1831, Postscript [and] remarks: Am. J. Sci., v. 20, no. 2, 419-420.
Praise of A. Eaton's research.

9702   1831, Principles of geology [review with excerpts of Geology of Yorkshire by John Phillips (see 8305)]: Am. J. Sci., v. 21, no. 1, 1-26.

9703   1831, Remarks by the editor: Am. J. Sci., v. 20, no. 1, 157-158.
In "Buffalo mineral spring" by M. Bristol (see 2013).

9704   1831, Remarks by the editor: Am. J. Sci., v. 20, no. 2, 371-372.
In "A collection of fossil bones, disinterred at Big Bone Lick, Kentucky" by W. Cooper et. al. (see 2636).

9705   1832, Acknowledgments to friends and correspondents: Am. J. Sci., v. 22, Addenda, 1-4.
Contains notice of recent geological literature.

9706   1832, Boring for water: Am. J. Sci., v. 23, no. 1, 206.

9707   1832, Museum of Gideon Mantell: Am. J. Sci., v. 23, no. 1, 162-179, fig.

9708   1832, Notice of a fountain of petroleum, called the oil spring [in Alleghany County, New York]: Am. J. Sci., v. 23, no. 1, 97-102.

9709   1832, Remarks: Am. J. Sci., v. 23, no. 1, 207.
Follows "Alluvial deposits of the Mohawk" by C. H. Tomlinson (see 10270).

9710   1833, Acknowledgments to friends and correspondents: Am. J. Sci., v. 23, Addenda, 1-2; v. 24, Addenda, 1-4.
Contains notice of recent geological literature.

9711   1833, The Cabinet of natural history and American rural sports, with illustrations; monthly; quarto; by J. & T. Doughty, Philadelphia [review of]: Am. J. Sci., v. 23, no. 1, 402-403.

9712   1833, Consistency of the discoveries of modern geology, with the sacred history of the Creation and the deluge: NHav., Hezekiah Howe and Co., 80 p.
Published as an appendix to Bakewell's Geology, second edition (see 1457).
For review see 4107 and 5984.

9713   1833, [Editor's note following "Vegetable origin of anthracite," by J. M. Bunker (see 2193)]: Am. J. Sci., v. 24, no. 1, 173 (10 lines).

9714   1833, Fossil vegetables [review of memoirs on fossil plants by H. Witham (see 11015)]: Am. J. Sci., v. 25, no. 1, 108-113.

9715   1834, Acknowledgments to friends and correspondents: Am. J. Sci., v. 25, Addenda, 1-4.
Contains notice of recent geological literature.

9716   1834, Mineralogical school at New Haven: Am. J. Sci., v. 25, no. 2, 431-432.
Notice of mineralogical school begun at Yale by C. U. Shepard.

9717   1834, Remarks: Am. J. Sci., v. 27, no. 1, 38.
Follows "Report on the projected survey of the state of Maryland" by J. T. Ducatel and J. H. Alexander (see 3232).

9718   1835, Acknowledgments to friends, correspondents and strangers: Am. J. Sci., v. 27, Addenda no. 1, 1-6; Addenda no. 2, 1-5; v. 28, Addenda no. 1, 4-6; Addenda no. 2, 5-8.
Contains notice of recent geological literature.

9719   1835, Lowell [Massachusetts]. Geological facts: Am. J. Sci., v. 27, no. 2, 340-347, plate.

9720   1835, Memoires geologiques et paleontologiques, published by A. Boué [notice of]: Am. J. Sci., v. 28, no. 2, 366-367 (9 lines).

9721   1835, Mr. Witham's transparent sections of fossil wood: Am. J. Sci., v. 27, no. 2, 415-416.
See also article by H. Witham (11014).

9722   1835, New publications: Am. J. Sci., v. 28, no. 2, 373-374.
Notice of Bos. J. Nat. History.

9723   1835, New scientific journals in Great Britain [notice of West of England journal of science and literature and of Records of general science]: Am. J. Sci., v. 28, no. 2, 365.

9724   1835, Notice of the work of T. Hawkins: Am. J. Sci., v. 27, no. 2, 413-415.
In review of Memoirs of Ichthyosauri and Plesiosauri by T. Hawkins, by G. A. Mantell (see 6727).

9725   1835, Professor Silliman's second lecture on geology, abridged. Internal fire or volcanoes: Farm's Reg., v. 3, 310.

9726   1835, Remark: Am. J. Sci., v. 27, no. 2, 399 (8 lines).
In "Progressive increase of the internal heat of the crust of the earth" (see 606).

9727   1836, Acknowledgments to friends, correspondents and strangers: Am. J. Sci., v. 29, Addenda no. 2, 3-8; v. 30, Addenda, 1-4.
Contains notice of recent geological literature.

9728   1836, [Crinoid from Schoharie, New York]: Am. J. Sci., v. 31, no. 1, 165-167, illus.

9729   1836, Notices of scientific works [for sale by Herrick and Noyes, New Haven]: Am. J. Sci., v. 29, Addenda no. 1, 1-5.

9730   1836, Professor Griscom's cabinet for sale: Am. J. Sci., v. 29, Addenda, no. 1, 5 (6 lines).

9731   1836, Report of an examination of the gold districts of the Virginia and New-England mining company with An act to incorporate: Fredericksburgh, n.p., 8, 4 p.

9732   1836, Report on the gold mine in Culpeper County [Virginia]: Fredericksburg, Va., 16 p.

9733   1836, [Report] to the President and Directors of the Richmond Mining Company: n.p., 16 p.
With a letter by F. Shepherd (see 9568).

9734   1836, Richmond Mining Company: Richmond, Va., 16 p.
On the Goochland County gold mine.

9735   1836, Valuable cabinet of minerals for sale: Am. J. Sci., v. 29, no. 2, 392-393.
Notice of Baron L. Lederer's mineral cabinet for sale.

9736   1836, Walton Mining Company. (A report): Fredericksburg, Va., 11 p.
Includes report by C. U. Shepard (see 9569). On the Walton Mine, Louisa County, Virginia.

9737   1837, Acknowledgments to correspondents, strangers and friends: Am. J. Sci., v. 31, Addenda no. 1, 1-5; Addenda no. 2, 1-3; v. 32, Addenda, 1-4.
Contains notices of recent geological literature.

Silliman, B. (cont.)
9738   1837, Culpeper gold mine, Virginia: Am. J. Sci., v. 32, no. 1, 185 (14 lines).
9739   1837, Geological drawings and illustrations: Am. J. Sci., v. 31, Addenda no. 1, 6.
       Geological drawings of R. Bakewell, jr. reviewed and recommended.
9740   1837, Geological reports [reviews of reports on geology of Ohio by W. W. Mather (see 5057), Maryland by J. T. Ducatel (see 3240), Virginia by W. B. Rogers (see 8968 and 8969), Pennsylvania by H. D. Rogers (see 8848), Maine by C. T. Jackson (see 5550), and Coteau de Prairie by G. W. Featherstonhaugh (see 3781)]: Am. J. Sci., v. 32, no. 1, 185-194.
9741   1837, Gold mines of Virginia: Am. J. Sci., v. 31, no. 2, 413 (11 lines).
9742   1837, Notice of "A report on the geological survey of the state of Connecticut; by Prof. Charles Upham Shepard" [see 9512]: Am. J. Sci., v. 33, no. 1, 151-175.
9743   1837, On the elevation of mountain ranges: Am. J. Sci., v. 31, no. 2, 290-291.
9744   1837, Remarks on some of the gold mines, and on parts of the gold region of Virginia, founded on personal observations, made in the months of August and September, 1836: Am. J. Sci., v. 32, no. 1, 98-130.
9745   1837, United States' gold mine at Fredericksburgh, Va.: Am. J. Sci., v. 32, no. 1, 183-184.
9746   1838, Acknowledgments to correspondents, friends and strangers: Am. J. Sci., v. 33, Addenda, 1-6.
       Contains notices of recent geological literature.
9747   1838, and B. Silliman, jr., Acknowledgments to correspondents, friends and strangers: Am. J. Sci., v. 34, Addenda no. 1, 1-8; Addenda no. 2, 1-7.
9748   1838, and B. Silliman, jr., Extracts from a report made to the Maryland Mining Company: NY, Scathard and Adams (printers), 27 p.
       Contains chemical analysis of coal by C. U. Shepard (see 9521) and experiments on coal by B. H. Latrobe (see 6069) with several letters of endorsement.
9749   1838, [Review with excerpts of First annual report on the geological survey of the state of Ohio by W. W. Mather (see 6891)]: Am. J. Sci., v. 34, no. 2, 347-364.
9750   1838, Notice of rocks, minerals, &c. from the Rev. Mr. Robertson, missionary in Greece: Am. J. Sci., v. 33, no. 2, 255-256.
9751   1838, Remarks by the editor: Am. J. Sci., v. 34, no. 1, 76-80.
       In "Popular notices of Mount Washington and the vicinity" by G. W. Nichols. Silliman's notes are geological additions to Nichols's non-scientific article.
9752   1838, [Remarks on "Lectures and remarks of Dr. Gideon Mantell" (see 6731)]: Am. J. Sci., v. 33, no. 2, 328-334.
9753   1838, Virginia and New England Mining Company: NY, James Narine (printer), 14 p.
       On gold mines of Louisa County, Virginia.
9754   1839, and B. Silliman, jr., Acknowledgments to correspondents, friends and strangers: Am. J. Sci., v. 35, Addenda, 1-7; n. 36, Addenda no. 1, 1-8; Addenda no. 2, 1-8.
9755   1839, Remarks introductory to the first American edition of Dr. Mantell's Wonders of geology: NHav., Hitchcock and Stafford, 31 p.
9756   1839, and B. Silliman, jr., [Extracts from a report on the geology of the Maryland Mining Company]: in 6812.
9757   1839, and B. Silliman, jr., New edition of Dr. Mantell's Wonders of geology [notice of New York edition (see 6732)]: Am. J. Sci., v. 37, Addenda, 8.
9758   1839, and B. Silliman, jr., Remarks by the editors: Am. J. Sci., v. 35, no. 2, 250-251.
       Following "Notes on American geology" by T. A. Conrad (see 2562).
9759   1839, Suggestions relative to the philosophy of geology as deduced from the facts, and to the consistency of both the facts and theory of this science with sacred history: NHav., B. and W. Noyes, 119 p.
       Also published as an appendix to Bakewell's Geology, third edition (see 1460), 461-579, 595-596 (index).
       For review with excerpts see 7818. For review see 1157.
9760   1840, and B. Silliman, jr., Acknowledgments to correspondents, friends and strangers: Am. J. Sci., v. 38, Addenda no. 1, 1-6; Addenda no. 2, 1-7; v. 39, Addenda, 1-12.
       Contains notice of recent geological literature.
9761   1840, and B. Silliman, jr., Notice of the Wonders of geology by Gideon Algernon Mantell [see 6732]: Am. J. Sci., v. 39, no. 1, 1-18.
9762   1840, Remarks at the agricultural meeting in Boston, Jan. 13, 1840: Farm's Mon. Vis., v. 2, 69-70.
9763   1841, and B. Silliman, jr., Acknowledgments to correspondents, friends and strangers: Am. J. Sci., v. 40, Addenda, 1-6; v. 41, Addenda, 1-8.
       Contains notice of recent geological literature.
9764   1841, [On the lack of erratic boulders near the equator]: Am. J. Sci., v. 41, no. 1, 174 (4 lines).
       In Assoc. Am. Geol's Nat's Proc.
9765   1842, and B. Silliman, jr., Acknowledgments to correspondents, friends and strangers: Am. J. Sci., v. 42, Addenda, 1-10; v. 43, Addenda, 1-10.
       Contains notice of recent geological literature.
9766   1842, Address before the Association of American Geologists and Naturalists, assembled at Boston, April 24, 1842: Am. J. Sci., v. 43, no. 2, 217-250.
9767   1842, Address ... [as above] ... With an abstract of the proceedings at the meeting: NY, Wiley and Putnam, 36 p. (address), 39 p. (proceedings).
9768   1842, and O. P. Hubbard, Chemical examination of bituminous coal from the pits of the Mid Lothian Coal Mining Company, south side of James River, fourteen miles from Richmond, Virginia, in Chesterfield County: Am. J. Sci., v. 42, no. 2, 369-374.
9769   1842, Mr. Lyell and Mr. Murchison: Am. J. Sci., v. 42, no. 1, 213-214.
       Notice of Boston lectures by C. Lyell and Russian tour by R. I. Murchison.
9770   1842, Remark by the senior editor: Am. J. Sci., v. 43, no. 1, 14 (11 lines).
       In "Geological and statistical notice of the coal mines in the vicinity of Richmond, Va." by A. S. Wooldridge (see 11050).
9771   1842, Tubular concretions of iron and sand from Florida [at Pensacola]: Am. J. Sci., v. 42, no. 1, 207-209.
       With observations by J. T. Gerry (see 4339).
9772   1842, [Volcanic islands]: Farm's Mon. Vis., v. 4, 50.

Silliman, B. (cont.)

9773 1843, and B. Silliman, jr., Acknowledgments to correspondents, friends and strangers: Am. J. Sci., v. 44, Addenda, 1-8; v. 45, Addenda no. 1, 1-6; Addenda no. 2, 1-6.
Contains notices of recent geological literature.

9774 1843, Lecture VII. Coal, its origin and organic remains: Pittsburgh, Penn.
Not seen. Lecture delivered before the Wirt Institute and citizens of Pittsburgh. See also 9800.

9775 1843, Vegetable origin of coal: Bos. Cult., v. 5, 234. Also in NY State Mech., v. 2, 204.

9776 1844, and B. Silliman, jr., Acknowledgments to correspondents, friends and strangers: Am. J. Sci., v. 46, Addenda, 1-4; v. 47, Addenda, 1-6.
Contains notices of recent geological literature.

9777 1844, and B. Silliman, jr., Editorial remark: Am. J. Sci., v. 46, no. 2, Appendix, 1.
On the controversy between J. D. Dana (see 2810, 2811; and 2813) and J. P. Couthouy (see 2671; 2673 and 2674).

9778 1844, and B. Silliman, jr., Acknowledgments to correspondents, friends and strangers: Am. J. Sci., v. 48, Addenda, 1-8.

9779 1845, [Change in elevation of Northeast America's coastline]: Am. Q. J. Ag. Sci., v. 2, 136-138.
In Assoc. Am. Geol's Nat's Proc.

9780 1845, Coal: Dwight's Am. Mag., v. 1, 252.

9781 1845, Extraordinary fossil animal: Dwight's Am. Mag., v. 1, 518.
Notice of Koch's fossil display.

9782 1845, [Native copper from Wallingford, Connecticut]: Am. Q. J. Ag. Sci., v. 2, 151 (4 lines).
In Assoc. Am. Geol's Nat's Proc.

9783 1845, Notice of The medals of Creation, of first lessons in geology, and in the study of organic remains; by G. A. Mantell [review with excerpts of 6744]: Am. J. Sci., v. 48, no. 1, 105-137.

9784 1845, [On New England drift]: Am. Q. J. Ag. Sci., v. 2, 142-144.
In Assoc. Am. Geol's Nat's Proc.

9785 1845, [On rhomboidal cleavage in the New Red Sandstone]: Am. Q. J. Ag. Sci., v. 2, 149-150.

9786 1847, Preface: Am. J. Sci., v. 50, iii-xviii.
History of the first 50 volumes of the American Journal of Science.

9787 1848, Death of Alexander Brongniart: Am. J. Sci., v. 5 (ns), no. 1, 141-144.

9788 1848, Gold in Canada: Niles' Wk. Reg., v. 74, 336.

9789 1848, [Opinion on the glacial theory]: Lit. World, v. 2, 229 (12 lines).
In Assoc. Am. Geol's Nat's Proc.

9790 1849, A cabinet of minerals for sale, collected by the late Dr. Crawe, of Watertown, Jefferson Co., N. Y.: Am. J. Sci., v. 7 (ns), no. 1, 138.

9791 1849, [Review of The dodo and its kindred ... , by H. E. Strickland and A. G. Melville (see 10038)]: Am. J. Sci., v. 7 (ns), no. 1, 52-67, fig., plate.

9792 1849, and F. C. Kropff, A plan and prospectus and geological survey of the Amherst Copper Mining and Smelting Company, Amherst County, Virginia: Phila., R. Wilson de Silver, 24 p., map.

9793 1850, and O. P. Hubbard, Connecticut copper mines [at Bristol and Litchfield]: Ann. Scien. Disc., v. 1, 252-253 (15 lines).
Abstracted from Farm. and Mech. and Bridgeport Farm.

Silliman, Benjamin, jr. (1816-1885)
See also American Journal of Science, volumes 34 to 10, new series (1838-1850). Silliman served as co-editor with B. Silliman, sr. (1838-1850), and J. D. Dana (1846-1850).
See also articles 9747, 9748, 9754, 9756 to 9758, 9760, 9761, 9763, 9765, 9773, 9776 to 9778 with B. Silliman, and discussion of article by J. Hall (4685).

9794 1841, [Diluvial boulders at 40° South latitude]: Am. J. Sci., v. 41, no. 1, 174 (4 lines).
Also in Assoc. Am. Geol's Nat's Proc., v. 2, 19 (4 lines). Also in Assoc. Am. Geol's Nat's Rept., v. 1, 27 (4 lines), 1843.

9795 1841, Report on the geological and agricultural survey of the state of Rhode Island, in 1839; by Dr. Charles T. Jackson [review of 5571]: Am. J. Sci., v. 40, no. 1, 182-194.

9796 1841, [Rhombic structure in Connecticut Valley Triassic]: Am. J. Sci., v. 41, no. 1, 173 (5 lines).
Also in Assoc. Am. Geol's Nat's Proc., v. 2, 17 (5 lines). Also in Assoc. Am. Geol's Nat's Rept., v. 1, 26 (5 lines), 1843.

9797 1841, [Soil of the Nile]: Am. J. Sci., v. 41, no. 1, 160 (5 lines).
Also in Assoc. Am. Geol's Nat's Proc., v. 2, 4 (5 lines). Also in Assoc. Am. Geol's Nat's Rept., v. 1, 13 (5 lines), 1843.

9798 1842, [Cause of dip in the Triassic of the Connecticut River Valley]: Am. J. Sci., v. 43, no. 1, 171 (4 lines).
Also in Assoc. Am. Geol's Nat's Proc., v. 3, 26 (4 lines). Also in Assoc. Am. Geol's Nat's Rept., v. 1, 64 (4 lines), 1843.

9799 1842, [Glacial versus man-made mounds]: Am. J. Sci., v. 43, no. 1, 154.
Also in Assoc. Am. Geol's Nat's Proc., v. 3, 7-8. Also in Assoc. Am. Geol's Nat's Rept., v. 1, 47, 1843.

9800 1843, Lectures on geology, delivered before the Wirt Institute, and citizens of Pittsburgh: Pittsburgh.
Not seen. See also 9774.

9801 1843, Abstract of proceedings of the fourth session of the Association of American Geologists and Naturalists: Am. J. Sci., v. 45, no. 1, 135-165, no. 2, 310-353.
B. Silliman, jr. served as Secretary of the Association.

9802 1843, [Copper at Bristol, Connecticut and Lake Superior]: Am. J. Sci., v. 45, no. 2, 331 (6 lines); 332 (11 lines).
In Assoc. Am. Geol's Nat's Proc.

9803 1843, [On crescent-shaped dikes in New Jersey and Massachusetts]: Am. J. Sci., v. 45, no. 2, 334 (8 lines).
In Assoc. Am. Geol's Nat's Proc.

Silliman, B., jr. (cont.)

9804   1844, Analysis of meteoric iron from Burlington, Otsega Co., N. Y.: Am. J. Sci., v. 46, no. 2, 401-403.

9805   1844, and D. Houghton, Connection of metallic copper with the trap of Connecticut and Michigan: Am. J. Sci., v. 47, no. 1, 132.
Also in Assoc. Am. Geol's Nat's Proc., v. 5, 40-41.

9806   1844, Natural polariscope: Am. J. Sci., v. 47, no. 2, 418 (18 lines).
Tourmalines from Grafton, New Hampshire used to polarize light.

9807   1844, Report on the intrusive trap of the New Red Sandstone of Connecticut: Am. J. Sci., v. 47, no. 1, 107-108.
Also in Assoc. Am. Geol's Nat's Proc., v. 5, 14.

9808   1844, [Review with excerpts of System of mineralogy by J. D. Dana (see 2812)]: Am. J. Sci., v. 46, no. 2, 362-387, 29 figs.

9809   1844, Abstract of the proceedings of the fifth session of the Association of American Geologists and Naturalists: Am. J. Sci., v. 47, no. 1, 94-134.
See also Assoc. Am. Geol's Nat's Proc., v. 5, 1-43 p.
B. Silliman, jr. was recording secretary of the Association.

9810   1844, Dr. Percival, the original observer of the crescent-formed dykes of trap in the New Red Sandstone of Connecticut: Am. J. Sci., v. 46, no. 1, 205-206.

9811   1844, Sillimanite and monazite [from Connecticut]: Am. J. Sci., v. 46, no. 1, 207-208.

9812   1845, Report on the chemical examination of several waters for the city of Boston: Bos., xxxii.

9813   1845, [Letter of endorsement]: in Catechism of agricultural chemistry and geology by J. F. W. Johnston (see 5825).

9814   1845, Notice of a mass of meteoric iron found at Cambria, near Lockport, in the state of New York: Am. J. Sci., v. 48, no. 2, 388-392, 2 figs.

9815   1845, [Notes in "Reply to a notice of Shepard's Mineralogy" by C. U. Shepard (see 9546)]: Am. J. Sci., v. 48, no. 1, 172-173; 179-180.

9816   1845, [Ores of Mine La Motte, Missouri]: Assoc. Am. Geol's Nat's Proc., v. 6, 82 (9 lines).

9817   1845, [Analysis of water of the Dead Sea]: Am. J. Sci., v. 48, no. 1, 10-12.
In J. D. Sherwood's "Observations upon the valley of the Jordan and the Dead Sea" (see 9588).

9818   1846, Chemical examination of several natural waters: Am. J. Sci., v. 2 (ns), no. 1, 218-224.

9819   1846, On the chemical compostion of the calcareous corals: Am. J. Sci., v. 1 (ns), no. 2, 189-199.

9820   1846, and T. S. Hunt, On the meteoric iron of Texas and Lockport: Am. J. Sci., v. 2 (ns), no. 3, 370-376, fig., photograph.

9821   1847, Description of a meteoric stone which fell in Concord, New Hampshire, in October, 1846: Am. J. Sci., v. 4 (ns), no. 3, 353-356, fig.

9822   1847, Hydrate of nickel, a new mineral: Am. J. Sci., v. 3 (ns), no. 3, 407-409.
On nickel hydroxide from Texas, Lancaster County, Pennsylvania.

9823   1847, On fossil trees found at Bristol, Conn., in the New Red Sandstone: Am. J. Sci., v. 4 (ns), no. 1, 116-118, fig.

9824   1848, Analysis of the Quebec waters: Mining J., v. 2, 54-55.

9825   1848, Analysis of the Quebec waters: NHav., 20 p.
"Appended to Mr. Baldwin's report."

9826   1848, Gold in Canada [at Chaudière]: Am. J. Sci., v. 6 (ns), no. 2, 274-275.

9827   1848, On emerald nickel from Texas, Lancaster County, Pa.: Am. J. Sci., v. 6 (ns), no. 2, 248-249.

9828   1849, Descriptions and analyses of several American minerals: Am. J. Sci., v. 8 (ns), no. 3, 377-394.
Also reprint edition.

9829   1849, On gibbsite and allophane from Richmond, Mass.: Am. J. Sci., v. 7 (ns), no. 3, 411-417.

9830   1850, Analyses of "green picrolite," and "slaty serpentine," so called, from Texas, Lancaster County, Pa.: Am. Assoc. Adv. Sci. Proc., v. 2, 134.

9831   1850, Gadolinite in New York [at Monroe, Orange County]: Am. Assoc. Adv. Sci. Proc., v. 2, 134 (5 lines).

9832   1850, Identity of several differently named American minerals: Ann. Scien. Disc., v. 1, 272-273.
Abstracted from Am. J. Sci.
On the equivalence of boltonite and sphene, and of sillimanite, bulcholzite and fibrolite.

9833   1850, Identity of sillimanite of Bowen, of Bucholzite of Brandes, and of fibrolite of Buoman with the species kyanite: Am. Assoc. Adv. Sci. Proc., v. 2, 111-113.

9834   1850, On boltonite of Shepard and bisilicate of magnesia of Dr. Thomson: Am. Assoc. Adv. Sci. Proc., v. 2, 109-110.

9835   1850, On "indianite" of Count Bournon, and on the American mineral which has been distributed under the same name: Am. Assoc. Adv. Sci. Proc., v. 2, 131-134.

9836   1850, On the new American mineral, lancasterite [from Lancaster County, Pennsylvania]: Am. J. Sci., v. 9 (ns), no. 2, 216-217.

9837   1850, Optical examination of several American micas: Am. J. Sci., v. 10 (ns), no. 3, 372-383, tables.
Also reprint edition.

Silliman, Benjamin Douglas (1805-1901; see 5091 with E. Hitchcock)

Simpson, Henry I.

9838   1848, The emigrant's guide to the gold mines; three weeks in the gold mines ... : NY, Joyce and Co., 30 p., map.

Simpson, James Hervey (1813-1883)

9839   1850, Journal of a military reconnaissance from Santa Fe, New Mexico, to the Navajo country in 1849: US SED 64, 31-1, v. 14 (562), 56-168, 75 plates, 2 maps.

9840   1850, Report and map of the route from Fort Smith to Santa Fe, New Mexico: US SED 12, 31-1, v. 6 (554), 2-25, map.

9841   1850, with R. B. Marcy, Report on a route from Fort Smith to Santa Fe: US HED 45, 31-1, v. 8 (577), 2-89, map.

Sing, Edward F.
9842   1835, and L. Clary, I. W. Hanchett, R. R. Davis and H. Hoyt, Medical topographical report of the county of Onondaga: Med. Soc. NY Trans., v. 2, 228-240.

Siskowit Mining Company
9843   1850, Charter and by-laws of the Siskowit Mining Company of Michigan, together with the treasurer's report, &c.: Phila., 23 p.
       Contains reports by J. Hancock (see 4755) and S. Richardson (see 8773).

Six weeks in Fauquier, being the substance of a series of familiar letters illustrating the scenery, localities, medical virtues, and general characteristics of the White Sulphur Springs at Warrenton, Fauquier County, Virginia:
9844   1839, NY, Samuel Colman, 67 p., front.

Skell, M.
9845   1808, [Analysis of Russian mineral waters from Lipetzk]: Am. Reg. or Gen. Repos., v. 2, 359.

Skinner, John Stuart (1788-1851)
9846   1835, The importance of geological and mineral surveys: Am. Farm., v. 2 (2s), 124.
9847   1845, Maryland coal: Mon. J. Ag., v. 1, no. 3, 162.
9848   1845, Memoir of the late Stephen Van Rensselaer, as the friend and patron of agriculture: Mon. J. Ag., v. 1, no. 1, 1-7.
9849   1846, The coal mines of Pennsylvania. The Reading Railroad and Bear Mountain coal fields: Mon. J. Ag., v. 2, no. 1, 42-45.
9850   1846, Randall's geology [review of Incentives for geology (see 8684)]: Mon. J. Ag., v. 2, no. 2, 85-86.

Skinner, Richard Cortland (d. 1834)
9851   1796, Petrifactions, &c. &c. collected in the state of Virginia, and presented to the Tammany Museum, ... : NY Mag., v. 1 (ns), 175-176.

Sloat, L. W.
9852   1847, The mines of Upper California: Hunt's Merch. Mag., v. 16, 365-367.

Smith, Alfred
9853   1832, Map showing the extent of Primitive and of Secondary formation in the Connecticut River Valley: Am. J. Sci., v. 22, no. 2, colored geological map, 48.5 x 23.9 cm.
       In 9854.
9854   1832, On the water courses, and the alluvial and rock formations of the Connecticut River Valley: Am. J. Sci., v. 22, no. 2, 205-231, colored map.
       Contains map 9853.

Smith, Benjamin Mosby (1811-1893)
9855   1850, The testimony of science to the truth of the Bible: Charlottesville, SC, O. S. Allen, 28 p.
       For review see 1683.

Smith, Buckingham (1810-1871)
9856   1848, Report on reconnoissance of the Everglades made to the Secretary of the Treasury:
       Not seen.

Smith, Charles Hamilton (1776-1859)
9857   1844, On the magnesian limestone: Franklin Inst. J., v. 8 (3s), 110-111.
       Abstracted from London Ath.

Smith, Daniel B. (1792-1883)
9858   1826, Preparation of sulphate of soda from sea-water: N. Am. Med. Surgical J., v. 1, no. 1, 238-239.
       Abstracted from Phila. College Pharmacy J.
9859   1831, On borax: Am. J. Pharmacy, v. 3, 119-123.
9860   1831, On the preparation of glauber and epsom salt, and magnesia from sea water: Am. J. Sci., v. 21, no. 1, 175.
       Abstracted from Am. J. Pharmacy.

Smith, Edward Darrell (1777-1819)
9861   1818, On the changes which have taken place in the wells of water situated in Columbia, South-Carolina, since the earthquakes of 1811-12: Am. J. Sci., v. 1, no. 1, 93-95.
9862   1819, Analysis of the Harrodsburg salts: Am. J. Sci., v. 1, no. 4, 403-405.
9863   1821, An account of the warm springs in Buncome County, state of North-Carolina: Am. J. Sci., v. 3, no. 1, 117-125.

Smith, Franklin L.
9864   1837, Notice of some facts connected with the gold of a portion of North Carolina: Am. J. Sci., v. 32, no. 1, 130-133.

Smith, George Hand
9865   1850, Analysis of the Byron Acid Spring, near Batavia [New York]: Med. Soc. NY Trans., v. 8, 161.

Smith, Hamilton Lanphere (1819-1903)
9866   1848, The world; or, first lessons in astronomy and geology, in connection with the present and past condition of our globe: Cleveland, Ohio, M. C. Younglove and Co., 324 p., illus.
9867   1849, Indiana - her resources and prospects: De Bow's Rev., v. 7, 246-261.

Smith, J. (of Jordan Hill, England F. R. S.)
9868 1837, [Abstract of "On the indications of change in the relative level of land and water in the estuary of the Clyde"] : Mus. For. Liter. Sci., v. 31, 47-48.
       In London Geol. Soc. Proc.
9869 1838, Changes of level: Am. J. Sci., v. 34, no. 1, 27 (15 lines).
       In British Assoc. Proc.
9870 1845, On the subsidence of the land at Puzzuoli: Eclec. Mag., v. 6, 287.
       Abstracted from London Ath.
9871 1846, Subsidence of the land at Puzzuoli: Am. J. Sci., v. 2 (ns), no. 2, 269-270.
       Abstracted from Jameson's J.

Smith, James
       See also 10894 with S. Whisler.
9872 1832, Geology of Natchez: Atl. J. and Friend Know., v. 1, no. 4, 135.

Smith, John Augustine (1782-1865)
       See also articles with W. Cooper (2635 and 2636) and S. L. Mitchill (7317).
9873 1840, [On the origin and formation of trap and basaltic rocks] : Am. Rep. Arts Sci's Manu's, v. 1, 420 (9 lines).
       Ascribed to "Dr. Smith." In NY Lyc. Nat. History Proc.
9874 1846, Central cavity of the Mastodon: Am. Q. J. Ag. Sci., v. 3, 19-22.
9875 1846, The mutations of the earth; or, an outline of the more remarkable physical changes, of which, in the progress of time, this earth has been the subject, and the theatre; including an examination into the scientific errors of the author of the "Vestiges of Creation" being the anniversary discourse for 1846, delivered in the chapel of the University before the Lyceum of Natural History of New York: NY, Bartlett and Welford, 64 p.
       Includes review of Vestiges of the natural history of Creation by R. Chambers (see 2292 et seq.).

Smith, John C.
9876 1810, An account of the whitten plaster [of Connecticut] : Ct. Acad. Arts Sci's Mem., v. 1, pt. 1, 81-82.
9877 1812, An account of whitten plaster: Am. Med. Philos. Reg., v. 3, no. 2, 209 (10 lines).
       Abstracted from Ct. Acad. Arts Sci's Mem.

Smith, John Lawrence (1818-1883)
9878 1839, Chromate of potassa - a reagent for distinguishing between the salts of baryta and strontia: Am. J. Sci., v. 36, no. 1, 183-184.
       Abstracted from London and Edinburgh Philos. Mag.
9879 1843, A new instrument for estimating the quantity of carbonate of lime in calcareous substances: Am. J. Sci., v. 45, no. 2, 262-266, 2 figs.
9880 1844, On fossil bones from the vicinity of Charleston, S. C.: Am. J. Sci., v. 47, no. 1, 116-117.
       Also in Assoc. Am. Geol's Nat's Proc., v. 5, 23-24.
9881 1844, Oxide of cobalt near Silver Bluff, South Carolina: Am. J. Sci., v. 47, no. 1, 131 (17 lines).
       Also in Assoc. Am. Geol's Nat's Proc., v. 5, 40 (17 lines).
9882 1845, Action of solutions of the neutral phosphates of the alkalis upon the carbonate of lime, &c.: Am. J. Sci., v. 48, no. 1, 97-98.
9883 1845, Composition of the marl from Ashley River, S. C.: Am. J. Sci., v. 48, no. 1, 101-103.
9884 1845, Oxide of cobalt from Silver Bluff, South Carolina: Am. J. Sci., v. 48, no. 1, 103-104.
9885 1845, Source of fluorine in fossil bones: Am. J. Sci., v. 48, no. 1, 99-101.
9886 1846, Observations on the more recent researches concerning the operations of the blast furnaces in the manufacture of iron: Am. J. Sci., v. 1 (ns), no. 2, 170-178, figs.; v. 2 (ns), no. 1, 95-102, 2 figs.
9887 1846, Report to the Black Oak Agricultural Society, on the ashes of the cotton stalk, the compostion of the cotton soils, and the nature of rust in cotton: Charleston, Miller & Browne, 14 p.
9888 1846, Water from the artesian well of Grenville, at Paris: Am. J. Sci., v. 1 (ns), no. 2, 264-265 (10 lines).
       Abstracted from Chemist.
9889 1848, Analysis of the cotton lands at the head waters of Cooper River: in Report on the geology of South Carolina by M. Tuomey (see 10438), xliii-xlviii.
9890 1848, Two new minerals, - medjibite (sulphate of uranium and lime) - liebigite (carbonate of uranium and lime): Am. J. Sci., v. 5 (ns), no. 3, 336-338.
       On minerals from Adrianople, Turkey. J. L. Smith was "geologist to the Sultan of Turkey."
9891 1849, Chrome and meerschaum of Asia Minor: Am. J. Sci., v. 7 (ns), no. 2, 285-286.
       Abstracted in Ann. Scien. Disc., v. 1, 265-266 (1850).
9892 1849, Emery formation of Asia Minor: Am. J. Sci., v. 7 (ns), no. 2, 283.
       Abstracted in Ann. Scien. Disc., v. 1, 265 (1850).
9893 1850, A list of the minerals associated with the emery of Asia Minor: Am. J. Sci., v. 9 (ns), no. 2, 289.
9894 1850, Memoir on emery. - First part. - On the geology and mineralogy of emery from observations made in Asia Minor: Am. J. Sci., v. 10 (ns), no. 3, 354-369.

Smith, John Pye (1774-1851)
9895 1840, On the relation between the Holy Scriptures and some parts of geological science: NY, D. Appleton, 364 p.
       For review see 1679, 1687, 3408 and 7149.
9896 1840, Introductory notice: in Elementary geology by E. Hitchcock (see 5132, 5134, 5142, 5157 and 5177).
       For excerpts see 229 and 744.
9897 1850, The relation between the Holy Scriptures and some parts of geological science: Phila., R. E. Paterson, xvi, 400 p.

Smith, Josiah (1704-1781)
9898   1730, The greatest sufferers not always the greatest sinners. A sermon delivered in Charleston, in the Province of South-Carolina, February 4th, 1727,8. Then occasioned by the terrible earthquake in New-England ... : Bos., ii, 21 p.
       Evans number 3355.

Smith, Nathan (1770-1835)
9899   1820, Fossil bones found in red sandstone [of East Windsor, Connecticut]: Am. J. Sci., v. 2, no. 1, 146-147.
       See also "Note" (345).

Smith, Nathan D.
9900   1835, Some account of the prairie soils of Arkansas: Farm's Reg., v. 3, 273-274.
9901   1836, Further notice of the prairie soils of Arkansas, with specimens and their localities particularly described: Farm's Reg., v. 3, 556-557.

Smith, Robert Angus (1817-1884)
9902   1850, On the air and water in towns, and the action of porous strata on water and organic matter: Am. J. Sci., v. 10 (ns), no. 3, 411-412.
       In British Assoc. Proc.

Smith, Robert H.
9903   1812, Geological and philosophical disquisition: Cin., John W. Browne and Co. (printers), 24 p.

Smith, Stephen
9904   1828, Notice of the salt springs and manufacture of salt at Salina, Syracuse, &c. N. Y. made at the request of the editor: Am. J. Sci., v. 15, no. 1, 6-12.

Smith, Thomas (D. D.)
9905   1806/1807, The wonders of nature and art; or a concise account of whatever is most curious and remarkable in the world, compiled from historical and geographical works of established celebrity, and illustrated with the discoveries of modern travellers: Phila., Carr for Birch and Small, 14 v., illus.
       "Revised by James Mease."

Smith, Thomas Peters
9906   1799, Account of chrystallized basaltes found in Pennslyvania: Am. Philos. Soc. Trans., v. 4, 445-446.
9907   1800, Geological remarks [on southern New Jersey]: Med. Repos., v. 3, 151-154.
9908   1800, On crystallization: Med. Repos., v. 3, 253-257.
9909   1801, Heterogeneous formation of garnet crystals: Am. Rev. and Lit. J., v. 1, 259-260.
       Abstracted from Med. Repos.

Smith, William
9910   1814, Observations on the Saint Maurice iron-works, near the town Three Rivers, in the Province of Lower Canada: Am. Min. J., v. 1, no. 4, 198-199.

Smith, William
9911   1834, Geological essays: Farm's Reg., v. 1, 473-474; 529-530; 605-606.

Smith, William B. (M. D.)
9912   1841, Essay on calcareous earth, and remarks on quicklime as an indirect manure: Farm's Reg., v. 9, 202-207.

Smithson, James (1765-1829)
9913   1812, On the composition of zeolite: Am. Min. J., v. 1, no. 3, 182-187.
       Abstracted from Philos. Trans.
9914   1825, Blowpipe experiments: Am. J. Sci., v. 9, no. 1, 201-202.
       Abstracted by J. Griscom from "Tech. Repository."

Smithsonian Institution
9915   1848, Building material: Am. J. Sci., v. 6 (ns), no. 2, 285-287.
9916   1848, [Reports on building stones for the Smithsonian Institution]: Smithsonian Institution Ann. Rept., v. 2, 4-74; 105-107; 109-114; 119; 121-122.
9917   1848, Smithsonian Institution: Am. J. Sci., v. 6 (ns), no. 2, 288-292.
       Description of organization and formation.

Snow, Edwin, M.
9918   1845, Letter from Mr. Edwin M. Snow [on Pomfret, Windsor County, Vermont]: in First annual report on the geology of the state of Vermont by C. B. Adams (see 23), 81-82.

Sobolewski, P.
9919   1835, Extraction of platina in Russia: Am. J. Pharmacy, v. 1 (ns), 237-241.
       Ascribed to "P. Sobolewskoy."
9920   1836, On the improvements lately introduced into the iron foundries of Russia: Am. J. Sci., v. 30, no. 1, 181.
       Abstracted and translated by "D." from Ann. des Mines.

Some rude and indigested thoughts on the terrible majesty of God in the works of nature particularly in the phenomena of earthquakes; occasioned by that memorable earthquake Octob. 29th, 1727. Wherein earthquakes in their causes, kinds, and astonishing effects, are briefly hinted, enumerated and described:
9921   1730, N. London, Timothy Green, [2], [12] p.
       Evans number 3359.

Somers, Benjamin
9922   1829, Invention of certain improvements on furnaces for smelting different kinds of metal ores and slaggs: Franklin Inst. J., v. 3 (ns), 332.

Somerville
9923   1801, On the subject of manures: Mass. Agri. Soc. Papers, for 1801, 34-80.

Something (Boston, 1809)
9924   1809, Meteoric stones: v. 1, no. 20, 309-311.

Sonden, M.
9925   1836, Brevicite, a new mineral: Am. J. Sci., v. 30, no. 1, 178-179.
       Abstracted and translated by "D." from "Neues Jahrbuch fur Min. Geog."

South Carolina. State of
9926   1824, [Appropriation for a state geological survey]: SC Gen. Assembly, 1824.
9927   1825, [On the advantages of a geological survey]: SC Gen. Assembly, December, 1825.
9928   1842, [Resolution for an agricultural survey]: SC Gen. Assembly, 1842.
9929   1842, [Resolution confirming opinion of above]: SC Gen. Assembly, December, 1842.
9930   1843, [Extracts from "Report of the Committee on Agriculture, on a geological and
       agricultural survey of the state, to the legislature of South Carolina"]: S. Q.
       Rev., v. 3, no. 2, 449-467.
       For review see 9960.

South Carolina Society For Promoting and Improving Agriculture
9931   1785, South Carolina. Society For Promoting and Improving Agriculture. Address, and
       rules of the South Carolina Society for Promoting and Improving Agriculture: Char.,
       Printed by John Miller?
       Evans number 19254.

South Carolina Weekly Museum (Charleston, South Carolina, 1797)
9932   1797, On the supposed virtues of the rarest stones that are usually set in rings: v. 1,
       165-167.

Southern Agriculturist, Horticulturist, and Register of Rural Affairs (Charleston, South
       Carolina, 1828-1846)
9933   1834, Notice of the climate, soil and production of Florida: v. 7, 23-25.
       Signed by "X. Y. Z."
9934   1834, Economy of fuel: v. 7, 607-610.
9935   1835, On chemistry, as connected with the development and growth of plants: v. 8,
       361-367.
9936   1836-1845, [Notes on mines and mining]: v. 9, 222; 448; 502; v. 10, 55; 110; 166;
       180-181; 391; v. 11, 109; v. 12, 168; v. 5 (ns), 77-78.
9937   1837-1846, [Notes on soils and soil analysis]: v. 10, 23-24; 603; 609-611; v. 11,
       161-163; 417-422; 442-443; 473-475; 529-531; v. 1 (ns), 7-9; 405-411; 467-471;
       513-518; v. 6 (ns), 7-8.
9938   1837, Heat of the earth's centre: v. 10, 277-278.
9939   1838, Origin of amber: v. 11, 390.
9940   1843, [Review of Muck manual ... , by S. L. Dana (see 2700) and Essay on calcareous
       manures, by E. Ruffin (see 9088)]: v. 3 (ns), 407-400 [i.e. 408].
9941   1844, Remarks on Mr. Ruffin's agricultural survey [review of 9087]: v. 4 (ns), 166-169;
       218-220; 241-246.

Southern and Western Monthly Magazine and Review (Charleston, South Carolina, 1845)
9942   1845, [Review of Geology of South-Carolina by M. Tuomey (see 10438)]: v. 1, no. 2,
       146-147.
9943   1845, [Notice of Vestiges of the natural history of Creation by R. Chambers (see 2292
       et. seq.)]: v. 1, no. 5, 295; no. 6, 369-378.

The Southern Literary Messenger (Richmond, Virginia, 1834-1850)
9944   1835, Visit to the Virginia Springs: v. 1, 474-477; 544-547; 613-616.
9945   1835, Observations on the importance of mineral possessions, and the cultivation of
       geological inquiry: v. 1, 484-487.
       Signed by "Gamma."
9946   1835, Dagger's Springs, in the County of Botetourt, Virginia: v. 1, 518-519.
9947   1835, The Red Sulphur Springs: v. 1, 519.
9948   1835, The Virginia Springs [review with excerpts of Letters ... , by P. Nicklin (see
       7865)]: v. 2, 592-593.
9949   1839, [Review with excerpts of Geology and mineralogy by W. Buckland (see 2160)]: v. 5,
       548-558.
       Signed "Δ."
9950   1846, A resume of geology [reviews of Address by H. D. Rogers (see 8903), Proceedings
       of the Association of American Geologists and Naturalists (see 1329), and reviews
       with excerpts of Travels in North America, Principles of Geology, and Elements of
       Geology by C. Lyell (see 6533, 6543, 6557)]: v. 12, 658-671.

Southern Medical and Surgical Journal (Augusta, Georgia, 1836-1850)
9951   1836, [Review with excerpts of Medical properties of the Grey Sulphur Springs, Virginia
       by A. E. Miller (see 7184)]: v. 1, 162-166.
9952   1847, [Review of The Virginia Springs by J. J. Moormann (see 7422)]: v. 3 (ns), 624 (5
       lines).
9953   1849, Notice of the mineral springs in the state of Georgia: v. 5 (ns), 442-444.

The Southern Planter (Richmond, Virginia, 1841-1850)
9954   1841-1850, [Notes on soil analysis and improvement]: v. 1, 193; v. 2, 42-43; v. 3, 233,
       woodcut; v. 4, 262-263; v. 6, 12-13; v. 7, 211-212; v. 10, 221-222, 330-341.
9955   1844, Green sand [from New Jersey and Delaware]: v. 4, 18-20.
       Contains excerpts from New Jersey report by H. D. Rogers (see 8863) and Delaware
       report by J. C. Booth (1769).
9956   1844, Lime: v. 4, 101-104.
       Contains excerpts from Lectures ... , by J. F. W. Johnston (see 5819).
9957   1845, Artesian well: v. 5, 29-30, woodcut.
9958   1850, Professor Johnston's lectures: v. 10, 362-363.

The Southern Quarterly Review (New Orleans, 1842-1850)
9959 1842, Geology of Louisiana: v. 1, no. 1, 268-270.
       Includes reviews of papers by W. M. Carpenter (see 2264) and C. Forshey.
9960 1843, [Review of Report on a geological and agricultural survey of South Carolina (see 9930)]: v. 3, no. 2, 449-467.
9961 1843, [Review of Muck manual ..., by S. L. Dana (see 2885) and Lectures ..., by J. F. W. Johnston (see 5819)]: v. 3, no. 1, 182-199.
9962 1849, [Review of Monograph of the fossil Squalidae of the United States by R. W. Gibbes (see 4352)]: v. 15, 526-531.
9963 1849, [Review of Report on the geology of South-Carolina by M. Tuomey (see 10438)]: v. 16, no. 1, 161-178.
9964 1850, [Review of California and its gold regions by F. Robinson (see 8820)]: v. 17, no. 1, 19-37.
9965 1850, [Review of An address on the sphere, interest and importance of geology by R. T. Brumby (see 2131)]: v. 17, no. 1, 265-266.
9966 1850, [Review with excerpts of A second visit to the United States by C. Lyell (see 6571)]: v. 17, no. 2, 406-426.
9967 1850, [Review of First biennial report on the geology of Alabama by M. Tuomey (see 10440)]: v. 18, no. 2, 542.

The Southern Review (Charleston, 1828-1832)
9968 1828, [Review with excerpts of Report on the geology of North-Carolina by D. Olmsted (see 8029 and 8030)]: v. 1, no. 1, 235-261.
9969 1830, [Reviews of Introduction to geology by R. Bakewell (see 1460), Geological lectures by B. Silliman (see 9687), New system of Geology by A. Ure, Outlines of geology by W. T. Brande, and rock classification schemes of H. T. De La Beche]: v. 6, no. 12, 287-307, 2 tables.
9970 1831, [Review of A discourse on the revolutions of the surface of the globe by G. Cuvier (see 2757)]: v. 8, no. 15, 69-88.

Southron, or Lily of the Valley (Gallatin, Tennessee, 1841)
9971 1841, Geology: v. 1, 136-137.
       Signed by "Tyro".

Spackman, George (see 2251 and 2252 with G. W. Carpenter)

Spafford, Horatio Gates (1778-1832)
9972 1813, A gazetteer of the state of New York: Alb., H. C. Southwick, 334, ii p., map.
9973 1816, Cabinet of mineralogy: Am. Mag., a Mon. Misc., v. 1, 360-361.
9974 1816, Valuable lead mine [at Ancram, New York]: Am. Mag., a Mon. Misc., v. 1, 329.
9975 1824, A gazetteer of the state of New York: embracing an ample survey and description of its counties, towns, cities, villages, canals, mountains, lakes, rivers, creeks, and natural topography: Alb., B. D. Packard; Troy, NY, the author, 620 p., map.
9976 1824, A pocket guide for the tourist and traveller, along the line of the canals, and the interior commerce of the state of New York: NY, T. and J. Swords, 72 p.
9977 1825, The school boy's introduction to the geography and statistics of the state of New York ...: Troy, NY, Platt, 46 p.

Spallanzani, Lazzaro (1729-1799)
9978 1800, Comparison between Vesuvius and Etna: Mon. Mag. and Am. Rev., v. 2, 461-462.
9979 1800, A visit to Vesuvius: Mon. Mag. and Am. Rev., v. 2, 459-461.

Sparks, Jared (1789-1866)
9980 1824, [Review with excerpts of An abstract of a new theory of the formation of the earth by I. Hill (see 5060)]: N. Am. Rev., v. 18, 266-279.
9981 1824, [Review of Universal geography, ancient and modern by W. Woodbridge (see 11038) and E. Williard (see 10962)]: N. Am. Rev., v. 19, 460-364 [i.e. 463].
9982 1825, Baltimore [copper mines]: N. Am. Rev., v. 20, 99-138.

Spencer, Jesse Ames (1816-1898)
9983 1850, The East; sketches of travel in Egypt and the Holy Land: NY, G. P. Putnam, 503 p.

The Spirit of the Pilgrims (Boston, 1828-1833)
9984 1833, [Review of Iceland by E. Henderson (see 4981)]: v. 6, 212-213.

Spoon, John Jones (Reverend)
9985 1794, A topographical description of the County of Prince George, in Virginia, 1793: Mass. Hist. Soc. Coll., v. 3, 85-92.

Sprague, Timothy Dwight (1819?-1849)
9986 1848, Benjamin Silliman: Am. Lit. Mag., v. 3, no. 1, 1-15, plate.
       Biographical sketch.
9987 1848, California: Am. Lit. Mag., v. 3, no. 6, 337-344.

Squier, Ephraim George (1821-1888)
9988 1846, Pipestone of the ancient pipes in the Indian mounds: Am. J. Sci., v. 2 (ns), no. 2, 287.
9989 1847, Hieroglyphical mica plates from the mounds [Indian burial mounds near the Missouri River]: Am. J. Sci., v. 4 (ns), no. 1, 145.
9990 1848, On the fossils, minerals, organic remains, &c., found in the mounds of the West: Lit. World, v. 2, 229-230.
       In Assoc. Am. Geol's Nat's Proc.
9991 1849, Gold hunting in California in the sixteenth century: Am. Whig Rev., v. 9, 84-88.
9992 1850, The volcanoes of Central America, and the geographical and topographical features of Nicaragua, as connected with the proposed inter-oceanic canal: NY, 20 p.

Staaf
9993   1846, Sillimanite: Am. J. Sci., v. 2 (ns), no. 3, 418 (6 lines).
        Abstracted from Berzelius Jahresbericht.

Stambaugh, S. C.
9994   1837, Iowa copper mines [from the Wisconsin Territory]: Niles' Wk. Reg., v. 51, 307.

Stanley, John Thomas (see 1727 with J. Black)

Stearns, John (1770-1848)
9995   1819, Address by the president [on geology]: Med. Soc. NY Trans., for 1819, 11-24.
9996   1819, Annual address to the Medical Society of the state of New-York: Am. Mon. Mag. and
        Crit. Rev., v. 4, no. 6, 459-465.
        Includes notice of the progress of geology.

Stearns, Samuel (1747-1819)
        See also 8111 with T. Paine and W. Young.
9997   1791, The American oracle. Comprehending an account of recent discoveries in the arts
        and sciences ... : NY, Hodge and Campbell, Berry and Rogers, and T. Allen, [2],
        [5], 627, xvii p.
        Evans number 23795.

Steel [or Steele], John Honeywood (1780-1838)
9998   1817, An analysis of the mineral waters of Saratoga and Ballston: Alb., E. & E.
        Hosford, 94 p.
        For review and excerpts see 4593 and 7681.
9999   1819, An analysis of the mineral waters of Saratoga and Ballston, with practical
        remarks on their use in various diseases containing observations on the geology and
        mineralogy of the surrounding country with a geological map: Saratoga, NY, G. M.
        Davison, 118 p.
        "2nd edition, enlarged" of 9998. Also 1828 printing.
10000   1821, New locality of chrysoberyl [at Saratoga Springs, New York]: Am. J. Sci., v. 4,
        no. 1, 37-38.
10001   1822, On the geological structure of the county of Saratoga, in the state of New-York;
        together with remarks on the nature and properties of the various soils and modes
        of culture ... : Plough Boy, v. 3, 354-355; 366-367; 373-374; 377-379; 385;
        401-402; 409; v. 4, 1-3; 9.
        Also reprint edition.
10002   1823, A report of the geological structure of the county of Saratoga [New York]: NY
        Board Ag. Mem., v. 2, 44-84; 155-161.
        For review see 386. For notice see 4247.
10003   1825, A description of the oolite formation lately discovered in the county of
        Saratoga, and the state of New-York: Am. J. Sci., v. 9, no. 1, 16-19, fig.
10004   1825, Notice of Snake Hill and Saratoga Lake and its environs: Am. J. Sci., v. 9, no.
        1, 1-4, plate with 2 figs.
10005   1829, Description of the High Rock Spring at Saratoga Springs, in the county of
        Saratoga, and state of New York, with a drawing: Am. J. Sci., v. 16, no. 2,
        341-345, plate.
10006   1829, Iodine in the mineral waters of Saratoga [New York]: Am. J. Sci., v. 16, no. 2,
        242-246. Also in: Am. J. Pharmacy, v. 1, 222-227.
10007   1831, Soil, produce, and modes of cultivation, of Saratoga County: Genesee Farm. and
        Gardener's J., v. 2, 17-18; 29-31.
10008   1831, An analysis of the Mineral Water of Saratoga and Ballston, with ... observations
        on the geology and mineralogy of the surrounding country. Saratoga Springs, NY, G.
        M. Davison, i-xii, [13]-208.
        Also 1838 "second edition."
10009   1847, Analysis of the Congress Spring, with remarks on its medical properties. Rev. by
        John L. Perry: Saratoga Springs, Davison, 35 p.

Steele, Richard
10010   1839, Black river marble, serpentine and soapstone quarries [from Plymouth, Windsor
        County, Vermont]: Am. Inst. NY J., v. 4, 222-224.

Steele's western guide book, and emigrant's directory; containing general descriptions of
        different routes through the states of New-York, Ohio, Indiana, Illinois, and the
        Territory of Michigan, with short descriptions of the climate, soil, productions,
        prospects, etc.:
10011   1834, Buffalo, NY, O. G. Steele, 90, 2 p.
        Numerous later editions and printings. See also 10318.

Stein, Albert
10012   1850, Mississippi Valley. Remarks on the improvement of the River Mississippi: De Bow's
        Rev., v. 8, 105-111; 335-338, fig.; v. 9, 55-66; 304-306, fig.; 352-357; 594-601.

Stein, Heinrich Wilhelm (1811-1889)
10013   1845, Lithia - blowpipe test when mixed with soda: Am. J. Sci., v. 48, no. 1, 193.
        Abstracted by J. L. Smith from "Journ. fur Prakt. Chem."

Steinhauer, Henry (Reverend)
10014   1818, On fossil reliquia of unknown vegetables in the coal strata: Am. Philos. Soc.
        Trans., v. 1 (ns), 265-297, 4 plates.

Stephenson, Samuel L.
10015   1839, Geology of the Androscoggin and Magalloway Rivers [in Maine]: in third Maine
        report, by C. T. Jackson (see 5566), 191-205.

Stevens, J. W.
10016   1799, On the creation and convulsions of the earth; the inequality of its surface, and
        the diminution of the sea: Wk. Mag., v. 4, 225-232.
        Contains notice of caves in Kentucky, and observations on the structural geology
        of the Alleghany Mountains. Author states that the oceans are gradually receding.

Stewart, Charles Samuel (1795-1870)
10017 1826, Volcano of Kirauea: Am. J. Sci., v. 11, no. 2, 362-376.
10018 1828, Journal of a residence in the Sandwich Islands during the years 1823, 1824, and 1825: NY, J. P. Haven, xviii, [19]-320 p., front.
  Also other editions 1828, 1839.
10019 1831, Hawaii, (Owyhee,) and its volcanic regions and productions; with some notices of the moral and civil progress of its inhabitants, and of those of Oahu: Am. J. Sci., v. 20, no. 2, 228-248.
  Contains "Notices" by J. Goodrich (see 4440).

Stewart, Thomas T.
10020 1828, Teeth of the mastodon: Am. J. Sci., v. 14, no. 1, 187-189.
  Description of specimens from Cheshire, Connecticut and Schooley's Mountain, New Jersey.

Stickney, B. F.
10021 1820, American cinnabar and native lead [from the Michigan Territory]: Am. J. Sci., v. 2, no. 1, 170-171.
  See also "Correction," 10022. Also in Saturday Mag., v. 3, 343-344.
10022 1839, Correction of an error - cinnabar not found in Michigan: Am. J. Sci., v. 37, no. 1, 185-187.
  See also 10021.

Stickney, J.
10023 1834, G. B. Palmer's gold washing machine: Am. Rr. J., v. 3, 262-263, woodcut.

Stilson, W. B.
10024 1818, Sketch of the geology and mineralogy of a part of the state of Indiana: Am. J. Sci., v. 1, no. 2, 131-133.

Stockton, John
10025 1845, Report on the mineral lands of Lake Superior: US SED 175, 28-2, v. 11 (461), 22 p., map.

Stoddard, Amos (1762-1813)
10026 1804, Description of the lead mines in upper Louisiana: US Congress, 8-2, Am. State Papers, v. 1, no. 103, 188-191.
10027 1807, Observations on the native salt, bearded Indians, earthquakes, and boundaries of Louisiana: Med. Repos., v. 10, 44-50.
10028 1812, Sketches, historical and descriptive, of Louisiana: Phila., M. Carey, viii, 488 p.

Stodder, Charles (d. 1884)
  See also discussion of 3131 by E. Desor.
10029 1844, [Review of Mineralogy of New York, by L. C. Beck (see 1577)]: Bos. Soc. Nat. History Proc., v. 1, 173.
10030 1846, [On drift in a ridge at South Boston, Massachusetts]: Bos. Soc. Nat. History Proc., v. 2, 131-132.
10031 1847, [Review of "On the origin of continents" by J. D. Dana (see 2829)]: Bos. Soc. Nat. History Proc., v. 2, 213 (7 lines).

Storer, David Humphreys (1804-1891)
10032 1837, Annual report of the curators of the Boston Society of Natural History: Am. J. Sci., v. 32, no. 2, 364-371.

Strachey, B. (Lieutenant)
10033 1850, On the motion of the glacier of the Pindur, in Kumaon: Ann. Scien. Disc., v. 1, 251 (12 lines).
10034 1850, The snow-line in the Himalaya: Ann. Scien. Disc., v. 1, 250-251.

Strain, Isaac G. (1821-1857)
10035 1844, [Fossils from Minas Geras, Brazil]: Phila. Acad. Nat. Sci's Proc., v. 2, no. 1, 11-14.

Stranger, John
10036 1818, Supplement to Dr. Mitchill's "Observations on the geology of North-America," just published by Messrs. Kirk & Mercein, in the description of a fossil elephant, discovered in Wythe County, southwest of the River Ihanhawa, in Virginia: Am. Mon. Mag. and Crit. Rev., v. 3, no. 1, 60.
  See also "Observations" by S. L. Mitchill (7307).

Strickland, Hugh Edwin (1811-1853)
  See also 4742 with W. I. Hamilton.
10037 1847, On the history of the dodo and other allied species of birds: Am. J. Sci., v. 4 (ns), no. 3, 422-423.
  Abstracted from London Ath.
10038 1849, and A. G. Melville, [Excerpts from The dodo and its kindred, or the history, affinities and osteology of the dodo, solitaire and other extinct birds of the Islands Mauritius, Rodriguez, and Bourbon]: Am. J. Sci., v. 7 (ns), no. 1, 52-67, plate, fig.
  In review by B. Silliman (see 9791).

Strode, Thomas
10039 1824, Notice of a siliceous petrifaction, from N. Carolina: Am. J. Sci., v. 7, no. 2, 249-250.
  On petrified wood from Fayetteville, North Carolina.

Stromeyer, Friedrich (1776-1835)
10040   1818, Analyse des sels de strontiane et de quelques mineraux: J. Sci. Arts, v. 3, no. 1, 215-216.
        Abstracted from Annales de Chimie.
10041   1823, Dichroite [from Greenland]: Bos. J. Phil. Arts, v. 1, no. 2, 188 (7 lines).
10042   1823, Eudialite [from Greenland]: Bos. J. Phil. Arts, v. 1, no. 2, 187 (6 lines).
10043   1823, Sapphirine [from Greenland]: Bos. J. Phil. Arts, v. 1, no. 2, 187 (7 lines).
10044   1824, M. Stromeyer: Am. J. Sci., v. 7, no. 2, 368-371.
        Abstracted by J. Griscom. On the chemical analysis of twelve minerals.
10045   1825, and J. F. L. Hausmann, Account of a new ore of lead, containing selenium [from the Hartz Mountain]: Bos. J. Phil. Arts, v. 2, no. 6, 597-601.
        Abstracted from the Learned Intelligencer by Frederick H. Hedge.
10046   1826, Selenium in the sulphur of the Lipari Islands: Bos. J. Phil. Arts, v. 3, no. 5, 499 (5 lines).
10047   1827, Comparative analysis of olivine and chrysolite: Am. J. Sci., v. 13, no. 1, 184-185.
        Abstracted by C. U. Shepard from Annales des Mines.
10048   1836, A new antimonuret of nickel [from Andreasberg]: Am. J. Sci., v. 30, no. 1, 177.
        Abstracted by "D." from Annales des Mines.

Strong, Edward (1813-1898)
10049   1846, Vestiges of Creation and its reviewers [extracts from reviews with excerpts of Vestiges ... , by R. Chambers (see 2292 et seq.)]: New Englander, v. 4, no. 1, 113-127.
        Extracted from N. Am. Rev., Am. Whig Rev., and N. British Am. Rev.

Strong, Henry King (1798-1860)
10050   1839, Report to the legislature of Pennsylvania, containing a description of the Swatara mining district, illustrated by diagrams: Harrisburg, Penn., Boas and Coplan, 61 p., plates, folded map.
10051   1839, Map of the Swatara Coal Region: Harrisburg, map.

Struder, B. (Professor)
10052   1839, On the recent explanations of the phenomenon of erratic blocks: Am. J. Sci., v. 36, no. 2, 325-332.
        Translated by W. A. Larned from Neues Jahrbuch für Mineralogie.

Strzelecki, Sir Paul Edmond de (1796-1873)
10053   1838, Sandwich Islands. - crater of Kirauea, Hawaii: Hawaiian Spectator, v. 1, no. 4, 434-437.
10054   1839, Crater of Kirauea, Hawaii. - Sandwich Islands: The Friend, v. 12, 258-259.
        Abstracted from Hawaiian Spectator.
10055   1846, Geology of New South Wales, New Holland and Van Dieman's Land: Am. J. Sci., v. 1 (ns), no. 2, 278.

Stuart, J.
10056   1823, By Mr. J. Stuart, of Peacham, Vermont: Am. J. Sci., v. 6, no. 2, 249.
        On mineral localities in Vermont.

Stuart, Moses (1780-1852)
10057   1836, Critical examination of some passages in Gen. I.; with remarks on difficulties that afford some of the present modes of geological reasoning: Bib. Repos., v. 7, 46-106.
        For review with excerpts see 655. For review see 5984. See also article by E. Hitchcock (5117).

Stuart, W.
10058   1839, On the limestone, the lime cement, and method of blasting, in the neighbourhood of Plymouth: Franklin Inst. J., v. 23 (ns), 212-213.
        Abstracted from J. Sci. Arts.

Sturm, Christoph Christian (1740-1786)
10059   1814, Reflections for every day in the year, on the works of God; and of his providence, throughout all nature: Hudson, NY, Ashbel Stoddard, 2 v.
        First American edition. Other editions include "revised edition," NY, A. Paul, 612 p., plates, translated by A. Clarke; and, as Reflections on the works of God, Phila., J. J. Woodward, 486 p., 1832 and 1836; NY, B. Waugh and T. Mason, 624 p., plates, 1834.

Stutchbury, Samuel (1798-1859)
10060   1837, and Riley, [Newly discovered saurian remains from England]: Am. J. Sci., v. 31, no. 2, 364-365.
        In British Assoc. Proc.
10061   1847, Plesiosaurus megacephalus [from Somersetshire, England]: Am. J. Sci., v. 3 (ns), no. 2, 276 (12 lines).
        Abstracted from Q. J. Geol. Soc.

The Subscribers to the annual support of the Cabinet of Natural History are informed that said cabinet ... will open from the first of November next ... :
10062   1804, NY, Broadside.

Suffern, Edward
10063   1817, [Mammoth bones from Rockland County, New York]: Am. Mon. Mag. and Crit. Rev., v. 2, no. 1, 46-47.

Sullivan, John Langdon (1777-1865)
10064 1828, Report, on the origin and increase of the Paterson manufactories ... : Paterson, NJ, Day & Burnett, 60, [2] p., map.
Includes data on NJ coal and iron trade.
10065 1830, Proposition for an anthracite coal steam power boat and barge company, for passengers on the North River: NY, Clayton and Van Norden, 12 p.
10066 1835, Extracting precious metals from their ores: Franklin Inst. J., v. 15 (ns), 195 (11 lines).

Sullivan, Jonathan
10067 1848, Improvement in washing gold ore: Franklin Inst. J., v. 15 (3s), 347 (7 lines).

Sullivant, Joseph (1809-1884)
10068 1838, An alphabetical catalogue of shells, fossils, minerals and zoophytes, in the cabinet of Joseph Sullivant: Columbus, Ohio, Cutler and Pilsbury (printers), 38 p.

Summers, B.
10069 1850, Bog iron ore [from Ohio, Michigan and Indiana]: Ohio Cult., v. 6, 84.

Summey, Peter
10070 1830, Machine for washing gold out of gravel and sand: Franklin Inst. J., v. 5 (ns), 230 (9 lines).

Sumter, T. W. A.
10071 1830, Machine for washing, separating, and saving alluvial gold: Franklin Inst. J., v. 6 (ns), 150-151.

Surveyor (pseudonym; see 455 and 465)

Sutherland, Joel Barlow (1792-1861)
10072 1813, Speculations on lime: Colum. Chemical Soc. Phila. Mem., v. 1, 58-62.

Svanberg, Lars Fredrick (1805-1878)
10073 1846, Discovery of some new earths in zircons: Am. J. Sci., v. 1 (ns), no. 2, 257 (8 lines).
Abstracted from "Chem. Gaz." by J. L. Smith.
10074 1847, Groppite, a new mineral: Am. J. Sci., v. 3 (ns), no. 3, 429 (9 lines).
Abstracted from Berzelius´ Jahresbericht.

Sweitzer, Lewis
10075 1808, Observations on the waters of Schooley´s Mountain (N. J.): Phila. Med. Mus., v. 6, 250-252.
Also reprint edition, 3 p.

Swift, William Henry (1800-1879; see 6648 with W. G. McNeill and G. W. Whistler)

Swinburne, C. H.
10076 1831, Volcano in the Mediterranean: Mon. Am. J. Geology, v. 1, no. 5, 229-232.
Abstracted in The Friend, v. 5, 13.

Symmes, John Cleves (1780-1829)
Passing references have been found to several works by this author. For example, on page 20 of Symzonia (see 10078), the narrator says "least of all did I omit Symmes´ Memoirs, and printed lectures." However, copies of pre-1826 memoirs or lectures have not been found.
10077 1818, [Circular sent to learned institutions of Europe and the United States, explaining Symmes´s theory of the earth]:
Not seen.
10078 1820, Symzonia; a voyage of discovery. By Captain Adam Seaborn: NY, J. Seymour (printer), xii, [13]-248, plate.
Adam Seaborn is apparently a pseudonym for J. C. Symmes.
10079 1826, Symmes theory of concentric spheres: demonstrating that the earth is hollow, habitable within, and widely open about the Poles. By a citizen of the United States: Cin., Morgan, Lodge and Fisher, 168 p.
For review see 6921. See also 8757.

Taber, Thomas
10080 1840, Practical remarks on gems, especially on some of those found in the United States: Am. J. Sci., v. 38, no. 1, 61-68.

Taddei, Giovacchino (1792-1860)
10081 1817, [Native caustic lime in Italy at Santa Gonda]: Am. Reg. or Summary Rev., v. 1 (ns), 449-450 (16 lines).
Abstracted from Brandes J.

Talcott, George
10082 1843, Lead mines: Army and Navy Chron. and Scien. Repos., v. 2, 72.
10083 1843, Lead mines of the Upper Mississippi: Army and Navy Chron. and Scien. Repos., v. 2, 70-72.
10084 1843, Mines in the south part of Illinois: Army and Navy Chron. and Scien. Repos., v. 2, 72.
10085 1843, Mines in the state of Arkansas: Army and Navy Chron. and Scien. Repos., v. 2, 72.
10086 1846, [Map of mine locations in the Lake Superior mining district]: in report by A. B. Gray (see 4499).
10087 1846, [Report on mineral lands in Illinois, Arkansas, Michigan, and adjacent districts]: US SED 160, 29-1, v. 4 (473), 9-19.

Tallmadge, James (1778-1853)
10088 1846, Geological meeting: NY J. Medicine and Collateral Sci´s, v. 7, 276 (11 lines).
Notice of the 1846 meeting of the Assoc. Am. Geol´s Nat´s.

Tanner, Henry Schenk (1786-1858)
    See also 4387 with J. B. Gibson and R. Harlan.
10089 1832, View of the Valley of the Mississippi, or traveller's guide to the West: Phila.
    Not seen.
10090 1834, The American traveller; or, guide through the United States ... : Phila., the
    author, 144 p., map.
    Numerous other printings published in most years through 1846.

Tappan, Benjamin (1773-1857)
10091 1828, On the boulders of Primitive rocks found in Ohio, and other western states and
    territories: Am. J. Sci., v. 14, no. 2, 291-297.
10092 c.1830, Circular. The Historical and Philosophical Society of Ohio: n.p., n.d., 16 mo,
    12 p.
10093 1833, A discourse delivered before the Historical and Philosophical Society of Ohio:
    Columbus, Ohio, 16 p., 12 mo.

Tarlor (Mr., jr.)
10094 1838, [Practical mining techniques for lead]: Am. J. Sci., v. 34, no. 1, 9.
    In British Assoc. Proc.

Tarver, Micajah
10095 1848, Manufacture of iron in St. Louis, Iron Mountain, and iron ore in Missouri, -
    causes hindering the production of iron in Missouri: West. J. Ag., v. 1, no. 1,
    36-39.

Tatham, William (1752-1819)
10096 1799, Supplement to Mr. Latrobe's memoir [see 6063]: Am. Philos. Soc. Trans., v. 4,
    444.
    Contains notes on fossils from Cape Henry, Virginia.

Tauscher (Dr., of Germany)
10097 1839, The salt lake of Inderskoi, and its environs, in the Kirghis Steppe, in Asia:
    Burton's Gentleman's Mag., v. 5, 92-95.

Taylor, Bayard (1825-1878)
10098 1850, Eldorado: adventures in the path of empire; comprising a voyage to California,
    via Panama; life in San Francisco and Monterey; pictures of the gold region and
    experiences of Mexican travel: NY, G. P. Putnam, 2 v. in 1, 251, 247 p.
    Contains appendix report by T. King (see 5968) on p. 201-247 of v. 2.
    For review with excerpts see 3421. For review see 5489.

Taylor, George
10099 1833, Effect of incorporated coal companies upon the anthracite coal trade of
    Pennsylvania: Pottsville, Penn., B. Bannan, 34 p.
    Another edition in 1834: Pottsville, Penn., 7 p.
10100 1847, Theories of creation and the universe: De Bow's Rev., v. 4, 177-194.
    Discussion of theories by J. Herschel and R. Chambers (see 2292 et seq.)

Taylor, J.
10101 1822, [Review of Narrative journal of travels ... , by H. R. Schoolcraft (see 9261)]:
    NY Med. Phys. J., v. 1, 124-125.

Taylor, John (1752-1835)
10102 1811, On gypsum: Phila. Soc. Promoting Ag. Mem., v. 2, 51-62; 75-78.

Taylor, John
10103 1835, Account of the depth of mines: Franklin Inst. J., v. 15 (ns), 280-283. Also in:
    Mech's Mag. and Reg. Inventions, v. 5, 92-94.
    Abstracted from British Assoc. Proc.
10104 1838, New process for extracting gold: Franklin Inst. J., v. 21 (ns), 429-430.
    Abstracted from Mining J.
10105 1846, Improvements in separating metals from each other, and from certain combinations
    with other substances: Franklin Inst. J., v. 11 (3s), 51-56.
    Abstracted from "Report. Pat. Inv."

Taylor, Julius S.
10106 1848, Notice of fragments of trilobites of gigantic size in the cabinet of Dr. Julius
    S. Taylor, Carrollton, Montgomery County, Ohio: Am. J. Sci., v. 6 (ns), no. 3, 431.
10107 1850, Notice of trilobites in the cabinet of Dr. Julius S. Taylor: Am. J. Sci., v. 10
    (ns), no. 1, 113-114.

Taylor, Richard Cowling (1789-1851)
    See also 8875 with others, and 4414 with P. B. Goddard and H. D. Rogers.
10108 1831, Mr. R. C. Taylor's fossils: Mon. Am. J. Geology, v. 1, no. 5, 239-240.
    Notice of fossils for sale.
10109 [1831], Phillipsburg Rail Road: Phila.?, 21 p., map.
10110 1832, Geology of the Alleghany Mountains and Moshannon Valley [Pennsylvania]: Hazard's
    Reg. Penn., v. 10, 305-306.
10111 1832, Section of the Alleghany Mountain, and Moshannon Valley, in Centre County, Penn.:
    Mon. Am. J. Geology, v. 1, no. 10, 433-438, plate.
10112 1833, Report on the surveys, undertaken with a view to the establishment of a rail
    road, from the coal and iron mines near Blossburg or Peter's camp to the state line
    at Lawrenceville, in the county of Tioga and the state of Pennsylvania, and
    mineralogical report on the coal region in the environs of Blossburg: Phila.,
    Mifflin and Parry (printers), 56 p., map.
10113 1833, Reports on the Woodville gold mine [in Orange County, Virginia]: Wash.?, 15 p.
10114 1834, On the geological position of certain beds which contain numerous fossil marine
    plants of the family Fucoides, near Lewistown, Mifflin County, Pennsylvania: Geol.
    Soc. Penn. Trans., v. 1, 5-15, 4 colored plates with 11 figs, map.
    Contains map 10116. See also notice 4325. Abstracted in Hazard's Reg. Penn., v.
    14, 168-170. Also reprint edition.

Taylor, R. C. (cont.)

10115 1834, Richmond coal basin and its coal trade: Penn. State J., v. 2, 567.
Not seen.

10116 1834, Rough sketch of the position of the transition beds near Lewistown, Mifflin County, Pennsylvania containing various species of fossil Fucoides: Geol. Soc. Penn. Trans., v. 1, plate 4, colored geological map, 23 x 27 cm.
In 10114.

10117 1835, Fossil marine plants: Am. J. Sci., v. 27, no. 2, 347-348.
In "Notice of the transactions of the Geological Society of Pennsylvania."

10118 1835, Map and section of Kishacoquillas Valley [in Mifflin County, Pennsylvania]: Geol. Soc. Penn. Trans., v. 1, plate 9, colored geological map, 34 x 23 cm.
In 10119.

10119 1835, Notice as to the evidences of the existence of an ancient lake, which appears to have formerly filled the limestone valley of Kishacoquillas, in Mifflin County, Pennsylvania: Geol. Soc. Penn. Trans., v. 1, 194-203, plate with 6 figs., map.
Contains map 10118.

10120 1835, Memoir of a section passing through the bituminous coal field near Richmond, in Virginia: Geol. Soc. Penn. Trans., v. 1, 275-294, folding plate.

10121 1835, On the mineral basin or coal field of Blossburg, on the Tioga River, Tioga County, Pennsylvania: Geol. Soc. Penn. Trans., v. 1, 204-219.
See also "Examination and analysis ...," by T. G. Clemson (2439).

10122 1835, On the relative position of the Transition and Secondary coal formations in Pennsylvania, and description of some Transition coal or bituminous anthracite, and iron ore beds near Broad Top Mountain, in Bedford County, and of a coal vein in Perry County, Pennsylvania. With sections: Geol. Soc. Penn. Trans., v. 1, 177-193, plate.

10123 1835, Review of geological phenomena, and the deductions derivable therefrom, in two hundred and fifty miles of sections in parts of Virginia and Maryland. Also notice of certain fossil acotyledonous plants in the Secondary strata of Fredericksburg: Geol. Soc. Penn. Trans., v. 1, 314-325, 2 plates.

10124 1835, and T. G. Clemson, Section from Fredericksburg through the gold region of Virginia to Winchester: Geol. Soc. Penn. Trans., v. 1, colored plate 17 (100 x 23 cm.)

10125 1836, Note on the occurrence of bituminous coal near the city of Havana, in Cuba: Franklin Inst. J., v. 18 (ns), 178.
Abstracted in Am. Rr. J., v. 5, 651.

10126 1838, Extracts from R. C. Taylor's report, on the surveys undertaken with a view to the establishment of a railroad ... : Alb., Packard & Van Benthuysen, 15 p.

10127 1839, and T. G. Clemson, Notice of a vein of bituminous coal, recently explored in the vicinity of the Havana, in the island of Cuba: Am. Philos. Soc. Trans., v. 6 (ns), 191-196.
Also reprint edition.

10128 1840, Two reports on the coal lands, mines and improvements of the Dauphin and Susquehanna Coal Company, and of the geological examinations, present condition, and prospectus of the Stony Creek Coal Estate, in the townships of Jackson, Rush, and Middle Paxtang, in the county of Dauphin, and of East Hanover Township, in the County of Lebanon, Pennsylvania: With an appendix, containing numerous tables and statistical information, and various maps, sections, and diagrams, chiefly an illustration of coal and iron: Phila., E. G. Dorsey, 113, 74 p., plate, 3 maps.
For review see Am. Rr. J., v. 11, 39.

10129 1841, [Geological sections and coloring]: Am. J. Sci., v. 41, no. 1, 177.
In Assoc. Am. Geol's Nat's Proc.

10130 1841, Notice of a model of the western portion of the Schuylkill or Southern Coal-field of Pennsylvania, in illustration of an address to the Association of American Geologists, on the most appropriate modes for representing geological phenomena: Am. J. Sci., v. 41, no. 1, 80-91, colored plate.
Also reprint edition, 14 p.

10131 1841, [The Dauphin and Lebanon coal region of Pennsylvania]: Am. J. Sci., v. 41, no. 1, 176.
Also in Assoc. Am. Geol's Nat's Proc., v. 2, 21-22 (11 lines). Also in Assoc. Am. Geol's Nat's Rept., v. 1, 29 (11 lines), 1843.

10132 1843, Notice of a model of the western portion of the Schuylkill or southern coal field of Pennsylvania, in illustration of an address to the Association of American Geologists, on the most appropriate modes for representing geological phenomena: Assoc. Am. Geol's Nat's Rept. v. 1, 81-94.

10133 1843, On fossil arborescent ferns of the family of Sigillaria, occurring in the roof and floor of a coal seam in Dauphin Co., Pa.: Am. Philos. Soc. Proc., v. 3, no. 27, 149-150.

10134 1843, On the geology of the northeast part of the Island of Cuba, and on the prospects of the copper region of Gibara: Am. Philos. Soc. Proc., v. 3, no. 27, 154-155.

10135 1845, Lead, silver, and gold mine in North Carolina: Am. Q. J. Ag. Sci., v.2, 130.
Extracted from Washington Silver Mine Reports (see 10138).

10136 1845, [Mr. Taylor's cabinet of fossils described]: Phila. Acad. Nat. Sci's Proc., v. 2, no. 10, 261.

10137 1845, On the anthracite and bituminous coal fields in China; the system of mining, and the prices of coal, and labour in its production, and transportation to Pekin: Franklin Inst. J., v. 10 (3s), 51-57.
Also reprint edition, Phila., 6 p.

10138 1845, Reports on the Washington Silver Mine in Davidson County, North Carolina. ... With appendix, containing assays of the ores ... : Phila., E. G. Dorsey (printer), 40 p., 4 folded color plates.

10139 1845, Silver mines of North Carolina: Hunt's Merch. Mag., v. 13, 171-173.
Extracted from Washington Silver Mine report (see 10138).

10140 1845, [Washington Silver Mine in Davidson County, North Carolina]: Am. Philos. Soc. Proc., v. 4, no. 32, 150-151.

10141 1846, Anthracite and bituminous coal in China: Am. J. Sci., v. 2 (ns), no. 1, 141-144.
Abstracted from Philos. Mag.

10142 1846, Memoir on the character and prospects of the copper region of Gibara, and a sketch of the geology of the northeast part of the Island of Cuba: Am. Philos. Soc. Trans., v. 9 (ns), 204-218, map.
Contains map 10144. Also reprint edition.

Taylor, R. C. (cont.)
10143   1846, Notice of fossil arborescent ferns, of the family of Sigillaria, and other coal plants, exhibited in the roof and floor of a coal seam, in Dauphin County, Pennsylvania: Am. Philos. Soc. Trans., v. 9 (ns), 219-227, folded plate.
10144   1846, Rough sketch or reconnoissance of the copper region and of the geology of the Savana region of Gibara in the island of Cuba: Am. Philos. Soc. Trans., v. 9 (ns), geological map, 33.1 x 47.1 cm.
         In 10142. Scale of 1" to the mile.
10145   1847, The coal mines and coal trade of Belgium - historical, statistical, and commercial: Hunt's Merch. Mag., v. 16, 235-256.
10146   1848, Report to the Dauphin and Susquehanna Coal Co.: Phila.
10147   1848, Statistics of coal: the geographical and geological distribution of mineral combustibles or fossil fuel, including also notices and localities of the various mineral bituminous substances employed in arts and manufactures ... embracing from official reports of the great coal producing countries, the respective amounts of their production in all parts of the world; together with their prices, tariffs, duties and international regulations: Phila., J. W. Moore, 148, 754 p., maps, diagram.
         For review see 938, 3689, 4085, 5446, and 5468. For notice see 5461.
10148   1848, [i.e. 1849], Map illustrative of the statistics of the coal trade of Pennsylvania ... : Phila., T. Sinclair (lithographer), geological map, 40 x 31 cm.
         In Report to the directors of the Pequa Railroad and Improvement Company, Phila., T. K. and P. G. Collins.
10149   1850, Reports on the Woodville gold mine: Wash., 15 p.
10150   1850, Report [on the Eagle Mine]: in Enterprise Mining Company report (see 3594).

Taylor, Stephen
10151   1841, Discovery in Virginia, of the regular mineral salt formation: Am. J. Sci., v. 41, no. 1, 214-215.

Taylor, Steuben
10152   1823, By Mr. Steuben Taylor [on mineral localities in Rhode Island and Connecticut]: Am. J. Sci., v. 6, no. 2, 245-246.
10153   1824, Miscellaneous localities [of minerals from northeastern United States]: Am. J. Sci., v. 7, no. 2, 253-254.
10154   1824, Notice of a rocking stone in Warwick, R. I.: Am. J. Sci., v. 7, no. 2, 201-203, plate.
10155   1824, and T. H. Webb, Notice of miscellaneous localities of minerals [in Rhode Island]: Am. J. Sci., v. 8, no. 2, 225-236.
10156   1825, Notice of pebbles from Cape Horn: Am. J. Sci., v. 9, no. 1, 48-49.

Tchihatcheff, M.
10157   1848, Emery in Asia Minor: Am. J. Sci., v. 6 (ns), no. 2, 272 (4 lines).

Teague, Moses
10158   1832, Improvement in the making and smelting pig iron: Franklin Inst. J., v. 10 (ns), 337-338.

Teissier, J. A.
10159   1828, Projected salt works in Boston Bay: Bos., 8 p.

Temascaltepec Mining Company, Baltimore
10160   1827, Documents laid before a meeting of the Temascaltepec Mining Company, of Baltimore, convened by public notice, August 24, 1827: Balt., Thomas Murphy, 33 p.

Tench, Watkin (1759-1833)
10161   1789, A narrative of the expedition to Botany Bay; with an account of New South Wales, its productions, inhabitants, &c. To which is subjoined, a list of the civil and military establishments at Port Jackson. By Captain Watkin Tench, of the Marines: NY, T. and J. Swords (printers).
         Evans number 22176. Not seen.

Tennessee. State of
10162   1831, [Act forming the geological survey of Tennessee]: Tenn. Gen. Assembly, December 21, 1831.
10163   1833, [Act appointing G. Troost geologist of the state]: Tenn. Gen. Assembly, November 15, 1833.
10164   1836, [Act continuing G. Troost's appointment as state geologist]: Tenn. Gen. Assembly, January 28, 1836.
10165   1836, [Resolution requesting G. Troost to make a geological survey of the Cherokee Nation]: Tenn. Gen. Assembly, October 24, 1836.

Tenney, Samuel (1748-1816)
10166   1793, An account of a number of medicinal springs at Saratoga, in the state of Newyork: Am. Acad. Arts Sci's Mem., v. 2, pt. 2, 43-61.
10167   1814, An account of an earthquake in New-England [on November 10, 1810]: Am. Acad. Arts Sci's Mem., v. 3, pt. 2, 346-349.

Teptoff, M.
10168   1836, [The mines of Russia]: Zodiac, v. 1, no. 7, 112.

Teschemacher, James Englebert (1790-1853)
10169   1841, Address delivered at the annual meeting of the Boston Natural History Society, May 5th, 1841: Bos., 46 p.
         For review with excerpts see 751.
10170   1841, [Analysis of tin ore, from New Hampshire]: Hazard's US Reg., v. 5, 63.
         In report by C. T. Jackson (see 5587).
10171   1841, [Mineralogical notes from Jackson's first New Hampshire report (see 5576)]: Farm's Mon. Vis., v. 3, 164-165, 2 figs.
10172   1842, and A. A. Hayes, On the identity of pyrochlore with the microlite of Prof. Shepard: Am. J. Sci., v. 43, no. 1, 33-35.

Teschemacher, J. E. (cont.)

10173 1842, On the occurrence of the phosphate of uranium in the tourmaline locality at Chesterfield [Massachusetts]: Bos. J. Nat. History, v. 4, no. 1, 35-37.
    With quantitative analyses by A. A. Hayes (see 4895).

10174 1843, [Artificial dendrites]: Bos. Soc. Nat. History Proc., v. 1, 102.

10175 1843, On the occurrence of phosphate of uranium in the tourmaline locality of Chesterfield [Massachusetts]: Bos. Soc. Nat. History Proc., v. 1, 15.

10176 1843, Description of the oxide of tin found at the tourmaline locality, Chesterfield, Mass.: Assoc. Am. Geol's Nat's Rept., v. 1, 296-297, fig.

10177 1843, [Fossil ferns from the anthracite coal mines at Mansfield, Pennsylvania]: Bos. Soc. Nat. History Proc., v. 1, 61-62.
    With discussion by C. T. Jackson.

10178 1843, [On slate with dendritic markings at Newton, Massachusetts]: Bos. Soc. Nat. History Proc., v. 1, 96.

10179 1843, [Review of the transactions of the Imperial Mineralogical Society of St. Petersburg]: Bos. Soc. Nat. History Proc., v. 1, 108-110.

10180 1843, [Silicified wood from Manilla]: Bos. Soc. Nat. History Proc., v. 1, 22.

10181 1843, [Minerals from Mount Vesuvius]: Bos. Soc. Nat. History Proc., v. 1, 3-4.

10182 1844, Blowpipe characters of the supposed pyrrhite of the Azores: Am. J. Sci., v. 47, no. 2, 418 (11 lines).
    Abstracted from Bos. J. Nat. History.

10183 1844, Mineralogical notices: Bos. J. Nat. History, v. 4, no. 4, 498-504.
    Summary of recent world-wide discoveries.

10184 1844, [On beryls from Acworth, New Hampshire]: Bos. Soc. Nat. History Proc., v. 1, 191-192.

10185 1844, [On the Jura limestone and its dendrites]: Bos. Soc. Nat. History Proc., v. 1, 170.

10186 1844, Proceedings of the Boston Society of Natural History: Am. J. Sci., v. 46, no. 1, 203-204.
    On activities of the Society in 1843.

10187 1844, Singular crystal of lead from Rossie, N. Y.: Am. J. Sci., v. 47, no. 2, 417.
    Abstracted from Bos. J. Nat. History.

10188 1845, [Notice on pyrochlore from Chesterfield, Massachusetts]: Am. J. Sci., v. 48, no. 1, 177-178.

10189 1845, On the occurrence of uranium in the beryl locality at Acworth, N. H.: Bos. J. Nat. History, v. 5, no. 1, 87-89.

10190 1845, Remarks on uranium and pyrochlore, in reply to Prof. Shepard: Am. J. Sci., v. 48, no. 2, 395-397.
    See also notice by C. U. Shepard (9546).

10191 1846, Damourite in the United States [from Chesterfield, Massachusetts and Leiperville, Pennsylvania]: Am. J. Sci., v. 2 (ns), no. 1, 119 (7 lines).
    Abstracted from Bos. Soc. Nat. History Proc.

10192 1846, [Notes on damourite and pyrrhite]: Bos. Soc. Nat. History Proc., v. 2, 107-109.

10193 1846, On the fossil vegetation of America: Bos. J. Nat. History, v. 5, no. 3, 370-385, 4 plates with 14 figs.

10194 1846, On the fossil vegetation of America: Bos. Soc. Nat. History Proc., v. 2, 146-147.

10195 1847, [Coal showing wood structure]: Bos. Soc. Nat. History Proc., v. 2, 260.

10196 1847, [Fossil plants of anthracite coal]: Am. Q. J. Ag. Sci., v. 6, 212 (9 lines).
    In Assoc. Am. Geol's Nat's Proc.

10197 1847, A new mineral from the Azores [pyrrhite]: Am. J. Sci., v. 3 (ns), no. 1, 32-34, fig.

10198 1847, On the fossil vegetation of America: Am. J. Sci., v. 3 (ns), no. 1, 86-90, 5 figs.

10199 1847, On the fossil vegetation of anthracite coal: Am. J. Sci., v. 4 (ns), no. 3, 420-421.
    Abstracted from Assoc. Am. Geol's Nat's Proc. Abstract in Lit. World, v. 2, 228 (1848).

10200 1848, Mr. Teschemacher: Am. Acad. Arts Sci's Mem., v. 1, 179-180.
    Notice of fossil seeds, and slickensides, in coal.

10201 1848, [Note on metamorphism, cleavage, and bedding in rocks]: Bos. Soc. Nat. History Proc., v. 3, 28-30.

10202 1848, [On the study of minerals]: Bos. Soc. Nat. History Proc., v. 3, 18-19.

10203 1848, [On the vegetation of anthracite coal]: Bos. Soc. Nat. History Proc., v. 3, 8-9; 35.

10204 1849, [Dendrites and their origins]: Bos. Soc. Nat. History Proc., v. 3, 99-100.

10205 1849, [Harmotome from Isle Royale, Lake Superior]: Bos. Soc. Nat. History Proc., v. 3, 105-106.

10206 1849, [On the identity of arkansite with brookite, and the measurement of their interfacial angles]: Bos. Soc. Nat. History Proc., v. 2, 131-132, fig.
    Abstracted in Am. J. Sci., v. 8 (ns), no. 2, 274 (7 lines).

10207 1850, [California gold and platina]: Bos. Soc. Nat. History Proc., v. 3, 280; 287 (11 lines).

10208 1850, [Fossil plants and the origin of coal]: Bos. Soc. Nat. History Proc., v. 3, 260-262.

10209 1850, [Fossil rain drops]: Bos. Soc. Nat. History Proc., v. 3, 201 (5 lines).

10210 1850, [On black oxide of copper from Copper Harbor, Michigan]: Ann. Scien. Disc., v. 1, 262 (13 lines).
    Abstracted from Am. Assoc. Adv. Sci. Proc.

10211 1850, On the existence of vanadium in the copper ores of Lake Superior: Ann. Scien. Disc., v. 1, 262-263.

10212 1850, [On the identity of vermiculite and pyrophyllite]: Bos. Soc. Nat. History Proc., v. 3, 245-246.
    Discussion of article by C. T. Jackson (see 5664).

10213 1850, On the occurrence of rare minerals in the United States: Ann. Scien. Disc., v. 1, 273 (11 lines).
    On tellurides from Virginia, and arkansite.

10214 1850, Platinum of California: Am. J. Sci., v. 10, no. 1, 121 (6 lines).

10215 1850, [Vanadium from Isle Royale, Lake Superior]: Bos. Soc. Nat. History Proc., v. 3, 229.

Testut, Charles (1818-1892)
10216　1843, Earthquake in the West Indies: Army and Navy Chron. and Scien. Repos., v. 1, 423-424.

Texier, Charles Félix Marie (1802-1871)
10217　1837, Asia Minor [geological notes]: Am. J. Sci., v. 32, no. 2, 399-400 (13 lines). Abstracted from London Ath.

Thacher, James (1754-1844)
10218　1804, Observations upon the art of extracting marine salt from sea water, by evaporation, produced by the sun's heat; with a description of the works, and the several processes used in preparing medicinal salts, and magnesia alba: Am. Acad. Arts Sci's Mem., v. 2, pt. 2, 107-121.

Thackeray, William Makepiece (1811-1863)
10219　1844, Visit to the Giant's Causeway: Arthur's Mag., v. 2, 41-44.
　　　Signed by "M. A. Titmarsh" (pseudonym).

Thayer, S. W., jr.
10220　1845, Letter from S. W. Thayer, M. D. [on the geology of Thetford]: in First annual report on the geology of the state of Vermont by C. B. Adams (see 23), 77-79.

Thomas, David (1776-1859)
10221　1819, Travels through the western country in the summer of 1816, including notices of the natural history, antiquities, topography, agriculture, commerce, and manufactures; with a map of Wabash county, now settling: Auburn, NY, David Rumsey, ii, 320 p., folding map.
10222　1830, Diluvial furrows and scratches: Am. J. Sci., v. 17, no. 2, 408.
10223　1830, Geological facts, by David Thomas: Am. J. Sci., v. 18, no. 2, 375-376.
　　　With notice by A. Eaton (see 3360). On stratigraphy at Cayuga Lake, New York.
10224　1831, Remarks on Professor Eaton's "Observations on the coal formations in the state of New York" [see 3351]: Am. J. Sci., v. 19, no. 2, 326-328.
10225　1832, Geological remarks between Buffaloe in New York and Pittsburgh, in Pennsylvania: Atl. J. and Friend Know., v. 1, no. 4, 135-137.

Thomas, Francis Hay
10226　1850, Improvement in smelting copper ores: Hunt's Merch. Mag., v. 22, 239.

Thomas, Isaac
10227　1830, Chester County Cabinet of Natural Science: Hazard's Reg. Penn., v. 5, 377-378.
　　　Notice of officers for 1830.

Thomas, S. H.
10228　1841, Abstract of a lecture on the continental method of assaying copper: Franklin Inst. J., v. 2 (3s), 105-111.

Thompson (Dr., of Naples)
10229　1797, Catalogue of certain productions discovered in the last eruption of Mount Vesuvius: NY Mag., v. 2 (ns), 623-624.

Thompson, Andrew
10230　1813, To separate copper from silver: Arch. Use. Know., v. 3, no. 3, 254-256.

Thompson, Lewis
10231　1840, On the separation of lime from magnesia, and on the assay of gold: Franklin Inst. J., v. 25 (ns), 262.
　　　Abstracted from London and Edinburgh Philos. Mag.
10232　1840, Assay of gold: Franklin Inst. J., v. 25 (ns), 379-380.
　　　Abstracted by "U. S. M."
10233　1840, To assay gold: Am. Rep. Arts Sci's Manu's, v. 1, 300 (11 lines).

Thompson, Thomas (see Thomson, Thomas; 1773-1852)

Thompson, William A. (Judge)
10234　1831, Scratches on elevated strata of horizontal graywacke in the Alleghany range; probably diluvial: Am. J. Sci., v. 20, no. 1, 124-125.
　　　With letter by A. Eaton (see 3370).
10235　1833, Facts relating to diluvial action: Am. J. Sci., v. 23, no. 2, 243-249.
　　　On glacial scouring and scratches in Sullivan County, New York.

Thompson, Zadock (1796-1856)
　　　Thompson was editor of the Green Mountain Repository and probably extracted the short unsigned notes in that periodical (see 4521 and 4522). See also 4734 with S. R. Hall.
10236　1824, A gazetteer of the state of Vermont; containing a brief general view of the state, a historical and topographical description of all the counties, towns, rivers, etc.: Montpelier, Vt., E. P. Walton and author, vi, [9]-310, [2] p., plate, map.
10237　1842, History of Vermont, natural, civil and statistical, in three parts, with a new map of the state, and 200 engravings: Burlington, Vt., C. Goodrich, 3 pts. in 1, illus., plates, map.
　　　For review see 925.
10238　1846, Report of Mr. Thompson [on Chittenden County]: in Second annual report on the geology of the state of Vermont by C. B. Adams (see 26), 259-261.
10239　1848, Geography and geology of Vermont, with state and county outline maps ... for the use of schools and families: Burlington, Vt., printed by Chauncey Goodrich for the author, [9]-219, [1] p., illus., maps.
10240　1849, Report ... on the geological survey of the state: Vt. House J., 1849, 316-317.

Thompson, Z. (cont.)
10241 1850, An account of some fossil bones found in Vermont in making excavations for the Rutland and Burlington Railroad: Am. J. Sci., v. 9 (ns), no. 2, 256-263, 2 plates with 13 figs.
Abstracted in Ann. Scien. Disc., v. 1, 285, and in Bos. Soc. Nat. History Proc., v. 3, 205-206.
10242 1850, Natural history of Vermont: an address delivered at Boston before the Boston Society of Natural History, June 5, 1850: Burlington, Vt., C. Goodrich, 32 p.

Thomson, Robert Dundas (1810-1864)
10243 1837, On some silicates of alumina: Franklin Inst. J., v. 20 (ns), 171-175.
Abstracted from Record Gen. Sci.

Thomson, Thomas (1773-1852)
10244 1811, Allanite: Am. Min. J., v. 1, no. 2, 121 (8 lines).
10245 1811, and M. Ekeberg, Sodalite: Am. Min. J., v. 1, no. 2, 122.
10246 1818, A system of chemistry: Phila., Abraham Small, 4 v.
First American from the fifth London edition. With notes by Thomas Cooper.
10247 1825, Notice of Dr. Thomson's first principles of chemistry; and extracts from foreign letters on various subjects: Am. J. Sci., v. 10, no. 1, 162-164.
Includes chemical analysis of Glasgow limestone by Thomson.
10248 1828, and J. Torrey, Chemical analysis of some minerals, chiefly from America: NY Lyc. Nat. History Annals, v. 3, no. 1/2, 9-86, 3 figs.
On the chemical analysis and physical properties of 32 minerals.
10249 1831, [Analyses of some American minerals]: Am. J. Sci., v. 19, no. 2, 355 (7 lines).
In NY Lyc. Nat. History Proc.
10250 1831, Analysis of the supposed anthophyllite of New York: Am. J. Sci., v. 19, no. 2, 359-360.
With note by A. F. Holmes (see 5236).
10251 1836, Analysis of North American minerals: Am. J. Sci., v. 31, no. 1, 172-174.
Includes analyses of holmesite from Amity, New York and feldspar from Bytown, Lower Canada.
10252 1836, Analysis of tabasheer from India: Am. J. Sci., v. 30, no. 2, 381 (7 lines).
Abstracted from Rec. Gen. Sci.
10253 1836, Emmonite, a new mineral species: Am. J. Sci., v. 31, no. 1, 171.
10254 1836, Method of determining the value of black oxide of manganese for manufacturing purposes: Franklin Inst. J., v. 18 (ns), 343-345.
10255 1839, On native diarseniate of lead: Am. J. Sci., v. 35, no. 2, 297.
In British Assoc. Proc.
10256 1841, On the most important chemical manufactures, carried on in Glasgow and the neighborhood: Am. J. Pharmacy, v. 7 (ns), 206-215.
Abstracted from British Assoc. Trans.
10257 1843, Sketch of the progress of physical science: NY, Greeley and McElrath, Cin., W. H. Moore and Co., 96 p.
10258 1845, Sillimanite: Am. J. Sci., v. 48, no. 1, 219 (9 lines); v. 49, no. 2, 396 (4 lines).
Sillimanite and bucholzite are said to be equivalent.

Thornton, Jessy Quinn (1810-1888)
10259 1849, Oregon and California in 1848, with an appendix including recent and authentic information on the subject of the gold mines of California and other valuable matter of interest to the emigrant, etc.: NY, Harper and Brothers, 2 v. (393; 379 p.), plates, map.
For review with excerpts see 5226, 5479 and 9178.

Thornton, William
10260 1806, North Carolina Gold Mine Company: Wash., 20 p.
10261 1806?, Letter to the members of the North Carolina Gold Mine Company:
Not seen.

Thury, Louis Etienne François Héricart-Ferrand, vicomte de (1776-1854)
10262 1830, Springs and artificial fountains: Mech's Mag., v. 1, 358-362, plate.
10263 1830, Springs and artificial fountains. - From the Annales de la societe d'Horticulture de Paris. - Considerations, geological and physical on the resevoirs of subterranean water, relative to the spouting fountains or wells, obtained by boring: Am. J. Sci., v. 18, no. 2, 267-276, 2 plates.
10264 1831, Bored wells; Am. J. Sci., v. 20, no. 2, 392-393.
Abstracted from Bulletin d'Encouragement by J. Griscom.

Tibelman, M.
10265 1846, Artificial quartz: Franklin Inst. J., v. 11 (3s), 70 (9 lines).
Abstracted from London Ath.

Tighlman, Richard Albert (1824-1899)
10266 1850, New process for the manufacture of alkalies: Ann. Scien. Disc., v. 1, 210.

Tilden, Bryant Parrot (1817?-1859)
10267 1847, Notes on the upper Rio Grande, explored in the months of October and November, 1846, on board the U. S. Steamer Major Brown, commanded by Capt. Mark Sterling, of Pittsburgh, by order of Major-General Patterson, U. S. A.: Phila., Lindsay and Blakiston, 32 p., 9 maps.

Tindall, J. G.
10268 1840, Soils and agriculture of Monroe County, Mississippi: S. Cult., v. 2, 209-210.

Titmarsh, M. A. (pseudonym; see Thackeray, William Makepiece)

Tod, David (1805-1868, see 1510 with W. Bartlet)

Tolman, Thomas (1756-1842)
10269 1795, Salt-springs at Bridport and Orwell: Rural Mag. or Vt. Repos., v. 1, 130.

Tomlinson, C. H.
10270 1832, Alluvial deposits of the Mohawk: Am. J. Sci., v. 23, no. 1, 207.
With remarks by B. Silliman (see 9709).

Torrey, John (1796-1873)
See also 10248 with T. Thomson.
10271 1819, Staurotide [from Manhattan Island, New York]: Am. J. Sci., v. 1, no. 4, 435-436, fig.
10272 1820, [Analysis of datolite from Patterson, New Jersey]: Am. J. Sci., v. 2, no. 2, 369.
In NY Lyc. Nat. History Proc.
10273 1820, Fibrous sulphate of barytes: Am. J. Sci., v. 2, no. 2, 368.
On the strontium content of barite from Carlisle, New York. In NY Lyc. Nat. History Proc.
10274 1820, Sidero-graphite [from Schooley's Mountain, New Jersey]: Am. J. Sci., v. 2, no. 1, 176; no. 2, 370 (7 lines).
10275 1821, Original scientific intelligence, or discoveries and remarks on natural sciences: West. Minerva, v. 1, 38.
On mineral localities in New Jersey, New York, and Massachusetts.
10276 1822, Description and analysis of a new ore of zinc: Am. J. Sci., v. 5, no. 2, 235-238.
Also in: NY Med. Phys. J., v. 1, 191-194.
On zinc ore from Ancram, New York.
10277 1822, Description and analysis of gibbsite, a new mineral [from Richmond, Massachusetts]: NY Med. Phys. J., v. 1, 68-73.
10278 1822, Franklinite [from Franklin, New Jersey]: Am. J. Sci., v. 5, no. 2, 400 (10 lines).
10279 1822, Green zinc ore of Ancram, (ancramite): Am. J. Sci., v. 5, no. 2, 399 (11 lines).
10280 1822, Jeffersonite [from Franklin, New Jersey]: Am. J. Sci., v. 5, no. 2, 402.
10281 1822, Siliceous oxyd of zinc [from Franklin, New Jersey]: Am. J. Sci., v. 5, no. 2, 400.
On willemite and franklinite.
10282 1822, Stilbite [from West Point, New York]: Am. J. Sci., v. 5, no. 2, 399-400.
10283 1822, Sulphuret of molybdena [from Hamburg, New Jersey]: Am. J. Sci., v. 5, no. 2, 401 (10 lines).
10284 1822, Summerville copper-mine: Am. J. Sci., v. 5, no. 2, 401-402.
10285 1822, Zircon from N. Carolina: Am. J. Sci., v. 5, no. 2, 401 (8 lines).
10286 1823, Notice of a locality of yenite in the United States [at Cumberland, Rhode Island]: NY Lyc. Nat. History Annals, v. 1, pt. 1, 51.
Abstracted in NY Med. Phys. J., v. 2, 512-513.
10287 1823, [On mineral localities in Massachusetts and New York]: Am. J. Sci., v. 6, no. 2, 364-365.
In NY Lyc. Nat. History Proc.
10288 1824, An account of the columbite of Haddam, (Connecticut,) with notices of several other North American minerals: NY Lyc. Nat. History Annals, v. 1, pt. 1, no. 3, 89-93, 2 figs.
10289 1825, West Point minerals: Am. J. Sci., v. 9, no. 2, 402 (13 lines).
On stilbite, sphene, and scapolite from West Point, New York.
10290 1829, Collections in natural history: Am. J. Sci., v. 16, no. 2, 368-369.
10291 1834, [Remarks on fossil wood]: Am. J. Sci., v. 27, no. 1, 155 (8 lines).
In NY Lyc. Nat. History Proc.
10292 1837, Discovery of vauquelinite, a rare ore of chromium, in the United States: NY Lyc. Nat. History Annals, v. 4, no. 1/2, 76-79.
On lead ores from Singsing, New York.

Torrie, Thomas Jefferson
10293 1832, Notice of a recent eruption of Vesuvius: The Friend, v. 5, 259.

Totten, Joseph Gilbert (1788-1864)
See also 1633 with S. Bernard.
10294 1824, Notes on some new supports for minerals subjected to the action of the common blow-pipe: NY Lyc. Nat. History Annals, v. 1, pt. 1, no. 4, 109.
10295 1838, Essays on hydraulic and common mortars and on lime burning: NY, Wiley & Putnam, 256 p., 2 folding plates, table.
10296 1842?, List of the cabinet of minerals presented to the National Institution in the city of Washington, by Col. Joseph G. Totten: Wash., 30 p.

Toulmin, Henry
10297 1793, A description of Kentucky, in North America. To which are prefixed miscellaneous observations respecting the United States: Lexington, Kentucky, John Bradford, 8vo., map.
Evans number 26268. Not seen.

A tour through part of Virginia, in the summer of 1808. In a series of letters, including an account of Harper's Ferry, the natural bridge, the new discovery called Weir's Cave, Monticello, and the different medicinal springs, hot and cold baths, visited by the author:
10298 1809, NY, H. C. Southwick (printer), for the author, 31 p.
This work has been variously ascribed to S. L. Mitchill, Joseph Caldwell (1773-1835), and T. Caldwell.

Towers, Jonas
10299 1845, Improvement in the mode of preparing, applying, and using certain fluxes for the reduction of ores, in the blast furnace: Franklin Inst. J., v. 10 (3s), 98.

Townsend
10300 1820, [Organic remains at Corlaer's Hook, New York]: Am. J. Sci., v. 2, no. 2, 371 (13 lines).
In NY Lyc. Nat. History Proc.

Townsend, John Kirk (1809-1851)
10301   1839, Narrative of a journey across the Rocky Mountains to the Columbia River, and a visit to the Sandwich Islands, Chile, &c. With a scientific appendix: Phila., Perkins; Bos., Perkins and Marvin, 352 p.
         For review with excerpts see 2737.

Townsend, Peter Solomon (1796-1849)
10302   1826, Memoir on the topography, weather, and diseases of the Bahama Islands: NY, J. Seymour, 80 p.

Townsend, William P.
10303   [1839], A report on the minerals of Chester County, that are used in the arts, read before the West Chester Lyceum, March 30, 1839: West Chester, Penn., 10 p.

Trail, Dr. (see Traill, Thomas Stewart)

Traill, Thomas Stewart (1781-1862)
10304   1818, [Baryto-strontianite, a new mineral from the Orkney Islands]: J. Sci. Arts, v. 3, no. 2, 375-376.
         Abstracted from Edinburgh Royal Soc. Proc.
10305   1820, Stromnite - a new mineral: Am. J. Sci., v. 2, no. 1, 177.
10306   1843, On the introduction into Scotland of granite, for ornamental purposes, by Messrs. Macdonald and Leslie, of Aberdeen: Franklin Inst. J., v. 5 (3s), 327-331.
         Abstracted from Edinburgh New Philos. J.
10307   1847, Volcanic dust of Hecla: Am. J. Sci., v. 3 (ns), no. 2, 272-273.
         Abstracted from Edinburgh Royal Soc. Proc.

The Transylvania Journal of Medicine and the Associated Sciences (Lexington, Kentucky, 1828-1839)
10308   1832, Analysis of the sulphur springs at Nashville: v. 5, 136.
10309   1832, [Review on On baths and mineral waters by J. Bell (see 1622)]: v. 5, 202-235; 402-433.
         Signed by "C." With extensive excerpts.

The Transylvanian, or Lexington Literary Journal (Lexington, Kentucky, 1829)
10310   1829, The gold region [of North Carolina]: v. 1, no. 7, 237-238.
10311   1829, Volcanoes: v. 1, no. 7, 262-264.
10312   1829, Town lyceums: v. 1, no. 7, 273-278.
10313   1829, Remains of the mammoth [from Bering Strait, Alaska]: v. 1, no. 8, 316-317.

Trautwine, John Cresson (1810-1883)
10314   1839, Description of a natural bridge in Claiborne County, Tennessee: Franklin Inst. J., v. 24 (ns), 82-85, woodcut.

Traveller, Thomas (pseudonym?)
10315   1833, Ascent of Vesuvius: Juv. Rambler, v. 2, no. 50, 197, woodcut.
10316   1833, Cone and crater of Vesuvius: Juv. Rambler, v. 2, no. 51, 201-202.
10317   1833, Eruption of Vesuvius in 1794: Juv. Rambler, v. 2, no. 7, 25, woodcut. Also in Youth's Companion, v. 6, 191-192, woodcut.

The traveller's directory, and emigrant's guide; containing general descriptions of different routes through the states of New-York, Ohio, Indiana, Illinois, and the territory of Michigan, with short descriptions of the climate, soil, productions, prospects, &c.:
10318   1832, Buffalo, NY, Steele and Faxon, 82, 2 p.
         See also 10011.

Traveller's directory and statistical view of the United States:
10319   1834, NHav., C. S. Williams, 1 leaf, folded to 24mo..

Travels through Upper and Lower Canada, with an accurate description of Niagara Falls
10320   1821, NY, 2v.

A Treatise on soils and manures, as founded on actual experience, and as combined with the leading principles of agriculture in which the theory and doctrines of Sir Humphrey Davy, and other agricultural chemists, are rendered familiar to the experienced farmer.
10321   1821, Phila., B. Warner, 92 p.

Tredgold, Thomas (1788-1829)
10322   1825, On the flexibility and strength of marble: Am. Mech's Mag., v. 1, no. 24, 371.
         Abstracted from London Mech's J.
10323   1826, On the flexibility and strength of Portland stone: Am. Mech's Mag., v. 2, no. 33, 112.
         Abstracted from London Mech's J.

Trego, Charles B. (1794-1874)
         See also 5193 with J. T. Hodge.
10324   1836, Report of the committee appointed on so much of the Governor's Message as related to a geological and mineralogical survey of the state of Pennsylvania: Harrisburg, Penn., 12 p.
10325   1843, A geography of Pennsylvania, containing an account of the history, geographical features, soil, climate, geology, botany ... with a separate description of each county ... : Phila., Biddle, 384 p., illus., plate, maps.

Tremper, J.
10326   1846, Synopsis of natural history: Geneva, NY, 16 p.
10327   1848, The coal mines at Blossburg [New York]: Am. Q. J. Ag. Sci., v. 7, no. 3, 113-116.

Trevelyan, Sir Walter Calverley (1797-1879)
10328   1839, Notice regarding the stone used in constucting the Temples at Paestum: Am. J. Sci., v. 37, no. 2, 366.
          Abstracted from Jameson's J.

Trimmer, Joshua (1795-1857)
10329   1839, On the occurrence of marine shells over the remains of terrestrial mammalia in Cefn Cave, in Denbighshire: Am. J. Sci., v. 35, no. 2, 306-307 (6 lines).
          In British Assoc. Proc.
10330   1842, Practical geology and mineralogy, with instructions for the quantitative analysis of minerals: Phila., Lea and Blanchard, 527 p., 200+ figs.
          For review see 769 and 11109. For notice see 1025.

Tripier
10331   1848, also F. A. Walchner, On the existence of copper, arsenic, antimony, and tin in mineral waters: Am. J. Sci., v. 5 (ns), no. 1, 120.
          Abstracted by T. S. Hunt from Chem. Gaz.

Troost, Gerard (1776-1850)
10332   1821, [Announcement of Troost's chemical lectures]: Saturday Mag., v. 5, 208.
10333   1821, Description of a variety of amber, and of a fossil substance supposed to be the nest of an insect discovered at Cape Sable, Magothy River, Ann-Arundel Co., Maryland: Am. J. Sci., v. 3, no. 1, 8-15.
10334   1821, Description of some new crystalline forms of the minerals of the United States: Phila. Acad. Nat. Sci's J., v. 2, pt. 1, 55-58, plate with 5 figs.
10335   1822, Description of a new crystalline form of quartz [from Lake George, New York]: Phila. Acad. Nat. Sci's J., v. 2, pt. 2, 212-214, 2 figs.
10336   1822, Description of some crystals of sulphate of strontian, from Lake Erie: Phila. Acad. Nat. Sci's J., v. 2, pt. 2, 300-302, 3 figs.
10337   1822, Description of some new crystalline forms of the minerals of the United States: Phila. J. Med. Phys. Sci's, v. 4, 304-386.
10338   1823, Account of the pyroxene of the United States, and descriptions of some new varieties of its crystalline forms: Phila. Acad. Nat. Sci's J., v. 3, pt. 1, 105-124, plate with 10 figs.
          Also reprint edition.
10339   1823, Notice of the yenite of Rhode Island, and several other American minerals: Phila. Acad. Nat. Sci's J., v. 3, pt. 1, 222-224.
10340   1824, Description of a new crystalline form of the andalusite [from Litchfield, Connecticut]: Phila. Acad. Nat. Sci's J., v. 4, pt. 1, 122-123.
10341   1824, Description of a new crystalline form of the chrysoberyl [from Saratoga, New York]: Phila. Acad. Nat. Sci's J., v. 3, pt. 2, 293-295, 2 figs.
10342   1824, Description of the American petalite from Lake Ontario: Phila. Acad. Nat. Sci's J., v. 3, pt. 2, 234-237.
10343   1825, Description and chemical analysis of the retinasphalt discovered at Cape Sable, Magothy River, Ann Arundel Co., Maryland: Am. Philos. Soc. Trans., v. 2 (ns), 110-115.
          Also reprint edition.
10344   1825, Description of a new crystalline form of apophyllite, laumonite, and amphibole, and of a variety of pearlstone [from Point Marmoaze, Lake Superior]: Phila. Acad. Nat. Sci's J., v. 5, pt. 1, 51-56, 4 figs.
10345   1825, Notice of a new crystalline form of the yenite of Rhode Island: Am. Philos. Soc. Trans., v. 2 (ns), 478-480.
          Also reprint edition.
10346   1825, Observations on the zinc ores of Franklin and Sterling, Sussex County, New Jersey: Phila. Acad. Nat. Sci's J., v. 4, pt. 2, 220-231.
10347   1826, Geological survey of the environs of Philadelphia, performed by order of the Philadelphia Society for Promoting Agriculture ... : Phila., H. S. Tanner, 40 p.
          See also map (10348).
10348   1826, [Geological map of the region near Philadelphia]:
          Not seen. Colored map to accompany 10347.
10349   1827, and C. A. Le Sueur, Calamine in Missouri: Am. J. Sci., v. 12, no. 2, 376-378.
10350   1827, and C. A. Le Sueur, Cobalt in Missouri: Am. J. Sci., v. 12, no. 2, 378 (5 lines).
          Also in: West. Med. Phys. J., v. 1, 188 (5 lines).
10351   1827, and C. A. Le Sueur, Description of some valuable zinc ore hitherto rejected as useless: New Harmony Gaz., v. 3, 197.
10352   1827, and C. A. Le Sueur, Description of the lead ores of the mining district of Missouri: New Harmony Gazette, v. 3, 228-229.
10353   1827, and C. A. Le Sueur, Description of an ochraceous clay, formerly called earth of Lemnos [from New Harmony, Indiana]: New-Harmony Gaz., v. 2, 204-205.
10354   1827, and C. A. Le Sueur, Lead ores of Missouri: Am. J. Sci., v. 12, no. 2, 379-380.
10355   1827, On pyroxene: Maclurian Lyc. Contributions, v. 1, 51-66.
          "To be continued" but no more found.
10356   1831, Address delivered before the legislature of Tennessee, at Nashville, October 19, 1831: Transylvania J. Medicine and Associated Sci's, v. 4, no. 4, 491-507.
          Also reprint edition. From Nat'l Banner and Nashville Whig.
10357   1832, First geological report to the 19th General Assembly of the state of Tennessee: "Does not appear to have been published" (M. Meisel, 1926, v. 2, 533).
10358   1833, Analysis of the geological description of Davidson, Williamson, and Maury Counties ... : in Tenn. House of Representatives J., 20th Gen. Assembly, 303-305.
10359   1833, Bones of the gigantic mastodon. Improperly called mammoth, found in the vicinity of Nashville: Cin. Mirror, v. 3, 61-62.
          Abstracted from Nashville Banner.
10360   1833, Bones of the gigantic mastodon, improperly called mammoth, found in the vicinity of Nashville [Williamson County, Tennessee]: NE Farm., v. 12, 168.
10361   1833, The great mastodon; improperly called 'mammoth,' found in the vicinity of Nashville - communicated for the Nashville Republican: Lit. J. Wk. Reg. Sci. Arts, v. 1, no. 30, 234.
10362   1833, Second geological report to the 20th General Assembly of the state of Tennessee: "Does not appear to have been published" (M. Meisel, 1926, v. 2, 533).
10363   1834, Important work [notice of Fossil organic remains by Prof. A. Goldfuss]: Am. J. Sci., v. 25, no. 2, 430-431.

Troost, G. (cont.)

10364 1834/1835, On the localities in Tennessee in which bones of the gigantic Mastodon and Megalonyx jeffersonii are found: Geol. Soc. Penn. Trans., v. 1, 139-146, 236-243. With footnote by R. H. (Harlan?), (see 4797). Abstracted in Am. J. Sci., v. 27, no. 2, 354-355.

10365 1835, Description of a new species of fossil Asterias (Asterias antiqua) [from Davidson County, Tennessee]: Geol. Soc. Penn. Trans., v. 1, 232-235, plate.

10366 1835, Description of some organic remains characterizing the strata of the upper Transition which composes middle Tennessee: Geol. Soc. Penn. Trans., v. 1, 244-247.

10367 1835, Map of the coal formation in Tennessee: NY, Curriers (lithographer), geological map, 35 x 30 cm.
    In third Tennessee report (see 10370).

10368 1835, On the organic remains which characterize the Transition series of the valley of the Mississippi, &c.: Geol. Soc. Penn. Trans., v. 1, 248-250.

10369 1835, On the Pentremites reinwardtii, a new fossil; with remarks on the genus Pentremites (Say), and its geognostic position in the states of Tennessee, Alabama and Kentucky: Geol. Soc. Penn. Trans., v. 1, 224-231, plate with 11 figs.

10370 1835, Third geological report of the state of Tennessee: Nashville, S. Nye and Co. (printers), 32 p., map.
    Contains map 10367. For excerpts see 10378.

10371 1836, Extracts from the third report on the geological survey of Tennessee. On the marl of Tennessee: Farm's Reg., v. 3, 696-699.

10372 1836, Third geological report to the 21st Assembly of the state of Tennessee, made Oct. 1835: Am. J. Sci., v. 30, no. 2, 391-392.
    Description of the limits of the Tennessee coal formation.

10373 1837, Fourth geological report to the 22nd General Assembly of the state of Tennessee: Nashville, Tenn., S. Nye and Co. (printer), 36, [2] p., folded color map.
    Contains map 10376.

10374 1837, Fourth report of the geological survey of the state of Tennessee by the state geologist: Nashville, Tenn., 24 p., map.
    Contains map 10376. Not seen.

10375 1837, Fourth geological report to the 22nd General Assembly of the state of Tennessee: Tenn. House J., 1837-1838, Appendix, 628-652.

10376 1837, Map of the Ocoee District to elucidate the 4th report of G. Troost, geologist of the state of Tennessee: Phila., Lehman and Duval (printers), geological map, 53 x 47 cm.
    In fourth Tennessee report (see 10373 and 10374).

10377 1839, [On marbles from East Tennessee]: S. Cult., v. 1, 60-61.

10378 1840, On marl, and its application and effect upon the soil: Agriculturist, v. 1, 6-11.
    From Third Geological Report (see 10370).

10379 1840, Cocke County: Agriculturist, v. 1, 31-33.
    From Fifth Geological Report (see 10382).

10380 1840, Geological description of the state of Tennessee: Agriculturist, v. 1, 77-79, 102-103.
    From Fifth Geological Report (see 10382).

10381 1840, Description and analysis of a meteoric mass, found in Tennessee, composed of metallic iron, graphite, hydroxide of iron and pyrites: Am. J. Sci., v. 38, no. 2, 250-254.

10382 1840, Fifth geological report to the 23rd General Assembly of Tennessee: Nashville, Tenn., J. George Harris (printer), 75 p., folded plate, 2 folded maps, folded table.
    Contains maps (see 10383 and 10384) and a list of Tennessee fossils (pp. 45-75). Also reprinted in Tenn. House J.? Not found. For excerpts see 10379 and 10380.

10383 1840, Geological map of the state of Tennessee by G. Troost, geologist of the state: Phila., P. S. Duval (lithographer), colored geological map, 63 x 25 cm.
    In fifth Tennessee report (see 10382).

10384 1840, Geological map of Cocke County. East Tenn. by G. Troost: Phila., P. S. Duval (lithographers), geological map, 37 x 27 cm.
    In fifth Tennessee report (see 10382).

10385 1841, Geological map of Sevier County Tenn. by G. Troost: in sixth Tennessee report (see 10386), geological map, 25 x 30 cm.

10386 1841, Sixth geological report to the 24th General Assembly of the state of Tennessee: Nashville, Tenn., B. R. M'Kennie, 48 p., folded map.
    Contains map (see 10385). For excerpts see 10389.

10387 1841, Sixth geological report to the 24th General Assembly of the state of Tennessee: Tenn. House J., 1841-1842, Appendix, 171-199.

10388 1841, Sixth geological report to the 24th General Assembly of the state of Tennessee: Tenn. Senate J., 1841-1842, Appendix, 155-183.

10389 1842, Sevier County: Agriculturist, v. 3, 6-8, 21-31.
    From Sixth Geological Report (see 10386).

10390 1843, [Analysis of mineral spring near Nashville]: Agriculturist, v. 4, 128.

10391 1843, Seventh geological report to the 25th General Assembly of the state of Tennessee: Nashville, Tenn., W. F. Bang and Co. and B. R. McKennis (printers), 45 p., colored map.
    Contains map (see 10393).

10392 1843, Seventh geological report to the 25th General Assembly of the state of Tennessee: Tenn. House and Senate Journals, 1843-1844, Appendix, 133-163.

10393 1843, Geological map of Davidson, Williamson & Maury Counties by G. Troost. Geologist of the state: in seventh Tennessee report (see 10391), colored geological map, 26 x 30 cm.

10394 1844, Greensand: Agriculturist, v. 5, 4-5.

10395 1844, Natural history of the elephant: Agriculturist, v. 5, 57-58.

10396 1845, An account of some ancient (human or other) remains in Tennessee: Trans. Am. Ethnological Soc., v. 1, 355-365.

10397 1845, (1.) Description of a mass of meteoric iron, which fell near Charlotte, Dickson County, Tennessee, in 1835; (2.) Of a mass of meteoric iron discovered in De Kalb County, Tenn.; (3.) Of a mass discovered in Green County, Tenn.; (4.) Of a mass discovered in Walker County, Alabama: Am. J. Sci., v. 49, no. 2, 336-346, 3 figs.

Troost, G. (cont.)
10398   1845, Eighth geological report to the 26th General Assembly of the state of Tennessee: Nashville, Tenn., W. F. Bang and Co. (printers), 20 p.
10399   1845, Eighth geological report to the 26th General Assembly of the state of Tennessee: Tenn. House and Senate Journals, 1845-1846, Appendix, 65-76.
10400   1846, Description of three varieties of meteoric iron. - 1. from near Carthage, Smith County, Tennessee; 2. from Jackson County, Tennessee; 3. from near Smithland, Livingston County, Kentucky: Am. J. Sci., v. 2 (ns), no. 2, 356-358.
10401   1846, The state of Tennessee: Fisher's Nat'l Mag., v. 2, no. 8, 702-707.
10402   1848, Description of a mass of meteoric iron, discovered near Murfreesboro, Rutherford County, Tenn.: Am. J. Sci., v. 5 (ns), no. 3, 351-352.
10403   1848, Kraurite and cacoxene in Tennessee [from Brush Creek, Cocke County]: Am. J. Sci., v. 5 (ns) no. 3, 421.
10404   1848, Ninth geological report to the 27th General Assembly of the state of Tennessee: Nashville, Tenn., W. F. Bang and Co. (printer), 39 p., 2 folded plates.
10405   1848, Ninth geological report to the 27th General Assembly of the state of Tennessee: Tenn. House J., 1847-1848, Appendix, 143-168, 2 plates.
10406   1848, Ninth geological report to the 27th General Assembly of the state of Tennessee: Tenn. Senate J., 1847-1848, Appendix, 315-341, 2 plates.
10407   1849, [List of the fossil crinoids of Tennessee]: Am. J. Sci., v. 8 (ns), no. 3, 419-420.
         In review of Ninth geological report (see 988).
10408   1850, Fossil crinoids of the United States: Ann. Scien. Disc., v. 1, 282-283.
10409   1850, A list of the fossil crinoids of Tennessee: Am. Assoc. Adv. Sci. Proc., v. 2, 59-64.
         With discussion by L. Agassiz and J. Hall. For notice see 10833.
10410   1850, Tenth geological report to the 28th General Assembly of the state of Tennessee, made Jan. 12, 1850 ... :
         "Does not appear to have been published" (M. Meisel, 1926, v. 2, 533).

Troost, Lewis
10411   1850, Mineral resources of Alabama: De Bow's Rev., v. 9, 330-331.
         Abstracted from Troost's "Railroad report" (see 10412).
10412   1850, [Survey of the route of a railroad in Alabama]:
         Not seen. For notice see 10411.

Troughton, Nicolas
10413   1845, An improvement in washing ores: Franklin Inst. J., v. 9 (3s), 176.

Troy Lyceum of Natural History (Troy, New York)
10414   1819, Troy Lyceum of Natural History: Plough Boy, v. 1, no. 27, 210-211.
         On current activities. Geological studies of J. D. Dickenson, J. Dekay, A. Eaton and T. R. Beck are noted.
10415   1819-1820, Extracts from the transactions of the Troy Lyceum of Natural History: Plough Boy, v. 1, no. 30, 233; no. 31, 250; no. 32, 257-258; no. 36, 282; no. 42, 337; no. 44, 353; v. 2, 65.
10416   1820, Troy Lyceum: Am. J. Sci., v 2, no. 1, 173.
         Notice of founding and goals.

True, T. N. (Dr.)
10417   1849, Axinite in Maine [from Wales, Maine]: Am. J. Sci., v. 7 (ns), no. 2, 286 (4 lines).

Trumbull, Benjamin (1735-1820)
10418   1797, A complete history of Connecticut, civil and ecclesiastical, from the emigration of its first planters from England in MDCXXX, to MDCCXIII: Hart., Hudson and Goodwin, xix, 587 p., illus., map.
         Evans number 32942. Contains notes on early earthquakes. Also reprint edition in 1818, NHav., Maltby, Goldsmith and Co., 2 v., plate.

Tschudi, Johann Jakob von (1818-1889)
10419   1847, Travels in Peru, during the years 1838-1842, on the coast, in the Sierra, across the Cordilleras and the Andes, into the primeval forests. Translated from German by Thomasin Ross: NY, Wiley and Putnam, 2 pts. in 1.
         For extract on "Peruvian silver mines" see Hunt's Merch. Mag., v. 17, 434-436.

Tucker, George (1775-1861)
10420   1850, [California gold mines]: Am. Philos. Soc. Proc., v. 5, no. 45, 148-150.
10421   1850, The gold mines of California: Hunt's Merch. Mag., v. 23, 19-28.

Tuckey, James Kingston (1776-1816)
10422   1818, Narrative of an expedition to explore the River Zaire, usually called the Congo, in South Africa in 1816: NY, W. B. Gilley, 410 p., illus., map.
         Also a printing by NY, Kirk and Mercein, 410 p., illus., map, 1818.

Tufts, Cotton (1732-1815)
10423   1801, Observations on a species of earth found in Weymouth, presented to the Agricultural Society of Massachusetts: Mass. Agri. Soc. Papers, for 1801, 19-20.

Tuomey, Michael (1805-1857)
         See also discussion of article by R. W. Gibbes (4358).
10424   1840, Calcareous earth for manure in Loudon County [Virginia]: Farm's Reg., v. 8, 570-571.
10425   1841, Analysis of coal ashes [of coal from the Appomattox pits]: Farm's Reg., v. 9, 707-708.
10426   1842, Discovery of a chambered univalve fossil in the Eocene Tertiary of James River, Virginia: Am. J. Sci., v. 43, no. 2, 187.
10427   1842, Notice of the Appomattox coal pits: Farm's Reg., v. 10, 449-450.
10428   1843, Notice of the discovery of a new locality of the "infusorial strata [at Petersburg, Virginia]: Am. J. Sci., v. 44, no. 2, 339-341.

Tuomey, M. (cont.)
10429 1844, [Age of Tertiary rocks of South Carolina]: Am. J. Sci., v. 47, no. 1, 117 (10 lines).
        Also in Assoc. Am. Geol's Nat's Proc., v. 5, 24 (10 lines).
10430 1844, Report on the geological and agricultural survey of South Carolina, 1844: Columbia, SC, A. S. Johnston (printer), iv, [5]-63 p., illus.
        Contains "Supplementary report" by E. Ruffin (see 9093). For extract on "Mineral resources of Alabama" see De Bow's Rev., v. 4, 404-405 (1847).
10431 1847, Minerals of South Carolina for building purposes, etc.: De Bow's Rev., v. 3, 343-344.
10432 1847, Notice of the discovery of a cranium of the Zeuglodon, (Basilosaurus): Phila. Acad. Nat. Sci's J., v. 1 (ns), 16-17, plate.
10433 1847, Notice of the discovery of a cranium of the Zeuglodon: Phila. Acad. Nat. Sci's Proc., v. 3, no. 7, 151-153, 2 figs.
        Abstracted in Am. J. Sci., v. 4 (ns), no. 2, 283-285, 2 figs.
10434 1847, Report on artesian wells, published agreeably to a resolution of the city of Charleston: Char., Miller and Browne, 21 p.
10435 1848, Geological map of South Carolina: NY, G. and W. Endicott (lithographers), colored geological map, 47 x 38.5 cm.
        In Report on the geology of South Carolina (see 10438).
10436 1848, Interesting cave [in Alabama]: Scien. Am., v. 3, no. 23, 179.
10437 1848, Map of the iron ore and limestone region of York and Spartanburg Districts: NY, G. and W. Endicott (lithographers), colored geological map, 22 x 17.5 cm.
        In Report on the geology of South Carolina (see 10438).
10438 1848, Report on the geology of South Carolina: Columbia, SC, A. S. Johnston (printer), vi, 293, lvi p., illus., plate, 2 colored maps, 47 figs.
        Contains reports by C. U. Shepard (see 9556), J. L. Smith (see 9889), L. Vanuxem (see 10593), and maps (see 10435 and 10437). For review see 9942, 1893, 9963, and 2722.
10439 1849, Results of observations in the Tertiary region of South Carolina: Am. Assoc. Adv. Sci. Proc., v. 1, 32-33.
10440 1850, First biennial report on the geology of Alabama: Tuscaloosa, Alabama, M. D. J. Slade (printer), 176 p., illus., map.
        Contains map (see 10441), and a geological glossary (pp. 171-176). For review see 1021, 2482 and 9967.
10441 1850, Geological map of Alabama: NY, Ackerman (lithographer), colored geological map, 42 x 55 cm.
        In First biennial report ..., (see 10440).

Turnbull, Robert (Reverend, of Hartford)
10442 1850, Scientific observations: Bib. Repos., v. 6 (3s), 283-303.
        On the methodology of science.

Turner, Charles
10443 1826, A description of Natardin or Catardin Mountain: Mass. Hist. Soc. Coll., v. 8 (2s), 112-116.

Turner, Edward (1798-1837)
10444 1831, On the mode of ascertaining the commercial value of ores of manganese: Am. J. Pharmacy, v. 3, 153-157.
        Abstracted from Royal Inst. J.
10445 1832, Manganese - mode of ascertaining the commercial value of its ores: Am. J. Sci., v. 21, no. 2, 364-365.
        Abstracted from Royal Inst. J.
10446 1839, Chemical examination of the fire-damp from the coal mines near Newcastle: Am. J. Sci., v. 37, no. 2, 201-210.
        Abstracted from London and Edinburgh New Philos. Mag.

Turner, George
10447 1787, Description of the chalybeate springs, near Saratoga, with a perspective view of the main spring, taken on the spot: Colum. Mag., v. 1, no. 7, 306-309, plate.
10448 1799, Memoir on the extraneous fossils, denominated mammoth bones: principally designed to shew, that they are the remains of more than one species of non-descript animal: Am. Philos. Soc. Trans., v. 4, 510-518.
        Also reprint edition, 11 p. (Evans number 35459).

Turner, Richard (1753-1788)
10449 1796, An abridgement of the arts and sciences: being a short, but comprehensive system, of useful and polite learning. Divided into lessons. Adapted to the use of schools and academies: New-London, Ct., James Springer, 167 p.
        Evans number 31332. Also 1802 NY edition, James Oran, 128 p., and later editions.

Turner, Sharon (1768-1847)
10450 1836, The deluge. Causes and objects of the general deluge, its history, and the traditional evidence corroborating the Mosaic account: Meth. Mag., v. 18 (v. 7, ns), 231-240.

Turpin, Pierre Jean François (1775-1840)
10451 1839, Cause of the red color of agates: Am. J. Sci., v. 36, no. 1, 207-208.
        Abstracted from Comptes Rendus.

Twining, Alexander Catlin (1801-1884)
10452 1838, Remarks upon Sherwood's magnetic discoveries: Am. Rr. J., v. 7, 105-108.

Tyro (pseudonym, see 9971)

Tyson, Phillip Thomas (1799-1877)
See also 126 with J. H. Alexander
10453   1830, Notice of some localities of minerals in the counties of Baltimore and Harford, Md.: Am. J. Sci., v. 18, no. 1, 78-81.
With an appendix by C. U. Shepard (see 9483).
10454   1837, A descriptive catalogue of the principal minerals of the state of Maryland: Md. Acad. Sci. Liter. Trans., v. 1, 102-117, fig.
10455   1837, A description of the Frostburg coal formation of Alleghany County, Maryland, with an account of its geological position: Md. Acad. Sci. Liter. Trans., v. 1, 92-98, plate.
10456   1837, George's Creek coal and iron company: Balt., n.p.
10457   1850, ... information in relation to the geology and topography of California: US SED 47, 31-1, v. 10, pt. 1 (558), 3-74, 10 plates and maps.
Contains report by J. F. Frazer (see 4101).
10458   1850, Geology and industrial resources of California ... : Wash.
Not seen.

Ulex, George Ludwig
10459   1846, Struvite [from Hamburg, Germany]: Am. J. Sci., v. 2 (ns), no. 2, 268 (5 lines).
Abstracted from l'Institut.

Underhill
10460   1846, Marl [from New Jersey]: NY State Agric. Soc. Trans., v. 5, 512.

Unger, Franz Joseph Andreas Nicolas (1800-1870)
10461   1846, Conspectus of the fossil flora: Am. J. Sci., v. 2 (ns), no. 1, 136.
Abstracted by A. Gray from Synopsis plantarum fossilium.
10462   1846, Distribution of the vestiges of palms in the geological formations: Am. J. Sci., v. 2 (ns), no. 1, 133-135.

Union Agriculturist and Western Prairie Farmer (Chicago, Illinois, 1841-1842)
10463   1841, Analysis of soil: v. 1, 5; 41; 77.
Includes notice of C. T. Jackson's offer to analyze soils.
10464   1841, Salt in Iowa [from St. Peters]: v. 1, 86 (3 lines).
10465   1842, Lime in prairie soils: v. 2, 2; 13; 49.
10466   1842, La Salle Co. salt [from Illinois]: v. 2, 20.
10467   1842, Analyzing soils: v. 2, 65.
Notice of C. T. Jackson's progress in analyzing prairie soils.

Union Potomac Company and the Union Company
10468   1840, Charters of the Union Potomac Company and the Union Company, with a description of their coal and iron mines, situate in Hampshire County, Virginia, and in Alleghany County, Maryland: Balt., 52 p., map.
10469   1840, The coal and iron mines, of the Union Potomac Company and the Union Company: Balt., 23 p.

United States
10470   1806, Discoveries made in exploring the Missouri, Red River and Washita, by Captains Lewis and Clark ... : Natchez, Mississippi, Andrew Marschalk (printer), 177 p.
Message from the President of the United States (Thomas Jefferson).
10471   1806, Message from the President of the United States communicating discoveries made in exploring the Missouri, Red River and Washita, by Captains Lewis and Clark ... : Wash., A. and G. Way (printers), 178 p.
Also another printing: NY, Hopkins and Seymour, 178 p.
Contains report by W. Dunbar and Dr. Hunter (see 3272).
10472   1826, Lead mines - Illinois and Missouri. Message from the President of the United States ... in relation to the lead mines belonging to the United States, in the state of Illinois and Missouri: Wash., Gales and Seaton, 13 p.
10473   1826, Statement in relation to the lead mines and salt springs in Missouri ... : Wash., Gales and Seaton, 6 p.
10474   1827, Joint resolution directing experiments to be made to ascertain the length of a pendulum vibrating sixty times in a minute ... : Wash., 1 p.
10475   1827, Salt springs in Illinois. Feb. 10, 1827. Mr. Scott, from the Committee on the Public Land to which was referred the memorial of the Legislature of the state of Illinois, upon the subject of lands reserved for the use of salt springs ... : Wash., 10 p.
10476   1828, Agent -- lead mines, Missouri and Illinois etc. Message from the President of the United States ... : Wash., Gales and Seaton, 8 p. (SED 30).
10477   1828, Lead Mines. Letter from the Secretary of War ... in relation to the lead mines of the United States: Wash., Gales & Seaton, 10 p. (SED 45).
10478   1828, Sales of land -- lead mines Missouri: Wash., Gales and Seaton, 2 p.
10479   1829, Lead ore imported. Letter from the Secretary of the Treasury in reply to a resolution of the House of Representatives, requiring information in relation to the quantity of lead ore imported into the United States: Wash., Gales and Seaton, 1 p. (SED 83).
10480   1829, Letter from the Secretary of the Treasury, transmitting further information on the subject of lead ore imported: Wash., Gales and Seaton, 1 p. (SED 101).
10481   1830, A bill authorizing the legislature of the territory of Arkansas to lease the salt springs in said territory ... : Wash., 2 p.
10482   1830, A bill to authorize the President to appoint a superintendent and receiver at the Fever River lead mines ... : Wash., 1 p.
10483   1830, A bill to authorize the state of Indiana to sell her reserved salt springs ... : Wash., 2 p.
10484   1830, Memorial of the manufacturers of salt in the county of Kenhawa, Virginia, against the repeal of the duty on imported salt ... : Wash., Duff Green, 2 p.
10485   1830, Salt-works -- United States. Letter from the Secretary of the Treasury, transmitting a report of the number and nature of the salt works established in the United States ... : Wash., Duff Green, 77 p. (HED 55).
10486   1830, School lands, salines, etc. in Missouri ... : Wash., Duff Green, 3 p.
10487   1838, [Report to accompany] a bill ... to occupy the Oregon Territory: US SED 470, 25-2, v. 5 (318), 1-23, 2 maps.

United States (cont.)
10488 1846, A brief account of the process employed in the assay of gold and silver coins at the mint of the United States: Franklin Inst. J., v. 11 (3s), 131-135.

The United States Christian Magazine (New York, 1796)
10489 1796, Essay on the consistency of the Holy Scriptures with the actual state and history of the world: v. 1, no. 1, 25-35; no. 2, 79-85; no. 3, 165-171.
Signed by "S."

The United States Literary Gazette (Boston, 1824-1826)
10490 1824, [Review of New theory of the earth by I. Hill (see 5060)]: v. 1, 7.
10491 1824-1826, [Notes on fossil bones from New Jersey, England, Russia and France]: v. 1, 77, 110; v. 2, 77, 436; v. 4, 312, 463.
10492 1824, [Review of A familiar introduction to crystallography by H. J. Brooke]: v. 1, 200.
10493 1824, Cabinet of minerals at Cambridge [at Harvard College]: v. 1, 204.
10494 1824, Heat produced by the compound blow-pipe: v. 1, 222.
10495 1825, Lehigh River and coal mine [in Pennsylvania]: v. 2, 36-37.
10496 1825, [Review of Travels in the central portions of the Mississippi Valley by H. R. Schoolcraft (see 9270)]: v. 2, 201-209.
10497 1825, Subsidence of the Baltic: v. 2, 235-236.
10498 1825, [Review of Lectures on geology by J. Van Rensselaer (see 10551)]: v. 2, 287-293.
10499 1825, Account of the earthquake at Lisbon: v. 3, 233-234.
10500 1826, [Review of Journal of a tour around Hawaii, the largest of the Sandwich Islands]: v. 3, 401-414.
10501 1826, [Notice of Manual of mineralogy by J. L. Comstock (see 2505)]: v. 4, 87 (4 lines).
10502 1826, [Review of Manual of mineralogy and geology by E. Emmons (see 3491)]: v. 4, 383.
10503 1826, Lead on the Mississippi: v. 4, 391 (7 lines).
10504 1826, Variation of the magnetic needle: v. 4, 464 (9 lines).
On the effect of Aurora on magnetic variation.

The United States Magazine and Democratic Review (Washington, 1837-1850)
10505 1840, The mineral lands of the United States: v. 8, 30-42.
10506 1849, California gold: v. 24, 3-13.

United States Medical and Surgical Journal (New York and Philadelphia, 1834-1836)
10507 1835, Geological descriptions, &c.: v. 2, 124 (14 lines).
Notice of geological reports sponsored by NY State Med. Soc. on counties in NY.

United States Mining Company
10508 1835, First annual report of the Board of Directors of the United States' Mining Company, August 1835: Fredericksburg, Va., Arena Publishing Co., 17 p., 5 plates.
Report signed by J. L. Marye, J. Metcalf, W. Jackson, and M. F. Maury.

United States papers, relative to an application to Congress, for an exclusive right of searching for and working mines, in the North-West and South-West Territory. By N[icholas] I. Roosevelt & J[acob] Mark, and their associates. February 10, 1796:
10509 1796, Phila., Samuel Harrison Smith (printer), [28] p.
Evans number 31473. See also 6062.

United States Review and Literary Gazette (Boston, 1827)
10510 1827, [Notes on fossil bones]: v. 1, 73, 315; v. 2, 73.
10511 1827, American mines: v. 1, 465.
Production statistics since 1800.
10512 1827, New mines of platinum [at Colombia]: v. 1, 465-466.

The Universal Asylum and Columbian Magazine (Philadelphia, 1790-1792)
10513 1790, Of the enormous bones found in America [in Kentucky and New York]: v. 5, 349-350.
10514 1791, [Review of Magnetic atlas and Explanation by W. Churchman (see 2370 and 2371)]: v. 6, 101-105, fig.
10515 1791, Mineral nitre discovered at Molfetta in the Kingdom of Naples [Italy]: v. 7, 237-238.
10516 1791, Account of some extraordinary petrifaction of Padua: v. 7, 326 (9 lines).
On fossil fish.
10517 1792, [Review with excerpts of Travels through North and South Carolina by W. Bartram (see 1544)]: v. 8, 195-197, 255-267.
10518 1792, [Review with excerpts of Treatise on mineral waters of Virginia by J. Rouelle (see 9040)]: v. 8, 267-269.

The Universal Magazine (Philadelphia, 1797-1798)
10519 1797, Curious particulars relating to the Island of Malta: v. 2, 98-99.
10520 1797, The loadstone: v. 2, 319-321.
10521 1797, Account of an earthquake at Naples: v. 3, 85-88.
Signed by "Petrarch."
10522 1797, Account of the earthquake at Messina in 1783: v. 3, 187-189.
10523 1797, Description of the copper-mine at Fahlun in Sweden, by a Dutch officer: v. 4, 17-23.

The Universalist Quarterly and General Review (Boston, 1844-1850)
10524 1845, Geology - its facts and its inferences: v. 2, no. 1, 5-21.
Signed by "G. W. M." On the difference between geological fact and speculation.
10525 1845, Geology and scripture: v. 2, no. 4, 349-384.
Signed by "W. F."
10526 1846, The White Mountains [review of Final report on New Hampshire by C. T. Jackson (see 5611)]: v. 3, no. 2, 113-143.
10527 1847, [Review with excerpts of Ancient world by D. T. Ansted (see 1280)]: v. 4, no. 3, 289-317.
Signed by "M. S."

The Universalist Quarterly (cont.)
10528   1850, [Review of Lake Superior by L. Agassiz (see 83)]: v. 7, no. 3, 291-292.
10529   1850, [Review of Footprints of the Creator by H. Miller (see 7186)]: v. 7, no. 4, 424-425.

Ure, Andrew (1778-1857)
10530   1821, A dictionary of chemistry, on the basis of Mr. Nicholson's, in which the principles of the science are investigated anew, and its applications to the phenomena of nature, medicine, mineralogy, agriculture, and manufactures detailed ... edited by Prof. Hare and assisted by Dr. Bache: Phila, Robert De Silver, 2 v. First American edition.
10531   1840, Report ... upon the asphalte rocks of Val-de-Travers, Seyssel, Pyrimont, &c., and their application as a mastich, in footpavements, roofs, aqueducts, cisterns, &c.: Franklin Inst. J., v. 25 (ns), 409-413.
10532   1842, A dictionary of arts, manufactures, and mines; containing a clear exposition of their principles and practise: NY, Le Roy Sunderland; and NY, D. Appleton and Co., 340 p., 1000+ figs.
        Numerous other printings, including "11 edition," 1848. For review see 4485, 5425.
10533   1844, Recent improvements in arts, manufactures, and mines; being a supplement to his dictionary: NY, Appleton, 304 p.
10534   1844, Apparatus for the analysis of carbonates by ascertaining the loss of weight from the disengagement of carbonic gas: Franklin Inst. J., v. 8 (3s), 354-355, fig.
        Abstracted from London J. Arts Sci's.
10535   1847, [Extract from a report on the geology of the lands of the Barrelville Mining Company]: see 1502.
10536   1847, [Report on the lands of the Phoenix Mining Company]: in their Documents (see 8328), p. 30.

The Useful Cabinet (Boston, 1808)
10537   1808, Stone, equal to the Turkey oil-stone: v. 1, 11 (5 lines).

Usher, F. C.
10538   1837, On the elevation of the banks of the Mississippi in 1811: Am. J. Sci., v. 31, no. 2, 294-296.

Usher, William
10539   1829, Iodine in Saratoga water: Am. J. Pharmacy, v. 1, 160.
        Abstracted from Am. J. Sci.

Valchner
10540   1847, Arsenic in mineral waters: Am. J. Sci., v. 3 (ns), no. 2, 260 (7 lines).
        Abstracted from l'Institut.

Valk, William W.
10541   1847, Remarks on the science of gardening - No. II: Hort. and J. Rural Arts, v. 2, 12-15.
        On the chemical analysis of soils.

Van Beuren, Peter
10542   1835, and R. G. Frary and S. P. White, Medical topographical report of the county of Columbia: Med. Soc. NY Trans., v. 2, 30-35.

Van Cleve, John W. (d. 1858)
10543   1849, Fossil zoophytes of western Ohio, with a few additions from other western localities: Am. Assoc. Adv. Sci. Proc., v. 1, 19-24.

Vancouver, Charles (fl. 1785-1813)
10544   1785, A general compendium of chemical, experimental, and natural philosophy: with a complete system of commerce: Phila.
        Evans number 19337. Not seen.

Vandewater, Robert J.
10545   1830, The tourist; or, pocket manual for travelers on the Hudson River, the western canal, and stage road; comprising also the routes to Lebanon, Ballston, and Saratoga Springs: NY, J. and J. Harper, 59 p.
        Also numerous later editions.

Van Rensselaer, Jeremiah (Dr.)
        See also 3021 with J. E. Dekay and W. Cooper.
10546   1820, Account of a journey to the summit of Mount Blanc: Am. J. Sci., v. 2, no. 1, 1-11.
10547   1822, On the natural history of the ocean, with two sea journals: Am. J. Sci., v. 5, no. 1, 128-140.
10548   1823, An essay on salt, containing notices of its origin, formation, geological position, and principal localities, embracing a particular description of the American salines, with a view to its uses in the arts, manufactures, and agriculture. Delivered as a lecture before the New York Lyceum of Natural History: NY, O. Wilder and J. M. Campbell, 80 p.
        For review see 7810.
10549   1823, Fresh water formation: Am. J. Sci., v. 6, no. 2, 381-382.
        On correlation of strata at Paris and Rome by G. Cuvier.
10550   1825, History of the diamond: Merrimack Mag., v. 1, 74-75.
        Extracted from Lectures on Mineralogy (see 10551).
10551   1825, Lectures on geology; being outlines of the science, delivered in the New-York Athenaeum: NY, E. Bliss and E. White, xiv, 15-358 p., illus.
        For review see 7823, 10498. For notice see 426.
10552   1825, Notice of fossil crustacea, from New-Jersey: NY Lyc. Nat. History Annals, v. 1, pt. 2, 195-198, plate with 4 figs.
10553   1825, Supplement to a notice of fossil crustacea: NY Lyc. Nat. History Annals, v. 1, pt. 2, 249.
        See also 10552.

Van Rensselaer, J. (cont.)
10554   1826, Notice of a recent discovery of the fossil remains of the Mastodon: Am. J. Sci., v. 11, no. 2, 246-250.
Fossil bones discovered at Long Branch, New Jersey.
10555   1827, On the fossil remains of the Mastodon lately found in Ontario County, New-York: Am. J. Sci., v. 12, no. 2, 380-381.
10556   1828, [Geological specimens from New York Island]: Am. J. Sci., v. 14, no. 1, 192 (16 lines).
In NY Lyc. Nat. History Proc.
10557   1828, On the fossil tooth of an elephant, found near the shore of Lake Erie, and on the skeleton of a Mastodon, lately discovered on the Delaware and Hudson Canal: Am. J. Sci., v. 14, no. 1, 31-33.

Van Rensselaer, Stephen (1765-1839)
10558   1820, Address of the general committee of the Board of Agriculture to the officers and members of the county agricultural societies: Alb., S. Southwick, 32 p. (address is found on pp. 13-32).
On the importance of geology to agriculture.
Also in NY State Board Ag. Mem., v. 1, xix-xlvi.
10559   1831, Gen. Van Rensselaer's note: Am. J. Sci., v. 20, no. 2, 419-420.
On the quality of Amos Eaton's field work.

Vanuxem, Lardner (1792-1848)
See also 8875 with others.
10560   1821, Analysis of the blue iron earth of New Jersey, made at the school of mines at Paris, in the year 1819: Phila. Acad. Nat. Sci's J., v. 2, pt. 1, 82.
For abstract see Phila. J. Med. Phys. Sci's, v. 4, 386-387 (1822).
10561   1821, Description and analysis of the table spar from the vicinity of Willsborough, Lake Champlain: Phila. Acad. Nat. Sci's J., v. 2, 182-185.
For abstract see Phila. J. Med. Phys. Sci's, v. 4, 388-389 (1822).
10562   1821, On two veins of pyroxene or augite in granite [in Columbia, South Carolina]: Phila. Acad. Nat. Sci's J., v. 2, 146-149.
For abstract see Phila. J. Med. Phys. Sci's, v. 4, 388 (1822).
10563   1822, and W. H. Keating, Account of the jeffersonite, a new mineral discovered at the Franklin Iron Works, near Sparta in New-Jersey: Phila. Acad. Nat. Sci's J., v. 2, pt. 2, 194-204.
Also reprint edition, 12 p.
10564   1822, On a new locality of automalite [at Franklin, New Jersey]: Phila. Acad. Nat. Sci's J., v. 2, pt. 2, 249-251.
For abstract see Am. J. Sci., v. 5, no. 2, 402 (2 lines).
10565   1822, and W. H. Keating, On the geology and mineralogy of Franklin, in Sussex County, New Jersey: Phila. Acad. Nat. Sci's J., v. 2, pt. 2, 277-288, plate.
10566   1823, Description, analysis, etc. of a lamellar pyroxene [from West Point, New York]: Phila. Acad. Nat. Sci's J., v. 3, pt. 1, 68-73.
10567   1823, Description and analysis of the zirconite of Buncombe County, North Carolina: Phila. Acad. Nat. Sci's J., v. 3, pt. 1, 59-64.
10568   1823, On the marmolite of Mr. Nuttall [from Bare Hills, Baltimore, Maryland]: Phila. Acad. Nat. Sci's J., v. 3, pt. 1, 129.
10569   1824, Account of an examination of fused charcoal: Am. J. Sci., v. 8, no. 2, 292-294.
See also articles by R. Hare (4767 to 4771).
10570   1824, Account of an examination of fused charcoal: Phila. Acad. Nat. Sci's J., v. 3, pt. 2, 371-374.
See also article by B. Silliman (9659).
10571   1824, and W. H. Keating, Observations upon some of the minerals discovered at Franklin, Sussex County, New Jersey: Phila. Acad. Nat. Sci's J., v. 4, pt. 1, 3-11.
For abstract see Bos. J. Phil. Arts, v. 2, no. 2, 133-138.
10572   1825, Experiments on anthracite, plumbago, &c.: Phila. Acad. Nat. Sci's J., v. 5, pt. 1, 17-27. Also in Am. J. Sci., v. 10, no. 1, 102-109. Also reprint edition, Phila., J. Harding (printer), 13 p.
10573   1825, Fusion of charcoal: Bos. J. Phil. Arts, v. 2, no. 6, 602.
Abstracted from Phila. Acad. Nat. Sci's J.
10574   1826, Report on the geological survey of the state of South Carolina: in Statistics of South Carolina by R. Mills (see 7194), 25-30.
10575   1827, Proofs, drawn from geology, of the abstraction of nitrogen from the atmosphere, by organization: Am. J. Sci., v. 12, no. 1, 84-93.
10576   1829, Analysis of cyanite and fibrolite, and their union in one species, under the name of disthene: Phila. Acad. Nat. Sci's J., v. 6, pt. 1, 41-45.
Also reprint editions.
10577   1829, Geological observations on the Secondary, Tertiary, and Alluvial formations of the Atlantic coast of the United States of America: Phila. Acad. Nat. Sci's J., v. 6, pt. 1, 59-71.
Arranged by S. G. Morton. Also reprint edition, 44 p.
10578   1829, Remarks on the characters and classification of certain American rock formations: Am. J. Sci., v. 16, no. 2, 254-256.
10579   1837, and J. Eights, First annual report of the geological survey of the fourth district of the state of New-York: NY Geol. Survey Ann. Rept., v. 1, 187-212.
See also report by T. A. Conrad (2557).
10580   1837, Geological survey of the state: Genesee Farm. and Gardener's J., v. 7, 82.
On the reasons for the New York survey.
10581   1838, The marl deposits of South Carolina: Farm's Reg., v. 6, 111.
10582   1838, Second annual report of so much of the geological survey of the third district of the state of New-York as relates to the objects of immediate utility: NY Geol. Survey Ann. Rept., v. 2, 253-286.
10583   1839, Analyses of sundry specimens of calcareous rock, or marl, in South Carolina: Farm's Reg., v. 7, 78-79.
10584   1839, Third annual report of the geological survey of the third district: NY Geol. Survey Ann. Rept., v. 3, 241-285.
10585   1840, Fourth annual report of the geological survey of the third district: NY Geol. Survey Ann. Rept., v. 4, 355-383.

Vanuxem, L. (cont.)
10586 1841, [Analogy between fossils in eastern and western New York]: Am. J. Sci., v. 41, no. 1, 164 (2 lines).
Also in Assoc. Am. Geol's Nat's Proc., v. 2, 9 (2 lines). Also in Assoc. Am. Geol's Nat's Rept., v. 1, 17 (2 lines), 1843.
10587 1841, and E. S. Carr, Fifth annual report of the geological survey of the third district: NY Geol. Survey Ann. Rept., v. 5, 137-147.
10588 1841, On the ancient oyster shell deposits observed near the Atlantic coast of the United States: Am. J. Sci., v. 41, no. 1, 168-170.
Also in Assoc. Am. Geol's Nat's Proc., v. 2, 13-15. Also in Assoc. Am. Geol's Nat's Rept., v. 1, 21-23, 1843.
10589 1842, Geology of New York. Part III. comprising the survey of the third geological district: Alb., W. and A. White and J. Visscher, 306, [1] p., numerous figs.
For review see 8066.
10590 1843, Marl and soft limestone of South Carolina. "Chalk Hills" improperly so-called. New Jersey green sand: Farm's Reg., v. 10, 486-488.
10591 1843, On the origin of mineral springs: Assoc. Am. Geol's Nat's Rept., v. 1, 224-229.
10592 1847, [Report on the Taconic system]: Am. Q. J. Ag. Sci., v. 6, 213 [i.e. 261] (3 lines).
In Assoc. Am. Geol's Nat's Proc.
10593 1848, Report [on minerals of South Carolina]: in Report on the geology and mineralogy of South Carolina by M. Tuomey (see 10438), xxxi-xxxii.

Van Zandt, Nicholas Biddle
10594 1818, A full description of the soil, water, timber, and prairies of each lot, or quarter section of the military lands between the Mississippi and Illinois rivers ... : Wash., P. Force, 127 p.

Varley, Delvalle Lowry (see Lowry, Delvalle)

Vattemare, Alexandre (1796-1864)
10595 1847, Instructions on the best mode of collecting, preserving and transporting objects of natural history: Dwight's Am. Mag., v. 3, 445-446; 462-463; 494-495; 590-591; 654-655.

Vaucluse Mine (Orange County, Virginia)
10596 1847, Plan and description of the Vaucluse Mine, Orange County, Virginia: Phila., J. H. Schwack (printer), 13 p.
10597 1847, Vaucluse gold mine. Description of the mine, machinery, and operations: Phila., 19 p.

Vauquelin, Louis Nicolas (1763-1829)
See also 3943 with A. F. Fourcroy.
10598 1808, [Iron and the formation of meteors]: Am. Reg. or Gen. Repos., v. 2, 378-379.
10599 1808, [Silver ore analyses]: Am. Reg. or Gen. Repos., v. 2, 365 (11 lines).
10600 1826, Discovery of iodine in the mineral kingdom [in Brazil]: N. Am. Med. Surgical J., v. 1, 239 (3 lines).
Abstracted from J. de Pharmacie.
10601 1827, Volcanic ashes [from Mount Etna]: Am. J. Sci., v. 12, no. 1, 194 (12 lines).
Abstracted by J. Griscom from Annales de Chimie.
10602 1830, Dictionary of chemistry, containing the principles and modern theories of the science, with its applications to the arts, manufactures and chemistry ... with additions and notes by Mrs. Almira H. Lincoln: NY, G. and C. Carvill, xxviii, [291-531 p.

Vaux, Frederic W.
10603 1849, Ultimate analysis of some varieties of coal: Franklin Inst. J., v. 17 (3s), 197-199.
Abstracted from "J. Chem. Soc."

Vaux, George
10604 1827, Bituminous coal near Harrisburg: Am. J. Sci., v. 12, no. 2, 378 (13 lines).

Vaux, James
10605 1826, Remarks upon the use of anthracite, and its application to the various purposes of domestic economy: Franklin Inst. J., v. 2, 292-296.

Vaux, Roberts (1786-1836)
10606 1830, and T. McEuen, Notice of the fall of a meteoric stone, at Deal, in New Jersey: Phila. Acad. Nat. Sci's J., v. 6, pt. 2, 181-182.

Vaux, William S.
10607 1850, Large crystals of sphene: Am. J. Sci., v. 9 (ns), no. 3, 430 (6 lines), 2 figs.
On sphene from Diana, Lewis County, New York.

Verheul, Captain
10608 1821, Account of a visit to the crater of the volcano Goenong Apie, one of the islands of Banda: The Pilot, v. 1, no. 2, [2].

Vermont. State of
10609 1836, [Recommendation for a geological survey of the state]: in Vt. Governor's Message for 1836.
10610 1838, Report and correspondence on the subject of a geological and topographical survey of the state of Vermont: Burlington, Vt., E. P. Walton and Son (printer), 20 p.
10611 1844, An act to provide for a geological survey of the state: Vt. Gen. Assembly, October 28, 1844.
10612 1844, [Geological survey, soil analyses, and agricultural development urged]: Vt. Governor's Message for 1844.
10613 1844, [On the Vermont geological survey]: Niles' Wk. Reg., v. 67, 175.
From Vt. Governor's Message, by Governor Slade.

Vermont (cont.)
10614 1845, Bill providing for a geological survey of Vermont: in First annual report on the geology of the state of Vermont by C. B. Adams (see 23), p. 63.
10615 1845, [Message from the Governor relating to the geological survey]: Vt. Senate J. for 1845, 170-171.
10616 1846, [The role of the geological survey in developing natural resouces]: Vt. Governor's Message for 1846, pp. 6-8.

Verneuil, Philippe Edouard Poulletierde de (1805-1873)
10617 1846, Correspondence: Am. Q. J. Ag. Sci., v. 4, 166.
   On the correlation of Carboniferous fossils in Ohio and Russia.
10618 1846, M. Verneuil on the Fusilina in the coal formation of Ohio: Am. J. Sci., v. 2 (ns), no. 2, 293.
   Abstracted from Am. Q. J. Ag. Sci.
10619 1847, A general review of the geology of Russia: Am. J. Sci., v. 3 (ns), 153-160.
   Translated by D. D. Owen.
10620 1848/1849, On the parallelism of the Paleozoic deposits of North America, with those of Europe; followed by a table of the species of fossils common to the two continents, with indication of the positions in which they occur, and terminated by a critical examination of each of these species: Am. J. Sci., v. 5 (ns), no. 2, 176-183; no. 3, 359-370; v. 7 (ns), no. 1, 45-51; no. 2, 218-231.
   Translated and condensed, with annotations, by J. Hall (see 4720).

Vespucius (pseudonym; see 7582)

Vestiges of the natural history of Creation (see Chambers, Robert)

Vetch, Captain
10621 1821, Mammoth [from Rochester, New York]: Saturday Mag., v. 1 (ns), 136-138.

Viator (pseudonym; see 4286)

Vicat, Louis Joseph (1786-1861)
10622 1838, New observations on magnesian hydraulic lime: Franklin Inst. J., v. 21 (ns), 338-339.
   Translated by J. Griscom, from J. des Mines.

The Villager, A Literary Paper (Greenwich Village, New York, 1819)
10623 1819, [Granite found at Kennebec County, Maine]: v. 1, 95 (6 lines).
10624 1819, [Incorporation of the Connecticut State Geological Society, with a list of members]: v. 1, 95.

Villefosse, Antoine Marie Héron de (1774-1852)
10625 1819, Some account of Werner: Port Folio, v. 7, 47-51.

Vincent, John Nicolas
10626 1825, Account of a journey to Monte Rosa, and of the first ascent of its southern summit: Bos. J. Phil. Arts, v. 2, no. 5, 413-423.
   Abstracted from Brewster's J. With geological notes on the Italian mountain.

Vine, James
10627 1823, Method of hardening gypsum: Bos. J. Phil. Arts, v. 1, no. 1, 93 (6 lines).
   Abstracted from Geol. Soc. London Trans.

Virginia. State of
10628 1831, [Memorial of colliers in Chesterfield County, Virginia]: n.p., 10 p.
10629 1833, [Recommendation for a geological survey of the state]: in Va. Governor's Message, session 1833-1834.
10630 1835, An act to authorize a geological reconnoissance of the state, with a view to the chemical composition of its soils, minerals, and mineral waters: Farm's Reg., v. 2, 692.
10631 1835, A bill to authorize a geological reconnoissance of the state, with a view to the chemical composition of its soils, minerals, and mineral waters: Va. Gen. Assembly, March 6, 1835.
10632 1835, Report ... from the select committee, to whom was referred certain materials from Morgan, Frederick and Shenandoah counties, praying for a geological survey of the state of Virginia, ... : Am. Farm., v. 16, 355-357.
10633 1836, An act to provide for a geological survey of the state, and for other purposes: Va. Gen. Assembly, February 29, 1836.
10634 1841, [An act repealing the geological survey, and funding completion of the reports]: Va. Gen. Assembly, 1841.

Virginia and New England Mining Company
10635 1835, An act to incorporate the Virginia and New England Mining Company: Fredericksburgh, 4 p.
10636 1838, Charter and description of the property of the Virginia and New England mining company: NY, J. Narine (printer), 17 p.

The Virginia Historical Register and Literary Companion (Richmond, Virginia, 1848-1850)
10637 1849, El Dorado. - The gold mines of California: v. 2, 41-44.
   Signed by "J. G. W."
10638 1849, The burning wells of Kanawha: v. 2, 50.
10639 1849, Lead ore in Nelson [Virginia]: v. 2, 114 (10 lines).
10640 1849, Gold in Virginia: v. 2, 114-115.
10641 1850, The gold mines [of Virginia]: v. 3, 117 (6 lines).

The Virginia Literary Museum and Journal of Belles Lettres, Arts, Sciences, &c. (Charlottesville, Virginia, 1829-1830)
10642   1829, Manufacture of diamonds: v. 1, 3-4, 107.
            Signed by "M." Notice of experiments by M. Gannal.
10643   1829, Great earthquake in Spain: v. 1, 207.
10644   1830, The new substance bromine in the salt of Kanawha County: v. 1, 736.
            Signed by "E."
10645   1830, Dr. John D. Godman [obituary]: v. 1, 770-772.
            Signed by "P."

Virlet d'Aoust, Pierre Theodore (1800-1894)
10646   1836, Progressive rise of a portion of the bottom of the Mediterranean: Am. J. Sci., v. 31, no. 1, 190-191.
            Abstracted from Edinburgh New Philos. Rev.

The Visitor (Richmond, Virginia, 1809)
10647   1809, A description of Mount Vesuvius: v. 1, 29-30.
10648   1809, On magnetism: v. 1, 117-118; 123-124; 132-133.
10649   1809, [On coal near Richmond, Virginia]: v. 1, 189-190.

Voelcker, R.
10650   1850, On the proportion of phosphoric acid in some natural water: Am. J. Sci., v. 10, no. 3, 412-413.
            In British Assoc. Proc.

Volney, Constantin François Chasseboeuf Boisgivais, comte de (1757-1820)
10651   1798, Travels through Syria and Egypt, in the years 1783, 1784 and 1785. Containing the present natural and political state of those countries, their productions, arts, manufactures, and commerce: NY, J. Tiebout, 2 v.
10652   1804, Of earthquakes and volcanoes of the United States: Lit. Mag. and Am. Reg., v. 2, 452-453.
10653   1804, A view of the soil and climate of the United States of America: with supplementary remarks upon Florida; on the French colonies on the Mississippi and Ohio, and in Canada; and on the aboriginal tribes of America. ... Translated with occasional remarks, by C. B. Brown: Phila., J. Conrad and Co., xxviii, 446 p., illus., 3 plates, 2 folded maps.
            Contains map 10654. For review see 1243, 3217, 6326 and 7047.
            Shaw and Shoemaker (volume for 1805, #9677) record an 1805 Philadelphia edition at the Quincy Free Public Library, Illinois. Not seen.
10654   1804, [Geological map of the United States]: in A view of the soil and climate of the United States (see 10653), colored geological map.
            While coloring guides to this map are given on the page of errata, most maps were never colored.

Von Buch, L. (see Buch, L. von)

Von Humboldt, A. (see Humboldt, A. von)

Von Kobell (see Kobell)

Von Leonhard, K. C. (see Leonhard, K. C. von)

W., J. G. (see 10637)

W., J. S. (see 1035)

Waagner, Joseph Guillaume
10655   1824, Splendid cabinet of minerals for sale: Bos. J. Phil. Arts, v. 2, no. 3, 298-299.
            Abstracted from Annals Phil. Notice of Jacques Frederick Von der Null's mineral cabinet for sale.

Wachmeister, M.
10656   1830, Analysis of a new mineral, from Hoboken [New Jersey]: Am. J. Sci., v. 18, no. 1, 167 (10 lines).
            Abstracted from Journal des Mines.

Wackenroder, Heinrich Wilhelm Ferdinand (1798-1854)
10657   1849, Arsenical nickel from Oelsnitz: Am. J. Sci., v. 8 (ns), no. 1, 128 (14 lines).
            Abstracted from "Annuaire de Chem."
10658   1849, and M. Ludwig, On an arsenio-sulphuret of nickel: Am. J. Sci., v. 8 (ns), no. 1, 128 (4 lines).
            Abstracted from "Jour. für Prakt. Ch."

Wadsworth, Daniel
10659   1830, Illustrations of a view taken from the Upper Falls of the Genesee River: Am. J. Sci., v. 18, no. 2, 209-211, plate.
            With "Remarks" by B. Silliman (see 9698).

Wadsworth, James Samuel (1807-1864)
10660   1836, Recent and curious facts and questions, respecting plaster beds in New York: Farm's Reg., v. 4, 187-188.

Wagner, William (1796-1885)
10661   1839, Description of five new fossils, of the older Pliocene formation of Maryland and North Carolina: Phila. Acad. Nat. Sci's J., v. 8, pt. 1, 51-53, plate with 5 figs.

Wagstaff, William
10662   1841, Report [on the geology of the Potomac and Alleghany Coal and Iron Company]: in the company's documents (8548).

Wailes, Benjamin Leonard Covington (1797-1862)
10663 1845, On the geology of Mississippi: Assoc. Am. Geol's Nat's Proc., v. 6, 80-81.
       Also in Am. Q. J. Ag. Sci., v. 2, 168-169.
10664 1846, Geological gleanings in Mississippi: Scien. Am., v. 1, no. 43, [3]; no. 44, [3]; no. 45, [3]; no. 46, [3]; no. 47, [3]; no. 48, [3]; no. 49, [3].
       Signed by "B. L. C. W." For notice see 9303.
10665 1847, On the formation of the Mississippi Bluff, near Natchez: Am. Q. J. Ag. Sci., v. 6, 208-209.
       In Assoc. Am. Geol's Nat's Proc. With discussion by A. Binney.
10666 1848, [On the Natchez bluff formations of Mississippi]: Lit. World, v. 2, 228.
       In Assoc. Am. Geol's Nat's Proc.

Walchner, Friedrich August (b. 1799; see 10331 with Tripier.)

Walckner (i.e. F. A. Walchner?)
10667 1824, Metallic titanium [from furnace slag]: Bos. J. Phil. Arts, v. 2, no. 3, 298 (12 lines).

Waldheim, Gotthelf Fischer von
10668 1846, Spondylosaurus [from Stichioukino, Russia]: Am. J. Sci., v. 1 (ns), no. 3, 440.
       Abstracted from l'Institut.
10669 1846, Thoracoceras, a new genus of the family of Orthoceratites: Am. J. Sci., v. 1 (ns), no. 3, 440 (9 lines).
       Abstracted from l'Institut.

Wales, B. L. C. (see Wailes, Benjamin Leonard Covington)

Walferdin
10670 1839, Subterranean temperatures: Am. J. Sci., v. 36, no. 1, 204-205.
       Abstracted from Comptes Rendus.

Walker (see 6518 with F. C. Lukis)

Walker
10671 1838, Discussion [of "Dipping needle for finding longitude" by H. H. Sherwood (see 9585)]: Franklin Inst. J., v. 22 (ns), 270-273.

Walker, James Scott (1793-1850)
10672 1830, Original account of the earthquake of Caracas, in South America, in 1812: Mus. For. Liter. Sci., v. 18, 121-129,
       From British Mag.

Walker, Robert
10673 1835, Analysis of the fossil tree seen at present imbedded in the sandstone at Craigleith Quarry [in Scotland]: Am. J. Sci., v. 28, no. 2, 390 (7 lines).
       Abstracted from Jameson's J.

Walker, Robert John (1801-1869)
10674 1845, Introductory address: Nat'l Inst. Prom. Sci. Bulletin, v. 3, 439-450.
       Includes notice of the progress of geology and mineralogy.

Walker, W.
10675 1795, Dissertation of the probable cause of the deluge: NY Mag., v. 6, 76-79.

Walker, W.
10676 1842, Geological changes produced by the Saxicava rugosa in Plymouth Sound: Am. J. Sci., v. 42, no. 2, 326 (6 lines).
       Abstracted from British Assoc. Proc.

Walpoole, George Augustus
10677 1795, Account of the hot-well at Bristol, England: NY Mag., v. 6, 734-735.
       Extracted from Walpoole's British traveller.
10678 1795, Some account of the baths, and of the nature of the mineral waters at the city of Bath, in England: NY Mag., v. 6, 735-737.
       Extracted from Walpoole's British traveller.

Walsh, Robert (1772-1852)
10679 1831, Notices of Brazil in 1828 and 1829: Bos., Richardson, Lord and Holbrook, 2 v.
10680 1831, Walsh's notices of Brazil: Mon. Am. J. Geology, v. 1, no. 4, 182-186.
10681 1833, Earthquake at Zante [Ionian Islands]: Dollar Mag., v. 1, no. 11, 170-172.

Waltershausen, Baron Wolfgang Sartorius von (1809-1876)
10682 1844, Extract of a letter: Am. J. Sci., v. 47, no. 1, 99-100.
       Also in Assoc. Am. Geol's Nat's Proc., v. 5, 7-8. Prospectus for his book Aetna and its convulsions.
10683 1850, Iceland, its climate and its geysers: Ann. Scien. Disc., v. 1, 231-232.

Ward, Nahum (1785-1860)
10684 1819, Description of the Mammoth Cave in the county of Warren, state of Kentucky: Methodist Mag., v. 2, 380-383; 415-418.
10685 1830, Fountains of fresh and salt water: Am. J. Sci., v. 18, no. 2, 379-380.

Warden, David Bailie (1778-1845)
10686 1803, Observations on the natural history of the village of Kinderhook, and its vicinity: Med. Repos., v. 6, 4-18.
10687 1809, Shower of stones in Italy [at Parma]: Med. Repos., v. 12, 183-184.
10688 1810, Analysis of the stone of the Islands of Fegee, employed by the savages to make their axes: Med. Repos., v. 13, 75-76.
10689 1810, Description and analysis of the meteorik stone, which fell at Weston, in North America, the 4th December, 1807: Mon. Anth., v. 9, 352-354.
       Abstracted from Annales de Chimie.

Warden, D. B. (cont.)
10690   1810, Sulphate of lime of Onondaga, state of New-York, analysed: Med. Repos., v. 13, 76-77.
10691   1811, Description and analysis of the meteoric stone, which fell at Weston, the 4th December, 1807: Med. Repos., v. 14, 194-196.
        Also in Am. Med. Philos. Reg., v. 3, no. 4, 413-416 (1813).

Warder, John Aston (1812-1883)
10692   1838, New trilobites (1.) Ceratocephala goniata [from Cincinnati, Ohio]: Am. J. Sci., v. 34, no. 2, 377-379, fig.

Warder, M. A.
10693   1839, Reclamation of M. A. Warder: Am. J. Sci., v. 36, no. 1, 187 (14 lines).
        On J. A. Warder's discovery of a trilobite.

Ware, John (1795-1864)
10694   1823, Some account of the discovery of the fossil bones of the mastodonte or great American mammoth, and of the anatomical character of that animal: Bos. J. Phil. Arts, v. 1, no. 3, 257-269; no. 4, 391-408.
        With extracts from "Researches on the fossil bones of quadrupeds" by G. Cuvier (see 2756) and "Historical disquisition on the mammoth" by R. Peale (see 8188).

Warren, John Collins (1778-1856)
        See also discussion of article by L. Agassiz (86).
10695   1846, [On mastodon bones of New York and New Jersey]: Am. Philos. Soc. Proc., v. 4, no. 35, 269 (9 lines).
10696   1847, On the Mastodon: Am. Q. J. Ag. Sci., v. 6, 200-201 [i.e. 248-249].
        In Assoc. Am. Geol's Nat's Proc. Abstracted in Lit. World, v. 2, 230.
10697   1848, [On the scientific collections of the Wilkes expedition]: Bos. Soc. Nat. History Proc., v. 3, 41-42.
10698   1849, [On the geological position of the Mastodon giganteus]: Bos. Soc. Nat. History Proc., v. 3, 111-117.
        With discussion by J. W. Foster (see 3934) and H. D. Rogers.
10699   1850, [American building sandstone]: Bos. Soc. Nat. History Proc., v. 3, 248.
10700   1850, [Fossil Pleseosaurus skeleton from England]: Bos. Soc. Nat. History Proc., v. 3, 205.
10701   1850, On the Mastodon angustidens: Am. Assoc. Adv. Sci. Proc., v. 2, 93-94.
        With discussion by H. D. Rogers, L. Agassiz, and R. W. Gibbes.

Washburne, Elihu Benjamin (1816-1887)
10702   1848, Lead region and lead trade of the Upper Mississippi: Hunt's Merch. Mag., v. 18, 285-293.

Washington Coal Company
10703   1847, An act to incorporate the Washington Coal Comapny, and for other purposes: NY, J. M. Elliott (printer), 8 p.

Waterhouse, Benjamin (1754-1846; see 11072 with J. B. Wyeth)

Waterhouse, John Fothergill
10704   1813, Description of certain minerals found in the state of Massachusetts: NE J. Medicine Surg., v. 2, 261-264.
10705   1814, Description of American minerals: NE J. Medicine Surg., v. 3, 126-128.

Watkins, John W. (Dr.)
10706   1800, On the disease called the lake-fever of the western counties of New-York: Med. Repos., v. 3, no. 4, 359-361.
        Contains notes on the geology of Seneca Lake.

Watson
10707   1795, Description of Iceland: NY Mag., v. 6, 752-756.
        Extracted from Watson's Universal Gazetteer.

Watson (Dr., of Philadelphia; i.e. John Fanning Watson?, 1779-1860)
10708   1821, [Review with excerpts of Geological essays by H. H. Hayden (see 4877)]: N. Am. Rev., v. 12, 134-149.

Watson, J. Y.
10709   1848/1849, The compendium of British mining: Mining J., v. 2, 116-117; 132; 146; 163.
        Abstracted from London Rr. Gaz.

Watson, William
10710   1819, Bedford springs: Saturday Mag., v. 2, 59-60.
10711   1822, Some account of the effects and use of the mineral springs of Bedford County, Pennsylvania: Am. Med. Rec'r, v. 5, 381-383.

Webb (Mr.)
10712   1839, Lunar volcanoes: Am. J. Sci., v. 35, no. 2, 305 (8 lines).

Webb, Thomas H.
        See also 8505 with J. Porter and H. U. Cambridge, and 10155 with S. Taylor.
10713   1822, Notice of the minerals in the vicinity of Providence, (R. I.): Am. J. Sci., v. 4, no. 2, 284-285.
10714   1822, Notice of mineral localities [in Rhode Island]: Am. J. Sci., v. 5, no. 2, 402-403.
10715   1823, Miscellaneous localities of minerals [in Massachusetts]: Am. J. Sci., v. 7, no. 1, 54-55.
10716   1825, By Thomas H. Webb: Am. J. Sci., v. 9, no. 2, 246-248.
        On mineral localities in Massachusetts and Rhode Island.

Webber, Charles Wilkins (1819-1856)
10717 1848, Old Hicks, the guide; or, adventures in the Comanche country in search of a gold mine: NY, Harper and Brothers, [ix]-x, [13]1-356 p.
10718 1849, Gold mines of the Gila. A sequel to Old Hicks the guide: NY, DeWitt and Davenport, 2 v. in 1.
    For review see 5484.

Webber, Samuel
10719 1842, Sketch of the great geological features of the Valley of the Connecticut River, at Charlestown, New Hampshire, and remarks on some crystals found in the slate-rock scattered in that region; with specimens: Nat'l Inst. Prom. Sci. Bulletin, v. 2, 197-200, woodcut.
10720 1844, Observations of some alluvial banks of the Connecticut River: Am. J. Sci., v. 47, no. 1, 98.
    Also in Assoc. Am. Geol's Nat's Proc., v. 5, 5.

Webster, Captain (Royal Navy)
10721 1830, Antarctic expedition: Am. J. Sci., v. 18, 188-190.

Webster, J. H. (Professor, i.e. John White Webster?)
10722 1848, New locality of idocrase, anorthite? and molybdenite: Am. J. Sci., v. 6 (ns), no. 3, 425.

Webster, John White (1793-1850)
    See also 10722 by J. H. Webster.
10723 1817, [Minerals from the Boston region]: J. Sci. Arts, v. 1, no. 1, 130-131.
10724 1818, Remarks on the structure of the Calton Hill, near Edinburgh, Scotland; and on the aqueous origin of wacke: Am. J. Sci., v. 1, no. 2, 230-234, fig.
10725 1819, Analysis of wacke: Am. J. Sci., v. 1, no. 3, 296-297.
10726 1819, Asbestos in anthracite. Extract of a letter from Dr. I. W. [i.e. J. W.] Webster: Am. J. Sci., v. 1, no. 3, 243.
    With "Remarks" (see 326).
10727 1820, Localities of minerals, observed principally in Haddam, in Connecticut, in Sept. 1819: Am. J. Sci., v. 2, no. 2, 239-240.
    With note by B. Silliman (see 9629).
10728 1820, Miscellaneous articles of foreign intellignece: Am. J. Sci., v. 2, no. 1, 166-167.
    Brief summary of scientific work of 14 researchers in Europe.
10729 1820, [Review with excerpts of An index to the geology of the Northern states by A. Eaton (see 3324)]: N. Am. Rev., v. 11, 225-239.
10730 1821, A description of the Island of St. Michael, comprising an account of its geological structure with remarks on the other Azores or Western Islands: Bos., R. P. and C. Williams, viii, 9-244 p., folded map.
    For review with excerpts see 377 and 7943.
10731 1821, Epidote [from Nahant, Massachusetts]: Am. J. Sci., v. 3, no. 2, 364 (3 lines).
10732 1821, Notice of some minerals from the New South Shetland Islands: Am. J. Sci., v. 4, no. 1, 25-27.
10733 1821, Notice from the Edinburgh Philosophical Journal: Am. J. Sci., v. 4, no. 1, 27-28.
    Mineralogical intelligence from Europe and South America.
10734 1821, Notice of vegetable remains in coal strata: Am. J. Sci., v. 3, no. 2, 389-390.
    With excerpt from Tilloch's Mag. On British fossil plants.
10735 1821, Siliceous sinter of the Azores: Am. J. Sci., v. 3, no. 2, 391-392.
10736 1822, Foreign notices in mineralogy, geology, ancient arts, &c.: Am. J. Sci., v. 4, no. 2, 243-245.
    Includes notice of the Geological Society of London, and the Wernerian Society of Edinburgh.
10737 1823, Chemical examination of a fragment of a meteor which fell in Maine, August, 1823, and of green feldspar from Beverly, Mass.: Bos. J. Phil. Arts, v. 1, no. 4, 386-390.
10738 1823, Extract of a letter [on the origin of bituminous marl]: Am. J. Sci., v. 6, no. 1, 75 (12 lines).
    In "Geology of the Connecticut ... " by E. Hitchcock (see 5081).
10739 1823, New American locality of rubellite and lepidolite: Bos. J. Phil. Arts, v. 1, no. 2, 190-191 (9 lines).
    On minerals of Mount Paris, Maine.
10740 1823, New locality of marble, &c. near Boston [at Stoneham, Massachusetts]: Bos. J. Phil. Arts, v. 1, no. 1, 95-96.
10741 1823, New periodical work [notice of NY Lyc. Nat. History Annals]: Bos. J. Phil. Arts, v. 1, no. 4, 406 (16 lines).
10742 1823, New work on crystallography [review of Introduction to crystallography by H. J. Brooke]: Bos. J. Phil. Arts, v. 1, no. 3, 298-299.
10743 1823, Notice of Messrs. Conybeare and Phillips' Outlines of the geology of England and Wales. - Part 1st. London, 1822 [notice with excerpts of 2597]: Bos. J. Phil. Arts, v. 1, no. 3, 239-250.
10744 1823, Phillip's Mineralogy [review of 8325]: Bos. J. Phil. Arts, v. 1, no. 3, 299-300 (8 lines).
10745 1823, Primitive boulders: Bos. J. Phil. Arts, v. 1, no. 1, 90-92.
    On boulders from Massachusetts, and a rocking stone in Roxbury, Massachusetts.
    With excerpts from Geol. Soc. London Trans. (see 1794).
10746 1824, Additions to the cabinet of minerals at Cambridge [Harvard College]: Bos. J. Phil. Arts, v. 2, no. 3, 299 (19 lines); no. 4, 394-395; no. 5, 505; no. 6, 610.
10747 1824, Cabinet of minerals at Cambridge: Bos. J. Phil. Arts, v. 2, no. 2, 201-203.
10748 1824, Chemical analysis of the sea-water of Boston Harbor: Bos. J. Phil. Arts, v. 2, no. 1, 96-97.
10749 1824, New localities of American minerals [from Massachusetts]: Bos. J. Phil. Arts, v. 1, no. 6, 599-600.
10750 1824, New locality of apatite [from Billerica, Massachusetts]: Bos. J. Phil. Arts, v. 2, no. 1, 104 (15 lines).

Webster, J. W. (cont.)
10751 1824, Notice of M. Brongniart's memoir on the lignite [with excerpts (see 2058)]: Bos. J. Phil. Arts, v. 2, no. 1, 88-96.
10752 1824, Obituary [of J. J. Conybeare]: Bos. J. Phil. Arts, v. 2, no. 2, 208 (8 lines).
10753 1824/1826, Remarks on the geology of Boston and its vicinity: Bos. J. Phil. Arts, v. 2, no. 3, 277-292; v. 3, no. 5, 486-489; to be continued but no more found.
With excerpts from "springs and wells ... of Boston" by J. Lathrop (see 6057).
10754 1825, Additions to the cabinet of minerals at Cambridge: Bos. J. Phil. Arts, v. 3, no. 1, 103.
10755 1825, Aerolite of Maine: Am. J. Sci., v. 9, no. 2, 400 (15 lines).
10756 1825, An elementary treatise on mineralogy [review of text by F. S. Beudant]: Bos. J. Phil. Arts, v. 3, no. 1, 71-81.
10757 1825, New scientific works: Bos. J. Phil. Arts, v. 2, no. 5, 502-503.
10758 1825, [Review of A catalogue of American minerals by S. Robinson (see 8825)]: N. Am. Rev., v. 21, 233-234.
10759 1825, Robinson's Catalogue of American minerals [review of 8825]: Bos. J. Phil. Arts, v. 2, no. 5, 505 (12 lines).
10760 1826, New work on mineralogy [review of Manual of mineralogy by E. Emmons (see 3491)]: Bos. J. Phil. Arts, v. 3, no. 5, 491-492 (14 lines).
10761 1826, New work on mineralogy [review of Manual of mineralogy by J. L. Comstock (see 2505)]: Bos. J. Phil. Arts, v. 3, no. 6, 599 (5 lines).
10762 1826, Notice of the mineralogy of Nova Scotia, and of several new localities of American minerals [in Massachusetts and Maine]: Bos. J. Phil. Arts, v. 3, no. 6, 594-599.
10763 1827, Garnet, (cinnamon stone?) &c [from Carlisle, Massachusetts]: Am. J. Sci., v. 12, no. 1, 176-177 (10 lines).
Abstracted from Bos. J. Phil. Arts.
10764 1827, [Review of Elements of mineralogy by J. L. Comstock (see 2506)]: N. Am. Rev., v. 24, 487-488.
10765 1829, Description of the polariscope, an instrument for observing some of the most interesting pehenomena of polarised light, invented by H. J. Brooke: Am. J. Sci., v. 15, no. 2, 369-373, plate with 4 figs.
10766 1838, [Review of Summary of geological observations made in a voyage to the island of Madeina, Porto Santo, and Azores by Count Varges de Bedemor]: N. Am. Rev., v. 46, 366-386.
10767 1842, A treatise on mineralogy, on the basis of Thompson's Outlines, with numerous additions; comprising the description of all new American and foreign minerals, their localities, &c. Designed as a text-book for students, travellers, and persons attending lectures on the science: Cambridge, Mass., J. Owen, 8vo.
This work is advertised for sale in an extra sheet of Bos. J. Nat. History, v. 4, no. 2, 1842. However, no other evidence that this text was ever published has been found.
10768 1844, [Review of An elementary treatise on mineralogy by W. Phillips (see 8325)]: N. Am. Rev., v. 58, 240-243.

Webster, Matthew Henry
10769 1824, Catalogue of the minerals which have been discovered in the state of New York, arranged under the heads of the respective counties and towns in which they are found: Alb., Websters and Skinners, 32 p.
10770 1841, Second meeting of the associated geologists of the United States: No. Light, v. 1, 44-45.

Webster, Noah (1758-1843)
10771 1801, A collection of phenomena, relative to the connection between earthquakes, tempests, and epidemic distempers: Med. Repos., v. 5, 25-31.
10772 1801, On the connection of earthquakes with epidemic diseases, and on the succession of epidemics: Med. Repos., v. 4, 340-344.
10773 1839, Critical interpretation of bara and asah, in a letter from Dr. Noah Webster to the Rev. William Buckland, Oxford, England: Am. J. Sci., v. 35, no. 2, 375-376.
Discussion of Mineralogy and geology by W. Buckland (see 2160). On the precise translation of Genesis.

Webster, Samuel (Reverend)
10774 1785, An account of an oil-stone found at Salisbury: Am. Acad. Arts Sci's Mem., v. 1, 380.

The Weekly Magazine (Philadelphia, 1798-1799)
10775 1798, Method of obtaining and preparing tin in the mines of Cornwall: v. 1, 78-79.
10776 1798, To the editor of the Weekly Magazine [on an earthquake in Canada]: v. 1, 332.
10777 1798, Geological observations: v. 2, 385-390.
On the organic nature of fossils, the Neptunian theory of rock formation, and "revolutions" of the globe. Signed by "T. P. S." For review see 10778.
10778 1799, Strictures on the author of "Geological observations" [review of 10777]: v. 3, 16-18.
Signed by "C."

The Weekly Register (see Niles' Weekly Register)

The Weekly Visitor, or, Ladies Miscellany (New York, 1802-1812)
10779 1803, [Gold mine in Cabarrus County, North Carolina]: v. 2, 46.
10780 1808, [Ballston spring, New York]: v. 7, 260-263, 278-281.
10781 1809, The earth [its formation and the origin of rocks]: v. 10, 41-43.
10782 1812, Tremendous earthquake [at Caracas]: v. 15, 13-14.
10783 1812, The earthquake: v. 15, 79-80.
A poem.

Weeks, John W.
10784   1839, [Letter on the importance of geology]: Farm's Mon. Vis., v. 1, 70.

Weisner, R. (Reverend)
10785   1848, Mineral lands of Wisconsin, Iowa and Illinois: Lit. Rec. and J. Linnaean Assoc. Penn. College, v. 4, 86-89.

Weld, Charles W. (1813-1869)
10786   1850, The first discovery (probably) of fossil bones and teeth in the state of New-York [in 1713]: NY State Mus. Ann. Rept., v. 3, 156.
        Extracted from Weld's History of the Royal Society.

Weld, Henry Thomas
10787   1839, A report made by Henry Thomas Weld, Esq., of the Maryland and New York Iron and Coal Company's land [in Alleghany Co., Maryland]: NY.
        Not seen. See also 6810.

Weld, Isaac (1774-1856)
10788   1841, [Letter to the National Institute of Science in Dublin on Irish progress in geology and mineralogy]: Nat'l Inst. Prom. Sci. Bulletin, v. 1, 14-17.

Wells, David Ames (1828-1896)
        See also discussion of article by G. J. Chase (2327).
10789   1850, and G. Bliss, jr., Louis Agassiz: Ann. Scien. Disc., v. 1, ix-x, plate.
        Brief biographical sketch.
10790   1850, [On a vein of phosphate of lime near Crown Point, New York]: Bos. Soc. Nat. History Proc., v. 3, 379.
10791   1850, [On the age of the sandstone of the Connecticut Valley]: Bos. Soc. Nat. History Proc., v. 3, 339-341.
10792   1850, and G. Bliss, jr., Prospectus [of Ann. Scien. Disc.]: Ann. Scien. Disc., v. 1, i-ii.
10793   1850, [Septaria from Connecticut, and their correlation with those of England]: Bos. Soc. Nat. History Proc., v. 3, 359 (5 lines).

Wells, Henry
10794   1832, and E. Lewis, I. Cooley, B. Chubbuk, and W. Russell, [Ores of Bradford County, Pennsylvania]: Mon. Am. J. Geology, v. 1, no. 11, 519-520.

Wells, I. M.
        See also 2178 with D. Buel, jr.
10795   1820, Observations on the geology of part of the counties of Albany and Greene [New York]: Plough Boy, v. 1, 257-258.

Wells, Nathaniel
10796   1794, A topographical description of Wells, in the County of York [Maine]: Mass. Hist. Soc. Coll., v. 3, 138-140.

Wells, Robert William (1795-1864)
10797   1819, On the origin of prairies: Am. J. Sci., v. 1, no. 3, 331-337.

Welsh, Jane Kilby
10798   1832/1833, Familiar lessons in mineralogy and geology designed for the use of young persons and lyceums: Bos., Clapp and Hull, 2 v. (v. 1, 404, 2 p., 77 woodcuts, front.; v. 2, 401, [1] p., 7 woodcuts, plate with 11 figs.).
        For review with excerpts see 1102 and 1813. For review see 3635.

Wesley
10799   1822, Of the formation of mountains after the flood: Meth. Mag., v. 5, 455-457.
        Extracted from Wesley's Natural Philosophy.

West Virginia Iron Mining and Manufacturing Company
10800   1838, Prospectus of the West Virginia Iron Mining and Manufacturing Company: NY, J. W. Bell (printer), 28 p.
        Contains report on geology by F. Shepherd (see 9570).

West, B.
10801   1849, On the presence of nitrogen in mineral waters: Scien. Am., v. 5, 67.

West, Charles Edwin (1809-1900)
10802   1843, Notice of certain siliceous tubes (fulgurites) formed in the earth [from Rome, New York]: Am. J. Sci., v. 45, no. 1, 220-222.

West, Samuel (1730-1807)
10803   1793, A letter concerning Gay Head [Martha's Vineyard, Massachusetts]: Am. Acad. Arts Sci's Mem., v. 2, pt. 1, 147-150.
        On geological features of Gay Head.

The Western Academician and Journal of Education and Science (Cincinnati, 1837-1838)
10804   1849, The science of geology: v. 1, no. 6, 323-329.
        Abstracted from Edinburgh Rev.

Western and Southern Medical Recorder (Lexington, Kentucky, 1841-1843)
10805   1841, Necrology [on Amos Eaton]: v. 1, 384.

The Western Farmer (see The Michigan Farmer)

The Western Gleaner, or Repository for Arts, Sciences and Literature (Pittsburgh, Pennsylvania, 1813-1814)
10806   1813, [Review with excerpts of Views of Louisiana by H. M. Brackenridge (see 1943)]: v. 1, 53-68.
10807   1814, Clay near Pittsburgh [at Wilkinsburg, Pennsylvania]: v. 1, 130.
10808   1814, On coal: v. 1, 151-167; 211-221; 273-283.

Western Gleaner (cont.)
10809   1814, On earthquakes: v. 1, 375-383.
10810   1814, On salt and salt-making: v. 2, 17-26; 72-78; 135-140.
10811   1814, Late fall of stones in France [at Agen]: v. 2, 249-250.

The western guide book ... (see Steele's western guide book, 10011)

The Western Journal and Civilian (see The Western Journal of Agriculture ...)

The Western Journal of Agriculture, Manufactures, Mechanic Arts, Internal Improvement, Commerce, and General Literature (St. Louis, Missouri, 1848-1850)
10812   1848, Arkansas lead: v. 1, no. 3, 169.
        Abstracted from Little Rock Gazette.
10813   1848, British North American Mining Company: v. 1, no. 3, 170.
        Abstracted from Am. Mining J.
10814   1848, Discovery of a gold mine in Michigan: v. 1, no. 3, 170 (10 lines).
        Abstracted from Buffalo Courier.
10815   1850, The Arkansas lead mines: v. 3, no. 4, 276-278.
        Abstracted from Arkansas State Democrat.
10816   1850, The precious metals, coins, and bank notes: v. 5, no. 1, 20-34.
        Abstracted from Hunt's Merch. Mag.

The Western Journal of Medicine and Surgery (Louisville, Kentucky, 1840-1850)
10817   1840, Clarendon Springs, Vermont: v. 1, 370.
        Signed by "D."
10818   1840, Harrodsburg Springs [in the Mississippi River Valley]: v. 1, 379 (8 lines).
        Signed by "D."
10819   1840, Parquet Springs [at Bullitt County, Kentucky]: v. 1, 395-397.
        Signed by "Y."
10820   1841, Geological voyage: v. 3, 399-400.
        Signed by "D." D. D. Owen's Ohio valley trip noted.
10821   1843, Tertiary formations [of the United States]: v. 7, 238 (7 lines).
10822   1843, Geological survey [of Alabama]: v. 7, 474.
10823   1845, [Review with excerpts of Vestiges of the natural history of Creation, by R. Chambers (see 2292 et seq.)]: v. 4 (ns), 233-244, 308-319.
10824   1845, Fossil remains: v. 4 (ns), 256-257.
        Notice of fossil birds from New Zealand, and A. Koch's collection of fossil bones from Alabama.
10825   1846, Vestiges of Creation [notice of the third American edition of text by R. Chambers (see 2390)]: v. 6 (ns), 182.
10826   1846, Analysis of Boston waters: v. 6 (ns), 362-363.
10827   1846, [Notice of Voyage of a naturalist by C. Darwin (see 2916)]: v. 6 (ns), 457.
10828   1847, [Review of The Virginia Springs by J. J. Moormann (see 7422)]: v. 8 (ns), 60-62.
10829   1847, Temperature of the Kenawha salt wells [in Virginia]: v. 8 (ns), 92 (9 lines).
10830   1847, Geological survey: v. 8 (ns), 180-181.
        Wisconsin and Minnesota surveys by D. D. Owen noted.
10831   1848, [Review with excerpts of Ancient world by D. T. Ansted (see 1280)]: v. 2 (3s), 119-131.
10832   1849, Fossil elephant [from Vermont]: v. 4 (3s), 363 (6 lines).
10833   1849, New work on natural history [notice of crinoid study by G. Troost (see 10409)]: v. 4 (3s), 365.

The Western Journal of the Medical and Physical Sciences (Cincinnati, 1827-1838)
10834   1829, Bath chalybeate spring [at Yellow Springs, Ohio and Oympian Springs, Kentucky]: v. 3, no. 1, 159-160.
10835   1831, Vanadium: v. 5, no. 3, 503 (8 lines).
        Abstracted from Annales de Chimie.
10836   1832, Analysis of the sulphur springs at Nashville [Tennessee]: v. 5, no. 4, 648.
        Abstracted from Transylvania J. Medicine.
10837   1833, [Review of On baths and mineral waters by J. Bell (see 1622)]: v. 6, no. 3, 409-422.
10838   1833, Geological survey of the state of Tennessee: v. 7, 319.
        Notice of appointment of G. Troost as state geologist.
10839   1835, Greenville epsom springs, Harrodsburg, Ky.: v. 5, 159.
10840   1835, Blue Sulphur Spring [in Virginia]: v. 5, 159-160.
10841   1835, Analysis of minerals and mineral waters: v. 5, 371 (8 lines).
        Notice of an offer to analyze specimens by Profs. W. B. Rogers and J. L. Riddell.
10842   1835, Geological surveys: v. 5, 376.
        Notice of United States progress.

The Western Literary Journal and Monthly Review (Cincinnati, 1836)
10843   1836, Geology of the West: v. 1, 140-141.
        On the need for a geological survey of Ohio.

The Western Literary Journal and Monthly Review (Cincinnati, 1844-1845)
10844   1844, [Fossils from Mississippi collected by B. L. C. Wailes]: v. 1, 251-252.
        Abstracted from the Macon Independent.

The Western Literary Magazine, and Journal of Education, Science, Arts, and Morals (Columbus, Ohio, 1849)
10845   1849, Volcanoes: v. 1, 348-352.

The Western Medical and Physical Journal, Original and Eclectic (Cincinnati, 1827-1828)
10846   1827, Meteoric stones [from Sumner County, Tennessee]: v. 1, 127 (14 lines).
10847   1827, Earthquakes [in the Mississippi River Valley]: v. 1, 247.

The Western Medical Gazette (Cincinnati, 1832-1835)
10848   1834, [Techniques of chemical] examination of ores, mineral waters, &c. &c.: v. 2, 142.
10849   1834, A well of alum water [in Columbus, Ohio]: v. 2, 192 (3 lines).
10850   1834, Copperas works in Kentucky [on the Green River]: v. 2, 192 (4 lines).

The Western Messenger; Devoted to Religion and Literature (Louisville, Kentucky, 1835-1841)
10851   1836, [Review of Principles of geology by C. Lyell (see 6532)]: v. 2, 1-9.
        Signed by "E. P."

The Western Miscellany (Dayton, Ohio, 1848-1849)
10852   1848, Cost of gems and diadems: v. 1, 99-100.
10853   1848, An exciting scene. An authentic narrative, by an eye witness: v. 1, 148-152.
        Account of a mine disaster at Bois-Monzil coal mine, France.
10854   1849, Salt [its natural occurrence and uses]: v. 1, 244.

The Western Monthly Magazine, a Continuation of the Illinois Monthly Magazine (Cincinnati, 1833)
10855   1833, Terrestrial magnetism [review of A new theory of terrestrial magnetism by S. L. Metcalf (see 7138)]: v. 1, pt. 2, 524-533.
        Signed by "J. E."
10856   1834, Gold mines and gold laws: v. 2, 617-625.
        Signed by "E. D. M."

Western Museum Society
10857   1818, An address to the people of the western country: Am. J. Sci., v. 1, no. 2, 203-207.
        Natural history specimens requested for museums.

The Western Quarterly Reporter of Medical, Surgical, and Natural Science (Cincinnati, 1822-1823)
10858   1822, [Gypsum from Ohio]: v. 1, no. 1, 101 (5 lines).
        Abstracted from London Med. Intell.

The Western Quarterly Review (Cincinnati, 1849)
10859   1849, Fossil footprints [from Greenfield, Massachusetts]: v. 1, 173-174.
10860   1849, Coal: v. 1, 184.
10861   1849, Copper [from Lake Superior]: v. 1, 186 (7 lines).
10862   1849, The age of the material universe: v. 1, 224-242.
10863   1849, Review of cosmogonies: v. 1, 257-275.
10864   1849, Ohio, her prospects and resources: v. 1, 335-348.
10865   1849, Magnetic stone: v. 1, 375-376.
        On the physical properties of magnetite.
10866   1849, Interesting geological fact: v. 1, 376-377 (10 lines).
        Notice of the absence of fossil flowers.

The Western Register, a Monthly Journal of Commerce, Navigation, Science and the Arts (St. Louis, Missouri, 1849)
10867   1849, Glance at Upper California - illustrated by a plat [sic] of the gold region: v. 1, no. 1, 26, folding map.
10868   1849, California gold: v. 1, no. 1, 32.
10869   1849, California - the salt lakes - immense beds of pure salt: v. 1, no. 2, 60-61.
10870   1849, Gold mines of Michigan: v. 1, no. 3, 85.
10871   1849, Gold in northern Texas: v. 1, no. 4, 123.
10872   1849, Texas silver [from Austin]: v. 1, no. 4, 124 (6 lines).

Western Review and Miscellaneous Magazine (Lexington, Kentucky, 1819-1821)
10873   1819, Geology of the western country: v. 1, 33-36.
        Signed by "B." On the geology of Kentucky.
10874   1820, Obituary [of John D. Clifford]: v. 2, 309-310.
        Signed by "E."
10875   1820, A letter touching Russell's Cave [in Lexington, Kentucky]: v. 3, 160-164.
10876   1820, [Review with excerpts of A short tour between Hartford and Quebec by B. Silliman (see 9631)]: v. 3, 193-234.

Wetherell
10877   1841, London clay: Am. Eclec., v. 1, no. 2, 388-390.
        Abstracted from Lit. Gaz. On the fossils of the London Clay in England.

Wetherill, J. P.
10878   1826, Geological remarks on the Perkioming Lead Mine: Phila.
        Not seen. This work is noted in Am. Philos. Soc. Trans., v. 3, 508.

Wetmore, Alphonso
10879   1837, Catalogue of minerals: in Gazetteer of the state of Missouri, St. Louis, C. Keemle, 382 p., illus., map, pp. 261-265.

Whalen, Seth
10880   1846, Saratoga County, N. Y.: The Cultivator, v. 3 (ns), 78.

Wheeler, Benjamin
10881   1837, Discovery of marl: Genesee Farm. and Gardener's J., v. 7, 317.

Wheelock, T. B.
10882   1834, [Journal of the campaign of the regiment of dragoons for the summer of 1834]: US SED 1, 23-2, v. 1 (266), 73-93.
        Also in US SED 209, 24-1, (281), not seen.

Whelpley, James Davenport (d. 1872)
10883   1845, [Classification of drift phenomena]: Assoc. Am. Geol's Nat's Proc., v. 6, 14-16.
        Also in Am. Q. J. Ag. Sci., v. 2, 144.
10884   1845, [On the relations of the trap and sandstones of the Connecticut Valley]: Assoc.
        Am. Geol's Nat's Proc., v. 6, 61-64.
        Also in Am. Q. J. Ag. Sci., v. 2, 163.

Whewell, William (1794-1866)
10885   1832, Whewell's written nomenclature for chemical compounds: Am. J. Sci., v. 21, no. 2, 369-370.
        Abstracted from Royal Inst. J.
10886   1836, Relative level of land and sea: Franklin Inst. J., v. 18 (ns), 417-419.
        Abstracted from British Assoc. Proc. On long term variation in sea level.
10887   1838, Presentation of the Wollaston Medal [to Richard Owen]: Am. J. Sci., v. 35, no. 1, 197.
        Abstracted from Edinburgh New Philos. J.
10888   1839, Extracts from the anniversary address of the Rev. Wm. Whewell, before the
        Geological Society of London: Am. J. Sci., v. 37, no. 2, 218-240.
        Review of progress in geology, especially in Paleozoic stratigraphy.
10889   1839, Notice of Prof. Ehrenberg's discoveries in relation to fossil animalcules: Am. J.
        Sci., v. 37, no. 1, 116-130.
        Includes list of officers of the Geol. Soc. London, notice of award of the
        Wollaston Medal to Ehrenberg, and obituaries of six geologists.
10890   1841, British Association for the Advancement of Science: Am. J. Sci., v. 41, no. 2, 391-399.
        Whewell's presidential address on the Society's history and progress.
10891   1845, Indications of the Creator: Phila.
        Not seen.
10892   1848, On the wave of translation in connexion with the Northern drift: Am. J. Sci., v. 6 (ns), no. 1, 115-119, fig.
        Abstracted from Q. J. Geol. Soc.

Whipple, S. H.
10893   1844, [Notice of Mastodon bones from the county of Benton, Missouri]: Am. Philos. Soc. Proc., v. 4, 35-36.

Whisler, Samuel
10894   1830, and J. Smith, Improvement in pounding, grinding, and separating gold from the earth, stone, or rock: Franklin Inst. J., v. 5 (ns), 220.

Whistler, George William (1822-1869)
        See also 6648 with W. G. McNeill and W. H. Swift.
10895   1849, Report on the use of anthracite coal in locomotive engines on the Reading Rail Road: Balt., J. D. Toy, 36 p.
10896   1850, Anthracite coal in locomotive engines: Ann. Scien. Disc., v. 1, 39-40.

White, George (1802-1887)
10897   1849, Statistics of the state of Georgia, including a record of its natural, civil, and
        ecclesiastical history; together with a particular description of each county ... :
        Savannah, Georgia, W. T. Williams, 624, 77 p., illus., plates, map.
        Contains map by W. G. Bonner (see 1752).

White, Joseph
10898   1850, Medical topography of the County of Montgomery: Med. Soc. NY Trans., v. 8, 155-160.

White, Samuel P. (see 10542 with P. Van Beuren and R. G. Frary)

White, Thomas
10899   1794, A short account of an excursion through the subterraneous cavern at Paris: Mass. Mag., v. 6, 325-327.
        Also in Univ. Mag., v. 3, 48-52 (1797).

Whiting, Henry (1788-1851)
10900   1838, Cursory remarks upon East Florida, in 1838: Am. J. Sci., v. 35, no. 1, 47-64.

Whiting, William Henry Chase (1824-1865)
10901   1849, Report ... of the exploration of a new route from San Antonio de Bexar to El
        Paso: US SED 1, 31-1, v. 3, pt. 3 (551), 281-293. Also in US HED 5, 31-1, v. 5, pt. 3 (571), 281-293.
10902   1850, Report ... on the reconnaissance of the western frontier of Texas: US SED 64, 31-1, v. 14 (562), 236-250.

Whitney, Asa (1797-1872)
10903   1838, Improvement in the method, or art, of carbonating and smelting iron ore: Franklin Inst. J., v. 21 (ns), 263-264.

Whitney, Josiah Dwight (1819-1896)
        See also 3928 to 3930, 3932, and 3936 to 3941 with J. W. Foster. Whitney was also
        translator of the American edition of Berzelius' treatise on the blowpipe (1668).
10904   1841, and M. B. Williams, Geology and topography of the northern corner of the state of
        New Hampshire: in First annual report on the geology of New Hampshire by C. T. Jackson (see 5576), 83-93.
        Also in Final report by C. T. Jackson (see 5611), 67-73, 1844.
10905   1841, and M. B. Williams, Report on the section from Portsmouth to Claremont through
        Concord: in First annual report on the geology of New Hampshire by C. T. Jackson (see 5576), 45-51.
        Also in Final Report by C. T. Jackson (see 5611), 49-52.
10906   1844, and M. B. Williams, Section from Concord to Wakefield: in Final report on the
        geology of the state of New Hampshire by C. T. Jackson (see 5611), 70-71.

Whitney, J. D. (cont.)

10907 1844, and M. B. Williams, Section from Wakefield to Haverill, and examination of the country adjacent to Lake Winnipissiogee: in Final report on the geology of the state of New Hampshire by C. T. Jackson (see 5611), 71-73.

10908 1847, Description and analysis of three minerals from Lake Superior: Bos. J. Nat. History, v. 5, no. 4, 486-489.
Abstracted in Am. J. Sci., v. 6 (ns), no. 2, 269-270 (1848).

10909 1847, [Report of work in the Upper Peninsula of Michigan]: in Michigan report of C. T. Jackson (see 5640), 223-230.

10910 1848, Chemical examination of some American minerals [from Lake Superior]: Bos. J. Nat. History, v. 6, no. 1, 36-42.
Abstracted in Am. J. Sci., v. 7 (ns), no. 3, 434-435 (1849).

10911 1848, [Chemical examination of some American minerals]: Bos. Soc. Nat. History Proc., v. 3, 78-79.

10912 1848, [Chlorastrolite from Isle Royale, Lake Superior]: Bos. Soc. Nat. History Proc., v. 3, 12 (4 lines).

10913 1848, Examination of three new mineralogical species proposed by Prof. C. U. Shepard: Bos. J. Nat. History, v. 6, no. 1, 42-48.
See also article by Shepard (9553). On arkansite, ozarkite and schorlomite.
Abstracted in Am. J. Sci., v. 7 (ns), no. 3, 433-434 (1849).

10914 1848, [New reflecting goniometer]: Bos. Soc. Nat. History Proc., v. 3, 78 (16 lines).

10915 1848, [On jacksonite, a new mineral from the Lake Superior region]: Bos. Soc. Nat. History Proc., v. 3, 5-6.

10916 1848, [Report of field work in the Lake Superior land district]: US SED 2, 30-2, v. 2 (530), 154-159.

10917 1849, [Analyses and descriptions of copper minerals from Lake Superior]: in Lake Superior report by C. T. Jackson (see 5655), 491-493.

10918 1849, Field notes for 1847 [on the Lake Superior region]: in Lake Superior report by C. T. Jackson (see 5655), 713-758.

10919 1849, Notes on the topography, soil, geology, etc., of the district between Portage Lake and the Ontonagon: in Lake Superior report by C. T. Jackson (see 5655), 649-701.

10920 1849, [On the composition of chloritoid or chlorite spar, and masonite; and on oxide of copper from Copper Harbor, Lake Superior]: Bos. Soc. Nat. History Proc., v. 3, 100-103.
Abstracted in Am. J. Sci., v. 8 (ns), no. 1, 272-274. Also abstracted in Ann. Scien. Disc., v. 1, 261-262 (1850).

10921 1849, On the constitution of some silicates containing carbonic acid, chlorine and sulphuric acid: Am. J. Sci., v. 7 (ns), no. 2, 435-436.
Abstracted from Poggendorff's Annalen.

10922 1849, [On three minerals from Arkansas]: Bos. Soc. Nat. History Proc., v. 3, 96.

10923 1849, [On the origin of California gold deposits]: Bos. Soc. Nat. History Proc., v. 3, 95 (6 lines).

10924 1850, [On fractured strata at Guilford, Vermont]: Bos. Soc. Nat. History Proc., v. 3, 226.

10925 1850, On the deposits of iron near Lake Superior: Ann. Scien. Disc., v. 1, 263 (14 lines).

10926 1850, [On the mineral lands of the Lake Superior regions]: Bos. Soc. Nat. History Proc., v. 3, 210-212.

Whitten, J. S.

10927 1841, Green-sand in Georgia: Farm's Reg., v. 9, 86-87.

Whittlesey, Charles (1808-1886)
See also discussion of article by E. Desor (3122).

10928 1838, Bituminous coal. Suggestions in opposition to the theory of the vegetable origin of mineral coal: Hesperian, v. 1, 111-114.

10929 1838 [i.e. 1839], Mr. Whittlesey's report [on Ohio]: in second annual Ohio report by W. W. Mather (see 6896), 41-71, 2 plates.

10930 1839, Analysis of soils: Hesperian, v. 3, 287-292.

10931 1839, Origin of bituminous coal: Hesperian, v. 3, 199-204.

10932 1843, Missouri and its resources: Hunt's Merch. Mag., v. 8, 535-544.

10933 1843, A statement of elevations in Ohio, with reference to the geological formations, and also the heights of various points in this state and elsewhere: Am. J. Sci., v. 45, no. 1, 12-18.

10934 1844, Analysis of soils: Am. Farm., v. 6 (3s), 151-152.
Abstracted from Plough Boy.

10935 1846, A dissertation upon the origin of mineral coal: Cleveland, M. C. Younglove (printer), 24 p.

10936 1846, Copper regions of Lake Superior: n.p., 64 p.
Not seen.

10937 1846, Two months in the copper region [of Lake Superior]: Fisher's Nat'l Mag., v. 2, no. 9, 814-846.

10938 1847, Coal and iron trade of the Ohio Valley: Hunt's Merch. Mag., v. 16, 450-455.

10939 1848, Notes upon the drift and alluvium of Ohio and the West: Am. J. Sci., v. 5 (ns), no. 2, 205-217, fig.

10940 1848, Outline sketch of the geology of Ohio: in Historical collections of Ohio ... , by H. Howe, Cin., 577-589, illus.

10941 1849, Description of a coal plant supposed to be new [from Tallmadge, Summit County, Ohio]: Am. J. Sci., v. 8 (ns), 375-377, fig. Also in Fam. Vis., v. 1, no. 2, 9, fig. (1850).

10942 1850, [Extract from reports on the Piscataqua Mining Company holdings]: in Piscataqua Mining Company Charter (see 3937), 17-19.
In 8380.

10943 1850, On the natural terraces and ridges of the country bordering Lake Erie: Am. J. Sci., v. 10 (ns), no. 1, 31-39.

Wickersham, C. P.
10944   1846, [Fossil tracks in the red sandstone of the Connecticut Valley] : Phila. Acad. Nat. Sci's Proc., v. 3, no. 6, 120-121.

Widder (Mr.)
10945   1848, Observations on Mr. Phillips' process of separating silver from ores: Am. Rr. J., v. 21, 41-42.
        See also article by R. Phillips (8318).

Wigglesworth, Samuel (1688-1768)
10946   1729, A religious fear of God's tokens explained and urged; in a sermon preached at Ipswich, November 1. 1727. Being a day of humiliation on account of the terrible earthquake, October 29, 1727 ... : Bos., Printed for C. Henchman and T. Hancock, [4], iii, 42 p.
        Evans number 3121.

Wilcox, Carlos (1794-1827)
10947   1826, Account of the late slides from the White Mountains: Bowen's Bos. News-letter, v. 2, 122-125.
10948   1829, Letter of the Rev. Carlos Wilcox: Am. J. Sci., v. 15, no. 2, 222-228.
        In article by B. Silliman (see 9684).

Wilde, James
10949   1834, [Report on the mines of the Delaware Coal Company] : in Charter and by-laws of the Delaware Coal Company (see 3066).

Wilkes, Charles (1798-1877)
10950   1841, Exploring expedition: Am. J. Sci., v. 40, no. 2, 394-399.
        Progress report.
10951   1843, United States exploring expedition: Am. J. Sci., v. 44, no. 2, 393-408.
        Includes note of magnetic observations.
10952   1844, On the formation of the Antarctic ice: Am. J. Sci., v. 47, no. 1, 114 (10 lines).
        Also in Assoc. Am. Geol's Nat's Proc., v. 5, 21 (10 lines).
10953   1844, Narrative of the United States exploring expedition during the years 1838, 1839, 1840, 1841, 1842 ... in five volumes, and an atlas: Phila., C. Sherman (printer), 5 v., plates, maps.
        There were nine reprint editions of this work by 1850. For references to these editions see Haskell (1942).
10954   1845, Narrative of the United States exploring expedition during the years 1838, 1839, 1840, 1841, 1842 ... : Am. J. Sci., v.49, no. 1, 149-166.
        Abstracted from the report (see 10953).
10955   1847, [Observations on the transport of drift by icebergs] : Am. Q. J. Ag. Sci., v. 6, 215.
        In Assoc. Am. Geol's Nat's Proc.
10956   1847, On the depth and saltness of the ocean: Am. Q. J. Ag. Sci., v. 6, 210-211.
        In Assoc. Am. Geol's Nat's Proc.
10957   1848, The depth and saltness of the ocean: Am. Q. Reg. Mag., v. 1, 215-217.
10958   1848, [Nature of icebergs in drift transfer]: Lit. World, v. 2, 229 (5 lines).
        Abstracted from Assoc. Am. Geol's Nat's Proc.
10959   1848, On the depth and saltness of the ocean: Am. J. Sci., v. 5 (ns), no. 1, 41-48.
10960   1849, Western America, including California and Oregon, with maps of those regions, and of "the Sacramento Valley": Phila., Lea and Blanchard, x, 13-130 p., colored map.

Wilkins, Edmund
10961   1836, Some of the geological features of the region surrounding Gaston, N. C., on the Roanoke: Farm's Reg., v. 4, 30-31.

Willard, Emma (Hart) (1787-1870)
10962   1829, Ancient geography, as connected with chronology ... : Hartford, Oliver D. Cooke and Co., W. E. Dean (printer), viii, 9-90 p.
        Third edition, improved. First and second editions not seen. Published with Universal geography by W. C. Woodbridge (see 11037).
        For review see 415 and 9981.

Willard, Joseph (1738-1804)
10963   1785, Observations made at Beverly [Massachusetts], Lat. 42°36'N. Long. 70°45'W. to determine the variation of the magnetical needle: Am. Acad. Arts Sci's Mem., v. 1, 318-321.

Williams, Eliphalet (1727-1803)
10964   1756, The duty of a people, under dark providences, or symptoms of approaching evils, to prepare to meet their God. A discourse delivered in East Hartford, November 23, 1755. The next Sabbath after the late terrible earthquake ... : New-London, Conn., Printed and sold by Timothy and John Green, 68 p.
        Evans number 7819.

Williams, John (1730-1795)
10965   1812, Of the indications of coal, and methods of searching for it: Am. Min. J., v. 1, no. 3, 166-182.
        Extracted from Natural history of the mineral kingdom.
10966   1814, [Extracts from The mineral kingdom]: Am. Wk. Mess., v. 2, no. 1, 9.
10967   1814, [Extracts from The mineral kingdom] : in North-American Coal and Mining Company Reports (see 7921), 13-14.

Williams, John Lee
10968   1827, A view of West Florida, embracing its geography, topography, etc. ... : Phila., H. S. Tanner, 178 p., folding map.
10969   1837, The territory of Florida: or, sketches of the topography and of the civil and natural history ... of Florida: NY, A. T. Goodrich, 304 p., plate, map, front.

Williams, Jonathan (1750-1815)
10970   1799, Barometrical measurement of the Blue-Ridge, Warm-Spring, and Alleghany Mountains, in Virginia, taken in the summer of the year 1791: Am. Philos. Soc. Trans., v. 4, 216-223, plate.
10971   1811, On the height of the mountains in Virginia and New York, with observations on the formation of rivers: Am. Med. Philos. Reg., v. 1, no. 3, 336-346.
10972   1811, Observations on the falls of the Ohio: Am. Med. Philos. Reg., v. 1, no. 3, 330-336.

Williams, Moses B. (see 10904 to 10907)

Williams, Samuel (1743-1817)
10973   1785, A memoir on the latitude of the University of Cambridge: with observations of the variation and dip of the magnetic needle: Am. Acad. Arts Sci's Mem., v. 1, 62-69.
        On the magnetic field at Cambridge, Massachusetts. Also in Am. Mus. or Univ. Mag., v. 3, 292-306, 567-579.
10974   1785, Observations and conjectures on the earthquakes of New-England: Am. Acad. Arts Sci's Trans., v. 1, 260-311.
        Abstracted in NY Mag., v. 3, 135-136; 208-210 (1792).
10975   1793, Magnetic observations: Am. Philos. Soc. Trans., v. 3, 115 (table only).
        On magnetic observations at Cambridge, Massachusetts.
10976   1794, The natural and civil history of Vermont: Walpole, NH., Isaiah Thomas and David Carlisle, 416 p., folded map.
        Evans number 28094. Also second edition (Burlington, Vt., S. Mills, 1809, 2 v., folded map.
10977   1795, Description of a curious subterranean cave, at Clarendon in Vermont: Mass. Mag., v. 7, 414-415.
        Extracted from his History of Vermont (see 10976).

Williams, Stephen C.
10978   1825, New locality of rubellite, beryl, tourmaline, &c. [at Middletown, Connecticut]: Am. J. Sci., v. 10, no. 1, 206-208.
        With "Remarks by the Editor" by B. Silliman (see 9665).

Williams, William (1688-1760)
10979   1729, Divine warnings to be received with faith & fear, and improved to excite to all proper methods for our own safety and our families. Shew'd in a discourse on Heb. XI. 7, on the publick fast, December 21, 1727, on occasion of the terrible earthquake October 29, 30, & frequently since repeated. To which is added, A discourse on Prov. 2, 1-6: Bos., Printed for Samuel Gerrish, [2], xii, 72, 132 p.
        Evans number 3125.

Williamson [William Crawford?]
10980   1838, Fossils with coal: Am. J. Sci., v. 34, no. 1, 26.
        Abstracted from British Assoc. Arts Sci's Proc.

Williamson, Hugh (1735-1819)
10981   1812, Of the soil and general state of health, in different parts of North Carolina: Eclec. Rep. and Analytic Rev., v. 3, 1-16.

Williamson, Robert Stockton (1824-1882)
10982   1850, [Report on limestone collected at Mount Diablo, California]: US SED 47, 31-1, v. 10, pt. 2 (558), 34-37.

Williamson, William Crawford (1816-1895)
10983   1837, [Abstract of "On the distribution of organic remains in the oolitic formation on the coast of Yorkshire"]: Mus. For. Liter. Sci., v. 31, 48.
        Abstracted from Geol. Soc. London Proc.

Williamson, William Durkee (1779-1846)
10984   1832, History of the state of Maine from its first discovery A. D. 1602, to the separation, A. D. 1820, inclusive: Hallowell, Me., 2 v.
        Also an 1839 printing: Hallowell, Me., Glazier, Masters and Smith, 2 v. Contains section on mineralogy, p. 174-182.

Willis, Oscar
10985   1832, Machine for washing gold: Franklin Inst. J., v. 9 (ns), 394.

Wilson, George (1818-1859)
10986   1846, On the solubility of fluoride of calcium in water, and its relation to the occurrence of fluorine in minerals, and in recent fossil plants and animals: Am. J. Sci., v. 2 (ns), no. 1, 114-115.
        Abstracted from "Chem. Gazette".
10987   1850, On the presence of fluorine in the waters of the firths of Fourth and Clyde, and the German Ocean: Am. J. Sci., v. 9 (ns), no. 1, 118-120. Also noted in Ann. Scien. Disc., v. 1, 201.
        Abstracted from London Ath.

Wilson, J. G. (Reverend)
10988   1850, The harmony of science and revelation: Bib. Repos., v. 6 (3s), 704-719.
10989   1850, Consistency of geology and scripture: Fam. Vis., v. 1, 121.
        Abstracted from Bib. Repos., and ascribed to "White" (i.e. J. G. Wilson).

Wilson, J. W.
10990   1821, Bursting of lakes through mountains: Am. J. Sci., v. 3, no. 2, 252-253.

Wilson, James (see 5711 by R. Jameson)

Windsor County Natural History Association (Vermont)
10991   1848, Windsor County Natural History Association: Sch. J. and Vt. Agriculturist, v. 2, 25.
        Includes notice of the progress of geology.
10992   1848, Windsor County Natural History Association: Sch. J. and Vt. Agriculturist, v. 2, 46.
        Notice of geology and mineralogy sessions in June, 1848.

Wines, Enoch Cobb (1806-1879)
10993   1841, Ascent of Vesuvius: Young People's Book, v. 1, 84-85.

Winifrede Mining and Manufacturing Company
10994   1850, An act to incorporate the Winifrede Mining and Manufacturing Company: n.p., 20 p.
        Mines situated in Kanawha and Boone Counties, Virginia.

Winkler, C. A.
10995   1843, On the ores used at Frieberg: Franklin Inst. J., v. 6 (3s), 109-110.
        Translated by T. F. Moss.

Winslow, Miron (1789-1864)
10996   1823, Mineralogy of the Island of Ceylon: Am. J. Sci., v. 6, no. 1, 192-195.
10997   1824, Miscellaneous notices from the Island of Ceylon: Am. J. Sci., v. 8, no. 1, 186-187.

Winterbotham, William (1763-1829)
10998   1795, An historical, geographical, commercial, and philosophical view of the American United States, and of the European settlements in America and the West Indies: Newark, New Jersey, D. Holt, 2 v.
        Evans number 29912. Also 1796 edition: NY, Tiebout and O'Brien for J. Reid, 4 v., plates, maps.
10999   1796, An historical, geographical and philosophical view of the Chinese empire, comprehending a description of the fifteen provinces of China, Chinese Tartary, tributary states; natural history of China; government, religion, laws, manners and customs, literature, arts, sciences, manufactures, etc.: London printed: Phila., Re-printed for Richard Lee, Dunning, Hyer and Palmer, 2 v.
        Evans number 31648.

Winthrop, James (1752-1821)
11000   1804, Account of an inscribed rock, at Dighton, in the Commonwealth of Massachusetts, accompanied with a copy of the inscription: Am. Acad. Arts Sci's Mem., v. 2, pt. 2, 126-129, plate.

Winthrop, John (d. 1747)
11001   1844, Selections from an ancient catalogue of objects of natural history, formed in New England more than one hundred years ago: Am. J. Sci., v. 47, no. 2, 282-290.
        Notice of New England fossil and mineral specimens collected in 1735.

Winthrop, John (1714-1779)
11002   1755, A lecture on earthquakes; read in the chapel of Harvard-College, in Cambridge, N. E. November 26th 1755. On occasion of the great earthquake which shook New-England the week before ... : Bos., Edes and Gill, 38 p.
        Evans number 7597. For review see 6270. For notice see 9139.
11003   1756, A letter to the publisher of the Boston Gazette, containing an answer to Mr. Prince's letter [upon earthquakes] inserted in said Gazette of January 26, 1756: Bos., Edes and Gill, 7 p.
        Evans number 7820.
11004   1757, On the causes of earthquakes: Am. Mag. and Mon. Chron., v. 1, no. 3, 111-116.
        Extracted from Lecture on earthquakes (see 11002).
11005   1789, Theory of earthquakes: Am. Mus. or Univ. Mag., v. 6, 64-67.
        Extracted from Lecture on earthquakes (see 11002).

Wirt, William
11006   1843, Red Sulphur Springs [Virginia]: S. Lit. Mess., v. 10, 422-424.

Wiseman, Nicholas Patrick Stephen, Cardinal (1802-1865)
11007   1837, Twelve lectures on the connexion between science and revealed religion: Andover, Mass. and NY, Gould and Newman, 404 p., front., folded map.
        For review see 676 and 3408.

Wislizenus, Frederick Adolphus (1810-1889)
11008   1848, Geological sketch [of northern Mexico]: in Memoir of a tour ..., (see 11009), geological map, 27 x 30 cm.
11009   1848, Memoir of a tour to northern Mexico, connected with Doniphan's expedition in 1846 and 1847: US SMD 26, 30-1, v. 1 (511), 141 p., plate, 2 maps.
        For review with excerpts see 949. Contains map (see 11008).

Wistar, Caspar (1761-1818)
11010   1799, A description of the bones deposited, by the President, in the Museum of the Society, and represented in the annexed plates: Am. Philos. Soc. Trans., v. 4, 526-531, 2 plates.
11011   1818, An account of two heads found in the morass, called the Big Bone Lick, and presented to the Society, by Mr. Jefferson: Am. Philos. Soc. Trans., v. 1 (ns), 375-380, 2 plates.

Wister, Charles I.
11012   1810, Description of melanite from Pennsylvania, and amber from New Jersey: Am. Min. J., v. 1, no. 1, 31.

Witham, Henry Thornton Maire (1779-1844)
11013   1830, On the vegetation of the first period of an ancient world: Am. J. Sci., v. 18, no. 1, 110-117, 3 figs.
11014   1832, [Thin sections of fossil plants]: Mon. Am. J. Geology, v. 1, no. 10, 471-472 (7 lines).
         In British Assoc. Proc. See also note by B. Silliman (9721).
11015   1833, [On fossil vegetation]: Am. J. Sci., v. 25, no. 1, 108-112.
         With discussion by B. Silliman (9714).

Withers, Robert W. (of Erie, Greene County, Alabama)
11016   1833, Geological notices respecting a part of Greene County, Alabama: Am. J. Sci., v. 24, no. 1, 187-189.
11017   1835, Description of certain remarkable prairie and woodland soils of Alabama: Farm's Reg., v. 3, 498.
11018   1835, Some account of the calcareous region of Alabama: Farm's Reg., v. 2, 637.

Witherspoon, J. D.
11019   1847, The use of marl in South Carolina: Mon. J. Ag., v. 3, no. 4, 156.

Wittstein, George Christoph (1810-1887)
11020   1846, On the quantitative determination of soda, and its separation from potash: Am. J. Sci., v. 1 (ns), no. 1, 110-111.
         Abstracted from Chemist, by J. L. Smith.

Wohler, Friedrich (1800-1882)
11021   1834, A method of obtaining iridium and osmium from the platinum residue: Am. J. Sci., v. 26, no. 2, 371.
         Translated by J. C. Booth.
11022   1835, Crystals of oxide of chrome: Am. J. Sci., v. 28, no. 2, 396 (5 lines).
         Abstracted from Jameson's J.
11023   1837, Artificial formation of crystallized pyrites: Franklin Inst. J., v. 19 (ns), 310 (8 lines).
11024   1839, On two new cobalt minerals, from Modum in Norway: Am. J. Sci., v. 36, no. 1, 332-335.
         With notes by C. U. Shepard (see 9525). Translated by W. A. Larned from Neues Jahrbuch fur Mineralogie.
11025   1846, Cryptolite [from Arendal, Norway]: Am. J. Sci., v. 2 (ns), no. 2, 268.
         Abstracted from Poggendorff's Annalen.
11026   1846, On the estimation of fluorine: Am. J. Sci., v. 1 (ns), no. 2, 261.
         Abstracted by J. L. Smith from "Chem. Gaz."

Wollaston, William Hyde (1766-1828)
11027   1807, [Notice of a portable blowpipe]: Am. Reg. or Gen. Repos., v. 1, pt. 2, 52.
11028   1810, On the identity of columbium and tantalum: Am. Min. J., v. 1, no. 1, 57-62.
         See also note by J. J. Berzelius (1647). For notice of this article see Am. Min. J., v. 1, no. 1, 53 (3 lines) and no. 2, 126 (10 lines), 1811.
11029   1817, [Native iron from Brazil]: J. Sci. Arts, v. 2, no. 1, 205 (6 lines).
11030   1821, Potash in sea water: Am. J. Sci., v. 3, no. 2, 371 (3 lines).
11031   1823, Metallic titanium: Am. J. Sci., v. 7, no. 1, 192 (4 lines).
         Abstracted by J. Griscom. On crystals from slag.

Wood, J.
11032   1825, Remarks on the moving of rocks by ice: Am. J. Sci., v. 9, no. 1, 144-145.

Wood, John (1775-1822?)
11033   1809, A new theory of the diurnal rotation of the Earth: Richmond, n.p., 89 p.

Wood, W.
11034   1809, Diamonds: Select Reviews, v. 2, 61-66.

Wood, William
11035   1834/1835, Notes on geology: Cin. Mirror, v. 4, 6-7; 17; 25; 41-42; 50; 66; 81; 90; 105; 127; 153-154; 184.
         See also articles by W. B. Powell (8559 to 8561) and J. L. Riddell (8781) on the origin of coal. Also in Eclec. J. Sci., v. 2, 763-765; 770-774; 793-795; 803-806; 826-829.
11036   1835, [Letter on the origin of coal]: Cin. Mirror, v. 4, 106.

Wood, William W. W.
11037   1848, Manufacture of iron in Alabama: Hunt's Merch. Mag., v. 18, 656-657.

Woodbridge, William Channing (1794-1845)
11038   1824, A system of universal geography, on the principles of comparison and classification: Hart., Oliver D. Cooke and Sons, xxx, 336 p., illus.
         Usually bound with Ancient geography by E. Willard (see 10962). For review see 415 and 9981.
         Includes section on geology, with a woodcut version of W. Maclure's geological map of the United States (see 6619).
         Numerous other editions and printings, including 1827, second edition, Hart., and 1829, third edition, Hart.
11039   1838, Meteoric iron in France [from Auvergne]: Am. J. Sci., v. 33, no. 2, 257-258.
         In article by R. Cox (see 2675).

Woodhouse, James (1770-1809)
11040   1802, Remarks on a letter of the Rev. Zechariah Lewis, relating to a subterranean wall, discovered in North Carolina: Med. Repos., v. 5, 21-24.
         Debate on the origin of a dike in Salisbury, North Carolina. See also article by J. Hall (4663).
11041   1804, Additional observations on the subterranean minerals near the Yadkin, and on their basaltic nature: Med. Repos., v. 7, 26-27.
11042   1805, Experiments and observations on the Lehigh coal: Phila. Med. Mus., v. 1, 441-444.

Woodhouse, James (cont.)
11043   1805, [On manganese ore from Northampton County, Pennsylvania]: Phila. Med. Mus., v. 1, 449-451.
11044   1808, An account of the Perkiomen zinc mines, with an analysis of the ore: Balt. Med. Phys. Rec'r, v. 1, 154-157. Also in Phila. Med. Mus., v. 5, 133-135.
      See also article by A. Seybert (9434).
11045   1808, Account of the meteor which was seen at Weston, in the state of Connecticut, on the 14th of December, 1807; with an analysis of the stone: Phila. Med. Mus., v. 5, 131-133.
11046   1808, Analysis of an iron ore, containing titanium [from Franklin, New Jersey]: Phila. Med. Mus., v. 4, 206-207.
11047   1808, [Molybdena from Chester County, Pennsylvania]: Phila. Med. Mus., v. 4, cxi (5 lines).
11048   1809, Reply to Seybert's strictures on his essay concerning the Perkiomen Zinc Mine: Phila. Med. Mus., v. 6, 44-54.
      See also "An account ... " by Woodhouse (11044) and article by A. Seybert (9434).
11049   1835, Lehigh coal [from Northampton County, Pennsylvania]: Hazard's Reg. Penn., v. 16, 225-226.

Wooldridge, A. S.
11050   1842, Geological and statistical notice of the coal mines in the vicinity of Richmond, Va.: Am. J. Sci., v. 43, no. 1, 1-14.
      Contains "Remarks" by B. Silliman (see 9770) and coal analysis by G. W. Andrews (see 1247).

Woolford, Frederick
11051   1848, Clays and minerals of Missouri: West. J. Ag., v. 1, no. 4, 193-195.
11052   1848, Mines and mining [in Washington County, Missouri]: West. J. Ag., v. 1, no. 3, 168-169.

Woolworth, Samuel
11053   1847, Description of a tooth of the Elephas americanus: Am. Q. J. Ag. Sci., v. 6, 31-37, 2 figs.
      On a fossil tooth from Homer, Cortlandt County, New York.

The Worcester Magazine and Historical Journal, Containing Articles Original and Selected (Worcester, Massachusetts, 1825-1826)
      Volume 2 contains historical sketches of towns which include topographical notes.
11054   1825, Trenton Falls [in Utica, New York]: v. 1, 46-49.
      Signed by "B."
11055   1825, The coral islands: v. 1, 70-73.
      On the development and geological importance of coral islands.
11056   1825, The Brazils: v. 1, 99-101.
      Abstracted from Q. Rev. On gold and diamond mines.
11057   1826, Topographical view of Templeton [Massachusetts]: v. 1, 116-122.
      Signed by "B." Includes notes on minerals, erosion patterns, and caves.
11058   1826, History of Sterling [Massachusetts]: v. 2, 37-52.
      Includes notice of silver mines.

Worcester, Samuel Melanchthon (1801-1886)
11059   1849, Address before the Naumkeag Mutual Trading and Mining Company: n.p., 11 p.
      On the theological aspects of mining.

Workman, Benjamin
11060   1789, Elements of geography, designed for young students in that science. In six sections ... : Phila., John M'Culloch, c. 140 p., maps.
      Evans Number 22291. Not seen. Numerous later editions and printings.

Wörth, Fr. von
11061   1847, Chiolite, a new mineral from Miask: Am. J. Sci., v. 3 (ns), no. 2, 267.
      Abstracted from "Verhandl. Min. Ges. zu. St. Petersburg."

Worthington, Wilmer (1804-1873)
11062   1831, Chester County Cabinet of natural science [in Pennsylvania]: Hazard's Reg. Penn., v. 7, 244-246.
      Contains a list of officers for 1831.

Wraxall, Sir Nathaniel William (1751-1831)
11063   1805, Description of the salt mines of Poland: Evening Fireside, v. 1, no. 28, 223-224.
      Extracted from Wraxall's Memoirs.

Wray, Thomas I.
11064   1814, Minerals from Tennessee: Am. Min. J., v. 1, no. 4, 265-266.

Wright, A. H. (Reverend)
11065   1847, Notices of Koordistan. Hot sulphur spring - manna - mines of lead - sulphur and orpiment - rock salt and saline springs - ruins, &c. [from Koordistan, Turkey]: Am. J. Sci., v. 3 (ns), no. 2, 347-354.

Wright, Benjamin (1770-1842)
11066   1821, Lime for water cement [in New York]: Am. J. Sci., v. 3, no. 2, 230-231.
      With note by Myron Holley (see 5234).

Wright, W. H. (Lieutenant)
11067   1845, A brief practical treatise on mortars: with an account of the processes employed at the public works in Boston Harbor: Bos., W. D. Ticknor and Co., 148 p.
      For review see 820.

Wurtz, Henry (1828-1910)
11068 1850, On a new method of decomposing silicates in the process of analysis: Am. J. Sci., v. 10, no. 3, 323-326.
11069 1850, On a supposed new mineral species [melanolite, from Cambridge, Massachusetts]: Am. J. Sci., v. 10, no. 1, 80-83.
11070 1850, On the availability of the greensand of New Jersey as a source of potash and its compounds: Am. J. Sci., v. 10, no. 3, 326-329.

Wyatt, Thomas
11071 1839, A synopsis of Natural History: Phila., Thomas Wardle, 191 p.

Wyeth, John B.
11072 1833, and B. Waterhouse, Oregon; or, a short history of a long journey from the Atlantic Ocean to the region of the Pacific by land ... : Cambridge, Mass., J. B. Wyeth, 87 p.

Wyman, Jeffries (1814-1874)
11073 1840, Notice of the tooth of a mastodon [from Burmah]: Am. J. Sci., v. 39, no. 1, 53-54.
11074 1840, Proceedings of the Boston Society of Natural History: Am. J. Sci., v. 38, no. 1, 193-197; no. 2, 391-396; v. 39, no. 1, 182-189; no. 2, 373-380.
        Includes short geological notes.
11075 1845, [On Hydrarchos sillimani]: Bos. Soc. Nat. History Proc., v. 2, 65-68.
11076 1846, Anatomical description of the cranium [of the Castoroides ohioensis]: Bos. J. Nat. History, v. 5, no. 3, 391-401.
11077 1846, [On a cranium and lower jaw of an extinct rodent from Wayne County, New York]: Bos. Soc. Nat. History Proc., v. 2, 138-139.
11078 1847, [Lignite from Richmond, Virginia]: Bos. Soc. Nat. History Proc., v. 2, 213 (7 lines).
11079 1850, [Fossil bones of seal from Richmond, Virginia]: Bos. Soc. Nat. History Proc., v. 3, 241-242.
11080 1850, Notice of fossil bones from the neighborhood of Memphis, Tennessee: Am. J. Sci., v. 10 (ns), no. 1, 56-64, 5 figs.
11081 1850, Notice of remains of vertebrate animals found at Richmond, Virginia: Am. J. Sci., v. 10 (ns), no. 2, 228-235, 8 figs.
11082 1850, [On bones of Zeuglodon from Washington County, Alabama]: Bos. Soc. Nat. History Proc., v. 3, 328-329.
11083 1850, [On boulder accumulations on the coast of Labrador]: Bos. Soc. Nat. History Proc., v. 3, 182-183.
11084 1850, [On teeth of fishes from the Tertiary deposit of Richmond, Virginia]: Bos. Soc. Nat. History Proc., v. 3, 246-247.
        With discussion by E. Desor and H. D. Rogers.
11085 1850, [On the remains of seals and a coprolite from Richmond, Virginia]: Bos. Soc. Nat. History Proc., v. 3, 323.
11086 1850, [Subsidence of Newfoundland's shore]: Bos. Soc. Nat. History Proc., v. 3, 375.
11087 1850, [Vertebrate fossils from the Mississippi alluvium at Memphis, Tennessee]: Bos. Soc. Nat. History Proc., v. 3, 280-281.

Wyoming Coal Association
11088 1847, Articles of agreement and association of the Wyoming Coal Association: NY, J. M. Elliott (printer), 18 p.

Yale Literary Magazine (New Haven, 1836-1850)
11089 1839, Science and religion: v. 5, 1-8.
        Usefulness of geology noted.

Yale Natural History Society
11090 1838, Yale Natural History Society: Am. J. Sci., v. 34, no. 2, 397-398.
        Notice of activities and cabinet.

Yandell, Lunsford Pitts (1805-1878)
11091 1832, An account of some of the principal mineral springs of Kentucky: Transylvania J. Medicine and Associated Sci's, v. 5, 375-401.
11092 1833, A notice of Drenon's Lick [in Kentucky]: Transylvania J. Medicine and Associated Sci's, v. 6, 399-400.
11093 1845, Lectures on geology: West. J. Medicine Surg., v. 3 (ns), 361-362.
        Notice and review of lectures by B. Lawrence.
11094 1847, and B. F. Shumard, Contributions to the geology of Kentucky: Louisville, Kentucky, Prentice and Weisinger, 36 p., plate.
11095 1847, and B. F. Shumard, Contributions to the geology of Kentucky: West. J. Medicine Surg., v. 8 (ns), 310-342, plate with 9 figs., table.
11096 1847, A notice of Grayson Springs [Kentucky]: West. J. Medicine Surg., v. 7 (ns), 142-143.
        Abstracted in Med. Exam., v. 3 (ns), 205-206.
11097 1850, Death of Dr. Troost [obituary]: West. J. Medicine Surg., v. 6 (3s), 269-271.

The Yankee (Boston, 1828-1829)
11098 1828, [Account of Mammoth Cave]: v. 1, 5.
11099 1829, [Review of Introduction to geology by R. Bakewell (see 1455)]: v. 2, 97-99, 110.

Yates (Reverend)
11100 1838, Fossil plants [from Worcestershire, England]: Am. J. Sci., v. 33, no. 2, 270-271.
        In British Assoc. Proc.

Yates, William
11101 1841, Blossburg coal: The Cultivator, v. 8, 55.

Yorkshire Philosophical Society
11102   1831, Yorkshire Philosophical Society: Am. J. Sci., v. 21, no. 1, 168-171.
        Abstracted by J. Griscom from the Society's Ann. Rept. Includes description of the cabinet.

The Young American's Magazine of Self-Improvement (Boston, 1847)
11103   1847, American Association of Naturalists and Geologists: v. 1, 355.
        Notice of 8th meeting, in Boston.

The Young Ladies' Journal of Literature and Science (Baltimore, 1830-1831)
11104   1830, [Lead and tin ores and their processing]: v. 1, 267-270.
        Signed by "O."
11105   1831, Crystallization: v. 1, 351-352.
        Signed by "O."
11106   1831, [On the salt mines of Cracow in Poland]: v. 1, 399-400.

The Young Mechanic (Boston, 1832-1834)
11107   1833, The burning springs [of Florida County, Kentucky]: v. 2, 66-67.
        Abstracted from Sunday School J.
11108   1833, Anthracite coal for steam: v. 2, 99 (6 lines).

The Young People's Book; or, Magazine of Useful and Entertaining Knowledge (Philadelphia, 1841-1842)
11109   1842, Practical geology and mineralogy, with instructions for the qualitative analysis of minerals. By Joshua Trimmer [review of 10330]: v. 2, 136.

Young, Arthur (1741-1820)
11110   1822, An essay on manures: Am. Farm., v. 3, 385-389.
        Extracted from Bath and West of England Soc. Papers.

Young, J. H. (Reverend)
11111   1839, A visit to Weyer's Cave, Virginia: Meth. Mag., v. 21, 349-356.

Young, J. P. (Dr., of Edenville, New York)
11112   1831, Oolite in situ, in Edenville, Orange Co., N. Y.: Am. J. Sci., v. 19, no. 2, 398 (10 lines).
11113   1832, and J. Heron, Geological and mineralogical map of a part of Orange Co. N. York. By Doct. J. P. Young of Edenville & Doct. J. Heron of Warwick. June 1st 1831: Am. J. Sci., v. 21, no. 2, geological map, 17.2 x 20.8 cm., facing p. 231 [i.e. 321].
11114   1833, Phosphate of lime in Edenville, NY: Am. J. Sci., v. 23, no. 2, 403-404.
11115   1839, Cabinet of minerals for sale: Am. J. Sci., v. 36, no. 2, 393-394 (9 lines).
11116   1843, and J. Heron, Geological and mineralogical map of part of Orange County N. York: in Geology of New York, first district, by W. W. Mather (see 6907), plate 41, geological map, 16 x 20 cm.
        Lithographed by Endicott of New York.

Young, William (see 8111 with T. Paine and S. Stearns)

Youth's Companion (Boston, 1827-1850)
11117   1827, The mineral kingdom: v. 1, 59.
11118   1828, Mining: v. 1, 191.
        On the deepest mine, in Bohemia.
11119   1828, Diamonds: v. 1, 191.
11120   1837, Subaqueous volcano: v. 11, 73, woodcut.
        Abstracted from Am. Mag. On the formation of Graham's Island.
11121   1845, Remarkable rock: v. 19, 88.
        On a native copper boulder from Lake Superior.
11122   1847, The principal crater of Vesuvius: v. 20, 169, woodcut.
11123   1847, Fingal's Cave: v. 20, 173, woodcut.
        Abstracted from Penny Mag.
11124   1848, Pencillings of a rambler "Saratoga Springs": v. 22, 134-135.
        Signed by "J. H. B."

The Youth's Medallion (Boston, 1841-1842)
11125   1841, Earthquake at Lisbon: v. 1, no. 3, 19.
11126   1841, Formation of the volcano Jorullo [in Mexico]: v. 1, no. 7, 50.

Z., X. Y. (see 9933)

Zabriskie, John B.
11127   1833, and N. A. Garrison and J. C. Fanning, Medical topographical account of the county of Kings [New York]: Med. Soc. NY Trans., v. 1, 174-182.

Zantzinger, William S.
11128   1847-1850, Report of the librarian for the year: Phila. Acad. Nat. Sci's Proc., v. 3, no. 12, 357-359; v. 4, no. 6, 130-132; no. 12, 252-254; v. 5, no. 6, 127-129.
        Contains list of recent geological literature acquired by the Academy.

Zennich
11129   1835-1836, The best method of assaying the ores of manganese: Am. J. Sci., v. 28, no. 1, 146-147; v. 29, no. 2, 374-375.
        Abstracted by C. U. Shepard from Annales des Mines.

Zinkin
11130 1834, A new ore of antimony [from the Hartz Mountains]: Am. J. Sci., v. 26, no. 2, 386 (4 lines).
　　　　Abstracted from Berzelius' Jahresbericht by L. Feuchtwanger.

The Zodiac, a Monthly Periodical, Devoted to Science, Literature and the Arts (Albany, 1835-1837)
11131 1835/1836, The study of natural history: v. 1, 80; 91-93; 110-111; 128.
　　　　On the importance of the study of natural history.
11132 1837, Land slides: v. 2, 97-98.
　　　　On the causes of land slides, with a list of major disasters.

Zollikoffer, William (1793-1853)
11133 1819, A materia medica of the United States, systematically arranged: Balt., Richard J. Matchett (printer), 120 p.
　　　　First edition. Also second edition: Balt., J. Lovegrove, 245 p. (1826).

INDEX

Index

Academy of Natural Sciences of Philadelphia: 11, 7467
    Fossil Collections: 8097
Accum, F.: 6258
Aden
    Mercury Mines and Mining: 2233
Africa
    Exploration: 1505
    Geology: 6585, 9048
    Gold Mines and Mining: 54, 971
    Meteorites: 728, 746
    Minerals, Localities: 7201
    Mines and Mining: 9139
    Natural History: 8117-8119
    Paleontology, Vertebrate: 5246
Agassiz, L.: 617, 643, 700, 721, 749, 765, 775, 786, 853, 870, 873, 908, 913, 1144, 1458, 2164, 3319, 3690, 4171, 4308, 4507, 4637, 5531, 5712, 6347, 6611, 6862, 7511, 7586, 10789
Alabama
    Caves: 10436
    Coal Mines and Mining: 5837, 6558, 6561, 5837, 8439, 8443
    Deltas: 4638, 4639
    Geological Survey: 264, 2127, 2128, 10440, 10822
    Geology: 803, 1189, 2125-2130, 2648, 4638, 4639, 5836, 8527, 9490, 10440, 11016-11018
    Iron, Ore Processing: 11037
    Limestone: 5914
    Maps, Geological: 10441
    Marl, Agricultural: 5698, 5699
    Medical Topography: 4662, 5029, 5030
    Mercury Mines and Mining: 5729
    Meteorites: 5555, 5589, 5617, 10397
    Mineral Springs: 2125, 2126, 2129, 2147
    Minerals, Localities: 2125, 2130
    Mines and Mining: 1180, 5449, 6565, 10411, 10412
    Paleobotany: 2190
    Paleontology, Invertebrate: 2546, 2579, 6084, 6093, 6569
    Paleontology, Vertebrate: 82, 812, 2171-2174, 2908, 2909, 4803, 4810, 6002, 6003, 6235, 6564, 11082
    Prairies: 5913, 9532, 9572
    Sandstone: 1694
    Soils: 3484, 3519, 6603, 6604, 9572
Alaska
    Paleontology, Vertebrate: 10313
Albany Institute: 1807
Algeria
    Iron Mines and Mining: 3946
    Salt Mines and Mining: 3945
Algiers
    Diamonds: 631
    Earthquakes: 211
    Minerals, Localities: 9164
    Mines and Mining: 2220
Alkalies
    Manufacture of: 10266
Allan, R.: 612
Alleghany Mountains
    Geology: 10970, 10971
Alluvial Formation (see also Glaciers)
    United States: 3827
Alps
    Elevation: 4613
    Glaciers: 5509
    Natural History: 4738
    Paleontology, Invertebrate: 6572
    Structure of: 4613
Alum Mines and Mining: 8496
    Scotland: 6323
Amber: 330, 433, 1631, 1998, 3799, 4016, 4450, 6682, 7108, 7544, 8695
    Fossils in: 6596
    Maryland: 10333, 10343
    New Jersey: 11012
    Origin of: 121, 9939
    Russia: 7672
American Academy of Arts and Sciences
    Donations: 177, 178, 179, 182
    Libraries, Scientific: 180
    Officers: 181
American Association for the Advancement of Science: 199, 202-206
American Association for the Promotion of Science: 207-210, 1276

American Geological Society: 282, 283, 284, 285, 8514, 9198
    Mineral Collections: 6627
American Institute
    Mineral Collections: 300
American Journal of Science: 434
American Mineralogical Society: 7015, 7016, 7270, 7274
American Museum of Natural History: 1110, 3285
American Philosophical Society: 1117-1139
Amherst College
    Mineral Collections: 943
Amianthus (see Asbestos)
Analysis, Chemical: 566, 1637, 1765, 2147, 2317, 2335, 2501, 3226
    Bone: 9885
    Bromine: 5927
    Cadmium: 5929
    Cerium: 7808
    Coal: 314, 2342, 5275, 5547, 5771, 6952, 8842, 9217, 10425, 10603
    Cobalt: 2878
    Columbium: 4850
    Copper Ore: 5784, 8397
    Dead Sea: 6771, 6780
    Dolomite: 8946, 8951, 8952, 8998
    Fluorine: 11026
    Fossil bones: 7180, 7181
    Glass: 8199
    Gypsum: 2630
    Iron: 3581, 6774
    Iron Ore: 2615, 4010, 5771
    Jordan River: 5683
    Lead ore: 5922
    Lime, Agricultural: 8958-8960, 8963
    Limestone: 2743, 5757
    Lithium: 8677, 10013
    Manganese: 2699
    Manganese Ores: 3384
    Mercury: 7193, 8602
    Metals: 2463, 7408, 9356, 9606
    Meteorites: 3259, 3260, 3841, 4896, 4901, 6071, 6072, 8947, 9526, 9674, 10689, 10691, 10737, 10755, 11045
    Mineral Springs: 1366, 1934, 1935, 1956, 1959, 2239, 2474, 2613, 2868, 2880, 2899, 3097, 3486, 4975, 4985, 4991, 5412-5414, 5546, 5674, 6947, 6948, 6951, 6965, 7109, 7111, 8841, 9862, 10331, 10539, 10540, 10650, 10801, 10826
    Minerals: 14, 15, 16, 17, 379, 521, 1644, 1763, 1876, 2343-2349, 2957-2960, 3676, 3850, 4408, 4409, 4642, 4643, 5408-5411, 6075, 6076, 8954, 8955, 8957, 8966, 9435-9453, 9528, 9828, 9832, 9878, 9879, 9882, 10040, 10044, 10073, 10247-10252, 10534, 10908, 10910-10912, 10915, 10917, 10920, 10921, 11020, 11068
    Noble Metals: 6316, 8288, 11021
    Notes: 3385, 4121, 4231, 4259, 4384, 5814, 5816, 5817, 6887, 8221, 10848, 10885
    Ocean: 2973, 2976, 5399, 6184, 6770, 7279, 8602, 9192, 9291, 9858, 9860, 10748, 10956, 10957, 10959, 11030
    Ohio: 2388, 2391
    Ores: 2491, 6397
    Precious Metals: 6112
    Rocks: 8844
    Sand: 5627
    Silica: 6015
    Silver: 4230
    Soils: 183
    Textbooks: 5804, 8123, 8204, 9138, 10246, 10330, 10530, 10532, 10544, 10602
    Uranium: 4895, 8159
    Volcanic Rocks: 2642, 2643, 2926, 2931
    Water: 9812, 9817, 9818, 9824, 9825
Annals of Natural History: 711
Annual of Scientific Discovery: 90, 1002, 1004
Ansted, D. T.: 1015
Antarctic
    Exploring Expeditions: 736
    Geology: 10952
    Paleobotany: 5252
Anthracite (see Coal)
Antigua
    Geology: 5323
    Minerals, Localities: 9468

Antigua (continued)
    Paleobotany: 318
Antimony: 11130
    Uses of: 8021
Antimony Mines and Mining
    United States: 7086, 7357
Apennines
    Geology: 4851
Appalachians
    Oolite: 9000
    Structure of: 8880, 8899, 8907, 8914, 8941, 8997
Arago, M.: 1256
Arctic
    Exploring Expeditions: 53, 1506
    Geology: 9333
    Natural History: 8772, 9622
Argentina
    Paleontology, Vertebrate: 7339
Arkansas
    Geology: 8564, 8600
    Lead Mines and Mining: 10812, 10815
    Meteorites: 945, 5034
    Minerals, Localities: 5533, 7120, 8679, 9549, 10922
    Mines and Mining: 3007, 3008, 5991, 10085, 10087
    Natural History: 9256, 9258
    Oil Stone: 5858
    Paleontology: 7501
    Prairie: 9900, 9901
    Salt Mines and Mining: 10481
    Silver Mines and Mining: 1270
    Springs, Hot: 688, 3685, 6269, 7135, 7655, 7685, 7692, 7776
Arkansas Territory
    Exploration: 7993
Arran, Isle of (see Scotland)
Arsenic: 7625
    In sea water: 533
    Mineral Springs: 4177, 6056
Artesian Wells: 1060, 2166, 2132, 2235, 2318, 3190, 3606, 3734, 7766, 9351, 9706, 9957, 10262-10264, 10434, 10685
    France: 9888
    Italy: 7005
    Massachusetts: 9812
    Temperature of: 8166, 8167
Aruba
    Gold Mines and Mining: 8376
Asbestos: 3981, 3984, 8149
    Connecticut: 468
    Uses of: 520
Asia
    Geology: 5150, 5163
Asphaltite: 10531
    France: 4036
Asphaltum (see also Pitch)
    Mexico: 7213
    New Brunswick: 5670
Assaying
    Copper: 10228
    Gold: 10231-10233, 10488
    Manganese: 10444, 10445, 11129
    Silver: 10488
Association of American Geologists: 1322-1342, 1569, 1572, 1576, 1603, 3172, 5138, 5139, 5594, 7570, 7610, 7838, 7897, 7899, 8902, 8903, 9308, 9766, 9767, 9801, 9809, 10088, 10770, 11103
Atomic Theory: 2338, 9310
    Weights: 2866, 5274
Auburn Theological Seminary: 4691
Austin, T.: 874
Australia (see also New South Wales)
    Caves: 2236
    Copper Mines and Mining: 866, 935, 2769, 6352
    Geology: 6365, 6592, 10055
    Gold Mines and Mining: 879, 983, 880
    Mines and Mining: 958
Austria
    Earthquakes: 8281
    Salt Mines and Mining: 7846, 8529
    Silver Mines and Mining: 8148
Azores
    Earthquakes: 1074, 7300, 8295
    Geology: 3314, 9316, 10730, 10735
    Mineral Springs: 5546

Azores (continued)
    Minerals, Localities: 10182, 10730, 10735
    Springs, Hot: 4155
    Volcanoes: 1074, 2763, 2764, 5032, 8463

Babbage, C.: 939
Bache, A. D.: 9321
Bahamas
    Natural History: 10302
Bakewell, R.: 787, 9739
Banda Island
    Volcanoes: 10608
Banka Island (Malaysia)
    Minerals, Localities: 5270
Barbados
    Geology: 9165, 9250
Barbary Coast
    Geology: 9047
Barnes, D. H.: 9686
Barrande, J.: 978
Barton, M.: 1532
Basalt: 7383
    Columnar: 923, 3659, 8872
    Iceland: 1062
    Lake Superior District: 3287
    Massachusetts: 5130
    Minerals: 2817, 2818
    Notes: 8145
    Origin of: 9873
    Pennsylvania: 9906
Batavia
    Volcanoes: 1236
Bavaria
    Minerals, Localities: 982
    Paleontology, Vertebrate: 4429
Beaumont, E. de: 869
Beck, T. R.: 10414
    Mineral Collections: 1591
Bedemor, Count: 10766
Belfast Natural History Society: 1615, 1616
Belgium
    Coal Mines and Mining: 10145
    Mines and Mining: 4017, 4080
    Shale: 1205
Berghmehl, Prof.: 227
Berkshire Medical Institution Lyceum: 1082
Berzelius, J. J.: 391, 956, 4089, 5860
Beudant, F. S.: 389, 6629, 10756
Big Bone Lick (see Kentucky)
Bitumen (see also Coal): 7216
Black Sea
    Geology: 4996
Blasting
    Rocks: 483, 4774, 4775, 10058
Blowpipe: 360, 918, 1351, 1411, 1561, 1666, 1818, 2508, 2890, 2894, 2895, 2966, 3078, 3079, 3104, 3136, 3678, 3983, 3995, 4216, 4529, 4577, 4762-4772, 4776, 4778-4780, 4905, 6915, 7010, 7019, 7114, 7211, 7228, 7499, 7533, 7859, 8469, 8471, 8474, 9145, 9146, 9318, 9332, 9337, 9612, 9614, 9616, 9617, 9623, 9642, 9644, 9647, 9648, 9650, 9651, 9659, 9662, 9663, 9914, 10182, 10294, 10569, 10570, 10572, 10573, 11027
    Hare, R.: 2620
    Newman's: 2410
    Treatise on: 1668
    Use by miners: 8109
Boase, Mr.: 6985
Bohemia
    Meteorites: 525, 922
    Minerals, Localities: 8800
    Mines and Mining: 11118
    Paleontology, Invertebrate: 1494
    Selenium: 1810
Bolivia
    Meteorites: 5887
    Mines and Mining: 7129
Bone Caves (see Caves)
Bonpland, M.: 1803
Borax Lagoons
    Tuscany: 4031
Boring
    Paleontology: 6342
    Rocks: 4995
Boston Journal of Natural History: 277, 278, 279
Boston Journal of Philosophy and the Arts: 395, 402

Boston Society of Natural History: 1710,
    1844-1848, 2182, 4459, 4460, 10169,
    10186, 11074
  Mineral Collections: 1846, 10032
Boué, A.: 844, 865, 9720
Boulders (see also Glaciers)
  Massachusetts: 1794
  Origin of: 1402, 3898
Bowen, G. T.: 9686
Brande, W. T.: 213, 1235
Brard, M.: 400
Brazil
  Diamond Mines and Mining:3264, 3265,
    3395, 4066, 4588, 5026, 5459, 5882,
    5897, 7561, 7565, 8162
  Geology: 2913, 6519, 6521
  Meteorites: 699, 7429
  Minerals, Localities: 2788, 10600, 11029
  Mines and Mining: 3855, 4093, 6929, 6930,
    9122, 11056
  Natural History: 10679, 10680
  Paleontology: 10035
Breccia
  France: 354
Breislak, S.: 453
Brewster, D.: 374
Bridgewater, Earl of: 3743
British Annual: 737
British Association for the Advancement of
  Science: 1678, 2014-2041, 3752, 5529,
  7513, 7514, 7523, 8843, 10890
British Columbia
  Coal Mines and Mining: 970
  Volcanoes: 4195
British India
  Geology: 7526
British Mining Museum: 7902
Brodie, P.: 833
Bromine: 9690, 10644
  Analysis, Chemical: 5927
  Discovery of: 1462
  From Brines: 174
  Mineral Springs: 4888, 4975
Brongniart, A.: 381, 390, 423, 1553, 7883,
  8224, 8225, 8645, 9649, 9787, 10751
Bronn, H. G.: 689, 1016, 6882
Brooke, H. J.: 10492, 10742, 10765
Bruce, A.: 327, 7089
  Mineral Collections: 7050
Buckland, W.: 228, 5532, 6422, 7134, 7210,
  7537, 7998, 9656
Building Stone (see also types of stone)
  Durability: 2176, 5808
  England: 5866
  Granite: 6518
  Greece: 10328
  Pennsylvania: 4747
  Properties of: 1204, 1316, 1453, 8105,
    9915, 9916
  Sandstone: 149, 10699
  Specific Gravity: 1626
  Strength of: 4109, 5211, 5681, 8945
  Thermal Expansion: 1511
  Vermont: 5170
  Virginia: 6067
Burma
  Earthquakes: 5954, 7374
  Geology: 6814
  Paleontology, Vertebrate: 11073
Burnet, T: 4431
Burning Springs (see Natural Gas; Springs;
  Petroleum)

Cadmium
  Analysis, Chemical: 5929
  Discovery of: 8414
Calabria
  Landslide: 2222
California
  Earthquakes: 9170
  Gazetteer: 7260
  Geology: 2837, 2838, 3098, 3099, 3420,
    4703, 5967-5970, 6576, 6577
  Gold Mines and Mining: 151, 990, 1152,
    1153, 1174, 1176, 1264, 1268, 2137,
    2138, 2180, 2360, 2725, 2835, 3014,
    3267, 3421, 3682, 3683, 3891, 3919,
    4519, 4821, 4919, 5475, 5760, 5761,
    5967-5970, 6201, 6361, 6402, 6579,
    6580, 6590, 6817, 6818, 7260, 7262,
    7985, 8044, 8335, 8574, 8756, 8807,

California, Gold Mining (continued)
    8820, 9147, 9283, 9293, 9455, 9838,
    9852, 9987, 9991, 10207, 10214, 10259,
    10420, 10421, 10506, 10637, 10867,
    10868, 10923, 10960
  Guide: 4848, 4849
  Limestone: 10982
  Mercury Mines and Mining: 6575
  Minerals, Localities: 985, 4101
  Mines and Mining: 4391
  Natural History: 2137, 2138, 4111-4119,
    4187, 4556-4558, 10259, 10457, 10458,
    10960
  Paleontology: 4700, 4701
  Platinum: 985
  Platinum Mines and Mining: 10207, 10214
  Salt Lake: 5680, 10869
  Volcanoes: 4195
Cambridge University: 6327
Canada (see also names of provinces)
  Caves: 1705, 6249
  Earthquakes: 7792, 10776
  Geological Survey: 805, 9338
  Geology: 1700, 3552, 4293, 5415, 6546,
    8020, 8547, 8878, 8992, 9631, 9658
  Gold Mines and Mining: 1173, 1401, 9788,
    9826
  Landslide: 1797
  Mineral Springs: 298, 5412-5414
  Minerals, Localities: 429, 1396, 1700,
    1702, 1831, 4151, 8821, 8822, 9529,
    10342
  Mines and Mining: 1171, 8502, 10813
  Natural History: 9174, 10324
  Oilstone: 5330
  Paleontology: 1702
  Paleontology, Vertebrate: 1705
  Scientific Societies: 489, 531
  Spring, Salt: 6282
  Terraces: 4723
  Waterfalls: 6272
Canal Surveys: 9591
Cape Horn
  Geology: 10156
Cape of Good Hope
  Meteorites: 1310
  Natural History: 2304
  Paleontology: 1549
Cape Verde
  Volcanoes: 896, 3140
Caracas
  Earthquakes: 8343, 8365
Carboniferous Formation: 10617
Carpenter, G. W.
  Mineral Collections: 2253
Catalonia
  Salt Mines and Mining: 2644
Catskill Lyceum: 8352, 8425
Caves: 4161
  Alabama: 10436
  Australia: 2236
  Canada: 1705, 6249
  Cuba: 2383
  England: 1919, 2152-2154, 2158, 2159,
    4033, 6469, 8151
  France: 6304, 6311, 9424, 9425, 10899
  Illinois: 8137
  India: 7774, 7847
  Indiana: 20, 8483
  Ireland: 5909, 6378, 7386, 7395, 7723
  Italy: 565, 4000
  Jamaica: 6328
  Java: 4477, 7711
  Kentucky: 1035, 1053, 1294, 1721, 1784,
    1871, 2080, 2219, 2337, 2378, 2379,
    2387, 2601, 2689, 3422, 3706, 3731,
    3815, 4173, 4658, 5020, 5273, 6438,
    6498, 7385, 7686, 8336, 8342, 8391,
    8579, 8643, 10684, 10875, 11098
  Missouri: 1788
  Moravia: 1858
  New Holland: 3741, 5513
  New York: 850, 3472, 3763, 4122, 4248,
    4264, 4526, 5188, 7041, 8018
  Notes: 1050, 3300, 3636, 3652, 3658,
    4136, 4145, 4161, 5372, 5386, 5388,
    6302, 6856, 6861, 6982, 7225, 7720,
    7834, 7882, 9129, 9141, 9295, 9402,
    9406, 3403
  Pennsylvania: 1450, 4222, 4935, 5419,
    5905, 7188, 7690, 8644
  Tennessee: 1053, 5911, 5912

Caves (continued)
    United States: 4961, 7053
    Vermont: 1111, 3895, 4527, 7876, 10977
    Virginia: 321, 1053, 1517, 2076, 2495,
        4793, 4847, 5421, 5505, 5721, 5838,
        5907, 6228, 6321, 7053, 7057, 7062,
        7102, 7382, 8139, 8431, 9119, 9366,
        11111
Cement: 1403, 1639, 2449, 3697, 3978, 4185,
    4186, 4576, 4869, 5234, 6017, 6513,
    6989, 7666, 7890, 8157, 8632, 9110,
    9452, 10058, 10295, 10622, 11066, 11067
    New York: 1597
    Strength of: 10323
    Water Line: 1399, 1403
Central America
    Geology: 1636
    Volcanoes: 9992
Cerium
    Chemical Analysis: 7808
Ceylon
    Geology: 2977-2979
    Mercury Mines and Mining: 2528
    Minerals, Localities: 6073, 10996, 10997
Chalk
    Nature of: 3453
    Chalk Formation: 2060, 2144, 2892, 3811,
        6527, 6535, 7132, 7468
    England: 7316
    Paleontology: 6191
Chalybeate Springs (see Mineral Springs)
Chapman, E. J.: 1013
Chemical Analysis (see Analysis, Chemical)
Chemical Society of Philadelphia
    Mineral Collections: 7018, 7024, 7031
Chemistry
    Encyclopedia of: 10530, 10532, 10544,
        10602
    Textbook: 2317
Chenevix, Dr.: 535
Chester County Cabinet: 2333, 2334, 4757,
    4937, 10227, 11062
    Mineral Collections: 485
Chile
    Coal Mines and Mining: 937, 5775, 5785
    Earthquakes: 621, 654, 5886, 7242, 7566,
        8134, 8383, 8415, 8825
    Geology: 4472, 4562, 9132
    Minerals, Localities: 1971, 3817, 8777,
        8778
    Mines and Mining: 3199, 8803
    Natural History: 7332, 10301
    Silver Mines and Mining: 6280, 9683,
        9694
China
    Coal Mines and Mining: 10137, 10141
    Geology: 8649, 8736
    Natural History: 10999
    Tin Mines and Mining: 984
Christie, T.: 7338
Clay
    Missouri: 11051
    Ochraceous: 10353
    Origin of: 1509
    Pennsylvania: 6243, 10807
Clay, Porcelain: 319, 590, 2776, 8125, 8501
    Connecticut: 9696
    Georgia: 6240
    Maryland: 1521
    Vermont: 1825, 3159, 7360, 7592
Claystone: 30
    Vermont: 33
Cleavage: 8929, 8935, 9460, 9461, 10201
Connecticut River Valley: 9785, 9796
    Mountains: 8914
    Origin of: 5187
Clemson, S. G.: 6293
Clemson, T. G.: 4316
Clifford, J. D.: 10874
Coal (see also Coal Mines and Mining)
    Age of: 5127, 8923, 8986, 8993
    Alabama: 5837
    Analysis, Chemical: 314, 2342, 5275,
        5547, 5571, 6952, 8842, 9217, 10425,
        10603
    Connecticut: 4732
    Experiments on: 2185-2188
    Footprints: 6100, 6101
    Gas: 4194
    Jet Ware: 1824
    New Hampshire: 9362

Coal (continued)
    New York: 6298
    Notes: 1819, 5992, 7953, 7962, 8032, 8144,
        10808, 10860
    Ohio: 5053
    Origin of: 877, 1792, 2163, 2167, 2193,
        2316, 3926, 4029, 4044, 4233, 4451, 5261,
        6545, 7671, 8408, 8557, 8558-8561, 8563,
        8781, 8876, 8889, 8909, 9774, 9775, 9780,
        10928, 10931, 10935, 11035, 11036
    Paleobotany: 587, 1435, 1438, 1555, 1606,
        1711, 2055, 2058, 2190, 2191, 2192,
        2399, 4492, 5253, 6558, 6561, 8934,
        10014, 10133, 10143, 10177, 10195,
        10196, 10199, 10200, 10203, 10208,
        10734, 10941, 10980
    Paleontology: 573
    Paleontology, Vertebrate: 5956-5959
    Pennsylvania: 6562
    Properties of: 5789, 5791, 5794-5798,
        5846, 6069, 8619
    Sea Coal: 3970
    Structure of: 5782, 5783
    Types of: 350, 514
    United States, Western regions: 4709
    Uses of: 891, 2163, 2167, 2685, 3740,
        4476, 4828, 5777, 5786, 6022, 6966,
        7651, 7772, 9404, 10605, 10895, 10896,
        11108
    Volcanoes: 6262
Coal Mines and Mining: 197, 313, 2596, 5846,
    8202, 10965-10967
    Alabama: 6558, 6561, 6927, 8439, 8443
    Belgium: 10145
    British Columbia: 970
    Chile: 937, 5775, 5785
    China: 10137, 10141
    Costa Rica: 8793
    Cuba: 1729, 2282, 10125
    Delaware: 3065, 3066, 10949
    East Indies: 921
    Egypt: 991, 2762
    England: 526, 1390, 1459, 1860, 2175,
        2202, 2232, 3268, 3377, 4025, 4087,
        4095, 4473, 5880, 7195, 7554
    Europe: 3856, 3861, 4077, 5787, 5807
    Explosions: 735
    Fires: 2910
    France: 1185, 3621, 3993, 4026, 4051,
        4070, 10853
    Great Britain: 4076, 6077, 8491
    Greece: 6200
    Illinois: 4001, 8493, 8576
    India: 895
    Maryland: 125, 126, 301, 1762, 2728,
        3031, 3137, 3214, 3598, 3857, 4180,
        4330, 4331, 5354, 5355, 5477, 5776,
        8451, 8548, 8558, 9847, 10455, 10456
    Massachusetts: 560, 1826, 3795, 3797,
        5144, 5554, 8792
    Michigan: 4292
    Mississippi River Valley: 4179, 4297,
        8390
    Missouri: 4170, 9748, 9756
    New England: 7631
    New Jersey: 4199, 5774
    New York: 113, 2104, 2287, 2476, 3054,
        3351, 3354, 3363, 3564, 8394, 10327
    North Carolina: 5725
    Notes: 194, 233, 237, 1077, 1160, 1195,
        2774, 3262, 4056, 4065, 4256, 4606,
        4823, 5362, 6313, 6372, 6858, 6954,
        6995, 7130, 7231, 7362, 7912, 9340,
        9395, 10446
    Nova Scotia: 4042, 4343, 6546
    Ohio: 515, 1028, 3010, 5482, 6912
    Pennsylvania: 299, 428, 529, 561, 1052,
        1078, 1148, 1167, 1214, 1215, 1596,
        1598, 1670, 1841, 1899, 1916, 1932,
        2148, 2149, 2189, 2273, 2274, 2339,
        2385, 2389, 2392, 2396, 2397,
        2437-2440, 2594, 2734-2736, 2942, 2943,
        3002, 3006, 3011, 3291-3296, 3351,
        3354, 3363, 3377, 3624, 3695, 3787,
        3788, 3794, 3833, 3849, 3852, 3869,
        4073, 4088, 4123, 4146, 4190, 4191,
        4480, 4589, 4702, 4707, 4932, 4933,
        4943, 5352, 5353, 5439, 5442, 5447,
        5456, 5460, 5462, 5700, 5701, 5746,
        5764-5766, 5768, 5769, 5772,
        5778-5783, 5802, 5806, 5988, 6004,
        6041, 6096, 6102, 6104-6106,

Coal Mines and Mining (continued)
    6143-6170, 6331, 6337, 6338, 6373,
    6522, 6574, 6612, 6704, 6708, 6932,
    6952, 6972, 7020, 7117, 7196, 7296,
    7369, 7601, 7688, 7761, 7762,
    7921-7924, 8205, 8206, 8209, 8210,
    8216, 8218, 8220, 8362, 8363, 8385,
    8429, 8434, 8437, 8451, 8814-8816,
    8943, 9288, 9289, 9363, 9457-9458,
    9597, 9615, 9666, 9667, 9692, 9695,
    9700, 9849, 10050, 10051, 10065, 10099,
    10110-10112, 10121, 10122, 10128,
    10130-10132, 10146-10148, 10495, 10604,
    10895, 10896, 11042, 11049, 11088,
    11101
  Rhode Island: 45, 1178, 5773, 6942, 7824,
    8284, 8760, 8763-8766, 8925, 9389,
    9667
  Safety: 49
  Scotland: 1302, 4561
  Straits of Magellan: 967
  United States: 1273, 1522, 1736, 1740,
    1778, 2340, 3474, 4587, 5807, 6077,
    6305, 7036, 10703
  Virginia: 98, 3214, 3870, 4491, 5365,
    6570, 7493, 7496, 8035, 8036, 8361,
    8981, 8986, 8989, 8993, 9768, 10115,
    10120, 10427, 10628, 10649, 11050,
    11078
  Wales: 5778, 5809
Coast (see also Tides; Ocean)
  Ancient: 2299, 2947
  Erosion: 5357
  Geology: 2225, 2226, 2946, 2947
  Origin of: 1385
Cobalt
  Analysis, Chemical: 2878
  Connecticut: 9185
  Ore Processing: 1503, 6503
Cobalt Mines and Mining
  Connecticut: 4403
Coke
  Virginia: 5788
Collections (see also Fossils; Minerals)
  Natural History: 3046, 3047
  Preservation Techniques: 1449, 2056,
    2150, 3878, 4020, 4098, 5067, 5068,
    6199, 6765, 6806, 10092, 10595
Colombia
  Earthquakes: 1915
  Exploration: 1387
  Geology: 1882, 1885
  Meteorites: 1883, 3975, 5379
  Minerals, Localities: 1352, 1884
  Platinum Mines and Mining: 10512
Colthurst, Miss: 1007
Columbium: 7032, 7051, 11028
  Analysis, Chemical: 4850
  Discovery of: 1851
Columnar Structure: 3514
Concord Natural History Society: 2521
Concretions: 30
  Florida: 4339, 9771
  Origin of: 4631
Conglomerate: 3523, 8879
  Massachusetts: 5083, 5657
  Ohio: 6455
Congo
  Exploration: 10422
Connecticut (see also Connecticut River Valley)
  Asbestos: 468
  Clay, Porcelain: 9696
  Coal: 4732
  Cobalt: 9185
  Cobalt Mines and Mining: 4403
  Copper Mines and Mining: 7668, 8259
  Earthquakes: 684, 743, 1162
  Geological Survey: 628, 2523-2527, 3438,
    4267, 9512, 10624
  Geology: 6197, 6880, 8230, 9630-9632,
    9658
  Gypsum: 5250
  Iron Mines and Mining: 361, 9485
  Maps, Geological: 6197, 6879, 8229
  Marble: 346, 8478
  Marl, Agricultural: 5129
  Meteorites: 2203, 3396, 4969, 5235, 6124,
    9599-9602, 10689, 10691, 11045
  Mineral Springs: 1854, 5955, 7714, 8353
  Minerals, Localities: 341, 342, 364,
    365, 1347, 1905, 1906, 1910,
    1939-1941, 2490, 2504, 2794, 2801,

Connecticut, Minerals (continued)
    2855, 3050, 3326, 4374, 4890, 5088,
    5831-5833, 6028, 6120, 6121, 6123,
    6263, 6717, 6865, 6872, 6873, 6880,
    6950, 8223, 8230, 8231, 8506, 8508,
    8509, 8524-8526, 8694, 8831, 9398,
    9467, 9470-9473, 9486, 9508, 9511,
    9512, 9517, 9518, 9538, 9605, 9607,
    9609, 9611, 9620, 9621, 9625,
    9629-9632, 9636, 9638, 9654, 9657,
    9664, 9669, 9782, 9793, 9802, 9805,
    9811, 9876, 9877, 10152, 10288, 10340,
    10727, 10978
  Mines and Mining: 8328
  Moving Rocks: 6118, 6122
  Natural History: 2522, 10418
  Oilstone: 10774
  Ornithichnology: 25, 669, 2311, 2312
  Paleobotany: 9823
  Paleontology: 472
  Paleontology, Vertebrate: 593, 687, 696,
    4733, 6289
  Sandstone: 84, 2310
  Septarian Nodules: 10793
  Slate: 540
  Vitriol: 7082
Connecticut River Valley (see also
    Connecticut; Massachusetts)
  Cleavage, Sandstone: 9785, 9796
  Coprolites: 5167
  Dikes: 8885
  Erosion: 3687
  Geology: 5071, 5072, 5074, 5079, 5081,
    5102, 5179, 6538, 6542, 9854, 10719,
    10720, 10884
  Joints: 5141
  Maps, Geological: 5071, 5079, 9853
  Ornithichnology: 1496, 1499, 2406,
    2984-2997, 4910, 4912, 5114-5116, 5120,
    5146, 5148, 5162-5164, 5166, 5176,
    5178, 5182-5184, 6743, 6785, 7517,
    8091, 8875, 10859, 10944
  Paleobotany: 5147
  Paleontology, Vertebrate: 8705, 9635, 9899
  Sandstone: 8900, 8912, 8987, 9798, 10791
  Rain Drop Impression: 3129
  Terraces: 3687, 5180, 5186
  Trap: 9807, 9810
  Volcanoes: 5165, 5181, 9680, 9807, 9810
Continents
  Origin of: 2829, 2830
Conybeare, J. J.: 10752
Cooper, T.: 8479
Copper Mines and Mining: 615, 649, 901,
    902, 987, 1169, 1408, 2074, 2614, 3016,
    3017, 3191, 3192, 7621, 8315, 8316,
    8319, 8921
  Assaying: 3908, 5784, 8397, 10228
  Australia: 866, 935, 2769, 6352
  Connecticut: 7668, 8259
  Cuba: 1645, 5437, 5544, 10134, 10142,
    10144
  England: 4069, 7140
  Europe: 3863
  Great Britain: 413
  Iowa: 9994
  Lake Superior Region: 600, 1165, 1172,
    1177, 1179, 1269, 1390, 1492, 1510,
    1733, 1790, 2277, 2279, 2308, 2309,
    2336, 2640, 2693, 3004, 3174, 3175,
    3207, 3470, 3529, 3608, 3609, 3684,
    3935, 3937, 3939, 4164, 4498-4500,
    4511-4513, 4598, 4755, 4904, 5061,
    5196, 5197, 5204-5206, 5208, 5291,
    5292-5304, 5306, 5319, 5321, 5322,
    5339-5341, 5361, 5491, 5492, 5618,
    5621, 5623, 5628, 5635, 5637,
    5639-5641, 5646, 5650-5653, 5655, 5675,
    5679, 5851, 5874, 5950, 6021, 6040,
    6358, 6385, 6458, 6461-6463, 6606,
    7105, 7171, 7198, 7235, 7760, 7957,
    8037, 8253, 8274, 8381, 8477, 8482,
    8518, 8586, 8739, 8773, 9098, 9154,
    9155, 9166, 9259, 9265-9267, 9276,
    9278, 9281, 9552, 9574-9580, 9802,
    9843, 10025, 10086, 10087, 10210, 10861,
    10908-10912, 10915-10920, 10926, 10936,
    10937, 10942, 11121
  Maryland: 9982
  Missouri: 306
  New Jersey: 1567, 1579, 2434, 5259, 6062-
    6065, 6287, 7236, 8275, 9360, 10284

Copper Mines and Mining (continued)
  Notes: 5919, 5924, 7402, 7621, 8805,
  Ore Processing: 615, 901, 902, 987, 1169,
    3016, 3017, 3191, 3192, 3269, 5843,
    5919, 5924, 7594-7596, 10226, 10230
  Saxony: 4090
  Spain: 614
  Sweden: 5422, 8150, 10523
  Vermont: 8770
  Virginia: 7354, 9792
Copperas Mines and Mining
  Kentucky: 10850
  Maryland: 3282
  Vermont: 6419
Coprolites: 6097, 8888
  Connecticut River Valley: 5167
  England: 4992
  Virginia: 6419
Coral Islands: 1384, 3452
  Formation of: 1038, 2668, 2670-2674
  Origin of: 2802, 2804, 2805, 2808-2811,
    2813, 4435, 4461, 4641, 4911, 11055
Correlation of Strata: 73, 3366, 8606,
    10617
  New York: 4687, 4695, 4696, 4704
Corsica
  Geology: 6284
  Minerals, Localities: 3081
Costa Rica
  Coal Mines and Mining: 8793
Cozzens, I.: 1354
Crawe, I. B.: 903, 9790
Creation: 1280
Cretaceous Formation: 7470, 7472, 7474,
    7476-7478, 7485
  New Jersey: 6551, 8838
  Paleontology, Invertebrate: 2574, 2593
  United States: 8845
Cross, A.: 4007, 4276
Crystallization: 9906-9909, 11105
Crystallography: 130, 1671, 1987, 1993,
    2792-2796, 2798, 2828, 2833, 2834,
    3369, 3643, 4496, 4528, 10914
  Models: 1381
Crystals, Twin: 2793
Cuba
  Caves: 2383
  Coal Mines and Mining: 1729, 2282, 10125
  Copper Mines and Mining: 1645, 5437,
    5544, 10134, 10142, 10144
  Geology: 10134, 10142, 10144
  Maps, Geological: 10144
  Minerals, Localities: 1765, 2445
  Mines and Mining: 1470, 5470, 9153
  Natural History: 7986
  Soils: 1641
Curacao
  Salt: 487
Cuvier, G.: 381, 1835, 2238, 4595, 5980,
    6271, 6278, 6977, 7028, 7074, 7227,
    7567, 7572, 7575, 7676, 9399, 9411,
    10549

D´Archiac, V.: 936
D´Orbigny, A.: 843, 868, 1437
Dana, J. F.: 7704
Darwin, C.: 882, 8087
Daubeny, C.: 926, 932
Davy, H.: 8418
Dead Sea
  Analysis, Chemical: 6771, 6780
  Jordan: 4410, 6581, 6656, 6771, 6780
    9588
Dekay, J.: 10414
Del Rio, A. M.: 3769, 7066
Delaware
  Coal Mines and Mining: 3065, 3066, 10949
  Gazetteer: 9370
  Geological Survey: 1764, 1769, 3059-3063
  Greensand: 9955
  Marl, Agricultural: 1768
  Minerals, Localities: 1929, 2251, 2252,
    2255, 2257, 2542, 4541
  Paleontology, Invertebrate: 7463, 7464,
    7476, 7485
Delaware Academy of Natural Sciences: 562
Delaware Chemical and Geological Society: 3064
Deltas
  Formation of: 4137
  Guiana: 4608
  Mississippi River Valley: 4638, 4639,
    6559, 6563, 6567, 6568

De Luc, Dr.: 7407, 9355
Deluge (see also Religion and Geology;
    Glaciers): 10675, 10799
Dendrites: 10174, 10178, 10185, 10204
Desmarest, N.: 4582
Devonian Formation
  United States: 8895
Devonian System
  Paleontology, Invertebrate: 2573, 2576
DeWitt, B.
  Mineral Collections: 3169
Diamond Mines and Mining
  Brazil: 3264, 3265, 4066, 4588, 5026,
    5459, 5882, 5897, 7561, 7565, 8162
  India: 1254, 1979, 4181, 6370, 8340,
    8344
  Notes: 8386
Dickenson, J. D.: 10414
Dickeson, M. W.: 9307
Dictionary of Science: 1958
Dighton Rock (Massachusetts): 5941,
    11000
Dikes: 2141, 5350, 6341, 8466, 8872
  Connecticut River Valley: 8885
  Massachusetts: 9803
  New Jersey: 9803
  North Carolina: 1608, 6219, 6220,
    8034, 11040, 11041
  Vermont: 2401
Diluvial Evidence (see Glaciers)
Dolomite: 2609, 5582, 7428, 9857
  Analysis, Chemical: 8946, 8951, 8952,
    8998
  Fetid: 3148
  Vermont: 2325
  Weathering: 2933
Dominican Republic
  Earthquakes: 8881
Drift
  Glaciers, Evidence for: 8720-8723
Ducatel, J. T.: 250, 251, 975

Earth
  Age of: 1477, 8160, 10862
  Cooling of: 2825, 2826
  Density of: 191, 2285, 4344, 6970, 9115
  Diameter of: 48, 2684
  Electrical Conductivity: 6495, 6919
  Figure of: 47, 190, 1048, 1311, 1714,
    1715, 1717, 1776, 1896, 2417, 2684,
    3638, 4106, 4223, 5876, 6052, 6722,
    7215, 7223, 7927, 8552, 8747, 9191,
    9193, 9320, 10474
  Gravity: 47
  Origin of: 176, 274, 1055, 1669, 1856,
    1868, 1950, 1951, 2305, 3319, 3844,
    3969, 3971-3973, 4431, 5060, 5395,
    5936, 6059, 6109, 6500, 6667, 7604,
    8039, 8040, 9875, 10016, 10100,
    10781, 10863
  Rotation of: 11033
  Structure of: 1861, 3309, 3312, 3637,
    3647, 3713, 3759, 3760, 5278, 5729,
    6389, 6921, 7210, 7221, 7627, 7885,
    8305, 8310, 8757, 10077-10079
  Temperature of: 58, 192, 244, 555, 606,
    693, 1051, 1249, 1300, 1373, 1840, 2124,
    2645, 2825, 2826, 3193, 3415, 3599,
    3691, 3899-3901, 3944, 3954, 3958,
    3960, 3999, 4012, 4013, 4130, 4270,
    5093, 5210, 5401, 5868, 6053, 6637,
    6638, 6666, 6971, 7552, 7961, 8153,
    8165-8167, 8306, 8372, 8375, 8417,
    8450, 8983, 8991, 9298, 9302, 9938,
    10670
Earth, Magnetic studies: 1, 2, 465, 522,
    578, 620, 640, 730, 741, 825, 826,
    840, 1076, 1175, 1227, 1297, 1298,
    1315, 1376-1380, 1382, 1383, 1386,
    1445, 1487, 1728, 1750, 1895, 1897,
    1898, 2011, 2070, 2201, 2218, 2370,
    2371, 2654, 2655, 2898, 3279, 3280,
    3286, 3343, 3426, 3582, 3600, 3601,
    3640, 3655, 3972, 4005, 4021, 4045,
    4052-4054, 4091, 4192, 4389, 4390,
    4392-4394, 4470, 4522, 4740, 4758,
    4759, 4822, 4987, 5014, 5039, 5040,
    5172, 5305, 5342, 5451, 5854, 5966,
    6018, 6023, 6058, 6253, 6261, 6266,
    6380, 6398, 6410-6416, 6423, 7138,
    6427-6430, 6432, 6435-6437, 6441,

Earth, Magnetic Studies (continued)
    6449-6452, 6454, 6458-6460, 6465,
    6482-6494, 6501-6507, 6979, 6992,
    7217, 7263, 7322, 7821, 7870, 7872,
    7976-7978, 8161, 8239, 8240, 8416,
    8424, 8614-8616, 8798, 9042, 9286,
    9148-9151, 9287, 9297, 9364, 9365,
    9430, 9585-9587, 10452, 10504,
    10648, 10671, 10963, 10973, 10975
Earthquake Sermons: 1489, 1490, 2207, 2216,
    2330, 2331, 2472, 2486, 2641, 2659,
    2660, 2889, 3201, 3322, 3877, 3961,
    3962, 4447, 4448, 6230, 6866-6868,
    6718, 6871, 6939-6941, 7330, 7430,
    7858, 8110, 8171, 8319, 8448, 8580,
    8591-8593, 8738, 8948, 8949, 9428,
    9429, 9581, 9898, 9921, 10946, 10964,
    10979
Earthquakes: 292
    Algiers: 211
    Austria: 8281
    Azores: 1074, 7300, 8295
    Burma: 5954, 7374
    California: 9170
    Canada: 7792, 10776
    Caracas: 8343, 8365
    Chile: 621, 654, 5886, 7242, 7566, 8134,
        8383, 8415, 8825
    Colombia: 1915
    Connecticut: 684, 743, 1162
    Dominican Republic: 8881
    Ecuador: 7796
    Effects on disease: 10771, 10772
    Effects on vegetation: 420
    England: 2652, 6976
    Europe: 5870
    Georgia: 7094
    Great Britain: 222
    Greece: 2275
    Illinois: 7932
    In Fiction: 4208
    India: 378, 864, 929, 9109
    Ionian Islands: 10681
    Italy: 622, 3321, 4104, 4745, 5730, 6317,
        6406, 6601, 7126, 7342, 7398, 7557,
        8279, 9107, 9415, 10521, 10522
    Jamaica: 2641, 5690, 8738
    Lecture on: 11002-11005
    Martinique: 304
    Massachusetts: 164, 1489, 1490, 2207,
        2216, 2276, 2330, 2331, 2472, 2480,
        2486, 2659, 2660, 2889, 3201, 3322,
        3433, 3605, 3679, 3680, 3877, 3961,
        3962, 4447, 4448, 6230, 6718,
        6866-6868, 6871, 6939-6941, 7112, 7330,
        7430, 7858, 8110, 8171, 8319, 8580,
        8591-8593, 8948, 8949, 9428, 9429,
        9898, 9921, 10946, 10964, 10979,
        11002-11005
    Mississippi River Valley: 3885, 3886,
        10847
    Missouri: 6991
    Motion of: 8886, 8888, 8901, 8905
    New England: 2313, 3258, 6247, 10167,
        10974
    New Granada: 6288
    New Hampshire: 118
    New York: 1780, 7044
    North Carolina: 2448
    Notes: 119, 1045, 1069, 1191, 1233, 1305,
        1314, 1367, 1389, 1486, 1723, 1852,
        1859, 2384, 2530, 3299, 3305, 3405,
        3424, 3633, 3644, 3645, 3704, 4014,
        4127, 4475, 4514, 5033, 5689, 6230,
        6237, 6238, 6276, 6301, 6325, 6707,
        6720, 6859, 7121, 7122, 7205, 7363,
        7376, 7393, 7640, 7717, 7827, 7841,
        7877, 7960, 8312, 8378, 8379, 8464,
        9248, 9296, 9346, 9350, 10809
    Ohio: 4825
    Origin of: 1031, 1291, 1716, 3255, 6092,
        8567, 8886, 8888, 8901, 8905
    Pennsylvania: 4936, 5265, 7011, 7026,
        8300
    Peru: 1406, 6516
    Poems: 5989, 6230, 10783
    Portugal: 13, 1063, 2514, 8141, 10499,
        11125
    Scotland: 4062
    Sicily: 3792, 3825
    South America: 4211, 4255, 4741, 5374,
        5387, 5393, 6290

Earthquakes (continued)
    South Carolina: 8299, 9589, 9861
    Spain: 3069, 10643
    United States: 12, 783, 1516, 2069, 4252,
        4469, 4596, 4967, 5530, 6231, 7299,
        8894, 8898, 9175, 10652
    United States, Western regions: 8364
    Venezuela: 1223, 1229, 2221, 5504, 6392,
        7100, 7301, 9414, 10672, 10782
    West Indies: 784, 2245, 10216
East Indies
    Coal Mines and Mining: 921
    Geology: 3180, 4301
Eaton, A.: 3744, 7832, 9701, 10414, 10559,
    10805
Eaton, T. D.: 9686
Ecuador
    Earthquakes: 7796
Education, Geological: 1955, 1957
    France: 3058
Egypt
    Coal Mines and Mining: 991, 2762
    Geology: 613, 691, 2122, 2123, 4175
    Gold Mines and Mining: 3427
    Mines and Mining: 4037
    Natural History: 2122, 2123, 9983, 10651
    Paleobotany: 2181, 6335
    Paleontology, Invertebrate: 1420
    Salt Mines and Mining: 5982
    Soils: 5799, 5801, 6055, 9797
Ehrenberg, C. G.: 1418, 1424, 1427, 1428,
    4154, 10889
Elba Island
    Minerals, Localities: 1974, 8289, 9202
Elevation
    Alps: 4613
    Changes in: 1373, 1551, 1741,
        2669, 2914, 4263, 4338, 4434, 4472,
        4562, 5256, 5506, 5507, 5516, 5597,
        5649, 6525, 6529, 6595, 6911, 6914,
        7133, 7212, 7366, 7833, 8236, 8377,
        8882, 9309, 9407, 9779, 9868-9871,
        8680, 10497, 10538, 10886, 11086
    Mountain Ranges: 9743
    Newfoundland: 3134
    Of Continents: 5392
    Scandinavia: 1962
Encyclopedias: 3440-3442, 3584-3588, 4567,
    4835, 7863
England (see also Great Britain)
    Building Stones: 5866
    Caves: 1919, 2152-2154, 2158, 2159, 4033,
        6469, 8151
    Chalk Formation: 7316
    Coal Mines and Mining: 526, 1390, 1459,
        1860, 2175, 2202, 2232, 3268, 3377,
        4025, 4087, 4095, 4473, 5880, 7195,
        7554
    Coal, Origin of: 2316
    Copper Mines and Mining: 4069, 7140
    Coprolites: 4992
    Earthquakes: 2652, 6976
    Footprints: 2162, 2183, 2731, 4866
    Geology: 823, 828, 1922, 2489, 2597,
        3447, 5359, 6537, 6761, 7400,
        7506-7510, 7551, 7614, 9397, 10983
    Glaciers, Evidence for: 9394
    Graphite: 266
    Gravel: 8307
    Greensand: 5827
    Iron Mines and Mining: 3268, 3853, 3858
    Lead: 435
    Mineral Springs: 5707, 10677, 10678
    Minerals, Localities: 1182, 1975, 2538,
        3226, 4980, 5883, 6931, 8839, 8918
    Mines and Mining: 1372, 3183, 3729, 3864,
        4008, 4011, 4015, 4993, 5358, 6268,
        6277, 7237, 7710, 7794
    Museum of Economic Geology: 953
    Natural Gas: 1072
    Paleobotany: 1606, 1711, 2048, 4041,
        4184, 4865, 7584, 7641, 11100
    Paleontology: 8176, 9396
    Paleontology, Invertebrate: 573, 838,
        1921, 5889, 6745, 6747, 6763, 10877
    Paleontology, Vertebrate: 460, 630, 644,
        646, 852, 1095, 1184, 1451, 1919,
        2152-2154, 2158, 2159, 2322, 3425,
        4040, 4055, 4819, 5244, 5245, 6330,
        6469, 6725, 6726, 6728, 6730, 6733,
        6734, 6746, 6748, 6750, 6754, 6755,
        6759, 6999, 7541, 8096, 8101, 8102,
        8126, 9459, 10060, 10700

England (continued)
  Petroleum: 969
  Rocking stones: 7394, 7397
  Salt Mines and Mining: 1781, 2024, 7064
  Sandstone: 8116
  Septarian Nodules: 589, 10793
  Slate: 9393
  Springs, Hot: 2531, 6357
  Tin Mines and Mining: 1107, 8152, 10775
  Whin Sill: 5497
Eocene
  Formations: 7529, 9095
  Paleontology: 6084
  Paleontology, Invertebrate: 2575, 2583, 2584, 2585, 2588, 2590, 2593
Erosion: 70, 441, 1049, 1204, 3896, 5350, 8043, 8985, 10990
  By Frost: 5073
  By Water: 9005
  Coastal: 8484
  Connecticut River Valley: 5357, 3687
  Mississippi River: 3056
  Mississippi River Valley: 3913-3914
  New York: 9598
  Rivers: 4299, 8246
  Vermont: 4648
Erratics (see Boulders; Rocking Stones; Glaciers)
Esmark, J.: 748
Essex County Natural History Society: 3607
Etna (see Volcanoes, Sicily)
Europe
  Coal Mines and Mining: 3856, 3861, 4077, 5787, 5807
  Copper Mines and Mining: 3863
  Earthquakes: 5870
  Geology: 1874, 1875, 3034
  Glaciers, Evidence for: 2956, 3110, 3903, 6539
  Iron Mines and Mining: 3856, 3861, 3863, 4657
  Lead Mines and Mining: 4035
  Minerals, Localities: 7077, 8631, 9479
  Natural History: 4534, 4585, 9603, 9628
  Paleontology, Invertebrate: 1887
  Paleontology, Vertebrate: 2753, 5875
  Platinum Mines and Mining: 4067
  Salt Mines and Mining: 4574, 4824
  Societies, Scientific: 7548
Evolution: 85, 2291-2298, 9875
Excavating Machine: 7831
Exploring Expeditions
  Africa: 5711
  Antarctic: 736, 9036, 10721
  Arctic: 1506, 3899
  Beagle: 2916
  Dead Sea: 6581-6584, 9313
  French: 380, 1299, 3284
  Great Lakes: 349
  Northwest Passage: 5709, 5710
  Pacific: 109, 8736
  Polar: 480
  Siberia: 505
  United States: 5737, 5805
  Wilkes: 2824, 2831, 2836, 2841-2844, 2847, 10697, 10950
Extinction: 5708

Falkland Islands
  Geology: 2915
  Natural History: 9046
Faroe Islands
  Geology: 160
  Minerals, Localities: 3603
Featherstonhaugh, G. W.: 8, 2278, 5249, 6651, 6886, 6888, 7630
  Mineral Collections: 467
Ferrara, F.: 8228
Feuchtwanger, L.
  Mineral Collections: 551
Figure of the Earth (see Earth, Figure of)
Fiji Islands
  Meteorites: 10691
Fingals Cave: 11123
Finland
  Geology: 4141
  Minerals, Localities: 5871, 7914, 7916, 7917, 9204
Fire Stone
  Pennsylvania: 8252

Flame Test
  Strontium: 1397
Flint: 6254, 8389
  New Jersey: 7293
  New York: 7271
Flood (see also Religion and Geology; Glaciers): 3619
  Evidence for: 3703, 3718
Florida
  Concretions: 4339, 9771
  Geology: 167, 2547, 2586, 8673, 9490
  Marl, Agricultural: 2690
  Medical Topography: 8237
  Natural History: 1544, 2905, 3905, 8359, 9012, 9013, 9856, 9933, 10900, 10968, 10969
  Oolite: 8915
  Paleontology, Invertebrate: 1350, 1442, 2547, 2586, 3186, 9208
  Petroleum: 2361
  Soils: 6654
Fluid Inclusions: 399, 1990, 1992, 1996, 2002, 2003, 9214
Fluorine
  Analysis, Chemical: 11026
  Distribution of: 10986, 10987
  In fossils: 1188
Folding: 8892, 8908
Footprints (see also Ornithichnology)
  Alligator: 3178
  Coal: 6100, 6101
  England: 2162, 2183, 2731, 4866
  Germany: 2653
  Human: 44, 1245, 1630, 1746, 2479, 2948, 4414, 5842, 6193, 6738, 8051, 9264, 9270, 9275, 9278
  Mammalia: 6729
  Missouri: 692, 4224, 4414
  Notes: 6374
  Pennsylvania: 5956-5959
  Saxony: 668
  Scotland: 3275, 6103
Forbes, J. D.: 790
Fossil Collections
  Academy of Natural Sciences of Philadelphia: 8097
  Munster, Herr: 854
  Taylor, R. C.: 10108, 10136
Foster, J. W.: 196, 198, 5648
Fox, R. W.: 4006, 9171
France
  Artesian wells: 9888
  Asphaltite: 4036
  Breccia: 354
  Caves: 6304, 6311, 9424, 9425, 10899
  Coal Mines and Mining: 1185, 3621, 3993, 4026, 4051, 4070, 10853
  Education, Geological: 3058
  Geology: 2052, 4032, 4599, 8332, 8392, 10546
  Lithographic stones: 822
  Maps, Geological: 7587
  Meteorites: 763, 767, 2675, 3430, 6246, 7087, 7208, 7406, 9203, 9341, 9412, 10811, 11039
  Minerals, Localities: 747, 1545, 1878, 2783, 4583, 5565, 5873, 5879
  Mines and Mining: 1870, 7963, 8501, 8503
  Natural History: 6599
  Oil Stone: 9176
  Paleobotany: 8536
  Paleontology: 536, 6322
  Paleontology, Vertebrate: 1804, 1835, 2155, 2156, 2536, 2752, 4579, 5864, 6271, 6278, 8528, 9424, 9425
  Petroleum: 7839
  Platinum: 957, 1663
  Salt Mines and Mining: 8409, 9116
  Soils: 1642
  Springs, Hot: 4228
  Volcanoes: 1073, 7423
Franklin Institute: 3974
  Mineral Collections: 4945, 4959
Franklin Institute of New Haven: 4097
Franklin Institute of Philadelphia: 4959
Freestone
  Georgia: 94
  Scotland: 7534
French Institute: 884
Fresenius, C. R.: 884
Friendly Islands
  Volcanoes: 928

Fuels
    Comparisons of: 2185-2188
    Heating power of: 5800
    Types of: 9934
    United States: 1027
Fulgurites
    Origin of: 792, 10802

Gadolin, M.,: 9204
Gannel, M.: 10642
Gas (see Natural Gas)
Gay, M.: 5665
Gay-Lussac, J.: 7675
Gazetteers
    California: 7260
    Delaware: 9370
    Georgia: 7452
    Maryland: 5364, 9370
    New Jersey: 4454, 4456
    New York: 4455, 9972, 9975-9977
    Pennsylvania: 4453, 9369
    Tennessee: 7431
    United States: 7451, 9368
    Vermont: 10236
    Virginia: 6791
Gemstones: 5893, 9932
    Cost of: 10852
    Treatise on: 3805
    United States: 10080
Geodes: 8660
    Formation of: 408
    Maryland: 6273
Geography
    Books: 7446-7450
    Physical: 4609, 4610
    Textbooks: 1504, 4863, 5936, 8170, 8312, 10962, 11038, 11060
Geological Field Trips
    England: 2026
Geological Maps (see Maps, Geological)
Geological Society of Bradford County (Pennsylvania): 4950
Geological Society of Connecticut: 9187
Geological Society of Dublin: 3230
Geological Society of France: 4303, 4304
Geological Society of Fredericksburg: 1307
Geological Society of London: 63, 772, 1143, 1232, 3550, 4306-4311, 7516, 7580, 10889
Geological Society of Pennsylvania: 2087, 3758, 4312-4314, 4948
Geological Survey
    Importance of: 3773, 5161, 5217
    United States: 307
Geology (see also geographical listings)
    Agencies of Change: 38, 39, 70
    Importance of: 9945, 9995, 9996, 10784
    Journal: 3746
    Lectures: 8846, 9799
    Museums: 4019
    Notes: 384, 1056, 1057, 1075, 1147, 1250, 1258, 1292, 1346, 1718, 1751, 1812, 1938, 1945, 2241, 2242, 2402, 2429, 2433, 2494, 2607, 2618, 2651, 2917, 2918, 2936, 3127, 3130, 3135, 3189, 3298, 3358, 3360, 3630, 3632, 3649, 3656, 3661, 3662, 3667, 3669, 3670, 3681, 3726, 3749, 3753-3755, 3761, 3842, 3904, 3989, 4047, 4110, 4129, 4140, 4142, 4338, 4371-4373, 4411, 4518, 4737, 4742, 4749, 4753, 4872, 4908, 4977, 5031, 5221, 5223, 5224, 5247, 5248, 5604, 5763, 5878, 5908, 5931, 6291, 6309, 6334, 6345, 6499, 6509, 6510, 6524, 6530, 6623, 6624, 6632, 6680, 6685, 6689, 6691, 6695, 6710, 6712, 6716, 6735, 6744, 6854, 6876, 6925, 6926, 7104, 7110, 7116, 7139, 7206, 7310, 7318, 7377, 7378, 7440, 7442, 7454, 7555, 7606, 7638, 7645, 7647, 7844, 7850, 7936, 8022, 8071, 8072, 8333, 8338, 8339, 8366, 8402, 8541, 9043, 9243, 9247, 9311, 9343, 9390, 9391, 9494, 9846, 9902, 9903, 9911, 9971, 10524, 10728, 10804, 10990
    Popularity of: 254
    Progress of: 10888
    Relationship to Geography: 1047
    Study of: 255
    Teaching: 452, 511, 596, 4260

Geology (continued)
    Textbooks: 1280, 1455, 1457, 1460, 2083, 2092, 2131, 2160, 2165, 2511, 2513, 2420, 2422, 2426, 2515, 2517, 2518, 2656, 2754, 2757, 3036-3039, 3324, 3325, 3361, 3374, 3382, 3833, 3920, 3921, 4415, 4442, 4445, 4609, 4610, 4805, 4877, 5044, 5218, 5219, 5222, 5132, 5134, 5142, 5157, 5177, 5804, 6125, 6192, 6426, 6500, 6532-6534, 6536, 6540, 6543, 6544, 6732, 6875, 6890, 7256, 7443, 8024, 8107, 8108, 8256, 8320, 8684, 8740, 8751, 8771, 9135, 9292, 9687, 9755, 9866, 10330, 10551, 10798
    Uses of: 1282
    Vermont: 22, 23
Georgia
    Clay, Porcelain: 6240
    Earthquakes: 7094
    Freestone: 94
    Gazetteer: 7452
    Geological Survey: 2657, 2658, 4332-4334
    Geology: 1752, 2657, 2658, 9490
    Gold Mines and Mining: 706, 1161, 3182, 5731, 6881, 7926, 8324
    Greensand: 10927
    Meteorites: 1550, 1936, 3070
    Mill Stone: 2665
    Mineral Springs: 9953
    Minerals, Localities: 145, 147, 2603, 5934
    Mines and Mining: 3009, 5194, 5202
    Natural History: 1544, 4761, 10897
    Paleontology: 712, 4615, 7304
    Paleontology, Invertebrate: 1892, 2662-2664, 7061, 7075, 7125, 7465
    Paleontology, Vertebrate: 69, 1353, 2662-2664, 4814, 7312
    Soils: 2658
    Waterfalls: 3918
Germany
    Footprints: 2653
    Geology: 5097, 7515, 7519
    Glaciers, Evidence for: 5976
    Gold Mines and Mining: 2939, 5898
    Lava: 4407
    Mercury Mines and Mining: 3018
    Meteorites: 7040
    Minerals, Localities: 5920, 5949, 6936, 10459, 11130
    Paleontology, Vertebrate: 2653, 9349
    Salt Mines and Mining: 638
    Scientific Societies: 532
    Silver Mines and Mining: 4868
Gesner, A.: 4059
Geysers
    Guatemala: 8666
    Iceland: 1064, 1066, 1727, 3103, 4752, 5255, 5268, 5269, 6607, 7145, 7238, 7389, 8140, 10683
Giant's Causeway, Ireland: 3418, 3657, 4139, 5398, 5508, 7143, 10219
Gibbons, H.: 562
Gibbs, G.: 574, 7070, 7088
    Mineral Collections: 348, 427, 1829, 7345
Gibralter
    Geology: 1293, 1446, 3409
Gilmor, R.: 959
Glacial Theory: 64, 232, 2302, 3319, 3712, 4178, 4183, 4494, 5257, 6611, 9789
    Iceberg Theory of Drift: 3113, 3197, 4913, 4915, 10955, 10958
Glaciers
    Action of: 3117, 3118
    Alps: 5509
    Europe: 2956, 3903
    Growth of: 5869
    Himalayas: 10033, 10034
    Motion of: 6719
    Notes: 7209, 7435
    Structure of: 5658, 8930, 8935

Glaciers (continued)
  Switzerland: 7380
  Temperature of: 6794
Glaciers, Evidence for: 1402
  Boulders: 416, 2874, 7632, 8911, 8916, 9604, 10091
  Drift: 3527, 3558, 3566, 5143, 5151, 5152, 5156, 5169, 5171, 5174, 6049, 8341, 8720-8723, 8867, 8884, 8897, 8911, 8922, 8931, 8932, 8939, 9579, 9764, 9784, 9794, 9799, 10030, 10883, 10892
  England: 9394
  Erratics: 221, 2363, 2365, 4611, 4612, 7583, 7589, 10052
  Europe: 3110, 6539
  Furrows: 28
  Germany: 5976
  Labrador: 11083
  Lake Superior District: 81, 3124, 9579
  Massachusetts: 88, 1059, 3898, 4501, 8911, 8916, 10030, 10745
  New Brunswick: 5803
  New England: 1498, 3107-3109, 3111, 3114, 3115, 3120, 5580, 5590, 5591, 5602, 5620, 5632, 5638, 5643, 5672, 5758
  New Hampshire: 5669, 5672
  New York: 3106, 3164, 4280, 8373, 8709, 8722
  Ohio: 6042, 6431, 8718, 9371, 10091
  Pennsylvania: 8867, 8879
  Polishing: 29
  Potholes: 5185, 8345
  Scotland: 61
  Scratches and Polishing: 37, 1466, 3196, 10222, 10234, 10235
  Switzerland: 62, 71, 74
  United States: 2223, 2224, 6903, 8714, 8715
  Vermont: 35
  Virginia: 8867
  Wales: 6605
Glass: 8501
  Analysis, Chemical: 8199
  Manufacture of: 3922
  Spun: 4218
Glossary: 262, 2425, 4083, 4571, 4988, 5189, 6528, 6646, 7237, 7738, 7742, 7999, 9136
Godman, J. D.: 10645
Godon, G.: 7346
Gold
  Ore Processing: 1773, 1774, 5645, 7590
  Assaying: 10231-10233, 10488
Gold Mines and Mining: 195, 1150, 1155, 1283, 3000, 3227, 3848, 7616, 10098
  Africa: 54, 971
  Aruba: 8376
  Australia: 879, 983
  California: 151, 990, 1152, 1153, 1174, 1176, 1264, 1268, 2137, 2138, 2180, 2360, 2725, 2835, 3014, 3267, 3421, 3682, 3683, 3891, 3919, 4519, 4821, 4919, 5475, 5760, 5761, 5967-5970, 6201, 6361, 6402, 6579, 6580, 6590, 6817, 6818, 7260, 7262, 7985, 8044, 8335, 8574, 8756, 8807, 8820, 9147, 9283, 9293, 9455, 9838, 9852, 9987, 9991, 10207, 10214, 10259, 10420, 10421, 10506, 10637, 10867, 10868, 10923, 10960
  Canada: 1173, 1401, 9788, 9826
  Egypt: 3427
  Fiction: 10717, 10718
  Georgia: 706, 1161, 3182, 5731, 6881, 7926, 8324
  Germany: 2939, 5898
  Indiana: 4100
  Maryland: 498, 993, 3575, 5892
  Mexico: 3362, 7023, 7370
  Michigan: 10814, 10870
  New England: 7611
  North Carolina: 290, 537, 1039, 1349, 1371, 1782, 1914, 1925, 2374, 2457, 3182, 3362, 4124, 4368, 4400, 5436, 5844, 6344, 7000, 7001, 7003, 7043, 7052, 7079, 7177, 7251-7253, 7348, 7952, 8031, 8197, 8276, 8304, 8374, 8489, 8735, 9038, 9039, 9864, 10260, 10261, 10310, 10779

Gold Mines and Mining (continued)
  Notes: 175, 286, 1151, 1267, 1722, 3801, 3824, 4256, 5389, 5852, 5939, 6024, 6139, 6300, 6967, 6997, 7136, 7396, 7525, 7527, 7616, 8114, 9011, 10104, 10856
  Pennsylvania: 4954
  Refining: 159
  Russia: 419, 509, 634, 942, 1252, 2234, 4034, 4086, 4094, 4918, 5383, 5384, 5390, 5440, 5443, 5445, 5478, 5483, 6008, 7550, 7560, 7571, 7724, 9252, 9358
  St. Domingo: 4402
  Southern United States: 245
  Texas: 10871
  United States: 3617, 5726, 6286, 6922, 8650, 9359
  Vermont: 152, 446, 3976
  Virginia: 966, 1159, 2045, 2441, 3088, 3182, 3397, 4272, 4603, 5500, 5656, 6208, 6920, 6923, 8393, 8970, 9565-9569, 9731-9734, 9736, 9738, 9741, 9744, 9745, 9753, 10113, 10124, 10149, 10596, 10640, 10641
  Washing: 310, 1468, 1557, 1793, 1872, 1880, 2184, 2955, 3195, 3439, 3593, 4555, 5739, 5741, 5742, 5840, 6019, 8154, 8556, 8599, 8804, 9173, 9236, 9328, 9387, 9584, 10023, 10067, 10070, 10071, 10985
Goldfuss, A.: 960, 3767, 10363
Goniometer: 3369, 3643, 4496, 10914
Graham Island
  Volcanoes: 107
Granite (see also Building Stone): 474, 6279, 6598, 6683, 9388
  Building Stone: 6518
  India: 376
  Italy: 398
  Maine: 3727, 7892, 10623
  Massachusetts: 3489, 3711
  New England: 293
  New York: 2350, 7837
  Norway: 697
  Notes: 3843
  Polishing of: 404, 3698
  Russia: 411
  Scotland: 10306
  South Carolina: 3012
  Types of: 297
Gravel
  England: 8307
Gravity (see Earth, Gravity)
Great Britain (see also England; Ireland; Scotland; Wales)
  Coal Mines and Mining: 4076, 6077, 8491
  Copper Mines and Mining: 413
  Earthquakes: 222
  Geological Survey: 4061, 6610
  Geology: 2597, 7891, 8406, 9460, 9461
  Landslide: 222
  Lead Mines and Mining: 103, 161, 951, 5405
  Mineral Collections: 5859
  Mines and Mining: 5403, 7573, 8309, 8504, 8538, 10709
  Museum, Economic geology: 5260
  Natural Gas: 219
  Paleobotany: 5253
  Paleontology, Invertebrate: 10329
  Paleontology, Vertebrate: 55, 7186, 7187, 7508
  Rocking stones: 5510
  Salt Mines and Mining: 104
Great Lakes Region
  Exploring Expedition: 349
  Geology: 2452
  Origin of: 7285, 7307
  Paleontology, Invertebrate: 3048
  Silver Mines and Mining: 1354
Greece
  Building stones: 10328
  Coal Mines and Mining: 6200
  Earthquakes: 2275
  Geology: 8022, 9750
  Mines and Mining: 8818
Greenland
  Geology: 8377
  Minerals, Localities: 632, 10041-10043
  Natural History: 3446

Greensand: 10394
   Delaware: 9955
   England: 5827
   Georgia: 10927
   New Jersey: 5678, 8940, 9084, 9085, 9453, 9955, 11070
   Uses of: 1963, 1964, 2268
   Virginia: 8962, 8964, 9076, 9077, 9081
Greville, C. F.: 1084
   Mineral Collections: 2114
Griscom, J.
   Mineral Collections: 4568, 9730
Grottos
   Antiparos: 1108, 4432, 9129
   Russia: 214
Guadaloupe
   Paleontology, Vertebrate: 7502
   Volcanoes: 7017, 8254
Guatemala
   Geyser: 8666
Guiana
   Deltas: 4608
Guiton, M.: 7019
Gurney, Dr.: 7211
Gypsum
   Analysis, Chemical: 2630
   Connecticut: 5250
   New York: 2394, 2395
   South Carolina: 4347
   Uses of: 2064, 2065

Hall, F.: 788
   Mineral Collections: 673
Hammer, Geological: 5080
Hare, R.: 360, 8471, 8474, 9337, 9612, 9614, 9616, 9617, 9623, 9642, 9644, 9647, 9648, 9650, 9651, 9659, 9662, 9663
   Blowpipe: 2620, 7228
Harlan, R.: 793, 8092
Hartz Mountains (see Germany)
Harvard College: 892, 4713, 6079, 7677
   Mineral Collections: 7046, 8620, 9131, 10493, 10746, 10747, 10754
   Natural History Society: 4844, 4845
Haughton, M. A.: 9318
Haüy, R. J.: 410, 5857, 7545, 7546, 9201
   Mineral Collections: 7540, 7543
Hawaii
   Geology: 4196
   Meteorites: 1708, 5501
   Minerals, Localities: 481
   Natural History: 3482, 5937, 10301
   Volcanoes: 1034, 1202, 1253, 2465, 2466, 2666, 2845, 2848, 2859, 2861, 2862, 3316, 4438-4441, 4634, 4860-4862, 5598, 5633, 5888, 5937, 6578, 7381, 7581, 8174, 8530, 8727, 9169, 10017-10019, 10053, 10054
Hawkins, T.: 646, 6727, 9724
Heidelberg
   Mineral Collections: 6189, 7331
Helmersen, Prof.: 890
Helvetic Society of Natural Sciences: 4978
Heuland, H.
   Mineral Collections: 450, 1806
Hibbert, S.: 623, 7538
Higgins, S.: 8438
Himalayas
   Geology: 5212
   Glaciers: 10033, 10034
   Paleontology, Vertebrate: 2284
Historical and Philosophical Society of Ohio: 10092, 10093
History of Geology: 4370, 9342
History of Mineralogy: 7419
History of Science: 3745, 5716, 7189, 8398, 8430
Hitchcock, E.: 2090
Hoeffer, F.: 231
Holbrook, J.
   Mineral Collections: 7642
Hooker, Dr.: 4508
Horsford, E. N.: 892
Horton, W.: 834
Hosack, D.
   Mineral Collections: 4531, 5285
Houghton, D.: 917, 3311, 4321, 5631
Hudson Association: 3
Hudson River
   Sediments: 4298
Hughes, V.: 7017

Hugi, F. J.: 9498
Humboldt, A.: 1801, 6333, 6602, 7214, 9409
Humor
   Geological Puns: 5727, 8908
Hungary
   Meteorites: 777, 1785
   Minerals, Localities: 4405, 5537, 5890
   Opals: 1672
   Paleontology, Vertebrate: 4584
   Steatite: 3080
Hutton, J.: 2026, 2607
Hydraulic Limestone (see Cement)

Iceberg Theory of Drift: 2667, 3113, 4913, 4915, 10955, 10958
Iceland
   Basalt: 1062
   Geology: 1181, 6329, 8413
   Geysers: 1064, 1066, 1727, 3103, 4752, 5255, 5268, 5269, 6607, 7145, 7238, 7389, 8140, 10683
   Natural History: 4981, 8810, 8811, 9405, 10707
   Springs, Hot: 1054
   Volcanoes: 2499, 3102, 4300, 6267, 6998, 7222, 7387, 10307
Igneous Processes: 1582, 1586, 2803, 3029
   Notes: 1965
Igneous Rocks: 6637, 6638
Illinois
   Caves: 8137
   Coal Mines and Mining: 4001, 8493, 8576
   Earthquakes: 7932
   Geological Survey: 8570
   Geology: 5503, 8050, 8061, 8065, 8068-8070, 8073, 8074, 9282, 9515
   Iron Mines and Mining: 8572
   Lead Mines and Mining: 242, 6032
   Maps, Geological: 8059-8062, 8067
   Minerals, Localities: 369, 1464, 5744, 8832
   Mines and Mining: 8194, 9531
   Natural Bridge: 9315
   Natural History: 1558, 7261
   Paleontology, Vertebrate: 6116
   Salt Mines and Mining: 10475
Ilmenium
   Discovery of: 876
India
   Caves: 7774, 7847
   Coal Mines and Mining: 895
   Diamond Mines and Mining: 1254, 1979, 4181, 6370, 8340, 8344
   Earthquakes: 378, 864, 929, 9109
   Geology: 496, 941, 7338, 7855
   Granite: 376
   Iron Mines and Mining: 895
   Lithographic Stone: 952
   Mercury Mines and Mining: 1853
   Meteorites: 4
   Minerals, Localities: 920, 8596, 10252
   Paleobotany: 5251
   Paleontology, Vertebrate: 8597
Indiana
   Caves: 20, 8483
   Geological Survey: 5502, 5523, 5524, 8046-8048
   Geology: 6591, 8046-8048, 8063, 8447, 10024
   Gold Mines and Mining: 4100
   Minerals, Localities: 2503
   Mines and Mining: 9867
   Oolite: 2404
   Paleobotany: 8052, 8055
   Paleontology, Vertebrate: 2281, 7980, 7982, 7983
   Salt Mines and Mining: 10483
   Silver Mines and Mining: 9117
Indiana Historical Society: 5525
Iodine
   In Minerals: 316, 10600
   Mineral Springs: 1799
Ionian Islands
   Earthquakes: 10681
Iowa
   Copper Mines and Mining: 9994
   Geology: 165, 8050, 8061, 8065, 8068-8070, 8073, 8074
   Guidebook: 7856, 7857
   Maps, Geological: 8061, 8062, 8068

Iowa (continued)
  Marble: 9099
  Meteorites: 1493, 4220
  Mines and Mining: 9867
  Natural History: 5, 6082, 7856, 7857
  Paleontology, Invertebrate: 3544, 8083
  Paleontology, Vertebrate: 3123
  Salt Mines and Mining: 10464, 10466
  Sulphur: 1039
Iran
  Geology: 8022
Ireland (see also Great Britain)
  Caves: 5909, 6378, 7386, 7395, 7723, 11123
  Geology: 2765, 10788
  Giant's Causeway: 923, 1476, 3418, 3657, 4139, 5398, 5508, 7143, 10219
  Lead Mines and Mining: 824
  Mineral Springs: 7532, 7654
  Minerals, Localities: 648, 7229
  Paleontology, Invertebrate: 3228
Iridium: 1659
Iron
  Analysis, Chemical: 3581, 6774
  Ore Processing: 102, 127, 128, 129, 1302, 1760, 1796, 1873, 1918, 2042, 2043, 2405, 2416, 2602, 2608, 2685, 2789, 3475, 3581, 3625, 3627, 3851, 3860, 3987, 3990, 3991, 4379, 4465, 4636, 4746, 5767, 5774, 5777, 5786, 6371, 7889, 8045, 8232, 8445, 8488, 8617, 8769, 8790, 8791, 9037, 9886, 9910, 9920, 10158, 10903, 11037
  Soils: 8318
Iron Mines and Mining: 2608, 3626, 7620
  Alabama: 11037
  Algeria: 3946
  Connecticut: 361, 9485
  England: 3268, 3853, 3858
  Europe: 3856, 3861, 3863, 4657
  Illinois: 8572
  India: 895
  Lake Superior District: 10925
  Maryland: 125, 1762, 3031, 3137, 5354
  Michigan: 5297
  Mississippi: 3015
  Missouri: 1347, 3266, 3782, 3809, 3810, 3812, 3813, 4167, 4660, 7244, 7245, 8950, 9516
  New England: 4920
  New Hampshire: 4375, 6968, 7705, 10170
  New York: 308, 500, 1141, 2679, 3503, 3504, 3526, 4294, 4670, 5770, 5929, 8748
  Norway: 9372
  Notes: 2303, 3808, 4859, 5660, 7233, 7275, 7401, 7620
  Ohio: 2400
  Pennsylvania: 1149, 1670, 1770, 2339, 4149, 4160, 4510, 4633, 4956, 5667, 5767-5769, 5779, 6523, 6612, 6704, 6708, 8214, 8218, 8441, 8944
  Portugal: 6285
  Scotland: 5756
  Sweden: 508, 3398, 5420, 9372
  Turkey: 3625
  United States: 550, 2341, 2393, 4168, 5201, 6037, 8435, 8443, 10069
  Vermont: 4376
  Virginia: 268, 5365, 9570, 9571
  Wales: 3917
  West Virginia: 10800
Iron Ores
  Analysis, Chemical: 2615, 4010, 5771
Islands
  Origin of: 7791
Isle of Man
  Paleontology, Vertebrate: 375
Isle of Skye
  Minerals, Localities: 4563
Isomorphism: 4336, 5815, 7628, 9225, 9227
Isostacy (see Elevation, Changes in)
Italy
  Artesian Wells: 7005
  Caves: 565, 4000
  Earthquakes: 622, 3321, 4104, 4745, 5730, 6317, 6406, 6601, 7126, 7342, 7398, 7557, 8279, 9107, 9415, 10521, 10522
  Geology: 162, 1923
  Granite: 398
  Lagoons: 1003
  Marble: 1255, 4148

Italy (continued)
  Mercury Mines and Mining: 9032
  Meteorites: 334, 878, 1800, 9417, 10687
  Mineral Springs: 4586
  Minerals, Localities: 4162, 6202, 7367, 7411, 7412, 8754, 8795, 9210, 10081, 10181, 10515
  Paleobotany: 5900, 6348
  Paleontology, Vertebrate: 999
  Scientific Meetings: 924
  Scientific Societies: 827
  Spring: 1368
  Sulphur Mines and Mining: 1869
  Volcanoes: 394, 451, 527, 565, 619, 972, 1029, 1085, 1089, 1106, 1154, 1183, 1207, 1480, 2054, 2329, 2408, 2791, 3277, 3278, 3318, 3580, 3618, 3874, 4495, 4573, 4607, 5521, 5697, 5714, 5715, 6317, 6406, 7142, 7146, 7182, 7358, 7361, 7388, 7390, 7391, 7409-7412, 7427, 7563, 7579, 7591, 7706, 7708, 7793, 7913, 8085, 8129, 8132, 8294, 8384, 8400, 8487, 8490, 8531, 8556, 9373, 9978, 9979, 10229, 10293, 10315-10317, 10647, 10993, 11122

Jackson, C. T.: 804, 4965, 5017, 5018, 6698, 7658, 10467
Jackson, J. R.: 1017
Jacquelin, V. A.: 9219
Jamaica
  Caves: 6328
  Earthquakes: 2641, 5690, 8738
  Geology: 24
  Mines and Mining: 4028
  Platinum Mines and Mining: 7033
Jameson, R.: 7798, 8411, 9195
Japan
  Geology: 8621
  Mines and Mining: 1272
Java
  Caves: 4477, 7711
  Sulphur: 320
  Vitriol Lake: 5904, 6196, 6281
  Volcanoes: 1230, 1265, 4521, 5862, 7384
Jefferson, T.: 7309
Jet (see Coal)
Jobert, A. C. G.: 961
Johnston, J. F. W.: 275, 2724, 2726, 2727, 5536, 5751, 7669, 9958
Joints: 5583, 6904
  Connecticut River Valley: 5141
  New York: 5343
Jordan
  Dead Sea: 1261, 4410, 6581, 6656, 6771, 6780, 9588
  Geology: 1068, 6656
  Natural History: 6581, 6656
Jordan River
  Analysis, Chemical: 5683
Jura Mountains
  Geology: 7553, 10185

Kansas (see also Arkansas)
  Geology: 5810, 9189
  Springs, Hot: 8423
Kentucky
  Caves: 1035, 1053, 1294, 1721, 1784, 1871, 2080, 2219, 2337, 2378, 2379, 2387, 2601, 2689, 3422, 3706, 3731, 3815, 4173, 4658, 5020, 5273, 6438, 6498, 7385, 7686, 8336, 8342, 8391, 8579, 8643, 10684, 10875, 11098
  Copperas: 10850
  Descriptive Account: 4482
  Geological Survey: 5943, 6898
  Geology: 1116, 3001, 8075, 8076, 8080, 8081, 8647, 10873, 11094, 11095
  Marble: 3443
  Marl, Agricultural: 8243, 8244
  Medical Topography: 3256, 5747
  Meteorites: 8634, 10400
  Mineral Springs: 3224, 10819, 10834, 10839, 11091, 11096
  Minerals, Localities: 3095, 3432, 8628
  Mines and Mining: 5944
  Natural Gas: 7056, 7098, 8297
  Natural History: 2215, 2483, 3822, 3823, 5511, 5512, 6645, 6665, 6787, 6788, 7336, 7790, 8627, 8635, 8636

Index

Kentucky (continued)
  Paleobotany: 9221
  Paleontology, Invertebrate: 704, 6198, 7075, 8630, 8633, 8654
  Paleontology, Vertebrate: 1115, 1817, 2484, 2485, 2493, 2634-2636, 2638, 4232, 4781, 4790-4792, 6552, 7073, 7684, 8155, 8664, 9689, 10513, 11010, 11011
  Petroleum: 260, 742, 4128, 5855, 5903, 7886, 7887, 11107
  Salt Licks: 11092
  Salt Mines and Mining: 8651
  Soils: 6209, 6792, 9423
  Springs, Hot: 5209
  Volcanoes: 7931
Kerguelin Island
  Geology: 6353
Kirwan, R.: 2119, 8280
Klaproth, M. H.: 3839, 3840, 8287
Knox, G.: 6956
Koch, A.: 3708, 4813, 4926, 4929, 5266, 5267, 7901, 8017, 8913, 9781, 10824
Koninck, L. de: 955
Kupffer, A. T.: 738

Laboratory Safety: 9610, 9618
Labrador
  Glaciers, Evidence for: 11083
Lake Huron
  Geology: 1698, 1701
Lake Superior District (see also Michigan)
  Basalt: 3287
  Geological Survey: 3404, 3927-3933, 3936-3941, 7903, 7905-7907
  Geology: 78, 80, 1308, 1510, 2204-2206, 2208-2212, 2230, 3287, 4383, 5675, 5679, 6606, 8371, 8920, 9155
  Glaciers, Evidence for: 81, 3124, 9579
  Iron Mines and Mining: 10925
  Maps, Geological: 2209, 2210, 3928-3930, 3938, 3941, 4500, 5652, 5653
  Minerals, Localities: 1972, 3053, 3174, 3175, 5615, 6113, 9553, 10205, 10210, 10211, 10215, 10344, 10908, 10910-10912, 10915, 10917, 10920
  Mines and Mining: 5618, 5621, 5623, 5628, 5635, 5637, 5639-5641, 5646, 5650-5653, 5655
  Natural History: 83, 6033, 9155
  Ornithichnology: 3133
  Salt Mines and Mining: 7172
  Silver Mines and Mining: 2375
  Vanadium: 10211, 10215
Lake Superior Region
  Copper Mines and Mining: 600, 1165, 1172, 1177, 1179, 1269, 1390, 1492, 1510, 1733, 1790, 2277, 2279, 2308, 2309, 2336, 2640, 2693, 3004, 3174, 3175, 3207, 3470, 3529, 3608, 3609, 3684, 3935, 3937, 3939, 4164, 4498-4500, 4511-4513, 4598, 4755, 4904, 5061, 5196, 5197, 5204-5206, 5208, 5291, 5292-5304, 5306, 5312, 5319, 5321, 5322, 5339-5341, 5361, 5491, 5492, 5618, 5621, 5623, 5628, 5635, 5637, 5639-5641, 5646, 5650-5653, 5655, 5675, 5679, 5851, 5874, 5950, 6021, 6040, 6358, 6385, 6458, 6461-6463, 6606, 7105, 7171, 7198, 7235, 7760, 7957, 8037, 8253, 8274, 8381, 8477, 8482, 8518, 8586, 8739, 8773, 9098, 9154, 9155, 9166, 9259, 9265-9267, 9276, 9278, 9281, 9552, 9574-9580, 9802, 9843, 10025, 10086, 10087, 10210, 10861, 10908-10912, 10915-10920, 10926, 10936, 10937, 10942, 11121
Lamouroux, Dr.: 8410
Landslides: 9296, 9348
  Calabria: 2222
  Canada: 1797
  Causes of: 11132
  Great Britain: 222
  Maine: 1607
  New Hampshire: 1465, 9684, 10947
  New York: 4126
  Russia: 223
  South Carolina: 5952
  Switzerland: 5881
  Timor: 930

Landt, G.: 9410
Lane, E.: 6881
Lanthanum: 7497
Laplace, P. S.: 9190
Latanium: 1665
Lava (see also Volcanoes)
  Germany: 4407
Lawrence, B.: 11093
Lawrence, J.: 907
Lead Mines and Mining: 2612, 7622, 11104
  Arkansas: 10812, 10815
  Europe: 4035
  Great Britain: 103, 161, 435, 951, 5405
  Illinois: 242, 6032
  Ireland: 824
  Louisiana: 1369, 6943, 7054
  Massachusetts: 3323, 5070, 5331, 6945, 7597, 7598, 9607
  Mexico: 3084
  Mississippi River Valley: 5696, 6442, 6453, 10082-10084, 10087, 10503, 10702
  Missouri: 269, 1197, 1312, 1912, 1947, 2306, 2468, 2578, 2580, 3243, 5195, 5316, 5960-5965, 6232, 6442, 6453, 6647, 7498, 8064, 8238, 8480, 8953, 9254, 9255, 9257, 9260, 10352, 10354, 10472, 10473, 10476-10480, 10482
  New Hampshire: 2475
  New Jersey: 4138
  New York: 261, 3796, 6248, 9974
  North Carolina: 1370, 6119
  Notes: 6026, 7405, 7622, 8573, 10094
  Ore Processing: 3590
  Pennsylvania: 4949, 5755, 7058, 10878
  Portugal: 1186
  Scotland: 2066
  United States: 439, 1563, 3688, 3909, 5200, 5207, 6973
  United States, Western regions: 1166
  Virginia: 8444, 10639
  Wisconsin: 5195, 6700
Lead ore
  Analysis, Chemical: 5922
Lecomte, C.: 9218
Lectures
  Mineralogy: 4420
  Silliman, B: 4261, 4262
Lederer, L.: 782
  Mineral Collections: 9735
Lehigh Coal and Navigation Company: 4146, 6574
Leibnitz, G. W. von: 7254
Leonhard, G.: 808, 6187
Leonhard, K. C. von: 885
  Mineral Collections: 2157
Leske, Prof.
  Mineral Collections: 7078
Liberia
  Gold Mines and Mining: 54
Libraries, Scientific: 117
  American Academy of Arts and Sciences: 180
Lignite (see Coal)
Lime, Agricultural: 27, 664, 1344, 1345, 1350, 1414, 1768, 2135, 2136, 2229, 2604, 2683, 2690, 2691, 2692, 3067, 3465, 3476, 3485, 3562, 4446, 4627, 4751, 4846, 5113, 5129, 5215, 5225, 5528, 5698, 5699, 6029, 6030, 6260, 6407, 6408, 6508-6510, 6784, 6964, 7468, 7487-7491, 7495, 8243, 8244, 8357, 8361, 8555, 8612, 8729, 8849, 8859, 8871, 8917, 8961, 9050, 9051, 9053-9056, 9058, 9060, 9061, 9063, 9065, 9066, 9068-9075, 9078-9080, 9082, 9083, 9086, 9088, 9251, 9548, 9556, 9557, 9912, 9923, 9956, 10072, 10378, 10424, 10460, 10581, 10583, 10590, 10881, 11019
  Analysis, Chemical: 8958-8960, 8963
  European: 469
  Notes: 538, 3615, 3616, 6215, 6216
  South Carolina: 9883
  Virginia: 1707
Limestone: 2141, 2609, 4462, 9330
  Alabama: 5914
  Analysis, Chemical: 2743, 5757
  California: 10982
  Folded: 3157
  Jura Mountains: 10185
  Massachusetts: 3157
  Missouri: 4224, 9260
  New Hampshire: 7425

Limestone (continued)
  New York: 3163, 3380, 4685, 7368, 8937
  Notes: 4462
  Ohio: 2390
  Pennsylvania: 2628, 2629, 3380, 8937
  South Carolina: 5396, 5397
  Vermont: 2324, 4645
  Virginia: 4466
Linnaean Association of New England: 7679
Lipari Islands
  Minerals, Localities: 10046
  Volcanoes: 7241, 7556
Literary and Philosophical Society of Quebec
  Mineral Collections: 6236
Lithification: 8043
Lithium
  Analysis, Chemical: 8677, 10013
  Discovery of: 333, 1304, 1656, 9245
Lithographic Stone
  France: 822
  India: 952
  Spain: 7413
Logan, W. E.: 977
London Geological Journal: 663, 883, 910
London University: 1013
Lonsdale, W.: 4310
Loss, C.: 7022
Louisiana: 2134
  Geology: 2748, 2900, 2901, 3915, 9594, 9595
  Lead Mines and Mining: 1369, 6943, 7054
  Medical Topography: 5027, 5028
  Meteorites: 407, 601, 4378, 6984, 7096, 7099, 9476
  Minerals, Localities: 1942, 2115, 2900, 2901
  Mines and Mining: 5438, 10026
  Natural History: 1943, 1944, 2900, 2901, 5027, 5028, 6194, 9593, 10027, 10028
  Paleobotany: 2262
  Paleontology: 1808
  Paleontology, Vertebrate: 2248, 2264, 2265, 4798, 6178, 7930
  Salt Mines and Mining: 4275
  Springs, Hot: 6655
Luxemburg
  Artesian Well: 879
Lyceum of Natural History
  New York: 10741
Lyell, C.: 629, 657, 759, 1039, 1143, 2164, 6934, 8327, 8882, 9319, 9769

Maclure, W.: 478, 3664, 7291, 7355, 7473, 7475, 7479, 7480, 7935, 7939, 8049, 8268, 8270, 7084
  Mineral Collections: 6643, 6644
Madagascar
  Geology: 6283
  Meteorites: 9547
Madeira
  Geology: 236
Maine
  Gazetteer: 4931
  Geological Survey: 5549-5552, 5556, 5558-5561, 5566, 6668-6677, 6687, 6697, 6711
  Geology: 1409, 2775, 4697, 5112, 5118, 5190, 5237, 5549-5552, 5556-5561, 5563, 5564, 5566, 6696, 8546, 10015, 10443
  Granite: 3727, 7892, 10623
  Landslide: 1607
  Meteorites: 2424, 5043, 9555, 10737, 10755
  Minerals, Localities: 4748, 5619, 5626, 5885, 6686, 7656, 8146, 10417, 10739
  Mines and Mining: 4559, 4560, 5553, 5570, 6039, 6684, 6715
  Natural History: 1946, 2775, 4108, 10984
  Paleontology, Invertebrate: 3125, 5199, 5581, 7183
  Paleontology, Vertebrate: 4463, 5636
  Slate: 294, 5348, 5586, 6107, 6108
  Tertiary Formation: 5642
Malaya
  Geology: 934
Malta
  Geology: 10519
Manchester Geological Society: 6721
Manganese
  Analysis, Chemical: 2699, 3384
  Assaying: 10444, 10445, 11129

Manganese Mines and Mining: 1638, 10254
  New York: 1789
  Ore Processing: 1503
  Pennsylvania: 11043
  Vermont: 4421, 4646
  Virginia: 1525
Mantell, G. A.: 588, 642, 756, 768, 773, 774, 811, 830, 861, 898, 962, 963, 998, 1458, 6526, 9707
  Mineral Collections: 618
Maps, Geological
  Alabama: 10441
  Coloring of: 8053, 8054, 10129
  Connecticut: 6197, 6879, 8229
  Connecticut River Valley: 5071, 5079, 9853
  Construction of: 5584
  Cuba: 10144
  France: 7587
  Illinois: 8059-8062, 8067
  Iowa: 8061, 8062, 8068
  Lake Superior District: 2209, 2210, 3928-3930, 3938, 3941, 4500, 5652, 5653
  Massachusetts: 2870, 3142, 3155, 5085, 5099, 5103, 5106, 5140, 5160, 7597
  Mexico: 11008
  Michigan: 4302, 5042, 5320, 5340
  Missouri: 2306
  New Hampshire: 5612
  New Jersey: 8864
  New South Wales: 2836
  New York: 2681, 3359, 3496, 3507, 3508, 3511, 3518, 3540, 6227, 6556, 6906, 6908-6910, 7749, 11113, 11116
  North Carolina: 7251, 7257, 8198
  Nova Scotia: 5539, 5541, 5578
  Ohio: 5054, 6424
  Pennsylvania: 4702, 6005, 10116, 10118, 10148, 10348
  Peru: 8802
  Rhode Island: 5569
  Sicily: 2919
  South Carolina: 10435, 10437
  Tennessee: 8198, 10367, 10376, 10383-10385, 10393
  United States: 2421, 2423, 5063, 6555, 6614, 6619, 8257, 10654
  United States, Western regions: 4682, 6081
  Virginia: 8776
  Wisconsin: 8061, 8062, 8068, 8078
Marble (see also Building Stones): 474
  Artificial: 1877
  Color of: 6855, 7372, 9199, 9301
  Connecticut: 346, 8478
  Flexible: 3158
  Iowa: 9099
  Italy: 1255, 4148
  Kentucky: 3443
  Markings on: 2075
  Massachusetts: 287, 3158, 6944, 6946, 9127, 10740
  New England: 293, 9246
  New York: 258, 6324, 9653
  North Carolina: 3724
  Pennsylvania: 4946
  Persia: 5894
  Strength of: 10322
  Tennessee: 99, 7607, 10377
  Tuscany: 2770
  Vermont: 719, 1592, 4644, 10010
  Wisconsin: 8575
Markoe, F.
  Mineral Collections: 944
Marl, Agricultural: 27, 469, 538, 664, 1344, 1345, 1414, 1707, 2136, 2229, 3067, 3476, 3562, 3615, 3616, 4446, 4751, 4846, 5215, 6029, 6030, 6215, 6216, 6407, 6408, 6508-6510, 8555, 8612, 8849, 8859, 8871, 8958-8960, 8961, 8963, 9050, 9051, 9053-9056, 9058, 9060, 9061, 9063, 9065, 9066, 9068-9075, 9078-9080, 9082, 9083, 9086, 9088, 9251, 9548, 9556, 9557, 9883, 9912, 9923, 9956, 10072, 10378, 10881
  Alabama: 5698, 5699
  Connecticut: 5129
  Delaware: 1768
  Florida: 2690
  Kentucky: 8243, 8244
  Maryland: 6260, 8361

Index

Marl, Agricultural (continued)
  Massachusetts: 5113
  Mississippi: 5528
  New Jersey: 1350, 2604, 2683, 3320, 4627, 5225, 7468, 8357, 8729, 8917, 10460
  New York: 3465
  North Carolina: 2135, 2691, 2692
  South Carolina: 3485, 10581, 10583, 10590, 11019
  Vermont: 6784
  Virginia: 6260, 6964, 7487-7491, 7495, 8361, 10424
Martinique
  Earthquakes: 304
Maryland
  Amber: 10333, 10343
  Clay, Porcelain: 1521
  Coal Mines and Mining: 125, 126, 301, 1762, 2728, 3031, 3137, 3214, 3598, 3857, 4180, 4330, 4331, 5354, 5355, 5477, 5776, 8451, 8548, 8558, 9847, 10455, 10456
  Copper Mines and Mining: 9982
  Copperas Mining: 3282
  Gazetteer: 5364, 9370
  Geodes: 6273
  Geological Survey: 250, 251, 267, 2243, 2244, 3231-3242, 3244-3252, 4283, 4317, 4326, 5738, 5951, 6609, 6797-6804, 8438
  Geology: 105, 106, 4874, 4884, 7433, 10123
  Gold Mines and Mining: 498, 993, 3575, 5892
  Iron Mines and Mining: 125, 1762, 3031, 3137, 5354
  Marl, Agricultural: 6260, 8361
  Meteorites: 2271, 2272, 2347, 4838
  Mineral Springs: 3845
  Minerals, Localities: 1474, 1478, 1528, 1640, 2257, 2265, 2428, 3073, 4396, 4423, 4426, 4875, 4879, 6205, 7060, 9440, 10453, 10454, 10568
  Mines and Mining: 1164, 3282, 4102, 4525, 5848, 5849, 6573, 6812, 8750, 8752, 9535, 10662, 10787
  Natural History: 5364
  Paleobotany: 2192, 7069
  Paleontology: 8084, 9590
  Paleontology, Invertebrate: 1423, 1431, 1440, 2544, 2546, 2570, 2571, 3458, 6093, 9207, 10661
  Paleontology, Vertebrate: 4812, 8196
  Slate: 7049
  Tertiary Formation: 6781
Maryland Academy of Science and Literature: 6587, 6805-6809
Massachusetts (see also Connecticut River Valley)
  Artesian Wells: 9812
  Basalt: 5130
  Boulders: 1794
  Coal Mines and Mining: 560, 1826, 3795, 3797, 5144, 5554, 8792
  Conglomerate: 5083, 5657
  Dighton Rock: 2953, 11000
  Dikes: 9803
  Earthquakes: 164, 1489, 1490, 2207, 2216, 2276, 2330, 2331, 2472, 2480, 2486, 2659, 2660, 2889, 3201, 3322, 3433, 3605, 3679, 3680, 3877, 3961, 3962, 4447, 4448, 6230, 6718, 6866-6868, 6871, 6939-6941, 7112, 7330, 7430, 7858, 8110, 8171, 8319, 8580, 8591-8593, 8948, 8949, 9428, 9429, 9898, 9921, 10946, 10964, 10979, 11002-11005
  Geological Survey: 2090, 4249, 4318, 4458, 5099-5101, 5103, 5105, 5106, 5110, 5123, 5128, 5131, 5135-5137, 5140, 5158, 5160, 5556, 5558, 5559, 6819-6843, 7644, 7646, 7660
  Geology: 1548, 2871, 2872, 3141, 3143, 3154, 3156, 3160, 3323, 4970, 5073, 5084, 5099-5101, 5105-5111, 5126, 5158, 5160, 5558, 5559, 7644, 7646, 7660, 8111, 8516, 8583, 8726, 8840, 8873, 9639, 9719, 10753, 10803, 11057
  Glaciers, Evidence for: 88, 1059, 3898, 4501, 8911, 8916, 10030, 10745
  Granite: 3489, 3711
  Graphite: 516

Massachusetts (continued)
  Lead Mines and Mining: 3323, 5070, 5331, 6945, 7597, 7598, 9607
  Limestone: 3157
  Maps, Geological: 2870, 3142, 3155, 5085, 5099, 5103, 5106, 5140, 5160, 7597
  Marble: 287, 3158, 6944, 6946, 9127, 10740
  Marl, Agricultural: 5113
  Meteorites: 682
  Mineral Collections: 7643
  Mineral Springs: 4888, 5884, 7636
  Minerals, Localities: 473, 710, 1365, 1732, 1811, 1908, 1911, 2067, 2113, 2799, 2863, 2864, 2949, 2950, 3145, 3153, 3326, 3488, 3490, 3537, 3791, 4382, 4422, 4457, 4842, 4843, 4895, 5078, 5087, 5089-5091, 5095, 5149, 5154, 5155, 5543, 5606-5608, 5607, 5610, 5613, 5616, 5664, 6074, 6121, 6245, 6553, 6950, 8507, 8509, 8510, 8512, 8513, 8515, 8581, 8583, 9300, 9463, 9465, 9495, 9503, 9507, 9517, 9829, 10173, 10175, 10176, 10191, 10277, 10287, 10704, 10715, 10716, 10723, 10731, 10749, 10750, 10762, 10763, 11069
  Mines and Mining: 5104, 5119, 5122, 5128, 5198, 6853, 7715
  Natural History: 5069, 6206
  Paint: 7670
  Paleontology: 5076, 5109, 5677, 8143
  Paleontology, Invertebrate: 1416
  Paleontology, Vertebrate: 3257, 8138
  Phyllite: 872
  Rocking Stones: 1058, 8505, 8511, 10745
  Salt Mines and Mining: 10159
  Silver Mines and Mining: 11058
  Soapstone: 2477
  Soils: 6847, 6848
  Springs: 6057
  Steatite: 432, 602, 1820
  Tertiary Formation: 6547
  Volcanoes: 5111
Materia Medica: 11133
  United States: 1515, 1533
Mather, W. W.
  Mineral Collections: 673, 5249
McClelland, J.: 679
Meade, W.: 575
  Mineral Collections: 582
Mechanic's Institute of New York: 7763
Medical Repository: 1210
Medical Topography: 1022
  Alabama: 4662, 5029, 5030
  Cholera: 3642, 4253, 4254, 5654
  Earthquakes: 10771, 10772
  Florida: 8237
  Kentucky: 3256, 5747
  Louisiana: 5027, 5028
  New York: 1394, 1395, 2646, 2954, 3892, 4841, 4982, 8499, 9014, 9015, 9842, 9996, 10507, 10542, 10705, 10898, 11127
  North Carolina: 10981
  Notes: 6050, 7813
  Ohio: 3216, 3218-3220, 3916
  Pennsylvania: 4566
  Portugal: 2766
  South Carolina: 8691
  Tennessee: 1747, 2145, 5035, 8682
  Theory: 7276, 7277
Medicine
  Minerals: 5975
Mediterranean
  Natural History: 5835
  Volcanoes: 10076
Mercury
  Analysis, Chemical: 7193, 8602
  Uses of: 4320
Mercury Mines and Mining: 3800, 7619
  Aden: 2233
  Alabama: 5729
  California: 6575
  Ceylon: 2528
  Germany: 3018
  India: 1853
  Italy: 9032
  Michigan: 10021, 10022
  Notes: 7403, 7619
  Pennsylvania: 4143
  Spain: 637, 3614, 3979, 4436, 6295, 7713, 7716, 7933

Metamorphic Rocks
  New Hampshire: 5600
  Rhode Island: 5647
Metamorphism: 2141, 2803, 3517, 6190, 8890, 8891, 8893, 8928, 10201
Meteorites
  Africa: 728, 746
  Alabama: 5555, 5589, 5617, 10397
  Analysis, Chemical: 3259, 3260, 3841, 4896, 4901, 6071, 6072, 8947, 9526, 9674, 10689, 10691, 10737, 10755, 11045
  Arkansas: 945, 5034
  Arva: 989
  Bohemia: 525, 922
  Bolivia: 5887
  Brazil: 699, 7429
  Cape of Good Hope: 1310
  Colombia: 1883, 3975, 5379
  Connecticut: 2203, 3396, 4969, 5235, 6124, 9599-9602, 10689, 10691, 11045
  Fiji Islands: 10691
  France: 763, 767, 2675, 3430, 6246, 7087, 7208, 7406, 9203, 9341, 9412, 10811, 11039
  Georgia: 1550, 1936, 3070
  Germany: 7040
  Hawaii: 1708, 5501
  Hungary: 777, 1785
  India: 4
  Iowa: 1493, 4220
  Italy: 334, 878, 1800, 9417, 10687
  Kentucky: 8634, 10400
  Louisiana: 407, 601, 4378, 6984, 7096, 7099, 9476
  Madagascar: 9547
  Maine: 2424, 5043, 9555, 10737, 10755
  Maryland: 2271, 2272, 2347, 4838
  Massachusetts: 682
  Mexico: 1632
  Missouri: 5010
  New Hampshire: 7661, 9821
  New Jersey: 10606
  New York: 9533, 9804, 9814, 9820
  North Carolina: 4369, 6661, 9524
  Notes: 658, 1046, 1070, 1225, 1226, 1251, 1625, 1667, 1849, 1867, 1960, 1965, 2099, 2102, 2116, 3308, 3392, 3400, 3641, 3677, 4038, 4135, 4217, 4377, 4497, 4944, 5856, 5877, 5896, 5897, 5902, 6299, 6303, 6318, 6678, 7226, 7542, 7673, 7853, 7880, 7911, 7915, 7929, 8156, 8458, 8461, 8492, 8532, 9240, 9324, 9345, 9492, 9544, 9550, 9551, 9558, 9560, 9564, 9924
  Origin of: 193, 2005, 2982, 3890, 8758, 10598
  Pennsylvania: 7026, 9674
  Red River Valley: 1086
  Russia: 7802
  Silesia: 764
  South America: 2286
  South Carolina: 1012, 9329, 9562
  Tennessee: 510, 517, 1036, 1839, 9454, 9536, 10381, 10397, 10400, 10402, 10846
  Texas: 2675, 9820
  Virginia: 2471, 8982, 9477, 9541
Meteorology
  Textbook: 3911
Mexico
  Asphaltum: 7213
  General: 4481
  Geology: 8826, 11009
  Gold Mines and Mining: 3362, 7023, 7370
  Lead Mines and Mining: 3084
  Maps, Geological: 11008
  Meteorites: 1632
  Minerals, Localities: 1441, 7066
  Mines and Mining: 2377, 3986, 6874, 6877, 6937, 7547, 8672
  Natural History: 4391, 6935, 7860, 8195
  Paleontology: 2415
  Silver Mines and Mining: 1791, 5450, 8494, 9682
  Volcanoes: 670, 4597, 6400, 6636, 6639, 6986, 7712, 11126
Michelotti, G.: 829
Michigan (see also Lake Superior)
  Coal Mines and Mining: 4292
  Geological Survey: 3209-3211, 5294-5314, 5318-5320, 5332-5341, 7157-7170

Michigan (continued)
  Geology: 3209-3211, 5038, 5041, 5042, 5296, 5299, 5301-5304, 5307, 5309, 5311, 5314, 5318, 5332, 5333, 5335-5337, 5339, 9097
  Gold Mines and Mining: 10814, 10870
  Iron Mines and Mining: 5297
  Maps, Geological: 4302, 5042, 5320, 5340
  Mercury Mines and Mining: 10021, 10022
  Minerals, Localities: 6586
  Mines and Mining: 8380
  Natural History: 6035, 6036, 8360
  Salt Mines and Mining: 1783, 5298, 5300, 5308, 5310, 5313, 5315
  Silver Mines and Mining: 9269
Microscope: 1391, 2200, 2237, 2260, 3413, 3454, 4048, 4397, 6737, 8367
Microscopical Society of London: 7178
Mill Stone
  Georgia: 2665
Miller, H.: 87
Miller, J. S.: 535
Millstone
  Ohio: 8387
  Pennsylvania: 8293
Mine Safety: 1953, 1980, 3942
  Accidents: 9395
  Coal Gas: 6372
  Fire Extinguisher: 4606
  Safety Lamp: 2965
  Ventilation: 4056, 6110
Mineral Black
  Pennsylvania: 95
Mineral Collections
  American Geological Society: 6627
  American Institute: 300
  American Museum: 1110
  Amherst College: 943
  Arrangement of: 7352
  Beck, T. R.: 1591
  Boston Society of Natural History: 1846, 10032
  Bruce, A.: 7050
  Carpenter, G. W.: 2253
  Catalogue of: 3105
  Chemical Society of Philadelphia: 7018, 7024, 7031
  Chester County Cabinet: 485
  Cozzens, I.: 1354
  Crawe, Dr.: 9790
  DeWitt, B.: 3169
  Dunn, N.: 720
  Featherstonhaugh, G. W.: 467
  Feuchtwanger, L.: 551
  Foreign: 442
  Franklin Institute: 4945, 4959
  Gibbs, G.: 348, 427, 1829, 7345
  Great Britain: 5859
  Greville, M.: 2114
  Griscom, J.: 4568, 9730
  Hall, F.: 673
  Harvard College: 7046, 8620, 9131, 10493, 10746, 10747, 10754
  Hauy, R. J.: 7540, 7543
  Heidelberg: 6189, 7331
  Heuland, H.: 450, 1806
  Holbrook, J.: 7642
  Hosack, D.: 4531, 5285
  Lederer, L.: 9735
  Leonhard, K. C. von: 2157
  Leske, Prof.: 7078
  Literary and Philosophical Society of Quebec: 6236
  London Mineral Prices: 382
  Maclure, W.: 6643, 6644
  Mantell, G. A.: 618
  Markoe, F.: 944
  Massachusetts: 7643
  Mather, W. W.: 673
  Meade, W.: 582
  Montgomery County, Pennsylvania: 7505
  Morton, S. G.: 430
  Mount Sylvan Academy: 7439
  New England: 519, 11001
  New York: 7755-7757, 9972, 10062
  New York Historical Society: 1742, 4380, 4381
  New York Lyceum of Natural History: 7314
  Notes: 10290
  Nuttall, T.: 841
  Perkins, B. D.: 7050

Mineral Collections (continued)
    Philadelphia Academy of Natural Sciences:
        6174, 6176, 6177, 6182
    Phillips, W.: 490, 798
    Rafinesque, C. S.: 8638
    Robinson, S.: 463
    Sale of: 462, 954
    Sullivant, J.: 10068
    Totten, J. G.: 10296
    Trading Notices: 134
    United States: 339, 444, 5356
    United States, Western regions: 10857
    Vanuxem, L.: 994
    Von der Null, J. F.: 10655
    Webster, J. W.: 329
    Werner, A. G.: 1192
    Yale College: 2111, 2120, 11090
    Young, J. P.: 635, 842
    Zimmern, L. W.: 5133
Mineral Springs
    Alabama: 2125, 2126, 2129, 2147
    Analysis, Chemical: 1366, 1934, 1935,
        1956, 1959, 2239, 2474, 2613, 2868,
        2880, 2899, 3097, 3486, 4975, 4985,
        4991, 5412-5414, 5546, 5674, 6947,
        6948, 6951, 7109, 7111, 8841,
        9862, 10331, 10539, 10540, 10650, 10801,
        10826
    Arsenic: 4177, 6056
    Artificial: 7426
    Azores: 5546
    Bromine: 4888, 4975
    Canada: 298, 5412-5414
    Connecticut: 1854, 5955, 7714, 8353
    England: 5707, 10677, 10678
    General: 1621, 1622
    Georgia: 9953
    Iodine: 1799
    Ireland: 7532, 7654
    Italy: 4586
    Kentucky: 3224, 10819, 10834, 10839,
        11091, 11096
    Localities: 235
    Maryland: 3845
    Massachusetts: 4888, 5884, 7636
    Mississippi River Valley: 1083, 10818
    New Jersey: 6275, 6649, 7034, 7282, 7287,
        7290, 7297, 8486, 10075
    New South Wales: 6221
    New York: 169, 170, 171, 239, 243, 482,
        1040, 1113, 1114, 1523, 1560, 1562,
        1574, 1827, 1828, 2013, 2346, 2349,
        2353, 2474, 2868, 2880, 3057, 3542,
        3559, 3561, 3574, 3604, 4569, 4570,
        4592, 4756, 4887, 4962, 5282-5284,
        5327, 5978, 6012-6014, 6078, 6409,
        6860, 6947, 6948, 6951, 6965, 7071,
        7109, 7111, 7289, 7343, 7709,
        7954-7956, 8354, 8696, 8733, 9124,
        9125, 9158-9163, 9244, 9383-9386, 9484,
        9690, 9865, 9998, 9999, 10005, 10006,
        10008, 10009, 10166, 10447, 10539,
        10545, 10780, 11124
    Notes: 1044, 1248, 2723, 2922, 2923,
        2925, 2927, 3302, 3639, 3648, 4002,
        7007, 7829, 7884, 8455, 8551, 9006,
        9010, 9305
    Ohio: 414, 2380, 3224, 3225, 6913, 10834,
        10849
    Origin of: 10591
    Pennsylvania: 1241, 1934, 1935,
        2366-2369, 2610, 2742, 2744, 3097,
        4881, 4882, 4938, 6857, 6955, 6957,
        7324, 8283, 8286, 8459, 8462, 8472,
        8497, 8498, 10710, 10711
    Peru: 8801
    Portugal: 8742
    Russia: 9845
    Scotland: 2534
    Tennessee: 10308, 10390, 10836
    United States: 2256, 3966, 3967, 6811
    Vermont: 163, 9016, 9128, 10817
    Virginia: 1452, 1481, 1537, 1623, 3486,
        3733, 3776, 3777, 4003, 4523, 4524,
        4870, 4871, 4881, 4882, 5490, 6140,
        6141, 6657, 6660, 6662, 6857, 7004,
        7048, 7184, 7420-7422, 7865-7867, 8203,
        8245, 8971, 8988, 8995, 9040, 9504,
        9844, 9944, 9946, 9947, 10840, 11006
    West Virginia: 2196-2198
Mineralogical Society of Virginia: 7199

Mineralogy (see also Chemical Analysis;
    Minerals, Species; Minerals, Localities)
    Beauty of the Science: 5526, 5527
    History of: 6091
    Lectures on: 2617, 2619, 2623, 2624,
        4420
    New York: 1577
    Mineralogy, Notes: 1109, 1147, 1187, 1201,
        1206, 1207, 1392, 1393, 1398, 1599,
        1646, 1648, 1651, 1662, 1666, 1671,
        1673, 2169, 2335, 2501, 2846, 2851,
        2857, 2891, 3082, 3092, 3139, 3215,
        3221, 3222, 3301, 3304, 3353, 3651,
        3798, 3806, 3850, 4213, 4604, 4617,
        4659, 4760, 4856, 4857, 4898, 4989,
        4990, 5702, 5706, 5763, 5893, 5932,
        6917, 6953, 6994, 6996, 7037, 7131,
        7283, 7462, 7535, 7617, 7633, 7920,
        8292, 8456, 8465, 8550, 8598, 9118,
        9246, 9354, 9357, 9932, 10202, 10674,
        10733, 10736, 11117
    Mineralogy, Textbooks: 101, 136, 137, 1577,
        1666, 1775, 1958, 2107, 2160, 2165,
        2420, 2422, 2426, 2502, 2505, 2506,
        2508-2510, 2512, 2516, 2624, 2798,
        2812, 2832, 2865, 3085, 3491, 3492,
        4424, 4426, 4655, 5804, 5930, 6514,
        7419, 7443, 8320-8323, 8325, 9489,
        9502, 9542, 10330, 10767, 10798
Minerals
    Analysis, Chemical: 14, 15, 16, 17, 379,
        521, 1644, 1763, 1876, 2343-2349,
        2957-2960, 3676, 3850, 4408, 4409,
        4642, 4643, 5408-5411, 6075, 6076,
        8954, 8955, 8957, 8966, 9435-9453,
        9528, 9828, 9832, 9878, 9879, 9882,
        10040, 10044, 10073, 10247-10252, 10534,
        10908, 10910-10912, 10915, 10917, 10920,
        10921, 11020, 11068
    Basalt: 2817, 2818
    Color of: 1802, 6379, 6517
    Crystallization: 9906, 9908, 9909
    Electrical Properties: 1609, 1612
    Fluid Inclusions: 3138
    Isomorphism: 2849, 2850, 2853, 2856
    Magnetic properties: 7014, 8446
    Medicine: 5975
    Optical properties: 153, 445, 475, 534,
        1193, 1237, 1987, 1989, 1991, 1994,
        1995, 1996, 1997, 2004, 2143, 2621,
        2777, 3902, 4622, 4624, 4853, 5713,
        5872, 8554, 9806, 9837, 10765
    Pseudomorphism: 1737, 1738, 2815, 2816
    Solubility in steam: 5733
    Specific Gravity: 1400, 1674, 5845, 6007,
        6621
    Specific Heat: 5762, 8241
    Synthetic: 393, 494, 1610, 1735, 2695,
        2940, 3386-3390, 4120, 4209, 4210,
        4214, 4276, 5407, 5663, 5923, 6387,
        4007, 7329, 7659, 9220,
        9420, 9421, 10265, 11023
    Thermal Conductivity: 7113, 9419
    Volcanoes: 451
Minerals, Localities
    Africa: 7201
    Alabama: 2125, 2130
    Algiers: 9164
    Antigua: 9468
    Arkansas: 5533, 7120, 8679, 9549, 10922
    Azores: 10182, 10730, 10735
    Banka Island: 5270
    Bavaria: 982
    Bohemia: 8800
    Brazil: 2788, 10600, 11029
    California: 985, 4101
    Canada: 429, 1396, 1700, 1702, 1831,
        4151, 8821, 8822, 9529, 10342
    Ceylon: 6073, 10996, 10997
    Chile: 1971, 3817, 8777, 8778
    Colombia: 1352, 1884
    Connecticut: 341, 342, 364, 1347, 1905,
        1906, 1910, 1939-1941, 2490, 2504,
        2794, 2801, 2855, 3050, 3326, 4374,
        4890, 5088, 5831-5833, 6028, 6120,
        6121, 6123, 6263, 6717, 6865, 6872,
        6873, 6880, 6950, 8223, 8230, 8231,
        8506, 8508, 8509, 8524-8526, 8694,
        8831, 9398, 9467, 9470-9473, 9486,
        9508, 9511, 9512, 9517, 9518, 9538,
        9605, 9607, 9609, 9611, 9620, 9621,

Minerals, Localities--Connecticut (continued)
9625, 9629-9632, 9636, 9638, 9654,
9657, 9664, 9669, 9782, 9793, 9802,
9805, 9811, 9876, 9877, 10152, 10288,
10340, 10727, 10978
Corsica: 3081
Cuba: 1765, 2445
Delaware: 1929, 2251, 2252, 2255, 2257, 2542, 4541
Elba Island: 1974, 8289, 9202
England: 1182, 1975, 2538, 3226, 4980, 5883, 6931, 8839, 8918
Europe: 458, 7077, 8631, 9479
Faroe Islands: 3603
Finland: 5871, 7914, 7916, 7917, 9204
France: 747, 1545, 1878, 2783, 4583, 5565, 5873, 5879
Georgia: 145, 147, 2603, 5934
Germany: 5920, 5949, 6936, 10459, 11130
Greenland: 632, 10041-10043
Hawaii: 481
Hungary: 4405, 5537, 5890
Illinois: 369, 1464, 5744, 8832
India: 920, 8596, 10252
Indiana: 2503
Ireland: 648, 7229
Isle of Skye: 4563
Italy: 4162, 6202, 7367, 7411, 7412, 8754, 8795, 9210, 10081, 10181, 10515
Kentucky: 3095, 3432, 8628
Lake Superior District: 1972, 3053, 3174, 3175, 5615, 6113, 9553, 10205, 10210, 10211, 10215, 10344, 10908, 10910, 10912, 10915, 10917, 10920
Lipari Islands: 10046
Louisiana: 1942, 2115, 2900, 2901
Maine: 4748, 5619, 5626, 5885, 6686, 7656, 8146, 10417, 10739
Maryland: 1474, 1478, 1528, 1640, 2257, 2265, 2428, 3073, 4396, 4423, 4426, 4875, 4879, 6205, 7060, 9440, 9453, 10454, 10568
Massachusetts: 473, 710, 1365, 1811, 1732, 1908, 1911, 2067, 2113, 2799, 2863, 2864, 2949, 2950, 3145, 3153, 3326, 3488, 3490, 3537, 3791, 4382, 4422, 4457, 4842, 4843, 4895, 5078, 5087, 5089-5091, 5095, 5149, 5154, 5155, 5543, 5606-5608, 5607, 5610, 5613, 5616, 5664, 6074, 6121, 6245, 6553, 6950, 8507, 8509, 8510, 8512, 8513, 8515, 8581, 8583, 9300, 9463, 9465, 9495, 9503, 9507, 9517, 9538, 9829, 10173, 10175, 10176, 10191, 10277, 10287, 10704, 10715, 10716, 10723, 10731, 10749, 10750, 10762, 10763, 11069
Mexico: 1441, 7066
Michigan: 6586
Missouri: 1947, 6114, 7777, 10349, 10879, 11051
Moldavia: 4406
New England: 6850
New Hampshire: 2478, 2873, 2875, 2876, 2879, 4886, 4892, 5077, 5349, 5601, 7313, 10171, 10184, 10189
New Holland: 133
New Jersey: 146, 148, 154, 387, 396, 1209, 1580, 1637, 1881, 1908, 1909, 2100, 2103, 2109, 2112, 2251, 2343, 2430, 2434, 2540, 2541, 2622, 2745, 3145, 3836, 3949, 4193, 4532, 4900, 4906, 5666, 6031, 7045, 7060, 7067, 7350, 7995, 7996, 8346, 8350, 8744, 8746, 9336, 9437, 9441, 9445, 9446, 9447, 10272, 10274, 10278, 10280, 10281, 10283, 10560, 10563, 10564, 10565, 10571, 10656
New South Shetland Islands: 9469, 10732
New York: 111, 114, 143, 356, 357, 362, 912, 1081, 1208, 1321, 1410, 1495, 1577, 1580, 1581, 1583, 1584, 1627, 1628, 1753, 1907, 1928, 1948, 1973, 2106, 2121, 2431, 2442, 2453, 2686, 2687, 2747, 3144, 3145, 3146, 3165, 3168, 3379, 3530, 3531, 3548, 3789, 3790, 3835, 3837, 3893, 3947, 3948, 4235-4237, 4245, 4274, 4616, 4640, 4647, 4649, 5098, 5236, 5280, 5288-5290, 5400, 5745, 5758, 6121, 6320, 6872, 6873, 6958, 7063, 7076,

Minerals, Localities--New York (continued)
7093, 7095, 7097, 7273, 7278, 7351, 7879, 8347, 8348, 8349, 8351, 8355, 8357, 8420, 9627, 9831, 10000, 10187, 10250, 10271, 10273, 10276, 10279, 10282, 10287, 10289, 10292, 10335, 10336, 10341, 10561, 10566, 10660, 10690, 10769, 10790, 11114
North Carolina: 2447, 3184, 3185, 4329, 5891, 7248, 9067, 9475, 10285, 10567
Norway: 1650, 3596, 3602, 6203, 9226, 9228, 11024
Nova Scotia: 131, 132, 150, 449, 2981, 4891, 4902, 5538-5542, 5539, 5541, 5574, 10762
Ohio: 1361, 1699, 3213, 3923, 10858
Ontario: 1703
Oregon: 4101
Orkney Islands: 10304
Palestine: 1719, 1720, 4651, 9661
Pennsylvania: 100, 1355, 1526, 1930-1933, 2249-2252, 2254, 2255, 2257, 2398, 3595, 3829, 3830, 3834, 4158, 4426, 4958, 5765, 5790, 6089, 6244, 7068, 7325, 7455, 7835, 8566, 8768, 8779, 9380, 9432, 9438, 9474, 9830, 10191, 10304, 11012, 11047
Peru: 1731, 4893, 4894, 4897, 7012, 8799
Rhode Island: 431, 1285, 1904, 3080, 6815, 6872, 8830, 10152, 10155, 10286, 10339, 10345, 10713, 10714, 10716
Russia: 492, 871, 950, 1605, 3416, 5000-5009, 5023, 5916, 6009, 6010, 7918, 8686, 11061
Saxony: 1969, 1970, 2246, 5921, 5926
Scotland: 5748, 5749, 6593, 6594, 8201, 9352
Silesia: 4404
South Carolina: 3185, 7203, 9881, 9884, 10562
Spain: 1739, 7919
Sweden: 391, 1654, 7497
Tennessee: 4661, 4878, 10403, 11064
Texas: 9559, 9822, 9827
Turkey: 2409, 9322, 9890, 9891-9894, 10157
United States: 141, 325, 388, 2118, 2133, 2420, 2422, 2426, 2432, 2498, 3381, 4650, 4652, 5094, 6127, 6130, 7027, 7281, 7306, 7797, 8506, 8507, 8821, 8822, 9216, 9326, 9433, 9435-9445, 9480-9482, 9489, 9561, 9645, 9832, 10153, 10213, 10275, 10288, 10334, 10337, 10338, 10705, 10722, 10739
United States, Western regions: 6080, 6081
Vermont: 155, 156, 157, 2430, 2875, 3149-3151, 3229, 3818, 3819, 3820, 3821, 4646, 4647, 4650, 4652, 4885, 6241, 8106, 8517, 9637, 9678, 10056
Virginia: 1535, 1538, 1540, 1541, 1542, 3847, 5644, 5676, 8285, 8984
World-wide: 10183
Minerals, Species
Acadiolite (see Chabasite)
Acmite: 1650
Agate: 3042, 3807, 5078, 6517, 9196, 8404, 10451
Albite: 6074
Algerite: 146, 2696, 5411, 5659
Allanite: 632, 10244
Alum: 364, 540, 1361, 1593, 1948, 4274, 8779, 9478, 9488
Alumine: 338
Aluminosilicates: 10243
Amethyst: 41, 431
Amoibite: 5995
Amphodellite: 7916
Anatase: 2778, 9019
Andalusite (see also Chiastolite): 2853, 3050, 4621, 5089, 6027, 10340
Anthophyllite: 5236, 10250
Antimonial nickel: 4852
Antimony: 6320, 7063
Antimony arsenide: 8674
Antimony ore: 11130
Antimony sulphide: 2115
Apatite: 2821, 3043, 3530, 3531, 4426, 6931, 9541, 10750, 10790, 11114
Aragonite: 1983, 2147, 3735, 3836
Argentine: 3150, 3152, 3153

Index 413

Minerals, Species (continued)
Arkansite: 7120, 10206
Arsenical nickel: 10657, 10658
Asbestos: 468, 848, 1071, 5885, 8130, 8566, 10726
Aspasiolite: 9226
Augite: 335, 2849, 4876, 6006
Aurichalcite: 2539
Automalite: 140, 10564
Aventurine: 4120
Axinite: 2938, 10417
Azurite: 8311
Bagrationite: 6009
Baierine: 982
Barite: 150, 1907, 2314, 2343, 2504, 2741, 3212, 3790, 3943, 7045, 7060, 8628, 9300, 9627, 10273
Barium phosphate: 4457
Barytocalcite: 3100
Beaumontite: 135, 3073, 5565, 6205
Beryl: 5832, 5833, 6865, 9380, 9657, 10184
Bismuth cobalt ore: 5921
Bismuth silver (see Childrenite)
Bismuth sulphide: 6878
Bismuth telluride: 3847
Blue Iron Earth (see Vivianite)
Bodenite: 1969, 5948
Bog iron ore: 10069
Boltonite: 9152, 9832, 9834
Boracite: 5925
Borax: 9859
Brandesite: 4625, 5996
Brevicite: 9925
Brewsterite: 2533
Brochantite: 8675
Bronzite: 2430, 3835
Brookite: 9019, 10206
Brucite: 148, 343, 1985, 1986, 2109, 4193, 8346
Bucholzite: 3595, 9832, 9833, 10258
Bucklandite: 6203
Buratite: 3077
Cacoxene: 10403
Cadmium Sulphide: 3101
Calamine (see Smithsonite)
Calcite: 393, 445, 1321, 1547, 1981-1983, 3735, 7328
Calcium carbonate: 366, 1611
Calcium lead phosphate: 1545
Calstronbarite: 9513, 9538
Cassiterite: 2940, 8984, 10176
Castor: 1974
Catlinite: 4968, 5562
Caustic Lime: 3676, 8795, 10081
Celestite: 1699
Ceramohalite: 5890
Cerrusite: 4616
Chabasite: 4902, 5214
Chalcedony: 486, 3998, 6608
Chiastolite: 5543, 6027
Childrenite: 2068
Chiolite: 11061
Chloanthite: 1970
Chlorastrolite: 10912
Chloritoid: 3597
Chlorophacite: 6593, 6594
Chlorophoeite: 5090
Chondrodite: 2821, 9437, 9441, 9446, 9447
Chromite: 1478, 1765, 5075, 5410, 8517, 9440
Chromium oxide: 11022
Chrysoberyl: 9449, 9629, 10000, 10341
Chrysocolla: 1908, 1909
Chrysoprase: 3818
Cinnabar: 2503, 5974, 10021, 10022
Cleavelandite: 403, 2067
Cobalt: 2877, 10350
Cobalt oxide: 9881, 9884
Cocolite: 6241
Columbite: 909, 2783, 2797, 5831, 9020, 9024, 9028, 9030, 9514, 10288
Conite: 6593, 6594
Copper ore: 2879
Copper phosphate: 5007
Copper Sulphato-chloride: 2538
Coracite: 6113
Corundum (see also Emery): 2447, 3184, 3699, 3949, 4214, 4864, 6515
Couzeranite: 2323

Minerals, Species (continued)
Crichtonite: 4428, 8830
Cryptolite: 11025
Cuban: 9232
Cuprite: 1610
Cuproplumbite: 1971
Cyanite (see Kyanite)
Damourite: 3076, 10191, 10192
Danalite: 4892
Danburite: 2855, 9518
Datholite: 9486
Datolite: 2411, 3602, 10272
Davidsonite: 8774
Deweylite: 9483
Diallage: 4618, 6006
Diamond: 503, 652, 985, 1026, 1351, 1999, 494, 524, 631, 858, 4209, 4210, 6251, 2000, 2444, 2603, 2966, 3200, 3264, 3265, 3395, 4027, 4329, 4479, 5895, 7010, 7019, 7659, 8340, 8344, 8533, 8537, 8737, 8796, 8956, 9121, 9123, 9167, 9317, 9347, 9403, 9545, 10550, 11034, 11119
Diaspore: 2779
Dichroite: 10041
Digenite: 1971
Dioptase: 2778
Diphanite: 7918
Dipyre (see also Scapolite): 3074
Disterrite: 5997
Dolomite: 2140
Dreelite: 3263
Dufrenoysite: 2784
Dysclasite: 1972
Dysluite: 140
Edwardsite: 9018, 9508, 9509, 9530
Ehlite: 8767
Elaterite: 747
Emerald: 2118, 6245, 7012, 8412
Emerald Nickel: 9822, 9827
Emery (see also Corundum): 9891-9894, 10157
Emmonite: 10253
Epidote: 4337, 10731
Epistilbite: 3804, 6204, 9022
Epsom Salt: 3432
Eremite: 2794, 2801
Erlanite: 2246
Ermite: 9511
Euclase: 9538
Eudialite: 10042
Eukolite: 9230
Euxenite: 9231
Feldspar: 710, 1654, 1823, 1929, 1930, 2542, 2712, 3081, 4188, 5923, 7683, 8242, 9224, 9464
Fer Titanne: 1640, 2428
Fibrolite (see also Sillimanite): 3819, 9832, 9833, 10576
Fischerite: 5001
Flint: 3998
Fluorite: 369, 1464, 1535, 1538, 1542, 1928, 2106, 2108, 2821, 3789, 4878, 5077, 5078, 5744, 6016, 7013, 8832, 8918, 9637
Forsterite (see also Olivine): 6202
Fowlerite: 1011
Franklinite: 158, 10278, 10281
Gadolinite: 1649, 9831
Galena: 710, 1270, 3946, 4886, 9638, 10187
Garnet: 365, 3473, 6073, 9398, 10763
Gaylusite: 1884
Gehlenite: 2411
Geochronite: 5947
Gibbsite: 9829, 10277
Gismondine: 6779
Glaucophane: 4854
Glinkite: 1605
Gmelinite: 2535
Gold (see also Gold Mines and Mining): 1732, 3820, 3821
Graphite: 266, 516, 671, 814, 2453, 2596, 3229, 4271, 4649, 4998, 4999, 5793, 6553, 7313, 7842, 8124, 8768, 9067
Grey Antimony: 5349
Groppite: 10074
Gypsum: 488, 580, 1209, 1284, 1540, 1541, 2314, 2315, 2394, 2395, 2611, 2911, 2981, 3165, 3487, 3790, 5740, 6466, 6467, 7002, 7093, 7432, 7879, 8163,

Minerals, Species--Gypsum (continued)
    8247-8251, 8255, 8258, 8519, 9052,
    9057, 9059, 9876, 9877, 10102, 10627,
    10660, 10690, 10858
Halite (see also Salt): 487
Halloysite: 2785, 2786
Harmotome: 2780, 10205
Hauerite: 4623
Haydenite: 6205
Hematite: 1931, 3537, 5407
Herschelite: 2782
Hetepozite: 3261
Heulandite: 135, 2787, 6204
Hopeite: 1988
Humboltine: 8800
Humite: 6778
Huraulite: 3261
Hyalite: 1410, 3386-3390
Hydrarchos: 6235
Hydrate of silica: 9237
Hydroboracite: 5024
Hydrophane: 3386-3390
Hydrous aluminosilicate: 5873, 5879
Hydrous zinc oxide: 7864
Hypersthene: 1588
Iberite: 7919
Ice: 1984, 3147, 4528, 9633
Idocrase: 1758, 4563, 8775
Ilmenite (see also Fer Titanne): 5940
Indianite: 9835
Indicolite: 2113
Iolite: 3923, 4619, 9486, 9534, 9554
Iron (see also Meteorites): 407
Iron silicate: 5993
Iron sulphate: 1643
Itacolumbite: 5891, 6099, 9545
Jacksonite: 10915
Jade: 2781
Jeffersonite: 1349, 9445, 10280, 10563
Jenite: 8289
Kaliphite: 5537
Kaolinite: 8350
Keilhauite: 3596
Kraurite: 10403
Krisuvigite: 8675
Kyanite: 153, 9466, 10576
Kyrosite: 9233
Lancaster: 9836
Lapis Lazuli: 950, 3416, 5847, 9290
Lardite: 5926
Lazulite: 1652
Lead: 435
Lead antimonate: 5000
Lead antimony sulphide: 1878
Lead arsenide: 10255
Lead ore: 10045
Lead phosphate: 4426
Lead vanadate: 5748, 5749
Ledererite: 595, 4891, 5540
Lepidolite: 732, 1713, 10739
Libethenite: 8767
Liebenerite: 6777
Liebigite: 9890
Lincolnite: 135
Loadstone (see Magnetite)
Loxoclase: 1973
Maclureite: 9437, 9441, 9446, 9447
Magnesian Minerals: 2418
Magnesite: 3283, 8347, 8348, 8596
Magnesium sulphate: 356, 8285
Magnetite: 2431, 3640, 5975, 6346, 10520, 10865
Malachite: 7325, 8311
Malacrone: 9223
Manganese: 2398, 9678
Manganese peroxide: 6085
Manganocalcite: 919
Margarodite: 9222
Marmolite: 10568
Martinsite: 5920
Masonite: 138
Medjibite: 9890
Megalonyx: 2639
Melanite: 11012
Melanolite: 11069
Mendipite: 9238
Mesitine: 1976
Miargyrite: 1757, 1759, 2849
Mica: 157, 566, 950, 1196, 1823, 1991,
    2852, 3151, 3791, 3902, 4022, 4777,
    6468, 6959, 8730, 8731, 9017, 9837,
    9989

Minerals, Species (continued)
Microlite: 4899, 9495, 9507, 9537, 10172, 10188, 10190
Molybdenite: 1474, 7068, 8524, 9438, 10283, 11047
Monazite: 5009, 9018, 9530, 9811
Mowenite: 2780
Mullicite: 2100, 2103
Muramontite: 5949
Naptha: 9596
Native Copper: 3817, 5149, 5154, 7341, 9782, 9793, 9802, 9805
Native Iron: 1352, 2286, 2442, 5013, 8694, 9474, 11029
Native Lead: 2503
Native Tin: 5002
Native Titanium: 8839
Natrolite: 4605
Needle Ore: 633
Nemalite: 2537
Neocronite: 331, 4875
Neolite: 9228
Nephrite: 1904
Nickel antimonide: 10048
Nickel arsenide: 1761
Nickel sulphide: 5289, 5758
Niobite: 9021
Nitrates: 4997
Nitre: 2082, 2413, 4175, 4893, 4894, 4897, 10515
Novaculite: 5934
Ochroit Earth: 8287
Oerstedite: 3907
Olivine (see also Forsterite): 424, 10047
Opal: 317, 1672, 2001
Orthite: 1654
Osmelite: 19
Ottrelite: 142
Oxide of chromium: 1735
Ozarkite: 2860
Ozokerite: 4406
Palladium selenide: 3803
Pargasite: 5871
Parisite: 2194
Pectolite: 19
Pelokonite: 8777, 8778
Penine: 6775
Periclase: 9210
Perovskite: 3101, 9027
Petalite: 1703, 9245, 10342
Phacolite: 142, 143, 8676
Phenakite: 9517, 9539
Philippsite: 6779
Phlogopite: 2686
Phosphorite: 154
Piauzite: 4620
Pickeringite: 4893, 4894, 4897
Pimelite: 3818
Pinguite: 1966
Pisolite: 8420
Pistomesite: 1976
Pitchblende: 9229
Platinum: 957, 985, 1657, 1663, 2772, 3802
Pliniane: 1975
Plumbacalcite: 5748, 5749
Plumbago (see Graphite)
Plumbic Ochre: 1441
Pollux: 1974
Polyadelphite: 142
Polycrase: 9223
Prehnite: 424, 2044, 5662, 5668
Prunnerite: 3603
Pyrargillite: 7917
Pyrite: 1661, 4050, 4063, 4397, 6007, 6586, 7653, 9500, 11023
Pyrochlore: 4899, 9537, 10172, 10188, 10190
Pyrolusite: 1612, 5075, 6244, 8581
Pyrophyllite: 8678, 10212
Pyroxene: 2121, 3548, 7329, 10338, 10355, 10562, 10566
Pyrrhite: 4905, 10182, 10192, 10197
Quartz: 155, 362, 1355, 1731, 1832, 2478,
    2940, 3959, 4153, 4976, 5351, 5713,
    6387, 7095, 7097, 7203, 8023, 8754,
    9214, 10265, 10335
Randanite: 9164
Rubellite (see also Tourmaline): 10739
Ruby: 4023, 7207, 8432
Rutilated Quartz: 156

Index

Minerals, Species (continued)
  Rutile: 2940, 5077, 5351, 5406, 7067, 9019
  Saccharite: 9235
  Sahlite: 1906
  Sal Ammoniac: 1528, 4583
  Samarskite: 9031
  Sapphire: 145
  Sapphirine: 10043
  Scapolite (Dipyre): 3074, 5607, 5610, 5613, 5616
  Scheelite: 1905
  Schorlomite: 8679
  Scolezite: 4605
  Scorodite: 336
  Serpentine: 719, 1365, 5935, 7995, 8106, 8242, 9830, 10047
  Siderite: 2430
  Sidero-graphite: 10274
  Silica: 918
  Sillimanite (see also Fibrolite): 1347, 1910, 8525, 9466, 9811, 9832, 9833, 9993, 10258
  Sismondine: 3075
  Smaelite: 4405
  Smithsonite: 4050, 10349
  Soda Nitre: 8799
  Sodalite: 2054, 10245
  Somervillite: 2068
  Sordawalite: 7914
  Spadaite: 5994
  Sphene: 9832, 9834, 10607
  Spodumene: 1908, 1911, 2133, 2863
  Stannine: 1975
  Staurolite: 2853, 2864, 5689, 10271
  Steinheilite: 9204
  Stilbite: 1739, 10282
  Stroganowite: 5003
  Stromnite: 10305
  Strontianite: 1396, 1907, 3049, 3213, 3493, 3943, 9491, 9499, 10304
  Strontium sulphate: 10336
  Struvite: 10459
  Sulphate of Iron: 499
  Sulphate of Strontian: 1081
  Sulphides: 6183
  Sulphur: 315, 320, 1039, 4239, 7327, 8158, 10046
  Sulphuret of Cobalt: 920
  Tabasheer: 10252
  Table Spar: 10561
  Talc: 3080, 6776
  Tantalite: 876, 9030, 9033
  Tautolite: 9553
  Telluric bismuth: 2788, 5676
  Tellurium: 341, 342, 5644
  Tenantite: 4980
  Terra di Sienna: 4158, 5790, 7835
  Thomaite: 6936
  Tin: 5095
  Titanium: 2117
  Topaz: 648, 1985, 1989, 2002, 2003, 2853, 5088, 5091, 5098, 5703, 6028, 6263, 9467, 9517, 9664, 9669
  Torrelite: 2797, 8746
  Tourmaline: 473, 698, 1609, 4382, 4409, 4858, 6931, 7072, 8168, 9806
  Tremolite: 2781
  Triphylme: 4189
  Tungstate of lead: 336
  Tungstates: 6773
  Tungsten: 341, 342
  Turgite: 5004
  Turquoise: 1652, 4404, 5005
  Uralorthite: 6010
  Uranite: 9503
  Uranium: 10189, 10190
  Uranium Ore: 8313, 8314
  Uranium phosphate: 10173, 10175
  Vanadate of Copper: 871
  Vauquelinite: 10292
  Vermiculite: 2200, 3091, 5664, 10212
  Vesuvianite (see Idocrase)
  Vivianite: 2409, 2622, 2745, 5671, 10560
  Volborthite: 8686
  Volknerite: 5008
  Warwickite: 9519
  Washingtonite: 5940, 9538
  Wavellite: 3144, 3146
  Wax: 4176
  Willemite: 10281
  Williamsite: 9559

Minerals, Species (continued)
  Wolchoskoit: 5916
  Wolfram: 9239
  Worthite: 5023
  Xanthite: 6872
  Xylite: 5006
  Yenite: 9464, 10286, 10339, 10345
  Yttrocerite: 142, 144, 396, 5155, 5606-5608, 5616
  Zeolite: 1526, 1881, 3906, 4855, 8831, 9913
  Zinc ore: 10276, 10279, 10351
  Zincite: 158, 2112, 4900, 4906
  Zircon: 1655, 2540, 2541, 5704, 5883, 6958, 9475, 10073, 10285, 10286
  Zirconite: 10567
  Zygadite: 1977
Mines and Mining (see also specific materials, e.g. Gold): 289, 3189, 3824, 5684, 6110
  Africa: 9139
  Alabama: 1180, 5449, 6565, 10411, 10412
  Algiers: 2220
  Arkansas: 3007, 3008, 5991, 10085, 10087
  Australia: 958
  Belgium: 4017, 4080
  Bohemia: 11118
  Bolivia: 7129
  Brazil: 3855, 4093, 6929, 6930, 9122, 11056
  California: 4391
  Canada: 1171, 8502, 10813
  Child Labor: 2231
  Chile: 3199, 8803
  Connecticut: 8328
  Cuba: 1470, 5470, 9153
  Egypt: 4037
  England: 1372, 3183, 3729, 3864, 4008, 4011, 4015, 4993, 5358, 6268, 6277, 7237, 7710, 7794
  France: 1870, 7963, 8501, 8503
  Georgia: 3009, 5194, 5202
  Great Britain: 5403, 7573, 8309, 8504, 8538, 10709
  Greece: 8818
  Illinois: 8194, 9531
  India: 9867
  Jamaica: 4028
  Japan: 1272
  Kentucky: 5944
  Lake Superior District: 5618, 5621, 5623, 5628, 5635, 5637, 5639-5641, 5646, 5650-5653, 5655
  Louisiana: 5438, 10026
  Maine: 4559, 4560, 5553, 5570, 6039, 6684, 6715
  Maryland: 1164, 3282, 4102, 4525, 5848, 5849, 6573, 6812, 8750, 8752, 9535, 10662, 10787
  Massachusetts: 5104, 5119, 5122, 5128, 5198, 6853, 7715
  Mexico: 2377, 3986, 6874, 6877, 6937, 7547, 8672
  Michigan: 8380
  Mississippi River Valley: 1624, 5960, 5963, 10785
  Missouri: 3187, 3188, 3812, 3813, 8194, 8605, 8607, 9120, 9531, 9816, 10932, 11050
  New England: 9234
  New Hampshire: 5587, 5596, 5603, 5605
  New Jersey: 4182, 4614, 9285, 10064, 11046
  New York: 1926, 1927, 2450, 3553, 3563, 5466
  North Carolina: 1767, 6869, 6870, 9563, 10135
  Norway: 3867
  Notes: 647, 1042, 1076, 1266, 1281, 1866, 2195, 3407, 3634, 3646, 3700, 3701, 3780, 3854, 3862, 3866, 3868, 3872, 3873, 4058, 4068, 4074, 4133, 4575, 6183, 6376, 6975, 7202, 7230, 7232, 7234, 7635, 7826, 7830, 7878, 8115, 8334, 8725, 9186, 9293, 9382, 9936, 10103, 10995, 11059
  Nova Scotia: 7727
  Ohio: 9367, 10864, 10938
  Pennsylvania: 8362, 10794
  Peru: 1238, 8803
  Portugal: 1952
  Potassium: 6813

Mines and Mining (continued)
    Precious metals: 10816
    Russia: 454, 1272, 3181, 3865, 10168
    Saxony: 567
    Scotland: 10256
    South America: 3198
    Spain: 5427, 7364, 7814
    Sweden: 2412, 3867, 9106
    Textbooks: 1483, 1484, 1775, 2502, 2520, 5928, 8041, 8395, 8396, 9381
    United States: 309, 1170, 1500-1502, 1546, 1556, 1777, 1778, 1836, 1924, 2012, 2217, 2240, 2444, 2464, 2469, 2470, 2498, 2678, 2907, 3003, 3068, 3814, 4207, 4960, 4983, 4984, 5203, 5424, 5433, 5750, 6038, 6127, 6130, 6550, 6557, 6571, 6810, 6961, 6962, 6974, 8329, 9242, 10150, 10160, 10505, 10508, 10509, 10511
    United States, Western regions: 247, 7042, 8823
    Virginia: 1978, 3715, 3732, 3775, 3889, 4174, 5454, 5473, 10468, 10469, 10635, 10994
    Wales: 4072
Mining Safety: 49
    Lighting: 51
Minnesota
    Exploration: 8453
Miocene
    Paleontology, Invertebrate: 2575, 2581
Mississippi
    Geological Survey: 7243
    Geology: 1745, 2648, 3130, 3176, 3671, 4203, 4204, 7438, 10538
    Iron Mines and Mining: 3015
    Marl, Agricultural: 5528
    Paleontology, Invertebrate: 1712, 2587, 2588
    Paleontology, Vertebrate: 1508, 3177, 6566
    Soils: 10268
Mississippi River
    Erosion: 3056
Mississippi River Valley
    Coal Mines and Mining: 4179, 4297, 8390
    Delta: 6559, 6563, 6567, 6568
    Earthquakes: 3885, 3886, 10847
    Erosion: 3913-3914
    Exploration: 3884, 8368, 8369
    Geology: 1447, 1448, 1633, 1949, 2010, 2446, 3270, 3273, 3274, 3313, 3883-3887, 3913, 3914, 4388, 4467, 5692, 5695, 5696, 5960, 5963, 7875, 7991, 7994, 8562, 8578, 8608, 8629, 9261, 9268, 9271-9274, 10663-10666
    Guidebooks: 10089
    Lead Mines and Mining: 5696, 6442, 6453, 10082-10084, 10087, 10503, 10702
    Mineral Springs: 1083, 10818
    Mines and Mining: 1624, 5960, 5963, 10785
    Natural History: 2247, 2729, 2748, 3479, 3480, 4664, 5493, 5496, 6033, 7334, 7335, 8578, 8608, 9261, 9268, 9271-9274, 10012
    Paleobotany: 1837
    Paleontology: 10368
    Paleontology, Invertebrate: 7061, 8637
    Paleontology, Vertebrate: 1834, 3071, 4788
    Pumice: 367
    Sedimentation: 3179, 6782, 6928, 8788, 8789
    Sediments: 2071
    Soils: 3557
Missouri
    Caves: 1788
    Clay: 11051
    Coal Mines and Mining: 4170, 9748, 9756
    Copper Mines and Mining: 306
    Earthquakes: 6991
    Footprints: 692, 4224, 4414
    Footprints, Human: 1245, 1630, 1746, 2479, 2948
    Geological Survey: 6790, 8797
    Geology: 3231-3242, 3244-3252, 3592, 5960-5965, 7871, 7873, 7874, 8605, 8607
    Iron Mines and Mining: 1347, 3266, 3782, 3809, 3810, 3812, 3813, 4167, 4660, 7244, 7245, 8950, 9516

Missouri (continued)
    Lead Mines and Mining: 269, 1197, 1312, 1912, 1947, 2306, 2468, 2578, 2580, 3243, 5195, 5316, 5960-5965, 6232, 6442, 6453, 6647, 7498, 8064, 8238, 8480, 8953, 9254, 9255, 9257, 9260, 10352, 10354, 10472, 10473, 10476, 10480, 10482
    Limestone: 4224, 9260
    Maps, Geological: 2306
    Meteorites: 5010
    Minerals, Localities: 1947, 6114, 7777, 10349, 10879, 11051
    Mines and Mining: 3187, 3188, 3812, 3813, 8194, 8605, 8607, 9120, 9531, 9816, 10932, 11050
    Natural History: 1558, 9256, 9258
    Paleobotany: 2697, 2698, 7379
    Paleontology, Invertebrate: 1413, 2578, 2580, 3591, 7981
    Paleontology, Vertebrate: 726, 2290, 3126, 3708, 4150, 4515, 4799, 4811, 4813, 4927, 5266, 5267, 5998, 5999, 6000, 6001, 6179, 6181, 6382, 6709, 8603, 8604, 10893
    Pumice: 8910
    Salt Mines and Mining: 6647
    Zinc Mines and Mining: 7928
Mohs, F.: 748, 7539 7807
Moldavia
    Minerals, Localities: 4406
    Paleobotany: 659
Monticelli, T.: 1846, 5262
Moon
    Geology: 8819
    Volcanoes: 2822, 5015, 5016, 8019, 10712
Moravia
    Caves: 1858
Morton, S. G.
    Mineral Collections: 430
Mount Ararat: 9
Mount Blanc
    Natural History: 5325
Mount Sylvan Academy
    Mineral Collections: 7439
Mountains
    Age of: 8659
    Cleavage Patterns: 8914
    Elevation: 9743
    European: 541
    Heights of: 347
    Map of: 9634
    Notes: 2455
    Origin of: 6307, 7648, 8880, 8888, 8899, 8941, 10799
    Parallelism: 3128
    Persia: 6011
    Structure of: 3554
Moving Rocks: 2352, 6949
    Connecticut: 6118, 6122
Munster, Herr
    Fossil Collection: 854
Murchison, R. I.: 220, 608, 780, 831, 855, 998, 1142, 1145, 2164, 2823, 3423, 4049, 4307, 5145, 9769
Murray, J.: 353, 8469
Museums: 50
    British: 953, 4043, 5260
    Geology: 4019
    Peale's: 1211, 6852, 7030, 8454, 8177, 8178, 8181-8183, 8189, 8191

National Institution for the Promotion of Science: 7602, 8449
Natural Bridge
    Tennessee: 10314
    Virginia: 1041, 1242, 2496, 3483, 3766, 4395, 4474, 4478, 4516, 5723, 6306, 7240, 7375, 7849, 8133
Natural Gas
    England: 1072
    From coal: 6022
    Great Britain: 219
    Kentucky: 7056, 7098, 8297
    New York: 4922, 7288
    Notes: 7246
    Production of: 9416
    Scotland: 4159
    Uses of: 1954
    Vermont: 3675
    Virginia: 172, 1065, 6207, 6388

Index

Natural History: 9905
  Africa: 8117-8119
  Alps: 4738
  Arctic: 8772, 9622
  Bahamas: 10302
  Brazil: 10679, 10680
  California: 4111-4119, 4187, 4556-4558, 10259, 10457, 10458, 10960
  Canada: 9174, 10324
  Chile: 7332, 10301
  China: 10999
  Collections: 3046, 3047
  Connecticut: 10418
  Cuba: 7986
  Egypt: 9983, 10651
  Europe: 4534, 4585, 9603, 9628
  Faulkland Island: 9046
  Florida: 2905, 3905, 8359, 9012, 9013, 9856, 9933, 10900, 10968, 10969
  France: 6599
  Georgia: 4761, 10897
  Greenland: 3446
  Hawaii: 3482, 5937, 10301
  Iceland: 4981, 8810, 8811, 9405, 10707
  Illinois: 7261
  Importance of: 5082, 11131
  Iowa: 5, 6082, 7856, 7857
  Jordan: 6581, 6656
  Kentucky: 3822, 3823, 5511, 5512, 6645, 6665, 6787, 6788, 7336, 7790, 8627, 8635, 8636
  Lake Superior District: 83, 6033, 9155
  Lectures: 7284, 8179, 8180, 8184
  Louisiana: 2900, 2901, 5027, 5028, 6194, 9593, 10027, 10028
  Lyceums: 3631
  Maine: 2775, 4108, 10984
  Maryland: 5364
  Massachusetts: 5069, 6206
  Mediterranean: 5835
  Mexico: 4391, 6935, 7860, 8195
  Michigan: 6035, 6036, 8360
  Mississippi River Valley: 2729, 2748, 3479, 3480, 4664, 5493, 5496, 6033, 7334, 7335, 8578, 8608, 9261, 9268, 9271-9274, 10012
  Mount Blanc: 5325
  New England: 3297
  New Hampshire: 9462
  New Jersey: 6171, 7321
  New South Wales: 10161
  New York: 3288, 3297, 3464, 3469, 3471, 3506, 6588, 10686
  Newfoundland: 2647
  Notes: 1629
  Nova Scotia: 7989
  Ocean: 10547
  Ohio: 3216, 3218-3220, 4206, 4666-4668, 4836
  Ohio River Valley: 5495
  Oregon: 4111-4119, 4556-4558, 10259, 10960, 11072
  Panama: 6417
  Pennsylvania: 2906, 3478, 5494, 6963, 9100-9109, 10325
  Peru: 10419
  Rocky Mountains: 4111-4119, 5534, 5535, 8121, 9213, 10301
  Russia: 3601, 8113
  South Africa: 10422
  South Carolina: 7194, 8681
  Sumatra: 6020
  Syria: 10651
  Tennessee: 4930
  Texas: 5232, 5233, 5942, 7416, 7417, 7860
  Textbooks: 68, 2595, 2605, 3888, 4099, 4430, 4433, 4443, 4444, 4809, 5394, 5717, 5811, 6051, 6652, 6793, 7268, 7424, 7434, 8584, 8585, 8669, 8671, 9133, 9134, 9137, 9253, 9997, 10326, 11070
  United States: 2678, 2732, 3610-3612, 5064-5066, 5977, 6034, 6070, 8588, 8589, 9280, 9284, 10998
  United States, Western regions: 2081, 4666-4668, 4972, 5062, 8235, 8623, 8624, 8823, 9213, 10221, 10318, 10319, 10594
  Vermont: 10237, 10242, 10976
  Virginia: 4973, 4974, 5719, 5945, 5946, 6851, 9985, 10298

Natural History (continued)
  West Indies: 10998
  Wisconsin: 5, 6046-6048, 6613
Natural Science
  Textbooks: 10449
Natural Tunnel
  Virginia: 6478
Natural Wall: 1608, 8466
  North Carolina: 4663, 6219, 6220, 8034, 11040, 11041
  Origin of: 6341
New Brunswick
  Asphaltum: 5670
  Geology: 8808
  Glaciers, Evidence for: 5803
New England
  Coal Mines and Mining: 7631
  Earthquakes: 2313, 3258, 6247
  Glaciers, Evidence for: 1498, 3107-3109, 3111, 3114, 3115, 3120, 5580, 5590, 5591, 5602, 5620, 5632, 5638, 5643, 5672, 5758
  Gold Mines and Mining: 7611
  Granite: 293
  Guide: 3291-3296
  Iron Mines and Mining: 4920
  Marble: 293, 9246
  Mineral Collections: 11001
  Minerals, Localities: 6850
  Mines and Mining: 9234
  Natural History: 3297
  Slate: 293
New Granada
  Earthquakes: 6288
  Salt Mines and Mining: 4278, 4367
  Volcanoes: 18
New Hampshire
  Coal: 9362
  Earthquakes: 118
  Geological Survey: 2711, 3705, 4965, 5576, 5577, 5579, 5592, 5593, 5599-5601, 5611, 5624, 5634, 7695-7703, 7895, 10904-10907
  Geology: 3703, 4165, 4653, 4656, 4914, 5344, 5350, 5576, 5577, 5579, 5592, 5593, 5599-5601, 5611, 5624, 5634, 5792, 8358, 8582, 8919, 10904-10907
  Glaciers, Evidence for: 5669, 5672
  Iron Mines and Mining: 4375, 6968, 7705, 10170
  Landslides: 1465, 9684, 10947
  Lead Mines and Mining: 2475
  Limestone: 7425
  Maps, Geological: 5612
  Metamorphic Rocks: 5600
  Meteorites: 7661, 9821
  Minerals, Localities: 2478, 2873, 2875, 2876, 2879, 4886, 4892, 5077, 5349, 5601, 7313, 10171, 10184, 10189
  Mines and Mining: 5587, 5596, 5603, 5605
  Natural History: 1617, 1619, 1695, 1696, 9462
  Paleontology: 8926
  Petroleum: 1618
  Pot Holes: 8345
  Rocking Stones: 4532, 7418, 8549
  Tourmaline: 698
  Volcanoes: 123, 168, 5839
New Holland
  Caves: 3741, 5513
  Minerals, Localities: 133
  Paleontology, Vertebrate: 5153
New Jersey
  Amber: 11012
  Coal Mines and Mining: 4199, 5774
  Copper Mines and Mining: 1567, 1579, 2434, 5259, 6062, 6064, 6065, 6287, 7236, 8275, 9360, 10284
  Cretaceous Formation: 6551, 8838
  Dikes: 9803
  Flint: 7293
  Gazetteer: 4454, 4456
  Geological Survey: 4322, 7721, 7722, 8850-8854, 8863, 8864
  Geology: 1385, 1485, 7282, 7287, 7290, 7297, 8349, 8351, 8355, 8357, 8484, 8850-8854, 8863, 8864, 9487, 9907, 10565, 10571
  Greensand: 5678, 8940, 9084, 9085, 9453, 9955, 11070
  Lead Mines and Mining: 4138

New Jersey (continued)
  Maps, Geological: 8864
  Marl, Agricultural: 1350, 2604, 2683,
    3320, 4627, 5225, 7468, 8357, 8729,
    8917, 10460
  Meteorites: 10606
  Mineral Springs: 6275, 6649, 7034, 7282,
    7287, 7290, 7297, 8486, 10075
  Minerals, Localities: 146, 148, 154,
    387, 396, 1209, 1580, 1637, 1881, 1908,
    1909, 2100, 2103, 2109, 2112, 2251,
    2343, 2430, 2434, 2540, 2541, 2622,
    2745, 3145, 3836, 3949, 4193, 4532,
    4900, 4906, 5666, 6031, 7045, 7060,
    7067, 7350, 7995, 7996, 8346, 8350,
    8744, 8746, 9336, 9437, 9441, 9445,
    9446, 9447, 10272, 10274, 10278, 10280,
    10281, 10283, 10560, 10563, 10564,
    10565, 10571, 10656
  Mines and Mining: 4182, 4614, 9285, 10064,
    11046
  Natural History: 6171, 7321
  Paint, Minerals: 7081
  Paleontology: 7286
  Paleontology, Invertebrate: 1415, 1888,
    2103, 6093, 7463, 7464, 7476, 7485,
    10552, 10553
  Paleontology, Vertebrate: 79, 1348, 2373,
    3021, 3028, 3030, 3589, 4197, 4739,
    4785, 4787, 4924, 5514, 6933, 7319,
    7320, 7478, 7481, 7482, 7483, 8710,
    8713, 8717, 10555, 10557
  Raindrop Impressions: 8710, 8713, 8717
  Silver Mines and Mining: 9360
  Zinc Mines and Mining: 139, 158, 1271,
    1349, 8442, 10346
New Mexico
  Exploration: 9839-9841
  Geology: 6772
New South Shetland Islands
  Minerals, Localities: 9469, 10732
New South Wales
  Maps, Geological: 2836
  Mineral Springs: 6221
  Natural History: 10161
  Platinum Mines and Mining: 6714
New York
  Caves: 850, 3472, 3763, 4122, 4248, 4264,
    4526, 5188, 7041, 8018
  Cement: 1597
  Coal Mines and Mining: 113, 2104, 2287,
    2476, 3054, 3351, 3354, 3363, 3564,
    6298, 8394, 10327
  Correlation of Strata: 4687, 4695, 4696,
    4704
  Earthquakes: 1780, 7044
  Erie Canal Survey: 3338, 3341
  Erosion: 9598
  Flint: 7271
  Gazetteer: 4455, 9972, 9975-9977
  General Description: 4483
  Geological Survey: 288, 291, 296, 302,
    305, 785, 796, 846, 1565, 1566, 1568,
    1570, 1571, 1573, 1575, 1577, 1587,
    1589, 2267, 2557, 2559, 2563, 2567,
    2569, 2709, 2720, 3051, 3052, 3317,
    3495-3501, 3507, 3508, 3510-3513, 3515,
    3518, 3520, 3573, 3723, 3728, 4266,
    4269, 4277, 4287, 4670-4675,
    4680-4686, 4690, 4692, 4693, 4712,
    4716-4719, 4726-4728, 5271, 5277, 5280,
    5281, 5345, 5347, 6883, 6889, 6895,
    6899-6902, 6906-6910, 6916, 6987, 7123,
    7729-7759, 7896, 8066, 8388, 9431,
    10579, 10580, 10582, 10584-10587, 10589
  Geology: 112-115, 1096, 1317, 1318,
    1491, 1539, 1697, 1926, 1927, 2072,
    2177-2179, 2450, 2451, 2650, 2680,
    3027, 3044, 3162, 3327, 3328,
    3330-3336, 3338, 3339, 3341, 3344,
    3349, 3351, 3355, 3463-3468,
    3495-3501, 3507, 3508, 3510-3513, 3515,
    3518, 3520, 3549, 3573, 3838, 3875,
    3876, 3893, 4198, 4200-4202, 5938,
    4235-4237, 4245, 4286, 4670-4675, 4690,
    4680-4686, 4692, 4693, 4712, 4716-4719,
    4726-4728, 4907, 5220, 5271, 5277,
    5281, 5287, 5345, 5347, 5736, 5745,
    6226, 7224, 7264, 7267, 7269, 7272,
    7273, 7280, 7291, 7294, 7303, 7444,
    7445, 7734-7740, 7743, 7746, 7773,
    8356, 8702, 8745, 8873, 9172, 9487,

New York, Geology (continued)
    9631, 9658, 9999, 10001, 10008,
    10223-10225, 10556, 10579, 10580, 10582,
    10584-10587, 10589, 10795, 10796, 10880,
    11113, 11116
  Glaciers, Evidence for: 3106, 3164, 4280,
    8373, 8709, 8722
  Granite: 2350, 7837
  Guidebook: 10545
  Gypsum: 488, 2394, 2395
  Iron Mines and Mining: 308, 500, 1141,
    2679, 3503, 3504, 3526, 4294, 4670,
    5770, 5929, 8748
  Joints: 5343
  Landslides: 4126
  Lead Mines and Mining: 261, 3796, 6248,
    9974
  Limestone: 3163, 3380, 4685, 7368
  Lyceum of Natural History: 10741
  Manganese Mines and Mining: 1789
  Maps, Geological: 2681, 3496, 3507, 3508,
    3511, 3518, 3540, 6227, 6556, 6906,
    6908-6910, 7749, 11113, 11116
  Marble: 258, 6324, 9653
  Marl, Agricultural: 3465
  Medical Topography: 1394, 1395, 2646,
    2954, 3892, 4841, 4982, 8499, 9014,
    9015, 9842, 9996, 10507, 10542, 10705,
    10898, 11127
  Meteorites: 9533, 9804, 9814, 9820
  Mineral Collections: 7755-7757, 9972,
    10062
  Mineral Springs: 169, 170, 171, 239,
    243, 482, 1040, 1113, 1114, 1523, 1560,
    1562, 1574, 1827, 1828, 2013, 2346,
    2349, 2353, 2474, 2868, 2880, 3057,
    3542, 3559, 3561, 3574, 3604, 4569,
    4570, 4592, 4756, 4887, 4962,
    5282-5284, 5327, 5978, 6012-6014, 6078,
    6409, 6860, 6947, 6948, 6951, 6965,
    7071, 7109, 7111, 7289, 7343, 7709,
    7954-7956, 8354, 8696, 8733, 9124,
    9125, 9158-9163, 9244, 9383-9386, 9484,
    9690, 9865, 9998, 9999, 10005, 10006,
    10008, 10009, 10166, 10447, 10539,
    10545, 10780, 11124
  Minerals, Localities: 111, 114, 143,
    356, 357, 362, 912, 1081, 1208, 1321,
    1410, 1495, 1577, 1580, 1581, 1583,
    1584, 1627, 1628, 1753, 1907, 1928,
    1948, 1973, 2106, 2121, 2431, 2442,
    2453, 2686, 2687, 2747, 3144, 3145,
    3146, 3165, 3168, 3379, 3530, 3531,
    3548, 3789, 3790, 3835, 3837, 3893,
    3947, 3948, 4235-4237, 4245, 4274,
    4616, 4640, 4647, 4649, 5098, 5236,
    5280, 5288-5290, 5400, 5745, 5758,
    6121, 6320, 6872, 6873, 6958, 7063,
    7076, 7093, 7095, 7097, 7273, 7278,
    7351, 7879, 8347, 8348, 8349, 8351,
    8355, 8357, 8420, 9487, 9491, 9499,
    9513, 9519, 9529, 9538, 9627, 9831,
    10000, 10187, 10250, 10271, 10273,
    10276, 10279, 10282, 10287, 10289,
    10292, 10335, 10336, 10341, 10561,
    10566, 10660, 10690, 10769, 10790,
    11114
  Mines and Mining: 1926, 1927, 2450, 3553,
    3563, 5466
  Natural Gas: 4922, 7288
  Natural History: 1693, 1734, 3288, 3297,
    3464, 3469, 3471, 3506, 6588, 10686
  Niagara Falls: 1461, 2454, 3291-3296,
    3477, 3572, 3748, 4565, 4600, 4677,
    4684, 5366, 6549, 6554, 6589, 7582,
    8460, 8817, 8847, 10324
  Oil Stone: 1519
  Oolite: 11112
  Paleobotany: 4532
  Paleontology: 131, 1112, 4705, 4712,
    4724, 4727, 4731, 7266, 7302, 7304,
    7305, 7308, 8303, 8657, 10300, 10586
  Paleontology, Invertebrate: 704, 1412,
    1434, 1585, 1704, 3020, 3022, 3023,
    3025, 3106, 3121, 3451, 3468, 4547,
    4681, 4719, 6093, 6198, 7035, 8641,
    8707, 8711, 8719, 8720, 8724, 9044,
    9727

Index

New York (continued)
  Paleontology, Vertebrate: 594, 837, 1198, 1260, 3024, 3032, 3033, 3161, 3170, 3171, 4105, 4419, 4471, 4602, 5685-5687, 7025, 7073, 7190-7192, 8590, 9592, 10063, 10513, 10621, 10786, 11053, 11076, 11077
  Petroleum: 5286, 6252, 6978, 7652, 8147, 8470, 9693, 9708
  Potash Mining: 1564
  Rocking Stone: 2649, 3660
  Salt Mines and Mining: 166, 295, 1246, 1559, 3166, 3167, 3331, 3334, 3336, 3910, 4335, 5435, 6250, 6349, 7115, 7118, 7124, 7728, 7811, 8290, 9422, 9904
  Slate: 1212, 3556, 7039
  Soils: 3518, 3521, 3522, 3524, 3539-3541, 3555, 3573, 8399, 8693, 9157, 9215, 10007
  Sulphur Mines and Mining: 7055
  Terraces: 3122, 3131
  Tertiary Formation: 3832
  Topography: 4986
  Trenton Falls: 9582, 9583, 11054
  Upper Falls: 10659
  Vitriol Spring: 7326
  Waterfalls: 4698
New York Historical Society: 7764
  Mineral Collections: 1742, 4380, 4381
New York Literary and Philosophical Society: 7770, 7771
New York Lyceum of Natural History: 1829, 3055, 7780-7788, 7809, 7812, 8697
  Mineral Collections: 7314
New York University: 596
New Zealand
  Paleontology, Vertebrate: 813, 2326, 2327, 6740, 6743, 6751, 6756, 6758, 6764, 8094, 8095, 8099, 10824
  Volcanoes: 6229
Newfoundland
  Elevation, Changes in: 3134
  Geology: 849, 1146, 8236, 9309, 11086
  Natural History: 2647
Niagara Falls: 2454, 3291-3296, 3477, 3572, 3748, 4565, 4600, 4677, 4684, 5366, 6549, 6554, 6589, 7582, 8460, 8817, 8847, 10324
Nicaragua
  Volcanoes: 4205
Nickel
  Ore Processing: 6503
Nicol, J.: 979
Nicollet, J. N.: 789
Niobium: 9025, 9026, 9029
Nitrogen
  From atmosphere: 10575
Noble Metals
  Analysis, Chemical: 6316, 8288, 11021
North America
  Paleontology, Vertebrate: 4786, 4796, 4804, 4805
North Carolina
  Coal Mines and Mining: 5725
  Diamonds: 858
  Dikes: 1608, 6219, 6220, 11040, 11041 8034
  Earthquakes: 2448
  Geological Survey: 4323, 7247, 7249-7253, 7255-7259, 7530, 7951, 8026, 8029, 8030
  Geology: 2549, 2556, 2564, 2577, 7247, 7249-7253, 7255-7259, 8025, 8027-8030, 8197, 8522
  Gold Mines and Mining: 290, 537, 1039, 1349, 1371, 1782, 1914, 1925, 2374, 2457, 3182, 3362, 4124, 4368, 4400, 5436, 5844, 6344, 7000, 7001, 7003, 7043, 7052, 7079, 7177, 7251-7253, 7348, 7952, 8031, 8197, 8276, 8304, 8374, 8489, 8735, 9038, 9039, 9864, 10260, 10261, 10310, 10779
  Graphite: 266
  Lead Mines and Mining: 1370, 6119
  Maps, Geological: 7251, 7257, 8198
  Marble: 3724
  Marl, Agricultural: 2135, 2691, 2692
  Medical Topography: 10981
  Meteorites: 4369, 6661, 9524

North Carolina (continued)
  Minerals, Localities: 2447, 3184, 3185, 4329, 5891, 7248, 9067, 9475, 10285, 10567
  Mines and Mining: 1767, 6869, 6870, 9563, 10135
  Natural History: 1544
  Natural Wall: 4663, 6219, 6220, 8034, 11040, 11041
  Paleobotany: 10039
  Paleontology: 3716, 5191, 5192, 6224
  Paleontology, Invertebrate: 2549, 2556, 2564, 2577, 8706, 10661
  Silver Mines and Mining: 4157, 4169, 10138-10140
  Soils: 6869, 6870, 7247, 7259
  Springs, Hot: 9863
  Volcanoes: 1140
Northern Academy of Arts and Sciences: 7959
Norway
  Granite: 697
  Iron Mines and Mining: 9372
  Minerals, Localities: 1650, 3596, 3602, 6203, 9226, 9228, 11024
  Mines and Mining: 3867
  Silver Mines and Mining: 968, 2773
Nott, Mr.: 3426
Nova Scotia
  Coal Mines and Mining: 4042, 4343, 6546
  Geology: 2073, 2077-2079, 3494, 4342, 5538-5542, 5539, 5541, 5574
  Maps, Geological: 5539, 5541, 5578
  Minerals, Localities: 131, 132, 150, 449, 2981, 4891, 4902, 5538-5542, 5539, 5541, 5574, 10762
  Mines and Mining: 7727
  Natural History: 7989
  Ornithichnology: 4921
  Paleobotany: 2078, 2079, 2980, 6548
  Paleontology, Invertebrate: 4539
  Sandstone: 8933
Nuttall, T.
  Mineral Collections: 841

Ocean (see also Tides; Coast)
  Analysis, Chemical: 2973, 2976, 5399, 6184, 6770, 7279, 8602, 9192, 9291, 9858, 9860, 10748, 10956, 10957, 10959, 11030
  Currents: 6925, 8716
  Natural History: 10547
  Salt: 2930, 10159, 10218
Ohio
  Analysis, Chemical: 2388, 2391
  Coal Mines and Mining: 515, 1028, 3010, 5053, 5482, 6912
  Conglomerate: 6455
  Earthquakes: 4825
  Exploration: 5059
  Geological Survey: 3924, 4279, 5057, 5058, 6420, 6424, 6425, 6891-6894, 6896, 6905, 8000-8013, 8785, 10843, 10929, 10933, 10940
  Geology: 1358, 1362, 1364, 2006-2008, 2403, 2748, 3916, 4493, 5045-5056, 5058, 6044, 6045, 6420, 6421, 6425, 6891-6894, 6896, 6905, 8647, 8780, 8782-8784, 8878, 8992, 10929, 10933, 10940, 10972
  Glaciers, Evidence for: 6042, 6431, 8718, 9371, 10091
  Iron Mines and Mining: 2400
  Limestone: 2390
  Maps, Geological: 5054, 6424
  Medical Topography: 3216, 3218-3220, 3916
  Millstone: 8387
  Mineral Springs: 414, 2380, 3224, 3225, 6913, 10834, 10849
  Minerals, Localities: 1361, 1699, 3213, 3923, 10858
  Mines and Mining: 9367, 10864, 10938
  Natural History: 3216, 3218-3220, 4206, 4666-4668, 4836
  Oil Stone: 311, 484, 6343
  Paleobotany: 1361, 4234, 4492, 5049, 6446, 10941
  Paleontology: 6456, 6464, 6917, 10543

Ohio (continued)
  Paleontology, Invertebrate: 1286-1290,
    1891, 2682, 4468, 4544, 4699, 6433,
    6439, 6440, 6443, 6445, 6447, 6448,
    6457, 10617, 10618, 10692, 10693
  Paleontology, Vertebrate: 303, 1360,
    2009, 4282, 5417, 8296
  Petroleum: 4251, 9111
  Prairies: 690
  Salt Mines and Mining: 8476
  Silver Mines and Mining: 92, 7779
  Terraces: 10943
Ohio River Valley
  Geology: 3223, 3882, 5360
  Natural History: 5495
  Paleontology, Vertebrate: 9045
Oil of Vitriol (see Copperas)
Oil Stone: 10537
  Arkansas: 5858
  Canada: 5330
  Connecticut: 10774
  France: 9176
  New York: 1519
  Ohio: 311, 484, 6343
  Pennsylvania: 9114
Oil Wells (see Petroleum)
Olmsted, D.: 867, 2999
Ontario
  Geology: 1755
  Minerals, Localities: 1703
Oolite Formation: 8658, 10903
  Appalachians: 9000
  Florida: 8915
  Indiana: 2404
  New York: 11112
  United States: 6094, 6095
  Uses of: 7220
Optical Properties (see Minerals)
Ore Processing: 1408, 1791, 2074, 2612,
    2614, 2616, 2944, 2955, 3982, 5939,
    6139, 6874, 6877, 8315, 8316, 8319,
    9314, 10105
  Cobalt: 1503, 6503
  Copper: 615, 901, 902, 987, 1169, 3016,
    3017, 3191, 3192, 3269, 5843, 5919,
    5924, 7594-7596, 8805, 8806, 8813,
    10226, 10230
  Furnaces: 9922, 10299
  Gold: 159, 1773, 1774, 5645, 7590, 10894
  Iron: 102, 127, 128, 129, 1302, 1760,
    1796, 1873, 1918, 2042, 2043, 2405,
    2416, 2602, 2608, 2685, 2789, 3475,
    3581, 3625, 3627, 3851, 3860, 3987,
    3990, 3991, 4379, 4465, 4636, 4746,
    5767, 5774, 5777, 5786, 6371, 7889,
    8045, 8232, 8445, 8488, 8617, 8769,
    8790, 8791, 9037, 9886, 9910, 9920,
    10158, 10903
  Lead: 3590
  Manganese: 1503
  Metals: 1168, 1169, 1482, 1503, 2463,
    2600, 4009, 4078, 4103, 4601, 5328,
    5329, 5734, 5735, 5739, 5843, 6897,
    8122
  Nickel: 6503
  Precious Metals: 10066
  Silver: 1200, 1613, 3087, 3089, 3194,
    10945
  Titanium: 10667
  Washing: 4839, 10413
  Zinc: 8827
Oregon
  Geology: 4703
  Guide: 4848, 4849
  Minerals, Localities: 4101
  Natural History: 4111-4119, 4556-4558,
    10259, 10960, 11072
  Paleontology: 4700, 4701, 7898
  Paleontology, Invertebrate: 1430, 1433,
    3462
Oregon Territory: 10487
  Paleontology, Vertebrate: 8233, 8234
Ores (see also specific metals)
  Analysis, Chemical: 6397
Organic Remains (see Paleontology)
Orkney Islands
  Minerals, Localities: 10304
Ornithichnology
  Connecticut: 25, 669, 2311, 2312

Ornithichnology (continued)
  Connecticut River Valley: 1496, 1499,
    2406, 2984-2997, 4910, 4912,
    5114-5116, 5120, 5146, 5148,
    5162-5164, 5166, 5176, 5178,
    5182-5184, 6743, 6785, 7517, 8091,
    8875, 10859, 10944
  Lake Superior District: 3133
  Nova Scotia: 4921
  Recent: 5688
Osmium: 1659
Owen, D. D.: 196, 198, 3664, 4964, 5502,
    6635, 7904, 7908, 10820, 10830
Owen, R.: 809, 860, 964, 995, 10887

Pacific Ocean
  Geology: 2841-2844, 2855, 2861
Paint and Pigments
  Manufacture of: 122, 2998
  Massachusetts: 7670
  New Jersey: 7081
  Pennsylvania: 7080, 7349
  Tennessee: 4166
Paleobotany
  Alabama: 2190
  Antarctica: 5252
  Antigua: 318
  Coal: 587, 1435, 1438, 1555, 1606, 1711,
    2055, 2058, 2190, 2191, 2192, 2399,
    4492, 4806, 5253, 6558, 6561, 8934,
    10014, 10133, 10143, 10177, 10195,
    10196, 10199, 10200, 10203, 10208,
    10734, 10941, 10980
  Conifers: 1432
  Connecticut: 9823
  Connecticut River Valley: 5147
  Corn: 5812, 7156
  Egypt: 2181, 6335
  England: 1606, 1711, 2048, 4041, 4184,
    4865, 7584, 7641, 11100
  Equisetum: 4802
  France: 8536
  Fruits: 6741
  Fucoides: 4789, 4795, 7584, 10114, 10116,
    10117
  Great Britain: 5253
  India: 5251
  Indiana: 8052, 8055
  Italy: 5900, 6348
  Kentucky: 9221
  Louisiana: 2262
  Maryland: 2192, 7069
  Mississippi River Valley: 1837
  Missouri: 2697, 2698, 7379
  Moldavia: 659
  New York: 4532
  North Carolina: 10039
  Notes: 67, 1920, 2049-2051, 2053, 2057,
    4131, 4449, 4505, 5910, 6347, 6795,
    6796, 10461, 10462, 10866
  Nova Scotia: 2078, 2079, 2980, 6548
  Nuts: 9095
  Ohio: 1361, 4234, 4492, 5049, 6446, 10941
  Palm Tree: 256, 8052, 8055, 9168
  Pennsylvania: 2399, 4806, 7156, 10114,
    10116, 10117, 10133, 10143, 10177
  Philippines: 10180
  Russia: 4581
  Scotland: 368, 639, 733, 9200, 10673
  Stigmaria: 2980
  Texas: 4517, 6225
  Thin Sections: 11013-11015
  Tree: 42, 368, 639, 733, 1726, 2181, 2697,
    2698, 4041, 4184, 4517, 4532, 4581,
    4865, 5049, 5251, 5724, 5900, 6225,
    6335, 6446, 6548, 6663, 6988, 7379,
    7641, 8536, 9112, 9221, 9262, 9263, 9823,
    10673
  United States: 1432, 10193, 10194, 10198
  Virginia: 2191, 10123
  Wax: 659, 6983
  West Virginia: 5812
  Wood: 318, 624, 4234, 7069, 7585, 7869,
    9221, 10039, 10291
Paleoclimates: 1359, 1578, 3911, 6131, 7558,
    8893, 8906
Paleontological Society of London: 8112
Paleontology
  Arkansas: 7501
  Brazil: 10035
  California: 4700, 4701

Paleontology (continued)
　Canada: 1702
　Cape of Good Hope: 1549
　Chalk Formation: 6191
　Coal: 573
　Connecticut: 472
　England: 8176, 9396
　Eocene: 6084
　Extinction: 7667
　Fossil Collections: 2061-2063
　France: 536, 6322
　From Borings: 6342
　Georgia: 712, 4615, 7304
　Louisiana: 1808
　Maryland: 8084, 9590
　Massachusetts: 5076, 5109, 5677, 8143
　Mexico: 2415
　Mississippi River Valley: 10368
　New Hampshire: 8926
　New Jersey: 7286
　New York: 131, 1112, 4705, 4712, 4724, 4727, 4731, 7266, 7302, 7304, 7305, 7308, 8303, 8657, 10300, 10586
　North Carolina: 3716, 5191, 5192, 6224
　Notes: 56, 66, 77, 1032, 1036, 1043, 1857, 2161, 2170, 2419, 2750, 2755, 2756, 2758, 2760, 2761, 3202, 3303, 3366, 3393, 3394, 3406, 3650, 3663, 3673, 3693, 4358, 5019, 5708, 5853, 5971, 5985, 6054, 6264, 6315, 6339, 6531, 6681, 7200, 7239, 7311, 8565, 9241, 9294, 9353, 10620
　Ohio: 6456, 6464, 6917, 10543
　Oregon: 4700, 4701, 7898
　Origin of fossils: 272
　Pennsylvania: 5755, 6090, 8644, 8646
　Plaster Casts: 642
　Scotland: 8164
　South America: 3203
　Tennessee: 6497, 7304, 7608, 10366
　Tertiary: 6086-6088, 6093
　Textbooks: 75, 2047, 2507, 2519, 4398, 4357, 4399
　United States: 1536, 4923, 6094, 6095, 8662, 8663, 8665
　United States, Western regions: 7854
　Vermont: 4710
　Virginia: 8169, 8426, 9094, 9851
　Wisconsin: 5630
Paleontology, Invertebrate: 10428
　Alabama: 2546, 2579, 6084, 6093, 6569
　Alps: 6572
　Amber: 6596
　Ammonites: 8192
　Animalcules: 2983
　Asterias: 4468, 6457, 10365
　Belemnites: 2060, 3111, 6739, 6760, 7465, 8098, 8685
　Belemnosepia: 2103
　Bilobites: 3020
　Bohemia: 1494
　Brachiopoda: 4694, 4730
　Conolites: 1886
　Conularia: 3544
　Conus: 4533
　Corals: 2820, 4461, 6481, 6688, 8663, 9206, 9819
　Crab: 2755
　Cretaceous Formation: 2574, 2593
　Crinoids: 91, 1288, 1754, 3591, 4681, 4719, 8083, 9209, 9727, 10407-10409
　Crustacea: 8308, 10552, 10553
　Delaware: 7463, 7464, 7476, 7485
　Devonian System: 2573, 2576
　Echinodermata: 59, 7484, 7981
　Echinoids: 7075
　Ecinidae: 8690
　Egypt: 1420
　Encrinites: 8641, 9044
　England: 573, 838, 1921, 5889, 6745, 6747, 6763, 10877
　Eocene: 2575, 2583, 2584, 2585, 2588, 2590, 2593
　Europe: 1887
　Eurypterus: 3023, 3025, 7305
　Florida: 1350, 1442, 2547, 2586, 3186, 9208
　Georgia: 1892, 2662-2664, 7061, 7075, 7125, 7465
　Glomeris: 8622
　Graptolites: 1585, 4729
　Great Britain: 10329

Paleontology, Invertebrate (continued)
　Great Lakes: 3048
　Infusoria: 766, 1412, 1413, 1415, 1416, 1417-1419, 1420, 1422, 1423, 1426, 1429, 1430, 1431, 1433, 1434, 1436, 1440, 1442, 1443, 1921, 3228, 4699, 3448-3462, 4154, 5125, 5581, 6138, 6745, 6747, 8755, 8904, 8927, 8996, 8999, 9023, 9250, 9426
　Iowa: 3544, 8083
　Ireland: 3228
　Kentucky: 704, 6198, 7075, 8630, 8633, 8654
　Limniadai: 4628
　Lucilites: 8652, 8653
　Maclurites: 6198
　Maine: 3125, 5199, 5581, 7183
　Maryland: 1423, 1431, 1440, 2544, 2546, 2570, 2571, 3458, 6093, 9207, 10661
　Massachusetts: 1416
　Microfossils (see Infusoria)
　Miocene: 2575, 2581
　Mississippi: 1712, 2587, 2588
　Mississippi River Valley: 7061, 8637
　Missouri: 1413, 2578, 2580, 3591, 7981
　Molluskite: 6742
　Myliobates: 4360, 4361
　New Jersey: 1415, 1888, 2103, 6093, 7463, 7464, 7476, 7485, 10552, 10553
　New York: 704, 1412, 1434, 1585, 1704, 3020, 3022, 3023, 3025, 3106, 3121, 3451, 3468, 4547, 4681, 4719, 6093, 6198, 7035, 8641, 8707, 8711, 8719, 8720, 8724, 9044, 9727
　North Carolina: 2549, 2556, 2564, 2577, 8706, 10661
　Notes: 1520, 2046, 4687, 4695, 4696, 4704, 6098, 8637
　Nova Scotia: 4539
　Nummulina: 2261
　Nummulites: 6569, 6572
　Ohio: 1286-1290, 1891, 2682, 4468, 4544, 4699, 6433, 6439, 6440, 6443, 6445, 6447, 6448, 6457, 10617, 10618, 10692, 10693
　Orbitolina: 3204
　Oregon: 1430, 1433, 3462
　Orthoceras: 1290
　Orthocerata: 4694
　Ostrea: 7456
　Oysters: 1771
　Paleozoic Formation: 8308
　Pennsylvania: 1886, 4544-4546, 4628, 4630, 4632, 7315, 8652, 8653, 8870
　Pentremites: 10369
　Prussia: 766
　Pygorhynchus: 1892
　Saxicava: 10676
　Silurian System: 2566, 2573, 2578
　South Carolina: 1426, 1429, 1838, 2579, 4360, 4361, 8690, 8904, 8927
　Sphaeroma: 3468
　Sponge: 6763
　Tennessee: 91, 10365, 10369, 10407, 10409
　Tertiary: 1887, 1890, 2543-2548, 2550-2556, 2558, 2560-2564, 2568, 2570-2572, 2579, 2581-2593
　Thoracoceras: 10669
　Trianisities: 8630
　Trilobites: 604, 704, 1286, 1287, 1494, 1704, 2199, 2578,, 3022, 3048, 3372, 3375, 3576, 4535, 4537, 4538, 4539, 4540, 4542-4554, 4582, 4630, 4632, 4669, 4705, 4722, 4800, 4807, 5889, 6195, 6433, 6439, 6440, 6443, 6445-6448, 7315, 8633, 8656, 9099, 10106, 10107, 10692, 10693
　United States: 1417-1419, 1426, 1431, 1614, 1743, 1771, 2543-2565, 2567-2593, 4533, 4535, 4537, 4540, 7457-7461, 7466, 7469, 7470, 7474, 7477, 7484, 8663, 9206, 10408
　United States, Western regions: 2589, 2592, 2824, 2831
　Vermont: 1436
　Virginia: 1422, 1423, 2546, 2561, 2577, 2579, 3458, 6086-6088, 6138, 8996, 8999, 10426, 10428

Paleontology, Vertebrate: 2134
  Africa: 5246
  Alabama: 82, 812, 2171-2174, 2908, 2909, 4803, 4810, 6002, 6003, 6235, 6564, 11082
  Alaska: 10313
  Ape: 4340
  Argentina: 7339
  Atlantochelys: 79
  Basilosaurus: 2173, 2565, 4345, 4346, 4350, 4351, 4353, 4354, 4803, 4810, 7503, 7504
  Bat: 1804
  Bavaria: 4429
  Birds: 60, 813, 1756, 2326, 2327, 2599, 5153, 6730, 6740, 6743, 6748, 6751, 6756, 6758, 6764, 8094, 8095, 8099, 10037, 10038
  Bone Analysis, Chemical: 7180, 7181
  Bone Caves: 2152-2154, 2158, 2159, 3741
  Burma: 11073
  Canada: 1705
  Castoroides: 4708, 11076, 11077
  Cheirotherium: 6374, 6562
  Coal: 5956-5959
  Connecticut: 593, 687, 696, 4733, 6289
  Connecticut River Valley: 8705, 9635, 9899
  Coprolites: 2888
  Crocodile: 79, 1095, 4785, 7478, 7482, 7483
  Dinorthis: 8091
  Dinotherium: 226, 4810, 6336, 9349
  Dodo: 1275
  Dogs: 7486
  Dolphin: 4812
  Dorudon: 72, 4345, 4346, 4350, 4351, 4353, 4354, 8099, 8100
  Early discovery: 3257
  Edentata: 4801
  Elephant: 86, 680, 1527, 1530, 1534, 4341, 4782, 5417, 5864, 7107, 8298, 8528, 8734, 10036, 10395, 10832
  Elephas: 4366, 11053
  Elk: 212, 375
  Enaliosaurus: 4784
  England: 460, 630, 644, 646, 852, 1095, 1184, 1451, 1919, 2152-2154, 2158, 2159, 2322, 3425, 4040, 4055, 4819, 5244, 5245, 6330, 6469, 6725, 6726, 6728, 6730, 6733, 6734, 6746, 6748, 6750, 6754, 6755, 6759, 6999, 7541, 8096, 8101, 8102, 8126, 9459, 10060, 10700
  Equus (see Horse)
  Eucrotaphus: 6180
  Europe: 2753, 5875
  Fish: 55, 57, 65, 76, 617, 687, 696, 725, 2536, 3032, 3365, 4197, 4355, 4463, 5712, 6330, 7186, 7187, 7472, 7508, 7980, 7982, 7983, 8138, 8698-8701, 8703, 8705, 8708, 8710, 8713, 8717, 9635, 10516, 11084
  Footprints (see also Ornithichnites): 668
  France: 1804, 1835, 2155, 2156, 2536, 2752, 4579, 5864, 6271, 6278, 8528, 9424, 9425
  Frozen animals: 7337
  Gavial: 3030
  Georgia: 69, 1353, 2662-2664, 4814, 7312
  Geosaurus: 3028
  Germany: 2653, 9349
  Glyptodon: 226
  Great Britain: 55, 7186, 7187, 7508
  Guadaloupe: 7502
  Harlanus: 8103
  Himalayas: 2284
  Horse: 4365, 5918, 6115, 6173
  Human: 383, 1360, 1830, 6520, 7333, 7502, 8405, 10396
  Hungary: 4584
  Hydrarchos (Hydrargos): 3306-3307, 6002, 6003, 7436, 7503, 7504, 8913, 11075
  Hylaesaurus: 6733, 6754
  Ichthyosaurus: 624, 4783, 4799, 4800, 8193, 8427, 9459
  Iguanodon: 6725, 6726, 6733, 6746, 6750, 6754, 6755, 6759
  Illinois: 6116
  India: 8597
  Indiana: 2281, 7980, 7982, 7983
  Iowa: 3123

Paleontology, Vertebrate (continued)
  Isle of Man: 375
  Italy: 999
  Kentucky: 1115, 1817, 2484, 2485, 2493, 2634-2636, 2638, 4232, 4781, 4790-4792, 6552, 7073, 7684, 8155, 8664, 9689, 10513, 11010, 11011
  Leiodon: 82
  Louisiana: 2248, 2264, 2265, 4798, 6178, 7930
  Maine: 4463, 5636
  Mammals: 3033
  Mammoth: 43, 246, 491, 528, 1095, 1203, 1518, 1534, 2373, 3072, 3161, 3583, 3613, 3708, 4040, 4055, 4105, 4257, 4282, 6289, 6658, 6659, 6664, 7025, 7127, 7190-7192, 7930, 8101, 8102, 8185-8188, 8190, 8296, 8301, 8302, 8457, 10313, 10448, 10621
  Maryland: 4812, 8196
  Massachusetts: 3257, 8138
  Mastodon: 226, 303, 477, 593, 726, 837, 999, 1198, 1348, 1360, 2009, 2228, 2381, 3021, 3116, 3126, 3589, 3925, 3934, 4341, 4364, 4416, 4464, 4504, 4506, 4602, 4739, 4926, 4929, 5263, 5264, 5266, 5267, 5685-5687, 5998, 5999, 6479, 6552, 6709, 6933, 7319, 7320, 8096, 8126, 8185-8188, 8190, 8330, 8590, 9592, 9874, 10020, 10359-10361, 10364, 10555, 10557, 10694-10696, 10698, 10701, 10893, 11073
  Megalonyx: 4781, 4790-4792, 4808, 4814, 4815, 5720, 7588, 8092
  Megalosaurus: 5514
  Megatherium: 1353, 2168, 2631-2633, 2664, 2751, 3757, 5213, 6566, 7312, 7339, 8088
  Merycoidodon: 6175
  Mississippi: 1508, 6566
  Mississippi River Valley: 1834, 3071, 4788
  Missouri: 726, 2290, 3126, 3708, 4150, 4515, 4799, 4811, 4813, 4927, 5266, 5267, 5998, 5999, 6000, 6001, 6179, 6181, 6382, 6709, 8603, 8604, 10893
  Missourium: 3730, 4150, 4412, 4413, 6000, 6001
  Mosasaurus: 3028, 4359, 4362, 4363, 7481
  Mylodon: 4814, 4815, 8092
  New Holland: 5153
  New Jersey: 79, 1348, 2373, 3021, 3028, 3030, 3589, 4197, 4739, 4785, 4787, 4924, 5514, 6933, 7319, 7320, 7478, 7481, 7482, 7483, 8710, 8713, 8717, 10555, 10557
  New York: 594, 837, 1198, 1260, 3024, 3032, 3033, 3161, 3170, 3171, 4105, 4419, 4471, 4602, 5685-5687, 7025, 7073, 7190-7192, 8590, 9592, 10063, 10513, 10621, 10786, 11053, 11076, 11077
  New Zealand: 813, 2326, 2327, 6740, 6743, 6751, 6756, 6758, 6764, 8094, 8095, 8099, 10824
  North America: 4786, 4796, 4804, 4805
  Notes: 344, 1188, 1274, 1524, 2688, 3045, 3762, 4030, 4043, 4125, 4635, 5728, 6762, 7414, 7415, 7437, 7639, 7828, 7881, 8089, 8090, 8093, 8097, 8104, 8127, 8135, 8428, 8433, 8485, 8534, 9885, 10491, 10510
  Ohio: 303, 1360, 2009, 4282, 5417, 8296
  Ohio River Valley: 9045
  Opossum: 6728
  Oregon Territory: 8233, 8234
  Orycterotherium: 4811, 4813
  Paleotherium: 2752, 8603, 8604
  Pelorosaurus: 4819, 6759
  Pennsylvania: 1450, 3072, 3365, 3445, 3751, 4222, 4939, 6223, 6980
  Platygonus: 6117
  Plesiosaurus: 644, 4787, 10061, 10700
  Poebrotherium: 6172
  Proboscidae: 4928
  Pterodactyl: 852, 4429
  Pygorhynchus: 69
  Reptiles: 110, 630, 2908, 2909, 6723, 6757
  Rhinoceroides: 3751
  Rhinoceros: 1961, 5244, 5246, 6181, 7541, 8655

Index

Paleontology, Vertebrate (continued)
  Russia: 217, 491, 1203, 1961, 3583, 5246, 6354, 7337, 8457, 10668
  Saurodon: 4924
  Saxony: 383
  Scandinavia: 7909
  Scotland: 8403, 9194
  Scutella: 8687-8689
  Seal: 11079, 11085
  Sharks: 3123, 5499
  Snakes: 4530
  South America: 2749, 2751, 7588
  South Carolina: 4345, 4346, 4350, 4351, 4353, 4354, 5242, 8687-8689, 8692, 9126, 9880
  South Dakota: 6180, 8082
  Spondylosaurus: 10668
  Squalidae: 4348, 4349, 4352, 4356, 8692
  Tapirus: 6178, 6724
  Teleosaurus: 3771
  Tennessee: 4808, 6479, 10359-10361, 10364, 11080, 11087
  Tertiary: 2134
  Tetracaulodon: 4418, 4925
  Textbooks: 4417
  Toxodon: 8086
  Turtle: 6734
  United States: 1527, 1530, 1534, 2500, 3271, 3281, 4348, 4349, 4352, 4356, 4359, 4362, 4363, 4804, 4963, 7472, 8698-8701, 8703, 8705, 8708, 8710, 8713, 8717, 8734, 10694-10696, 10698, 10701
  United States, Western regions: 1199, 4355
  Vermont: 86, 10241, 10832
  Virginia: 1517, 2639, 5263, 5264, 5718-5720, 5722, 7317, 8298, 8701, 10036, 11079, 11081, 11084, 11085
  Wales: 2529
  Walrus: 7317
  West Indies: 8405
  Whale: 5242, 5636, 8403, 9194
  Zeuglodon: 72, 812, 1508, 2174, 2565, 3528, 3547, 4345, 4346, 4350, 4351, 4353, 4354, 6564, 7503, 7504, 8099, 8100, 8913, 10432, 10433, 11082
  Zygodon: 1444, 2173
Paleozoic Formation: 10, 8075-8077, 10620
Paleontology, Invertebrate: 8308
  United States: 8907
Paleozoic System: 4695, 4696, 4711, 4720, 4721, 4727
Palestine
  Geology: 1719, 1720, 6129
  Minerals, Localities: 1719, 1720, 4651, 9661
Palladium: 4064
  Separation of: 2456, 2458
Panama
  Geology: 5258
  Natural History: 6417
Paris School of Mining: 4079
Parkinson, J.: 1798
Peale, C. W.: 2497, 6852, 7030
Peale's Museum (see Museums, Peale's)
Pelopium: 9025, 9026, 9029
Pennsylvania
  Basalt: 9906
  Building Stone: 4747
  Caves: 1450, 4222, 4935, 5419, 5905, 7188, 7690, 8644
  Clay: 6243, 10807
  Coal Mines and Mining: 299, 428, 529, 561, 1052, 1078, 1148, 1167, 1214, 1215, 1596, 1598, 1670, 1841, 1899, 1916, 1932, 2148, 2149, 2189, 2273, 2274, 2339, 2385, 2389, 2392, 2396, 2397, 2437-2440, 2594, 2734-2736, 2942, 2943, 3002, 3006, 3011, 3291-3296, 3351, 3354, 3363, 3377, 3624, 3695, 3787, 3788, 3794, 3833, 3849, 3852, 3869, 4073, 4088, 4123, 4146, 4190, 4480, 4589, 4702, 4707, 4932, 4933, 4943, 5352, 5353, 5439, 5442, 5447, 5456, 5460, 5462, 5700, 5701, 5746, 5764-5766, 5768, 5769, 5772, 5778-5783, 5802, 5806, 5988, 6004, 6041, 6096, 6102, 6104-6106, 6562, 6143-6170, 6331, 6337, 6338, 6373, 6522, 6574, 6612, 6704, 6708, 6932,

Pennsylvania, Coal Mining (continued)
  6952, 6972, 7020, 7117, 7196, 7296, 7369, 7601, 7688, 7761, 7762, 7921-7924, 8205, 8206, 8209, 8210, 8216, 8218, 8220, 8362, 8363, 8385, 8429, 8434, 8437, 8451, 8814-8816, 8943, 9288, 9289, 9363, 9457-9458, 9597, 9615, 9666, 9667, 9692, 9695, 9700, 9849, 10050, 10051, 10065, 10099, 10110-10112, 10121, 10122, 10128, 10130-10132, 10146-10148, 10495, 10604, 10895, 10896, 11042, 11049, 11088, 11101
  Earthquakes: 4936, 5265, 7011, 7026, 8300
  Fire Stone: 8252
  Footprints: 5956-5959
  Gazetteer: 4453, 9369
  General: 120, 4480
  Geological Survey: 448, 2084, 3977, 4152, 4319, 4387, 7900, 8207, 8208, 8211-8213, 8215, 8217, 8219, 8601, 8848, 8855-8858, 8861-8863, 8865, 8866, 8868, 8869, 8883, 10324
  Geology: 1512, 1514, 2086, 2088, 2089, 2091, 2098, 2435, 2436, 2443, 2492, 2951, 2952, 3829, 3830, 3834, 4385, 4386, 4940-4942, 4953, 4955, 4994, 5193, 5423, 6600, 6631, 6634, 7185, 8362, 8648, 8683, 8848, 8855-8858, 8861-8863, 8865, 8866, 8868, 8869, 8874, 8883, 8924, 9335, 10110, 10111, 10112, 10119, 10121, 10122, 10347
  Glaciers, Evidence for: 8867, 8879
  Gold Mines and Mining: 4954
  Iron Mines and Mining: 1149, 1670, 1770, 2339, 4149, 4160, 4510, 4633, 4956, 5667, 5767-5769, 5779, 6523, 6612, 6704, 6708, 8214, 8218, 8441, 8944
  Lead Mines and Mining: 4949, 5755, 7058, 10878
  Limestone: 2628, 2629, 3380
  Manganese Mines and Mining: 11043
  Maps, Geological: 4702, 6005, 10116, 10118, 10148, 10348
  Marble: 4946
  Medical Topography: 4566
  Mercury Mines and Mining: 4143
  Meteorites: 7026, 9674
  Millstone: 8293
  Mineral Black: 95
  Mineral Springs: 1241, 1934, 1935, 2366-2369, 2610, 2742, 2744, 3097, 4881, 4882, 4938, 6857, 6955, 6957, 7324, 8283, 8286, 8459, 8462, 8472, 8497, 8498, 10710, 10711
  Minerals, Localities: 100, 1355, 1526, 1930-1933, 2249-2252, 2254, 2255, 2257, 2398, 3595, 3829, 3830, 3834, 4158, 4426, 4958, 5765, 5790, 6089, 6244, 7068, 7325, 7455, 7835, 8566, 8768, 8779, 9380, 9432, 9438, 9474, 9830, 10191, 10304, 11012, 11047
  Mines and Mining: 8362, 10794
  Natural History: 2906, 3478, 5494, 6963, 9100-9109, 10325
  Oil Stone: 9114
  Paint, Ochres: 7080, 7349
  Paleobotany: 2399, 4806, 7156, 10114, 10116, 10117, 10133, 10143, 10177
  Paleontology: 5755, 6090, 8644, 8646
  Paleontology, Invertebrate: 1886, 4544-4546, 4628, 4630, 4632, 7315, 8652, 8653, 8870
  Paleontology, Vertebrate: 1450, 3072, 3365, 3445, 3751, 4222, 4939, 6223, 6980
  Perkiomen Mine: 96, 1479, 2741
  Petroleum: 4281, 4536, 5953
  Rocking Stone: 4947
  Salt Mines and Mining: 4934, 7800
  School for Geology: 4273
  Silver Mines and Mining: 1917, 4957
  Slate: 7049, 7958
  Springs: 3444
  Springs, Hot: 5418
  Volcanoes: 215
  Zinc Mines and Mining: 1772, 9434, 11044, 11048
Pennsylvania College of Mines: 6993
Perkins, B. D.: 2110
  Mineral Collections: 7050

Perkiomen Mine (Pennsylvania): 2741, 9434, 10878, 11044, 11048
Persia
  Geology: 6011
  Marble: 5894
  Mountains: 6011
Peru
  Earthquakes: 1406, 6516
  Geology: 1730, 5382, 8618
  Maps, Geological: 8802
  Mineral Springs: 8801
  Minerals, Localities: 1731, 4893, 4894, 4897, 7012, 8799
  Mines and Mining: 1238, 8803
  Natural History: 10419
  Silver Mines and Mining: 1257, 9683, 9694
  Volcanoes: 905, 2733, 8801
Petroleum
  England: 969
  Florida: 2361
  France: 7839
  Kentucky: 260, 742, 4128, 5855, 5903, 7886, 7887, 11107
  New Hampshire: 1618
  New York: 5286, 6252, 6978, 7652, 8147, 8470, 9693, 9708
  Ohio: 4251, 9111
  Pennsylvania: 4281, 4536, 5953
  Poland: 8282
  Tennessee: 2481
  Virginia: 4172, 7718, 8611, 10638
Philadelphia Academy of Natural Sciences: 8260-8273, 11128
  Mineral Collections: 6174, 6176, 6177, 6182
Philadelphia College
  Courses: 8277
Philadelphia Linnaen Society: 8278
Philadelphia Museum: 6310
Philippines
  Paleobotany: 10180
  Volcanoes: 3429, 8326
Phillips, W.: 228, 4590
  Mineral Collections: 490, 798
Phosphorite Rock
  Spain: 2935
Phyllite
  Massachusetts: 872
Physical and Medical Journal of Cincinnati: 456
Pipestone
  United States: 9988
Pitch Lake (see also Asphaltum)
  Trinidad: 124, 3399, 4004, 4163, 6296
Pittsfield Lyceum: 8382
Platinum Mines and Mining: 1744, 2146, 5025, 6967, 7615, 8794
  California: 985, 10207, 10214
  Colombia: 10512
  Europe: 4067
  France: 957, 1663
  Jamaica: 7033
  New South Wales: 6714
  Notes: 4082
  Russia: 634, 1657, 1822, 3996, 5380, 7550, 7560, 7571, 7664, 7724, 9919
Playfair, J.: 1805
Poems
  Coal and Diamond: 7137
  Earthquakes: 5989, 6230, 10783
  Fingal's Cave: 7386
  Fossil Fern: 7578
  Geological: 9211, 9212
  Geologist's Wife: 6395
  Gold vs. Agriculture: 7174
  Mammoth Caves: 8579
  New York Mineral Springs: 9386
  On a Mineralogist: 4979
  Rhode Island Coal: 7824
  The World, or Instability: 8670
Poland
  Petroleum: 8282
  Salt Mines and Mining: 1224, 1816, 2706, 5901, 6653, 7141, 7719, 7840, 8136, 8291, 9108, 9140, 9408, 11063, 11106
Polymorphism: 7327, 9019
Porcelain Clay (see Clay)
Portland (Maine) Mineral Society: 1833

Portugal
  Earthquakes: 13, 1063, 2514, 8141, 10499, 11125
  Iron Mines and Mining: 6285
  Lead Mines and Mining: 1186
  Medical Topography: 2766
  Mineral Springs: 8742
  Mines and Mining: 1952
  Springs, Hot: 1186
Pot Holes (see also Glaciers, Evidence for)
  New Hampshire: 8345
Potassium
  Manufacture of: 4816
  Mines and Mining: 1564, 6813
  Production of: 2606
Potsdam Sandstone Formation: 3121, 3132
Prairies: 8787
  Alabama: 5913, 9532, 9572
  Arkansas: 9900, 9901
  Ohio: 690
  Origin of: 7992, 10797
  United States, Western regions: 1356, 1357, 1879, 2263, 5841
Precious Metals (see Gold; Silver; Platinum)
Providence Franklin Society: 8609, 8610
Prussia
  Paleontology, Invertebrate: 766
Pseudomorphism: 4619
Publications, Scientific (lists, catalogues, notices): 352, 371, 374, 400, 412, 422, 497, 579, 605, 653, 662, 707, 729, 771, 776, 795, 819, 856, 863, 886, 900, 916, 940, 965, 973, 981, 997, 1019, 1079, 1080, 1795, 6256, 8642, 9496, 9497, 9619, 9626, 9705, 9710, 9711, 9715, 9718, 9722, 9723, 9727, 9729, 9737, 9746, 9747, 9754, 9760, 9763, 9765, 9773, 9776, 9778, 9786, 10757, 11128
  English: 564
  Foreign: 4572
  United States: 572, 577, 591, 605, 636, 1277, 1278, 1279
Pumice: 4889, 5693, 7218
  Mississippi River Valley: 367
  Missouri: 8910

Quebec
  Geology: 1706, 3502, 3509
Queries, Geological: 4305, 9205
Quicksilver (see Mercury)

Rafinesque, C. S.: 340, 651, 3768, 4794
  Mineral Collections: 8638
Railroad Surveys
  Bear Mountain: 5753
  Norwich and Worcester: 7979
  Pennsylvania: 10126
  Phillipsburg: 10109
  Tennessee: 6083
  Western Corporation: 6648
Raindrop Impressions (see Sedimentary Features)
Ranking, J.: 1156
Red River Valley
  Exploration: 7176
  Meteorites: 1086
Religion and Geology: 224, 1259, 1677, 2328, 2358, 2386, 2627, 2759, 3408, 3619, 3622, 3623, 3628, 3629, 3672, 3880, 3881, 3968, 4134, 4156, 4830-4834, 4877, 5107, 5117, 5121, 5972, 5973, 5984, 6332, 6938, 7147, 7219, 7441, 7531, 7910, 7987, 7988, 8039, 8172, 8331, 8452, 8588, 8589, 8741, 9392, 9400, 9427, 9712, 9759, 9855, 9895, 9897, 10057, 10059, 10450, 10489, 10525, 10675, 10773, 10891, 10988, 10989, 11007, 11059, 11089
Rensselaer Polytechnic Institute: 511, 3364
Rhode Island
  Coal Mines and Mining: 45, 1178, 5773, 6942, 7824, 8284, 8760, 8763-8766, 8925, 9389, 9667
  Geological Survey: 5571, 8761, 8762
  Geology: 5571, 8840
  Maps, Geological: 5569
  Metamorphic Rocks: 5647
  Mineralogy: 41

Rhode Island (continued)
  Minerals, Localities: 431, 1285, 1904,
    3080, 6815, 6872, 8830, 10152, 10155,
    10286, 10339, 10345, 10713, 10714,
    10716
  Rocking Stones: 6816, 10154
Rhodium
  Alloyed with gold: 3083
  Separation of: 2458
Richardson, J.: 947
Riddell, J. L.: 10841
Rionium (see Vanadium)
Ripple Marks (see Sedimentary Features)
Rivers
  Erosion: 4299
Road Survey
  Washington to New York: 6477
Robinson, S.
  Mineral Collections: 463
Rock Harmonium: 9299
Rocking Stones
  England: 7394, 7397
  Great Britain: 5510
  Massachusetts: 1058, 8505, 8511, 10745
  New Hampshire: 4532, 7418, 8549
  New York: 2649, 3660
  Pennsylvania: 4947
  Rhode Island: 6816, 10154
  United States: 3828
Rocks
  Analysis, Chemical: 8844
  Blasting: 4774, 4775
  Classification of: 3034, 3035, 3337,
    3340, 3345-3348, 3352, 3356, 3357,
    3367, 3373, 3747, 6188, 6618, 6622,
    6628, 6630, 6633, 6640, 6642, 7512,
    8606
  Formation of: 5377, 6716, 9306
  Moving: 40, 3026, 11032
  Origin of: 4144
  Strength of: 6597
Rocky Mountains
  Exploration: 5534, 5535, 5694, 8121
  Geology: 9323
  Natural History: 4111-4119, 5534, 5535,
    8121, 9213, 10301
Rogers, H. D.: 4152
Rogers, M. M.: 281
Rogers, W. B.: 1864, 10841
Royal Institution: 5863, 5867
Royal Society Transactions: 501
Russia
  Amber: 7672
  Diamonds: 524
  Geology: 495, 862, 1142, 2823, 7179,
    7518-7521, 7528, 7851, 10617, 10619
  Gold Mines and Mining: 419, 509, 634,
    942, 1252, 2234, 4034, 4086, 4094,
    4918, 5383, 5384, 5390, 5440, 5443,
    5445, 5478, 5483, 6008, 7550, 7560,
    7571, 7724, 9252, 9358
  Granite: 411
  Grotto: 214
  Landslide: 223
  Meteorites: 7802
  Mineral Springs: 9845
  Minerals, Localities: 476, 492, 871, 950,
    1605, 3416, 5000-5009, 5023, 5916,
    6009, 6010, 7918, 8686, 11061
  Mines and Mining: 454, 1272, 3181, 3865,
    10168
  Natural History: 3601, 8113
  Paleobotany: 4581
  Paleontology, Vertebrate: 43, 217, 491,
    1203, 1961, 3583, 5246, 6354, 7337,
    8457, 10668
  Platinum Mines and Mining: 634, 1657,
    1822, 3996, 5380, 7550, 7560, 7571,
    7664, 7724, 9919
  Salt Lake: 10097
  Silver Mines and Mining: 5383, 5384,
    5390, 5440, 5443, 5445, 5478, 5483
  Volcanoes: 10

St. Croix
  Geology: 5324
St. Domingo
  Gold Mines and Mining: 4402
St. Fond, M. F. de: 351
St. Lucia
  Volcanoes: 2280

St. Petersburg Academy: 9156
St. Petersburg Mineralogical Society: 10179
St. Vincent
  Volcanoes: 6242, 7562, 7674, 7298
Salt
  Ocean: 10218
  Origin of: 8936, 8938
Salt Lake
  California: 5680, 10869
  Russia: 10097
Salt Licks
  Kentucky: 11092
Salt Mines and Mining: 7889
  Algeria: 3945
  Arkansas: 10481
  Austria: 7846, 8529
  Catalonia: 2644
  Egypt: 5982
  England: 1781, 2024, 7064
  Europe: 4574, 4824
  France: 8409, 9116
  Germany: 638
  Great Britain: 104
  Illinois: 10475
  Indiana: 10483
  Iowa: 10464, 10466
  Kentucky: 8651
  Lake Superior District: 7172
  Louisiana: 4275
  Massachusetts: 10159
  Michigan: 1783, 5298, 5300, 5308, 5310,
    5313, 5315
  Missouri: 6647
  New Granada: 4278, 4367
  New York: 166, 295, 1246, 1559, 3166,
    3167, 3331, 3334, 3336, 3910, 4335,
    5435, 6250, 6349, 7115, 7118, 7124,
    7728, 7811, 8290, 9422, 9904
  Notes: 1343, 1454, 1456, 1601, 3871,
    4132, 4754, 4867, 6292, 6308, 7650,
    7663, 8639, 10218, 10810, 10854
  Ohio: 8476
  Pennsylvania: 4934, 7800
  Poland: 1224, 1816, 2706, 5901, 6653,
    7141, 7719, 7840, 8136, 8291, 9108,
    9140, 9408, 11063, 11106
  Spain: 5416
  Tennessee: 8475
  Textbook: 1319, 10548
  United States: 5429, 10485, 10486
  Virginia: 97, 248, 1036, 1061, 1787,
    4873, 5917, 7119, 9096, 10151, 10484,
    10644, 10829
Saltpetre
  Virginia: 7085, 7356
Salses (see Volcanoes)
Sand
  Analysis, Chemical: 5627
  For Molding: 4084
  Notes: 1391
  Thermal Conductivity: 7600
Sandstone (see also Building Stone): 2176,
    6279
  Alabama: 1694
  Building Stone: 149, 10699
  Connecticut: 2310
  Connecticut River Valley: 8900, 8912,
    8987, 9798, 10791
  England: 8116
  Flexible: 1894
  Nova Scotia: 8933
  Scotland: 10724, 10725
  South Carolina: 6099
  United States: 5673
  Variegated: 3831
  Virginia: 8994
Saxony
  Bitumen: 3431
  Copper Mines and Mining: 4090
  Footprints: 668
  Minerals, Localities: 1969, 1970, 2246,
    5921, 5926
  Mines and Mining: 567
  Paleontology, Vertebrate: 383
  School of mines: 7500
Say, T.: 888
Scafe, J.: 6257
Scandinavia
  Elevation, Changes in: 1962
  Geology: 7522
  Paleontology, Vertebrate: 7909

School for Geology
  Pennsylvania: 4273
Schoolcraft, H. R.: 4626, 8421
Schöpf, J. D.: 2800
Science
  Methodology: 10442
  Progress of: 10257
Scientific Instruments: 8367
Scientific Meetings
  Berlin: 493
  Italy: 924
Scientific Societies (see also individual societies)
  Canada: 489, 531
  Germany: 532
  Italy: 827
  Lyceums: 3631, 10312
  Massachusetts: 504
  United States: 436, 543, 544
Scotland (see also Great Britain)
  Alum Mines and Mining: 6323
  Coal Mines and Mining: 1302, 4561
  Earthquakes: 4062
  Footprints: 3275, 6103
  Freestone: 7534
  Glaciers, Evidence for: 61
  Granite: 10306
  Iron Mines and Mining: 5756
  Lead Mines and Mining: 2066
  Mineral Springs: 2534
  Minerals, Localities: 5748, 5749, 6593, 6594, 8201, 9352
  Mines and Mining: 10256
  Natural Gas: 4159
  Paleobotany: 368, 639, 733, 9200, 10673
  Paleontology: 8164
  Paleontology, Vertebrate: 8403, 9194
  Rock, Moving: 3173
  Sandstone: 10724, 10725
  Wales: 5756
Scriptural Natural History: 2258, 2259
Scudder, Dr.: 7091
Sedgwick, A.: 542
Sedimentary Features
  Fluting: 4493
  Mud Furrows: 8896
  Rain Drop Impressions: 89, 2730, 3129, 8710, 8713, 8717, 8942, 10209
  Ripple Marks: 2227
Sedimentation (see also Deltas): 10270
  In rivers: 3056, 3096, 8704
  Mississippi River Valley: 3179, 6782, 6928, 8788, 8789
  Potomac River: 6066, 6068
  Red River: 6233
Sediments
  Furrows: 4688, 4689
  Hudson River: 4298
  Wave mark: 4688, 4689
Selenium
  Bohemia: 1810
  Discovery of: 337
Septarian Nodules
  Connecticut: 10793
  England: 589, 10793
Shale
  Belgium: 1205
Shepard, C. U.: 586, 1012, 1684, 1913, 9716
Siberia (see Russia)
Sicily
  Earthquakes: 3792, 3825
  Geology: 2920, 3793, 8228
  Maps, Geological: 2919
  Natural History: 2142
  Volcanoes: 660, 1190, 1313, 1507, 1552, 2740, 3825, 4238, 4818, 4917, 5759, 6969, 7789, 8535, 8824, 9144, 10601, 10682
Sideroscope: 475
Silesia
  Meteorites: 764
  Minerals, Localities: 4404
Silica
  Analysis, Chemical: 6015
Silicon: 1653
Silliman, B.: 253, 875, 3402, 6690, 6692, 7373, 7657, 7659, 7940, 9986
  Lectures on Geology: 4261, 4262
Silliman, B., jr.: 892
Silurian System: 4730, 7524
  Paleontology, Invertebrate: 2566, 2573, 2578

Silver Mines and Mining: 195, 1150, 1155, 1200, 3000, 3087, 3089, 3227, 3848, 7618
  Analysis, Chemical: 4230
  Arkansas: 1270
  Assaying: 10488
  Austria: 8148
  Chile: 6280, 9683, 9694
  Germany: 4868
  Great Lakes: 1354
  Indiana: 9117
  Lake Superior District: 2375
  Massachusetts: 11058
  Mexico: 1791, 5450, 8494, 9682
  Michigan: 9269
  New Jersey: 9360
  North Carolina: 4157, 4169, 10138-10140
  Norway: 968, 2773
  Notes: 4075, 6025, 6300, 6997, 7399, 7618, 10599
  Ohio: 92, 7779
  Ore Processing: 1200, 1613, 3087, 3089, 3194, 10945
  Pennsylvania: 1917, 4957
  Peru: 1257, 9683, 9694
  Russia: 5383, 5384, 5390, 5440, 5443, 5445, 5478, 5483
  Spain: 4081, 5481
  Texas: 10872
  United States: 9359
Slate (see also Building Stone): 1937, 5682, 6990, 7577, 7726
  Connecticut: 540
  England: 9393
  Maine: 294, 5348, 5586, 6107, 6108
  Maryland: 7049
  New England: 293
  New York: 1212, 3556, 7039
  Pennsylvania: 7049, 7958
  Vermont: 2325
Smith, J. L.: 9322
Smith, T. P.: 4850
Smith, W.: 748, 3739, 6390, 8481
Smithson, J.: 535
Smithsonian Institution: 906, 9917
Soapstone (see Steatite)
Societies, Scientific: 5813, 6849
  Europe: 7548
  Paris: 7204
Society for Promoting Agriculture
  South Carolina: 9931
Soda Lake, South America: 3736
Sodium
  Manufacture of: 4816
Soils: 218, 580, 664, 1320, 1344, 1345, 1540, 2064, 2065, 2314, 2315, 4541, 4971, 5175, 5216, 5220, 5221, 5740, 5977, 7002, 7432, 8163, 8247-8251, 8255, 8258, 9506, 9527, 9532, 9539, 9540, 9543, 9548, 9556, 9557, 9762, 9887, 9889, 9912, 10102, 10378, 10463, 10465, 10467, 10930, 10934, 11110
  Alabama: 3484, 3519, 6603, 6604, 9572
  Analysis, Chemical: 183
  Cuba: 1641
  Egypt: 5799, 5801, 6055, 9797
  Florida: 6654
  France: 1642
  Georgia: 2658
  Iron: 8318
  Kentucky: 6209, 6792, 9423
  Massachusetts: 6847, 6848
  Mississippi River Valley: 3557, 10268
  New York: 3518, 3521, 3522, 3524, 3539-3541, 3555, 3573, 8399, 8693, 9157, 9215, 10007
  North Carolina: 6869, 6870, 7247, 7259
  Notes: 187, 189, 238, 241, 1295, 1296, 1301, 1404, 1405, 1779, 1786, 1963, 1964, 2268, 2459-2462, 2611, 2661, 2700, 2881-2887, 2934, 2935, 2961, 2967-2971, 2975, 3019, 3276, 3342, 3371, 3376, 3531, 3532, 3567, 3571, 3674, 3692, 3694, 3696, 3714, 3717, 3738, 3774, 3816, 4212, 4221, 4225-4227, 4244, 4247, 4290, 4401, 4452, 4714, 4715, 4837, 4840, 4952, 5036, 5037, 5092, 5272, 5334, 5338, 5498, 5567, 5568, 5575, 5582, 5585, 5588, 5595, 5609, 5625, 5743, 5752, 5754, 5820-5822, 5824, 5826-5830, 5915, 6043, 6126, 6132-6137, 6217, 6222,

Soils, Notes (continued)
    6294, 6407, 6408, 6418, 6471, 6472,
    6508-6510, 6679, 6685, 6767, 6768,
    6786, 6844-6848, 6885, 7173, 7175,
    7197, 7365, 7492, 7536, 7603, 7613,
    7634, 7665, 7825, 7861, 7862, 7868,
    7879, 7968-7972, 7974, 7975,
    8014-8016, 8200, 8401, 8436, 8568,
    8571, 8759, 8812, 9034, 9035, 9041,
    9049-9066, 9113, 9312, 9331, 9935,
    9937, 9954, 10541, 10558
  Peat: 7265, 7323, 8743
  South Carolina: 1388, 9378, 9379
  Texas: 5276
  Textbooks: 5092, 5818, 5819, 5823, 5825,
    6060, 6466, 6789, 7973, 8828, 8829
  Treatises: 2962, 2963
  Virginia: 3894, 7487-7491, 7495
  Wisconsin: 3535
Sounds, Subterranean: 437, 5381,
  7549, 8407, 9197, 9304
South Africa: 5013
  Natural History: 10422
South America (see also individual countries)
  Earthquakes: 4211, 4255, 4741, 5374,
    5387, 5393, 6290
  Geology: 1865, 3653, 5378, 6274
  Meteorites: 2286
  Mines and Mining: 3198
  Paleontology: 3203
  Paleontology, Vertebrate: 2749, 2751,
    7588
  Soda Lake: 3736
  Volcanoes: 568, 5367-5371, 5375, 5376,
    5385, 5391
South Carolina
  Agricultural Survey: 9087, 9089-9093
  Earthquakes: 8299, 9589, 9861
  Geological Survey: 4323, 4750, 6142,
    9926-9930, 10429-10431, 10438, 10439
  Geology: 3668, 5238-5243, 8522,
    10429-10431, 10438, 10439, 10574, 10593
  Granite: 3012
  Gypsum: 4347
  Landslide: 5952
  Lime, Agricultural: 9883
  Limestone: 5396, 5397
  Maps, Geological: 10435, 10437
  Marl, Agricultural: 3485, 10581, 10583,
    10590, 11019
  Medical Topography: 8691
  Meteorites: 1012, 9329, 9562
  Minerals, Localities: 3185, 7203, 9881,
    9884, 10562
  Natural History: 1544, 2289, 7194, 8681
  Paleontology, Invertebrate: 1426, 1429,
    1838, 2579, 4360, 4361, 8690, 8904,
    8927
  Paleontology, Vertebrate: 4345, 4346,
    4350, 4351, 4353, 4354, 5242,
    8687-8689, 8692, 9126, 9880
  Sandstone: 6099
  Society for Promoting Agriculture: 9931
  Soils: 1388, 9378, 9379
South Dakota
  Paleontology, Vertebrate: 6180, 8082
Spain
  Copper Mines and Mining: 614
  Earthquakes: 3069, 10643
  Geology: 523, 6626
  Lithographic Stone: 7413
  Mercury Mines and Mining: 637, 3614,
    3979, 4436, 6295, 7713, 7716, 7933
  Minerals, Localities: 1739, 7919
  Mines and Mining: 5427, 7364, 7814
  Phosphorite Rock: 2935
  Salt Mines and Mining: 5416
  Silver Mines and Mining: 4081, 5481
Specific Gravity: 1821, 2746, 4773
  Building Stones: 1626
  Minerals: 1400, 1674, 1968, 2467, 2893,
    3078, 3079
Specific Heat: 5762, 8241
Spencer, J. A.: 1008
Springs: 607
  Boring for: 750
  Intermittant: 7392
  Italy: 1368
  Massachusetts: 6057
  Notes: 1060
  Origin of: 438, 3289, 6057, 6470
  Pennsylvania: 3444

Springs, Hot: 216, 1724, 7453, 7829, 7884,
  9305
  Arkansas: 688, 3685, 6269, 7655, 7685,
    7692, 7776, 7135
  Azores: 4155
  England: 2531, 6357
  France: 4228
  Iceland: 1054
  Kansas: 8423
  Kentucky: 5209
  Louisiana: 6655
  North Carolina: 9863
  Notes: 5661
  Pennsylvania: 5418
  Portugal: 1186
  Switzerland: 2266
  Turkey: 4743
  United States, Western regions: 6473
  Virginia: 1024, 2085, 4437, 6657, 6660,
    6662, 7848, 8971, 8988, 8995
  Volcanoes: 2923, 2924, 2928, 2929
Springs, Salt (see also Salt Mines and Mining)
  Canada: 6282
  Vermont: 10269
Steatite: 3150, 3152
  Hungary: 3080
  Massachusetts: 432, 602, 1820, 2477
  Uses of: 312, 471, 1407, 1809, 3383,
    3980, 3985, 4024
Stone (see also Building Stones; types
  of stone)
  Artificial Coloring: 1262
  Cutting machine: 9325
  Thermal Expansion: 46, 3992, 3994
Straits of Magellan
  Coal Mines and Mining: 967
Stratigraphy: 10888
Strontium
  Flame Test: 1397
  Scotland: 8201
Sullivant, J.
  Mineral Collections: 10068
Sulphur Mines and Mining: 1602
  Italy: 1869
  New York: 7055
Sulphur Springs (see Mineral Springs)
Sumatra
  Natural History: 6020
Survey
  Ohio to Missouri: 6650
Sweden
  Copper Mines and Mining: 5422, 8150,
    10523
  Geology: 629, 6525, 6529
  Iron Mines and Mining: 508, 3398, 5420,
    9372
  Minerals, Localities: 391, 1654, 7497
  Mines and Mining: 2412, 3867, 9106
Switzerland
  Glaciers: 62, 71, 74,7380
  Landslide: 5881
  Springs, Hot: 2266
Synthetic Minerals (see Minerals, Synthetic)
Syria
  Natural History: 10651

Taconic System: 31, 3520, 3551, 3568, 4629,
  4706, 10592
  Vermont: 36
Talus Slopes
  Angle of repose: 6111
Tantalum: 1647, 11028
Taylor, R. C.
  Fossil Collections: 10108, 10136
Tchihatcheff, M. P. de: 817
Tellurium: 1664
Tennessee
  Caves: 1053, 5911, 5912
  Gazetteer: 7431
  Geological Survey: 10162-10165,
    10356-10358, 10362, 10367, 10370-10376,
    10378-10380, 10382-10393, 10398, 10399,
    10401, 10404-10406, 10410
  Geology: 2648, 5911, 5912, 6083, 6496,
    7593, 7605, 7612, 8197, 10356-10358,
    10362, 10367, 10370-10376, 10378-10380,
    10382-10393, 10398, 10399, 10401,
    10404-10406, 10410
  Maps, Geological: 8198, 10367, 10376,
    10383-10385, 10393
  Marble: 99, 7607, 10377

Tennessee (continued)
  Medical Topography: 1747, 2145, 5035, 8682
  Meteorites: 510, 517, 1036, 1839, 9454, 9536, 10381, 10397, 10400, 10402, 10846
  Mineral Springs: 10308, 10390, 10836
  Minerals, Localities: 4661, 4878, 10403, 11064
  Natural Bridge: 10314
  Natural History: 4930
  Paint Rock: 4166
  Paleontology: 6497, 7304, 7608, 10366
  Paleontology, Invertebrate: 91, 10365, 10369, 10407, 10409
  Paleontology, Vertebrate: 4808, 6479, 10359-10361, 10364, 11080, 11087
  Petroleum: 2481
  Railroad Survey: 6083
  Salt Mines and Mining: 8475
  Terraces: 2300, 2839, 8939, 9002
    Canada: 4723
    Connecticut River Valley: 3687, 5180, 5186
    New York: 3122, 3131
    Ohio: 10943
  Tertiary Formation: 2321, 5677
    Maine: 5642
    Maryland: 6781
    Massachusetts: 6547
    New York: 3832
    Notes: 3119
    Paleontology: 6086-6088, 6093
    Paleontology, Invertebrate: 1887, 1890, 2543-2548, 2550-2556, 2558, 2560-2564, 2568, 2570-2572, 2579, 2581-2593
    Paleontology, Vertebrate: 2134
    United States: 3826, 6924, 10821
    Virginia: 8972, 8977, 9094
Texas
  Exploration: 5834, 10267, 10901, 10902
  General Description: 4481
  Geology: 7836, 8786, 8833-8837
  Gold Mines and Mining: 10871
  Meteorites: 2675, 9820
  Minerals, Localities: 9559, 9822, 9827
  Natural History: 5232, 5233, 5942, 7416, 7417, 7860
  Paleobotany: 4517, 6225
  Silver Mines and Mining: 10872
  Soils: 5276
The Naturalist and Journal of Agriculture: 859
Thermal Conductivity
  Sand: 7600
Thermal Expansion
  Building Stone: 1511
  Stone: 46, 3992, 3994
Thomas, J.: 4147
Thorium
  Discovery of: 1658
Tibet
  Geology: 5254
Tides
  Geological Action of: 2946, 2947
Timor
  Landslide: 930
Tin Mines and Mining: 2616, 2937, 7623, 11104
  China: 984
  England: 1107, 8152, 10775
Titanium: 11031
  Ore Processing: 10667
Totten, J. G.
  Mineral Collections: 10296
Trap (see also Basalt)
  Connecticut River Valley: 9807, 9810
Travertine
  Virginia: 3777
Triassic Formation
  United States: 8877, 8900, 8987
Trinidad
  Pitch Lake: 124, 3399, 4004, 4163, 6296
Tristan de Chuna
  Volcanoes: 6319
Troost, G.: 1020, 4315, 6293, 7707, 10163-10165, 10838, 11097
  Lectures: 10332
Troy Lyceum: 8402, 10414-10416
Tuomey, M.: 116
Turkey
  Geology: 3897, 4742, 8022, 10217, 11065
  Iron Mines and Mining: 3625

Turkey (continued)
  Minerals, Localities: 2409, 9322, 9890, 9891-9894, 10157
  Springs, Hot: 4743
  Volcanoes: 4744
Tuscany
  Borax Lagoons: 4031
  Marble: 2770

Unger, Prof.: 4505
United States
  Alluvial Formation: 3827
  Antimony Mines and Mining: 7086, 7357
  Caves: 4961, 7053
  Coal Mines and Mining: 1273, 1522, 1736, 1740, 1778, 2340, 3474, 4587, 5807, 6077, 6305, 7036, 10703
  Cretaceous Formation: 8845
  Devonian Formation: 8895
  Earthquakes: 12, 783, 1516, 2069, 4252, 4469, 4596, 4967, 5530, 6231, 7299, 8894, 8898, 9175, 10652
  Exploring Expeditions: 5737, 5805
  Fuel: 1027
  Gazetteers: 7451, 9368
  Gemstones: 10080
  Geological Survey: 201, 252, 257, 695, 714, 734, 755, 1194, 1263, 2101, 3208, 3401, 5317, 5545, 6981, 7966, 10842
  Geology: 234, 1554, 2151, 2363-2365, 2932, 3324, 3325, 3361, 3374, 3382, 3665, 3666, 3784, 4827, 4877, 4880, 5326, 5572, 5573, 6550, 6557, 6571, 6615, 6616, 6620, 6625, 6960, 7285, 7292, 7307, 8668, 8902, 8903, 9007, 9872, 10577, 10578, 10588, 10653
  Glaciers, Evidence for: 2223, 2224, 6903, 8714, 8715
  Gold Mines and Mining: 3617, 5726, 6286, 6922, 8650, 9359
  Guidebooks: 10090
  Iron Mines and Mining: 550, 2341, 2393, 4168, 5201, 6037, 8435, 8443, 10069
  Land Surveys: 1634, 1635
  Lead Mines and Mining: 439, 1563, 3688, 3909, 5200, 5207, 6973
  Maps, Geological: 2421, 2423, 5063, 6555, 6614, 6619, 8257, 10654
  Materia Medica: 1515, 1533
  Mineral Collections: 5356
  Mineral Springs: 2256, 3966, 3967, 6811
  Minerals, Localities: 141, 325, 388, 2118, 2133, 2420, 2422, 2426, 2432, 2498, 3381, 4650, 4652, 5094, 6127, 6130, 7027, 7281, 7306, 7797, 8506, 8507, 8821, 8822, 9216, 9326, 9433, 9435-9445, 9480-9482, 9489, 9561, 9645, 9832, 10153, 10213, 10275, 10288, 10334, 10337, 10338, 10705, 10722, 10739
  Mines and Mining: 309, 1170, 1500-1502, 1546, 1556, 1777, 1778, 1836, 1855, 1863, 1924, 2012, 2217, 2240, 2444, 2464, 2469, 2470, 2498, 2678, 2907, 3003, 3068, 3814, 4207, 4960, 4983, 4984, 5203, 5424, 5433, 5750, 6038, 6127, 6130, 6550, 6557, 6571, 6810, 6961, 6962, 6974, 8329, 9242, 10150, 10160, 10505, 10508, 10509, 10511
  Natural History: 1529, 2464, 2469, 2470, 2678, 2732, 3610-3612, 5064-5066, 5977, 6034, 6070, 8588, 8589, 9280, 9284, 10998
  Oolitic Formation: 6094, 6095
  Paleobotany: 1432, 10193, 10194, 10198
  Paleontology: 1536, 4923, 6094, 6095, 8662, 8663, 8665
  Paleontology, Invertebrate: 1417-1419, 1426, 1431, 1614, 1743, 1771, 2543-2565, 2567-2593, 4533, 4535, 4537, 4540, 7457-7461, 7466, 7469, 7470, 7474, 7477, 7484, 8663, 9206, 10408
  Paleontology, Vertebrate: 1527, 1530, 1534, 2500, 3271, 3281, 4348, 4349, 4352, 4356, 4359, 4362, 4363, 4804, 4963, 7472, 8698-8701, 8703, 8705, 8708, 8710, 8713, 8717, 8734, 10694-10696, 10698, 10701
  Paleozoic Formation: 8907
  Pipestone: 9988
  Rocking Stones: 3828
  Salt Mines and Mining: 5429, 10485, 10486

Index 429

United States (continued)
  Sandstone: 5673
  Scientific Societies: 544
  Silver Mines and Mining: 9359
  Tertiary Formation: 3826, 6924, 10821
  Triassic Formation: 8877, 8900, 8987
  Volcanoes: 10652
United States, Western regions: 165
  Coal: 4709
  Earthquakes: 8364
  Exploration: 6, 7, 1320, 1363, 2269,
    2270, 2676, 2694, 2729, 3479, 3480,
    3578, 3579, 4111-4119, 4215, 4470,
    4817, 4836, 5614, 5834, 5933, 6210-6214,
    6772, 6368, 8369, 9839-9841, 10470,
    10471, 10882
  Gazetteer: 2081
  Geology: 1306, 1467, 2283, 2841-2844,
    3378, 3778, 3779, 3781, 3912, 4676,
    4678, 4909, 6080, 6081, 6434,
    6473-6476, 6480, 6502, 8056-8058, 8887,
    8910, 9990
  Guidebooks: 2790, 2902, 2904, 10011, 10318,
    10319
  Lead Mines and Mining: 1166
  Maps, Geological: 4682, 6081
  Mineral Collections: 10857
  Minerals, Localities: 6080, 6081
  Mines and Mining: 247, 7042, 8823
  Natural History: 2081, 4666-4668, 4972,
    5062, 8235, 8623, 8624, 8823, 9213,
    10221, 10318, 10319, 10594
  Paleontology: 7854
  Paleontology, Invertebrate: 2589, 2592,
    2824, 2831
  Paleontology, Vertebrate: 1199, 4355
  Prairies: 1356, 1357, 1879, 2263, 5841
  Resources: 548
  Scientific Societies: 543
  Springs, Hot: 6473
University of Pennsylvania: 2617
Uranium
  Analysis, Chemical: 4895, 8159

Van Rensselaer, J.: 8226, 9675
Van Rensselaer, S.: 724, 1488, 9848
Vanadium
  Discovery of: 1660, 3742, 3769, 10835
  Lake Superior District: 10211, 10215
Vanuxem, L.: 931, 4725
  Mineral Collections: 994
Veins: 7920, 9171
  Nature of: 3764
  Origin of: 1725, 2937, 2941, 3950-3953,
    3955, 3956, 3957, 4006, 5402, 5404,
    8732, 9493
Venezuela
  Earthquakes: 1223, 1229, 2221, 5504,
    6392, 7100, 7301, 9414, 10672, 10782
Verd Antique (see Marble)
Vermont
  Building stones: 5170
  Caves: 1111, 3895, 4527, 7876, 10977
  Clay, Porcelain: 1825, 3159, 7360, 7592
  Claystone: 33
  Copper Mines and Mining: 8770
  Copperas Mines and Mining: 6419
  Dike: 2401
  Dolomite: 2325
  Erosion: 4648
  Gazetteer: 10236
  Geological Survey: 200, 806, 851, 4243,
    4734-4736, 4966, 8033, 10609-10616
  Geology: 22, 23, 26, 32, 34, 3290, 4654,
    5691, 6061, 8809, 9918, 10220,
    10238-10240, 10924
  Glaciers, Evidence for: 35
  Gold Mines and Mining: 152, 446, 3976
  Iron Mines and Mining: 4376
  Limestone: 2324, 4645
  Manganese Mines and Mining: 4421, 4646
  Marble: 719, 1592, 4644, 10010
  Marl, Agricultural: 6784
  Mineral Springs: 163, 9016, 9128, 10817
  Minerals, Localities: 155, 156, 157,
    366, 2430, 2875, 3149-3151, 3229, 3818,
    3819, 3820, 3821, 4646, 4647, 4650,
    4652, 4885, 6241, 8106, 8517, 9637,
    9678, 10056
  Natural Gas: 3675

Vermont (continued)
  Natural History: 10237, 10242, 10976
  Paleontology: 4710
  Paleontology, Invertebrate: 1436
  Paleontology, Vertebrate: 86, 10241,
    10832
  Slate: 2325
  Springs, Salt: 10269
  Taconic System: 36
  Waterfalls: 4648
Vesuvius (see Italy; Volcanoes)
Virginia
  Building Stone: 6067
  Caves: 321, 1053, 1517, 2076, 2495, 4793,
    4847, 5421, 5505, 5721, 5838, 5907,
    6228, 6321, 7053, 7057, 7062, 7102,
    7382, 8139, 8431, 9119, 9366, 11111
  Coal Mines and Mining: 98, 3214, 3870,
    4491, 5365, 6570, 7493, 7496, 8035,
    8036, 8361, 8981, 8986, 8989, 8993,
    9768, 10115, 10120, 10427, 10628, 10649,
    11050, 11078
  Coke: 5788
  Copper Mines and Mining: 7354, 9792
  Coprolite: 11085
  Gazetteer: 6791
  General: 4484
  Geological Survey: 2738, 2739, 3720,
    7893, 7894, 8965, 8967-8969,
    8973-8976, 8978-8980, 8990, 10629-
    10634
  Geology: 265, 2093-2096, 2648, 2738,
    2739, 3770, 6063, 6228, 7038, 7295,
    7494, 8965, 8967-8969, 8973-8976,
    8978-8980, 8990, 10123
  Glaciers, Evidence for: 8867
  Gold Mines and Mining: 966, 1159, 2045,
    2441, 3088, 3182, 3397, 4272, 4603,
    5500, 5656, 6208, 6920, 6923, 8393,
    8970, 9565-9569, 9731-9734, 9736, 9738,
    9741, 9744, 9745, 9753, 10113, 10124,
    10149, 10596, 10640, 10641
  Greensand: 8962, 8964, 9076, 9077, 9081
  Gypsum: 1540
  Iron Mines and Mining: 268, 5365, 9570,
    9571
  Lead Mines and Mining: 8444, 10639
  Lime, Agricultural: 1707
  Limestone: 4466
  Manganese Mining: 1525
  Maps, Geological: 8776
  Marl, Agricultural: 6260, 6964,
    7487-7491, 7495, 8361, 10424
  Meteorites: 2471, 8982, 9477, 9541
  Mineral Springs: 1452, 1481, 1537, 1623,
    3486, 3733, 3776, 3777, 4003, 4523,
    4524, 4870, 4871, 4881, 4882, 5490,
    6140, 6141, 6657, 6660, 6662, 6857,
    7004, 7048, 7184, 7420-7422,
    7865-7867, 8203, 8245, 8971, 8988,
    8995, 9040, 9504, 9844, 9944, 9946,
    9947, 10840, 11006
  Minerals, Localities: 1535, 1538, 1540,
    1541, 1542, 3847, 5644, 5676, 8285,
    8984
  Mines and Mining: 1978, 3715, 3732, 3775,
    3889, 4174, 5454, 5473, 10468, 10469,
    10635, 10994
  Natural Bridge: 1041, 1242, 2496, 3483,
    3766, 4395, 4474, 4478, 4516, 5723,
    6306, 7240, 7375, 7849, 8133
  Natural Gas: 172, 1065, 6207, 6388
  Natural History: 1978, 4973, 4974, 5719,
    5945, 5946, 6851, 9985, 10298
  Natural Tunnel: 6478
  Paleobotany: 2191, 10123
  Paleontology: 8169, 8426, 9094, 9851
  Paleontology, Invertebrate: 1422, 1423,
    2546, 2561, 2577, 2579, 3458,
    6086-6088, 6138, 8996, 8999, 10426,
    10428
  Paleontology, Vertebrate: 1517, 2639,
    5263, 5264, 5718-5720, 5722, 7317,
    8298, 8701, 10036, 11079, 11081, 11084,
    11085
  Petroleum: 4172, 7718, 8611, 10638
  Salt Mines and Mining: 97, 248, 1036,
    1061, 1787, 4873, 5917, 7119, 9096,
    10151, 10484, 10644, 10829
  Saltpetre: 7085, 7356
  Sandstone: 8994
  Soils: 3894, 7487-7491, 7495

Virginia (continued)
  Springs, Hot: 1024, 2085, 4437, 6657, 6660, 6662, 7848, 8971, 8988, 8995
  Tertiary Formation: 8972, 8977, 9094
  Travertine: 3777
Vitriol (see also Copperas)
  Connecticut: 7082
  Java: 5904, 6196, 6281
  New York: 7326
Volcanic Rocks
  Analysis, Chemical: 2642, 2643, 2926, 2931
  Minerals of: 3093, 3094
Volcanoes
  At Sea: 4046, 4829, 7144, 7795, 8120, 8131, 11120
  Azores: 1074, 2763, 2764, 5032, 8463
  Banda Island: 10608
  Batavia: 1236
  British Columbia: 4195
  California: 4195
  Cape Verde: 896, 3140
  Central America: 9992
  Coal: 6262
  Connecticut River Valley: 5165, 5181, 9680, 9807, 9810
  Cotopaxi (see South America)
  France: 1073, 7423
  Friendly Islands: 928
  General Descriptions: 292, 866
  Graham Island: 107
  Guadaloupe: 7017, 8254
  Hawaii: 1034, 1202, 1253, 2465, 2466, 2666, 2845, 2848, 2859, 2861, 2862, 3316, 4438-4441, 4634, 4860-4862, 5598, 5633, 5888, 5937, 6578, 7381, 7581, 8174, 8530, 8727, 9169, 10017-10019, 10053, 10054
  Iceland: 2499, 3102, 4300, 6267, 6998, 7222, 7387, 10307
  Island: 108, 225, 645, 675, 1067, 1469, 2362, 3772, 8370
  Italy: 394, 451, 527, 565, 619, 972, 1029, 1085, 1089, 1106, 1154, 1183, 1207, 1480, 2054, 2329, 2408, 2791, 3277, 3278, 3318, 3580, 3618, 3874, 4495, 4573, 4607, 5521, 5697, 5714, 5715, 6317, 6406, 7142, 7146, 7182, 7358, 7361, 7388, 7390, 7391, 7409-7412, 7427, 7563, 7579, 7591, 7706, 7708, 7793, 7913, 8085, 8129, 8132, 8294, 8384, 8400, 8487, 8490, 8531, 8556, 9373, 9978, 9979, 10229, 10293, 10315-10317, 10647, 10993, 11122
  Java: 1230, 1265, 4521, 5862, 7384
  Kentucky: 7931
  Lipari Islands: 7241, 7556
  Localities: 240
  Massachusetts: 5111
  Mediterranean: 10076
  Mexico: 670, 4597, 6400, 6636, 6639, 6986, 7712, 11126
  Minerals: 451
  Moon: 2822, 5015, 5016, 8019, 10712
  New Granada: 18
  New Hampshire: 123, 168, 5839
  New Zealand: 6229
  Nicaragua: 4205
  North Carolina: 1140
  Notes: 1030, 1032, 1045, 1088, 1305, 1723, 1859, 2213, 2384, 2407, 2859, 2867, 2921, 2972, 2974, 3299, 3305, 3315, 3405, 3424, 3633, 3645, 3704, 4127, 4219, 4229, 4240, 4475, 4916, 5033, 5367-5371, 5375, 5376, 5385, 5391, 5899, 5906, 6705, 6706, 7122, 7205, 7363, 7404, 7626, 7640, 7717, 7827, 7841, 7877, 8128, 8142, 8312, 8337, 8523, 8640, 9142, 9188, 9296, 9346, 9374-9377, 9725, 9772, 10311, 10845
  Origin of: 2625, 3255, 5011, 5012, 6265, 6312, 9673, 9676
  Pennsylvania: 215
  Peru: 905, 2733, 8801
  Philippines: 3429, 8326
  Russia: 10
  St. Lucia: 2280
  St. Vincent: 6242, 7298, 7562, 7674

Volcanoes (continued)
  Sicily: 660, 1190, 1313, 1507, 1552, 2740, 3825, 4238, 4818, 4917, 5759, 6969, 7789, 8535, 8824, 9144, 10601, 10682
  South America: 568, 5367-5371, 5375, 5376, 5385, 5391
  Springs, Hot: 2923, 2924, 2928, 2929
  Submarine: 887
  Tristan de Chuna: 6319
  Turkey: 4744
  United States: 10652
Volney, C. F. C.: 2912
Von der Null, J. F.
  Mineral Collections: 10655

Wailes, B. L. C.: 10844
Wales (see also Great Britain)
  Coal Mines and Mining: 5778, 5809
  Geology: 2597, 8680
  Glaciers, Evidence for: 6605
  Iron Mines and Mining: 3917
  Mines and Mining: 4072
  Paleontology, Vertebrate: 2529
  Scotland: 5756
Water
  Analysis, Chemical: 9812, 9817, 9818, 9824, 9825
Waterfalls (see also Niagara Falls): 10659, 11054
  Canada: 6272
  Georgia: 3918
  New York: 4698
  Vermont: 4648
Watson, W.: 2532
Weathering (see also Soils; Erosion): 1850, 1862, 2140, 3391
  Carbonates: 9003, 9004, 9008, 9009
  Dolomite: 2933
Webster, J. W.: 328
  Mineral Collections: 329
Werner, A. G.: 1092, 2607, 3436, 6186, 6622, 10625
  Mineral Collections: 1192
West Chester County Cabinet: 554
West Indies
  Earthquakes: 784, 2245, 10216
  Geology: 6617, 6641, 7990
  Natural History: 10998
  Paleontology, Vertebrate: 8405
West Virginia
  Geology: 2427
  Iron Mines and Mining: 10800
  Mineral Springs: 2196-2198
  Paleobotany: 5812
Whin Sill
  England: 5497
Whitney, J. D.: 196, 198, 5648
Williams, C. W.: 230
Windsor County Natural History Association: 10991, 10992
Winthrop, J.: 9130
Wisconsin
  Geology: 1749, 6046-6048, 7984, 8050, 8061, 8065, 8068-8070, 8073, 8074, 8078, 8079
  Lead Mines and Mining: 5195, 6700
  Maps, Geological: 8061, 8062, 8068, 8078
  Marble: 8575
  Natural History: 5, 6046-6048, 6613
  Paleontology: 5630
  Soils: 3535
Witham, H.: 587, 9721
Wollaston, W. H.: 370, 4591, 9679
Woodbridge, W. C.: 835

Yale College: 892, 1463, 9610, 9618, 9716
  Mineral Collections: 2111, 2120, 11090
Yorkshire Philosophical Society: 11102
Young, J. P.
  Mineral Collections: 635, 842
Zimmern, L. W.
  Mineral Collections: 5133
Zinc Mines and Mining: 7624
  Missouri: 7928
  New Jersey: 139, 158, 1271, 1349, 8442, 10346
  Notes: 4246, 8440, 8827
  Ore Processing: 8827
  Pennsylvania: 1772, 9434, 11044, 11048

ABOUT THE AUTHORS

Robert M. Hazen, experimental mineralogist at the Carnegie Institution of Washington's Geophysical Laboratory, received the B.S. and S.M. in geology at the Massachusetts Institute of Technology in 1971 and the Ph.D. in mineralogy at Harvard University in 1975. Prior to joining the Carnegie Institution, he was a NATO Fellow at Cambridge University in England. His research interests include high-temperature and high-pressure crystallography and the crystal chemistry of rock-forming minerals, as well as the history of North American geology. In addition, as a part-time professional trumpeter, Dr. Hazen has performed with numerous ensembles, including the Boston and National Symphony orchestras. He is editor of *North American Geology: Early Writings*, published by Dowden, Hutchinson & Ross.

Margaret Hindle Hazen received the B.A. at Wellesley College in 1970 and the M.A. in history at Boston University. She also holds an M.L.S. degree from Simmons College. Mrs. Hazen was a senior cataloger at the library of Boston University from 1972 to 1974 and subsequently became the manuscript and rare book librarian of the New England Historic Genealogical Society. Her research interests include aspects of American social history and the scientific exploring expeditions of nineteenth-century America.

The authors reside in Bethesda, Maryland, with their two children, Benjamin and Elizabeth.